Higher Engineering Mathematics

Now in its eighth edition, *Higher Engineering Mathematics* has helped thousands of students succeed in their exams. Theory is kept to a minimum, with the emphasis firmly placed on problem-solving skills, making this a thoroughly practical introduction to the advanced engineering mathematics that students need to master. The extensive and thorough topic coverage makes this an ideal text for upper-level vocational courses and for undergraduate degree courses. It is also supported by a fully updated companion website with resources for both students and lecturers. It has full solutions to all 2,000 further questions contained in the 277 practice exercises.

John Bird, BSc (Hons), CMath, CEng, CSci, FITE, FIMA, FCollT, is the former Head of Applied Electronics in the Faculty of Technology at Highbury College, Portsmouth, UK. More recently he has combined freelance lecturing and examining, and is the author of over 130 textbooks on engineering and mathematical subjects with worldwide sales of over one million copies. He is currently lecturing at the Defence School of Marine Engineering in the Defence College of Technical Training at HMS Sultan, Gosport, Hampshire, UK.

Why is knowledge of mathematics important in engineering?

A career in any engineering or scientific field will require both basic and advanced mathematics. Without mathematics to determine principles, calculate dimensions and limits, explore variations, prove concepts, and so on, there would be no mobile telephones, televisions, stereo systems, video games, microwave ovens, computers, or virtually anything electronic. There would be no bridges, tunnels, roads, skyscrapers, automobiles, ships, planes, rockets or most things mechanical. There would be no metals beyond the common ones, such as iron and copper, no plastics, no synthetics. In fact, society would most certainly be less advanced without the use of mathematics throughout the centuries and into the future.

Electrical engineers require mathematics to design, develop, test, or supervise the manufacturing and installation of electrical equipment, components, or systems for commercial, industrial, military, or scientific use.

Mechanical engineers require mathematics to perform engineering duties in planning and designing tools, engines, machines, and other mechanically functioning equipment; they oversee installation, operation, maintenance, and repair of such equipment as centralised heat, gas, water, and steam systems.

Aerospace engineers require mathematics to perform a variety of engineering work in designing, constructing, and testing aircraft, missiles, and spacecraft; they conduct basic and applied research to evaluate adaptability of materials and equipment to aircraft design and manufacture and recommend improvements in testing equipment and techniques.

Nuclear engineers require mathematics to conduct research on nuclear engineering problems or apply principles and theory of nuclear science to problems concerned with release, control, and utilisation of nuclear energy and nuclear waste disposal.

Petroleum engineers require mathematics to devise methods to improve oil and gas well production and determine the need for new or modified tool designs; they oversee drilling and offer technical advice to achieve economical and satisfactory progress.

Industrial engineers require mathematics to design, develop, test, and evaluate integrated systems for managing industrial production processes, including human work factors, quality control, inventory control, logistics and material flow, cost analysis, and production co-ordination.

Environmental engineers require mathematics to design, plan, or perform engineering duties in the prevention, control, and remediation of environmental health hazards, using various engineering disciplines; their work may include waste treatment, site remediation, or pollution control technology.

Civil engineers require mathematics in all levels in civil engineering – structural engineering, hydraulics and geotechnical engineering are all fields that employ mathematical tools such as differential equations, tensor analysis, field theory, numerical methods and operations research.

Knowledge of mathematics is therefore needed by each of the engineering disciplines listed above.

It is intended that this text – *Higher Engineering Mathematics* – will provide a step-by-step approach to learning fundamental mathematics needed for your engineering studies.

Higher Engineering Mathematics

Eighth Edition

John Bird

Routledge
Taylor & Francis Group

LONDON AND NEW YORK

Eighth edition published 2017
by Routledge
2 Park Square, Milton Park, Abingdon, Oxon OX14 4RN

and by Routledge
711 Third Avenue, New York, NY 10017

Routledge is an imprint of the Taylor & Francis Group, an informa business

First edition published by Elsevier 1993
Seventh edition published by Routledge 2014

British Library Cataloguing in Publication Data
A catalogue record for this book is available from the British Library

Library of Congress Cataloging in Publication Data
Names: Bird, J. O.Title: Higher engineering mathematics / John Bird.
Description: 8th ed. | Boca Raton : CRC Press, 2017. | Includes index.
Identifiers: LCCN 2016038671 | ISBN 9781138673571
Subjects: LCSH: Engineering mathematics.
Classification: LCC TA330 .B52 2017 | DDC 620.001/51–dc23LC
record available at https://lccn.loc.gov/2016038671

ISBN: 978-1-138-67357-1 (pbk)
ISBN: 978-1-315-26502-5 (ebk)

Typeset in Times by
Servis Filmsetting Ltd, Stockport, Cheshire
Printed and bound in Great Britain by Bell & Bain Ltd, Glasgow

Visit the companion website: www.routledge.com/cw/bird

To Sue

Contents

Section B   Geometry and trigonometry   121

Section C   Graphs   217

Preface

This **eighth edition of** *Higher Engineering Mathematics* covers essential mathematical material suitable for students studying **Degrees, Foundation Degrees, and Higher National Certificate and Diploma courses in Engineering disciplines.**

The text has been conveniently divided into the following **fourteen convenient categories**: number and algebra, geometry and trigonometry, graphs, complex numbers, matrices and determinants, vector geometry, introduction to calculus, further differential calculus, further integral calculus, further differential equations, statistics and probability, Laplace transforms, Fourier series and z-transforms.

Increasingly, **difficulty in understanding algebra** is proving a problem for many students as they commence studying engineering courses. Inevitably there are a lot of formulae and calculations involved with engineering studies that require a sound grasp of algebra. On the website www.routledge.com/cw/bird/ is a document which offers **a quick revision of the main areas of algebra** essential for further study, i.e. basic algebra, simple equations, transposition of formulae, simultaneous equations and quadratic equations.

In this new edition, all of the chapters of the previous edition are included, plus one extra, but the order of presenting some of the calculus chapters has been changed. **New material** has been added on the introduction to numbering systems, Bayes' theorem in probability, the comparison of numerical methods and z-transforms.

The **primary aim of the material in this text** is to provide the fundamental analytical and underpinning knowledge and techniques needed to successfully complete scientific and engineering principles modules of Degree, Foundation Degree and Higher National Engineering programmes. The material has been designed to enable students to use techniques learned for the analysis, modelling and solution of realistic engineering problems at Degree and Higher National level. It also aims to provide some of the more advanced knowledge required for those wishing to pursue careers in mechanical engineering, aeronautical engineering, electrical and electronic engineering, communications engineering, systems engineering and all variants of control engineering.

In *Higher Engineering Mathematics 8th Edition*, theory is introduced in each chapter by a full outline of essential definitions, formulae, laws, procedures, etc; **problem solving** is extensively used to establish and exemplify the theory. It is intended that readers will gain real understanding through seeing problems solved and then through solving similar problems themselves.

Access to **software packages** such as Maple, Mathematica and Derive, or a graphics calculator, will enhance understanding of some of the topics in this text.

Each topic considered in the text is presented in a way that assumes in the reader only knowledge attained in BTEC National Certificate/Diploma, or similar, in an Engineering discipline.

Higher Engineering Mathematics 8th Edition provides a follow-up to *Engineering Mathematics 8th Edition*.

This textbook contains over **1050 worked problems**, followed by nearly **2000 further problems (with answers),** arranged within **277 Practice Exercises**. Some **552 line diagrams** further enhance understanding.

Worked solutions to all 2000 of the further problems have been prepared and can be **accessed free by students and staff via the website** www.routledge.com/cw/bird/

At the end of the text, a list of **Essential Formulae** is included for convenience of reference.

At intervals throughout the text are some **21 Revision Tests** to check understanding. For example, Revision Test 1 covers the material in chapters 1 to 5, Revision Test 2 covers the material in chapters 6 to 8, Revision Test 3 covers the material in chapters 9 to 11, and so on. An **Instructor's Manual**, containing full solutions to the Revision Tests, is available free to lecturers/instructors via the website (see below).

'Learning by example' is at the heart of *Higher Engineering Mathematics 8th Edition*.

JOHN BIRD
Royal Naval Defence College of Marine and Air
Engineering, HMS Sultan,
formerly University of Portsmouth
and Highbury College, Portsmouth

John Bird is the former Head of Applied Electronics in the Faculty of Technology at Highbury College, Portsmouth, UK. More recently, he has combined freelance lecturing at the University of Portsmouth with examiner responsibilities for Advanced Mathematics with City and Guilds, and examining for International Baccalaureate. He is the author of some 130 textbooks on engineering and mathematical subjects with worldwide sales of one million copies. He is a chartered engineer, a chartered mathematician, a chartered scientist and a Fellow of three professional institutions, and is currently lecturing at the Defence College of Marine and Air Engineering in the Defence College of Technical Training at HMS Sultan, Gosport, Hampshire, UK.

Free Web downloads

The following support material is available from www.routledge.com/cw/bird/

For Students:

1. Full solutions to all 2000 further questions contained in the 277 Practice Exercises

2. Revision of some important algebra topics

3. A list of Essential Formulae

4. Information on 32 Mathematicians/Engineers mentioned in the text

For Lecturers/Instructors:

1. Full solutions to all 2000 further questions contained in the 277 Practice Exercises

2. Revision of some important algebra topics

3. Full solutions and marking scheme for each of the 21 Revision Tests; also, each test may be downloaded for distribution to students. In addition, solutions to the Revision Test given in the 'Revision of Algebra Topics' is also included.

4. A list of Essential Formulae

5. Information on 32 Mathematicians/Engineers mentioned in the text

6. All 552 illustrations used in the text may be downloaded for use in PowerPoint presentations

Syllabus guidance

This textbook is written for **undergraduate engineering degree and foundation degree courses**; however, it is also most appropriate for **BTEC levels 4 and 5 HNC/D studies in engineering** and three syllabuses are covered. The appropriate chapters for these three syllabuses are shown in the table below.

Chapter		Analytical Methods for Engineers	Further Analytical Methods for Engineers	Advanced Mathematics for Engineering
1.	Algebra	×		
2.	Partial fractions	×		
3.	Logarithms	×		
4.	Exponential functions	×		
5.	Inequalities			
6.	Arithmetic and geometric progressions	×		
7.	The binomial series	×		
8.	Maclaurin's series	×		
9.	Solving equations by iterative methods		×	
10.	Binary, octal and hexadecimal		×	
11.	Boolean algebra and logic circuits		×	
12.	Introduction to trigonometry	×		
13.	Cartesian and polar co-ordinates	×		
14.	The circle and its properties	×		
15.	Trigonometric waveforms	×		
16.	Hyperbolic functions	×		
17.	Trigonometric identities and equations	×		
18.	The relationship between trigonometric and hyperbolic functions	×		
19.	Compound angles	×		
20.	Functions and their curves		×	
21.	Irregular areas, volumes and mean value of waveforms		×	
22.	Complex numbers		×	
23.	De Moivre's theorem		×	
24.	The theory of matrices and determinants		×	
25.	Applications of matrices and determinants		×	

(*Continued*)

Chapter		Analytical Methods for Engineers	Further Analytical Methods for Engineers	Advanced Mathematics for Engineering
26.	Vectors		×	
27.	Methods of adding alternating waveforms		×	
28.	Scalar and vector products		×	
29.	Methods of differentiation	×		
30.	Some applications of differentiation	×		
31.	Standard integration	×		
32.	Some applications of integration	×		
33.	Introduction to differential equations		×	
34.	Differentiation of parametric equations			
35.	Differentiation of implicit functions	×		
36.	Logarithmic differentiation	×		
37.	Differentiation of hyperbolic functions	×		
38.	Differentiation of inverse trigonometric and hyperbolic functions	×		
39.	Partial differentiation			×
40.	Total differential, rates of change and small changes			×
41.	Maxima, minima and saddle points for functions of two variables			×
42.	Integration using algebraic substitutions	×		
43.	Integration using trigonometric and hyperbolic substitutions	×		
44.	Integration using partial fractions	×		
45.	The t = tan θ/2 substitution			
46.	Integration by parts	×		
47.	Reduction formulae	×		
48.	Double and triple integrals			
49.	Numerical integration		×	
50.	Homogeneous first-order differential equations			
51.	Linear first-order differential equations		×	
52.	Numerical methods for first-order differential equations		×	×
53.	Second-order differential equations of the form $a\dfrac{d^2 y}{dx^2} + b\dfrac{dy}{dx} + cy = 0$		×	

(Continued)

Chapter		Analytical Methods for Engineers	Further Analytical Methods for Engineers	Advanced Mathematics for Engineering
54.	Second-order differential equations of the form $a\dfrac{d^2 y}{dx^2} + b\dfrac{dy}{dx} + cy = f(x)$		×	
55.	Power series methods of solving ordinary differential equations			×
56.	An introduction to partial differential equations			×
57.	Presentation of statistical data	×		
58.	Measures of central tendency and dispersion	×		
59.	Probability	×		
60.	The binomial and Poisson distributions	×		
61.	The normal distribution	×		
62.	Linear correlation	×		
63.	Linear regression	×		
64.	Sampling and estimation theories	×		
65.	Significance testing	×		
66.	Chi-square and distribution-free tests	×		
67.	Introduction to Laplace transforms			×
68.	Properties of Laplace transforms			×
69.	Inverse Laplace transforms			×
70.	The Laplace transform of the Heaviside function			
71.	Solution of differential equations using Laplace transforms			×
72.	The solution of simultaneous differential equations using Laplace transforms			×
73.	Fourier series for periodic functions of period 2π			×
74.	Fourier series for non-periodic functions over range 2π			×
75.	Even and odd functions and half-range Fourier series			×
76.	Fourier series over any range			×
77.	A numerical method of harmonic analysis			×
78.	The complex or exponential form of a Fourier series			×
79.	An introduction to z-transforms			

Section A

Number and algebra

Chapter 1

Algebra

Why it is important to understand: **Algebra, polynomial division and the factor and remainder theorems**

It is probably true to say that there is no branch of engineering, physics, economics, chemistry or computer science which does not require the understanding of the basic laws of algebra, the laws of indices, the manipulation of brackets, the ability to factorise and the laws of precedence. This then leads to the ability to solve simple, simultaneous and quadratic equations which occur so often. The study of algebra also revolves around using and manipulating polynomials. Polynomials are used in engineering, computer programming, software engineering, in management, and in business. Mathematicians, statisticians and engineers of all sciences employ the use of polynomials to solve problems; among them are aerospace engineers, chemical engineers, civil engineers, electrical engineers, environmental engineers, industrial engineers, materials engineers, mechanical engineers and nuclear engineers. The factor and remainder theorems are also employed in engineering software and electronic mathematical applications, through which polynomials of higher degrees and longer arithmetic structures are divided without any complexity. The study of algebra, equations, polynomial division and the factor and remainder theorems is therefore of some considerable importance in engineering.

At the end of this chapter, you should be able to:

- understand and apply the laws of indices
- understand brackets, factorisation and precedence
- transpose formulae and solve simple, simultaneous and quadratic equations
- divide algebraic expressions using polynomial division
- factorise expressions using the factor theorem
- use the remainder theorem to factorise algebraic expressions

---

1.1 Introduction

In this chapter, polynomial division and the factor and remainder theorems are explained (in Sections 1.4 to 1.6). However, before this, some essential algebra revision on basic laws and equations is included.

For further algebra revision, go to the website:
www.routledge.com/cw/bird

1.2 Revision of basic laws

(a) Basic operations and laws of indices

The **laws of indices** are:

(i) $a^m \times a^n = a^{m+n}$ (ii) $\dfrac{a^m}{a^n} = a^{m-n}$

(iii) $(a^m)^n = a^{m \times n}$ (iv) $a^{\frac{m}{n}} = \sqrt[n]{a^m}$

(v) $a^{-n} = \dfrac{1}{a^n}$ (vi) $a^0 = 1$

Problem 1. Evaluate $4a^2bc^3 - 2ac$ when $a = 2$, $b = \frac{1}{2}$ and $c = 1\frac{1}{2}$

$$4a^2bc^3 - 2ac = 4(2)^2\left(\frac{1}{2}\right)\left(\frac{3}{2}\right)^3 - 2(2)\left(\frac{3}{2}\right)$$

$$= \frac{4 \times 2 \times 2 \times 3 \times 3 \times 3}{2 \times 2 \times 2 \times 2} - \frac{12}{2}$$

$$= 27 - 6 = \mathbf{21}$$

Problem 2. Multiply $3x + 2y$ by $x - y$

$$3x + 2y$$
$$x - y$$

Multiply by x $\rightarrow$ $\quad 3x^2 + 2xy$

Multiply by $-y$ $\rightarrow$ $\quad\quad\quad -3xy - 2y^2$

Adding gives: $\quad \mathbf{3x^2 - xy - 2y^2}$

Alternatively,

$$(3x + 2y)(x - y) = 3x^2 - 3xy + 2xy - 2y^2$$

$$= \mathbf{3x^2 - xy - 2y^2}$$

Problem 3. Simplify $\dfrac{a^3b^2c^4}{abc^{-2}}$ and evaluate when $a = 3$, $b = \frac{1}{8}$ and $c = 2$

$$\frac{a^3b^2c^4}{abc^{-2}} = a^{3-1}b^{2-1}c^{4-(-2)} = \mathbf{a^2bc^6}$$

When $a = 3$, $b = \frac{1}{8}$ and $c = 2$,

$$a^2bc^6 = (3)^2\left(\frac{1}{8}\right)(2)^6 = (9)\left(\frac{1}{8}\right)(64) = \mathbf{72}$$

Problem 4. Simplify $\dfrac{x^2y^3 + xy^2}{xy}$

$$\frac{x^2y^3 + xy^2}{xy} = \frac{x^2y^3}{xy} + \frac{xy^2}{xy}$$

$$= x^{2-1}y^{3-1} + x^{1-1}y^{2-1}$$

$$= \mathbf{xy^2 + y} \quad \text{or} \quad \mathbf{y(xy + 1)}$$

Problem 5. Simplify $\dfrac{(x^2\sqrt{y})(\sqrt{x}\,\sqrt[3]{y^2})}{(x^5y^3)^{\frac{1}{2}}}$

$$\frac{(x^2\sqrt{y})(\sqrt{x}\,\sqrt[3]{y^2})}{(x^5y^3)^{\frac{1}{2}}} = \frac{x^2y^{\frac{1}{2}}x^{\frac{1}{2}}y^{\frac{2}{3}}}{x^{\frac{5}{2}}y^{\frac{3}{2}}}$$

$$= x^{2+\frac{1}{2}-\frac{5}{2}}y^{\frac{1}{2}+\frac{2}{3}-\frac{3}{2}}$$

$$= x^0y^{-\frac{1}{3}}$$

$$= \mathbf{y^{-\frac{1}{3}}} \quad \text{or} \quad \mathbf{\frac{1}{y^{\frac{1}{3}}}} \quad \text{or} \quad \mathbf{\frac{1}{\sqrt[3]{y}}}$$

Now try the following Practice Exercise

Practice Exercise 1   Basic algebraic operations and laws of indices (Answers on page 856)

1. Evaluate $2ab + 3bc - abc$ when $a = 2$, $b = -2$ and $c = 4$

2. Find the value of $5pq^2r^3$ when $p = \frac{2}{5}$, $q = -2$ and $r = -1$

3. From $4x - 3y + 2z$ subtract $x + 2y - 3z$.

4. Multiply $2a - 5b + c$ by $3a + b$

5. Simplify $(x^2y^3z)(x^3yz^2)$ and evaluate when $x = \frac{1}{2}$, $y = 2$ and $z = 3$

6. Evaluate $(a^{\frac{3}{2}}bc^{-3})(a^{\frac{1}{2}}b^{-\frac{1}{2}}c)$ when $a = 3$, $b = 4$ and $c = 2$

7. Simplify $\dfrac{a^2b + a^3b}{a^2b^2}$

8. Simplify $\dfrac{(a^3b^{\frac{1}{2}}c^{-\frac{1}{2}})(ab)^{\frac{1}{3}}}{(\sqrt{a^3}\sqrt{b}\,c)}$

(b) Brackets, factorisation and precedence

Problem 6. Simplify $a^2 - (2a - ab) - a(3b + a)$

$$a^2 - (2a - ab) - a(3b + a)$$

$$= a^2 - 2a + ab - 3ab - a^2$$

$$= \mathbf{-2a - 2ab} \quad \text{or} \quad \mathbf{-2a(1 + b)}$$

Problem 7. Remove the brackets and simplify the expression:

$$2a - [3\{2(4a - b) - 5(a + 2b)\} + 4a]$$

Removing the innermost brackets gives:

$$2a - [3\{8a - 2b - 5a - 10b\} + 4a]$$

Collecting together similar terms gives:

$$2a - [3\{3a - 12b\} + 4a]$$

Removing the 'curly' brackets gives:

$$2a - [9a - 36b + 4a]$$

Collecting together similar terms gives:

$$2a - [13a - 36b]$$

Removing the square brackets gives:

$$2a - 13a + 36b = \mathbf{-11a + 36b} \quad \text{or}$$

$$\mathbf{36b - 11a}$$

Problem 8. Factorise (a) $xy - 3xz$
(b) $4a^2 + 16ab^3$ (c) $3a^2b - 6ab^2 + 15ab$

(a) $xy - 3xz = \mathbf{x(y - 3z)}$

(b) $4a^2 + 16ab^3 = \mathbf{4a(a + 4b^3)}$

(c) $3a^2b - 6ab^2 + 15ab = \mathbf{3ab(a - 2b + 5)}$

Problem 9. Simplify $3c + 2c \times 4c + c \div 5c - 8c$

The order of precedence is division, multiplication, addition, and subtraction (sometimes remembered by BODMAS). Hence

$$3c + 2c \times 4c + c \div 5c - 8c$$

$$= 3c + 2c \times 4c + \left(\frac{c}{5c}\right) - 8c$$

$$= 3c + 8c^2 + \frac{1}{5} - 8c$$

$$= \mathbf{8c^2 - 5c + \frac{1}{5}} \quad \text{or} \quad \mathbf{c(8c - 5) + \frac{1}{5}}$$

Problem 10. Simplify
$(2a - 3) \div 4a + 5 \times 6 - 3a$

$$(2a - 3) \div 4a + 5 \times 6 - 3a$$

$$= \frac{2a - 3}{4a} + 5 \times 6 - 3a$$

$$= \frac{2a - 3}{4a} + 30 - 3a$$

$$= \frac{2a}{4a} - \frac{3}{4a} + 30 - 3a$$

$$= \frac{1}{2} - \frac{3}{4a} + 30 - 3a = \mathbf{30\frac{1}{2} - \frac{3}{4a} - 3a}$$

Now try the following Practice Exercise

Practice Exercise 2   Brackets, factorisation and precedence (Answers on page 856)

1. Simplify $2(p + 3q - r) - 4(r - q + 2p) + p$

2. Expand and simplify $(x + y)(x - 2y)$

3. Remove the brackets and simplify:

   $$24p - [2\{3(5p - q) - 2(p + 2q)\} + 3q]$$

4. Factorise $21a^2b^2 - 28ab$

5. Factorise $2xy^2 + 6x^2y + 8x^3y$

6. Simplify $2y + 4 \div 6y + 3 \times 4 - 5y$

7. Simplify $3 \div y + 2 \div y - 1$

8. Simplify $a^2 - 3ab \times 2a \div 6b + ab$

1.3   Revision of equations

(a) Simple equations

Problem 11. Solve $4 - 3x = 2x - 11$

Since $4 - 3x = 2x - 11$ then $4 + 11 = 2x + 3x$

i.e. $15 = 5x$ from which, $x = \frac{15}{5} = \mathbf{3}$

Problem 12. Solve

$$4(2a - 3) - 2(a - 4) = 3(a - 3) - 1$$

Removing the brackets gives:

$$8a - 12 - 2a + 8 = 3a - 9 - 1$$

Rearranging gives:

$$8a - 2a - 3a = -9 - 1 + 12 - 8$$

i.e. $\qquad 3a = -6$

and $\qquad a = \dfrac{-6}{3} = -2$

Problem 13. Solve $\dfrac{3}{x-2} = \dfrac{4}{3x+4}$

By 'cross-multiplying': $\quad 3(3x + 4) = 4(x - 2)$

Removing brackets gives: $\quad 9x + 12 = 4x - 8$

Rearranging gives: $\quad 9x - 4x = -8 - 12$

i.e. $\qquad\qquad\qquad\qquad 5x = -20$

and $\qquad\qquad\qquad\qquad x = \dfrac{-20}{5}$

$$= -4$$

Problem 14. Solve $\left(\dfrac{\sqrt{t} + 3}{\sqrt{t}} \right) = 2$

$$\sqrt{t} \left(\dfrac{\sqrt{t} + 3}{\sqrt{t}} \right) = 2\sqrt{t}$$

i.e. $\qquad \sqrt{t} + 3 = 2\sqrt{t}$

and $\qquad 3 = 2\sqrt{t} - \sqrt{t}$

i.e. $\qquad 3 = \sqrt{t}$

and $\qquad \mathbf{9 = t}$

(c) Transposition of formulae

Problem 15. Transpose the formula $v = u + \dfrac{ft}{m}$ to make f the subject.

$u + \dfrac{ft}{m} = v$ from which, $\dfrac{ft}{m} = v - u$

and $\qquad m \left(\dfrac{ft}{m} \right) = m(v - u)$

i.e. $\qquad ft = m(v - u)$

and $\qquad f = \dfrac{m}{t}(v - u)$

Problem 16. The impedance of an a.c. circuit is given by $Z = \sqrt{R^2 + X^2}$. Make the reactance X the subject.

$\sqrt{R^2 + X^2} = Z$ and squaring both sides gives

$R^2 + X^2 = Z^2$, from which,

$$X^2 = Z^2 - R^2 \text{ and } \textbf{reactance } X = \sqrt{Z^2 - R^2}$$

Problem 17. Given that $\dfrac{D}{d} = \sqrt{\left(\dfrac{f+p}{f-p} \right)}$ express p in terms of D, d and f.

Rearranging gives: $\qquad \sqrt{\left(\dfrac{f+p}{f-p} \right)} = \dfrac{D}{d}$

Squaring both sides gives: $\qquad \dfrac{f+p}{f-p} = \dfrac{D^2}{d^2}$

'Cross-multiplying' gives:

$$d^2(f + p) = D^2(f - p)$$

Removing brackets gives:

$$d^2 f + d^2 p = D^2 f - D^2 p$$

Rearranging gives: $\quad d^2 p + D^2 p = D^2 f - d^2 f$

Factorising gives: $\quad p(d^2 + D^2) = f(D^2 - d^2)$

and $\qquad\qquad p = \dfrac{f(D^2 - d^2)}{(d^2 + D^2)}$

Now try the following Practice Exercise

Practice Exercise 3   Simple equations and transposition of formulae (Answers on page 856)

In problems 1 to 4 solve the equations

1. $3x - 2 - 5x = 2x - 4$

2. $8 + 4(x - 1) - 5(x - 3) = 2(5 - 2x)$

3. $\dfrac{1}{3a - 2} + \dfrac{1}{5a + 3} = 0$

4. $\dfrac{3\sqrt{t}}{1 - \sqrt{t}} = -6$

5. Transpose $y = \dfrac{3(F - f)}{L}$ for f

6. Make l the subject of $t = 2\pi\sqrt{\dfrac{l}{g}}$

7. Transpose $m = \dfrac{\mu L}{L + rCR}$ for L

8. Make r the subject of the formula
$$\frac{x}{y} = \frac{1 + r^2}{1 - r^2}$$

(d) Simultaneous equations

Problem 18. Solve the simultaneous equations:

$$7x - 2y = 26 \qquad (1)$$

$$6x + 5y = 29 \qquad (2)$$

$5 \times$ equation (1) gives:

$$35x - 10y = 130 \qquad (3)$$

$2 \times$ equation (2) gives:

$$12x + 10y = 58 \qquad (4)$$

Equation (3) + equation (4) gives:

$$47x + 0 = 188$$

from which, $\qquad x = \dfrac{188}{47} = 4$

Substituting $x = 4$ in equation (1) gives:

$$28 - 2y = 26$$

from which, $28 - 26 = 2y$ and $y = 1$

Problem 19. Solve

$$\frac{x}{8} + \frac{5}{2} = y \qquad (1)$$

$$11 + \frac{y}{3} = 3x \qquad (2)$$

$8 \times$ equation (1) gives: $\qquad x + 20 = 8y \qquad (3)$

$3 \times$ equation (2) gives: $\qquad 33 + y = 9x \qquad (4)$

i.e. $\qquad\qquad\qquad\qquad x - 8y = -20 \qquad (5)$

and $\qquad\qquad\qquad\qquad 9x - y = 33 \qquad (6)$

$8 \times$ equation (6) gives: $\quad 72x - 8y = 264 \qquad (7)$

Equation (7) − equation (5) gives:

$$71x = 284$$

from which, $\qquad\qquad x = \dfrac{284}{71} = 4$

Substituting $x = 4$ in equation (5) gives:

$$4 - 8y = -20$$

from which, $\qquad 4 + 20 = 8y$ and $y = 3$

(e) Quadratic equations

Problem 20. Solve the following equations by factorisation:
(a) $3x^2 - 11x - 4 = 0$

(b) $4x^2 + 8x + 3 = 0$

(a) The factors of $3x^2$ are $3x$ and x and these are placed in brackets thus:

$$(3x \quad)(x \quad)$$

The factors of -4 are $+1$ and -4 or -1 and $+4$, or -2 and $+2$. Remembering that the product of the two inner terms added to the product of the two outer terms must equal $-11x$, the only combination to give this is $+1$ and -4, i.e.,

$$3x^2 - 11x - 4 = (3x + 1)(x - 4)$$

Thus $\quad (3x + 1)(x - 4) = 0$ hence

either $\qquad (3x + 1) = 0$ i.e. $x = -\dfrac{1}{3}$

or $\qquad (x - 4) = 0$ i.e. $x = 4$

(b) $4x^2 + 8x + 3 = (2x + 3)(2x + 1)$

Thus $\quad (2x + 3)(2x + 1) = 0$ hence

either $\qquad (2x + 3) = 0$ i.e. $x = -\dfrac{3}{2}$

or $\qquad (2x + 1) = 0$ i.e. $x = -\dfrac{1}{2}$

Problem 21. The roots of a quadratic equation are $\frac{1}{3}$ and -2. Determine the equation in x.

If $\dfrac{1}{3}$ and -2 are the roots of a quadratic equation then,

$$(x - \frac{1}{3})(x + 2) = 0$$

i.e. $\quad x^2 + 2x - \dfrac{1}{3}x - \dfrac{2}{3} = 0$

i.e. $\qquad x^2 + \dfrac{5}{3}x - \dfrac{2}{3} = 0$

or $\qquad 3x^2 + 5x - 2 = 0$

Problem 22. Solve $4x^2 + 7x + 2 = 0$ giving the answer correct to 2 decimal places.

From the quadratic formula if $ax^2 + bx + c = 0$ then,

$$x = \frac{-b \pm \sqrt{b^2 - 4ac}}{2a}$$

Hence if $4x^2 + 7x + 2 = 0$

then $x = \dfrac{-7 \pm \sqrt{7^2 - 4(4)(2)}}{2(4)}$

$\quad = \dfrac{-7 \pm \sqrt{17}}{8}$

$\quad = \dfrac{-7 \pm 4.123}{8}$

$\quad = \dfrac{-7 + 4.123}{8}$ or $\dfrac{-7 - 4.123}{8}$

i.e. $\quad x = -0.36$ or -1.39

Now try the following Practice Exercise

Practice Exercise 4   Simultaneous and quadratic equations (Answers on page 856)

In problems 1 to 3, solve the simultaneous equations

1. $8x - 3y = 51$

$\quad 3x + 4y = 14$

2. $5a = 1 - 3b$

$\quad 2b + a + 4 = 0$

3. $\dfrac{x}{5} + \dfrac{2y}{3} = \dfrac{49}{15}$

$\quad \dfrac{3x}{7} - \dfrac{y}{2} + \dfrac{5}{7} = 0$

4. Solve the following quadratic equations by factorisation:

(a) $x^2 + 4x - 32 = 0$

(b) $8x^2 + 2x - 15 = 0$

5. Determine the quadratic equation in x whose roots are 2 and -5

6. Solve the following quadratic equations, correct to 3 decimal places:

(a) $2x^2 + 5x - 4 = 0$

(b) $4t^2 - 11t + 3 = 0$

1.4 Polynomial division

Before looking at long division in algebra let us revise long division with numbers (we may have forgotten, since calculators do the job for us!).

For example, $\dfrac{208}{16}$ is achieved as follows:

$$
\begin{array}{r}
13 \\
16 \overline{)\ 208} \\
16 \\
\hline
48 \\
48 \\
\hline
\cdot\,\cdot \\
\hline
\end{array}
$$

(1)   16 divided into 2 won't go

(2)   16 divided into 20 goes 1

(3)   Put 1 above the zero

(4)   Multiply 16 by 1 giving 16

(5)   Subtract 16 from 20 giving 4

(6)   Bring down the 8

(7)   16 divided into 48 goes 3 times

(8)   Put the 3 above the 8

(9) $3 \times 16 = 48$

(10) $48 - 48 = 0$

Hence $\dfrac{208}{16} = \mathbf{13}$ exactly

Similarly, $\dfrac{172}{15}$ is laid out as follows:

$$
\begin{array}{r}
11 \\
15 \overline{)\ 172} \\
15 \\
\hline
22 \\
15 \\
\hline
7 \\
\hline
\end{array}
$$

Hence $\dfrac{172}{15} = 11$ remainder 7 or $11 + \dfrac{7}{15} = \mathbf{11\dfrac{7}{15}}$

Below are some examples of division in algebra, which in some respects is similar to long division with numbers.

(Note that a **polynomial** is an expression of the form

$$f(x) = a + bx + cx^2 + dx^3 + \cdots$$

and **polynomial division** is sometimes required when resolving into partial fractions – see Chapter 2.)

Problem 23.   Divide $2x^2 + x - 3$ by $x - 1$

$2x^2 + x - 3$ is called the **dividend** and $x - 1$ the **divisor**. The usual layout is shown below with the dividend and divisor both arranged in descending powers of the symbols.

$$
\begin{array}{r}
2x + 3 \\
x - 1 \overline{\smash{\big)}\, 2x^2 + x - 3} \\
\underline{2x^2 - 2x} \\
3x - 3 \\
\underline{3x - 3} \\
\cdot \quad \cdot
\end{array}
$$

Dividing the first term of the dividend by the first term of the divisor, i.e. $\dfrac{2x^2}{x}$ gives $2x$, which is put above the first term of the dividend as shown. The divisor is then multiplied by $2x$, i.e. $2x(x - 1) = 2x^2 - 2x$, which is placed under the dividend as shown. Subtracting gives $3x - 3$. The process is then repeated, i.e. the first term of the divisor, x, is divided into $3x$, giving $+3$, which is placed above the dividend as shown. Then $3(x - 1) = 3x - 3$, which is placed under the $3x - 3$. The remainder, on subtraction, is zero, which completes the process.

Thus $\mathbf{(2x^2 + x - 3) \div (x - 1) = (2x + 3)}$

[A check can be made on this answer by multiplying $(2x + 3)$ by $(x - 1)$ which equals $2x^2 + x - 3$.]

Problem 24.   Divide $3x^3 + x^2 + 3x + 5$ by $x + 1$

$$
\begin{array}{r}
(1) \quad (4) \quad (7) \\
3x^2 - 2x \ + 5 \\
x + 1 \overline{\smash{\big)}\, 3x^3 + \ x^2 + 3x + 5} \\
\underline{3x^3 + 3x^2} \\
-2x^2 + 3x + 5 \\
\underline{-2x^2 - 2x} \\
5x + 5 \\
\underline{5x + 5} \\
\cdot \quad \cdot
\end{array}
$$

(1) x into $3x^3$ goes $3x^2$. Put $3x^2$ above $3x^3$

(2) $3x^2(x + 1) = 3x^3 + 3x^2$

(3)   Subtract

(4) x into $-2x^2$ goes $-2x$. Put $-2x$ above the dividend

(5) $-2x(x + 1) = -2x^2 - 2x$

(6)   Subtract

(7) x into $5x$ goes 5. Put 5 above the dividend

(8) $5(x + 1) = 5x + 5$

(9)   Subtract

Thus $\dfrac{3x^3 + x^2 + 3x + 5}{x + 1} = \mathbf{3x^2 - 2x + 5}$

Problem 25.   Simplify $\dfrac{x^3 + y^3}{x + y}$

$$
\begin{array}{r}
(1) \quad (4) \quad (7) \\
x^2 - \ xy \ + y^2 \\
x + y \overline{\smash{\big)}\, x^3 + \ 0 \ + \ 0 \ + y^3} \\
\underline{x^3 + x^2 y} \\
-x^2 y \qquad + y^3 \\
\underline{-x^2 y - xy^2} \\
xy^2 + y^3 \\
\underline{xy^2 + y^3} \\
\cdot \quad \cdot
\end{array}
$$

(1) x into x^3 goes x^2. Put x^2 above x^3 of dividend

(2) $x^2(x + y) = x^3 + x^2 y$

(3)   Subtract

(4) x into $-x^2 y$ goes $-xy$. Put $-xy$ above dividend

(5) $-xy(x + y) = -x^2 y - xy^2$

(6)   Subtract

(7) x into xy^2 goes y^2. Put y^2 above dividend

(8) $y^2(x + y) = xy^2 + y^3$

(9)   Subtract

Thus

$$
\frac{x^3 + y^3}{x + y} = \mathbf{x^2 - xy + y^2}
$$

The zeros shown in the dividend are not normally shown, but are included to clarify the subtraction process and to keep similar terms in their respective columns.

Problem 26. Divide $(x^2 + 3x - 2)$ by $(x - 2)$

$$
\begin{array}{r}
x + 5 \\
x - 2 \overline{\smash{)}\ x^2 + 3x - 2} \\
\underline{x^2 - 2x} \\
5x - 2 \\
\underline{5x - 10} \\
8
\end{array}
$$

Hence

$$\frac{x^2 + 3x - 2}{x - 2} = x + 5 + \frac{8}{x - 2}$$

Problem 27. Divide $4a^3 - 6a^2b + 5b^3$ by $2a - b$

$$
\begin{array}{r}
2a^2 - 2ab - b^2 \\
2a - b \overline{\smash{)}\ 4a^3 - 6a^2b \qquad + 5b^3} \\
\underline{4a^3 - 2a^2b} \\
-4a^2b \qquad + 5b^3 \\
\underline{-4a^2b + 2ab^2} \\
-2ab^2 + 5b^3 \\
\underline{-2ab^2 + b^3} \\
4b^3
\end{array}
$$

Thus

$$\frac{4a^3 - 6a^2b + 5b^3}{2a - b}$$

$$= 2a^2 - 2ab - b^2 + \frac{4b^3}{2a - b}$$

Now try the following Practice Exercise

Practice Exercise 5   Polynomial division (Answers on page 856)

1. Divide $(2x^2 + xy - y^2)$ by $(x + y)$

2. Divide $(3x^2 + 5x - 2)$ by $(x + 2)$

3. Determine $(10x^2 + 11x - 6) \div (2x + 3)$

4. Find $\dfrac{14x^2 - 19x - 3}{2x - 3}$

5. Divide $(x^3 + 3x^2y + 3xy^2 + y^3)$ by $(x + y)$

6. Find $(5x^2 - x + 4) \div (x - 1)$

7. Divide $(3x^3 + 2x^2 - 5x + 4)$ by $(x + 2)$

8. Determine $(5x^4 + 3x^3 - 2x + 1)/(x - 3)$

1.5   The factor theorem

There is a simple relationship between the factors of a quadratic expression and the roots of the equation obtained by equating the expression to zero.
For example, consider the quadratic equation $x^2 + 2x - 8 = 0$
To solve this we may factorise the quadratic expression $x^2 + 2x - 8$ giving $(x - 2)(x + 4)$
Hence $(x - 2)(x + 4) = 0$
Then, if the product of two numbers is zero, one or both of those numbers must equal zero. Therefore,

either $(x - 2) = 0$, from which, $x = 2$

or $\quad (x + 4) = 0$, from which, $x = -4$

It is clear, then, that a factor of $(x - 2)$ indicates a root of $+2$, while a factor of $(x + 4)$ indicates a root of -4
In general, we can therefore say that:

a factor of $(x - a)$ corresponds to a root of $x = a$

In practice, we always deduce the roots of a simple quadratic equation from the factors of the quadratic expression, as in the above example. However, we could reverse this process. If, by trial and error, we could determine that $x = 2$ is a root of the equation $x^2 + 2x - 8 = 0$ we could deduce at once that $(x - 2)$ is a factor of the expression $x^2 + 2x - 8$. We wouldn't normally solve quadratic equations this way – but suppose we have to factorise a cubic expression (i.e. one in which the highest power of the variable is 3). A cubic equation might have three simple linear factors and the difficulty of discovering all these factors by trial and error would be considerable. It is to deal with this kind of case that we use the **factor theorem**. This is just a generalised version of what we established above for the quadratic expression. The factor theorem provides a method of factorising any polynomial, $f(x)$, which has simple factors.
A statement of the **factor theorem** says:

'if $x = a$ is a root of the equation

$$f(x) = 0, \text{ then } (x - a) \text{ is a factor of } f(x)'$$

The following worked problems show the use of the factor theorem.

Problem 28. Factorise $x^3 - 7x - 6$ and use it to solve the cubic equation $x^3 - 7x - 6 = 0$.

Let $f(x) = x^3 - 7x - 6$

If $x = 1$, then $f(1) = 1^3 - 7(1) - 6 = -12$

If $x = 2$, then $f(2) = 2^3 - 7(2) - 6 = -12$

If $x = 3$, then $f(3) = 3^3 - 7(3) - 6 = 0$

If $f(3) = 0$, then $(x - 3)$ is a factor – from the factor theorem.

We have a choice now. We can divide $x^3 - 7x - 6$ by $(x - 3)$ or we could continue our 'trial and error' by substituting further values for x in the given expression – and hope to arrive at $f(x) = 0$

Let us do both ways. Firstly, dividing out gives:

$$
\begin{array}{r}
x^2 + 3x + 2 \\
x - 3 \overline{)\ x^3 - 0\ \ - 7x\ - 6} \\
\underline{x^3 - 3x^2} \\
3x^2 - 7x - 6 \\
\underline{3x^2 - 9x} \\
2x - 6 \\
\underline{2x - 6} \\
\cdot\ \ \cdot
\end{array}
$$

Hence $\dfrac{x^3 - 7x - 6}{x - 3} = x^2 + 3x + 2$

i.e. $x^3 - 7x - 6 = (x - 3)(x^2 + 3x + 2)$

$x^2 + 3x + 2$ factorises 'on sight' as $(x + 1)(x + 2)$.

Therefore

$$x^3 - 7x - 6 = (x - 3)(x + 1)(x + 2)$$

A second method is to continue to substitute values of x into $f(x)$.

Our expression for $f(3)$ was $3^3 - 7(3) - 6$. We can see that if we continue with positive values of x the first term will predominate such that $f(x)$ will not be zero.

Therefore let us try some negative values for x. Therefore $f(-1) = (-1)^3 - 7(-1) - 6 = 0$; hence $(x + 1)$ is a factor (as shown above). Also $f(-2) = (-2)^3 - 7(-2) - 6 = 0$; hence $(x + 2)$ is a factor (also as shown above).

To solve $x^3 - 7x - 6 = 0$, we substitute the factors, i.e.

$$(x - 3)(x + 1)(x + 2) = 0$$

from which, $x = 3$, $x = -1$ and $x = -2$

Note that the values of x, i.e. 3, -1 and -2, are all factors of the constant term, i.e. 6. This can give us a clue as to what values of x we should consider.

Problem 29. Solve the cubic equation $x^3 - 2x^2 - 5x + 6 = 0$ by using the factor theorem.

Let $f(x) = x^3 - 2x^2 - 5x + 6$ and let us substitute simple values of x like 1, 2, 3, -1, -2, and so on.

$$f(1) = 1^3 - 2(1)^2 - 5(1) + 6 = 0,$$
$$\text{hence } (x - 1) \text{ is a factor}$$

$$f(2) = 2^3 - 2(2)^2 - 5(2) + 6 \neq 0$$
$$f(3) = 3^3 - 2(3)^2 - 5(3) + 6 = 0,$$
$$\text{hence } (x - 3) \text{ is a factor}$$

$$f(-1) = (-1)^3 - 2(-1)^2 - 5(-1) + 6 \neq 0$$
$$f(-2) = (-2)^3 - 2(-2)^2 - 5(-2) + 6 = 0,$$
$$\text{hence } (x + 2) \text{ is a factor}$$

Hence $x^3 - 2x^2 - 5x + 6 = (x - 1)(x - 3)(x + 2)$

Therefore if $x^3 - 2x^2 - 5x + 6 = 0$

then $(x - 1)(x - 3)(x + 2) = 0$

from which, $x = 1$, $x = 3$ and $x = -2$

Alternatively, having obtained one factor, i.e. $(x - 1)$ we could divide this into $(x^3 - 2x^2 - 5x + 6)$ as follows:

$$
\begin{array}{r}
x^2 - x - 6 \\
x - 1 \overline{)\ x^3 - 2x^2 - 5x + 6} \\
\underline{x^3 - x^2} \\
-x^2 - 5x + 6 \\
\underline{-x^2 + x} \\
-6x + 6 \\
\underline{-6x + 6} \\
\cdot\ \ \cdot
\end{array}
$$

Hence $x^3 - 2x^2 - 5x + 6$

$$= (x - 1)(x^2 - x - 6)$$

$$= (x - 1)(x - 3)(x + 2)$$

Summarising, the factor theorem provides us with a method of factorising simple expressions, and an alternative, in certain circumstances, to polynomial division.

Now try the following Practice Exercise

> **Practice Exercise 6   The factor theorem (Answers on page 856)**
>
> Use the factor theorem to factorise the expressions given in problems 1 to 4.
>
> 1. $x^2 + 2x - 3$
>
> 2. $x^3 + x^2 - 4x - 4$
>
> 3. $2x^3 + 5x^2 - 4x - 7$
>
> 4. $2x^3 - x^2 - 16x + 15$
>
> 5. Use the factor theorem to factorise $x^3 + 4x^2 + x - 6$ and hence solve the cubic equation $x^3 + 4x^2 + x - 6 = 0$
>
> 6. Solve the equation $x^3 - 2x^2 - x + 2 = 0$

1.6 The remainder theorem

Dividing a general quadratic expression $(ax^2 + bx + c)$ by $(x - p)$, where p is any whole number, by long division (see Section 1.4) gives:

$$
\begin{array}{r}
ax + (b + ap) \\
x - p \overline{\smash{)}\; ax^2 + bx \qquad\quad + c} \\
\underline{ax^2 - apx} \\
(b + ap)x + c \\
\underline{(b + ap)x - (b + ap)p} \\
c + (b + ap)p
\end{array}
$$

The remainder, $c + (b + ap)p = c + bp + ap^2$ or $ap^2 + bp + c$. This is, in fact, what the **remainder theorem** states, i.e.

> **'if $(ax^2 + bx + c)$ is divided by $(x - p)$, the remainder will be $ap^2 + bp + c$'**

If, in the dividend $(ax^2 + bx + c)$, we substitute p for x we get the remainder $ap^2 + bp + c$

For example, when $(3x^2 - 4x + 5)$ is divided by $(x - 2)$ the remainder is $ap^2 + bp + c$ (where $a = 3$, $b = -4$, $c = 5$ and $p = 2$),

i.e. the remainder is

$$3(2)^2 + (-4)(2) + 5 = 12 - 8 + 5 = \mathbf{9}$$

We can check this by dividing $(3x^2 - 4x + 5)$ by $(x - 2)$ by long division:

$$
\begin{array}{r}
3x + 2 \\
x - 2 \overline{\smash{)}\; 3x^2 - 4x + 5} \\
\underline{3x^2 - 6x} \\
2x + 5 \\
\underline{2x - 4} \\
9
\end{array}
$$

Similarly, when $(4x^2 - 7x + 9)$ is divided by $(x + 3)$, the remainder is $ap^2 + bp + c$ (where $a = 4$, $b = -7$, $c = 9$ and $p = -3$), i.e. the remainder is
$4(-3)^2 + (-7)(-3) + 9 = 36 + 21 + 9 = \mathbf{66}$
Also, when $(x^2 + 3x - 2)$ is divided by $(x - 1)$, the remainder is $1(1)^2 + 3(1) - 2 = \mathbf{2}$
It is not particularly useful, on its own, to know the remainder of an algebraic division. However, if the remainder should be zero then $(x - p)$ is a factor. This is very useful therefore when factorising expressions.
For example, when $(2x^2 + x - 3)$ is divided by $(x - 1)$, the remainder is $2(1)^2 + 1(1) - 3 = 0$, which means that $(x - 1)$ is a factor of $(2x^2 + x - 3)$.
In this case the other factor is $(2x + 3)$, i.e.

$$(2x^2 + x - 3) = (x - 1)(2x - 3)$$

The **remainder theorem** may also be stated for a **cubic equation** as:

> **'if $(ax^3 + bx^2 + cx + d)$ is divided by $(x - p)$, the remainder will be $ap^3 + bp^2 + cp + d$'**

As before, the remainder may be obtained by substituting p for x in the dividend.
For example, when $(3x^3 + 2x^2 - x + 4)$ is divided by $(x - 1)$, the remainder is $ap^3 + bp^2 + cp + d$ (where $a = 3$, $b = 2$, $c = -1$, $d = 4$ and $p = 1$), i.e. the remainder is $3(1)^3 + 2(1)^2 + (-1)(1) + 4 = 3 + 2 - 1 + 4 = \mathbf{8}$
Similarly, when $(x^3 - 7x - 6)$ is divided by $(x - 3)$, the remainder is $1(3)^3 + 0(3)^2 - 7(3) - 6 = 0$, which means that $(x - 3)$ is a factor of $(x^3 - 7x - 6)$
Here are some more examples on the remainder theorem.

> **Problem 30.** Without dividing out, find the remainder when $2x^2 - 3x + 4$ is divided by $(x - 2)$

By the remainder theorem, the remainder is given by $ap^2 + bp + c$, where $a = 2$, $b = -3$, $c = 4$ and $p = 2$. Hence **the remainder is:**

$$2(2)^2 + (-3)(2) + 4 = 8 - 6 + 4 = \mathbf{6}$$

Problem 31. Use the remainder theorem to determine the remainder when $(3x^3 - 2x^2 + x - 5)$ is divided by $(x + 2)$

By the remainder theorem, the remainder is given by $ap^3 + bp^2 + cp + d$, where $a = 3$, $b = -2$, $c = 1$, $d = -5$ and $p = -2$
Hence **the remainder is:**

$$3(-2)^3 + (-2)(-2)^2 + (1)(-2) + (-5)$$
$$= -24 - 8 - 2 - 5$$
$$= \mathbf{-39}$$

Problem 32. Determine the remainder when $(x^3 - 2x^2 - 5x + 6)$ is divided by (a) $(x - 1)$ and (b) $(x + 2)$. Hence factorise the cubic expression.

(a) When $(x^3 - 2x^2 - 5x + 6)$ is divided by $(x - 1)$, the remainder is given by $ap^3 + bp^2 + cp + d$, where $a = 1$, $b = -2$, $c = -5$, $d = 6$ and $p = 1$,

i.e. **the remainder** $= (1)(1)^3 + (-2)(1)^2$
$$+ (-5)(1) + 6$$
$$= 1 - 2 - 5 + 6 = \mathbf{0}$$

Hence $(x - 1)$ is a factor of $(x^3 - 2x^2 - 5x + 6)$.

(b) When $(x^3 - 2x^2 - 5x + 6)$ is divided by $(x + 2)$, **the remainder is** given by

$$(1)(-2)^3 + (-2)(-2)^2 + (-5)(-2) + 6$$
$$= -8 - 8 + 10 + 6 = \mathbf{0}$$

Hence $(x + 2)$ is also a factor of $(x^3 - 2x^2 - 5x + 6)$. Therefore $(x - 1)(x + 2)(x \quad) = x^3 - 2x^2 - 5x + 6$. To determine the third factor (shown blank) we could

 (i) divide $(x^3 - 2x^2 - 5x + 6)$ by $(x - 1)(x + 2)$
or (ii) use the factor theorem where $f(x) = x^3 - 2x^2 - 5x + 6$ and hoping to choose a value of x which makes $f(x) = 0$
or (iii) use the remainder theorem, again hoping to choose a factor $(x - p)$ which makes the remainder zero.

(i) Dividing $(x^3 - 2x^2 - 5x + 6)$ by $(x^2 + x - 2)$ gives:

$$
\begin{array}{r}
x - 3 \\
x^2 + x - 2 \overline{\smash{)}\; x^3 - 2x^2 - 5x + 6} \\
\underline{x^3 + x^2 - 2x} \\
-3x^2 - 3x + 6 \\
\underline{-3x^2 - 3x + 6} \\
\cdot \qquad \cdot \qquad \cdot
\end{array}
$$

Thus $(x^3 - 2x^2 - 5x + 6)$
$$= (x - 1)(x + 2)(x - 3)$$

(ii) Using the factor theorem, we let
$$f(x) = x^3 - 2x^2 - 5x + 6$$

Then $f(3) = 3^3 - 2(3)^2 - 5(3) + 6$
$$= 27 - 18 - 15 + 6 = 0$$

Hence $(x - 3)$ is a factor.

(iii) Using the remainder theorem, when $(x^3 - 2x^2 - 5x + 6)$ is divided by $(x - 3)$, the remainder is given by $ap^3 + bp^2 + cp + d$, where $a = 1$, $b = -2$, $c = -5$, $d = 6$ and $p = 3$
Hence the remainder is:

$$1(3)^3 + (-2)(3)^2 + (-5)(3) + 6$$
$$= 27 - 18 - 15 + 6 = 0$$

Hence $(x - 3)$ is a factor.

Thus $(x^3 - 2x^2 - 5x + 6)$
$$= (x - 1)(x + 2)(x - 3)$$

Now try the following Practice Exercise

Practice Exercise 7   The remainder theorem (Answers on page 856)

1. Find the remainder when $3x^2 - 4x + 2$ is divided by

   (a) $(x - 2)$ (b) $(x + 1)$

2. Determine the remainder when $x^3 - 6x^2 + x - 5$ is divided by

   (a) $(x + 2)$ (b) $(x - 3)$

3. Use the remainder theorem to find the factors of $x^3 - 6x^2 + 11x - 6$

4. Determine the factors of $x^3 + 7x^2 + 14x + 8$ and hence solve the cubic equation $x^3 + 7x^2 + 14x + 8 = 0$

5. Determine the value of 'a' if $(x+2)$ is a factor of $(x^3 - ax^2 + 7x + 10)$

6. Using the remainder theorem, solve the equation $2x^3 - x^2 - 7x + 6 = 0$

For more help on basic algebra, simple equations, transposition of formulae, simultaneous equations and quadratic equations, go to the website:
www.routledge.com/cw/bird

For fully worked solutions to each of the problems in Practice Exercises 1 to 17 in this chapter, go to website:
www.routledge.com/cw/bird

Chapter 2

Partial fractions

Why it is important to understand: **Partial fractions**

The algebraic technique of resolving a complicated fraction into partial fractions is often needed by electrical and mechanical engineers for not only determining certain integrals in calculus, but for determining inverse Laplace transforms, and for analysing linear differential equations with resonant circuits and feedback control systems.

At the end of this chapter, you should be able to:

- understand the term 'partial fraction'
- appreciate the conditions needed to resolve a fraction into partial fractions
- resolve into partial fractions a fraction containing linear factors in the denominator
- resolve into partial fractions a fraction containing repeated linear factors in the denominator
- resolve into partial fractions a fraction containing quadratic factors in the denominator

2.1 Introduction to partial fractions

By algebraic addition,

$$\frac{1}{x-2} + \frac{3}{x+1} = \frac{(x+1)+3(x-2)}{(x-2)(x+1)}$$

$$= \frac{4x-5}{x^2-x-2}$$

The reverse process of moving from $\dfrac{4x-5}{x^2-x-2}$

to $\dfrac{1}{x-2} + \dfrac{3}{x+1}$ is called resolving into **partial fractions**.

In order to resolve an algebraic expression into partial fractions:

(i) the denominator must factorise (in the above example, x^2-x-2 factorises as $(x-2)(x+1)$), and

(ii) the numerator must be at least one degree less than the denominator (in the above example $(4x-5)$ is of degree 1 since the highest powered x term is x^1 and (x^2-x-2) is of degree 2).

When the degree of the numerator is equal to or higher than the degree of the denominator, the numerator must be divided by the denominator until the remainder is of less degree than the denominator (see Problems 3 and 4). There are basically three types of partial fraction and the form of partial fraction used is summarised in Table 2.1, where $f(x)$ is assumed to be of less degree than the relevant denominator and A, B and C are constants to be determined.

(In the latter type in Table 2.1, ax^2+bx+c is a quadratic expression which does not factorise without containing surds or imaginary terms.)

Resolving an algebraic expression into partial fractions is used as a preliminary to integrating certain functions (see Chapter 44) and in determining inverse Laplace transforms (see Chapter 69).

Table 2.1

Type	Denominator containing	Expression	Form of partial fraction
1	Linear factors (see Problems 1 to 4)	$\dfrac{f(x)}{(x+a)(x-b)(x+c)}$	$\dfrac{A}{(x+a)} + \dfrac{B}{(x-b)} + \dfrac{C}{(x+c)}$
2	Repeated linear factors (see Problems 5 to 7)	$\dfrac{f(x)}{(x+a)^3}$	$\dfrac{A}{(x+a)} + \dfrac{B}{(x+a)^2} + \dfrac{C}{(x+a)^3}$
3	Quadratic factors (see Problems 8 and 9)	$\dfrac{f(x)}{(ax^2+bx+c)(x+d)}$	$\dfrac{Ax+B}{(ax^2+bx+c)} + \dfrac{C}{(x+d)}$

2.2 Worked problems on partial fractions with linear factors

Problem 1. Resolve $\dfrac{11-3x}{x^2+2x-3}$ into partial fractions.

The denominator factorises as $(x-1)(x+3)$ and the numerator is of less degree than the denominator. Thus $\dfrac{11-3x}{x^2+2x-3}$ may be resolved into partial fractions.

Let $\dfrac{11-3x}{x^2+2x-3} \equiv \dfrac{11-3x}{(x-1)(x+3)}$

$$\equiv \dfrac{A}{(x-1)} + \dfrac{B}{(x+3)}$$

where A and B are constants to be determined,

i.e. $\dfrac{11-3x}{(x-1)(x+3)} \equiv \dfrac{A(x+3)+B(x-1)}{(x-1)(x+3)}$

by algebraic addition.

Since the denominators are the same on each side of the identity then the numerators are equal to each other.

Thus, $11-3x \equiv A(x+3)+B(x-1)$

To determine constants A and B, values of x are chosen to make the term in A or B equal to zero.

When $x=1$, then

$$11-3(1) \equiv A(1+3)+B(0)$$

i.e. $\qquad 8 = 4A$

i.e. $\qquad \boldsymbol{A=2}$

When $x=-3$, then

$$11-3(-3) \equiv A(0)+B(-3-1)$$

i.e. $\qquad 20 = -4B$

i.e. $\qquad \boldsymbol{B=-5}$

Thus $\dfrac{\boldsymbol{11-3x}}{\boldsymbol{x^2+2x-3}} \equiv \dfrac{2}{(x-1)} + \dfrac{-5}{(x+3)}$

$$\equiv \dfrac{\boldsymbol{2}}{\boldsymbol{(x-1)}} - \dfrac{\boldsymbol{5}}{\boldsymbol{(x+3)}}$$

$$\left[\text{Check: } \dfrac{2}{(x-1)} - \dfrac{5}{(x+3)} = \dfrac{2(x+3)-5(x-1)}{(x-1)(x+3)}\right.$$

$$\left. = \dfrac{11-3x}{x^2+2x-3}\right]$$

Problem 2. Convert $\dfrac{2x^2-9x-35}{(x+1)(x-2)(x+3)}$ into the sum of three partial fractions.

Let $\dfrac{2x^2-9x-35}{(x+1)(x-2)(x+3)}$

$$\equiv \dfrac{A}{(x+1)} + \dfrac{B}{(x-2)} + \dfrac{C}{(x+3)}$$

$$\equiv \dfrac{\left(\begin{array}{c} A(x-2)(x+3)+B(x+1)(x+3) \\ +C(x+1)(x-2) \end{array}\right)}{(x+1)(x-2)(x+3)}$$

by algebraic addition.

Equating the numerators gives:

$$2x^2-9x-35 \equiv A(x-2)(x+3)$$

$$+B(x+1)(x+3)+C(x+1)(x-2)$$

Let $x=-1$. Then

$$2(-1)^2-9(-1)-35 \equiv A(-3)(2)$$

$$+B(0)(2)+C(0)(-3)$$

i.e. $\qquad -24 = -6A$

i.e. $\qquad A = \dfrac{-24}{-6} = 4$

Let $x = 2$. Then

$$2(2)^2 - 9(2) - 35 \equiv A(0)(5) + B(3)(5) + C(3)(0)$$

i.e. $-45 = 15B$

i.e. $B = \dfrac{-45}{15} = -3$

Let $x = -3$. Then

$$2(-3)^2 - 9(-3) - 35 \equiv A(-5)(0) + B(-2)(0)$$
$$+ C(-2)(-5)$$

i.e. $10 = 10C$

i.e. $C = 1$

Thus $\dfrac{2x^2 - 9x - 35}{(x+1)(x-2)(x+3)}$

$$\equiv \frac{4}{(x+1)} - \frac{3}{(x-2)} + \frac{1}{(x+3)}$$

Problem 3.   Resolve $\dfrac{x^2+1}{x^2-3x+2}$ into partial fractions.

The denominator is of the same degree as the numerator. Thus dividing out gives:

$$
\begin{array}{r}
1 \\
x^2 - 3x + 2 \overline{)\, x^2 + 1} \\
\underline{x^2 - 3x + 2} \\
3x - 1
\end{array}
$$

For more on polynomial division, see Section 1.4, page 8.

Hence $\dfrac{x^2+1}{x^2-3x+2} \equiv 1 + \dfrac{3x-1}{x^2-3x+2}$

$$\equiv 1 + \frac{3x-1}{(x-1)(x-2)}$$

Let $\dfrac{3x-1}{(x-1)(x-2)} \equiv \dfrac{A}{(x-1)} + \dfrac{B}{(x-2)}$

$$\equiv \frac{A(x-2) + B(x-1)}{(x-1)(x-2)}$$

Equating numerators gives:
$$3x - 1 \equiv A(x-2) + B(x-1)$$

Let $x = 1$. Then $2 = -A$

i.e. $A = -2$

Let $x = 2$. Then $5 = B$

Hence $\dfrac{3x-1}{(x-1)(x-2)} \equiv \dfrac{-2}{(x-1)} + \dfrac{5}{(x-2)}$

Thus $\dfrac{x^2+1}{x^2-3x+2} \equiv 1 - \dfrac{2}{(x-1)} + \dfrac{5}{(x-2)}$

Problem 4.   Express $\dfrac{x^3 - 2x^2 - 4x - 4}{x^2 + x - 2}$ in partial fractions.

The numerator is of higher degree than the denominator. Thus dividing out gives:

$$
\begin{array}{r}
x - 3 \\
x^2 + x - 2 \overline{)\, x^3 - 2x^2 - 4x - 4} \\
\underline{x^3 + x^2 - 2x} \\
-3x^2 - 2x - 4 \\
\underline{-3x^2 - 3x + 6} \\
x - 10
\end{array}
$$

Thus $\dfrac{x^3 - 2x^2 - 4x - 4}{x^2 + x - 2} \equiv x - 3 + \dfrac{x - 10}{x^2 + x - 2}$

$$\equiv x - 3 + \frac{x-10}{(x+2)(x-1)}$$

Let $\dfrac{x-10}{(x+2)(x-1)} \equiv \dfrac{A}{(x+2)} + \dfrac{B}{(x-1)}$

$$\equiv \frac{A(x-1) + B(x+2)}{(x+2)(x-1)}$$

Equating the numerators gives:

$$x - 10 \equiv A(x-1) + B(x+2)$$

Let $x = -2$. Then $-12 = -3A$

i.e. $A = 4$

Let $x = 1$. Then $-9 = 3B$

i.e. $B = -3$

Hence $\dfrac{x-10}{(x+2)(x-1)} \equiv \dfrac{4}{(x+2)} - \dfrac{3}{(x-1)}$

Thus $\dfrac{x^3 - 2x^2 - 4x - 4}{x^2 + x - 2}$

$$\equiv x - 3 + \frac{4}{(x+2)} - \frac{3}{(x-1)}$$

Now try the following Practice Exercise

Practice Exercise 8   Partial fractions with linear factors (Answers on page 856)

Resolve the following into partial fractions.

1. $\dfrac{12}{x^2 - 9}$

2. $\dfrac{4(x - 4)}{x^2 - 2x - 3}$

3. $\dfrac{x^2 - 3x + 6}{x(x - 2)(x - 1)}$

4. $\dfrac{3(2x^2 - 8x - 1)}{(x + 4)(x + 1)(2x - 1)}$

5. $\dfrac{x^2 + 9x + 8}{x^2 + x - 6}$

6. $\dfrac{x^2 - x - 14}{x^2 - 2x - 3}$

7. $\dfrac{3x^3 - 2x^2 - 16x + 20}{(x - 2)(x + 2)}$

2.3 Worked problems on partial fractions with repeated linear factors

Problem 5. Resolve $\dfrac{2x + 3}{(x - 2)^2}$ into partial fractions.

The denominator contains a repeated linear factor, $(x - 2)^2$.

Let $\dfrac{2x + 3}{(x - 2)^2} \equiv \dfrac{A}{(x - 2)} + \dfrac{B}{(x - 2)^2}$

$\equiv \dfrac{A(x - 2) + B}{(x - 2)^2}$

Equating the numerators gives:

$$2x + 3 \equiv A(x - 2) + B$$

Let $x = 2$. Then $\qquad 7 = A(0) + B$

i.e. $\qquad \boldsymbol{B = 7}$

$2x + 3 \equiv A(x - 2) + B \equiv Ax - 2A + B$

Since an identity is true for all values of the unknown, the coefficients of similar terms may be equated.
Hence, equating the coefficients of x gives: $\boldsymbol{2 = A}$

[Also, as a check, equating the constant terms gives:

$$3 = -2A + B$$

When $A = 2$ and $B = 7$,

$$\text{RHS} = -2(2) + 7 = 3 = \text{LHS}]$$

Hence $\dfrac{\boldsymbol{2x + 3}}{\boldsymbol{(x - 2)^2}} \equiv \dfrac{\boldsymbol{2}}{\boldsymbol{(x - 2)}} + \dfrac{\boldsymbol{7}}{\boldsymbol{(x - 2)^2}}$

Problem 6. Express $\dfrac{5x^2 - 2x - 19}{(x + 3)(x - 1)^2}$ as the sum of three partial fractions.

The denominator is a combination of a linear factor and a repeated linear factor.

Let $\dfrac{5x^2 - 2x - 19}{(x + 3)(x - 1)^2}$

$\equiv \dfrac{A}{(x + 3)} + \dfrac{B}{(x - 1)} + \dfrac{C}{(x - 1)^2}$

$\equiv \dfrac{A(x - 1)^2 + B(x + 3)(x - 1) + C(x + 3)}{(x + 3)(x - 1)^2}$

by algebraic addition.
Equating the numerators gives:

$$5x^2 - 2x - 19 \equiv A(x - 1)^2 + B(x + 3)(x - 1)$$
$$+ C(x + 3) \qquad (1)$$

Let $x = -3$. Then

$5(-3)^2 - 2(-3) - 19 \equiv A(-4)^2 + B(0)(-4) + C(0)$

i.e. $32 = 16A$

i.e. $\boldsymbol{A = 2}$

Let $x = 1$. Then

$5(1)^2 - 2(1) - 19 \equiv A(0)^2 + B(4)(0) + C(4)$

i.e. $-16 = 4C$

i.e. $\boldsymbol{C = -4}$

Without expanding the RHS of equation (1) it can be seen that equating the coefficients of x^2 gives: $5 = A + B$, and since $A = 2$, $\boldsymbol{B = 3}$

[Check: Identity (1) may be expressed as:

$5x^2 - 2x - 19 \equiv A(x^2 - 2x + 1)$
$$+ B(x^2 + 2x - 3) + C(x + 3)$$

i.e. $5x^2 - 2x - 19 \equiv Ax^2 - 2Ax + A + Bx^2 + 2Bx$
$$- 3B + Cx + 3C$$

Equating the x term coefficients gives:

$-2 \equiv -2A + 2B + C$

When $A = 2$, $B = 3$ and $C = -4$ then

$-2A + 2B + C = -2(2) + 2(3) - 4$
$$= -2 = \text{LHS}$$

Equating the constant term gives:

$-19 \equiv A - 3B + 3C$

$\text{RHS} = 2 - 3(3) + 3(-4) = 2 - 9 - 12$
$$= -19 = \text{LHS}]$$

Hence $\dfrac{5x^2 - 2x - 19}{(x+3)(x-1)^2}$

$$\equiv \frac{2}{(x+3)} + \frac{3}{(x-1)} - \frac{4}{(x-1)^2}$$

Problem 7. Resolve $\dfrac{3x^2 + 16x + 15}{(x+3)^3}$ into partial fractions.

Let $\dfrac{3x^2 + 16x + 15}{(x+3)^3}$

$$\equiv \frac{A}{(x+3)} + \frac{B}{(x+3)^2} + \frac{C}{(x+3)^3}$$

$$\equiv \frac{A(x+3)^2 + B(x+3) + C}{(x+3)^3}$$

Equating the numerators gives:

$3x^2 + 16x + 15 \equiv A(x+3)^2 + B(x+3) + C$ (1)

Let $x = -3$. Then

$3(-3)^2 + 16(-3) + 15 \equiv A(0)^2 + B(0) + C$

i.e. $\boldsymbol{-6 = C}$

Identity (1) may be expanded as:

$3x^2 + 16x + 15 \equiv A(x^2 + 6x + 9) + B(x + 3) + C$

i.e. $3x^2 + 16x + 15 \equiv Ax^2 + 6Ax + 9A$
$$+ Bx + 3B + C$$

Equating the coefficients of x^2 terms gives: $\boldsymbol{3 = A}$

Equating the coefficients of x terms gives:

$16 = 6A + B$

Since $A = 3$, $\boldsymbol{B = -2}$

[Check: equating the constant terms gives:

$15 = 9A + 3B + C$

When $A = 3$, $B = -2$ and $C = -6$,

$9A + 3B + C = 9(3) + 3(-2) + (-6)$
$$= 27 - 6 - 6 = 15 = \text{LHS}]$$

Thus $\dfrac{3x^2 + 16x + 15}{(x+3)^3}$

$$\equiv \frac{3}{(x+3)} - \frac{2}{(x+3)^2} - \frac{6}{(x+3)^3}$$

Now try the following Practice Exercise

Practice Exercise 9   Partial fractions with repeated linear factors (Answers on page 857)

1. $\dfrac{4x-3}{(x+1)^2}$

2. $\dfrac{x^2+7x+3}{x^2(x+3)}$

3. $\dfrac{5x^2-30x+44}{(x-2)^3}$

4. $\dfrac{18+21x-x^2}{(x-5)(x+2)^2}$

2.4 Worked problems on partial fractions with quadratic factors

Problem 8. Express $\dfrac{7x^2+5x+13}{(x^2+2)(x+1)}$ in partial fractions.

The denominator is a combination of a quadratic factor, (x^2+2), which does not factorise without introducing imaginary surd terms, and a linear factor, $(x+1)$. Let,

$$\frac{7x^2+5x+13}{(x^2+2)(x+1)}\equiv\frac{Ax+B}{(x^2+2)}+\frac{C}{(x+1)}$$

$$\equiv\frac{(Ax+B)(x+1)+C(x^2+2)}{(x^2+2)(x+1)}$$

Equating numerators gives:

$$7x^2+5x+13\equiv(Ax+B)(x+1)+C(x^2+2)\quad(1)$$

Let $x=-1$. Then

$$7(-1)^2+5(-1)+13\equiv(Ax+B)(0)+C(1+2)$$

i.e. $\qquad 15=3C$

i.e. $\qquad\quad C=5$

Identity (1) may be expanded as:

$$7x^2+5x+13\equiv Ax^2+Ax+Bx+B+Cx^2+2C$$

Equating the coefficients of x^2 terms gives:

$$7=A+C,\text{ and since }C=5,A=2$$

Equating the coefficients of x terms gives:

$$5=A+B,\text{ and since }A=2,B=3$$

[Check: equating the constant terms gives:

$$13=B+2C$$

When $B=3$ and $C=5$,

$$B+2C=3+10=13=\text{LHS}]$$

Hence $\dfrac{7x^2+5x+13}{(x^2+2)(x+1)}\equiv\dfrac{2x+3}{(x^2+2)}+\dfrac{5}{(x+1)}$

Problem 9. Resolve $\dfrac{3+6x+4x^2-2x^3}{x^2(x^2+3)}$ into partial fractions.

Terms such as x^2 may be treated as $(x+0)^2$, i.e. they are repeated linear factors.

Let $\dfrac{3+6x+4x^2-2x^3}{x^2(x^2+3)}\equiv\dfrac{A}{x}+\dfrac{B}{x^2}+\dfrac{Cx+D}{(x^2+3)}$

$$\equiv\frac{Ax(x^2+3)+B(x^2+3)+(Cx+D)x^2}{x^2(x^2+3)}$$

Equating the numerators gives:

$$3+6x+4x^2-2x^3\equiv Ax(x^2+3)+B(x^2+3)$$
$$+(Cx+D)x^2$$
$$\equiv Ax^3+3Ax+Bx^2+3B$$
$$+Cx^3+Dx^2$$

Let $x=0$. Then $3=3B$

i.e. $\qquad\quad\boldsymbol{B=1}$

Equating the coefficients of x^3 terms gives:

$$-2=A+C\quad(1)$$

Equating the coefficients of x^2 terms gives:

$$4=B+D$$

Since $B=1,D=3$

Equating the coefficients of x terms gives:

$$6=3A$$

i.e. $\qquad A = 2$

From equation (1), since $A = 2, C = -4$

Hence $\dfrac{3 + 6x + 4x^2 - 2x^3}{x^2(x^2 + 3)} \equiv \dfrac{2}{x} + \dfrac{1}{x^2} + \dfrac{-4x + 3}{x^2 + 3}$

$\qquad\qquad\qquad\qquad = \dfrac{2}{x} + \dfrac{1}{x^2} + \dfrac{3 - 4x}{x^2 + 3}$

Now try the following Practice Exercise

Practice Exercise 10 Partial fractions with quadratic factors (Answers on page 857)

1. $\dfrac{x^2 - x - 13}{(x^2 + 7)(x - 2)}$

2. $\dfrac{6x - 5}{(x - 4)(x^2 + 3)}$

3. $\dfrac{15 + 5x + 5x^2 - 4x^3}{x^2(x^2 + 5)}$

4. $\dfrac{x^3 + 4x^2 + 20x - 7}{(x - 1)^2(x^2 + 8)}$

5. When solving the differential equation $\dfrac{d^2\theta}{dt^2} - 6\dfrac{d\theta}{dt} - 10\theta = 20 - e^{2t}$ by Laplace transforms, for given boundary conditions, the following expression for $\mathcal{L}\{\theta\}$ results:

$$\mathcal{L}\{\theta\} = \dfrac{4s^3 - \dfrac{39}{2}s^2 + 42s - 40}{s(s - 2)(s^2 - 6s + 10)}$$

Show that the expression can be resolved into partial fractions to give:

$$\mathcal{L}\{\theta\} = \dfrac{2}{s} - \dfrac{1}{2(s - 2)} + \dfrac{5s - 3}{2(s^2 - 6s + 10)}$$

For fully worked solutions to each of the problems in Practice Exercises 8 to 10 in this chapter, go to the website:
www.routledge.com/cw/bird

Chapter 3

Logarithms

Why it is important to understand: **Logarithms**

All types of engineers use natural and common logarithms. Chemical engineers use them to measure radioactive decay and pH solutions, both of which are measured on a logarithmic scale. The Richter scale, which measures earthquake intensity, is a logarithmic scale. Biomedical engineers use logarithms to measure cell decay and growth, and also to measure light intensity for bone mineral density measurements. In electrical engineering, a dB (decibel) scale is very useful for expressing attenuations in radio propagation and circuit gains, and logarithms are used for implementing arithmetic operations in digital circuits. Logarithms are especially useful when dealing with the graphical analysis of non-linear relationships and logarithmic scales are used to linearise data to make data analysis simpler. Understanding and using logarithms is clearly important in all branches of engineering.

At the end of this chapter, you should be able to:

- define base, power, exponent and index
- define a logarithm
- distinguish between common and Napierian (i.e. hyperbolic or natural) logarithms
- evaluate logarithms to any base
- state the laws of logarithms
- simplify logarithmic expressions
- solve equations involving logarithms
- solve indicial equations
- sketch graphs of $\log_{10} x$ and $\log_e x$

3.1 Introduction to logarithms

With the use of calculators firmly established, logarithmic tables are no longer used for calculations. However, the theory of logarithms is important, for there are several scientific and engineering laws that involve the rules of logarithms.

From the laws of indices: $\qquad 16 = 2^4$

The number 4 is called the **power** or the **exponent** or the **index**. In the expression 2^4, the number 2 is called the **base**.

In another example: $\qquad 64 = 8^2$

In this example, 2 is the power, or exponent, or index. The number 8 is the base.

What is a logarithm?

Consider the expression $16 = 2^4$.

An alternative, yet equivalent, way of writing this expression is: $\log_2 16 = 4$.

This is stated as 'log to the base 2 of 16 equals 4'.

We see that the logarithm is the same as the power or index in the original expression. It is the base in the original expression which becomes the base of the logarithm.

The two statements: $16 = 2^4$ and $\log_2 16 = 4$ are equivalent.

If we write either of them, we are automatically implying the other.

In general, if a number y can be written in the form a^x, then the index 'x' is called the 'logarithm of y to the base of a',

i.e.   if $y = a^x$ then $x = \log_a y$

In another example, if we write that $64 = 8^2$ then the equivalent statement using logarithms is:

$$\log_8 64 = 2$$

In another example, if we write that $\log_3 81 = 4$ then the equivalent statement using powers is:

$$3^4 = 81$$

So the two sets of statements, one involving powers and one involving logarithms, are equivalent.

Common logarithms

From above, if we write that $1000 = 10^3$, then $3 = \log_{10} 1000$

This may be checked using the 'log' button on your calculator.

Logarithms having a base of 10 are called **common logarithms** and $\log_{10}$ is often abbreviated to lg.

The following values may be checked by using a calculator:

$$\lg 27.5 = 1.4393\ldots, \quad \lg 378.1 = 2.5776\ldots$$
$$\text{and} \quad \lg 0.0204 = -1.6903\ldots$$

Napierian logarithms

Logarithms having a base of e (where 'e' is a mathematical constant approximately equal to 2.7183) are called **hyperbolic**, **Napierian** or **natural logarithms**, and $\log_e$ is usually abbreviated to ln.

The following values may be checked by using a calculator:

$$\ln 3.65 = 1.2947\ldots, \quad \ln 417.3 = 6.0338\ldots$$
$$\text{and} \quad \ln 0.182 = -1.7037\ldots$$

More on Napierian logarithms is explained in Chapter 4.

Here are some worked problems to help understanding of logarithms.

Problem 1. Evaluate $\log_3 9$

Let $x = \log_3 9$ then $3^x = 9$ from the definition
of a logarithm,

i.e. $3^x = 3^2$ from which, $x = 2$

Hence, $\log_3 9 = 2$

Problem 2. Evaluate $\log_{10} 10$

Let $x = \log_{10} 10$ then $10^x = 10$ from the
definition of a logarithm,

i.e. $10^x = 10^1$ from which, $x = 1$

Hence, $\log_{10} 10 = 1$ (which may be checked
using a calculator)

Problem 3. Evaluate $\log_{16} 8$

Let $x = \log_{16} 8$ then $16^x = 8$ from the definition
of a logarithm,

i.e. $(2^4)^x = 2^3$ i.e. $2^{4x} = 2^3$ from the laws of indices,

from which, $4x = 3$ and $x = \dfrac{3}{4}$

Hence, $\log_{16} 8 = \dfrac{3}{4}$

Problem 4. Evaluate $\lg 0.001$

Let $x = \lg 0.001 = \log_{10} 0.001$ then $10^x = 0.001$

i.e. $10^x = 10^{-3}$ from which, $x = -3$

Hence, $\lg 0.001 = -3$ (which may be checked
using a calculator)

Problem 5. Evaluate $\ln e$

Let $x = \ln e = \log_e e$ then $e^x = e$

i.e. $e^x = e^1$

from which, $x = 1$

Hence, $\ln e = 1$ (which may be checked
using a calculator)

Problem 6. Evaluate $\log_3 \dfrac{1}{81}$

Let $x = \log_3 \dfrac{1}{81}$ then $3^x = \dfrac{1}{81} = \dfrac{1}{3^4} = 3^{-4}$

from which, $x = -4$

Hence, $\log_3 \dfrac{1}{81} = -4$

Problem 7. Solve the equation: $\lg x = 3$

If $\lg x = 3$ then $\log_{10} x = 3$

and $x = 10^3$ i.e. $x = 1000$

Problem 8. Solve the equation: $\log_2 x = 5$

If $\log_2 x = 5$ then $x = 2^5 = 32$

Problem 9. Solve the equation: $\log_5 x = -2$

If $\log_5 x = -2$ then $x = 5^{-2} = \dfrac{1}{5^2} = \dfrac{1}{25}$

Now try the following Practice Exercise

Practice Exercise 11   Introduction to logarithms (Answers on page 857)

In Problems 1 to 11, evaluate the given expressions:

1. $\log_{10} 10000$ 2. $\log_2 16$

3. $\log_5 125$ 4. $\log_2 \frac{1}{8}$

5. $\log_8 2$ 6. $\log_7 343$

7. $\lg 100$ 8. $\lg 0.01$

9. $\log_4 8$ 10. $\log_{27} 3$

11. $\ln e^2$

In Problems 12 to 18 solve the equations:

12. $\log_{10} x = 4$ 13. $\lg x = 5$

14. $\log_3 x = 2$ 15. $\log_4 x = -2\dfrac{1}{2}$

16. $\lg x = -2$ 17. $\log_8 x = -\dfrac{4}{3}$

18. $\ln x = 3$

3.2 Laws of logarithms

There are three laws of logarithms, which apply to any base:

(i) To multiply two numbers:

$$\log (A \times B) = \log A + \log B$$

The following may be checked by using a calculator:

$$\lg 10 = 1$$

Also, $\lg 5 + \lg 2 = 0.69897\ldots$
$$+ 0.301029\ldots = 1$$
Hence, $\lg (5 \times 2) = \lg 10 = \lg 5 + \lg 2$

(ii) To divide two numbers:

$$\log \left(\frac{A}{B} \right) = \log A - \log B$$

The following may be checked using a calculator:

$$\ln \left(\frac{5}{2} \right) = \ln 2.5 = 0.91629\ldots$$

Also, $\ln 5 - \ln 2 = 1.60943\ldots - 0.69314\ldots$
$$= 0.91629\ldots$$
Hence, $\ln \left(\dfrac{5}{2} \right) = \ln 5 - \ln 2$

(iii) To raise a number to a power:

$$\log A^n = n \log A$$

The following may be checked using a calculator:

$$\lg 5^2 = \lg 25 = 1.39794\ldots$$

Also, $2 \lg 5 = 2 \times 0.69897\ldots = 1.39794\ldots$

Hence, $\lg 5^2 = 2 \lg 5$

Here are some worked problems to help understanding of the laws of logarithms.

Problem 10. Write $\log 4 + \log 7$ as the logarithm of a single number.

$\log 4 + \log 7 = \log(7 \times 4)$

by the first law of logarithms

$= \log 28$

Problem 11. Write $\log 16 - \log 2$ as the logarithm of a single number.

$\log 16 - \log 2 = \log\left(\dfrac{16}{2}\right)$

by the second law of logarithms

$= \log 8$

Problem 12. Write $2\log 3$ as the logarithm of a single number.

$2\log 3 = \log 3^2$ by the third law of logarithms

$= \log 9$

Problem 13. Write $\dfrac{1}{2}\log 25$ as the logarithm of a single number.

$\dfrac{1}{2}\log 25 = \log 25^{\frac{1}{2}}$ by the third law of logarithms

$= \log\sqrt{25} = \log 5$

Problem 14. Simplify: $\log 64 - \log 128 + \log 32$.

$64 = 2^6, 128 = 2^7$ and $32 = 2^5$

Hence, $\log 64 - \log 128 + \log 32$

$= \log 2^6 - \log 2^7 + \log 2^5$

$= 6\log 2 - 7\log 2 + 5\log 2$

by the third law of logarithms

$= 4\log 2$

Problem 15. Write $\dfrac{1}{2}\log 16 + \dfrac{1}{3}\log 27 - 2\log 5$ as the logarithm of a single number.

$\dfrac{1}{2}\log 16 + \dfrac{1}{3}\log 27 - 2\log 5$

$= \log 16^{\frac{1}{2}} + \log 27^{\frac{1}{3}} - \log 5^2$

by the third law of logarithms

$= \log\sqrt{16} + \log\sqrt[3]{27} - \log 25$

by the laws of indices

$= \log 4 + \log 3 - \log 25$

$= \log\left(\dfrac{4 \times 3}{25}\right)$

by the first and second laws of logarithms

$= \log\left(\dfrac{12}{25}\right) = \log 0.48$

Problem 16. Write (a) $\log 30$ (b) $\log 450$ in terms of $\log 2, \log 3$ and $\log 5$ to any base.

(a) $\log 30 = \log(2 \times 15) = \log(2 \times 3 \times 5)$

$= \log 2 + \log 3 + \log 5$

by the first law of logarithms

(b) $\log 450 = \log(2 \times 225) = \log(2 \times 3 \times 75)$

$= \log(2 \times 3 \times 3 \times 25)$

$= \log(2 \times 3^2 \times 5^2)$

$= \log 2 + \log 3^2 + \log 5^2$

by the first law of logarithms

i.e. $\log 450 = \log 2 + 2\log 3 + 2\log 5$

by the third law of logarithms

Problem 17. Write $\log\left(\dfrac{8 \times \sqrt[4]{5}}{81}\right)$ in terms of $\log 2, \log 3$ and $\log 5$ to any base.

$\log\left(\dfrac{8 \times \sqrt[4]{5}}{81}\right) = \log 8 + \log\sqrt[4]{5} - \log 81$

by the first and second laws of logarithms

$= \log 2^3 + \log 5^{\frac{1}{4}} - \log 3^4$

by the laws of indices

i.e. $\log\left(\dfrac{8 \times \sqrt[4]{5}}{81}\right) = 3\log 2 + \dfrac{1}{4}\log 5 - 4\log 3$

by the third law of logarithms

Problem 18. Evaluate:

$$\frac{\log 25 - \log 125 + \frac{1}{2}\log 625}{3\log 5}$$

$$\frac{\log 25 - \log 125 + \frac{1}{2}\log 625}{3\log 5}$$

$$= \frac{\log 5^2 - \log 5^3 + \frac{1}{2}\log 5^4}{3\log 5}$$

$$= \frac{2\log 5 - 3\log 5 + \frac{4}{2}\log 5}{3\log 5} = \frac{1\log 5}{3\log 5} = \frac{1}{3}$$

Problem 19. Solve the equation:
$$\log(x-1) + \log(x+8) = 2\log(x+2)$$

$$\text{LHS} = \log(x-1) + \log(x+8)$$

$$= \log(x-1)(x+8)$$

from the first law of logarithms

$$= \log(x^2 + 7x - 8)$$

$$\text{RHS} = 2\log(x+2) = \log(x+2)^2$$

from the third law of logarithms

$$= \log(x^2 + 4x + 4)$$

Hence, $\log(x^2 + 7x - 8) = \log(x^2 + 4x + 4)$

from which, $x^2 + 7x - 8 = x^2 + 4x + 4$

i.e. $7x - 8 = 4x + 4$

i.e. $3x = 12$

and $x = 4$

Problem 20. Solve the equation: $\dfrac{1}{2}\log 4 = \log x$

$\dfrac{1}{2}\log 4 = \log 4^{\frac{1}{2}}$ from the third law of logarithms

$\qquad\quad = \log\sqrt{4}$ from the laws of indices

Hence, $\dfrac{1}{2}\log 4 = \log x$

becomes $\log\sqrt{4} = \log x$

i.e. $\log 2 = \log x$

from which, $2 = x$

i.e. the **solution of the equation is: $x = 2$**

Problem 21. Solve the equation:
$$\log(x^2 - 3) - \log x = \log 2$$

$$\log(x^2 - 3) - \log x = \log\left(\frac{x^2-3}{x}\right)$$

from the second law of logarithms

Hence, $\log\left(\dfrac{x^2-3}{x}\right) = \log 2$

from which, $\dfrac{x^2-3}{x} = 2$

Rearranging gives: $x^2 - 3 = 2x$

and $x^2 - 2x - 3 = 0$

Factorising gives: $(x-3)(x+1) = 0$

from which, $x = 3$ or $x = -1$

$x = -1$ is not a valid solution since the logarithm of a negative number has no real root.

Hence, **the solution of the equation is: $x = 3$**

Now try the following Practice Exercise

Practice Exercise 12   Laws of logarithms (Answers on page 857)

In Problems 1 to 11, write as the logarithm of a single number:

1. $\log 2 + \log 3$

2. $\log 3 + \log 5$

3. $\log 3 + \log 4 - \log 6$

4. $\log 7 + \log 21 - \log 49$

5. $2\log 2 + \log 3$

6. $2\log 2 + 3\log 5$

7. $2\log 5 - \dfrac{1}{2}\log 81 + \log 36$

8. $\dfrac{1}{3}\log 8 - \dfrac{1}{2}\log 81 + \log 27$

9. $\dfrac{1}{2}\log 4 - 2\log 3 + \log 45$

10. $\dfrac{1}{4}\log 16 + 2\log 3 - \log 18$

11. $2\log 2 + \log 5 - \log 10$

Simplify the expressions given in Problems 12 to 14:

12. $\log 27 - \log 9 + \log 81$

13. $\log 64 + \log 32 - \log 128$

14. $\log 8 - \log 4 + \log 32$

Evaluate the expressions given in Problems 15 and 16:

15. $\dfrac{\frac{1}{2}\log 16 - \frac{1}{3}\log 8}{\log 4}$

16. $\dfrac{\log 9 - \log 3 + \frac{1}{2}\log 81}{2\log 3}$

Solve the equations given in Problems 17 to 22:

17. $\log x^4 - \log x^3 = \log 5x - \log 2x$

18. $\log 2t^3 - \log t = \log 16 + \log t$

19. $2\log b^2 - 3\log b = \log 8b - \log 4b$

20. $\log(x+1) + \log(x-1) = \log 3$

21. $\dfrac{1}{3}\log 27 = \log(0.5a)$

22. $\log\left(x^2 - 5\right) - \log x = \log 4$

3.3   Indicial equations

The laws of logarithms may be used to solve certain equations involving powers – called **indicial equations**. For example, to solve, say, $3^x = 27$, logarithms to a base of 10 are taken of both sides, i.e. $\log_{10} 3^x = \log_{10} 27$

and $x\log_{10} 3 = \log_{10} 27$, by the third law of logarithms. Rearranging gives

$$x = \frac{\log_{10} 27}{\log_{10} 3} = \frac{1.43136\ldots}{0.4771\ldots} = 3$$

which may be readily checked

$$\left(\text{Note, } \left(\frac{\log 8}{\log 2}\right) \text{ is } \textbf{not} \text{ equal to } \lg\left(\frac{8}{2}\right)\right)$$

Problem 22.   Solve the equation $2^x = 3$, correct to 4 significant figures.

Taking logarithms to base 10 of both sides of $2^x = 3$ gives:

$$\log_{10} 2^x = \log_{10} 3$$

i.e. $x\log_{10} 2 = \log_{10} 3$

Rearranging gives:

$$x = \frac{\log_{10} 3}{\log_{10} 2} = \frac{0.47712125\ldots}{0.30102999\ldots}$$

$$= \textbf{1.585}, \text{ correct to 4 significant figures.}$$

Problem 23.   Solve the equation $2^{x+1} = 3^{2x-5}$ correct to 2 decimal places.

Taking logarithms to base 10 of both sides gives:

$$\log_{10} 2^{x+1} = \log_{10} 3^{2x-5}$$

i.e. $(x+1)\log_{10} 2 = (2x-5)\log_{10} 3$

$$x\log_{10} 2 + \log_{10} 2 = 2x\log_{10} 3 - 5\log_{10} 3$$

$$x(0.3010) + (0.3010) = 2x(0.4771) - 5(0.4771)$$

i.e. $0.3010x + 0.3010 = 0.9542x - 2.3855$

Hence

$$2.3855 + 0.3010 = 0.9542x - 0.3010x$$

$$2.6865 = 0.6532x$$

from which $x = \dfrac{2.6865}{0.6532} = \textbf{4.11}$, correct to

2 decimal places.

Problem 24.   Solve the equation $x^{3.2} = 41.15$, correct to 4 significant figures.

Taking logarithms to base 10 of both sides gives:

$$\log_{10} x^{3.2} = \log_{10} 41.15$$

$$3.2\log_{10} x = \log_{10} 41.15$$

Hence $\log_{10} x = \dfrac{\log_{10} 41.15}{3.2} = 0.50449$

Thus $x = \text{antilog}\, 0.50449 = 10^{0.50449} = \textbf{3.195}$ correct to 4 significant figures.

Now try the following Practice Exercise

Practice Exercise 13   Indicial equations (Answers on page 857)

Solve the following indicial equations for x, each correct to 4 significant figures:

1. $3^x = 6.4$

2. $2^x = 9$

3. $2^{x-1} = 3^{2x-1}$

4. $x^{1.5} = 14.91$

5. $25.28 = 4.2^x$

6. $4^{2x-1} = 5^{x+2}$

7. $x^{-0.25} = 0.792$

8. $0.027^x = 3.26$

9. The decibel gain n of an amplifier is given by:

$$n = 10 \log_{10} \left(\frac{P_2}{P_1} \right)$$

where P_1 is the power input and P_2 is the power output. Find the power gain $\dfrac{P_2}{P_1}$ when $n = 25$ decibels.

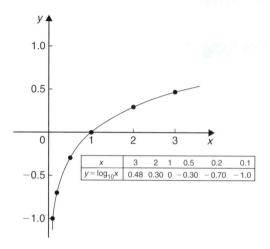

x	3	2	1	0.5	0.2	0.1
$y = \log_{10} x$	0.48	0.30	0	-0.30	-0.70	-1.0

Figure 3.1

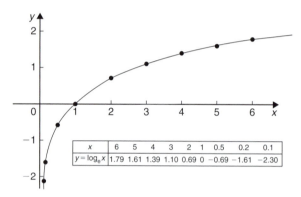

x	6	5	4	3	2	1	0.5	0.2	0.1
$y = \log_e x$	1.79	1.61	1.39	1.10	0.69	0	-0.69	-1.61	-2.30

Figure 3.2

3.4 Graphs of logarithmic functions

A graph of $y = \log_{10} x$ is shown in Fig. 3.1 and a graph of $y = \log_e x$ is shown in Fig. 3.2. Both are seen to be of similar shape; in fact, the same general shape occurs for a logarithm to any base.

In general, with a logarithm to any base a, it is noted that:

(i) **$\log_a 1 = 0$**

Let $\log_a = x$, then $a^x = 1$ from the definition of the logarithm.

If $a^x = 1$ then $x = 0$ from the laws of indices.
Hence $\log_a 1 = 0$. In the above graphs it is seen that $\log_{10} 1 = 0$ and $\log_e 1 = 0$

(ii) **$\log_a a = 1$**

Let $\log_a a = x$ then $a^x = a$ from the definition of a logarithm.

If $a^x = a$ then $x = 1$

Hence $\log_a a = 1$. (Check with a calculator that $\log_{10} 10 = 1$ and $\log_e e = 1$)

(iii) **$\log_a 0 \to -\infty$**

Let $\log_a 0 = x$ then $a^x = 0$ from the definition of a logarithm.

If $a^x = 0$, and a is a positive real number, then x must approach minus infinity. (For example, check with a calculator, $2^{-2} = 0.25$, $2^{-20} = 9.54 \times 10^{-7}$, $2^{-200} = 6.22 \times 10^{-61}$, and so on)

Hence $\log_a 0 \to -\infty$

For fully worked solutions to each of the problems in Practice Exercises 11 to 13 in this chapter, go to the website:
www.routledge.com/cw/bird

Chapter 4

Exponential functions

Why it is important to understand: Exponential functions

Exponential functions are used in engineering, physics, biology and economics. There are many quantities that grow exponentially; some examples are population, compound interest and charge in a capacitor. With exponential growth, the rate of growth increases as time increases. We also have exponential decay; some examples are radioactive decay, atmospheric pressure, Newton's law of cooling and linear expansion. Understanding and using exponential functions is important in many branches of engineering.

At the end of this chapter, you should be able to:

- evaluate exponential functions using a calculator
- state the exponential series for e^x
- plot graphs of exponential functions
- evaluate Napierian logarithms using a calculator
- solve equations involving Napierian logarithms
- appreciate the many examples of laws of growth and decay in engineering and science
- perform calculations involving the laws of growth and decay
- reduce exponential laws to linear form using log-linear graph paper

---

4.1 Introduction to exponential functions

An exponential function is one which contains e^x, e being a constant called the exponent and having an approximate value of 2.7183. The exponent arises from the natural laws of growth and decay and is used as a base for natural or Napierian logarithms.

The most common method of evaluating an exponential function is by using a scientific notation **calculator**. Use your calculator to check the following values:

$e^1 = 2.7182818$, correct to 8 significant figures,

$e^{-1.618} = 0.1982949$, each correct to 7 significant figures,

$e^{0.12} = 1.1275$, correct to 5 significant figures,

$e^{-1.47} = 0.22993$, correct to 5 decimal places,

$e^{-0.431} = 0.6499$, correct to 4 decimal places,

$e^{9.32} = 11\ 159$, correct to 5 significant figures,

$e^{-2.785} = 0.0617291$, correct to 7 decimal places.

Problem 1. Evaluate the following correct to 4 decimal places, using a calculator:

$$0.0256 \left(e^{5.21} - e^{2.49} \right)$$

$$0.0256 \left(e^{5.21} - e^{2.49} \right) = 0.0256 \, (183.094058\ldots$$
$$- 12.0612761\ldots)$$
$$= \mathbf{4.3784}, \text{ correct to 4}$$
$$\text{decimal places.}$$

Problem 2. Evaluate the following correct to 4 decimal places, using a calculator:

$$5 \left(\frac{e^{0.25} - e^{-0.25}}{e^{0.25} + e^{-0.25}} \right)$$

$$5 \left(\frac{e^{0.25} - e^{-0.25}}{e^{0.25} + e^{-0.25}} \right)$$

$$= 5 \left(\frac{1.28402541\ldots - 0.77880078\ldots}{1.28402541\ldots + 0.77880078\ldots} \right)$$

$$= 5 \left(\frac{0.5052246\ldots}{2.0628262\ldots} \right)$$

$$= \mathbf{1.2246}, \text{ correct to 4 decimal places.}$$

Problem 3. The instantaneous voltage v in a capacitive circuit is related to time t by the equation: $v = V e^{-t/CR}$ where V, C and R are constants. Determine v, correct to 4 significant figures, when $t = 50$ ms, $C = 10\,\mu$F, $R = 47$ kΩ and $V = 300$ volts.

$$v = V e^{-t/CR} = 300 e^{(-50 \times 10^{-3})/(10 \times 10^{-6} \times 47 \times 10^{3})}$$

Using a calculator,

$$v = 300 e^{-0.1063829\ldots} = 300(0.89908025\ldots)$$

$$= \mathbf{269.7 \text{ volts}}$$

Now try the following Practice Exercise

Practice Exercise 14   Evaluating exponential functions (Answers on page 857)

1. Evaluate the following, correct to 4 significant figures:
    (a) $e^{-1.8}$ (b) $e^{-0.78}$ (c) e^{10}

2. Evaluate the following, correct to 5 significant figures:
    (a) $e^{1.629}$ (b) $e^{-2.7483}$ (c) $0.62 e^{4.178}$

In Problems 3 and 4, evaluate correct to 5 decimal places:

3. (a) $\dfrac{1}{7} e^{3.4629}$ (b) $8.52 e^{-1.2651}$ (c) $\dfrac{5 e^{2.6921}}{3 e^{1.1171}}$

4. (a) $\dfrac{5.6823}{e^{-2.1347}}$ (b) $\dfrac{e^{2.1127} - e^{-2.1127}}{2}$
    (c) $\dfrac{4(e^{-1.7295} - 1)}{e^{3.6817}}$

5. The length of a bar l at a temperature θ is given by $l = l_0 e^{\alpha\theta}$, where l_0 and α are constants. Evaluate l, correct to 4 significant figures, where $l_0 = 2.587, \theta = 321.7$ and $\alpha = 1.771 \times 10^{-4}$

6. When a chain of length $2L$ is suspended from two points, $2D$ metres apart, on the same horizontal level: $D = k \left\{ \ln \left(\frac{L + \sqrt{L^2 + k^2}}{k} \right) \right\}$. Evaluate D when $k = 75$ m and $L = 180$ m.

4.2 The power series for e^x

The value of e^x can be calculated to any required degree of accuracy since it is defined in terms of the following **power series**:

$$e^x = 1 + x + \frac{x^2}{2!} + \frac{x^3}{3!} + \frac{x^4}{4!} + \cdots \qquad (1)$$

(where $3! = 3 \times 2 \times 1$ and is called 'factorial 3')
The series is valid for all values of x.
The series is said to **converge**, i.e. if all the terms are added, an actual value for e^x (where x is a real number) is obtained. The more terms that are taken, the closer will be the value of e^x to its actual value. The value of the exponent e, correct to, say, 4 decimal places, may be determined by substituting $x = 1$ in the power series of equation (1). Thus,

$$e^1 = 1 + 1 + \frac{(1)^2}{2!} + \frac{(1)^3}{3!} + \frac{(1)^4}{4!} + \frac{(1)^5}{5!}$$

$$+ \frac{(1)^6}{6!} + \frac{(1)^7}{7!} + \frac{(1)^8}{8!} + \cdots$$

$$= 1 + 1 + 0.5 + 0.16667 + 0.04167$$

$$+ 0.00833 + 0.00139 + 0.00020$$

$$+ 0.00002 + \cdots$$

i.e. $e = 2.71828 = 2.7183,$
        correct to 4 decimal places.

The value of $e^{0.05}$, correct to, say, 8 significant figures, is found by substituting $x = 0.05$ in the power series for e^x. Thus

$$e^{0.05} = 1 + 0.05 + \frac{(0.05)^2}{2!} + \frac{(0.05)^3}{3!}$$

$$+ \frac{(0.05)^4}{4!} + \frac{(0.05)^5}{5!} + \cdots$$

$$= 1 + 0.05 + 0.00125 + 0.000020833$$

$$+ 0.000000260 + 0.000000003$$

and by adding,

$e^{0.05} = 1.0512711$, correct to 8 significant figures.

In this example, successive terms in the series grow smaller very rapidly and it is relatively easy to determine the value of $e^{0.05}$ to a high degree of accuracy. However, when x is nearer to unity or larger than unity, a very large number of terms are required for an accurate result.

If in the series of equation (1), x is replaced by $-x$, then,

$$e^{-x} = 1 + (-x) + \frac{(-x)^2}{2!} + \frac{(-x)^3}{3!} + \cdots$$

i.e. $e^{-x} = 1 - x + \dfrac{x^2}{2!} - \dfrac{x^3}{3!} + \cdots$

In a similar manner the power series for e^x may be used to evaluate any exponential function of the form $a\,e^{kx}$, where a and k are constants. In the series of equation (1), let x be replaced by kx. Then,

$$a\,e^{kx} = a\left\{1 + (kx) + \frac{(kx)^2}{2!} + \frac{(kx)^3}{3!} + \cdots\right\}$$

Thus $5\,e^{2x} = 5\left\{1 + (2x) + \dfrac{(2x)^2}{2!} + \dfrac{(2x)^3}{3!} + \cdots\right\}$

$$= 5\left\{1 + 2x + \frac{4x^2}{2} + \frac{8x^3}{6} + \cdots\right\}$$

i.e. $5\,e^{2x} = 5\left\{1 + 2x + 2x^2 + \dfrac{4}{3}x^3 + \cdots\right\}$

> **Problem 4.** Determine the value of $5\,e^{0.5}$, correct to 5 significant figures, by using the power series for e^x

$$e^x = 1 + x + \frac{x^2}{2!} + \frac{x^3}{3!} + \frac{x^4}{4!} + \cdots$$

Hence $e^{0.5} = 1 + 0.5 + \dfrac{(0.5)^2}{(2)(1)} + \dfrac{(0.5)^3}{(3)(2)(1)}$

$$+ \frac{(0.5)^4}{(4)(3)(2)(1)} + \frac{(0.5)^5}{(5)(4)(3)(2)(1)}$$

$$+ \frac{(0.5)^6}{(6)(5)(4)(3)(2)(1)}$$

$$= 1 + 0.5 + 0.125 + 0.020833$$

$$+ 0.0026042 + 0.0002604$$

$$+ 0.0000217$$

i.e. $e^{0.5} = 1.64872$,

correct to 6 significant figures.

Hence $5e^{0.5} = 5(1.64872) = \mathbf{8.2436}$,

correct to 5 significant figures.

> **Problem 5.** Expand $e^x(x^2 - 1)$ as far as the term in x^5

The power series for e^x is,

$$e^x = 1 + x + \frac{x^2}{2!} + \frac{x^3}{3!} + \frac{x^4}{4!} + \frac{x^5}{5!} + \cdots$$

Hence $e^x(x^2 - 1)$

$$= \left(1 + x + \frac{x^2}{2!} + \frac{x^3}{3!} + \frac{x^4}{4!} + \frac{x^5}{5!} + \cdots\right)(x^2 - 1)$$

$$= \left(x^2 + x^3 + \frac{x^4}{2!} + \frac{x^5}{3!} + \cdots\right)$$

$$- \left(1 + x + \frac{x^2}{2!} + \frac{x^3}{3!} + \frac{x^4}{4!} + \frac{x^5}{5!} + \cdots\right)$$

Grouping like terms gives:

$e^x(x^2 - 1)$

$$= -1 - x + \left(x^2 - \frac{x^2}{2!}\right) + \left(x^3 - \frac{x^3}{3!}\right)$$

$$+ \left(\frac{x^4}{2!} - \frac{x^4}{4!}\right) + \left(\frac{x^5}{3!} - \frac{x^5}{5!}\right) + \cdots$$

$$= -1 - x + \frac{1}{2}x^2 + \frac{5}{6}x^3 + \frac{11}{24}x^4 + \frac{19}{120}x^5$$

when expanded as far as the term in x^5

Now try the following Practice Exercise

Practice Exercise 15   Power series for e^x (Answers on page 857)

1. Evaluate $5.6\,e^{-1}$, correct to 4 decimal places, using the power series for e^x

2. Use the power series for e^x to determine, correct to 4 significant figures, (a) e^2 (b) $e^{-0.3}$, and check your result by using a calculator.

3. Expand $(1-2x)\,e^{2x}$ as far as the term in x^4

4. Expand $\left(2\,e^{x^2}\right)\left(x^{\frac{1}{2}}\right)$ to six terms.

4.3   Graphs of exponential functions

Values of e^x and e^{-x} obtained from a calculator, correct to 2 decimal places, over a range $x=-3$ to $x=3$, are shown in the following table.

x	−3.0	−2.5	−2.0	−1.5	−1.0	−0.5	0
e^x	0.05	0.08	0.14	0.22	0.37	0.61	1.00
e^{-x}	20.09	12.18	7.39	4.48	2.72	1.65	1.00

x	0.5	1.0	1.5	2.0	2.5	3.0
e^x	1.65	2.72	4.48	7.39	12.18	20.09
e^{-x}	0.61	0.37	0.22	0.14	0.08	0.05

Fig. 4.1 shows graphs of $y=e^x$ and $y=e^{-x}$

Problem 6.   Plot a graph of $y=2\,e^{0.3x}$ over a range of $x=-2$ to $x=3$. Hence determine the value of y when $x=2.2$ and the value of x when $y=1.6$

A table of values is drawn up as shown below.

x	−3	−2	−1	0	1	2	3
$0.3x$	−0.9	−0.6	−0.3	0	0.3	0.6	0.9
$e^{0.3x}$	0.407	0.549	0.741	1.000	1.350	1.822	2.460
$2\,e^{0.3x}$	0.81	1.10	1.48	2.00	2.70	3.64	4.92

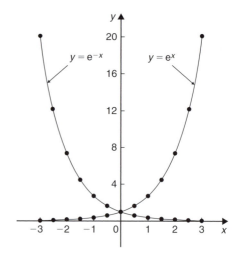

Figure 4.1

A graph of $y=2\,e^{0.3x}$ is shown plotted in Fig. 4.2.

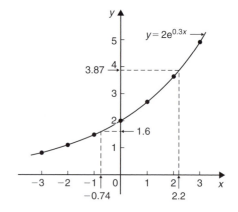

Figure 4.2

From the graph, **when $x=2.2$, $y=3.87$** and **when $y=1.6$, $x=-0.74$**

Problem 7.   Plot a graph of $y=\frac{1}{3}\,e^{-2x}$ over the range $x=-1.5$ to $x=1.5$. Determine from the graph the value of y when $x=-1.2$ and the value of x when $y=1.4$

A table of values is drawn up as shown below.

x	−1.5	−1.0	−0.5	0	0.5	1.0	1.5
$-2x$	3	2	1	0	−1	−2	−3
e^{-2x}	20.086	7.389	2.718	1.00	0.368	0.135	0.050
$\frac{1}{3}\,e^{-2x}$	6.70	2.46	0.91	0.33	0.12	0.05	0.02

A graph of $\frac{1}{3}e^{-2x}$ is shown in Fig. 4.3.

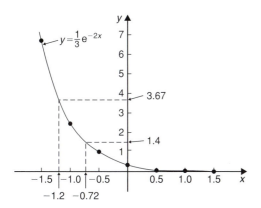

Figure 4.3

From the graph, **when $x = -1.2$, $y = 3.67$ and when $y = 1.4$, $x = -0.72$**

Problem 8. The decay of voltage, v volts, across a capacitor at time t seconds is given by $v = 250\,e^{\frac{-t}{3}}$. Draw a graph showing the natural decay curve over the first six seconds. From the graph, find (a) the voltage after 3.4 s, and (b) the time when the voltage is 150 V.

A table of values is drawn up as shown below.

t	0	1	2	3
$e^{\frac{-t}{3}}$	1.00	0.7165	0.5134	0.3679
$v = 250\,e^{\frac{-t}{3}}$	250.0	179.1	128.4	91.97

t	4	5	6
$e^{\frac{-t}{3}}$	0.2636	0.1889	0.1353
$v = 250\,e^{\frac{-t}{3}}$	65.90	47.22	33.83

The natural decay curve of $v = 250\,e^{\frac{-t}{3}}$ is shown in Fig. 4.4.

From the graph:

(a) **when time $t = 3.4$ s, voltage $v = 80$ V and**

(b) **when voltage $v = 150$ V, time $t = 1.5$ s**

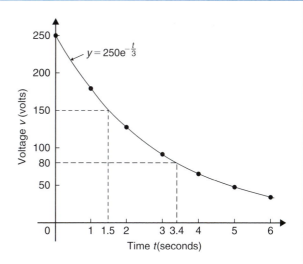

Figure 4.4

Now try the following Practice Exercise

Practice Exercise 16   Exponential graphs (Answers on page 857)

1. Plot a graph of $y = 3\,e^{0.2x}$ over the range $x = -3$ to $x = 3$. Hence determine the value of y when $x = 1.4$ and the value of x when $y = 4.5$

2. Plot a graph of $y = \frac{1}{2}e^{-1.5x}$ over a range $x = -1.5$ to $x = 1.5$ and hence determine the value of y when $x = -0.8$ and the value of x when $y = 3.5$

3. In a chemical reaction the amount of starting material $C\,\text{cm}^3$ left after t minutes is given by $C = 40\,e^{-0.006t}$. Plot a graph of C against t and determine (a) the concentration C after one hour, and (b) the time taken for the concentration to decrease by half.

4. The rate at which a body cools is given by $\theta = 250\,e^{-0.05t}$ where the excess of temperature of a body above its surroundings at time t minutes is $\theta\,^\circ\text{C}$. Plot a graph showing the natural decay curve for the first hour of cooling. Hence determine (a) the temperature after 25 minutes, and (b) the time when the temperature is $195\,^\circ\text{C}$.

4.4   Napierian logarithms

Logarithms having a base of 'e' are called **hyperbolic**, **Napierian** or **natural logarithms** and the Napierian logarithm of x is written as $\log_e x$, or more commonly as $\ln x$. Logarithms were invented by **John Napier***, a Scotsman (1550–1617).

The most common method of evaluating a Napierian logarithm is by a scientific notation **calculator**. Use your calculator to check the following values:

$\ln 4.328 = 1.46510554\ldots$
$\qquad = 1.4651$, correct to 4 decimal places

$\ln 1.812 = 0.59443$, correct to 5 significant figures

$\ln 1 = 0$

$\ln 527 = 6.2672$, correct to 5 significant figures

$\ln 0.17 = -1.772$, correct to 4 significant figures

$\ln 0.00042 = -7.77526$, correct to 6 significant figures

$\ln e^3 = 3$

$\ln e^1 = 1$

From the last two examples we can conclude that:

$$\log_e e^x = x$$

* **Who was Napier**? **John Napier** of Merchiston (1550–4 April 1617) is best known as the discoverer of logarithms. The inventor of the so-called 'Napier's bones', Napier also made common the use of the decimal point in arithmetic and mathematics. To find out more go to **www.routledge.com/cw/bird**

This is useful when **solving equations involving exponential functions**. For example, to solve $e^{3x} = 7$, take Napierian logarithms of both sides, which gives:

$$\ln e^{3x} = \ln 7$$

i.e. $\qquad 3x = \ln 7$

from which $\qquad x = \dfrac{1}{3}\ln 7 = \mathbf{0.6486,}$ correct to 4 decimal places.

Problem 9. Evaluate the following, each correct to 5 significant figures:

(a) $\dfrac{1}{2}\ln 4.7291$ (b) $\dfrac{\ln 7.8693}{7.8693}$ (c) $\dfrac{3.17\ln 24.07}{e^{-0.1762}}$

(a) $\dfrac{1}{2}\ln 4.7291 = \dfrac{1}{2}(1.5537349\ldots) = \mathbf{0.77687,}$

correct to 5 significant figures

(b) $\dfrac{\ln 7.8693}{7.8693} = \dfrac{2.06296911\ldots}{7.8693} = \mathbf{0.26215,}$

correct to 5 significant figures

(c) $\dfrac{3.17\ln 24.07}{e^{-0.1762}} = \dfrac{3.17(3.18096625\ldots)}{0.83845027\ldots}$

$= \mathbf{12.027,}$

correct to 5 significant figures.

Problem 10. Evaluate the following:

(a) $\dfrac{\ln e^{2.5}}{\lg 10^{0.5}}$ (b) $\dfrac{5e^{2.23}\lg 2.23}{\ln 2.23}$ correct to 3 decimal places.

(a) $\dfrac{\ln e^{2.5}}{\lg 10^{0.5}} = \dfrac{2.5}{0.5} = \mathbf{5}$

(b) $\dfrac{5e^{2.23}\lg 2.23}{\ln 2.23}$

$= \dfrac{5(9.29986607\ldots)(0.34830486\ldots)}{0.80200158\ldots}$

$= \mathbf{20.194}$, correct to 3 decimal places.

Problem 11. Solve the equation: $9 = 4e^{-3x}$ to find x, correct to 4 significant figures.

Rearranging $9 = 4e^{-3x}$ gives: $\qquad \dfrac{9}{4} = e^{-3x}$

Taking the reciprocal of both sides gives:

$$\dfrac{4}{9} = \dfrac{1}{e^{-3x}} = e^{3x}$$

Taking Napierian logarithms of both sides gives:

$$\ln\left(\frac{4}{9}\right) = \ln(e^{3x})$$

Since $\log_e e^{\alpha} = \alpha$, then $\ln\left(\frac{4}{9}\right) = 3x$

Hence, $x = \frac{1}{3}\ln\left(\frac{4}{9}\right) = \frac{1}{3}(-0.81093) = -\mathbf{0.2703}$,

correct to 4 significant figures.

Problem 12. Given $32 = 70\left(1 - e^{-\frac{t}{2}}\right)$ determine the value of t, correct to 3 significant figures.

Rearranging $32 = 70(1 - e^{-\frac{t}{2}})$ gives:

$$\frac{32}{70} = 1 - e^{-\frac{t}{2}}$$

and $\quad\quad e^{-\frac{t}{2}} = 1 - \frac{32}{70} = \frac{38}{70}$

Taking the reciprocal of both sides gives:

$$e^{\frac{t}{2}} = \frac{70}{38}$$

Taking Napierian logarithms of both sides gives:

$$\ln e^{\frac{t}{2}} = \ln\left(\frac{70}{38}\right)$$

i.e. $\quad\quad \frac{t}{2} = \ln\left(\frac{70}{38}\right)$

from which, $t = 2\ln\left(\frac{70}{38}\right) = \mathbf{1.22}$, correct to 3 significant figures.

Problem 13. Solve the equation: $2.68 = \ln\left(\frac{4.87}{x}\right)$ to find x.

From the definition of a logarithm, since

$2.68 = \ln\left(\frac{4.87}{x}\right)$ then $e^{2.68} = \frac{4.87}{x}$

Rearranging gives: $\quad x = \frac{4.87}{e^{2.68}} = 4.87e^{-2.68}$

i.e. $\quad\quad\quad\quad x = \mathbf{0.3339}$, correct to 4 significant figures.

Problem 14. Solve $\frac{7}{4} = e^{3x}$ correct to 4 significant figures.

Taking natural logarithms of both sides gives:

$$\ln\frac{7}{4} = \ln e^{3x}$$

$$\ln\frac{7}{4} = 3x\ln e$$

Since $\ln e = 1$ $\quad\quad \ln\frac{7}{4} = 3x$

i.e. $\quad\quad\quad\quad 0.55962 = 3x$

i.e. $\quad\quad\quad\quad x = \mathbf{0.1865}$, correct to 4 significant figures.

Problem 15. Solve: $e^{x-1} = 2e^{3x-4}$ correct to 4 significant figures.

Taking natural logarithms of both sides gives:

$$\ln\left(e^{x-1}\right) = \ln\left(2e^{3x-4}\right)$$

and by the first law of logarithms,

$$\ln\left(e^{x-1}\right) = \ln 2 + \ln\left(e^{3x-4}\right)$$

i.e. $\quad\quad x - 1 = \ln 2 + 3x - 4$

Rearranging gives: $\quad 4 - 1 - \ln 2 = 3x - x$

i.e. $\quad\quad\quad\quad 3 - \ln 2 = 2x$

from which, $\quad\quad x = \frac{3 - \ln 2}{2}$

$$= \mathbf{1.153}$$

Problem 16. Solve, correct to 4 significant figures: $\ln(x-2)^2 = \ln(x-2) - \ln(x+3) + 1.6$

Rearranging gives:

$$\ln(x-2)^2 - \ln(x-2) + \ln(x+3) = 1.6$$

and by the laws of logarithms,

$$\ln\left\{\frac{(x-2)^2(x+3)}{(x-2)}\right\} = 1.6$$

Cancelling gives: $\quad \ln\{(x-2)(x+3)\} = 1.6$

and $\quad\quad\quad\quad (x-2)(x+3) = e^{1.6}$

i.e. $\quad\quad\quad\quad x^2 + x - 6 = e^{1.6}$

or $\quad\quad\quad\quad x^2 + x - 6 - e^{1.6} = 0$

i.e. $\quad\quad\quad\quad x^2 + x - 10.953 = 0$

Using the quadratic formula,

$$x = \frac{-1 \pm \sqrt{1^2 - 4(1)(-10.953)}}{2}$$

$$= \frac{-1 \pm \sqrt{44.812}}{2} = \frac{-1 \pm 6.6942}{2}$$

i.e. $x = 2.847$ or -3.8471

$x = -3.8471$ is not valid since the logarithm of a negative number has no real root.

Hence, **the solution of the equation is: $x = 2.847$**

Now try the following Practice Exercise

Practice Exercise 17   Napierian logarithms (Answers on page 857)

In Problems 1 and 2, evaluate correct to 5 significant figures:

1.   (a) $\dfrac{1}{3}\ln 5.2932$ (b) $\dfrac{\ln 82.473}{4.829}$

     (c) $\dfrac{5.62\ln 321.62}{e^{1.2942}}$

2.   (a) $\dfrac{1.786\ln e^{1.76}}{\lg 10^{1.41}}$ (b) $\dfrac{5e^{-0.1629}}{2\ln 0.00165}$

     (c) $\dfrac{\ln 4.8629 - \ln 2.4711}{5.173}$

In Problems 3 to 7 solve the given equations, each correct to 4 significant figures.

3. $\ln x = 2.10$

4. $24 + e^{2x} = 45$

5. $5 = e^{x+1} - 7$

6. $1.5 = 4e^{2t}$

7. $7.83 = 2.91e^{-1.7x}$

8. $16 = 24\left(1 - e^{-\frac{t}{2}}\right)$

9. $5.17 = \ln\left(\dfrac{x}{4.64}\right)$

10. $3.72\ln\left(\dfrac{1.59}{x}\right) = 2.43$

11. $5 = 8\left(1 - e^{\frac{-x}{2}}\right)$

12. $\ln(x + 3) - \ln x = \ln(x - 1)$

13. $\ln(x - 1)^2 - \ln 3 = \ln(x - 1)$

14. $\ln(x + 3) + 2 = 12 - \ln(x - 2)$

15. $e^{(x+1)} = 3e^{(2x-5)}$

16. $\ln(x + 1)^2 = 1.5 - \ln(x - 2) + \ln(x + 1)$

17.   Transpose: $b = \ln t - a\ln D$ to make t the subject.

18.   If $\dfrac{P}{Q} = 10\log_{10}\left(\dfrac{R_1}{R_2}\right)$ find the value of R_1 when $P = 160$, $Q = 8$ and $R_2 = 5$

19.   If $U_2 = U_1 e^{\left(\frac{W}{PV}\right)}$ make W the subject of the formula.

20.   The work done in an isothermal expansion of a gas from pressure p_1 to p_2 is given by:

$$w = w_0 \ln\left(\frac{p_1}{p_2}\right)$$

If the initial pressure $p_1 = 7.0$ kPa, calculate the final pressure p_2 if $w = 3w_0$

21.   The velocity v_2 of a rocket is given by: $v_2 = v_1 + C\ln\left(\dfrac{m_1}{m_2}\right)$ where v_1 is the initial rocket velocity, C is the velocity of the jet exhaust gases, m_1 is the mass of the rocket before the jet engine is fired, and m_2 is the mass of the rocket after the jet engine is switched off. Calculate the velocity of the rocket given $v_1 = 600$ m/s, $C = 3500$ m/s, $m_1 = 8.50 \times 10^4$ kg and $m_2 = 7.60 \times 10^4$ kg.

4.5   Laws of growth and decay

The laws of exponential growth and decay are of the form $y = A e^{-kx}$ and $y = A(1 - e^{-kx})$, where A and k are constants. When plotted, the form of each of these equations is as shown in Fig. 4.5. The laws occur frequently in engineering and science and examples of quantities related by a natural law include.

(i)   linear expansion $l = l_0 e^{\alpha\theta}$

(ii)   change in electrical resistance with temperature $R_\theta = R_0 e^{\alpha\theta}$

(iii)   tension in belts $T_1 = T_0 e^{\mu\theta}$

(iv)   newton's law of cooling $\theta = \theta_0 e^{-kt}$

(a)

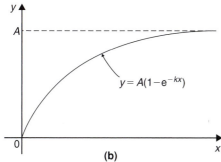

(b)

Figure 4.5

  (v) biological growth $y = y_0\,e^{kt}$

 (vi) discharge of a capacitor $q = Q\,e^{-t/CR}$

(vii) atmospheric pressure $p = p_0\,e^{-h/c}$

(viii) radioactive decay $N = N_0\,e^{-\lambda t}$

 (ix) decay of current in an
       inductive circuit $i = I\,e^{-Rt/L}$

  (x) growth of current in a
       capacitive circuit $i = I(1 - e^{-t/CR})$

Problem 17. The resistance R of an electrical conductor at temperature $\theta\,°C$ is given by $R = R_0\,e^{\alpha\theta}$, where α is a constant and $R_0 = 5 \times 10^3$ ohms. Determine the value of α, correct to 4 significant figures, when $R = 6 \times 10^3$ ohms and $\theta = 1500°C$. Also, find the temperature, correct to the nearest degree, when the resistance R is 5.4×10^3 ohms.

Transposing $R = R_0\,e^{\alpha\theta}$ gives $\dfrac{R}{R_0} = e^{\alpha\theta}$

Taking Napierian logarithms of both sides gives:

$$\ln\frac{R}{R_0} = \ln e^{\alpha\theta} = \alpha\theta$$

Hence $\alpha = \dfrac{1}{\theta}\ln\dfrac{R}{R_0} = \dfrac{1}{1500}\ln\left(\dfrac{6 \times 10^3}{5 \times 10^3}\right)$

$$= \frac{1}{1500}(0.1823215\ldots)$$

$$= 1.215477\cdots \times 10^{-4}$$

Hence $\boldsymbol{\alpha = 1.215 \times 10^{-4}}$,
                     correct to 4 significant figures.

From above, $\ln\dfrac{R}{R_0} = \alpha\theta$

hence $\theta = \dfrac{1}{\alpha}\ln\dfrac{R}{R_0}$

When $R = 5.4 \times 10^3$, $\alpha = 1.215477\ldots \times 10^{-4}$ and $R_0 = 5 \times 10^3$

$$\theta = \frac{1}{1.215477\ldots \times 10^{-4}}\ln\left(\frac{5.4 \times 10^3}{5 \times 10^3}\right)$$

$$= \frac{10^4}{1.215477\ldots}(7.696104\ldots \times 10^{-2})$$

$$= \boldsymbol{633°C},\text{ correct to the nearest degree.}$$

Problem 18. In an experiment involving Newton's law of cooling, the temperature $\theta\,(°C)$ is given by $\theta = \theta_0\,e^{-kt}$. Find the value of constant k when $\theta_0 = 56.6°C$, $\theta = 16.5°C$ and $t = 83.0$ seconds.

Transposing $\theta = \theta_0\,e^{-kt}$ gives

$$\frac{\theta}{\theta_0} = e^{-kt}$$

from which $\dfrac{\theta_0}{\theta} = \dfrac{1}{e^{-kt}} = e^{kt}$

Taking Napierian logarithms of both sides gives:

$$\ln\frac{\theta_0}{\theta} = kt$$

from which,

$$k = \frac{1}{t}\ln\frac{\theta_0}{\theta} = \frac{1}{83.0}\ln\left(\frac{56.6}{16.5}\right)$$

$$= \frac{1}{83.0}(1.2326486\ldots)$$

Hence $\boldsymbol{k = 1.485 \times 10^{-2}}$

Problem 19. The current i amperes flowing in a capacitor at time t seconds is given by $i = 8.0(1 - e^{\frac{-t}{CR}})$, where the circuit resistance R is 25×10^3 ohms and capacitance C is 16×10^{-6} farads. Determine (a) the current i after 0.5 seconds and (b) the time, to the nearest millisecond, for the current to reach 6.0 A. Sketch the graph of current against time.

(a) Current $i = 8.0(1 - e^{\frac{-t}{CR}})$

$$= 8.0[1 - e^{\frac{-0.5}{(16 \times 10^{-6})(25 \times 10^3)}}] = 8.0(1 - e^{-1.25})$$

$$= 8.0(1 - 0.2865047\ldots) = 8.0(0.7134952\ldots)$$

$$= \textbf{5.71 amperes}$$

(b) Transposing $i = 8.0(1 - e^{\frac{-t}{CR}})$

gives $\dfrac{i}{8.0} = 1 - e^{\frac{-t}{CR}}$

from which, $e^{\frac{-t}{CR}} = 1 - \dfrac{i}{8.0} = \dfrac{8.0 - i}{8.0}$

Taking the reciprocal of both sides gives:

$$e^{\frac{t}{CR}} = \frac{8.0}{8.0 - i}$$

Taking Napierian logarithms of both sides gives:

$$\frac{t}{CR} = \ln\left(\frac{8.0}{8.0 - i}\right)$$

Hence

$$t = CR\ln\left(\frac{8.0}{8.0 - i}\right)$$

$$= (16 \times 10^{-6})(25 \times 10^3)\ln\left(\frac{8.0}{8.0 - 6.0}\right)$$

when $i = 6.0$ amperes,

i.e. $t = \dfrac{400}{10^3}\ln\left(\dfrac{8.0}{2.0}\right) = 0.4\ln 4.0$

$$= 0.4(1.3862943\ldots) = 0.5545\,\text{s}$$

$$= \textbf{555 ms}, \text{ to the nearest millisecond.}$$

A graph of current against time is shown in Fig. 4.6.

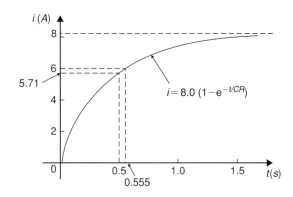

Figure 4.6

Problem 20. The temperature θ_2 of a winding which is being heated electrically at time t is given by: $\theta_2 = \theta_1(1 - e^{\frac{-t}{\tau}})$ where θ_1 is the temperature (in degrees Celsius) at time $t = 0$ and τ is a constant. Calculate,

(a) θ_1, correct to the nearest degree, when θ_2 is $50°C$, t is $30\,\text{s}$ and τ is $60\,\text{s}$

(b) the time t, correct to 1 decimal place, for θ_2 to be half the value of θ_1

(a) Transposing the formula to make θ_1 the subject gives:

$$\theta_1 = \frac{\theta_2}{(1 - e^{\frac{-t}{T}})} = \frac{50}{1 - e^{\frac{-30}{60}}}$$

$$= \frac{50}{1 - e^{-0.5}} = \frac{50}{0.393469\ldots}$$

i.e. $\boldsymbol{\theta_1 = 127°C}$, correct to the nearest degree.

(b) Transposing to make t the subject of the formula gives:

$$\frac{\theta_2}{\theta_1} = 1 - e^{\frac{-t}{\tau}}$$

from which, $e^{\frac{-t}{\tau}} = 1 - \dfrac{\theta_2}{\theta_1}$

Hence $-\dfrac{t}{\tau} = \ln\left(1 - \dfrac{\theta_2}{\theta_1}\right)$

i.e. $t = -\tau\ln\left(1 - \dfrac{\theta_2}{\theta_1}\right)$

Since $\theta_2 = \dfrac{1}{2}\theta_1$

$$t = -60\ln\left(1 - \frac{1}{2}\right)$$

$$= -60\ln 0.5 = 41.59\,\text{s}$$

Hence the time for the temperature θ_2 to be one half of the value of θ_1 is 41.6 s, correct to 1 decimal place.

Now try the following Practice Exercise

Practice Exercise 18   The laws of growth and decay (Answers on page 858)

1. The temperature, $T\,°C$, of a cooling object varies with time, t minutes, according to the equation: $T = 150e^{-0.04t}$. Determine the temperature when (a) $t = 0$, (b) $t = 10$ minutes.

2. The pressure p pascals at height h metres above ground level is given by $p = p_0 e^{\frac{-h}{C}}$, where p_0 is the pressure at ground level and C is a constant. Find pressure p when $p_0 = 1.012 \times 10^5$ Pa, height $h = 1420$ m, and $C = 71\,500$

3. The voltage drop, v volts, across an inductor L henrys at time t seconds is given by $v = 200e^{\frac{-Rt}{L}}$, where $R = 150\,\Omega$ and $L = 12.5 \times 10^{-3}$ H. Determine (a) the voltage when $t = 160 \times 10^{-6}$ s, and (b) the time for the voltage to reach 85 V.

4. The length l metres of a metal bar at temperature $t\,°C$ is given by $l = l_0 e^{\alpha t}$, where l_0 and α are constants. Determine (a) the value of α when $l = 1.993$ m, $l_0 = 1.894$ m and $t = 250\,°C$, and (b) the value of l_0 when $l = 2.416$, $t = 310\,°C$ and $\alpha = 1.682 \times 10^{-4}$

5. The temperature $\theta_2\,°C$ of an electrical conductor at time t seconds is given by: $\theta_2 = \theta_1(1 - e^{-t/T})$, where θ_1 is the initial temperature and T seconds is a constant. Determine:

    (a) θ_2 when $\theta_1 = 159.9\,°C$, $t = 30$ s and $T = 80$ s, and

    (b) the time t for θ_2 to fall to half the value of θ_1 if T remains at 80 s.

6. A belt is in contact with a pulley for a sector of $\theta = 1.12$ radians and the coefficient of friction between these two surfaces is $\mu = 0.26$. Determine the tension on the taut side of the belt, T newtons, when tension on the slack side $T_0 = 22.7$ newtons, given

that these quantities are related by the law $T = T_0 e^{\mu\theta}$. Determine also the value of θ when $T = 28.0$ newtons.

7. The instantaneous current i at time t is given by: $i = 10e^{\frac{-t}{CR}}$ when a capacitor is being charged. The capacitance C is 7×10^{-6} farads and the resistance R is 0.3×10^6 ohms. Determine:

    (a) the instantaneous current when t is 2.5 seconds, and

    (b) the time for the instantaneous current to fall to 5 amperes.

    Sketch a curve of current against time from $t = 0$ to $t = 6$ seconds.

8. The amount of product x (in mol/cm^3) found in a chemical reaction starting with 2.5 mol/cm^3 of reactant is given by $x = 2.5(1 - e^{-4t})$ where t is the time, in minutes, to form product x. Plot a graph at 30-second intervals up to 2.5 minutes and determine x after 1 minute.

9. The current i flowing in a capacitor at time t is given by:
    $$i = 12.5(1 - e^{\frac{-t}{CR}})$$
    where resistance R is 30 kilohms and the capacitance C is 20 micro-farads. Determine:

    (a) the current flowing after 0.5 seconds, and

    (b) the time for the current to reach 10 amperes.

10. The percentage concentration C of the starting material in a chemical reaction varies with time t according to the equation $C = 100e^{-0.004t}$. Determine the concentration when (a) $t = 0$, (b) $t = 100$ s, (c) $t = 1000$ s

11. The current i flowing through a diode at room temperature is given by: $i = i_S\left(e^{40V} - 1\right)$ amperes. Calculate the current flowing in a silicon diode when the reverse saturation current $i_S = 50$ nA and the forward voltage $V = 0.27$ V

12. A formula for chemical decomposition is given by: $C = A\left(1 - e^{-\frac{t}{10}}\right)$ where t is the time in seconds. Calculate the time, in milliseconds, for a compound to decompose to a value of $C = 0.12$ given $A = 8.5$

13. The mass, m, of pollutant in a water reservoir decreases according to the law $m = m_0 e^{-0.1t}$ where t is the time in days and m_0 is the initial mass. Calculate the percentage decrease in the mass after 60 days, correct to 3 decimal places.

14. A metal bar is cooled with water. Its temperature, in °C, is given by: $\theta = 15 + 1300e^{-0.2t}$ where t is the time in minutes. Calculate how long it will take for the temperature, θ, to decrease to 36°C, correct to the nearest second.

4.6 Reduction of exponential laws to linear form

Frequently, the relationship between two variables, say x and y, is not a linear one, i.e. when x is plotted against y a curve results. In such cases the non-linear equation may be modified to the linear form, $y = mx + c$, so that the constants, and thus the law relating the variables can be determined. This technique is called 'determination of law'.

Graph paper is available where the scale markings along the horizontal and vertical axes are proportional to the logarithms of the numbers. Such graph paper is called **log-log graph paper**.

A **logarithmic scale** is shown in Fig. 4.7 where the distance between, say, 1 and 2, is proportional to $\lg 2 - \lg 1$, i.e. 0.3010 of the total distance from 1 to 10. Similarly, the distance between 7 and 8 is proportional to $\lg 8 - \lg 7$, i.e. 0.05799 of the total distance from 1 to 10. Thus the distance between markings progressively decreases as the numbers increase from 1 to 10.

Figure 4.7

With log-log graph paper the scale markings are from 1 to 9, and this pattern can be repeated several times. The number of times the pattern of markings is repeated on an axis signifies the number of **cycles**. When the vertical axis has, say, three sets of values from 1 to 9, and the horizontal axis has, say, two sets of values from 1 to 9, then this log-log graph paper is called 'log 3 cycle × 2 cycle'. Many different arrangements are available ranging from 'log 1 cycle × 1 cycle' through to 'log 5 cycle × 5 cycle'.

To depict a set of values, say, from 0.4 to 161, on an axis of log-log graph paper, four cycles are required, from 0.1 to 1, 1 to 10, 10 to 100 and 100 to 1000.

Graphs of the form $y = a e^{kx}$

Taking logarithms to a base of e of both sides of $y = a e^{kx}$ gives:

$$\ln y = \ln(a e^{kx}) = \ln a + \ln e^{kx} = \ln a + kx \ln e$$

i.e. $\ln y = kx + \ln a$ (since $\ln e = 1$)

which compares with $Y = mX + c$

Thus, by plotting $\ln y$ vertically against x horizontally, a straight line results, i.e. the equation $y = a e^{kx}$ is reduced to linear form. In this case, graph paper having a linear horizontal scale and a logarithmic vertical scale may be used. This type of graph paper is called **log-linear graph paper**, and is specified by the number of cycles on the logarithmic scale.

Problem 21. The data given below are believed to be related by a law of the form $y = a e^{kx}$, where a and b are constants. Verify that the law is true and determine approximate values of a and b. Also determine the value of y when x is 3.8 and the value of x when y is 85

x	−1.2	0.38	1.2	2.5	3.4	4.2	5.3
y	9.3	22.2	34.8	71.2	117	181	332

Since $y = a e^{kx}$ then $\ln y = kx + \ln a$ (from above), which is of the form $Y = mX + c$, showing that to produce a straight line graph $\ln y$ is plotted vertically against x horizontally. The value of y ranges from 9.3 to 332, hence 'log 3 cycle × linear' graph paper is used. The plotted co-ordinates are shown in Fig. 4.8, and since a straight line passes through the points the law $y = a e^{kx}$ is verified.

Gradient of straight line,

$$k = \frac{AB}{BC} = \frac{\ln 100 - \ln 10}{3.12 - (-1.08)} = \frac{2.3026}{4.20}$$

$$= 0.55, \text{ correct to 2 significant figures.}$$

Since $\ln y = kx + \ln a$, when $x = 0$, $\ln y = \ln a$, i.e. $y = a$ The vertical axis intercept value at $x = 0$ is 18, hence $a = 18$

The law of the graph is thus $y = 18 e^{0.55x}$

When x is 3.8, $y = 18 e^{0.55(3.8)} = 18 e^{2.09}$

$$= 18(8.0849) = \mathbf{146}$$

Figure 4.8

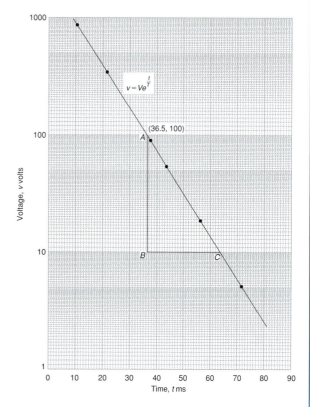

Figure 4.9

When y is 85, $85 = 18\,e^{0.55x}$

Hence, $e^{0.55x} = \dfrac{85}{18} = 4.7222$

and $0.55x = \ln 4.7222 = 1.5523$

Hence $x = \dfrac{1.5523}{0.55} = \mathbf{2.82}$

Problem 22.   The voltage, v volts, across an inductor is believed to be related to time, t ms, by the law $v = V\,e^{\frac{t}{T}}$, where V and T are constants. Experimental results obtained are:

v volts	883	347	90	55.5	18.6	5.2
t ms	10.4	21.6	37.8	43.6	56.7	72.0

Show that the law relating voltage and time is as stated and determine the approximate values of V and T. Find also the value of voltage after 25 ms and the time when the voltage is 30.0 V.

Since $v = V\,e^{\frac{t}{T}}$ then $\ln v = \frac{1}{T}t + \ln V$ which is of the form $Y = mX + c$.

Using 'log 3 cycle × linear' graph paper, the points are plotted as shown in Fig. 4.9.

Since the points are joined by a straight line the law $v = V\,e^{\frac{t}{T}}$ is verified.

Gradient of straight line,

$$\frac{1}{T} = \frac{AB}{BC}$$

$$= \frac{\ln 100 - \ln 10}{36.5 - 64.2}$$

$$= \frac{2.3026}{-27.7}$$

Hence $\boldsymbol{T} = \dfrac{-27.7}{2.3026}$

$$= \mathbf{-12.0}, \text{ correct to 3 significant figures.}$$

Since the straight line does not cross the vertical axis at $t = 0$ in Fig. 4.9, the value of V is determined by selecting any point, say A, having co-ordinates $(36.5, 100)$ and substituting these values into $v = V\,e^{\frac{t}{T}}$.

Thus $100 = V\,e^{\frac{36.5}{-12.0}}$

i.e. $V = \dfrac{100}{e^{\frac{-36.5}{12.0}}}$

$= \mathbf{2090\ volts}$,

correct to 3 significant figures.

Hence the law of the graph is $v = 2090\,e^{\frac{-t}{12.0}}$

When time $\quad t = 25$ ms,

voltage $\quad v = 2090\,e^{\frac{-25}{12.0}} = \mathbf{260\,V}$

When the voltage is 30.0 volts, $30.0 = 2090\,e^{\frac{-t}{12.0}}$,

hence $\quad e^{\frac{-t}{12.0}} = \dfrac{30.0}{2090}$

and $\quad e^{\frac{t}{12.0}} = \dfrac{2090}{30.0} = 69.67$

Taking Napierian logarithms gives:

$$\frac{t}{12.0} = \ln 69.67 = 4.2438$$

from which, time $t = (12.0)(4.2438) = \mathbf{50.9\,ms}$

Now try the following Practice Exercise

Practice Exercise 19 Reducing exponential laws to linear form (Answers on page 858)

1. Atmospheric pressure p is measured at varying altitudes h and the results are as shown:

Altitude, h m	pressure, p cm
500	73.39
1500	68.42
3000	61.60
5000	53.56
8000	43.41

Show that the quantities are related by the law $p = a\,e^{kh}$, where a and k are constants. Determine the values of a and k and state the law. Find also the atmospheric pressure at 10 000 m.

2. At particular times, t minutes, measurements are made of the temperature, $\theta\,^\circ C$, of a cooling liquid and the following results are obtained:

Temperature $\theta\,^\circ C$	Time t minutes
92.2	10
55.9	20
33.9	30
20.6	40
12.5	50

Prove that the quantities follow a law of the form $\theta = \theta_0\,e^{kt}$, where θ_0 and k are constants, and determine the approximate value of θ_0 and k.

For fully worked solutions to each of the problems in Practice Exercises 14 to 19 in this chapter, go to the website:
www.routledge.com/cw/bird

Chapter 5

Inequalities

Why it is important to understand: Inequalities

In mathematics, an inequality is a relation that holds between two values when they are different. A working knowledge of inequalities can be beneficial to the practising engineer, and inequalities are central to the definitions of all limiting processes, including differentiation and integration. When exact solutions are unavailable, inconvenient, or unnecessary, inequalities can be used to obtain error bounds for numerical approximation. Understanding and using inequalities is important in many branches of engineering.

At the end of this chapter, you should be able to:

- define an inequality
- state simple rules for inequalities
- solve simple inequalities
- solve inequalities involving a modulus
- solve inequalities involving quotients
- solve inequalities involving square functions
- solve quadratic inequalities

5.1 Introduction to inequalities

An **inequality** is any expression involving one of the symbols $<$, $>$, $\leq$ or $\geq$

$p < q$ means p is less than q

$p > q$ means p is greater than q

$p \leq q$ means p is less than or equal to q

$p \geq q$ means p is greater than or equal to q

Some simple rules

(i) When a quantity is **added or subtracted** to both sides of an inequality, the inequality still remains.

For example, if $p < 3$

then $p + 2 < 3 + 2$ (adding 2 to both sides)

and $p - 2 < 3 - 2$ (subtracting 2 from both sides)

(ii) When **multiplying or dividing** both sides of an inequality by a **positive** quantity, say 5, the inequality **remains the same**. For example,

if $p > 4$ then $5p > 20$ and $\dfrac{p}{5} > \dfrac{4}{5}$

(iii) When **multiplying or dividing** both sides of an inequality by a **negative** quantity, say -3, **the inequality is reversed**. For example,

if $p > 1$ then $-3p < -3$ and $\dfrac{p}{-3} < \dfrac{1}{-3}$

(Note $>$ has changed to $<$ in each example.)

To **solve an inequality** means finding all the values of the variable for which the inequality is true.

Knowledge of simple equations and quadratic equations is needed in this chapter.

5.2 Simple inequalities

The solution of some simple inequalities, using only the rules given in section 5.1, is demonstrated in the following worked problems.

Problem 1. Solve the following inequalities:
(a) $3 + x > 7$ (b) $3t < 6$
(c) $z - 2 \geq 5$ (d) $\frac{p}{3} \leq 2$

(a) Subtracting 3 from both sides of the inequality: $3 + x > 7$ gives:

$$3 + x - 3 > 7 - 3, \text{ i.e. } \mathbf{x > 4}$$

Hence, all values of x greater than 4 satisfy the inequality.

(b) Dividing both sides of the inequality: $3t < 6$ by 3 gives:

$$\frac{3t}{3} < \frac{6}{3}, \text{ i.e. } \mathbf{t < 2}$$

Hence, all values of t less than 2 satisfy the inequality.

(c) Adding 2 to both sides of the inequality $z - 2 \geq 5$ gives:

$$z - 2 + 2 \geq 5 + 2, \text{ i.e. } \mathbf{z \geq 7}$$

Hence, all values of z greater than or equal to 7 satisfy the inequality.

(d) Multiplying both sides of the inequality $\frac{p}{3} \leq 2$ by 3 gives:

$$(3)\frac{p}{3} \leq (3)2, \text{ i.e. } \mathbf{p \leq 6}$$

Hence, all values of p less than or equal to 6 satisfy the inequality.

Problem 2. Solve the inequality: $4x + 1 > x + 5$

Subtracting 1 from both sides of the inequality: $4x + 1 > x + 5$ gives:

$$4x > x + 4$$

Subtracting x from both sides of the inequality: $4x > x + 4$ gives:

$$3x > 4$$

Dividing both sides of the inequality: $3x > 4$ by 3 gives:

$$x > \frac{4}{3}$$

Hence all values of x greater than $\frac{4}{3}$ satisfy the inequality:

$$4x + 1 > x + 5$$

Problem 3. Solve the inequality: $3 - 4t \leq 8 + t$

Subtracting 3 from both sides of the inequality: $3 - 4t \leq 8 + t$ gives:

$$-4t \leq 5 + t$$

Subtracting t from both sides of the inequality: $-4t \leq 5 + t$ gives:

$$-5t \leq 5$$

Dividing both sides of the inequality $-5t \leq 5$ by -5 gives:

$$t \geq -1 \text{ (remembering to reverse the inequality)}$$

Hence, all values of t greater than or equal to -1 satisfy the inequality.

Now try the following Practice Exercise

Practice Exercise 20   Simple inequalities (Answers on page 858)

Solve the following inequalities:

1. (a) $3t > 6$ (b) $2x < 10$

2. (a) $\frac{x}{2} > 1.5$ (b) $x + 2 \geq 5$

3. (a) $4t - 1 \leq 3$ (b) $5 - x \geq -1$

4. (a) $\frac{7 - 2k}{4} \leq 1$ (b) $3z + 2 > z + 3$

5. (a) $5 - 2y \leq 9 + y$ (b) $1 - 6x \leq 5 + 2x$

5.3 Inequalities involving a modulus

The **modulus** of a number is the size of the number, regardless of sign. Vertical lines enclosing the number denote a modulus.

For example, $|4| = 4$ and $|-4| = 4$ (the modulus of a number is never negative),

The inequality: $|t| < 1$ means that all numbers whose actual size, regardless of sign, is less than 1, i.e. any value between -1 and $+1$

Thus $|t| < 1$ means $-1 < t < 1$

Similarly, $|x| > 3$ means all numbers whose actual size, regardless of sign, is greater than 3, i.e. any value greater than 3 and any value less than -3

Thus $|x| > 3$ means $x > 3$ and $x < -3$

Inequalities involving a modulus are demonstrated in the following worked problems.

Problem 4. Solve the following inequality:
$$|3x + 1| < 4$$

Since $|3x + 1| < 4$ then $-4 < 3x + 1 < 4$

Now $-4 < 3x + 1$ becomes $-5 < 3x$,

i.e. $-\dfrac{5}{3} < x$ and $3x + 1 < 4$ becomes $3x < 3$,

i.e. $x < 1$

Hence, these two results together become $-\dfrac{5}{3} < x < 1$ and mean that the inequality $|3x + 1| < 4$ is satisfied for any value of x greater than $-\dfrac{5}{3}$ but less than 1

Problem 5. Solve the inequality: $|1 + 2t| \le 5$

Since $|1 + 2t| \le 5$ then $-5 \le 1 + 2t \le 5$

Now $-5 \le 1 + 2t$ becomes $-6 \le 2t$, i.e. $-3 \le t$

and $1 + 2t \le 5$ becomes $2t \le 4$ i.e. $t \le 2$

Hence, these two results together become: $-3 \le t \le 2$

Problem 6. Solve the inequality: $|3z - 4| > 2$

$|3z - 4| > 2$ means $3z - 4 > 2$ and $3z - 4 < -2$,

i.e. $3z > 6$ and $3z < 2$,

i.e. the inequality: $|3z - 4| > 2$ is satisfied when

$$z > 2 \text{ and } z < \frac{2}{3}$$

Now try the following Practice Exercise

Practice Exercise 21 Inequalities involving a modulus (Answers on page 858)

Solve the following inequalities:

1. $|t + 1| < 4$
2. $|y + 3| \le 2$
3. $|2x - 1| < 4$
4. $|3t - 5| > 4$
5. $|1 - k| \ge 3$

5.4 Inequalities involving quotients

If $\dfrac{p}{q} > 0$ then $\dfrac{p}{q}$ must be a **positive** value.

For $\dfrac{p}{q}$ to be positive, **either** p is positive **and** q is positive **or** p is negative **and** q is negative.

i.e. $\dfrac{+}{+} = +$ and $\dfrac{-}{-} = +$

If $\dfrac{p}{q} < 0$ then $\dfrac{p}{q}$ must be a **negative** value.

For $\dfrac{p}{q}$ to be negative, **either** p is positive **and** q is negative **or** p is negative **and** q is positive.

i.e. $\dfrac{+}{-} = -$ and $\dfrac{-}{+} = -$

This reasoning is used when solving inequalities involving quotients, as demonstrated in the following worked problems.

Problem 7. Solve the inequality: $\dfrac{t + 1}{3t - 6} > 0$

Since $\dfrac{t + 1}{3t - 6} > 0$ then $\dfrac{t + 1}{3t - 6}$ must be **positive**.

For $\dfrac{t + 1}{3t - 6}$ to be positive,

either (i) $t + 1 > 0$ and $3t - 6 > 0$

or (ii) $t + 1 < 0$ and $3t - 6 < 0$

(i) If $t + 1 > 0$ then $t > -1$ and if $3t - 6 > 0$ then $3t > 6$ and $t > 2$

Both of the inequalities $t > -1$ and $t > 2$ are only true when $t > 2$,

i.e. the fraction $\dfrac{t+1}{3t-6}$ is positive when **$t > 2$**

(ii) If $t + 1 < 0$ then $t < -1$ and if $3t - 6 < 0$ then $3t < 6$ and $t < 2$

Both of the inequalities $t < -1$ and $t < 2$ are only true when $t < -1$,

i.e. the fraction $\dfrac{t+1}{3t-6}$ is positive when **$t < -1$**

Summarising, $\dfrac{t+1}{3t-6} > 0$ when **$t > 2$** or **$t < -1$**

Problem 8. Solve the inequality: $\dfrac{2x+3}{x+2} \le 1$

Since $\dfrac{2x+3}{x+2} \le 1$ then $\dfrac{2x+3}{x+2} - 1 \le 0$

i.e. $\dfrac{2x+3}{x+2} - \dfrac{x+2}{x+2} \le 0$,

i.e. $\dfrac{2x+3-(x+2)}{x+2} \le 0$ or $\dfrac{x+1}{x+2} \le 0$

For $\dfrac{x+1}{x+2}$ to be negative or zero,

either (i) $x + 1 \le 0$ **and** $x + 2 > 0$

or (ii) $x + 1 \ge 0$ **and** $x + 2 < 0$

(i) If $x + 1 \le 0$ then $x \le -1$ and if $x + 2 > 0$ then $x > -2$

(Note that $>$ is used for the denominator, not $\ge$; a zero denominator gives a value for the fraction which is impossible to evaluate.)

Hence, the inequality $\dfrac{x+1}{x+2} \le 0$ is true when x is greater than -2 and less than or equal to -1, which may be written as **$-2 < x \le -1$**

(ii) If $x + 1 \ge 0$ then $x \ge -1$ and if $x + 2 < 0$ then $x < -2$

It is not possible to satisfy both $x \ge -1$ and $x < -2$ thus no values of x satisfies (ii).

Summarising, $\dfrac{2x+3}{x+2} \le 1$ when **$-2 < x \le -1$**

Now try the following Practice Exercise

Practice Exercise 22   Inequalities involving quotients (Answers on page 858)

Solve the following inequalities:

1. $\dfrac{x+4}{6-2x} \ge 0$

2. $\dfrac{2t+4}{t-5} > 1$

3. $\dfrac{3z-4}{z+5} \le 2$

4. $\dfrac{2-x}{x+3} \ge 4$

5.5 Inequalities involving square functions

The following two general rules apply when inequalities involve square functions:

(i) **if** $x^2 > k$ **then** $x > \sqrt{k}$ **or** $x < -\sqrt{k}$ (1)

(ii) **if** $x^2 < k$ **then** $-\sqrt{k} < x < \sqrt{k}$ (2)

These rules are demonstrated in the following worked problems.

Problem 9. Solve the inequality: $t^2 > 9$

Since $t^2 > 9$ then $t^2 - 9 > 0$, i.e. $(t+3)(t-3) > 0$ by factorising

For $(t+3)(t-3)$ to be positive,

either (i) $(t+3) > 0$ **and** $(t-3) > 0$

or (ii) $(t+3) < 0$ **and** $(t-3) < 0$

(i) If $(t+3) > 0$ then $t > -3$ and if $(t-3) > 0$ then $t > 3$

Both of these are true only when **$t > 3$**

(ii) If $(t+3) < 0$ then $t < -3$ and if $(t-3) < 0$ then $t < 3$

Both of these are true only when **$t < -3$**

Summarising, $t^2 > 9$ when **$t > 3$** or **$t < -3$**

This demonstrates the general rule:

if $x^2 > k$ **then** $x > \sqrt{k}$ **or** $x < -\sqrt{k}$ (1)

Problem 10.   Solve the inequality: $x^2 > 4$

From the general rule stated in equation (1):
if $x^2 > 4$ then $x > \sqrt{4}$ or $x < -\sqrt{4}$

i.e. the inequality: $x^2 > 4$ is satisfied when **$x > 2$** or **$x < -2$**

Problem 11.   Solve the inequality: $(2z + 1)^2 > 9$

From equation (1), if $(2z + 1)^2 > 9$ then

$$2z + 1 > \sqrt{9} \quad \text{or} \quad 2z + 1 < -\sqrt{9}$$
$$\text{i.e.} \quad 2z + 1 > 3 \quad \text{or} \quad 2z + 1 < -3$$
$$\text{i.e.} \quad 2z > 2 \quad \text{or} \quad 2z < -4$$
$$\text{i.e.} \quad \boldsymbol{z > 1} \quad \text{or} \quad \boldsymbol{z < -2}$$

Problem 12.   Solve the inequality: $t^2 < 9$

Since $t^2 < 9$ then $t^2 - 9 < 0$, i.e. $(t + 3)(t - 3) < 0$ by factorising. For $(t + 3)(t - 3)$ to be negative,

either   (i) $(t + 3) > 0$ **and** $(t - 3) < 0$
or   (ii) $(t + 3) < 0$ **and** $(t - 3) > 0$

(i)   If $(t + 3) > 0$ then $t > -3$ and if $(t - 3) < 0$ then $t < 3$
Hence (i) is satisfied when $t > -3$ and $t < 3$ which may be written as: $\boldsymbol{-3 < t < 3}$

(ii)   If $(t + 3) < 0$ then $t < -3$ and if $(t - 3) > 0$ then $t > 3$
It is not possible to satisfy both $t < -3$ and $t > 3$, thus no values of t satisfies (ii).

Summarising, $t^2 < 9$ when $\boldsymbol{-3 < t < 3}$ which means that all values of t between -3 and $+3$ will satisfy the inequality.
This demonstrates the general rule:

$$\textbf{if } x^2 < k \textbf{ then } -\sqrt{k} < x < \sqrt{k} \qquad (2)$$

Problem 13.   Solve the inequality: $x^2 < 4$

From the general rule stated in equation (2):
if $x^2 < 4$ then $-\sqrt{4} < x < \sqrt{4}$
i.e. the inequality: $x^2 < 4$ is satisfied when:

$$-2 < x < 2$$

Problem 14.   Solve the inequality: $(y - 3)^2 \leq 16$

From equation (2), $-\sqrt{16} \leq (y - 3) \leq \sqrt{16}$

i.e. $-4 \leq (y - 3) \leq 4$
from which, $3 - 4 \leq y \leq 4 + 3$,
i.e. $\boldsymbol{-1 \leq y \leq 7}$

Now try the following Practice Exercise

Practice Exercise 23   Inequalities involving square functions (Answers on page 858)

Solve the following inequalities:

1. $z^2 > 16$

2. $z^2 < 16$

3. $2x^2 \geq 6$

4. $3k^2 - 2 \leq 10$

5. $(t - 1)^2 \leq 36$

6. $(t - 1)^2 \geq 36$

7. $7 - 3y^2 \leq -5$

8. $(4k + 5)^2 > 9$

5.6   Quadratic inequalities

Inequalities involving quadratic expressions are solved using either **factorisation** or '**completing the square**'. For example,

$$x^2 - 2x - 3 \text{ is factorised as } (x + 1)(x - 3)$$

and $6x^2 + 7x - 5$ is factorised as $(2x - 1)(3x + 5)$

If a quadratic expression does not factorise, then the technique of 'completing the square' is used. In general, the procedure for $x^2 + bx + c$ is:

$$x^2 + bx + c \equiv \left(x + \frac{b}{2}\right)^2 + c - \left(\frac{b}{2}\right)^2$$

For example, $x^2 + 4x - 7$ does not factorise; completing the square gives:

$$x^2 + 4x - 7 \equiv (x + 2)^2 - 7 - 2^2 \equiv (x + 2)^2 - 11$$

Similarly,

$$x^2 - 6x - 5 \equiv (x - 3)^2 - 5 - 3^2 \equiv (x - 3)^2 - 14$$

Solving quadratic inequalities is demonstrated in the following worked problems.

Problem 15. Solve the inequality:
$$x^2 + 2x - 3 > 0$$

Since $x^2 + 2x - 3 > 0$ then $(x - 1)(x + 3) > 0$ by factorising. For the product $(x - 1)(x + 3)$ to be positive,

either (i) $(x - 1) > 0$ **and** $(x + 3) > 0$

or (ii) $(x - 1) < 0$ **and** $(x + 3) < 0$

(i) Since $(x - 1) > 0$ then $x > 1$ and since $(x + 3) > 0$ then $x > -3$
Both of these inequalities are satisfied only when **$x > 1$**

(ii) Since $(x - 1) < 0$ then $x < 1$ and since $(x + 3) < 0$ then $x < -3$
Both of these inequalities are satisfied only when **$x < -3$**

Summarising, $x^2 + 2x - 3 > 0$ is satisfied when either **$x > 1$ or $x < -3$**

Problem 16. Solve the inequality:
$$t^2 - 2t - 8 < 0$$

Since $t^2 - 2t - 8 < 0$ then $(t - 4)(t + 2) < 0$ by factorising.
For the product $(t - 4)(t + 2)$ to be negative,

either (i) $(t - 4) > 0$ **and** $(t + 2) < 0$

or (ii) $(t - 4) < 0$ **and** $(t + 2) > 0$

(i) Since $(t - 4) > 0$ then $t > 4$ and since $(t + 2) < 0$ then $t < -2$
It is not possible to satisfy both $t > 4$ and $t < -2$, thus no values of t satisfies the inequality (i)

(ii) Since $(t - 4) < 0$ then $t < 4$ and since $(t + 2) > 0$ then $t > -2$
Hence, (ii) is satisfied when $-2 < t < 4$

Summarising, $t^2 - 2t - 8 < 0$ is satisfied when **$-2 < t < 4$**

Problem 17. Solve the inequality:
$$x^2 + 6x + 3 < 0$$

$x^2 + 6x + 3$ does not factorise; completing the square gives:

$$x^2 + 6x + 3 \equiv (x + 3)^2 + 3 - 3^2$$
$$\equiv (x + 3)^2 - 6$$

The inequality thus becomes: $(x + 3)^2 - 6 < 0$ or $(x + 3)^2 < 6$
From equation (2), $-\sqrt{6} < (x + 3) < \sqrt{6}$

from which, $(-\sqrt{6} - 3) < x < (\sqrt{6} - 3)$

Hence, $x^2 + 6x + 3 < 0$ is satisfied when

$-5.45 < x < -0.55$ correct to 2 decimal places.

Problem 18. Solve the inequality:
$$y^2 - 8y - 10 \geq 0$$

$y^2 - 8y - 10$ does not factorise; completing the square gives:

$$y^2 - 8y - 10 \equiv (y - 4)^2 - 10 - 4^2$$
$$\equiv (y - 4)^2 - 26$$

The inequality thus becomes: $(y - 4)^2 - 26 \geq 0$ or $(y - 4)^2 \geq 26$

From equation (1), $(y - 4) \geq \sqrt{26}$ or $(y - 4) \leq -\sqrt{26}$

from which, **$y \geq 4 + \sqrt{26}$ or $y \leq 4 - \sqrt{26}$**

Hence, $y^2 - 8y - 10 \geq 0$ is satisfied when **$y \geq 9.10$** or **$y \leq -1.10$** correct to 2 decimal places.

Now try the following Practice Exercise

Practice Exercise 24 Quadratic inequalities (Answers on page 858)

Solve the following inequalities:

1. $x^2 - x - 6 > 0$

2. $t^2 + 2t - 8 \leq 0$

3. $2x^2 + 3x - 2 < 0$

4. $y^2 - y - 20 \geq 0$

5. $z^2 + 4z + 4 \leq 4$

6. $x^2 + 6x + 6 \leq 0$

7. $t^2 - 4t - 7 \geq 0$

8. $k^2 + k - 3 \geq 0$

**For fully worked solutions to each of the problems in Practice Exercises 20 to 24 in this chapter,
go to the website:
www.routledge.com/cw/bird**

Revision Test 1 Algebra, partial fractions, logarithms, exponentials and inequalities

This Revision Test covers the material contained in Chapters 1 to 5. *The marks for each question are shown in brackets at the end of each question.*

1. Factorise $x^3 + 4x^2 + x - 6$ using the factor theorem. Hence solve the equation

$$x^3 + 4x^2 + x - 6 = 0 \qquad (6)$$

2. Use the remainder theorem to find the remainder when $2x^3 + x^2 - 7x - 6$ is divided by

   (a) $(x - 2)$ (b) $(x + 1)$

   Hence factorise the cubic expression $\qquad (8)$

3. Simplify $\dfrac{6x^2 + 7x - 5}{2x - 1}$ by dividing out $\qquad (4)$

4. Resolve the following into partial fractions

   (a) $\dfrac{x - 11}{x^2 - x - 2}$ (b) $\dfrac{3 - x}{(x^2 + 3)(x + 3)}$

   (c) $\dfrac{x^3 - 6x + 9}{x^2 + x - 2}$ $\qquad (25)$

5. Evaluate, correct to 3 decimal places,

$$\dfrac{5\,e^{-0.982}}{3\ln 0.0173} \qquad (2)$$

6. Solve the following equations, each correct to 4 significant figures:

   (a) $\ln x = 2.40$ (b) $3^{x-1} = 5^{x-2}$

   (c) $5 = 8(1 - e^{-\frac{x}{2}})$ $\qquad (10)$

7. (a) The pressure p at height h above ground level is given by: $p = p_0 e^{-kh}$ where p_0 is the pressure at ground level and k is a constant. When p_0 is 101 kilopascals and the pressure at a height of 1500 m is 100 kilopascals, determine the value of k.

   (b) Sketch a graph of p against h (p the vertical axis and h the horizontal axis) for values of height from zero to 12 000 m when p_0 is 101 kilopascals.

   (c) If pressure $p = 95$ kPa, ground-level pressure $p_0 = 101$ kPa, constant $k = 5 \times 10^{-6}$, determine the height above ground level, h, in kilometres correct to 2 decimal places. $\qquad (13)$

8. Solve the following equations:

   (a) $\log(x^2 + 8) - \log(2x) = \log 3$

   (b) $\ln x + \ln(x - 3) = \ln 6x - \ln(x - 2)$ $\qquad (13)$

9. If $\theta_f - \theta_i = \dfrac{R}{J}\ln\left(\dfrac{U_2}{U_1}\right)$ find the value of U_2 given that $\theta_f = 3.5$, $\theta_i = 2.5$, $R = 0.315$, $J = 0.4$, $U_1 = 50$ $\qquad (6)$

10. Solve, correct to 4 significant figures:

    (a) $13e^{2x-1} = 7e^x$

    (b) $\ln(x + 1)^2 = \ln(x + 1) - \ln(x + 2) + 2$ $\qquad (15)$

11. Solve the following inequalities:

    (a) $2 - 5x \le 9 + 2x$ (b) $|3 + 2t| \le 6$

    (c) $\dfrac{x - 1}{3x + 5} > 0$ (d) $(3t + 2)^2 > 16$

    (e) $2x^2 - x - 3 < 0$ $\qquad (18)$

For lecturers/instructors/teachers, fully worked solutions to each of the problems in Revision Test 1, together with a full marking scheme, are available at the website:
www.routledge.com/cw/bird

Chapter 6

Arithmetic and geometric progressions

Why it is important to understand: Arithmetic and geometric progressions

Number sequences are widely used in engineering applications, including computer data structure and sorting algorithms, financial engineering, audio compression, and architectural engineering. Thanks to engineers, robots have migrated from factory shop floors as industrial manipulators, to outer space as interplanetary explorers, hospitals as minimally invasive surgical assistants, homes as vacuum cleaners and lawn mowers, and battlefields as unmanned air, underwater, and ground vehicles. Arithmetic progressions are used in simulation engineering and in the reproductive cycle of bacteria. Some uses of APs in daily life include uniform increase in speed at regular intervals, completing patterns of objects, calculating simple interest, speed of an aircraft, increase or decrease in the costs of goods, sales and production, and so on. GPs are used in compound interest and the range of speeds on a drilling machine. In fact, GPs are used throughout mathematics, and they have many important applications in physics, engineering, biology, economics, computer science, queuing theory, and finance.

At the end of this chapter, you should be able to:

- calculate the nth term of an AP
- calculate the sum of n terms of an AP
- calculate the nth term of a GP
- calculate the sum of n terms of a GP
- calculate the sum to infinity of a GP

6.1 Arithmetic progressions

When a sequence has a constant difference between successive terms it is called an **arithmetic progression** (often abbreviated to AP).
Examples include:

(i) 1, 4, 7, 10, 13,... where the **common difference** is 3 and

(ii) a, $a+d$, $a+2d$, $a+3d$,... where the common difference is d.

General expression for the nth term of an AP

If the first term of an AP is 'a' and the common difference is 'd' then

the nth term is: $a + (n - 1)d$

In example (i) above, the seventh term is given by $1 + (7 - 1)3 = \mathbf{19}$, which may be readily checked.

Sum of *n* terms of an AP

The sum S of an AP can be obtained by multiplying the average of all the terms by the number of terms.

The average of all the terms $= \dfrac{a+l}{2}$, where a is the first term and l is the last term, i.e. $l = a + (n-1)d$, for n terms.

Hence the sum of n terms,

$$S_n = n\left(\frac{a+l}{2}\right)$$

$$= \frac{n}{2}\{a + [a + (n-1)d]\}$$

i.e. $$S_n = \frac{n}{2}[2a + (n-1)d]$$

For example, the sum of the first seven terms of the series $1, 4, 7, 10, 13, \ldots$ is given by

$$S_7 = \frac{7}{2}[2(1) + (7-1)3], \text{ since } a = 1 \text{ and } d = 3$$

$$= \frac{7}{2}[2 + 18] = \frac{7}{2}[20] = \mathbf{70}$$

6.2 Worked problems on arithmetic progressions

Problem 1. Determine (a) the ninth and (b) the sixteenth term of the series $2, 7, 12, 17, \ldots$

$2, 7, 12, 17, \ldots$ is an arithmetic progression with a common difference, d, of 5

(a) The nth term of an AP is given by $a + (n-1)d$
Since the first term $a = 2$, $d = 5$ and $n = 9$ then the ninth term is:
$2 + (9-1)5 = 2 + (8)(5) = 2 + 40 = \mathbf{42}$

(b) The sixteenth term is:
$2 + (16-1)5 = 2 + (15)(5) = 2 + 75 = \mathbf{77}$

Problem 2. The sixth term of an AP is 17 and the thirteenth term is 38. Determine the nineteenth term.

The nth term of an AP is $a + (n-1)d$

The sixth term is: $\quad a + 5d = 17$ \hfill (1)

The nineteenth term is: $\quad a + 12d = 38$ \hfill (2)

Equation (2) $-$ equation (1) gives: $7d = 21$, from which,
$$d = \frac{21}{7} = 3$$

Substituting in equation (1) gives: $a + 15 = 17$, from which, $a = 2$

Hence the nineteenth term is:
$a + (n-1)d = 2 + (19-1)3 = 2 + (18)(3)$
$$= 2 + 54 = \mathbf{56}$$

Problem 3. Determine the number of the term whose value is 22 in the series $2\frac{1}{2}, 4, 5\frac{1}{2}, 7, \ldots$

$2\frac{1}{2}, 4, 5\frac{1}{2}, 7, \ldots$ is an AP where $a = 2\frac{1}{2}$ and $d = 1\frac{1}{2}$

Hence if the nth term is 22 then: $a + (n-1)d = 22$
i.e. $2\frac{1}{2} + (n-1)\left(1\frac{1}{2}\right) = 22$
$(n-1)\left(1\frac{1}{2}\right) = 22 - 2\frac{1}{2} = 19\frac{1}{2}$

$$n - 1 = \frac{19\frac{1}{2}}{1\frac{1}{2}} = 13 \text{ and } n = 13 + 1 = 14$$

i.e. **the fourteenth term of the AP is 22**

Problem 4. Find the sum of the first 12 terms of the series $5, 9, 13, 17, \ldots$

$5, 9, 13, 17, \ldots$ is an AP where $a = 5$ and $d = 4$. The sum of n terms of an AP,

$$S_n = \frac{n}{2}[2a + (n-1)d]$$

Hence the sum of the first 12 terms,

$$S_{12} = \frac{12}{2}[2(5) + (12-1)4]$$

$$= 6[10 + 44] = 6(54) = \mathbf{324}$$

Problem 5. Find the sum of the first 21 terms of the series $3.5, 4.1, 4.7, 5.3, \ldots$

$3.5, 4.1, 4.7, 5.3, \ldots$ is an AP where $a = 3.5$ and $d = 0.6$
The sum of the first 21 terms,

$$S_{21} = \frac{21}{2}[2a + (n-1)d]$$

$$= \frac{21}{2}[2(3.5) + (21-1)0.6] = \frac{21}{2}[7 + 12]$$

$$= \frac{21}{2}(19) = \frac{399}{2} = \mathbf{199.5}$$

Now try the following Practice Exercise

Practice Exercise 25   Arithmetic progressions (Answers on page 858)

1. Find the eleventh term of the series 8, 14, 20, 26, …

2. Find the seventeenth term of the series 11, 10.7, 10.4, 10.1, …

3. The seventh term of a series is 29 and the eleventh term is 54. Determine the sixteenth term.

4. Find the fifteenth term of an arithmetic progression of which the first term is 2.5 and the tenth term is 16.

5. Determine the number of the term which is 29 in the series 7, 9.2, 11.4, 13.6, …

6. Find the sum of the first 11 terms of the series 4, 7, 10, 13, …

7. Determine the sum of the series 6.5, 8.0, 9.5, 11.0, …, 32

6.3 Further worked problems on arithmetic progressions

Problem 6. The sum of seven terms of an AP is 35 and the common difference is 1.2. Determine the first term of the series.

$n = 7$, $d = 1.2$ and $S_7 = 35$
Since the sum of n terms of an AP is given by

$$S_n = \frac{n}{2}[2a + (n-1)d], \text{ then}$$

$$35 = \frac{7}{2}[2a + (7-1)1.2] = \frac{7}{2}[2a + 7.2]$$

Hence $\dfrac{35 \times 2}{7} = 2a + 7.2$

$$10 = 2a + 7.2$$

Thus $2a = 10 - 7.2 = 2.8$,

from which $a = \dfrac{2.8}{2} = 1.4$

i.e. **the first term, $a = 1.4$**

Problem 7. Three numbers are in arithmetic progression. Their sum is 15 and their product is 80. Determine the three numbers.

Let the three numbers be $(a-d)$, a and $(a+d)$

Then $(a-d) + a + (a+d) = 15$, i.e. $3a = 15$, from which, $a = 5$

Also, $a(a-d)(a+d) = 80$, i.e. $a(a^2 - d^2) = 80$

Since $a = 5, 5(5^2 - d^2) = 80$

$$125 - 5d^2 = 80$$

$$125 - 80 = 5d^2$$

$$45 = 5d^2$$

from which, $d^2 = \dfrac{45}{5} = 9$. Hence $d = \sqrt{9} = \pm 3$.
The three numbers are thus $(5-3)$, 5 and $(5+3)$, i.e. **2, 5 and 8**.

Problem 8. Find the sum of all the numbers between 0 and 207 which are exactly divisible by 3.

The series 3, 6, 9, 12, …, 207 is an AP whose first term $a = 3$ and common difference $d = 3$

The last term is $a + (n-1)d = 207$

i.e. $3 + (n-1)3 = 207$,

from which $(n-1) = \dfrac{207 - 3}{3} = 68$

Hence $n = 68 + 1 = 69$

The sum of all 69 terms is given by

$$S_{69} = \frac{n}{2}[2a + (n-1)d]$$

$$= \frac{69}{2}[2(3) + (69-1)3]$$

$$= \frac{69}{2}[6 + 204] = \frac{69}{2}(210) = \mathbf{7245}$$

Problem 9. The first, twelfth and last term of an arithmetic progression are 4, $31\frac{1}{2}$ and $376\frac{1}{2}$, respectively. Determine (a) the number of terms in the series, (b) the sum of all the terms and (c) the 80th term.

(a) Let the AP be $a, a+d, a+2d, \ldots, a+(n-1)d$, where $a = 4$

The twelfth term is: $a + (12 - 1)d = 31\frac{1}{2}$

i.e. $4 + 11d = 31\frac{1}{2}$

from which, $11d = 31\frac{1}{2} - 4 = 27\frac{1}{2}$

Hence $d = \dfrac{27\frac{1}{2}}{11} = 2\frac{1}{2}$

The last term is $a + (n - 1)d$

i.e. $4 + (n - 1)\left(2\frac{1}{2}\right) = 376\frac{1}{2}$

$$(n - 1) = \frac{376\frac{1}{2} - 4}{2\frac{1}{2}}$$

$$= \frac{372\frac{1}{2}}{2\frac{1}{2}} = 149$$

Hence the number of terms in the series, $n = 149 + 1 = 150$

(b)   Sum of all the terms,

$$S_{150} = \frac{n}{2}[2a + (n - 1)d]$$

$$= \frac{150}{2}\left[2(4) + (150 - 1)\left(2\frac{1}{2}\right)\right]$$

$$= 75\left[8 + (149)\left(2\frac{1}{2}\right)\right]$$

$$= 85[8 + 372.5]$$

$$= 75(380.5) = \mathbf{28\,537}\frac{\mathbf{1}}{\mathbf{2}}$$

(c)   The 80th term is:

$$a + (n - 1)d = 4 + (80 - 1)\left(2\frac{1}{2}\right)$$

$$= 4 + (79)\left(2\frac{1}{2}\right)$$

$$= 4 + 197.5 = \mathbf{201}\frac{\mathbf{1}}{\mathbf{2}}$$

Problem 10.   An oil company bores a hole 80 m deep. Estimate the cost of boring if the cost is £30 for drilling the first metre with an increase in cost of £2 per metre for each succeeding metre.

The series is: $30, 32, 34, \ldots$ to 80 terms, i.e. $a = 30$, $d = 2$ and $n = 80$

Thus, **total cost,**

$$S_n = \frac{n}{2}\left[2a + (n - 1)d\right]$$

$$= \frac{80}{2}[2(30) + (80 - 1)(2)]$$

$$= 40[60 + 158] = 40(218) = \mathbf{£8720}$$

Now try the following Practice Exercise

Practice Exercise 26   Arithmetic progressions (Answers on page 858)

1. The sum of 15 terms of an arithmetic progression is 202.5 and the common difference is 2. Find the first term of the series.

2. Three numbers are in arithmetic progression. Their sum is 9 and their product is 20.25. Determine the three numbers.

3. Find the sum of all the numbers between 5 and 250 which are exactly divisible by 4

4. Find the number of terms of the series 5, 8, 11, … of which the sum is 1025

5. Insert four terms between 5 and 22.5 to form an arithmetic progression.

6. The first, tenth and last terms of an arithmetic progression are 9, 40.5 and 425.5, respectively. Find (a) the number of terms, (b) the sum of all the terms and (c) the 70th term.

7. On commencing employment a man is paid a salary of £16 000 per annum and receives annual increments of £480. Determine his salary in the ninth year and calculate the total he will have received in the first 12 years.

8. An oil company bores a hole 120 m deep. Estimate the cost of boring if the cost is £70 for drilling the first metre with an increase in cost of £3 per metre for each succeeding metre.

6.4   Geometric progressions

When a sequence has a constant ratio between successive terms it is called a **geometric progression** (often abbreviated to GP). The constant is called the **common ratio, r**.

Examples include

(i) $1, 2, 4, 8, \ldots$ where the common ratio is 2 and

(ii) $a, ar, ar^2, ar^3, \ldots$ where the common ratio is r.

General expression for the nth term of a GP

If the first term of a GP is a and the common ratio is r, then

$$\text{the } n\text{th term is: } ar^{n-1}$$

which can be readily checked from the above examples.

For example, the eighth term of the GP $1, 2, 4, 8, \ldots$ is $(1)(2)^7 = \mathbf{128}$, since $a = 1$ and $r = 2$

Sum of n terms of a GP

Let a GP be $a, ar, ar^2, ar^3, \ldots, ar^{n-1}$
then the sum of n terms,

$$S_n = a + ar + ar^2 + ar^3 + \cdots + ar^{n-1} \cdots \quad (1)$$

Multiplying throughout by r gives:

$$rS_n = ar + ar^2 + ar^3 + ar^4$$
$$+ \cdots + ar^{n-1} + ar^n + \cdots \quad (2)$$

Subtracting equation (2) from equation (1) gives:

$$S_n - rS_n = a - ar^n$$

i.e. $S_n(1 - r) = a(1 - r^n)$

Thus the sum of n terms, $S_n = \dfrac{a(1 - r^n)}{(1 - r)}$, which is valid when $r < 1$

Subtracting equation (1) from equation (2) gives

$$S_n = \frac{a(r^n - 1)}{(r - 1)} \text{ which is valid when } r > 1$$

For example, the sum of the first eight terms of the GP $1, 2, 4, 8, 16, \ldots$ is given by $S_8 = \dfrac{1(2^8 - 1)}{(2 - 1)}$, since $a = 1$ and $r = 2$

i.e. $S_8 = \dfrac{1(256 - 1)}{1} = \mathbf{255}$

Sum to infinity of a GP

When the common ratio r of a GP is less than unity, the sum of n terms, $S_n = \dfrac{a(1 - r^n)}{(1 - r)}$, which may be written

as $S_n = \dfrac{a}{(1 - r)} - \dfrac{ar^n}{(1 - r)}$

Since $r < 1$, r^n becomes less as n increases, i.e. $r^n \to 0$ as $n \to \infty$

Hence $\dfrac{ar^n}{(1 - r)} \to 0$ as $n \to \infty$. Thus $S_n \to \dfrac{a}{(1 - r)}$ as $n \to \infty$

The quantity $\dfrac{a}{(1 - r)}$ is called the **sum to infinity**, S_∞, and is the limiting value of the sum of an infinite number of terms,

i.e. $S_\infty = \dfrac{a}{(1 - r)}$ which is valid when $-1 < r < 1$

Convergence means that the values of the terms must get progressively smaller and the sum of the terms must reach a limiting value.

For example, the function $y = \dfrac{1}{x}$ converges to zero as x increases.

Similarly, the series $1 + \dfrac{1}{2} + \dfrac{1}{4} + \dfrac{1}{8} + \ldots$ is convergent since the value of the terms is getting smaller and the sum of the terms is approaching a limiting value of 2, i.e. the sum to infinity, $S_\infty = \dfrac{a}{1 - r} = \dfrac{1}{1 - \dfrac{1}{2}} = 2$

6.5 Worked problems on geometric progressions

Problem 11. Determine the tenth term of the series $3, 6, 12, 24, \ldots$

$3, 6, 12, 24, \ldots$ is a geometric progression with a common ratio r of 2. The nth term of a GP is ar^{n-1}, where a is the first term. Hence the tenth term is: $(3)(2)^{10-1} = (3)(2)^9 = 3(512) = \mathbf{1536}$

Problem 12. Find the sum of the first seven terms of the series, $\dfrac{1}{2}, 1\dfrac{1}{2}, 4\dfrac{1}{2}, 13\dfrac{1}{2}, \ldots$

$\dfrac{1}{2}, 1\dfrac{1}{2}, 4\dfrac{1}{2}, 13\dfrac{1}{2}, \ldots$ is a GP with a common ratio $r = 3$

The sum of n terms, $S_n = \dfrac{a(r^n - 1)}{(r - 1)}$

Hence $S_7 = \dfrac{\frac{1}{2}(3^7 - 1)}{(3 - 1)} = \dfrac{\frac{1}{2}(2187 - 1)}{2} = \mathbf{546\dfrac{1}{2}}$

Problem 13. The first term of a geometric progression is 12 and the fifth term is 55. Determine the eighth term and the eleventh term.

The fifth term is given by $ar^4 = 55$, where the first term $a = 12$

Hence $r^4 = \dfrac{55}{a} = \dfrac{55}{12}$

and $r = \sqrt[4]{\left(\dfrac{55}{12}\right)} = 1.4631719\ldots$

The eighth term is $ar^7 = (12)(1.4631719\ldots)^7 = \mathbf{172.3}$
The eleventh term is $ar^{10} = (12)(1.4631719\ldots)^{10}$
$= \mathbf{539.7}$

Problem 14. Which term of the series 2187, 729, 243, … is $\frac{1}{9}$?

2187, 729, 243, … is a GP with a common ratio $r = \frac{1}{3}$ and first term $a = 2187$

The nth term of a GP is given by: ar^{n-1}

Hence $\dfrac{1}{9} = (2187)\left(\dfrac{1}{3}\right)^{n-1}$

from which $\left(\dfrac{1}{3}\right)^{n-1} = \dfrac{1}{(9)(2187)} = \dfrac{1}{3^2 3^7}$

$= \dfrac{1}{3^9} = \left(\dfrac{1}{3}\right)^9$

Thus $(n-1) = 9$, from which, $n = 9 + 1 = 10$
i.e. $\frac{1}{9}$ **is the tenth term of the GP.**

Problem 15. Find the sum of the first nine terms of the series 72.0, 57.6, 46.08, …

The common ratio, $r = \dfrac{ar}{a} = \dfrac{57.6}{72.0} = 0.8$

$\left(\text{also } \dfrac{ar^2}{ar} = \dfrac{46.08}{57.6} = 0.8\right)$

The sum of nine terms,

$S_9 = \dfrac{a(1-r^n)}{(1-r)} = \dfrac{72.0(1-0.8^9)}{(1-0.8)}$

$= \dfrac{72.0(1-0.1342)}{0.2} = \mathbf{311.7}$

Problem 16. Find the sum to infinity of the series 3, 1, $\frac{1}{3}$, …

3, 1, $\frac{1}{3}$, … is a GP of common ratio, $r = \frac{1}{3}$

The sum to infinity,

$$S_\infty = \frac{a}{1-r} = \frac{3}{1-\frac{1}{3}} = \frac{3}{\frac{2}{3}} = \frac{9}{2} = \mathbf{4\frac{1}{2}}$$

Now try the following Practice Exercise

Practice Exercise 27   Geometric progressions (Answers on page 858)

1. Find the tenth term of the series 5, 10, 20, 40, …

2. Determine the sum of the first seven terms of the series $\frac{1}{4}$, $\frac{3}{4}$, $2\frac{1}{4}$, $6\frac{3}{4}$, …

3. The first term of a geometric progression is 4 and the sixth term is 128. Determine the eighth and eleventh terms.

4. Find the sum of the first seven terms of the series 2, 5, $12\frac{1}{2}$, … (correct to 4 significant figures).

5. Determine the sum to infinity of the series 4, 2, 1, …

6. Find the sum to infinity of the series $2\frac{1}{2}$, $-1\frac{1}{4}$, $\frac{5}{8}$, …

6.6 Further worked problems on geometric progressions

Problem 17. In a geometric progression the sixth term is eight times the third term and the sum of the seventh and eighth terms is 192. Determine (a) the common ratio, (b) the first term, and (c) the sum of the fifth to eleventh terms, inclusive.

(a) Let the GP be a, ar, ar^2, ar^3, …, ar^{n-1}
The third term $= ar^2$ and the sixth term $= ar^5$
The sixth term is eight times the third.
Hence $ar^5 = 8ar^2$ from which, $r^3 = 8$, $r = \sqrt[3]{8}$
i.e. **the common ratio $r = 2$**

(b) The sum of the seventh and eighth terms is 192.
Hence $ar^6 + ar^7 = 192$

Since $r = 2$, then $64a + 128a = 192$

$192a = 192,$

from which, a, the first term, $\mathbf{= 1}$

(c) The sum of the fifth to eleventh terms (inclusive) is given by:

$$S_{11} - S_4 = \frac{a(r^{11} - 1)}{(r - 1)} - \frac{a(r^4 - 1)}{(r - 1)}$$

$$= \frac{1(2^{11} - 1)}{(2 - 1)} - \frac{1(2^4 - 1)}{(2 - 1)}$$

$$= (2^{11} - 1) - (2^4 - 1)$$

$$= 2^{11} - 2^4 = 2048 - 16 = \mathbf{2032}$$

Problem 18. A hire tool firm finds that their net return from hiring tools is decreasing by 10% per annum. If their net gain on a certain tool this year is £400, find the possible total of all future profits from this tool (assuming the tool lasts forever).

The net gain forms a series:

$$£400 + £400 \times 0.9 + £400 \times 0.9^2 + \cdots,$$

which is a GP with $a = 400$ and $r = 0.9$.
The sum to infinity,

$$S_\infty = \frac{a}{(1 - r)} = \frac{400}{(1 - 0.9)}$$

$$= \mathbf{£4000} = \textbf{total future profits}$$

Problem 19. If £100 is invested at compound interest of 8% per annum, determine (a) the value after ten years, (b) the time, correct to the nearest year, it takes to reach more than £300.

(a) Let the GP be $a, ar, ar^2, \ldots, ar^n$
The first term $a = £100$
The common ratio $r = 1.08$
Hence the second term is

$$ar = (100)(1.08) = £108,$$

which is the value after one year,
the third term is

$$ar^2 = (100)(1.08)^2 = £116.64,$$

which is the value after two years, and so on.
Thus the value after ten years

$$= ar^{10} = (100)(1.08)^{10} = \mathbf{£215.89}$$

(b) When £300 has been reached, $300 = ar^n$

i.e. $300 = 100(1.08)^n$

and $3 = (1.08)^n$

Taking logarithms to base 10 of both sides gives:

$$\lg 3 = \lg(1.08)^n = n \lg(1.08),$$

by the laws of logarithms
from which, $n = \dfrac{\lg 3}{\lg 1.08} = 14.3$

Hence it will take 15 years to reach more than £300.

Problem 20. A drilling machine is to have six speeds ranging from 50 rev/min to 750 rev/min. If the speeds form a geometric progression determine their values, each correct to the nearest whole number.

Let the GP of n terms be given by $a, ar, ar^2, \ldots, ar^{n-1}$.
The first term $a = 50$ rev/min
The sixth term is given by ar^{6-1}, which is 750 rev/min,

i.e. $ar^5 = 750$

from which $r^5 = \dfrac{750}{a} = \dfrac{750}{50} = 15$

Thus the common ratio, $r = \sqrt[5]{15} = 1.7188$
The first term is $a = 50$ rev/min
the second term is $ar = (50)(1.7188) = 85.94$,
the third term is $ar^2 = (50)(1.7188)^2 = 147.71$,
the fourth term is $ar^3 = (50)(1.7188)^3 = 253.89$,
the fifth term is $ar^4 = (50)(1.7188)^4 = 436.39$,
the sixth term is $ar^5 = (50)(1.7188)^5 = 750.06$
Hence, correct to the nearest whole number, the six speeds of the drilling machine are **50, 86, 148, 254, 436 and 750 rev/min.**

Now try the following Practice Exercise

Practice Exercise 28   Geometric progressions (Answers on page 858)

1. In a geometric progression the fifth term is nine times the third term and the sum of the sixth and seventh terms is 1944. Determine (a) the common ratio, (b) the first term and

(c) the sum of the fourth to tenth terms inclusive.

2. Which term of the series 3, 9, 27,... is 59 049?

3. The value of a lathe originally valued at £3000 depreciates 15% per annum. Calculate its value after 4 years. The machine is sold when its value is less than £550. After how many years is the lathe sold?

4. If the population of Great Britain is 55 million and is decreasing at 2.4% per annum, what will be the population in five years' time?

5. 100 g of a radioactive substance disintegrates at a rate of 3% per annum. How much of the substance is left after 11 years?

6. If £250 is invested at compound interest of 6% per annum, determine (a) the value after 15 years, (b) the time, correct to the nearest year, it takes to reach £750

7. A drilling machine is to have eight speeds ranging from 100 rev/min to 1000 rev/min. If the speeds form a geometric progression determine their values, each correct to the nearest whole number.

Section A

For fully worked solutions to each of the problems in Practice Exercises 25 to 28 in this chapter, go to the website:
www.routledge.com/cw/bird

Chapter 7

The binomial series

Why it is important to understand: **The binomial series**

There are many applications of the binomial theorem in every part of algebra, and in general with permutations, combinations and probability. It is also used in atomic physics, where it is used to count s, p, d and f orbitals. There are applications of the binomial series in financial mathematics to determine the number of stock price paths that leads to a particular stock price at maturity.

At the end of this chapter, you should be able to:

- define a binomial expression
- use Pascal's triangle to expand a binomial expression
- state the general binomial expansion of $(a + x)^n$ and $(1 + x)^n$
- use the binomial series to expand expressions of the form $(a + x)^n$ for positive, negative and fractional values of n
- determine the rth term of a binomial expansion
- use the binomial expansion with practical applications

7.1 Pascal's triangle

A **binomial expression** is one which contains two terms connected by a plus or minus sign. Thus $(p + q)$, $(a + x)^2$, $(2x + y)^3$ are examples of binomial expressions. Expanding $(a + x)^n$ for integer values of n from 0 to 6 gives the results as shown at the top of page 59.

From these results the following patterns emerge:

(i) 'a' decreases in power moving from left to right.

(ii) 'x' increases in power moving from left to right.

(iii) The coefficients of each term of the expansions are symmetrical about the middle coefficient when n is even and symmetrical about the two middle coefficients when n is odd.

$$
\begin{aligned}
(a+x)^0 &= && 1\\
(a+x)^1 &= a+x && a+x\\
(a+x)^2 &= (a+x)(a+x) && a^2+2ax+x^2\\
(a+x)^3 &= (a+x)^2(a+x) = && a^3+3a^2x+3ax^2+x^3\\
(a+x)^4 &= (a+x)^3(a+x) = && a^4+4a^3x+6a^2x^2+4ax^3+x^4\\
(a+x)^5 &= (a+x)^4(a+x) = && a^5+5a^4x+10a^3x^2+10a^2x^3+5ax^4+x^5\\
(a+x)^6 &= (a+x)^5(a+x) = a^6+&&6a^5x+15a^4x^2+20a^3x^3+15a^2x^4+6ax^5+x^6
\end{aligned}
$$

(iv) The coefficients are shown separately in Table 7.1 and this arrangement is known as **Pascal's triangle**.* A coefficient of a term may be obtained by adding the two adjacent coefficients immediately above in the previous row. This is shown by the triangles in Table 7.1, where, for example, $1+3=4$, $10+5=15$, and so on.

(v) Pascal's triangle method is used for expansions of the form $(a+x)^n$ for integer values of n less than about 8.

* **Who was Pascal?** **Blaise Pascal** (19 June 1623–19 August 1662) was a French polymath. A child prodigy, he wrote a significant treatise on the subject of projective geometry at the age of 16, and later corresponded with Pierre de Fermat on probability theory, strongly influencing the development of modern economics and social science. To find out more go to **www.routledge.com/cw/bird**

Table 7.1

$(a+x)^0$						1					
$(a+x)^1$					1		1				
$(a+x)^2$				1		2		1			
$(a+x)^3$			1		3		3		1		
$(a+x)^4$		1		4		6		4		1	
$(a+x)^5$	1		5		10		10		5		1
$(a+x)^6$	1		6	15		20		15		6	1

Problem 1. Use Pascal's triangle method to determine the expansion of $(a+x)^7$

From Table 7.1, the row of Pascal's triangle corresponding to $(a+x)^6$ is as shown in (1) below. Adding adjacent coefficients gives the coefficients of $(a+x)^7$ as shown in (2) below.

$$
\begin{array}{cccccccc}
1 & 6 & 15 & 20 & 15 & 6 & 1 & \quad(1)\\
1 & 7 & 21 & 35 & 35 & 21 & 7 & 1 \quad(2)
\end{array}
$$

The first and last terms of the expansion of $(a+x)^7$ are a^7 and x^7, respectively. The powers of a decrease and the powers of x increase moving from left to right. Hence

$$
(a+x)^7 = a^7 + 7a^6x + 21a^5x^2 + 35a^4x^3
$$
$$
+ 35a^3x^4 + 21a^2x^5 + 7ax^6 + x^7
$$

Problem 2. Determine, using Pascal's triangle method, the expansion of $(2p-3q)^5$

Comparing $(2p-3q)^5$ with $(a+x)^5$ shows that $a=2p$ and $x=-3q$

Using Pascal's triangle method:

$$
(a+x)^5 = a^5 + 5a^4x + 10a^3x^2 + 10a^2x^3 + \cdots
$$

Hence

$$(2p - 3q)^5 = (2p)^5 + 5(2p)^4(-3q)$$

$$+ 10(2p)^3(-3q)^2$$

$$+ 10(2p)^2(-3q)^3$$

$$+ 5(2p)(-3q)^4 + (-3q)^5$$

i.e. $(2p - 3q)^5 = 32p^5 - 240p^4q + 720p^3q^2$

$$- 1080p^2q^3 + 810pq^4 - 243q^5$$

Now try the following Practice Exercise

Practice Exercise 29   Pascal's triangle (Answers on page 858)

1. Use Pascal's triangle to expand $(x - y)^7$

2. Expand $(2a + 3b)^5$ using Pascal's triangle.

7.2   The binomial series

The **binomial series** or **binomial theorem** is a formula for raising a binomial expression to any power without lengthy multiplication. The general binomial expansion of $(a + x)^n$ is given by:

$$(a + x)^n = a^n + na^{n-1}x + \frac{n(n - 1)}{2!}a^{n-2}x^2$$

$$+ \frac{n(n - 1)(n - 2)}{3!}a^{n-3}x^3$$

$$+ \cdots$$

where 3! denotes $3 \times 2 \times 1$ and is termed 'factorial 3'. With the binomial theorem, n may be a fraction, a decimal fraction, or a positive or negative integer.
When n is a positive integer, the series is finite, i.e., it comes to an end; when n is a negative integer, or a fraction, the series is infinite.
In the general expansion of $(a + x)^n$ it is noted that the fourth term is: $\frac{n(n-1)(n-2)}{3!}a^{n-3}x^3$. The number 3 is very evident in this expression.

For any term in a binomial expansion, say the rth term, $(r - 1)$ is very evident. It may therefore be reasoned that **the rth term of the expansion $(a + x)^n$** is:

$$\frac{n(n - 1)(n - 2)\ldots \text{ to } (r - 1) \text{ terms}}{(r - 1)!}a^{n-(r-1)}x^{r-1}$$

If $a = 1$ in the binomial expansion of $(a + x)^n$ then:

$$(1 + x)^n = 1 + nx + \frac{n(n - 1)}{2!}x^2$$

$$+ \frac{n(n - 1)(n - 2)}{3!}x^3 + \cdots$$

which is valid for $-1 < x < 1$
When x is small compared with 1 then:

$$(1 + x)^n \approx 1 + nx$$

7.3   Worked problems on the binomial series

Problem 3. Use the binomial series to determine the expansion of $(2 + x)^7$

The binomial expansion is given by:

$$(a + x)^n = a^n + na^{n-1}x + \frac{n(n - 1)}{2!}a^{n-2}x^2$$

$$+ \frac{n(n - 1)(n - 2)}{3!}a^{n-3}x^3 + \cdots$$

When $a = 2$ and $n = 7$:

$$(2 + x)^7 = 2^7 + 7(2)^6x + \frac{(7)(6)}{(2)(1)}(2)^5x^2$$

$$+ \frac{(7)(6)(5)}{(3)(2)(1)}(2)^4x^3 + \frac{(7)(6)(5)(4)}{(4)(3)(2)(1)}(2)^3x^4$$

$$+ \frac{(7)(6)(5)(4)(3)}{(5)(4)(3)(2)(1)}(2)^2x^5$$

$$+ \frac{(7)(6)(5)(4)(3)(2)}{(6)(5)(4)(3)(2)(1)}(2)x^6$$

$$+ \frac{(7)(6)(5)(4)(3)(2)(1)}{(7)(6)(5)(4)(3)(2)(1)}x^7$$

i.e. $(2 + x)^7 = 128 + 448x + 672x^2 + 560x^3$

$$+ 280x^4 + 84x^5 + 14x^6 + x^7$$

Problem 4. Use the binomial series to determine the expansion of $(2a - 3b)^5$

From equation (1), the binomial expansion is given by:

$$(a + x)^n = a^n + na^{n-1}x + \frac{n(n - 1)}{2!}a^{n-2}x^2$$

$$+ \frac{n(n - 1)(n - 2)}{3!}a^{n-3}x^3 + \cdots$$

When $a = 2a$, $x = -3b$ and $n = 5$:

$$(2a - 3b)^5 = (2a)^5 + 5(2a)^4(-3b)$$

$$+ \frac{(5)(4)}{(2)(1)}(2a)^3(-3b)^2$$

$$+ \frac{(5)(4)(3)}{(3)(2)(1)}(2a)^2(-3b)^3$$

$$+ \frac{(5)(4)(3)(2)}{(4)(3)(2)(1)}(2a)(-3b)^4$$

$$+ \frac{(5)(4)(3)(2)(1)}{(5)(4)(3)(2)(1)}(-3b)^5$$

i.e. $(2a - 3b)^5 = 32a^5 - 240a^4b + 720a^3b^2$

$$- 1080a^2b^3 + 810ab^4 - 243b^5$$

Problem 5.   Expand $\left(c - \dfrac{1}{c}\right)^5$ using the binomial series.

$$\left(c - \frac{1}{c}\right)^5 = c^5 + 5c^4\left(-\frac{1}{c}\right)$$

$$+ \frac{(5)(4)}{(2)(1)}c^3\left(-\frac{1}{c}\right)^2$$

$$+ \frac{(5)(4)(3)}{(3)(2)(1)}c^2\left(-\frac{1}{c}\right)^3$$

$$+ \frac{(5)(4)(3)(2)}{(4)(3)(2)(1)}c\left(-\frac{1}{c}\right)^4$$

$$+ \frac{(5)(4)(3)(2)(1)}{(5)(4)(3)(2)(1)}\left(-\frac{1}{c}\right)^5$$

i.e. $\left(c - \dfrac{1}{c}\right)^5 = c^5 - 5c^3 + 10c - \dfrac{10}{c} + \dfrac{5}{c^3} - \dfrac{1}{c^5}$

Problem 6.   Without fully expanding $(3 + x)^7$, determine the fifth term.

The rth term of the expansion $(a + x)^n$ is given by:

$$\frac{n(n - 1)(n - 2)\ldots\text{to }(r - 1)\text{ terms}}{(r - 1)!}a^{n-(r-1)}x^{r-1}$$

Substituting $n = 7$, $a = 3$ and $r - 1 = 5 - 1 = 4$ gives:

$$\frac{(7)(6)(5)(4)}{(4)(3)(2)(1)}(3)^{7-4}x^4$$

i.e. the fifth term of $(3 + x)^7 = 35(3)^3x^4 = 945x^4$

Problem 7.   Find the middle term of

$$\left(2p - \frac{1}{2q}\right)^{10}$$

In the expansion of $(a + x)^{10}$ there are $10 + 1$, i.e. 11 terms. Hence the middle term is the sixth. Using the general expression for the rth term where $a = 2p$, $x = -\dfrac{1}{2q}$, $n = 10$ and $r - 1 = 5$ gives:

$$\frac{(10)(9)(8)(7)(6)}{(5)(4)(3)(2)(1)}(2p)^{10-5}\left(-\frac{1}{2q}\right)^5$$

$$= 252(32p^5)\left(-\frac{1}{32q^5}\right)$$

Hence the middle term of $\left(2p - \dfrac{1}{2q}\right)^{10}$ is $-252\dfrac{p^5}{q^5}$

Problem 8.   Evaluate $(1.002)^9$ using the binomial theorem correct to (a) 3 decimal places and (b) 7 significant figures.

$$(1 + x)^n = 1 + nx + \frac{n(n - 1)}{2!}x^2$$

$$+ \frac{n(n - 1)(n - 2)}{3!}x^3 + \cdots$$

$$(1.002)^9 = (1 + 0.002)^9$$

Substituting $x = 0.002$ and $n = 9$ in the general expansion for $(1 + x)^n$ gives:

$$(1 + 0.002)^9 = 1 + 9(0.002) + \frac{(9)(8)}{(2)(1)}(0.002)^2$$

$$+ \frac{(9)(8)(7)}{(3)(2)(1)}(0.002)^3 + \cdots$$

$$= 1 + 0.018 + 0.000144$$

$$+ 0.000000672 + \cdots$$

$$= 1.018144672\ldots$$

Hence $(1.002)^9 = \mathbf{1.018,}$ **correct to 3 decimal places**

$$= \mathbf{1.018145,}$ **correct to 7 significant figures.**

Problem 9.   Evaluate $(0.97)^6$ correct to 4 significant figures using the binomial expansion.

Section A

$(0.97)^6$ is written as $(1-0.03)^6$

Using the expansion of $(1+x)^n$ where $n=6$ and $x=-0.03$ gives:

$$(1-0.03)^6 = 1 + 6(-0.03) + \frac{(6)(5)}{(2)(1)}(-0.03)^2$$
$$+ \frac{(6)(5)(4)}{(3)(2)(1)}(-0.03)^3$$
$$+ \frac{(6)(5)(4)(3)}{(4)(3)(2)(1)}(-0.03)^4 + \cdots$$
$$= 1 - 0.18 + 0.0135 - 0.00054$$
$$+ 0.00001215 - \cdots$$
$$\approx 0.83297215$$

i.e. $(0.97)^6 = 0.8330$, **correct to 4 significant figures.**

Problem 10. Determine the value of $(3.039)^4$, correct to 6 significant figures using the binomial theorem.

$(3.039)^4$ may be written in the form $(1+x)^n$ as:

$$(3.039)^4 = (3+0.039)^4$$
$$= \left[3\left(1+\frac{0.039}{3}\right)\right]^4$$
$$= 3^4(1+0.013)^4$$

$$(1+0.013)^4 = 1 + 4(0.013)$$
$$+ \frac{(4)(3)}{(2)(1)}(0.013)^2$$
$$+ \frac{(4)(3)(2)}{(3)(2)(1)}(0.013)^3 + \cdots$$
$$= 1 + 0.052 + 0.001014$$
$$+ 0.000008788 + \cdots$$
$$= 1.0530228$$

correct to 8 significant figures

Hence $(3.039)^4 = 3^4(1.0530228)$

$= 85.2948$, **correct to 6 significant figures.**

Now try the following Practice Exercise

Practice Exercise 30   The binomial series (Answers on page 859)

1. Use the binomial theorem to expand $(a+2x)^4$

2. Use the binomial theorem to expand $(2-x)^6$

3. Expand $(2x-3y)^4$

4. Determine the expansion of $\left(2x+\frac{2}{x}\right)^5$

5. Expand $(p+2q)^{11}$ as far as the fifth term.

6. Determine the sixth term of $\left(3p+\frac{q}{3}\right)^{13}$

7. Determine the middle term of $(2a-5b)^8$

8. Use the binomial theorem to determine, correct to 4 decimal places:
   (a) $(1.003)^8$ (b) $(1.042)^7$

9. Use the binomial theorem to determine, correct to 5 significant figures:
   (a) $(0.98)^7$ (b) $(2.01)^9$

10. Evaluate $(4.044)^6$ correct to 2 decimal places.

7.4 Further worked problems on the binomial series

Problem 11.

(a) Expand $\frac{1}{(1+2x)^3}$ in ascending powers of x as far as the term in x^3, using the binomial series.

(b) State the limits of x for which the expansion is valid.

(a) Using the binomial expansion of $(1+x)^n$, where $n=-3$ and x is replaced by $2x$ gives:

$$\frac{1}{(1+2x)^3} = (1+2x)^{-3}$$
$$= 1 + (-3)(2x) + \frac{(-3)(-4)}{2!}(2x)^2$$
$$+ \frac{(-3)(-4)(-5)}{3!}(2x)^3 + \cdots$$
$$= 1 - 6x + 24x^2 - 80x^3 + \cdots$$

(b) The expansion is valid provided $|2x| < 1$,

i.e. $|x| < \dfrac{1}{2}$ or $-\dfrac{1}{2} < x < \dfrac{1}{2}$

Problem 12.
(a) Expand $\dfrac{1}{(4-x)^2}$ in ascending powers of x as far as the term in x^3, using the binomial theorem.

(b) What are the limits of x for which the expansion in (a) is true?

(a) $\dfrac{1}{(4-x)^2} = \dfrac{1}{\left[4\left(1-\dfrac{x}{4}\right)\right]^2} = \dfrac{1}{4^2\left(1-\dfrac{x}{4}\right)^2}$

$= \dfrac{1}{16}\left(1-\dfrac{x}{4}\right)^{-2}$

Using the expansion of $(1+x)^n$

$\dfrac{1}{(4-x)^2} = \dfrac{1}{16}\left(1-\dfrac{x}{4}\right)^{-2}$

$= \dfrac{1}{16}\left[1 + (-2)\left(-\dfrac{x}{4}\right)\right.$

$\left. + \dfrac{(-2)(-3)}{2!}\left(-\dfrac{x}{4}\right)^2 \right.$

$\left. + \dfrac{(-2)(-3)(-4)}{3!}\left(-\dfrac{x}{4}\right)^3 + \cdots \right]$

$= \dfrac{1}{16}\left(1 + \dfrac{x}{2} + \dfrac{3x^2}{16} + \dfrac{x^3}{16} + \cdots\right)$

(b) The expansion in (a) is true provided $\left|\dfrac{x}{4}\right| < 1$,

i.e. $|x| < 4$ or $-4 < x < 4$

Problem 13. Use the binomial theorem to expand $\sqrt{4+x}$ in ascending powers of x to four terms. Give the limits of x for which the expansion is valid.

$\sqrt{4+x} = \sqrt{\left[4\left(1+\dfrac{x}{4}\right)\right]}$

$= \sqrt{4}\sqrt{\left(1+\dfrac{x}{4}\right)} = 2\left(1+\dfrac{x}{4}\right)^{\frac{1}{2}}$

Using the expansion of $(1+x)^n$,

$2\left(1+\dfrac{x}{4}\right)^{\frac{1}{2}}$

$= 2\left[1 + \left(\dfrac{1}{2}\right)\left(\dfrac{x}{4}\right) + \dfrac{(1/2)(-1/2)}{2!}\left(\dfrac{x}{4}\right)^2\right.$

$\left. + \dfrac{(1/2)(-1/2)(-3/2)}{3!}\left(\dfrac{x}{4}\right)^3 + \cdots\right]$

$= 2\left(1 + \dfrac{x}{8} - \dfrac{x^2}{128} + \dfrac{x^3}{1024} - \cdots\right)$

$= 2 + \dfrac{x}{4} - \dfrac{x^2}{64} + \dfrac{x^3}{512} - \cdots$

This is valid when $\left|\dfrac{x}{4}\right| < 1$,

i.e. $|x| < 4$ or $-4 < x < 4$

Problem 14. Expand $\dfrac{1}{\sqrt{(1-2t)}}$ in ascending powers of t as far as the term in t^3.

State the limits of t for which the expression is valid.

$\dfrac{1}{\sqrt{(1-2t)}}$

$= (1-2t)^{-\frac{1}{2}}$

$= 1 + \left(-\dfrac{1}{2}\right)(-2t) + \dfrac{(-1/2)(-3/2)}{2!}(-2t)^2$

$+ \dfrac{(-1/2)(-3/2)(-5/2)}{3!}(-2t)^3 + \cdots,$

using the expansion for $(1+x)^n$

$= 1 + t + \dfrac{3}{2}t^2 + \dfrac{5}{2}t^3 + \cdots$

The expression is valid when $|2t| < 1$,

i.e. $|t| < \dfrac{1}{2}$ or $-\dfrac{1}{2} < t < \dfrac{1}{2}$

Problem 15. Simplify $\dfrac{\sqrt[3]{(1-3x)}\sqrt{(1+x)}}{\left(1+\dfrac{x}{2}\right)^3}$

given that powers of x above the first may be neglected.

$\dfrac{\sqrt[3]{(1-3x)}\sqrt{(1+x)}}{\left(1+\dfrac{x}{2}\right)^3}$

$= (1-3x)^{\frac{1}{3}}(1+x)^{\frac{1}{2}}\left(1+\dfrac{x}{2}\right)^{-3}$

$\approx \left[1 + \left(\dfrac{1}{3}\right)(-3x)\right]\left[1 + \left(\dfrac{1}{2}\right)(x)\right]\left[1 + (-3)\left(\dfrac{x}{2}\right)\right]$

when expanded by the binomial theorem as far as the x term only,

$$= (1-x)\left(1+\frac{x}{2}\right)\left(1-\frac{3x}{2}\right)$$

$$= \left(1 - x + \frac{x}{2} - \frac{3x}{2}\right) \quad \begin{array}{l}\text{when powers of } x \text{ higher than}\\ \text{unity are neglected}\end{array}$$

$$= (1-2x)$$

Problem 16. Express $\dfrac{\sqrt{(1+2x)}}{\sqrt[3]{(1-3x)}}$ as a power series as far as the term in x^2. State the range of values of x for which the series is convergent.

$$\frac{\sqrt{(1+2x)}}{\sqrt[3]{(1-3x)}} = (1+2x)^{\frac{1}{2}}(1-3x)^{-\frac{1}{3}}$$

$$(1+2x)^{\frac{1}{2}} = 1 + \left(\frac{1}{2}\right)(2x)$$

$$+ \frac{(1/2)(-1/2)}{2!}(2x)^2 + \cdots$$

$$= 1 + x - \frac{x^2}{2} + \cdots \text{ which is valid for}$$

$$|2x| < 1, \text{ i.e. } |x| < \frac{1}{2}$$

$$(1-3x)^{-\frac{1}{3}} = 1 + (-1/3)(-3x)$$

$$+ \frac{(-1/3)(-4/3)}{2!}(-3x)^2 + \cdots$$

$$= 1 + x + 2x^2 + \cdots \text{ which is valid for}$$

$$|3x| < 1, \text{ i.e. } |x| < \frac{1}{3}$$

Hence

$$\frac{\sqrt{(1+2x)}}{\sqrt[3]{(1-3x)}} = (1+2x)^{\frac{1}{2}}(1-3x)^{-\frac{1}{3}}$$

$$= \left(1 + x - \frac{x^2}{2} + \cdots\right)(1 + x + 2x^2 + \cdots)$$

$$= 1 + x + 2x^2 + x + x^2 - \frac{x^2}{2},$$

neglecting terms of higher power than 2,

$$= 1 + 2x + \frac{5}{2}x^2$$

The series is convergent if $-\dfrac{1}{3} < x < \dfrac{1}{3}$

Now try the following Practice Exercise

Practice Exercise 31   The binomial series (Answers on page 859)

In problems 1 to 5 expand in ascending powers of x as far as the term in x^3, using the binomial theorem. State in each case the limits of x for which the series is valid.

1. $\dfrac{1}{(1-x)}$

2. $\dfrac{1}{(1+x)^2}$

3. $\dfrac{1}{(2+x)^3}$

4. $\sqrt{2+x}$

5. $\dfrac{1}{\sqrt{1+3x}}$

6. Expand $(2+3x)^{-6}$ to three terms. For what values of x is the expansion valid?

7. When x is very small show that:

   (a) $\dfrac{1}{(1-x)^2\sqrt{(1-x)}} \approx 1 + \dfrac{5}{2}x$

   (b) $\dfrac{(1-2x)}{(1-3x)^4} \approx 1 + 10x$

   (c) $\dfrac{\sqrt{1+5x}}{\sqrt[3]{1-2x}} \approx 1 + \dfrac{19}{6}x$

8. If x is very small such that x^2 and higher powers may be neglected, determine the power series for $\dfrac{\sqrt{x+4}\sqrt[3]{8-x}}{\sqrt[5]{(1+x)^3}}$

9. Express the following as power series in ascending powers of x as far as the term in x^2. State in each case the range of x for which the series is valid.

   (a) $\sqrt{\left(\dfrac{1-x}{1+x}\right)}$ (b) $\dfrac{(1+x)\sqrt[3]{(1-3x)^2}}{\sqrt{(1+x^2)}}$

7.5 Practical problems involving the binomial theorem

Binomial expansions may be used for numerical approximations, for calculations with small variations and in probability theory (see Chapter 60).

Problem 17.   The radius of a cylinder is reduced by 4% and its height is increased by 2%. Determine the approximate percentage change in (a) its volume and (b) its curved surface area, (neglecting the products of small quantities).

Volume of cylinder $= \pi r^2 h$.
Let r and h be the original values of radius and height.
The new values are $0.96r$ or $(1 - 0.04)r$ and $1.02h$ or $(1 + 0.02)h$.

(a) New volume $= \pi[(1 - 0.04)r]^2[(1 + 0.02)h]$

$$= \pi r^2 h(1 - 0.04)^2(1 + 0.02)$$

Now $(1 - 0.04)^2 = 1 - 2(0.04) + (0.04)^2$

$$= (1 - 0.08),$$

neglecting powers of small terms.

Hence new volume

$$\approx \pi r^2 h(1 - 0.08)(1 + 0.02)$$

$$\approx \pi r^2 h(1 - 0.08 + 0.02), \text{ neglecting}$$

products of small terms

$$\approx \pi r^2 h(1 - 0.06) \text{ or } 0.94\pi r^2 h, \text{ i.e. } 94\%$$

of the original volume

Hence the volume is reduced by approximately 6%

(b)   Curved surface area of cylinder $= 2\pi r h$.

New surface area

$$= 2\pi[(1 - 0.04)r][(1 + 0.02)h]$$

$$= 2\pi r h(1 - 0.04)(1 + 0.02)$$

$$\approx 2\pi r h(1 - 0.04 + 0.02), \text{ neglecting}$$

products of small terms

$$\approx 2\pi r h(1 - 0.02) \text{ or } 0.98(2\pi r h),$$

i.e. 98% of the original surface area

Hence the curved surface area is reduced by approximately 2%

Problem 18.   The second moment of area of a rectangle through its centroid is given by $\dfrac{bl^3}{12}$

Determine the approximate change in the second moment of area if b is increased by 3.5% and l is reduced by 2.5%

New values of b and l are $(1 + 0.035)b$ and $(1 - 0.025)l$ respectively.

New second moment of area

$$= \frac{1}{12}[(1 + 0.035)b][(1 - 0.025)l]^3$$

$$= \frac{bl^3}{12}(1 + 0.035)(1 - 0.025)^3$$

$$\approx \frac{bl^3}{12}(1 + 0.035)(1 - 0.075), \text{ neglecting}$$

powers of small terms

$$\approx \frac{bl^3}{12}(1 + 0.035 - 0.075), \text{ neglecting}$$

products of small terms

$$\approx \frac{bl^3}{12}(1 - 0.040) \text{ or } (0.96)\frac{bl^3}{12}, \text{ i.e. } 96\%$$

of the original second moment of area

Hence the second moment of area is reduced by approximately 4%

Problem 19.   The resonant frequency of a vibrating shaft is given by: $f = \dfrac{1}{2\pi}\sqrt{\dfrac{k}{I}}$, where k is the stiffness and I is the inertia of the shaft. Use the binomial theorem to determine the approximate percentage error in determining the frequency using the measured values of k and I when the measured value of k is 4% too large and the measured value of I is 2% too small.

Let f, k and I be the true values of frequency, stiffness and inertia respectively. Since the measured value of stiffness, k_1, is 4% too large, then

$$k_1 = \frac{104}{100}k = (1 + 0.04)k$$

The measured value of inertia, I_1, is 2% too small, hence

$$I_1 = \frac{98}{100}I = (1 - 0.02)I$$

The measured value of frequency,

$$f_1 = \frac{1}{2\pi}\sqrt{\frac{k_1}{I_1}} = \frac{1}{2\pi}k_1^{\frac{1}{2}}I_1^{-\frac{1}{2}}$$

$$= \frac{1}{2\pi}[(1+0.04)k]^{\frac{1}{2}}[(1-0.02)I]^{-\frac{1}{2}}$$

$$= \frac{1}{2\pi}(1+0.04)^{\frac{1}{2}}k^{\frac{1}{2}}(1-0.02)^{-\frac{1}{2}}I^{-\frac{1}{2}}$$

$$= \frac{1}{2\pi}k^{\frac{1}{2}}I^{-\frac{1}{2}}(1+0.04)^{\frac{1}{2}}(1-0.02)^{-\frac{1}{2}}$$

i.e. $f_1 = f(1+0.04)^{\frac{1}{2}}(1-0.02)^{-\frac{1}{2}}$

$$\approx f\left[1+\left(\frac{1}{2}\right)(0.04)\right]\left[1+\left(-\frac{1}{2}\right)(-0.02)\right]$$

$$\approx f(1+0.02)(1+0.01)$$

Neglecting the products of small terms,

$$f_1 \approx (1+0.02+0.01)f \approx 1.03f$$

Thus the percentage error in f based on the measured values of k and I is approximately $[(1.03)(100) - 100]$, i.e. **3% too large**.

Now try the following Practice Exercise

Practice Exercise 32   Practical problems involving the binomial theorem (Answers on page 859)

1. Pressure p and volume v are related by $pv^3 = c$, where c is a constant. Determine the approximate percentage change in c when p is increased by 3% and v decreased by 1.2%

2. Kinetic energy is given by $\frac{1}{2}mv^2$. Determine the approximate change in the kinetic energy when mass m is increased by 2.5% and the velocity v is reduced by 3%

3. An error of $+1.5\%$ was made when measuring the radius of a sphere. Ignoring the products of small quantities, determine the approximate error in calculating (a) the volume, and (b) the surface area.

4. The power developed by an engine is given by $I = k$ PLAN, where k is a constant. Determine the approximate percentage change in the power when P and A are each increased by 2.5% and L and N are each decreased by 1.4%

5. The radius of a cone is increased by 2.7% and its height reduced by 0.9%. Determine the approximate percentage change in its volume, neglecting the products of small terms.

6. The electric field strength H due to a magnet of length $2l$ and moment M at a point on its axis distance x from the centre is given by
$$H = \frac{M}{2l}\left\{\frac{1}{(x-l)^2} - \frac{1}{(x+l)^2}\right\}$$
Show that if l is very small compared with x, then $H \approx \frac{2M}{x^3}$

7. The shear stress τ in a shaft of diameter D under a torque T is given by: $\tau = \frac{kT}{\pi D^3}$. Determine the approximate percentage error in calculating τ if T is measured 3% too small and D 1.5% too large.

8. The energy W stored in a flywheel is given by: $W = kr^5N^2$, where k is a constant, r is the radius and N the number of revolutions. Determine the approximate percentage change in W when r is increased by 1.3% and N is decreased by 2%

9. In a series electrical circuit containing inductance L and capacitance C the resonant frequency is given by: $f_r = \frac{1}{2\pi\sqrt{LC}}$. If the values of L and C used in the calculation are 2.6% too large and 0.8% too small respectively, determine the approximate percentage error in the frequency.

10. The viscosity η of a liquid is given by: $\eta = \frac{kr^4}{vl}$, where k is a constant. If there is an error in r of $+2\%$, in v of $+4\%$ and l of -3%, what is the resultant error in η?

11. A magnetic pole, distance x from the plane of a coil of radius r, and on the axis of the coil, is subject to a force F when a current flows in the coil. The force is given by: $F = \frac{kx}{\sqrt{(r^2+x^2)^5}}$, where k is a constant. Use

the binomial theorem to show that when x is small compared to r, then

$$F \approx \frac{kx}{r^5} - \frac{5kx^3}{2r^7}$$

12. The flow of water through a pipe is given by: $G = \sqrt{\dfrac{(3d)^5 H}{L}}$. If d decreases by 2% and H by 1%, use the binomial theorem to estimate the decrease in G.

For fully worked solutions to each of the problems in Practice Exercises 29 to 32 in this chapter, go to the website:
www.routledge.com/cw/bird

Chapter 8

Maclaurin's series

> **_Why it is important to understand:_** **Maclaurin's series**
>
> One of the simplest kinds of function to deal with, in either algebra or calculus, is a polynomial (i.e. an expression of the form $a + bx + cx^2 + dx^3 + \ldots$). Polynomials are easy to substitute numerical values into, and they are easy to differentiate. One useful application of Maclaurin's series is to approximate, to a polynomial, functions which are not already in polynomial form. In the simple theory of flexure of beams, the slope, bending moment, shearing force, load and other quantities are functions of a derivative of y with respect to x. The elastic curve of a transversely loaded beam can be represented by the Maclaurin series. Substitution of the values of the derivatives gives a direct solution of beam problems. Another application of Maclaurin series is in relating inter-atomic potential functions. At this stage, not all of the above applications would have been met or understood; however, sufficient to say that Maclaurin's series has a number of applications in engineering and science.

At the end of this chapter, you should be able to:

- determine simple derivatives
- derive Maclaurin's theorem
- appreciate the conditions of Maclaurin's series
- use Maclaurin's series to determine the power series for simple trigonometric, logarithmic and exponential functions
- evaluate a definite integral using Maclaurin's series
- state L'Hôpital's rule
- determine limiting values of functions

8.1   Introduction

Some mathematical functions may be represented as power series, containing terms in ascending powers of the variable. For example,

$$e^x = 1 + x + \frac{x^2}{2!} + \frac{x^3}{3!} + \cdots$$

$$\sin x = x - \frac{x^3}{3!} + \frac{x^5}{5!} - \frac{x^7}{7!} + \cdots$$

$$\text{and } \cosh x = 1 + \frac{x^2}{2!} + \frac{x^4}{4!} + \cdots$$

(as shown in Chapter 16)

Using a series, called **Maclaurin's**[*] **series**, mixed functions containing, say, algebraic, trigonometric and exponential functions, may be expressed solely as

Page Lordp
MACLAURIN,
London Pub.ᵈ as the Act directs Oct.ᵗ 1ᵗ 1815. by Gilason.

[*] **Who was Maclaurin**? **Colin Maclaurin** (February 1698–14 June 1746) was a Scottish mathematician who made important contributions to geometry and algebra. **The Maclaurin series** are named after him. To find out more go to **www.routledge.com/cw/bird**

algebraic functions, and differentiation and integration can often be more readily performed.

To expand a function using Maclaurin's theorem, some knowledge of differentiation is needed (more on differentiation is given in Chapter 29). Here is a revision of derivatives of the main functions needed in this chapter.

y or $f(x)$	$\frac{dy}{dx}$ or $f'(x)$
ax^n	anx^{n-1}
$\sin ax$	$a \cos ax$
$\cos ax$	$-a \sin ax$
e^{ax}	ae^{ax}
$\ln ax$	$\frac{1}{x}$
$\sinh ax$	$a \cosh ax$
$\cosh ax$	$a \sinh ax$

Given a general function $f(x)$, then $f'(x)$ is the first derivative, $f''(x)$ is the second derivative, and so on. Also, $f(0)$ means the value of the function when $x = 0$, $f'(0)$ means the value of the first derivative when $x = 0$, and so on.

8.2   Derivation of Maclaurin's theorem

Let the power series for $f(x)$ be

$$f(x) = a_0 + a_1 x + a_2 x^2 + a_3 x^3 + a_4 x^4 \\ + a_5 x^5 + \cdots \quad (1)$$

where $a_0, a_1, a_2, \ldots$ are constants.

When $x = 0, \boldsymbol{f(0) = a_0}$
Differentiating equation (1) with respect to x gives:

$$f'(x) = a_1 + 2a_2 x + 3a_3 x^2 + 4a_4 x^3 \\ + 5a_5 x^4 + \cdots \quad (2)$$

When $x = 0, \boldsymbol{f'(0) = a_1}$
Differentiating equation (2) with respect to x gives:

$$f''(x) = 2a_2 + (3)(2)a_3 x + (4)(3)a_4 x^2 \\ + (5)(4)a_5 x^3 + \cdots \quad (3)$$

When $x = 0, \ f''(0) = 2a_2 = 2! a_2$, i.e. $\boldsymbol{a_2 = \dfrac{f''(0)}{2!}}$

Differentiating equation (3) with respect to x gives:

$$f'''(x) = (3)(2)a_3 + (4)(3)(2)a_4 x$$
$$+ (5)(4)(3)a_5 x^2 + \cdots \quad (4)$$

When $x = 0$, $f'''(0) = (3)(2)a_3 = 3! a_3$, i.e. $a_3 = \dfrac{f'''(0)}{3!}$

Continuing the same procedure gives $a_4 = \dfrac{f^{iv}(0)}{4!}$,

$a_5 = \dfrac{f^{v}(0)}{5!}$, and so on.

Substituting for $a_0, a_1, a_2, \ldots$ in equation (1) gives:

$$f(x) = f(0) + f'(0)x + \frac{f''(0)}{2!}x^2$$
$$+ \frac{f'''(0)}{3!}x^3 + \cdots$$

i.e. $\boldsymbol{f(x) = f(0) + xf'(0) + \dfrac{x^2}{2!}f''(0)}$

$$\boldsymbol{+ \dfrac{x^3}{3!}f'''(0) + \cdots} \quad (5)$$

Equation (5) is a mathematical statement called **Maclaurin's theorem** or **Maclaurin's series**.

8.3 Conditions of Maclaurin's series

Maclaurin's series may be used to represent any function, say $f(x)$, as a power series provided that at $x = 0$ the following three conditions are met:

(a) $\boldsymbol{f(0) \neq \infty}$

For example, for the function $f(x) = \cos x$, $f(0) = \cos 0 = 1$, thus $\cos x$ meets the condition. However, if $f(x) = \ln x$, $f(0) = \ln 0 = -\infty$, thus $\ln x$ does not meet this condition.

(b) $\boldsymbol{f'(0), f''(0), f'''(0), \ldots \neq \infty}$

For example, for the function $f(x) = \cos x$, $f'(0) = -\sin 0 = 0$, $f''(0) = -\cos 0 = -1$, and so on; thus $\cos x$ meets this condition. However, if $f(x) = \ln x$, $f'(0) = \frac{1}{0} = \infty$, thus $\ln x$ does not meet this condition

(c) **The resultant Maclaurin's series must be convergent**.

In general, this means that the values of the terms, or groups of terms, must get progressively smaller

and the sum of the terms must reach a limiting value.

For example, the series $1 + \frac{1}{2} + \frac{1}{4} + \frac{1}{8} + \cdots$ is convergent since the value of the terms is getting smaller and the sum of the terms is approaching a limiting value of 2.

8.4 Worked problems on Maclaurin's series

Problem 1. Determine the first four terms of the power series for $\cos x$

The values of $f(0)$, $f'(0)$, $f''(0)$, $\ldots$ in Maclaurin's series are obtained as follows:

$f(x) = \cos x \qquad f(0) = \cos 0 = 1$

$f'(x) = -\sin x \qquad f'(0) = -\sin 0 = 0$

$f''(x) = -\cos x \qquad f''(0) = -\cos 0 = -1$

$f'''(x) = \sin x \qquad f'''(0) = \sin 0 = 0$

$f^{iv}(x) = \cos x \qquad f^{iv}(0) = \cos 0 = 1$

$f^{v}(x) = -\sin x \qquad f^{v}(0) = -\sin 0 = 0$

$f^{vi}(x) = -\cos x \qquad f^{vi}(0) = -\cos 0 = -1$

Substituting these values into equation (5) gives:

$$f(x) = \cos x = 1 + x(0) + \frac{x^2}{2!}(-1) + \frac{x^3}{3!}(0)$$
$$+ \frac{x^4}{4!}(1) + \frac{x^5}{5!}(0) + \frac{x^6}{6!}(-1) + \cdots$$

i.e. $\boldsymbol{\cos x = 1 - \dfrac{x^2}{2!} + \dfrac{x^4}{4!} - \dfrac{x^6}{6!} + \cdots}$

Problem 2. Determine the power series for $\cos 2\theta$

Replacing x with 2θ in the series obtained in Problem 1 gives:

$$\cos 2\theta = 1 - \frac{(2\theta)^2}{2!} + \frac{(2\theta)^4}{4!} - \frac{(2\theta)^6}{6!} + \cdots$$
$$= 1 - \frac{4\theta^2}{2} + \frac{16\theta^4}{24} - \frac{64\theta^6}{720} + \cdots$$

i.e. $\boldsymbol{\cos 2\theta = 1 - 2\theta^2 + \dfrac{2}{3}\theta^4 - \dfrac{4}{45}\theta^6 + \cdots}$

Problem 3.   Using Maclaurin's series, find the first 4 (non zero) terms for the function $f(x) = \sin x$

$$f(x) = \sin x \qquad f(0) = \sin 0 = 0$$

$$f''(x) = -\sin x \qquad f''(0) = -\sin 0 = 0$$

$$f'''(x) = -\cos x \qquad f'''(0) = -\cos 0 = -1$$

$$f^{iv}(x) = \sin x \qquad f^{iv}(0) = \sin 0 = 0$$

$$f^{v}(x) = \cos x \qquad f^{v}(0) = \cos 0 = 1$$

$$f^{vi}(x) = -\sin x \qquad f^{vi}(0) = -\sin 0 = 0$$

$$f^{vii}(x) = -\cos x \qquad f^{vii}(0) = -\cos 0 = -1$$

Substituting the above values into Maclaurin's series of equation (5) gives:

$$\sin x = 0 + x\,(1) + \frac{x^2}{2!}\,(0) + \frac{x^3}{3!}\,(-1) + \frac{x^4}{4!}\,(0)$$

$$+ \frac{x^5}{5!}\,(1) + \frac{x^6}{6!}\,(0) + \frac{x^7}{7!}\,(-1) + \cdots$$

i.e. $\mathbf{\sin x = x - \dfrac{x^3}{3!} + \dfrac{x^5}{5!} - \dfrac{x^7}{7!} + \cdots}$

Problem 4.   Using Maclaurin's series, find the first five terms for the expansion of the function $f(x) = e^{3x}$

$$f(x) = e^{3x} \qquad f(0) = e^0 = 1$$

$$f'(x) = 3e^{3x} \qquad f'(0) = 3e^0 = 3$$

$$f''(x) = 9e^{3x} \qquad f''(0) = 9e^0 = 9$$

$$f'''(x) = 27e^{3x} \qquad f'''(0) = 27e^0 = 27$$

$$f^{iv}(x) = 81e^{3x} \qquad f^{iv}(0) = 81e^0 = 81$$

Substituting the above values into Maclaurin's series of equation (5) gives:

$$e^{3x} = 1 + x\,(3) + \frac{x^2}{2!}\,(9) + \frac{x^3}{3!}\,(27)$$

$$+ \frac{x^4}{4!}\,(81) + \cdots$$

$$e^{3x} = 1 + 3x + \frac{9x^2}{2!} + \frac{27x^3}{3!} + \frac{81x^4}{4!} + \cdots$$

i.e. $e^{3x} = 1 + 3x + \dfrac{9x^2}{2} + \dfrac{9x^3}{2} + \dfrac{27x^4}{8} + \cdots$

Problem 5.   Determine the power series for $\tan x$ as far as the term in x^3

$$f(x) = \tan x$$

$$f(0) = \tan 0 = 0$$

$$f'(x) = \sec^2 x$$

$$f'(0) = \sec^2 0 = \frac{1}{\cos^2 0} = 1$$

$$f''(x) = (2\sec x)(\sec x \tan x)$$

$$= 2\sec^2 x \tan x$$

$$f''(0) = 2\sec^2 0 \tan 0 = 0$$

$$f'''(x) = (2\sec^2 x)(\sec^2 x)$$

$$\qquad + (\tan x)(4\sec x \sec x \tan x), \text{ by the}$$

$$\text{product rule,}$$

$$= 2\sec^4 x + 4\sec^2 x \tan^2 x$$

$$f'''(0) = 2\sec^4 0 + 4\sec^2 0 \tan^2 0 = 2$$

Substituting these values into equation (5) gives:

$$f(x) = \tan x = 0 + (x)(1) + \frac{x^2}{2!}\,(0) + \frac{x^3}{3!}\,(2)$$

i.e. $\mathbf{\tan x = x + \dfrac{1}{3}x^3}$

Problem 6.   Expand $\ln(1+x)$ to five terms.

$$f(x) = \ln(1+x) \qquad f(0) = \ln(1+0) = 0$$

$$f'(x) = \frac{1}{(1+x)} \qquad f'(0) = \frac{1}{1+0} = 1$$

$$f''(x) = \frac{-1}{(1+x)^2} \qquad f''(0) = \frac{-1}{(1+0)^2} = -1$$

$$f'''(x) = \frac{2}{(1+x)^3} \qquad f'''(0) = \frac{2}{(1+0)^3} = 2$$

$$f^{iv}(x) = \frac{-6}{(1+x)^4} \qquad f^{iv}(0) = \frac{-6}{(1+0)^4} = -6$$

$$f^{v}(x) = \frac{24}{(1+x)^5} \qquad f^{v}(0) = \frac{24}{(1+0)^5} = 24$$

Substituting these values into equation (5) gives:

$$f(x) = \ln(1+x) = 0 + x(1) + \frac{x^2}{2!}(-1)$$

$$+ \frac{x^3}{3!}(2) + \frac{x^4}{4!}(-6) + \frac{x^5}{5!}(24)$$

i.e. $\ln(1+x) = x - \frac{x^2}{2} + \frac{x^3}{3} - \frac{x^4}{4} + \frac{x^5}{5} - \cdots$

Problem 7. Expand $\ln(1-x)$ to five terms.

Replacing x by $-x$ in the series for $\ln(1+x)$ in Problem 6 gives:

$$\ln(1-x) = (-x) - \frac{(-x)^2}{2} + \frac{(-x)^3}{3}$$

$$- \frac{(-x)^4}{4} + \frac{(-x)^5}{5} - \cdots$$

i.e. $\ln(1-x) = -x - \frac{x^2}{2} - \frac{x^3}{3} - \frac{x^4}{4} - \frac{x^5}{5} - \cdots$

Problem 8. Determine the power series for $\ln\left(\frac{1+x}{1-x}\right)$

$\ln\left(\frac{1+x}{1-x}\right) = \ln(1+x) - \ln(1-x)$ by the laws of logarithms, and from Problems 6 and 7,

$$\ln\left(\frac{1+x}{1-x}\right) = \left(x - \frac{x^2}{2} + \frac{x^3}{3} - \frac{x^4}{4} + \frac{x^5}{5} - \cdots\right)$$

$$- \left(-x - \frac{x^2}{2} - \frac{x^3}{3} - \frac{x^4}{4} - \frac{x^5}{5} - \cdots\right)$$

$$= 2x + \frac{2}{3}x^3 + \frac{2}{5}x^5 + \cdots$$

i.e. $\ln\left(\frac{1+x}{1-x}\right) = 2\left(x + \frac{x^3}{3} + \frac{x^5}{5} + \cdots\right)$

Problem 9. Use Maclaurin's series to find the expansion of $(2+x)^4$

$$f(x) = (2+x)^4 \qquad f(0) = 2^4 = 16$$

$$f'(x) = 4(2+x)^3 \qquad f'(0) = 4(2)^3 = 32$$

$$f''(x) = 12(2+x)^2 \qquad f''(0) = 12(2)^2 = 48$$

$$f'''(x) = 24(2+x)^1 \qquad f'''(0) = 24(2) = 48$$

$$f^{iv}(x) = 24 \qquad f^{iv}(0) = 24$$

Substituting in equation (5) gives:

$(2+x)^4$

$$= f(0) + xf'(0) + \frac{x^2}{2!}f''(0) + \frac{x^3}{3!}f'''(0) + \frac{x^4}{4!}f^{iv}(0)$$

$$= 16 + (x)(32) + \frac{x^2}{2!}(48) + \frac{x^3}{3!}(48) + \frac{x^4}{4!}(24)$$

$$= 16 + 32x + 24x^2 + 8x^3 + x^4$$

(This expression could have been obtained by applying the binomial theorem.)

Problem 10. Expand $e^{\frac{x}{2}}$ as far as the term in x^4

$$f(x) = e^{\frac{x}{2}} \qquad f(0) = e^0 = 1$$

$$f'(x) = \frac{1}{2}e^{\frac{x}{2}} \qquad f'(0) = \frac{1}{2}e^0 = \frac{1}{2}$$

$$f''(x) = \frac{1}{4}e^{\frac{x}{2}} \qquad f''(0) = \frac{1}{4}e^0 = \frac{1}{4}$$

$$f'''(x) = \frac{1}{8}e^{\frac{x}{2}} \qquad f'''(0) = \frac{1}{8}e^0 = \frac{1}{8}$$

$$f^{iv}(x) = \frac{1}{16}e^{\frac{x}{2}} \qquad f^{iv}(0) = \frac{1}{16}e^0 = \frac{1}{16}$$

Substituting in equation (5) gives:

$$e^{\frac{x}{2}} = f(0) + xf'(0) + \frac{x^2}{2!}f''(0)$$

$$+ \frac{x^3}{3!}f'''(0) + \frac{x^4}{4!}f^{iv}(0) + \cdots$$

$$= 1 + (x)\left(\frac{1}{2}\right) + \frac{x^2}{2!}\left(\frac{1}{4}\right) + \frac{x^3}{3!}\left(\frac{1}{8}\right)$$

$$+ \frac{x^4}{4!}\left(\frac{1}{16}\right) + \cdots$$

i.e. $e^{\frac{x}{2}} = 1 + \frac{1}{2}x + \frac{1}{8}x^2 + \frac{1}{48}x^3 + \frac{1}{384}x^4 + \cdots$

Problem 11. Develop a series for $\sinh x$ using Maclaurin's series.

$$f(x) = \sinh x \qquad f(0) = \sinh 0 = \frac{e^0 - e^{-0}}{2} = 0$$

$$f'(x) = \cosh x \qquad f'(0) = \cosh 0 = \frac{e^0 + e^{-0}}{2} = 1$$

$$f''(x) = \sinh x \qquad f''(0) = \sinh 0 = 0$$

$$f'''(x) = \cosh x \qquad f'''(0) = \cosh 0 = 1$$

$$f^{iv}(x) = \sinh x \qquad f^{iv}(0) = \sinh 0 = 0$$

$$f^{v}(x) = \cosh x \qquad f^{v}(0) = \cosh 0 = 1$$

Substituting in equation (5) gives:

$$\sinh x = f(0) + x f'(0) + \frac{x^2}{2!} f''(0) + \frac{x^3}{3!} f'''(0)$$

$$+ \frac{x^4}{4!} f^{iv}(0) + \frac{x^5}{5!} f^{v}(0) + \cdots$$

$$= 0 + (x)(1) + \frac{x^2}{2!}(0) + \frac{x^3}{3!}(1) + \frac{x^4}{4!}(0)$$

$$+ \frac{x^5}{5!}(1) + \cdots$$

i.e. $\mathbf{\sinh x = x + \dfrac{x^3}{3!} + \dfrac{x^5}{5!} + \cdots}$

(as shown in Chapter 16)

Problem 12. Produce a power series for $\cos^2 2x$ as far as the term in x^6

From double angle formulae, $\cos 2A = 2\cos^2 A - 1$ (see Chapter 19).

from which, $\qquad \cos^2 A = \dfrac{1}{2}(1 + \cos 2A)$

and $\qquad\qquad \cos^2 2x = \dfrac{1}{2}(1 + \cos 4x)$

From Problem 1,

$$\cos x = 1 - \frac{x^2}{2!} + \frac{x^4}{4!} - \frac{x^6}{6!} + \cdots$$

hence $\qquad \cos 4x = 1 - \dfrac{(4x)^2}{2!} + \dfrac{(4x)^4}{4!} - \dfrac{(4x)^6}{6!} + \cdots$

$$= 1 - 8x^2 + \frac{32}{3}x^4 - \frac{256}{45}x^6 + \cdots$$

Thus $\quad \cos^2 2x = \dfrac{1}{2}(1 + \cos 4x)$

$$= \frac{1}{2}\left(1 + 1 - 8x^2 + \frac{32}{3}x^4 - \frac{256}{45}x^6 + \cdots\right)$$

i.e. $\mathbf{\cos^2 2x = 1 - 4x^2 + \dfrac{16}{3}x^4 - \dfrac{128}{45}x^6 + \cdots}$

Now try the following Practice Exercise

Practice Exercise 33   Maclaurin's series (Answers on page 859)

1. Determine the first four terms of the power series for $\sin 2x$ using Maclaurin's series.

2. Use Maclaurin's series to produce a power series for $\cosh 3x$ as far as the term in x^6

3. Use Maclaurin's theorem to determine the first three terms of the power series for $\ln(1 + e^x)$

4. Determine the power series for $\cos 4t$ as far as the term in t^6

5. Expand $e^{\frac{3}{2}x}$ in a power series as far as the term in x^3

6. Develop, as far as the term in x^4, the power series for $\sec 2x$

7. Expand $e^{2\theta}\cos 3\theta$ as far as the term in θ^2 using Maclaurin's series.

8. Determine the first three terms of the series for $\sin^2 x$ by applying Maclaurin's theorem.

9. Use Maclaurin's series to determine the expansion of $(3 + 2t)^4$

8.5 Numerical integration using Maclaurin's series

The value of many integrals cannot be determined using the various analytical methods. In Chapter 49, the trapezoidal, mid-ordinate and Simpson's rules are used to numerically evaluate such integrals. Another method of finding the approximate value of a definite integral is to express the function as a power series using Maclaurin's series, and then integrating each algebraic term in turn. This is demonstrated in the following worked problems.

As a reminder, the general solution of integrals of the form $\int a x^n \mathrm{d}x$, where a and n are constants, is given by:

$$\int a x^n \, \mathrm{d}x = \frac{a x^{n+1}}{n+1} + c$$

Problem 13. Evaluate $\int_{0.1}^{0.4} 2\,e^{\sin\theta}\,d\theta$, correct to 3 significant figures.

A power series for $e^{\sin\theta}$ is firstly obtained using Maclaurin's series.

$$f(\theta) = e^{\sin\theta} \qquad f(0) = e^{\sin 0} = e^0 = 1$$

$$f'(\theta) = \cos\theta\,e^{\sin\theta} \qquad f'(0) = \cos 0\,e^{\sin 0} = (1)e^0 = 1$$

$$f''(\theta) = (\cos\theta)(\cos\theta\,e^{\sin\theta}) + (e^{\sin\theta})(-\sin\theta),$$
$$\text{by the product rule}$$

$$= e^{\sin\theta}(\cos^2\theta - \sin\theta);$$

$$f''(0) = e^0(\cos^2 0 - \sin 0) = 1$$

$$f'''(\theta) = (e^{\sin\theta})[(2\cos\theta(-\sin\theta) - \cos\theta)]$$
$$+ (\cos^2\theta - \sin\theta)(\cos\theta\,e^{\sin\theta})$$

$$= e^{\sin\theta}\cos\theta[-2\sin\theta - 1 + \cos^2\theta - \sin\theta]$$

$$f'''(0) = e^0\cos 0[(0 - 1 + 1 - 0)] = 0$$

Hence from equation (5):

$$e^{\sin\theta} = f(0) + \theta f'(0) + \frac{\theta^2}{2!}f''(0) + \frac{\theta^3}{3!}f'''(0) + \cdots$$

$$= 1 + \theta + \frac{\theta^2}{2} + 0$$

Thus $\int_{0.1}^{0.4} 2\,e^{\sin\theta}\,d\theta = \int_{0.1}^{0.4} 2\left(1 + \theta + \frac{\theta^2}{2}\right)d\theta$

$$= \int_{0.1}^{0.4} (2 + 2\theta + \theta^2)\,d\theta$$

$$= \left[2\theta + \frac{2\theta^2}{2} + \frac{\theta^3}{3}\right]_{0.1}^{0.4}$$

$$= \left(0.8 + (0.4)^2 + \frac{(0.4)^3}{3}\right)$$

$$\qquad - \left(0.2 + (0.1)^2 + \frac{(0.1)^3}{3}\right)$$

$$= 0.98133 - 0.21033$$

$$= \mathbf{0.771}, \text{ correct to 3 significant figures.}$$

Problem 14. Evaluate $\int_0^1 \frac{\sin\theta}{\theta}\,d\theta$ using Maclaurin's series, correct to 3 significant figures.

Let $f(\theta) = \sin\theta \qquad f(0) = 0$

$$f'(\theta) = \cos\theta \qquad f'(0) = 1$$

$$f''(\theta) = -\sin\theta \qquad f''(0) = 0$$

$$f'''(\theta) = -\cos\theta \qquad f'''(0) = -1$$

$$f^{\text{iv}}(\theta) = \sin\theta \qquad f^{\text{iv}}(0) = 0$$

$$f^{\text{v}}(\theta) = \cos\theta \qquad f^{\text{v}}(0) = 1$$

Hence from equation (5):

$$\sin\theta = f(0) + \theta f'(0) + \frac{\theta^2}{2!}f''(0) + \frac{\theta^3}{3!}f'''(0)$$

$$+ \frac{\theta^4}{4!}f^{\text{iv}}(0) + \frac{\theta^5}{5!}f^{\text{v}}(0) + \cdots$$

$$= 0 + \theta(1) + \frac{\theta^2}{2!}(0) + \frac{\theta^3}{3!}(-1)$$

$$+ \frac{\theta^4}{4!}(0) + \frac{\theta^5}{5!}(1) + \cdots$$

i.e. $\sin\theta = \theta - \frac{\theta^3}{3!} + \frac{\theta^5}{5!} - \cdots$

Hence

$$\int_0^1 \frac{\sin\theta}{\theta}\,d\theta$$

$$= \int_0^1 \frac{\left(\theta - \dfrac{\theta^3}{3!} + \dfrac{\theta^5}{5!} - \dfrac{\theta^7}{7!} + \cdots\right)}{\theta}\,d\theta$$

$$= \int_0^1 \left(1 - \frac{\theta^2}{6} + \frac{\theta^4}{120} - \frac{\theta^6}{5040} + \cdots\right)d\theta$$

$$= \left[\theta - \frac{\theta^3}{18} + \frac{\theta^5}{600} - \frac{\theta^7}{7(5040)} + \cdots\right]_0^1$$

$$= 1 - \frac{1}{18} + \frac{1}{600} - \frac{1}{7(5040)} + \cdots$$

$$= \mathbf{0.946}, \text{ correct to 3 significant figures.}$$

Problem 15. Evaluate $\int_0^{0.4} x\ln(1+x)\,dx$ using Maclaurin's theorem, correct to 3 decimal places.

From Problem 6,

$$\ln(1+x) = x - \frac{x^2}{2} + \frac{x^3}{3} - \frac{x^4}{4} + \frac{x^5}{5} - \cdots$$

Hence $\int_0^{0.4} x\ln(1+x)\,dx$

$$= \int_0^{0.4} x\left(x - \frac{x^2}{2} + \frac{x^3}{3} - \frac{x^4}{4} + \frac{x^5}{5} - \cdots\right) dx$$

$$= \int_0^{0.4} \left(x^2 - \frac{x^3}{2} + \frac{x^4}{3} - \frac{x^5}{4} + \frac{x^6}{5} - \cdots\right) dx$$

$$= \left[\frac{x^3}{3} - \frac{x^4}{8} + \frac{x^5}{15} - \frac{x^6}{24} + \frac{x^7}{35} - \cdots\right]_0^{0.4}$$

$$= \left(\frac{(0.4)^3}{3} - \frac{(0.4)^4}{8} + \frac{(0.4)^5}{15} - \frac{(0.4)^6}{24}\right.$$

$$\left. + \frac{(0.4)^7}{35} - \cdots\right) - (0)$$

$$= 0.02133 - 0.0032 + 0.0006827 - \cdots$$

$$= \mathbf{0.019}, \text{ correct to 3 decimal places.}$$

Now try the following Practice Exercise

4. Use Maclaurin's theorem to expand $\sqrt{x}\ln(x+1)$ as a power series. Hence evaluate, correct to 3 decimal places, $\int_0^{0.5} \sqrt{x}\ln(x+1)\,dx$.

8.6 Limiting values

It is sometimes necessary to find limits of the form $\lim\limits_{x\to a}\left\{\frac{f(x)}{g(x)}\right\}$, where $f(a)=0$ and $g(a)=0$

For example,

$$\lim_{x\to 1}\left\{\frac{x^2+3x-4}{x^2-7x+6}\right\} = \frac{1+3-4}{1-7+6} = \frac{0}{0}$$

and $\frac{0}{0}$ is generally referred to as indeterminate. For certain limits a knowledge of series can sometimes help.

For example,

$$\lim_{x\to 0}\left\{\frac{\tan x - x}{x^3}\right\}$$

$$\equiv \lim_{x\to 0}\left\{\frac{x + \frac{1}{3}x^3 + \cdots - x}{x^3}\right\} \text{ from Problem 5}$$

$$= \lim_{x\to 0}\left\{\frac{\frac{1}{3}x^3 + \cdots}{x^3}\right\} = \lim_{x\to 0}\left\{\frac{1}{3}\right\} = \mathbf{\frac{1}{3}}$$

Similarly,

$$\lim_{x\to 0}\left\{\frac{\sinh x}{x}\right\}$$

$$\equiv \lim_{x\to 0}\left\{\frac{x + \frac{x^3}{3!} + \frac{x^5}{5!} +}{x}\right\} \text{ from Problem 11}$$

$$= \lim_{x\to 0}\left\{1 + \frac{x^2}{3!} + \frac{x^4}{5!} + \cdots\right\} = \mathbf{1}$$

However, a knowledge of series does not help with examples such as $\lim\limits_{x\to 1}\left\{\frac{x^2+3x-4}{x^2-7x+6}\right\}$

Section A

L'Hôpital's* **rule** will enable us to determine such limits when the differential coefficients of the numerator and denominator can be found.

L'Hôpital's rule states:

$$\lim_{x \to a} \left\{ \frac{f(x)}{g(x)} \right\} = \lim_{x \to a} \left\{ \frac{f'(x)}{g'(x)} \right\}$$

provided $g'(a) \neq 0$

It can happen that $\lim_{x \to a} \left\{ \frac{f'(x)}{g'(x)} \right\}$ is still $\frac{0}{0}$; if so, the numerator and denominator are differentiated again (and again) until a non-zero value is obtained for the denominator.

The following worked problems demonstrate how L'Hôpital's rule is used. Refer to Chapter 29 for methods of differentiation.

Problem 16. Determine $\lim_{x \to 1} \left\{ \frac{x^2 + 3x - 4}{x^2 - 7x + 6} \right\}$

The first step is to substitute $x = 1$ into both numerator and denominator. In this case we obtain $\frac{0}{0}$. It is

* **Who was L'Hôpital?** **Guillaume François Antoine, Marquis de l'Hôpital** (1661–2 February 1704, Paris) was a French mathematician. His name is firmly associated with l'Hôpital's rule for calculating limits involving indeterminate forms 0/0 and ∞/∞. To find out more go to **www.routledge.com/cw/bird**

only when we obtain such a result that we then use L'Hôpital's rule. Hence applying L'Hôpital's rule,

$$\lim_{x \to 1} \left\{ \frac{x^2 + 3x - 4}{x^2 - 7x + 6} \right\} = \lim_{x \to 1} \left\{ \frac{2x + 3}{2x - 7} \right\}$$

i.e. both numerator and denominator have been differentiated

$$= \frac{5}{-5} = -1$$

Problem 17. Determine $\lim_{x \to 0} \left\{ \frac{\sin x - x}{x^2} \right\}$

Substituting $x = 0$ gives

$$\lim_{x \to 0} \left\{ \frac{\sin x - x}{x^2} \right\} = \frac{\sin 0 - 0}{0} = \frac{0}{0}$$

Applying L'Hôpital's rule gives

$$\lim_{x \to 0} \left\{ \frac{\sin x - x}{x^2} \right\} = \lim_{x \to 0} \left\{ \frac{\cos x - 1}{2x} \right\}$$

Substituting $x = 0$ gives

$$\frac{\cos 0 - 1}{0} = \frac{1 - 1}{0} = \frac{0}{0} \quad \text{again}$$

Applying L'Hôpital's rule again gives

$$\lim_{x \to 0} \left\{ \frac{\cos x - 1}{2x} \right\} = \lim_{x \to 0} \left\{ \frac{-\sin x}{2} \right\} = 0$$

Problem 18. Determine $\lim_{x \to 0} \left\{ \frac{x - \sin x}{x - \tan x} \right\}$

Substituting $x = 0$ gives

$$\lim_{x \to 0} \left\{ \frac{x - \sin x}{x - \tan x} \right\} = \frac{0 - \sin 0}{0 - \tan 0} = \frac{0}{0}$$

Applying L'Hôpital's rule gives

$$\lim_{x \to 0} \left\{ \frac{x - \sin x}{x - \tan x} \right\} = \lim_{x \to 0} \left\{ \frac{1 - \cos x}{1 - \sec^2 x} \right\}$$

Substituting $x = 0$ gives

$$\lim_{x \to 0} \left\{ \frac{1 - \cos x}{1 - \sec^2 x} \right\} = \frac{1 - \cos 0}{1 - \sec^2 0} = \frac{1 - 1}{1 - 1} = \frac{0}{0} \quad \text{again}$$

Applying L'Hôpital's rule gives

$$\lim_{x\to 0}\left\{\frac{1-\cos x}{1-\sec^2 x}\right\}=\lim_{x\to 0}\left\{\frac{\sin x}{(-2\sec x)(\sec x\tan x)}\right\}$$

$$=\lim_{x\to 0}\left\{\frac{\sin x}{-2\sec^2 x\tan x}\right\}$$

Substituting $x=0$ gives

$$\frac{\sin 0}{-2\sec^2 0\tan 0}=\frac{0}{0}\quad\text{again}$$

Applying L'Hôpital's rule gives

$$\lim_{x\to 0}\left\{\frac{\sin x}{-2\sec^2 x\tan x}\right\}$$

$$=\lim_{x\to 0}\left\{\frac{\cos x}{(-2\sec^2 x)(\sec^2 x)+(\tan x)(-4\sec^2 x\tan x)}\right\}$$

$$\text{using the product rule}$$

Substituting $x=0$ gives

$$\frac{\cos 0}{-2\sec^4 0-4\sec^2 0\tan^2 0}=\frac{1}{-2-0}$$

$$=-\frac{1}{2}$$

Hence $\lim_{x\to 0}\left\{\dfrac{x-\sin x}{x-\tan x}\right\}=-\dfrac{1}{2}$

Now try the following Practice Exercise

Practice Exercise 35　Limiting values (Answers on page 859)

Determine the following limiting values

1. $\lim_{x\to 1}\left\{\dfrac{x^3-2x+1}{2x^3+3x-5}\right\}$

2. $\lim_{x\to 0}\left\{\dfrac{\sin x}{x}\right\}$

3. $\lim_{x\to 0}\left\{\dfrac{\ln(1+x)}{x}\right\}$

4. $\lim_{x\to 0}\left\{\dfrac{x^2-\sin 3x}{3x+x^2}\right\}$

5. $\lim_{\theta\to 0}\left\{\dfrac{\sin\theta-\theta\cos\theta}{\theta^3}\right\}$

6. $\lim_{t\to 1}\left\{\dfrac{\ln t}{t^2-1}\right\}$

7. $\lim_{x\to 0}\left\{\dfrac{\sinh x-\sin x}{x^3}\right\}$

8. $\lim_{\theta\to\frac{\pi}{2}}\left\{\dfrac{\sin\theta-1}{\ln\sin\theta}\right\}$

9. $\lim_{t\to 0}\left\{\dfrac{\sec t-1}{t\sin t}\right\}$

Section A

For fully worked solutions to each of the problems in Practice Exercises 33 to 35 in this chapter,
go to the website:
www.routledge.com/cw/bird

Revision Test 2 APs, GPs and binomial and Maclaurin's series

This Revision Test covers the material contained in Chapters 6 to 8. *The marks for each question are shown in brackets at the end of each question.*

1. Determine the twentieth term of the series $15.6, 15, 14.4, 13.8, \ldots$ (3)

2. The sum of 13 terms of an arithmetic progression is 286 and the common difference is 3. Determine the first term of the series. (4)

3. An engineer earns £21 000 per annum and receives annual increments of £600. Determine the salary in the ninth year and calculate the total earnings in the first eleven years. (5)

4. Determine the eleventh term of the series $1.5, 3, 6, 12, \ldots$ (2)

5. Find the sum of the first eight terms of the series $1, 2.5, 6.25, \ldots$, correct to 1 decimal place. (4)

6. Determine the sum to infinity of the series $5, 1, \frac{1}{5}, \ldots$ (3)

7. A machine is to have seven speeds ranging from 25 rev/min to 500 rev/min. If the speeds form a geometric progression, determine their value, each correct to the nearest whole number. (8)

8. Use the binomial series to expand $(2a - 3b)^6$ (7)

9. Determine the middle term of $\left(3x - \dfrac{1}{3y}\right)^{18}$ (6)

10. Expand the following in ascending powers of t as far as the term in t^3

    (a) $\dfrac{1}{1 + t}$ (b) $\dfrac{1}{\sqrt{(1 - 3t)}}$

    For each case, state the limits for which the expansion is valid. (11)

11. When x is very small show that:

$$\frac{1}{(1 + x)^2 \sqrt{(1 - x)}} \approx 1 - \frac{3}{2}x \tag{5}$$

12. The modulus of rigidity G is given by $G = \dfrac{R^4 \theta}{L}$ where R is the radius, θ the angle of twist and L the length. Find the approximate percentage error in G when R is measured 1.5% too large, θ is measured 3% too small and L is measured 1% too small. (8)

13. Use Maclaurin's series to determine a power series for $e^{2x} \cos 3x$ as far as the term in x^2. (10)

14. Show, using Maclaurin's series, that the first four terms of the power series for $\cosh 2x$ is given by:

$$\cosh 2x = 1 + 2x^2 + \frac{2}{3}x^4 + \frac{4}{45}x^6 \tag{11}$$

15. Expand the function $x^2 \ln(1 + \sin x)$ using Maclaurin's series and hence evaluate:

    $$\int_0^{\frac{1}{2}} x^2 \ln(1 + \sin x) \, dx \text{ correct to 2 significant figures.} \tag{13}$$

Chapter 9

Solving equations by iterative methods

Why it is important to understand: **Solving equations by iterative methods**

There are many, many different types of equations used in every branch of engineering and science. There are straightforward methods for solving simple, quadratic and simultaneous equations; however, there are many other types of equations than these three. Great progress has been made in the engineering and scientific disciplines regarding the use of iterative methods for linear systems. In engineering it is important that we can solve any equation; iterative methods, such as the bisection method, an algebraic method of successive approximations, and the Newton-Raphson method, help us do that.

At the end of this chapter, you should be able to:

- define iterative methods
- use the method of bisection to solve equations
- use an algebraic method of successive approximations to solve equations
- state the Newton–Raphson formula
- use Newton's method to solve equations

9.1 Introduction to iterative methods

Many equations can only be solved graphically or by methods of successive approximations to the roots, called **iterative methods**. Three methods of successive approximations are (i) bisection method, introduced in Section 9.2, (ii) an algebraic method, introduced in Section 9.3, and (iii) by using the Newton–Raphson formula, given in Section 9.4.

Each successive approximation method relies on a reasonably good first estimate of the value of a root being made. One way of determining this is to sketch a graph of the function, say $y = f(x)$, and determine the approximate values of roots from the points where the graph cuts the x-axis. Another way is by using a functional notation method. This method uses the property that the value of the graph of $f(x)=0$ changes sign for values of x just before and just after the value of a root. For example, one root of the equation $x^2 - x - 6 = 0$ is $x = 3$. Using functional notation:

$$f(x) = x^2 - x - 6$$
$$f(2) = 2^2 - 2 - 6 = -4$$
$$f(4) = 4^2 - 4 - 6 = +6$$

It can be seen from these results that the value of $f(x)$ changes from -4 at $f(2)$ to $+6$ at $f(4)$, indicating that a root lies between 2 and 4. This is shown more clearly in Fig. 9.1.

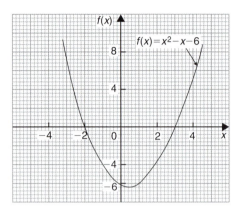

Figure 9.1

9.2 The bisection method

As shown above, by using functional notation it is possible to determine the vicinity of a root of an equation by the occurrence of a change of sign, i.e. if x_1 and x_2 are such that $f(x_1)$ and $f(x_2)$ have opposite signs, there is at least one root of the equation $f(x)=0$ in the interval between x_1 and x_2 (provided $f(x)$ is a continuous function). In the **method of bisection** the mid-point of the interval, i.e. $x_3 = \dfrac{x_1 + x_2}{2}$, is taken, and from the sign of $f(x_3)$ it can be deduced whether a root lies in the half interval to the left or right of x_3. Whichever half interval is indicated, its mid-point is then taken and the procedure repeated. The method often requires many iterations and is therefore slow, but never fails to eventually produce the root. The procedure stops when two successive values of x are equal—to the required degree of accuracy.

The method of bisection is demonstrated in Problems 1 to 3 following.

Problem 1. Use the method of bisection to find the positive root of the equation $5x^2 + 11x - 17 = 0$ correct to 3 significant figures.

Let $f(x) = 5x^2 + 11x - 17$

then, using functional notation:

$$f(0) = -17$$

$$f(1) = 5(1)^2 + 11(1) - 17 = -1$$

$$f(2) = 5(2)^2 + 11(2) - 17 = +25$$

Since there is a change of sign from negative to positive there must be a root of the equation between $x = 1$ and $x = 2$. This is shown graphically in Fig. 9.2.

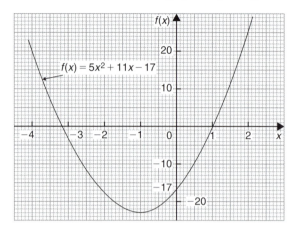

Figure 9.2

The method of bisection suggests that the root is at $\dfrac{1+2}{2} = 1.5$, i.e. the interval between 1 and 2 has been bisected.
Hence

$$f(1.5) = 5(1.5)^2 + 11(1.5) - 17$$

$$= +10.75$$

Since $f(1)$ is negative, $f(1.5)$ is positive, and $f(2)$ is also positive, a root of the equation must lie between $x = 1$ and $x = 1.5$, since a **sign change** has occurred between $f(1)$ and $f(1.5)$
Bisecting this interval gives $\dfrac{1+1.5}{2}$ i.e. 1.25 as the next root.
Hence

$$f(1.25) = 5(1.25)^2 + 11x - 17$$

$$= +4.5625$$

Since $f(1)$ is negative and $f(1.25)$ is positive, a root lies between $x = 1$ and $x = 1.25$
Bisecting this interval gives $\dfrac{1+1.25}{2}$ i.e. 1.125
Hence

$$f(1.125) = 5(1.125)^2 + 11(1.125) - 17$$

$$= +1.703125$$

Since $f(1)$ is negative and $f(1.125)$ is positive, a root lies between $x = 1$ and $x = 1.125$
Bisecting this interval gives $\dfrac{1+1.125}{2}$ i.e. 1.0625

Hence

$$f(1.0625) = 5(1.0625)^2 + 11(1.0625) - 17$$
$$= +0.33203125$$

Since $f(1)$ is negative and $f(1.0625)$ is positive, a root lies between $x = 1$ and $x = 1.0625$

Bisecting this interval gives $\dfrac{1 + 1.0625}{2}$ i.e. 1.03125
Hence

$$f(1.03125) = 5(1.03125)^2 + 11(1.03125) - 17$$
$$= -0.338867\ldots$$

Since $f(1.03125)$ is negative and $f(1.0625)$ is positive, a root lies between $x = 1.03125$ and $x = 1.0625$

Bisecting this interval gives

$$\frac{1.03125 + 1.0625}{2} \text{ i.e. } 1.046875$$

Hence

$$f(1.046875) = 5(1.046875)^2 + 11(1.046875) - 17$$
$$= -0.0046386\ldots$$

Since $f(1.046875)$ is negative and $f(1.0625)$ is positive, a root lies between $x = 1.046875$ and $x = 1.0625$

Bisecting this interval gives

$$\frac{1.046875 + 1.0625}{2} \text{ i.e. } \mathbf{1.0546875}$$

The last three values obtained for the root are 1.03125, 1.046875 and 1.0546875. The last two values are both 1.05, correct to 3 significant figure. We therefore stop the iterations here.

Thus, correct to 3 significant figures, the positive root of $5x^2 + 11x - 17 = 0$ is 1.05

> **Problem 2.** Use the bisection method to determine the positive root of the equation $x + 3 = e^x$, correct to 3 decimal places.

Let $f(x) = x + 3 - e^x$
  then, using functional notation:
$$f(0) = 0 + 3 - e^0 = \mathbf{+2}$$
$$f(1) = 1 + 3 - e^1 = \mathbf{+1.2817\ldots}$$
$$f(2) = 2 + 3 - e^2 = \mathbf{-2.3890\ldots}$$

Since $f(1)$ is positive and $f(2)$ is negative, a root lies between $x = 1$ and $x = 2$. A sketch of $f(x) = x + 3 - e^x$, i.e. $x + 3 = e^x$ is shown in Fig. 9.3.

Bisecting the interval between $x = 1$ and $x = 2$ gives $\dfrac{1 + 2}{2}$ i.e. 1.5

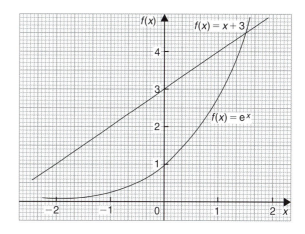

Figure 9.3

Hence

$$f(1.5) = 1.5 + 3 - e^{1.5}$$
$$= \mathbf{+0.01831\ldots}$$

Since $f(1.5)$ is positive and $f(2)$ is negative, a root lies between $x = 1.5$ and $x = 2$.
Bisecting this interval gives $\dfrac{1.5 + 2}{2}$ i.e. 1.75
Hence

$$f(1.75) = 1.75 + 3 - e^{1.75}$$
$$= \mathbf{-1.00460\ldots}$$

Since $f(1.75)$ is negative and $f(1.5)$ is positive, a root lies between $x = 1.75$ and $x = 1.5$

Bisecting this interval gives $\dfrac{1.75 + 1.5}{2}$ i.e. 1.625
Hence

$$f(1.625) = 1.625 + 3 - e^{1.625}$$
$$= \mathbf{-0.45341\ldots}$$

Since $f(1.625)$ is negative and $f(1.5)$ is positive, a root lies between $x = 1.625$ and $x = 1.5$
Bisecting this interval gives $\dfrac{1.625 + 1.5}{2}$ i.e. 1.5625
Hence

$$f(1.5625) = 1.5625 + 3 - e^{1.5625}$$
$$= \mathbf{-0.20823\ldots}$$

Since $f(1.5625)$ is negative and $f(1.5)$ is positive, a root lies between $x = 1.5625$ and $x = 1.5$

Bisecting this interval gives

$$\frac{1.5625 + 1.5}{2} \text{ i.e. } 1.53125$$

Hence

$$f(1.53125) = 1.53125 + 3 - e^{1.53125}$$
$$= \mathbf{-0.09270\ldots}$$

Since $f(1.53125)$ is negative and $f(1.5)$ is positive, a root lies between $x = 1.53125$ and $x = 1.5$

Bisecting this interval gives

$$\frac{1.53125 + 1.5}{2} \text{ i.e. } 1.515625$$

Hence

$$f(1.515625) = 1.515625 + 3 - e^{1.515625}$$

$$= -0.03664\ldots$$

Since $f(1.515625)$ is negative and $f(1.5)$ is positive, a root lies between $x = 1.515625$ and $x = 1.5$

Bisecting this interval gives

$$\frac{1.515625 + 1.5}{2} \text{ i.e. } 1.5078125$$

Hence

$$f(1.5078125) = 1.5078125 + 3 - e^{1.5078125}$$

$$= -0.009026\ldots$$

Since $f(1.5078125)$ is negative and $f(1.5)$ is positive, a root lies between $x = 1.5078125$ and $x = 1.5$

Bisecting this interval gives

$$\frac{1.5078125 + 1.5}{2} \text{ i.e. } 1.50390625$$

Hence

$$f(1.50390625) = 1.50390625 + 3 - e^{1.50390625}$$

$$= +0.004676\ldots$$

Since $f(1.50390625)$ is positive and $f(1.5078125)$ is negative, a root lies between $x = 1.50390625$ and $x = 1.5078125$

Bisecting this interval gives

$$\frac{1.50390625 + 1.5078125}{2} \text{ i.e. } 1.505859375$$

Hence

$$f(1.505859375) = 1.505859375 + 3 - e^{1.505859375}$$

$$= -0.0021666\ldots$$

Since $f(1.50589375)$ is negative and $f(1.50390625)$ is positive, a root lies between $x = 1.50589375$ and $x = 1.50390625$

Bisecting this interval gives

$$\frac{1.505859375 + 1.50390625}{2} \text{ i.e. } 1.504882813$$

Hence

$$f(1.504882813) = 1.504882813 + 3 - e^{1.504882813}$$

$$= +0.001256\ldots$$

Since $f(1.504882813)$ is positive and $f(1.505859375)$ is negative,

a root lies between $x = 1.504882813$ and $x = 1.505859375$

Bisecting this interval gives

$$\frac{1.504882813 + 1.50589375}{2} \text{ i.e. } \mathbf{1.505388282}$$

The last two values of x are 1.504882813 and 1.505388282, i.e. both are equal to 1.505, correct to 3 decimal places.

Hence the root of $x + 3 = e^x$ is $x = 1.505$, correct to 3 decimal places.

The above is a lengthy procedure and it is probably easier to present the data in a table.

x_1	x_2	$x_3 = \dfrac{x_1 + x_2}{2}$	$f(x_3)$
		0	+2
		1	+1.2817...
		2	−2.3890...
1	2	1.5	+0.0183...
1.5	2	1.75	−1.0046...
1.5	1.75	1.625	−0.4534...
1.5	1.625	1.5625	−0.2082...
1.5	1.5625	1.53125	−0.0927...
1.5	1.53125	1.515625	−0.0366...
1.5	1.515625	1.5078125	−0.0090...
1.5	1.5078125	1.50390625	+0.0046...
1.50390625	1.5078125	1.505859375	−0.0021...
1.50390625	1.505859375	**1.504882813**	+0.0012...
1.504882813	1.505859375	**1.505388282**	

Problem 3. Solve, correct to 2 decimal places, the equation $2\ln x + x = 2$ using the method of bisection.

Let $f(x) = 2\ln x + x - 2$

$\quad f(0.1) = 2\ln(0.1) + 0.1 - 2 = -6.5051\ldots$

$\qquad\qquad$ (Note that $\ln 0$ is infinite that is why $x = 0$ was not chosen)

$\quad f(1) = 2\ln 1 + 1 - 2 = -1$

$\quad f(2) = 2\ln 2 + 2 - 2 = +1.3862\ldots$

A change of sign indicates a root lies between $x = 1$ and $x = 2$

Since $2\ln x + x = 2$ then $2\ln x = -x + 2$; sketches of $2\ln x$ and $-x + 2$ are shown in Fig. 9.4.

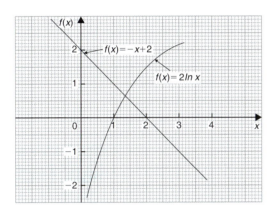

Figure 9.4

As shown in Problem 2, a table of values is produced to reduce space needed.

x_1	x_2	$x_3 = \dfrac{x_1 + x_2}{2}$	$f(x_3)$
		0.1	$-6.6051\ldots$
		1	-1
		2	$+1.3862\ldots$
1	2	1.5	$+0.3109\ldots$
1	1.5	1.25	$-0.3037\ldots$
1.25	1.5	1.375	$+0.0119\ldots$
1.25	1.375	1.3125	$-0.1436\ldots$
1.3125	1.375	1.34375	$-0.0653\ldots$
1.34375	1.375	1.359375	$-0.0265\ldots$
1.359375	1.375	**1.3671875**	$-0.0073\ldots$
1.3671875	1.375	**1.37109375**	$+0.0023\ldots$

The last two values of x_3 are both equal to 1.37 when expressed to 2 decimal places. We therefore stop the iterations.

Hence, the solution of $2\ln x + x = 2$ is $x = 1.37$, correct to 2 decimal places.

Now try the following Practice Exercise

Use the method of bisection to solve the following equations to the accuracy stated.

1. Find the positive root of the equation $x^2 + 3x - 5 = 0$, correct to 3 significant figures, using the method of bisection.

2. Using the bisection method, solve $e^x - x = 2$, correct to 4 significant figures.

3. Determine the positive root of $x^2 = 4\cos x$, correct to 2 decimal places using the method of bisection.

4. Solve $x - 2 - \ln x = 0$ for the root near to 3, correct to 3 decimal places using the bisection method.

5. Solve, correct to 4 significant figures, $x - 2\sin^2 x = 0$ using the bisection method.

9.3 An algebraic method of successive approximations

This method can be used to solve equations of the form:

$$a + bx + cx^2 + dx^3 + \cdots = 0,$$

where $a, b, c, d, \ldots$ are constants.

Procedure:

First approximation

(a) Using a graphical or the functional notation method (see Section 9.1) determine an approximate value of the root required, say x_1

Second approximation

(b) Let the true value of the root be $(x_1 + \delta_1)$

(c) Determine x_2 the approximate value of $(x_1 + \delta_1)$ by determining the value of $f(x_1 + \delta_1) = 0$, but neglecting terms containing products of δ_1

Third approximation

(d) Let the true value of the root be $(x_2 + \delta_2)$

(e) Determine x_3, the approximate value of $(x_2 + \delta_2)$ by determining the value of $f(x_2 + \delta_2) = 0$, but neglecting terms containing products of δ_2

(f) The fourth and higher approximations are obtained in a similar way.

Using the techniques given in paragraphs (b) to (f), it is possible to continue getting values nearer and nearer to the required root. The procedure is repeated until the value of the required root does not change on two consecutive approximations, when expressed to the required degree of accuracy.

> **Problem 4.** Use an algebraic method of successive approximations to determine the value of the negative root of the quadratic equation: $4x^2 - 6x - 7 = 0$ correct to 3 significant figures. Check the value of the root by using the quadratic formula.

A first estimate of the values of the roots is made by using the functional notation method

$$f(x) = 4x^2 - 6x - 7$$
$$f(0) = 4(0)^2 - 6(0) - 7 = -7$$
$$f(-1) = 4(-1)^2 - 6(-1) - 7 = 3$$

These results show that the negative root lies between 0 and -1, since the value of $f(x)$ changes sign between $f(0)$ and $f(-1)$ (see Section 9.1). The procedure given above for the root lying between 0 and -1 is followed.

First approximation

(a) Let a first approximation be such that it divides the interval 0 to -1 in the ratio of -7 to 3, i.e. let $x_1 = -0.7$

Second approximation

(b) Let the true value of the root, x_2, be $(x_1 + \delta_1)$.

(c) Let $f(x_1 + \delta_1) = 0$, then, since $x_1 = -0.7$,

$$4(-0.7 + \delta_1)^2 - 6(-0.7 + \delta_1) - 7 = 0$$

Hence, $4[(-0.7)^2 + (2)(-0.7)(\delta_1) + \delta_1^2]$

$$- (6)(-0.7) - 6\delta_1 - 7 = 0$$

Neglecting terms containing products of δ_1 gives:

$$1.96 - 5.6\delta_1 + 4.2 - 6\delta_1 - 7 \approx 0$$

i.e. $\quad -5.6\delta_1 - 6\delta_1 = -1.96 - 4.2 + 7$

i.e. $\quad \delta_1 \approx \dfrac{-1.96 - 4.2 + 7}{-5.6 - 6}$

$$\approx \dfrac{0.84}{-11.6}$$

$$\approx -0.0724$$

Thus, x_2, a second approximation to the root is $[-0.7 + (-0.0724)]$

i.e. $x_2 = -0.7724$, correct to 4 significant figures. (Since the question asked for 3 significant figure accuracy, it is usual to work to one figure greater than this.)

The procedure given in (b) and (c) is now repeated for $x_2 = -0.7724$

Third approximation

(d) Let the true value of the root, x_3, be $(x_2 + \delta_2)$

(e) Let $f(x_2 + \delta_2) = 0$, then, since $x_2 = -0.7724$,

$$4(-0.7724 + \delta_2)^2 - 6(-0.7724 + \delta_2) - 7 = 0$$

$$4[(-0.7724)^2 + (2)(-0.7724)(\delta_2) + \delta_2^2]$$

$$- (6)(-0.7724) - 6\delta_2 - 7 = 0$$

Neglecting terms containing products of δ_2 gives:

$$2.3864 - 6.1792\delta_2 + 4.6344 - 6\delta_2 - 7 \approx 0$$

i.e. $\delta_2 \approx \dfrac{-2.3864 - 4.6344 + 7}{-6.1792 - 6}$

$$\approx \dfrac{-0.0208}{-12.1792}$$

$$\approx +0.001708$$

Thus x_3, the third approximation to the root is $(-0.7724 + 0.001708)$

i.e. $x_3 = -0.7707$, correct to 4 significant figures (or -0.771 correct to 3 significant figures).

Fourth approximation

(f) The procedure given for the second and third approximations is now repeated for

$$x_3 = -0.7707$$

Let the true value of the root, x_4, be $(x_3 + \delta_3)$.
Let $f(x_3 + \delta_3) = 0$, then since $x_3 = -0.7707$,

$$4(-0.7707 + \delta_3)^2 - 6(-0.7707$$
$$+ \delta_3) - 7 = 0$$

$$4[(-0.7707)^2 + (2)(-0.7707)\delta_3 + \delta_3^2]$$
$$- 6(-0.7707) - 6\delta_3 - 7 = 0$$

Neglecting terms containing products of δ_3 gives:

$$2.3759 - 6.1656\delta_3 + 4.6242 - 6\delta_3 - 7 \approx 0$$

i.e. $\delta_3 \approx \dfrac{-2.3759 - 4.6242 + 7}{-6.1656 - 6}$

$$\approx \dfrac{-0.0001}{-12.156}$$

$$\approx +0.00000822$$

Thus, x_4, the fourth approximation to the root is $(-0.7707 + 0.00000822)$, i.e. $x_4 = -0.7707$, correct to 4 significant figures, and -0.771, correct to 3 significant figures.

Since the values of the roots are the same on two consecutive approximations, when stated to the required degree of accuracy, then the negative root of $4x^2 - 6x - 7 = 0$ is **−0.771, correct to 3 significant figures**.

[Checking, using the quadratic formula:

$$x = \dfrac{-(-6) \pm \sqrt{[(-6)^2 - (4)(4)(-7)]}}{(2)(4)}$$

$$= \dfrac{6 \pm 12.166}{8} = \textbf{−0.771 and 2.27},$$

correct to 3 significant figures]

[**Note on accuracy and errors.** Depending on the accuracy of evaluating the $f(x + \delta)$ terms, one or two iterations (i.e. successive approximations) might be saved. However, it is not usual to work to more than about 4 significant figures accuracy in this type of calculation. If a small error is made in calculations, the only likely effect is to increase the number of iterations.]

Problem 5. Determine the value of the smallest positive root of the equation $3x^3 - 10x^2 + 4x + 7 = 0$, correct to 3 significant figures, using an algebraic method of successive approximations.

The functional notation method is used to find the value of the first approximation.

$$f(x) = 3x^3 - 10x^2 + 4x + 7$$

$$f(0) = 3(0)^3 - 10(0)^2 + 4(0) + 7 = 7$$

$$f(1) = 3(1)^3 - 10(1)^2 + 4(1) + 7 = 4$$

$$f(2) = 3(2)^3 - 10(2)^2 + 4(2) + 7 = -1$$

Following the above procedure:

First approximation

(a) Let the first approximation be such that it divides the interval 1 to 2 in the ratio of 4 to -1, i.e. let x_1 be 1.8

Second approximation

(b) Let the true value of the root, x_2, be $(x_1 + \delta_1)$

(c) Let $f(x_1 + \delta_1) = 0$, then since $x_1 = 1.8$

$$3(1.8 + \delta_1)^3 - 10(1.8 + \delta_1)^2$$
$$+ 4(1.8 + \delta_1) + 7 = 0$$

Neglecting terms containing products of δ_1 and using the binomial series gives:

$$3[1.8^3 + 3(1.8)^2\delta_1] - 10[1.8^2 + (2)(1.8)\delta_1]$$
$$+ 4(1.8 + \delta_1) + 7 \approx 0$$

$$3(5.832 + 9.720\delta_1) - 32.4 - 36\delta_1$$
$$+ 7.2 + 4\delta_1 + 7 \approx 0$$

$$17.496 + 29.16\delta_1 - 32.4 - 36\delta_1$$
$$+ 7.2 + 4\delta_1 + 7 \approx 0$$

$$\delta_1 \approx \dfrac{-17.496 + 32.4 - 7.2 - 7}{29.16 - 36 + 4}$$

$$\approx -\dfrac{0.704}{2.84} \approx -0.2479$$

Thus $x_2 \approx 1.8 - 0.2479 = 1.5521$

Third approximation

(d) Let the true value of the root, x_3, be $(x_2 + \delta_2)$

(e) Let $f(x_2 + \delta_2) = 0$, then since $x_2 = 1.5521$

$$3(1.5521 + \delta_2)^3 - 10(1.5521 + \delta_2)^2$$
$$+ 4(1.5521 + \delta_2) + 7 = 0$$

Neglecting terms containing products of δ_2 gives:

$$11.217 + 21.681\delta_2 - 24.090 - 31.042\delta_2$$
$$+ 6.2084 + 4\delta_2 + 7 \approx 0$$

$$\delta_2 \approx \frac{-11.217 + 24.090 - 6.2084 - 7}{21.681 - 31.042 + 4}$$

$$\approx \frac{-0.3354}{-5.361}$$

$$\approx 0.06256$$

Thus $x_3 \approx 1.5521 + 0.06256 \approx 1.6147$

(f) Values of x_4 and x_5 are found in a similar way.

$$f(x_3 + \delta_3) = 3(1.6147 + \delta_3)^3 - 10(1.6147$$
$$+ \delta_3)^2 + 4(1.6147 + \delta_3) + 7 = 0$$

giving $\delta_3 \approx 0.003175$ and $x_4 \approx 1.618$, i.e. 1.62 correct to 3 significant figures.

$$f(x_4 + \delta_4) = 3(1.618 + \delta_4)^3 - 10(1.618 + \delta_4)^2$$
$$+ 4(1.618 + \delta_4) + 7 = 0$$

giving $\delta_4 \approx 0.0000417$, and $x_5 \approx 1.62$, correct to 3 significant figures.

Since x_4 and x_5 are the same when expressed to the required degree of accuracy, then the required root is **1.62**, correct to 3 significant figures.

Now try the following Practice Exercise

Practice Exercise 37   Solving equations by an algebraic method of successive approximations (Answers on page 859)

Use an algebraic method of successive approximation to solve the following equations to the accuracy stated.

1. $3x^2 + 5x - 17 = 0$, correct to 3 significant figures.

2. $x^3 - 2x + 14 = 0$, correct to 3 decimal places.

3. $x^4 - 3x^3 + 7x - 5.5 = 0$, correct to 3 significant figures.

4. $x^4 + 12x^3 - 13 = 0$, correct to 4 significant figures.

9.4   The Newton–Raphson method

The Newton–Raphson formula,* often just referred to as **Newton's method**, may be stated as follows:

If r_1 is the approximate value of a real root of the equation $f(x) = 0$, then a closer approximation to the root r_2 is given by:

$$r_2 = r_1 - \frac{f(r_1)}{f'(r_1)}$$

The advantages of Newton's method over the algebraic method of successive approximations is that it can be used for any type of mathematical equation (i.e. ones containing trigonometric, exponential, logarithmic, hyperbolic and algebraic functions), and it is usually easier to apply than the algebraic method.

* **Who were Newton and Raphson**? **Sir Isaac Newton** PRS MP (25 December 1642–20 March 1727) was an English polymath. Newton showed that the motions of objects are governed by the same set of natural laws, by demonstrating the consistency between Kepler's laws of planetary motion and his theory of gravitation. To find out more go to **www.routledge.com/cw/bird**

Joseph Raphson was an English mathematician known best for the Newton–Raphson method for approximating the roots of an equation. To find out more go to **www.routledge.com/cw/bird**

Problem 6.　Use Newton's method to determine the positive root of the quadratic equation $5x^2 + 11x - 17 = 0$, correct to 3 significant figures. Check the value of the root by using the quadratic formula.

The functional notation method is used to determine the first approximation to the root.

$$f(x) = 5x^2 + 11x - 17$$

$$f(0) = 5(0)^2 + 11(0) - 17 = -17$$

$$f(1) = 5(1)^2 + 11(1) - 17 = -1$$

$$f(2) = 5(2)^2 + 11(2) - 17 = 25$$

This shows that the value of the root is close to $x = 1$

Let the first approximation to the root, r_1, be 1

Newton's formula states that a closer approximation,

$$r_2 = r_1 - \frac{f(r_1)}{f'(r_1)}$$
$$f(x) = 5x^2 + 11x - 17$$

thus,　$f(r_1) = 5(r_1)^2 + 11(r_1) - 17$

$$= 5(1)^2 + 11(1) - 17 = -1$$

$f'(x)$ is the differential coefficient of $f(x)$,

i.e.　$f'(x) = 10x + 11$

Thus $f'(r_1) = 10(r_1) + 11$

$$= 10(1) + 11 = 21$$

By Newton's formula, a better approximation to the root is:

$$r_2 = 1 - \frac{-1}{21} = 1 - (-0.048) = 1.05,$$

correct to 3 significant figures.

A still better approximation to the root, r_3, is given by:

$$r_3 = r_2 - \frac{f(r_2)}{f'(r_2)}$$

$$= 1.05 - \frac{[5(1.05)^2 + 11(1.05) - 17]}{[10(1.05) + 11]}$$

$$= 1.05 - \frac{0.0625}{21.5}$$

$$= 1.05 - 0.003 = 1.047,$$

i.e. 1.05, correct to 3 significant figures.

Since the values of r_2 and r_3 are the same when expressed to the required degree of accuracy, the required root is **1.05**, correct to 3 significant figures. Checking, using the quadratic equation formula,

$$x = \frac{-11 \pm \sqrt{[121 - 4(5)(-17)]}}{(2)(5)}$$

$$= \frac{-11 \pm 21.47}{10}$$

The positive root is 1.047, i.e. **1.05**, correct to 3 significant figures (This root was determined in Problem 1 using the bisection method; Newton's method is clearly quicker).

Problem 7.　Taking the first approximation as 2, determine the root of the equation $x^2 - 3\sin x + 2\ln(x+1) = 3.5$, correct to 3 significant figures, by using Newton's method.

Newton's formula states that $r_2 = r_1 - \dfrac{f(r_1)}{f'(r_1)}$, where r_1 is a first approximation to the root and r_2 is a better approximation to the root.

Since　$f(x) = x^2 - 3\sin x + 2\ln(x+1) - 3.5$

$$f(r_1) = f(2) = 2^2 - 3\sin 2 + 2\ln 3 - 3.5$$

where $\sin 2$ means the sine of 2 radians

$$= 4 - 2.7279 + 2.1972 - 3.5$$

$$= -0.0307$$

$$f'(x) = 2x - 3\cos x + \frac{2}{x+1}$$

$$f'(r_1) = f'(2) = 2(2) - 3\cos 2 + \frac{2}{3}$$

$$= 4 + 1.2484 + 0.6667$$

$$= 5.9151$$

Hence,　$r_2 = r_1 - \dfrac{f(r_1)}{f'(r_1)}$

$$= 2 - \frac{-0.0307}{5.9151}$$

$$= 2.005 \text{ or } 2.01, \text{ correct to}$$

3 significant figures.

A still better approximation to the root, r_3, is given by:

$$r_3 = r_2 - \frac{f(r_2)}{f'(r_2)}$$

$$= 2.005 - \frac{[(2.005)^2 - 3\sin 2.005 + 2\ln 3.005 - 3.5]}{\left[2(2.005) - 3\cos 2.005 + \dfrac{2}{2.005 + 1}\right]}$$

$$= 2.005 - \frac{(-0.00104)}{5.9376} = 2.005 + 0.000175$$

i.e. $r_3 = 2.01$, correct to 3 significant figures.

Since the values of r_2 and r_3 are the same when expressed to the required degree of accuracy, then the required root is **2.01**, correct to 3 significant figures.

Problem 8. Use Newton's method to find the positive root of:

$$(x + 4)^3 - e^{1.92x} + 5\cos\frac{x}{3} = 9,$$

correct to 3 significant figures.

The functional notational method is used to determine the approximate value of the root.

$$f(x) = (x + 4)^3 - e^{1.92x} + 5\cos\frac{x}{3} - 9$$

$$f(0) = (0 + 4)^3 - e^0 + 5\cos 0 - 9 = 59$$

$$f(1) = 5^3 - e^{1.92} + 5\cos\frac{1}{3} - 9 \approx 114$$

$$f(2) = 6^3 - e^{3.84} + 5\cos\frac{2}{3} - 9 \approx 164$$

$$f(3) = 7^3 - e^{5.76} + 5\cos 1 - 9 \approx 19$$

$$f(4) = 8^3 - e^{7.68} + 5\cos\frac{4}{3} - 9 \approx -1660$$

From these results, let a first approximation to the root be $r_1 = 3$

Newton's formula states that a better approximation to the root,

$$r_2 = r_1 - \frac{f(r_1)}{f'(r_1)}$$

$$f(r_1) = f(3) = 7^3 - e^{5.76} + 5\cos 1 - 9$$

$$= 19.35$$

$$f'(x) = 3(x + 4)^2 - 1.92e^{1.92x} - \frac{5}{3}\sin\frac{x}{3}$$

$$f'(r_1) = f'(3) = 3(7)^2 - 1.92e^{5.76} - \frac{5}{3}\sin 1$$

$$= -463.7$$

Thus, $r_2 = 3 - \dfrac{19.35}{-463.7} = 3 + 0.042$

$$= 3.042 = 3.04,$$
correct to 3 significant figures.

Similarly, $r_3 = 3.042 - \dfrac{f(3.042)}{f'(3.042)}$

$$= 3.042 - \frac{(-1.146)}{(-513.1)}$$

$$= 3.042 - 0.0022 = 3.0398 = 3.04,$$
correct to 3 significant figures.

Since r_2 and r_3 are the same when expressed to the required degree of accuracy, then the required root is **3.04**, correct to 3 significant figures.

Now try the following Practice Exercise

Practice Exercise 38   Newton's method (Answers on page 859)

In Problems 1 to 7, use **Newton's method** to solve the equations given to the accuracy stated.

1. $x^2 - 2x - 13 = 0$, correct to 3 decimal places.

2. $3x^3 - 10x = 14$, correct to 4 significant figures.

3. $x^4 - 3x^3 + 7x = 12$, correct to 3 decimal places.

4. $3x^4 - 4x^3 + 7x - 12 = 0$, correct to 3 decimal places.

5. $3\ln x + 4x = 5$, correct to 3 decimal places.

6. $x^3 = 5\cos 2x$, correct to 3 significant figures.

7. $300e^{-2\theta} + \dfrac{\theta}{2} = 6$, correct to 3 significant figures.

8. Solve the equations in Problems 1 to 5, Practice Exercise 36, page 83 and Problems 1 to 4, Practice Exercise 37, page 86 using Newton's method.

9. A Fourier analysis of the instantaneous value of a waveform can be represented by:

$$y = \left(t + \frac{\pi}{4}\right) + \sin t + \frac{1}{8}\sin 3t$$

Use Newton's method to determine the value of t near to 0.04, correct to 4 decimal places, when the amplitude, y, is 0.880

10. A damped oscillation of a system is given by the equation:

$$y = -7.4e^{0.5t} \sin 3t$$

Determine the value of t near to 4.2, correct to 3 significant figures, when the magnitude y of the oscillation is zero.

11. The critical speeds of oscillation, λ, of a loaded beam are given by the equation:

$$\lambda^3 - 3.250\lambda^2 + \lambda - 0.063 = 0$$

Determine the value of λ which is approximately equal to 3.0 by Newton's method, correct to 4 decimal places.

For fully worked solutions to each of the problems in Practice Exercises 36 to 38 in this chapter, go to the website:
www.routledge.com/cw/bird

Chapter 10

Binary, octal and hexadecimal numbers

Why it is important to understand: Binary, octal and hexadecimal numbers

There are infinite ways to represent a number. The four commonly associated with modern computers and digital electronics are decimal, binary, octal, and hexadecimal. All four number systems are equally capable of representing any number. Furthermore, a number can be perfectly converted between the various number systems without any loss of numeric value. At first look, it seems like using any number system other than decimal is complicated and unnecessary. However, since the job of electrical and software engineers is to work with digital circuits, engineers require number systems that can best transfer information between the human world and the digital circuit world. Thus the way in which a number is represented can make it easier for the engineer to perceive the meaning of the number as it applies to a digital circuit, i.e. the appropriate number system can actually make things less complicated. Binary, octal and hexadecimal numbers are explained in this chapter.

At the end of this chapter, you should be able to:

- recognise a binary number
- convert binary to decimal and vice-versa
- add binary numbers
- recognise an octal number
- convert decimal to binary via octal and vice-versa
- recognise a hexadecimal number
- convert from hexadecimal to decimal and vice-versa
- convert from binary to hexadecimal and vice-versa

10.1 Introduction

Man's earliest number or counting system was probably developed to help determine how many possessions a person had. As daily activities became more complex, numbers became more important in trade, time, distance, and all other phases of human life. Ever since people discovered that it was necessary to count objects, they have been looking for easier ways to do so. The **abacus**, developed by the Chinese, is one of the earliest known calculators; it is still in use in some parts of the world.

Blaise Pascal* invented the first adding machine in 1642. Twenty years later, an Englishman, **Sir Samuel Morland***, developed a more compact device that could multiply, add, and subtract. About 1672, **Gottfried Wilhelm von Leibniz*** perfected a machine that could perform all the basic operations (add, subtract, multiply, divide), as well as extract the square root. Modern electronic digital computers still use von Leibniz's principles.

Computers are now employed wherever repeated calculations or the processing of huge amounts of data are needed. The greatest applications are found in the military, scientific, and commercial fields. They have applications that range from mail sorting, and engineering design, to the identification and destruction of enemy targets. The advantages of digital computers include speed, accuracy, and man-power savings. Often computers are able to take over routine jobs and release personnel for more important work that cannot be handled by a computer. People and computers do not normally speak the same language. Methods of translating information into forms that are understandable and usable to both are necessary. Humans generally speak in words and numbers expressed in the decimal number system, while computers only understand coded electronic pulses that represent digital information.

All data in modern computers is stored as series of **bits**, a bit being a **b**inary dig**it**, and can have one of two values, the numbers 0 and 1. The most basic form of representing computer data is to represent a piece of data as a string of 1s and 0s, one for each bit. This is called a **binary** or base-2 number.

Because binary notation requires so many bits to represent relatively small numbers, two further compact notations are often used, called **octal** and **hexadecimal**. Computer programmers who design sequences of number codes instructing a computer what to do would have a very difficult task if they were forced to work with nothing but long strings of 1s and 0s, the 'native language' of any digital circuit.

Octal notation represents data as base-8 numbers with each digit in an octal number representing three bits. Similarly, hexadecimal notation uses base-16 numbers, representing four bits with each digit. Octal numbers use only the digits 0–7, while hexadecimal numbers use all ten base-10 digits (0–9) and the letters A–F (representing the numbers 10–15).

This chapter explains how to convert between the decimal, binary, octal and hexadecimal systems.

* Who was Leibniz? **Gottfried Wilhelm Leibniz** (sometimes von Leibniz theorem*) (1 July 1646–14 November 1716) was a German mathematician and philosopher. Leibniz developed infinitesimal calculus and invented the Leibniz wheel. To find out more go to **www.routledge.com/cw/bird**

* **Who was Morland**? **Sir Samuel Morland, 1st Baronet** (1625–1695), was an English academic, diplomat, spy, inventor and mathematician of the 17th century, a polymath credited with early developments in relation to computing, hydraulics and steam power. To find out more go to **www.routledge.com/cw/bird**

* **Who was Pascal**? For image and resume of Pascal, see page 59. To find out more go to **www.routledge.com/cw/bird**

10.2 Binary numbers

The system of numbers in everyday use is the **denary** or **decimal** system of numbers, using the digits 0 to 9. It has ten different digits (0, 1, 2, 3, 4, 5, 6, 7, 8 and 9) and is said to have a **radix** or **base** of 10.

The **binary** system of numbers has a radix of 2 and uses only the digits 0 and 1.

(a) Conversion of binary to decimal

The decimal number 234.5 is equivalent to

$$2 \times 10^2 + 3 \times 10^1 + 4 \times 10^0 + 5 \times 10^{-1}$$

i.e. the sum of terms comprising: (a digit) multiplied by (the base raised to some power).

In the binary system of numbers, the base is 2, so 1101.1 is equivalent to:

$$1 \times 2^3 + 1 \times 2^2 + 0 \times 2^1 + 1 \times 2^0 + 1 \times 2^{-1}$$

Thus the decimal number equivalent to the binary number 1101.1 is $8 + 4 + 0 + 1 + \frac{1}{2}$, that is 13.5 i.e. **$1101.1_2 = 13.5_{10}$**, the suffixes 2 and 10 denoting binary and decimal systems of numbers respectively.

Problem 1. Convert 11011_2 to a decimal number.

From above: $11011_2 = 1 \times 2^4 + 1 \times 2^3 + 0 \times 2^2$

$$+ 1 \times 2^1 + 1 \times 2^0$$

$$= 16 + 8 + 0 + 2 + 1$$

$$= \mathbf{27_{10}}$$

Problem 2. Convert 0.1011_2 to a decimal fraction.

$0.1011_2 = 1 \times 2^{-1} + 0 \times 2^{-2} + 1 \times 2^{-3} + 1 \times 2^{-4}$

$$= 1 \times \frac{1}{2} + 0 \times \frac{1}{2^2} + 1 \times \frac{1}{2^3} + 1 \times \frac{1}{2^4}$$

$$= \frac{1}{2} + \frac{1}{8} + \frac{1}{16}$$

$$= 0.5 + 0.125 + 0.0625$$

$$= \mathbf{0.6875_{10}}$$

Problem 3. Convert 101.0101_2 to a decimal number.

$101.0101_2 = 1 \times 2^2 + 0 \times 2^1 + 1 \times 2^0 + 0 \times 2^{-1}$

$$+ 1 \times 2^{-2} + 0 \times 2^{-3} + 1 \times 2^{-4}$$

$$= 4 + 0 + 1 + 0 + 0.25 + 0 + 0.0625$$

$$= \mathbf{5.3125_{10}}$$

Now try the following Practice Exercise

Practice Exercise 39 Conversion of binary to decimal numbers (Answers on page 859)

In Problems 1 to 5, convert the binary numbers given to decimal numbers.

1. (a) 110 (b) 1011 (c) 1110 (d) 1001

2. (a) 10101 (b) 11001 (c) 101101 (d) 110011

3. (a) 101010 (b) 111000 (c) 1000001 (d) 10111000

4. (a) 0.1101 (b) 0.11001 (c) 0.00111 (d) 0.01011

5. (a) 11010.11 (b) 10111.011 (c) 110101.0111 (d) 11010101.10111

(b) Conversion of decimal to binary

An integer decimal number can be converted to a corresponding binary number by repeatedly dividing by 2 and noting the remainder at each stage, as shown below for 39_{10}.

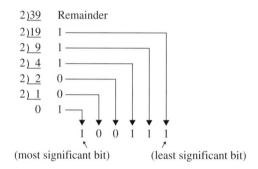

The result is obtained by writing the top digit of the remainder as the least significant bit (a bit is a **b**inary dig**it** and the least significant bit is the one on the right). The bottom bit of the remainder is the most significant bit, i.e. the bit on the left.

Thus   **$39_{10} = 100111_2$**

The fractional part of a decimal number can be converted to a binary number by repeatedly multiplying by 2, as shown below for the fraction 0.625

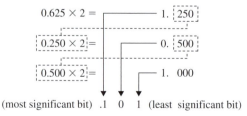

For fractions, the most significant bit of the result is the top bit obtained from the integer part of multiplication by 2. The least significant bit of the result is the bottom bit obtained from the integer part of multiplication by 2.

Thus   **$0.625_{10} = 0.101_2$**

Problem 4.   Convert 47_{10} to a binary number.

From above, repeatedly dividing by 2 and noting the remainder gives:

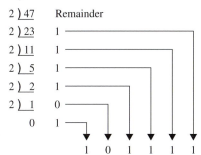

Thus $\mathbf{47_{10} = 101111_2}$

Problem 5.   Convert 0.40625_{10} to a binary number.

From above, repeatedly multiplying by 2 gives:

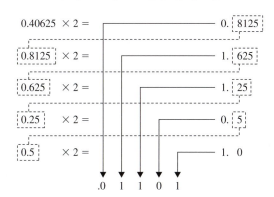

i.e. $\mathbf{0.40625_{10} = 0.01101_2}$

Problem 6.   Convert 58.3125_{10} to a binary number.

The integer part is repeatedly divided by 2, giving:

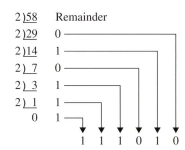

The fractional part is repeatedly multiplied by 2 giving:

$$
\begin{array}{ll}
0.3125 \times 2 = & 0.625 \\
0.625 \ \times 2 = & 1.25 \\
0.25 \ \ \times 2 = & 0.5 \\
0.5 \ \ \ \times 2 = & 1.0
\end{array}
$$

.0 1 0 1

Thus $\mathbf{58.3125_{10} = 111010.0101_2}$

Now try the following Practice Exercise

Practice Exercise 40   Conversion of decimal to binary numbers (Answers on page 860)

In Problems 1 to 5, convert the decimal numbers given to binary numbers.

1.   (a) 5 (b) 15 (c) 19 (d) 29

2.   (a) 31 (b) 42 (c) 57 (d) 63

3.   (a) 47 (b) 60 (c) 73 (d) 84

4.   (a) 0.25 (b) 0.21875 (c) 0.28125 (d) 0.59375

5.   (a) 47.40625 (b) 30.8125 (c) 53.90625 (d) 61.65625

(c)   Binary addition

Binary addition of two/three bits is achieved according to the following rules:

sum	carry	sum	carry
$0 + 0 = 0$	0	$0 + 0 + 0 = 0$	0
$0 + 1 = 1$	0	$0 + 0 + 1 = 1$	0
$1 + 0 = 1$	0	$0 + 1 + 0 = 1$	0
$1 + 1 = 0$	1	$0 + 1 + 1 = 0$	1
		$1 + 0 + 0 = 1$	0
		$1 + 0 + 1 = 0$	1
		$1 + 1 + 0 = 0$	1
		$1 + 1 + 1 = 1$	1

These rules are demonstrated in the following worked problems.

Problem 7.   Perform the binary addition:
$$1001 + 10110$$

$$
\begin{array}{r}
1001 \\
+\,10110 \\
\hline
\mathbf{11111} \\
\hline
\end{array}
$$

Problem 8. Perform the binary addition:
$$11111 + 10101$$

$$
\begin{array}{r}
11111 \\
+10101 \\
\hline
\text{sum} \quad \mathbf{110100} \\
\text{carry} \quad 11111
\end{array}
$$

Problem 9. Perform the binary addition:
$$1101001 + 1110101$$

$$
\begin{array}{r}
1101001 \\
+1110101 \\
\hline
\text{sum} \quad \mathbf{11011110} \\
\text{carry} \quad 11 \quad 1
\end{array}
$$

Problem 10. Perform the binary addition:
$$1011101 + 1100001 + 110101$$

$$
\begin{array}{r}
1011101 \\
1100001 \\
+ \quad 110101 \\
\hline
\text{sum} \quad \mathbf{11110011} \\
\text{carry} \quad 11111 \; 1
\end{array}
$$

Now try the following Practice Exercise

Practice Exercise 41   Binary addition (Answers on page 860)

Perform the following binary additions:
1. $10 + 11$
2. $101 + 110$
3. $1101 + 111$
4. $1111 + 11101$
5. $110111 + 10001$
6. $10000101 + 10000101$
7. $11101100 + 111001011$
8. $110011010 + 11100011$
9. $10110 + 1011 + 11011$
10. $111 + 10101 + 11011$
11. $1101 + 1001 + 11101$
12. $100011 + 11101 + 101110$

10.3 Octal numbers

For decimal integers containing several digits, repeatedly dividing by 2 can be a lengthy process. In this case, it is usually easier to convert a decimal number to a binary number via the octal system of numbers. This system has a radix of 8, using the digits 0, 1, 2, 3, 4, 5, 6 and 7. The decimal number equivalent to the octal number 4317_8 is:

$$4 \times 8^3 + 3 \times 8^2 + 1 \times 8^1 + 7 \times 8^0$$

i.e. $4 \times 512 + 3 \times 64 + 1 \times 8 + 7 \times 1$ or 2255_{10}

An integer decimal number can be converted to a corresponding octal number by repeatedly dividing by 8 and noting the remainder at each stage, as shown below for 493_{10}

$$
\begin{array}{lll}
8\,)\underline{493} & \text{Remainder} \\
8\,)\underline{\ 61} & 5 \\
8\,)\underline{\ \ 7} & 5 \\
\quad\ 0 & 7 \\
\end{array}
$$
$$7 \quad 5 \quad 5$$

Thus $493_{10} = 755_8$

The fractional part of a decimal number can be converted to an octal number by repeatedly multiplying by 8, as shown below for the fraction 0.4375_{10}

$$
\begin{array}{ll}
0.4375 \times 8 = & 3.\boxed{5} \\
\boxed{0.5} \times 8 = & 4.\ 0 \\
& .3 \quad 4
\end{array}
$$

For fractions, the most significant bit is the top integer obtained by multiplication of the decimal fraction by 8, thus,
$$0.4375_{10} = 0.34_8$$

The natural binary code for digits 0 to 7 is shown in Table 10.1, and an octal number can be converted to a binary number by writing down the three bits corresponding to the octal digit.

Thus $437_8 = 100\ 011\ 111_2$

and $26.35_8 = 010\ 110.011\ 101_2$

The '0' on the extreme left does not signify anything, thus $26.35_8 = 10\ 110.011\ 101_2$
Conversion of decimal to binary via octal is demonstrated in the following worked problems.

Table 10.1

Octal digit	Natural binary number
0	000
1	001
2	010
3	011
4	100
5	101
6	110
7	111

Problem 11.   Convert 3714_{10} to a binary number, via octal.

Dividing repeatedly by 8, and noting the remainder gives:

$$
\begin{array}{r|l}
8\,)\underline{3714} & \text{Remainder} \\
8\,)\underline{\ 464} & 2 \\
8\,)\underline{\ \ 58} & 0 \\
8\,)\underline{\ \ \ 7} & 2 \\
0 & 7
\end{array}
$$

$$7\ \ 2\ \ 0\ \ 2$$

From Table 10.1, $7202_8 = 111\ 010\ 000\ 010_2$

i.e. $\mathbf{3714_{10} = 111\ 010\ 000\ 010_2}$

Problem 12.   Convert 0.59375_{10} to a binary number, via octal.

Multiplying repeatedly by 8, and noting the integer values, gives:

$$0.59375 \times 8 = 4.75$$
$$0.75 \times 8 = 6.00$$

$$.4\ \ 6$$

Thus $0.59375_{10} = 0.46_8$

From Table 10.1, $0.46_8 = 0.100\ 110_2$

i.e. $\mathbf{0.59375_{10} = 0.100\ 11_2}$

Problem 13.   Convert 5613.90625_{10} to a binary number, via octal.

The integer part is repeatedly divided by 8, noting the remainder, giving:

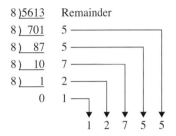

$$
\begin{array}{r|l}
8\,)\underline{5613} & \text{Remainder} \\
8\,)\underline{\ 701} & 5 \\
8\,)\underline{\ \ 87} & 5 \\
8\,)\underline{\ \ 10} & 7 \\
8\,)\underline{\ \ \ 1} & 2 \\
0 & 1
\end{array}
$$

$$1\ \ 2\ \ 7\ \ 5\ \ 5$$

This octal number is converted to a binary number, (see Table 10.1).

$$12755_8 = 001\ 010\ 111\ 101\ 101_2$$

i.e. $5613_{10} = 1\ 010\ 111\ 101\ 101_2$

The fractional part is repeatedly multiplied by 8, and noting the integer part, giving:

$$0.90625 \times 8 = 7.25$$
$$0.25 \times 8 = 2.00$$

$$.7\ \ 2$$

This octal fraction is converted to a binary number, (see Table 10.1).

$$0.72_8 = 0.111\ 010_2$$

i.e. $0.90625_{10} = 0.111\ 01_2$

Thus, $\mathbf{5613.90625_{10} = 1\ 010\ 111\ 101\ 101.111\ 01_2}$

Problem 14.   Convert $11\ 110\ 011.100\ 01_2$ to a decimal number via octal.

Grouping the binary number in three's from the binary point gives: $011\ 110\ 011.100\ 010_2$

Using Table 10.1 to convert this binary number to an octal number gives 363.42_8 and 363.42_8

$$= 3 \times 8^2 + 6 \times 8^1 + 3 \times 8^0 + 4 \times 8^{-1} + 2 \times 8^{-2}$$

$$= 192 + 48 + 3 + 0.5 + 0.03125$$

$$= \mathbf{243.53125_{10}}$$

Now try the following Practice Exercise

Practice Exercise 42   Conversion between decimal and binary numbers via octal (Answers on page 860)

In Problems 1 to 3, convert the decimal numbers given to binary numbers, via octal.

1. (a) 343 (b) 572 (c) 1265

2. (a) 0.46875 (b) 0.6875 (c) 0.71875

3. (a) 247.09375 (b) 514.4375 (c) 1716.78125

4. Convert the binary numbers given to decimal numbers via octal.

    (a) 111.011 1 (b) 101 001.01
    (c) 1 110 011 011 010.001 1

10.4   Hexadecimal numbers

The hexadecimal system is particularly important in computer programming, since four bits (each consisting of a one or zero) can be succinctly expressed using a single hexadecimal digit. Two hexadecimal digits represent numbers from 0 to 255, a common range used, for example, to specify colours. Thus, in the HTML language of the web, colours are specified using three pairs of hexadecimal digits RRGGBB, where RR is the amount of red, GG the amount of green, and BB the amount of blue.

A **hexadecimal numbering system** has a radix of 16 and uses the following 16 distinct digits:

$$0, 1, 2, 3, 4, 5, 6, 7, 8, 9, \text{A}, \text{B}, \text{C}, \text{D}, \text{E and F}$$

'A' corresponds to 10 in the decimal system, B to 11, C to 12, and so on.

(a)   Converting from hexadecimal to decimal

For example

$$1\text{A}_{16} = 1 \times 16^1 + \text{A} \times 16^0$$
$$= 1 \times 16^1 + 10 \times 1$$
$$= 16 + 10 = 26$$

i.e. $1\text{A}_{16} = 26_{10}$

Similarly, $2\text{E}_{16} = 2 \times 16^1 + \text{E} \times 16^0$
$$= 2 \times 16^1 + 14 \times 16^0$$
$$= 32 + 14 = 46_{10}$$

and $1\text{BF}_{16} = 1 \times 16^2 + \text{B} \times 16^1 + \text{F} \times 16^0$
$$= 1 \times 16^2 + 11 \times 16^1 + 15 \times 16^0$$
$$= 256 + 176 + 15 = 447_{10}$$

Table 10.2 compares decimal, binary, octal and hexadecimal numbers and shows, for example, that $23_{10} = 10111_2 = 27_8 = 17_{16}$

Problem 15.   Convert the following hexadecimal numbers into their decimal equivalents:
(a) 7A_{16} (b) 3F_{16}

(a) $7\text{A}_{16} = 7 \times 16^1 + \text{A} \times 16^0 = 7 \times 16 + 10 \times 1$
$$= 112 + 10 = 122$$

Thus $7\text{A}_{16} = 122_{10}$

(b) $3\text{F}_{16} = 3 \times 16^1 + \text{F} \times 16^0 = 3 \times 16 + 15 \times 1$
$$= 48 + 15 = 63$$

Thus $3\text{F}_{16} = 63_{10}$

Problem 16.   Convert the following hexadecimal numbers into their decimal equivalents:
(a) C9_{16} (b) BD_{16}

(a) $\text{C9}_{16} = \text{C} \times 16^1 + 9 \times 16^0 = 12 \times 16 + 9 \times 1$
$$= 192 + 9 = 201$$

Thus $\text{C9}_{16} = 201_{10}$

(b) $\text{BD}_{16} = \text{B} \times 16^1 + \text{D} \times 16^0$
$$= 11 \times 16 + 13 \times 1 = 176 + 13 = 189$$

Thus $\text{BD}_{16} = 189_{10}$

Problem 17.   Convert 1A4E_{16} into a decimal number.

$1\text{A4E}_{16} = 1 \times 16^3 + \text{A} \times 16^2 + 4 \times 16^1 + \text{E} \times 16^0$
$$= 1 \times 16^3 + 10 \times 16^2 + 4 \times 16^1 + 14 \times 16^0$$
$$= 1 \times 4096 + 10 \times 256 + 4 \times 16 + 14 \times 1$$
$$= 4096 + 2560 + 64 + 14 = 6734$$

Thus $1\text{A4E}_{16} = 6734_{10}$

Table 10.2

Decimal	Binary	Octal	Hexadecimal
0	0000	0	0
1	0001	1	1
2	0010	2	2
3	0011	3	3
4	0100	4	4
5	0101	5	5
6	0110	6	6
7	0111	7	7
8	1000	10	8
9	1001	11	9
10	1010	12	A
11	1011	13	B
12	1100	14	C
13	1101	15	D
14	1110	16	E
15	1111	17	F
16	10000	20	10
17	10001	21	11
18	10010	22	12
19	10011	23	13
20	10100	24	14
21	10101	25	15
22	10110	26	16
23	10111	27	17
24	11000	30	18
25	11001	31	19
26	11010	32	1A
27	11011	33	1B
28	11100	34	1C
29	11101	35	1D
30	11110	36	1E
31	11111	37	1F
32	100000	40	20

(b) Converting from decimal to hexadecimal

This is achieved by repeatedly dividing by 16 and noting the remainder at each stage, as shown below for 26_{10}

$$
\begin{array}{rl}
16\,)\,26 & \text{Remainder} \\
16\,)\,\underline{1} & 10 \equiv A_{16} \\
0 & 1 \equiv 1_{16}
\end{array}
$$

most significant bit $\rightarrow$ 1   A $\leftarrow$ least significant bit

Hence $\mathbf{26_{10} = 1A_{16}}$

Similarly, for 447_{10}

$$
\begin{array}{rl}
16\,)\,447 & \text{Remainder} \\
16\,)\,\underline{27} & 15 \equiv F_{16} \\
16\,)\,\underline{1} & 11 \equiv B_{16} \\
0 & 1 \equiv 1_{16}
\end{array}
$$

1 B F

Thus $\mathbf{447_{10} = 1BF_{16}}$

Problem 18. Convert the following decimal numbers into their hexadecimal equivalents:
(a) 37_{10} (b) 108_{10}

(a)
$$
\begin{array}{rl}
16\,)\,37 & \text{Remainder} \\
16\,)\,\underline{2} & 5 = 5_{16} \\
0 & 2 = 2_{16}
\end{array}
$$

2 5

most significant bit $\rfloor$ $\lfloor$ least significant bit

Hence $\mathbf{37_{10} = 25_{16}}$

(b)
$$
\begin{array}{rl}
16\,)\,108 & \text{Remainder} \\
16\,)\,\underline{6} & 12 = C_{16} \\
0 & 6 = 6_{16}
\end{array}
$$

6 C

Hence $\mathbf{108_{10} = 6C_{16}}$

Problem 19. Convert the following decimal numbers into their hexadecimal equivalents:
(a) 162_{10} (b) 239_{10}

(a) $16 \overline{)162}$ Remainder

     $16 \overline{)10}$ $2 = 2_{16}$

        0 $10 = A_{16}$

                 A   2

Hence $162_{10} = A2_{16}$

(b) $16 \overline{)239}$ Remainder

     $16 \overline{)14}$ $15 = F_{16}$

        0 $14 = E_{16}$

                 E   F

Hence $239_{10} = EF_{16}$

Now try the following Practice Exercise

Practice Exercise 43   Hexadecimal numbers (Answers on page 860)

In Problems 1 to 4, convert the given hexadecimal numbers into their decimal equivalents.

1. $E7_{16}$ 2. $2C_{16}$

3. 98_{16} 4. $2F1_{16}$

In Problems 5 to 8, convert the given decimal numbers into their hexadecimal equivalents.

5. 54_{10} 6. 200_{10}

7. 91_{10} 8. 238_{10}

(c) Converting from binary to hexadecimal

The binary bits are arranged in groups of four, starting from right to left, and a hexadecimal symbol is assigned to each group. For example, the binary number 1110011110101001 is initially grouped in fours as:

$\underbrace{1110}_{E} \ \underbrace{0111}_{7} \ \underbrace{1010}_{A} \ \underbrace{1001}_{9}$ and a hexadecimal symbol assigned to each group as above, from Table 10.2.

Hence $1110011110101001_2 = E7A9_{16}$

Problem 20. Convert the following binary numbers into their hexadecimal equivalents:
(a) 11010110_2 (b) 1100111_2

(a) Grouping bits in fours from the right gives:

$\underbrace{1101}_{D} \ \underbrace{0110}_{6}$ and assigning hexadecimal symbols

to each group gives as above, from Table 10.2.

Thus, $11010110_2 = D6_{16}$

(b) Grouping bits in fours from the right gives:

$\underbrace{0110}_{6} \ \underbrace{0111}_{7}$ and assigning hexadecimal symbols
to each group gives as above, from Table 10.2.

Thus, $1100111_2 = 67_{16}$

Problem 21. Convert the following binary numbers into their hexadecimal equivalents:
(a) 11001111_2 (b) 110011110_2

(a) Grouping bits in fours from the right gives:

$\underbrace{1100}_{C} \ \underbrace{1111}_{F}$ and assigning hexadecimal symbols
to each group gives as above, from Table 10.2.

Thus, $11001111_2 = CF_{16}$

(b) Grouping bits in fours from the right gives:

$\underbrace{0001}_{1} \ \underbrace{1001}_{9} \ \underbrace{1110}_{E}$ and assigning hexadecimal
symbols to each group gives as above, from Table 10.2.

Thus, $110011110_2 = 19E_{16}$

(d) Converting from hexadecimal to binary

The above procedure is reversed; thus, for example,

$6CF3_{16} = 0110 \ 1100 \ 1111 \ 0011$

                      from Table 10.2

i.e. $6CF3_{16} = 110110011110011_2$

Problem 22. Convert the following hexadecimal numbers into their binary equivalents:
(a) $3F_{16}$ (b) $A6_{16}$

(a) Spacing out hexadecimal digits gives:

$\underbrace{3}_{0011} \ \underbrace{F}_{1111}$ and converting each into binary
gives as above, from Table 10.2.

Thus, $3F_{16} = 111111_2$

(b) Spacing out hexadecimal digits gives:

$\underbrace{A}_{1010} \ \underbrace{6}_{0110}$ and converting each into binary
gives as above, from Table 10.2.

Thus, $A6_{16} = 10100110_2$

Problem 23.   Convert the following hexadecimal numbers into their binary equivalents:
(a) $7B_{16}$ (b) $17D_{16}$

(a)   Spacing   out   hexadecimal   digits   gives:
   7   B
$\overbrace{0111}$ $\overbrace{1011}$ and converting each into binary gives as above, from Table 10.2.

   Thus, $7B_{16} = 1111011_2$

(b)   Spacing   out   hexadecimal   digits   gives:
   1   7   D
$\overbrace{0001}$ $\overbrace{0111}$ $\overbrace{1101}$ and converting each into binary gives as above, from Table 10.2.

   Thus, $17D_{16} = 101111101_2$

3. 10001011_2

4. 10100101_2

In Problems 5 to 8, convert the given hexadecimal numbers into their binary equivalents.

5. 37_{16}

6. ED_{16}

7. $9F_{16}$

8. $A21_{16}$

Now try the following Practice Exercise

Practice Exercise 44   Hexadecimal numbers (Answers on page 860)

In Problems 1 to 4, convert the given binary numbers into their hexadecimal equivalents.

1. 11010111_2

2. 11101010_2

For fully worked solutions to each of the problems in Practice Exercises 39 to 44 in this chapter, go to the website:
www.routledge.com/cw/bird

Chapter 11

Boolean algebra and logic circuits

Why it is important to understand: **Boolean algebra and logic circuits**

Logic circuits are the basis for modern digital computer systems; to appreciate how computer systems operate an understanding of digital logic and Boolean algebra is needed. Boolean algebra (named after its developer, George Boole), is the algebra of digital logic circuits all computers use. Boolean algebra is the algebra of binary systems. A logic gate is a physical device implementing a Boolean function, performing a logical operation on one or more logic inputs, and produces a single logic output. Logic gates are implemented using diodes or transistors acting as electronic switches, but can also be constructed using electromagnetic relays, fluidic relays, pneumatic relays, optics, molecules or even mechanical elements. Learning Boolean algebra for logic analysis, learning about gates that process logic signals and learning how to design some smaller logic circuits is clearly of importance to computer engineers.

At the end of this chapter, you should be able to:

- draw a switching circuit and truth table for a two-input and three-input or-function and state its Boolean expression
- draw a switching circuit and truth table for a two-input and three-input and-function and state its Boolean expression
- produce the truth table for a two-input not-function and state its Boolean expression
- simplify Boolean expressions using the laws and rules of Boolean algebra
- simplify Boolean expressions using de Morgan's laws
- simplify Boolean expressions using Karnaugh maps
- draw a circuit diagram symbol and truth table for a three-input and-gate and state its Boolean expression
- draw a circuit diagram symbol and truth table for a three-input or-gate and state its Boolean expression
- draw a circuit diagram symbol and truth table for a three-input invert (or nor)-gate and state its Boolean expression
- draw a circuit diagram symbol and truth table for a three-input nand-gate and state its Boolean expression
- draw a circuit diagram symbol and truth table for a three-input nor-gate and state its Boolean expression
- devise logic systems for particular Boolean expressions
- use universal gates to devise logic circuits for particular Boolean expressions

11.1 Boolean algebra and switching circuits

A **two-state device** is one whose basic elements can only have one of two conditions. Thus, two-way switches, which can either be on or off, and the binary numbering system, having the digits 0 and 1 only, are two-state devices. In Boolean algebra (named after **George Boole**),* if A represents one state, then $\overline{A}$, called 'not-A', represents the second state.

The or-function

In Boolean algebra, the **or**-function for two elements A and B is written as $A + B$, and is defined as 'A, or B, or both A and B'. The equivalent electrical circuit for a two-input **or**-function is given by two switches connected in parallel. With reference to Fig. 11.1(a), the lamp will be on when A is on, when B is on, or when both A and B are on. In the table shown in Fig. 11.1(b), all the possible switch combinations are shown in columns 1 and 2, in which a 0 represents a switch being off and a 1 represents the switch being on, these columns being called the inputs. Column 3 is called the output and a 0 represents the lamp being off and a 1 represents the lamp being on. Such a table is called a **truth table**.

1 Input (switches)	2	3 Output (lamp)
A	B	$Z = A + B$
0	0	0
0	1	1
1	0	1
1	1	1

(a) Switching circuit for or-function   **(b)** Truth table for or-function

Figure 11.1

The and-function

In Boolean algebra, the **and**-function for two elements A and B is written as $A \cdot B$ and is defined as 'both A and B'. The equivalent electrical circuit for a two-input **and**-function is given by two switches connected in series. With reference to Fig. 11.2(a) the lamp will be on only when both A and B are on. The truth table for a two-input **and**-function is shown in Fig. 11.2(b).

Input (switches)		Output (lamp)
A	B	$Z = A \cdot B$
0	0	0
0	1	0
1	0	0
1	1	1

(a) Switching circuit for and-function   **(b)** Truth table for and-function

Figure 11.2

The not-function

In Boolean algebra, the **not**-function for element A is written as $\overline{A}$, and is defined as 'the opposite to A'. Thus if A means switch A is on, $\overline{A}$ means that switch A is off. The truth table for the **not**-function is shown in Table 11.1.

In the above, the Boolean expressions, equivalent switching circuits and truth tables for the three functions used in Boolean algebra are given for a two-input

* **Who was Boole**? **George Boole** (2 November 1815– 8 December 1864) was an English mathematician, philosopher and logician that worked in the fields of differential equations and algebraic logic. Best known as the author of *The Laws of Thought*, Boole is also the inventor of the prototype of what is now called Boolean logic, which became the basis of the modern digital computer. To find out more go to **www.routledge.com/cw/bird**

Table 11.1

Input	Output
A	$Z = \overline{A}$
0	1
1	0

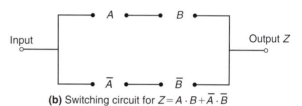

1	2	3	4	5
A	B	$A \cdot B$	$\overline{A} \cdot \overline{B}$	$Z = AB + \overline{A} \cdot \overline{B}$
0	0	0	1	1
0	1	0	0	0
1	0	0	0	0
1	1	1	0	1

(a) Truth table for $Z = A \cdot B + \overline{A} \cdot \overline{B}$

(b) Switching circuit for $Z = A \cdot B + \overline{A} \cdot \overline{B}$

Figure 11.4

system. A system may have more than two inputs and the Boolean expression for a three-input **or**-function having elements A, B and C is $A + B + C$. Similarly, a three-input **and**-function is written as $A \cdot B \cdot C$. The equivalent electrical circuits and truth tables for three-input **or** and **and**-functions are shown in Figs 11.3(a) and (b) respectively.

To achieve a given output, it is often necessary to use combinations of switches connected both in series and in parallel. If the output from a switching circuit is given by the Boolean expression $Z = A \cdot B + \overline{A} \cdot \overline{B}$, the truth table is as shown in Fig. 11.4(a). In this table, columns 1 and 2 give all the possible combinations of A and B. Column 3 corresponds to $A \cdot B$ and column 4 to $\overline{A} \cdot \overline{B}$, i.e. a 1 output is obtained when $A = 0$ and when $B = 0$. Column 5 is the **or**-function applied to columns 3 and 4 giving an output of $Z = A \cdot B + \overline{A} \cdot \overline{B}$. The corresponding switching circuit is shown in Fig. 11.4(b) in which A and B are connected in series to give $A \cdot B$, $\overline{A}$ and $\overline{B}$ are connected in series to give $\overline{A} \cdot \overline{B}$, and $A \cdot B$ and $\overline{A} \cdot \overline{B}$ are connected in parallel to give $A \cdot B + \overline{A} \cdot \overline{B}$. The circuit symbols used are such that A means the switch is on

when A is 1, $\overline{A}$ means the switch is on when A is 0, and so on.

Problem 1. Derive the Boolean expression and construct a truth table for the switching circuit shown in Fig. 11.5.

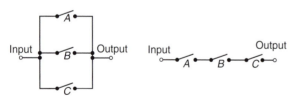

Figure 11.5

The switches between 1 and 2 in Fig. 11.5 are in series and have a Boolean expression of $B \cdot A$. The parallel circuit 1 to 2 and 3 to 4 have a Boolean expression of $(B \cdot A + \overline{B})$. The parallel circuit can be treated as a single switching unit, giving the equivalent of switches 5 to 6, 6 to 7 and 7 to 8 in series. Thus the output is given by:

$$Z = \overline{A} \cdot (B \cdot A + \overline{B}) \cdot \overline{B}$$

The truth table is as shown in Table 11.2. Columns 1 and 2 give all the possible combinations of switches A and B. Column 3 is the **and**-function applied to columns 1 and 2, giving $B \cdot A$. Column 4 is $\overline{B}$, i.e. the opposite to column 2. Column 5 is the **or**-function applied to columns 3 and 4. Column 6 is $\overline{A}$, i.e. the opposite to column 1. The output is column 7 and is obtained by applying the **and**-function to columns 4, 5 and 6.

Input	Output
A B C	$Z = A + B + C$
0 0 0	0
0 0 1	1
0 1 0	1
0 1 1	1
1 0 0	1
1 0 1	1
1 1 0	1
1 1 1	1

(a) The or-function electrical circuit and truth table

Input	Output
A B C	$Z = A \cdot B \cdot C$
0 0 0	0
0 0 1	0
0 1 0	0
0 1 1	0
1 0 0	0
1 0 1	0
1 1 0	0
1 1 1	1

(b) The and-function electrical circuit and truth table

Figure 11.3

Table 11.2

1	2	3	4	5	6	7
A	B	$B \cdot A$	$\overline{B}$	$B \cdot A + \overline{B}$	$\overline{A}$	$Z = \overline{A} \cdot (B \cdot A + \overline{B}) \cdot \overline{B}$
0	0	0	1	1	1	1
0	1	0	0	0	1	0
1	0	0	1	1	0	0
1	1	1	0	1	0	0

Problem 2. Derive the Boolean expression and construct a truth table for the switching circuit shown in Fig. 11.6.

Figure 11.6

The parallel circuit 1 to 2 and 3 to 4 gives $(A + \overline{B})$ and this is equivalent to a single switching unit between 7 and 2. The parallel circuit 5 to 6 and 7 to 2 gives $C + (A + \overline{B})$ and this is equivalent to a single switching unit between 8 and 2. The series circuit 9 to 8 and 8 to 2 gives the output

$$Z = B \cdot [C + (A + \overline{B})]$$

The truth table is shown in Table 11.3. Columns 1, 2 and 3 give all the possible combinations of A, B and C. Column 4 is $\overline{B}$ and is the opposite to column 2. Column 5 is the **or**-function applied to columns 1 and 4, giving $(A + \overline{B})$. Column 6 is the **or**-function applied to columns 3 and 5 giving $C + (A + \overline{B})$. The output is given in column 7 and is obtained by applying the **and**-function to columns 2 and 6, giving $Z = B \cdot [C + (A + \overline{B})]$.

Problem 3. Construct a switching circuit to meet the requirements of the Boolean expression: $Z = A \cdot \overline{C} + \overline{A} \cdot B + \overline{A} \cdot B \cdot \overline{C}$. Construct the truth table for this circuit.

The three terms joined by **or**-functions, $(+)$, indicate three parallel branches,

Table 11.3

1	2	3	4	5	6	7
A	B	C	$\overline{B}$	$A + \overline{B}$	$C + (A + \overline{B})$	$Z = B \cdot [C + (A + \overline{B})]$
0	0	0	1	1	1	0
0	0	1	1	1	1	0
0	1	0	0	0	0	0
0	1	1	0	0	1	1
1	0	0	1	1	1	0
1	0	1	1	1	1	0
1	1	0	0	1	1	1
1	1	1	0	1	1	1

having: branch 1 A **and** $\overline{C}$ in series
branch 2 $\overline{A}$ **and** B in series
and   branch 3 $\overline{A}$ **and** B **and** $\overline{C}$ in series

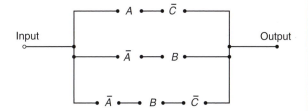

Figure 11.7

Hence the required switching circuit is as shown in Fig. 11.7. The corresponding truth table is shown in Table 11.4.

Table 11.4

1	2	3	4	5	6	7	8	9
A	B	C	$\overline{C}$	$A \cdot \overline{C}$	$\overline{A}$	$\overline{A} \cdot B$	$\overline{A} \cdot V \cdot \overline{C}$	$Z = A \cdot \overline{C} + \overline{A} \cdot B + \overline{A} \cdot B \cdot \overline{C}$
0	0	0	1	0	1	0	0	0
0	0	1	0	0	1	0	0	0
0	1	0	1	0	1	1	1	1
0	1	1	0	0	1	1	0	1
1	0	0	1	1	0	0	0	1
1	0	1	0	0	0	0	0	0
1	1	0	1	1	0	0	0	1
1	1	1	0	0	0	0	0	0

Column 4 is $\overline{C}$, i.e. the opposite to column 3

Column 5 is $A \cdot \overline{C}$, obtained by applying the **and**-function to columns 1 and 4

Column 6 is $\overline{A}$, the opposite to column 1

Column 7 is $\overline{A} \cdot B$, obtained by applying the **and**-function to columns 2 and 6

Column 8 is $\overline{A} \cdot B \cdot \overline{C}$, obtained by applying the **and**-function to columns 4 and 7

Column 9 is the output, obtained by applying the **or**-function to columns 5, 7 and 8

Problem 4. Derive the Boolean expression and construct the switching circuit for the truth table given in Table 11.5.

Table 11.5

	A	B	C	Z
1	0	0	0	1
2	0	0	1	0
3	0	1	0	1
4	0	1	1	1
5	1	0	0	0
6	1	0	1	1
7	1	1	0	0
8	1	1	1	0

Examination of the truth table shown in Table 11.5 shows that there is a 1 output in the Z-column in rows 1, 3, 4 and 6. Thus, the Boolean expression and switching circuit should be such that a 1 output is obtained for row 1 **or** row 3 **or** row 4 **or** row 6. In row 1, A is 0 **and** B is 0 **and** C is 0 and this corresponds to the Boolean expression $\overline{A} \cdot \overline{B} \cdot \overline{C}$. In row 3, A is 0 **and** B is 1 **and** C is 0, i.e. the Boolean expression in $\overline{A} \cdot B \cdot \overline{C}$. Similarly in rows 4 and 6, the Boolean expressions are $\overline{A} \cdot B \cdot C$ and $A \cdot \overline{B} \cdot C$ respectively. Hence the Boolean expression is:

$$Z = \overline{A} \cdot \overline{B} \cdot \overline{C} + \overline{A} \cdot B \cdot \overline{C}$$
$$+ \overline{A} \cdot B \cdot C + A \cdot \overline{B} \cdot C$$

The corresponding switching circuit is shown in Fig. 11.8. The four terms are joined by **or**-functions, $(+)$, and are represented by four parallel circuits. Each

term has three elements joined by an **and**-function, and is represented by three elements connected in series.

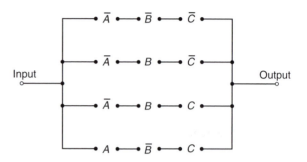

Figure 11.8

Now try the following Practice Exercise

Practice Exercise 45 Boolean algebra and switching circuits (Answers on page 860)

In Problems 1 to 4, determine the Boolean expressions and construct truth tables for the switching circuits given.

1. The circuit shown in Fig. 11.9.

Figure 11.9

2. The circuit shown in Fig. 11.10.

Figure 11.10

3. The circuit show in Fig. 11.11.

Figure 11.11

4. The circuit shown in Fig. 11.12.

Figure 11.12

In Problems 5 to 7, construct switching circuits to meet the requirements of the Boolean expressions given.

5. $A \cdot C + A \cdot \overline{B} \cdot C + A \cdot B$

6. $A \cdot B \cdot C \cdot (A + B + C)$.

7. $A \cdot (A \cdot \overline{B} \cdot C + B \cdot (A + \overline{C}))$.

In Problems 8 to 10, derive the Boolean expressions and construct the switching circuits for the truth table stated.

8. Table 11.6, column 4.

Table 11.6

1	2	3	4	5	6
A	B	C			
0	0	0	0	1	1
0	0	1	1	0	0
0	1	0	0	0	1
0	1	1	0	1	0
1	0	0	0	1	1
1	0	1	0	0	1
1	1	0	1	0	0
1	1	1	0	0	0

9. Table 11.6, column 5.

10. Table 11.6, column 6.

11.2 Simplifying Boolean expressions

A Boolean expression may be used to describe a complex switching circuit or logic system. If the Boolean expression can be simplified, then the number of switches or logic elements can be reduced, resulting

in a saving in cost. Three principal ways of simplifying Boolean expressions are:

(a) by using the laws and rules of Boolean algebra (see Section 11.3),

(b) by applying de Morgan's laws (see Section 11.4), and

(c) by using Karnaugh maps (see Section 11.5).

11.3 Laws and rules of Boolean algebra

A summary of the principal laws and rules of Boolean algebra are given in Table 11.7. The way in which these laws and rules may be used to simplify Boolean expressions is shown in Problems 5 to 10.

Table 11.7

Ref.	Name	Rule or law
1	Commutative laws	$A + B = B + A$
2		$A \cdot B = B \cdot A$
3	Associative laws	$(A + B) + C = A + (B + C)$
4		$(A \cdot B) \cdot C = A \cdot (B \cdot C)$
5	Distributive laws	$A \cdot (B + C) = A \cdot B + A \cdot C$
6		$A + (B \cdot C)$
		$\quad = (A + B) \cdot (A + C)$
7	Sum rules	$A + 0 = A$
8		$A + 1 = 1$
9		$A + A = A$
10		$A + \overline{A} = 1$
11	Product	$A \cdot 0 = 0$
12	rules	$A \cdot 1 = A$
13		$A \cdot A = A$
14		$A \cdot \overline{A} = 0$
15	Absorption	$A + A \cdot B = A$
16	rules	$A \cdot (A + B) = A$
17		$A + \overline{A} \cdot B = A + B$

Problem 5. Simplify the Boolean expression: $\overline{P} \cdot \overline{Q} + \overline{P} \cdot Q + P \cdot \overline{Q}$

With reference to Table 11.7: *Reference*

$$\overline{P} \cdot \overline{Q} + \overline{P} \cdot Q + P \cdot \overline{Q}$$
$$= \overline{P} \cdot (\overline{Q} + Q) + P \cdot \overline{Q} \qquad\qquad 5$$
$$= \overline{P} \cdot 1 + P \cdot \overline{Q} \qquad\qquad 10$$
$$= \boldsymbol{\overline{P} + P \cdot \overline{Q}} \qquad\qquad 12$$

Problem 6. Simplify
$(P + \overline{P} \cdot Q) \cdot (Q + \overline{Q} \cdot P)$

With reference to Table 11.7: *Reference*

$$(P + \overline{P} \cdot Q) \cdot (Q + \overline{Q} \cdot P)$$
$$= P \cdot (Q + \overline{Q} \cdot P)$$
$$\quad + \overline{P} \cdot Q \cdot (Q + \overline{Q} \cdot P) \qquad\qquad 5$$
$$= P \cdot Q + P \cdot \overline{Q} \cdot P + \overline{P} \cdot Q \cdot Q$$
$$\quad + \overline{P} \cdot Q \cdot \overline{Q} \cdot P \qquad\qquad 5$$
$$= P \cdot Q + P \cdot \overline{Q} + \overline{P} \cdot Q$$
$$\quad + \overline{P} \cdot Q \cdot \overline{Q} \cdot P \qquad\qquad 13$$
$$= P \cdot Q + P \cdot \overline{Q} + \overline{P} \cdot Q + 0 \qquad\qquad 14$$
$$= P \cdot Q + P \cdot \overline{Q} + \overline{P} \cdot Q \qquad\qquad 7$$
$$= P \cdot (Q + \overline{Q}) + \overline{P} \cdot Q \qquad\qquad 5$$
$$= P \cdot 1 + \overline{P} \cdot Q \qquad\qquad 10$$
$$= \boldsymbol{P + \overline{P} \cdot Q} \qquad\qquad 12$$

Problem 7. Simplify
$F \cdot G \cdot \overline{H} + F \cdot G \cdot H + \overline{F} \cdot G \cdot H$

With reference to Table 11.7: *Reference*

$$F \cdot G \cdot \overline{H} + F \cdot G \cdot H + \overline{F} \cdot G \cdot H$$
$$= F \cdot G \cdot (\overline{H} + H) + \overline{F} \cdot G \cdot H \qquad\qquad 5$$
$$= F \cdot G \cdot 1 + \overline{F} \cdot G \cdot H \qquad\qquad 10$$
$$= F \cdot G + \overline{F} \cdot G \cdot H \qquad\qquad 12$$
$$= \boldsymbol{G \cdot (F + \overline{F} \cdot H)} \qquad\qquad 5$$

Problem 8. Simplify
$\overline{F} \cdot \overline{G} \cdot H + \overline{F} \cdot G \cdot H + F \cdot \overline{G} \cdot H + F \cdot G \cdot H$

With reference to Table 11.7: *Reference*

$$\overline{F} \cdot \overline{G} \cdot H + \overline{F} \cdot G \cdot H + F \cdot \overline{G} \cdot H + F \cdot G \cdot H$$
$$= \overline{G} \cdot H \cdot (\overline{F} + F) + G \cdot H \cdot (\overline{F} + F) \qquad\qquad 5$$
$$= \overline{G} \cdot H \cdot 1 + G \cdot H \cdot 1 \qquad\qquad 10$$
$$= \overline{G} \cdot H + G \cdot H \qquad\qquad 12$$
$$= H \cdot (\overline{G} + G) \qquad\qquad 5$$
$$= H \cdot 1 = \boldsymbol{H} \qquad\qquad 10 \text{ and } 12$$

Problem 9. Simplify
$A \cdot \overline{C} + \overline{A} \cdot (B + C) + A \cdot B \cdot (C + \overline{B})$

using the rules of Boolean algebra.

With reference to Table 11.7: *Reference*

$$A \cdot \overline{C} + \overline{A} \cdot (B + C)$$
$$\quad + A \cdot B \cdot (C + \overline{B})$$
$$= A \cdot \overline{C} + \overline{A} \cdot B + \overline{A} \cdot C$$
$$\quad + A \cdot B \cdot C + A \cdot B \cdot \overline{B} \qquad\qquad 5$$
$$= A \cdot \overline{C} + \overline{A} \cdot B + \overline{A} \cdot C$$
$$\quad + A \cdot B \cdot C + A \cdot 0 \qquad\qquad 14$$
$$= A \cdot \overline{C} + \overline{A} \cdot B + \overline{A} \cdot C + A \cdot B \cdot C \qquad\qquad 11$$
$$= A \cdot (\overline{C} + B \cdot C) + \overline{A} \cdot B + \overline{A} \cdot C \qquad\qquad 5$$
$$= A \cdot (\overline{C} + B) + \overline{A} \cdot B + \overline{A} \cdot C \qquad\qquad 17$$
$$= A \cdot \overline{C} + A \cdot B + \overline{A} \cdot B + \overline{A} \cdot C \qquad\qquad 5$$
$$= A \cdot \overline{C} + B \cdot (A + \overline{A}) + \overline{A} \cdot C \qquad\qquad 5$$
$$= A \cdot \overline{C} + B \cdot 1 + \overline{A} \cdot C \qquad\qquad 10$$
$$= \boldsymbol{A \cdot \overline{C} + B + \overline{A} \cdot C} \qquad\qquad 12$$

Problem 10. Simplify the expression
$P \cdot \overline{Q} \cdot R + P \cdot Q \cdot (\overline{P} + R) + Q \cdot R \cdot (\overline{Q} + P)$,
using the rules of Boolean algebra.

With reference to Table 11.7: *Reference*

$$P \cdot \overline{Q} \cdot R + P \cdot Q \cdot (\overline{P} + R)$$
$$\quad + Q \cdot R \cdot (\overline{Q} + P)$$
$$= P \cdot \overline{Q} \cdot R + P \cdot Q \cdot \overline{P} + P \cdot Q \cdot R$$
$$\quad + Q \cdot R \cdot \overline{Q} + Q \cdot R \cdot P \qquad\qquad 5$$
$$= P \cdot \overline{Q} \cdot R + 0 \cdot Q + P \cdot Q \cdot R$$
$$\quad + 0 \cdot R + P \cdot Q \cdot R \qquad\qquad 14$$
$$= P \cdot \overline{Q} \cdot R + P \cdot Q \cdot R + P \cdot Q \cdot R \qquad\qquad 7 \text{ and } 11$$
$$= P \cdot \overline{Q} \cdot R + P \cdot Q \cdot R \qquad\qquad 9$$
$$= P \cdot R \cdot (Q + \overline{Q}) \qquad\qquad 5$$
$$= P \cdot R \cdot 1 \qquad\qquad 10$$
$$= \boldsymbol{P \cdot R} \qquad\qquad 12$$

Now try the following Practice Exercise

Practice Exercise 46 Laws and rules of Boolean algebra (Answers on page 861)

Use the laws and rules of Boolean algebra given in Table 11.7 to simplify the following expressions:

1. $\overline{P} \cdot \overline{Q} + \overline{P} \cdot Q$

2. $\overline{P} \cdot Q + P \cdot Q + \overline{P} \cdot \overline{Q}$

3. $\overline{F} \cdot \overline{G} + F \cdot \overline{G} + \overline{G} \cdot (F + \overline{F})$

4. $F \cdot \overline{G} + F \cdot (G + \overline{G}) + F \cdot G$

5. $(P + P \cdot Q) \cdot (Q + Q \cdot P)$

6. $\overline{F} \cdot \overline{G} \cdot H + \overline{F} \cdot G \cdot H + F \cdot \overline{G} \cdot H$

7. $F \cdot \overline{G} \cdot \overline{H} + F \cdot G \cdot H + \overline{F} \cdot G \cdot H$

8. $\overline{P} \cdot \overline{Q} \cdot \overline{R} + \overline{P} \cdot Q \cdot R + P \cdot \overline{Q} \cdot \overline{R}$

9. $\overline{F} \cdot \overline{G} \cdot \overline{H} + \overline{F} \cdot \overline{G} \cdot H + F \cdot \overline{G} \cdot \overline{H} + F \cdot \overline{G} \cdot H$

10. $F \cdot \overline{G} \cdot H + F \cdot G \cdot H + F \cdot G \cdot \overline{H} + \overline{F} \cdot G \cdot \overline{H}$

11. $R \cdot (P \cdot Q + P \cdot \overline{Q}) + \overline{R} \cdot (\overline{P} \cdot \overline{Q} + \overline{P} \cdot Q)$

12. $\overline{R} \cdot (\overline{P} \cdot \overline{Q} + P \cdot Q + P \cdot \overline{Q})$
$+ P \cdot (Q \cdot R + \overline{Q} \cdot R)$

11.4 De Morgan's laws

De Morgan's **laws** may be used to simplify **not**-functions having two or more elements. The laws state that:

*Who was de Morgan? **Augustus de Morgan** (27 June 1806–18 March 1871) was a British mathematician and logician. He formulated **de Morgan's laws** and introduced the term mathematical induction. To find out more go to www.routledge.com/cw/bird

$$\overline{A + B} = \overline{A} \cdot \overline{B} \quad \text{and} \quad \overline{A \cdot B} = \overline{A} + \overline{B}$$

and may be verified by using a truth table (see Problem 11). The application of de Morgan's laws in simplifying Boolean expressions is shown in Problems 12 and 13.

Problem 11. Verify that $\overline{A + B} = \overline{A} \cdot \overline{B}$

A Boolean expression may be verified by using a truth table. In Table 11.8, columns 1 and 2 give all the possible arrangements of the inputs A and B. Column 3 is the **or**-function applied to columns 1 and 2 and column 4 is the **not**-function applied to column 3. Columns 5 and 6 are the **not**-function applied to columns 1 and 2, respectively, and column 7 is the **and**-function applied to columns 5 and 6.

Table 11.8

1	2	3	4	5	6	7
A	B	$A+B$	$\overline{A+B}$	$\overline{A}$	$\overline{B}$	$\overline{A} \cdot \overline{B}$
0	0	0	1	1	1	1
0	1	1	0	1	0	0
1	0	1	0	0	1	0
1	1	1	0	0	0	0

Since columns 4 and 7 have the same pattern of 0's and 1's this verifies that $\overline{A + B} = \overline{A} \cdot \overline{B}$

Problem 12. Simplify the Boolean expression $(\overline{\overline{A} \cdot B}) + (\overline{\overline{A} + B})$ by using de Morgan's laws and the rules of Boolean algebra.

Applying de Morgan's law to the first term gives:

$$\overline{\overline{A} \cdot B} = \overline{\overline{A}} + \overline{B} = A + \overline{B} \quad \text{since} \ \overline{\overline{A}} = A$$

Applying de Morgan's law to the second term gives:

$$\overline{\overline{A} + B} = \overline{\overline{A}} \cdot \overline{B} = A \cdot \overline{B}$$

Thus, $(\overline{\overline{A} \cdot B}) + (\overline{\overline{A} + B}) = (A + \overline{B}) + A \cdot \overline{B}$

Removing the bracket and reordering gives: $A + A \cdot \overline{B} + \overline{B}$

But, by rule 15, Table 11.7, $A + A \cdot B = A$. It follows that: $A + A \cdot \overline{B} = A$

Thus: $(\overline{\overline{A} \cdot B}) + (\overline{\overline{A} + B}) = A + \overline{B}$

Problem 13. Simplify the Boolean expression $\overline{(A \cdot \overline{B} + C)} \cdot \overline{(\overline{A} + B \cdot \overline{C})}$ by using de Morgan's laws and the rules of Boolean algebra.

Applying de Morgan's laws to the first term gives:

$$\overline{A \cdot \overline{B} + C} = \overline{A \cdot \overline{B}} \cdot \overline{C} = (\overline{A} + \overline{\overline{B}}) \cdot \overline{C}$$

$$= (\overline{A} + B) \cdot \overline{C} = \overline{A} \cdot \overline{C} + B \cdot \overline{C}$$

Applying de Morgan's law to the second term gives:

$$\overline{\overline{A} + B \cdot \overline{C}} = \overline{\overline{A}} \cdot \overline{(B + \overline{\overline{C}})} = \overline{\overline{A}} + (\overline{B} + \overline{C})$$

Thus $\overline{(A \cdot \overline{B} + C)} \cdot \overline{(\overline{A} + B \cdot \overline{C})}$

$$= (\overline{A} \cdot \overline{C} + B \cdot \overline{C}) \cdot (\overline{A} + \overline{B} + C)$$

$$= \overline{A} \cdot \overline{A} \cdot \overline{C} + \overline{A} \cdot \overline{B} \cdot \overline{C} + \overline{A} \cdot \overline{C} \cdot C$$

$$+ \overline{A} \cdot B \cdot \overline{C} + B \cdot \overline{B} \cdot \overline{C} + B \cdot \overline{C} \cdot C$$

But from Table 11.7, $\overline{A} \cdot \overline{A} = \overline{A}$ and $\overline{C} \cdot C = B \cdot \overline{B} = 0$
Hence the Boolean expression becomes:

$$\overline{A} \cdot \overline{C} + \overline{A} \cdot \overline{B} \cdot \overline{C} + \overline{A} \cdot B \cdot \overline{C}$$

$$= \overline{A} \cdot \overline{C} (1 + \overline{B} + B)$$

$$= \overline{A} \cdot \overline{C} (1 + B)$$

$$= \overline{A} \cdot \overline{C}$$

Thus: $\overline{(A \cdot \overline{B} + C)} \cdot \overline{(\overline{A} + B \cdot \overline{C})} = \overline{A} \cdot \overline{C}$

Now try the following Practice Exercise

Practice Exercise 47   Simplifying Boolean expressions using de Morgan's laws (Answers on page 861)

Use de Morgan's laws and the rules of Boolean algebra given in Table 11.7 to simplify the following expressions.

1. $(\overline{A} \cdot \overline{B}) \cdot (\overline{A} \cdot B)$

2. $(A + \overline{B \cdot C}) + (\overline{A \cdot B} + C)$

3. $(\overline{\overline{A} \cdot B + B \cdot \overline{C}}) \cdot A \cdot \overline{B}$

4. $(A \cdot \overline{B} + B \cdot \overline{C}) + (\overline{A} \cdot B)$

5. $(\overline{P \cdot \overline{Q} + \overline{P} \cdot R}) \cdot (\overline{P \cdot \overline{Q} \cdot R})$

11.5   Karnaugh maps

(a) Two-variable Karnaugh maps

A truth table for a two-variable expression is shown in Table 11.9(a), the '1' in the third row output showing that $Z = A \cdot \overline{B}$. Each of the four possible Boolean expressions associated with a two-variable function can be depicted as shown in Table 11.9(b), in which one cell is allocated to each row of the truth table. A matrix similar to that shown in Table 11.9(b) can be used to depict $Z = A \cdot \overline{B}$, by putting a 1 in the cell corresponding to $A \cdot \overline{B}$ and 0s in the remaining cells. This method of depicting a Boolean expression is called a two-variable **Karnaugh* map**, and is shown in Table 11.9(c).

To simplify a two-variable Boolean expression, the Boolean expression is depicted on a Karnaugh map, as outlined above. Any cells on the map having either a common vertical side or a common horizontal side are grouped together to form a **couple**. (This is a coupling

***Who is Karnaugh**? **Maurice Karnaugh** (4 October 1924 in New York City) is an American physicist, famous for the Karnaugh map used in Boolean algebra. To find out more go to **www.routledge.com/cw/bird**

Table 11.9

Inputs		Output	Boolean
A	B	Z	expression
0	0	0	$\overline{A} \cdot \overline{B}$
0	1	0	$\overline{A} \cdot B$
1	0	1	$A \cdot \overline{B}$
1	1	0	$A \cdot B$

(a)

A $\searrow$ B	0 ($\overline{A}$)	1 (A)		A $\searrow$ B	0	1
0($\overline{B}$)	$\overline{A}\cdot\overline{B}$	$A\cdot\overline{B}$		0	0	1
1(B)	$\overline{A}\cdot B$	$A\cdot B$		1	0	0

(b)   **(c)**

together of cells, just combining two together). The simplified Boolean expression for a couple is given by those variables common to all cells in the couple. See Problem 14.

(b) Three-variable Karnaugh maps

A truth table for a three-variable expression is shown in Table 11.10(a), the 1s in the output column showing that:

$$Z = \overline{A} \cdot \overline{B} \cdot C + \overline{A} \cdot B \cdot C + A \cdot B \cdot \overline{C}$$

Each of the eight possible Boolean expressions associated with a three-variable function can be depicted as shown in Table 11.10(b), in which one cell is allocated to each row of the truth table. A matrix similar to that shown in Table 11.10(b) can be used to depict: $Z = \overline{A} \cdot \overline{B} \cdot C + \overline{A} \cdot B \cdot C + A \cdot B \cdot \overline{C}$, by putting 1s in the cells corresponding to the Boolean terms on the right of the Boolean equation and 0s in the remaining cells. This method of depicting a three-variable Boolean expression is called a three-variable Karnaugh map, and is shown in Table 11.10(c).

To simplify a three-variable Boolean expression, the Boolean expression is depicted on a Karnaugh map as outlined above. Any cells on the map having common edges either vertically or horizontally are grouped together to form couples of four cells or two cells. During coupling the horizontal lines at the top and bottom of the cells are taken as a common edge, as are the vertical lines on the left and right of the cells. The simplified Boolean expression for a couple is given by those variables common to all cells in the couple. See Problems 15 to 17.

Table 11.10

Inputs			Output	Boolean
A	B	C	Z	expression
0	0	0	0	$\overline{A} \cdot \overline{B} \cdot \overline{C}$
0	0	1	1	$\overline{A} \cdot \overline{B} \cdot C$
0	1	0	0	$\overline{A} \cdot B \cdot \overline{C}$
0	1	1	1	$\overline{A} \cdot B \cdot C$
1	0	0	0	$A \cdot \overline{B} \cdot \overline{C}$
1	0	1	0	$A \cdot \overline{B} \cdot C$
1	1	0	1	$A \cdot B \cdot \overline{C}$
1	1	1	0	$A \cdot B \cdot C$

(a)

C $\searrow$ A·B	00 ($\overline{A}\cdot\overline{B}$)	01 ($\overline{A}\cdot B$)	11 ($A\cdot B$)	10 ($A\cdot\overline{B}$)
0($\overline{C}$)	$\overline{A}\cdot\overline{B}\cdot\overline{C}$	$\overline{A}\cdot B\cdot\overline{C}$	$A\cdot B\cdot\overline{C}$	$A\cdot\overline{B}\cdot\overline{C}$
1(C)	$\overline{A}\cdot\overline{B}\cdot C$	$\overline{A}\cdot B\cdot C$	$A\cdot B\cdot C$	$A\cdot\overline{B}\cdot C$

(b)

C $\searrow$ A.B	00	01	11	10
0	0	0	1	0
1	1	1	0	0

(c)

(c) Four-variable Karnaugh maps

A truth table for a four-variable expression is shown in Table 11.11(a), the 1's in the output column showing that:

$$Z = \overline{A} \cdot \overline{B} \cdot C \cdot \overline{D} + \overline{A} \cdot B \cdot C \cdot \overline{D}$$
$$+ A \cdot \overline{B}. C. \overline{D} + A. B. C. \overline{D}$$

Each of the 16 possible Boolean expressions associated with a four-variable function can be depicted as shown in Table 11.11(b), in which one cell is allocated to each row of the truth table. A matrix similar to that shown in Table 11.11(b) can be used to depict

$$Z = \overline{A} \cdot \overline{B} \cdot C \cdot \overline{D} + \overline{A} \cdot B \cdot C \cdot \overline{D}$$
$$+ A \cdot \overline{B} \cdot C \cdot \overline{D} + A \cdot B \cdot C \cdot \overline{D}$$

by putting 1s in the cells corresponding to the Boolean terms on the right of the Boolean equation and 0s in the remaining cells. This method of depicting a four-variable expression is called a four-variable Karnaugh map, and is shown in Table 11.11(c).

To simplify a four-variable Boolean expression, the Boolean expression is depicted on a Karnaugh map as outlined above. Any cells on the map having common edges either vertically or horizontally are grouped together to form couples of eight cells, four cells or two cells. During coupling, the horizontal lines at the top and bottom of the cells may be considered to be common edges, as are the vertical lines on the left and the right of the cells. The simplified Boolean expression for a couple is given by those variables common to all cells in the couple. See Problems 18 and 19.

Summary of procedure when simplifying a Boolean expression using a Karnaugh map

(a) Draw a four-, eight- or sixteen-cell matrix, depending on whether there are two, three or four variables.

(b) Mark in the Boolean expression by putting 1s in the appropriate cells.

(c) Form couples of 8, 4 or 2 cells having common edges, forming the largest groups of cells possible. (Note that a cell containing a 1 may be used more than once when forming a couple. Also note that each cell containing a 1 must be used at least once.)

(d) The Boolean expression for the couple is given by the variables which are common to all cells in the couple.

> **Problem 14.** Use the Karnaugh map techniques to simplify the expression $\overline{P} \cdot \overline{Q} + \overline{P} \cdot Q$

Using the above procedure:

(a) The two-variable matrix is drawn and is shown in Table 11.12.

(b) The term $\overline{P} \cdot \overline{Q}$ is marked with a 1 in the top left-hand cell, corresponding to $P = 0$ and $Q = 0$; $\overline{P} \cdot Q$ is marked with a 1 in the bottom left-hand cell corresponding to $P = 0$ and $Q = 1$

(c) The two cells containing 1s have a common horizontal edge and thus a vertical couple can be formed.

Table 11.11

Inputs				Output	Boolean
A	B	C	D	Z	expression
0	0	0	0	0	$\overline{A} \cdot \overline{B} \cdot \overline{C} \cdot \overline{D}$
0	0	0	1	0	$\overline{A} \cdot \overline{B} \cdot \overline{C} \cdot D$
0	0	1	0	1	$\overline{A} \cdot \overline{B} \cdot C \cdot \overline{D}$
0	0	1	1	0	$\overline{A} \cdot \overline{B} \cdot C \cdot D$
0	1	0	0	0	$\overline{A} \cdot B \cdot \overline{C} \cdot \overline{D}$
0	1	0	1	0	$\overline{A} \cdot B \cdot \overline{C} \cdot D$
0	1	1	0	1	$\overline{A} \cdot B \cdot C \cdot \overline{D}$
0	1	1	1	0	$\overline{A} \cdot B \cdot C \cdot D$
1	0	0	0	0	$A \cdot \overline{B} \cdot \overline{C} \cdot \overline{D}$
1	0	0	1	0	$A \cdot \overline{B} \cdot \overline{C} \cdot D$
1	0	1	0	1	$A \cdot \overline{B} \cdot C \cdot \overline{D}$
1	0	1	1	0	$A \cdot \overline{B} \cdot C \cdot D$
1	1	0	0	0	$A \cdot B \cdot \overline{C} \cdot \overline{D}$
1	1	0	1	0	$A \cdot B \cdot \overline{C} \cdot D$
1	1	1	0	1	$A \cdot B \cdot C \cdot \overline{D}$
1	1	1	1	0	$A \cdot B \cdot C \cdot D$

(a)

$\begin{array}{c} \\ C \cdot D \end{array}$ $\begin{array}{c} A \cdot B \end{array}$	00 $(\overline{A} \cdot \overline{B})$	01 $(\overline{A} \cdot B)$	11 $(A \cdot B)$	10 $(A \cdot \overline{B})$
00 $(\overline{C} \cdot \overline{D})$	$\overline{A} \cdot \overline{B} \cdot \overline{C} \cdot \overline{D}$	$\overline{A} \cdot B \cdot \overline{C} \cdot \overline{D}$	$A \cdot B \cdot \overline{C} \cdot \overline{D}$	$A \cdot \overline{B} \cdot \overline{C} \cdot \overline{D}$
01 $(\overline{C} \cdot D)$	$\overline{A} \cdot \overline{B} \cdot \overline{C} \cdot D$	$\overline{A} \cdot B \cdot \overline{C} \cdot D$	$A \cdot B \cdot \overline{C} \cdot D$	$A \cdot \overline{B} \cdot \overline{C} \cdot D$
11 $(C \cdot D)$	$\overline{A} \cdot \overline{B} \cdot C \cdot D$	$\overline{A} \cdot B \cdot C \cdot D$	$A \cdot B \cdot C \cdot D$	$A \cdot \overline{B} \cdot C \cdot D$
10 $(C \cdot \overline{D})$	$\overline{A} \cdot \overline{B} \cdot C \cdot \overline{D}$	$\overline{A} \cdot B \cdot C \cdot \overline{D}$	$A \cdot B \cdot C \cdot \overline{D}$	$A \cdot \overline{B} \cdot C \cdot \overline{D}$

(b)

$\begin{array}{c} \\ C \cdot D \end{array}$ $\begin{array}{c} A \cdot B \end{array}$	0.0	0.1	1.1	1.0
0.0	0	0	0	0
0.1	0	0	0	0
1.1	0	0	0	0
1.0	1	1	1	1

(c)

(d) The variable common to both cells in the couple is $P = 0$, i.e. $\overline{P}$ thus

$$\overline{P} \cdot \overline{Q} + \overline{P} \cdot Q = \overline{P}$$

Table 11.12

Problem 15. Simplify the expression
$\overline{X} \cdot Y \cdot \overline{Z} + \overline{X} \cdot \overline{Y} \cdot Z + X \cdot Y \cdot \overline{Z} + X \cdot \overline{Y} \cdot Z$
by using Karnaugh map techniques.

Using the above procedure:

(a)   A three-variable matrix is drawn and is shown in Table 11.13

Table 11.13

Z \ X·Y	0·0	0·1	1·1	1·0
0	0	1	1	0
1	1	0	0	1

(b)   The 1s on the matrix correspond to the expression given, i.e. for $\overline{X} \cdot Y \cdot \overline{Z}$, $X = 0$, $Y = 1$ and $Z = 0$ and hence corresponds to the cell in the two rows and second column, and so on.

(c)   Two couples can be formed as shown. The couple in the bottom row may be formed since the vertical lines on the left and right of the cells are taken as a common edge.

(d)   The variables common to the couple in the top row are $Y = 1$ and $Z = 0$, that is, $Y \cdot \overline{Z}$ and the variables common to the couple in the bottom row are $Y = 0$, $Z = 1$, that is, $\overline{Y} \cdot Z$. Hence:

$$\overline{X} \cdot Y \cdot \overline{Z} + \overline{X} \cdot \overline{Y} \cdot Z + X \cdot Y \cdot \overline{Z}$$
$$+ X \cdot \overline{Y} \cdot Z = Y \cdot \overline{Z} + \overline{Y} \cdot Z$$

Problem 16. Use a Karnaugh map technique to simplify the expression $(\overline{A} \cdot B) \cdot (\overline{A} + B)$.

Using the procedure, a two-variable matrix is drawn and is shown in Table 11.14.

$\overline{A} \cdot B$ corresponds to the bottom left-hand cell and $(\overline{A} \cdot B)$ must therefore be all cells except this one, marked with a 1 in Table 11.14. $(\overline{A} + B)$ corresponds to all the cells except the top right-hand cell marked with a

Table 11.14

B \ A	0	1
0	1	1 2
1		1

2 in Table 11.14. Hence $(\overline{A} + B)$ must correspond to the cell marked with a 2. The expression $(\overline{A} \cdot B) \cdot (\overline{A} + B)$ corresponds to the cell having both 1 and 2 in it, i.e.

$$\overline{(\overline{A} \cdot B)} \cdot \overline{(\overline{A} + B)} = A \cdot \overline{B}$$

Problem 17. Simplify $\overline{(P + \overline{Q} \cdot R)} + \overline{(P \cdot Q + \overline{R})}$ using a Karnaugh map technique.

The term $(P + \overline{Q} \cdot R)$ corresponds to the cells marked 1 on the matrix in Table 11.15(a), hence $\overline{(P + \overline{Q} \cdot R)}$ corresponds to the cells marked 2. Similarly, $(P \cdot Q + \overline{R})$ corresponds to the cells marked 3 in Table 11.15(a), hence $\overline{(P \cdot Q + \overline{R})}$ corresponds to the cells marked 4. The expression $\overline{(P + \overline{Q} \cdot R)} + \overline{(P \cdot Q + \overline{R})}$ corresponds to cells marked with either a 2 or with a 4 and is shown in Table 11.15(b) by Xs. These cells may be coupled as shown. The variables common to the group of four cells is $P = 0$, i.e. $\overline{P}$, and those common to the group of two cells are $Q = 0$, $R = 1$, i.e. $\overline{Q} \cdot R$
Thus: $\overline{(P + \overline{Q} \cdot R)} + \overline{(P \cdot Q + \overline{R})} = \overline{P} + \overline{Q} \cdot R$

Table 11.15

R \ P·Q	0·0	0·1	1·1	1·0
0	3 2	3 2	3 1	3 1
1	4 1	4 2	3 1	4 1

(a)

R \ P·Q	0·0	0·1	1·1	1·0
0	X	X		
1	X	X		X

(b)

Problem 18. Use Karnaugh map techniques to simplify the expression: $A \cdot B \cdot \overline{C} \cdot \overline{D} + A \cdot B \cdot C \cdot D + \overline{A} \cdot B \cdot C \cdot D + A \cdot B \cdot C \cdot \overline{D} + \overline{A} \cdot B \cdot C \cdot \overline{D}$.

Using the procedure, a four-variable matrix is drawn and is shown in Table 11.16. The 1s marked on the matrix correspond to the expression given. Two couples can be formed as shown. The four-cell couple has $B = 1$, $C = 1$, i.e. $B \cdot C$ as the common variables to all four cells and the two-cell couple has $A \cdot B \cdot \overline{D}$ as the common variables to both cells. Hence, the expression simplifies to:

$$B \cdot C + A \cdot B \cdot \overline{D} \quad \text{i.e.} \quad B \cdot (C + A \cdot \overline{D})$$

Table 11.16

C·D \ A·B	0.0	0.1	1.1	1.0
0.0			1	
0.1				
1.1		1	1	
1.0		1	1	

Problem 19. Simplify the expression $\overline{A} \cdot \overline{B} \cdot \overline{C} \cdot D + A \cdot \overline{B} \cdot \overline{C} \cdot D + \overline{A} \cdot \overline{B} \cdot C \cdot \overline{D} + A \cdot \overline{B} \cdot C \cdot \overline{D} + A \cdot B \cdot C \cdot D$ by using Karnaugh map techniques.

The Karnaugh map for the expression is shown in Table 11.17. Since the top and bottom horizontal lines are common edges and the vertical lines on the left and right of the cells are common, then the four corner cells form a couple, $\overline{B} \cdot \overline{D}$ (the cells can be considered as if they are stretched to completely corner a sphere, as far as common edges are concerned). The cell $A \cdot B \cdot C \cdot D$ cannot be coupled with any other. Hence the expression simplifies to

$$\overline{B} \cdot \overline{D} + A \cdot B \cdot C \cdot D$$

Table 11.17

C·D \ A·B	0.0	0.1	1.1	1.0
0.0	1			1
0.1				
1.1			1	
1.0	1			1

Now try the following Practice Exercise

Practice Exercise 48 Simplifying Boolean expressions using Karnaugh maps (Answers on page 861)

In Problems 1 to 12 use Karnaugh map techniques to simplify the expressions given.

1. $\overline{X} \cdot Y + X \cdot Y$

2. $\overline{X} \cdot \overline{Y} + \overline{X} \cdot Y + X \cdot Y$

3. $(\overline{P} \cdot \overline{Q}) \cdot (\overline{\overline{P} \cdot Q})$

4. $A \cdot \overline{C} + \overline{A} \cdot (B + C) + A \cdot B \cdot (C + \overline{B})$

5. $\overline{P} \cdot \overline{Q} \cdot \overline{R} + \overline{P} \cdot Q \cdot \overline{R} + P \cdot Q \cdot \overline{R}$

6. $\overline{P} \cdot \overline{Q} \cdot \overline{R} + P \cdot Q \cdot \overline{R} + P \cdot Q \cdot R + P \cdot \overline{Q} \cdot R$

7. $\overline{A} \cdot \overline{B} \cdot \overline{C} \cdot \overline{D} + \overline{A} \cdot B \cdot \overline{C} \cdot \overline{D} + \overline{A} \cdot B \cdot \overline{C} \cdot D$

8. $\overline{A} \cdot \overline{B} \cdot C \cdot D + \overline{A} \cdot \overline{B} \cdot C \cdot \overline{D} + A \cdot \overline{B} \cdot C \cdot \overline{D}$

9. $\overline{A} \cdot B \cdot \overline{C} \cdot D + A \cdot B \cdot \overline{C} \cdot D + A \cdot B \cdot C \cdot D + A \cdot \overline{B} \cdot \overline{C} \cdot D + A \cdot \overline{B} \cdot C \cdot D$

10. $\overline{A} \cdot \overline{B} \cdot \overline{C} \cdot D + A \cdot B \cdot \overline{C} \cdot \overline{D} + A \cdot \overline{B} \cdot \overline{C} \cdot \overline{D} + A \cdot B \cdot C \cdot \overline{D} + A \cdot \overline{B} \cdot C \cdot \overline{D}$

11. $A \cdot B \cdot \overline{C} \cdot \overline{D} + \overline{A} \cdot \overline{B} \cdot \overline{C} \cdot \overline{D} + \overline{A} \cdot B \cdot C \cdot D + \overline{A} \cdot \overline{B} \cdot C \cdot D + A \cdot \overline{B} \cdot \overline{C} \cdot D + \overline{A} \cdot B \cdot C \cdot D + \overline{A} \cdot B \cdot C \cdot \overline{D}$

11.6 Logic circuits

In practice, logic gates are used to perform the **and**, **or** and **not**-functions introduced in Section 11.1. Logic gates can be made from switches, magnetic devices or fluidic devices but most logic gates in use are electronic devices. Various logic gates are available. For example, the Boolean expression $(A \cdot B \cdot C)$ can be produced using a three-input **and**-gate and $(C + D)$ by using a two-input **or**-gate. The principal gates in common use are introduced below. The term 'gate' is used in the same sense as a normal gate, the open state being indicated by a binary '1' and the closed state by a binary '0'. A gate will only open when the requirements of the gate are met and, for example, there will only be a '1' output on a two-input **and**-gate when both the inputs to the gate are at a '1' state.

The and-gate

The different symbols used for a three-input, **and**-gate are shown in Fig. 11.13(a) and the truth table is shown in Fig. 11.13(b). This shows that there will only be a '1' output when A is 1 and B is 1 and C is 1, written as:

$$Z = A \cdot B \cdot C$$

BRITISH   AMERICAN

(a)

INPUTS			OUTPUT
A	B	C	$Z = A \cdot B \cdot C$
0	0	0	0
0	0	1	0
0	1	0	0
0	1	1	0
1	0	0	0
1	0	1	0
1	1	0	0
1	1	1	1

(b)

Figure 11.13

The or-gate

The different symbols used for a three-input or-gate are shown in Fig. 11.14(a) and the truth table is shown in Fig. 11.14(b). This shows that there will be a '1' output when A is 1, or B is 1, or C is 1, or any combination of A, B or C is 1, written as:

$$Z = A + B + C$$

The invert-gate or not-gate

The different symbols used for an **invert**-gate are shown in Fig. 11.15 (a) and the truth table is shown in Fig. 11.15(b). This shows that a '0' input gives a '1' output and vice-versa, i.e. it is an 'opposite to' function. The invert of A is written $\overline{A}$ and is called 'not-A'.

BRITISH   AMERICAN

(a)

INPUTS			OUTPUT
A	B	C	$Z = A + B + C$
0	0	0	0
0	0	1	1
0	1	0	1
0	1	1	1
1	0	0	1
1	0	1	1
1	1	0	1
1	1	1	1

(b)

Figure 11.14

BRITISH   AMERICAN

(a)

INPUT A	OUTPUT $Z = \overline{A}$
0	1
1	0

(b)

Figure 11.15

The nand-gate

The different symbols used for a **nand**-gate are shown in Fig. 11.16(a) and the truth table is shown in Fig. 11.16(b). This gate is equivalent to an **and**-gate and an **invert**-gate in series (not-and=nand) and the output is written as:

$$Z = \overline{A \cdot B \cdot C}$$

The nor-gate

The different symbols used for a **nor**-gate are shown in Fig. 11.17(a) and the truth table is shown in

BRITISH · AMERICAN

(a)

| INPUTS | | | | OUTPUT |
A	B	C	$A.B.C.$	$Z=\overline{A.B.C.}$
0	0	0	0	1
0	0	1	0	1
0	1	0	0	1
0	1	1	0	1
1	0	0	0	1
1	0	1	0	1
1	1	0	0	1
1	1	1	1	0

(b)

Figure 11.16

Fig. 11.17(b). This gate is equivalent to an **or**-gate and an **invert**-gate in series (not-or=nor), and the output is written as:

$$Z = \overline{A + B + C}$$

BRITISH · AMERICAN

(a)

| INPUTS | | | | OUTPUT |
A	B	C	$A+B+C$	$Z=\overline{A+B+C}$
0	0	0	0	1
0	0	1	1	0
0	1	0	1	0
0	1	1	1	0
1	0	0	1	0
1	0	1	1	0
1	1	0	1	0
1	1	1	1	0

(b)

Figure 11.17

Combinational logic networks

In most logic circuits, more than one gate is needed to give the required output. Except for the **invert**-gate, logic gates generally have two, three or four inputs and are confined to one function only. Thus, for example, a two-input **or**-gate or a four-input **and**-gate can be used when designing a logic circuit. The way in which logic gates are used to generate a given output is shown in Problems 20 to 23.

> **Problem 20.** Devise a logic system to meet the requirements of: $Z = A \cdot \overline{B} + C$

With reference to Fig. 11.18 an **invert**-gate, shown as (1), gives $\overline{B}$. The **and**-gate, shown as (2), has inputs of A and $\overline{B}$, giving $A \cdot \overline{B}$. The **or**-gate, shown as (3), has inputs of $A \cdot \overline{B}$ and C, giving:

$$Z = A \cdot \overline{B} + C$$

Figure 11.18

> **Problem 21.** Devise a logic system to meet the requirements of $(P + \overline{Q}) \cdot (\overline{R} + S)$

The logic system is shown in Fig. 11.19. The given expression shows that two **invert**-functions are needed to give $\overline{Q}$ and $\overline{R}$ and these are shown as gates (1) and (2). Two **or**-gates, shown as (3) and (4), give $(P + \overline{Q})$ and $(\overline{R} + S)$ respectively. Finally, an **and**-gate, shown as (5), gives the required output,

$$Z = (P + \overline{Q}) \cdot (\overline{R} + S)$$

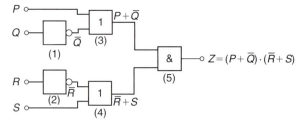

Figure 11.19

Problem 22. Devise a logic circuit to meet the requirements of the output given in Table 11.18, using as few gates as possible.

Table 11.18

Inputs			Output
A	B	C	Z
0	0	0	0
0	0	1	0
0	1	0	0
0	1	1	0
1	0	0	0
1	0	1	1
1	1	0	1
1	1	1	1

The '1' outputs in rows 6, 7 and 8 of Table 11.18 show that the Boolean expression is:

$$Z = A \cdot \overline{B} \cdot C + A \cdot B \cdot \overline{C} + A \cdot B \cdot C$$

The logic circuit for this expression can be built using three, three-input **and**-gates and one, three-input **or**-gate, together with two **invert**-gates. However, the number of gates required can be reduced by using the techniques introduced in Sections 11.3 to 11.5, resulting in the cost of the circuit being reduced. Any of the techniques can be used, and in this case, the rules of Boolean algebra (see Table 11.7) are used.

$$Z = A \cdot \overline{B} \cdot C + A \cdot B \cdot \overline{C} + A \cdot B \cdot C$$
$$= A \cdot [\overline{B} \cdot C + B \cdot \overline{C} + B \cdot C]$$
$$= A \cdot [\overline{B} \cdot C + B(\overline{C} + C)] = A \cdot [\overline{B} \cdot C + B]$$
$$= A \cdot [B + \overline{B} \cdot C] = A \cdot [B + C]$$

The logic circuit to give this simplified expression is shown in Fig. 11.20.

Figure 11.20

Problem 23. Simplify the expression:

$$Z = \overline{P} \cdot \overline{Q} \cdot \overline{R} \cdot \overline{S} + \overline{P} \cdot \overline{Q} \cdot \overline{R} \cdot S + \overline{P} \cdot Q \cdot \overline{R} \cdot \overline{S}$$
$$+ \overline{P} \cdot Q \cdot \overline{R} \cdot S + P \cdot \overline{Q} \cdot \overline{R} \cdot \overline{S}$$

and devise a logic circuit to give this output.

The given expression is simplified using the Karnaugh map techniques introduced in Section 11.5. Two couples are formed as shown in Fig. 11.21(a) and the simplified expression becomes:

$$Z = \overline{Q} \cdot \overline{R} \cdot \overline{S} + \overline{P} \cdot \overline{R}$$

i.e $$\boldsymbol{Z = \overline{R} \cdot (\overline{P} + \overline{Q} \cdot \overline{S})}$$

The logic circuit to produce this expression is shown in Fig. 11.21(b).

(b)

Figure 11.21

Now try the following Practice Exercise

Practice Exercise 49 Logic circuits (Answers on page 862)

In Problems 1 to 4, devise logic systems to meet the requirements of the Boolean expressions given.

1. $Z = \overline{A} + B \cdot C$

2. $Z = A \cdot \overline{B} + B \cdot \overline{C}$

3. $Z = A \cdot B \cdot \overline{C} + \overline{A} \cdot \overline{B} \cdot C$

4. $Z = (\overline{A} + B) \cdot (\overline{C} + D)$

In Problems 5 to 7, simplify the expression given in the truth table and devise a logic circuit to meet the requirements stated.

5. Column 4 of Table 11.19.

6. Column 5 of Table 11.19.

7. Column 6 of Table 11.19.

Table 11.19

1	2	3	4	5	6
A	B	C	Z_1	Z_2	Z_3
0	0	0	0	0	0
0	0	1	1	0	0
0	1	0	0	0	1
0	1	1	1	1	1
1	0	0	0	1	0
1	0	1	1	1	1
1	1	0	1	0	1
1	1	1	1	1	1

In Problems 8 to 12, simplify the Boolean expressions given and devise logic circuits to give the requirements of the simplified expressions.

8. $\overline{P} \cdot \overline{Q} + \overline{P} \cdot Q + P \cdot Q$

9. $\overline{P} \cdot \overline{Q} \cdot \overline{R} + P \cdot Q \cdot \overline{R} + P \cdot \overline{Q} \cdot \overline{R}$

10. $P \cdot \overline{Q} \cdot R + P \cdot \overline{Q} \cdot \overline{R} + \overline{P} \cdot \overline{Q} \cdot \overline{R}$

11. $\overline{A} \cdot \overline{B} \cdot \overline{C} \cdot \overline{D} + A \cdot \overline{B} \cdot \overline{C} \cdot \overline{D} + \overline{A} \cdot \overline{B} \cdot C \cdot \overline{D}$
    $+ \overline{A} \cdot B \cdot C \cdot \overline{D} + A \cdot \overline{B} \cdot C \cdot \overline{D}$

12. $\overline{(\overline{P} \cdot Q \cdot R)} \cdot \overline{(P + \overline{Q} \cdot R)}$

11.7   Universal logic gates

The function of any of the five logic gates in common use can be obtained by using either **nand**-gates or **nor**-gates and when used in this manner, the gate selected is called a **universal gate**. The way in which a universal **nand**-gate is used to produce the **invert**, **and**, **or** and **nor**-functions is shown in Problem 24. The way in which a universal **nor**-gate is used to produce the **invert**, **or**, **and** and **nand**-functions is shown in Problem 25.

> **Problem 24.** Show how **invert**, **and**, **or** and **nor**-functions can be produced using nand-gates only.

A single input to a **nand**-gate gives the **invert**-function, as shown in Fig. 11.22(a). When two **nand**-gates are connected, as shown in Fig. 11.22(b), the output from the first gate is $\overline{A \cdot B \cdot C}$ and this is inverted by the second gate, giving $Z = \overline{\overline{A \cdot B \cdot C}} = A \cdot B \cdot C$, i.e. the **and**-function is produced. When $\overline{A}$, $\overline{B}$ and $\overline{C}$ are the inputs to a **nand**-gate, the output is $\overline{\overline{A} \cdot \overline{B} \cdot \overline{C}}$

(a)

(b)

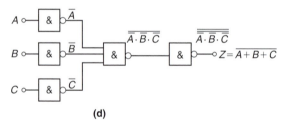

(c)

(d)

Figure 11.22

By de Morgan's law, $\overline{\overline{A} \cdot \overline{B} \cdot \overline{C}} = \overline{\overline{A}} + \overline{\overline{B}} + \overline{\overline{C}} = A + B + C$, i.e. a **nand**-gate is used to produce the or-function. The logic circuit is shown in Fig. 11.22(c). If the output from the logic circuit in Fig. 11.22(c) is inverted by adding an additional **nand**-gate, the output becomes the invert of an **or**-function, i.e. the **nor**-function, as shown in Fig. 11.22(d).

> **Problem 25.** Show how **invert, or, and** and **nand**-functions can be produced by using **nor**-gates only.

A single input to a **nor**-gate gives the **invert**-function, as shown in Fig. 11.23(a). When two **nor**-gates are connected, as shown in Fig. 11.23(b), the output from the first gate is $\overline{A + B + C}$ and this is inverted by the second gate, giving $Z = \overline{\overline{A + B + C}} = A + B + C$, i.e. the **or**-function is produced. Inputs of $\overline{A}$, $\overline{B}$, and $\overline{C}$ to a **nor**-gate give an output of $\overline{A} + \overline{B} + \overline{C}$

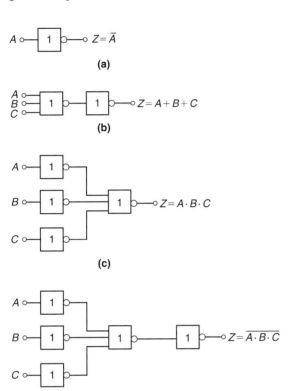

(a)

(b)

(c)

(d)

Figure 11.23

By de Morgan's law, $\overline{A + B + C} = \overline{\overline{A} \cdot \overline{B} \cdot \overline{C}} = A \cdot B \cdot C$, i.e. the **nor**-gate can be used to produce the **and**-function. The logic circuit is shown in Fig. 11.23(c). When the output of the logic circuit, shown in Fig. 11.23(c), is inverted by adding an additional **nor**-gate, the output then becomes the invert of an **or**-function, i.e. the **nor**-function as shown in Fig. 11.23(d).

Problem 26. Design a logic circuit, using **nand**-gates having not more than three inputs, to meet the requirements of the Boolean expression $Z = \overline{A} + \overline{B} + C + \overline{D}$

When designing logic circuits, it is often easier to start at the output of the circuit. The given expression shows there are four variables joined by **or**-functions. From the principles introduced in Problem 24, if a four-input **nand**-gate is used to give the expression given, the inputs are $\overline{\overline{A}}$, $\overline{\overline{B}}$, $\overline{C}$ and $\overline{\overline{D}}$ that is A, B, $\overline{C}$ and D.

However, the problem states that three-inputs are not to be exceeded so two of the variables are joined, i.e. the inputs to the three-input **nand**-gate, shown as gate (1) in Fig. 11.24, is A, B, $\overline{C}$ and D. From Problem 24, the **and**-function is generated by using two **nand**-gates connected in series, as shown by gates (2) and (3) in Fig. 11.24. The logic circuit required to produce the given expression is as shown in Fig. 11.24.

Figure 11.24

Problem 27. Use **nor**-gates only to design a logic circuit to meet the requirements of the expression $Z = \overline{D} \cdot (\overline{A} + B + \overline{C})$

It is usual in logic circuit design to start the design at the output. From Problem 25, the **and**-function between $\overline{D}$ and the terms in the bracket can be produced by using inputs of $\overline{\overline{D}}$ and $\overline{\overline{A} + B + \overline{C}}$ to a **nor**-gate, i.e. by de Morgan's law, inputs of D and $A \cdot \overline{B} \cdot C$. Again, with reference to Problem 25, inputs of $\overline{A} \cdot B$ and $\overline{C}$ to a **nor**-gate give an output of $\overline{A} + B + \overline{C}$, which by de Morgan's law is $A \cdot \overline{B} \cdot C$. The logic circuit to produce the required expression is as shown in Fig. 11.25.

Figure 11.25

Problem 28. An alarm indicator in a grinding mill complex should be activated if (a) the power supply to all mills is off, (b) the hopper feeding the mills is less than 10% full, and (c) if less than two of the three grinding mills are in action. Devise a logic system to meet these requirements.

Let variable A represent the power supply on to all the mills, then $\overline{A}$ represents the power supply off. Let B represent the hopper feeding the mills being more than 10%

full, then $\overline{B}$ represents the hopper being less than 10% full. Let C, D and E represent the three mills respectively being in action, then $\overline{C}$, $\overline{D}$ and $\overline{E}$ represent the three mills respectively not being in action. The required expression to activate the alarm is:

$$Z = \overline{A} \cdot \overline{B} \cdot (\overline{C} + \overline{D} + \overline{E})$$

There are three variables joined by **and**-functions in the output, indicating that a three-input **and**-gate is required, having inputs of $\overline{A}$, $\overline{B}$ and $(\overline{C} + \overline{D} + \overline{E})$. The term $(\overline{C} + \overline{D} + \overline{E})$ is produce by a three-input **nand**-gate. When variables C, D and E are the inputs to a **nand**-gate, the output is $\overline{C \cdot D \cdot E}$ which, by de Morgan's law, is $\overline{C} + \overline{D} + \overline{E}$. Hence the required logic circuit is as shown in Fig. 11.26.

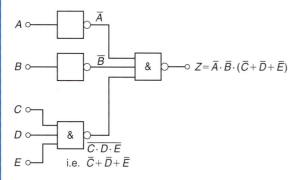

Figure 11.26

Now try the following Practice Exercise

Practice Exercise 50 Universal logic circuits (Answers on page 862)

In Problems 1 to 3, use **nand**-gates only to devise the logic systems stated.

1. $Z = A + B \cdot C$

2. $Z = A \cdot \overline{B} + B \cdot \overline{C}$

3. $Z = A \cdot B \cdot \overline{C} + \overline{A} \cdot \overline{B} \cdot C$

In Problems 4 to 6, use **nor**-gates only to devise the logic systems stated.

4. $Z = (\overline{A} + B) \cdot (\overline{C} + D)$

5. $Z = A \cdot \overline{B} + B \cdot \overline{C} + C \cdot \overline{D}$

6. $Z = \overline{P} \cdot Q + P \cdot (Q + R)$

7. In a chemical process, three of the transducers used are P, Q and R, giving output signals of either 0 or 1. Devise a logic system to give a 1 output when:
   (a) P and Q and R all have 0 outputs, or when:
   (b) P is 0 and (Q is 1 or R is 0)

8. Lift doors should close (Z) if:
   (a) the master switch (A) is on and either
   (b) a call (B) is received from any other floor, or
   (c) the doors (C) have been open for more than 10 seconds, or
   (d) the selector push within the lift (D) is pressed for another floor.
   Devise a logic circuit to meet these requirements.

9. A water tank feeds three separate processes. When any two of the processes are in operation at the same time, a signal is required to start a pump to maintain the head of water in the tank.
   Devise a logic circuit using **nor**-gates only to give the required signal.

10. A logic signal is required to give an indication when:
    (a) the supply to an oven is on, and
    (b) the temperature of the oven exceeds 210 °C, or
    (c) the temperature of the oven is less than 190 °C.
    Devise a logic circuit using **nand**-gates only to meet these requirements.

For fully worked solutions to each of the problems in Practice Exercises 45 to 50 in this chapter, go to the website:
www.routledge.com/cw/bird

Revision Test 3 Equations, numbering systems, Boolean algebra and logic circuits

This Revision Test covers the material contained in Chapters 9 to 11. *The marks for each question are shown in brackets at the end of each question.*

1. Use the method of bisection to evaluate the root of the equation: $x^3 + 5x = 11$ in the range $x = 1$ to $x = 2$, correct to 3 significant figures. (11)

2. Repeat question 1 using an algebraic method of successive approximations. (16)

3. The solution to a differential equation associated with the path taken by a projectile for which the resistance to motion is proportional to the velocity is given by:

$$y = 2.5(e^x - e^{-x}) + x - 25$$

   Use Newton's method to determine the value of x, correct to 2 decimal places, for which the value of y is zero. (11)

4. Convert the following binary numbers to decimal form:

   (a) 1101 (b) 101101.0101 (5)

5. Convert the following decimal numbers to binary form:

   (a) 27 (b) 44.1875 (9)

6. Convert the following decimal numbers to binary, via octal:

   (a) 479 (b) 185.2890625 (9)

7. Convert

   (a) $5F_{16}$ into its decimal equivalent

   (b) 132_{10} into its hexadecimal equivalent

   (c) 110101011_2 into its hexadecimal equivalent (8)

8. Use the laws and rules of Boolean algebra to simplify the following expressions:

   (a) $B \cdot (A + \overline{B}) + A \cdot \overline{B}$

   (b) $\overline{A} \cdot \overline{B} \cdot \overline{C} + \overline{A} \cdot B \cdot \overline{C} + \overline{A} \cdot B \cdot C + \overline{A} \cdot \overline{B} \cdot C$ (9)

9. Simplify the Boolean expression

   $\overline{A \cdot \overline{B} + A \cdot B \cdot \overline{C}}$ using de Morgan's laws. (6)

10. Use a Karnaugh map to simplify the Boolean expression:

    $\overline{A} \cdot \overline{B} \cdot \overline{C} + \overline{A} \cdot B \cdot \overline{C} + \overline{A}. B. C + A \cdot \overline{B} \cdot C$ (8)

11. A clean room has two entrances, each having two doors, as shown in Fig. RT3.1. A warning bell must sound if both doors A and B or doors C and D are open at the same time. Write down the Boolean expression depicting this occurrence, and devise a logic network to operate the bell using nand-gates only. (8)

Figure RT3.1

Geometry and trigonometry

Chapter 12

Introduction to trigonometry

Why it is important to understand: **Introduction to trigonometry**

There are an enormous number of uses of trigonometry and trigonometric functions. Fields that use trigonometry or trigonometric functions include astronomy (especially for locating apparent positions of celestial objects, in which spherical trigonometry is essential) and hence navigation (on the oceans, in aircraft, and in space), music theory, acoustics, optics, analysis of financial markets, electronics, probability theory, statistics, biology, medical imaging (CAT scans and ultrasound), pharmacy, chemistry, number theory (and hence cryptology), seismology, meteorology, oceanography, many physical sciences, land surveying and geodesy (a branch of earth sciences), architecture, phonetics, economics, electrical engineering, mechanical engineering, civil engineering, computer graphics, cartography, crystallography and game development. It is clear that a good knowledge of trigonometry is essential in many fields of engineering.

At the end of this chapter, you should be able to:

- state the theorem of Pythagoras and use it to find the unknown side of a right-angled triangle
- define sine, cosine, tangent, secant, cosecant and cotangent of an angle in a right-angled triangle
- evaluate trigonometric ratios of angles
- solve right-angled triangles
- understand angles of elevation and depression
- use the sine and cosine rules to solve non-right-angled triangles
- calculate the area of any triangle
- solve practical problems involving trigonometry

12.1 Trigonometry

Trigonometry is the branch of mathematics which deals with the measurement of sides and angles of triangles, and their relationship with each other. There are many applications in engineering where knowledge of trigonometry is needed.

12.2 The theorem of Pythagoras

With reference to Fig. 12.1, the side opposite the right angle (i.e. side b) is called the **hypotenuse**. The **theorem of Pythagoras*** states:

'In any right-angled triangle, the square on the hypotenuse is equal to the sum of the squares on the other two sides.'

Hence $b^2 = a^2 + c^2$

Problem 1. In Fig. 12.2, find the length of EF.

Figure 12.2

By Pythagoras' theorem:

$$e^2 = d^2 + f^2$$

Hence

$$13^2 = d^2 + 5^2$$
$$169 = d^2 + 25$$
$$d^2 = 169 - 25 = 144$$

Thus

$$d = \sqrt{144} = 12\,\text{cm}$$

i.e.

$$\boldsymbol{EF = 12\,\text{cm}}$$

Figure 12.1

* **Who was Pythagoras?** **Pythagoras of Samos** (Born c. 570 BC and died about 495 BC) was an Ionian Greek philosopher and mathematician. He is best known for the Pythagorean theorem, which states that in a right-angled triangle $a^2 + b^2 = c^2$. To find out more go to **www.routledge.com/cw/bird**

Problem 2. Two aircraft leave an airfield at the same time. One travels due north at an average speed of 300 km/h and the other due west at an average speed of 220 km/h. Calculate their distance apart after four hours.

After four hours, the first aircraft has travelled $4 \times 300 = 1200$ km, due north, and the second aircraft has travelled $4 \times 220 = 880$ km due west, as shown in Fig. 12.3. Distance apart after four hours $= BC$.

Figure 12.3

From Pythagoras' theorem:

$$BC^2 = 1200^2 + 880^2 = 1\,440\,000 + 774\,400$$

and $BC = \sqrt{(2\,214\,400)}$

Hence distance apart after four hours = 1488 km

Now try the following Practice Exercise

Practice Exercise 51   The theorem of Pythagoras (Answers on page 863)

1. In a triangle CDE, $D = 90°$, $CD = 14.83$ mm and $CE = 28.31$ mm. Determine the length of DE.

2. Triangle PQR is isosceles, Q being a right angle. If the hypotenuse is 38.47 cm find (a) the lengths of sides PQ and QR, and (b) the value of $\angle QPR$.

3. A man cycles 24 km due south and then 20 km due east. Another man, starting at the same time as the first man, cycles 32 km due east and then 7 km due south. Find the distance between the two men.

4. A ladder 3.5 m long is placed against a perpendicular wall with its foot 1.0 m from the wall. How far up the wall (to the nearest centimetre) does the ladder reach? If the foot of the ladder is now moved 30 cm further away from the wall, how far does the top of the ladder fall?

5. Two ships leave a port at the same time. One travels due west at 18.4 km/h and the other due south at 27.6 km/h. Calculate how far apart the two ships are after four hours.

6. Figure 12.4 shows a bolt rounded off at one end. Determine the dimension h.

Figure 12.4

7. Figure 12.5 shows a cross-section of a component that is to be made from a round bar. If the diameter of the bar is 74 mm, calculate the dimension x.

Figure 12.5

12.3 Trigonometric ratios of acute angles

(a) With reference to the right-angled triangle shown in Fig. 12.6:

(i) $\quad$ sine $\theta = \dfrac{\text{opposite side}}{\text{hypotenuse}}$

i.e. $\quad \sin\theta = \dfrac{b}{c}$

(ii) $\quad$ cosine $\theta = \dfrac{\text{adjacent side}}{\text{hypotenuse}}$

i.e. $\quad \cos\theta = \dfrac{a}{c}$

(iii) $\quad$ tangent $\theta = \dfrac{\text{opposite side}}{\text{adjacent side}}$

i.e. $\quad \tan\theta = \dfrac{b}{a}$

(iv) $\quad$ secant $\theta = \dfrac{\text{hypotenuse}}{\text{adjacent side}}$

i.e. $\quad \sec\theta = \dfrac{c}{a}$

(v) $\quad$ cosecant $\theta = \dfrac{\text{hypotenuse}}{\text{opposite side}}$

i.e. $\quad \operatorname{cosec}\theta = \dfrac{c}{b}$

(vi) $\text{cotangent } \theta = \dfrac{\text{adjacent side}}{\text{opposite side}}$

i.e. $\cot \theta = \dfrac{a}{b}$

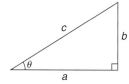

Figure 12.6

(b)   From above,

(i) $\dfrac{\sin \theta}{\cos \theta} = \dfrac{\dfrac{b}{c}}{\dfrac{a}{c}} = \dfrac{b}{a} = \tan \theta,$

i.e. $\tan \theta = \dfrac{\sin \theta}{\cos \theta}$

(ii) $\dfrac{\cos \theta}{\sin \theta} = \dfrac{\dfrac{a}{c}}{\dfrac{b}{c}} = \dfrac{a}{b} = \cot \theta,$

i.e. $\cot \theta = \dfrac{\cos \theta}{\sin \theta}$

(iii) $\sec \theta = \dfrac{1}{\cos \theta}$

(iv) $\csc \theta = \dfrac{1}{\sin \theta}$

(Note, the first letter of each ratio, 's' and 'c' go together)

(v) $\cot \theta = \dfrac{1}{\tan \theta}$

Secants, cosecants and cotangents are called the **reciprocal ratios**.

Problem 3.   If $\cos X = \dfrac{9}{41}$ determine the value of the other five trigonometry ratios.

Fig. 12.7 shows a right-angled triangle XYZ.

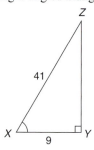

Figure 12.7

Since $\cos X = \dfrac{9}{41}$, then $XY = 9$ units and $XZ = 41$ units.
Using Pythagoras' theorem: $41^2 = 9^2 + YZ^2$ from which $YZ = \sqrt{(41^2 - 9^2)} = 40$ units.
Thus

$$\sin X = \frac{40}{41}, \tan X = \frac{40}{9} = 4\frac{4}{9},$$

$$\csc X = \frac{41}{40} = 1\frac{1}{40},$$

$$\sec X = \frac{41}{9} = 4\frac{5}{9} \text{ and } \cot X = \frac{9}{40}$$

Problem 4.   If $\sin \theta = 0.625$ and $\cos \theta = 0.500$ determine, without using trigonometric tables or calculators, the values of $\csc \theta, \sec \theta, \tan \theta$ and $\cot \theta$.

$$\csc \theta = \frac{1}{\sin \theta} = \frac{1}{0.625} = \mathbf{1.60}$$

$$\sec \theta = \frac{1}{\cos \theta} = \frac{1}{0.500} = \mathbf{2.00}$$

$$\tan \theta = \frac{\sin \theta}{\cos \theta} = \frac{0.625}{0.500} = \mathbf{1.25}$$

$$\cot \theta = \frac{\cos \theta}{\sin \theta} = \frac{0.500}{0.625} = \mathbf{0.80}$$

Problem 5.   Point A lies at co-ordinate $(2, 3)$ and point B at $(8, 7)$. Determine (a) the distance AB, (b) the gradient of the straight line AB, and (c) the angle AB makes with the horizontal.

(a)   Points A and B are shown in Fig. 12.8(a).

In Fig. 12.8(b), the horizontal and vertical lines AC and BC are constructed.

Since ABC is a right-angled triangle, and $AC = (8 - 2) = 6$ and $BC = (7 - 3) = 4$, then by Pythagoras' theorem

$$AB^2 = AC^2 + BC^2 = 6^2 + 4^2$$

$$\text{and } AB = \sqrt{(6^2 + 4^2)} = \sqrt{52} = \mathbf{7.211,}$$

correct to 3 decimal places.

(b)   The gradient of AB is given by $\tan A$,

i.e. **gradient** $= \tan A = \dfrac{BC}{AC} = \dfrac{4}{6} = \dfrac{\mathbf{2}}{\mathbf{3}}$

(c)   **The angle AB makes with the horizontal** is given by $\tan^{-1} \frac{2}{3} = \mathbf{33.69°}$

(a)

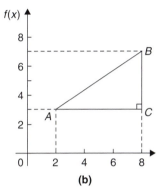

(b)

Figure 12.8

Now try the following Practice Exercise

Practice Exercise 52   Trigonometric ratios of acute angles (Answers on page 863)

1. In triangle ABC shown in Fig. 12.9, find $\sin A$, $\cos A$, $\tan A$, $\sin B$, $\cos B$ and $\tan B$.

    Figure 12.9

2. If $\cos A = \dfrac{15}{17}$ find $\sin A$ and $\tan A$, in fraction form.

3. For the right-angled triangle shown in Fig. 12.10, find:
    (a) $\sin \alpha$ (b) $\cos \theta$ (c) $\tan \theta$

    Figure 12.10

4. Point P lies at co-ordinate $(-3, 1)$ and point Q at $(5, -4)$. Determine
    (a)   the distance PQ
    (b)   the gradient of the straight line PQ and
    (c)   the angle PQ makes with the horizontal.

12.4   Evaluating trigonometric ratios

The easiest method of evaluating trigonometric functions of any angle is by using a **calculator**.
The following values, correct to 4 decimal places, may be checked:

sine $18° = 0.3090$,   cosine $56° = 0.5592$

sine $172° = 0.1392$ cosine $115° = -0.4226$,

sine $241.63° = -0.8799$, cosine $331.78° = 0.8811$

tangent $29° = 0.5543$,

tangent $178° = -0.0349$

tangent $296.42° = -2.0127$

To evaluate, say, sine $42°23'$ using a calculator means finding sine $42\dfrac{23°}{60}$ since there are 60 minutes in 1 degree.

$$\frac{23}{60} = 0.383\dot{3} \text{ thus } 42°23' = 42.38\dot{3}°$$

Thus sine $42°23' = $ sine $42.38\dot{3}° = 0.6741$, correct to 4 decimal places.

Similarly, cosine $72°38' = $ cosine $72\dfrac{38°}{60} = 0.2985$, correct to 4 decimal places.

Most calculators contain only sine, cosine and tangent functions. Thus to evaluate secants, cosecants and cotangents, reciprocals need to be used. The following values, correct to 4 decimal places, may be checked:

$$\text{secant } 32° = \frac{1}{\cos 32°} = 1.1792$$

$$\text{cosecant } 75° = \frac{1}{\sin 75°} = 1.0353$$

$$\text{cotangent } 41° = \frac{1}{\tan 41°} = 1.1504$$

$$\text{secant } 215.12° = \frac{1}{\cos 215.12°} = -1.2226$$

$$\text{cosecant}\,321.62° = \frac{1}{\sin 321.62°} = -1.6106$$

$$\text{cotangent}\,263.59° = \frac{1}{\tan 263.59°} = 0.1123$$

If we know the value of a trigonometric ratio and need to find the angle we use the **inverse function** on our calculators.

For example, using shift and sin on our calculator gives $\sin^{-1}($

If, for example, we know the sine of an angle is 0.5 then the value of the angle is given by:

$$\sin^{-1} 0.5 = \mathbf{30°} \quad (\text{Check that } \sin 30° = 0.5)$$

(Note that $\sin^{-1} x$ does not mean $\dfrac{1}{\sin x}$; also, $\sin^{-1} x$ may also be written as $\arcsin x$)

Similarly, if $\cos\theta = 0.4371$ then

$$\theta = \cos^{-1} 0.4371 = \mathbf{64.08°}$$

and   if $\tan A = 3.5984$ then $A = \tan^{-1} 3.5984$

$$= \mathbf{74.47°}$$

each correct to 2 decimal places.

Use your calculator to check the following worked examples.

Problem 6.   Determine, correct to 4 decimal places, $\sin 43°39'$

$$\sin 43°39' = \sin 43\frac{39°}{60} = \sin 43.65°$$

$$= \mathbf{0.6903}$$

This answer can be obtained using the **calculator** as follows:

1. Press sin   2. Enter 43   3. Press °'''
4. Enter 39   5. Press °'''   6. Press)
7. Press =   Answer = **0.6902512**....

Problem 7.   Determine, correct to 3 decimal places, $6\cos 62°12'$

$$6\cos 62°12' = 6\cos 62\frac{12°}{60} = 6\cos 62.20°$$

$$= \mathbf{2.798}$$

This answer can be obtained using the **calculator** as follows:

1. Enter 6   2. Press cos   3. Enter 62
4. Press °'''   5. Enter 12   6. Press °'''
7. Press)   8. Press =   Answer = **2.798319**....

Problem 8.   Evaluate, correct to 4 decimal places:
(a) sine $168°14'$ (b) cosine $271.41°$
(c) tangent $98°4'$

(a) $\sin 168°14' = \sin 168\dfrac{14°}{60} = \mathbf{0.2039}$

(b) $\text{cosine}\,271.41° = \mathbf{0.0246}$

(c) $\text{tangent}\,98°4' = \tan 98\dfrac{4°}{60} = \mathbf{-7.0558}$

Problem 9.   Evaluate, correct to 4 decimal places:
(a) secant $161°$ (b) secant $302°29'$

(a) $\sec 161° = \dfrac{1}{\cos 161°} = \mathbf{-1.0576}$

(b) $\sec 302°29' = \dfrac{1}{\cos 302°29'} = \dfrac{1}{\cos 302\dfrac{29°}{60}}$

$$= \mathbf{1.8620}$$

Problem 10.   Evaluate, correct to 4 significant figures:
(a) cosecant $279.16°$ (b) cosecant $49°7'$

(a) $\text{cosec}\,279.16° = \dfrac{1}{\sin 279.16°} = \mathbf{-1.013}$

(b) $\text{cosec}\,49°7' = \dfrac{1}{\sin 49°7'} = \dfrac{1}{\sin 49\dfrac{7°}{60}}$

$$= \mathbf{1.323}$$

Problem 11.   Evaluate, correct to 4 decimal places:
(a) cotangent $17.49°$ (b) cotangent $163°52'$

(a) $\cot 17.49° = \dfrac{1}{\tan 17.49°} = \mathbf{3.1735}$

(b) $\cot 163°52' = \dfrac{1}{\tan 163°52'} = \dfrac{1}{\tan 163\dfrac{52°}{60}}$

$$= \mathbf{-3.4570}$$

Problem 12.   Evaluate, correct to 4 significant figures:
(a) $\sin 1.481$ (b) $\cos(3\pi/5)$ (c) $\tan 2.93$

(a) $\sin 1.481$ means the sine of 1.481 radians. Hence a calculator needs to be on the radian function. Hence $\sin 1.481 = \mathbf{0.9960}$

(b) $\cos(3\pi/5) = \cos 1.884955\ldots = \mathbf{-0.3090}$

(c) $\tan 2.93 = \mathbf{-0.2148}$

> **Problem 13.** Evaluate, correct to 4 decimal places:
> (a) secant 5.37 (b) cosecant $\pi/4$
> (c) cotangent $\pi/24$

(a) Again, with no degrees sign, it is assumed that 5.37 means 5.37 radians.

Hence $\sec 5.37 = \dfrac{1}{\cos 5.37} = \mathbf{1.6361}$

(b) $\operatorname{cosec}(\pi/4) = \dfrac{1}{\sin(\pi/4)} = \dfrac{1}{\sin 0.785398\ldots}$
$= \mathbf{1.4142}$

(c) $\cot(5\pi/24) = \dfrac{1}{\tan(5\pi/24)} = \dfrac{1}{\tan 0.654498\ldots}$
$= \mathbf{1.3032}$

> **Problem 14.** Find, in degrees, the acute angle $\sin^{-1} 0.4128$ correct to 2 decimal places.

$\sin^{-1} 0.4128$ means 'the angle whose sine is 0.4128'

Using a calculator:

1. Press shift 2. Press sin 3. Enter 0.4128

4. Press) 5. Press = The answer $24.380848\ldots$ is displayed

Hence, $\mathbf{\sin^{-1} 0.4128 = 24.38^\circ}$

> **Problem 15.** Find the acute angle $\cos^{-1} 0.2437$ in degrees and minutes.

$\cos^{-1} 0.2437$ means 'the angle whose cosine is 0.2437'

Using a calculator:

1. Press shift 2. Press cos 3. Enter 0.2437

4. Press) 5. Press = The answer $75.894979\ldots$ is displayed

6. Press $^\circ$ '" and $75^\circ 53' 41.93''$ is displayed

Hence, $\mathbf{\cos^{-1} 0.2437 = 75.89^\circ = 77^\circ 54'}$ correct to the nearest minute.

> **Problem 16.** Find the acute angle $\tan^{-1} 7.4523$ in degrees and minutes.

$\tan^{-1} 7.4523$ means 'the angle whose tangent is 7.4523'

Using a calculator:

1. Press shift 2. Press tan 3. Enter 7.4523

4. Press) 5. Press = The answer $82.357318\ldots$ is displayed

6. Press $^\circ$ '" and $82^\circ 21' 26.35''$ is displayed

Hence, $\mathbf{\tan^{-1} 7.4523 = 82.36^\circ = 82^\circ 21'}$ correct to the nearest minute.

> **Problem 17.** Determine the acute angles:
> (a) $\sec^{-1} 2.3164$ (b) $\operatorname{cosec}^{-1} 1.1784$
> (c) $\cot^{-1} 2.1273$

(a) $\sec^{-1} 2.3164 = \cos^{-1}\left(\dfrac{1}{2.3164}\right)$
$= \cos^{-1} 0.4317\ldots$
$= \mathbf{64.42^\circ}$ or $\mathbf{64^\circ 25'}$
or $\mathbf{1.124\,radians}$

(b) $\operatorname{cosec}^{-1} 1.1784 = \sin^{-1}\left(\dfrac{1}{1.1784}\right)$
$= \sin^{-1} 0.8486\ldots$
$= \mathbf{58.06^\circ}$ or $\mathbf{58^\circ 4'}$
or $\mathbf{1.013\,radians}$

(c) $\cot^{-1} 2.1273 = \tan^{-1}\left(\dfrac{1}{2.1273}\right)$
$= \tan^{-1} 0.4700\ldots$
$= \mathbf{25.18^\circ}$ or $\mathbf{25^\circ 11'}$
or $\mathbf{0.439\,radians}$

> **Problem 18.** Evaluate the following expression, correct to 4 significant figures:
> $$\frac{4\sec 32^\circ 10' - 2\cot 15^\circ 19'}{3\operatorname{cosec} 63^\circ 8' \tan 14^\circ 57'}$$

Section B

By calculator:

$$\sec 32°10' = 1.1813, \cot 15°19' = 3.6512$$

$$\operatorname{cosec} 63°8' = 1.1210, \tan 14°57' = 0.2670$$

Hence
$$\frac{4\sec 32°10' - 2\cot 15°19'}{3\operatorname{cosec} 63°8' \tan 14°57'}$$

$$= \frac{4(1.1813) - 2(3.6512)}{3(1.1210)(0.2670)}$$

$$= \frac{4.7252 - 7.3024}{0.8979}$$

$$= \frac{-2.5772}{0.8979} = \mathbf{-2.870}$$

correct to 4 significant figures.

Problem 19. Evaluate correct to 4 decimal places:
(a) $\sec(-115°)$ (b) $\operatorname{cosec}(-95°47')$

(a) Positive angles are considered by convention to be anticlockwise and negative angles as clockwise.

Hence $-115°$ is actually the same as $245°$ (i.e. $360° - 115°$)

Hence $\sec(-115°) = \sec 245° = \dfrac{1}{\cos 245°}$

$$= \mathbf{-2.3662}$$

(b) $\operatorname{cosec}(-95°47') = \dfrac{1}{\sin\left(-95\dfrac{47°}{60}\right)} = \mathbf{-1.0051}$

Problem 20. In triangle EFG in Fig. 12.11, calculate angle G.

Figure 12.11

With reference to $\angle G$, the two sides of the triangle given are the opposite side EF and the hypotenuse EG; hence, sine is used,

i.e. $$\sin G = \frac{2.30}{8.71} = 0.26406429\ldots$$

from which, $G = \sin^{-1} 0.26406429\ldots$

i.e. $$G = 15.311360\ldots$$

Hence, $$\angle \mathbf{G} = \mathbf{15.31°} \text{ or } \mathbf{15°19'}$$

Now try the following Practice Exercise

Practice Exercise 53   Evaluating trigonometric ratios (Answers on page 863)

In Problems 1 to 8, evaluate correct to 4 decimal places:

1. (a) sine $27°$ (b) sine $172.41°$
    (c) sine $302°52'$

2. (a) cosine $124°$ (b) cosine $21.46°$
    (c) cosine $284°10'$

3. (a) tangent $145°$ (b) tangent $310.59°$
    (c) tangent $49°16'$

4. (a) secant $73°$ (b) secant $286.45°$
    (c) secant $155°41'$

5. (a) cosecant $213°$ (b) cosecant $15.62°$
    (c) cosecant $311°50'$

6. (a) cotangent $71°$ (b) cotangent $151.62°$
    (c) cotangent $321°23'$

7. (a) sine $\dfrac{2\pi}{3}$ (b) cos 1.681 (c) tan 3.672

8. (a) sec $\dfrac{\pi}{8}$ (b) cosec 2.961 (c) cot 2.612

In Problems 9 to 14, determine the acute angle in degrees (correct to 2 decimal places), degrees and minutes, and in radians (correct to 3 decimal places).

9. $\sin^{-1} 0.2341$

10. $\cos^{-1} 0.8271$

11. $\tan^{-1} 0.8106$

12. $\sec^{-1} 1.6214$

13. $\csc^{-1} 2.4891$

14. $\cot^{-1} 1.9614$

15. In the triangle shown in Fig. 12.12, determine angle θ, correct to 2 decimal places.

Figure 12.12

16. In the triangle shown in Fig. 12.13, determine angle θ in degrees and minutes.

Figure 12.13

In Problems 17 to 20, evaluate correct to 4 significant figures:

17. $4\cos 56°19' - 3\sin 21°57'$

18. $\dfrac{11.5\tan 49°11' - \sin 90°}{3\cos 45°}$

19. $\dfrac{5\sin 86°3'}{3\tan 14°29' - 2\cos 31°9'}$

20. $\dfrac{6.4\csc 29°5' - \sec 81°}{2\cot 12°}$

21. Determine the acute angle, in degrees and minutes, correct to the nearest minute, given by $\sin^{-1}\left(\dfrac{4.32\sin 42°16'}{7.86}\right)$

22. If $\tan x = 1.5276$, determine $\sec x$, $\csc x$, and $\cot x$. (Assume x is an acute angle.)

In Problems 23 to 25 evaluate correct to 4 significant figures:

23. $\dfrac{(\sin 34°27')(\cos 69°2')}{(2\tan 53°39')}$

24. $3\cot 14°15'\sec 23°9'$

25. $\dfrac{\csc 27°19' + \sec 45°29'}{1 - \csc 27°19'\sec 45°29'}$

26. Evaluate correct to 4 decimal places:
    (a) sine $(-125°)$ (b) $\tan(-241°)$
    (c) $\cos(-49°15')$

27. Evaluate correct to 5 significant figures:
    (a) $\csc(-143°)$ (b) $\cot(-252°)$
    (c) $\sec(-67°22')$

12.5 Solution of right-angled triangles

To 'solve a right-angled triangle' means 'to find the unknown sides and angles'. This is achieved by using (i) the theorem of Pythagoras, and/or (ii) trigonometric ratios. This is demonstrated in the following problems.

Problem 21. In triangle PQR shown in Fig. 12.14, find the lengths of PQ and PR.

Figure 12.14

$$\tan 38° = \frac{PQ}{QR} = \frac{PQ}{7.5}$$

hence $PQ = 7.5\tan 38° = 7.5(0.7813)$

$$= \mathbf{5.860\,cm}$$

$$\cos 38° = \frac{QR}{PR} = \frac{7.5}{PR}$$

hence $PR = \dfrac{7.5}{\cos 38°} = \dfrac{7.5}{0.7880} = \mathbf{9.518\,cm}$

[Check: Using Pythagoras' theorem

$$(7.5)^2 + (5.860)^2 = 90.59 = (9.518)^2]$$

Problem 22. Solve the triangle ABC shown in Fig. 12.15.

Figure 12.15

To 'solve triangle ABC' means 'to find the length AC and angles B and C'

$$\sin C = \frac{35}{37} = 0.94595$$

hence $\angle C = \sin^{-1} 0.94595 = 71.08° = \mathbf{71°5'}$
$\angle B = 180° - 90° - 71°5' = \mathbf{18°55'}$ (since angles in a triangle add up to $180°$)

$$\sin B = \frac{AC}{37}$$

hence $\quad AC = 37 \sin 18°55' = 37(0.3242)$

$$= \mathbf{12.0\,mm}$$

or, using Pythagoras' theorem, $37^2 = 35^2 + AC^2$, from which, $AC = \sqrt{(37^2 - 35^2)} = \mathbf{12.0\,mm}$.

Problem 23. Solve triangle XYZ given $\angle X = 90°$, $\angle Y = 23°17'$ and $YZ = 20.0\,$mm. Determine also its area.

It is always advisable to make a reasonably accurate sketch so as to visualise the expected magnitudes of unknown sides and angles. Such a sketch is shown in Fig. 12.16.

$$\angle Z = 180° - 90° - 23°17' = \mathbf{66°43'}$$

$$\sin 23°17' = \frac{XZ}{20.0}$$

Figure 12.16

hence $\quad XZ = 20.0 \sin 23°17'$

$$= 20.0(0.3953) = \mathbf{7.906\,mm}$$

$$\cos 23°17' = \frac{XY}{20.0}$$

hence $\quad XY = 20.0 \cos 23°17'$

$$= 20.0(0.9186) = \mathbf{18.37\,mm}$$

[Check: Using Pythagoras' theorem

$$(18.37)^2 + (7.906)^2 = 400.0 = (20.0)^2]$$

Area of triangle XYZ

$$= \tfrac{1}{2}\,(\text{base})\,(\text{perpendicular height})$$

$$= \tfrac{1}{2}(XY)(XZ) = \tfrac{1}{2}(18.37)(7.906)$$

$$= \mathbf{72.62\,mm^2}$$

Now try the following Practice Exercise

Practice Exercise 54   The solution of right-angled triangles (Answers on page 863)

1. Solve triangle ABC in Fig. 12.17(i).

Figure 12.17

2. Solve triangle DEF in Fig. 12.17(ii).

3. Solve triangle GHI in Fig. 12.17(iii).

4. Solve the triangle JKL in Fig. 12.18(i) and find its area.

5. Solve the triangle MNO in Fig. 12.18(ii) and find its area.

Figure 12.18

6. Solve the triangle PQR in Fig. 12.18(iii) and find its area.

7. A ladder rests against the top of the perpendicular wall of a building and makes an angle of $73°$ with the ground. If the foot of the ladder is 2 m from the wall, calculate the height of the building.

12.6 Angles of elevation and depression

(a) If, in Fig. 12.19, BC represents horizontal ground and AB a vertical flagpole, then the **angle of elevation** of the top of the flagpole, A, from the point C is the angle that the imaginary straight line AC must be raised (or elevated) from the horizontal CB, i.e. angle θ.

Figure 12.19

(b) If, in Fig. 12.20, PQ represents a vertical cliff and R a ship at sea, then the **angle of depression** of the ship from point P is the angle through which the imaginary straight line PR must be lowered (or depressed) from the horizontal to the ship, i.e. angle ϕ.

Figure 12.20

(Note, $\angle PRQ$ is also ϕ – alternate angles between parallel lines.)

Problem 24.   An electricity pylon stands on horizontal ground. At a point 80 m from the base of the pylon, the angle of elevation of the top of the pylon is $23°$. Calculate the height of the pylon to the nearest metre.

Figure 12.21 shows the pylon AB and the angle of elevation of A from point C is $23°$

$$\tan 23° = \frac{AB}{BC} = \frac{AB}{80}$$

Hence height of pylon AB

$$= 80 \tan 23° = 80(0.4245) = 33.96 \, \text{m}$$

$$= \textbf{34 m to the nearest metre}.$$

Figure 12.21

Problem 25.   A surveyor measures the angle of elevation of the top of a perpendicular building as $19°$. He moves 120 m nearer the building and finds the angle of elevation is now $47°$. Determine the height of the building.

The building PQ and the angles of elevation are shown in Fig. 12.22.

In triangle PQS,

$$\tan 19° = \frac{h}{x + 120}$$

hence $h = \tan 19°(x + 120)$,

i.e. $h = 0.3443(x + 120)$ (1)

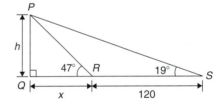

Figure 12.22

In triangle PQR, $\tan 47° = \dfrac{h}{x}$

hence $h = \tan 47°(x)$, i.e. $h = 1.0724x$ (2)

Equating equations (1) and (2) gives:

$$0.3443(x + 120) = 1.0724x$$

$$0.3443x + (0.3443)(120) = 1.0724x$$

$$(0.3443)(120) = (1.0724 - 0.3443)x$$

$$41.316 = 0.7281x$$

$$x = \frac{41.316}{0.7281} = 56.74 \, \text{m}$$

From equation (2), **height of building,**

$h = 1.0724x = 1.0724(56.74) = \textbf{60.85 m}$

Problem 26. The angle of depression of a ship viewed at a particular instant from the top of a 75 m vertical cliff is 30°. Find the distance of the ship from the base of the cliff at this instant. The ship is sailing away from the cliff at constant speed and one minute later its angle of depression from the top of the cliff is 20°. Determine the speed of the ship in km/h.

Fig. 12.23 shows the cliff AB, the initial position of the ship at C and the final position at D. Since the angle of depression is initially 30° then $\angle ACB = 30°$ (alternate angles between parallel lines).

$$\tan 30° = \frac{AB}{BC} = \frac{75}{BC}$$

hence $\quad BC = \frac{75}{\tan 30°} = \frac{75}{0.5774} = \textbf{129.9 m}$

$$= \textbf{initial position of ship from}$$
$$\textbf{base of cliff}$$

Figure 12.23

In triangle ABD,

$$\tan 20° = \frac{AB}{BD} = \frac{75}{BC + CD}$$

$$= \frac{75}{129.9 + x}$$

Hence $\quad 129.9 + x = \frac{75}{\tan 20°} = \frac{75}{0.3640}$

$$= 206.0\,\text{m}$$

from which $\quad x = 206.0 - 129.9 = 76.1\,\text{m}$

Thus the ship sails 76.1 m in one minute, i.e. 60 s, hence speed of ship

$$= \frac{\text{distance}}{\text{time}} = \frac{76.1}{60}\ \text{m/s}$$

$$= \frac{76.1 \times 60 \times 60}{60 \times 1000}\ \text{km/h} = \textbf{4.57 km/h}$$

Now try the following Practice Exercise

Practice Exercise 55   Angles of elevation and depression (Answers on page 864)

1. If the angle of elevation of the top of a vertical 30 m high aerial is 32°, how far is it to the aerial?

2. From the top of a vertical cliff 80.0 m high the angles of depression of two buoys lying due west of the cliff are 23° and 15°, respectively. How far are the buoys apart?

3. From a point on horizontal ground a surveyor measures the angle of elevation of the top of a flagpole as 18°40′. He moves 50 m nearer to the flagpole and measures the angle of elevation as 26°22′. Determine the height of the flagpole.

4. A flagpole stands on the edge of the top of a building. At a point 200 m from the building the angles of elevation of the top and bottom of the pole are 32° and 30° respectively. Calculate the height of the flagpole.

5. From a ship at sea, the angles of elevation of the top and bottom of a vertical lighthouse standing on the edge of a vertical cliff are 31° and 26°, respectively. If the lighthouse is 25.0 m high, calculate the height of the cliff.

6. From a window 4.2 m above horizontal ground the angle of depression of the foot of a building across the road is 24° and the angle of elevation of the top of the building is 34°. Determine, correct to the nearest centimetre, the width of the road and the height of the building.

7. The elevation of a tower from two points, one due east of the tower and the other due west of it are 20° and 24°, respectively, and the two points of observation are 300 m apart. Find the height of the tower to the nearest metre.

12.7 Sine and cosine rules

To 'solve a triangle' means 'to find the values of unknown sides and angles'. If a triangle is **right angled**, trigonometric ratios and the theorem of Pythagoras may be used for its solution, as shown in Section 12.5.

However, for a **non-right-angled triangle**, trigonometric ratios and Pythagoras' theorem **cannot** be used. Instead, two rules, called the **sine rule** and the **cosine rule**, are used.

Sine rule

With reference to triangle ABC of Fig. 12.24, the **sine rule** states:

$$\frac{a}{\sin A} = \frac{b}{\sin B} = \frac{c}{\sin C}$$

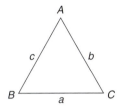

Figure 12.24

The rule may be used only when:

(i)   one side and any two angles are initially given, or

(ii) two sides and an angle (not the included angle) are initially given.

Cosine rule

With reference to triangle ABC of Fig. 12.24, the **cosine rule** states:

$$a^2 = b^2 + c^2 - 2bc\cos A$$
$$\text{or}\quad b^2 = a^2 + c^2 - 2ac\cos B$$
$$\text{or}\quad c^2 = a^2 + b^2 - 2ab\cos C$$

The rule may be used only when:

(i)   two sides and the included angle are initially given, or

(ii) three sides are initially given.

12.8   Area of any triangle

The **area of any triangle** such as ABC of Fig. 12.24 is given by:

(i) $\frac{1}{2}$ × **base × perpendicular height, or**

(ii) $\frac{1}{2}ab\sin C$ or $\frac{1}{2}ac\sin B$ or $\frac{1}{2}bc\sin A$, **or**

(iii) $\sqrt{[s(s-a)(s-b)(s-c)]}$, **where**

$$s = \frac{a+b+c}{2}$$

12.9   Worked problems on the solution of triangles and finding their areas

Problem 27.   In a triangle XYZ, $\angle X = 51°$, $\angle Y = 67°$ and $YZ = 15.2$ cm. Solve the triangle and find its area.

The triangle XYZ is shown in Fig. 12.25. Since the angles in a triangle add up to $180°$, then $Z = 180° - 51° - 67° = \mathbf{62°}$. Applying the sine rule:

$$\frac{15.2}{\sin 51°} = \frac{y}{\sin 67°} = \frac{z}{\sin 62°}$$

Using $\dfrac{15.2}{\sin 51°} = \dfrac{y}{\sin 67°}$ and transposing gives:

$$y = \frac{15.2\ \sin 67°}{\sin 51°} = \mathbf{18.00\,cm} = XZ$$

Using $\dfrac{15.2}{\sin 51°} = \dfrac{z}{\sin 62°}$ and transposing gives:

$$z = \frac{15.2\sin 62°}{\sin 51°} = \mathbf{17.27\,cm} = XY$$

Figure 12.25

Area of triangle $XYZ = \frac{1}{2}xy\sin Z$
$= \frac{1}{2}(15.2)(18.00)\sin 62° = \mathbf{120.8\,cm^2}$ (or area
$= \frac{1}{2}xz\sin Y = \frac{1}{2}(15.2)(17.27)\sin 67° = \mathbf{120.8\,cm^2}$).

   It is always worth checking with triangle problems that the longest side is opposite the largest angle, and vice-versa. In this problem, Y is the largest angle and XZ is the longest of the three sides.

Problem 28. Solve the triangle PQR and find its area given that $QR = 36.5$ mm, $PR = 29.6$ mm and $\angle Q = 36°$.

Triangle PQR is shown in Fig. 12.26.

Figure 12.26

Applying the sine rule:

$$\frac{29.6}{\sin 36°} = \frac{36.5}{\sin P}$$

from which,

$$\sin P = \frac{36.5 \sin 36°}{29.6} = 0.7248$$

Hence $P = \sin^{-1} 0.7248 = 46°27'$ or $133°33'$

When $P = 46°27'$ and $Q = 36°$ then
$R = 180° - 46°27' - 36° = 97°33'$
When $P = 133°33'$ and $Q = 36°$ then
$R = 180° - 133°33' - 36° = 10°27'$

Thus, in this problem there are **two** separate sets of results and both are feasible solutions. Such a situation is called the **ambiguous case**.

Case 1. $P = 46°27'$, $Q = 36°$, $R = 97°33'$, $p = 36.5$ mm and $q = 29.6$ mm.
From the sine rule:

$$\frac{r}{\sin 97°33'} = \frac{29.6}{\sin 36°}$$

from which,

$$r = \frac{29.6 \sin 97°33'}{\sin 36°} = \textbf{49.92 mm}$$

Area $= \frac{1}{2} pq \sin R = \frac{1}{2}(36.5)(29.6) \sin 97°33'$

$$= \textbf{535.5 mm}^2$$

Case 2. $P = 133°33'$, $Q = 36°$, $R = 10°27'$, $p = 36.5$ mm and $q = 29.6$ mm.
From the sine rule:

$$\frac{r}{\sin 10°27'} = \frac{29.6}{\sin 36°}$$

from which,

$$r = \frac{29.6 \sin 10°27'}{\sin 36°} = \textbf{9.134 mm}$$

Area $= \frac{1}{2} pq \sin R = \frac{1}{2}(36.5)(29.6) \sin 10°27'$

$$= \textbf{97.98 mm}^2$$

Triangle PQR for case 2 is shown in Fig. 12.27.

Figure 12.27

Now try the following Practice Exercise

Practice Exercise 56   Solving triangles and finding their areas (Answers on page 864)

In Problems 1 and 2, use the sine rule to solve the triangles ABC and find their areas.

1. $A = 29°$, $B = 68°$, $b = 27$ mm.

2. $B = 71°26'$, $C = 56°32'$, $b = 8.60$ cm.

In Problems 3 and 4, use the sine rule to solve the triangles DEF and find their areas.

3. $d = 17$ cm, $f = 22$ cm, $F = 26°$

4. $d = 32.6$ mm, $e = 25.4$ mm, $D = 104°22'$

In Problems 5 and 6, use the sine rule to solve the triangles JKL and find their areas.

5. $j = 3.85$ cm, $k = 3.23$ cm, $K = 36°$

6. $k = 46$ mm, $l = 36$ mm, $L = 35°$

12.10   Further worked problems on solving triangles and finding their areas

Problem 29. Solve triangle DEF and find its area given that $EF = 35.0$ mm, $DE = 25.0$ mm and $\angle E = 64°$

Triangle *DEF* is shown in Fig. 12.28.

Figure 12.28

Applying the cosine rule:

$$e^2 = d^2 + f^2 - 2d\,f\cos E$$

i.e. $e^2 = (35.0)^2 + (25.0)^2$

$$- [2(35.0)(25.0)\cos 64^\circ]$$

$$= 1225 + 625 - 767.1 = 1083$$

from which, $e = \sqrt{1083} = \textbf{32.91 mm}$

Applying the sine rule:

$$\frac{32.91}{\sin 64^\circ} = \frac{25.0}{\sin F}$$

from which, $\sin F = \dfrac{25.0\sin 64^\circ}{32.91} = 0.6828$

Thus $\angle F = \sin^{-1} 0.6828$

$$= 43^\circ 4' \text{ or } 136^\circ 56'$$

$F = 136^\circ 56'$ is not possible in this case since $136^\circ 56' + 64^\circ$ is greater than 180°. Thus only $\textbf{F} = \textbf{43}^\circ\textbf{4}'$ is valid

$$\angle \textbf{D} = 180^\circ - 64^\circ - 43^\circ 4' = \textbf{72}^\circ\textbf{56}'$$

Area of triangle $DEF = \frac{1}{2}\,d\,f\sin E$
$$= \tfrac{1}{2}(35.0)(25.0)\sin 64^\circ = \textbf{393.2 mm}^2$$

> **Problem 30.**   A triangle *ABC* has sides $a = 9.0$ cm, $b = 7.5$ cm and $c = 6.5$ cm. Determine its three angles and its area.

Triangle *ABC* is shown in Fig. 12.29. It is usual first to calculate the largest angle to determine whether the triangle is acute or obtuse. In this case the largest angle is *A* (i.e. opposite the longest side).

Applying the cosine rule:

$$a^2 = b^2 + c^2 - 2bc\cos A$$

from which, $2bc\cos A = b^2 + c^2 - a^2$

and $\cos A = \dfrac{b^2 + c^2 - a^2}{2bc} = \dfrac{7.5^2 + 6.5^2 - 9.0^2}{2(7.5)(6.5)}$

$$= 0.1795$$

Figure 12.29

Hence $A = \cos^{-1} 0.1795 = \textbf{79}^\circ\textbf{40}'$ (or $280^\circ 20'$, which is obviously impossible). The triangle is thus acute angled since $\cos A$ is positive. (If $\cos A$ had been negative, angle *A* would be obtuse, i.e. lie between 90° and 180°.) Applying the sine rule:

$$\frac{9.0}{\sin 79^\circ 40'} = \frac{7.5}{\sin B}$$

from which,

$$\sin B = \frac{7.5\sin 79^\circ 40'}{9.0} = 0.8198$$

Hence $\textbf{B} = \sin^{-1} 0.8198 = \textbf{55}^\circ\textbf{4}'$

and $\textbf{C} = 180^\circ - 79^\circ 40' - 55^\circ 4' = \textbf{45}^\circ\textbf{16}'$

$$\text{Area} = \sqrt{[s(s-a)(s-b)(s-c)]}$$

where $s = \dfrac{a+b+c}{2} = \dfrac{9.0+7.5+6.5}{2}$

$$= 11.5\,\text{cm}$$

Hence **area**

$$= \sqrt{[11.5(11.5-9.0)(11.5-7.5)(11.5-6.5)]}$$

$$= \sqrt{[11.5(2.5)(4.0)(5.0)]} = \textbf{23.98 cm}^2$$

Alternatively, area $= \frac{1}{2}ab\sin C$

$$= \tfrac{1}{2}(9.0)(7.5)\sin 45^\circ 16' = \textbf{23.98 cm}^2$$

Now try the following Practice Exercise

Practice Exercise 57   Solving triangles and finding their areas (Answers on page 864)

In Problems 1 and 2, use the cosine and sine rules to solve the triangles *PQR* and find their areas.

1. $q = 12$ cm, $r = 16$ cm, $P = 54^\circ$

2. $q = 3.25$ m, $r = 4.42$ m, $P = 105^\circ$

In problems 3 and 4, use the cosine and sine rules to solve the triangles *XYZ* and find their areas.

Section B

3. $x = 10.0\,\text{cm}, y = 8.0\,\text{cm}, z = 7.0\,\text{cm}.$

4. $x = 21\,\text{mm}, y = 34\,\text{mm}, z = 42\,\text{mm}.$

12.11 Practical situations involving trigonometry

There are a number of **practical situations** where the use of trigonometry is needed to find unknown sides and angles of triangles. This is demonstrated in the following problems.

Problem 31. A room 8.0 m wide has a span roof which slopes at $33°$ on one side and $40°$ on the other. Find the length of the roof slopes, correct to the nearest centimetre.

A section of the roof is shown in Fig. 12.30.

Figure 12.30

Angle at ridge, $B = 180° - 33° - 40° = 107°$
From the sine rule:

$$\frac{8.0}{\sin 107°} = \frac{a}{\sin 33°}$$

from which,

$$a = \frac{8.0 \sin 33°}{\sin 107°} = 4.556\,\text{m}$$

Also from the sine rule:

$$\frac{8.0}{\sin 107°} = \frac{c}{\sin 40°}$$

from which,

$$c = \frac{8.0 \sin 40°}{\sin 107°} = 5.377\,\text{m}$$

Hence the roof slopes are 4.56 m and 5.38 m, correct to the nearest centimetre.

Problem 32. Two voltage phasors are shown in Fig. 12.31. If $V_1 = 40\,\text{V}$ and $V_2 = 100\,\text{V}$ determine the value of their resultant (i.e. length OA) and the angle the resultant makes with V_1

Figure 12.31

Angle $OBA = 180° - 45° = 135°$

Applying the cosine rule:

$$OA^2 = V_1^2 + V_2^2 - 2V_1V_2 \cos OBA$$

$$= 40^2 + 100^2 - \{2(40)(100)\cos 135°\}$$

$$= 1600 + 10\,000 - \{-5657\}$$

$$= 1600 + 10\,000 + 5657 = 17\,257$$

The resultant

$$OA = \sqrt{(17257)} = 131.4\,\text{V}$$

Applying the sine rule:

$$\frac{131.4}{\sin 135°} = \frac{100}{\sin AOB}$$

from which, $\sin AOB = \dfrac{100 \sin 135°}{131.4}$

$$= 0.5381$$

Hence   angle $AOB = \sin^{-1} 0.5381 = 32°33'$ (or $147°27'$, which is impossible in this case).

Hence the resultant voltage is 131.4 volts at $32°33'$ to V_1

Problem 33. In Fig. 12.32, PR represents the inclined jib of a crane and is 10.0 long. PQ is 4.0 m long. Determine the inclination of the jib to the vertical and the length of tie QR.

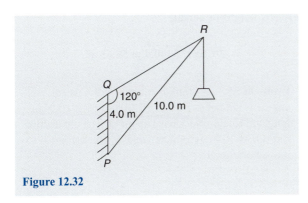

Figure 12.32

Applying the sine rule:

$$\frac{PR}{\sin 120°} = \frac{PQ}{\sin R}$$

from which,

$$\sin R = \frac{PQ \sin 120°}{PR} = \frac{(4.0)\sin 120°}{10.0}$$
$$= 0.3464$$

Hence $\angle R = \sin^{-1} 0.3464 = 20°16'$ (or $159°44'$, which is impossible in this case).
$\angle P = 180° - 120° - 20°16' = \mathbf{39°44'}$, **which is the inclination of the jib to the vertical**.

Applying the sine rule:

$$\frac{10.0}{\sin 120°} = \frac{QR}{\sin 39°44'}$$

from which, **length of tie**,

$$QR = \frac{10.0 \sin 39°44'}{\sin 120°} = \mathbf{7.38\,m}$$

Now try the following Practice Exercise

Practice Exercise 58   Practical situations involving trigonometry (Answers on page 864)

1. A ship P sails at a steady speed of 45 km/h in a direction of W 32° N (i.e. a bearing of 302°) from a port. At the same time another ship Q leaves the port at a steady speed of 35 km/h in a direction N 15° E (i.e. a bearing of 015°). Determine their distance apart after four hours.

2. Two sides of a triangular plot of land are 52.0 m and 34.0 m, respectively. If the area

of the plot is 620 m² find (a) the length of fencing required to enclose the plot and (b) the angles of the triangular plot.

3. A jib crane is shown in Fig. 12.33. If the tie rod PR is 8.0 long and PQ is 4.5 m long, determine (a) the length of jib RQ and (b) the angle between the jib and the tie rod.

Figure 12.33

4. A building site is in the form of a quadrilateral, as shown in Fig. 12.34, and its area is 1510 m². Determine the length of the perimeter of the site.

Figure 12.34

5. Determine the length of members BF and EB in the roof truss shown in Fig. 12.35.

Figure 12.35

6. A laboratory 9.0 m wide has a span roof which slopes at 36° on one side and 44° on the other. Determine the lengths of the roof slopes.

12.12 Further practical situations involving trigonometry

Problem 34. A vertical aerial stands on horizontal ground. A surveyor positioned due east of the aerial measures the elevation of the top as 48°. He moves due south 30.0 m and measures the elevation as 44°. Determine the height of the aerial.

In Fig. 12.36, DC represents the aerial, A is the initial position of the surveyor and B his final position.

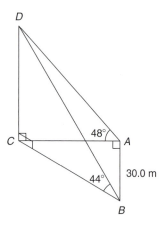

Figure 12.36

From triangle ACD, $\tan 48° = \dfrac{DC}{AC}$

from which $\qquad AC = \dfrac{DC}{\tan 48°}$

Similarly, from triangle BCD,

$$BC = \frac{DC}{\tan 44°}$$

For triangle ABC, using Pythagoras' theorem:

$$BC^2 = AB^2 + AC^2$$

$$\left(\frac{DC}{\tan 44°}\right)^2 = (30.0)^2 + \left(\frac{DC}{\tan 48°}\right)^2$$

$$DC^2\left(\frac{1}{\tan^2 44°} - \frac{1}{\tan^2 48°}\right) = 30.0^2$$

$$DC^2(1.072323 - 0.810727) = 30.0^2$$

$$DC^2 = \frac{30.0^2}{0.261596} = 3440.4$$

Hence, height of aerial,

$$DC = \sqrt{3440.4} = 58.65\,\text{m}$$

Problem 35. A crank mechanism of a petrol engine is shown in Fig. 12.37. Arm OA is 10.0 cm long and rotates clockwise about O. The connecting rod AB is 30.0 cm long and end B is constrained to move horizontally.

Figure 12.37

(a) For the position shown in Fig. 12.37 determine the angle between the connecting rod AB and the horizontal and the length of OB.

(b) How far does B move when angle AOB changes from 50° to 120°?

(a) Applying the sine rule:

$$\frac{AB}{\sin 50°} = \frac{AO}{\sin B}$$

from which,

$$\sin B = \frac{AO \sin 50°}{AB} = \frac{10.0 \sin 50°}{30.0}$$

$$= 0.2553$$

Hence $B = \sin^{-1} 0.2553 = 14°47'$ (or $165°13'$, which is impossible in this case).

Hence the connecting rod AB makes an angle of 14°47′ with the horizontal.

Angle $OAB = 180° - 50° - 14°47' = 115°13'$.

Applying the sine rule:

$$\frac{30.0}{\sin 50°} = \frac{OB}{\sin 115°13'}$$

from which,

$$OB = \frac{30.0 \sin 115°13'}{\sin 50°} = 35.43\,\text{cm}$$

(b) Fig. 12.38 shows the initial and final positions of the crank mechanism. In triangle $OA'B'$, applying the sine rule:

$$\frac{30.0}{\sin 120°} = \frac{10.0}{\sin A'B'O}$$

from which,

$$\sin A'B'O = \frac{10.0 \sin 120°}{30.0} = 0.2887$$

Figure 12.38

Hence $A'B'O = \sin^{-1} 0.2887 = 16°47'$ (or $163°13'$ which is impossible in this case).

Angle $OA'B' = 180° - 120° - 16°47' = 43°13'$

Applying the sine rule:

$$\frac{30.0}{\sin 120°} = \frac{OB'}{\sin 43°13'}$$

from which,

$$OB' = \frac{30.0 \sin 43°13'}{\sin 120°} = 23.72 \, \text{cm}$$

Since $OB = 35.43 \, \text{cm}$ and $OB' = 23.72 \, \text{cm}$ then $BB' = 35.43 - 23.72 = 11.71 \, \text{cm}$.

Hence B moves 11.71 cm when angle AOB changes from 50° to 120°

Problem 36. The area of a field is in the form of a quadrilateral $ABCD$ as shown in Fig. 12.39. Determine its area.

Figure 12.39

A diagonal drawn from B to D divides the quadrilateral into two triangles.

Area of quadrilateral $ABCD$

$$= \text{area of triangle } ABD + \text{area of triangle } BCD$$

$$= \tfrac{1}{2}(39.8)(21.4)\sin 114° + \tfrac{1}{2}(42.5)(62.3)\sin 56°$$

$$= 389.04 + 1097.5 = \mathbf{1487 \, m^2}$$

Now try the following Practice Exercise

Practice Exercise 59   Practical situations involving trigonometry (Answers on page 864)

1. PQ and QR are the phasors representing the alternating currents in two branches of a circuit. Phasor PQ is 20.0 A and is horizontal. Phasor QR (which is joined to the end of PQ to form triangle PQR) is 14.0 A and is at an angle of 35° to the horizontal. Determine the resultant phasor PR and the angle it makes with phasor PQ.

2. Three forces acting on a fixed point are represented by the sides of a triangle of dimensions 7.2 cm, 9.6 cm and 11.0 cm. Determine the angles between the lines of action and the three forces.

3. Calculate, correct to 3 significant figures, the co-ordinates x and y to locate the hole centre at P shown in Fig. 12.40.

Figure 12.40

4. An idler gear, 30 mm in diameter, has to be fitted between a 70 mm diameter driving gear and a 90 mm diameter driven gear as shown in Fig. 12.41. Determine the value of angle θ between the centre lines.

Section B

Figure 12.41

5. A reciprocating engine mechanism is shown in Fig. 12.42. The crank AB is 12.0 cm long and the connecting rod BC is 32.0 cm long. For the position shown determine the length of AC and the angle between the crank and the connecting rod.

Figure 12.42

6. From Fig. 12.42, determine how far C moves, correct to the nearest millimetre, when angle CAB changes from $40°$ to $160°$, B moving in an anticlockwise direction.

7. A surveyor, standing W $25°$ S of a tower, measures the angle of elevation of the top of the tower as $46°30'$. From a position E $23°$ S from the tower the elevation of the top is $37°15'$. Determine the height of the tower if the distance between the two observations is 75 m.

8. An aeroplane is sighted due east from a radar station at an elevation of $40°$ and a height of 8000 m and later at an elevation of $35°$ and height 5500 m in a direction E $70°$ S. If it is descending uniformly, find the angle of descent. Determine also the speed of the aeroplane in km/h if the time between the two observations is 45 s.

9. Sixteen holes are equally spaced on a pitch circle of 70 mm diameter. Determine the length of the chord joining the centres of two adjacent holes.

Chapter 13

Cartesian and polar co-ordinates

Why it is important to understand: Cartesian and polar co-ordinates

Applications where polar co-ordinates would be used include terrestrial navigation with sonar-like devices, and those in engineering and science involving energy radiation patterns. Applications where Cartesian co-ordinates would be used include any navigation on a grid and anything involving raster graphics (e.g. bitmap – a dot matrix data structure representing a generally rectangular grid of pixels). The ability to change from Cartesian to polar co-ordinates is vitally important when using complex numbers and their use in a.c. electrical circuit theory and with vector geometry.

At the end of this chapter, you should be able to:

- change from Cartesian to polar co-ordinates
- change from polar to Cartesian co-ordinates
- use a scientific notation calculator to change from Cartesian to polar co-ordinates and vice-versa

13.1 Introduction

There are two ways in which the position of a point in a plane can be represented. These are

(a) by **Cartesian co-ordinates**, (named after **Descartes**[*]), i.e. (x, y), and

(b) by **polar co-ordinates**, i.e. (r, θ), where r is a 'radius' from a fixed point and θ is an angle from a fixed point.

13.2 Changing from Cartesian into polar co-ordinates

In Fig. 13.1, if lengths x and y are known, then the length of r can be obtained from Pythagoras' theorem (see Chapter 12) since OPQ is a right-angled triangle. Hence $r^2 = (x^2 + y^2)$

[*] **Who was Descartes**? **René Descartes** (31 March 1596– 11 February 1650) was a French philosopher, mathematician, and writer. He wrote many influential texts including *Meditations on First Philosophy*. Descartes is best known for the philosophical statement 'Cogito ergo sum' (I think, therefore I am), found in part IV of *Discourse on the Method*. To find out more go to **www.routledge.com/cw/bird**

from which, $r = \sqrt{x^2 + y^2}$

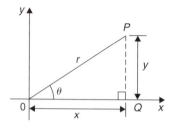

Figure 13.1

From trigonometric ratios (see Chapter 12),

$$\tan \theta = \frac{y}{x}$$

from which $\theta = \tan^{-1} \dfrac{y}{x}$

$r = \sqrt{x^2 + y^2}$ and $\theta = \tan^{-1} \dfrac{y}{x}$ are the two formulae we need to change from Cartesian to polar co-ordinates. The angle θ, which may be expressed in degrees or radians, must **always** be measured from the positive x-axis, i.e., measured from the line OQ in Fig. 13.1. It is suggested that when changing from Cartesian to polar co-ordinates a diagram should always be sketched.

Problem 1. Change the Cartesian co-ordinates (3, 4) into polar co-ordinates.

A diagram representing the point (3, 4) is shown in Fig. 13.2.

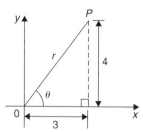

Figure 13.2

From Pythagoras' theorem, $r = \sqrt{3^2 + 4^2} = 5$ (note that -5 has no meaning in this context). By trigonometric ratios, $\theta = \tan^{-1} \frac{4}{3} = 53.13°$ or 0.927 rad.

[Note that $53.13° = 53.13 \times (\pi/180)$ rad $= 0.927$ rad.]

Hence (3, 4) in Cartesian co-ordinates corresponds to (5, 53.13°) or (5, 0.927 rad) in polar co-ordinates.

Problem 2.   Express in polar co-ordinates the position $(-4, 3)$

A diagram representing the point using the Cartesian co-ordinates $(-4, 3)$ is shown in Fig. 13.3.

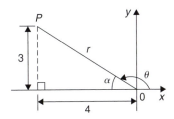

Figure 13.3

From Pythagoras' theorem, $r = \sqrt{4^2 + 3^2} = 5$

By trigonometric ratios, $\alpha = \tan^{-1}\frac{3}{4} = 36.87°$ or 0.644 rad.

Hence $\theta = 180° - 36.87° = 143.13°$ or $\theta = \pi - 0.644 = 2.498$ rad.

Hence the position of point P in polar co-ordinate form is (5, 143.13°) or (5, 2.498 rad).

Problem 3.   Express $(-5, -12)$ in polar co-ordinates.

A sketch showing the position $(-5, -12)$ is shown in Fig. 13.4.

$$r = \sqrt{5^2 + 12^2} = 13$$

and $\alpha = \tan^{-1}\dfrac{12}{5}$

$\qquad = 67.38°$ or 1.176 rad

Hence $\theta = 180° + 67.38° = 247.38°$ or

$\qquad \theta = \pi + 1.176 = 4.318$ rad

Thus $(-5, -12)$ in Cartesian co-ordinates corresponds to (13, 247.38°) or (13, 4.318 rad) in polar co-ordinates.

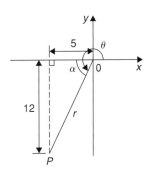

Figure 13.4

Problem 4.   Express $(2, -5)$ in polar co-ordinates.

A sketch showing the position $(2, -5)$ is shown in Fig. 13.5.

$$r = \sqrt{2^2 + 5^2} = \sqrt{29} = 5.385 \text{ correct to}$$
$$3 \text{ decimal places.}$$

$$\alpha = \tan^{-1}\frac{5}{2} = 68.20° \text{ or } 1.190 \text{ rad}$$

Hence $\theta = 360° - 68.20° = 291.80°$ or

$$\theta = 2\pi - 1.190 = 5.093 \text{ rad}$$

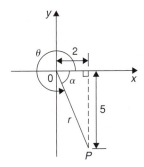

Figure 13.5

Thus $(2, -5)$ in Cartesian co-ordinates corresponds to (5.385, 291.80°) or (5.385, 5.093 rad) in polar co-ordinates.

Section B

Now try the following Practice Exercise

Practice Exercise 60   Changing Cartesian into polar co-ordinates (Answers on page 864)

In Problems 1 to 8, express the given Cartesian co-ordinates as polar co-ordinates, correct to 2 decimal places, in both degrees and in radians.

1. $(3, 5)$

2. $(6.18, 2.35)$

3. $(-2, 4)$

4. $(-5.4, 3.7)$

5. $(-7, -3)$

6. $(-2.4, -3.6)$

7. $(5, -3)$

8. $(9.6, -12.4)$

13.3 Changing from polar into Cartesian co-ordinates

From the right-angled triangle OPQ in Fig. 13.6,

$$\cos\theta = \frac{x}{r} \text{ and } \sin\theta = \frac{y}{r}, \text{ from}$$

trigonometric ratios

Hence $x = r\cos\theta$ and $y = r\sin\theta$

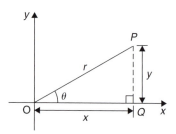

Figure 13.6

If lengths r and angle θ are known then $x = r\cos\theta$ and $y = r\sin\theta$ are the two formulae we need to change from polar to Cartesian co-ordinates.

Problem 5.   Change $(4, 32°)$ into Cartesian co-ordinates.

A sketch showing the position $(4, 32°)$ is shown in Fig. 13.7.

Now $x = r\cos\theta = 4\cos 32° = 3.39$
and $y = r\sin\theta = 4\sin 32° = 2.12$

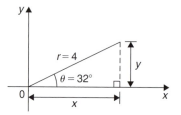

Figure 13.7

Hence (4, 32°) in polar co-ordinates corresponds to (3.39, 2.12) in Cartesian co-ordinates.

Problem 6.   Express $(6, 137°)$ in Cartesian co-ordinates.

A sketch showing the position $(6, 137°)$ is shown in Fig. 13.8.

$$x = r\cos\theta = 6\cos 137° = -4.388$$

which corresponds to length OA in Fig. 13.8.

$$y = r\sin\theta = 6\sin 137° = 4.092$$

which corresponds to length AB in Fig. 13.8.

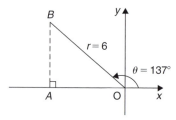

Figure 13.8

Thus (6, 137°) in polar co-ordinates corresponds to (−4.388, 4.092) in Cartesian co-ordinates.

Section B

(Note that when changing from polar to Cartesian co-ordinates it is not quite so essential to draw a sketch. Use of $x = r\cos\theta$ and $y = r\sin\theta$ automatically produces the correct signs.)

> **Problem 7.** Express $(4.5, 5.16\,\text{rad})$ in Cartesian co-ordinates.

A sketch showing the position $(4.5, 5.16\,\text{rad})$ is shown in Fig. 13.9.

$$x = r\cos\theta = 4.5\cos 5.16 = 1.948$$

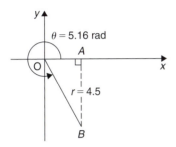

Figure 13.9

which corresponds to length OA in Fig. 13.9.

$$y = r\sin\theta = 4.5\sin 5.16 = -4.057$$

which corresponds to length AB in Fig. 13.9.

Thus $(1.948, -4.057)$ in Cartesian co-ordinates corresponds to $(4.5, 5.16\,\text{rad})$ in polar co-ordinates.

Now try the following Practice Exercise

> **Practice Exercise 61   Changing polar into Cartesian co-ordinates (Answers on page 864)**
>
> In Problems 1 to 8, express the given polar co-ordinates as Cartesian co-ordinates, correct to 3 decimal places.
>
> 1. $(5, 75°)$
>
> 2. $(4.4, 1.12\,\text{rad})$
>
> 3. $(7, 140°)$
>
> 4. $(3.6, 2.5\,\text{rad})$
>
> 5. $(10.8, 210°)$
>
> 6. $(4, 4\,\text{rad})$

7. $(1.5, 300°)$

8. $(6, 5.5\,\text{rad})$

9. Fig. 13.10 shows five equally spaced holes on an 80 mm pitch circle diameter. Calculate their co-ordinates relative to axes Ox and Oy in (a) polar form, (b) Cartesian form.
(c) Calculate also the shortest distance between the centres of two adjacent holes.

Figure 13.10

13.4   Use of Pol/Rec functions on calculators

Another name for Cartesian co-ordinates is **rectangular** co-ordinates. Many scientific notation calculators possess **Pol** and **Rec** functions. 'Rec' is an abbreviation of 'rectangular' (i.e. Cartesian) and 'Pol' is an abbreviation of 'polar'. Check the operation manual for your particular calculator to determine how to use these two functions. They make changing from Cartesian to polar co-ordinates, and vice-versa, so much quicker and easier.

For example, with the Casio fx-991ES PLUS calculator, or similar, to change the Cartesian number $(3, 4)$ into polar form, the following procedure is adopted:

1. Press 'shift'   2. Press 'Pol'   3. Enter 3
4. Enter 'comma' (obtained by 'shift' then))
5. Enter 4   6. Press)
7. Press $=$ The answer is: $r = 5, \theta = 53.13°$

Hence, (3, 4) in Cartesian form is the same as (5, 53.13°) in polar form.

If the angle is required in **radians**, then before repeating the above procedure press 'shift', 'mode' and then 4 to change your calculator to radian mode.

Similarly, to change the polar form number $(7, 126°)$ into Cartesian or rectangular form, adopt the following procedure:

1. Press 'shift'   2. Press 'Rec'   3. Enter 7
4. Enter 'comma'

5. Enter 126 (assuming your calculator is in degrees mode)
6. Press)   7. Press =

The answer is: $X = -4.11$, **and scrolling across**, $Y = 5.66$, correct to 2 decimal places.

Hence, (7, 126°) in polar form is the same as (−4.11, 5.66) in rectangular or Cartesian form.

Now return to Practice Exercises 60 and 61 in this chapter and use your calculator to determine the answers, and see how much more quickly they may be obtained.

For fully worked solutions to each of the problems in Practice Exercises 60 and 61 in this chapter, go to the website:
www.routledge.com/cw/bird

Chapter 14

The circle and its properties

Why it is important to understand: **The circle and its properties**

A circle is one of the fundamental shapes of geometry; it consists of all the points that are equidistant from a central point. Knowledge of calculations involving circles is needed with crank mechanisms, with determinations of latitude and longitude, with pendulums, and even in the design of paper clips. The floodlit area at a football ground, the area an automatic garden sprayer sprays and the angle of lap of a belt drive all rely on calculations involving the arc of a circle. The ability to handle calculations involving circles and its properties is clearly essential in several branches of engineering design.

At the end of this chapter, you should be able to:

- define a circle
- state some properties of a circle – including radius, circumference, diameter, semicircle, quadrant, tangent, sector, chord, segment and arc
- appreciate the angle in a semicircle is a right angle
- define a radian, change radians to degrees, and vice-versa
- determine arc length, area of a circle and area of a sector of a circle
- state the equation of a circle
- sketch a circle given its equation
- understand linear and angular velocity
- understand centripetal force

14.1 Introduction

A **circle** is a plain figure enclosed by a curved line, every point on which is equidistant from a point within, called the **centre**.

14.2 Properties of circles

(i) The distance from the centre to the curve is called the **radius**, r, of the circle (see OP in Fig. 14.1).

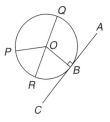

Figure 14.1

(ii) The boundary of a circle is called the **circumference**, c.

(iii) Any straight line passing through the centre and touching the circumference at each end is called the **diameter**, d (see QR in Fig. 14.1). Thus $d = 2r$.

(iv) The ratio $\dfrac{\text{circumference}}{\text{diameter}} = $ a constant for any circle.

This constant is denoted by the Greek letter π (pronounced 'pie'), where $\pi = 3.14159$, correct to 5 decimal places.

Hence $c/d = \pi$ or $c = \pi d$ or $c = 2\pi r$.

(v) A **semicircle** is one half of the whole circle.

(vi) A **quadrant** is one quarter of a whole circle.

(vii) A **tangent** to a circle is a straight line which meets the circle in one point only and does not cut the circle when produced. AC in Fig. 14.1 is a tangent to the circle since it touches the curve at point B only. If radius OB is drawn, then angle ABO is a right angle.

(viii) A **sector** of a circle is the part of a circle between radii (for example, the portion OXY of Fig. 14.2 is a sector). If a sector is less than a semicircle it is called a **minor sector**, if greater than a semicircle it is called a **major sector**.

Figure 14.2

(ix) A **chord** of a circle is any straight line which divides the circle into two parts and is terminated at each end by the circumference. ST, in Fig. 14.2 is a chord.

(x) A **segment** is the name given to the parts into which a circle is divided by a chord. If the segment is less than a semicircle it is called a **minor segment** (see shaded area in Fig. 14.2). If the segment is greater than a semicircle it is called a **major segment** (see the unshaded area in Fig. 14.2).

(xi) An **arc** is a portion of the circumference of a circle. The distance SRT in Fig. 14.2 is called a **minor arc** and the distance $SXYT$ is called a **major arc**.

(xii) The angle at the centre of a circle, subtended by an arc, is double the angle at the circumference subtended by the same arc. With reference to Fig. 14.3, **Angle $AOC = 2 \times$ angle ABC**.

(xiii) The angle in a semicircle is a right angle (see angle BQP in Fig. 14.3).

Figure 14.3

Problem 1. If the diameter of a circle is 75 mm, find its circumference.

Circumference, $c = \pi \times \text{diameter} = \pi d$
$$= \pi(75) = \textbf{235.6 mm}$$

Problem 2. In Fig. 14.4, AB is a tangent to the circle at B. If the circle radius is 40 mm and $AB = 150$ mm, calculate the length AO.

Figure 14.4

A tangent to a circle is at right angles to a radius drawn from the point of contact, i.e. $ABO = 90°$. Hence, using Pythagoras' theorem:

$$AO^2 = AB^2 + OB^2$$

$$AO = \sqrt{(AB^2 + OB^2)} = \sqrt{[(150)^2 + (40)^2]}$$

$$= \textbf{155.2 mm}$$

Now try the following Practice Exercise

Practice Exercise 62   Properties of circles (Answers on page 864)

1. If the radius of a circle is 41.3 mm, calculate the circumference of the circle.

2. Find the diameter of a circle whose perimeter is 149.8 cm.

3. A crank mechanism is shown in Fig. 14.5, where XY is a tangent to the circle at point X. If

the circle radius OX is 10 cm and length OY is 40 cm, determine the length of the connecting rod XY.

Figure 14.5

4. If the circumference of the Earth is 40 000 km at the equator, calculate its diameter.

5. Calculate the length of wire in the paper clip shown in Fig. 14.6. The dimensions are in millimetres.

Figure 14.6

14.3 Radians and degrees

One **radian** is defined as the angle subtended at the centre of a circle by an arc equal in length to the radius.

Figure 14.7

With reference to Fig. 14.7, for arc length s,

$$\theta \text{ radians} = \frac{s}{r}$$

When $s =$ whole circumference $(= 2\pi r)$ then

$$\theta = \frac{s}{r} = \frac{2\pi r}{r} = 2\pi$$

i.e. **2π radians $= 360°$ or π radians $= 180°$**

Thus, **$1 \text{ rad} = \dfrac{180°}{\pi} = 57.30°$**, correct to 2 decimal places.

Since π rad $= 180°$, then $\dfrac{\pi}{2} = 90°$, $\dfrac{\pi}{3} = 60°$, $\dfrac{\pi}{4} = 45°$, and so on.

Problem 3. Convert to radians: (a) 125° (b) 69°47′

(a) Since $180° = \pi$ rad then $1° = \pi/180$ rad, therefore

$$125° = 125\left(\frac{\pi}{180}\right)^c = \mathbf{2.182 \text{ rad}}$$

(Note that ^{c} means 'circular measure' and indicates radian measure.)

(b) $69°47′ = 69\dfrac{47°}{60} = 69.783°$

$$69.783° = 69.783\left(\frac{\pi}{180}\right)^c = \mathbf{1.218 \text{ rad}}$$

Problem 4. Convert to degrees and minutes: (a) 0.749 rad (b) $3\pi/4$ rad.

(a) Since π rad $= 180°$ then $1 \text{ rad} = 180°/\pi$, therefore

$$0.749 = 0.749\left(\frac{180}{\pi}\right)^° = 42.915°$$

$0.915° = (0.915 \times 60)′ = 55′$, correct to the nearest minute, hence

$0.749 \text{ rad} = 42°55′$

(b) Since $1 \text{ rad} = \left(\dfrac{180}{\pi}\right)^°$ then

$$\frac{3\pi}{4} \text{ rad} = \frac{3\pi}{4}\left(\frac{180}{\pi}\right)^° = \frac{3}{4}(180)^° = \mathbf{135°}$$

Problem 5. Express in radians, in terms of π, (a) 150° (b) 270° (c) 37.5°

Since $180° = \pi$ rad then $1° = \pi/180$, hence

(a) $\quad 150° = 150\left(\dfrac{\pi}{180}\right)$ rad $= \dfrac{5\pi}{6}$ **rad**

(b) $\quad 270° = 270\left(\dfrac{\pi}{180}\right)$ rad $= \dfrac{3\pi}{2}$ **rad**

(c) $\quad 37.5° = 37.5\left(\dfrac{\pi}{180}\right)$ rad $= \dfrac{75\pi}{360}$ rad $= \dfrac{5\pi}{24}$ **rad**

Now try the following Practice Exercise

Practice Exercise 63   Radians and degrees (Answers on page 865)

1. Convert to radians in terms of π: (a) $30°$ (b) $75°$ (c) $225°$

2. Convert to radians: (a) $48°$ (b) $84°51'$ (c) $232°15'$

3. Convert to degrees: (a) $\dfrac{7\pi}{6}$ rad (b) $\dfrac{4\pi}{9}$ rad (c) $\dfrac{7\pi}{12}$ rad.

4. Convert to degrees and minutes: (a) 0.0125 rad (b) 2.69 rad (c) 7.241 rad.

5. A car engine speed is 1000 rev/min. Convert this speed into rad/s.

14.4   Arc length and area of circles and sectors

Arc length

From the definition of the radian in the previous section and Fig. 14.7,

$\quad$ **arc length, $s = r\theta$**   where θ is in radians

Area of circle

For any circle, area $= \pi \times$ (radius)^2

i.e. $\quad\quad\quad\quad\quad$ **area $= \pi r^2$**

Since $r = \dfrac{d}{2}$, then **area $= \pi r^2$** or $\dfrac{\pi d^2}{4}$

Area of sector

Area of a sector $= \dfrac{\theta}{360}(\pi r^2)$ when θ is in degrees

$\quad\quad\quad\quad\quad = \dfrac{\theta}{2\pi}(\pi r^2) = \dfrac{1}{2}r^2\theta$

$\quad\quad\quad\quad\quad\quad\quad\quad$ when θ is in radians

Problem 6. A hockey pitch has a semicircle of radius 14.63 m around each goal net. Find the area enclosed by the semicircle, correct to the nearest square metre.

Area of a semicircle $= \dfrac{1}{2}\pi r^2$

When $r = 14.63$ m, $\quad$ area $= \dfrac{1}{2}\pi(14.63)^2$

i.e. $\quad$ **area of semicircle $= 336$ m^2**

Problem 7. Find the area of a circular metal plate, correct to the nearest square millimetre, having a diameter of 35.0 mm.

Area of a circle $= \pi r^2 = \dfrac{\pi d^2}{4}$

When $d = 35.0$ mm, $\quad$ area $= \dfrac{\pi(35.0)^2}{4}$

i.e. $\quad$ **area of circular plate $= 962$ mm^2**

Problem 8. Find the area of a circle having a circumference of 60.0 mm.

Circumference, $c = 2\pi r$

from which radius $r = \dfrac{c}{2\pi} = \dfrac{60.0}{2\pi} = \dfrac{30.0}{\pi}$

Area of a circle $= \pi r^2$

i.e. $\quad$ **area $= \pi\left(\dfrac{30.0}{\pi}\right)^2 = 286.5$ mm^2**

Problem 9. Find the length of arc of a circle of radius 5.5 cm when the angle subtended at the centre is 1.20 rad.

Length of arc, $s = r\theta$, where θ is in radians, hence

$\quad s = (5.5)(1.20) = 6.60$ **cm**

Problem 10. Determine the diameter and circumference of a circle if an arc of length 4.75 cm subtends an angle of 0.91 rad.

Since $s = r\theta$ then $r = \dfrac{s}{\theta} = \dfrac{4.75}{0.91} = 5.22\,\text{cm}$

Diameter $= 2 \times$ radius $= 2 \times 5.22 = \mathbf{10.44\,cm}$

Circumference, $c = \pi d = \pi(10.44) = \mathbf{32.80\,cm}$

Problem 11. If an angle of $125°$ is subtended by an arc of a circle of radius 8.4 cm, find the length of (a) the minor arc, and (b) the major arc, correct to 3 significant figures.

(a) Since $180° = \pi$ rad then $1° = \left(\dfrac{\pi}{180}\right)$ rad and $125° = 125\left(\dfrac{\pi}{180}\right)$ rad.

Length of minor arc,

$$s = r\theta = (8.4)(125)\left(\frac{\pi}{180}\right) = \mathbf{18.3\,cm},$$

correct to 3 significant figures.

(b) Length of major arc

$= $ (circumference $-$ minor arc)

$= 2\pi(8.4) - 18.3 = 34.5\,\text{cm}$,

correct to 3 significant figures.

(Alternatively, major arc $= r\theta$
$= 8.4(360 - 125)(\pi/180) = \mathbf{34.5\,cm}.$)

Problem 12. Determine the angle, in degrees and minutes, subtended at the centre of a circle of diameter 42 mm by an arc of length 36 mm. Calculate also the area of the minor sector formed.

Since length of arc, $s = r\theta$ then $\theta = s/r$

Radius, $r = \dfrac{\text{diameter}}{2} = \dfrac{42}{2} = 21\,\text{mm}$

hence $\theta = \dfrac{s}{r} = \dfrac{36}{21} = 1.7143\,\text{rad}$

$1.7143\,\text{rad} = 1.7143 \times (180/\pi)° = 98.22° = \mathbf{98°13'} =$ angle subtended at centre of circle.

Area of sector
$$= \tfrac{1}{2}r^2\theta = \tfrac{1}{2}(21)^2(1.7143) = \mathbf{378\,mm^2}$$

Problem 13. A football stadium floodlight can spread its illumination over an angle of $45°$ to a distance of 55 m. Determine the maximum area that is floodlit.

Floodlit area $=$ area of sector

$$= \frac{1}{2}r^2\theta = \frac{1}{2}(55)^2\left(45 \times \frac{\pi}{180}\right)$$

$$= \mathbf{1188\,m^2}$$

Problem 14. An automatic garden spray produces a spray to a distance of 1.8 m and revolves through an angle α which may be varied. If the desired spray catchment area is to be $2.5\,\text{m}^2$, to what should angle α be set, correct to the nearest degree.

Area of sector $= \tfrac{1}{2}r^2\theta$, hence $2.5 = \tfrac{1}{2}(1.8)^2\alpha$

from which, $\alpha = \dfrac{2.5 \times 2}{1.8^2} = 1.5432\,\text{rad}$

$1.5432\,\text{rad} = \left(1.5432 \times \dfrac{180°}{\pi}\right) = 88.42°$

Hence **angle $\alpha = 88°$**, correct to the nearest degree.

Problem 15. The angle of a tapered groove is checked using a 20 mm diameter roller as shown in Fig. 14.8. If the roller lies 2.12 mm below the top of the groove, determine the value of angle θ.

Figure 14.8

In Fig. 14.9, triangle ABC is right-angled at C (see Section 14.2 (vii)).

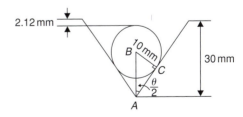

Figure 14.9

Length $BC = 10\,\text{mm}$ (i.e. the radius of the circle), and $AB = 30 - 10 - 2.12 = 17.88\,\text{mm}$ from Fig. 14.9.

Hence, $\sin\dfrac{\theta}{2} = \dfrac{10}{17.88}$ and $\dfrac{\theta}{2} = \sin^{-1}\left(\dfrac{10}{17.88}\right) = 34°$

and **angle $\theta = 68°$**

Now try the following Practice Exercise

Practice Exercise 64   Arc length and area of circles and sectors (Answers on page 865)

1. Calculate the area of a circle of radius 6.0 cm, correct to the nearest square centimetre.

2. The diameter of a circle is 55.0 mm. Determine its area, correct to the nearest square millimetre.

3. The perimeter of a circle is 150 mm. Find its area, correct to the nearest square millimetre.

4. Find the area of the sector, correct to the nearest square millimetre, of a circle having a radius of 35 mm, with angle subtended at centre of 75°

5. An annulus has an outside diameter of 49.0 mm and an inside diameter of 15.0 mm. Find its area correct to 4 significant figures.

6. Find the area, correct to the nearest square metre, of a 2 m wide path surrounding a circular plot of land 200 m in diameter.

7. A rectangular park measures 50 m by 40 m. A 3 m flower bed is made round the two longer sides and one short side. A circular fish pond of diameter 8.0 m is constructed in the centre of the park. It is planned to grass the remaining area. Find, correct to the nearest square metre, the area of grass.

8. Find the length of an arc of a circle of radius 8.32 cm when the angle subtended at the centre is 2.14 rad. Calculate also the area of the minor sector formed.

9. If the angle subtended at the centre of a circle of diameter 82 mm is 1.46 rad, find the lengths of the (a) minor arc (b) major arc.

10. A pendulum of length 1.5 m swings through an angle of 10° in a single swing. Find, in centimetres, the length of the arc traced by the pendulum bob.

11. Determine the length of the radius and circumference of a circle if an arc length of 32.6 cm subtends an angle of 3.76 rad.

12. Determine the angle of lap, in degrees and minutes, if 180 mm of a belt drive are in contact with a pulley of diameter 250 mm.

13. Determine the number of complete revolutions a motorcycle wheel will make in travelling 2 km, if the wheel's diameter is 85.1 cm.

14. The floodlight at a sports ground spreads its illumination over an angle of 40° to a distance of 48 m. Determine (a) the angle in radians, and (b) the maximum area that is floodlit.

15. Determine (a) the shaded area in Fig. 14.10 (b) the percentage of the whole sector that the area of the shaded portion represents.

Figure 14.10

16. Determine the length of steel strip required to make the clip shown in Fig. 14.11.

Figure 14.11

17. A 50° tapered hole is checked with a 40 mm diameter ball as shown in Fig. 14.12. Determine the length shown as x.

Figure 14.12

14.5   The equation of a circle

The simplest equation of a circle, centre at the origin, radius r, is given by:

$$x^2 + y^2 = r^2$$

For example, Fig. 14.13 shows a circle $x^2 + y^2 = 9$. More generally, the equation of a circle, centre (a, b), radius r, is given by:

$$(x - a)^2 + (y - b)^2 = r^2 \qquad (1)$$

Figure 14.14 shows a circle $(x - 2)^2 + (y - 3)^2 = 4$. The general equation of a circle is:

$$x^2 + y^2 + 2ex + 2fy + c = 0 \qquad (2)$$

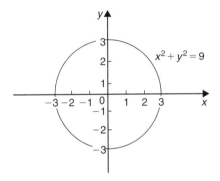

Figure 14.13

Multiplying out the bracketed terms in equation (1) gives:

$$x^2 - 2ax + a^2 + y^2 - 2by + b^2 = r^2$$

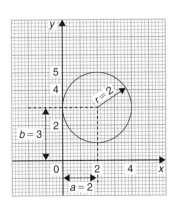

Figure 14.14

Comparing this with equation (2) gives:

$$2e = -2a, \text{ i.e. } \boldsymbol{a = -\dfrac{2e}{2}}$$

and $2f = -2b$, i.e. $\boldsymbol{b = -\dfrac{2f}{2}}$

and $c = a^2 + b^2 - r^2$,

i.e., $\boldsymbol{r = \sqrt{(a^2 + b^2 - c)}}$

Thus, for example, the equation

$$x^2 + y^2 - 4x - 6y + 9 = 0$$

represents a circle with centre $a = -\left(\dfrac{-4}{2}\right)$, $b = -\left(\dfrac{-6}{2}\right)$, i.e. at (2, 3) and radius $r = \sqrt{(2^2 + 3^2 - 9)} = 2$

Hence $x^2 + y^2 - 4x - 6y + 9 = 0$ is the circle shown in Fig. 14.14 (which may be checked by multiplying out the brackets in the equation

$$(x - 2)^2 + (y - 3)^2 = 4)$$

> **Problem 16.**   Determine (a) the radius and (b) the co-ordinates of the centre of the circle given by the equation: $x^2 + y^2 + 8x - 2y + 8 = 0$

$x^2 + y^2 + 8x - 2y + 8 = 0$ is of the form shown in equation (2),

where $a = -\left(\dfrac{8}{2}\right) = -4, b = -\left(\dfrac{-2}{2}\right) = 1$

and $r = \sqrt{[(-4)^2 + (1)^2 - 8]} = \sqrt{9} = 3$

Hence $x^2 + y^2 + 8x - 2y + 8 = 0$ represents a circle **centre (−4, 1)** and **radius 3**, as shown in Fig. 14.15.

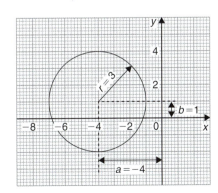

Figure 14.15

Section B

Alternatively, $x^2 + y^2 + 8x - 2y + 8 = 0$ may be re-arranged as:

$$(x + 4)^2 + (y - 1)^2 - 9 = 0$$

i.e. $$(x + 4)^2 + (y - 1)^2 = 3^2$$

which represents a circle, **centre $(-4, 1)$** and **radius 3**, as stated above.

Problem 17. Sketch the circle given by the equation: $x^2 + y^2 - 4x + 6y - 3 = 0$

The equation of a circle, centre (a, b), radius r is given by:

$$(x - a)^2 + (y - b)^2 = r^2$$

The general equation of a circle is

$$x^2 + y^2 + 2ex + 2fy + c = 0$$

From above, $a = -\dfrac{2e}{2}$, $b = -\dfrac{2f}{2}$ and

$$r = \sqrt{(a^2 + b^2 - c)}$$

Hence if $x^2 + y^2 - 4x + 6y - 3 = 0$

then $a = -\left(\dfrac{-4}{2}\right) = 2$, $b = -\left(\dfrac{6}{2}\right) = -3$

and $r = \sqrt{[(2)^2 + (-3)^2 - (-3)]}$
$= \sqrt{16} = 4$

Thus **the circle has centre $(2, -3)$ and radius 4**, as shown in Fig. 14.16.

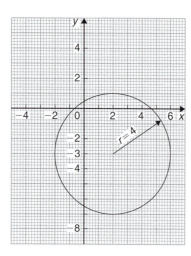

Figure 14.16

Alternatively, $x^2 + y^2 - 4x + 6y - 3 = 0$ may be re-arranged as:

$$(x - 2)^2 + (y + 3)^2 - 3 - 13 = 0$$

i.e. $$(x - 2)^2 + (y + 3)^2 = 4^2$$

which represents a circle, **centre $(2, -3)$** and **radius 4**, as stated above.

Now try the following Practice Exercise

Practice Exercise 65   The equation of a circle (Answers on page 865)

1. Determine (a) the radius, and (b) the co-ordinates of the centre of the circle given by the equation $x^2 + y^2 + 6x - 2y - 26 = 0$

2. Sketch the circle given by the equation $x^2 + y^2 - 6x + 4y - 3 = 0$

3. Sketch the curve $x^2 + (y - 1)^2 - 25 = 0$

4. Sketch the curve $x = 6\sqrt{\left[1 - (y/6)^2\right]}$

14.6 Linear and angular velocity

Linear velocity

Linear velocity v is defined as the rate of change of linear displacement s with respect to time t. For motion in a straight line:

$$\text{linear velocity} = \frac{\text{change of displacement}}{\text{change of time}}$$

i.e. $$v = \frac{s}{t} \qquad (1)$$

The unit of linear velocity is metres per second (m/s).

Angular velocity

The speed of revolution of a wheel or a shaft is usually measured in revolutions per minute or revolutions per second but these units do not form part of a coherent system of units. The basis in SI units is the angle turned through in one second.

Angular velocity is defined as the rate of change of angular displacement θ, with respect to time t. For an object rotating about a fixed axis at a constant speed:

$$\text{angular velocity} = \frac{\text{angle turned through}}{\text{time taken}}$$

i.e.
$$\omega = \frac{\theta}{t} \qquad (2)$$

The unit of angular velocity is radians per second (rad/s). An object rotating at a constant speed of n revolutions per second subtends an angle of $2\pi n$ radians in one second, i.e. its angular velocity ω is given by:

$$\omega = 2\pi n \, \text{rad/s} \qquad (3)$$

From page 152, $s = r\theta$ and from equation (2) above, $\theta = \omega t$

hence
$$s = r(\omega t)$$

from which
$$\frac{s}{t} = \omega r$$

However, from equation (1) $v = \frac{s}{t}$

hence
$$v = \omega r \qquad (4)$$

Equation (4) gives the relationship between linear velocity v and angular velocity ω.

Problem 18. A wheel of diameter 540 mm is rotating at $\dfrac{1500}{\pi}$ rev/min. Calculate the angular velocity of the wheel and the linear velocity of a point on the rim of the wheel.

From equation (3), angular velocity $\omega = 2\pi n$ where n is the speed of revolution in rev/s. Since in this case $n = \dfrac{1500}{\pi}$ rev/min $= \dfrac{1500}{60\pi} = $ rev/s, then

$$\textbf{angular velocity}\,\omega = 2\pi \left(\frac{1500}{60\pi}\right) = 50 \, \text{rad/s}$$

The linear velocity of a point on the rim, $v = \omega r$, where r is the radius of the wheel, i.e.

$$\frac{540}{2} \, \text{mm} = \frac{0.54}{2} \, \text{m} = 0.27 \, \text{m}$$

Thus **linear velocity** $v = \omega r = (50)(0.27)$
$$= 13.5 \, \text{m/s}$$

Problem 19. A car is travelling at 64.8 km/h and has wheels of diameter 600 mm.
(a) Find the angular velocity of the wheels in both rad/s and rev/min.

(b) If the speed remains constant for 1.44 km, determine the number of revolutions made by the wheel, assuming no slipping occurs.

(a) Linear velocity $v = 64.8$ km/h
$$= 64.8 \, \frac{\text{km}}{\text{h}} \times 1000 \, \frac{\text{m}}{\text{km}} \times \frac{1}{3600} \, \frac{\text{h}}{\text{s}} = 18 \, \text{m/s}$$

The radius of a wheel $= \dfrac{600}{2} = 300 \, \text{mm}$
$$= 0.3 \, \text{m}$$

From equation (5), $v = \omega r$, from which,

$$\textbf{angular velocity}\,\omega = \frac{v}{r} = \frac{18}{0.3}$$
$$= 60 \, \text{rad/s}$$

From equation (4), angular velocity, $\omega = 2\pi n$, where n is in rev/s.

Hence angular speed $n = \dfrac{\omega}{2\pi} = \dfrac{60}{2\pi}$ rev/s

$$= 60 \times \frac{60}{2\pi} \, \text{rev/min}$$

$$= 573 \, \text{rev/min}$$

(b) From equation (1), since $v = s/t$ then the time taken to travel 1.44 km, i.e. 1440 m at a constant speed of 18 m/s is given by:

$$\text{time } t = \frac{s}{v} = \frac{1440 \, \text{m}}{18 \, \text{m/s}} = 80 \, \text{s}$$

Since a wheel is rotating at 573 rev/min, then in 80/60 minutes it makes

$$573 \, \text{rev/min} \times \frac{80}{60} \, \text{min} = \textbf{764 revolutions}$$

Now try the following Practice Exercise

Practice Exercise 66   Linear and angular velocity (Answers on page 865)

1. A pulley driving a belt has a diameter of 300 mm and is turning at $2700/\pi$ revolutions per minute. Find the angular velocity of the pulley and the linear velocity of the belt assuming that no slip occurs.

2. A bicycle is travelling at 36 km/h and the diameter of the wheels of the bicycle is 500 mm. Determine the linear velocity of a point on the rim of one of the wheels of the bicycle, and the angular velocity of the wheels.

3. A train is travelling at 108 km/h and has wheels of diameter 800 mm.
   (a) Determine the angular velocity of the wheels in both rad/s and rev/min.
   (b) If the speed remains constant for 2.70 km, determine the number of revolutions made by a wheel, assuming no slipping occurs.

14.7 Centripetal force

When an object moves in a circular path at constant speed, its direction of motion is continually changing and hence its velocity (which depends on both magnitude and direction) is also continually changing. Since acceleration is the (change in velocity)/(time taken), the object has an acceleration. Let the object be moving with a constant angular velocity of ω and a tangential velocity of magnitude v and let the change of velocity for a small change of angle of θ ($=\omega t$) be V in Fig. 14.17. Then $v_2 - v_1 = V$. The vector diagram is shown in Fig. 14.17(b) and since the magnitudes of v_1 and v_2 are the same, i.e. v, the vector diagram is an isosceles triangle.

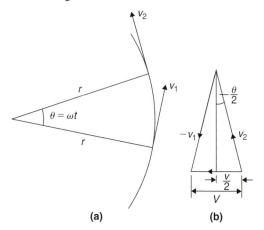

Figure 14.17

Bisecting the angle between v_2 and v_1 gives:

$$\sin\frac{\theta}{2} = \frac{V/2}{v_2} = \frac{V}{2v}$$

i.e. $V = 2v\sin\dfrac{\theta}{2}$ (1)

Since $\theta = \omega t$ then

$$t = \frac{\theta}{\omega}$$ (2)

Dividing equation (1) by equation (2) gives:

$$\frac{V}{t} = \frac{2v\sin(\theta/2)}{(\theta/\omega)} = \frac{v\omega\sin(\theta/2)}{(\theta/2)}$$

For small angles $\dfrac{\sin(\theta/2)}{(\theta/2)} \approx 1$,

hence $\dfrac{V}{t} = \dfrac{\text{change of velocity}}{\text{change of time}}$

$= $ acceleration $a = v\omega$

However, $\omega = \dfrac{v}{r}$ (from Section 14.6)

thus $v\omega = v \cdot \dfrac{v}{r} = \dfrac{v^2}{r}$

i.e. the acceleration a is $\dfrac{v^2}{r}$ and is towards the centre of the circle of motion (along V). It is called the **centripetal acceleration**. If the mass of the rotating object is m, then by Newton's second law, the **centripetal force is $\dfrac{mv^2}{r}$** and its direction is towards the centre of the circle of motion.

Problem 20. A vehicle of mass 750 kg travels around a bend of radius 150 m, at 50.4 km/h. Determine the centripetal force acting on the vehicle.

The centripetal force is given by $\dfrac{mv^2}{r}$ and its direction is towards the centre of the circle.

Mass $m = 750$ kg, $v = 50.4$ km/h

$$= \frac{50.4 \times 1000}{60 \times 60}\,\text{m/s}$$

$$= 14\,\text{m/s}$$

and radius $r = 150$ m,

thus **centripetal force** $= \dfrac{750(14)^2}{150} = \mathbf{980\,N}$

Problem 21. An object is suspended by a thread 250 mm long and both object and thread move in a horizontal circle with a constant angular velocity of 2.0 rad/s. If the tension in the thread is 12.5 N, determine the mass of the object.

Centripetal force (i.e. tension in thread),

$$F = \frac{mv^2}{r} = 12.5\,\text{N}$$

Angular velocity $\omega = 2.0$ rad/s and radius $r = 250$ mm $= 0.25$ m.

Since linear velocity $v = \omega r$, $v = (2.0)(0.25)$
$$= 0.5 \text{ m/s}.$$

Since $F = \dfrac{mv^2}{r}$, then mass $m = \dfrac{Fr}{v^2}$,

i.e. **mass of object, $m = \dfrac{(12.5)(0.25)}{0.5^2} = 12.5 \text{ kg}$**

Problem 22. An aircraft is turning at constant altitude, the turn following the arc of a circle of radius 1.5 km. If the maximum allowable acceleration of the aircraft is $2.5\,g$, determine the maximum speed of the turn in km/h. Take g as 9.8 m/s^2.

The acceleration of an object turning in a circle is $\dfrac{v^2}{r}$. Thus, to determine the maximum speed of turn, $\dfrac{v^2}{r} = 2.5\,g$, from which,

$$\textbf{velocity, } v = \sqrt{(2.5\,gr)} = \sqrt{(2.5)(9.8)(1500)}$$
$$= \sqrt{36\,750} = 191.7 \text{ m/s}$$

and $191.7 \text{ m/s} = 191.7 \times \dfrac{60 \times 60}{1000}$ km/h $= \textbf{690 km/h}$

Now try the following Practice Exercise

Practice Exercise 67   Centripetal force (Answers on page 865)

1. Calculate the tension in a string when it is used to whirl a stone of mass 200 g round in a horizontal circle of radius 90 cm with a constant speed of 3 m/s.

2. Calculate the centripetal force acting on a vehicle of mass 1 tonne when travelling around a bend of radius 125 m at 40 km/h. If this force should not exceed 750 N, determine the reduction in speed of the vehicle to meet this requirement.

3. A speed-boat negotiates an S-bend consisting of two circular arcs of radii 100 m and 150 m. If the speed of the boat is constant at 34 km/h, determine the change in acceleration when leaving one arc and entering the other.

Section B

For fully worked solutions to each of the problems in Practice Exercises 62 to 67 in this chapter, go to the website:
www.routledge.com/cw/bird

Revision Test 4 Trigonometry, Cartesian and polar co-ordinates and properties of circles

This Revision Test covers the material contained in Chapters 12 to 14. *The marks for each question are shown in brackets at the end of each question.*

1. A 2.0 m long ladder is placed against a perpendicular pylon with its foot 52 cm from the pylon. (a) Find how far up the pylon (correct to the nearest mm) the ladder reaches. (b) If the foot of the ladder is moved 10 cm towards the pylon, how far does the top of the ladder rise? (7)

2. Evaluate correct to 4 significant figures:
    (a) $\cos 124°13'$ (b) $\cot 72.68°$ (4)

3. From a point on horizontal ground a surveyor measures the angle of elevation of a church spire as 15°. He moves 30 m nearer to the church and measures the angle of elevation as 20°. Calculate the height of the spire. (9)

4. If secant $\theta = 2.4613$, determine the acute angle θ (4)

5. Evaluate, correct to 3 significant figures:
$$\frac{3.5 \csc 31°17' - \cot(-12°)}{3 \sec 79°41'}$$ (5)

6. A man leaves a point walking at 6.5 km/h in a direction E 20°N (i.e. a bearing of 70°). A cyclist leaves the same point at the same time in a direction E 40°S (i.e. a bearing of 130°) travelling at a constant speed. Find the average speed of the cyclist if the walker and cyclist are 80 km apart after five hours. (8)

7. A crank mechanism shown in Fig. RT4.1 comprises arm OP, of length 0.90 m, which rotates anticlockwise about the fixed point O, and connecting rod PQ of length 4.20 m. End Q moves horizontally in a straight line OR.

    (a) If $\angle POR$ is initially zero, how far does end Q travel in $\frac{1}{4}$ revolution.

    (b) If $\angle POR$ is initially 40° find the angle between the connecting rod and the horizontal and the length OQ.

    (c) Find the distance Q moves (correct to the nearest cm) when $\angle POR$ changes from 40° to 140°. (16)

Figure RT4.1

8. Change the following Cartesian co-ordinates into polar co-ordinates, correct to 2 decimal places, in both degrees and in radians:

    (a) $(-2.3, 5.4)$ (b) $(7.6, -9.2)$ (10)

9. Change the following polar co-ordinates into Cartesian co-ordinates, correct to 3 decimal places: (a) $(6.5, 132°)$ (b) $(3, 3\,\text{rad})$ (6)

10. (a) Convert 2.154 radians into degrees and minutes.

    (b) Change $71°17'$ into radians. (4)

11. 140 mm of a belt drive is in contact with a pulley of diameter 180 mm which is turning at 300 revolutions per minute. Determine (a) the angle of lap, (b) the angular velocity of the pulley, and (c) the linear velocity of the belt, assuming that no slipping occurs. (9)

12. Figure RT4.2 shows a cross-section through a circular water container where the shaded area represents the water in the container. Determine: (a) the depth, h, (b) the area of the shaded portion, and (c) the area of the unshaded area. (11)

Figure RT4.2

13. Determine (a) the co-ordinates of the centre of the circle and (b) the radius, given the equation

$$x^2 + y^2 - 2x + 6y + 6 = 0 \qquad (7)$$

Section B

Chapter 15

Trigonometric waveforms

Why it is important to understand: **Trigonometric waveforms**

Trigonometric graphs are commonly used in all areas of science and engineering for modelling many different natural and mechanical phenomena such as waves, engines, acoustics, electronics, populations, UV intensity, growth of plants and animals, and so on. Periodic trigonometric graphs mean that the shape repeats itself exactly after a certain amount of time. Anything that has a regular cycle, like the tides, temperatures, rotation of the Earth, and so on, can be modelled using a sine or cosine curve. The most common periodic signal waveform that is used in electrical and electronic engineering is the sinusoidal waveform. However, an alternating a.c. waveform may not always take a smooth shape based around the sine and cosine function; a.c. waveforms can also take the shape of square or triangular waves, i.e. complex waves. In engineering, it is therefore important to have a clear understanding of sine and cosine waveforms.

At the end of this chapter, you should be able to:

- sketch sine, cosine and tangent waveforms
- determine angles of any magnitude
- understand cycle, amplitude, period, periodic time, frequency, lagging/leading angles with reference to sine and cosine waves
- perform calculations involving sinusoidal form $A \sin(\omega t \pm \alpha)$
- define a complex wave and harmonic analysis
- use harmonic synthesis to construct a complex waveform

15.1 Graphs of trigonometric functions

By drawing up tables of values from $0°$ to $360°$, graphs of $y = \sin A$, $y = \cos A$ and $y = \tan A$ may be plotted. Values obtained with a calculator (correct to 3 decimal places – which is more than sufficient for plotting graphs), using $30°$ intervals, are shown below, with the respective graphs shown in Fig. 15.1.

(a) $y = \sin A$

A	0	30°	60°	90°	120°	150°	180°
$\sin A$	0	0.500	0.866	1.000	0.866	0.500	0

A	210°	240°	270°	300°	330°	360°
$\sin A$	−0.500	−0.866	−1.000	−0.866	−0.500	0

(b) $y = \cos A$

A	0	30°	60°	90°	120°	150°	180°
$\cos A$	1.000	0.866	0.500	0	−0.500	−0.866	−1.000

A	210°	240°	270°	300°	330°	360°
$\cos A$	−0.866	−0.500	0	0.500	0.866	1.000

(c) $y = \tan A$

A	0	30°	60°	90°	120°	150°	180°
$\tan A$	0	0.577	1.732	∞	−1.732	−0.577	0

A	210°	240°	270°	300°	330°	360°
$\tan A$	0.577	1.732	∞	−1.732	−0.577	0

From Fig. 15.1 it is seen that:

(i) Sine and cosine graphs oscillate between peak values of ±1.

(ii) The cosine curve is the same shape as the sine curve but displaced by 90°.

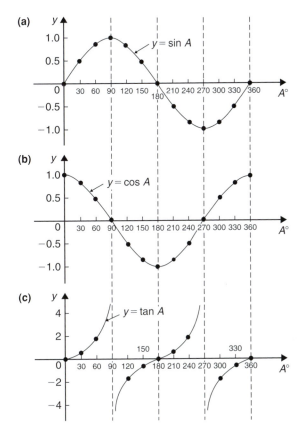

(a) $y = \sin A$

(b) $y = \cos A$

(c) $y = \tan A$

Figure 15.1

(iii) The sine and cosine curves are continuous and they repeat at intervals of 360°; the tangent curve appears to be discontinuous and repeats at intervals of 180°.

15.2 Angles of any magnitude

(i) Fig. 15.2 shows rectangular axes XX' and YY' intersecting at origin O. As with graphical work, measurements made to the right and above O are positive while those to the left and downwards are negative. Let OA be free to rotate about O. By convention, when OA moves anticlockwise angular measurement is considered positive, and vice-versa.

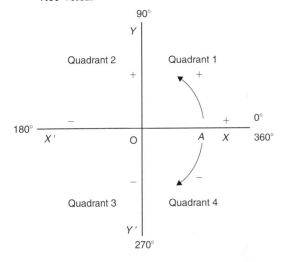

Figure 15.2

(ii) Let OA be rotated anticlockwise so that θ_1 is any angle in the first quadrant and let perpendicular AB be constructed to form the right-angled triangle OAB (see Fig. 15.3). Since all three sides of the triangle are positive, all six trigonometric ratios are positive in the first quadrant. (Note: OA is always positive since it is the radius of a circle.)

(iii) Let OA be further rotated so that θ_2 is any angle in the second quadrant and let AC be constructed to form the right-angled triangle OAC. Then:

$$\sin\theta_2 = \frac{+}{+} = + \qquad \cos\theta_2 = \frac{-}{+} = -$$

$$\tan\theta_2 = \frac{+}{-} = - \qquad \operatorname{cosec}\theta_2 = \frac{+}{+} = +$$

$$\sec\theta_2 = \frac{+}{-} = - \qquad \cot\theta_2 = \frac{-}{+} = -$$

(iv) Let OA be further rotated so that θ_3 is any angle in the third quadrant and let AD be constructed to form the right-angled triangle OAD. Then:

Section B

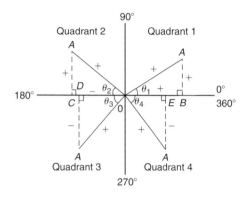

Figure 15.3

$$\sin\theta_3 = \frac{-}{+} = -\ (\text{and hence } \operatorname{cosec}\theta_3 \text{ is } -)$$

$$\cos\theta_3 = \frac{-}{+} = -\ (\text{and hence } \sec\theta_3 \text{ is } +)$$

$$\tan\theta_3 = \frac{-}{-} = +\ (\text{and hence } \cot\theta_3 \text{ is } -)$$

(v)   Let OA be further rotated so that θ_4 is any angle in the fourth quadrant and let AE be constructed to form the right-angled triangle OAE. Then:

$$\sin\theta_4 = \frac{-}{+} = -\ (\text{and hence } \operatorname{cosec}\theta_4 \text{ is } -)$$

$$\cos\theta_4 = \frac{+}{+} = +\ (\text{and hence } \sec\theta_4 \text{ is } +)$$

$$\tan\theta_4 = \frac{-}{+} = -\ (\text{and hence } \cot\theta_4 \text{ is } -)$$

(vi)   The results obtained in (ii) to (v) are summarised in Fig. 15.4. The letters underlined spell the word CAST when starting in the fourth quadrant and moving in an anticlockwise direction.

(vii)   In the first quadrant of Fig. 15.1 all the curves have positive values; in the second only sine is positive; in the third only tangent is positive; in the fourth only cosine is positive (exactly as summarised in Fig. 15.4).

A knowledge of angles of any magnitude is needed when finding, for example, all the angles between $0°$ and $360°$ whose sine is, say, 0.3261. If 0.3261 is entered into a calculator and then the inverse sine key pressed (or $\sin^{-1}$ key) the answer $19.03°$ appears. However, there is a second angle between $0°$ and $360°$ which the calculator does not give. Sine is also positive in the second quadrant (either from CAST or from Fig. 15.1(a)). The other angle is shown in Fig. 15.5 as angle θ where $\theta = 180° - 19.03° = 160.97°$. Thus $19.03°$ **and** $160.97°$ are the angles between $0°$ and $360°$ whose sine is 0.3261 (check that $\sin 160.97° = 0.3261$ on your calculator).

Figure 15.4

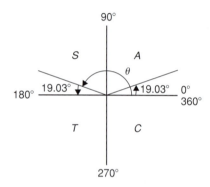

Figure 15.5

Be careful! Your calculator only gives you one of these answers. The second answer needs to be deduced from a knowledge of angles of any magnitude, as shown in the following problems.

Problem 1.   Determine all the angles between $0°$ and $360°$ whose sine is -0.4638

The angles whose sine is -0.4638 occurs in the third and fourth quadrants since sine is negative in these quadrants (see Fig. 15.6(a)). From Fig. 15.6(b), $\theta = \sin^{-1} 0.4638 = 27°38'$

Measured from $0°$, the two angles between $0°$ and $360°$ whose sine is -0.4638 are $180° + 27°38'$, i.e. **207°38'** and $360° - 27°38'$, i.e. **332°22'**. (Note that a calculator generally only gives one answer, i.e. $-27.632588°$)

Problem 2.   Determine all the angles between $0°$ and $360°$ whose tangent is 1.7629

A tangent is positive in the first and third quadrants (see Fig. 15.7(a)). From Fig. 15.7(b), $\theta = \tan^{-1} 1.7629 = 60°26'$. Measured from $0°$, the two

(a)

(b)

Figure 15.6

(a)

(b)

Figure 15.7

angles between $0°$ and $360°$ whose tangent is 1.7629 are **$60°26'$** and $180° + 60°26'$, i.e. **$240°26'$**

Problem 3.   Solve $\sec^{-1}(-2.1499) = \alpha$ for angles of α between $0°$ and $360°$.

Secant is negative in the second and third quadrants (i.e. the same as for cosine). From Fig. 15.8,

$$\theta = \sec^{-1} 2.1499 = \cos^{-1}\left(\frac{1}{2.1499}\right) = 62°17'$$

Measured from $0°$, the two angles between $0°$ and $360°$ whose secant is -2.1499 are

$$\alpha = 180° - 62°17' = \mathbf{117°43'} \quad \text{and}$$

$$\alpha = 180° + 62°17' = \mathbf{242°17'}$$

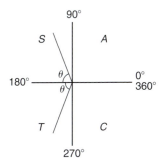

Figure 15.8

Problem 4.   Solve $\cot^{-1} 1.3111 = \alpha$ for angles of α between $0°$ and $360°$.

Cotangent is positive in the first and third quadrants (i.e. same as for tangent). From Fig. 15.9,

$$\theta = \cot^{-1} 1.3111 = \tan^{-1}\left(\frac{1}{1.3111}\right) = 37°20'$$

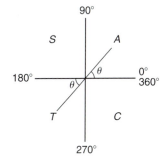

Figure 15.9

Hence $\alpha = \mathbf{37°20'}$

and $\alpha = 180° + 37°20' = \mathbf{217°20'}$

Now try the following Practice Exercise

Practice Exercise 68   Evaluating trigonometric ratios of any magnitude (Answers on page 865)

1.   Find all the angles between $0°$ and $360°$ whose sine is -0.7321

2. Determine the angles between 0° and 360° whose cosecant is 2.5317

3. If cotangent $x = -0.6312$, determine the values of x in the range $0° \leq x \leq 360°$

In Problems 4 to 6 solve the given equations.

4. $\cos^{-1}(-0.5316) = t$

5. $\sec^{-1} 2.3162 = x$

6. $\tan^{-1} 0.8314 = \theta$

15.3 The production of a sine and cosine wave

In Fig. 15.10, let OR be a vector 1 unit long and free to rotate anticlockwise about O. In one revolution a circle is produced and is shown with 15° sectors. Each

radius arm has a vertical and a horizontal component. For example, at 30°, the vertical component is TS and the horizontal component is OS.
From trigonometric ratios,

$$\sin 30° = \frac{TS}{TO} = \frac{TS}{1}, \text{ i.e. } TS = \sin 30°$$

$$\text{and} \cos 30° = \frac{OS}{TO} = \frac{OS}{1}, \text{ i.e. } OS = \cos 30°$$

The vertical component TS may be projected across to $T'S'$, which is the corresponding value of 30° on the graph of y against angle $x°$. If all such vertical components as TS are projected onto the graph, then a **sine wave** is produced as shown in Fig. 15.10.

If all horizontal components such as OS are projected onto a graph of y against angle $x°$, then a **cosine wave** is produced. It is easier to visualise these projections by redrawing the circle with the radius arm OR initially in a vertical position, as shown in Fig. 15.11.

Figure 15.10

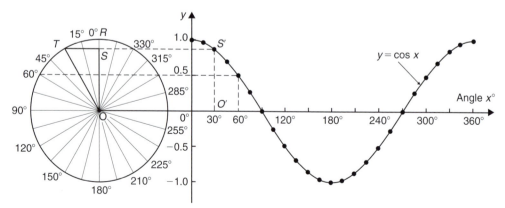

Figure 15.11

From Figs. 15.10 and 15.11 it is seen that a cosine curve is of the same form as the sine curve but is displaced by $90°$ (or $\pi/2$ radians).

15.4 Sine and cosine curves

Graphs of sine and cosine waveforms

(i) A graph of $y = \sin A$ is shown by the broken line in Fig. 15.12 and is obtained by drawing up a table of values as in Section 15.1. A similar table may be produced for $y = \sin 2A$.

$A°$	$2A$	$\sin 2A$	$A°$	$2A$	$\sin 2A$
0	0	0	225	450	1.0
30	60	0.866	240	480	0.866
45	90	1.0	270	540	0
60	120	0.866	300	600	−0.866
90	180	0	315	630	−1.0
120	240	−0.866	330	660	−0.866
135	270	−1.0	360	720	0
150	300	−0.866			
180	360	0			
210	420	0.866			

A graph of $y = \sin 2A$ is shown in Fig. 15.12.

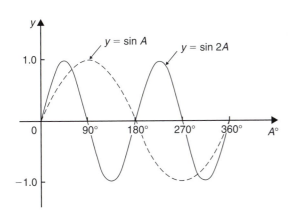

Figure 15.12

(ii) A graph of $y = \sin \frac{1}{2}A$ is shown in Fig. 15.13 using the following table of values.

$A°$	$\frac{1}{2}A$	$\sin \frac{1}{2}A$
0	0	0
30	15	0.259
60	30	0.500
90	45	0.707
120	60	0.866
150	75	0.966
180	90	1.00
210	105	0.966
240	120	0.866
270	135	0.707
300	150	0.500
330	165	0.259
360	180	0

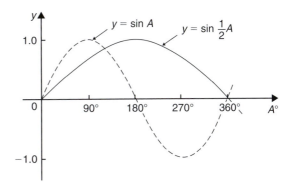

Figure 15.13

(iii) A graph of $y = \cos A$ is shown by the broken line in Fig. 15.14 and is obtained by drawing up a

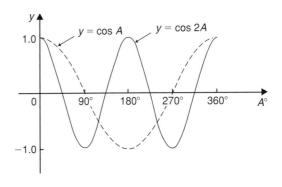

Figure 15.14

table of values. A similar table may be produced for $y = \cos 2A$ with the result as shown.

(iv) A graph of $y = \cos \frac{1}{2} A$ is shown in Fig. 15.15, which may be produced by drawing up a table of values, similar to above.

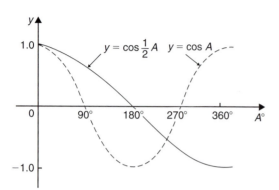

Figure 15.15

Periodic functions and period

(i) Each of the graphs shown in Figs. 15.12 to 15.15 will repeat themselves as angle A increases and are thus called **periodic functions**.

(ii) $y = \sin A$ and $y = \cos A$ repeat themselves every $360°$ (or 2π radians); thus $360°$ is called the **period** of these waveforms. $y = \sin 2A$ and $y = \cos 2A$ repeat themselves every $180°$ (or π radians); thus $180°$ is the period of these waveforms.

(iii) In general, if $y = \sin pA$ or $y = \cos pA$ (where p is a constant) then the period of the waveform is $360°/p$ (or $2\pi/p$ rad). Hence if $y = \sin 3A$ then the period is $360/3$, i.e. $120°$, and if $y = \cos 4A$ then the period is $360/4$, i.e. $90°$

Amplitude

Amplitude is the name given to the maximum or peak value of a sine wave. Each of the graphs shown in Figs. 15.12 to 15.15 has an amplitude of $+1$ (i.e. they oscillate between $+1$ and -1). However, if $y = 4 \sin A$, each of the values in the table is multiplied by 4 and the maximum value, and thus amplitude, is 4. Similarly, if $y = 5 \cos 2A$, the amplitude is 5 and the period is $360°/2$, i.e. $180°$.

Problem 5. Sketch $y = \sin 3A$ between $A = 0°$ and $A = 360°$.

Amplitude $= 1$; period $= 360°/3 = 120°$.

A sketch of $y = \sin 3A$ is shown in Fig. 15.16.

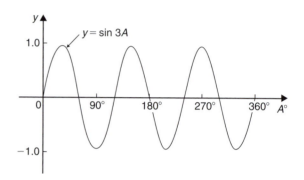

Figure 15.16

Problem 6. Sketch $y = 3 \sin 2A$ from $A = 0$ to $A = 2\pi$ radians.

Amplitude $= 3$, period $= 2\pi/2 = \pi$ rads (or $180°$)

A sketch of $y = 3 \sin 2A$ is shown in Fig. 15.17.

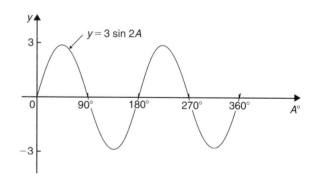

Figure 15.17

Problem 7. Sketch $y = 4 \cos 2x$ from $x = 0°$ to $x = 360°$.

Amplitude $= 4$; period $= 360°/2 = 180°$

A sketch of $y = 4 \cos 2x$ is shown in Fig. 15.18.

Figure 15.18

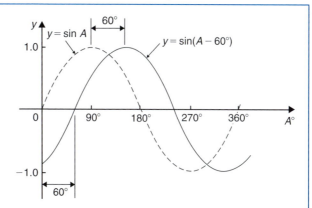

Figure 15.20

Problem 8. Sketch $y = 2 \sin \dfrac{3}{5} A$ over one cycle.

Amplitude $= 2$; period $= \dfrac{360°}{\dfrac{3}{5}} = \dfrac{360° \times 5}{3} = 600°$

A sketch of $y = 2 \sin \dfrac{3}{5} A$ is shown in Fig. 15.19.

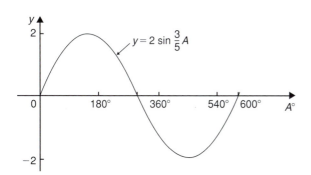

Figure 15.19

Lagging and leading angles

(i)   A sine or cosine curve may not always start at $0°$. To show this a periodic function is represented by $y = \sin(A \pm \alpha)$ or $y = \cos(A \pm \alpha)$ where α is a phase displacement compared with $y = \sin A$ or $y = \cos A$.

(ii) By drawing up a table of values, a graph of $y = \sin(A - 60°)$ may be plotted as shown in Fig. 15.20. If $y = \sin A$ is assumed to start at $0°$ then $y = \sin(A - 60°)$ starts $60°$ later (i.e. has a zero value $60°$ later). Thus $y = \sin(A - 60°)$ is said to **lag** $y = \sin A$ by $60°$.

(iii) By drawing up a table of values, a graph of $y = \cos(A + 45°)$ may be plotted as shown in Fig. 15.21. If $y = \cos A$ is assumed to start at $0°$ then $y = \cos(A + 45°)$ starts $45°$ earlier (i.e. has a zero value $45°$ earlier). Thus $y = \cos(A + 45°)$ is said to **lead** $y = \cos A$ by $45°$

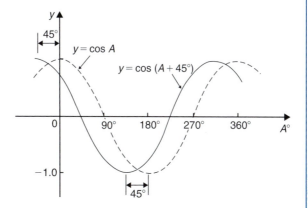

Figure 15.21

(iv) Generally, a graph of $y = \sin(A - \alpha)$ lags $y = \sin A$ by angle α, and a graph of $y = \sin(A + \alpha)$ leads $y = \sin A$ by angle α.

(v)   A cosine curve is the same shape as a sine curve but starts $90°$ earlier, i.e. leads by $90°$. Hence $\cos A = \sin(A + 90°)$

Problem 9. Sketch $y = 5 \sin(A + 30°)$ from $A = 0°$ to $A = 360°$.

Amplitude $= 5$; period $= 360°/1 = 360°$.

$5 \sin(A + 30°)$ leads $5 \sin A$ by $30°$ (i.e. starts $30°$ earlier).

Section B

A sketch of $y = 5 \sin(A + 30°)$ is shown in Fig. 15.22.

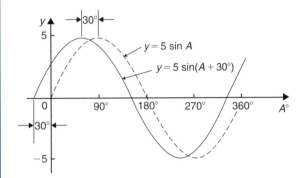

Figure 15.22

Problem 10. Sketch $y = 7 \sin(2A - \pi/3)$ in the range $0 \le A \le 2\pi$

Amplitude $= 7$; period $= 2\pi/2 = \pi$ radians.

In general, $y = \sin(pt - \alpha)$ **lags** $y = \sin pt$ **by** α/p, hence $7 \sin(2A - \pi/3)$ lags $7 \sin 2A$ by $(\pi/3)/2$, i.e. $\pi/6$ rad or $30°$.

A sketch of $y = 7 \sin(2A - \pi/3)$ is shown in Fig. 15.23.

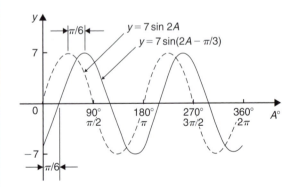

Figure 15.23

Problem 11. Sketch $y = 2 \cos(\omega t - 3\pi/10)$ over one cycle.

Amplitude $= 2$; period $= 2\pi/\omega$ rad.

$2 \cos(\omega t - 3\pi/10)$ lags $2 \cos \omega t$ by $3\pi/10\omega$ seconds.

A sketch of $y = 2 \cos(\omega t - 3\pi/10)$ is shown in Fig. 15.24.

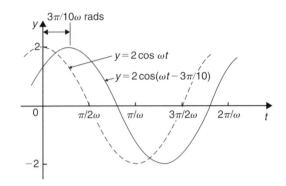

Figure 15.24

Graphs of $\sin^2 A$ and $\cos^2 A$

(i) A graph of $y = \sin^2 A$ is shown in Fig. 15.25 using the following table of values.

$A°$	$\sin A$	$(\sin A)^2 = \sin^2 A$
0	0	0
30	0.50	0.25
60	0.866	0.75
90	1.0	1.0
120	0.866	0.75
150	0.50	0.25
180	0	0
210	−0.50	0.25
240	−0.866	0.75
270	−1.0	1.0
300	−0.866	0.75
330	−0.50	0.25
360	0	0

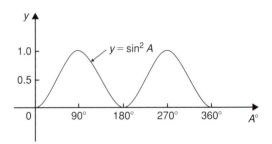

Figure 15.25

(ii) A graph of $y = \cos^2 A$ is shown in Fig. 15.26, obtained by drawing up a table of values, similar to above.

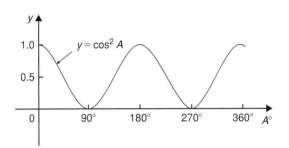

Figure 15.26

(iii) $y = \sin^2 A$ and $y = \cos^2 A$ are both periodic functions of period $180°$ (or π rad) and both contain only positive values. Thus a graph of $y = \sin^2 2A$ has a period $180°/2$, i.e. $90°$. Similarly, a graph of $y = 4\cos^2 3A$ has a maximum value of 4 and a period of $180°/3$, i.e. $60°$.

> **Problem 12.** Sketch $y = 3\sin^2 \frac{1}{2} A$ in the range $0 < A < 360°$.

Maximum value $= 3$; period $= 180°/(1/2) = 360°$
A sketch of $3\sin^2 \frac{1}{2} A$ is shown in Fig. 15.27.

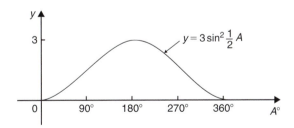

Figure 15.27

> **Problem 13.** Sketch $y = 7\cos^2 2A$ between $A = 0°$ and $A = 360°$

Maximum value $= 7$; period $= 180°/2 = 90°$
A sketch of $y = 7\cos^2 2A$ is shown in Fig. 15.28.

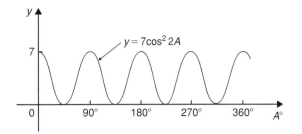

Figure 15.28

Now try the following Practice Exercise

> **Practice Exercise 69   Sine and cosine curves (Answers on page 865)**
>
> In Problems 1 to 9 state the amplitude and period of the waveform and sketch the curve between $0°$ and $360°$.
>
> 1. $y = \cos 3A$
>
> 2. $y = 2\sin \dfrac{5x}{2}$
>
> 3. $y = 3\sin 4t$
>
> 4. $y = 3\cos \dfrac{\theta}{2}$
>
> 5. $y = \dfrac{7}{2}\sin \dfrac{3x}{8}$
>
> 6. $y = 6\sin(t - 45°)$
>
> 7. $y = 4\cos(2\theta + 30°)$
>
> 8. $y = 2\sin^2 2t$
>
> 9. $y = 5\cos^2 \dfrac{3}{2}\theta$

15.5 Sinusoidal form $A \sin(\omega t \pm \alpha)$

In Fig. 15.29, let OR represent a vector that is free to rotate anticlockwise about O at a velocity of ω rad/s. A rotating vector is called a **phasor**. After a time t seconds OR will have turned through an angle ωt radians (shown as angle TOR in Fig. 15.29). If ST is constructed perpendicular to OR, then $\sin \omega t = ST/TO$, i.e. $ST = TO \sin \omega t$.

If all such vertical components are projected onto a graph of y against ωt, a sine wave results of amplitude OR (as shown in Section 15.3).

If phasor OR makes one revolution (i.e. 2π radians) in T seconds, then the angular velocity,

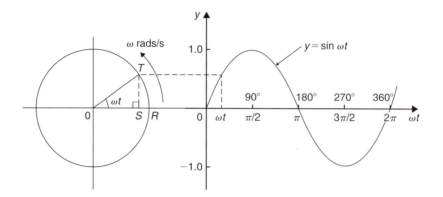

Figure 15.29

$\omega = 2\pi/T$ rad/s, from which, $\boldsymbol{T = 2\pi/\omega \textbf{ seconds}}$.
T is known as the **periodic time**.
The number of complete cycles occurring per second is called the **frequency, f**

$$\text{Frequency} = \frac{\text{number of cycles}}{\text{second}} = \frac{1}{T}$$

$$= \frac{\omega}{2\pi} \text{ i.e. } \boldsymbol{f = \frac{\omega}{2\pi} \textbf{ Hz}}$$

Hence **angular velocity**, $\boldsymbol{\omega = 2\pi f \textbf{ rad/s}}$

Amplitude is the name given to the maximum or peak value of a sine wave, as explained in Section 15.3. The amplitude of the sine wave shown in Fig. 15.29 is 1.

A sine or cosine wave may not always start at $0°$. To show this a periodic function is represented by $y = \sin(\omega t \pm \alpha)$ or $y = \cos(\omega t \pm \alpha)$, where α is a phase displacement compared with $y = \sin A$ or $y = \cos A$. A graph of $y = \sin(\omega t - \alpha)$ **lags** $y = \sin \omega t$ by angle α, and a graph of $y = \sin(\omega t + \alpha)$ **leads** $y = \sin \omega t$ by angle α.

The angle ωt is measured in **radians** (i.e. $\left(\omega \dfrac{\text{rad}}{\text{s}}\right)(t\,\text{s}) = \omega t \text{ radians}$) hence angle α should also be in radians.

The relationship between degrees and radians is:

$$360° = 2\pi \text{ radians or } \boldsymbol{180° = \pi \textbf{ radians}}$$

Hence 1 rad $= \dfrac{180}{\pi} = 57.30°$ and, for example,

$71° = 71 \times \dfrac{\pi}{180} = 1.239$ rad.

Given a general sinusoidal function $\boldsymbol{y = A\sin(\omega t \pm \alpha)}$, then

(i) $A = $ amplitude

(ii) $\omega = $ angular velocity $= 2\pi f$ rad/s

(iii) $\dfrac{2\pi}{\omega} = $ periodic time T seconds

(iv) $\dfrac{\omega}{2\pi} = $ frequency, f hertz

(v) $\alpha = $ angle of lead or lag (compared with $y = A\sin \omega t$)

Problem 14. An alternating current is given by $i = 30\sin(100\pi t + 0.27)$ amperes. Find the amplitude, periodic time, frequency and phase angle (in degrees and minutes).

$i = 30\sin(100\pi t + 0.27)$ A, hence **amplitude = 30 A**
Angular velocity $\omega = 100\pi$, hence

$$\textbf{periodic time, } \boldsymbol{T} = \frac{2\pi}{\omega} = \frac{2\pi}{100\pi} = \frac{1}{50}$$

$$= \textbf{0.02 s or 20 ms}$$

$$\textbf{Frequency, } \boldsymbol{f} = \frac{1}{T} = \frac{1}{0.02} = \textbf{50 Hz}$$

$$\textbf{Phase angle, } \boldsymbol{\alpha} = 0.27\,\text{rad} = \left(0.27 \times \frac{180}{\pi}\right)^{\circ}$$

$$= \textbf{15.47° or 15°28' leading}$$

$$\boldsymbol{i = 30\sin(100\pi t)}$$

Problem 15. An oscillating mechanism has a maximum displacement of 2.5 m and a frequency of 60 Hz. At time $t = 0$ the displacement is 90 cm. Express the displacement in the general form $A\sin(\omega t \pm \alpha)$.

Amplitude = maximum displacement = 2.5 m.
Angular velocity, $\omega = 2\pi f = 2\pi(60) = 120\pi$ rad/s.
Hence displacement = $2.5 \sin(120\pi t + \alpha)$ m.
When $t = 0$, displacement = 90 cm = 0.90 m.

Hence $\qquad 0.90 = 2.\sin(0 + \alpha)$

i.e. $\qquad \sin\alpha = \dfrac{0.90}{2.5} = 0.36$

Hence $\qquad \alpha = \arcsin 0.36 = 21.10° = 21°6'$

$\qquad\qquad\qquad = 0.368\,\text{rad}$

Thus **displacement = 2.5 sin(120πt + 0.368) m**

Problem 16. The instantaneous value of voltage in an a.c. circuit at any time t seconds is given by $v = 340\sin(50\pi t - 0.541)$ volts. Determine:

(a) the amplitude, periodic time, frequency and phase angle (in degrees)

(b) the value of the voltage when $t = 0$

(c) the value of the voltage when $t = 10$ ms

(d) the time when the voltage first reaches 200 V

(e) the time when the voltage is a maximum.

Sketch one cycle of the waveform.

(a) **Amplitude = 340 V**
Angular velocity, $\omega = 50\pi$

Hence **periodic time**, $T = \dfrac{2\pi}{\omega} = \dfrac{2\pi}{50\pi} = \dfrac{1}{25}$

$\qquad\qquad\qquad = \mathbf{0.04\,s}$ or **40 ms**

Frequency, $f = \dfrac{1}{T} = \dfrac{1}{0.04} = \mathbf{25\,Hz}$

Phase angle $= 0.541\,\text{rad} = \left(0.541 \times \dfrac{180}{\pi}\right)$

$\qquad\qquad = \mathbf{31°\ lagging}\ v = 340\sin(50\pi t)$

(b) **When $t = 0$,**

$\qquad v = 340\sin(0 - 0.541) = 340\sin(-31°)$

$\qquad\qquad = \mathbf{-175.1\,V}$

(c) **When $t = 10$ ms**

then $v = 340\sin\left(50\pi\dfrac{10}{10^3} - 0.541\right)$

$\qquad = 340\sin(1.0298) = 340\sin 59°$

$\qquad = \mathbf{291.4\,V}$

(d) When $v = 200$ volts

then $200 = 340\sin(50\pi t - 0.541)$

$\qquad \dfrac{200}{340} = \sin(50\pi t - 0.541)$

Hence $(50\pi t - 0.541) = \sin^{-1}\dfrac{200}{340}$

$\qquad\qquad = 36.03°$ or $0.6288\,\text{rad}$

$\qquad 50\pi t = 0.6288 + 0.541$

$\qquad\qquad = 1.1698$

Hence when $v = 200$ V,

$\qquad$ **time,** $t = \dfrac{1.1698}{50\pi} = \mathbf{7.447\,ms}$

(e) When the voltage is a maximum, $v = 340$ V.

Hence $\qquad 340 = 340\sin(50\pi t - 0.541)$

$\qquad\qquad 1 = \sin(50\pi t - 0.541)$

$\qquad 50\pi t - 0.541 = \sin^{-1} 1$

$\qquad\qquad = 90°$ or $1.5708\,\text{rad}$

$\qquad 50\pi t = 1.5708 + 0.541 = 2.1118$

Hence time, $t = \dfrac{2.1118}{50\pi} = \mathbf{13.44\,ms}$

A sketch of $v = 340\sin(50\pi t - 0.541)$ volts is shown in Fig. 15.30.

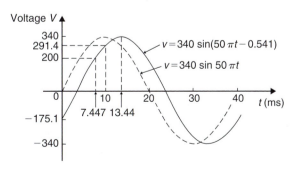

Figure 15.30

Section B

Now try the following Practice Exercise

Practice Exercise 70   The sinusoidal form
$A \sin(\omega t \pm \alpha)$ **(Answers on page 865)**

In Problems 1 to 3, find (a) the amplitude, (b) the frequency, (c) the periodic time, and (d) the phase angle (stating whether it is leading or lagging $A \sin \omega t$) of the alternating quantities given.

1. $i = 40 \sin(50\pi t + 0.29)$ mA

2. $y = 75 \sin(40t - 0.54)$ cm

3. $v = 300 \sin(200\pi t - 0.412)$ V

4. A sinusoidal voltage has a maximum value of 120 V and a frequency of 50 Hz. At time $t = 0$, the voltage is (a) zero and (b) 50 V.
    Express the instantaneous voltage v in the form $v = A \sin(\omega t \pm \alpha)$

5. An alternating current has a periodic time of 25 ms and a maximum value of 20 A. When time $t = 0$, current $i = -10$ amperes. Express the current i in the form $i = A \sin(\omega t \pm \alpha)$

6. An oscillating mechanism has a maximum displacement of 3.2 m and a frequency of 50 Hz. At time $t = 0$ the displacement is 150 cm. Express the displacement in the general form $A \sin(\omega t \pm \alpha)$

7. The current in an a.c. circuit at any time t seconds is given by:

    $i = 5 \sin(100\pi t - 0.432)$ amperes

    Determine (a) the amplitude, frequency, periodic time and phase angle (in degrees), (b) the value of current at $t = 0$, (c) the value of current at $t = 8$ ms, (d) the time when the current is first a maximum, (e) the time when the current first reaches 3A. Sketch one cycle of the waveform showing relevant points.

15.6 Harmonic synthesis with complex waveforms

A waveform that is not sinusoidal is called a **complex wave**. **Harmonic analysis** is the process of resolving a complex periodic waveform into a series of sinusoidal components of ascending order of frequency. Many of the waveforms met in practice can be represented by the following mathematical expression.

$$v = V_{1m} \sin(\omega t + \alpha_1) + V_{2m} \sin(2\omega t + \alpha_2)$$
$$+ \cdots + V_{nm} \sin(n\omega t + \alpha_n)$$

and the magnitude of their harmonic components together with their phase may be calculated using **Fourier series** (see Chapters 73 to 76). **Numerical methods** are used to analyse waveforms for which simple mathematical expressions cannot be obtained. A numerical method of harmonic analysis is explained in Chapter 77 on page 801. In a laboratory, waveform analysis may be performed using a **waveform analyser** which produces a direct readout of the component waves present in a complex wave.

By adding the instantaneous values of the fundamental and progressive harmonics of a complex wave for given instants in time, the shape of a complex waveform can be gradually built up. This graphical procedure is known as **harmonic synthesis** (synthesis meaning 'the putting together of parts or elements so as to make up a complex whole'). For more on the addition of alternating waveforms, see Chapter 27.

Some examples of harmonic synthesis are considered in the following worked problems.

Problem 17. Use harmonic synthesis to construct the complex voltage given by:

$$v_1 = 100 \sin \omega t + 30 \sin 3\omega t \text{ volts.}$$

The waveform is made up of a fundamental wave of maximum value 100 V and frequency, $f = \omega/2\pi$ hertz and a third harmonic component of maximum value 30 V and frequency $= 3\omega/2\pi (= 3f)$, the fundamental and third harmonics being initially in phase with each other.

In Fig. 15.31, the fundamental waveform is shown by the broken line plotted over one cycle, the periodic time T being $2\pi/\omega$ seconds. On the same axis is plotted $30 \sin 3\omega t$, shown by the dotted line, having a maximum value of 30 V and for which three cycles are completed in time T seconds. At zero time, $30 \sin 3\omega t$ is in phase with $100 \sin \omega t$.

The fundamental and third harmonic are combined by adding ordinates at intervals to produce the waveform for v_1, as shown. For example, at time $T/12$ seconds, the fundamental has a value of 50 V and the third harmonic a value of 30 V. Adding gives a value of 80 V for waveform v_1 at time $T/12$ seconds. Similarly, at time

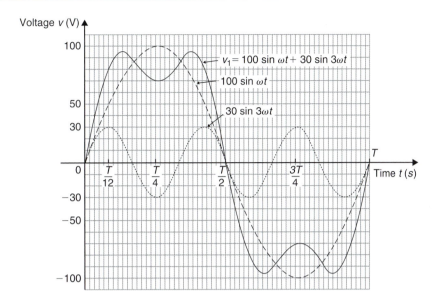

Figure 15.31

$T/4$ seconds, the fundamental has a value of 100 V and the third harmonic a value of -30 V. After addition, the resultant waveform v_1 is 70 V at $T/4$. The procedure is continued between $t = 0$ and $t = T$ to produce the complex waveform for v_1. The negative half-cycle of waveform v_1 is seen to be identical in shape to the positive half-cycle.

If further odd harmonics of the appropriate amplitude and phase were added to v_1 a good approximation to a **square wave** would result.

Problem 18. Construct the complex voltage given by:

$$v_2 = 100\sin\omega t + 30\sin\left(3\omega t + \frac{\pi}{2}\right) \text{ volts.}$$

The peak value of the fundamental is 100 volts and the peak value of the third harmonic is 30 V. However, the third harmonic has a phase displacement of $\frac{\pi}{2}$ radian leading (i.e. leading $30\sin 3\omega t$ by $\frac{\pi}{2}$ radian). Note that, since the periodic time of the fundamental is T seconds, the periodic time of the third harmonic is $T/3$ seconds, and a phase displacement of $\frac{\pi}{2}$ radian or $\frac{1}{4}$ cycle of the third harmonic represents a time interval of $(T/3) \div 4$, i.e. $T/12$ seconds.

Fig. 15.32 shows graphs of $100\sin\omega t$ and $30\sin\left(3\omega t + \frac{\pi}{2}\right)$ over the time for one cycle of the fundamental. When ordinates of the two graphs are added at intervals, the resultant waveform v_2 is

as shown. If the negative half-cycle in Fig. 15.32 is reversed it can be seen that the shape of the positive and negative half-cycles are identical.

Problems 17 and 18 demonstrate that whenever odd harmonics are added to a fundamental waveform, whether initially in phase with each other or not, the positive and negative half-cycles of the resultant complex wave are identical in shape. This is a feature of waveforms containing the fundamental and odd harmonics.

Problem 19. Use harmonic synthesis to construct the complex current given by:

$$i_1 = 10\sin\omega t + 4\sin 2\omega t \text{ amperes.}$$

Current i_1 consists of a fundamental component, $10\sin\omega t$, and a second harmonic component, $4\sin 2\omega t$, the components being initially in phase with each other. The fundamental and second harmonic are shown plotted separately in Fig. 15.33. By adding ordinates at intervals, the complex waveform representing i_1 is produced as shown. It is noted that if all the values in the negative half-cycle were reversed then this half-cycle would appear as a mirror image of the positive half-cycle about a vertical line drawn through time, $t = T/2$.

Problem 20. Construct the complex current given by:

$$i_2 = 10\sin\omega t + 4\sin\left(2\omega t + \frac{\pi}{2}\right) \text{ amperes.}$$

Section B

Figure 15.32

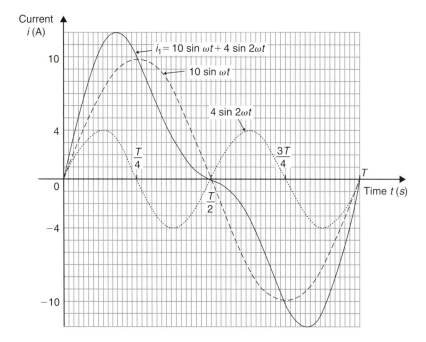

Figure 15.33

The fundamental component, $10\sin\omega t$, and the second harmonic component, having an amplitude of 4 A and a phase displacement of $\dfrac{\pi}{2}$ radian leading (i.e. leading $4\sin 2\omega t$ by $\dfrac{\pi}{2}$ radian or $T/8$ seconds), are shown plotted separately in Fig. 15.34. By adding ordinates at intervals, the complex waveform for i_2 is produced as shown. The positive and negative half-cycles of the resultant waveform are seen to be quite dissimilar.

From Problems 18 and 19 it is seen that whenever even harmonics are added to a fundamental component:

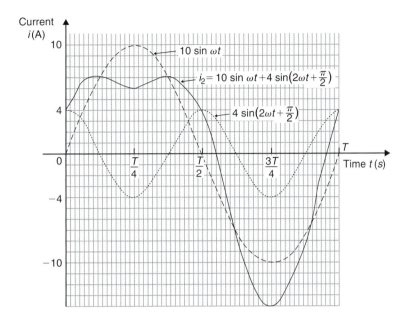

Figure 15.34

(a) if the harmonics are initially in phase, the negative half-cycle, when reversed, is a mirror image of the positive half-cycle about a vertical line drawn through time, $t = T/2$.

(b) if the harmonics are initially out of phase with each other, the positive and negative half-cycles are dissimilar.

These are features of waveforms containing the fundamental and even harmonics.

Problem 21. Use harmonic synthesis to construct the complex current expression given by:

$$i = 32 + 50\sin\omega t + 20\sin\left(2\omega t - \frac{\pi}{2}\right) \text{ mA}$$

The current i comprises three components – a 32 mA d.c. component, a fundamental of amplitude 50 mA and a second harmonic of amplitude 20 mA, lagging by $\frac{\pi}{2}$ radian. The fundamental and second harmonic are shown separately in Fig. 15.35. Adding ordinates at intervals gives the complex waveform $50\sin\omega t + 20\sin\left(2\omega t - \frac{\pi}{2}\right)$

This waveform is then added to the 32 mA d.c. component to produce the waveform i as shown.

The effect of the d.c. component is to shift the whole wave 32 mA upward. The waveform approaches that expected from a **half-wave rectifier**.

Problem 22. A complex waveform v comprises a fundamental voltage of 240 V rms and frequency 50 Hz, together with a 20% third harmonic which has a phase angle lagging by $3\pi/4$ rad at time $t = 0$. (a) Write down an expression to represent voltage v. (b) Use harmonic synthesis to sketch the complex waveform representing voltage v over one cycle of the fundamental component.

(a) A fundamental voltage having an rms value of 240 V has a maximum value, or amplitude of $\sqrt{2}\,(240)$, i.e. 339.4 V.

If the fundamental frequency is 50 Hz then angular velocity, $\omega = 2\pi f = 2\pi (50) = 100\pi$ rad/s. Hence the fundamental voltage is represented by $339.4\sin 100\pi t$ volts. Since the fundamental frequency is 50 Hz, the time for one cycle of the fundamental is given by $T = 1/f = 1/50$ s or 20 ms.

The third harmonic has an amplitude equal to 20% of 339.4 V, i.e. 67.9 V. The frequency of the third harmonic component is $3 \times 50 = 150$ Hz, thus the angular velocity is $2\pi(150)$, i.e. 300π rad/s. Hence

Section B

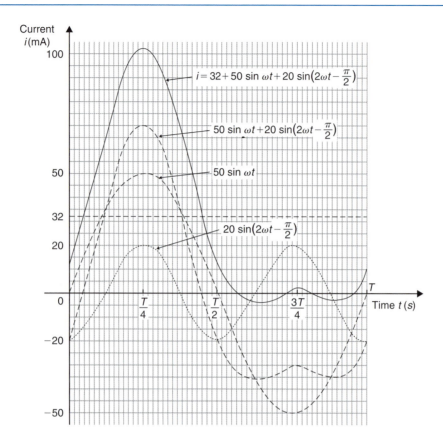

Figure 15.35

the third harmonic voltage is represented by $67.9\sin(300\pi t - 3\pi/4)$ volts. Thus

voltage, $v = 339.4\sin 100\pi t$

$$+ 67.9\sin(300\pi t - 3\pi/4)\,\textbf{volts}$$

(b) One cycle of the fundamental, $339.4\sin 100\pi t$, is shown sketched in Fig. 15.36, together with three cycles of the third harmonic component, $67.9\sin(300\pi t - 3\pi/4)$ initially lagging by $3\pi/4$ rad. By adding ordinates at intervals,

Figure 15.36

the complex waveform representing voltage is produced as shown.

Now try the following Practice Exercise

Practice Exercise 71   Harmonic synthesis with complex waveforms (Answers on page 865)

1. A complex current waveform i comprises a fundamental current of 50 A rms and frequency 100 Hz, together with a 24% third harmonic, both being in phase with each other at zero time. (a) Write down an expression to represent current i. (b) Sketch the complex waveform of current using harmonic synthesis over one cycle of the fundamental.

2. A complex voltage waveform v comprises of a 212.1 V rms fundamental voltage at a frequency of 50 Hz, a 30% second harmonic component lagging by $\pi/2$ rad, and a 10% fourth harmonic component leading by $\pi/3$ rad. (a) Write down an expression to represent voltage v. (b) Sketch the complex voltage waveform using harmonic synthesis over one cycle of the fundamental waveform.

3. A voltage waveform is represented by:

    $$v = 20 + 50\sin\omega t$$
    $$+\, 20\sin(2\omega t - \pi/2)\,\text{volts.}$$

    Draw the complex waveform over one cycle of the fundamental by using harmonic synthesis.

4. Write down an expression representing a current i having a fundamental component of amplitude 16 A and frequency 1 kHz, together with its third and fifth harmonics being respectively one-fifth and one-tenth the amplitude of the fundamental, all components being in phase at zero time. Sketch the complex current waveform for one cycle of the fundamental using harmonic synthesis.

5. A voltage waveform is described by

    $$v = 200\sin 377t + 80\sin\left(1131t + \frac{\pi}{4}\right)$$
    $$+\, 20\sin\left(1885t - \frac{\pi}{3}\right)\text{volts}$$

    Determine (a) the fundamental and harmonic frequencies of the waveform, (b) the percentage third harmonic and (c) the percentage fifth harmonic. Sketch the voltage waveform using harmonic synthesis over one cycle of the fundamental.

For fully worked solutions to each of the problems in Practice Exercises 68 to 71 in this chapter, go to the website:
www.routledge.com/cw/bird

Chapter 16

Hyperbolic functions

Why it is important to understand: **Hyperbolic functions**

There are two combinations of e^x and e^{-x} which are used so often in engineering that they are given their own name. They are the hyperbolic sine, sinh, and the hyperbolic cosine, cosh. They are of interest because they have many properties analogous to those of trigonometric functions and because they arise in the study of falling bodies, hanging cables, ocean waves, and many other phenomena in science and engineering. The shape of a chain hanging under gravity is well described by cosh x and the deformation of uniform beams can be expressed in terms of hyperbolic tangents. Other applications of hyperbolic functions are found in fluid dynamics, optics, heat, mechanical engineering, and in astronomy when dealing with the curvature of light in the presence of black holes.

At the end of this chapter, you should be able to:

- define a hyperbolic function
- state practical applications of hyperbolic functions
- define sinh x, cosh x, tanh x, cosech x, sech x and coth x
- evaluate hyperbolic functions
- sketch graphs of hyperbolic functions
- state Osborne's rule
- prove simple hyperbolic identities
- solve equations involving hyperbolic functions
- derive the series expansions for cosh x and sinh x

16.1 Introduction to hyperbolic functions

Functions which are associated with the geometry of the conic section called a hyperbola are called **hyperbolic functions**. Applications of hyperbolic functions include transmission line theory and catenary problems. By definition:

(i) Hyperbolic sine of x,

$$\sinh x = \frac{e^x - e^{-x}}{2} \qquad (1)$$

'sinh x' is often abbreviated to 'sh x' and is pronounced as 'shine x'

(ii) Hyperbolic cosine of x,

$$\cosh x = \frac{e^x + e^{-x}}{2} \qquad (2)$$

'cosh x' is often abbreviated to 'ch x' and is pronounced as 'kosh x'

(iii) Hyperbolic tangent of x,

$$\tanh x = \frac{\sinh x}{\cosh x} = \frac{e^x - e^{-x}}{e^x + e^{-x}} \qquad (3)$$

'tanh x' is often abbreviated to 'th x' and is pronounced as 'than x'

(iv) Hyperbolic cosecant of x,

$$\textbf{cosech } x = \frac{1}{\sinh x} = \frac{2}{e^x - e^{-x}} \qquad (4)$$

'cosech x' is pronounced as 'coshec x'

(v) Hyperbolic secant of x,

$$\textbf{sech } x = \frac{1}{\cosh x} = \frac{2}{e^x + e^{-x}} \qquad (5)$$

'sech x' is pronounced as 'shec x'

(vi) Hyperbolic cotangent of x,

$$\textbf{coth} x = \frac{1}{\tanh x} = \frac{e^x + e^{-x}}{e^x - e^{-x}} \qquad (6)$$

'coth x' is pronounced as 'koth x'

Some properties of hyperbolic functions

Replacing x by 0 in equation (1) gives:

$$\sinh 0 = \frac{e^0 - e^{-0}}{2} = \frac{1-1}{2} = 0$$

Replacing x by 0 in equation (2) gives:

$$\cosh 0 = \frac{e^0 + e^{-0}}{2} = \frac{1+1}{2} = 1$$

If a function of x, $f(-x) = -f(x)$, then $f(x)$ is called an **odd function** of x. Replacing x by $-x$ in equation (1) gives:

$$\sinh(-x) = \frac{e^{-x} - e^{-(-x)}}{2} = \frac{e^{-x} - e^x}{2}$$

$$= -\left(\frac{e^x - e^{-x}}{2}\right) = -\sinh x$$

Replacing x by $-x$ in equation (3) gives:

$$\tanh(-x) = \frac{e^{-x} - e^{-(-x)}}{e^{-x} + e^{-(-x)}} = \frac{e^{-x} - e^x}{e^{-x} + e^x}$$

$$= -\left(\frac{e^x - e^{-x}}{e^x + e^{-x}}\right) = -\tanh x$$

Hence **sinh x and tanh x are both odd functions** (see Section 16.1), as also are cosech $x\left(=\dfrac{1}{\sinh x}\right)$ and coth $x\left(=\dfrac{1}{\tanh x}\right)$

If a function of x, $f(-x) = f(x)$, then $f(x)$ is called an **even function** of x. Replacing x by $-x$ in equation (2) gives:

$$\cosh(-x) = \frac{e^{-x} + e^{-(-x)}}{2} = \frac{e^{-x} + e^x}{2}$$

$$= \cosh x$$

Hence **cosh x is an even function** (see Section 16.2), as also is sech $x\left(=\dfrac{1}{\cosh x}\right)$

Hyperbolic functions may be evaluated most easily using a calculator. Many scientific notation calculators actually possess sinh and cosh functions; however, if a calculator does not contain these functions, then the definitions given above may be used.

> **Problem 1.** Evaluate sinh 5.4, correct to 4 significant figures.

Using a calculator,

(i) press hyp

(ii) press 1 and sinh(appears

(iii) type in 5.4

(iv) press) to close the brackets

(v) press = and 110.7009498 appears

Hence, **sinh 5.4 = 110.7**, correct to 4 significant figures.

Alternatively, $\sinh 5.4 = \dfrac{1}{2}(e^{5.4} - e^{-5.4})$

$$= \frac{1}{2}(221.406416\ldots - 0.00451658\ldots)$$

$$= \frac{1}{2}(221.401899\ldots)$$

$$= \textbf{110.7}, \text{ correct to 4 significant figures.}$$

> **Problem 2.** Evaluate cosh 1.86, correct to 3 decimal places.

Using a calculator with the procedure similar to that used in Problem 1,

$\quad$ **cosh 1.86 = 3.290**, correct to 3 decimal places.

Problem 3.   Evaluate th 0.52, correct to 4 significant figures.

Using a calculator with the procedure similar to that used in Problem 1,

th 0.52 = 0.4777, correct to 4 significant figures.

Problem 4.   Evaluate cosech 1.4, correct to 4 significant figures.

$$\text{cosech} 1.4 = \frac{1}{\sinh 1.4}$$

Using a calculator,

(i)   press hyp

(ii)   press 1 and sinh(appears

(iii)   type in 1.4

(iv)   press) to close the brackets

(v)   press = and 1.904301501 appears

(vi)   press x^{-1}

(vii)   press = and 0.5251269293 appears

Hence, **cosech 1.4 = 0.5251**, correct to 4 significant figures.

Problem 5.   Evaluate sech 0.86, correct to 4 significant figures.

$$\text{sech} 0.86 = \frac{1}{\cosh 0.86}$$

Using a calculator with the procedure similar to that used in Problem 4,

sech 0.86 = 0.7178, correct to 4 significant figures.

Problem 6.   Evaluate coth 0.38, correct to 3 decimal places.

$$\coth 0.38 = \frac{1}{\tanh 0.38}$$

Using a calculator with the procedure similar to that used in Problem 4,

coth 0.38 = 2.757, correct to 3 decimal places.

Now try the following Practice Exercise

Practice Exercise 72   Evaluating hyperbolic functions (Answers on page 866)

In Problems 1 to 6, evaluate correct to 4 significant figures.

1.   (a) sh 0.64   (b) sh 2.182

2.   (a) ch 0.72   (b) ch 2.4625

3.   (a) th 0.65   (b) th 1.81

4.   (a) cosech 0.543   (b) cosech 3.12

5.   (a) sech 0.39   (b) sech 2.367

6.   (a) coth 0.444   (b) coth 1.843

7.   A telegraph wire hangs so that its shape is described by $y = 50 \text{ch} \dfrac{x}{50}$. Evaluate, correct to 4 significant figures, the value of y when $x = 25$

8.   The length l of a heavy cable hanging under gravity is given by $l = 2c \text{ sh}(L/2c)$. Find the value of l when $c = 40$ and $L = 30$

9. $V^2 = 0.55L \tanh(6.3d/L)$ is a formula for velocity V of waves over the bottom of shallow water, where d is the depth and L is the wavelength. If $d = 8.0$ and $L = 96$, calculate the value of V.

16.2   Graphs of hyperbolic functions

A graph of $y = \sinh x$ may be plotted using calculator values of hyperbolic functions. The curve is shown in Fig. 16.1. Since the graph is symmetrical about the origin, $\sinh x$ is an **odd function** (as stated in Section 16.1).

A graph of $y = \cosh x$ may be plotted using calculator values of hyperbolic functions. The curve is shown in Fig. 16.2. Since the graph is symmetrical about the y-axis, $\cosh x$ is an **even function** (as stated in Section 16.1). The shape of $y = \cosh x$ is that of a heavy rope or chain hanging freely under gravity and is called a **catenary**. Examples include transmission lines, a telegraph wire or a fisherman's line, and is used in the design of roofs and arches. Graphs of $y = \tanh x$,

Figure 16.1

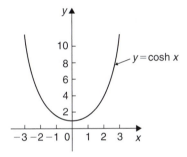

Figure 16.2

$y=\text{cosech}\,x$, $y=\text{sech}\,x$ and $y=\coth x$ are deduced in Problems 7 and 8.

Problem 7. Sketch graphs of (a) $y=\tanh x$ and (b) $y=\coth x$ for values of x between -3 and 3

A table of values is drawn up as shown below.

x	-3	-2	-1
sh$x$	-10.02	-3.63	-1.18
ch$x$	10.07	3.76	1.54
$y=\text{th}\,x=\dfrac{\text{sh}\,x}{\text{ch}\,x}$	-0.995	-0.97	-0.77
$y=\coth x=\dfrac{\text{ch}\,x}{\text{sh}\,x}$	-1.005	-1.04	-1.31

x	0	1	2	3
sh$x$	0	1.18	3.63	10.02
ch$x$	1	1.54	3.76	10.07
$y=\text{th}\,x=\dfrac{\text{sh}\,x}{\text{ch}\,x}$	0	0.77	0.97	0.995
$y=\coth x=\dfrac{\text{ch}\,x}{\text{sh}\,x}$	$\pm\infty$	1.31	1.04	1.005

(a)   A graph of $y=\tanh x$ is shown in Fig. 16.3(a)

(b)   A graph of $y=\coth x$ is shown in Fig. 16.3(b)

Both graphs are symmetrical about the origin, thus $\tanh x$ and $\coth x$ are odd functions.

Problem 8. Sketch graphs of (a) $y=\text{cosech}\,x$ and (b) $y=\text{sech}\,x$ from $x=-4$ to $x=4$, and, from the graphs, determine whether they are odd or even functions.

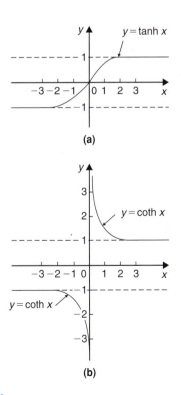

Figure 16.3

A table of values is drawn up as shown below.

x	-4	-3	-2	-1
$\operatorname{sh} x$	-22.29	-10.02	-3.63	-1.18
$\operatorname{cosech} x = \dfrac{1}{\operatorname{sh} x}$	-0.04	-0.10	-0.28	-0.85
$\operatorname{ch} x$	27.31	10.07	3.76	1.54
$\operatorname{sech} x = \dfrac{1}{\operatorname{ch} x}$	0.04	0.10	0.27	0.65

x	0	1	2	3	4
$\operatorname{sh} x$	0	1.18	3.63	10.02	27.29
$\operatorname{cosech} x = \dfrac{1}{\operatorname{sh} x}$	$\pm\infty$	0.85	0.28	0.10	0.04
$\operatorname{ch} x$	1	1.54	3.76	10.07	27.31
$\operatorname{sech} x = \dfrac{1}{\operatorname{ch} x}$	1	0.65	0.27	0.10	0.04

(a) A graph of $y = \operatorname{cosech} x$ is shown in Fig. 16.4(a). The graph is symmetrical about the origin and is thus an **odd function**.

(b) A graph of $y = \operatorname{sech} x$ is shown in Fig. 16.4(b). The graph is symmetrical about the y-axis and is thus an **even function**.

(a)

(b)

Figure 16.4

16.3 Hyperbolic identities

For every trigonometric identity there is a corresponding hyperbolic identity. **Hyperbolic identities** may be proved by either

(i) replacing $\operatorname{sh} x$ by $\dfrac{e^x - e^{-x}}{2}$ and $\operatorname{ch} x$ by $\dfrac{e^x + e^{-x}}{2}$, or

(ii) by using **Osborne's rule**, which states: *'the six trigonometric ratios used in trigonometrical identities relating general angles may be replaced by their corresponding hyperbolic functions, but the sign of any direct or implied product of two sines must be changed'.*

For example, since $\cos^2 x + \sin^2 x = 1$ then, by Osborne's rule, $\operatorname{ch}^2 x - \operatorname{sh}^2 x = 1$, i.e. the trigonometric functions have been changed to their corresponding hyperbolic functions and since $\sin^2 x$ is a product of two sines the sign is changed from $+$ to $-$. Table 16.1 shows some trigonometric identities and their corresponding hyperbolic identities.

> **Problem 9.** Prove the hyperbolic identities
> (a) $\operatorname{ch}^2 x - \operatorname{sh}^2 x = 1$ (b) $1 - \operatorname{th}^2 x = \operatorname{sech}^2 x$
> (c) $\operatorname{coth}^2 x - 1 = \operatorname{cosech}^2 x$

(a) $\operatorname{ch} x + \operatorname{sh} x = \left(\dfrac{e^x + e^{-x}}{2}\right) + \left(\dfrac{e^x - e^{-x}}{2}\right) = e^x$

$\operatorname{ch} x - \operatorname{sh} x = \left(\dfrac{e^x + e^{-x}}{2}\right) - \left(\dfrac{e^x - e^{-x}}{2}\right)$

$\qquad = e^{-x}$

$(\operatorname{ch} x + \operatorname{sh} x)(\operatorname{ch} x - \operatorname{sh} x) = (e^x)(e^{-x}) = e^0 = 1$

i.e. $\mathbf{ch^2 x - sh^2 x = 1}$ (1)

(b) Dividing each term in equation (1) by $\operatorname{ch}^2 x$ gives:

$$\frac{\operatorname{ch}^2 x}{\operatorname{ch}^2 x} - \frac{\operatorname{sh}^2 x}{\operatorname{ch}^2 x} = \frac{1}{\operatorname{ch}^2 x}$$

i.e. $\mathbf{1 - th^2 x = sech^2 x}$

Table 16.1

Trigonometric identity	Corresponding hyperbolic identity
$\cos^2 x + \sin^2 x = 1$	$\text{ch}^2 x - \text{sh}^2 x = 1$
$1 + \tan^2 x = \sec^2 x$	$1 - \text{th}^2 x = \text{sech}^2 x$
$\cot^2 x + 1 = \text{cosec}^2 x$	$\coth^2 x - 1 = \text{cosech}^2 x$
Compound angle formulae	
$\sin(A \pm B) = \sin A \cos B \pm \cos A \sin B$	$\text{sh}(A \pm B) = \text{sh}\,A\,\text{ch}\,B \pm \text{ch}\,A\,\text{sh}\,B$
$\cos(A \pm B) = \cos A \cos B \mp \sin A \sin B$	$\text{ch}(A \pm B) = \text{ch}\,A\,\text{ch}\,B \pm \text{sh}\,A\,\text{sh}\,B$
$\tan(A \pm B) = \dfrac{\tan A \pm \tan B}{1 \mp \tan A \tan B}$	$\text{th}(A \pm B) = \dfrac{\text{th}\,A \pm \text{th}\,B}{1 \pm \text{th}\,A\,\text{th}\,B}$
Double angles	
$\sin 2x = 2 \sin x \cos x$	$\text{sh}\,2x = 2\,\text{sh}\,x\,\text{ch}\,x$
$\cos 2x = \cos^2 x - \sin^2 x$	$\text{ch}\,2x = \text{ch}^2 x + \text{sh}^2 x$
$\quad = 2\cos^2 x - 1$	$\quad = 2\,\text{ch}^2 x - 1$
$\quad = 1 - 2\sin^2 x$	$\quad = 1 + 2\text{sh}^2 x$
$\tan 2x = \dfrac{2\tan x}{1 - \tan^2 x}$	$\text{th}\,2x = \dfrac{2\,\text{th}\,x}{1 + \text{th}^2 x}$

(c) Dividing each term in equation (1) by $\text{sh}^2 x$ gives:

$$\frac{\text{ch}^2 x}{\text{sh}^2 x} - \frac{\text{sh}^2 x}{\text{sh}^2 x} = \frac{1}{\text{sh}^2 x}$$

i.e. **$\coth^2 x - 1 = \text{cosech}^2 x$**

Problem 10. Prove, using Osborne's rule
(a) $\text{ch}\,2A = \text{ch}^2 A + \text{sh}^2 A$
(b) $1 - \text{th}^2 x = \text{sech}^2 x$

(a) From trigonometric ratios,

$$\cos 2A = \cos^2 A - \sin^2 A \tag{1}$$

Osborne's rule states that trigonometric ratios may be replaced by their corresponding hyperbolic functions but the sign of any product of two sines has to be changed. In this case, $\sin^2 A = (\sin A)(\sin A)$, i.e. a product of two sines, thus the sign of the corresponding hyperbolic function, $\text{sh}^2 A$, is changed from $+$ to $-$. Hence, from (1), **$\text{ch}\,2A = \text{ch}^2 A + \text{sh}^2 A$**

(b) From trigonometric ratios,

$$1 + \tan^2 x = \sec^2 x \tag{2}$$

and $\tan^2 x = \dfrac{\sin^2 x}{\cos^2 x} = \dfrac{(\sin x)(\sin x)}{\cos^2 x}$

i.e. a product of two sines.

Hence, in equation (2), the trigonometric ratios are changed to their equivalent hyperbolic function and the sign of $\text{th}^2 x$ changed $+$ to $-$, i.e.
$1 - \text{th}^2 x = \text{sech}^2 x$

Problem 11. Prove that $1 + 2\text{sh}^2 x = \text{ch}\,2x$

Left hand side (LHS)

$$= 1 + 2\text{sh}^2 x = 1 + 2\left(\frac{e^x - e^{-x}}{2}\right)^2$$

$$= 1 + 2\left(\frac{e^{2x} - 2e^x e^{-x} + e^{-2x}}{4}\right)$$

$$= 1 + \frac{e^{2x} - 2 + e^{-2x}}{2}$$

$$= 1 + \left(\frac{e^{2x} + e^{-2x}}{2}\right) - \frac{2}{2}$$

$$= \frac{e^{2x} + e^{-2x}}{2} = \text{ch}\,2x = \text{RHS}$$

Problem 12. Show that $\text{th}^2 x + \text{sech}^2 x = 1$

$$\text{LHS} = \text{th}^2 x + \text{sech}^2 x = \frac{\text{sh}^2 x}{\text{ch}^2 x} + \frac{1}{\text{ch}^2 x}$$

$$= \frac{\text{sh}^2 x + 1}{\text{ch}^2 x}$$

Since $\text{ch}^2 x - \text{sh}^2 x = 1$ then $1 + \text{sh}^2 x = \text{ch}^2 x$

Thus $\dfrac{\text{sh}^2 x + 1}{\text{ch}^2 x} = \dfrac{\text{ch}^2 x}{\text{ch}^2 x} = 1 = \text{RHS}$

Problem 13. Given $Ae^x + Be^{-x} \equiv 4\text{ch}\,x - 5\text{sh}\,x$, determine the values of A and B.

$$Ae^x + Be^{-x} \equiv 4\,\text{ch}\,x - 5\,\text{sh}\,x$$

$$= 4\left(\frac{e^x + e^{-x}}{2}\right) - 5\left(\frac{e^x - e^{-x}}{2}\right)$$

$$= 2e^x + 2e^{-x} - \frac{5}{2}e^x + \frac{5}{2}e^{-x}$$

$$= -\frac{1}{2}e^x + \frac{9}{2}e^{-x}$$

Equating coefficients gives: $A = -\dfrac{1}{2}$ and $B = 4\dfrac{1}{2}$

Problem 14. If $4e^x - 3e^{-x} \equiv P\text{sh}\,x + Q\text{ch}\,x$, determine the values of P and Q

$$4e^x - 3e^{-x} \equiv P\,\text{sh}\,x + Q\,\text{ch}\,x$$

$$= P\left(\frac{e^x - e^{-x}}{2}\right) + Q\left(\frac{e^x + e^{-x}}{2}\right)$$

$$= \frac{P}{2}e^x - \frac{P}{2}e^{-x} + \frac{Q}{2}e^x + \frac{Q}{2}e^{-x}$$

$$= \left(\frac{P+Q}{2}\right)e^x + \left(\frac{Q-P}{2}\right)e^{-x}$$

Equating coefficients gives:

$$4 = \frac{P+Q}{2} \quad \text{and} \quad -3 = \frac{Q-P}{2}$$

i.e. $P + Q = 8$ \hfill (1)

$-P + Q = -6$ \hfill (2)

Adding equations (1) and (2) gives: $2Q = 2$, i.e. $Q = 1$
Substituting in equation (1) gives: $P = 7$

Now try the following Practice Exercise

Practice Exercise 73   Hyperbolic identities (Answers on page 866)

In Problems 1 to 4, prove the given identities.

1. (a) $\text{ch}(P - Q) \equiv \text{ch}\,P\,\text{ch}\,Q - \text{sh}\,P\,\text{sh}\,Q$

    (b) $\text{ch}\,2x \equiv \text{ch}^2 x + \text{sh}^2 x$

2. (a) $\coth x \equiv 2\,\text{cosech}\,2x + \text{th}\,x$

    (b) $\text{ch}\,2\theta - 1 \equiv 2\,\text{sh}^2 \theta$

3. (a) $\text{th}(A - B) \equiv \dfrac{\text{th}\,A - \text{th}\,B}{1 - \text{th}\,A\,\text{th}\,B}$

    (b) $\text{sh}\,2A \equiv 2\,\text{sh}\,A\,\text{ch}\,A$

4. (a) $\text{sh}(A + B) \equiv \text{sh}\,A\,\text{ch}\,B + \text{ch}\,A\,\text{sh}\,B$

    (b) $\dfrac{\text{sh}^2 x + \text{ch}^2 x - 1}{2\,\text{ch}^2 x\,\coth^2 x} \equiv \tanh^4 x$

5. Given $Pe^x - Qe^{-x} \equiv 6\text{ch}\,x - 2\text{sh}\,x$, find P and Q

6. If $5e^x - 4e^{-x} \equiv A\,\text{sh}\,x + B\,\text{ch}\,x$, find A and B.

16.4 Solving equations involving hyperbolic functions

Equations such as $\sinh x = 3.25$ or $\coth x = 3.478$ may be determined using a calculator. This is demonstrated in Problems 15 to 21.

Problem 15. Solve the equation $\text{sh}\,x = 3$, correct to 4 significant figures.

If $\sinh x = 3$, then $x = \sinh^{-1} 3$
This can be determined by calculator.

(i) Press hyp

(ii) Choose 4, which is $\sinh^{-1}$

(iii) Type in 3

(iv) Close bracket)

(v) Press $=$ and the answer is 1.818448459

i.e. the solution of $\text{sh}\,x = 3$ is: $x = 1.818$, correct to 4 significant figures.

Problem 16. Solve the equation $\text{ch}\,x = 1.52$, correct to 3 decimal places.

Using a calculator with a similar procedure as in Problem 15, check that:

$$x = 0.980, \text{ correct to 3 decimal places.}$$

With reference to Fig. 16.2, it can be seen that there will be two values corresponding to $y = \cosh x = 1.52$. Hence, $x = \pm 0.980$

Problem 17. Solve the equation $\tanh\theta = 0.256$, correct to 4 significant figures.

Using a calculator with a similar procedure as in Problem 15, check that

$$\theta = 0.2618, \text{ correct to 4 significant figures.}$$

Problem 18. Solve the equation $\operatorname{sech} x = 0.4562$, correct to 3 decimal places.

If $\operatorname{sech} x = 0.4562$, then $x = \operatorname{sech}^{-1} 0.4562 = \cosh^{-1}\left(\dfrac{1}{0.4562}\right)$ since $\cosh = \dfrac{1}{\operatorname{sech}}$

i.e. $x = 1.421$, correct to 3 decimal places.

With reference to the graph of $y = \operatorname{sech} x$ in Fig. 16.4, it can be seen that there will be two values corresponding to $y = \operatorname{sech} x = 0.4562$
Hence, $x = \pm 1.421$

Problem 19. Solve the equation $\operatorname{cosech} y = -0.4458$, correct to 4 significant figures.

If $\operatorname{cosech} y = -0.4458$, then $y = \operatorname{cosech}^{-1}(-0.4458) = \sinh^{-1}\left(\dfrac{1}{-0.4458}\right)$ since $\sinh = \dfrac{1}{\operatorname{cosech}}$
i.e. $y = -1.547$, correct to 4 significant figures.

Problem 20. Solve the equation $\coth A = 2.431$, correct to 3 decimal places.

If $\coth A = 2.431$, then $A = \coth^{-1} 2.431 = \tanh^{-1}\left(\dfrac{1}{2.431}\right)$ since $\tanh = \dfrac{1}{\coth}$
i.e. $A = 0.437$, correct to 3 decimal places.

Problem 21. A chain hangs in the form given by $y = 40\,\mathrm{ch}\,\dfrac{x}{40}$. Determine, correct to 4 significant figures, (a) the value of y when x is 25, and (b) the value of x when $y = 54.30$

(a) $y = 40\,\mathrm{ch}\,\dfrac{x}{40}$, and when $x = 25$,

$$y = 40\,\mathrm{ch}\,\frac{25}{40} = 40\,\mathrm{ch}\,0.625$$

$$= 40(1.2017536\ldots) = \mathbf{48.07}$$

(b) When $y = 54.30$, $54.30 = 40\,\mathrm{ch}\,\dfrac{x}{40}$, from which

$$\mathrm{ch}\,\frac{x}{40} = \frac{54.30}{40} = 1.3575$$

Hence, $\dfrac{x}{40} = \cosh^{-1} 1.3575 = \pm 0.822219\ldots$

(see Fig. 16.2 for the reason as to why the answer is $\pm$) from which, $x = 40(\pm 0.822219\ldots.) = \pm 32.89$

Equations of the form $a\,\mathrm{ch}\,x + b\,\mathrm{sh}\,x = c$, where a, b and c are constants may be solved either by:

(a) plotting graphs of $y = a\,\mathrm{ch}\,x + b\,\mathrm{sh}\,x$ and $y = c$ and noting the points of intersection, or more accurately,

(b) by adopting the following procedure:

(i) Change $\mathrm{sh}\,x$ to $\left(\dfrac{e^x - e^{-x}}{2}\right)$ and $\mathrm{ch}\,x$ to $\left(\dfrac{e^x + e^{-x}}{2}\right)$

(ii) Rearrange the equation into the form $pe^x + qe^{-x} + r = 0$, where p, q and r are constants.

(iii) Multiply each term by e^x, which produces an equation of the form $p(e^x)^2 + re^x + q = 0$ (since $(e^{-x})(e^x) = e^0 = 1$)

(iv) Solve the quadratic equation $p(e^x)^2 + re^x + q = 0$ for e^x by factorising or by using the quadratic formula.

(v) Given $e^x = a$ constant (obtained by solving the equation in (iv)), take Napierian logarithms of both sides to give $x = \ln(\text{constant})$

This procedure is demonstrated in Problem 22.

Problem 22. Solve the equation $2.6\,\mathrm{ch}\,x + 5.1\,\mathrm{sh}\,x = 8.73$, correct to 4 decimal places.

Following the above procedure:

(i) $2.6\,\mathrm{ch}\,x + 5.1\,\mathrm{sh}\,x = 8.73$

   i.e. $2.6\left(\dfrac{\mathrm{e}^x + \mathrm{e}^{-x}}{2}\right) + 5.1\left(\dfrac{\mathrm{e}^x - \mathrm{e}^{-x}}{2}\right) = 8.73$

(ii) $1.3\mathrm{e}^x + 1.3\mathrm{e}^{-x} + 2.55\mathrm{e}^x - 2.55\mathrm{e}^{-x} = 8.73$

   i.e. $3.85\mathrm{e}^x - 1.25\mathrm{e}^{-x} - 8.73 = 0$

(iii) $3.85(\mathrm{e}^x)^2 - 8.73\mathrm{e}^x - 1.25 = 0$

(iv) e^x

   $= \dfrac{-(-8.73) \pm \sqrt{[(-8.73)^2 - 4(3.85)(-1.25)]}}{2(3.85)}$

   $= \dfrac{8.73 \pm \sqrt{95.463}}{7.70} = \dfrac{8.73 \pm 9.7705}{7.70}$

   Hence $\mathrm{e}^x = 2.4027$ or $\mathrm{e}^x = -0.1351$

(v) $x = \ln 2.4027$ or $x = \ln(-0.1351)$ which has no real solution.
   Hence $x = \mathbf{0.8766}$, correct to 4 decimal places.

Now try the following Practice Exercise

Practice Exercise 74   Hyperbolic equations (Answers on page 866)

In Problems 1 to 8, solve the given equations correct to 4 decimal places.

1. (a) $\sinh x = 1$ (b) $\mathrm{sh}\,A = -2.43$

2. (a) $\cosh B = 1.87$ (b) $2\,\mathrm{ch}\,x = 3$

3. (a) $\tanh y = -0.76$ (b) $3\,\mathrm{th}\,x = 2.4$

4. (a) $\mathrm{sech}\,B = 0.235$ (b) $\mathrm{sech}\,Z = 0.889$

5. (a) $\mathrm{cosech}\,\theta = 1.45$ (b) $5\,\mathrm{cosech}\,x = 4.35$

6. (a) $\coth x = 2.54$ (b) $2\coth y = -3.64$

7. $3.5\,\mathrm{sh}\,x + 2.5\,\mathrm{ch}\,x = 0$

8. $2\,\mathrm{sh}\,x + 3\,\mathrm{ch}\,x = 5$

9. $4\,\mathrm{th}\,x - 1 = 0$

10. A chain hangs so that its shape is of the form $y = 56\cosh\left(\dfrac{x}{56}\right)$. Determine, correct to 4 significant figures, (a) the value of y when x is 35, and (b) the value of x when y is 62.35

16.5 Series expansions for cosh x and sinh x

By definition,

$$\mathrm{e}^x = 1 + x + \frac{x^2}{2!} + \frac{x^3}{3!} + \frac{x^4}{4!} + \frac{x^5}{5!} + \cdots$$

from Chapter 4.
Replacing x by $-x$ gives:

$$\mathrm{e}^{-x} = 1 - x + \frac{x^2}{2!} - \frac{x^3}{3!} + \frac{x^4}{4!} - \frac{x^5}{5!} + \cdots$$

$$\cosh x = \frac{1}{2}(\mathrm{e}^x + \mathrm{e}^{-x})$$

$$= \frac{1}{2}\left[\left(1 + x + \frac{x^2}{2!} + \frac{x^3}{3!} + \frac{x^4}{4!} + \frac{x^5}{5!} + \cdots\right)\right.$$

$$\left. + \left(1 - x + \frac{x^2}{2!} - \frac{x^3}{3!} + \frac{x^4}{4!} - \frac{x^5}{5!} + \cdots\right)\right]$$

$$= \frac{1}{2}\left[\left(2 + \frac{2x^2}{2!} + \frac{2x^4}{4!} + \cdots\right)\right]$$

i.e. $\cosh x = 1 + \dfrac{x^2}{2!} + \dfrac{x^4}{4!} + \cdots$ (which is valid for all values of x). $\cosh x$ is an even function and contains only even powers of x in its expansion.

$$\sinh x = \frac{1}{2}(\mathrm{e}^x - \mathrm{e}^{-x})$$

$$= \frac{1}{2}\left[\left(1 + x + \frac{x^2}{2!} + \frac{x^3}{3!} + \frac{x^4}{4!} + \frac{x^5}{5!} + \cdots\right)\right.$$

$$\left. - \left(1 - x + \frac{x^2}{2!} - \frac{x^3}{3!} + \frac{x^4}{4!} - \frac{x^5}{5!} + \cdots\right)\right]$$

$$= \frac{1}{2}\left[2x + \frac{2x^3}{3!} + \frac{2x^5}{5!} + \cdots\right]$$

i.e. $\sinh x = x + \dfrac{x^3}{3!} + \dfrac{x^5}{5!} + \cdots$ (which is valid for all values of x). $\sinh x$ is an odd function and contains only odd powers of x in its series expansion.

Problem 23. Using the series expansion for $\mathrm{ch}\,x$, evaluate $\mathrm{ch}\,1$ correct to 4 decimal places.

$$\operatorname{ch} x = 1 + \frac{x^2}{2!} + \frac{x^4}{4!} + \cdots \text{ from above}$$

Let $x = 1$,

then $\operatorname{ch} 1 = 1 + \dfrac{1^2}{2 \times 1} + \dfrac{1^4}{4 \times 3 \times 2 \times 1}$

$$+ \frac{1^6}{6 \times 5 \times 4 \times 3 \times 2 \times 1} + \cdots$$

$$= 1 + 0.5 + 0.04167 + 0.001389 + \cdots$$

i.e. **ch 1 = 1.5431**, correct to 4 decimal places, which may be checked by using a calculator.

Problem 24. Determine, correct to 3 decimal places, the value of sh 3 using the series expansion for sh x.

$$\operatorname{sh} x = x + \frac{x^3}{3!} + \frac{x^5}{5!} + \cdots \text{ from above}$$

Let $x = 3$, then

$$\operatorname{sh} 3 = 3 + \frac{3^3}{3!} + \frac{3^5}{5!} + \frac{3^7}{7!} + \frac{3^9}{9!} + \frac{3^{11}}{11!} + \cdots$$

$$= 3 + 4.5 + 2.025 + 0.43393 + 0.05424$$

$$+ 0.00444 + \cdots$$

i.e. **sh 3 = 10.018**, correct to 3 decimal places.

Problem 25. Determine the power series for $2\operatorname{ch}\left(\dfrac{\theta}{2}\right) - \operatorname{sh} 2\theta$ as far as the term in θ^5

In the series expansion for ch x, let $x = \dfrac{\theta}{2}$ then:

$$2\operatorname{ch}\left(\frac{\theta}{2}\right) = 2\left[1 + \frac{(\theta/2)^2}{2!} + \frac{(\theta/2)^4}{4!} + \cdots \right]$$

$$= 2 + \frac{\theta^2}{4} + \frac{\theta^4}{192} + \cdots$$

In the series expansion for sh x, let $x = 2\theta$, then:

$$\operatorname{sh} 2\theta = 2\theta + \frac{(2\theta)^3}{3!} + \frac{(2\theta)^5}{5!} + \cdots$$

$$= 2\theta + \frac{4}{3}\theta^3 + \frac{4}{15}\theta^5 + \cdots$$

Hence

$$\operatorname{ch}\left(\frac{\theta}{2}\right) - \operatorname{sh} 2\theta = \left(2 + \frac{\theta^2}{4} + \frac{\theta^4}{192} + \cdots \right)$$

$$- \left(2\theta + \frac{4}{3}\theta^3 + \frac{4}{15}\theta^5 + \cdots \right)$$

$$= 2 - 2\theta + \frac{\theta^2}{4} - \frac{4}{3}\theta^3 + \frac{\theta^4}{192}$$

$$- \frac{4}{15}\theta^5 + \cdots \text{ as far the term in } \theta^5$$

Now try the following Practice Exercise

Practice Exercise 75   Series expansions for cosh x and sinh x (Answers on page 866)

1. Use the series expansion for ch x to evaluate, correct to 4 decimal places: (a) ch 1.5 (b) ch 0.8

2. Use the series expansion for sh x to evaluate, correct to 4 decimal places: (a) sh 0.5 (b) sh 2

3. Expand the following as a power series as far as the term in x^5: (a) sh 3x (b) ch 2x

In Problems 4 and 5, prove the given identities, the series being taken as far as the term in θ^5 only.

4. $\operatorname{sh} 2\theta - \operatorname{sh}\theta \equiv \theta + \dfrac{7}{6}\theta^3 + \dfrac{31}{120}\theta^5$

5. $2\operatorname{sh}\dfrac{\theta}{2} - \operatorname{ch}\dfrac{\theta}{2} \equiv -1 + \theta - \dfrac{\theta^2}{8} + \dfrac{\theta^3}{24} - \dfrac{\theta^4}{384}$
$$+ \frac{\theta^5}{1920}$$

For fully worked solutions to each of the problems in Practice Exercises 72 to 75 in this chapter,
go to the website:
www.routledge.com/cw/bird

Chapter 17

Trigonometric identities and equations

> **Why it is important to understand: Trigonometric identities and equations**
>
> In engineering, trigonometric identities occur often, examples being in the more advanced areas of calculus to generate derivatives and integrals, with tensors/vectors, and with differential equations and partial differential equations. One of the skills required for more advanced work in mathematics, especially in calculus, is the ability to use identities to write expressions in alternative forms. In software engineering, working, say, on the next big blockbuster film, trigonometric identities are needed for computer graphics; an RF engineer working on the next-generation mobile phone will also need trigonometric identities. In addition, identities are needed in electrical engineering when dealing with a.c. power, and wave addition/subtraction and the solutions of trigonometric equations often require knowledge of trigonometric identities.

At the end of this chapter, you should be able to:

- state simple trigonometric identities
- prove simple identities
- solve equations of the form $b \sin A + c = 0$
- solve equations of the form $a \sin^2 A + c = 0$
- solve equations of the form $a \sin^2 A + b \sin A + c = 0$
- solve equations requiring trigonometric identities

17.1 Trigonometric identities

A **trigonometric identity** is a relationship that is true for all values of the unknown variable.

$$\tan \theta = \frac{\sin \theta}{\cos \theta}, \cot \theta = \frac{\cos \theta}{\sin \theta}, \sec \theta = \frac{1}{\cos \theta}$$

$$\operatorname{cosec} \theta = \frac{1}{\sin \theta} \text{ and } \cot \theta = \frac{1}{\tan \theta}$$

are examples of trigonometric identities from Chapter 12.

Applying Pythagoras' theorem to the right-angled triangle shown in Fig. 17.1 gives:

$$a^2 + b^2 = c^2 \qquad (1)$$

Dividing each term of equation (1) by c^2 gives:

$$\frac{a^2}{c^2} + \frac{b^2}{c^2} = \frac{c^2}{c^2}$$

i.e. $\left(\dfrac{a}{c}\right)^2 + \left(\dfrac{b}{c}\right)^2 = 1$

$(\cos\theta)^2 + (\sin\theta)^2 = 1$

Hence $\mathbf{cos^2\,\theta + sin^2\,\theta = 1}$ (2)

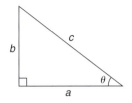

Figure 17.1

Dividing each term of equation (1) by a^2 gives:

$\dfrac{a^2}{a^2} + \dfrac{b^2}{a^2} = \dfrac{c^2}{a^2}$

i.e. $1 + \left(\dfrac{b}{a}\right)^2 = \left(\dfrac{c}{a}\right)^2$

Hence $\mathbf{1 + tan^2\,\theta = sec^2\,\theta}$ (3)

Dividing each term of equation (1) by b^2 gives:

$\dfrac{a^2}{b^2} + \dfrac{b^2}{b^2} = \dfrac{c^2}{b^2}$

i.e. $\left(\dfrac{a}{b}\right)^2 + 1 = \left(\dfrac{c}{b}\right)^2$

Hence $\mathbf{cot^2\,\theta + 1 = cosec^2\,\theta}$ (4)

Equations (2), (3) and (4) are three further examples of trigonometric identities. For the proof of further trigonometric identities, see Section 17.2.

17.2 Worked problems on trigonometric identities

Problem 1.   Prove the identity
$\sin^2\theta\cot\theta\sec\theta = \sin\theta$

With trigonometric identities it is necessary to start with the left-hand side (LHS) and attempt to make it equal to the right-hand side (RHS) or vice-versa. It is often useful to change all of the trigonometric ratios into sines and cosines where possible. Thus,

$\text{LHS} = \sin^2\theta\cot\theta\sec\theta$

$= \sin^2\theta\left(\dfrac{\cos\theta}{\sin\theta}\right)\left(\dfrac{1}{\cos\theta}\right)$

$= \sin\theta\,(\text{by cancelling}) = \text{RHS}$

Problem 2.   Prove that

$$\dfrac{\tan x + \sec x}{\sec x\left(1 + \dfrac{\tan x}{\sec x}\right)} = 1$$

$\text{LHS} = \dfrac{\tan x + \sec x}{\sec x\left(1 + \dfrac{\tan x}{\sec x}\right)}$

$= \dfrac{\dfrac{\sin x}{\cos x} + \dfrac{1}{\cos x}}{\left(\dfrac{1}{\cos x}\right)\left(1 + \dfrac{\dfrac{\sin x}{\cos x}}{\dfrac{1}{\cos x}}\right)}$

$= \dfrac{\dfrac{\sin x + 1}{\cos x}}{\left(\dfrac{1}{\cos x}\right)\left[1 + \left(\dfrac{\sin x}{\cos x}\right)\left(\dfrac{\cos x}{1}\right)\right]}$

$= \dfrac{\dfrac{\sin x + 1}{\cos x}}{\left(\dfrac{1}{\cos x}\right)[1 + \sin x]}$

$= \left(\dfrac{\sin x + 1}{\cos x}\right)\left(\dfrac{\cos x}{1 + \sin x}\right)$

$= 1\,(\text{by cancelling}) = \text{RHS}$

Problem 3.   Prove that $\dfrac{1 + \cot\theta}{1 + \tan\theta} = \cot\theta$

$\text{LHS} = \dfrac{1 + \cot\theta}{1 + \tan\theta}$

$= \dfrac{1 + \dfrac{\cos\theta}{\sin\theta}}{1 + \dfrac{\sin\theta}{\cos\theta}} = \dfrac{\dfrac{\sin\theta + \cos\theta}{\sin\theta}}{\dfrac{\cos\theta + \sin\theta}{\cos\theta}}$

$$= \left(\frac{\sin \theta + \cos \theta}{\sin \theta} \right) \left(\frac{\cos \theta}{\cos \theta + \sin \theta} \right)$$

$$= \frac{\cos \theta}{\sin \theta} = \cot \theta = \text{RHS}$$

Problem 4. Show that
$$\cos^2 \theta - \sin^2 \theta = 1 - 2 \sin^2 \theta$$

From equation (2), $\cos^2 \theta + \sin^2 \theta = 1$, from which, $\cos^2 \theta = 1 - \sin^2 \theta$

Hence, LHS

$$= \cos^2 \theta - \sin^2 \theta = (1 - \sin^2 \theta) - \sin^2 \theta$$

$$= 1 - \sin^2 \theta - \sin^2 \theta = 1 - 2 \sin^2 \theta = \text{RHS}$$

Problem 5. Prove that
$$\sqrt{\left(\frac{1 - \sin x}{1 + \sin x} \right)} = \sec x - \tan x$$

$$\text{LHS} = \sqrt{\left(\frac{1 - \sin x}{1 + \sin x} \right)} = \sqrt{\left\{ \frac{(1 - \sin x)(1 - \sin x)}{(1 + \sin x)(1 - \sin x)} \right\}}$$

$$= \sqrt{\left\{ \frac{(1 - \sin x)^2}{(1 - \sin^2 x)} \right\}}$$

Since $\cos^2 x + \sin^2 x = 1$ then $1 - \sin^2 x = \cos^2 x$

$$\text{LHS} = \sqrt{\left\{ \frac{(1 - \sin x)^2}{(1 - \sin^2 x)} \right\}} = \sqrt{\left\{ \frac{(1 - \sin x)^2}{\cos^2 x} \right\}}$$

$$= \frac{1 - \sin x}{\cos x} = \frac{1}{\cos x} - \frac{\sin x}{\cos x}$$

$$= \sec x - \tan x = \text{RHS}$$

Now try the following Practice Exercise

Practice Exercise 76   Trigonometric identities (Answers on page 866)

In Problems 1 to 6 prove the trigonometric identities.

1. $\sin x \cot x = \cos x$

2. $\dfrac{1}{\sqrt{(1 - \cos^2 \theta)}} = \text{cosec} \, \theta$

3. $2 \cos^2 A - 1 = \cos^2 A - \sin^2 A$

4. $\dfrac{\cos x - \cos^3 x}{\sin x} = \sin x \cos x$

5. $(1 + \cot \theta)^2 + (1 - \cot \theta)^2 = 2 \, \text{cosec}^2 \theta$

6. $\dfrac{\sin^2 x (\sec x + \text{cosec} \, x)}{\cos x \tan x} = 1 + \tan x$

17.3   Trigonometric equations

Equations which contain trigonometric ratios are called **trigonometric equations**. There are usually an infinite number of solutions to such equations; however, solutions are often restricted to those between $0°$ and $360°$.

A knowledge of angles of any magnitude is essential in the solution of trigonometric equations and calculators cannot be relied upon to give all the solutions (as shown in Chapter 15). Fig. 17.2 shows a summary for angles of any magnitude.

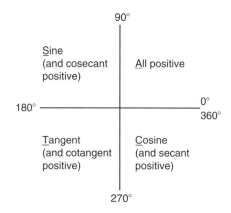

Figure 17.2

Equations of the type $a \sin^2 A + b \sin A + c = 0$

(i) **When $a = 0$,** $b \sin A + c = 0$, hence
$$\sin A = -\frac{c}{b} \quad \text{and} \quad A = \sin^{-1} \left(-\frac{c}{b} \right)$$
There are two values of A between $0°$ and $360°$ which satisfy such an equation, provided $-1 \leq \dfrac{c}{b} \leq 1$ (see Problems 6 to 9).

(ii) **When $b = 0$,** $a \sin^2 A + c = 0$, hence
$$\sin^2 A = -\frac{c}{a}, \ \sin A = \sqrt{\left(-\frac{c}{a} \right)}$$
$$\text{and} \ A = \sin^{-1} \sqrt{\left(-\frac{c}{a} \right)}$$

If either a or c is a negative number, then the value within the square root sign is positive. Since when a square root is taken there is a positive and negative answer there are four values of A between $0°$ and $360°$ which satisfy such an equation, provided $-1 \leq \dfrac{c}{a} \leq 1$ (see Problems 10 and 11).

(iii) **When a, b and c are all non-zero:**
$a\sin^2 A + b\sin A + c = 0$ is a quadratic equation in which the unknown is $\sin A$. The solution of a quadratic equation is obtained either by factorising (if possible) or by using the quadratic formula:

$$\sin A = \frac{-b \pm \sqrt{(b^2 - 4ac)}}{2a}$$

(see Problems 12 and 13).

(iv) Often the trigonometric identities
$\cos^2 A + \sin^2 A = 1$, $\quad 1 + \tan^2 A = \sec^2 A$ and $\cot^2 A + 1 = \csc^2 A$ need to be used to reduce equations to one of the above forms (see Problems 14 to 16).

17.4 Worked problems (i) on trigonometric equations

Problem 6. Solve the trigonometric equation $5\sin\theta + 3 = 0$ for values of θ from $0°$ to $360°$

$5\sin\theta + 3 = 0$, from which $\sin\theta = -\frac{3}{5} = -0.6000$

Hence $\theta = \sin^{-1}(-0.6000)$. Sine is negative in the third and fourth quadrants (see Fig. 17.3). The acute angle $\sin^{-1}(0.6000) = 36.87°$ (shown as α in Fig. 17.3(b)). Hence,

$$\theta = 180° + 36.87°, \text{ i.e. } \mathbf{216.87°} \quad \text{or}$$

$$\theta = 360° - 36.87°, \text{ i.e. } \mathbf{323.13°}$$

Problem 7. Solve $1.5\tan x - 1.8 = 0$ for $0° \leq x \leq 360°$

$1.5\tan x - 1.8 = 0$, from which
$\tan x = \dfrac{1.8}{1.5} = 1.2000$
Hence $x = \tan^{-1} 1.2000$

(a)

(b)

Figure 17.3

Tangent is positive in the first and third quadrants (see Fig. 17.4).
The acute angle $\tan^{-1} 1.2000 = 50.19°$. Hence,

$$x = \mathbf{50.19°} \text{ or } 180° + 50.19° = \mathbf{230.19°}$$

(a)

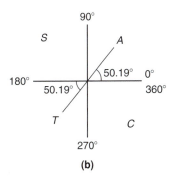

(b)

Figure 17.4

Problem 8. Solve for θ in the range $0° \leq \theta \leq 360°$ for $2\sin\theta = \cos\theta$

Dividing both sides by $\cos\theta$ gives: $\dfrac{2\sin\theta}{\cos\theta} = 1$

From Section 17.1, $\tan\theta = \dfrac{\sin\theta}{\cos\theta}$

hence $2\tan\theta = 1$

Dividing by 2 gives: $\tan\theta = \frac{1}{2}$

from which, $\theta = \tan^{-1}\frac{1}{2}$

Since tangent is positive in the first and third quadrants,

$\theta = \mathbf{26.57°}$ and $\mathbf{206.57°}$

Problem 9. Solve $4\sec t = 5$ for values of t between $0°$ and $360°$

$4\sec t = 5$, from which $\sec t = \frac{5}{4} = 1.2500$

Hence $t = \sec^{-1} 1.2500$

Secant $= (1/\text{cosine})$ is positive in the first and fourth quadrants (see Fig. 17.5) The acute angle $\sec^{-1} 1.2500 = 36.87°$. Hence,

$$t = \mathbf{36.87°} \text{ or } 360° - 36.87° = \mathbf{323.13°}$$

Figure 17.5

___

Now try the following Practice Exercise

Practice Exercise 77   Trigonometric equations (Answers on page 866)

In Problems 1 to 3 solve the equations for angles between $0°$ and $360°$.

1. $4 - 7\sin\theta = 0$

2. $3\operatorname{cosec} A + 5.5 = 0$

3. $4(2.32 - 5.4\cot t) = 0$

In Problems 4 to 6, solve for θ in the range $0° \leq \theta \leq 360°$.

4. $\sec\theta = 2$

5. $\cot\theta = 0.6$

6. $\operatorname{cosec}\theta = 1.5$

In Problems 7 to 9, solve for x in the range $-180° \leq x \leq 180°$.

7. $\sec x = -1.5$

8. $\cot x = 1.2$

9. $\operatorname{cosec} x = -2$

In Problem 10 and 11, solve for θ in the range $0° \leq \theta \leq 360°$.

10. $3\sin\theta = 2\cos\theta$

11. $5\cos\theta = -\sin\theta$

17.5 Worked problems (ii) on trigonometric equations

Problem 10. Solve $2 - 4\cos^2 A = 0$ for values of A in the range $0° < A < 360°$.

$2 - 4\cos^2 A = 0$, from which $\cos^2 A = \frac{2}{4} = 0.5000$

Hence $\cos A = \sqrt{(0.5000)} = \pm 0.7071$ and

$A = \cos^{-1}(\pm 0.7071)$

Cosine is positive in quadrants one and four and negative in quadrants two and three. Thus in this case there are four solutions, one in each quadrant (see Fig. 17.6). The acute angle $\cos^{-1} 0.7071 = 45°$.

Hence, $\mathbf{A = 45°, 135°, 225°}$ or $\mathbf{315°}$

Problem 11. Solve $\frac{1}{2}\cot^2 y = 1.3$ for $0° < y < 360°$

$\frac{1}{2}\cot^2 y = 1.3$, from which, $\cot^2 y = 2(1.3) = 2.6$

Hence $\cot y = \sqrt{2.6} = \pm 1.6125$, and $y = \cot^{-1}(\pm 1.6125)$. There are four solutions, one in each quadrant. The acute angle $\cot^{-1} 1.6125 = 31.81°$

Hence $y = \mathbf{31.81°, 148.19°, 211.81°}$ or $\mathbf{328.19°}$

Figure 17.6

Now try the following Practice Exercise

Practice Exercise 78   Trigonometric equations (Answers on page 866)

In Problems 1 to 5 solve the equations for angles between $0°$ and $360°$.

1. $5\sin^2 y = 3$

2. $\cos^2 \theta = 0.25$

3. $\tan^2 x = 3$

4. $5 + 3\operatorname{cosec}^2 D = 8$

5. $2\cot^2 \theta = 5$

17.6 Worked problems (iii) on trigonometric equations

Problem 12.   Solve the equation
$$8\sin^2 \theta + 2\sin\theta - 1 = 0,$$
for all values of θ between $0°$ and $360°$.

Factorising $8\sin^2 \theta + 2\sin\theta - 1 = 0$ gives
$(4\sin\theta - 1)(2\sin\theta + 1) = 0$
Hence $4\sin\theta - 1 = 0$, from which, $\sin\theta = \frac{1}{4} = 0.2500$
or $2\sin\theta + 1 = 0$, from which, $\sin\theta = -\frac{1}{2} = -0.5000$
(Instead of factorising, the quadratic formula can, of course, be used.)
$\theta = \sin^{-1} 0.2500 = 14.48°$ or $165.52°$, since sine is positive in the first and second quadrants, or $\theta = \sin^{-1}(-0.5000) = 210°$ or $330°$, since sine is negative in the third and fourth quadrants.

Hence $\boldsymbol{\theta = 14.48°, 165.52°, 210°}$ **or** $\boldsymbol{330°}$

Problem 13.   Solve $6\cos^2 \theta + 5\cos\theta - 6 = 0$ for values of θ from $0°$ to $360°$

Factorising $6\cos^2 \theta + 5\cos\theta - 6 = 0$ gives
$(3\cos\theta - 2)(2\cos\theta + 3) = 0$
Hence $3\cos\theta - 2 = 0$, from which, $\cos\theta = \frac{2}{3} = 0.6667$
or $2\cos\theta + 3 = 0$, from which, $\cos\theta = -\frac{3}{2} = -1.5000$
The minimum value of a cosine is -1, hence the latter expression has no solution and is thus neglected. Hence,

$$\theta = \cos^{-1} 0.6667 = \boldsymbol{48.18°}\ \textbf{or}\ \boldsymbol{311.82°}$$

since cosine is positive in the first and fourth quadrants.

Now try the following Practice Exercise

Practice Exercise 79   Trigonometric equations (Answers on page 866)

In Problems 1 to 4 solve the equations for angles between $0°$ and $360°$.

1. $15\sin^2 A + \sin A - 2 = 0$

2. $8\tan^2 \theta + 2\tan\theta = 15$

3. $2\operatorname{cosec}^2 t - 5\operatorname{cosec} t = 12$

4. $2\cos^2 \theta + 9\cos\theta - 5 = 0$

17.7 Worked problems (iv) on trigonometric equations

Problem 14.   Solve $5\cos^2 t + 3\sin t - 3 = 0$ for values of t from $0°$ to $360°$.

Since $\cos^2 t + \sin^2 t = 1$, $\cos^2 t = 1 - \sin^2 t$. Substituting for $\cos^2 t$ in $5\cos^2 t + 3\sin t - 3 = 0$ gives:

$$5(1 - \sin^2 t) + 3\sin t - 3 = 0$$

$$5 - 5\sin^2 t + 3\sin t - 3 = 0$$

$$-5\sin^2 t + 3\sin t + 2 = 0$$

$$5\sin^2 t - 3\sin t - 2 = 0$$

Factorising gives $(5\sin t + 2)(\sin t - 1) = 0$. Hence $5\sin t + 2 = 0$, from which, $\sin t = -\frac{2}{5} = -0.4000$, or $\sin t - 1 = 0$, from which, $\sin t = 1$
$t = \sin^{-1}(-0.4000) = 203.58°$ or $336.42°$, since sine is negative in the third and fourth quadrants, or $t = \sin^{-1} 1 = 90°$. Hence $t = 90°, 203.58°$ **or** $336.42°$ as shown in Fig. 17.7.

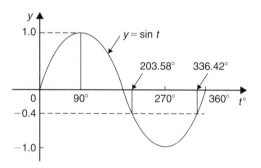

Figure 17.7

Problem 15. Solve $18\sec^2 A - 3\tan A = 21$ for values of A between $0°$ and $360°$.

$1 + \tan^2 A = \sec^2 A$. Substituting for $\sec^2 A$ in $18\sec^2 A - 3\tan A = 21$ gives
$18(1 + \tan^2 A) - 3\tan A = 21$,

i.e. $18 + 18\tan^2 A - 3\tan A - 21 = 0$

$$18\tan^2 A - 3\tan A - 3 = 0$$

Factorising gives $(6\tan A - 3)(3\tan A + 1) = 0$

Hence $6\tan A - 3 = 0$, from which, $\tan A = \frac{3}{6} = 0.5000$ or $3\tan A + 1 = 0$, from which, $\tan A = -\frac{1}{3} = -0.3333$
Thus $A = \tan^{-1}(0.5000) = 26.57°$ or $206.57°$, since tangent is positive in the first and third quadrants, or

$A = \tan^{-1}(-0.3333) = 161.57°$ or $341.57°$, since tangent is negative in the second and fourth quadrants.

Hence, $A = \mathbf{26.57°, 161.57°, 206.57°}$ **or** $\mathbf{341.57°}$

Problem 16. Solve $3\csc^2 \theta - 5 = 4\cot \theta$ in the range $0 < \theta < 360°$.

$\cot^2 \theta + 1 = \csc^2 \theta$. Substituting for $\csc^2 \theta$ in $3\csc^2 \theta - 5 = 4\cot \theta$ gives:

$$3(\cot^2 \theta + 1) - 5 = 4\cot \theta$$

$$3\cot^2 \theta + 3 - 5 = 4\cot \theta$$

$$3\cot^2 \theta - 4\cot \theta - 2 = 0$$

Since the LHS does not factorise the quadratic formula is used. Thus,

$$\cot \theta = \frac{-(-4) \pm \sqrt{[(-4)^2 - 4(3)(-2)]}}{2(3)}$$

$$= \frac{4 \pm \sqrt{(16 + 24)}}{6} = \frac{4 \pm \sqrt{40}}{6}$$

$$= \frac{10.3246}{6} \text{ or } -\frac{2.3246}{6}$$

Hence $\cot \theta = 1.7208$ or -0.3874, $\theta = \cot^{-1} 1.7208 = 30.17°$ or $210.17°$, since cotangent is positive in the first and third quadrants, or $\theta = \cot^{-1}(-0.3874) = 111.18°$ or $291.18°$, since cotangent is negative in the second and fourth quadrants.

Hence, $\theta = \mathbf{30.17°, 111.18°, 210.17°}$ **or** $\mathbf{291.18°}$

Now try the following Practice Exercise

Practice Exercise 80   **Trigonometric equations (Answers on page 866)**

In Problems 1 to 12 solve the equations for angles between $0°$ and $360°$.

1. $2\cos^2 \theta + \sin \theta = 1$

2. $4\cos^2 t + 5\sin t = 3$

3. $2\cos\theta - 4\sin^2\theta = 0$

4. $3\cos\theta + 2\sin^2\theta = 3$

5. $12\sin^2\theta - 6 = \cos\theta$

6. $16\sec x - 2 = 14\tan^2 x$

7. $4\cot^2 A - 6\csc A + 6 = 0$

8. $5\sec t + 2\tan^2 t = 3$

9. $2.9\cos^2 a - 7\sin a + 1 = 0$

10. $3\csc^2\beta = 8 - 7\cot\beta$

11. $\cot\theta = \sin\theta$

12. $\tan\theta + 3\cot\theta = 5\sec\theta$

For fully worked solutions to each of the problems in Practice Exercises 76 to 80 in this chapter,
go to the website:
www.routledge.com/cw/bird

Section B

Chapter 18

The relationship between trigonometric and hyperbolic functions

Why it is important to understand: **The relationship between trigonometric and hyperbolic functions**

There are similarities between notations used for hyperbolic and trigonometric functions. Both trigonometric and hyperbolic functions have applications in many areas of engineering. For example, the shape of a chain hanging under gravity (known as a catenary curve) is described by the hyperbolic cosine, $\cosh x$, and the deformation of uniform beams can be expressed in terms of hyperbolic tangents. Hyperbolic functions are also used in electrical engineering applications such as transmission line theory. Einstein's special theory of relativity used hyperbolic functions and both trigonometric and hyperbolic functions are needed for certain areas of more advanced integral calculus. There are many identities showing relationships between hyperbolic and trigonometric functions; these may be used to evaluate trigonometric and hyperbolic functions of complex numbers.

At the end of this chapter, you should be able to:

- identify the relationship between trigonometric and hyperbolic functions
- state hyperbolic identities

18.1 The relationship between trigonometric and hyperbolic functions

In Chapter 23 it is shown that

$$\cos\theta + j\sin\theta = e^{j\theta} \qquad (1)$$

and $\quad \cos\theta - j\sin\theta = e^{-j\theta} \qquad (2)$

Adding equations (1) and (2) gives:

$$\cos\theta = \frac{1}{2}(e^{j\theta} + e^{-j\theta}) \qquad (3)$$

Subtracting equation (2) from equation (1) gives:

$$\sin\theta = \frac{1}{2j}(e^{j\theta} - e^{-j\theta}) \qquad (4)$$

Substituting $j\theta$ for θ in equations (3) and (4) gives:

$$\cos j\theta = \frac{1}{2}(e^{j(j\theta)} + e^{-j(j\theta)})$$

and $\sin j\theta = \frac{1}{2j}(e^{j(j\theta)} - e^{-j(j\theta)})$

Since $j^2 = -1$, $\cos j\theta = \frac{1}{2}(e^{-\theta} + e^{\theta}) = \frac{1}{2}(e^{\theta} + e^{-\theta})$

Hence from Chapter 16, $\quad \cos j\theta = \cosh\theta \qquad (5)$

Similarly, $\sin j\theta = \frac{1}{2j}(e^{-\theta} - e^{\theta}) = -\frac{1}{2j}(e^{\theta} - e^{-\theta})$

$$= \frac{-1}{j}\left[\frac{1}{2}(e^{\theta} - e^{-\theta})\right]$$

$$= -\frac{1}{j}\sinh\theta \quad \text{(see Chapter 16)}$$

But $\quad -\frac{1}{j} = -\frac{1}{j} \times \frac{j}{j} = -\frac{j}{j^2} = j,$

hence $\quad \sin j\theta = j\sinh\theta \qquad (6)$

Equations (5) and (6) may be used to verify that in all standard trigonometric identities, $j\theta$ may be written for θ and the identity still remains true.

> **Problem 1.** Verify that $\cos^2 j\theta + \sin^2 j\theta = 1$

From equation (5), $\cos j\theta = \cosh\theta$, and from equation (6), $\sin j\theta = j\sinh\theta$

Thus, $\cos^2 j\theta + \sin^2 j\theta = \cosh^2\theta + j^2\sinh^2\theta$, and since $j^2 = -1$,

$$\cos^2 j\theta + \sin^2 j\theta = \cosh^2\theta - \sinh^2\theta$$

But from Chapter 16, Problem 6,

$$\cosh^2\theta - \sinh^2\theta = 1,$$

hence $\quad \cos^2 j\theta + \sin^2 j\theta = 1$

> **Problem 2.** Verify that $\sin j2A = 2\sin jA\cos jA$

From equation (6), writing $2A$ for θ, $\sin j2A = j\sinh 2A$, and from Chapter 16, Table 16.1, page 185, $\sinh 2A = 2\sinh A\cosh A$

Hence, $\quad \sin j2A = j(2\sinh A\cosh A)$

But, $\sinh A = \frac{1}{2}(e^A - e^{-A})$ and $\cosh A = \frac{1}{2}(e^A + e^{-A})$

Hence, $\sin j2A = j2\left(\dfrac{e^A - e^{-A}}{2}\right)\left(\dfrac{e^A + e^{-A}}{2}\right)$

$$= -\frac{2}{j}\left(\frac{e^A - e^{-A}}{2}\right)\left(\frac{e^A + e^{-A}}{2}\right)$$

$$= -\frac{2}{j}\left(\frac{\sin j\theta}{j}\right)(\cos j\theta)$$

$$= 2\sin jA\cos jA \quad \text{since } j^2 = -1$$

i.e. $\quad \sin j2A = 2\sin jA\cos jA$

Now try the following Practice Exercise

> **Practice Exercise 81 The relationship between trigonometric and hyperbolic functions (Answers on page 866)**
>
> Verify the following identities by expressing in exponential form.
>
> 1. $\sin j(A+B) = \sin jA\cos jB + \cos jA\sin jB$
> 2. $\cos j(A-B) = \cos jA\cos jB + \sin jA\sin jB$
> 3. $\cos j2A = 1 - 2\sin^2 jA$
> 4. $\sin jA\cos jB = \frac{1}{2}[\sin j(A+B) + \sin j(A-B)]$
> 5. $\sin jA - \sin jB$
> $$= 2\cos j\left(\frac{A+B}{2}\right)\sin j\left(\frac{A-B}{2}\right)$$

18.2 Hyperbolic identities

From Chapter 16, $\cosh\theta = \frac{1}{2}(e^{\theta} + e^{-\theta})$

Substituting $j\theta$ for θ gives:

$\cosh j\theta = \frac{1}{2}(e^{j\theta} + e^{-j\theta}) = \cos\theta$, from equation (3),

i.e. $\quad \cosh j\theta = \cos\theta \qquad (7)$

Similarly, from Chapter 16,

$$\sinh\theta = \frac{1}{2}(e^{\theta} - e^{-\theta})$$

Substituting $j\theta$ for θ gives:

$\sinh j\theta = \frac{1}{2}(e^{j\theta} - e^{-j\theta}) = j\sin\theta$, from equation (4).

Hence $\sinh j\theta = j\sin\theta$ (8)

$$\tan j\theta = \frac{\sin j\theta}{\cosh j\theta}$$

From equations (5) and (6),

$$\frac{\sin j\theta}{\cos j\theta} = \frac{j\sinh\theta}{\cosh\theta} = j\tanh\theta$$

Hence $\tan j\theta = j\tanh\theta$ (9)

Similarly, $\tanh j\theta = \dfrac{\sinh j\theta}{\cosh j\theta}$

From equations (7) and (8),

$$\frac{\sinh j\theta}{\cosh j\theta} = \frac{j\sin\theta}{\cos\theta} = j\tan\theta$$

Hence $\tanh j\theta = j\tan\theta$ (10)

Two methods are commonly used to verify hyperbolic identities. These are (a) by substituting $j\theta$ (and $j\phi$) in the corresponding trigonometric identity and using the relationships given in equations (5) to (10) (see Problems 3 to 5) and (b) by applying Osborne's rule given in Chapter 16, page 184.

Problem 3.   By writing jA for θ in $\cot^2\theta + 1 = \mathrm{cosec}^2\theta$, determine the corresponding hyperbolic identity.

Substituting jA for θ gives:

$$\cot^2 jA + 1 = \mathrm{cosec}^2 jA,$$

i.e. $\dfrac{\cos^2 jA}{\sin^2 jA} + 1 = \dfrac{1}{\sin^2 jA}$

But from equation (5), $\cos jA = \cosh A$

and from equation (6), $\sin jA = j\sinh A$

Hence $\dfrac{\cosh^2 A}{j^2\sinh^2 A} + 1 = \dfrac{1}{j^2\sinh^2 A}$

and since $j^2 = -1$, $-\dfrac{\cosh^2 A}{\sinh^2 A} + 1 = -\dfrac{1}{\sinh^2 A}$

Multiplying throughout by -1, gives:

$$\frac{\cosh^2 A}{\sinh^2 A} - 1 = \frac{1}{\sinh^2 A}$$

i.e. $\coth^2 A - 1 = \mathrm{cosech}^2 A$

Problem 4.   By substituting jA and jB for θ and ϕ, respectively, in the trigonometric identity for $\cos\theta - \cos\phi$, show that

$$\cosh A - \cosh B = 2\sinh\left(\frac{A+B}{2}\right)\sinh\left(\frac{A-B}{2}\right)$$

$$\cos\theta - \cos\phi = -2\sin\left(\frac{\theta+\phi}{2}\right)\sin\left(\frac{\theta-\phi}{2}\right)$$

(see Chapter 19, page 211)

thus $\cos jA - \cos jB$

$$= -2\sin j\left(\frac{A+B}{2}\right)\sin j\left(\frac{A-B}{2}\right)$$

But from equation (5), $\cos jA = \cosh A$

and from equation (6), $\sin jA = j\sinh A$

Hence, $\cosh A - \cosh B$

$$= -2\,j\sinh\left(\frac{A+B}{2}\right) j\sinh\left(\frac{A-B}{2}\right)$$

$$= -2\,j^2\sinh\left(\frac{A+B}{2}\right)\sinh\left(\frac{A-B}{2}\right)$$

But $j^2 = -1$, hence

$$\cosh A - \cosh B = 2\sinh\left(\frac{A+B}{2}\right)\sinh\left(\frac{A-B}{2}\right)$$

Problem 5.   Develop the hyperbolic identity corresponding to $\sin 3\theta = 3\sin\theta - 4\sin^3\theta$ by writing jA for θ

Substituting jA for θ gives:

$$\sin 3\,jA = 3\sin jA - 4\sin^3 jA$$

and since from equation (6),

$$\sin jA = j\sinh A,$$

$$j\sinh 3A = 3\,j\sinh A - 4\,j^3\sinh^3 A$$

Dividing throughout by j gives:

$$\sinh 3A = 3\sinh A - j^2 4\sinh^3 A$$

But $j^2 = -1$, hence

$$\sinh 3A = 3\sinh A + 4\sinh^3 A$$

[An examination of Problems 3 to 5 shows that whenever the trigonometric identity contains a term which is the product of two sines, or the implied product of two sines (e.g. $\tan^2\theta = \sin^2\theta/\cos^2\theta$, thus $\tan^2\theta$ is the implied product of two sines), the sign of the corresponding term in the hyperbolic function changes. This relationship between trigonometric and hyperbolic functions is known as Osborne's rule, as discussed in Chapter 16, page 184].

Now try the following Practice Exercise

Practice Exercise 82   Hyperbolic identities (Answers on page 866)

In Problems 1 to 9, use the substitution $A = j\theta$ (and $B = j\phi$) to obtain the hyperbolic identities corresponding to the trigonometric identities given.

1. $1 + \tan^2 A = \sec^2 A$

2. $\cos(A+B) = \cos A \cos B - \sin A \sin B$

3. $\sin(A-B) = \sin A \cos B - \cos A \sin B$

4. $\tan 2A = \dfrac{2\tan A}{1 - \tan^2 A}$

5. $\cos A \sin B = \dfrac{1}{2}[\sin(A+B) - \sin(A-B)]$

6. $\sin^3 A = \dfrac{3}{4}\sin A - \dfrac{1}{4}\sin 3A$

7. $\cot^2 A(\sec^2 A - 1) = 1$

For fully worked solutions to each of the problems in Practice Exercises 81 and 82 in this chapter, go to the website:
www.routledge.com/cw/bird

Chapter 19

Compound angles

Why it is important to understand: **Compound angles**

It is often necessary to rewrite expressions involving sines, cosines and tangents in alternative forms. To do this formulae known as trigonometric identities are used as explained previously. Compound angle (or sum and difference) formulae, and double angles are further commonly used identities. Compound angles are required, for example, in the analysis of acoustics (where a beat is an interference between two sounds of slightly different frequencies), and with phase detectors (which is a frequency mixer, analogue multiplier, or logic circuit that generates a voltage signal which represents the difference in phase between two signal inputs). Many rational functions of sine and cosine are difficult to integrate without compound angle formulae.

At the end of this chapter, you should be able to:

- state compound angle formulae for $\sin(A \pm B)$, $\cos(A \pm B)$ and $\tan(A \pm B)$
- convert $a \sin \omega t + b \cos \omega t$ into $R \sin(\omega t + \alpha)$
- derive double angle formulae
- change products of sines and cosines into sums or differences
- change sums or differences of sines and cosines into products
- develop expressions for power in a.c. circuits – purely resistive, inductive and capacitive circuits, R–L and R–C circuits

19.1 Compound angle formulae

An electric current i may be expressed as $i = 5\sin(\omega t - 0.33)$ amperes. Similarly, the displacement x of a body from a fixed point can be expressed as $x = 10\sin(2t + 0.67)$ metres. The angles $(\omega t - 0.33)$ and $(2t + 0.67)$ are called **compound angles** because they are the sum or difference of two angles. The **compound angle formulae** for sines and cosines of the sum and difference of two angles A and B are:

$$\sin(A + B) = \sin A \cos B + \cos A \sin B$$
$$\sin(A - B) = \sin A \cos B - \cos A \sin B$$
$$\cos(A + B) = \cos A \cos B - \sin A \sin B$$
$$\cos(A - B) = \cos A \cos B + \sin A \sin B$$

(Note, $\sin(A + B)$ is **not** equal to $(\sin A + \sin B)$, and so on.)

The formulae stated above may be used to derive two further compound angle formulae:

$$\tan(A + B) = \frac{\tan A + \tan B}{1 - \tan A \tan B}$$

$$\tan(A - B) = \frac{\tan A - \tan B}{1 + \tan A \tan B}$$

The compound angle formulae are true for all values of A and B, and by substituting values of A and B into the formulae they may be shown to be true.

Problem 1. Expand and simplify the following expressions:
(a) $\sin(\pi + \alpha)$ (b) $-\cos(90° + \beta)$
(c) $\sin(A - B) - \sin(A + B)$

(a) $\sin(\pi + \alpha) = \sin\pi \cos\alpha + \cos\pi \sin\alpha$ (from

the formula for $\sin(A + B)$)

$$= (0)(\cos\alpha) + (-1)\sin\alpha = -\sin\alpha$$

(b) $-\cos(90° + \beta)$

$$= -[\cos 90° \cos\beta - \sin 90° \sin\beta]$$

$$= -[(0)(\cos\beta) - (1)\sin\beta] = \sin\beta$$

(c) $\sin(A - B) - \sin(A + B)$

$$= [\sin A \cos B - \cos A \sin B]$$

$$- [\sin A \cos B + \cos A \sin B]$$

$$= -2\cos A \sin B$$

Problem 2. Prove that

$$\cos(y - \pi) + \sin\left(y + \frac{\pi}{2}\right) = 0$$

$$\cos(y - \pi) = \cos y \cos\pi + \sin y \sin\pi$$

$$= (\cos y)(-1) + (\sin y)(0)$$

$$= -\cos y$$

$$\sin\left(y + \frac{\pi}{2}\right) = \sin y \cos\frac{\pi}{2} + \cos y \sin\frac{\pi}{2}$$

$$= (\sin y)(0) + (\cos y)(1) = \cos y$$

Hence $\cos(y - \pi) + \sin\left(y + \frac{\pi}{2}\right)$

$$= (-\cos y) + (\cos y) = 0$$

Problem 3. Show that

$$\tan\left(x + \frac{\pi}{4}\right)\tan\left(x - \frac{\pi}{4}\right) = -1$$

$$\tan\left(x + \frac{\pi}{4}\right) = \frac{\tan x + \tan\frac{\pi}{4}}{1 - \tan x \tan\frac{\pi}{4}}$$

from the formula for $\tan(A + B)$

$$= \frac{\tan x + 1}{1 - (\tan x)(1)} = \left(\frac{1 + \tan x}{1 - \tan x}\right)$$

since $\tan\frac{\pi}{4} = 1$

$$\tan\left(x - \frac{\pi}{4}\right) = \frac{\tan x - \tan\frac{\pi}{4}}{1 + \tan x \tan\frac{\pi}{4}} = \left(\frac{\tan x - 1}{1 + \tan x}\right)$$

Hence $\tan\left(x + \frac{\pi}{4}\right)\tan\left(x - \frac{\pi}{4}\right)$

$$= \left(\frac{1 + \tan x}{1 - \tan x}\right)\left(\frac{\tan x - 1}{1 + \tan x}\right)$$

$$= \frac{\tan x - 1}{1 - \tan x} = \frac{-(1 - \tan x)}{1 - \tan x} = -1$$

Problem 4. If $\sin P = 0.8142$ and $\cos Q = 0.4432$ evaluate, correct to 3 decimal places:

(a) $\sin(P - Q)$ (b) $\cos(P + Q)$
(c) $\tan(P + Q)$, using the compound angle formulae.

Since $\sin P = 0.8142$ then
$P = \sin^{-1} 0.8142 = 54.51°$

Thus $\cos P = \cos 54.51° = 0.5806$ and
$\tan P = \tan 54.51° = 1.4025$

Since $\cos Q = 0.4432$, $Q = \cos^{-1} 0.4432 = 63.69°$.
Thus $\sin Q = \sin 63.69° = 0.8964$ and
$\tan Q = \tan 63.69° = 2.0225$

(a) $\sin(P - Q)$

$$= \sin P \cos Q - \cos P \sin Q$$

$$= (0.8142)(0.4432) - (0.5806)(0.8964)$$

$$= 0.3609 - 0.5204 = -0.160$$

(b) $\cos(P + Q)$

$$= \cos P \cos Q - \sin P \sin Q$$

$$= (0.5806)(0.4432) - (0.8142)(0.8964)$$

$$= 0.2573 - 0.7298 = -0.473$$

(c) $\tan(P + Q)$

$$= \frac{\tan P + \tan Q}{1 - \tan P \tan Q} = \frac{(1.4025) + (2.0225)}{1 - (1.4025)(2.0225)}$$

$$= \frac{3.4250}{-1.8366} = -1.865$$

Problem 5. Solve the equation

$$4\sin(x - 20°) = 5\cos x$$

for values of x between $0°$ and $90°$

$4\sin(x - 20°) = 4[\sin x \cos 20° - \cos x \sin 20°]$

from the formula for $\sin(A - B)$

$$= 4[\sin x (0.9397) - \cos x (0.3420)]$$

$$= 3.7588\sin x - 1.3680\cos x$$

Since $4\sin(x - 20°) = 5\cos x$ then

$3.7588\sin x - 1.3680\cos x = 5\cos x$

Rearranging gives:

$3.7588\sin x = 5\cos x + 1.3680\cos x$

$$= 6.3680\cos x$$

and $\dfrac{\sin x}{\cos x} = \dfrac{6.3680}{3.7588} = 1.6942$

i.e. $\tan x = 1.6942$, and $x = \tan^{-1} 1.6942 = 59.449°$ or **59°27'**

[Check: LHS $= 4\sin(59.449° - 20°)$

$$= 4\sin 39.449° = 2.542$$

RHS $= 5\cos x = 5\cos 59.449° = 2.542]$

Now try the following Practice Exercise

Practice Exercise 83   Compound angle formulae (Answers on page 867)

1. Reduce the following to the sine of one angle:

   (a) $\sin 37° \cos 21° + \cos 37° \sin 21°$
   (b) $\sin 7t \cos 3t - \cos 7t \sin 3t$

2. Reduce the following to the cosine of one angle:

   (a) $\cos 71° \cos 33° - \sin 71° \sin 33°$
   (b) $\cos\dfrac{\pi}{3}\cos\dfrac{\pi}{4} + \sin\dfrac{\pi}{3}\sin\dfrac{\pi}{4}$

3. Show that:

   (a) $\sin\left(x + \dfrac{\pi}{3}\right) + \sin\left(x + \dfrac{2\pi}{3}\right) = \sqrt{3}\cos x$

   (b) $-\sin\left(\dfrac{3\pi}{2} - \phi\right) = \cos\phi$

4. Prove that:

   (a) $\sin\left(\theta + \dfrac{\pi}{4}\right) - \sin\left(\theta - \dfrac{3\pi}{4}\right)$

   $$= \sqrt{2}(\sin\theta + \cos\theta)$$

   (b) $\dfrac{\cos(270° + \theta)}{\cos(360° - \theta)} = \tan\theta$

5. Given $\cos A = 0.42$ and $\sin B = 0.73$, evaluate (a) $\sin(A - B)$ (b) $\cos(A - B)$ (c) $\tan(A+B)$, correct to 4 decimal places.

In Problems 6 and 7, solve the equations for values of θ between $0°$ and $360°$.

6. $3\sin(\theta + 30°) = 7\cos\theta$

7. $4\sin(\theta - 40°) = 2\sin\theta$

19.2 Conversion of $a\sin\omega t + b\cos\omega t$ into $R\sin(\omega t + \alpha)$

(i) $R\sin(\omega t + \alpha)$ represents a sine wave of maximum value R, periodic time $2\pi/\omega$, frequency $\omega/2\pi$ and leading $R\sin\omega t$ by angle α (see Chapter 15).

(ii) $R\sin(\omega t + \alpha)$ may be expanded using the compound angle formula for $\sin(A + B)$, where $A = \omega t$ and $B = \alpha$. Hence,

$R\sin(\omega t + \alpha)$

$$= R[\sin\omega t \cos\alpha + \cos\omega t \sin\alpha]$$

$$= R\sin\omega t \cos\alpha + R\cos\omega t \sin\alpha$$

$$= (R\cos\alpha)\sin\omega t + (R\sin\alpha)\cos\omega t$$

(iii) If $a = R\cos\alpha$ and $b = R\sin\alpha$, where a and b are constants, then $R\sin(\omega t + \alpha) = a\sin\omega t + b\cos\omega t$, i.e. a sine and cosine function of the same frequency when added produce a sine wave of the same frequency (which is further demonstrated in Chapter 27).

(iv) Since $a = R\cos\alpha$, then $\cos\alpha = a/R$, and since $b = R\sin\alpha$, then $\sin\alpha = b/R$

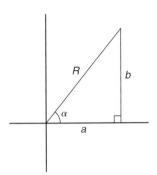

Figure 19.1

If the values of a and b are known then the values of R and α may be calculated. The relationship between constants a, b, R and α are shown in Fig. 19.1.

From Fig. 19.1, by Pythagoras' theorem:

$$R = \sqrt{a^2 + b^2}$$

and from trigonometric ratios:

$$\alpha = \tan^{-1} b/a$$

Problem 6. Find an expression for $3\sin\omega t + 4\cos\omega t$ in the form $R\sin(\omega t + \alpha)$ and sketch graphs of $3\sin\omega t$, $4\cos\omega t$ and $R\sin(\omega t + \alpha)$ on the same axes.

Let $3\sin\omega t + 4\cos\omega t = R\sin(\omega t + \alpha)$

then $3\sin\omega t + 4\cos\omega t$

$$= R[\sin\omega t \cos\alpha + \cos\omega t \sin\alpha]$$

$$= (R\cos\alpha)\sin\omega t + (R\sin\alpha)\cos\omega t$$

Equating coefficients of $\sin\omega t$ gives:

$$3 = R\cos\alpha, \text{ from which, } \cos\alpha = \frac{3}{R}$$

Equating coefficients of $\cos\omega t$ gives:

$$4 = R\sin\alpha, \text{ from which, } \sin\alpha = \frac{4}{R}$$

There is only one quadrant where both $\sin\alpha$ **and** $\cos\alpha$ are positive, and this is the first, as shown in Fig. 19.2. From Fig. 19.2, by Pythagoras' theorem:

$$R = \sqrt{(3^2 + 4^2)} = 5$$

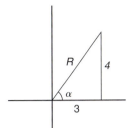

Figure 19.2

From trigonometric ratios: $\alpha = \tan^{-1}\frac{4}{3} = 53.13°$ or 0.927 radians.
Hence $3\sin\omega t + 4\cos\omega t = 5\sin(\omega t + 0.927)$

A sketch of $3\sin\omega t$, $4\cos\omega t$ and $5\sin(\omega t + 0.927)$ is shown in Fig. 19.3 on page 206.

Two periodic functions of the same frequency may be combined by

(a) plotting the functions graphically and combining ordinates at intervals, or

(b) resolution of phasors by drawing or calculation.

Problem 6, together with Problems 7 and 8 following, demonstrate a third method of combining waveforms.

Problem 7. Express $4.6\sin\omega t - 7.3\cos\omega t$ in the form $R\sin(\omega t + \alpha)$

Let $4.6\sin\omega t - 7.3\cos\omega t = R\sin(\omega t + \alpha)$

then $4.6\sin\omega t - 7.3\cos\omega t$

$$= R[\sin\omega t \cos\alpha + \cos\omega t \sin\alpha]$$

$$= (R\cos\alpha)\sin\omega t + (R\sin\alpha)\cos\omega t$$

Equating coefficients of $\sin\omega t$ gives:

$$4.6 = R\cos\alpha, \text{ from which, } \cos\alpha = \frac{4.6}{R}$$

Equating coefficients of $\cos\omega t$ gives:

$$-7.3 = R\sin\alpha, \text{ from which, } \sin\alpha = \frac{-7.3}{R}$$

There is only one quadrant where cosine is positive **and** sine is negative, i.e. the fourth quadrant, as shown in Fig. 19.4. By Pythagoras' theorem:

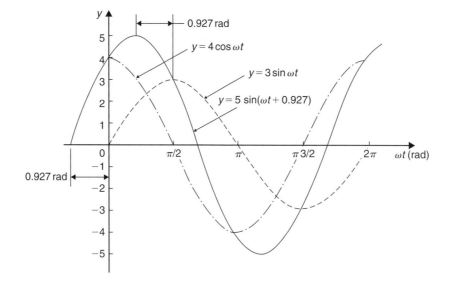

Figure 19.3

$$R = \sqrt{[(4.6)^2 + (-7.3)^2]} = 8.628$$

By trigonometric ratios:

$$\alpha = \tan^{-1}\left(\frac{-7.3}{4.6}\right)$$

$$= -57.78° \text{ or } -1.008 \text{ radians.}$$

Hence

$$\mathbf{4.6 \sin \omega t - 7.3 \cos \omega t = 8.628 \sin(\omega t - 1.008)}$$

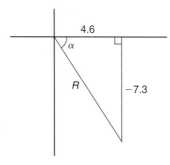

Figure 19.4

Problem 8. Express $-2.7 \sin \omega t - 4.1 \cos \omega t$ in the form $R \sin(\omega t + \alpha)$

Let $-2.7 \sin \omega t - 4.1 \cos \omega t = R \sin(\omega t + \alpha)$

$$= R[\sin \omega t \cos \alpha + \cos \omega t \sin \alpha]$$

$$= (R \cos \alpha) \sin \omega t + (R \sin \alpha) \cos \omega t$$

Equating coefficients gives:

$$-2.7 = R \cos \alpha, \text{ from which, } \cos \alpha = \frac{-2.7}{R}$$

and $\quad -4.1 = R \sin \alpha, \text{ from which, } \sin \alpha = \dfrac{-4.1}{R}$

There is only one quadrant in which both cosine **and** sine are negative, i.e. the third quadrant, as shown in Fig. 19.5. From Fig. 19.5,

$$R = \sqrt{[(-2.7)^2 + (-4.1)^2]} = 4.909$$

and $\quad \theta = \tan^{-1}\dfrac{4.1}{2.7} = 56.63°$

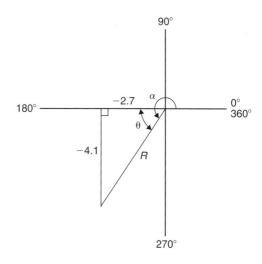

Figure 19.5

Hence $\alpha = 180° + 56.63° = 236.63°$ or 4.130 radians.
Thus,

$$-2.7\sin\omega t - 4.1\cos\omega t = 4.909\sin(\omega t + 4.130)$$

An angle of $236.63°$ is the same as $-123.37°$ or -2.153 radians. Hence $-2.7\sin\omega t - 4.1\cos\omega t$ may be expressed also as **$4.909\sin(\omega t - 2.153)$**, which is preferred since it is the **principal value** (i.e. $-\pi \le \alpha \le \pi$)

Problem 9. Express $3\sin\theta + 5\cos\theta$ in the form $R\sin(\theta + \alpha)$, and hence solve the equation $3\sin\theta + 5\cos\theta = 4$, for values of θ between $0°$ and $360°$

Let $3\sin\theta + 5\cos\theta = R\sin(\theta + \alpha)$

$$= R[\sin\theta\cos\alpha + \cos\theta\sin\alpha]$$

$$= (R\cos\alpha)\sin\theta + (R\sin\alpha)\cos\theta$$

Equating coefficients gives:

$$3 = R\cos\alpha, \text{ from which, } \cos\alpha = \frac{3}{R}$$

and $5 = R\sin\alpha$, from which, $\sin\alpha = \dfrac{5}{R}$

Since both $\sin\alpha$ and $\cos\alpha$ are positive, R lies in the first quadrant, as shown in Fig. 19.6.

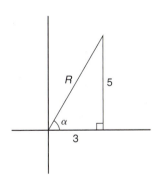

Figure 19.6

From Fig. 19.6, $R = \sqrt{(3^2 + 5^2)} = 5.831$ and
$\alpha = \tan^{-1}\frac{5}{3} = 59.03°$
Hence $3\sin\theta + 5\cos\theta = 5.831\sin(\theta + 59.03°)$

However $\qquad 3\sin\theta + 5\cos\theta = 4$

Thus $5.831\sin(\theta + 59.03°) = 4$, from which

$$(\theta + 59.03°) = \sin^{-1}\left(\frac{4}{5.831}\right)$$

i.e. $\qquad \theta + 59.03° = 43.32°$ or $136.68°$

Hence $\theta = 43.32° - 59.03° = -15.71°$

or $\qquad \theta = 136.68° - 59.03° = 77.65°$

Since $-15.71°$ is the same as $-15.71° + 360°$, i.e. $344.29°$, then the solutions are $\theta = 77.65°$ or $344.29°$, which may be checked by substituting into the original equation.

Problem 10. Solve the equation $3.5\cos A - 5.8\sin A = 6.5$ for $0° \le A \le 360°$

Let $3.5\cos A - 5.8\sin A = R\sin(A + \alpha)$

$$= R[\sin A\cos\alpha + \cos A\sin\alpha]$$

$$= (R\cos\alpha)\sin A + (R\sin\alpha)\cos A$$

Equating coefficients gives:

$$3.5 = R\sin\alpha, \text{ from which, } \sin\alpha = \frac{3.5}{R}$$

and $-5.8 = R\cos\alpha$, from which, $\cos\alpha = \dfrac{-5.8}{R}$

There is only one quadrant in which both sine is positive **and** cosine is negative, i.e. the second, as shown in Fig. 19.7.

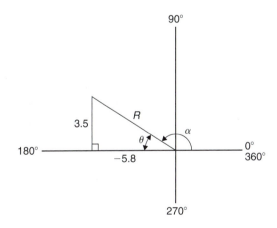

Figure 19.7

From Fig. 19.7, $R = \sqrt{[(3.5)^2 + (-5.8)^2]} = 6.774$ and
$\theta = \tan^{-1}\dfrac{3.5}{5.8} = 31.12°$
Hence $\alpha = 180° - 31.12° = 148.88°$
Thus

$$3.5\cos A - 5.8\sin A = 6.774\sin(A + 144.88°) = 6.5$$

Hence $\sin(A + 148.88°) = \dfrac{6.5}{6.774}$, from which,

$$(A + 148.88°) = \sin^{-1}\dfrac{6.5}{6.774}$$

$$= 73.65° \text{ or } 106.35°$$

Thus $A = 73.65° - 148.88° = -75.23°$

$$\equiv (-75.23° + 360°) = 284.77°$$

or $A = 106.35° - 148.88° = -42.53°$

$$\equiv (-42.53° + 360°) = 317.47°$$

The solutions are thus $A = \mathbf{284.77°}$ **or** $\mathbf{317.47°}$, which may be checked in the original equation.

Now try the following Practice Exercise

Practice Exercise 84   The conversion of
$a\,\sin\omega t + b\cos\omega t$ **into** $R\sin(\omega t \pm \alpha)$
(Answers on page 867)

In Problems 1 to 4, change the functions into the form $R\sin(\omega t \pm \alpha)$.

1. $5\sin\omega t + 8\cos\omega t$

2. $4\sin\omega t - 3\cos\omega t$

3. $-7\sin\omega t + 4\cos\omega t$

4. $-3\sin\omega t - 6\cos\omega t$

5. Solve the following equations for values of θ between $0°$ and $360°$: (a) $2\sin\theta + 4\cos\theta = 3$ (b) $12\sin\theta - 9\cos\theta = 7$

6. Solve the following equations for $0° < A < 360°$:   (a) $3\cos A + 2\sin A = 2.8$ (b) $12\cos A - 4\sin A = 11$

7. Solve the following equations for values of θ between $0°$ and $360°$: (a) $3\sin\theta + 4\cos\theta = 3$ (b) $2\cos\theta + \sin\theta = 2$

8. Solve the following equations for values of θ between $0°$ and $360°$: (a) $6\cos\theta + \sin\theta = \sqrt{3}$ (b) $2\sin 3\theta + 8\cos 3\theta = 1$

9. The third harmonic of a wave motion is given by $4.3\cos 3\theta - 6.9\sin 3\theta$. Express this in the form $R\sin(3\theta \pm \alpha)$

10. The displacement x metres of a mass from a fixed point about which it is oscillating is given by $x = 2.4\sin\omega t + 3.2\cos\omega t$, where t is the time in seconds. Express x in the form $R\sin(\omega t + \alpha)$.

11. Two voltages, $v_1 = 5\cos\omega t$ and $v_2 = -8\sin\omega t$ are inputs to an analogue circuit. Determine an expression for the output voltage if this is given by $(v_1 + v_2)$.

12. The motion of a piston moving in a cylinder can be described by:
$x = (5\cos 2t + 5\sin 2t)$cm.
Express x in the form $R\,\sin(\omega t + \alpha)$

19.3   Double angles

(i) If, in the compound angle formula for $\sin(A + B)$, we let $B = A$ then

$$\mathbf{\sin 2A = 2\sin A \cos A}$$

Also, for example,

$$\sin 4A = 2\sin 2A \cos 2A$$

and $\sin 8A = 2\sin 4A \cos 4A$, and so on.

(ii) If, in the compound angle formula for $\cos(A + B)$, we let $B = A$ then

$$\mathbf{\cos 2A = \cos^2 A - \sin^2 A}$$

Since $\cos^2 A + \sin^2 A = 1$, then
$\cos^2 A = 1 - \sin^2 A$, and $\sin^2 A = 1 - \cos^2 A$, and two further formula for $\cos 2A$ can be produced.

Thus $\cos 2A = \cos^2 A - \sin^2 A$

$$= (1 - \sin^2 A) - \sin^2 A$$

i.e. $\mathbf{\cos 2A = 1 - 2\sin^2 A}$

and $\cos 2A = \cos^2 A - \sin^2 A$

$$= \cos^2 A - (1 - \cos^2 A)$$

i.e. $\mathbf{\cos 2A = 2\cos^2 A - 1}$

Also, for example,

$$\cos 4A = \cos^2 2A - \sin^2 2A \quad \text{or}$$

$$1 - 2\sin^2 2A \quad \text{or}$$

$$2\cos^2 2A - 1$$

and $\quad \cos 6A = \cos^2 3A - \sin^2 3A \quad \text{or}$

$$1 - 2\sin^2 3A \quad \text{or}$$

$$2\cos^2 3A - 1,$$

and so on.

(iii) If, in the compound angle formula for $\tan(A+B)$, we let $B = A$ then

$$\tan 2A = \frac{2\tan A}{1 - \tan^2 A}$$

Also, for example,

$$\tan 4A = \frac{2\tan 2A}{1 - \tan^2 2A}$$

and $\quad \tan 5A = \dfrac{2\tan \frac{5}{2}A}{1 - \tan^2 \frac{5}{2}A}$ and so on.

Problem 11. $I_3 \sin 3\theta$ is the third harmonic of a waveform. Express the third harmonic in terms of the first harmonic $\sin\theta$, when $I_3 = 1$

When $I_3 = 1$,

$$I_3 \sin 3\theta = \sin 3\theta = \sin(2\theta + \theta)$$

$$= \sin 2\theta \cos\theta + \cos 2\theta \sin\theta,$$

$$\text{from the } \sin(A+B) \text{ formula}$$

$$= (2\sin\theta \cos\theta)\cos\theta + (1 - 2\sin^2\theta)\sin\theta,$$

$$\text{from the double angle expansions}$$

$$= 2\sin\theta \cos^2\theta + \sin\theta - 2\sin^3\theta$$

$$= 2\sin\theta(1 - \sin^2\theta) + \sin\theta - 2\sin^3\theta,$$

$$(\text{since } \cos^2\theta = 1 - \sin^2\theta)$$

$$= 2\sin\theta - 2\sin^3\theta + \sin\theta - 2\sin^3\theta$$

i.e. $\sin 3\theta = 3\sin\theta - 4\sin^3\theta$

Problem 12. Prove that $\dfrac{1 - \cos 2\theta}{\sin 2\theta} = \tan\theta$

$$\text{LHS} = \frac{1 - \cos 2\theta}{\sin 2\theta} = \frac{1 - (1 - 2\sin^2\theta)}{2\sin\theta \cos\theta}$$

$$= \frac{2\sin^2\theta}{2\sin\theta \cos\theta} = \frac{\sin\theta}{\cos\theta}$$

$$= \tan\theta = \text{RHS}$$

Problem 13. Prove that

$$\cot 2x + \operatorname{cosec} 2x = \cot x$$

$$\text{LHS} = \cot 2x + \operatorname{cosec} 2x = \frac{\cos 2x}{\sin 2x} + \frac{1}{\sin 2x}$$

$$= \frac{\cos 2x + 1}{\sin 2x}$$

$$= \frac{(2\cos^2 x - 1) + 1}{\sin 2x}$$

$$= \frac{2\cos^2 x}{\sin 2x} = \frac{2\cos^2 x}{2\sin x \cos x}$$

$$= \frac{\cos x}{\sin x} = \cot x = \text{RHS}$$

Problem 14. Solve the equation $\cos 2\theta + 3\sin\theta = 2$ for θ in the range $0° \leq \theta \leq 360°$

Replacing the double angle term with the relationship $\cos 2\theta = 1 - 2\sin^2\theta$ gives:

$$1 - 2\sin^2\theta + 3\sin\theta = 2$$

Rearranging gives: $\quad\quad -2\sin^2\theta + 3\sin\theta - 1 = 0$
or $\quad\quad\quad\quad\quad\quad\quad 2\sin^2\theta - 3\sin\theta + 1 = 0$
which is a quadratic in $\sin\theta$
Using the quadratic formula or by factorising gives:

$$(2\sin\theta - 1)(\sin\theta - 1) = 0$$

from which, $\quad 2\sin\theta - 1 = 0 \quad \text{or} \quad \sin\theta - 1 = 0$
and $\quad\quad\quad\quad\quad \sin\theta = \frac{1}{2} \quad \text{or} \quad \sin\theta = 1$
from which, $\quad\quad \boldsymbol{\theta = 30° \text{ or } 150° \text{ or } 90°}$

Now try the following Practice Exercise

Practice Exercise 85   Double angles (Answers on page 867)

1. The power p in an electrical circuit is given by $p = \dfrac{v^2}{R}$. Determine the power in terms of V, R and $\cos 2t$ when $v = V\cos t$.

2. Prove the following identities:

(a) $1 - \dfrac{\cos 2\phi}{\cos^2 \phi} = \tan^2 \phi$

(b) $\dfrac{1 + \cos 2t}{\sin^2 t} = 2\cot^2 t$

(c) $\dfrac{(\tan 2x)(1 + \tan x)}{\tan x} = \dfrac{2}{1 - \tan x}$

(d) $2\operatorname{cosec} 2\theta \cos 2\theta = \cot \theta - \tan \theta$

3. If the third harmonic of a waveform is given by $V_3 \cos 3\theta$, express the third harmonic in terms of the first harmonic $\cos \theta$, when $V_3 = 1$

In Problems 4 to 8, solve for θ in the range $-180° \le \theta \le 180°$

4. $\cos 2\theta = \sin \theta$

5. $3\sin 2\theta + 2\cos \theta = 0$

6. $\sin 2\theta + \cos \theta = 0$

7. $\cos 2\theta + 2\sin \theta = -3$

8. $\tan \theta + \cot \theta = 2$

19.4 Changing products of sines and cosines into sums or differences

(i) $\sin(A + B) + \sin(A - B) = 2\sin A \cos B$ (from the formulae in Section 19.1)

i.e. $\mathbf{sin\,A\,cos\,B}$

$$= \tfrac{1}{2}[\sin(A + B) + \sin(A - B)] \qquad (1)$$

(ii) $\sin(A + B) - \sin(A - B) = 2\cos A \sin B$

i.e. $\mathbf{cos\,A\,sin\,B}$

$$= \tfrac{1}{2}[\sin(A + B) - \sin(A - B)] \qquad (2)$$

(iii) $\cos(A + B) + \cos(A - B) = 2\cos A \cos B$

i.e. $\mathbf{cos\,A\,cos\,B}$

$$= \tfrac{1}{2}[\cos(A + B) + \cos(A - B)] \qquad (3)$$

(iv) $\cos(A + B) - \cos(A - B) = -2\sin A \sin B$

i.e. $\mathbf{sin\,A\,sin\,B}$

$$= -\tfrac{1}{2}[\cos(A + B) - \cos(A - B)] \qquad (4)$$

Problem 15. Express $\sin 4x \cos 3x$ as a sum or difference of sines and cosines.

From equation (1),

$\sin 4x \cos 3x = \tfrac{1}{2}[\sin(4x + 3x) + \sin(4x - 3x)]$

$\qquad\qquad = \tfrac{1}{2}(\sin 7x + \sin x)$

Problem 16. Express $2\cos 5\theta \sin 2\theta$ as a sum or difference of sines or cosines.

From equation (2),

$2\cos 5\theta \sin 2\theta = 2\left\{ \dfrac{1}{2}[\sin(5\theta + 2\theta) - \sin(5\theta - 2\theta)]\right\}$

$\qquad\qquad = \mathbf{sin\,7\theta - sin\,3\theta}$

Problem 17. Express $3\cos 4t \cos t$ as a sum or difference of sines or cosines.

From equation (3),

$3\cos 4t \cos t = 3\left\{ \dfrac{1}{2}[\cos(4t + t) + \cos(4t - t)]\right\}$

$\qquad\qquad = \dfrac{3}{2}(\cos 5t + \cos 3t)$

Thus, if the integral $\int 3\cos 4t \cos t\, dt$ was required (for integration see Chapter 31), then

$\displaystyle\int 3\cos 4t \cos t\, dt = \int \dfrac{3}{2}(\cos 5t + \cos 3t)\, dt$

$\qquad\qquad = \dfrac{3}{2}\left[\dfrac{\sin 5t}{5} + \dfrac{\sin 3t}{3}\right] + c$

Problem 18. In an alternating current circuit, voltage $v = 5\sin \omega t$ and current $i = 10\sin(\omega t - \pi/6)$. Find an expression for the instantaneous power p at time t given that $p = vi$, expressing the answer as a sum or difference of sines and cosines.

$p = vi = (5\sin \omega t)[10\sin(\omega t - \pi/6)]$

$\qquad = 50\sin \omega t \sin(\omega t - \pi/6)$

From equation (4),

$$50 \sin \omega t \sin(\omega t - \pi/6)$$

$$= (50)\left[-\tfrac{1}{2}\left\{\cos(\omega t + \omega t - \pi/6)\right.\right.$$

$$\left.\left. - \cos\left[\omega t - (\omega t - \pi/6)\right]\right\}\right]$$

$$= -25\{\cos(2\omega t - \pi/6) - \cos \pi/6\}$$

i.e. instantaneous power,

$$p = 25[\cos \pi/6 - \cos(2\omega t - \pi/6)]$$

Now try the following Practice Exercise

Practice Exercise 86   Changing products of sines and cosines into sums or differences (Answers on page 867)

In Problems 1 to 5, express as sums or differences:

1. $\sin 7t \cos 2t$

2. $\cos 8x \sin 2x$

3. $2 \sin 7t \sin 3t$

4. $4 \cos 3\theta \cos \theta$

5. $3 \sin \dfrac{\pi}{3} \cos \dfrac{\pi}{6}$

6. Determine $\int 2 \sin 3t \cos t \, dt$

7. Evaluate $\displaystyle\int_0^{\frac{\pi}{2}} 4 \cos 5x \cos 2x \, dx$

8. Solve the equation: $2 \sin 2\phi \sin \phi = \cos \phi$ in the range $\phi = 0$ to $\phi = 180°$

19.5 Changing sums or differences of sines and cosines into products

In the compound angle formula let,

$$(A + B) = X$$

and

$$(A - B) = Y$$

Solving the simultaneous equations gives:

$$A = \frac{X+Y}{2} \text{ and } B = \frac{X-Y}{2}$$

Thus $\sin(A+B) + \sin(A-B) = 2 \sin A \cos B$ becomes,

$$\sin X + \sin Y = 2 \sin\left(\frac{X+Y}{2}\right)\cos\left(\frac{X-Y}{2}\right) \quad (5)$$

Similarly,

$$\sin X - \sin Y = 2 \cos\left(\frac{X+Y}{2}\right)\sin\left(\frac{X-Y}{2}\right) \quad (6)$$

$$\cos X + \cos Y = 2 \cos\left(\frac{X+Y}{2}\right)\cos\left(\frac{X-Y}{2}\right) \quad (7)$$

$$\cos X - \cos Y = -2 \sin\left(\frac{X+Y}{2}\right)\sin\left(\frac{X-Y}{2}\right) \quad (8)$$

Problem 19. Express $\sin 5\theta + \sin 3\theta$ as a product.

From equation (5),

$$\sin 5\theta + \sin 3\theta = 2 \sin\left(\frac{5\theta+3\theta}{2}\right)\cos\left(\frac{5\theta-3\theta}{2}\right)$$

$$= 2 \sin 4\theta \cos \theta$$

Problem 20. Express $\sin 7x - \sin x$ as a product.

From equation (6),

$$\sin 7x - \sin x = 2 \cos\left(\frac{7x+x}{2}\right)\sin\left(\frac{7x-x}{2}\right)$$

$$= 2 \cos 4x \sin 3x$$

Problem 21. Express $\cos 2t - \cos 5t$ as a product.

From equation (8),

$$\cos 2t - \cos 5t = -2 \sin\left(\frac{2t+5t}{2}\right)\sin\left(\frac{2t-5t}{2}\right)$$

$$= -2 \sin\frac{7}{2}t \sin\left(-\frac{3}{2}t\right) = 2 \sin\frac{7}{2}t \sin\frac{3}{2}t$$

$$\left(\text{since } \sin\left(-\frac{3}{2}t\right) = -\sin\frac{3}{2}t\right)$$

Problem 22. Show that $\dfrac{\cos 6x + \cos 2x}{\sin 6x + \sin 2x} = \cot 4x$

From equation (7),

$$\cos 6x + \cos 2x = 2\cos 4x \cos 2x$$

From equation (5),

$$\sin 6x + \sin 2x = 2\sin 4x \cos 2x$$

Hence

$$\frac{\cos 6x + \cos 2x}{\sin 6x + \sin 2x} = \frac{2\cos 4x \cos 2x}{2\sin 4x \cos 2x}$$

$$= \frac{\cos 4x}{\sin 4x} = \cot 4x$$

Problem 23. Solve the equation $\cos 4\theta + \cos 2\theta = 0$ for θ in the range $0° \leq \theta \leq 360°$

From equation (7),

$$\cos 4\theta + \cos 2\theta = 2\cos\left(\frac{4\theta + 2\theta}{2}\right)\cos\left(\frac{4\theta - 2\theta}{2}\right)$$

Hence, $$2\cos 3\theta \cos\theta = 0$$

Dividing by 2 gives: $$\cos 3\theta \cos\theta = 0$$

Hence, either $$\cos 3\theta = 0 \quad \text{or} \quad \cos\theta = 0$$

Thus, $$3\theta = \cos^{-1} 0 \quad \text{or} \quad \theta = \cos^{-1} 0$$

from which, $3\theta = 90°$ or $270°$ or $450°$ or $630°$ or $810°$ or $990°$

and $\theta = 30°, 90°, 150°, 210°, 270°$ or $330°$

Now try the following Practice Exercise

Practice Exercise 87   Changing sums or differences of sines and cosines into products (Answers on page 867)

In Problems 1 to 5, express as products:

1. $\sin 3x + \sin x$

2. $\frac{1}{2}(\sin 9\theta - \sin 7\theta)$

3. $\cos 5t + \cos 3t$

4. $\frac{1}{8}(\cos 5t - \cos t)$

5. $\frac{1}{2}\left(\cos\frac{\pi}{3} + \cos\frac{\pi}{4}\right)$

6. Show that:

    (a) $\dfrac{\sin 4x - \sin 2x}{\cos 4x + \cos 2x} = \tan x$

    (b) $\frac{1}{2}\{\sin(5x - \alpha) - \sin(x + \alpha)\}$
    $$= \cos 3x \sin(2x - \alpha)$$

In Problems 7 and 8, solve for θ in the range $0° \leq \theta \leq 180°$

7. $\cos 6\theta + \cos 2\theta = 0$

8. $\sin 3\theta - \sin\theta = 0$

In Problems 9 and 10, solve in the range $0°$ to $360°$

9. $\cos 2x = 2\sin x$

10. $\sin 4t + \sin 2t = 0$

19.6 Power waveforms in a.c. circuits

(a) Purely resistive a.c. circuits

Let a voltage $v = V_m \sin\omega t$ be applied to a circuit comprising resistance only. The resulting current is $i = I_m \sin\omega t$, and the corresponding instantaneous power, p, is given by:

$$p = vi = (V_m \sin\omega t)(I_m \sin\omega t)$$

i.e. $p = V_m I_m \sin^2 \omega t$

From double angle formulae of Section 19.3,

$$\cos 2A = 1 - 2\sin^2 A, \text{ from which,}$$

$$\sin^2 A = \tfrac{1}{2}(1 - \cos 2A) \text{ thus}$$

$$\sin^2 \omega t = \tfrac{1}{2}(1 - \cos 2\omega t)$$

Then power $p = V_m I_m\left[\tfrac{1}{2}(1 - \cos 2\omega t)\right]$

i.e. $$p = \tfrac{1}{2}V_m I_m(1 - \cos 2\omega t)$$

The waveforms of v, i and p are shown in Fig. 19.8. The waveform of power repeats itself after π/ω seconds and hence the power has a frequency twice that of voltage and current. The power is always positive, having a maximum value of $V_m I_m$. The average or mean value of the power is $\frac{1}{2}V_m I_m$

The rms value of voltage $V = 0.707 V_m$, i.e. $V = \dfrac{V_m}{\sqrt{2}}$, from which, $V_m = \sqrt{2}\,V$

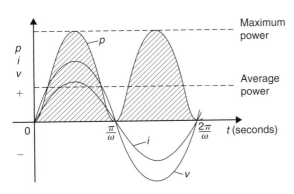

Figure 19.8

Similarly, the rms value of current, $I = \dfrac{I_m}{\sqrt{2}}$, from which, $I_m = \sqrt{2}\,I$. Hence the average power, P, developed in a purely resistive a.c. circuit is given by $P = \frac{1}{2} V_m I_m = \frac{1}{2}(\sqrt{2}V)(\sqrt{2}I) = VI$ watts.

Also, power $P = I^2 R$ or V^2/R as for a d.c. circuit, since $V = IR$.

Summarising, the average power P in a purely resistive a.c. circuit given by

$$P = VI = I^2 R = \frac{V^2}{R}$$

where V and I are rms values.

(b) Purely inductive a.c. circuits

Let a voltage $v = V_m \sin \omega t$ be applied to a circuit containing pure inductance (theoretical case). The resulting current is $i = I_m \sin\left(\omega t - \dfrac{\pi}{2}\right)$ since current lags voltage by $\dfrac{\pi}{2}$ radians or $90°$ in a purely inductive circuit, and the corresponding instantaneous power, p, is given by:

$$p = vi = (V_m \sin \omega t) I_m \sin\left(\omega t - \frac{\pi}{2}\right)$$

i.e. $p = V_m I_m \sin \omega t \sin\left(\omega t - \dfrac{\pi}{2}\right)$

However,

$$\sin\left(\omega t - \frac{\pi}{2}\right) = -\cos \omega t \text{ thus}$$
$$p = -V_m I_m \sin \omega t \cos \omega t$$

Rearranging gives:

$$p = -\tfrac{1}{2} V_m I_m (2 \sin \omega t \cos \omega t)$$

However, from double angle formulae,

$$2 \sin \omega t \cos \omega t = \sin 2\omega t.$$

Thus **power, $p = -\frac{1}{2} V_m I_m \sin 2\omega t$**

The waveforms of v, i and p are shown in Fig. 19.9. The frequency of power is twice that of voltage and current. For the power curve shown in Fig. 19.9, the area above the horizontal axis is equal to the area below, thus over a complete cycle the average power P is zero. It is noted that when v and i are both positive, power p is positive and energy is delivered from the source to the inductance; when v and i have opposite signs, power p is negative and energy is returned from the inductance to the source.

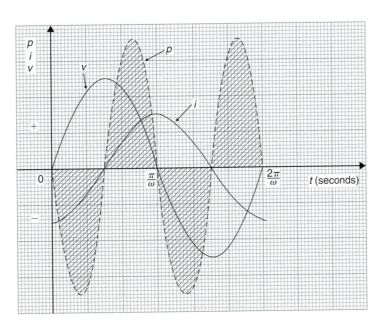

Figure 19.9

Section B

In general, when the current through an inductance is increasing, energy is transferred from the circuit to the magnetic field, but this energy is returned when the current is decreasing.

Summarising, **the average power P in a purely inductive a.c. circuit is zero**.

(c) Purely capacitive a.c. circuits

Let a voltage $v = V_m \sin \omega t$ be applied to a circuit containing pure capacitance. The resulting current is $i = I_m \sin\left(\omega t + \frac{\pi}{2}\right)$, since current leads voltage by $90°$ in a purely capacitive circuit, and the corresponding instantaneous power, p, is given by:

$$p = vi = (V_m \sin \omega t) I_m \sin\left(\omega t + \frac{\pi}{2}\right)$$

i.e. $\quad p = V_m I_m \sin \omega t \sin\left(\omega t + \frac{\pi}{2}\right)$

However, $\quad \sin\left(\omega t + \frac{\pi}{2}\right) = \cos \omega t$

thus $\qquad p = V_m I_m \sin \omega t \cos \omega t$

Rearranging gives

$$p = \tfrac{1}{2} V_m I_m (2 \sin \omega t \cos \omega t)$$

Thus **power**, $p = \tfrac{1}{2} V_m I_m \sin 2\omega t$.

The waveforms of v, i and p are shown in Fig. 19.10. Over a complete cycle the average power P is zero. When the voltage across a capacitor is increasing, energy is transferred from the circuit to the electric field, but this energy is returned when the voltage is decreasing.

Summarising, **the average power P in a purely capacitive a.c. circuit is zero**.

(d) R–L or R–C a.c. circuits

Let a voltage $v = V_m \sin \omega t$ be applied to a circuit containing resistance and inductance or resistance and capacitance. Let the resulting current be $i = I_m \sin(\omega t + \phi)$, where phase angle ϕ will be positive for an R–C circuit and negative for an R–L circuit. The corresponding instantaneous power, p, is given by:

$$p = vi = (V_m \sin \omega t) I_m \sin(\omega t + \phi)$$

i.e. $\quad p = V_m I_m \sin \omega t \sin(\omega t + \phi)$

Products of sine functions may be changed into differences of cosine functions as shown in Section 19.4,

i.e. $\quad \sin A \sin B = -\tfrac{1}{2}[\cos(A+B) - \cos(A-B)]$

Substituting $\omega t = A$ and $(\omega t + \phi) = B$ gives:

power, $\quad p = V_m I_m \{ -\tfrac{1}{2} [\cos(\omega t + \omega t + \phi)$
$$- \cos(\omega t - (\omega t + \phi))] \}$$

i.e. $\qquad p = \tfrac{1}{2} V_m I_m [\cos(-\phi) - \cos(2\omega t + \phi)]$

However, $\cos(-\phi) = \cos \phi$

Thus $p = \tfrac{1}{2} V_m I_m [\cos \phi - \cos(2\omega t + \phi)]$

The instantaneous power p thus consists of

Figure 19.10

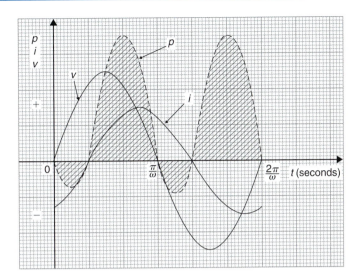

Figure 19.11

(i)   a sinusoidal term, $-\frac{1}{2}V_m I_m \cos(2\omega t + \phi)$ which has a mean value over a cycle of zero, and

(ii)   a constant term, $\frac{1}{2}V_m I_m \cos\phi$ (since ϕ is constant for a particular circuit).

Thus the average value of power $P = \frac{1}{2}V_m I_m \cos\phi$. Since $V_m = \sqrt{2}\,V$ and $I_m = \sqrt{2}\,I$, average power $P = \frac{1}{2}(\sqrt{2}\,V)(\sqrt{2}\,I)\cos\phi$

i.e. $\boldsymbol{P = VI\cos\phi}$

The waveforms of v, i and p, are shown in Fig. 19.11 for an R–L circuit. The waveform of power is seen to pulsate at twice the supply frequency. The areas of the power curve (shown shaded) above the horizontal time axis represent power supplied to the load; the small areas below the axis represent power being returned to the supply from the inductance as the magnetic field collapses.

A similar shape of power curve is obtained for an R–C circuit, the small areas below the horizontal axis representing power being returned to the supply from the charged capacitor. The difference between the areas above and below the horizontal axis represents the heat loss due to the circuit resistance. Since power is dissipated only in a pure resistance, the alternative equations for power, $P = I_R^2 R$, may be used, where I_R is the rms current flowing through the resistance.

Summarising, **the average power P in a circuit containing resistance and inductance and/or capacitance, whether in series or in parallel, is given by $P = VI\cos\phi$ or $P = I_R^2 R$** (V, I and I_R being rms values).

For fully worked solutions to each of the problems in Practice Exercises 83 to 87 in this chapter, go to the website:
www.routledge.com/cw/bird

Revision Test 5 Further trigonometry and hyperbolic functions

This Revision Test covers the material contained in Chapters 15 to 19. *The marks for each question are shown in brackets at the end of each question.*

1. Solve the following equations in the range $0°$ to $360°$.

   (a) $\sin^{-1}(-0.4161) = x$
   (b) $\cot^{-1}(2.4198) = \theta$ (8)

2. Sketch the following curves, labelling relevant points:

   (a) $y = 4\cos(\theta + 45°)$
   (b) $y = 5\sin(2t - 60°)$ (8)

3. The current in an alternating current circuit at any time t seconds is given by:

   $$i = 120\sin(100\pi t + 0.274) \text{ amperes.}$$

   Determine

   (a) the amplitude, periodic time, frequency and phase angle (with reference to $120\sin 100\pi t$)

   (b) the value of current when $t = 0$

   (c) the value of current when $t = 6\,\text{ms}$

   (d) the time when the current first reaches $80\,\text{A}$

   Sketch one cycle of the oscillation. (19)

4. A complex voltage waveform v is comprised of a $141.4\,\text{V rms}$ fundamental voltage at a frequency of $100\,\text{Hz}$, a 35% third harmonic component leading the fundamental voltage at zero time by $\pi/3$ radians, and a 20% fifth harmonic component lagging the fundamental at zero time by $\pi/4$ radians.

   (a) Write down an expression to represent voltage v.

   (b) Draw the complex voltage waveform using harmonic synthesis over one cycle of the fundamental waveform using scales of $12\,\text{cm}$ for the time for one cycle horizontally and $1\,\text{cm} = 20\,\text{V}$ vertically. (15)

5. Evaluate correct to 4 significant figures:

   (a) $\sinh 2.47$ (b) $\tanh 0.6439$
   (c) $\operatorname{sech} 1.385$ (d) $\operatorname{cosech} 0.874$ (6)

6. The increase in resistance of strip conductors due to eddy currents at power frequencies is given by:

   $$\lambda = \frac{\alpha t}{2}\left[\frac{\sinh \alpha t + \sin \alpha t}{\cosh \alpha t - \cos \alpha t}\right]$$

   Calculate λ, correct to 5 significant figures, when $\alpha = 1.08$ and $t = 1$ (5)

7. If $A\operatorname{ch}x - B\operatorname{sh}x \equiv 4e^x - 3e^{-x}$ determine the values of A and B.   (6)

8. Solve the following equation:

   $$3.52\operatorname{ch}x + 8.42\operatorname{sh}x = 5.32$$

   correct to 4 decimal places.   (8)

9. Prove the following identities:

   (a) $\sqrt{\left[\dfrac{1 - \cos^2\theta}{\cos^2\theta}\right]} = \tan\theta$

   (b) $\cos\left(\dfrac{3\pi}{2} + \phi\right) = \sin\phi$

   (c) $\dfrac{\sin^2 x}{1 + \cos 2x} = \tfrac{1}{2}\tan^2 x$ (9)

10. Solve the following trigonometric equations in the range $0° \le x \le 360°$:

    (a) $4\cos x + 1 = 0$

    (b) $3.25\operatorname{cosec}x = 5.25$

    (c) $5\sin^2 x + 3\sin x = 4$

    (d) $2\sec^2\theta + 5\tan\theta = 3$ (18)

11. Solve the equation $5\sin(\theta - \pi/6) = 8\cos\theta$ for values $0 \le \theta \le 2\pi$ (8)

12. Express $5.3\cos t - 7.2\sin t$ in the form $R\sin(t + \alpha)$. Hence solve the equation $5.3\cos t - 7.2\sin t = 4.5$ in the range $0 \le t \le 2\pi$ (12)

13. Determine $\int 2\cos 3t \sin t\, dt$ (3)

For lecturers/instructors/teachers, fully worked solutions to each of the problems in Revision Test 5, together with a full marking scheme, are available at the website:

www.routledge.com/cw/bird

Section C

Graphs

Chapter 20

Functions and their curves

Why it is important to understand: Functions and their curves

Engineers use many basic mathematical functions to represent, say, the input/output of systems – linear, quadratic, exponential, sinusoidal, and so on, and knowledge of these is needed to determine how these are used to generate some of the more unusual input/output signals such as the square wave, saw-tooth wave and fully rectified sine wave. Periodic functions are used throughout engineering and science to describe oscillations, waves and other phenomena that exhibit periodicity. Graphs and diagrams provide a simple and powerful approach to a variety of problems that are typical to computer science in general, and software engineering in particular; graphical transformations have many applications in software engineering problems. Understanding of continuous and discontinuous functions, odd and even functions, and inverse functions are helpful in this – it's all part of the 'language of engineering'.

At the end of this chapter, you should be able to:

- recognise standard curves and their equations – straight line, quadratic, cubic, trigonometric, circle, ellipse, hyperbola, rectangular hyperbola, logarithmic function, exponential function and polar curves
- perform simple graphical transformations
- define a periodic function
- define continuous and discontinuous functions
- define odd and even functions
- define inverse functions
- determine asymptotes
- sketch curves

20.1 Standard curves

When a mathematical equation is known, co-ordinates may be calculated for a limited range of values, and the equation may be represented pictorially as a graph, within this range of calculated values. Sometimes it is useful to show all the characteristic features of an equation, and in this case a sketch depicting the equation can be drawn, in which all the important features are shown, but the accurate plotting of points is less important. This technique is called 'curve sketching' and can involve the use of differential calculus, with, for example, calculations involving turning points.

If, say, y depends on, say, x, then y is said to be a function of x and the relationship is expressed as $y = f(x)$; x is called the independent variable and y is the dependent variable.

In engineering and science, corresponding values are obtained as a result of tests or experiments.

Here is a brief resumé of standard curves, some of which have been met earlier in this text.

(i) Straight line

The general equation of a straight line is $y = mx + c$, where m is the gradient $\left(\text{i.e.} \dfrac{dy}{dx} \right)$ and c is the y-axis intercept.

Two examples are shown in Fig. 20.1

(a)

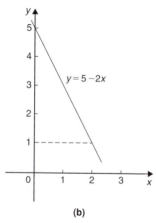

(b)

Figure 20.1

(ii) Quadratic graphs

The general equation of a quadratic graph is $y = ax^2 + bx + c$, and its shape is that of a parabola. The simplest example of a quadratic graph, $y = x^2$, is shown in Fig. 20.2.

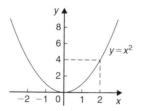

Figure 20.2

(iii) Cubic equations

The general equation of a cubic graph is $y = ax^3 + bx^2 + cx + d$.

The simplest example of a cubic graph, $y = x^3$, is shown in Fig. 20.3.

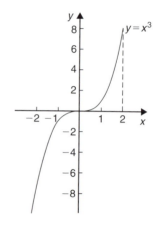

Figure 20.3

(iv) Trigonometric functions (see Chapter 15, page 162)

Graphs of $y = \sin\theta$, $y = \cos\theta$ and $y = \tan\theta$ are shown in Fig. 20.4.

(a)

(b)

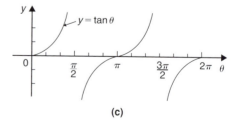

(c)

Figure 20.4

(v) Circle (see Chapter 14, page 155)

The simplest equation of a circle is $x^2 + y^2 = r^2$, with centre at the origin and radius r, as shown in Fig. 20.5.

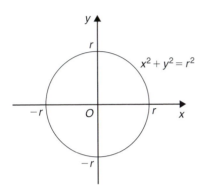

Figure 20.5

More generally, the equation of a circle, centre (a, b), radius r, is given by:

$$(x - a)^2 + (y - b)^2 = r^2$$

Figure 20.6 shows a circle

$$(x - 2)^2 + (y - 3)^2 = 4$$

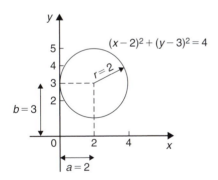

Figure 20.6

(vi) Ellipse

The equation of an ellipse is

$$\frac{x^2}{a^2} + \frac{y^2}{b^2} = 1$$

and the general shape is as shown in Fig. 20.7.
The length AB is called the **major axis** and CD the **minor axis**.
In the above equation, a is the semi-major axis and b is the semi-minor axis.

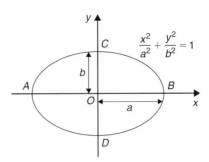

Figure 20.7

(Note that if $b = a$, the equation becomes $\frac{x^2}{a^2} + \frac{y^2}{a^2} = 1$, i.e. $x^2 + y^2 = a^2$, which is a circle of radius a).

(vii) Hyperbola

The equation of a hyperbola is

$$\frac{x^2}{a^2} - \frac{y^2}{b^2} = 1$$

and the general shape is shown in Fig. 20.8. The curve is seen to be symmetrical about both the x- and y-axes. The distance AB in Fig. 20.8 is given by $2a$.

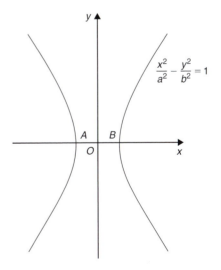

Figure 20.8

(viii) Rectangular hyperbola

The equation of a rectangular hyperbola is $xy = c$ or $y = \dfrac{c}{x}$ and the general shape is shown in Fig. 20.9.

(ix) Logarithmic function (see Chapter 3, page 28)

$y = \ln x$ and $y = \lg x$ are both of the general shape shown in Fig. 20.10.

Section C

Figure 20.9

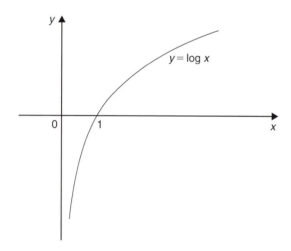

Figure 20.10

(x) Exponential functions (see Chapter 4, page 32)

$y = e^x$ is of the general shape shown in Fig. 20.11.

(xi) Polar curves

The equation of a polar curve is of the form $r = f(\theta)$. An example of a polar curve, $r = a \sin\theta$, is shown in Fig. 20.12.

20.2 Simple transformations

From the graph of $y = f(x)$ it is possible to deduce the graphs of other functions which are transformations of $y = f(x)$. For example, knowing the graph of $y = f(x)$, can help us draw the graphs of $y = af(x)$,

Figure 20.11

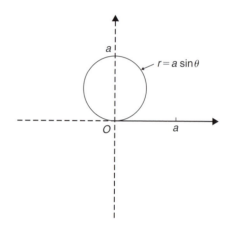

Figure 20.12

$y = f(x) + a$, $y = f(x+a)$, $y = f(ax)$, $y = -f(x)$ and $y = f(-x)$

(i) $y = af(x)$

For each point (x_1, y_1) on the graph of $y = f(x)$ there exists a point (x_1, ay_1) on the graph of $y = af(x)$. Thus the graph of $y = af(x)$ can be obtained by stretching $y = f(x)$ parallel to the y-axis by a scale factor 'a'

Graphs of $y = x + 1$ and $y = 3(x+1)$ are shown in Fig. 20.13(a) and graphs of $y = \sin\theta$ and $y = 2\sin\theta$ are shown in Fig. 20.13(b).

(ii) $y = f(x) + a$

The graph of $y = f(x)$ is translated by a units parallel to the y-axis to obtain $y = f(x) + a$. For example, if $f(x) = x$, $y = f(x) + 3$ becomes $y = x + 3$, as shown in Fig. 20.14(a). Similarly, if $f(\theta) = \cos\theta$, then $y = f(\theta) + 2$ becomes $y = \cos\theta + 2$, as shown in Fig. 20.14(b). Also, if $f(x) = x^2$, then $y = f(x) + 3$ becomes $y = x^2 + 3$, as shown in Fig. 20.14(c).

(a)

(b)

Figure 20.13

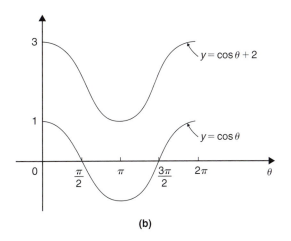

(b)

(iii) $y = f(x+a)$

The graph of $y = f(x)$ is translated by a units parallel to the x-axis to obtain $y = f(x+a)$. If $a > 0$ it moves $y = f(x)$ in the negative direction on the x-axis (i.e. to the left), and if $a < 0$ it moves $y = f(x)$ in the positive direction on the x-axis (i.e. to the right). For example, if $f(x) = \sin x$, $y = f\left(x - \dfrac{\pi}{3}\right)$ becomes $y = \sin\left(x - \dfrac{\pi}{3}\right)$ as shown in Fig. 20.15(a) and $y = \sin\left(x + \dfrac{\pi}{4}\right)$ is shown in Fig. 20.15(b).

Similarly graphs of $y = x^2$, $y = (x-1)^2$ and $y = (x+2)^2$ are shown in Fig. 20.16.

(iv) $y = f(ax)$

For each point (x_1, y_1) on the graph of $y = f(x)$, there exists a point $\left(\dfrac{x_1}{a}, y_1\right)$ on the graph of $y = f(ax)$. Thus the graph of $y = f(ax)$ can be obtained by stretching $y = f(x)$ parallel to the x-axis by a scale factor $\dfrac{1}{a}$

(c)

Figure 20.14

(a)

(b)

Figure 20.15

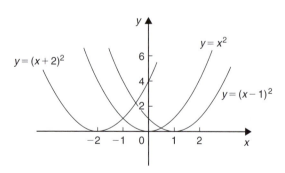

Figure 20.16

For example, if $f(x) = (x-1)^2$, and $a = \dfrac{1}{2}$, then

$$f(ax) = \left(\dfrac{x}{2} - 1\right)^2$$

Both of these curves are shown in Fig. 20.17(a).

Similarly, $y = \cos x$ and $y = \cos 2x$ are shown in Fig. 20.17(b).

(a)

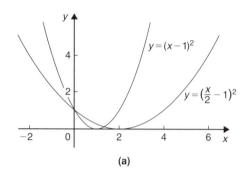

(b)

Figure 20.17

(v) $y = -f(x)$

The graph of $y = -f(x)$ is obtained by reflecting $y = f(x)$ in the x-axis. For example, graphs of $y = e^x$ and $y = -e^x$ are shown in Fig. 20.18(a) and graphs of $y = x^2 + 2$ and $y = -(x^2 + 2)$ are shown in Fig. 20.18(b).

(vi) $y = f(-x)$

The graph of $y = f(-x)$ is obtained by reflecting $y = f(x)$ in the y-axis. For example, graphs of $y = x^3$ and $y = (-x)^3 = -x^3$ are shown in Fig. 20.19(a) and graphs of $y = \ln x$ and $y = -\ln x$ are shown in Fig. 20.19(b).

(a)

Figure 20.18

(b)

Figure 20.18 (*Continued*)

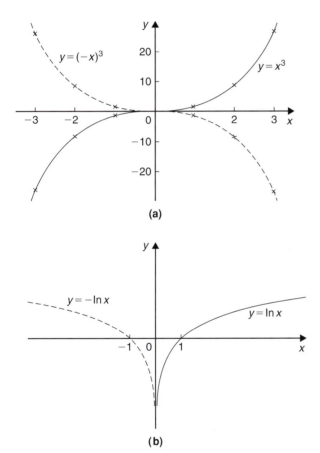

(a)

(b)

Figure 20.19

> **Problem 1.** Sketch the following graphs, showing relevant points:
>
> (a) $y = (x - 4)^2$ (b) $y = x^3 - 8$

(a) In Fig. 20.20 a graph of $y = x^2$ is shown by the broken line. The graph of $y = (x - 4)^2$ is of the form $y = f(x + a)$. Since $a = -4$, then $y = (x - 4)^2$ is translated 4 units to the right of $y = x^2$, parallel to the x-axis.

(See Section (iii) above).

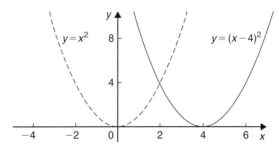

Figure 20.20

(b) In Fig. 20.21 a graph of $y = x^3$ is shown by the broken line. The graph of $y = x^3 - 8$ is of the form $y = f(x) + a$. Since $a = -8$, then $y = x^3 - 8$ is translated 8 units down from $y = x^3$, parallel to the y-axis.

(See Section (ii) above).

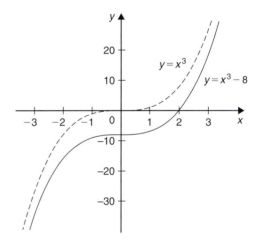

Figure 20.21

> **Problem 2.** Sketch the following graphs, showing relevant points:
>
> (a) $y = 5 - (x + 2)^3$ (b) $y = 1 + 3\sin 2x$

(a) Fig. 20.22(a) shows a graph of $y = x^3$. Fig. 20.22(b) shows a graph of $y = (x + 2)^3$ (see $f(x + a)$, Section (iii) above).

Fig. 20.22(c) shows a graph of $y = -(x + 2)^3$ (see $-f(x)$, Section (v) above). Figure 20.22(d)

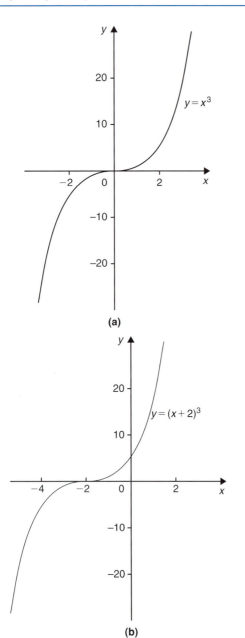

Figure 20.22

shows the graph of $y = 5 - (x+2)^3$ (see $f(x) + a$, Section (ii) above).

(b) Fig. 20.23(a) shows a graph of $y = \sin x$. Fig. 20.23(b) shows a graph of $y = \sin 2x$ (see $f(ax)$, Section (iv) above). Fig. 20.23(c) shows a graph of $y = 3 \sin 2x$ (see $a\,f(x)$, Section (i) above). Fig. 20.23(d) shows a graph of $y = 1 + 3 \sin 2x$ (see $f(x) + a$, Section (ii) above).

Now try the following Practice Exercise

Practice Exercise 88 Simple transformations with curve sketching (Answers on page 867)

Sketch the following graphs, showing relevant points:

1. $y = 3x - 5$

2. $y = -3x + 4$

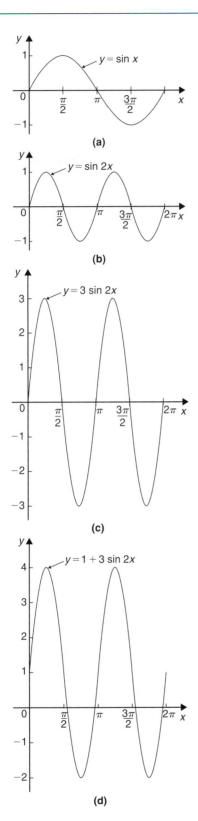

Figure 20.23

3. $y = x^2 + 3$

4. $y = (x - 3)^2$

5. $y = (x - 4)^2 + 2$

6. $y = x - x^2$

7. $y = x^3 + 2$

8. $y = 1 + 2\cos 3x$

9. $y = 3 - 2\sin\left(x + \dfrac{\pi}{4}\right)$

10. $y = 2\ln x$

20.3 Periodic functions

A function $f(x)$ is said to be **periodic** if $f(x+T) = f(x)$ for all values of x, where T is some positive number. T is the interval between two successive repetitions and is called the period of the function $f(x)$. For example, $y = \sin x$ is periodic in x with period 2π since $\sin x = \sin(x + 2\pi) = \sin(x + 4\pi)$, and so on. Similarly, $y = \cos x$ is a periodic function with period 2π since $\cos x = \cos(x + 2\pi) = \cos(x + 4\pi)$, and so on. In general, if $y = \sin \omega t$ or $y = \cos \omega t$ then the period of the waveform is $2\pi/\omega$. The function shown in Fig. 20.24 is also periodic of period 2π and is defined by:

$$f(x) = \begin{cases} -1, & \text{when } -\pi \le x \le 0 \\ 1, & \text{when } 0 \le x \le \pi \end{cases}$$

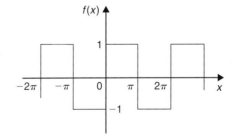

Figure 20.24

20.4 Continuous and discontinuous functions

If a graph of a function has no sudden jumps or breaks it is called a **continuous function**, examples being the graphs of sine and cosine functions. However, other graphs make finite jumps at a point or points in the

Section C

interval. The square wave shown in Fig. 20.24 has **finite discontinuities** as $x = \pi$, 2π, 3π, and so on, and is therefore a discontinuous function. $y = \tan x$ is another example of a discontinuous function.

20.5 Even and odd functions

Even functions

A function $y = f(x)$ is said to be even if $f(-x) = f(x)$ for all values of x. Graphs of even functions are always symmetrical about the y-axis (i.e. is a mirror image). Two examples of even functions are $y = x^2$ and $y = \cos x$ as shown in Fig. 20.25.

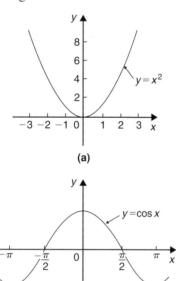

(a)

(b)

Figure 20.25

Odd functions

A function $y = f(x)$ is said to be odd if $f(-x) = -f(x)$ for all values of x. Graphs of odd functions are always symmetrical about the origin. Two examples of odd functions are $y = x^3$ and $y = \sin x$ as shown in Fig. 20.26.

Many functions are neither even nor odd, two such examples being shown in Fig. 20.27.

Problem 3. Sketch the following functions and state whether they are even or odd functions:
(a) $y = \tan x$

(a)

(b)

Figure 20.26

(a)

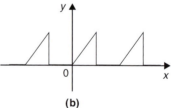

(b)

Figure 20.27

(b) $f(x) = \begin{cases} 2, & \text{when } 0 \leq x \leq \dfrac{\pi}{2} \\ -2, & \text{when } \dfrac{\pi}{2} \leq x \leq \dfrac{3\pi}{2}, \\ 2, & \text{when } \dfrac{3\pi}{2} \leq x \leq 2\pi \end{cases}$
and is periodic of period 2π

(a) A graph of $y = \tan x$ is shown in Fig. 20.28(a) and is symmetrical about the origin and is thus an **odd function** (i.e. $\tan(-x) = -\tan x$).

(b) A graph of $f(x)$ is shown in Fig. 20.28(b) and is symmetrical about the $f(x)$ axis hence the function is an **even** one, ($f(-x) = f(x)$).

(a)

(a)

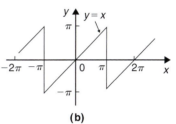

(b)

Figure 20.29

Now try the following Practice Exercise

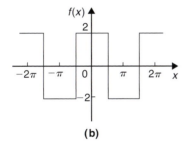

(b)

Figure 20.28

Problem 4. Sketch the following graphs and state whether the functions are even, odd or neither even nor odd:

(a) $y = \ln x$

(b) $f(x) = x$ in the range $-\pi$ to π and is periodic of period 2π.

(a) A graph of $y = \ln x$ is shown in Fig. 20.29(a) and the curve is neither symmetrical about the y-axis nor symmetrical about the origin and is thus **neither even nor odd**.

(b) A graph of $y = x$ in the range $-\pi$ to π is shown in Fig. 20.29(b) and is symmetrical about the origin and is thus an **odd function**.

Practice Exercise 89   Even and odd functions (Answers on page 868)

In Problems 1 and 2 determine whether the given functions are even, odd or neither even nor odd.

1. (a) x^4 (b) $\tan 3x$ (c) $2e^{3t}$ (d) $\sin^2 x$

2. (a) $5t^3$ (b) $e^x + e^{-x}$ (c) $\dfrac{\cos\theta}{\theta}$ (d) e^x

3. State whether the following functions, which are periodic of period 2π, are even or odd:

(a) $f(\theta) = \begin{cases} \theta, & \text{when } -\pi \leq \theta \leq 0 \\ -\theta, & \text{when } 0 \leq \theta \leq \pi \end{cases}$

(b) $f(x) = \begin{cases} x, & \text{when } -\dfrac{\pi}{2} \leq x \leq \dfrac{\pi}{2} \\ 0, & \text{when } \dfrac{\pi}{2} \leq x \leq \dfrac{3\pi}{2} \end{cases}$

20.6   Inverse functions

If y is a function of x, the graph of y against x can be used to find x when any value of y is given. Thus the

Section C

graph also expresses that x is a function of y. Two such functions are called **inverse functions**.

In general, given a function $y = f(x)$, its inverse may be obtained by interchanging the roles of x and y and then transposing for y. The inverse function is denoted by $y = f^{-1}(x)$.

For example, if $y = 2x + 1$, the inverse is obtained by

(i) transposing for x, i.e. $x = \dfrac{y-1}{2} = \dfrac{y}{2} - \dfrac{1}{2}$ and

(ii) interchanging x and y, giving the inverse as $y = \dfrac{x}{2} - \dfrac{1}{2}$

Thus if $f(x) = 2x + 1$, then $f^{-1}(x) = \dfrac{x}{2} - \dfrac{1}{2}$

A graph of $f(x) = 2x + 1$ and its inverse $f^{-1}(x) = \dfrac{x}{2} - \dfrac{1}{2}$ is shown in Fig. 20.30 and $f^{-1}(x)$ is seen to be a reflection of $f(x)$ in the line $y = x$

Similarly, if $y = x^2$, the inverse is obtained by

(i) transposing for x, i.e. $x = \pm\sqrt{y}$ and

(ii) interchanging x and y, giving the inverse $y = \pm\sqrt{x}$

Hence the inverse has two values for every value of x. Thus $f(x) = x^2$ does not have a single inverse. In such a case the domain of the original function may be restricted to $y = x^2$ for $x > 0$. Thus the inverse is then $y = +\sqrt{x}$. A graph of $f(x) = x^2$ and its inverse $f^{-1}(x) = \sqrt{x}$ for $x > 0$ is shown in Fig. 20.31 and, again, $f^{-1}(x)$ is seen to be a reflection of $f(x)$ in the line $y = x$.

It is noted from the latter example, that not all functions have an inverse. An inverse, however, can be determined if the range is restricted.

Figure 20.30

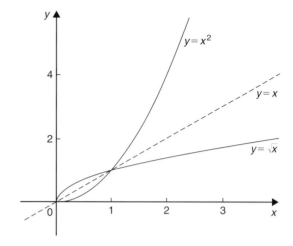

Figure 20.31

Problem 5. Determine the inverse for each of the following functions:

(a) $f(x) = x - 1$ (b) $f(x) = x^2 - 4$ $(x > 0)$

(c) $f(x) = x^2 + 1$

(a) If $y = f(x)$, then $y = x - 1$
Transposing for x gives $x = y + 1$
Interchanging x and y gives $y = x + 1$
Hence if $f(x) = x - 1$, then $\boldsymbol{f^{-1}(x) = x + 1}$

(b) If $y = f(x)$, then $y = x^2 - 4$ $(x > 0)$
Transposing for x gives $x = \sqrt{y + 4}$
Interchanging x and y gives $y = \sqrt{x + 4}$
Hence if $f(x) = x^2 - 4$ $(x > 0)$ then
$\boldsymbol{f^{-1}(x) = \sqrt{x + 4}}$ if $x > -4$

(c) If $y = f(x)$, then $y = x^2 + 1$
Transposing for x gives $x = \sqrt{y - 1}$
Interchanging x and y gives $y = \sqrt{x - 1}$, which has two values.
Hence there is no inverse of $f(x) = x^2 + 1$, since the domain of $f(x)$ is not restricted.

Inverse trigonometric functions

If $y = \sin x$, then x is the angle whose sine is y. Inverse trigonometrical functions are denoted by prefixing the function with 'arc' or, more commonly, $^{-1}$. Hence transposing $y = \sin x$ for x gives $x = \sin^{-1} y$. Interchanging x and y gives the inverse $y = \sin^{-1} x$.

Similarly, $y = \cos^{-1} x$, $y = \tan^{-1} x$, $y = \sec^{-1} x$, $y = \mathrm{cosec}^{-1} x$ and $y = \cot^{-1} x$ are all inverse trigonometric functions. The angle is always expressed in radians.

Section C

Inverse trigonometric functions are periodic so it is necessary to specify the smallest or principal value of the angle. For $\sin^{-1}x$, $\tan^{-1}x$, $\operatorname{cosec}^{-1}x$ and $\cot^{-1}x$, the principal value is in the range $-\dfrac{\pi}{2}<y<\dfrac{\pi}{2}$. For $\cos^{-1}x$ and $\sec^{-1}x$ the principal value is in the range $0<y<\pi$. Graphs of the six inverse trigonometric functions are shown in Fig. 38.1, page 436.

Problem 6. Determine the principal values of

(a) arcsin 0.5 (b) arctan(−1)

(c) $\arccos\left(-\dfrac{\sqrt{3}}{2}\right)$ (d) $\operatorname{arccosec}(\sqrt{2})$

Using a calculator,

(a) $\arcsin 0.5 \equiv \sin^{-1}0.5 = 30°$

$$= \frac{\pi}{6}\text{ rad or }\mathbf{0.5236\,rad}$$

(b) $\arctan(-1) \equiv \tan^{-1}(-1) = -45°$

$$= -\frac{\pi}{4}\text{ rad or }\mathbf{-0.7854\,rad}$$

(c) $\arccos\left(-\dfrac{\sqrt{3}}{2}\right) \equiv \cos^{-1}\left(-\dfrac{\sqrt{3}}{2}\right) = 150°$

$$= \frac{5\pi}{6}\text{ rad or }\mathbf{2.6180\,rad}$$

(d) $\operatorname{arccosec}(\sqrt{2}) = \arcsin\left(\dfrac{1}{\sqrt{2}}\right)$

$$\equiv \sin^{-1}\left(\frac{1}{\sqrt{2}}\right) = 45°$$

$$= \frac{\pi}{4}\text{ rad or }\mathbf{0.7854\,rad}$$

Problem 7. Evaluate (in radians), correct to 3 decimal places: $\sin^{-1}0.30 + \cos^{-1}0.65$

$\sin^{-1}0.30 = 17.4576° = 0.3047\,\text{rad}$

$\cos^{-1}0.65 = 49.4584° = 0.8632\,\text{rad}$

Hence $\sin^{-1}0.30 + \cos^{-1}0.65$
$= 0.3047 + 0.8632 = \mathbf{1.168}$, correct to 3 decimal places.

Now try the following Practice Exercise

Practice Exercise 90   Inverse functions (Answers on page 869)

Determine the inverse of the functions given in Problems 1 to 4.

1. $f(x) = x + 1$

2. $f(x) = 5x - 1$

3. $f(x) = x^3 + 1$

4. $f(x) = \dfrac{1}{x} + 2$

Determine the principal value of the inverse functions in Problems 5 to 11.

5. $\sin^{-1}(-1)$

6. $\cos^{-1}0.5$

7. $\tan^{-1}1$

8. $\cot^{-1}2$

9. $\operatorname{cosec}^{-1}2.5$

10. $\sec^{-1}1.5$

11. $\sin^{-1}\left(\dfrac{1}{\sqrt{2}}\right)$

12. Evaluate x, correct to 3 decimal places:
$$x = \sin^{-1}\frac{1}{3} + \cos^{-1}\frac{4}{5} - \tan^{-1}\frac{8}{9}$$

13. Evaluate y, correct to 4 significant figures:
$$y = 3\sec^{-1}\sqrt{2} - 4\operatorname{cosec}^{-1}\sqrt{2} + 5\cot^{-1}2$$

20.7 Asymptotes

If a table of values for the function $y = \dfrac{x+2}{x+1}$ is drawn up for various values of x and then y plotted against x, the graph would be as shown in Fig. 20.32. The straight lines AB, i.e. $x = -1$, and CD, i.e. $y = 1$, are known as **asymptotes**.

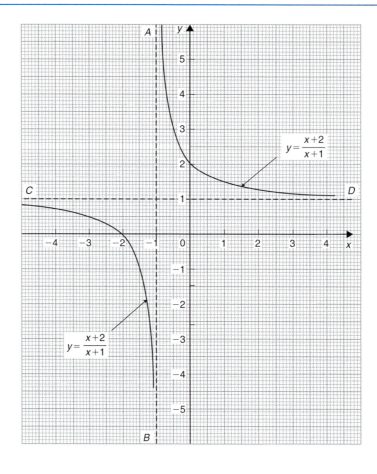

Figure 20.32

An asymptote to a curve is defined as a straight line to which the curve approaches as the distance from the origin increases. Alternatively, an asymptote can be considered as a tangent to the curve at infinity.

Asymptotes parallel to the *x*- and *y*-axes

There is a simple rule which enables asymptotes parallel to the *x*- and *y*-axes to be determined. For a curve $y = f(x)$:

(i) The asymptotes parallel to the *x*-axis are found by equating the coefficient of the highest power of *x* to zero.

(ii) The asymptotes parallel to the *y*-axis are found by equating the coefficient of the highest power of *y* to zero.

With the above example $y = \dfrac{x+2}{x+1}$, rearranging gives:

$$y(x+1) = x+2$$

i.e. $yx + y - x - 2 = 0$ (1)

and $x(y-1) + y - 2 = 0$

The coefficient of the highest power of *x* (in this case x^1) is $(y-1)$. Equating to zero gives: $y - 1 = 0$

From which, $y = 1$, which is an asymptote of $y = \dfrac{x+2}{x+1}$ as shown in Fig. 20.32.

Returning to equation (1): $yx + y - x - 2 = 0$

from which, $y(x+1) - x - 2 = 0$

The coefficient of the highest power of *y* (in this case y^1) is $(x+1)$. Equating to zero gives: $x + 1 = 0$ from which, $x = -1$, which is another asymptote of $y = \dfrac{x+2}{x+1}$ as shown in Fig. 20.32.

Problem 8. Determine the asymptotes for the function $y = \dfrac{x-3}{2x+1}$ and hence sketch the curve.

Rearranging $y = \dfrac{x-3}{2x+1}$ gives: $y(2x+1) = x-3$

i.e. $\qquad 2xy + y = x - 3$

or $\qquad 2xy + y - x + 3 = 0$

and $\quad x(2y-1) + y + 3 = 0$

Equating the coefficient of the highest power of x to zero gives: $2y - 1 = 0$ from which, $y = \frac{1}{2}$ which is an asymptote.

Since $y(2x+1) = x - 3$ then equating the coefficient of the highest power of y to zero gives: $2x + 1 = 0$ from which, $x = -\frac{1}{2}$ which is also an asymptote.

When $x = 0$, $y = \dfrac{x-3}{2x+1} = \dfrac{-3}{1} = -3$ and when $y = 0$,

$0 = \dfrac{x-3}{2x+1}$ from which, $x - 3 = 0$ and $x = 3$.

A sketch of $y = \dfrac{x-3}{2x+1}$ is shown in Fig. 20.33.

> **Problem 9.** Determine the asymptotes parallel to the x- and y-axes for the function
> $$x^2 y^2 = 9(x^2 + y^2)$$

Asymptotes parallel to the x-axis:

Rearranging $x^2 y^2 = 9(x^2 + y^2)$ gives

$$x^2 y^2 - 9x^2 - 9y^2 = 0$$

hence $\quad x^2(y^2 - 9) - 9y^2 = 0$

Equating the coefficient of the highest power of x to zero gives $y^2 - 9 = 0$ from which, $y^2 = 9$ and $y = \pm 3$

Asymptotes parallel to the y-axis:

Since $\quad x^2 y^2 - 9x^2 - 9y^2 = 0$

then $\qquad y^2(x^2 - 9) - 9x^2 = 0$

Equating the coefficient of the highest power of y to zero gives $x^2 - 9 = 0$ from which, $x^2 = 9$ and $x = \pm 3$.
Hence asymptotes occur at $y = \pm 3$ and $x = \pm 3$

Other asymptotes

To determine asymptotes other than those parallel to x- and y-axes a simple procedure is:

(i)   substitute $y = mx + c$ in the given equation

(ii) simplify the expression

(iii) equate the coefficients of the two highest powers of x to zero and determine the values of m and c. $y = mx + c$ gives the asymptote.

> **Problem 10.** Determine the asymptotes for the function: $y(x+1) = (x-3)(x+2)$ and sketch the curve.

Following the above procedure:

(i) Substituting $y = mx + c$ into
$y(x+1) = (x-3)(x+2)$ gives:

$$(mx + c)(x + 1) = (x - 3)(x + 2)$$

(ii) Simplifying gives

$$mx^2 + mx + cx + c = x^2 - x - 6$$

and $\quad (m-1)x^2 + (m+c+1)x + c + 6 = 0$

(iii) Equating the coefficient of the highest power of x to zero gives $m - 1 = 0$ from which, $m = 1$
Equating the coefficient of the next highest power of x to zero gives $m + c + 1 = 0$
and since $m = 1$, $1 + c + 1 = 0$ from which, $c = -2$
Hence $y = mx + c = 1x - 2$.

i.e. $y = x - 2$ is an asymptote.

To determine any asymptotes parallel to the x-axis:

Rearranging $\quad y(x+1) = (x-3)(x+2)$

gives $\qquad yx + y = x^2 - x - 6$

The coefficient of the highest power of x (i.e. x^2) is 1. Equating this to zero gives $1 = 0$ which is not an equation of a line. Hence there is no asymptote parallel to the x-axis.

To determine any asymptotes parallel to the y-axis:
Since $y(x+1) = (x-3)(x+2)$ the coefficient of the highest power of y is $x + 1$. Equating this to zero gives $x + 1 = 0$, from which, $x = -1$. Hence $x = -1$ is an asymptote.
When $x = 0$, $y(1) = (-3)(2)$, i.e. $y = -6$
When $y = 0$, $0 = (x-3)(x+2)$, i.e. $x = 3$ and $x = -2$
A sketch of the function $y(x+1) = (x-3)(x+2)$ is shown in Fig. 20.34.

> **Problem 11.** Determine the asymptotes for the function $x^3 - xy^2 + 2x - 9 = 0$

Following the procedure:

(i) Substituting $y = mx + c$ gives
$x^3 - x(mx + c)^2 + 2x - 9 = 0$

Figure 20.33

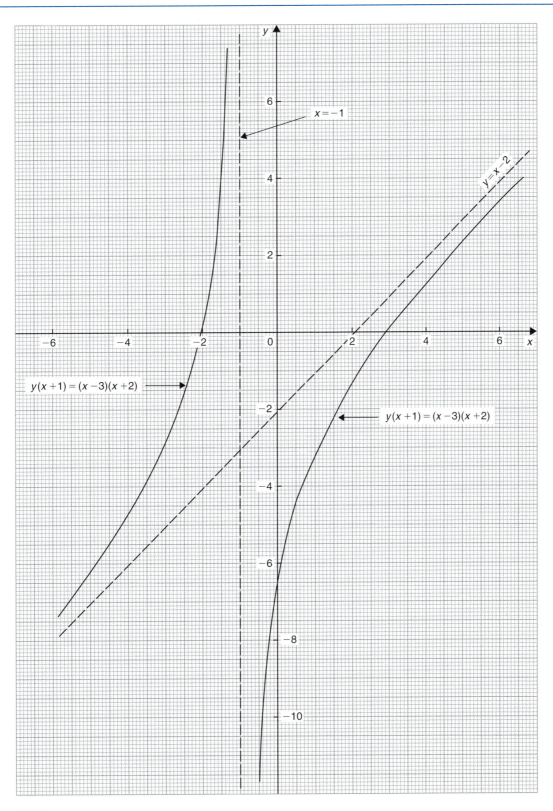

Figure 20.34

(ii) Simplifying gives

$$x^3 - x[m^2x^2 + 2mcx + c^2] + 2x - 9 = 0$$

i.e. $x^3 - m^2x^3 - 2mcx^2 - c^2x + 2x - 9 = 0$

and $x^3(1 - m^2) - 2mcx^2 - c^2x + 2x - 9 = 0$

(iii) Equating the coefficient of the highest power of x (i.e. x^3 in this case) to zero gives $1 - m^2 = 0$, from which, $m = \pm 1$

Equating the coefficient of the next highest power of x (i.e. x^2 in this case) to zero gives $-2mc = 0$, from which, $c = 0$

Hence $y = mx + c = \pm 1x + 0$, i.e. **$y = x$ and $y = -x$ are asymptotes**.

To determine any asymptotes parallel to the x- and y-axes for the function $x^3 - xy^2 + 2x - 9 = 0$:

Equating the coefficient of the highest power of x term to zero gives $1 = 0$ which is not an equation of a line. Hence there is no asymptote parallel with the x-axis.

Equating the coefficient of the highest power of y term to zero gives $-x = 0$ from which, $x = 0$

Hence $x = 0$, $y = x$ and $y = -x$ are asymptotes for the function $x^3 - xy^2 + 2x - 9 = 0$

> **Problem 12.** Find the asymptotes for the function $y = \dfrac{x^2 + 1}{x}$ and sketch a graph of the function.

Rearranging $y = \dfrac{x^2 + 1}{x}$ gives $yx = x^2 + 1$

Equating the coefficient of the highest power x term to zero gives $1 = 0$, hence there is no asymptote parallel to the x-axis.

Equating the coefficient of the highest power y term to zero gives $x = 0$

Hence there is an asymptote at $x = 0$ (i.e. the y-axis)

To determine any other asymptotes we substitute $y = mx + c$ into $yx = x^2 + 1$ which gives

$$(mx + c)x = x^2 + 1$$

i.e. $mx^2 + cx = x^2 + 1$

and $(m - 1)x^2 + cx - 1 = 0$

Equating the coefficient of the highest power x term to zero gives $m - 1 = 0$, from which $m = 1$

Equating the coefficient of the next highest power x term to zero gives $c = 0$. Hence $y = mx + c = 1x + 0$, i.e. **$y = x$ is an asymptote**.

A sketch of $y = \dfrac{x^2 + 1}{x}$ is shown in Fig. 20.35. It is possible to determine maximum/minimum points on the graph (see Chapter 30).

Since $y = \dfrac{x^2 + 1}{x} = \dfrac{x^2}{x} + \dfrac{1}{x} = x + x^{-1}$

then $\dfrac{dy}{dx} = 1 - x^{-2} = 1 - \dfrac{1}{x^2} = 0$

for a turning point.

Hence $1 = \dfrac{1}{x^2}$ and $x^2 = 1$, from which, $x = \pm 1$

When $x = 1$,

$$y = \dfrac{x^2 + 1}{x} = \dfrac{1 + 1}{1} = 2$$

and when $x = -1$,

$$y = \dfrac{(-1)^2 + 1}{-1} = -2$$

i.e. $(1, 2)$ and $(-1, -2)$ are the co-ordinates of the turning points. $\dfrac{d^2y}{dx^2} = 2x^{-3} = \dfrac{2}{x^3}$; when $x = 1$, $\dfrac{d^2y}{dx^2}$ is positive, which indicates a minimum point and when $x = -1$, $\dfrac{d^2y}{dx^2}$ is negative, which indicates a maximum point, as shown in Fig. 20.35.

Now try the following Practice Exercise

Practice Exercise 91   Asymptotes (Answers on page 869)

In Problems 1 to 3, determine the asymptotes parallel to the x- and y-axes.

1. $y = \dfrac{x - 2}{x + 1}$

2. $y^2 = \dfrac{x}{x - 3}$

3. $y = \dfrac{x(x + 3)}{(x + 2)(x + 1)}$

In Problems 4 and 5, determine all the asymptotes.

4. $8x - 10 + x^3 - xy^2 = 0$

5. $x^2(y^2 - 16) = y$

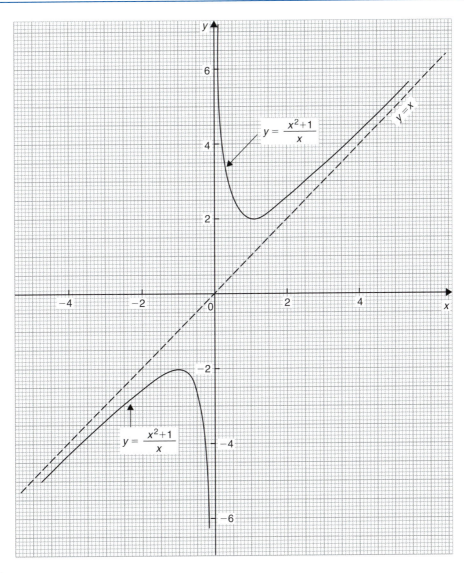

Figure 20.35

In Problems 6 and 7, determine the asymptotes and sketch the curves.

6. $y = \dfrac{x^2 - x - 4}{x + 1}$

7. $xy^2 - x^2y + 2x - y = 5$

20.8   Brief guide to curve sketching

The following steps will give information from which the graphs of many types of functions $y = f(x)$ can be sketched.

(i)   Use calculus to determine the location and nature of maximum and minimum points (see Chapter 30).

(ii)   Determine where the curve cuts the x- and y-axes.

(iii) Inspect the equation for symmetry.

(a)   If the equation is unchanged when $-x$ is substituted for x, the graph will be symmetrical about the y-axis (i.e. it is an **even function**).

(b)   If the equation is unchanged when $-y$ is substituted for y, the graph will be symmetrical about the x-axis.

(c)   If $f(-x) = -f(x)$, the graph is symmetrical about the origin (i.e. it is an **odd function**).

(iv)   Check for any asymptotes.

20.9 Worked problems on curve sketching

Problem 13. Sketch the graphs of

(a) $y = 2x^2 + 12x + 20$

(b) $y = -3x^2 + 12x - 15$

(a) $y = 2x^2 + 12x + 20$ is a parabola since the equation is a quadratic. To determine the turning point:

Gradient $= \dfrac{dy}{dx} = 4x + 12 = 0$ for a turning point.

Hence $4x = -12$ and $x = -3$

When $x = -3$, $y = 2(-3)^2 + 12(-3) + 20 = 2$

Hence $(-3, 2)$ are the co-ordinates of the turning point

$\dfrac{d^2y}{dx^2} = 4$, which is positive, hence $(-3, 2)$ is a minimum point.

When $x = 0$, $y = 20$, hence the curve cuts the y-axis at $y = 20$

Thus knowing the curve passes through $(-3, 2)$ and $(0, 20)$ and appreciating the general shape of a parabola results in the sketch given in Fig. 20.36.

(b) $y = -3x^2 + 12x - 15$ is also a parabola (but 'upside down' due to the minus sign in front of the x^2 term).

Gradient $= \dfrac{dy}{dx} = -6x + 12 = 0$ for a turning point.

Hence $6x = 12$ and $x = 2$

When $x = 2$, $y = -3(2)^2 + 12(2) - 15 = -3$

Hence $(2, -3)$ are the co-ordinates of the turning point

$\dfrac{d^2y}{dx^2} = -6$, which is negative, hence $(2, -3)$ is a maximum point.

When $x = 0$, $y = -15$, hence the curve cuts the axis at $y = -15$

The curve is shown sketched in Fig. 20.36.

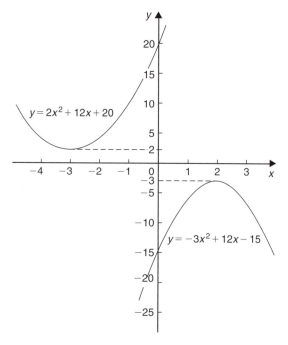

Figure 20.36

Problem 14. Sketch the curves depicting the following equations:

(a) $x = \sqrt{9 - y^2}$ (b) $y^2 = 16x$

(c) $xy = 5$

(a) Squaring both sides of the equation and transposing gives $x^2 + y^2 = 9$. Comparing this with the standard equation of a circle, centre origin and radius a, i.e. $x^2 + y^2 = a^2$, shows that $x^2 + y^2 = 9$ represents a circle, centre origin and radius 3. A sketch of this circle is shown in Fig. 20.37(a).

(b) The equation $y^2 = 16x$ is symmetrical about the x-axis and having its vertex at the origin $(0, 0)$. Also, when $x = 1$, $y = \pm 4$. A sketch of this parabola is shown in Fig. 20.37(b).

(c) The equation $y = \dfrac{a}{x}$ represents a rectangular hyperbola lying entirely within the first and third quadrants. Transposing $xy = 5$ gives $y = \dfrac{5}{x}$, and therefore represents the rectangular hyperbola shown in Fig. 20.37(c).

Section C

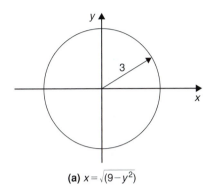

(a) $x = \sqrt{(9 - y^2)}$

(b) $y^2 = 16x$

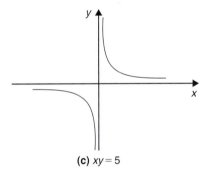

(c) $xy = 5$

Figure 20.37

Problem 15.   Sketch the curves depicting the following equations:

(a) $4x^2 = 36 - 9y^2$ (b) $3y^2 + 15 = 5x^2$

(a)   By dividing throughout by 36 and transposing, the equation $4x^2 = 36 - 9y^2$ can be written as $\dfrac{x^2}{9} + \dfrac{y^2}{4} = 1$. The equation of an ellipse is of the form $\dfrac{x^2}{a^2} + \dfrac{y^2}{b^2} = 1$, where $2a$ and $2b$ represent the length of the axes of the ellipse. Thus $\dfrac{x^2}{3^2} + \dfrac{y^2}{2^2} = 1$ represents an ellipse, having its axes coinciding

with the x- and y-axes of a rectangular co-ordinate system, the major axis being $2(3)$, i.e. 6 units long and the minor axis $2(2)$, i.e. 4 units long, as shown in Fig. 20.38(a).

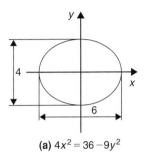

(a) $4x^2 = 36 - 9y^2$

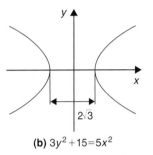

(b) $3y^2 + 15 = 5x^2$

Figure 20.38

(b)   Dividing $3y^2 + 15 = 5x^2$ throughout by 15 and transposing gives $\dfrac{x^2}{3} - \dfrac{y^2}{5} = 1$. The equation $\dfrac{x^2}{a^2} - \dfrac{y^2}{b^2} = 1$ represents a hyperbola which is symmetrical about both the x- and y-axes, the distance between the vertices being given by $2a$

Thus a sketch of $\dfrac{x^2}{3} - \dfrac{y^2}{5} = 1$ is as shown in Fig. 20.38(b), having a distance of $2\sqrt{3}$ between its vertices.

Problem 16.   Describe the shape of the curves represented by the following equations:

(a) $x = 2\sqrt{\left[1 - \left(\dfrac{y}{2}\right)^2\right]}$ (b) $\dfrac{y^2}{8} = 2x$

(c) $y = 6\left(1 - \dfrac{x^2}{16}\right)^{1/2}$

(a)   Squaring the equation gives $x^2 = 4\left[1 - \left(\dfrac{y}{2}\right)^2\right]$ and transposing gives $x^2 = 4 - y^2$, i.e.

Section C

$x^2 + y^2 = 4$. Comparing this equation with $x^2 + y^2 = a^2$ shows that $x^2 + y^2 = 4$ is the equation of a **circle** having centre at the origin $(0, 0)$ and of radius 2 units.

(b) Transposing $\dfrac{y^2}{8} = 2x$ gives $y = 4\sqrt{x}$. Thus $\dfrac{y^2}{8} = 2x$ is the equation of a **parabola** having its axis of symmetry coinciding with the x-axis and its vertex at the origin of a rectangular co-ordinate system.

(c) $y = 6\left(1 - \dfrac{x^2}{16}\right)^{1/2}$ can be transposed to $\dfrac{y}{6} = \left(1 - \dfrac{x^2}{16}\right)^{1/2}$ and squaring both sides gives

$$\dfrac{y^2}{36} = 1 - \dfrac{x^2}{16}, \text{ i.e. } \dfrac{x^2}{16} + \dfrac{y^2}{36} = 1$$

This is the equation of an **ellipse**, centre at the origin of a rectangular co-ordinate system, the major axis coinciding with the y-axis and being $2\sqrt{36}$, i.e. 12 units long. The minor axis coincides with the x-axis and is $2\sqrt{16}$, i.e. 8 units long.

Problem 17. Describe the shape of the curves represented by the following equations:

(a) $\dfrac{x}{5} = \sqrt{\left[1 + \left(\dfrac{y}{2}\right)^2\right]}$ (b) $\dfrac{y}{4} = \dfrac{15}{2x}$

(a) Since $\dfrac{x}{5} = \sqrt{\left[1 + \left(\dfrac{y}{2}\right)^2\right]}$

$\dfrac{x^2}{25} = 1 + \left(\dfrac{y}{2}\right)^2$

i.e. $\dfrac{x^2}{25} - \dfrac{y^2}{4} = 1$

This is a **hyperbola** which is symmetrical about both the x- and y-axes, the vertices being $2\sqrt{25}$, i.e. 10 units apart.

(With reference to Section 20.1 (vii), a is equal to ± 5)

(b) The equation $\dfrac{y}{4} = \dfrac{15}{2x}$ is of the form $y = \dfrac{a}{x}$,

$a = \dfrac{60}{2} = 30$

This represents a **rectangular hyperbola**, symmetrical about both the x- and y-axes, and lying entirely in the first and third quadrants, similar in shape to the curves shown in Fig. 20.9.

Now try the following Practice Exercise

Practice Exercise 92   Curve sketching (Answers on page 869)

1. Sketch the graphs of (a) $y = 3x^2 + 9x + \dfrac{7}{4}$ (b) $y = -5x^2 + 20x + 50$

In Problems 2 to 8, sketch the curves depicting the equations given.

2. $x = 4\sqrt{\left[1 - \left(\dfrac{y}{4}\right)^2\right]}$

3. $\sqrt{x} = \dfrac{y}{9}$

4. $y^2 = \dfrac{x^2 - 16}{4}$

5. $\dfrac{y^2}{5} = 5 - \dfrac{x^2}{2}$

6. $x = 3\sqrt{1 + y^2}$

7. $x^2 y^2 = 9$

8. $x = \dfrac{1}{3}\sqrt{(36 - 18y^2)}$

9. Sketch the circle given by the equation $x^2 + y^2 - 4x + 10y + 25 = 0$

In Problems 10 to 15 describe the shape of the curves represented by the equations given.

10. $y = \sqrt{[3(1 - x^2)]}$

11. $y = \sqrt{[3(x^2 - 1)]}$

12. $y = \sqrt{9 - x^2}$

13. $y = 7x^{-1}$

14. $y = (3x)^{1/2}$

15. $y^2 - 8 = -2x^2$

For fully worked solutions to each of the problems in Practice Exercises 88 to 92 in this chapter, go to the website:
www.routledge.com/cw/bird

Chapter 21

Irregular areas, volumes and mean values of waveforms

Why it is important to understand: Irregular areas, volumes and mean values of waveforms

Surveyors, farmers and landscapers often need to determine the area of irregularly shaped pieces of land to work with the land properly. There are many applications in business, economics and the sciences, including all aspects of engineering, where finding the areas of irregular shapes, the volumes of solids, and the lengths of irregular shaped curves are important applications. Typical earthworks include roads, railway beds, causeways, dams and canals. Other common earthworks are land grading to reconfigure the topography of a site, or to stabilise slopes. Engineers need to concern themselves with issues of geotechnical engineering (such as soil density and strength) and with quantity estimation to ensure that soil volumes in the cuts match those of the fills, while minimizing the distance of movement. Simpson's rule is a staple of scientific data analysis and engineering; it is widely used, for example, by naval architects to numerically determine hull offsets and cross-sectional areas to determine volumes and centroids of ships or lifeboats. There are therefore plenty of examples where irregular areas and volumes need to be determined by engineers.

At the end of this chapter, you should be able to:

- use the trapezoidal rule to determine irregular areas
- use the mid-ordinate rule to determine irregular areas
- use Simpson's rule to determine irregular areas
- estimate the volume of irregular solids
- determine the mean values of waveforms

21.1 Areas of irregular figures

Areas of irregular plane surfaces may be approximately determined by using (a) a planimeter, (b) the trapezoidal rule, (c) the mid-ordinate rule, and (d) Simpson's rule. Such methods may be used, for example, by engineers estimating areas of indicator diagrams of steam engines, surveyors estimating areas of plots of land or naval architects estimating areas of water planes or transverse sections of ships.

(a) **A planimeter** is an instrument for directly measuring small areas bounded by an irregular curve.

(b) **Trapezoidal rule**

To determine the areas $PQRS$ in Fig. 21.1:

(i) Divide base PS into any number of equal intervals, each of width d (the greater the number of intervals, the greater the accuracy).

(ii) Accurately measure ordinates y_1, y_2, y_3, etc.

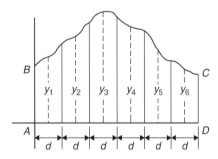

Figure 21.1

(iii) Areas $PQRS$

$$= d\left[\frac{y_1 + y_7}{2} + y_2 + y_3 + y_4 + y_5 + y_6\right]$$

In general, the trapezoidal rule states:

Area =

$$\begin{pmatrix}\textbf{width of}\\\textbf{interval}\end{pmatrix}\left[\frac{1}{2}\begin{pmatrix}\textbf{first +}\\\textbf{last}\\\textbf{ordinate}\end{pmatrix} + \begin{matrix}\textbf{sum of}\\\textbf{remaining}\\\textbf{ordinates}\end{matrix}\right]$$

(c) **Mid-ordinate rule**

To determine the area $ABCD$ of Fig. 21.2:

Figure 21.2

(i) Divide base AD into any number of equal intervals, each of width d (the greater the number of intervals, the greater the accuracy).

(ii) Erect ordinates in the middle of each interval (shown by broken lines in Fig. 21.2).

(iii) Accurately measure ordinates y_1, y_2, y_3, etc.

(iv) Area $ABCD = d(y_1 + y_2 + y_3 + y_4 + y_5 + y_6)$

In general, the mid-ordinate rule states:

$$\textbf{Area} = \begin{pmatrix}\textbf{width of}\\\textbf{interval}\end{pmatrix}\begin{pmatrix}\textbf{sum of}\\\textbf{mid-ordinates}\end{pmatrix}$$

(d) **Simpson's rule**[*]

To determine the area $PQRS$ of Fig. 21.1:

(i) Divide base PS into an **even** number of intervals, each of width d (the greater the number of intervals, the greater the accuracy).

(ii) Accurately measure ordinates y_1, y_2, y_3, etc.

(iii) Area $PQRS = \dfrac{d}{3}[(y_1 + y_7) + 4(y_2 + y_4 + y_6) + 2(y_3 + y_5)]$

In general, Simpson's rule states:

$$\textbf{Area} = \frac{1}{3}\begin{pmatrix}\textbf{width of}\\\textbf{interval}\end{pmatrix}\left[\begin{pmatrix}\textbf{first + last}\\\textbf{ordinate}\end{pmatrix}\right.$$

$$+ 4\begin{pmatrix}\textbf{sum of even}\\\textbf{ordinates}\end{pmatrix}$$

$$\left.+ 2\begin{pmatrix}\textbf{sum of remaining}\\\textbf{odd ordinates}\end{pmatrix}\right]$$

[*] **Who was Simpson?** **Thomas Simpson** FRS (20 August 1710–14 May 1761) was the British mathematician who invented Simpson's rule to approximate definite integrals. To find out more go to **www.routledge.com/cw/bird**

Problem 1. A car starts from rest and its speed is measured every second for 6 s:

Time $t(s)$	0	1	2	3	4	5	6
Speed v (m/s)	0	2.5	5.5	8.75	12.5	17.5	24.0

Determine the distance travelled in 6 seconds (i.e. the area under the v/t graph), by (a) the trapezoidal rule, (b) the mid-ordinate rule, and (c) Simpson's rule.

A graph of speed/time is shown in Fig. 21.3.

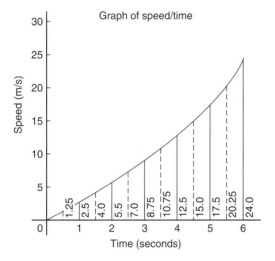

Figure 21.3

(a) **Trapezoidal rule** (see para. (b) on page 241)

The time base is divided into six strips each of width 1 s, and the length of the ordinates measured. Thus

$$\text{area} = (1)\left[\left(\frac{0+24.0}{2}\right) + 2.5 + 5.5\right.$$
$$\left. + 8.75 + 12.5 + 17.5\right]$$
$$= \mathbf{58.75\,m}$$

(b) **Mid-ordinate rule** (see para. (c) on page 242)

The time base is divided into six strips each of width 1 s.

Mid-ordinates are erected as shown in Fig. 21.3 by the broken lines. The length of each mid-ordinate is measured. Thus

$$\text{area} = (1)[1.25 + 4.0 + 7.0 + 10.75$$
$$+ 15.0 + 20.25]$$
$$= \mathbf{58.25\,m}$$

(c) **Simpson's rule** (see para. (d) on page 242)

The time base is divided into six strips each of width 1 s, and the length of the ordinates measured. Thus

$$\text{area} = \tfrac{1}{3}(1)[(0 + 24.0) + 4(2.5 + 8.75$$
$$+ 17.5) + 2(5.5 + 12.5)]$$
$$= \mathbf{58.33\,m}$$

Problem 2. A river is 15 m wide. Soundings of the depth are made at equal intervals of 3 m across the river and are as shown below.

Depth (m)	0	2.2	3.3	4.5	4.2	2.4	0

Calculate the cross-sectional area of the flow of water at this point using Simpson's rule.

From para. (d) on page 242,

$$\text{Area} = \tfrac{1}{3}(3)[(0 + 0) + 4(2.2 + 4.5 + 2.4)$$
$$+ 2(3.3 + 4.2)]$$
$$= (1)[0 + 36.4 + 15] = \mathbf{51.4\,m^2}$$

Now try the following Practice Exercise

Practice Exercise 93   Areas of irregular figures (Answers on page 869)

1. Plot a graph of $y = 3x - x^2$ by completing a table of values of y from $x = 0$ to $x = 3$. Determine the area enclosed by the curve, the x-axis and ordinate $x = 0$ and $x = 3$ by (a) the trapezoidal rule, (b) the mid-ordinate rule and (c) by Simpson's rule.

2. Plot the graph of $y = 2x^2 + 3$ between $x = 0$ and $x = 4$. Estimate the area enclosed by the curve, the ordinates $x = 0$ and $x = 4$, and the x-axis by an approximate method.

3. The velocity of a car at one second intervals is given in the following table:

time t (s)	0	1	2	3	4	5	6
velocity v (m/s)	0	2.0	4.5	8.0	14.0	21.0	29.0

Determine the distance travelled in six seconds (i.e. the area under the v/t graph) using Simpson's rule.

4. The shape of a piece of land is shown in Fig. 21.4. To estimate the area of the land, a surveyor takes measurements at intervals of 50 m, perpendicular to the straight portion with the results shown (the dimensions being in metres). Estimate the area of the land in hectares (1 ha $= 10^4$ m^2).

140	160	200	190	180	130
50	50	50	50	50	50

Figure 21.4

5. The deck of a ship is 35 m long. At equal intervals of 5 m the width is given by the following table:

Width (m)	0	2.8	5.2	6.5	5.8	4.1	3.0	2.3

Estimate the area of the deck.

21.2 Volumes of irregular solids

If the cross-sectional areas $A_1, A_2, A_3, \ldots$ of an irregular solid bounded by two parallel planes are known at equal intervals of width d (as shown in Fig. 21.5), then by Simpson's rule:

$$\textbf{volume, } V = \frac{d}{3}[(A_1 + A_7) + 4(A_2 + A_4 + A_6) + 2(A_3 + A_5)]$$

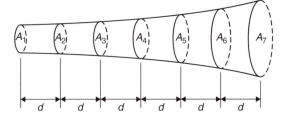

Figure 21.5

Problem 3. A tree trunk is 12 m in length and has a varying cross-section. The cross-sectional areas at intervals of 2 m measured from one end are:

0.52, 0.55, 0.59, 0.63, 0.72, 0.84, 0.97 m^2

Estimate the volume of the tree trunk.

A sketch of the tree trunk is similar to that shown in Fig. 21.5 above, where $d=2$ m, $A_1 = 0.52$ m^2, $A_2 = 0.55$ m^2, and so on.
Using Simpson's rule for volumes gives:

$$\text{Volume} = \tfrac{2}{3}[(0.52 + 0.97) + 4(0.55 + 0.63 + 0.84) + 2(0.59 + 0.72)]$$

$$= \tfrac{2}{3}[1.49 + 8.08 + 2.62] = \textbf{8.13 m}^3$$

Problem 4. The areas of seven horizontal cross-sections of a water reservoir at intervals of 10 m are:

210, 250, 320, 350, 290, 230, 170 m^2

Calculate the capacity of the reservoir in litres.

Using Simpson's rule for volumes gives:

$$\text{Volume} = \frac{10}{3}[(210 + 170) + 4(250 + 350 + 230) + 2(320 + 290)]$$

$$= \frac{10}{3}[380 + 3320 + 1220]$$

$$= \textbf{16 400 m}^3$$

16 400 m^3 = 16 400 $\times$ 10^6 cm^3 and since
1 litre = 1000 cm^3,

capacity of reservoir $= \dfrac{16\,400 \times 10^6}{1000}$ litres

$= 16\,400\,000$

$= \mathbf{1.64 \times 10^7}$ **litres**

Now try the following **Practice Exercise**

1. The areas of equidistantly spaced sections of the underwater form of a small boat are as follows:

    1.76, 2.78, 3.10, 3.12, 2.61, 1.24, 0.85 m^2

    Determine the underwater volume if the sections are 3 m apart.

2. To estimate the amount of earth to be removed when constructing a cutting, the cross-sectional area at intervals of 8 m were estimated as follows:

    0,   2.8,   3.7,   4.5,   4.1,   2.6,   0 m^3

    Estimate the volume of earth to be excavated.

3. The circumference of a 12 m long log of timber of varying circular cross-section is measured at intervals of 2 m along its length and the results are:

Distance from one end (m)	Circumference (m)
0	2.80
2	3.25
4	3.94
6	4.32
8	5.16
10	5.82
12	6.36

    Estimate the volume of the timber in cubic metres.

21.3   The mean or average value of a waveform

The mean or average value, y, of the waveform shown in Fig. 21.6 is given by:

$$y = \frac{\text{area under curve}}{\text{length of base, } b}$$

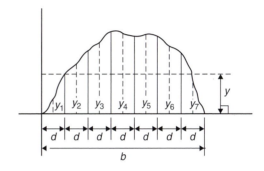

Figure 21.6

If the mid-ordinate rule is used to find the area under the curve, then:

$$y = \frac{\text{sum of mid-ordinates}}{\text{number of mid-ordinates}}$$

$$\left(= \frac{y_1 + y_2 + y_3 + y_4 + y_5 + y_6 + y_7}{7} \right.$$

$$\left. \text{for Fig. 21.6} \right)$$

For a **sine wave**, the mean or average value:

(i)   over one complete cycle is zero (see Fig. 21.7(a)),

(ii) over half a cycle is **0.637 × maximum value**, or **(2/π) × maximum value**,

(iii) of a full-wave rectified waveform (see Fig. 21.7(b)) is **0.637 × maximum value**,

(iv) of a half-wave rectified waveform (see Fig. 21.7(c)) is **0.318 × maximum value**, or **(1/π) maximum value**.

Problem 5.   Determine the average values over half a cycle of the periodic waveforms shown in Fig. 21.8.

Section C

Figure 21.7

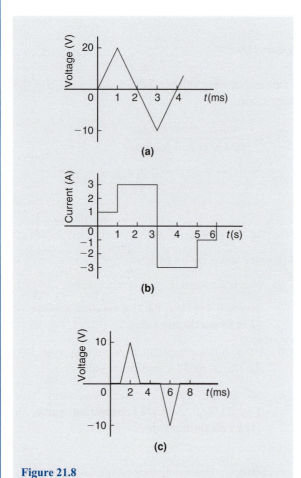

Figure 21.8

(a) Area under triangular waveform (a) for a half-cycle is given by:

$$\text{Area} = \tfrac{1}{2}\,(\text{base})\,(\text{perpendicular height})$$

$$= \tfrac{1}{2}(2 \times 10^{-3})(20)$$

$$= 20 \times 10^{-3}\,\text{Vs}$$

Average value of waveform

$$= \frac{\text{area under curve}}{\text{length of base}}$$

$$= \frac{20 \times 10^{-3}\,\text{Vs}}{2 \times 10^{-3}\,\text{s}}$$

$$= \mathbf{10\,V}$$

(b) Area under waveform (b) for a half-cycle =
$(1 \times 1) + (3 \times 2) = 7\,\text{As}$.

Average value of waveform

$$= \frac{\text{area under curve}}{\text{length of base}}$$

$$= \frac{7\,\text{As}}{3\,\text{s}}$$

$$= \mathbf{2.33\,A}$$

(c) A half-cycle of the voltage waveform (c) is completed in 4 ms.

$$\text{Area under curve} = \tfrac{1}{2}\{(3-1)10^{-3}\}(10)$$

$$= 10 \times 10^{-3}\,\text{Vs}$$

Average value of waveform

$$= \frac{\text{area under curve}}{\text{length of base}}$$

$$= \frac{10 \times 10^{-3}\,\text{Vs}}{4 \times 10^{-3}\,\text{s}}$$

$$= \mathbf{2.5\,V}$$

Problem 6. Determine the mean value of current over one complete cycle of the periodic waveforms shown in Fig. 21.9.

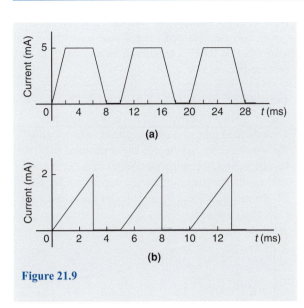

Figure 21.9

(a) One cycle of the trapezoidal waveform (a) is completed in 10 ms (i.e. the periodic time is 10 ms).

Area under curve = area of trapezium

$$= \tfrac{1}{2} \text{ (sum of parallel sides) (perpendicular}$$

distance between parallel sides)

$$= \tfrac{1}{2}\{(4+8) \times 10^{-3}\}(5 \times 10^{-3})$$

$$= 30 \times 10^{-6} \text{As}$$

Mean value over one cycle

$$= \frac{\text{area under curve}}{\text{length of base}} = \frac{30 \times 10^{-6} \text{As}}{10 \times 10^{-3} \text{s}}$$

$$= \mathbf{3\,mA}$$

(b) One cycle of the sawtooth waveform (b) is completed in 5 ms.

$$\text{Area under curve} = \tfrac{1}{2}(3 \times 10^{-3})(2)$$

$$= 3 \times 10^{-3} \text{As}$$

Mean value over one cycle

$$= \frac{\text{area under curve}}{\text{length of base}} = \frac{3 \times 10^{-3} \text{As}}{5 \times 10^{-3} \text{s}}$$

$$= \mathbf{0.6\,A}$$

Problem 7. The power used in a manufacturing process during a six hour period is recorded at intervals of one hour as shown below.

Time (h)	0	1	2	3	4	5	6
Power (kW)	0	14	29	51	45	23	0

Plot a graph of power against time and, by using the mid-ordinate rule, determine (a) the area under the curve and (b) the average value of the power.

The graph of power/time is shown in Fig. 21.10.

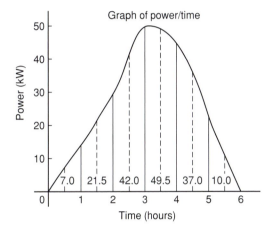

Figure 21.10

(a) The time base is divided into six equal intervals, each of width one hour. Mid-ordinates are erected (shown by broken lines in Fig. 21.10) and measured. The values are shown in Fig. 21.10.

Area under curve = (width of interval)

$$\times \text{(sum of mid-ordinates)}$$

$$= (1)[7.0 + 21.5 + 42.0$$

$$+ 49.5 + 37.0 + 10.0]$$

$$= \mathbf{167\,kWh}\,(\text{i.e. a measure}$$

of electrical energy)

(b) Average value of waveform

$$= \frac{\text{area under curve}}{\text{length of base}}$$

$$= \frac{167\,\text{kWh}}{6\,\text{h}} = \mathbf{27.83\,kW}$$

Section C

Alternatively, average value

$$= \frac{\text{sum of mid-ordinates}}{\text{number of mid-ordinates}}$$

Problem 8. Fig. 21.11 shows a sinusoidal output voltage of a full-wave rectifier. Determine, using the mid-ordinate rule with six intervals, the mean output voltage.

Figure 21.11

One cycle of the output voltage is completed in π radians or $180°$. The base is divided into six intervals, each of width $30°$. The mid-ordinate of each interval will lie at $15°$, $45°$, $75°$, etc.

At $15°$ the height of the mid-ordinate is $10\sin 15° = 2.588$ V.

At $45°$ the height of the mid-ordinate is $10\sin 45° = 7.071$ V, and so on.

The results are tabulated below:

Mid-ordinate	Height of mid-ordinate
$15°$	$10\sin 15° = 2.588$ V
$45°$	$10\sin 45° = 7.071$ V
$75°$	$10\sin 75° = 9.659$ V
$105°$	$10\sin 105° = 9.659$ V
$135°$	$10\sin 135° = 7.071$ V
$165°$	$10\sin 165° = 2.588$ V
	Sum of mid-ordinates $= 38.636$ V

Mean or average value of output voltage

$$= \frac{\text{sum of mid-ordinates}}{\text{number of mid-ordinates}}$$
$$= \frac{38.636}{6}$$
$$= \mathbf{6.439\,V}$$

(With a larger number of intervals a more accurate answer may be obtained.) For a sine wave the actual mean value is $0.637 \times$ maximum value, which in this problem gives 6.37 V.

Problem 9. An indicator diagram for a steam engine is shown in Fig. 21.12. The base line has been divided into six equally spaced intervals and the lengths of the seven ordinates measured with the results shown in centimetres. Determine (a) the area of the indicator diagram using Simpson's rule, and (b) the mean pressure in the cylinder given that 1 cm represents 100 kPa.

Figure 21.12

(a) The width of each interval is $\dfrac{12.0}{6}$ cm. Using Simpson's rule,

$$\text{area} = \tfrac{1}{3}(2.0)[(3.6 + 1.6) + 4(4.0$$
$$+ 2.9 + 1.7) + 2(3.5 + 2.2)]$$
$$= \tfrac{2}{3}[5.2 + 34.4 + 11.4]$$
$$= \mathbf{34\,cm^2}$$

(b) Mean height of ordinates

$$= \frac{\text{area of diagram}}{\text{length of base}} = \frac{34}{12}$$
$$= 2.83\,\text{cm}$$

Since 1 cm represents 100 kPa, the mean pressure in the cylinder

$$= 2.83\,\text{cm} \times 100\,\text{kPa/cm} = \mathbf{283\,kPa}$$

Now try the following Practice Exercise

Practice Exercise 95   Mean or average values of waveforms (Answers on page 869)

1. Determine the mean value of the periodic waveforms shown in Fig. 21.13 over a half-cycle.

Figure 21.13

2. Find the average value of the periodic wave-forms shown in Fig. 21.14 over one complete cycle.

Figure 21.14

3. An alternating current has the following values at equal intervals of 5 ms

Time (ms)	0	5	10	15	20	25	30
Current (A)	0	0.9	2.6	4.9	5.8	3.5	0

Plot a graph of current against time and estimate the area under the curve over the 30 ms period using the mid-ordinate rule and determine its mean value.

4. Determine, using an approximate method, the average value of a sine wave of maximum value 50 V for (a) a half-cycle and (b) a complete cycle.

5. An indicator diagram of a steam engine is 12 cm long. Seven evenly spaced ordinates, including the end ordinates, are measured as follows:

5.90, 5.52, 4.22, 3.63, 3.32, 3.24, 3.16 cm

Determine the area of the diagram and the mean pressure in the cylinder if 1 cm represents 90 kPa.

For fully worked solutions to each of the problems in Practice Exercises 93 to 95 in this chapter, go to the website:
www.routledge.com/cw/bird

Section C

Revision Test 6 Functions and their curves and irregular areas and volumes

This Revision Test covers the material contained in Chapters 20 and 21. *The marks for each question are shown in brackets at the end of each question.*

1. Sketch the following graphs, showing the relevant points:

   (a) $y=(x-2)^2$ (c) $x^2+y^2-2x+4y-4=0$

   (b) $y=3-\cos 2x$ (d) $9x^2-4y^2=36$

   (e) $f(x) = \begin{cases} -1 & -\pi \leq x \leq -\dfrac{\pi}{2} \\ x & -\dfrac{\pi}{2} \leq x \leq \dfrac{\pi}{2} \\ 1 & \dfrac{\pi}{2} \leq x \leq \pi \end{cases}$

   (15)

2. Determine the inverse of $f(x)=3x+1$ (3)

3. Evaluate, correct to 3 decimal places:
   $2\tan^{-1}1.64 + \sec^{-1}2.43 - 3\csc^{-1}3.85$ (4)

4. Determine the asymptotes for the following function and hence sketch the curve:
   $$y = \frac{(x-1)(x+4)}{(x-2)(x-5)}$$ (9)

5. Plot a graph of $y=3x^2+5$ from $x=1$ to $x=4$. Estimate, correct to 2 decimal places, using six intervals, the area enclosed by the curve, the ordinates $x=1$ and $x=4$, and the x-axis by (a) the trapezoidal rule, (b) the mid-ordinate rule, and (c) Simpson's rule. (16)

6. A circular cooling tower is 20 m high. The inside diameter of the tower at different heights is given in the following table:

Height (m)	0	5.0	10.0	15.0	20.0
Diameter (m)	16.0	13.3	10.7	8.6	8.0

   Determine the area corresponding to each diameter and hence estimate the capacity of the tower in cubic metres. (7)

7. A vehicle starts from rest and its velocity is measured every second for 6 seconds, with the following results:

Time t (s)	0	1	2	3	4	5	6
Velocity v (m/s)	0	1.2	2.4	3.7	5.2	6.0	9.2

   Using Simpson's rule, calculate (a) the distance travelled in 6 s (i.e. the area under the v/t graph) and (b) the average speed over this period. (6)

Section C

Complex numbers

Chapter 22

Complex numbers

Why it is important to understand: Complex numbers

Complex numbers are used in many scientific fields, including engineering, electromagnetism, quantum physics, and applied mathematics, such as chaos theory. Any physical motion which is periodic, such as an oscillating beam, string, wire, pendulum, electronic signal, or electromagnetic wave can be represented by a complex number function. This can make calculations with the various components simpler than with real numbers and sines and cosines. In control theory, systems are often transformed from the time domain to the frequency domain using the Laplace transform. In fluid dynamics, complex functions are used to describe potential flow in two dimensions. In electrical engineering, the Fourier transform is used to analyse varying voltages and currents. Complex numbers are used in signal analysis and other fields for a convenient description for periodically varying signals. This use is also extended into digital signal processing and digital image processing, which utilise digital versions of Fourier analysis (and wavelet analysis) to transmit, compress, restore, and otherwise process digital audio signals, still images, and video signals. Knowledge of complex numbers is clearly absolutely essential for further studies in so many engineering disciplines.

At the end of this chapter, you should be able to:

- define a complex number
- solve quadratic equations with imaginary roots
- use an Argand diagram to represent a complex number pictorially
- add, subtract, multiply and divide Cartesian complex numbers
- solve complex equations
- convert a Cartesian complex number into polar form, and vice-versa
- multiply and divide polar form complex numbers
- apply complex numbers to practical applications

22.1 Cartesian complex numbers

There are several applications of complex numbers in science and engineering, in particular in electrical alternating current theory and in mechanical vector analysis.

There are two main forms of complex number – **Cartesian form** (named after **Descartes**[*]) **and polar form** – and both are explained in this chapter.

If we can add, subtract, multiply and divide complex numbers in both forms and represent the numbers on an Argand diagram then a.c. theory and vector analysis become considerably easier.

(i) If the quadratic equation $x^2 + 2x + 5 = 0$ is solved using the quadratic formula then,

$$x = \frac{-2 \pm \sqrt{[(2)^2 - (4)(1)(5)]}}{2(1)}$$

$$= \frac{-2 \pm \sqrt{[-16]}}{2} = \frac{-2 \pm \sqrt{[(16)(-1)]}}{2}$$

$$= \frac{-2 \pm \sqrt{16}\sqrt{-1}}{2} = \frac{-2 \pm 4\sqrt{-1}}{2}$$

$$= -1 \pm 2\sqrt{-1}$$

It is not possible to evaluate $\sqrt{-1}$ in real terms. However, if an operator j is defined as $j = \sqrt{-1}$ then the solution may be expressed as $x = -1 \pm j2$.

(ii) $-1 + j2$ and $-1 - j2$ are known as **complex numbers**. Both solutions are of the form $a + jb$, 'a' being termed the **real part** and jb the **imaginary part**. A complex number of the form $a + jb$ is called **Cartesian complex number**.

(iii) In pure mathematics the symbol i is used to indicate $\sqrt{-1}$ (i being the first letter of the word imaginary). However i is the symbol of electric current in engineering, and to avoid possible confusion the next letter in the alphabet, j, is used to represent $\sqrt{-1}$

---

[*] **Who was Descartes**? For image and resume of Descartes, see page 144. To find out more go to **www.routledge.com/cw/bird**

Problem 1. Solve the quadratic equation $x^2 + 4 = 0$

Since $x^2 + 4 = 0$ then $x^2 = -4$ and $x = \sqrt{-4}$

i.e., $x = \sqrt{[(-1)(4)]} = \sqrt{(-1)}\sqrt{4} = j(\pm 2)$

$= \pm j2$, (since $j = \sqrt{-1}$)

(Note that $\pm j2$ may also be written $\pm 2j$)

Problem 2. Solve the quadratic equation $2x^2 + 3x + 5 = 0$

Using the quadratic formula,

$$x = \frac{-3 \pm \sqrt{[(3)^2 - 4(2)(5)]}}{2(2)}$$

$$= \frac{-3 \pm \sqrt{-31}}{4} = \frac{-3 \pm \sqrt{(-1)}\sqrt{31}}{4}$$

$$= \frac{-3 \pm j\sqrt{31}}{4}$$

Hence $x = -\dfrac{3}{4} \pm j\dfrac{\sqrt{31}}{4}$ or $-0.750 \pm j1.392$,

correct to 3 decimal places.

(Note, a graph of $y = 2x^2 + 3x + 5$ does not cross the x-axis and hence $2x^2 + 3x + 5 = 0$ has no real roots.)

Problem 3. Evaluate

(a) j^3 (b) j^4 (c) j^{23} (d) $\dfrac{-4}{j^9}$

(a) $j^3 = j^2 \times j = (-1) \times j = -j$, since $j^2 = -1$

(b) $j^4 = j^2 \times j^2 = (-1) \times (-1) = 1$

(c) $j^{23} = j \times j^{22} = j \times (j^2)^{11} = j \times (-1)^{11}$

$= j \times (-1) = -j$

(d) $j^9 = j \times j^8 = j \times (j^2)^4 = j \times (-1)^4$

$= j \times 1 = j$

Hence $\dfrac{-4}{j^9} = \dfrac{-4}{j} = \dfrac{-4}{j} \times \dfrac{-j}{-j} = \dfrac{4j}{-j^2}$

$= \dfrac{4j}{-(-1)} = 4j$ or $j4$

Now try the following Practice Exercise

Practice Exercise 96 Introduction to Cartesian complex numbers (Answers on page 869)

In Problems 1 to 9, solve the quadratic equations.

1. $x^2 + 25 = 0$
2. $x^2 - 2x + 2 = 0$
3. $x^2 - 4x + 5 = 0$
4. $x^2 - 6x + 10 = 0$
5. $2x^2 - 2x + 1 = 0$
6. $x^2 - 4x + 8 = 0$
7. $25x^2 - 10x + 2 = 0$
8. $2x^2 + 3x + 4 = 0$
9. $4t^2 - 5t + 7 = 0$
10. Evaluate (a) j^8 (b) $-\dfrac{1}{j^7}$ (c) $\dfrac{4}{2j^{13}}$

22.2 The Argand diagram

A complex number may be represented pictorially on rectangular or Cartesian axes. The horizontal (or x) axis is used to represent the real axis and the vertical (or y) axis is used to represent the imaginary axis. Such a diagram is called an **Argand diagram***. In Fig. 22.1, the point A represents the complex number $(3 + j2)$ and is obtained by plotting the co-ordinates $(3, \ j2)$ as in graphical work. Fig. 22.1 also shows the Argand points B, C and D representing the complex numbers $(-2 + j4), (-3 - j5)$ and $(1 - j3)$ respectively.

22.3 Addition and subtraction of complex numbers

Two complex numbers are added/subtracted by adding/subtracting separately the two real parts and the two imaginary parts.

For example, if $Z_1 = a + jb$ and $Z_2 = c + jd$,

$$\text{then} \quad Z_1 + Z_2 = (a + jb) + (c + jd)$$
$$= (a + c) + j(b + d)$$
$$\text{and} \quad Z_1 - Z_2 = (a + jb) - (c + jd)$$
$$= (a - c) + j(b - d)$$

Thus, for example,

$$(2 + j3) + (3 - j4) = 2 + j3 + 3 - j4$$
$$= \mathbf{5 - j1}$$

Figure 22.1

* **Who was Argand**? **Jean-Robert Argand** (18 July 1768–13 August 1822) was a highly influential mathematician. He privately published a landmark essay on the representation of imaginary quantities which became known as the **Argand diagram**. To find out more go to **www.routledge.com/cw/bird**

and $(2+j3)-(3-j4)=2+j3-3+j4$

$$=-1+j7$$

The addition and subtraction of complex numbers may be achieved graphically as shown in the Argand diagram of Fig. 22.2. $(2+j3)$ is represented by vector OP and $(3-j4)$ by vector OQ. In Fig. 22.2(a) by vector addition (i.e. the diagonal of the parallelogram) $OP+OQ=OR$. R is the point $(5,-j1)$

Hence $(2+j3)+(3-j4)=5-j1$

(a)

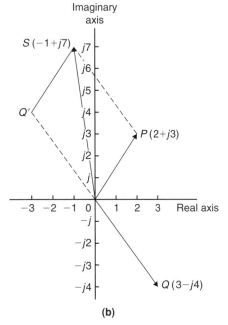

(b)

Figure 22.2

In Fig. 22.2(b), vector OQ is reversed (shown as OQ') since it is being subtracted. (Note $OQ=3-j4$ and $OQ'=-(3-j4)=-3+j4$) $OP-OQ=OP+OQ'=OS$ is found to be the Argand point $(-1, j7)$

Hence $(2+j3)-(3-j4)=-1+j7$

> **Problem 4.** Given $Z_1=2+j4$ and $Z_2=3-j$ determine (a) Z_1+Z_2, (b) Z_1-Z_2, (c) Z_2-Z_1 and show the results on an Argand diagram.

(a) $Z_1+Z_2=(2+j4)+(3-j)$

$$=(2+3)+j(4-1)=5+j3$$

(b) $Z_1-Z_2=(2+j4)-(3-j)$

$$=(2-3)+j(4-(-1))=-1+j5$$

(c) $Z_2-Z_1=(3-j)-(2+j4)$

$$=(3-2)+j(-1-4)=1-j5$$

Each result is shown in the Argand diagram of Fig. 22.3.

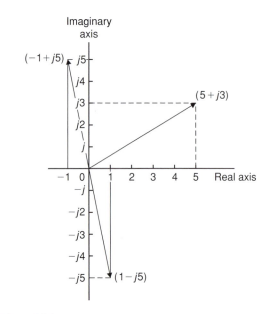

Figure 22.3

22.4 Multiplication and division of complex numbers

(i) **Multiplication of complex numbers** is achieved by assuming all quantities involved are real and then using $j^2=-1$ to simplify.

Hence $(a + jb)(c + jd)$

$$= ac + a(jd) + (jb)c + (jb)(jd)$$
$$= ac + jad + jbc + j^2bd$$
$$= (ac - bd) + j(ad + bc),$$
$$\text{since } j^2 = -1$$

Thus $(3 + j2)(4 - j5)$

$$= 12 - j15 + j8 - j^2 10$$
$$= (12 - (-10)) + j(-15 + 8)$$
$$= 22 - j7$$

(ii) The **complex conjugate** of a complex number is obtained by changing the sign of the imaginary part. Hence the complex conjugate of $a + jb$ is $a - jb$. The product of a complex number and its complex conjugate is always a real number.

For example,

$$(3 + j4)(3 - j4) = 9 - j12 + j12 - j^2 16$$
$$= 9 + 16 = 25$$

$[(a + jb)(a - jb)$ may be evaluated 'on sight' as $a^2 + b^2]$

(iii) **Division of complex numbers** is achieved by multiplying both numerator and denominator by the complex conjugate of the denominator.

For example,

$$\frac{2 - j5}{3 + j4} = \frac{2 - j5}{3 + j4} \times \frac{(3 - j4)}{(3 - j4)}$$

$$= \frac{6 - j8 - j15 + j^2 20}{3^2 + 4^2}$$

$$= \frac{-14 - j23}{25} = \frac{-14}{25} - j\frac{23}{25}$$

$$\text{or } -0.56 - j0.92$$

Problem 5. If $Z_1 = 1 - j3$, $Z_2 = -2 + j5$ and $Z_3 = -3 - j4$, determine in $a + jb$ form:

(a) $Z_1 Z_2$ (b) $\dfrac{Z_1}{Z_3}$

(c) $\dfrac{Z_1 Z_2}{Z_1 + Z_2}$ (d) $Z_1 Z_2 Z_3$

(a) $Z_1 Z_2 = (1 - j3)(-2 + j5)$

$$= -2 + j5 + j6 - j^2 15$$
$$= (-2 + 15) + j(5 + 6), \text{ since } j^2 = -1,$$
$$= 13 + j11$$

(b) $\dfrac{Z_1}{Z_3} = \dfrac{1 - j3}{-3 - j4} = \dfrac{1 - j3}{-3 - j4} \times \dfrac{-3 + j4}{-3 + j4}$

$$= \frac{-3 + j4 + j9 - j^2 12}{3^2 + 4^2}$$

$$= \frac{9 + j13}{25} = \frac{9}{25} + j\frac{13}{25}$$

$$\text{or } 0.36 + j0.52$$

(c) $\dfrac{Z_1 Z_2}{Z_1 + Z_2} = \dfrac{(1 - j3)(-2 + j5)}{(1 - j3) + (-2 + j5)}$

$$= \frac{13 + j11}{-1 + j2}, \text{ from part (a)},$$

$$= \frac{13 + j11}{-1 + j2} \times \frac{-1 - j2}{-1 - j2}$$

$$= \frac{-13 - j26 - j11 - j^2 22}{1^2 + 2^2}$$

$$= \frac{9 - j37}{5} = \frac{9}{5} - j\frac{37}{5} \text{ or } 1.8 - j7.4$$

(d) $Z_1 Z_2 Z_3 = (13 + j11)(-3 - j4)$, since

$$Z_1 Z_2 = 13 + j11, \text{ from part (a)}$$
$$= -39 - j52 - j33 - j^2 44$$
$$= (-39 + 44) - j(52 + 33)$$
$$= 5 - j85$$

Problem 6. Evaluate:

(a) $\dfrac{2}{(1 + j)^4}$ (b) $j\left(\dfrac{1 + j3}{1 - j2}\right)^2$

(a) $(1 + j)^2 = (1 + j)(1 + j) = 1 + j + j + j^2$

$$= 1 + j + j - 1 = j2$$

$(1 + j)^4 = [(1 + j)^2]^2 = (j2)^2 = j^2 4 = -4$

Hence $\dfrac{2}{(1 + j)^4} = \dfrac{2}{-4} = -\dfrac{1}{2}$

(b) $\dfrac{1+j3}{1-j2} = \dfrac{1+j3}{1-j2} \times \dfrac{1+j2}{1+j2}$

$$= \dfrac{1+j2+j3+j^2 6}{1^2+2^2} = \dfrac{-5+j5}{5}$$

$$= -1+j1 = -1+j$$

$$\left(\dfrac{1+j3}{1-j2}\right)^2 = (-1+j)^2 = (-1+j)(-1+j)$$

$$= 1-j-j+j^2 = -j2$$

Hence $j\left(\dfrac{1+j3}{1-j2}\right)^2 = j(-j2) = -j^2 2 = \mathbf{2},$

since $j^2 = -1$

Now try the following Practice Exercise

Practice Exercise 97   Operations involving Cartesian complex numbers (Answers on page 869)

1. Evaluate (a) $(3+j2)+(5-j)$ and (b) $(-2+j6)-(3-j2)$ and show the results on an Argand diagram.

2. Write down the complex conjugates of (a) $3+j4$, (b) $2-j$

3. If $z=2+j$ and $w=3-j$ evaluate
    (a) $z+w$ (b) $w-z$ (c) $3z-2w$
    (d) $5z+2w$ (e) $j(2w-3z)$ (f) $2jw-jz$

    In Problems 4 to 8 evaluate in $a+jb$ form given $Z_1 = 1+j2$, $Z_2 = 4-j3$, $Z_3 = -2+j3$ and $Z_4 = -5-j$.

4. (a) $Z_1+Z_2-Z_3$ (b) $Z_2-Z_1+Z_4$

5. (a) $Z_1 Z_2$ (b) $Z_3 Z_4$

6. (a) $Z_1 Z_3 + Z_4$ (b) $Z_1 Z_2 Z_3$

7. (a) $\dfrac{Z_1}{Z_2}$ (b) $\dfrac{Z_1+Z_3}{Z_2-Z_4}$

8. (a) $\dfrac{Z_1 Z_3}{Z_1+Z_3}$ (b) $Z_2 + \dfrac{Z_1}{Z_4} + Z_3$

9. Evaluate (a) $\dfrac{1-j}{1+j}$ (b) $\dfrac{1}{1+j}$

10. Show that $\dfrac{-25}{2}\left(\dfrac{1+j2}{3+j4} - \dfrac{2-j5}{-j}\right)$

    $= 57+j24$

22.5   Complex equations

If two complex numbers are equal, then their real parts are equal and their imaginary parts are equal. Hence if $a+jb = c+jd$, then $a=c$ and $b=d$

Problem 7.   Solve the complex equations:
(a) $2(x+jy) = 6-j3$
(b) $(1+j2)(-2-j3) = a+jb$

(a) $2(x+jy) = 6-j3$ hence $2x+j2y = 6-j3$

Equating the real parts gives:

$$2x=6, \text{ i.e. } \mathbf{\textit{x}=3}$$

Equating the imaginary parts gives:

$$2y=-3, \text{ i.e. } \mathbf{\textit{y}=-\tfrac{3}{2}}$$

(b) $(1+j2)(-2-j3) = a+jb$

$$-2-j3-j4-j^2 6 = a+jb$$

Hence $4-j7 = a+jb$

Equating real and imaginary terms gives:

$$\mathbf{\textit{a}=4} \text{ and } \mathbf{\textit{b}=-7}$$

Problem 8.   Solve the equations:

(a) $(2-j3) = \sqrt{(a+jb)}$
(b) $(x-j2y)+(y-j3x) = 2+j3$

(a) $(2-j3) = \sqrt{(a+jb)}$

Hence $\qquad (2-j3)^2 = a+jb,$

i.e. $\qquad (2-j3)(2-j3) = a+jb$

Hence $4-j6-j6+j^2 9 = a+jb$

and $\qquad -5-j12 = a+jb$

Thus $\mathbf{\textit{a}=-5}$ and $\mathbf{\textit{b}=-12}$

(b) $(x-j2y)+(y-j3x) = 2+j3$

Hence $(x+y)+j(-2y-3x) = 2+j3$

Equating real and imaginary parts gives:

$$x+y=2 \qquad\qquad (1)$$

and $-3x-2y=3 \qquad\qquad (2)$

i.e. two simultaneous equations to solve.

Multiplying equation (1) by 2 gives:

$$2x + 2y = 4 \qquad (3)$$

Adding equations (2) and (3) gives:

$$-x = 7, \text{ i.e., } x = -7$$

From equation (1), $y = 9$, which may be checked in equation (2).

Now try the following Practice Exercise

Practice Exercise 98   Complex equations (Answers on page 871)

In Problems 1 to 4 solve the complex equations.

1. $(2 + j)(3 - j2) = a + jb$

2. $\dfrac{2 + j}{1 - j} = j(x + jy)$

3. $(2 - j3) = \sqrt{(a + jb)}$

4. $(x - j2y) - (y - jx) = 2 + j$

5. If $Z = R + j\omega L + 1/j\omega C$, express Z in $(a + jb)$ form when $R = 10$, $L = 5$, $C = 0.04$ and $\omega = 4$

22.6 The polar form of a complex number

(i) Let a complex number z be $x + jy$ as shown in the Argand diagram of Fig. 22.4. Let distance OZ be r and the angle OZ makes with the positive real axis be θ

From trigonometry, $x = r \cos \theta$ and

$$y = r \sin \theta$$

Hence $Z = x + jy = r \cos \theta + jr \sin \theta$

$$= r(\cos \theta + j \sin \theta)$$

$Z = r(\cos \theta + j \sin \theta)$ is usually abbreviated to $Z = r\angle\theta$ which is known as the **polar form** of a complex number.

(ii) r is called the **modulus** (or magnitude) of Z and is written as mod Z or $|Z|$
r is determined using Pythagoras' theorem on triangle OAZ in Fig. 22.4,

Figure 22.4

i.e. $r = \sqrt{(x^2 + y^2)}$

(iii) θ is called the **argument** (or amplitude) of Z and is written as arg Z

By trigonometry on triangle OAZ,

$$\arg Z = \theta = \tan^{-1} \frac{y}{x}$$

(iv) Whenever changing from Cartesian form to polar form, or vice-versa, a sketch is invaluable for determining the quadrant in which the complex number occurs.

Problem 9.   Determine the modulus and argument of the complex number $Z = 2 + j3$, and express Z in polar form.

$Z = 2 + j3$ lies in the first quadrant as shown in Fig. 22.5.

Figure 22.5

Modulus, $|Z| = r = \sqrt{(2^2 + 3^2)} = \sqrt{13}$ or **3.606**, correct to 3 decimal places.

Argument, $\arg Z = \theta = \tan^{-1} \frac{3}{2}$

$$= \mathbf{56.31°} \text{ or } \mathbf{56°19'}$$

In polar form, $2 + j3$ is written as $\mathbf{3.606\angle 56.31°}$

Problem 10. Express the following complex numbers in polar form:

(a) $3+j4$ (b) $-3+j4$

(c) $-3-j4$ (d) $3-j4$

(a) $3+j4$ is shown in Fig. 22.6 and lies in the first quadrant.

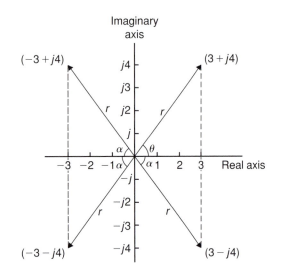

Figure 22.6

Modulus, $r=\sqrt{(3^2+4^2)}=5$ and argument $\theta=\tan^{-1}\frac{4}{3}=53.13°$

Hence $3+j4=5\angle53.13°$

(b) $-3+j4$ is shown in Fig. 22.6 and lies in the second quadrant.

Modulus, $r=5$ and angle $\alpha=53.13°$, from part (a).

Argument $=180°-53.13°=126.87°$ (i.e. the argument must be measured from the positive real axis).

Hence $-3+j4=5\angle126.87°$

(c) $-3-j4$ is shown in Fig. 22.6 and lies in the third quadrant.

Modulus, $r=5$ and $\alpha=53.13°$, as above.

Hence the argument $=180°+53.13°=233.13°$, which is the same as $-126.87°$

Hence $(-3-j4)=5\angle233.13°$ or $5\angle-126.87°$

(By convention the **principal value** is normally used, i.e. the numerically least value, such that $-\pi<\theta<\pi$)

(d) $3-j4$ is shown in Fig. 22.6 and lies in the fourth quadrant.

Modulus, $r=5$ and angle $\alpha=53.13°$, as above.

Hence $(3-j4)=5\angle-53.13°$

Problem 11. Convert (a) $4\angle30°$ (b) $7\angle-145°$ into $a+jb$ form, correct to 4 significant figures.

(a) $4\angle30°$ is shown in Fig. 22.7(a) and lies in the first quadrant.

(a)

(b)

Figure 22.7

Using trigonometric ratios, $x=4\cos30°=3.464$ and $y=4\sin30°=2.000$

Hence $4\angle30°=3.464+j2.000$

(b) $7\angle145°$ is shown in Fig. 22.7(b) and lies in the third quadrant.

Angle $\alpha=180°-145°=35°$

Hence $x=7\cos35°=5.734$
and $y=7\sin35°=4.015$

Hence $7\angle-145°=-5.734-j4.015$

Alternatively

$7\angle-145°=7\cos(-145°)+j7\sin(-145°)$

$\qquad\qquad=-5.734-j4.015$

Calculator

Using the 'Pol' and 'Rec' functions on a calculator enables changing from Cartesian to polar and vice-versa to be achieved more quickly.

Since complex numbers are used with vectors and with electrical engineering a.c. theory, it is essential that the calculator can be used quickly and accurately.

22.7 Multiplication and division in polar form

If $Z_1 = r_1 \angle \theta_1$ and $Z_2 = r_2 \angle \theta_2$ then:

(i) $Z_1 Z_2 = r_1 r_2 \angle (\theta_1 + \theta_2)$ and

(ii) $\dfrac{Z_1}{Z_2} = \dfrac{r_1}{r_2} \angle (\theta_1 - \theta_2)$

Problem 12. Determine, in polar form:

(a) $8 \angle 25° \times 4 \angle 60°$

(b) $3 \angle 16° \times 5 \angle -44° \times 2 \angle 80°$

(a) $8 \angle 25° \times 4 \angle 60° = (8 \times 4) \angle (25° + 60°) = \mathbf{32 \angle 85°}$

(b) $3 \angle 16° \times 5 \angle -44° \times 2 \angle 80°$

$= (3 \times 5 \times 2) \angle [16° + (-44°) + 80°] = \mathbf{30 \angle 52°}$

Problem 13. Evaluate in polar form

(a) $\dfrac{16 \angle 75°}{2 \angle 15°}$ (b) $\dfrac{10 \angle \frac{\pi}{4} \times 12 \angle \frac{\pi}{2}}{6 \angle -\frac{\pi}{3}}$

(a) $\dfrac{16 \angle 75°}{2 \angle 15°} = \dfrac{16}{2} \angle (75° - 15°) = \mathbf{8 \angle 60°}$

(b) $\dfrac{10 \angle \frac{\pi}{4} \times 12 \angle \frac{\pi}{2}}{6 \angle -\frac{\pi}{3}} = \dfrac{10 \times 12}{6} \angle \left(\dfrac{\pi}{4} + \dfrac{\pi}{2} - \left(-\dfrac{\pi}{3} \right) \right)$

$= 20 \angle \dfrac{13\pi}{12}$ or $20 \angle -\dfrac{11\pi}{12}$ or

$\mathbf{20 \angle 195°}$ or $\mathbf{20 \angle -165°}$

Problem 14. Evaluate, in polar form
$2 \angle 30° + 5 \angle -45° - 4 \angle 120°$

Addition and subtraction in polar form is not possible directly. Each complex number has to be converted into Cartesian form first.

$2 \angle 30° = 2(\cos 30° + j \sin 30°)$

$= 2 \cos 30° + j2 \sin 30° = 1.732 + j1.000$

$5 \angle -45° = 5(\cos(-45°) + j \sin(-45°))$

$= 5 \cos(-45°) + j5 \sin(-45°)$

$= 3.536 - j3.536$

$4 \angle 120° = 4(\cos 120° + j \sin 120°)$

$= 4 \cos 120° + j4 \sin 120°$

$= -2.000 + j3.464$

Hence $2 \angle 30° + 5 \angle -45° - 4 \angle 120°$

$= (1.732 + j1.000) + (3.536 - j3.536)$

$- (-2.000 + j3.464)$

$= 7.268 - j6.000$, which lies in the fourth quadrant

$= \sqrt{[(7.268)^2 + (6.000)^2]} \angle \tan^{-1} \left(\dfrac{-6.000}{7.268} \right)$

$= \mathbf{9.425 \angle -39.54°}$

Now try the following Practice Exercise

Practice Exercise 99 Polar form (Answers on page 871)

1. Determine the modulus and argument of (a) $2 + j4$ (b) $-5 - j2$ (c) $j(2 - j)$

In Problems 2 and 3 express the given Cartesian complex numbers in polar form, leaving answers in surd form.

2. (a) $2 + j3$ (b) -4 (c) $-6 + j$

3. (a) $-j3$ (b) $(-2 + j)^3$ (c) $j^3(1 - j)$

In Problems 4 and 5 convert the given polar complex numbers into $(a + jb)$ form giving answers correct to 4 significant figures.

4. (a) $5 \angle 30°$ (b) $3 \angle 60°$ (c) $7 \angle 45°$

5. (a) $6 \angle 125°$ (b) $4 \angle \pi$ (c) $3.5 \angle -120°$

In Problems 6 to 8, evaluate in polar form.

6. (a) $3\angle 20° \times 15\angle 45°$
    (b) $2.4\angle 65° \times 4.4\angle -21°$

7. (a) $6.4\angle 27° \div 2\angle -15°$
    (b) $5\angle 30° \times 4\angle 80° \div 10\angle -40°$

8. (a) $4\angle \dfrac{\pi}{6} + 3\angle \dfrac{\pi}{8}$
    (b) $2\angle 120° + 5.2\angle 58° - 1.6\angle -40°$

22.8 Applications of complex numbers

There are several applications of complex numbers in science and engineering, in particular in electrical alternating current theory and in mechanical vector analysis.

The effect of multiplying a phasor by j is to rotate it in a positive direction (i.e. anticlockwise) on an Argand diagram through $90°$ without altering its length. Similarly, multiplying a phasor by $-j$ rotates the phasor through $-90°$. These facts are used in a.c. theory since certain quantities in the phasor diagrams lie at $90°$ to each other. For example, in the $R-L$ series circuit shown in Fig. 22.8(a), V_L leads I by $90°$ (i.e. I lags V_L by $90°$) and may be written as jV_L, the vertical axis being regarded as the imaginary axis of an Argand diagram. Thus $V_R + jV_L = V$ and since $V_R = IR$, $V = IX_L$ (where X_L is the inductive reactance, $2\pi fL$ ohms) and $V = IZ$ (where Z is the impedance) then $R + jX_L = Z$

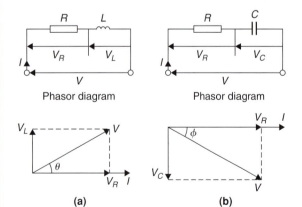

Phasor diagram (a)   Phasor diagram (b)

Figure 22.8

Similarly, for the $R-C$ circuit shown in Fig. 22.8(b), V_C lags I by $90°$ (i.e. I leads V_C by $90°$) and

$V_R - jV_C = V$, from which $R - jX_C = Z$ (where X_C is the capacitive reactance $\dfrac{1}{2\pi fC}$ ohms).

Problem 15. Determine the resistance and series inductance (or capacitance) for each of the following impedances, assuming a frequency of 50 Hz:

(a) $(4.0 + j7.0)\ \Omega$ (b) $-j20\ \Omega$

(c) $15\angle -60°\ \Omega$

(a) Impedance, $Z = (4.0 + j7.0)\ \Omega$ hence, **resistance = 4.0 Ω** and reactance $= 7.00\ \Omega$.

Since the imaginary part is positive, the reactance is inductive,

i.e. $X_L = 7.0\ \Omega$

Since $X_L = 2\pi fL$ then **inductance**,

$$L = \frac{X_L}{2\pi f} = \frac{7.0}{2\pi (50)} = \mathbf{0.0223\,H} \text{ or } \mathbf{22.3\,mH}$$

(b) Impedance, $Z = j20$, i.e. $Z = (0 - j20)\ \Omega$ hence **resistance = 0** and reactance $= 20\ \Omega$. Since the imaginary part is negative, the reactance is capacitive, i.e., $X_C = 20\ \Omega$ and since $X_C = \dfrac{1}{2\pi fC}$ then:

$$\textbf{capacitance, } C = \frac{1}{2\pi fX_C} = \frac{1}{2\pi (50)(20)}\ \text{F}$$

$$= \frac{10^6}{2\pi (50)(20)}\ \mu\text{F} = \mathbf{159.2\,\mu F}$$

(c) Impedance, Z

$$= 15\angle -60° = 15[\cos(-60°) + j\sin(-60°)]$$

$$= 7.50 - j12.99\ \Omega$$

Hence **resistance = 7.50 Ω** and capacitive reactance, $X_C = 12.99\ \Omega$

Since $X_C = \dfrac{1}{2\pi fC}$ then **capacitance**,

$$C = \frac{1}{2\pi fX_C} = \frac{10^6}{2\pi (50)(12.99)}\ \mu\text{F}$$

$$= \mathbf{245\,\mu F}$$

Problem 16. An alternating voltage of 240 V, 50 Hz is connected across an impedance of $(60 - j100)\ \Omega$. Determine (a) the resistance, (b) the capacitance, (c) the magnitude of the impedance and its phase angle and (d) the current flowing.

(a) Impedance $Z = (60 - j100)\,\Omega$.

Hence **resistance = 60 Ω**

(b) Capacitive reactance $X_C = 100\,\Omega$ and since $X_C = \dfrac{1}{2\pi f C}$ then

capacitance, $C = \dfrac{1}{2\pi f X_C} = \dfrac{1}{2\pi (50)(100)}$

$= \dfrac{10^6}{2\pi (50)(100)}\,\mu F$

$= \mathbf{31.83\,\mu F}$

(c) Magnitude of impedance,

$$|Z| = \sqrt{[(60)^2 + (-100)^2]} = \mathbf{116.6\,\Omega}$$

Phase angle, $\arg Z = \tan^{-1}\left(\dfrac{-100}{60}\right) = \mathbf{-59.04°}$

(d) Current flowing, $I = \dfrac{V}{Z} = \dfrac{240\angle 0°}{116.6\angle -59.04°}$

$= \mathbf{2.058\angle 59.04°\,A}$

The circuit and phasor diagrams are as shown in Fig. 22.8(b).

Problem 17. For the parallel circuit shown in Fig. 22.9, determine the value of current I and its phase relative to the 240 V supply, using complex numbers.

Figure 22.9

Current $I = \dfrac{V}{Z}$. Impedance Z for the three-branch parallel circuit is given by:

$$\frac{1}{Z} = \frac{1}{Z_1} + \frac{1}{Z_2} + \frac{1}{Z_3}$$

where $Z_1 = 4 + j3$, $Z_2 = 10$ and $Z_3 = 12 - j5$

Admittance, $Y_1 = \dfrac{1}{Z_1} = \dfrac{1}{4 + j3}$

$= \dfrac{1}{4 + j3} \times \dfrac{4 - j3}{4 - j3} = \dfrac{4 - j3}{4^2 + 3^2}$

$= 0.160 - j0.120$ siemens

Admittance, $Y_2 = \dfrac{1}{Z_2} = \dfrac{1}{10} = 0.10$ siemens

Admittance, $Y_3 = \dfrac{1}{Z_3} = \dfrac{1}{12 - j5}$

$= \dfrac{1}{12 - j5} \times \dfrac{12 + j5}{12 + j5} = \dfrac{12 + j5}{12^2 + 5^2}$

$= 0.0710 + j0.0296$ siemens

Total admittance, $Y = Y_1 + Y_2 + Y_3$

$= (0.160 - j0.120) + (0.10)$

$+ (0.0710 + j0.0296)$

$= 0.331 - j0.0904$

$= 0.343\angle -15.28°$ siemens

Current $I = \dfrac{V}{Z} = VY$

$= (240\angle 0°)(0.343\angle -15.28°)$

$= \mathbf{82.32\angle -15.28°\,A}$

Problem 18. Determine the magnitude and direction of the resultant of the three coplanar forces given below, when they act at a point.

Force A, 10 N acting at 45° from the positive horizontal axis.

Force B, 87 N acting at 120° from the positive horizontal axis.

Force C, 15 N acting at 210° from the positive horizontal axis.

The space diagram is shown in Fig. 22.10. The forces may be written as complex numbers.
Thus force A, $f_A = 10\angle 45°$, force B, $f_B = 8\angle 120°$ and force C, $f_C = 15\angle 210°$

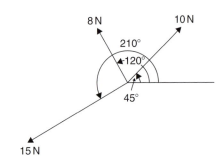

Figure 22.10

The resultant force

$$= f_A + f_B + f_C$$

$$= 10\angle 45° + 8\angle 120° + 15\angle 210°$$

$$= 10(\cos 45° + j\sin 45°) + 8(\cos 120°$$

$$+ j\sin 120°) + 15(\cos 210° + j\sin 210°)$$

$$= (7.071 + j7.071) + (-4.00 + j6.928)$$

$$+ (-12.99 - j7.50)$$

$$= -9.919 + j6.499$$

Magnitude of resultant force

$$= \sqrt{[(-9.919)^2 + (6.499)^2]} = \mathbf{11.86\,N}$$

Direction of resultant force

$$= \tan^{-1}\left(\frac{6.499}{-9.919}\right) = \mathbf{146.77°}$$

(since $-9.919 + j6.499$ lies in the second quadrant).

Now try the following Practice Exercise

Practice Exercise 100   Applications of complex numbers (Answers on page 871)

1. Determine the resistance R and series inductance L (or capacitance C) for each of the following impedances assuming the frequency to be 50 Hz.

   (a) $(3 + j8)\,\Omega$ (b) $(2 - j3)\,\Omega$
   (c) $j14\,\Omega$ (d) $8\angle -60°\,\Omega$

2. Two impedances, $Z_1 = (3 + j6)\,\Omega$ and $Z_2 = (4 - j3)\,\Omega$ are connected in series to a supply voltage of 120 V. Determine the magnitude of the current and its phase angle relative to the voltage.

3. If the two impedances in Problem 2 are connected in parallel determine the current flowing and its phase relative to the 120 V supply voltage.

4. A series circuit consists of a 12 Ω resistor, a coil of inductance 0.10 H and a capacitance of 160 μF. Calculate the current flowing and its phase relative to the supply voltage of 240 V, 50 Hz. Determine also the power factor of the circuit.

5. For the circuit shown in Fig. 22.11, determine the current I flowing and its phase relative to the applied voltage.

6. Determine, using complex numbers, the magnitude and direction of the resultant of the coplanar forces given below, which are acting at a point. Force A, 5 N acting horizontally, Force B, 9 N acting at an angle of 135° to force A, Force C, 12 N acting at an angle of 240° to force A.

Figure 22.11

7. A delta-connected impedance Z_A is given by:

$$Z_A = \frac{Z_1 Z_2 + Z_2 Z_3 + Z_3 Z_1}{Z_2}$$

Determine Z_A in both Cartesian and polar form given $Z_1 = (10 + j0)\,\Omega$, $Z_2 = (0 - j10)\,\Omega$ and $Z_3 = (10 + j10)\,\Omega$.

8. In the hydrogen atom, the angular momentum p of the de Broglie wave is given by: $p\psi = -\left(\dfrac{jh}{2\pi}\right)(\pm jm\psi)$. Determine an expression for p.

9. An aircraft P flying at a constant height has a velocity of $(400 + j300)$ km/h. Another aircraft Q at the same height has a velocity of $(200 - j600)$ km/h. Determine (a) the velocity of P relative to Q, and (b) the velocity of Q relative to P. Express the answers in polar form, correct to the nearest km/h.

10. Three vectors are represented by P, $2\angle 30°$, Q, $3\angle 90°$ and R, $4\angle -60°$. Determine in polar form the vectors represented by (a) $P + Q + R$, (b) $P - Q - R$

11. In a Schering bridge circuit,
$$Z_X = (R_X - jX_{C_X}), \quad Z_2 = -jX_{C_2},$$
$$Z_3 = \frac{(R_3)(-jX_{C_3})}{(R_3 - jX_{C_3})} \quad \text{and} \quad Z_4 = R_4$$
where $X_C = \dfrac{1}{2\pi fC}$

At balance: $(Z_X)(Z_3) = (Z_2)(Z_4)$.

Show that at balance $R_X = \dfrac{C_3 R_4}{C_2}$ and $C_X = \dfrac{C_2 R_3}{R_4}$

12. An amplifier has a transfer function T given by: $T = \dfrac{500}{1 + j\omega(5 \times 10^{-4})}$ where ω is the angular frequency. The gain of the amplifier is given by the modulus of T and the phase is given by the argument of T. If $\omega = 2000$ rad/s, determine the gain and the phase (in degrees).

13. The sending end current of a transmission line is given by: $I_S = \dfrac{V_S}{Z_0} \tanh PL$. Calculate the value of the sending current, in polar form, given $V_S = 200V$, $Z_0 = 560 + j420\,\Omega$, $P = 0.20$ and $L = 10$

With a calculator such as the CASIO 991ES PLUS it is possible, using the *complex mode*, to achieve many of the calculations in this chapter much more quickly.

For fully worked solutions to each of the problems in Practice Exercises 96 to 100 in this chapter, go to the website:
www.routledge.com/cw/bird

Chapter 23

De Moivre's theorem

Why it is important to understand: **De Moivre's theorem**

There are many, many examples of the use of complex numbers in engineering and science. De Moivre's theorem has several uses, including finding powers and roots of complex numbers, solving polynomial equations, calculating trigonometric identities, and for evaluating the sums of trigonometric series. The theorem is also used to calculate exponential and logarithmic functions of complex numbers. De Moivre's theorem has applications in electrical engineering and physics.

At the end of this chapter, you should be able to:

- state de Moivre's theorem
- calculate powers of complex numbers
- calculate roots of complex numbers
- state the exponential form of a complex number
- convert Cartesian/polar form into exponential form and vice-versa.
- determine loci in the complex plane

23.1 Introduction

From multiplication of complex numbers in polar form,

$$(r\angle\theta) \times (r\angle\theta) = r^2\angle 2\theta$$

Similarly, $(r\angle\theta) \times (r\angle\theta) \times (r\angle\theta) = r^3\angle 3\theta$, and so on. In general, **de Moivre's theorem***

$$[r\angle\theta]^n = r^n\angle n\theta$$

The theorem is true for all positive, negative and fractional values of n. The theorem is used to determine powers and roots of complex numbers.

23.2 Powers of complex numbers

For example $[3\angle 20°]^4 = 3^4\angle(4 \times 20°) = 81\angle 80°$ by de Moivre's theorem.

> **Problem 1.** Determine, in polar form
> (a) $[2\angle 35°]^5$ (b) $(-2 + j3)^6$

* **Who was de Moivre**? **Abraham de Moivre** (26 May 1667–27 November 1754) was a French mathematician famous for **de Moivre's formula**, which links complex numbers and trigonometry, and for his work on the normal distribution and probability theory. To find out more go to **www.routledge.com/cw/bird**

(a) $[2\angle 35°]^5 = 2^5\angle(5 \times 35°)$,

from de Moivre's theorem

$$= \mathbf{32\angle 175°}$$

(b) $(-2 + j3) = \sqrt{[(-2)^2 + (3)^2]}\angle\tan^{-1}\dfrac{3}{-2}$

$= \sqrt{13}\angle 123.69°$, since $-2 + j3$

lies in the second quadrant

$(-2 + j3)^6 = [\sqrt{13}\angle 123.69°]^6$

$= (\sqrt{13})^6\angle(6 \times 123.69°)$,

by de Moivre's theorem

$= 2197\angle 742.14°$

$= 2197\angle 382.14°$ (since 742.14

$\equiv 742.14° - 360° = 382.14°$)

$= \mathbf{2197\angle 22.14°}$ (since $382.14°$

$\equiv 382.14° - 360° = 22.14°$)

or $\mathbf{2197\angle 22°8'}$

> **Problem 2.** Determine the value of $(-7 + j5)^4$, expressing the result in polar and rectangular forms.

$(-7 + j5) = \sqrt{[(-7)^2 + 5^2]}\angle\tan^{-1}\dfrac{5}{-7}$

$= \sqrt{74}\angle 144.46°$

(Note, by considering the Argand diagram, $-7 + j5$ must represent an angle in the second quadrant and **not** in the fourth quadrant.)

Applying de Moivre's theorem:

$(-7 + j5)^4 = [\sqrt{74}\angle 144.46°]^4$

$= \sqrt{74}^4\angle 4 \times 144.46°$

$= 5476\angle 577.84°$

$= \mathbf{5476\angle 217.84°}$

or $\mathbf{5476\angle 217°50'}$ in polar form

Since $r\angle\theta = r\cos\theta + jr\sin\theta$,

$5476\angle 217.84° = 5476\cos 217.84°$

$+ j5476\sin 217.84°$

$= -4325 - j3359$

i.e. $\mathbf{(-7 + j5)^4 = -4325 - j3359}$

in rectangular form

Now try the following Practice Exercise

1. Determine in polar form (a) $[1.5\angle 15°]^5$ (b) $(1 + j2)^6$

2. Determine in polar and Cartesian forms (a) $[3\angle 41°]^4$ (b) $(-2 - j)^5$

3. Convert $(3 - j)$ into polar form and hence evaluate $(3 - j)^7$, giving the answer in polar form.

In problems 4 to 7, express in both polar and rectangular forms.

4. $(6 + j5)^3$

5. $(3 - j8)^5$

6. $(-2 + j7)^4$

7. $(-16 - j9)^6$

23.3 Roots of complex numbers

The **square root** of a complex number is determined by letting $n = 1/2$ in de Moivre's theorem,

i.e. $\sqrt{[r\angle\theta]} = [r\angle\theta]^{\frac{1}{2}} = r^{\frac{1}{2}}\angle\frac{1}{2}\theta = \sqrt{r}\angle\frac{\theta}{2}$

There are two square roots of a real number, equal in size but opposite in sign.

Problem 3. Determine the two square roots of the complex number $(5 + j12)$ in polar and Cartesian forms and show the roots on an Argand diagram.

$(5 + j12) = \sqrt{[5^2 + 12^2]}\angle\tan^{-1}\left(\frac{12}{5}\right)$

$= 13\angle 67.38°$

When determining square roots two solutions result. To obtain the second solution one way is to express $13\angle 67.38°$ also as $13\angle(67.38° + 360°)$, i.e. $13\angle 427.38°$. When the angle is divided by 2 an angle less than 360° is obtained.

Hence

$\sqrt{(5 + j12)} = \sqrt{[13\angle 67.38°]}$ and $\sqrt{[13\angle 427.38°]}$

$= [13\angle 67.38°]^{\frac{1}{2}}$ and $[13\angle 427.38°]^{\frac{1}{2}}$

$= 13^{\frac{1}{2}}\angle\left(\frac{1}{2} \times 67.38°\right)$ and

$13^{\frac{1}{2}}\angle\left(\frac{1}{2} \times 427.38°\right)$

$= \sqrt{13}\angle 33.69°$ and $\sqrt{13}\angle 213.69°$

$= 3.61\angle 33.69°$ and $3.61\angle 213.69°$

Thus, in polar form, the two roots are 3.61∠33.69° and 3.61∠−146.31°

$\sqrt{13}\angle 33.69° = \sqrt{13}(\cos 33.69° + j\sin 33.69°)$

$= 3.0 + j2.0$

$\sqrt{13}\angle 213.69° = \sqrt{13}(\cos 213.69° + j\sin 213.69°)$

$= -3.0 - j2.0$

Thus, in cartesian form the two roots are ±(3.0 + j2.0)

From the Argand diagram shown in Fig. 23.1 the two roots are seen to be 180° apart, which is always true when finding square roots of complex numbers.

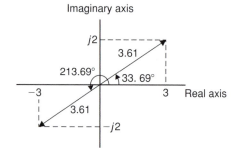

Figure 23.1

In general, **when finding the nth root of a complex number, there are n solutions**. For example, there are three solutions to a cube root, five solutions to a fifth root, and so on. In the solutions to the roots of a complex number, the modulus, r, is always the same, but the arguments, θ, are different. It is shown in Problem 3 that arguments are symmetrically spaced on an Argand diagram and are $(360/n)°$ apart, where n is the number of the roots required. Thus if one of the solutions to the cube root of a complex number is, say, $5\angle 20°$, the other two roots are symmetrically spaced $(360/3)°$, i.e. $120°$ from this root and the three roots are $5\angle 20°$, $5\angle 140°$ and $5\angle 260°$

Problem 4. Find the roots of $[(5+j3)]^{\frac{1}{2}}$ in rectangular form, correct to 4 significant figures.

$$(5+j3) = \sqrt{34}\angle 30.96°$$

Applying de Moivre's theorem:

$$(5+j3)^{\frac{1}{2}} = \sqrt{34}^{\frac{1}{2}}\angle \tfrac{1}{2} \times 30.96°$$

$$= 2.415\angle 15.48° \text{ or } 2.415\angle 15°29'$$

The second root may be obtained as shown above, i.e. having the same modulus but displaced $(360/2)°$ from the first root.

Thus, $(5+j3)^{\frac{1}{2}} = 2.415\angle(15.48° + 180°)$

$$= 2.415\angle 195.48°$$

In rectangular form:

$$2.415\angle 15.48° = 2.415\cos 15.48°$$
$$+ j2.415\sin 15.48°$$
$$= 2.327 + j0.6446$$

and $\quad 2.415\angle 195.48° = 2.415\cos 195.48°$
$$+ j2.415\sin 195.48°$$
$$= -2.327 - j0.6446$$

Hence $\quad [(5+j3)]^{\frac{1}{2}} = 2.415\angle 15.48° \text{ and}$

$$2.415\angle 195.48° \text{ or}$$
$$\pm(2.327 + j0.6446)$$

Problem 5. Express the roots of $(-14+j3)^{\frac{-2}{5}}$ in polar form.

$$(-14+j3) = \sqrt{205}\angle 167.905°$$

$$(-14+j3)^{\frac{-2}{5}} = \sqrt{205}^{\frac{-2}{5}}\angle\left[\left(-\frac{2}{5}\right) \times 167.905°\right]$$

$$= 0.3449\angle -67.164°$$

or $0.3449\angle -67°10'$

There are five roots to this complex number,

$$\left(x^{\frac{-2}{5}} = \frac{1}{x^{\frac{2}{5}}} = \frac{1}{\sqrt[5]{x^2}}\right)$$

The roots are symmetrically displaced from one another $(360/5)°$, i.e. $72°$ apart round an Argand diagram.

Thus the required roots are **$0.3449\angle -67°10'$, $0.3449\angle 4°50'$, $0.3449\angle 76°50'$, $0.3449\angle 148°50'$ and $0.3449\angle 220°50'$**

Now try the following Practice Exercise

Practice Exercise 102 The roots of complex numbers (Answers on page 871)

In Problems 1 to 3 determine the two square roots of the given complex numbers in Cartesian form and show the results on an Argand diagram.

1. (a) $1+j$ (b) j

2. (a) $3 - j4$ (b) $-1 - j2$

3. (a) $7\angle 60°$ (b) $12\angle\dfrac{3\pi}{2}$

In Problems 4 to 7, determine the moduli and arguments of the complex roots.

4. $(3+j4)^{\frac{1}{3}}$

5. $(-2+j)^{\frac{1}{4}}$

6. $(-6-j5)^{\frac{1}{2}}$

7. $(4-j3)^{\frac{-2}{3}}$

8. For a transmission line, the characteristic impedance Z_0 and the propagation coefficient

γ are given by:

$$Z_0 = \sqrt{\left(\frac{R + j\omega L}{G + j\omega C}\right)} \quad \text{and}$$

$$\gamma = \sqrt{[(R + j\omega L)(G + j\omega C)]}$$

Given $R = 25\,\Omega$, $L = 5 \times 10^{-3}\,\text{H}$, $G = 80 \times 10^{-6}$ siemens, $C = 0.04 \times 10^{-6}\,\text{F}$ and $\omega = 2000\pi$ rad/s, determine, in polar form, Z_0 and γ

23.4 The exponential form of a complex number

Certain mathematical functions may be expressed as power series (for example, by Maclaurin's series – see Chapter 8), three examples being:

$$\text{(i)} \quad e^x = 1 + x + \frac{x^2}{2!} + \frac{x^3}{3!} + \frac{x^4}{4!} + \frac{x^5}{5!} + \cdots \quad (1)$$

$$\text{(ii)} \quad \sin x = x - \frac{x^3}{3!} + \frac{x^5}{5!} - \frac{x^7}{7!} + \cdots \quad (2)$$

$$\text{(iii)} \quad \cos x = 1 - \frac{x^2}{2!} + \frac{x^4}{4!} - \frac{x^6}{6!} + \cdots \quad (3)$$

Replacing x in equation (1) by the imaginary number $j\theta$ gives:

$$e^{j\theta} = 1 + j\theta + \frac{(j\theta)^2}{2!} + \frac{(j\theta)^3}{3!} + \frac{(j\theta)^4}{4!} + \frac{(j\theta)^5}{5!} + \cdots$$

$$= 1 + j\theta + \frac{j^2\theta^2}{2!} + \frac{j^3\theta^3}{3!} + \frac{j^4\theta^4}{4!} + \frac{j^5\theta^5}{5!} + \cdots$$

By definition, $j = \sqrt{(-1)}$, hence $j^2 = -1$, $j^3 = -j$, $j^4 = 1$, $j^5 = j$, and so on.

Thus $e^{j\theta} = 1 + j\theta - \frac{\theta^2}{2!} - j\frac{\theta^3}{3!} + \frac{\theta^4}{4!} + j\frac{\theta^5}{5!} - \cdots$

Grouping real and imaginary terms gives:

$$e^{j\theta} = \left(1 - \frac{\theta^2}{2!} + \frac{\theta^4}{4!} - \cdots\right)$$
$$+ j\left(\theta - \frac{\theta^3}{3!} + \frac{\theta^5}{5!} - \cdots\right)$$

However, from equations (2) and (3):

$$\left(1 - \frac{\theta^2}{2!} + \frac{\theta^4}{4!} - \cdots\right) = \cos\theta$$

and

$$\left(\theta - \frac{\theta^3}{3!} + \frac{\theta^5}{5!} - \cdots\right) = \sin\theta$$

Thus $\qquad e^{j\theta} = \cos\theta + j\sin\theta \qquad (4)$

Writing $-\theta$ for θ in equation (4), gives:

$$e^{j(-\theta)} = \cos(-\theta) + j\sin(-\theta)$$

However, $\cos(-\theta) = \cos\theta$ and $\sin(-\theta) = -\sin\theta$

Thus $\qquad e^{-j\theta} = \cos\theta - j\sin\theta \qquad (5)$

The polar form of a complex number z is: $z = r(\cos\theta + j\sin\theta)$. But, from equation (4), $\cos\theta + j\sin\theta = e^{j\theta}$

Therefore $\qquad z = re^{j\theta}$

When a complex number is written in this way, it is said to be expressed in **exponential form**.

There are therefore three ways of expressing a complex number:

(i) $z = (a + jb)$, called **Cartesian** or **rectangular form**,

(ii) $z = r(\cos\theta + j\sin\theta)$ or $r\angle\theta$, called **polar form**, and

(iii) $z = re^{j\theta}$ called **exponential form**.

The exponential form is obtained from the polar form. For example, $4\angle 30°$ becomes $4e^{j\frac{\pi}{6}}$ in exponential form. (Note that in $re^{j\theta}$, θ must be in radians.)

Problem 6. Change $(3 - j4)$ into (a) polar form, (b) exponential form.

(a) $(3 - j4) = 5\angle-53.13°$ or $5\angle-0.927$

in polar form

(b) $(3 - j4) = 5\angle-0.927 = 5e^{-j0.927}$

in exponential form

Problem 7. Convert $7.2e^{j1.5}$ into rectangular form.

$7.2e^{j1.5} = 7.2\angle 1.5\,\text{rad}(= 7.2\angle 85.94°)$ in polar form

$= 7.2\cos 1.5 + j7.2\sin 1.5$

$= (0.509 + j7.182)$ in rectangular form

Problem 8. Express $z = 2e^{1+j\frac{\pi}{3}}$ in Cartesian form.

$z = (2e^1)\left(e^{j\frac{\pi}{3}}\right)$ by the laws of indices

$= (2e^1)\angle\dfrac{\pi}{3}$ (or $2e\angle 60°$)in polar form

$= 2e\left(\cos\dfrac{\pi}{3} + j\sin\dfrac{\pi}{3}\right)$

$= (2.718 + j4.708)$ in Cartesian form

Problem 9. Change $6e^{2-j3}$ into $(a + jb)$ form.

$6e^{2-j3} = (6e^2)(e^{-j3})$ by the laws of indices

$= 6e^2\angle -3$ rad   (or $6e^2\angle -171.89^0$) in polar form

$= 6e^2[\cos(-3) + j\sin(-3)]$

$= (-43.89 - j6.26)$ in $(a + jb)$ form

Problem 10. If $z = 4e^{j1.3}$, determine $\ln z$ (a) in Cartesian form, and (b) in polar form.

If $z = re^{j\theta}$ then $\ln z = \ln(re^{j\theta})$

$= \ln r + \ln e^{j\theta}$

i.e. $\ln z = \ln r + j\theta,$

by the laws of logarithms

(a) Thus if $z = 4e^{j1.3}$ then $\ln z = \ln(4e^{j1.3})$

$= \mathbf{\ln 4 + j1.3}$

(or $\mathbf{1.386 + j1.300}$) in Cartesian form.

(b) $(1.386 + j1.300) = \mathbf{1.90\angle 43.17°}$ or $\mathbf{1.90\angle 0.753}$ in polar form.

Problem 11. Given $z = 3e^{1-j}$, find $\ln z$ in polar form.

If $z = 3e^{1-j}$, then

$\ln$ $z = \ln(3e^{1-j})$

$= \ln 3 + \ln e^{1-j}$

$= \ln 3 + 1 - j$

$= (1 + \ln 3) - j$

$= 2.0986 - j1.0000$

$= \mathbf{2.325\angle -25.48°}$ or $\mathbf{2.325\angle -0.445}$

Problem 12. Determine, in polar form, $\ln(3 + j4)$

$\ln(3 + j4) = \ln[5\angle 0.927] = \ln[5e^{j0.927}]$

$= \ln 5 + \ln(e^{j0.927})$

$= \ln 5 + j0.927$

$= 1.609 + j0.927$

$= \mathbf{1.857\angle 29.95°}$ or $\mathbf{1.857\angle 0.523}$

Now try the following Practice Exercise

Practice Exercise 103   The exponential form of complex numbers (Answers on page 871)

1. Change $(5 + j3)$ into exponential form.

2. Convert $(-2.5 + j4.2)$ into exponential form.

3. Change $3.6e^{j2}$ into Cartesian form.

4. Express $2e^{3+j\frac{\pi}{6}}$ in $(a + jb)$ form.

5. Convert $1.7e^{1.2-j2.5}$ into rectangular form.

6. If $z = 7e^{j2.1}$, determine $\ln z$ (a) in Cartesian form, and (b) in polar form.

7. Given $z = 4e^{1.5-j2}$, determine $\ln z$ in polar form.

8. Determine in polar form (a) $\ln(2 + j5)$ (b) $\ln(-4 - j3)$

9. When displaced electrons oscillate about an equilibrium position the displacement x is given by the equation:

$$x = Ae^{\left\{-\frac{ht}{2m} + j\frac{\sqrt{(4mf - h^2)}}{2m - a}t\right\}}$$

Determine the real part of x in terms of t, assuming $(4mf - h^2)$ is positive.

23.5   Introduction to locus problems

The locus is a set of points that satisfy a certain condition. For example, the locus of points that are, say, 2 units from point C, refers to the set of all points that are

2 units from C; this would be a circle with centre C as shown in Fig. 23.2.

This circle is the locus of points 2 units from C

Figure 23.2

It is sometimes needed to find the locus of a point which moves in the Argand diagram according to some stated condition. Loci (the plural of locus) are illustrated by the following worked problems.

Problem 13. Determine the locus defined by $|z| = 4$, given that $z = x + jy$

If $z = x + jy$, then on an Argand diagram as shown in Figure 23.3, the **modulus z**,

$$|z| = \sqrt{x^2 + y^2}$$

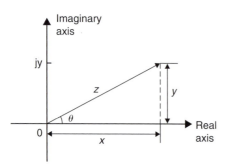

Figure 23.3

In this case, $\sqrt{x^2 + y^2} = 4$ from which, $x^2 + y^2 = 4^2$

From Chapter 14, $x^2 + y^2 = 4^2$ **is a circle, with centre at the origin and with radius 4**

The locus (or path) of $|z| = 4$ is shown in Fig. 23.4.

Problem 14. Determine the locus defined by arg $z = \dfrac{\pi}{4}$, given that $z = x + jy$

In Fig. 23.3 above, $\qquad \boldsymbol{\theta = \tan^{-1}\left(\dfrac{y}{x}\right)}$

where θ is called the **argument** and is written as $\boldsymbol{\arg z = \tan^{-1}\left(\dfrac{y}{x}\right)}$

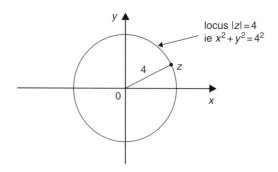

locus $|z| = 4$ ie $x^2 + y^2 = 4^2$

Figure 23.4

Hence, in this example,

$$\tan^{-1}\left(\frac{y}{x}\right) = \frac{\pi}{4} \text{ i.e. } \frac{y}{x} = \tan\frac{\pi}{4} = \tan 45° = 1$$

Thus, $\qquad$ if $\dfrac{y}{x} = 1$, then $\boldsymbol{y = x}$

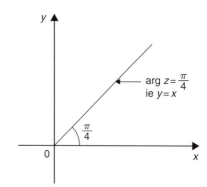

arg $z = \dfrac{\pi}{4}$ ie $y = x$

Figure 23.5

Hence, the locus (or path) **of arg $z = \dfrac{\pi}{4}$ is a straight line $y = x$** (with $y > 0$) as shown in Fig. 23.5.

Problem 15. If $z = x + jy$, determine the locus defined by arg $(z - 1) = \dfrac{\pi}{6}$

If arg $(z - 1) = \dfrac{\pi}{6}$, then arg $(x + jy - 1) = \dfrac{\pi}{6}$

i.e.
$$\arg[(x - 1) + jy] = \frac{\pi}{6}$$

In Fig. 25.3,

$$\theta = \tan^{-1}\left(\frac{y}{x}\right) \text{ i.e. arg } z = \tan^{-1}\left(\frac{y}{x}\right)$$

Hence, in this example,

$$\tan^{-1}\left(\frac{y}{x-1}\right) = \frac{\pi}{6} \text{ i.e. } \frac{y}{x-1} = \tan\frac{\pi}{6} = \tan 30° = \frac{1}{\sqrt{3}}$$

Thus,   if $\dfrac{y}{x-1} = \dfrac{1}{\sqrt{3}}$, then $y = \dfrac{1}{\sqrt{3}}(x-1)$

Hence, the locus of arg $(z-1) = \dfrac{\pi}{6}$ is a straight line

$$y = \dfrac{1}{\sqrt{3}}x - \dfrac{1}{\sqrt{3}}$$

Problem 16. Determine the locus defined by $|z-2| = 3$, given that $z = x + jy$

If $z = x + jy$,   then $|z-2| = |x + jy - 2|$
$$= |(x-2) + jy| = 3$$

On the Argand diagram shown in Figure 23.3, $|z| = \sqrt{x^2 + y^2}$

Hence, in this case, $|z-2| = \sqrt{(x-2)^2 + y^2} = 3$ from which, $(x-2)^2 + y^2 = 3^2$

From Chapter 14, $(x-a)^2 + (y-b)^2 = r^2$ is a circle, with centre (a, b) and radius r.

Hence, $(x-2)^2 + y^2 = 3^2$ is a circle, with centre $(2, 0)$ and radius 3

The locus of $|z-2| = 3$ is shown in Figure 23.6.

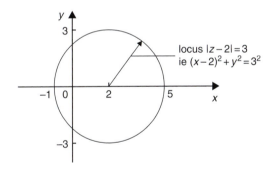

locus $|z-2| = 3$
ie $(x-2)^2 + y^2 = 3^2$

Figure 23.6

Problem 17. If $z = x + jy$, determine the locus defined by $\left|\dfrac{z-1}{z+1}\right| = 3$

$z - 1 = x + jy - 1 = (x-1) + jy$
$z + 1 = x + jy + 1 = (x+1) + jy$

Hence, $\left|\dfrac{z-1}{z+1}\right| = \left|\dfrac{(x-1) + jy}{(x+1) + jy}\right|$

$$= \dfrac{\sqrt{(x-1)^2 + y^2}}{\sqrt{(x+1)^2 + y^2}} = 3$$

and squaring both sides gives:

$$\dfrac{(x-1)^2 + y^2}{(x+1)^2 + y^2} = 9$$

from which,

$$(x-1)^2 + y^2 = 9[(x+1)^2 + y^2]$$
$$x^2 - 2x + 1 + y^2 = 9[x^2 + 2x + 1 + y^2]$$
$$x^2 - 2x + 1 + y^2 = 9x^2 + 18x + 9 + 9y^2$$
$$0 = 8x^2 + 20x + 8 + 8y^2$$

i.e.

$$8x^2 + 20x + 8 + 8y^2 = 0$$

and dividing by 4 gives:

$2x^2 + 5x + 2 + 2y^2 = 0$ which is the equation of the locus.

Rearranging gives: $x^2 + \dfrac{5}{2}x + y^2 = -1$

Completing the square gives:

$$\left(x + \dfrac{5}{4}\right)^2 - \dfrac{25}{16} + y^2 = -1$$

i.e. $\left(x + \dfrac{5}{4}\right)^2 + y^2 = -1 + \dfrac{25}{16}$

i.e. $\left(x + \dfrac{5}{4}\right)^2 + y^2 = \dfrac{9}{16}$

i.e. $\left(x + \dfrac{5}{4}\right)^2 + y^2 = \left(\dfrac{3}{4}\right)^2$ which is the equation of a circle.

Hence the locus defined by $\left|\dfrac{z-1}{z+1}\right| = 3$ is a circle of centre $\left(-\dfrac{5}{4}, 0\right)$ and radius $\dfrac{3}{4}$

Problem 18. If $z = x + jy$, determine the locus defined by $\arg\left(\dfrac{z+1}{z}\right) = \dfrac{\pi}{4}$

$$\left(\dfrac{z+1}{z}\right) = \left(\dfrac{(x+1) + jy}{x + jy}\right)$$

$$= \dfrac{[(x+1) + jy](x - jy)}{(x + jy)(x - jy)}$$

$$= \dfrac{x(x+1) - j(x+1)y + jxy + y^2}{x^2 + y^2}$$

$$= \dfrac{x^2 + x - jxy - jy + jxy + y^2}{x^2 + y^2}$$

$$= \frac{x^2 + x - jy + y^2}{x^2 + y^2}$$

$$= \frac{(x^2 + x + y^2) - jy}{x^2 + y^2}$$

$$= \frac{x^2 + x + y^2}{x^2 + y^2} - \frac{jy}{x^2 + y^2}$$

Since $\arg \left(\frac{z+1}{z} \right) = \frac{\pi}{4}$ then $\tan^{-1} \left(\dfrac{\dfrac{-y}{x^2 + y^2}}{\dfrac{x^2 + x + y^2}{x^2 + y^2}} \right) = \frac{\pi}{4}$

i.e.

$$\tan^{-1} \left(\frac{-y}{x^2 + x + y^2} \right) = \frac{\pi}{4}$$

from which,

$$\frac{-y}{x^2 + x + y^2} = \tan \frac{\pi}{4} = 1$$

Hence,

$$-y = x^2 + x + y^2$$

Hence, **the locus defined by** $\arg \left(\dfrac{z+1}{z} \right) = \dfrac{\pi}{4}$ **is:**
$x^2 + x + y + y^2 = 0$

Completing the square gives:

$$\left(x + \frac{1}{2} \right)^2 + \left(y + \frac{1}{2} \right)^2 = \frac{1}{2}$$

which is a circle, centre $\left(-\dfrac{1}{2}, -\dfrac{1}{2} \right)$ **and radius** $\dfrac{1}{\sqrt{2}}$

Problem 19. Determine the locus defined by $|z - j| = |z - 3|$, given that $z = x + jy$

Since $|z - j| = |z - 3|$

then $\qquad |x + j(y - 1)| = |(x - 3) + jy|$

and $\qquad \sqrt{x^2 + (y - 1)^2} = \sqrt{(x - 3)^2 + y^2}$

Squaring both sides gives:
$$x^2 + (y - 1)^2 = (x - 3)^2 + y^2$$

i.e. $\qquad x^2 + y^2 - 2y + 1 = x^2 - 6x + 9 + y^2$

from which, $\qquad -2y + 1 = -6x + 9$

i.e. $\qquad 6x - 8 = 2y$

or $\qquad y = 3x - 4$

Hence, the locus defined by $|z - j| = |z - 3|$ **is a straight line:** $y = 3x - 4$

Now try the following Practice Exercise

Practice Exercise 104 Locus problems (Answers on page 872)

For each of the following, if $z = x + jy$, (a) determine the equation of the locus, (b) sketch the locus.

1. $|z| = 2$

2. $|z| = 5$

3. $\arg(z - 2) = \dfrac{\pi}{3}$

4. $\arg(z + 1) = \dfrac{\pi}{6}$

5. $|z - 2| = 4$

6. $|z + 3| = 5$

7. $\left| \dfrac{z+1}{z-1} \right| = 3$

8. $\left| \dfrac{z-1}{z} \right| = \sqrt{2}$

9. $\arg \left| \dfrac{z-1}{z} \right| = \dfrac{\pi}{4}$

10. $\arg \left| \dfrac{z+2}{z} \right| = \dfrac{\pi}{4}$

11. $|z + j| = |z + 2|$

12. $|z - 4| = |z - 2j|$

13. $|z - 1| = |z|$

For fully worked solutions to each of the problems in Practice Exercises 101 to 104 in this chapter, go to the website:
www.routledge.com/cw/bird

Matrices and determinants

Chapter 24

The theory of matrices and determinants

Why it is important to understand: The theory of matrices and determinants

Matrices are used to solve problems in electronics, optics, quantum mechanics, statics, robotics, linear programming, optimisation, genetics, and much more. Matrix calculus is a mathematical tool used in connection with linear equations, linear transformations, systems of differential equations, and so on, and is vital for calculating forces, vectors, tensions, masses, loads and a lot of other factors that must be accounted for in engineering to ensure safe and resource-efficient structure. Electrical and mechanical engineers, chemists, biologists and scientists all need knowledge of matrices to solve problems. In computer graphics, matrices are used to project a three-dimensional image on to a two-dimensional screen, and to create realistic motion. Matrices are therefore very important in solving engineering problems.

At the end of this chapter, you should be able to:

- understand matrix notation
- add, subtract and multiply 2 by 2 and 3 by 3 matrices
- recognise the unit matrix
- calculate the determinant of a 2 by 2 matrix
- determine the inverse (or reciprocal) of a 2 by 2 matrix
- calculate the determinant of a 3 by 3 matrix
- determine the inverse (or reciprocal) of a 3 by 3 matrix

24.1 Matrix notation

Matrices and determinants are mainly used for the solution of linear simultaneous equations. The theory of matrices and determinants is dealt with in this chapter and this theory is then used in Chapter 25 to solve simultaneous equations.

The coefficients of the variables for linear simultaneous equations may be shown in matrix form.

The coefficients of x and y in the simultaneous equations

$$x + 2y = 3$$
$$4x - 5y = 6$$

become $\begin{pmatrix} 1 & 2 \\ 4 & -5 \end{pmatrix}$ in matrix notation.

Similarly, the coefficients of p, q and r in the equations

$$1.3p - 2.0q + r = 7$$

$$3.7p + 4.8q - 7r = 3$$

$$4.1p + 3.8q + 12r = -6$$

become $\begin{pmatrix} 1.3 & -2.0 & 1 \\ 3.7 & 4.8 & -7 \\ 4.1 & 3.8 & 12 \end{pmatrix}$ in matrix form.

The numbers within a matrix are called an **array** and the coefficients forming the array are called the **elements** of the matrix. The number of rows in a matrix is usually specified by m and the number of columns by n and a matrix referred to as an 'm by n' matrix. Thus, $\begin{pmatrix} 2 & 3 & 6 \\ 4 & 5 & 7 \end{pmatrix}$ is a '2 by 3' matrix. Matrices cannot be expressed as a single numerical value, but they can often be simplified or combined, and unknown element values can be determined by comparison methods. Just as there are rules for addition, subtraction, multiplication and division of numbers in arithmetic, rules for these operations can be applied to matrices and the rules of matrices are such that they obey most of those governing the algebra of numbers.

24.2 Addition, subtraction and multiplication of matrices

(i) Addition of matrices

Corresponding elements in two matrices may be added to form a single matrix.

Problem 1. Add the matrices

(a) $\begin{pmatrix} 2 & -1 \\ -7 & 4 \end{pmatrix}$ and $\begin{pmatrix} -3 & 0 \\ 7 & -4 \end{pmatrix}$

(b) $\begin{pmatrix} 3 & 1 & -4 \\ 4 & 3 & 1 \\ 1 & 4 & -3 \end{pmatrix}$ and $\begin{pmatrix} 2 & 7 & -5 \\ -2 & 1 & 0 \\ 6 & 3 & 4 \end{pmatrix}$

(a) Adding the corresponding elements gives:

$$\begin{pmatrix} 2 & -1 \\ -7 & 4 \end{pmatrix} + \begin{pmatrix} -3 & 0 \\ 7 & -4 \end{pmatrix}$$

$$= \begin{pmatrix} 2 + (-3) & -1 + 0 \\ -7 + 7 & 4 + (-4) \end{pmatrix}$$

$$= \begin{pmatrix} \mathbf{-1} & \mathbf{-1} \\ \mathbf{0} & \mathbf{0} \end{pmatrix}$$

(b) Adding the corresponding elements gives:

$$\begin{pmatrix} 3 & 1 & -4 \\ 4 & 3 & 1 \\ 1 & 4 & -3 \end{pmatrix} + \begin{pmatrix} 2 & 7 & -5 \\ -2 & 1 & 0 \\ 6 & 3 & 4 \end{pmatrix}$$

$$= \begin{pmatrix} 3+2 & 1+7 & -4+(-5) \\ 4+(-2) & 3+1 & 1+0 \\ 1+6 & 4+3 & -3+4 \end{pmatrix}$$

$$= \begin{pmatrix} \mathbf{5} & \mathbf{8} & \mathbf{-9} \\ \mathbf{2} & \mathbf{4} & \mathbf{1} \\ \mathbf{7} & \mathbf{7} & \mathbf{1} \end{pmatrix}$$

(ii) Subtraction of matrices

If A is a matrix and B is another matrix, then $(A - B)$ is a single matrix formed by subtracting the elements of B from the corresponding elements of A.

Problem 2. Subtract

(a) $\begin{pmatrix} -3 & 0 \\ 7 & -4 \end{pmatrix}$ from $\begin{pmatrix} 2 & -1 \\ -7 & 4 \end{pmatrix}$

(b) $\begin{pmatrix} 2 & 7 & -5 \\ -2 & 1 & 0 \\ 6 & 3 & 4 \end{pmatrix}$ from $\begin{pmatrix} 3 & 1 & -4 \\ 4 & 3 & 1 \\ 1 & 4 & -3 \end{pmatrix}$

To find matrix A minus matrix B, the elements of B are taken from the corresponding elements of A. Thus:

(a) $\begin{pmatrix} 2 & -1 \\ -7 & 4 \end{pmatrix} - \begin{pmatrix} -3 & 0 \\ 7 & -4 \end{pmatrix}$

$$= \begin{pmatrix} 2 - (-3) & -1 - 0 \\ -7 - 7 & 4 - (-4) \end{pmatrix}$$

$$= \begin{pmatrix} \mathbf{5} & \mathbf{-1} \\ \mathbf{-14} & \mathbf{8} \end{pmatrix}$$

(b) $\begin{pmatrix} 3 & 1 & -4 \\ 4 & 3 & 1 \\ 1 & 4 & -3 \end{pmatrix} - \begin{pmatrix} 2 & 7 & -5 \\ -2 & 1 & 0 \\ 6 & 3 & 4 \end{pmatrix}$

$$= \begin{pmatrix} 3-2 & 1-7 & -4-(-5) \\ 4-(-2) & 3-1 & 1-0 \\ 1-6 & 4-3 & -3-4 \end{pmatrix}$$

$$= \begin{pmatrix} \mathbf{1} & \mathbf{-6} & \mathbf{1} \\ \mathbf{6} & \mathbf{2} & \mathbf{1} \\ \mathbf{-5} & \mathbf{1} & \mathbf{-7} \end{pmatrix}$$

Problem 3. If

$A = \begin{pmatrix} -3 & 0 \\ 7 & -4 \end{pmatrix}, B = \begin{pmatrix} 2 & -1 \\ -7 & 4 \end{pmatrix}$ and

$C = \begin{pmatrix} 1 & 0 \\ -2 & -4 \end{pmatrix}$ find $A + B - C$

$$A + B = \begin{pmatrix} -1 & -1 \\ 0 & 0 \end{pmatrix}$$

(from Problem 1)

Hence, $A + B - C = \begin{pmatrix} -1 & -1 \\ 0 & 0 \end{pmatrix} - \begin{pmatrix} 1 & 0 \\ -2 & -4 \end{pmatrix}$

$$= \begin{pmatrix} -1-1 & -1-0 \\ 0-(-2) & 0-(-4) \end{pmatrix}$$

$$= \begin{pmatrix} -2 & -1 \\ 2 & 4 \end{pmatrix}$$

Alternatively $A + B - C$

$$= \begin{pmatrix} -3 & 0 \\ 7 & -4 \end{pmatrix} + \begin{pmatrix} 2 & -1 \\ -7 & 4 \end{pmatrix} - \begin{pmatrix} 1 & 0 \\ -2 & -4 \end{pmatrix}$$

$$= \begin{pmatrix} -3+2-1 & 0+(-1)-0 \\ 7+(-7)-(-2) & -4+4-(-4) \end{pmatrix}$$

$$= \begin{pmatrix} -2 & -1 \\ 2 & 4 \end{pmatrix} \text{ as obtained previously}$$

(iii) Multiplication

When a matrix is multiplied by a number, called **scalar multiplication**, a single matrix results in which each element of the original matrix has been multiplied by the number.

Problem 4. If $A = \begin{pmatrix} -3 & 0 \\ 7 & -4 \end{pmatrix}$,

$B = \begin{pmatrix} 2 & -1 \\ -7 & 4 \end{pmatrix}$ and $C = \begin{pmatrix} 1 & 0 \\ -2 & -4 \end{pmatrix}$ find

$2A - 3B + 4C$

For scalar multiplication, each element is multiplied by the scalar quantity, hence

$$2A = 2\begin{pmatrix} -3 & 0 \\ 7 & -4 \end{pmatrix} = \begin{pmatrix} -6 & 0 \\ 14 & -8 \end{pmatrix}$$

$$3B = 3\begin{pmatrix} 2 & -1 \\ -7 & 4 \end{pmatrix} = \begin{pmatrix} 6 & -3 \\ -21 & 12 \end{pmatrix}$$

and $4C = 4\begin{pmatrix} 1 & 0 \\ -2 & -4 \end{pmatrix} = \begin{pmatrix} 4 & 0 \\ -8 & -16 \end{pmatrix}$

Hence $2A - 3B + 4C$

$$= \begin{pmatrix} -6 & 0 \\ 14 & -8 \end{pmatrix} - \begin{pmatrix} 6 & -3 \\ -21 & 12 \end{pmatrix} + \begin{pmatrix} 4 & 0 \\ -8 & -16 \end{pmatrix}$$

$$= \begin{pmatrix} -6-6+4 & 0-(-3)+0 \\ 14-(-21)+(-8) & -8-12+(-16) \end{pmatrix}$$

$$= \begin{pmatrix} -8 & 3 \\ 27 & -36 \end{pmatrix}$$

When a matrix A is multiplied by another matrix B, a single matrix results in which elements are obtained from the sum of the products of the corresponding rows of A and the corresponding columns of B.

Two matrices A and B may be multiplied together, provided the number of elements in the rows of matrix A are equal to the number of elements in the columns of matrix B. In general terms, when multiplying a matrix of dimensions (m by n) by a matrix of dimensions (n by r), the resulting matrix has dimensions (m by r). Thus a 2 by 3 matrix multiplied by a 3 by 1 matrix gives a matrix of dimensions 2 by 1.

Problem 5. If $A = \begin{pmatrix} 2 & 3 \\ 1 & -4 \end{pmatrix}$ and $B = \begin{pmatrix} -5 & 7 \\ -3 & 4 \end{pmatrix}$ find $A \times B$

Let $A \times B = C$ where $C = \begin{pmatrix} C_{11} & C_{12} \\ C_{21} & C_{22} \end{pmatrix}$

C_{11} is the sum of the products of the first row elements of A and the first column elements of B taken one at a time,

i.e. $C_{11} = (2 \times (-5)) + (3 \times (-3)) = -19$

C_{12} is the sum of the products of the first row elements of A and the second column elements of B, taken one at a time,

i.e. $C_{12} = (2 \times 7) + (3 \times 4) = 26$

C_{21} is the sum of the products of the second row elements of A and the first column elements of B, taken one at a time,

i.e. $C_{21} = (1 \times (-5)) + (-4 \times (-3)) = 7$

Finally, C_{22} is the sum of the products of the second row elements of A and the second column elements of B, taken one at a time,

i.e. $C_{22} = (1 \times 7) + ((-4) \times 4) = -9$

Thus, $A \times B = \begin{pmatrix} -19 & 26 \\ 7 & -9 \end{pmatrix}$

Section E

Problem 6. Simplify

$$\begin{pmatrix} 3 & 4 & 0 \\ -2 & 6 & -3 \\ 7 & -4 & 1 \end{pmatrix} \times \begin{pmatrix} 2 \\ 5 \\ -1 \end{pmatrix}$$

The sum of the products of the elements of each row of the first matrix and the elements of the second matrix, (called a **column matrix**), are taken one at a time. Thus:

$$\begin{pmatrix} 3 & 4 & 0 \\ -2 & 6 & -3 \\ 7 & -4 & 1 \end{pmatrix} \times \begin{pmatrix} 2 \\ 5 \\ -1 \end{pmatrix}$$

$$= \begin{pmatrix} (3 \times 2) & + (4 \times 5) & + (0 \times (-1)) \\ (-2 \times 2) & + (6 \times 5) & + (-3 \times (-1)) \\ (7 \times 2) & + (-4 \times 5) & + (1 \times (-1)) \end{pmatrix}$$

$$= \begin{pmatrix} 26 \\ 29 \\ -7 \end{pmatrix}$$

Problem 7. If $A = \begin{pmatrix} 3 & 4 & 0 \\ -2 & 6 & -3 \\ 7 & -4 & 1 \end{pmatrix}$ and

$B = \begin{pmatrix} 2 & -5 \\ 5 & -6 \\ -1 & -7 \end{pmatrix}$, find $A \times B$

The sum of the products of the elements of each row of the first matrix and the elements of each column of the second matrix are taken one at a time. Thus:

$$\begin{pmatrix} 3 & 4 & 0 \\ -2 & 6 & -3 \\ 7 & -4 & 1 \end{pmatrix} \times \begin{pmatrix} 2 & -5 \\ 5 & -6 \\ -1 & -7 \end{pmatrix}$$

$$= \begin{pmatrix} \begin{array}{l} [(3 \times 2) \\ + (4 \times 5) \\ + (0 \times (-1))] \end{array} & \begin{array}{l} [(3 \times (-5)) \\ + (4 \times (-6)) \\ + (0 \times (-7))] \end{array} \\ \begin{array}{l} [(-2 \times 2) \\ + (6 \times 5) \\ + (-3 \times (-1))] \end{array} & \begin{array}{l} [(-2 \times (-5)) \\ + (6 \times (-6)) \\ + (-3 \times (-7))] \end{array} \\ \begin{array}{l} [(7 \times 2) \\ + (-4 \times 5) \\ + (1 \times (-1))] \end{array} & \begin{array}{l} [(7 \times (-5)) \\ + (-4 \times (-6)) \\ + (1 \times (-7))] \end{array} \end{pmatrix}$$

$$= \begin{pmatrix} 26 & -39 \\ 29 & -5 \\ -7 & -18 \end{pmatrix}$$

Problem 8. Determine

$$\begin{pmatrix} 1 & 0 & 3 \\ 2 & 1 & 2 \\ 1 & 3 & 1 \end{pmatrix} \times \begin{pmatrix} 2 & 2 & 0 \\ 1 & 3 & 2 \\ 3 & 2 & 0 \end{pmatrix}$$

The sum of the products of the elements of each row of the first matrix and the elements of each column of the second matrix are taken one at a time. Thus:

$$\begin{pmatrix} 1 & 0 & 3 \\ 2 & 1 & 2 \\ 1 & 3 & 1 \end{pmatrix} \times \begin{pmatrix} 2 & 2 & 0 \\ 1 & 3 & 2 \\ 3 & 2 & 0 \end{pmatrix}$$

$$= \begin{pmatrix} \begin{array}{l} [(1 \times 2) \\ + (0 \times 1) \\ + (3 \times 3)] \end{array} & \begin{array}{l} [(1 \times 2) \\ + (0 \times 3) \\ + (3 \times 2)] \end{array} & \begin{array}{l} [(1 \times 0) \\ + (0 \times 2) \\ + (3 \times 0)] \end{array} \\ \begin{array}{l} [(2 \times 2) \\ + (1 \times 1) \\ + (2 \times 3)] \end{array} & \begin{array}{l} [(2 \times 2) \\ + (1 \times 3) \\ + (2 \times 2)] \end{array} & \begin{array}{l} [(2 \times 0) \\ + (1 \times 2) \\ + (2 \times 0)] \end{array} \\ \begin{array}{l} [(1 \times 2) \\ + (3 \times 1) \\ + (1 \times 3)] \end{array} & \begin{array}{l} [(1 \times 2) \\ + (3 \times 3) \\ + (1 \times 2)] \end{array} & \begin{array}{l} [(1 \times 0) \\ + (3 \times 2) \\ + (1 \times 0)] \end{array} \end{pmatrix}$$

$$= \begin{pmatrix} 11 & 8 & 0 \\ 11 & 11 & 2 \\ 8 & 13 & 6 \end{pmatrix}$$

In algebra, the commutative law of multiplication states that $a \times b = b \times a$. For matrices, this law is only true in a few special cases, and in general $A \times B$ is **not** equal to $B \times A$

Problem 9. If $A = \begin{pmatrix} 2 & 3 \\ 1 & 0 \end{pmatrix}$ and

$B = \begin{pmatrix} 2 & 3 \\ 0 & 1 \end{pmatrix}$ show that $A \times B \neq B \times A$

$$A \times B = \begin{pmatrix} 2 & 3 \\ 1 & 0 \end{pmatrix} \times \begin{pmatrix} 2 & 3 \\ 0 & 1 \end{pmatrix}$$

$$= \begin{pmatrix} [(2 \times 2) + (3 \times 0)] & [(2 \times 3) + (3 \times 1)] \\ [(1 \times 2) + (0 \times 0)] & [(1 \times 3) + (0 \times 1)] \end{pmatrix}$$

$$= \begin{pmatrix} 4 & 9 \\ 2 & 3 \end{pmatrix}$$

$$B \times A = \begin{pmatrix} 2 & 3 \\ 0 & 1 \end{pmatrix} \times \begin{pmatrix} 2 & 3 \\ 1 & 0 \end{pmatrix}$$

$$= \begin{pmatrix} [(2 \times 2) + (3 \times 1)] & [(2 \times 3) + (3 \times 0)] \\ [(0 \times 2) + (1 \times 1)] & [(0 \times 3) + (1 \times 0)] \end{pmatrix}$$

$$= \begin{pmatrix} 7 & 6 \\ 1 & 0 \end{pmatrix}$$

Since $\begin{pmatrix} 4 & 9 \\ 2 & 3 \end{pmatrix} \neq \begin{pmatrix} 7 & 6 \\ 1 & 0 \end{pmatrix}$, then $A \times B \neq B \times A$

Now try the following Practice Exercise

Practice Exercise 105   Addition, subtraction and multiplication of matrices (Answers on page 872)

In Problems 1 to 13, the matrices A to K are:

$A = \begin{pmatrix} 3 & -1 \\ -4 & 7 \end{pmatrix}$ $B = \begin{pmatrix} 5 & 2 \\ -1 & 6 \end{pmatrix}$

$C = \begin{pmatrix} -1.3 & 7.4 \\ 2.5 & -3.9 \end{pmatrix}$

$D = \begin{pmatrix} 4 & -7 & 6 \\ -2 & 4 & 0 \\ 5 & 7 & -4 \end{pmatrix}$

$E = \begin{pmatrix} 3 & 6 & 2 \\ 5 & -3 & 7 \\ -1 & 0 & 2 \end{pmatrix}$

$F = \begin{pmatrix} 3.1 & 2.4 & 6.4 \\ -1.6 & 3.8 & -1.9 \\ 5.3 & 3.4 & -4.8 \end{pmatrix}$ $G = \begin{pmatrix} 6 \\ -2 \end{pmatrix}$

$H = \begin{pmatrix} -2 \\ 5 \end{pmatrix}$ $J = \begin{pmatrix} 4 \\ -11 \\ 7 \end{pmatrix}$ $K = \begin{pmatrix} 1 & 0 \\ 0 & 1 \\ 1 & 0 \end{pmatrix}$

In Problems 1 to 12, perform the matrix operation stated.

1. $A + B$

2. $D + E$

3. $A - B$

4. $A + B - C$

5. $5A + 6B$

6. $2D + 3E - 4F$

7. $A \times H$

8. $A \times B$

9. $A \times C$

10. $D \times J$

11. $E \times K$

12. $D \times F$

13. Show that $A \times C \neq C \times A$

24.3   The unit matrix

A **unit matrix, I**, is one in which all elements of the leading diagonal ($\backslash$) have a value of 1 and all other elements have a value of 0. Multiplication of a matrix by I is the equivalent of multiplying by 1 in arithmetic.

24.4   The determinant of a 2 by 2 matrix

The **determinant** of a 2 by 2 matrix $\begin{pmatrix} a & b \\ c & d \end{pmatrix}$ is defined as $(ad - bc)$

The elements of the determinant of a matrix are written between vertical lines. Thus, the determinant of $\begin{pmatrix} 3 & -4 \\ 1 & 6 \end{pmatrix}$ is written as $\begin{vmatrix} 3 & -4 \\ 1 & 6 \end{vmatrix}$ and is equal to $(3 \times 6) - (-4 \times 1)$, i.e. $18 - (-4)$ or 22. Hence the determinant of a matrix can be expressed as a single numerical value, i.e. $\begin{vmatrix} 3 & -4 \\ 1 & 6 \end{vmatrix} = 22$

Problem 10.   Determine the value of

$$\begin{vmatrix} 3 & -2 \\ 7 & 4 \end{vmatrix}$$

$$\begin{vmatrix} 3 & -2 \\ 7 & 4 \end{vmatrix} = (3 \times 4) - (-2 \times 7)$$

$$= 12 - (-14) = \mathbf{26}$$

Problem 11.   Evaluate $\begin{vmatrix} (1+j) & j2 \\ -j3 & (1-j4) \end{vmatrix}$

$$\begin{vmatrix} (1+j) & j2 \\ -j3 & (1-j4) \end{vmatrix} = (1+j)(1-j4) - (j2)(-j3)$$

$$= 1 - j4 + j - j^2 4 + j^2 6$$

$$= 1 - j4 + j - (-4) + (-6)$$

since from Chapter 22, $j^2 = -1$

$$= 1 - j4 + j + 4 - 6$$

$$= \mathbf{-1 - j\,3}$$

Problem 12. Evaluate $\begin{vmatrix} 5\angle 30° & 2\angle -60° \\ 3\angle 60° & 4\angle -90° \end{vmatrix}$

$$\begin{vmatrix} 5\angle 30° & 2\angle -60° \\ 3\angle 60° & 4\angle -90° \end{vmatrix} = (5\angle 30°)(4\angle -90°)$$

$$- (2\angle -60°)(3\angle 60°)$$

$$= (20\angle -60°) - (6\angle 0°)$$

$$= (10 - j17.32) - (6 + j0)$$

$$= (4 - j\,17.32) \text{ or } 17.78\angle -77°$$

Now try the following Practice Exercise

Practice Exercise 106   2 by 2 determinants
(Answers on page 872)

1. Calculate the determinant of $\begin{pmatrix} 3 & -1 \\ -4 & 7 \end{pmatrix}$

2. Calculate the determinant of
$\begin{pmatrix} -2 & 5 \\ 3 & -6 \end{pmatrix}$

3. Calculate the determinant of
$\begin{pmatrix} -1.3 & 7.4 \\ 2.5 & -3.9 \end{pmatrix}$

4. Evaluate $\begin{vmatrix} j2 & -j3 \\ (1+j) & j \end{vmatrix}$

5. Evaluate $\begin{vmatrix} 2\angle 40° & 5\angle -20° \\ 7\angle -32° & 4\angle -117° \end{vmatrix}$

6. Given matrix $A = \begin{pmatrix} (x-2) & 6 \\ 2 & (x-3) \end{pmatrix}$,
determine values of x for which $|A| = 0$

24.5 The inverse or reciprocal of a 2 by 2 matrix

The inverse of matrix A is A^{-1} such that $A \times A^{-1} = I$, the unit matrix.

Let matrix A be $\begin{pmatrix} 1 & 2 \\ 3 & 4 \end{pmatrix}$ and let the inverse matrix, A^{-1}

be $\begin{pmatrix} a & b \\ c & d \end{pmatrix}$

Then, since $A \times A^{-1} = I$,

$$\begin{pmatrix} 1 & 2 \\ 3 & 4 \end{pmatrix} \times \begin{pmatrix} a & b \\ c & d \end{pmatrix} = \begin{pmatrix} 1 & 0 \\ 0 & 1 \end{pmatrix}$$

Multiplying the matrices on the left-hand side, gives

$$\begin{pmatrix} a+2c & b+2d \\ 3a+4c & 3b+4d \end{pmatrix} = \begin{pmatrix} 1 & 0 \\ 0 & 1 \end{pmatrix}$$

Equating corresponding elements gives:

$$b + 2d = 0, \text{ i.e. } b = -2d$$

and $3a + 4c = 0$, i.e. $a = -\dfrac{4}{3}c$

Substituting for a and b gives:

$$\begin{pmatrix} -\dfrac{4}{3}c + 2c & -2d + 2d \\ 3\left(-\dfrac{4}{3}c\right) + 4c & 3(-2d) + 4d \end{pmatrix} = \begin{pmatrix} 1 & 0 \\ 0 & 1 \end{pmatrix}$$

i.e. $\begin{pmatrix} \dfrac{2}{3}c & 0 \\ 0 & -2d \end{pmatrix} = \begin{pmatrix} 1 & 0 \\ 0 & 1 \end{pmatrix}$

showing that $\dfrac{2}{3}c = 1$, i.e. $c = \dfrac{3}{2}$ and $-2d = 1$, i.e. $d = -\dfrac{1}{2}$

Since $b = -2d$, $b = 1$ and since $a = -\dfrac{4}{3}c$, $a = -2$

Thus the inverse of matrix $\begin{pmatrix} 1 & 2 \\ 3 & 4 \end{pmatrix}$ is $\begin{pmatrix} a & b \\ c & d \end{pmatrix}$ that is,

$$\begin{pmatrix} -2 & 1 \\ \dfrac{3}{2} & -\dfrac{1}{2} \end{pmatrix}$$

There is, however, **a quicker method of obtaining the inverse** of a 2 by 2 matrix.

For any matrix $\begin{pmatrix} p & q \\ r & s \end{pmatrix}$ the inverse may be obtained by:

(i)   interchanging the positions of p and s,

(ii) changing the signs of q and r, and

(iii) multiplying this new matrix by the reciprocal of the determinant of $\begin{pmatrix} p & q \\ r & s \end{pmatrix}$

Thus the inverse of matrix $\begin{pmatrix} 1 & 2 \\ 3 & 4 \end{pmatrix}$ is

$$\frac{1}{4-6}\begin{pmatrix} 4 & -2 \\ -3 & 1 \end{pmatrix} = \begin{pmatrix} -2 & 1 \\ \dfrac{3}{2} & -\dfrac{1}{2} \end{pmatrix}$$

as obtained previously.

Problem 13. Determine the inverse of
$\begin{pmatrix} 3 & -2 \\ 7 & 4 \end{pmatrix}$

The inverse of matrix $\begin{pmatrix} p & q \\ r & s \end{pmatrix}$ is obtained by interchanging the positions of p and s, changing the signs of q and r and multiplying by the reciprocal of the determinant $\begin{vmatrix} p & q \\ r & s \end{vmatrix}$. Thus, the inverse of

$$\begin{pmatrix} 3 & -2 \\ 7 & 4 \end{pmatrix} = \frac{1}{(3 \times 4) - (-2 \times 7)} \begin{pmatrix} 4 & 2 \\ -7 & 3 \end{pmatrix}$$

$$= \frac{1}{26} \begin{pmatrix} 4 & 2 \\ -7 & 3 \end{pmatrix} = \begin{pmatrix} \dfrac{2}{13} & \dfrac{1}{13} \\ \dfrac{-7}{26} & \dfrac{3}{26} \end{pmatrix}$$

Now try the following Practice Exercise

Practice Exercise 107   The inverse of 2 by 2 matrices (Answers on page 872)

1. Determine the inverse of $\begin{pmatrix} 3 & -1 \\ -4 & 7 \end{pmatrix}$

2. Determine the inverse of $\begin{pmatrix} \dfrac{1}{2} & \dfrac{2}{3} \\ -\dfrac{1}{3} & -\dfrac{3}{5} \end{pmatrix}$

3. Determine the inverse of $\begin{pmatrix} -1.3 & 7.4 \\ 2.5 & -3.9 \end{pmatrix}$

24.6 The determinant of a 3 by 3 matrix

(i) The **minor** of an element of a 3 by 3 matrix is the value of the 2 by 2 determinant obtained by covering up the row and column containing that element.

Thus for the matrix $\begin{pmatrix} 1 & 2 & 3 \\ 4 & 5 & 6 \\ 7 & 8 & 9 \end{pmatrix}$ the minor of element 4 is obtained by covering the row $(4\ 5\ 6)$ and the column $\begin{pmatrix} 1 \\ 4 \\ 7 \end{pmatrix}$, leaving the 2 by 2 determinant $\begin{vmatrix} 2 & 3 \\ 8 & 9 \end{vmatrix}$, i.e. the minor of element 4 is $(2 \times 9) - (3 \times 8) = -6$

(ii) The sign of a minor depends on its position within the matrix, the sign pattern being $\begin{pmatrix} + & - & + \\ - & + & - \\ + & - & + \end{pmatrix}$.

Thus the signed-minor of element 4 in the matrix $\begin{pmatrix} 1 & 2 & 3 \\ 4 & 5 & 6 \\ 7 & 8 & 9 \end{pmatrix}$ is $-\begin{vmatrix} 2 & 3 \\ 8 & 9 \end{vmatrix} = -(-6) = 6$

The signed-minor of an element is called the **cofactor** of the element.

(iii) **The value of a 3 by 3 determinant is the sum of the products of the elements and their cofactors of any row or any column of the corresponding 3 by 3 matrix.**

There are thus six different ways of evaluating a 3×3 determinant – and all should give the same value.

Problem 14.   Find the value of

$$\begin{vmatrix} 3 & 4 & -1 \\ 2 & 0 & 7 \\ 1 & -3 & -2 \end{vmatrix}$$

The value of this determinant is the sum of the products of the elements and their cofactors, of any row or of any column. If the second row or second column is selected, the element 0 will make the product of the element and its cofactor zero and reduce the amount of arithmetic to be done to a minimum.

Supposing a second row expansion is selected. The minor of 2 is the value of the determinant remaining when the row and column containing the 2 (i.e. the second row and the first column), is covered up.

Thus the cofactor of element 2 is $\begin{vmatrix} 4 & -1 \\ -3 & -2 \end{vmatrix}$ i.e. -11

The sign of element 2 is minus (see (ii) above), hence the cofactor of element 2, (the signed-minor) is $+11$

Similarly the minor of element 7 is $\begin{vmatrix} 3 & 4 \\ 1 & -3 \end{vmatrix}$ i.e. -13, and its cofactor is $+13$. Hence the value of the sum of the products of the elements and their cofactors is $2 \times 11 + 7 \times 13$, i.e.,

$$\begin{vmatrix} 3 & 4 & -1 \\ 2 & 0 & 7 \\ 1 & -3 & -2 \end{vmatrix} = 2(11) + 0 + 7(13) = \mathbf{113}$$

The same result will be obtained whichever row or column is selected. For example, the third column expansion is

$$(-1)\begin{vmatrix} 2 & 0 \\ 1 & -3 \end{vmatrix} - 7\begin{vmatrix} 3 & 4 \\ 1 & -3 \end{vmatrix} + (-2)\begin{vmatrix} 3 & 4 \\ 2 & 0 \end{vmatrix}$$

$$= 6 + 91 + 16 = \mathbf{113}, \text{ as obtained previously.}$$

Problem 15. Evaluate $\begin{vmatrix} 1 & 4 & -3 \\ -5 & 2 & 6 \\ -1 & -4 & 2 \end{vmatrix}$

Using the first row: $\begin{vmatrix} 1 & 4 & -3 \\ -5 & 2 & 6 \\ -1 & -4 & 2 \end{vmatrix}$

$$= 1\begin{vmatrix} 2 & 6 \\ -4 & 2 \end{vmatrix} - 4\begin{vmatrix} -5 & 6 \\ -1 & 2 \end{vmatrix} + (-3)\begin{vmatrix} -5 & 2 \\ -1 & -4 \end{vmatrix}$$

$$= (4 + 24) - 4(-10 + 6) - 3(20 + 2)$$

$$= 28 + 16 - 66 = \mathbf{-22}$$

Using the second column: $\begin{vmatrix} 1 & 4 & -3 \\ -5 & 2 & 6 \\ -1 & -4 & 2 \end{vmatrix}$

$$= -4\begin{vmatrix} -5 & 6 \\ -1 & 2 \end{vmatrix} + 2\begin{vmatrix} 1 & -3 \\ -1 & 2 \end{vmatrix} - (-4)\begin{vmatrix} 1 & -3 \\ -5 & 6 \end{vmatrix}$$

$$= -4(-10 + 6) + 2(2 - 3) + 4(6 - 15)$$

$$= 16 - 2 - 36 = \mathbf{-22}$$

Problem 16. Determine the value of
$$\begin{vmatrix} j2 & (1+j) & 3 \\ (1-j) & 1 & j \\ 0 & j4 & 5 \end{vmatrix}$$

Using the first column, the value of the determinant is:

$$(j2)\begin{vmatrix} 1 & j \\ j4 & 5 \end{vmatrix} - (1-j)\begin{vmatrix} (1+j) & 3 \\ j4 & 5 \end{vmatrix}$$

$$+ (0)\begin{vmatrix} (1+j) & 3 \\ 1 & j \end{vmatrix}$$

$$= j2(5 - j^2 4) - (1 - j)(5 + j5 - j12) + 0$$

$$= j2(9) - (1 - j)(5 - j7)$$

$$= j18 - [5 - j7 - j5 + j^2 7]$$

$$= j18 - [-2 - j12]$$

$$= j18 + 2 + j12 = \mathbf{2 + j\,30} \text{ or } \mathbf{30.07\angle 86.19°}$$

Now try the following Practice Exercise

Practice Exercise 108   3 by 3 determinants (Answers on page 873)

1. Find the matrix of minors of
$$\begin{pmatrix} 4 & -7 & 6 \\ -2 & 4 & 0 \\ 5 & 7 & -4 \end{pmatrix}$$

2. Find the matrix of cofactors of
$$\begin{pmatrix} 4 & -7 & 6 \\ -2 & 4 & 0 \\ 5 & 7 & -4 \end{pmatrix}$$

3. Calculate the determinant of
$$\begin{pmatrix} 4 & -7 & 6 \\ -2 & 4 & 0 \\ 5 & 7 & -4 \end{pmatrix}$$

4. Evaluate $\begin{vmatrix} 8 & -2 & -10 \\ 2 & -3 & -2 \\ 6 & 3 & 8 \end{vmatrix}$

5. Calculate the determinant of
$$\begin{pmatrix} 3.1 & 2.4 & 6.4 \\ -1.6 & 3.8 & -1.9 \\ 5.3 & 3.4 & -4.8 \end{pmatrix}$$

6. Evaluate $\begin{vmatrix} j2 & 2 & j \\ (1+j) & 1 & -3 \\ 5 & -j4 & 0 \end{vmatrix}$

7. Evaluate $\begin{vmatrix} 3\angle 60° & j2 & 1 \\ 0 & (1+j) & 2\angle 30° \\ 0 & 2 & j5 \end{vmatrix}$

8. Find the eigenvalues λ that satisfy the following equations:

   (a) $\begin{vmatrix} (2 - \lambda) & 2 \\ -1 & (5 - \lambda) \end{vmatrix} = 0$

   (b) $\begin{vmatrix} (5 - \lambda) & 7 & -5 \\ 0 & (4 - \lambda) & -1 \\ 2 & 8 & (-3 - \lambda) \end{vmatrix} = 0$

   (You may need to refer to Chapter 1, pages 10–14, for the solution of cubic equations).

24.7 The inverse or reciprocal of a 3 by 3 matrix

The **adjoint** of a matrix A is obtained by:

(i)   forming a matrix B of the cofactors of A, and

(ii) **transposing** matrix B to give B^T, where B^T is the matrix obtained by writing the rows of B as the columns of B^T. Then **adj $A = B^T$**

The **inverse of matrix** A, A^{-1} is given by

$$A^{-1} = \frac{\text{adj } A}{|A|}$$

where adj A is the adjoint of matrix A and $|A|$ is the determinant of matrix A.

Problem 17.   Determine the inverse of the matrix

$$\begin{pmatrix} 3 & 4 & -1 \\ 2 & 0 & 7 \\ 1 & -3 & -2 \end{pmatrix}$$

The inverse of matrix A, $A^{-1} = \dfrac{\text{adj } A}{|A|}$

The adjoint of A is found by:

(i)   obtaining the matrix of the cofactors of the elements, and

(ii) transposing this matrix.

The cofactor of element 3 is $+\begin{vmatrix} 0 & 7 \\ -3 & -2 \end{vmatrix} = 21$

The cofactor of element 4 is $-\begin{vmatrix} 2 & 7 \\ 1 & -2 \end{vmatrix} = 11$, and so on.

The matrix of cofactors is $\begin{pmatrix} 21 & 11 & -6 \\ 11 & -5 & 13 \\ 28 & -23 & -8 \end{pmatrix}$

The transpose of the matrix of cofactors, i.e. the adjoint of the matrix, is obtained by writing the rows as columns,

and is $\begin{pmatrix} 21 & 11 & 28 \\ 11 & -5 & -23 \\ -6 & 13 & -8 \end{pmatrix}$

From Problem 14, the determinant of $\begin{vmatrix} 3 & 4 & -1 \\ 2 & 0 & 7 \\ 1 & -3 & -2 \end{vmatrix}$

is 113

Hence the inverse of $\begin{pmatrix} 3 & 4 & -1 \\ 2 & 0 & 7 \\ 1 & -3 & -2 \end{pmatrix}$ is

$$\frac{\begin{pmatrix} 21 & 11 & 28 \\ 11 & -5 & -23 \\ -6 & 13 & -8 \end{pmatrix}}{113} \text{ or } \frac{1}{113}\begin{pmatrix} 21 & 11 & 28 \\ 11 & -5 & -23 \\ -6 & 13 & -8 \end{pmatrix}$$

Problem 18.   Find the inverse of

$$\begin{pmatrix} 1 & 5 & -2 \\ 3 & -1 & 4 \\ -3 & 6 & -7 \end{pmatrix}$$

$$\text{Inverse} = \frac{\text{adjoint}}{\text{determinant}}$$

The matrix of cofactors is $\begin{pmatrix} -17 & 9 & 15 \\ 23 & -13 & -21 \\ 18 & -10 & -16 \end{pmatrix}$

The transpose of the matrix of cofactors (i.e. the adjoint) is $\begin{pmatrix} -17 & 23 & 18 \\ 9 & -13 & -10 \\ 15 & -21 & -16 \end{pmatrix}$

The determinant of $\begin{pmatrix} 1 & 5 & -2 \\ 3 & -1 & 4 \\ -3 & 6 & -7 \end{pmatrix}$

$$= 1(7 - 24) - 5(-21 + 12) - 2(18 - 3)$$

$$= -17 + 45 - 30 = -2$$

Hence the inverse of $\begin{pmatrix} 1 & 5 & -2 \\ 3 & -1 & 4 \\ -3 & 6 & -7 \end{pmatrix}$

$$= \frac{\begin{pmatrix} -17 & 23 & 18 \\ 9 & -13 & -10 \\ 15 & -21 & -16 \end{pmatrix}}{-2}$$

$$= \begin{pmatrix} 8.5 & -11.5 & -9 \\ -4.5 & 6.5 & 5 \\ -7.5 & 10.5 & 8 \end{pmatrix}$$

Now try the following Practice Exercise

Practice Exercise 109   The inverse of a 3 by 3 matrix (Answers on page 873)

1. Write down the transpose of

$$\begin{pmatrix} 4 & -7 & 6 \\ -2 & 4 & 0 \\ 5 & 7 & -4 \end{pmatrix}$$

Section E

2. Write down the transpose of

$$\begin{pmatrix} 3 & 6 & \frac{1}{2} \\ 5 & -\frac{2}{3} & 7 \\ -1 & 0 & \frac{3}{5} \end{pmatrix}$$

3. Determine the adjoint of

$$\begin{pmatrix} 4 & -7 & 6 \\ -2 & 4 & 0 \\ 5 & 7 & -4 \end{pmatrix}$$

4. Determine the adjoint of

$$\begin{pmatrix} 3 & 6 & \frac{1}{2} \\ 5 & -\frac{2}{3} & 7 \\ -1 & 0 & \frac{3}{5} \end{pmatrix}$$

5. Find the inverse of

$$\begin{pmatrix} 4 & -7 & 6 \\ -2 & 4 & 0 \\ 5 & 7 & -4 \end{pmatrix}$$

6. Find the inverse of $\begin{pmatrix} 3 & 6 & \frac{1}{2} \\ 5 & -\frac{2}{3} & 7 \\ -1 & 0 & \frac{3}{5} \end{pmatrix}$

**For fully worked solutions to each of the problems in Practice Exercises 105 to 109 in this chapter,
go to the website:
www.routledge.com/cw/bird**

Chapter 25

Applications of matrices and determinants

Why it is important to understand: Applications of matrices and determinants

As mentioned previously, matrices are used to solve problems, for example, in electrical circuits, optics, quantum mechanics, statics, robotics, genetics, and much more, and for calculating forces, vectors, tensions, masses, loads and a lot of other factors that must be accounted for in engineering. In the main, matrices and determinants are used to solve a system of simultaneous linear equations. The simultaneous solution of multiple equations finds its way into many common engineering problems. In fact, modern structural engineering analysis techniques are all about solving systems of equations simultaneously. Eigenvalues and eigenvectors, which are based on matrix theory, are very important in engineering and science. For example, car designers analyse eigenvalues in order to damp out the noise in a car, eigenvalue analysis is used in the design of car stereo systems, eigenvalues can be used to test for cracks and deformities in a solid, and oil companies use eigenvalues analysis to explore land for oil.

At the end of this chapter, you should be able to:

- solve simultaneous equations in two and three unknowns using matrices
- solve simultaneous equations in two and three unknowns using determinants
- solve simultaneous equations using Cramer's rule
- solve simultaneous equations using Gaussian elimination
- determine the eigenvalues of a 2 by 2 and 3 by 3 matrix
- determine the eigenvectors of a 2 by 2 and 3 by 3 matrix

25.1 Solution of simultaneous equations by matrices

(a) The procedure for solving linear simultaneous equations in **two unknowns using matrices** is:
  (i) write the equations in the form

$$a_1 x + b_1 y = c_1$$
$$a_2 x + b_2 y = c_2$$

  (ii) write the matrix equation corresponding to these equations,

  i.e. $\begin{pmatrix} a_1 & b_1 \\ a_2 & b_2 \end{pmatrix} \times \begin{pmatrix} x \\ y \end{pmatrix} = \begin{pmatrix} c_1 \\ c_2 \end{pmatrix}$

  (iii) determine the inverse matrix of $\begin{pmatrix} a_1 & b_1 \\ a_2 & b_2 \end{pmatrix}$

  i.e. $\dfrac{1}{a_1 b_2 - b_1 a_2} \begin{pmatrix} b_2 & -b_1 \\ -a_2 & a_1 \end{pmatrix}$

  (from Chapter 24)
  (iv) multiply each side of (ii) by the inverse matrix, and
  (v) solve for x and y by equating corresponding elements.

Problem 1. Use matrices to solve the simultaneous equations:

$$3x + 5y - 7 = 0 \tag{1}$$
$$4x - 3y - 19 = 0 \tag{2}$$

(i) Writing the equations in the $a_1 x + b_1 y = c$ form gives:

$$3x + 5y = 7$$
$$4x - 3y = 19$$

(ii) The matrix equation is

$$\begin{pmatrix} 3 & 5 \\ 4 & -3 \end{pmatrix} \times \begin{pmatrix} x \\ y \end{pmatrix} = \begin{pmatrix} 7 \\ 19 \end{pmatrix}$$

(iii) The inverse of matrix $\begin{pmatrix} 3 & 5 \\ 4 & -3 \end{pmatrix}$ is

$$\frac{1}{3 \times (-3) - 5 \times 4} \begin{pmatrix} -3 & -5 \\ -4 & 3 \end{pmatrix}$$

i.e. $\begin{pmatrix} \dfrac{3}{29} & \dfrac{5}{29} \\ \dfrac{4}{29} & \dfrac{-3}{29} \end{pmatrix}$

(iv) Multiplying each side of (ii) by (iii) and remembering that $A \times A^{-1} = I$, the unit matrix, gives:

$$\begin{pmatrix} 1 & 0 \\ 0 & 1 \end{pmatrix} \begin{pmatrix} x \\ y \end{pmatrix} = \begin{pmatrix} \dfrac{3}{29} & \dfrac{5}{29} \\ \dfrac{4}{29} & \dfrac{-3}{29} \end{pmatrix} \times \begin{pmatrix} 7 \\ 19 \end{pmatrix}$$

Thus $\begin{pmatrix} x \\ y \end{pmatrix} = \begin{pmatrix} \dfrac{21}{29} + \dfrac{95}{29} \\ \dfrac{28}{29} - \dfrac{57}{29} \end{pmatrix}$

i.e. $\begin{pmatrix} x \\ y \end{pmatrix} = \begin{pmatrix} 4 \\ -1 \end{pmatrix}$

(v) By comparing corresponding elements:

$$x = 4 \quad \text{and} \quad y = -1$$

Checking:

equation (1),

$$3 \times 4 + 5 \times (-1) - 7 = 0 = \text{RHS}$$

equation (2),

$$4 \times 4 - 3 \times (-1) - 19 = 0 = \text{RHS}$$

(b) The procedure for solving linear simultaneous equations in **three unknowns using matrices** is:
  (i) write the equations in the form

$$a_1 x + b_1 y + c_1 z = d_1$$
$$a_2 x + b_2 y + c_2 z = d_2$$
$$a_3 x + b_3 y + c_3 z = d_3$$

  (ii) write the matrix equation corresponding to these equations, i.e.

$$\begin{pmatrix} a_1 & b_1 & c_1 \\ a_2 & b_2 & c_2 \\ a_3 & b_3 & c_3 \end{pmatrix} \times \begin{pmatrix} x \\ y \\ z \end{pmatrix} = \begin{pmatrix} d_1 \\ d_2 \\ d_3 \end{pmatrix}$$

  (iii) determine the inverse matrix of

$$\begin{pmatrix} a_1 & b_1 & c_1 \\ a_2 & b_2 & c_2 \\ a_3 & b_3 & c_3 \end{pmatrix} \text{ (see Chapter 24)}$$

  (iv) multiply each side of (ii) by the inverse matrix, and
  (v) solve for x, y and z by equating the corresponding elements.

Problem 2. Use matrices to solve the simultaneous equations:

$$x + y + z - 4 = 0 \qquad (1)$$
$$2x - 3y + 4z - 33 = 0 \qquad (2)$$
$$3x - 2y - 2z - 2 = 0 \qquad (3)$$

(i) Writing the equations in the $a_1x + b_1y + c_1z = d_1$ form gives:

$$x + y + z = 4$$
$$2x - 3y + 4z = 33$$
$$3x - 2y - 2z = 2$$

(ii) The matrix equation is

$$\begin{pmatrix} 1 & 1 & 1 \\ 2 & -3 & 4 \\ 3 & -2 & -2 \end{pmatrix} \times \begin{pmatrix} x \\ y \\ z \end{pmatrix} = \begin{pmatrix} 4 \\ 33 \\ 2 \end{pmatrix}$$

(iii) The inverse matrix of

$$A = \begin{pmatrix} 1 & 1 & 1 \\ 2 & -3 & 4 \\ 3 & -2 & -2 \end{pmatrix}$$

is given by

$$A^{-1} = \frac{\text{adj } A}{|A|}$$

The adjoint of A is the transpose of the matrix of the cofactors of the elements (see Chapter 24). The matrix of cofactors is

$$\begin{pmatrix} 14 & 16 & 5 \\ 0 & -5 & 5 \\ 7 & -2 & -5 \end{pmatrix}$$

and the transpose of this matrix gives

$$\text{adj } A = \begin{pmatrix} 14 & 0 & 7 \\ 16 & -5 & -2 \\ 5 & 5 & -5 \end{pmatrix}$$

The determinant of A, i.e. the sum of the products of elements and their cofactors, using a first row expansion is

$$1 \begin{vmatrix} -3 & 4 \\ -2 & -2 \end{vmatrix} - 1 \begin{vmatrix} 2 & 4 \\ 3 & -2 \end{vmatrix} + 1 \begin{vmatrix} 2 & -3 \\ 3 & -2 \end{vmatrix}$$

$$= (1 \times 14) - (1 \times (-16)) + (1 \times 5) = 35$$

Hence the inverse of A,

$$A^{-1} = \frac{1}{35} \begin{pmatrix} 14 & 0 & 7 \\ 16 & -5 & -2 \\ 5 & 5 & -5 \end{pmatrix}$$

(iv) Multiplying each side of (ii) by (iii), and remembering that $A \times A^{-1} = I$, the unit matrix, gives

$$\begin{pmatrix} 1 & 0 & 0 \\ 0 & 1 & 0 \\ 0 & 0 & 1 \end{pmatrix} \times \begin{pmatrix} x \\ y \\ z \end{pmatrix} = \frac{1}{35} \begin{pmatrix} 14 & 0 & 7 \\ 16 & -5 & -2 \\ 5 & 5 & -5 \end{pmatrix} \times \begin{pmatrix} 4 \\ 33 \\ 2 \end{pmatrix}$$

$$\begin{pmatrix} x \\ y \\ z \end{pmatrix} = \frac{1}{35} \begin{pmatrix} (14 \times 4) + (0 \times 33) + (7 \times 2) \\ (16 \times 4) + ((-5) \times 33) + ((-2) \times 2) \\ (5 \times 4) + (5 \times 33) + ((-5) \times 2) \end{pmatrix}$$

$$= \frac{1}{35} \begin{pmatrix} 70 \\ -105 \\ 175 \end{pmatrix}$$

$$= \begin{pmatrix} 2 \\ -3 \\ 5 \end{pmatrix}$$

(v) By comparing corresponding elements, $x = 2$, $y = -3$, $z = 5$, which can be checked in the original equations.

Now try the following Practice Exercise

Practice Exercise 110   Solving simultaneous equations using matrices (Answers on page 873)

In Problems 1 to 5 use **matrices** to solve the simultaneous equations given.

1. $3x + 4y = 0$
   $2x + 5y + 7 = 0$

2. $2p + 5q + 14.6 = 0$
   $3.1p + 1.7q + 2.06 = 0$

3. $x + 2y + 3z = 5$
   $2x - 3y - z = 3$
   $-3x + 4y + 5z = 3$

4. $3a + 4b - 3c = 2$
   $-2a + 2b + 2c = 15$
   $7a - 5b + 4c = 26$

5. $p + 2q + 3r + 7.8 = 0$
   $2p + 5q - r - 1.4 = 0$
   $5p - q + 7r - 3.5 = 0$

Section E

6. In two closed loops of an electrical circuit, the currents flowing are given by the simultaneous equations:
$$I_1 + 2I_2 + 4 = 0$$
$$5I_1 + 3I_2 - 1 = 0$$
Use matrices to solve for I_1 and I_2

7. The relationship between the displacement s, velocity v, and acceleration a, of a piston is given by the equations:
$$s + 2v + 2a = 4$$
$$3s - v + 4a = 25$$
$$3s + 2v - a = -4$$
Use matrices to determine the values of s, v and a

8. In a mechanical system, acceleration $\ddot{x}$, velocity $\dot{x}$ and distance x are related by the simultaneous equations:
$$3.4\ddot{x} + 7.0\dot{x} - 13.2x = -11.39$$
$$-6.0\ddot{x} + 4.0\dot{x} + 3.5x = 4.98$$
$$2.7\ddot{x} + 6.0\dot{x} + 7.1x = 15.91$$
Use matrices to find the values of $\ddot{x}$, $\dot{x}$ and x

25.2 Solution of simultaneous equations by determinants

(a) When solving linear simultaneous equations in **two unknowns using determinants:**

(i) write the equations in the form
$$a_1 x + b_1 y + c_1 = 0$$
$$a_2 x + b_2 y + c_2 = 0$$
and then

(ii) the solution is given by
$$\frac{x}{D_x} = \frac{-y}{D_y} = \frac{1}{D}$$

where $D_x = \begin{vmatrix} b_1 & c_1 \\ b_2 & c_2 \end{vmatrix}$

i.e. the determinant of the coefficients left when the x-column is covered up,

$$D_y = \begin{vmatrix} a_1 & c_1 \\ a_2 & c_2 \end{vmatrix}$$

i.e. the determinant of the coefficients left when the y-column is covered up,

and $\quad D = \begin{vmatrix} a_1 & b_1 \\ a_2 & b_2 \end{vmatrix}$

i.e. the determinant of the coefficients left when the constants-column is covered up.

Problem 3. Solve the following simultaneous equations using determinants:

$$3x - 4y = 12$$

$$7x + 5y = 6.5$$

Following the above procedure:

(i) $3x - 4y - 12 = 0$
$7x + 5y - 6.5 = 0$

(ii) $\dfrac{x}{\begin{vmatrix} -4 & -12 \\ 5 & -6.5 \end{vmatrix}} = \dfrac{-y}{\begin{vmatrix} 3 & -12 \\ 7 & -6.5 \end{vmatrix}} = \dfrac{1}{\begin{vmatrix} 3 & -4 \\ 7 & 5 \end{vmatrix}}$

i.e. $\dfrac{x}{(-4)(-6.5) - (-12)(5)}$

$= \dfrac{-y}{(3)(-6.5) - (-12)(7)}$

$= \dfrac{1}{(3)(5) - (-4)(7)}$

i.e. $\dfrac{x}{26 + 60} = \dfrac{-y}{-19.5 + 84} = \dfrac{1}{15 + 28}$

i.e. $\dfrac{x}{86} = \dfrac{-y}{64.5} = \dfrac{1}{43}$

Since $\dfrac{x}{86} = \dfrac{1}{43}$ then $x = \dfrac{86}{43} = \mathbf{2}$

and since

$\dfrac{-y}{64.5} = \dfrac{1}{43}$ then $y = -\dfrac{64.5}{43} = \mathbf{-1.5}$

Problem 4. The velocity of a car, accelerating at uniform acceleration a between two points, is given by $v = u + at$, where u is its velocity when passing the first point and t is the time taken to pass between the two points. If $v = 21$ m/s when $t = 3.5$ s and $v = 33$ m/s when $t = 6.1$ s, use determinants to find the values of u and a, each correct to 4 significant figures.

Substituting the given values in $v = u + at$ gives:

$$21 = u + 3.5a \tag{1}$$
$$33 = u + 6.1a \tag{2}$$

(i) The equations are written in the form

$$a_1 x + b_1 y + c_1 = 0$$

i.e. $u + 3.5a - 21 = 0$

and $u + 6.1a - 33 = 0$

(ii) The solution is given by

$$\frac{u}{D_u} = \frac{-a}{D_a} = \frac{1}{D}$$

where D_u is the determinant of coefficients left when the u column is covered up,

i.e. $D_u = \begin{vmatrix} 3.5 & -21 \\ 6.1 & -33 \end{vmatrix}$

$$= (3.5)(-33) - (-21)(6.1)$$
$$= 12.6$$

Similarly, $D_a = \begin{vmatrix} 1 & -21 \\ 1 & -33 \end{vmatrix}$

$$= (1)(-33) - (-21)(1)$$
$$= -12$$

and $D = \begin{vmatrix} 1 & 3.5 \\ 1 & 6.1 \end{vmatrix}$

$$= (1)(6.1) - (3.5)(1) = 2.6$$

Thus $\dfrac{u}{12.6} = \dfrac{-a}{-12} = \dfrac{1}{26}$

i.e. $u = \dfrac{12.6}{2.6} = \mathbf{4.846\,m/s}$

and $a = \dfrac{12}{2.6} = \mathbf{4.615\,m/s^2},$

each correct to 4 significant figures.

Problem 5.   Applying Kirchhoff's laws to an electric circuit results in the following equations:

$$(9 + j12)I_1 - (6 + j8)I_2 = 5$$
$$-(6 + j8)I_1 + (8 + j3)I_2 = (2 + j4)$$

Solve the equations for I_1 and I_2

Following the procedure:

(i) $(9 + j12)I_1 - (6 + j8)I_2 - 5 = 0$

$$-(6 + j8)I_1 + (8 + j3)I_2 - (2 + j4) = 0$$

(ii) $\dfrac{I_1}{\begin{vmatrix} -(6+j8) & -5 \\ (8+j3) & -(2+j4) \end{vmatrix}}$

$$= \frac{-I_2}{\begin{vmatrix} (9+j12) & -5 \\ -(6+j8) & -(2+j4) \end{vmatrix}}$$

$$= \frac{1}{\begin{vmatrix} (9+j12) & -(6+j8) \\ -(6+j8) & (8+j3) \end{vmatrix}}$$

$$\frac{I_1}{(-20 + j40) + (40 + j15)}$$

$$= \frac{-I_2}{(30 - j60) - (30 + j40)}$$

$$= \frac{1}{(36 + j123) - (-28 + j96)}$$

$$\frac{I_1}{20 + j55} = \frac{-I_2}{-j100} = \frac{1}{64 + j27}$$

Hence $I_1 = \dfrac{20 + j55}{64 + j27}$

$$= \frac{58.52\angle 70.02°}{69.46\angle 22.87°} = \mathbf{0.84\angle 47.15° A}$$

and $I_2 = \dfrac{100\angle 90°}{69.46\angle 22.87°}$

$$= \mathbf{1.44\angle 67.13° A}$$

(b) When solving simultaneous equations in **three unknowns using determinants:**

(i)   Write the equations in the form

$$a_1 x + b_1 y + c_1 z + d_1 = 0$$
$$a_2 x + b_2 y + c_2 z + d_2 = 0$$
$$a_3 x + b_3 y + c_3 z + d_3 = 0$$

and then

(ii)   the solution is given by

$$\frac{x}{D_x} = \frac{-y}{D_y} = \frac{z}{D_z} = \frac{-1}{D}$$

where D_x is $\begin{vmatrix} b_1 & c_1 & d_1 \\ b_2 & c_2 & d_2 \\ b_3 & c_3 & d_3 \end{vmatrix}$

i.e. the determinant of the coefficients obtained by covering up the x column.

D_y is $\begin{vmatrix} a_1 & c_1 & d_1 \\ a_2 & c_2 & d_2 \\ a_3 & c_3 & d_3 \end{vmatrix}$

i.e., the determinant of the coefficients obtained by covering up the y column.

D_z is $\begin{vmatrix} a_1 & b_1 & d_1 \\ a_2 & b_2 & d_2 \\ a_3 & b_3 & d_3 \end{vmatrix}$

i.e. the determinant of the coefficients obtained by covering up the z column.

and D is $\begin{vmatrix} a_1 & b_1 & c_1 \\ a_2 & b_2 & c_2 \\ a_3 & b_3 & c_3 \end{vmatrix}$

i.e. the determinant of the coefficients obtained by covering up the constants column.

Problem 6. A d.c. circuit comprises three closed loops. Applying Kirchhoff's laws to the closed loops gives the following equations for current flow in milliamperes:

$$2I_1 + 3I_2 - 4I_3 = 26$$

$$I_1 - 5I_2 - 3I_3 = -87$$

$$-7I_1 + 2I_2 + 6I_3 = 12$$

Use determinants to solve for I_1, I_2 and I_3

(i) Writing the equations in the $a_1 x + b_1 y + c_1 z + d_1 = 0$ form gives:

$$2I_1 + 3I_2 - 4I_3 - 26 = 0$$

$$I_1 - 5I_2 - 3I_3 + 87 = 0$$

$$-7I_1 + 2I_2 + 6I_3 - 12 = 0$$

(ii) the solution is given by

$$\frac{I_1}{D_{I_1}} = \frac{-I_2}{D_{I_2}} = \frac{I_3}{D_{I_3}} = \frac{-1}{D}$$

where D_{I_1} is the determinant of coefficients obtained by covering up the I_1 column, i.e.

$$D_{I_1} = \begin{vmatrix} 3 & -4 & -26 \\ -5 & -3 & 87 \\ 2 & 6 & -12 \end{vmatrix}$$

$$= (3)\begin{vmatrix} -3 & 87 \\ 6 & -12 \end{vmatrix} - (-4)\begin{vmatrix} -5 & 87 \\ 2 & -12 \end{vmatrix}$$

$$+ (-26)\begin{vmatrix} -5 & -3 \\ 2 & 6 \end{vmatrix}$$

$$= 3(-486) + 4(-114) - 26(-24)$$

$$= \mathbf{-1290}$$

$$D_{I_2} = \begin{vmatrix} 2 & -4 & -26 \\ 1 & -3 & 87 \\ -7 & 6 & -12 \end{vmatrix}$$

$$= (2)(36 - 522) - (-4)(-12 + 609)$$

$$+ (-26)(6 - 21)$$

$$= -972 + 2388 + 390$$

$$= \mathbf{1806}$$

$$D_{I_3} = \begin{vmatrix} 2 & 3 & -26 \\ 1 & -5 & 87 \\ -7 & 2 & -12 \end{vmatrix}$$

$$= (2)(60 - 174) - (3)(-12 + 609)$$

$$+ (-26)(2 - 35)$$

$$= -228 - 1791 + 858 = \mathbf{-1161}$$

and $D = \begin{vmatrix} 2 & 3 & -4 \\ 1 & -5 & -3 \\ -7 & 2 & 6 \end{vmatrix}$

$$= (2)(-30 + 6) - (3)(6 - 21)$$

$$+ (-4)(2 - 35)$$

$$= -48 + 45 + 132 = \mathbf{129}$$

Thus

$$\frac{I_1}{-1290} = \frac{-I_2}{1806} = \frac{I_3}{-1161} = \frac{-1}{129}$$

giving

$$I_1 = \frac{-1290}{-129} = \mathbf{10\,mA},$$

$$I_2 = \frac{1806}{129} = \mathbf{14\,mA}$$

and $I_3 = \dfrac{1161}{129} = \mathbf{9\,mA}$

Now try the following Practice Exercise

Practice Exercise 111   Solving simultaneous equations using determinants (Answers on page 873)

In Problems 1 to 5 use **determinants** to solve the simultaneous equations given.

1. $3x - 5y = -17.6$

   $7y - 2x - 22 = 0$

2. $2.3m - 4.4n = 6.84$

   $8.5n - 6.7m = 1.23$

3. $3x + 4y + z = 10$
   $2x - 3y + 5z + 9 = 0$

   $x + 2y - z = 6$

4. $1.2p - 2.3q - 3.1r + 10.1 = 0$

   $4.7p + 3.8q - 5.3r - 21.5 = 0$

   $3.7p - 8.3q + 7.4r + 28.1 = 0$

5. $\dfrac{x}{2} - \dfrac{y}{3} + \dfrac{2z}{5} = -\dfrac{1}{20}$

   $\dfrac{x}{4} + \dfrac{2y}{3} - \dfrac{z}{2} = \dfrac{19}{40}$

   $x + y - z = \dfrac{59}{60}$

6. In a system of forces, the relationship between two forces F_1 and F_2 is given by:

   $5F_1 + 3F_2 + 6 = 0$

   $3F_1 + 5F_2 + 18 = 0$

   Use determinants to solve for F_1 and F_2

7. Applying mesh-current analysis to an a.c. circuit results in the following equations:

   $(5 - j4)I_1 - (-j4)I_2 = 100\angle 0°$

   $(4 + j3 - j4)I_2 - (-j4)I_1 = 0$

   Solve the equations for I_1 and I_2

8. Kirchhoff's laws are used to determine the current equations in an electrical network and show that

   $i_1 + 8i_2 + 3i_3 = -31$

   $3i_1 - 2i_2 + i_3 = -5$

   $2i_1 - 3i_2 + 2i_3 = 6$

   Use determinants to find the values of i_1, i_2 and i_3

9. The forces in three members of a framework are F_1, F_2 and F_3. They are related by the simultaneous equations shown below.

   $1.4F_1 + 2.8F_2 + 2.8F_3 = 5.6$

   $4.2F_1 - 1.4F_2 + 5.6F_3 = 35.0$

   $4.2F_1 + 2.8F_2 - 1.4F_3 = -5.6$

   Find the values of F_1, F_2 and F_3 using determinants.

10. Mesh-current analysis produces the following three equations:

    $20\angle 0° = (5 + 3 - j4)I_1 - (3 - j4)I_2$

    $10\angle 90° = (3 - j4 + 2)I_2 - (3 - j4)I_1 - 2I_3$

    $-15\angle 0° - 10\angle 90° = (12 + 2)I_3 - 2I_2$

    Solve the equations for the loop currents I_1, I_2 and I_3

25.3   Solution of simultaneous equations using Cramer's rule

Cramer's* rule states that if

$a_{11}x + a_{12}y + a_{13}z = b_1$

$a_{21}x + a_{22}y + a_{23}z = b_2$

$a_{31}x + a_{32}y + a_{33}z = b_3$

then $\quad x = \dfrac{\boldsymbol{D_x}}{\boldsymbol{D}}, \quad y = \dfrac{\boldsymbol{D_y}}{\boldsymbol{D}}$ and $z = \dfrac{\boldsymbol{D_z}}{\boldsymbol{D}}$

* **Who was Cramer**? **Gabriel Cramer** (31 July 1704–4 January 1752) was a Swiss mathematician. His articles cover a wide range of subjects including the study of geometric problems, the history of mathematics, philosophy, and the date of Easter. Cramer's most famous book is a work which Cramer modelled on Newton's memoir on cubic curves. To find out more go to **www.routledge.com/cw/bird**

where
$$D = \begin{vmatrix} a_{11} & a_{12} & a_{13} \\ a_{21} & a_{22} & a_{23} \\ a_{31} & a_{32} & a_{33} \end{vmatrix}$$

$$D_x = \begin{vmatrix} b_1 & a_{12} & a_{13} \\ b_2 & a_{22} & a_{23} \\ b_3 & a_{32} & a_{33} \end{vmatrix}$$

i.e. the x-column has been replaced by the RHS b column,

$$D_y = \begin{vmatrix} a_{11} & b_1 & a_{13} \\ a_{21} & b_2 & a_{23} \\ a_{31} & b_3 & a_{33} \end{vmatrix}$$

i.e. the y-column has been replaced by the RHS b column,

$$D_z = \begin{vmatrix} a_{11} & a_{12} & b_1 \\ a_{21} & a_{22} & b_2 \\ a_{31} & a_{32} & b_3 \end{vmatrix}$$

i.e. the z-column has been replaced by the RHS b column.

Problem 7. Solve the following simultaneous equations using Cramer's rule.

$$x + y + z = 4$$
$$2x - 3y + 4z = 33$$
$$3x - 2y - 2z = 2$$

(This is the same as Problem 2 and a comparison of methods may be made). Following the above method:

$$D = \begin{vmatrix} 1 & 1 & 1 \\ 2 & -3 & 4 \\ 3 & -2 & -2 \end{vmatrix}$$

$$= 1(6 - (-8)) - 1((-4) - 12)$$

$$+ 1((-4) - (-9)) = 14 + 16 + 5 = \mathbf{35}$$

$$D_x = \begin{vmatrix} 4 & 1 & 1 \\ 33 & -3 & 4 \\ 2 & -2 & -2 \end{vmatrix}$$

$$= 4(6 - (-8)) - 1((-66) - 8)$$

$$+ 1((-66) - (-6)) = 56 + 74 - 60 = \mathbf{70}$$

$$D_y = \begin{vmatrix} 1 & 4 & 1 \\ 2 & 33 & 4 \\ 3 & 2 & -2 \end{vmatrix}$$

$$= 1((-66) - 8) - 4((-4) - 12) + 1(4 - 99)$$

$$= -74 + 64 - 95 = \mathbf{-105}$$

$$D_z = \begin{vmatrix} 1 & 1 & 4 \\ 2 & -3 & 33 \\ 3 & -2 & 2 \end{vmatrix}$$

$$= 1((-6) - (-66)) - 1(4 - 99)$$

$$+ 4((-4) - (-9)) = 60 + 95 + 20 = \mathbf{175}$$

Hence

$$x = \frac{D_x}{D} = \frac{70}{35} = \mathbf{2}, \quad y = \frac{D_y}{D} = \frac{-105}{35} = \mathbf{-3}$$

and $\quad z = \frac{D_z}{D} = \frac{175}{35} = \mathbf{5}$

Now try the following Practice Exercise

Practice Exercise 112 Solving simultaneous equations using Cramer's rule (Answers on page 873)

1. Repeat problems 3, 4, 5, 7 and 8 of Exercise 110 on page 289, using Cramer's rule.

2. Repeat problems 3, 4, 8 and 9 of Exercise 111 on page 293, using Cramer's rule.

25.4 Solution of simultaneous equations using the Gaussian elimination method

Consider the following simultaneous equations:

$$x + y + z = 4 \tag{1}$$
$$2x - 3y + 4z = 33 \tag{2}$$
$$3x - 2y - 2z = 2 \tag{3}$$

Leaving equation (1) as it is gives:

$$x + y + z = 4 \tag{1}$$

Equation $(2) - 2 \times$ equation (1) gives:

$$0 - 5y + 2z = 25 \tag{2'}$$

and equation $(3) - 3 \times$ equation (1) gives:

$$0 - 5y - 5z = -10 \tag{3'}$$

Leaving equations (1) and $(2')$ as they are gives:

$$x + y + z = 4 \tag{1}$$
$$0 - 5y + 2z = 25 \tag{2'}$$

Equation $(3')$ − equation $(2')$ gives:

$$0 + 0 - 7z = -35 \tag{3''}$$

By appropriately manipulating the three original equations we have deliberately obtained zeros in the positions shown in equations $(2')$ and $(3'')$.

Working backwards, from equation $(3'')$,

$$z = \frac{-35}{-7} = \mathbf{5}$$

from equation $(2')$,

$$-5y + 2(5) = 25,$$

from which,

$$y = \frac{25 - 10}{-5} = \mathbf{-3}$$

and from equation (1),

$$x + (-3) + 5 = 4$$

from which,

$$x = 4 + 3 - 5 = \mathbf{2}$$

(This is the same example as Problems 2 and 7, and a comparison of methods can be made). The

above method is known as the **Gaussian* elimination method**.

We conclude from the above example that if

$$a_{11}x + a_{12}y + a_{13}z = b_1$$
$$a_{21}x + a_{22}y + a_{23}z = b_2$$
$$a_{31}x + a_{32}y + a_{33}z = b_3$$

The three-step **procedure** to solve simultaneous equations in three unknowns using the **Gaussian elimination method** is:

(i)   Equation $(2) - \dfrac{a_{21}}{a_{11}} \times$ equation (1) to form equation $(2')$ and equation $(3) - \dfrac{a_{31}}{a_{11}} \times$ equation (1) to form equation $(3')$.

(ii) Equation $(3') - \dfrac{a_{32}}{a_{22}} \times$ equation $(2')$ to form equation $(3'')$.

(iii) Determine z from equation $(3'')$, then y from equation $(2')$ and finally, x from equation (1).

* **Who was Gauss**? **Johann Carl Friedrich Gauss** (30 April 1777–23 February 1855) was a German mathematician and physical scientist who contributed significantly to many fields, including number theory, statistics, electrostatics, astronomy and optics. To find out more go to **www.routledge.com/cw/bird**

Problem 8. A d.c. circuit comprises three closed loops. Applying Kirchhoff's laws to the closed loops gives the following equations for current flow in milliamperes:

$$2I_1 + 3I_2 - 4I_3 = 26 \tag{1}$$

$$I_1 - 5I_2 - 3I_3 = -87 \tag{2}$$

$$-7I_1 + 2I_2 + 6I_3 = 12 \tag{3}$$

Use the Gaussian elimination method to solve for I_1, I_2 and I_3

(This is the same example as Problem 6 on page 292, and a comparison of methods may be made)

Following the above procedure:

1. $2I_1 + 3I_2 - 4I_3 = 26$ (1)

   Equation (2) $- \dfrac{1}{2} \times$ equation (1) gives:

   $0 - 6.5I_2 - I_3 = -100$ (2′)

   Equation (3) $- \dfrac{-7}{2} \times$ equation (1) gives:

   $0 + 12.5I_2 - 8I_3 = 103$ (3′)

2. $2I_1 + 3I_2 - 4I_3 = 26$ (1)

   $0 - 6.5I_2 - I_3 = -100$ (2′)

   Equation (3′) $- \dfrac{12.5}{-6.5} \times$ equation (2′) gives:

   $0 + 0 - 9.923I_3 = -89.308$ (3″)

3. From equation (3″),

   $$I_3 = \frac{-89.308}{-9.923} = 9\,\text{mA}$$

   from equation (2′), $-6.5I_2 - 9 = -100$,

   from which, $I_2 = \dfrac{-100 + 9}{-6.5} = 14\,\text{mA}$

   and from equation (1), $2I_1 + 3(14) - 4(9) = 26$,

   from which, $I_1 = \dfrac{26 - 42 + 36}{2} = \dfrac{20}{2}$

   $$= 10\,\text{mA}$$

Now try the following Practice Exercise

Practice Exercise 113   Solving simultaneous equations using Gaussian elimination (Answers on page 873)

1. In a mass–spring–damper system, the acceleration $\ddot{x}\,\text{m/s}^2$, velocity $\dot{x}\,\text{m/s}$ and displacement x m are related by the following simultaneous equations:

   $$6.2\ddot{x} + 7.9\dot{x} + 12.6x = 18.0$$

   $$7.5\ddot{x} + 4.8\dot{x} + 4.8x = 6.39$$

   $$13.0\ddot{x} + 3.5\dot{x} - 13.0x = -17.4$$

   By using Gaussian elimination, determine the acceleration, velocity and displacement for the system, correct to 2 decimal places.

2. The tensions, T_1, T_2 and T_3 in a simple framework are given by the equations:

   $$5T_1 + 5T_2 + 5T_3 = 7.0$$

   $$T_1 + 2T_2 + 4T_3 = 2.4$$

   $$4T_1 + 2T_2 \qquad = 4.0$$

   Determine T_1, T_2 and T_3 using Gaussian elimination.

3. Repeat problems 3, 4, 5, 7 and 8 of Exercise 110 on page 289, using the Gaussian elimination method.

4. Repeat problems 3, 4, 8 and 9 of Exercise 111 on page 293, using the Gaussian elimination method.

25.5   Eigenvalues and eigenvectors

In practical applications, such as coupled oscillations and vibrations, equations of the form:

$$A\,x = \lambda\,x$$

occur, where A is a square matrix and λ (Greek lambda) is a number. Whenever $x \neq 0$, the values of λ are called the **eigenvalues** of the matrix A; the corresponding solutions of the equation $A x = \lambda x$ are called the **eigenvectors** of A.

Sometimes, instead of the term *eigenvalues*, **characteristic values** or **latent roots** are used. Also, instead of the term *eigenvectors*, **characteristic vectors** is used.

From above, if $A x = \lambda x$ then $A x - \lambda x = 0$ i.e. $(A - \lambda I) = 0$ where I is the unit matrix.

If $x \neq 0$ then $|A - \lambda I| = 0$

$|A - \lambda I|$ is called the **characteristic determinant** of A and $|A - \lambda I| = 0$ is called the **characteristic equation**. Solving the characteristic equation will give the value(s) of the eigenvalues, as demonstrated in the following worked problems.

Problem 9. Determine the eigenvalues of the matrix $A = \begin{pmatrix} 3 & 4 \\ 2 & 1 \end{pmatrix}$

The eigenvalue is determined by solving the characteristic equation $|A - \lambda I| = 0$

i.e. $\left| \begin{pmatrix} 3 & 4 \\ 2 & 1 \end{pmatrix} - \lambda \begin{pmatrix} 1 & 0 \\ 0 & 1 \end{pmatrix} \right| = 0$

i.e. $\left| \begin{pmatrix} 3 & 4 \\ 2 & 1 \end{pmatrix} - \begin{pmatrix} \lambda & 0 \\ 0 & \lambda \end{pmatrix} \right| = 0$

i.e. $\begin{vmatrix} 3 - \lambda & 4 \\ 2 & 1 - \lambda \end{vmatrix} = 0$

(Given a square matrix, we can get used to going straight to this characteristic equation)

Hence, $(3 - \lambda)(1 - \lambda) - (2)(4) = 0$

i.e. $3 - 3\lambda - \lambda + \lambda^2 - 8 = 0$

and $\lambda^2 - 4\lambda - 5 = 0$

i.e. $(\lambda - 5)(\lambda + 1) = 0$

from which, $\lambda - 5 = 0$ i.e. $\boldsymbol{\lambda = 5}$ or $\lambda + 1 = 0$ i.e. $\boldsymbol{\lambda = -1}$

(Instead of factorising, the quadratic formula could be used; even electronic calculators can solve quadratic equations.)

Hence, the eigenvalues of the matrix $\begin{pmatrix} 3 & 4 \\ 2 & 1 \end{pmatrix}$ are 5 and –1

Problem 10. Determine the eigenvectors of the matrix $A = \begin{pmatrix} 3 & 4 \\ 2 & 1 \end{pmatrix}$

From Problem 9, the eigenvalues of $\begin{pmatrix} 3 & 4 \\ 2 & 1 \end{pmatrix}$ are $\lambda_1 = 5$ and $\lambda_2 = -1$

Using the equation $(A - \lambda I)x = 0$ for $\lambda_1 = 5$

then $\begin{pmatrix} 3 - 5 & 4 \\ 2 & 1 - 5 \end{pmatrix} \begin{pmatrix} x_1 \\ x_2 \end{pmatrix} = \begin{pmatrix} 0 \\ 0 \end{pmatrix}$

i.e.

$\begin{pmatrix} -2 & 4 \\ 2 & -4 \end{pmatrix} \begin{pmatrix} x_1 \\ x_2 \end{pmatrix} = \begin{pmatrix} 0 \\ 0 \end{pmatrix}$

from which,

$$-2x_1 + 4x_2 = 0$$

and

$$2x_1 - 4x_2 = 0$$

From either of these two equations, $x_1 = 2x_2$

Hence, whatever value x_2 is, the value of x_1 will be two times greater. Hence the simplest eigenvector is:

$x_1 = \begin{pmatrix} 2 \\ 1 \end{pmatrix}$

Using the equation $(A - \lambda I)x = 0$ for $\lambda_2 = -1$

then $\begin{pmatrix} 3 - -1 & 4 \\ 2 & 1 - -1 \end{pmatrix} \begin{pmatrix} x_1 \\ x_2 \end{pmatrix} = \begin{pmatrix} 0 \\ 0 \end{pmatrix}$

i.e.

$\begin{pmatrix} 4 & 4 \\ 2 & 2 \end{pmatrix} \begin{pmatrix} x_1 \\ x_2 \end{pmatrix} = \begin{pmatrix} 0 \\ 0 \end{pmatrix}$

from which,

$$4x_1 + 4x_2 = 0$$

and

$$2x_1 + 2x_2 = 0$$

From either of these two equations, $x_1 = -x_2$ or $x_2 = -x_1$

Hence, whatever value x_1 is, the value of x_2 will be –1 times greater. Hence the simplest eigenvector is:

$x_2 = \begin{pmatrix} 1 \\ -1 \end{pmatrix}$

Summarising, $x_1 = \begin{pmatrix} 2 \\ 1 \end{pmatrix}$ is an eigenvector corresponding to $\boldsymbol{\lambda_1 = 5}$ and $x_2 = \begin{pmatrix} 1 \\ -1 \end{pmatrix}$ is an eigenvector corresponding to $\boldsymbol{\lambda_2 = -1}$

Problem 11. Determine the eigenvalues of the matrix $A = \begin{pmatrix} 5 & -2 \\ -9 & 2 \end{pmatrix}$

The eigenvalue is determined by solving the characteristic equation $|A - \lambda I| = 0$

i.e. $\begin{vmatrix} 5 - \lambda & -2 \\ -9 & 2 - \lambda \end{vmatrix} = 0$

Hence, $(5 - \lambda)(2 - \lambda) - (-9)(-2) = 0$

i.e. $10 - 5\lambda - 2\lambda + \lambda^2 - 18 = 0$

and $\lambda^2 - 7\lambda - 8 = 0$

i.e. $(\lambda - 8)(\lambda + 1) = 0$

from which, $\lambda - 8 = 0$ i.e. $\lambda = 8$ or $\lambda + 1 = 0$ i.e. $\lambda = -1$ (or use the quadratic formula).

Hence, the eigenvalues of the matrix $\begin{pmatrix} 5 & -2 \\ -9 & 2 \end{pmatrix}$ **are 8 and** -1

Problem 12. Determine the eigenvectors of the matrix $A = \begin{pmatrix} 5 & -2 \\ -9 & 2 \end{pmatrix}$

From Problem 11, the eigenvalues of $\begin{pmatrix} 5 & -2 \\ -9 & 2 \end{pmatrix}$ are $\lambda_1 = 8$ and $\lambda_2 = -1$

Using the equation $(A - \lambda I)x = 0$ for $\lambda_1 = 8$

then $\begin{pmatrix} 5-8 & -2 \\ -9 & 2-8 \end{pmatrix} \begin{pmatrix} x_1 \\ x_2 \end{pmatrix} = \begin{pmatrix} 0 \\ 0 \end{pmatrix}$

i.e.

$$\begin{pmatrix} -3 & -2 \\ -9 & -6 \end{pmatrix} \begin{pmatrix} x_1 \\ x_2 \end{pmatrix} = \begin{pmatrix} 0 \\ 0 \end{pmatrix}$$

from which,

$$-3x_1 - 2x_2 = 0$$

and

$$-9x_1 - 6x_2 = 0$$

From either of these two equations, $3x_1 = -2x_2$ or $x_1 = -\dfrac{2}{3}x_2$

Hence, if $x_2 = 3$, $x_1 = -2$. Hence the simplest eigenvector is: $x_1 = \begin{pmatrix} -2 \\ 3 \end{pmatrix}$

Using the equation $(A - \lambda I)x = 0$ for $\lambda_2 = -1$

then $\begin{pmatrix} 5--1 & -2 \\ -9 & 2--1 \end{pmatrix} \begin{pmatrix} x_1 \\ x_2 \end{pmatrix} = \begin{pmatrix} 0 \\ 0 \end{pmatrix}$

i.e.

$$\begin{pmatrix} 6 & -2 \\ -9 & 3 \end{pmatrix} \begin{pmatrix} x_1 \\ x_2 \end{pmatrix} = \begin{pmatrix} 0 \\ 0 \end{pmatrix}$$

from which,

$$6x_1 - 2x_2 = 0$$

and

$$-9x_1 + 3x_2 = 0$$

From either of these two equations, $x_2 = 3x_1$

Hence, if $x_1 = 1$, $x_2 = 3$. Hence the simplest eigenvector is: $x_2 = \begin{pmatrix} 1 \\ 3 \end{pmatrix}$

Summarising, $x_1 = \begin{pmatrix} -2 \\ 3 \end{pmatrix}$ is an eigenvector corresponding to $\lambda_1 = 8$ and $x_2 = \begin{pmatrix} 1 \\ 3 \end{pmatrix}$ is an eigenvector corresponding to $\lambda_2 = -1$

Problem 13. Determine the eigenvalues of the matrix $A = \begin{pmatrix} 1 & 2 & 1 \\ 6 & -1 & 0 \\ -1 & -2 & -1 \end{pmatrix}$

The eigenvalue is determined by solving the characteristic equation $|A - \lambda I| = 0$

i.e. $\begin{vmatrix} 1-\lambda & 2 & 1 \\ 6 & -1-\lambda & 0 \\ -1 & -2 & -1-\lambda \end{vmatrix} = 0$

Hence, using the top row:

$$(1-\lambda)[(-1-\lambda)(-1-\lambda) - (-2)(0)]$$
$$- 2[6(-1-\lambda) - (-1)(0)]$$
$$+ 1[(6)(-2) - (-1)(-1-\lambda) = 0$$

i.e. $(1-\lambda)[1+\lambda+\lambda+\lambda^2] - 2[-6-6\lambda]$
$$+ 1[-12-1-\lambda] = 0$$

i.e. $(1-\lambda)[\lambda^2 + 2\lambda + 1] + 12 + 12\lambda$
$$- 13 - \lambda = 0$$

and $\lambda^2 + 2\lambda + 1 - \lambda^3 - 2\lambda^2 - \lambda + 12 + 12\lambda$
$$- 13 - \lambda = 0$$

i.e. $-\lambda^3 - \lambda^2 + 12\lambda = 0$

or $\lambda^3 + \lambda^2 - 12\lambda = 0$

i.e. $\lambda(\lambda^2 + \lambda - 12) = 0$

i.e. $\lambda(\lambda - 3)(\lambda + 4) = 0$ by factorising

from which, $\lambda = 0$, $\lambda = 3$ or $\lambda = -4$

Hence, the eigenvalues of the matrix $\begin{pmatrix} 1 & 2 & 1 \\ 6 & -1 & 0 \\ -1 & -2 & -1 \end{pmatrix}$ **are 0, 3 and** -4

Problem 14. Determine the eigenvectors of the matrix $A = \begin{pmatrix} 1 & 2 & 1 \\ 6 & -1 & 0 \\ -1 & -2 & -1 \end{pmatrix}$

From Problem 13, the eigenvalues of $\begin{pmatrix} 1 & 2 & 1 \\ 6 & -1 & 0 \\ -1 & -2 & -1 \end{pmatrix}$ are $\lambda_1 = 0$, $\lambda_2 = 3$ and $\lambda_3 = -4$

Using the equation $(A - \lambda I)x = 0$ for $\lambda_1 = 0$

then $\begin{pmatrix} 1-0 & 2 & 1 \\ 6 & -1-0 & 0 \\ -1 & -2 & -1-0 \end{pmatrix} \begin{pmatrix} x_1 \\ x_2 \\ x_3 \end{pmatrix} = \begin{pmatrix} 0 \\ 0 \\ 0 \end{pmatrix}$

i.e.

$$\begin{pmatrix} 1 & 2 & 1 \\ 6 & -1 & 0 \\ -1 & -2 & -1 \end{pmatrix} \begin{pmatrix} x_1 \\ x_2 \\ x_3 \end{pmatrix} = \begin{pmatrix} 0 \\ 0 \\ 0 \end{pmatrix}$$

from which,

$$x_1 + 2x_2 + x_3 = 0$$

$$6x_1 - x_2 = 0$$

$$-x_1 - 2x_2 - x_3 = 0$$

From the second equation,

$$6x_1 = x_2$$

Substituting in the first equation,

$$x_1 + 12x_1 + x_3 = 0 \quad \text{i.e.} \ -13x_1 = x_3$$

Hence, when $x_1 = 1$, $x_2 = 6$ and $x_3 = -13$

Hence the simplest eigenvector corresponding to $\lambda_1 = 0$

is: $x_1 = \begin{pmatrix} 1 \\ 6 \\ -13 \end{pmatrix}$

Using the equation $(A - \lambda I)x = 0$ for $\lambda_2 = 3$

then $\begin{pmatrix} 1-3 & 2 & 1 \\ 6 & -1-3 & 0 \\ -1 & -2 & -1-3 \end{pmatrix} \begin{pmatrix} x_1 \\ x_2 \\ x_3 \end{pmatrix} = \begin{pmatrix} 0 \\ 0 \\ 0 \end{pmatrix}$

i.e.

$$\begin{pmatrix} -2 & 2 & 1 \\ 6 & -4 & 0 \\ -1 & -2 & -4 \end{pmatrix} \begin{pmatrix} x_1 \\ x_2 \\ x_3 \end{pmatrix} = \begin{pmatrix} 0 \\ 0 \\ 0 \end{pmatrix}$$

from which,

$$-2x_1 + 2x_2 + x_3 = 0$$

$$6x_1 - 4x_2 = 0$$

$$-x_1 - 2x_2 - 4x_3 = 0$$

From the second equation,

$$3x_1 = 2x_2$$

Substituting in the first equation,

$$-2x_1 + 3x_1 + x_3 = 0 \quad \text{i.e.} \ x_3 = -x_1$$

Hence, if $x_2 = 3$, then $x_1 = 2$ and $x_3 = -2$

Hence the simplest eigenvector corresponding to $\lambda_2 = 3$

is: $x_2 = \begin{pmatrix} 2 \\ 3 \\ -2 \end{pmatrix}$

Using the equation $(A - \lambda I)x = 0$ for $\lambda_2 = -4$

then $\begin{pmatrix} 1--4 & 2 & 1 \\ 6 & -1--4 & 0 \\ -1 & -2 & -1--4 \end{pmatrix} \begin{pmatrix} x_1 \\ x_2 \\ x_3 \end{pmatrix}$
$$= \begin{pmatrix} 0 \\ 0 \\ 0 \end{pmatrix}$$

i.e.

$$\begin{pmatrix} 5 & 2 & 1 \\ 6 & 3 & 0 \\ -1 & -2 & 3 \end{pmatrix} \begin{pmatrix} x_1 \\ x_2 \\ x_3 \end{pmatrix} = \begin{pmatrix} 0 \\ 0 \\ 0 \end{pmatrix}$$

from which,

$$5x_1 + 2x_2 + x_3 = 0$$

$$6x_1 + 3x_2 = 0$$

$$-x_1 - 2x_2 + 3x_3 = 0$$

From the second equation,

$$x_2 = -2x_1$$

Substituting in the first equation,

$$5x_1 - 4x_1 + x_3 = 0 \quad \text{i.e.} \ x_3 = -x_1$$

Hence, if $x_1 = -1$, then $x_2 = 2$ and $x_3 = 1$

Hence the simplest eigenvector corresponding to $\lambda_2 = -4$ is: $x_3 = \begin{pmatrix} -1 \\ 2 \\ 1 \end{pmatrix}$

Problem 15.   Determine the eigenvalues of the matrix $A = \begin{pmatrix} 1 & -4 & -2 \\ 0 & 3 & 1 \\ 1 & 2 & 4 \end{pmatrix}$

The eigenvalue is determined by solving the characteristic equation $|A - \lambda I| = 0$

i.e. $\begin{vmatrix} 1-\lambda & -4 & -2 \\ 0 & 3-\lambda & 1 \\ 1 & 2 & 4-\lambda \end{vmatrix} = 0$

Hence, using the top row:

$$(1-\lambda)[(3-\lambda)(4-\lambda)-(2)(1)]-(-4)[0-(1)(1)]$$
$$- 2[0-1(3-\lambda)] = 0$$

i.e. $(1-\lambda)[12-3\lambda-4\lambda+\lambda^2-2]+4[-1]$
$$-2[-3+\lambda] = 0$$

i.e. $(1-\lambda)[\lambda^2-7\lambda+10]-4+6-2\lambda = 0$

and $\lambda^2 - 7\lambda + 10 - \lambda^3 + 7\lambda^2 - 10\lambda - 4 + 6 - 2\lambda = 0$

Section E

i.e. $\qquad -\lambda^3 + 8\lambda^2 - 19\lambda + 12 = 0$
or $\qquad \lambda^3 - 8\lambda^2 + 19\lambda - 12 = 0$

To solve a cubic equation, the **factor theorem** (see Chapter 1) may be used. (Alternatively, electronic calculators can solve such cubic equations.)

Let $f(\lambda) = \lambda^3 - 8\lambda^2 + 19\lambda - 12$

If $\lambda = 1$, then $f(1) = 1^3 - 8(1)^2 + 19(1) - 12 = 0$

Since $f(1) = 0$ then $(\lambda - 1)$ is a factor.

If $\lambda = 2$, then $f(2) = 2^3 - 8(2)^2 + 19(2) - 12 \neq 0$

If $\lambda = 3$, then $f(3) = 3^3 - 8(3)^2 + 19(3) - 12 = 0$

Since $f(3) = 0$ then $(\lambda - 3)$ is a factor.

If $\lambda = 4$, then $f(4) = 4^3 - 8(4)^2 + 19(4) - 12 = 0$

Since $f(4) = 0$ then $(\lambda - 4)$ is a factor.

Thus, since $\lambda^3 - 8\lambda^2 + 19\lambda - 12 = 0$

then $(\lambda - 1)(\lambda - 3)(\lambda - 4) = 0$

from which, $\lambda = 1$ or $\lambda = 3$ or $\lambda = 4$

Hence, the eigenvalues of the matrix $\begin{pmatrix} 1 & -4 & -2 \\ 0 & 3 & 1 \\ 1 & 2 & 4 \end{pmatrix}$ are 1, 3 and 4

Problem 16. Determine the eigenvectors of the matrix $A = \begin{pmatrix} 1 & -4 & -2 \\ 0 & 3 & 1 \\ 1 & 2 & 4 \end{pmatrix}$

From Problem 15, the eigenvalues of $\begin{pmatrix} 1 & -4 & -2 \\ 0 & 3 & 1 \\ 1 & 2 & 4 \end{pmatrix}$

are $\lambda_1 = 1$, $\lambda_2 = 3$ and $\lambda_3 = 4$

Using the equation $(A - \lambda I)x = 0$ for $\lambda_1 = 1$

then $\begin{pmatrix} 1-1 & -4 & -2 \\ 0 & 3-1 & 1 \\ 1 & 2 & 4-1 \end{pmatrix} \begin{pmatrix} x_1 \\ x_2 \\ x_3 \end{pmatrix} = \begin{pmatrix} 0 \\ 0 \\ 0 \end{pmatrix}$

i.e.

$\begin{pmatrix} 0 & -4 & -2 \\ 0 & 2 & 1 \\ 1 & 2 & 3 \end{pmatrix} \begin{pmatrix} x_1 \\ x_2 \\ x_3 \end{pmatrix} = \begin{pmatrix} 0 \\ 0 \\ 0 \end{pmatrix}$

from which,

$$-4x_2 - 2x_3 = 0$$
$$2x_2 + x_3 = 0$$
$$x_1 + 2x_2 + 3x_3 = 0$$

From the first two equations,

$$x_3 = -2x_2 \quad \text{(i.e. if } x_2 = 1, x_3 = -2\text{)}$$

From the last equation,

$$x_1 = -2x_2 - 3x_3 \quad \text{i.e.} \quad x_1 = -2x_2 - 3(-2x_2)$$

i.e.

$$x_1 = 4x_2 \quad \text{(i.e. if } x_2 = 1, x_1 = 4\text{)}$$

Hence the simplest eigenvector corresponding to $\lambda_1 = 1$

is: $x_1 = \begin{pmatrix} 4 \\ 1 \\ -2 \end{pmatrix}$

Using the equation $(A - \lambda I)x = 0$ for $\lambda_2 = 3$

then $\begin{pmatrix} 1-3 & -4 & -2 \\ 0 & 3-3 & 1 \\ 1 & 2 & 4-3 \end{pmatrix} \begin{pmatrix} x_1 \\ x_2 \\ x_3 \end{pmatrix} = \begin{pmatrix} 0 \\ 0 \\ 0 \end{pmatrix}$

i.e.

$\begin{pmatrix} -2 & -4 & -2 \\ 0 & 0 & 1 \\ 1 & 2 & 1 \end{pmatrix} \begin{pmatrix} x_1 \\ x_2 \\ x_3 \end{pmatrix} = \begin{pmatrix} 0 \\ 0 \\ 0 \end{pmatrix}$

from which

$$-2x_1 - 4x_2 - 2x_3 = 0$$
$$x_3 = 0$$
$$x_1 + 2x_2 + x_3 = 0$$

Since $x_3 = 0$, $x_1 = -2x_2$ (i.e. if $x_2 = 1$, $x_1 = -2$)

Hence the simplest eigenvector corresponding to $\lambda_2 = 3$

is: $x_2 = \begin{pmatrix} -2 \\ 1 \\ 0 \end{pmatrix}$

Using the equation $(A - \lambda I)x = 0$ for $\lambda_3 = 4$

then $\begin{pmatrix} 1-4 & -4 & -2 \\ 0 & 3-4 & 1 \\ 1 & 2 & 4-4 \end{pmatrix} \begin{pmatrix} x_1 \\ x_2 \\ x_3 \end{pmatrix} = \begin{pmatrix} 0 \\ 0 \\ 0 \end{pmatrix}$

i.e.

$\begin{pmatrix} -3 & -4 & -2 \\ 0 & -1 & 1 \\ 1 & 2 & 0 \end{pmatrix} \begin{pmatrix} x_1 \\ x_2 \\ x_3 \end{pmatrix} = \begin{pmatrix} 0 \\ 0 \\ 0 \end{pmatrix}$

from which,

$$-3x_1 - 4x_2 - 2x_3 = 0$$
$$-x_2 + x_3 = 0$$
$$x_1 + 2x_2 = 0$$

from which, $x_3 = x_2$ and $x_1 = -2x_2$ (i.e. if $x_2 = 1$, $x_1 = -2$ and $x_3 = 1$)

Hence the simplest eigenvector corresponding to $\lambda_3 = 4$

is: $x_3 = \begin{pmatrix} -2 \\ 1 \\ 1 \end{pmatrix}$

Now try the following Practice Exercise

Practice Exercise 114   Eigenvalues and eigenvectors (Answers on page 873)

For each of the following matrices, determine their (a) eigenvalues (b) eigenvectors

1. $\begin{pmatrix} 2 & -4 \\ -1 & -1 \end{pmatrix}$
2. $\begin{pmatrix} 3 & 6 \\ 1 & 4 \end{pmatrix}$
3. $\begin{pmatrix} 3 & 1 \\ -2 & 0 \end{pmatrix}$

4. $\begin{pmatrix} -1 & -1 & 1 \\ -4 & 2 & 4 \\ -1 & 1 & 5 \end{pmatrix}$
5. $\begin{pmatrix} 1 & -1 & 0 \\ -1 & 2 & -1 \\ 0 & -1 & 1 \end{pmatrix}$

6. $\begin{pmatrix} 2 & 2 & -2 \\ 1 & 3 & 1 \\ 1 & 2 & 2 \end{pmatrix}$
7. $\begin{pmatrix} 1 & 1 & 2 \\ 0 & 2 & 2 \\ -1 & 1 & 3 \end{pmatrix}$

For fully worked solutions to each of the problems in Practice Exercises 110 to 114 in this chapter, go to the website:
www.routledge.com/cw/bird

Section E

COMPANION @ WEBSITE

Revision Test 7   Complex numbers, matrices and determinants

This Revision Test covers the material contained in Chapters 22 to 25. *The marks for each question are shown in brackets at the end of each question.*

1. Solve the quadratic equation $x^2 - 2x + 5 = 0$ and show the roots on an Argand diagram. (9)

2. If $Z_1 = 2 + j5$, $Z_2 = 1 - j3$ and $Z_3 = 4 - j$ determine, in both Cartesian and polar forms, the value of $\dfrac{Z_1 Z_2}{Z_1 + Z_2} + Z_3$, correct to 2 decimal places. (9)

3. Three vectors are represented by A, $4.2\angle 45°$, B, $5.5\angle -32°$ and C, $2.8\angle 75°$. Determine in polar form the resultant D, where $D = B + C - A$ (8)

4. Two impedances, $Z_1 = (2 + j7)$ ohms and $Z_2 = (3 - j4)$ ohms, are connected in series to a supply voltage V of $150\angle 0°$ V. Determine the magnitude of the current I and its phase angle relative to the voltage. (6)

5. Determine in both polar and rectangular forms:

   (a) $[2.37\angle 35°]^4$ (b) $[3.2 - j4.8]^5$

   (c) $\sqrt{[-1 - j3]}$ (18)

In questions 6 to 10, the matrices stated are:

$$A = \begin{pmatrix} -5 & 2 \\ 7 & -8 \end{pmatrix} \quad B = \begin{pmatrix} 1 & 6 \\ -3 & -4 \end{pmatrix}$$

$$C = \begin{pmatrix} j3 & (1 + j2) \\ (-1 - j4) & -j2 \end{pmatrix}$$

$$D = \begin{pmatrix} 2 & -1 & 3 \\ -5 & 1 & 0 \\ 4 & -6 & 2 \end{pmatrix} \quad E = \begin{pmatrix} -1 & 3 & 0 \\ 4 & -9 & 2 \\ -5 & 7 & 1 \end{pmatrix}$$

6. Determine $A \times B$ (4)

7. Calculate the determinant of matrix C (4)

8. Determine the inverse of matrix A (4)

9. Determine $E \times D$ (9)

10. Calculate the determinant of matrix D (6)

11. Solve the following simultaneous equations:

$$4x - 3y = 17$$
$$x + y + 1 = 0$$

using matrices. (7)

12. Use determinants to solve the following simultaneous equations:

$$4x + 9y + 2z = 21$$
$$-8x + 6y - 3z = 41$$
$$3x + y - 5z = -73$$ (11)

13. The simultaneous equations representing the currents flowing in an unbalanced, three-phase, star-connected, electrical network are as follows:

$$2.4I_1 + 3.6I_2 + 4.8I_3 = 1.2$$
$$-3.9I_1 + 1.3I_2 - 6.5I_3 = 2.6$$
$$1.7I_1 + 11.9I_2 + 8.5I_3 = 0$$

Using matrices, solve the equations for I_1, I_2 and I_3 (11)

14. For the matrix $\begin{pmatrix} 2 & -4 \\ -1 & -1 \end{pmatrix}$ find the (a) eigenvalues (b) eigenvectors. (14)

Section F

Vector geometry

Vectors

At the end of this chapter, you should be able to:

- distinguish between scalars and vectors
- recognise how vectors are represented
- add vectors using the nose-to-tail method
- add vectors using the parallelogram method
- resolve vectors into their horizontal and vertical components
- add vectors by calculation – horizontal and vertical components, complex numbers
- perform vector subtraction
- understand relative velocity
- understand i, j, k notation

26.1 Introduction

This chapter initially explains the difference between scalar and vector quantities and shows how a vector is drawn and represented.

Any object that is acted upon by an external force will respond to that force by moving in the line of the force. However, if two or more forces act simultaneously, the result is more difficult to predict; the ability to add two or more vectors then becomes important.

This chapter thus shows how vectors are added and subtracted, both by drawing and by calculation, and finding the resultant of two or more vectors has many uses in engineering. (Resultant means the single vector which would have the same effect as the individual vectors.) Relative velocities and vector i, j, k notation are also briefly explained.

26.2 Scalars and vectors

The time taken to fill a water tank may be measured as, say, 50 s. Similarly, the temperature in a room may be measured as, say, 16°C, or the mass of a bearing may be measured as, say, 3 kg.

Quantities such as time, temperature and mass are entirely defined by a numerical value and are called **scalars** or **scalar quantities**.

Not all quantities are like this. Some are defined by more than just size; some also have direction. For example, the velocity of a car is 90 km/h due west, or a force of 20 N acts vertically downwards, or an acceleration of $10\,\text{m/s}^2$ acts at 50° to the horizontal.

Quantities such as velocity, force and acceleration, which **have both a magnitude and a direction**, are called **vectors**.

Now try the following Practice Exercise

1. State the difference between scalar and vector quantities.

In Problems 2 to 9, state whether the quantities given are scalar (S) or vector (V)

2. A temperature of 70°C

3. $5\,\text{m}^3$ volume

4. A downward force of 20 N

5. 500 J of work

6. $30\,\text{cm}^2$ area

7. A south-westerly wind of 10 knots

8. 50 m distance

9. An acceleration of $15\,\text{m/s}^2$ at 60° to the horizontal

26.3 Drawing a vector

A vector quantity can be represented graphically by a line, drawn so that:

(a) the **length** of the line denotes the magnitude of the quantity, and

(b) the **direction** of the line denotes the direction in which the vector quantity acts.

An arrow is used to denote the sense, or direction, of the vector. The arrow end of a vector is called the 'nose' and the other end the 'tail'. For example, a force of 9 N acting at 45° to the horizontal is shown in Fig. 26.1.

Note that an angle of **+45°** is drawn from the horizontal and moves **anticlockwise**.

Figure 26.1

A velocity of 20 m/s at −**60°** is shown in Fig. 26.2. Note that an angle of −**60°** is drawn from the horizontal and moves **clockwise**.

Figure 26.2

Representing a vector

There are a number of ways of representing vector quantities. These include:

1. Using **bold print**

2. $\overrightarrow{AB}$ where an arrow above two capital letters denotes the sense of direction, where A is the starting point and B the end point of the vector

3. $\overline{AB}$ or $\overline{a}$ i.e. a line over the top of letters

4. $\underline{a}$ i.e. an underlined letter

The force of 9 N at 45° shown in Fig. 26.1 may be represented as:

$$\mathbf{0a} \quad \text{or} \quad \overrightarrow{0a} \quad \text{or} \quad \overline{0a}$$

The magnitude of the force is $0a$

Similarly, the velocity of 20 m/s at −60° shown in Fig. 26.2 may be represented as:

$$\mathbf{0b} \quad \text{or} \quad \overrightarrow{0b} \quad \text{or} \quad \overline{0b}$$

The magnitude of the velocity is $0b$

In this chapter a vector quantity is denoted by **bold print**.

26.4 Addition of vectors by drawing

Adding two or more vectors by drawing assumes that a ruler, pencil and protractor are available. Results obtained by drawing are naturally not as accurate as those obtained by calculation.

(a) Nose-to-tail method

Two force vectors, F_1 and F_2, are shown in Fig. 26.3. When an object is subjected to more than one force, the resultant of the forces is found by the addition of vectors.

Figure 26.3

To add forces F_1 and F_2:

(i)   Force F_1 is drawn to scale horizontally, shown as **0a** in Fig. 26.4.

(ii) From the nose of F_1, force F_2 is drawn at angle θ to the horizontal, shown as **ab**.

(iii) The resultant force is given by length **0b**, which may be measured.

This procedure is called the '**nose-to-tail**' or '**triangle**' method.

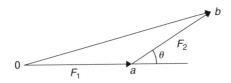

Figure 26.4

(b) Parallelogram method

To add the two force vectors, F_1 and F_2, of Fig. 26.3:

(i)   A line **cb** is constructed which is parallel to and equal in length to **0a** (see Fig. 26.5).

(ii) A line **ab** is constructed which is parallel to and equal in length to **0c**.

(iii) The resultant force is given by the diagonal of the parallelogram, i.e. length **0b**.

This procedure is called the '**parallelogram**' method.

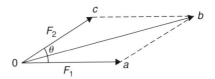

Figure 26.5

Problem 1. A force of 5 N is inclined at an angle of 45° to a second force of 8 N, both forces acting at a point. Find the magnitude of the resultant of these two forces and the direction of the resultant with respect to the 8 N force by:

(a) the 'nose-to-tail' method, and (b) the 'parallelogram' method.

The two forces are shown in Fig. 26.6. (Although the 8 N force is shown horizontal, it could have been drawn in any direction.)

Figure 26.6

(a) 'Nose-to tail' method

(i)   The 8 N force is drawn horizontally 8 units long, shown as **0a** in Fig. 26.7

(ii) From the nose of the 8 N force, the 5 N force is drawn 5 units long at an angle of 45° to the horizontal, shown as **ab**

(iii) The resultant force is given by length **0b** and is measured as **12 N** and angle θ is measured as **17°**

Figure 26.7

Section F

(b) 'Parallelogram' method

(i) In Fig. 26.8, a line is constructed which is parallel to and equal in length to the 8 N force

(ii) A line is constructed which is parallel to and equal in length to the 5 N force

(iii) The resultant force is given by the diagonal of the parallelogram, i.e. length 0*b*, and is measured as **12 N** and angle θ is measured as **17°**.

Figure 26.8

Thus, **the resultant of the two force vectors in Fig. 26.6 is 12 N at 17° to the 8 N force.**

Problem 2. Forces of 15 N and 10 N are at an angle of 90° to each other as shown in Fig. 26.9. Find, by drawing, the magnitude of the resultant of these two forces and the direction of the resultant with respect to the 15 N force.

Figure 26.9

Using the 'nose-to-tail' method:

(i) The 15 N force is drawn horizontally 15 units long as shown in Fig. 26.10

(ii) From the nose of the 15 N force, the 10 N force is drawn 10 units long at an angle of 90° to the horizontal as shown

(iii) The resultant force is shown as R and is measured as **18 N** and angle θ is measured as **34°**

Thus, **the resultant of the two force vectors is 18 N at 34° to the 15 N force.**

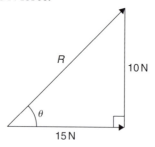

Figure 26.10

Problem 3. Velocities of 10 m/s, 20 m/s and 15 m/s act as shown in Fig. 26.11. Determine, by drawing, the magnitude of the resultant velocity and its direction relative to the horizontal.

Figure 26.11

When more than two vectors are being added the 'nose-to-tail' method is used.

The order in which the vectors are added does not matter. In this case the order taken is v_1, then v_2, then v_3. However, if a different order is taken the same result will occur.

(i) v_1 is drawn 10 units long at an angle of 30° to the horizontal, shown as **0*a*** in Fig. 26.12.

(ii) From the nose of v_1, v_2 is drawn 20 units long at an angle of 90° to the horizontal, shown as ***ab***.

(iii) From the nose of v_2, v_3 is drawn 15 units long at an angle of 195° to the horizontal, shown as ***br***.

(iv) The resultant velocity is given by length **0*r*** and is measured as **22 m/s** and the angle measured to the horizontal is **105°**.

Thus, **the resultant of the three velocities is 22 m/s at 105° to the horizontal.**

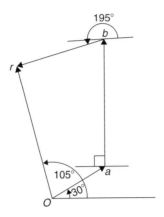

Figure 26.12

Worked Problems 1 to 3 have demonstrated how vectors are added to determine their resultant and their direction. However, drawing to scale is time-consuming and not highly accurate. The following sections demonstrate how to determine resultant vectors by calculation using horizontal and vertical components and, where possible, by Pythagoras' theorem.

26.5 Resolving vectors into horizontal and vertical components

A force vector F is shown in Fig. 26.13 at angle θ to the horizontal. Such a vector can be resolved into two components such that the vector addition of the components is equal to the original vector.

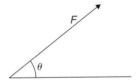

Figure 26.13

The two components usually taken are a **horizontal component** and a **vertical component**.

If a right-angled triangle is constructed as shown in Fig. 26.14, then $\mathbf{0a}$ is called the horizontal component of F and $\mathbf{ab}$ is called the vertical component of F.

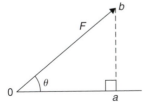

Figure 26.14

From trigonometry (see Chapter 12),

$$\cos\theta = \frac{0a}{0b} \quad \text{from which,} \quad \mathbf{0a} = \mathbf{0b}\cos\theta$$
$$= F\cos\theta$$

i.e. **the horizontal component of $F = F\cos\theta$**

and $\quad \sin\theta = \dfrac{ab}{0b} \quad$ from which, $\quad \mathbf{ab} = \mathbf{0b}\sin\theta$
$$= F\sin\theta$$

i.e. **the vertical component of $F = F\sin\theta$**

Problem 4. Resolve the force vector of 50 N at an angle of $35°$ to the horizontal into its horizontal and vertical components.

The **horizontal component** of the 50 N force,
$\mathbf{0a} = 50\cos35° = \mathbf{40.96\,N}$
The **vertical component** of the 50 N force,
$\mathbf{ab} = 50\sin35° = \mathbf{28.68\,N}$
The horizontal and vertical components are shown in Fig. 26.15.

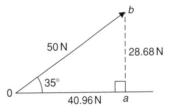

Figure 26.15

(Checking: by Pythagoras, $\mathbf{0b} = \sqrt{40.96^2 + 28.68^2}$
$$= 50\,N$$

and $\quad \theta = \tan^{-1}\left(\dfrac{28.68}{40.96}\right) = 35°$

Thus, the vector addition of components 40.96 N and 28.68 N is 50 N at $35°$)

Problem 5. Resolve the velocity vector of 20 m/s at an angle of $-30°$ to the horizontal into horizontal and vertical components.

The **horizontal component** of the 20 m/s velocity,
$\mathbf{0a} = 20\cos(-30°) = \mathbf{17.32\,m/s}$
The **vertical component** of the 20 m/s velocity,
$\mathbf{ab} = 20\sin(-30°) = \mathbf{-10\,m/s}$
The horizontal and vertical components are shown in Fig. 26.16.

Figure 26.16

Problem 6. Resolve the displacement vector of 40 m at an angle of $120°$ into horizontal and vertical components.

The **horizontal component** of the 40 m displacement, $0a = 40\cos 120° = -20.0\,\text{m}$
The **vertical component** of the 40 m displacement, $ab = 40\sin 120° = 34.64\,\text{m}$
The horizontal and vertical components are shown in Fig. 26.17.

Figure 26.17

26.6 Addition of vectors by calculation

Two force vectors, F_1 and F_2, are shown in Fig. 26.18, F_1 being at an angle of θ_1 and F_2 being at an angle of θ_2

Figure 26.18

A method of adding two vectors together is to use horizontal and vertical components.
The horizontal component of force F_1 is $F_1\cos\theta_1$ and the horizontal component of force F_2 is $F_2\cos\theta_2$
The total horizontal component of the two forces,
$$H = F_1\cos\theta_1 + F_2\cos\theta_2$$
The vertical component of force F_1 is $F_1\sin\theta_1$ and the vertical component of force F_2 is $F_2\sin\theta_2$
The total vertical component of the two forces,
$$V = F_1\sin\theta_1 + F_2\sin\theta_2$$
Since we have H and V, the resultant of F_1 and F_2 is obtained by using the theorem of Pythagoras. From Fig. 26.19, $\qquad 0b^2 = H^2 + V^2$
i.e. $\qquad \textbf{resultant} = \sqrt{H^2 + V^2}\qquad$ at an angle

given by $\theta = \tan^{-1}\left(\dfrac{V}{H}\right)$

Figure 26.19

Problem 7. A force of 5 N is inclined at an angle of $45°$ to a second force of 8 N, both forces acting at a point. Calculate the magnitude of the resultant of these two forces and the direction of the resultant with respect to the 8 N force.

The two forces are shown in Fig. 26.20.

Figure 26.20

The horizontal component of the 8 N force is $8\cos 0°$ and the horizontal component of the 5 N force is $5\cos 45°$
The total horizontal component of the two forces,

$$H = 8\cos 0° + 5\cos 45° = 8 + 3.5355$$
$$= \textbf{11.5355}$$

The vertical component of the 8 N force is $8 \sin 0°$ and the vertical component of the 5 N force is $5 \sin 45°$ The total vertical component of the two forces,

$$V = 8 \sin 0° + 5 \sin 45° = 0 + 3.5355$$

$$= \mathbf{3.5355}$$

From Fig. 26.21, magnitude of resultant vector

$$= \sqrt{H^2 + V^2}$$

$$= \sqrt{11.5355^2 + 3.5355^2} = \mathbf{12.07 \, N}$$

Figure 26.21

The direction of the resultant vector,

$$\theta = \tan^{-1}\left(\frac{V}{H}\right) = \tan^{-1}\left(\frac{3.5355}{11.5355}\right)$$

$$= \tan^{-1} 0.30648866\ldots = \mathbf{17.04°}$$

Thus, **the resultant of the two forces is a single vector of 12.07 N at 17.04° to the 8 N vector**.

Perhaps an easier and quicker method of calculating the magnitude and direction of the resultant is to use **complex numbers** (see Chapter 22).

In this example, the **resultant**

$$= 8\angle 0° + 5\angle 45°$$

$$= (8 \cos 0° + j8 \sin 0°) + (5 \cos 45° + j5 \sin 45°)$$

$$= (8 + j0) + (3.536 + j3.536)$$

$$= (11.536 + j3.536) \, N \quad \text{or} \quad \mathbf{12.07 \angle 17.04°N}$$

as obtained above using horizontal and vertical components.

Problem 8. Forces of 15 N and 10 N are at an angle of 90° to each other as shown in Fig. 26.22. Calculate the magnitude of the resultant of these two forces and its direction with respect to the 15 N force.

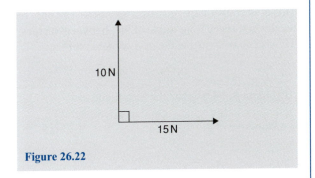

Figure 26.22

The horizontal component of the 15 N force is $15 \cos 0°$ and the horizontal component of the 10 N force is $10 \cos 90°$ The total horizontal component of the two velocities,

$$H = 15 \cos 0° + 10 \cos 90° = 15 + 0 = \mathbf{15}$$

The vertical component of the 15 N force is $15 \sin 0°$ and the vertical component of the 10 N force is $10 \sin 90°$ The total vertical component of the two velocities,

$$V = 15 \sin 0° + 10 \sin 90° = 0 + 10 = \mathbf{10}$$

Magnitude of resultant vector

$$= \sqrt{H^2 + V^2} = \sqrt{15^2 + 10^2} = \mathbf{18.03 \, N}$$

The direction of the resultant vector,

$$\theta = \tan^{-1}\left(\frac{V}{H}\right) = \tan^{-1}\left(\frac{10}{15}\right) = \mathbf{33.69°}$$

Thus, **the resultant of the two forces is a single vector of 18.03 N at 33.69° to the 15 N vector**.

There is an alternative method of calculating the resultant vector in this case. If we used the triangle method, then the diagram would be as shown in Fig. 26.23.

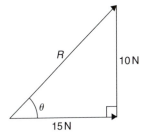

Figure 26.23

Since a right-angled triangle results then we could use **Pythagoras' theorem** without needing to go through the procedure for horizontal and vertical components. In fact, the horizontal and vertical components are 15 N and 10 N respectively.

Section F

This is, of course, a special case. Pythagoras **can only be used when there is an angle of 90° between vectors**. This is demonstrated in the next worked problem.

Problem 9. Calculate the magnitude and direction of the resultant of the two acceleration vectors shown in Fig. 26.24.

Figure 26.24

The $15\,\text{m/s}^2$ acceleration is drawn horizontally, shown as **0a** in Fig. 26.25.
From the nose of the $15\,\text{m/s}^2$ acceleration, the $28\,\text{m/s}^2$ acceleration is drawn at an angle of $90°$ to the horizontal, shown as **ab**.

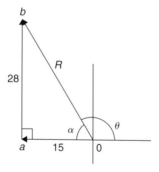

Figure 26.25

The resultant acceleration R is given by length **0b**.
Since a right-angled triangle results, the theorem of Pythagoras may be used.

$$0b = \sqrt{15^2 + 28^2} = \textbf{31.76 m/s}^2$$

and $\qquad \alpha = \tan^{-1}\left(\dfrac{28}{15}\right) = \textbf{61.82}°$

Measuring from the horizontal,
$\theta = 180° - 61.82° = \textbf{118.18}°$

Thus, **the resultant of the two accelerations is a single vector of 31.76 m/s^2 at 118.18° to the horizontal**.

Problem 10. Velocities of $10\,\text{m/s}$, $20\,\text{m/s}$ and $15\,\text{m/s}$ act as shown in Fig. 26.26. Calculate the magnitude of the resultant velocity and its direction relative to the horizontal.

Figure 26.26

The horizontal component of the $10\,\text{m/s}$ velocity is $10\cos 30° = 8.660\,\text{m/s}$,

the horizontal component of the $20\,\text{m/s}$ velocity is $20\cos 90° = 0\,\text{m/s}$,

and the horizontal component of the $15\,\text{m/s}$ velocity is $15\cos 195° = -14.489\,\text{m/s}$.

The total horizontal component of the three velocities,

$$H = 8.660 + 0 - 14.489 = \textbf{-5.829 m/s}$$

The vertical component of the $10\,\text{m/s}$ velocity is $10\sin 30° = 5\,\text{m/s}$,

the vertical component of the $20\,\text{m/s}$ velocity is $20\sin 90° = 20\,\text{m/s}$,

and the vertical component of the $15\,\text{m/s}$ velocity is $15\sin 195° = -3.882\,\text{m/s}$.

The total vertical component of the three forces,

$$V = 5 + 20 - 3.882 = \textbf{21.118 m/s}$$

From Fig. 26.27, magnitude of resultant vector,
$$R = \sqrt{H^2 + V^2} = \sqrt{5.829^2 + 21.118^2} = \textbf{21.91 m/s}$$

The direction of the resultant vector,
$$\alpha = \tan^{-1}\left(\frac{V}{H}\right) = \tan^{-1}\left(\frac{21.118}{5.829}\right) = \textbf{74.57}°$$

Measuring from the horizontal,
$\theta = 180° - 74.57° = \textbf{105.43}°$

Figure 26.27

Thus, **the resultant of the three velocities is a single vector of 21.91 m/s at 105.43° to the horizontal**.

Using **complex numbers**, from Fig. 26.26,

resultant $= 10\angle 30° + 20\angle 90° + 15\angle 195°$

$= (10\cos 30° + j10\sin 30°)$

$\qquad + (20\cos 90° + j20\sin 90°)$

$\qquad + (15\cos 195° + j15\sin 195°)$

$= (8.660 + j5.000) + (0 + j20.000)$

$\qquad + (-14.489 - j3.882)$

$= (-5.829 + j21.118)\,\text{N} \quad \text{or}$

$$\mathbf{21.91\angle 105.43°\,N}$$

as obtained above using horizontal and vertical components.

The method used to add vectors by calculation will not be specified – the choice is yours, but probably the quickest and easiest method is by using complex numbers.

Now try the following Practice Exercise

Practice Exercise 116   Addition of vectors by calculation (Answers on page 874)

1. A force of 7 N is inclined at an angle of 50° to a second force of 12 N, both forces acting at a point. Calculate magnitude of the resultant of the two forces, and the direction of the resultant with respect to the 12 N force.

2. Velocities of 5 m/s and 12 m/s act at a point at 90° to each other. Calculate the resultant velocity and its direction relative to the 12 m/s velocity.

3. Calculate the magnitude and direction of the resultant of the two force vectors shown in Fig. 26.28.

Figure 26.28

4. Calculate the magnitude and direction of the resultant of the two force vectors shown in Fig. 26.29.

Figure 26.29

5. A displacement vector s_1 is 30 m at 0°. A second displacement vector s_2 is 12 m at 90°. Calculate magnitude and direction of the resultant vector $s_1 + s_2$

6. Three forces of 5 N, 8 N and 13 N act as shown in Fig. 26.30. Calculate the magnitude and direction of the resultant force.

Figure 26.30

7. If velocity $v_1 = 25$ m/s at $60°$ and $v_2 = 15$ m/s at $-30°$, calculate the magnitude and direction of $v_1 + v_2$

8. Calculate the magnitude and direction of the resultant vector of the force system shown in Fig. 26.31.

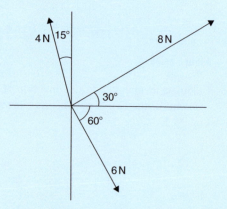

Figure 26.31

9. Calculate the magnitude and direction of the resultant vector of the system shown in Fig. 26.32.

Figure 26.32

10. An object is acted upon by two forces of magnitude 10 N and 8 N at an angle of $60°$ to each other. Determine the resultant force on the object.

11. A ship heads in a direction of E 20° S at a speed of 20 knots while the current is 4 knots in a direction of N 30° E. Determine the speed and actual direction of the ship.

26.7 Vector subtraction

In Fig. 26.33, a force vector F is represented by oa. The vector $(-oa)$ can be obtained by drawing a vector from o in the opposite sense to oa but having the same magnitude, shown as ob in Fig. 26.33, i.e. $ob = (-oa)$

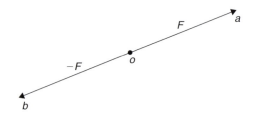

Figure 26.33

For two vectors acting at a point, as shown in Fig. 26.34(a), the resultant of vector addition is: $os = oa + ob$
Figure 26.33(b) shows vectors $ob + (-oa)$, that is, $ob - oa$ and the vector equation is $ob - oa = od$. Comparing od in Fig. 26.34(b) with the broken line ab in Fig. 26.34(a) shows that the second diagonal of the 'parallelogram' method of vector addition gives the magnitude and direction of vector subtraction of oa from ob.

Figure 26.34

Problem 11. Accelerations of $a_1 = 1.5$ m/s^2 at $90°$ and $a_2 = 2.6$ m/s^2 at $145°$ act at a point. Find $a_1 + a_2$ and $a_1 - a_2$ (a) by drawing a scale vector diagram, and (b) by calculation.

(a) The scale vector diagram is shown in Fig. 26.35. By measurement,

$$a_1 + a_2 = 3.7 \text{ m/s}^2 \text{ at } 126°$$

$$a_1 - a_2 = 2.1 \text{ m/s}^2 \text{ at } 0°$$

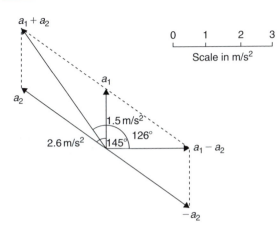

Figure 26.35

(b) Resolving horizontally and vertically gives:

Horizontal component of $a_1 + a_2$,
$$H = 1.5\cos 90° + 2.6\cos 145° = -2.13$$

Vertical component of $a_1 + a_2$,
$$V = 1.5\sin 90° + 2.6\sin 145° = 2.99$$

From Fig. 26.36, magnitude of $a_1 + a_2$,
$$R = \sqrt{(-2.13)^2 + 2.99^2} = 3.67 \, \text{m/s}^2$$

In Fig. 26.36, $\alpha = \tan^{-1}\left(\dfrac{2.99}{2.13}\right) = 54.53°$ and
$$\theta = 180° - 54.53° = 125.47°$$

Thus, $\quad a_1 + a_2 = 3.67 \, \text{m/s}^2 \text{ at } 125.47°$

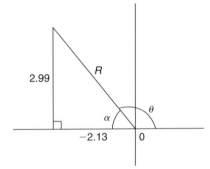

Figure 26.36

Horizontal component of $a_1 - a_2$
$$= 1.5\cos 90° - 2.6\cos 145° = 2.13$$

Vertical component of $a_1 - a_2$
$$= 1.5\sin 90° - 2.6\sin 145° = 0$$

Magnitude of $a_1 - a_2 = \sqrt{2.13^2 + 0^2}$
$$= 2.13 \, \text{m/s}^2$$

Direction of $a_1 - a_2 = \tan^{-1}\left(\dfrac{0}{2.13}\right) = 0°$

Thus, $\quad a_1 - a_2 = 2.13 \, \text{m/s}^2 \text{ at } 0°$

Problem 12. Calculate the resultant of (a) $v_1 - v_2 + v_3$ and (b) $v_2 - v_1 - v_3$ when $v_1 = 22$ units at $140°$, $v_2 = 40$ units at $190°$ and $v_3 = 15$ units at $290°$.

(a) The vectors are shown in Fig. 26.37.

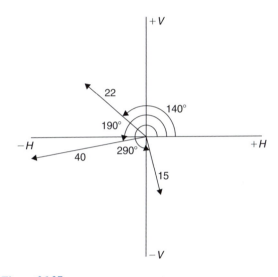

Figure 26.37

The horizontal component of

$$v_1 - v_2 + v_3 = (22\cos 140°) - (40\cos 190°)$$
$$+ (15\cos 290°)$$
$$= (-16.85) - (-39.39) + (5.13)$$
$$= 27.67 \, \text{units}$$

The vertical component of

$$v_1 - v_2 + v_3 = (22\sin 140°) - (40\sin 190°)$$
$$+ (15\sin 290°)$$
$$= (14.14) - (-6.95) + (-14.10)$$
$$= 6.99 \, \text{units}$$

The magnitude of the resultant,
$$R = \sqrt{27.67^2 + 6.99^2} = 28.54 \, \text{units}$$

The direction of the resultant $R = \tan^{-1}\left(\dfrac{6.99}{27.67}\right)$
$$= 14.18°$$

Thus, $v_1 - v_2 + v_3 = 28.54 \, \text{units at } 14.18°$

Section F

Using **complex numbers**,

$$v_1 - v_2 + v_3 = 22\angle140° - 40\angle190° + 15\angle290°$$

$$= (-16.853 + j14.141)$$

$$- (-39.392 - j6.946)$$

$$+ (5.130 - j14.095)$$

$$= 27.669 + j6.992 = \mathbf{28.54\angle14.18°}$$

(b) The horizontal component of

$$v_2 - v_1 - v_3 = (40\cos190°) - (22\cos140°)$$

$$- (15\cos290°)$$

$$= (-39.39) - (-16.85) - (5.13)$$

$$= \mathbf{-27.67\ units}$$

The vertical component of

$$v_2 - v_1 - v_3 = (40\sin190°) - (22\sin140°)$$

$$- (15\sin290°)$$

$$= (-6.95) - (14.14) - (-14.10)$$

$$= \mathbf{-6.99\ units}$$

From Fig. 26.38 the magnitude of the resultant,
$$R = \sqrt{(-27.67)^2 + (-6.99)^2} = \mathbf{28.54\ units}$$

and $\alpha = \tan^{-1}\left(\dfrac{6.99}{27.67}\right) = 14.18°$, from which,
$$\theta = 180° + 14.18° = 194.18°$$

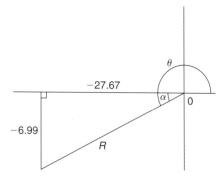

Figure 26.38

Thus, $v_2 - v_1 - v_3 = \mathbf{28.54\ units\ at\ 194.18°}$

This result is as expected, since $v_2 - v_1 - v_3 = -(v_1 - v_2 + v_3)$ and the vector 28.54 units at 194.18° is minus times (i.e. is 180° out of phase with) the vector 28.54 units at 14.18°

Using **complex numbers**,

$$v_2 - v_1 - v_3 = 40\angle190° - 22\angle140° - 15\angle290°$$

$$= (-39.392 - j6.946)$$

$$- (-16.853 + j14.141)$$

$$- (5.130 - j14.095)$$

$$= -27.669 - j6.992$$

$$= \mathbf{28.54\angle-165.82°}\ \text{or}$$

$$\mathbf{28.54\angle194.18°}$$

Now try the following Practice Exercise

26.8 Relative velocity

For relative velocity problems, some fixed datum point needs to be selected. This is often a fixed point on the Earth's surface. In any vector equation, only the start and finish points affect the resultant vector of a system. Two different systems are shown in Fig. 26.39, but in each of the systems, the resultant vector is ad.

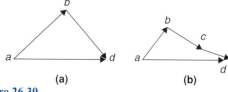

(a)   (b)

Figure 26.39

The vector equation of the system shown in Fig. 26.39(a) is:

$$ad = ab + bd$$

and that for the system shown in Fig. 26.39(b) is:

$$ad = ab + bc + cd$$

Thus in vector equations of this form, only the first and last letters, 'a' and 'd', respectively, fix the magnitude and direction of the resultant vector. This principle is used in relative velocity problems.

Problem 13. Two cars, P and Q, are travelling towards the junction of two roads which are at right angles to one another. Car P has a velocity of 45 km/h due east and car Q a velocity of 55 km/h due south. Calculate (a) the velocity of car P relative to car Q, and (b) the velocity of car Q relative to car P.

(a) The directions of the cars are shown in Fig. 26.40(a), called a **space diagram**. The velocity diagram is shown in Fig. 26.40(b), in which **pe** is taken as the velocity of car P relative to point e on the Earth's surface. The velocity of P relative to Q is vector **pq** and the vector equation is **pq** = **pe** + **eq**. Hence the vector directions are as shown, **eq** being in the opposite direction to **qe**.

From the geometry of the vector triangle, the magnitude of $pq = \sqrt{45^2 + 55^2} = 71.06$ km/h and the

direction of $pq = \tan^{-1}\left(\dfrac{55}{45}\right) = 50.71°$

i.e. **the velocity of car P relative to car Q is 71.06 km/h at 50.71°**

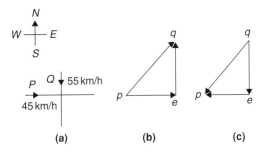

(a)   (b)   (c)

Figure 26.40

(b) The velocity of car Q relative to car P is given by the vector equation **qp** = **qe** + **ep** and the vector diagram is as shown in Fig. 26.40(c), having **ep** opposite in direction to **pe**.

From the geometry of this vector triangle, the magnitude of $qp = \sqrt{45^2 + 55^2} = 71.06$ m/s and

the direction of $qp = \tan^{-1}\left(\dfrac{55}{45}\right) = 50.71°$ but

must lie in the third quadrant, i.e. the required angle is: $180° + 50.71° = 230.71°$

i.e. **the velocity of car Q relative to car P is 71.06 m/s at 230.71°**

Now try the following Practice Exercise

26.9 i, j and k notation

A method of completely specifying the direction of a vector in space relative to some reference point is to use three unit vectors, i, j and k, mutually at right angles to each other, as shown in Fig. 26.41.

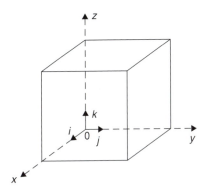

Figure 26.41

Calculations involving vectors given in i, jk notation are carried out in exactly the same way as standard algebraic calculations, as shown in the worked examples below.

Problem 14. Determine:
$(3i + 2j + 2k) - (4i - 3j + 2k)$

$$(3i + 2j + 2k) - (4i - 3j + 2k) = 3i + 2j + 2k$$
$$- 4i + 3j - 2k$$
$$= -i + 5j$$

Problem 15. Given $p = 3i + 2k$,
$q = 4i - 2j + 3k$ and $r = -3i + 5j - 4k$
determine:

(a) $-r$ (b) $3p$ (c) $2p + 3q$ (d) $-p + 2r$
(e) $0.2p + 0.6q - 3.2r$

(a) $-r = -(-3i + 5j - 4k) = +3i - 5j + 4k$

(b) $3p = 3(3i + 2k) = 9i + 6k$

(c) $2p + 3q = 2(3i + 2k) + 3(4i - 2j + 3k)$
$$= 6i + 4k + 12i - 6j + 9k$$
$$= 18i - 6j + 13k$$

(d) $-p + 2r = -(3i + 2k) + 2(-3i + 5j - 4k)$
$$= -3i - 2k + (-6i + 10j - 8k)$$
$$= -3i - 2k - 6i + 10j - 8k$$
$$= -9i + 10j - 10k$$

(e) $0.2p + 0.6q - 3.2r = 0.2(3i + 2k)$
$$+0.6(4i - 2j + 3k) - 3.2(-3i + 5j - 4k)$$
$$= 0.6i + 0.4k + 2.4i - 1.2j + 1.8k$$
$$+9.6i - 16j + 12.8k$$
$$= \mathbf{12.6i - 17.2j + 15k}$$

Now try the following Practice Exercise

Practice Exercise 119   *i, j, k* **notation**
(Answers on page 874)

Given that $p = 2i + 0.5j - 3k$, $q = -i + j + 4k$
and $r = 6j - 5k$, evaluate and simplify the following vectors in i, j, k form:

1. $-q$
2. $2p$
3. $q + r$
4. $-q + 2p$
5. $3q + 4r$
6. $q - 2p$
7. $p + q + r$
8. $p + 2q + 3r$
9. $2p + 0.4q + 0.5r$
10. $7r - 2q$

For fully worked solutions to each of the problems in Practice Exercises 115 to 119 in this chapter,
go to the website:
www.routledge.com/cw/bird

Section F

Chapter 27

Methods of adding alternating waveforms

Why it is important to understand: **Methods of adding alternating waveforms**

In electrical engineering, a phasor is a rotating vector representing a quantity such as an alternating current or voltage that varies sinusoidally. Sometimes it is necessary when studying sinusoidal quantities to add together two alternating waveforms, for example in an a.c. series circuit, where they are not in-phase with each other. Electrical engineers, electronics engineers, electronic engineering technicians and aircraft engineers all use phasor diagrams to visualise complex constants and variables. So, given oscillations to add and subtract, the required rotating vectors are constructed, called a phasor diagram, and graphically the resulting sum and/or difference oscillation are added or calculated. Phasors may be used to analyse the behaviour of electrical and mechanical systems that have reached a kind of equilibrium called sinusoidal steady state. Hence, discovering different methods of combining sinusoidal waveforms is of some importance in certain areas of engineering.

At the end of this chapter, you should be able to:

- determine the resultant of two phasors by graph plotting
- determine the resultant of two or more phasors by drawing
- determine the resultant of two phasors by the sine and cosine rules
- determine the resultant of two or more phasors by horizontal and vertical components
- determine the resultant of two or more phasors by complex numbers

27.1 Combination of two periodic functions

There are a number of instances in engineering and science where waveforms have to be combined and where it is required to determine the single phasor (called the resultant) that could replace two or more separate phasors. Uses are found in electrical alternating current theory, in mechanical vibrations, in the addition of forces and with sound waves.

There are a number of methods of determining the resultant waveform. These include:

(a) by drawing the waveforms and adding graphically

(b) by drawing the phasors and measuring the resultant

(c) by using the cosine and sine rules

(d) by using horizontal and vertical components

(e) by using complex numbers

27.2 Plotting periodic functions

This may be achieved by sketching the separate functions on the same axes and then adding (or subtracting) ordinates at regular intervals. This is demonstrated in the following worked Problems.

Problem 1. Plot the graph of $y_1 = 3\sin A$ from $A = 0°$ to $A = 360°$. On the same axes plot $y_2 = 2\cos A$. By adding ordinates, plot $y_R = 3\sin A + 2\cos A$ and obtain a sinusoidal expression for this resultant waveform.

$y_1 = 3\sin A$ and $y_2 = 2\cos A$ are shown plotted in Fig. 27.1. Ordinates may be added at, say, $15°$ intervals. For example,

at $0°$, $y_1 + y_2 = 0 + 2 = 2$

at $15°$, $y_1 + y_2 = 0.78 + 1.93 = 2.71$

at $120°$, $y_1 + y_2 = 2.60 + -1 = 1.6$

at $210°$, $y_1 + y_2 = -1.50 - 1.73 = -3.23$, and so on.

The resultant waveform, shown by the broken line, has the same period, i.e. $360°$, and thus the same frequency as the single phasors. The maximum value, or amplitude, of the resultant is 3.6. The resultant waveform **leads** $y_1 = 3\sin A$ by $34°$ or $34 \times \dfrac{\pi}{180}$ rad $= 0.593$ rad.

The sinusoidal expression for the resultant waveform is:

$$y_R = 3.6\sin(A + 34°) \quad \text{or}$$

$$y_R = 3.6\sin(A + 0.593)$$

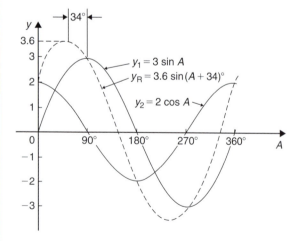

Figure 27.1

Problem 2. Plot the graphs of $y_1 = 4\sin\omega t$ and $y_2 = 3\sin(\omega t - \pi/3)$ on the same axes, over one cycle. By adding ordinates at intervals plot $y_R = y_1 + y_2$ and obtain a sinusoidal expression for the resultant waveform.

$y_1 = 4\sin\omega t$ and $y_2 = 3\sin(\omega t - \pi/3)$ are shown plotted in Fig. 27.2.

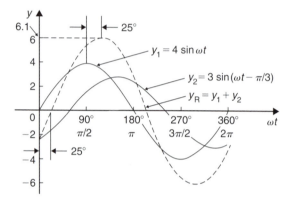

Figure 27.2

Ordinates are added at $15°$ intervals and the resultant is shown by the broken line. The amplitude of the resultant is 6.1 and it **lags** y_1 by $25°$ or 0.436 rad.

Hence, the sinusoidal expression for the resultant waveform is:

$$y_R = 6.1\sin(\omega t - 0.436)$$

Problem 3. Determine a sinusoidal expression for $y_1 - y_2$ when $y_1 = 4\sin\omega t$ and $y_2 = 3\sin(\omega t - \pi/3)$

y_1 and y_2 are shown plotted in Fig. 27.3. At $15°$ intervals y_2 is subtracted from y_1. For example:

at $0°$, $y_1 - y_2 = 0 - (-2.6) = +2.6$

at $30°$, $y_1 - y_2 = 2 - (-1.5) = +3.5$

at $150°$, $y_1 - y_2 = 2 - 3 = -1$, and so on.

The amplitude, or peak value of the resultant (shown by the broken line), is 3.6 and it leads y_1 by $45°$ or 0.79 rad. Hence,

$$y_1 - y_2 = 3.6\sin(\omega t + 0.79)$$

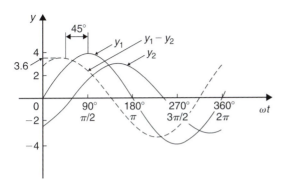

Figure 27.3

Problem 4. Two alternating currents are given by:
$i_1 = 20 \sin \omega t$ amperes and
$i_2 = 10 \sin \left(\omega t + \dfrac{\pi}{3} \right)$ amperes.
By drawing the waveforms on the same axes and adding, determine the sinusoidal expression for the resultant $i_1 + i_2$

i_1 and i_2 are shown plotted in Fig. 27.4. The resultant waveform for $i_1 + i_2$ is shown by the broken line. It has the same period, and hence frequency, as i_1 and i_2

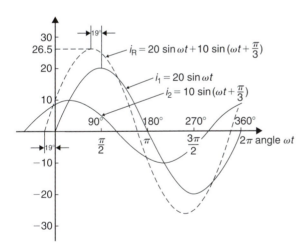

Figure 27.4

The amplitude or peak value is 26.5 A
The resultant waveform leads the waveform of $i_1 = 20 \sin \omega t$ by 19° or 0.33 rad
Hence, the sinusoidal expression for the resultant $i_1 + i_2$ is given by:

$$i_R = i_1 + i_2 = 26.5 \sin (\omega t + 0.33) \text{ A}$$

Now try the following Practice Exercise

Practice Exercise 120 Plotting periodic functions (Answers on page 874)

1. Plot the graph of $y = 2 \sin A$ from $A = 0°$ to $A = 360°$. On the same axes plot $y = 4 \cos A$. By adding ordinates at intervals plot $y = 2 \sin A + 4 \cos A$ and obtain a sinusoidal expression for the waveform.

2. Two alternating voltages are given by $v_1 = 10 \sin \omega t$ volts and $v_2 = 14 \sin(\omega t + \pi/3)$ volts. By plotting v_1 and v_2 on the same axes over one cycle obtain a sinusoidal expression for (a) $v_1 + v_2$ (b) $v_1 - v_2$

3. Express $12 \sin \omega t + 5 \cos \omega t$ in the form $A \sin(\omega t \pm \alpha)$ by drawing and measurement.

27.3 Determining resultant phasors by drawing

The resultant of two periodic functions may be found from their relative positions when the time is zero. For example, if $y_1 = 4 \sin \omega t$ and $y_2 = 3 \sin(\omega t - \pi/3)$ then each may be represented as phasors as shown in Fig. 27.5, y_1 being 4 units long and drawn horizontally and y_2 being 3 units long, lagging y_1 by $\pi/3$ radians or 60°. To determine the resultant of $y_1 + y_2$, y_1 is drawn horizontally as shown in Fig. 27.6 and y_2 is joined to the

Figure 27.5

Figure 27.6

Figure 27.7

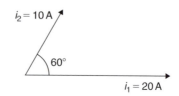

Figure 27.9

end of y_1 at $60°$ to the horizontal. The resultant is given by y_R. This is the same as the diagonal of a parallelogram that is shown completed in Fig. 27.7.

Resultant y_R, in Figs. 27.6 and 27.7, may be determined by drawing the phasors and their directions to scale and measuring using a ruler and protractor.

In this example, y_R is measured as 6 units long and angle ϕ is measured as $25°$

$$25° = 25 \times \frac{\pi}{180} \text{ radians } = 0.44 \text{ rad}$$

Hence, summarising, by drawing:
$$y_R = y_1 + y_2 = 4\sin\omega t + 3\sin(\omega t - \pi/3)$$
$$= 6\sin(\omega t - 0.44)$$

If the resultant phasor $y_R = y_1 - y_2$ is required, then y_2 is still 3 units long but is drawn in the opposite direction, as shown in Fig. 27.8.

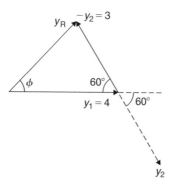

Figure 27.8

Problem 5. Two alternating currents are given by: $i_1 = 20\sin\omega t$ amperes and $i_2 = 10\sin\left(\omega t + \frac{\pi}{3}\right)$ amperes. Determine $i_1 + i_2$ by drawing phasors.

The relative positions of i_1 and i_2 at time $t = 0$ are shown as phasors in Fig. 27.9, where $\frac{\pi}{3}\text{rad} = 60°$

The phasor diagram in Fig. 27.10 is drawn to scale with a ruler and protractor.

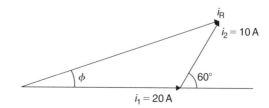

Figure 27.10

The resultant i_R is shown and is measured as 26 A and angle ϕ as $19°$ or 0.33 rad leading i_1. Hence, by drawing and measuring:

$$i_R = i_1 + i_2 = 26\sin(\omega t + 0.33)\text{A}$$

Problem 6. For the currents in Problem 5, determine $i_1 - i_2$ by drawing phasors.

At time $t = 0$, current i_1 is drawn 20 units long horizontally as shown by $0a$ in Fig. 27.11. Current i_2 is shown, drawn 10 units long in broken line and leading by $60°$. The current $-i_2$ is drawn in the opposite direction to the broken line of i_2, shown as ab in Fig. 27.11. The resultant i_R is given by $0b$ lagging by angle ϕ.

By measurement, $i_R = 17$A and $\phi = 30°$ or 0.52 rad

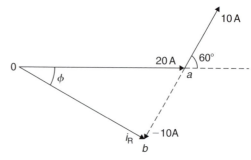

Figure 27.11

Hence, by drawing phasors:

$$i_R = i_1 - i_2 = 17\sin(\omega t - 0.52)$$

Now try the following Practice Exercise

Practice Exercise 121   Determining resultant phasors by drawing (Answers on page 874)

1. Determine a sinusoidal expression for $2\sin\theta + 4\cos\theta$ by drawing phasors.

2. If $v_1 = 10\sin\omega t$ volts and $v_2 = 14\sin(\omega t + \pi/3)$ volts, determine by drawing phasors sinusoidal expressions for (a) $v_1 + v_2$ (b) $v_1 - v_2$

3. Express $12\sin\omega t + 5\cos\omega t$ in the form $R\sin(\omega t \pm \alpha)$ by drawing phasors.

27.4   Determining resultant phasors by the sine and cosine rules

As stated earlier, the resultant of two periodic functions may be found from their relative positions when the time is zero. For example, if $y_1 = 5\sin\omega t$ and $y_2 = 4\sin(\omega t - \pi/6)$ then each may be represented by phasors as shown in Fig. 27.12, y_1 being 5 units long and drawn horizontally and y_2 being 4 units long, lagging y_1 by $\pi/6$ radians or $30°$. To determine the resultant of $y_1 + y_2$, y_1 is drawn horizontally as shown in Fig. 27.13 and y_2 is joined to the end of y_1 at $\pi/6$ radians, i.e. $30°$ to the horizontal. The resultant is given by y_R.
Using the cosine rule on triangle $0ab$ of Fig. 27.13 gives:

$$y_R^2 = 5^2 + 4^2 - [2(5)(4)\cos 150°]$$
$$= 25 + 16 - (-34.641)$$
$$= 75.641$$

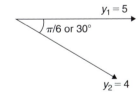

Figure 27.12

Figure 27.13

from which, $y_R = \sqrt{75.641} = 8.697$

Using the sine rule, $\dfrac{8.697}{\sin 150°} = \dfrac{4}{\sin\phi}$

from which, $\sin\phi = \dfrac{4\sin 150°}{8.697}$
$$= 0.22996$$

and $\phi = \sin^{-1} 0.22996$
$$= 13.29° \text{ or } 0.232 \text{ rad}$$

Hence, $y_R = y_1 + y_2 = 5\sin\omega t + 4\sin(\omega t - \pi/6)$
$$= 8.697\sin(\omega t - 0.232)$$

Problem 7.   Given $y_1 = 2\sin\omega t$ and $y_2 = 3\sin(\omega t + \pi/4)$, obtain an expression, by calculation, for the resultant, $y_R = y_1 + y_2$

When time $t = 0$, the position of phasors y_1 and y_2 are as shown in Fig. 27.14(a). To obtain the resultant, y_1 is drawn horizontally, 2 units long, y_2 is drawn 3 units long at an angle of $\pi/4$ rads or $45°$ and joined to the end of y_1 as shown in Fig. 27.14(b).
From Fig. 27.14(b), and using the cosine rule:

$$y_R^2 = 2^2 + 3^2 - [2(2)(3)\cos 135°]$$
$$= 4 + 9 - [-8.485] = 21.49$$

Hence, $y_R = \sqrt{21.49} = 4.6357$

Using the sine rule: $\dfrac{3}{\sin\phi} = \dfrac{4.6357}{\sin 135°}$

from which, $\sin\phi = \dfrac{3\sin 135°}{4.6357} = 0.45761$

Hence, $\phi = \sin^{-1} 0.45761$
$$= 27.23° \text{ or } 0.475 \text{ rad.}$$

Thus, by calculation, $y_R = 4.635\sin(\omega t + 0.475)$

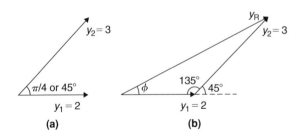

(a)   **(b)**

Figure 27.14

Problem 8. Determine
$20\sin\omega t + 10\sin\left(\omega t + \dfrac{\pi}{3}\right)$ using the cosine and sine rules.

From the phasor diagram of Fig. 27.15, and using the cosine rule:

$$i_R^2 = 20^2 + 10^2 - [2(20)(10)\cos 120°]$$
$$= 700$$

Hence, $i_R = \sqrt{700} = \mathbf{26.46\ A}$

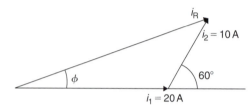

Figure 27.15

Using the sine rule gives : $\dfrac{10}{\sin\phi} = \dfrac{26.46}{\sin 120°}$

from which, $\quad \sin\phi = \dfrac{10\sin 120°}{26.46}$

$$= 0.327296$$

and $\quad \phi = \sin^{-1} 0.327296 = \mathbf{19.10°}$

$$= 19.10 \times \dfrac{\pi}{180} = \mathbf{0.333\ rad}$$

Hence, by cosine and sine rules,

$$i_R = i_1 + i_2 = \mathbf{26.46\sin(\omega t + 0.333)\ A}$$

Now try the following Practice Exercise

Practice Exercise 122   Resultant phasors by the sine and cosine rules (Answers on page 874)

1. Determine, using the cosine and sine rules, a sinusoidal expression for:

$$y = 2\sin A + 4\cos A$$

2. Given $v_1 = 10\sin\omega t$ volts and $v_2 = 14\sin(\omega t + \pi/3)$ volts use the cosine and sine rules to determine sinusoidal expressions for (a) $v_1 + v_2$ (b) $v_1 - v_2$

In Problems 3 to 5, express the given expressions in the form $A\sin(\omega t \pm \alpha)$ by using the cosine and sine rules.

3. $12\sin\omega t + 5\cos\omega t$

4. $7\sin\omega t + 5\sin\left(\omega t + \dfrac{\pi}{4}\right)$

5. $6\sin\omega t + 3\sin\left(\omega t - \dfrac{\pi}{6}\right)$

6. The sinusoidal currents in two parallel branches of an electrical network are $400\sin\omega t$ and $750\sin(\omega t - \pi/3)$, both measured in milliamperes. Determine the total current flowing into the parallel arrangement. Give the answer in sinusoidal form and in amperes.

27.5 Determining resultant phasors by horizontal and vertical components

If a right-angled triangle is constructed as shown in Fig. 27.16, then $\mathbf{0a}$ is called the horizontal component of F and $\mathbf{ab}$ is called the vertical component of F.

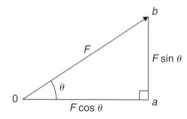

Figure 27.16

From trigonometry (see Chapter 12),

$$\cos\theta = \dfrac{\mathbf{0a}}{\mathbf{0b}} \quad \text{from which,}$$

$$\mathbf{0a} = \mathbf{0b}\cos\theta = F\cos\theta$$

i.e. **the horizontal component of F, $H = F\cos\theta$**

and $\sin\theta = \dfrac{\mathbf{ab}}{\mathbf{0b}}$ from which $\mathbf{ab} = \mathbf{0b}\sin\theta$

$$= F\sin\theta$$

i.e. **the vertical component of F, $V = F\sin\theta$**

Determining resultant phasors by horizontal and vertical components is demonstrated in the following worked Problems.

Problem 9. Two alternating voltages are given by $v_1 = 15 \sin \omega t$ volts and $v_2 = 25 \sin(\omega t - \pi/6)$ volts. Determine a sinusoidal expression for the resultant $v_R = v_1 + v_2$ by finding horizontal and vertical components.

The relative positions of v_1 and v_2 at time $t = 0$ are shown in Fig. 27.17(a) and the phasor diagram is shown in Fig. 27.17(b).

The horizontal component of v_R,
$H = 15 \cos 0° + 25 \cos(-30°) = 0a + ab = \mathbf{36.65\,V}$

The vertical component of v_R,
$V = 15 \sin 0° + 25 \sin(-30°) = bc = \mathbf{-12.50\,V}$

Hence, $v_R = 0c = \sqrt{36.65^2 + (-12.50)^2}$

$$\text{by Pythagoras' theorem}$$

$$= \mathbf{38.72\ volts}$$

$$\tan \phi = \frac{V}{H} = \frac{-12.50}{36.65} = -0.3411$$

from which, $\phi = \tan^{-1}(-0.3411) = -18.83°$

$$\text{or} \quad -0.329 \text{ radians.}$$

Hence, $v_R = v_1 + v_2 = \mathbf{38.72 \sin(\omega t - 0.329)V}$

Problem 10. For the voltages in Problem 9, determine the resultant $v_R = v_1 - v_2$ using horizontal and vertical components.

The horizontal component of v_R,
$H = 15 \cos 0° - 25 \cos(-30°) = \mathbf{-6.65V}$

The vertical component of v_R,
$V = 15 \sin 0° - 25 \sin(-30°) = \mathbf{12.50V}$

Hence, $v_R = \sqrt{(-6.65)^2 + (12.50)^2}$

$$\text{by Pythagoras' theorem}$$

$$= \mathbf{14.16\ volts}$$

$$\tan \phi = \frac{V}{H} = \frac{12.50}{-6.65} = -1.8797$$

from which, $\phi = \tan^{-1}(-1.8797) = 118.01°$

$$\text{or } 2.06 \text{ radians.}$$

Hence,

$$v_R = v_1 - v_2 = \mathbf{14.16 \sin(\omega t + 2.06)V}$$

The phasor diagram is shown in Fig. 27.18.

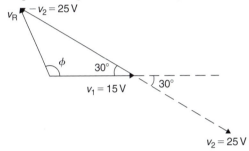

Figure 27.18

Problem 11. Determine $20 \sin \omega t + 10 \sin\left(\omega t + \dfrac{\pi}{3}\right)$ using horizontal and vertical components.

From the phasors shown in Fig. 27.19:

Figure 27.19

Total horizontal component,
$H = 20 \cos 0° + 10 \cos 60° = \mathbf{25.0}$
Total vertical component,
$V = 20 \sin 0° + 10 \sin 60° = \mathbf{8.66}$
By Pythagoras, the resultant, $i_R = \sqrt{[25.0^2 + 8.66^2]}$
$$= \mathbf{26.46\,A}$$

Phase angle, $\phi = \tan^{-1}\left(\dfrac{8.66}{25.0}\right) = \mathbf{19.11°}$

$$\text{or } \mathbf{0.333\ rad}$$

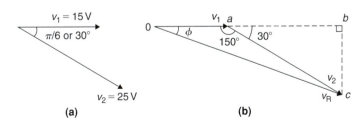

(a) (b)

Figure 27.17

Hence, by using horizontal and vertical components,

$$20\sin\omega t + 10\sin\left(\omega t + \frac{\pi}{3}\right) = 26.46\sin(\omega t + 0.333)$$

Now try the following Practice Exercise

Practice Exercise 123   Resultant phasors by horizontal and vertical components (Answers on page 874)

In Problems 1 to 5, express the combination of periodic functions in the form $A\sin(\omega t \pm \alpha)$ by horizontal and vertical components:

1. $7\sin\omega t + 5\sin\left(\omega t + \frac{\pi}{4}\right)$

2. $6\sin\omega t + 3\sin\left(\omega t - \frac{\pi}{6}\right)$

3. $i = 25\sin\omega t - 15\sin\left(\omega t + \frac{\pi}{3}\right)$

4. $v = 8\sin\omega t - 5\sin\left(\omega t - \frac{\pi}{4}\right)$

5. $x = 9\sin\left(\omega t + \frac{\pi}{3}\right) - 7\sin\left(\omega t - \frac{3\pi}{8}\right)$

6. The voltage drops across two components when connected in series across an a.c. supply are: $v_1 = 200\sin 314.2t$ and $v_2 = 120\sin(314.2t - \pi/5)$ volts respectively. Determine the:
    (a) voltage of the supply (given by $v_1 + v_2$) in the form $A\sin(\omega t \pm \alpha)$
    (b) frequency of the supply.

7. If the supply to a circuit is $v = 20\sin 628.3t$ volts and the voltage drop across one of the components is $v_1 = 15\sin(628.3t - 0.52)$ volts, calculate the:
    (a) voltage drop across the remainder of the circuit, given by $v - v_1$, in the form $A\sin(\omega t \pm \alpha)$
    (b) supply frequency
    (c) periodic time of the supply.

8. The voltages across three components in a series circuit when connected across an a.c.

supply are:

$$v_1 = 25\sin\left(300\pi t + \frac{\pi}{6}\right)\text{ volts,}$$

$$v_2 = 40\sin\left(300\pi t - \frac{\pi}{4}\right)\text{ volts, and}$$

$$v_3 = 50\sin\left(300\pi t + \frac{\pi}{3}\right)\text{ volts.}$$

Calculate the:
(a) supply voltage, in sinusoidal form, in the form $A\sin(\omega t \pm \alpha)$
(b) frequency of the supply
(c) periodic time.

9. In an electrical circuit, two components are connected in series. The voltage across the first component is given by $80\sin(\omega t + \pi/3)$ volts, and the voltage across the second component is given by $150\sin(\omega t - \pi/4)$ volts. Determine the total supply voltage to the two components. Give the answer in sinusoidal form.

27.6 Determining resultant phasors by complex numbers

As stated earlier, the resultant of two periodic functions may be found from their relative positions when the time is zero. For example, if $y_1 = 5\sin\omega t$ and $y_2 = 4\sin(\omega t - \pi/6)$ then each may be represented by phasors as shown in Fig. 27.20, y_1 being 5 units long and drawn horizontally and y_2 being 4 units long, lagging y_1 by $\pi/6$ radians or 30°. To determine the resultant of $y_1 + y_2$, y_1 is drawn horizontally as shown in Fig. 27.21

Figure 27.20

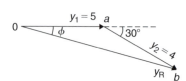

Figure 27.21

and y_2 is joined to the end of y_1 at $\pi/6$ radians, i.e. $30°$ to the horizontal. The resultant is given by y_R

In polar form, $y_R = 5\angle 0 + 4\angle -\dfrac{\pi}{6}$

$$= 5\angle 0° + 4\angle -30°$$

$$= (5 + j0) + (4.33 - j2.0)$$

$$= 9.33 - j2.0 = 9.54\angle -12.10°$$

$$= 9.54\angle -0.21\,\text{rad}$$

Hence, by using complex numbers, the resultant in sinusoidal form is:

$$y_1 + y_2 = 5\sin\omega t + 4\sin(\omega t - \pi/6)$$

$$= \mathbf{9.54\sin(\omega t - 0.21)}$$

Problem 12. Two alternating voltages are given by $v_1 = 15\sin\omega t$ volts and $v_2 = 25\sin(\omega t - \pi/6)$ volts. Determine a sinusoidal expression for the resultant $v_R = v_1 + v_2$ by using complex numbers.

The relative positions of v_1 and v_2 at time $t = 0$ are shown in Fig. 27.22(a) and the phasor diagram is shown in Fig. 27.22(b).

In polar form, $v_R = v_1 + v_2 = 15\angle 0 + 25\angle -\dfrac{\pi}{6}$

$$= 15\angle 0° + 25\angle -30°$$

$$= (15 + j0) + (21.65 - j12.5)$$

$$= 36.65 - j12.5 = 38.72\angle -18.83°$$

$$= 38.72\angle -0.329\,\text{rad}$$

Hence, by using complex numbers, the resultant in sinusoidal form is:

$$v_R = v_1 + v_2 = 15\sin\omega t + 25\sin(\omega t - \pi/6)$$

$$= \mathbf{38.72\sin(\omega t - 0.329)}$$

Problem 13. For the voltages in Problem 12, determine the resultant $v_R = v_1 - v_2$ using complex numbers.

In polar form, $y_R = v_1 - v_2 = 15\angle 0 - 25\angle -\dfrac{\pi}{6}$

$$= 15\angle 0° - 25\angle -30°$$

$$= (15 + j0) - (21.65 - j12.5)$$

$$= -6.65 + j12.5 = 14.16\angle 118.01°$$

$$= 14.16\angle 2.06\,\text{rad}$$

Hence, by using complex numbers, the resultant in sinusoidal form is:

$$y_1 - y_2 = 15\sin\omega t - 25\sin(\omega t - \pi/6)$$

$$= \mathbf{14.16\sin(\omega t - 2.06)}$$

Problem 14. Determine $20\sin\omega t + 10\sin\left(\omega t + \dfrac{\pi}{3}\right)$ using complex numbers.

From the phasors shown in Fig. 27.23, the resultant may be expressed in polar form as:

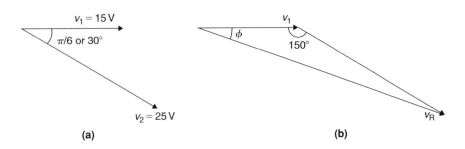

Figure 27.23

$$i_R = 20\angle 0° + 10\angle 60°$$

i.e. $i_R = (20 + j0) + (5 + j8.66)$

$$= (25 + j8.66) = \mathbf{26.46\angle 19.11°\,A}\quad\text{or}$$

$$\mathbf{26.46\angle 0.333\,rad\,A}$$

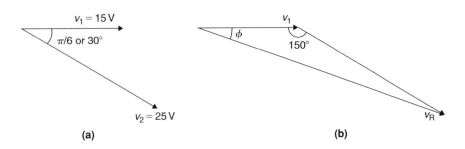

$v_1 = 15\,\text{V}$

$\pi/6$ or $30°$

$v_2 = 25\,\text{V}$

(a)

v_1

ϕ

$150°$

v_R

(b)

Figure 27.22

Hence, by using complex numbers, the resultant in sinusoidal form is:
$$i_R = i_1 + i_2 = 26.46 \sin(\omega t + 0.333) \text{A}$$

Problem 15. If the supply to a circuit is $v = 30 \sin 100\pi t$ volts and the voltage drop across one of the components is $v_1 = 20 \sin(100\pi t - 0.59)$ volts, calculate the:

(a) voltage drop across the remainder of the circuit, given by $v - v_1$, in the form $A \sin(\omega t \pm \alpha)$

(b) supply frequency

(c) periodic time of the supply

(d) rms value of the supply voltage.

(a) Supply voltage, $v = v_1 + v_2$ where v_2 is the voltage across the remainder of the circuit.

Hence, $v_2 = v - v_1 = 30 \sin 100\pi t$
$$- 20 \sin(100\pi t - 0.59)$$
$$= 30\angle 0 - 20\angle - 0.59 \,\text{rad}$$
$$= (30 + j0) - (16.619 - j11.127)$$
$$= 13.381 + j11.127$$
$$= 17.40\angle 0.694 \text{ rad}$$

Hence, by using complex numbers, the resultant in sinusoidal form is:

$$v - v_1 = 30 \sin 100\pi t - 20 \sin(100\pi t - 0.59)$$
$$= 17.40 \sin(\omega t + 0.694) \text{ volts}$$

(b) **Supply frequency,** $f = \dfrac{\omega}{2\pi} = \dfrac{100\pi}{2\pi} = 50 \text{ Hz}$

(c) **Periodic time,** $T = \dfrac{1}{f} = \dfrac{1}{50} = 0.02 \text{ s}$ or **20 ms**

(d) **Rms value of supply voltage,** $= 0.707 \times 30$
$$= 21.21 \text{ volts}$$

Now try the following Practice Exercise

Practice Exercise 124 Resultant phasors by complex numbers (Answers on page 875)

In Problems 1 to 4, express the combination of periodic functions in the form $A \sin(\omega t \pm \alpha)$ by using complex numbers:

1. $8 \sin \omega t + 5 \sin\left(\omega t + \dfrac{\pi}{4}\right)$

2. $6 \sin \omega t + 9 \sin\left(\omega t - \dfrac{\pi}{6}\right)$

3. $v = 12 \sin \omega t - 5 \sin\left(\omega t - \dfrac{\pi}{4}\right)$

4. $x = 10 \sin\left(\omega t + \dfrac{\pi}{3}\right) - 8 \sin\left(\omega t - \dfrac{3\pi}{8}\right)$

5. The voltage drops across two components when connected in series across an a.c. supply are: $v_1 = 240 \sin 314.2t$ and $v_2 = 150 \sin(314.2t - \pi/5)$ volts respectively. Determine the:
   (a) voltage of the supply (given by $v_1 + v_2$) in the form $A \sin(\omega t \pm \alpha)$
   (b) frequency of the supply.

6. If the supply to a circuit is $v = 25 \sin 200\pi t$ volts and the voltage drop across one of the components is $v_1 = 18 \sin(200\pi t - 0.43)$ volts, calculate the:
   (a) voltage drop across the remainder of the circuit, given by $v - v_1$, in the form $A \sin(\omega t \pm \alpha)$
   (b) supply frequency
   (c) periodic time of the supply.

7. The voltages across three components in a series circuit when connected across an a.c. supply are:

   $$v_1 = 20 \sin\left(300\pi t - \dfrac{\pi}{6}\right) \text{ volts,}$$

   $$v_2 = 30 \sin\left(300\pi t + \dfrac{\pi}{4}\right) \text{ volts, and}$$

   $$v_3 = 60 \sin\left(300\pi t - \dfrac{\pi}{3}\right) \text{ volts.}$$

   Calculate the:
   (a) supply voltage, in sinusoidal form, in the form $A \sin(\omega t \pm \alpha)$
   (b) frequency of the supply

(c) periodic time

(d) rms value of the supply voltage.

8. Measurements made at a substation at peak demand of the current in the red, yellow and blue phases of a transmission system are: $I_{red} = 1248\angle -15°A$, $I_{yellow} = 1120\angle -135°A$ and $I_{blue} = 1310\angle 95°A$. Determine the current in the neutral cable if the sum of the currents flows through it.

Section F

Chapter 28

Scalar and vector products

Why it is important to understand: Scalar and vector products

Common applications of the scalar product in engineering and physics is to test whether two vectors are perpendicular or to find the angle between two vectors when they are expressed in Cartesian form or to find the component of one vector in the direction of another. In mechanical engineering the scalar and vector product is used with forces, displacement, moments, velocities and the determination of work done. In electrical engineering, scalar and vector product calculations are important in electromagnetic calculations; most electromagnetic equations use vector calculus which is based on the scalar and vector product. Such applications include electric motors and generators and power transformers, and so on. Knowledge of scalar and vector products thus have important engineering applications.

At the end of this chapter, you should be able to:

- define a unit triad
- determine the scalar (or dot) product of two vectors
- calculate the angle between two vectors
- determine the direction cosines of a vector
- apply scalar products to practical situations
- determine the vector (or cross) product of two vectors
- apply vector products to practical situations
- determine the vector equation of a line

28.1 The unit triad

When a vector x of magnitude x units and direction $\theta°$ is divided by the magnitude of the vector, the result is a vector of unit length at angle $\theta°$. The unit vector for a velocity of 10 m/s at 50° is $\dfrac{10\,\text{m/s at } 50°}{10\,\text{m/s}}$, i.e. 1 at 50°.

In general, the unit vector for oa is $\dfrac{oa}{|oa|}$, the oa being a vector and having both magnitude and direction and $|oa|$ being the magnitude of the vector only.

One method of completely specifying the direction of a vector in space relative to some reference point is to use three unit vectors, mutually at right angles to each other, as shown in Fig. 28.1. Such a system is called a **unit triad**.

Figure 28.1

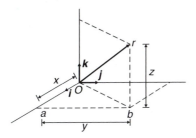

Figure 28.2

In Fig. 28.2, one way to get from o to r is to move x units along i to point a, then y units in direction j to get to b and finally z units in direction k to get to r. The vector or is specified as

$$or = xi + yj + zk$$

Problem 1. With reference to three axes drawn mutually at right angles, depict the vectors (a) $op = 4i + 3j - 2k$ and (b) $or = 5i - 2j + 2k$

The required vectors are depicted in Fig. 28.3, op being shown in Fig. 28.3(a) and or in Fig. 28.3(b).

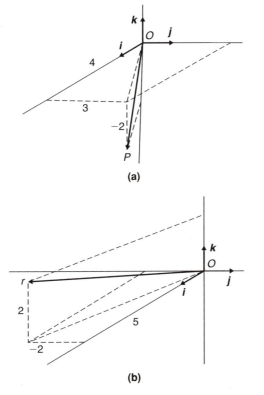

Figure 28.3

28.2 The scalar product of two vectors

When vector oa is multiplied by a scalar quantity, say k, the magnitude of the resultant vector will be k times the magnitude of oa and its direction will remain the same. Thus $2 \times (5\,\text{N at } 20°)$ results in a vector of magnitude $10\,\text{N at } 20°$.

One of the products of two vector quantities is called the **scalar** or **dot product** of two vectors and is defined as the product of their magnitudes multiplied by the cosine of the angle between them. The scalar product of oa and ob is shown as $oa \cdot ob$. For vectors $oa = oa$ at θ_1, and $ob = ob$ at θ_2 where $\theta_2 > \theta_1$, **the scalar product is:**

$$oa \cdot ob = oa\, ob \cos(\theta_2 - \theta_1)$$

For vectors v_1 and v_2 shown in Fig. 28.4, the scalar product is:

$$v_1 \cdot v_2 = v_1 v_2 \cos\theta$$

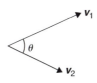

Figure 28.4

The commutative law of algebra, $a \times b = b \times a$ applies to scalar products. This is demonstrated in Fig. 28.5. Let oa represent vector v_1 and ob represent vector v_2. Then:

$$oa \cdot ob = v_1 v_2 \cos\theta \text{ (by definition of a scalar product)}$$

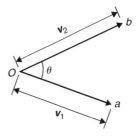

Figure 28.5

Similarly, $ob \cdot oa = v_2 v_1 \cos\theta = v_1 v_2 \cos\theta$ by the commutative law of algebra. Thus $oa \cdot ob = ob \cdot oa$

Section F

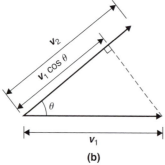

Figure 28.6

The projection of **ob** on **oa** is shown in Fig. 28.6(a) and by the geometry of triangle *obc*, it can be seen that the projection is $v_2 \cos\theta$. Since, by definition

$$\boldsymbol{oa} \cdot \boldsymbol{ob} = v_1(v_2 \cos\theta),$$

it follows that

$$\boldsymbol{oa} \cdot \boldsymbol{ob} = v_1 \text{ (the projection of } v_2 \text{ on } v_1)$$

Similarly the projection of **oa** on **ob** is shown in Fig. 28.6(b) and is $v_1 \cos\theta$. Since by definition

$$\boldsymbol{ob} \cdot \boldsymbol{oa} = v_2(v_1 \cos\theta),$$

it follows that

$$\boldsymbol{ob} \cdot \boldsymbol{oa} = v_2(\text{the projection of } v_1 \text{ on } v_2)$$

This shows that the scalar product of two vectors is the product of the magnitude of one vector and the magnitude of the projection of the other vector on it. The **angle between two vectors** can be expressed in terms of the vector constants as follows:
Because $\boldsymbol{a} \cdot \boldsymbol{b} = a\,b \cos\theta$,

then

$$\cos\theta = \frac{\boldsymbol{a} \cdot \boldsymbol{b}}{a\,b} \qquad (1)$$

Let $\quad \boldsymbol{a} = a_1\boldsymbol{i} + a_2\boldsymbol{j} + a_3\boldsymbol{k}$

and $\quad \boldsymbol{b} = b_1\boldsymbol{i} + b_2\boldsymbol{j} + b_3\boldsymbol{k}$

$$\boldsymbol{a} \cdot \boldsymbol{b} = (a_1\boldsymbol{i} + a_2\boldsymbol{j} + a_3\boldsymbol{k}) \cdot (b_1\boldsymbol{i} + b_2\boldsymbol{j} + b_3\boldsymbol{k})$$

Multiplying out the brackets gives:

$$\boldsymbol{a} \cdot \boldsymbol{b} = a_1b_1\boldsymbol{i} \cdot \boldsymbol{i} + a_1b_2\boldsymbol{i} \cdot \boldsymbol{j} + a_1b_3\boldsymbol{i} \cdot \boldsymbol{k}$$
$$+ a_2b_1\boldsymbol{j} \cdot \boldsymbol{i} + a_2b_2\boldsymbol{j} \cdot \boldsymbol{j} + a_2b_3\boldsymbol{j} \cdot \boldsymbol{k}$$
$$+ a_3b_1\boldsymbol{k} \cdot \boldsymbol{i} + a_3b_2\boldsymbol{k} \cdot \boldsymbol{j} + a_3b_3\boldsymbol{k} \cdot \boldsymbol{k}$$

However, the unit vectors **i**, **j** and **k** all have a magnitude of 1 and $\boldsymbol{i} \cdot \boldsymbol{i} = (1)(1)\cos 0° = 1$, $\boldsymbol{i} \cdot \boldsymbol{j} = (1)(1)\cos 90° = 0$, $\boldsymbol{i} \cdot \boldsymbol{k} = (1)(1)\cos 90° = 0$ and similarly $\boldsymbol{j} \cdot \boldsymbol{j} = 1$, $\boldsymbol{j} \cdot \boldsymbol{k} = 0$ and $\boldsymbol{k} \cdot \boldsymbol{k} = 1$. Thus, only terms containing $\boldsymbol{i} \cdot \boldsymbol{i}$, $\boldsymbol{j} \cdot \boldsymbol{j}$ or $\boldsymbol{k} \cdot \boldsymbol{k}$ in the expansion above will not be zero.
Thus, the scalar product

$$\boldsymbol{a} \cdot \boldsymbol{b} = a_1b_1 + a_2b_2 + a_3b_3 \qquad (2)$$

Both *a* and *b* in equation (1) can be expressed in terms of a_1, b_1, a_2, b_2, a_3 and b_3

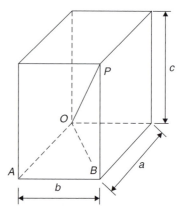

Figure 28.7

From the geometry of Fig. 28.7, the length of diagonal *OP* in terms of side lengths *a*, *b* and *c* can be obtained from Pythagoras' theorem as follows:

$$OP^2 = OB^2 + BP^2 \text{ and}$$
$$OB^2 = OA^2 + AB^2$$

Thus, $\quad OP^2 = OA^2 + AB^2 + BP^2$
$$= a^2 + b^2 + c^2,$$
$$\text{in terms of side lengths}$$

Thus, the **length** or **modulus** or **magnitude** or **norm of vector *OP*** is given by:

$$OP = \sqrt{(a^2 + b^2 + c^2)} \qquad (3)$$

Relating this result to the two vectors $a_1 i + a_2 j + a_3 k$ and $b_1 i + b_2 j + b_3 k$, gives:

$$a = \sqrt{(a_1^2 + a_2^2 + a_3^2)}$$

and $\quad b = \sqrt{(b_1^2 + b_2^2 + b_3^2)}$

That is, from equation (1),

$$\cos \theta = \frac{a_1 b_1 + a_2 b_2 + a_3 b_3}{\sqrt{(a_1^2 + a_2^2 + a_3^2)} \sqrt{(b_1^2 + b_2^2 + b_3^2)}} \qquad (4)$$

Problem 2. Find vector *a* joining points *P* and *Q* where point *P* has co-ordinates $(4, -1, 3)$ and point *Q* has co-ordinates $(2, 5, 0)$. Also, find $|a|$, the magnitude or norm of *a*

Let O be the origin, i.e. its co-ordinates are $(0, 0, 0)$. The position vector of *P* and *Q* are given by:

$$OP = 4i - j + 3k \text{ and } OQ = 2i + 5j$$

By the addition law of vectors $OP + PQ = OQ$

Hence $\quad a = PQ = OQ - OP$

i.e. $\quad a = PQ = (2i + 5j) - (4i - j + 3k)$

$$= -2i + 6j - 3k$$

From equation (3), the magnitude or norm of *a*,

$$|a| = \sqrt{(a^2 + b^2 + c^2)}$$

$$= \sqrt{[(-2)^2 + 6^2 + (-3)^2]} = \sqrt{49} = 7$$

Problem 3. If $p = 2i + j - k$ and $q = i - 3j + 2k$ determine:

(a) $p \cdot q$ (b) $p + q$

(c) $|p + q|$ (d) $|p| + |q|$

(a) From equation (2),

if $\qquad p = a_1 i + a_2 j + a_3 k$

and $\qquad q = b_1 i + b_2 j + b_3 k$

then $\quad p \cdot q = a_1 b_1 + a_2 b_2 + a_3 b_3$

When $\quad p = 2i + j - k$,

$$a_1 = 2, \ a_2 = 1 \text{ and } a_3 = -1$$

and when $\quad q = i - 3j + 2k$,

$$b_1 = 1, \ b_2 = -3 \text{ and } b_3 = 2$$

Hence $\quad p \cdot q = (2)(1) + (1)(-3) + (-1)(2)$

i.e. $\qquad p \cdot q = -3$

(b) $\quad p + q = (2i + j - k) + (i - 3j + 2k)$

$$= 3i - 2j + k$$

(c) $\quad |p + q| = |3i - 2j + k|$

From equation (3),

$$|p + q| = \sqrt{[3^2 + (-2)^2 + 1^2]} = \sqrt{14}$$

(d) From equation (3),

$$|p| = |2i + j - k|$$

$$= \sqrt{[2^2 + 1^2 + (-1)^2]} = \sqrt{6}$$

Similarly,

$$|q| = |i - 3j + 2k|$$

$$= \sqrt{[1^2 + (-3)^2 + 2^2]} = \sqrt{14}$$

Hence $|p| + |q| = \sqrt{6} + \sqrt{14} = \textbf{6.191}$, correct to 3 decimal places.

Problem 4. Determine the angle between vectors *oa* and *ob* when

$$oa = i + 2j - 3k$$

and $\quad ob = 2i - j + 4k$

An equation for $\cos \theta$ is given in equation (4)

$$\cos \theta = \frac{a_1 b_1 + a_2 b_2 + a_3 b_3}{\sqrt{(a_1^2 + a_2^2 + a_3^2)} \sqrt{(b_1^2 + b_2^2 + b_3^2)}}$$

Section F

Since $oa = i + 2j - 3k$,

$$a_1 = 1, a_2 = 2 \text{ and } a_3 = -3$$

Since $ob = 2i - j + 4k$,

$$b_1 = 2, b_2 = -1 \text{ and } b_3 = 4$$

Thus,

$$\cos\theta = \frac{(1 \times 2) + (2 \times -1) + (-3 \times 4)}{\sqrt{(1^2 + 2^2 + (-3)^2)}\sqrt{(2^2 + (-1)^2 + 4^2)}}$$

$$= \frac{-12}{\sqrt{14}\sqrt{21}} = -0.6999$$

i.e. $\theta = 134.4°$ or $225.6°$

By sketching the position of the two vectors as shown in Problem 1, it will be seen that $225.6°$ is not an acceptable answer.

Thus the angle between the vectors oa and ob, $\theta = 134.4°$

Direction cosines

From Fig. 28.2, $or = xi + yj + zk$ and from equation (3), $|or| = \sqrt{x^2 + y^2 + z^2}$

If or makes angles of α, β and γ with the co-ordinate axes i, j and k respectively, then:

The direction cosines are:

$$\cos\alpha = \frac{x}{\sqrt{x^2 + y^2 + z^2}}$$

$$\cos\beta = \frac{y}{\sqrt{x^2 + y^2 + z^2}}$$

and $\cos\gamma = \dfrac{y}{\sqrt{x^2 + y^2 + z^2}}$

such that $\cos^2\alpha + \cos^2\beta + \cos^2\gamma = 1$
The values of cos α, cos β and cos γ are called the **direction cosines** of or

Problem 5. Find the direction cosines of 3

$$i + 2j + k$$

.

$$\sqrt{x^2 + y^2 + z^2} = \sqrt{3^2 + 2^2 + 1^2} = \sqrt{14}$$

The direction cosines are:

$$\cos\alpha = \frac{x}{\sqrt{x^2 + y^2 + z^2}} = \frac{3}{\sqrt{14}} = 0.802$$

$$\cos\beta = \frac{y}{\sqrt{x^2 + y^2 + z^2}} = \frac{2}{\sqrt{14}} = 0.535$$

and $\cos\gamma = \dfrac{y}{\sqrt{x^2 + y^2 + z^2}} = \dfrac{1}{\sqrt{14}} = 0.267$

(and hence $\alpha = \cos^{-1} 0.802 = 36.7°$,
$\beta = \cos^{-1} 0.535 = 57.7°$ and $\gamma = \cos^{-1} 0.267 = 74.5°$)
Note that $\cos^2\alpha + \cos^2\beta + \cos^2\gamma$
$$= 0.802^2 + 0.535^2 + 0.267^2 = 1$$

Practical application of scalar product

Problem 6. A constant force of $F = 10i + 2j - k$ newtons displaces an object from $A = i + j + k$ to $B = 2i - j + 3k$ (in metres). Find the work done in newton metres.

One of the applications of scalar products is to the work done by a constant force when moving a body. The work done is the product of the applied force and the distance moved in the direction of the force.

i.e. **work done** $= F \cdot d$

The principles developed in Problem 13, page 317, apply equally to this problem when determining the displacement. From the sketch shown in Fig. 28.8,

$$AB = AO + OB = OB - OA$$

that is $AB = (2i - j + 3k) - (i + j + k)$

$$= i - 2j + 2k$$

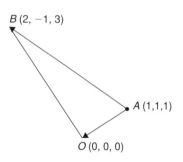

Figure 28.8

The work done is $\mathbf{F} \cdot \mathbf{d}$, that is $\mathbf{F} \cdot \mathbf{AB}$ in this case

i.e. **work done** $= (10\mathbf{i} + 2\mathbf{j} - \mathbf{k}) \cdot (\mathbf{i} - 2\mathbf{j} + 2\mathbf{k})$

But from equation (2),

$$\mathbf{a} \cdot \mathbf{b} = a_1 b_1 + a_2 b_2 + a_3 b_3$$

Hence **work done** $=$
$$(10 \times 1) + (2 \times (-2)) + ((-1) \times 2) = \mathbf{4\,N\,m}$$
(Theoretically, it is quite possible to get a negative answer to a 'work done' problem. This indicates that the force must be in the opposite sense to that given, in order to give the displacement stated.)

Now try the following Practice Exercise

Practice Exercise 125   Scalar products (Answers on page 875)

1. Find the scalar product $\mathbf{a} \cdot \mathbf{b}$ when
   (a) $\mathbf{a} = \mathbf{i} + 2\mathbf{j} - \mathbf{k}$ and $\mathbf{b} = 2\mathbf{i} + 3\mathbf{j} + \mathbf{k}$
   (b) $\mathbf{a} = \mathbf{i} - 3\mathbf{j} + \mathbf{k}$ and $\mathbf{b} = 2\mathbf{i} + \mathbf{j} + \mathbf{k}$

Given $\mathbf{p} = 2\mathbf{i} - 3\mathbf{j}$, $\mathbf{q} = 4\mathbf{j} - \mathbf{k}$ and $\mathbf{r} = \mathbf{i} + 2\mathbf{j} - 3\mathbf{k}$, determine the quantities stated in problems 2 to 8.

2. (a) $\mathbf{p} \cdot \mathbf{q}$ (b) $\mathbf{p} \cdot \mathbf{r}$

3. (a) $\mathbf{q} \cdot \mathbf{r}$ (b) $\mathbf{r} \cdot \mathbf{q}$

4. (a) $|\mathbf{p}|$ (b) $|\mathbf{r}|$

5. (a) $\mathbf{p} \cdot (\mathbf{q} + \mathbf{r})$ (b) $2\mathbf{r} \cdot (\mathbf{q} - 2\mathbf{p})$

6. (a) $|\mathbf{p} + \mathbf{r}|$ (b) $|\mathbf{p}| + |\mathbf{r}|$

7. Find the angle between (a) $\mathbf{p}$ and $\mathbf{q}$ (b) $\mathbf{q}$ and $\mathbf{r}$

8. Determine the direction cosines of (a) $\mathbf{p}$ (b) $\mathbf{q}$ (c) $\mathbf{r}$

9. Determine the angle between the forces:
   $$\mathbf{F_1} = 3\mathbf{i} + 4\mathbf{j} + 5\mathbf{k} \text{ and}$$
   $$\mathbf{F_2} = \mathbf{i} + \mathbf{j} + \mathbf{k}$$

10. Find the angle between the velocity vectors $\mathbf{v_1} = 5\mathbf{i} + 2\mathbf{j} + 7\mathbf{k}$ and $\mathbf{v_2} = 4\mathbf{i} + \mathbf{j} - \mathbf{k}$

11. Calculate the work done by a force $\mathbf{F} = (-5\mathbf{i} + \mathbf{j} + 7\mathbf{k})\,\text{N}$ when its point of application moves from point $(-2\mathbf{i} - 6\mathbf{j} + \mathbf{k})\,\text{m}$ to the point $(\mathbf{i} - \mathbf{j} + 10\mathbf{k})\,\text{m}$.

28.3   Vector products

A second product of two vectors is called the **vector** or **cross product** and is defined in terms of its modulus and the magnitudes of the two vectors and the sine of the angle between them. The vector product of vectors $\mathbf{oa}$ and $\mathbf{ob}$ is written as $\mathbf{oa \times ob}$ and is defined by:

$$|\mathbf{oa \times ob}| = oa\,ob\,\sin\theta$$

where θ is the angle between the two vectors. The direction of $\mathbf{oa \times ob}$ is perpendicular to both $\mathbf{oa}$ and $\mathbf{ob}$, as shown in Fig. 28.9.

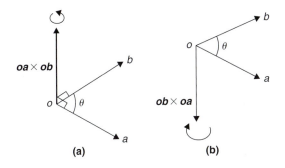

Figure 28.9

The direction is obtained by considering that a right-handed screw is screwed along $\mathbf{oa \times ob}$ with its head at the origin and if the direction of $\mathbf{oa \times ob}$ is correct, the head should rotate from $\mathbf{oa}$ to $\mathbf{ob}$, as shown in Fig. 28.9(a). It follows that the direction of $\mathbf{ob \times oa}$ is as shown in Fig. 28.9(b). Thus $\mathbf{oa \times ob}$ is not equal to $\mathbf{ob \times oa}$. The magnitudes of $oa\,ob\,\sin\theta$ are the same but their directions are $180°$ displaced, i.e.

$$\mathbf{oa \times ob} = -\mathbf{ob \times oa}$$

The vector product of two vectors may be expressed in terms of the unit vectors. Let two vectors, $\mathbf{a}$ and $\mathbf{b}$, be such that:

$$\mathbf{a} = a_1\mathbf{i} + a_2\mathbf{j} + a_3\mathbf{k} \text{ and}$$
$$\mathbf{b} = b_1\mathbf{i} + b_2\mathbf{j} + b_3\mathbf{k}$$

Then,

$$\mathbf{a \times b} = (a_1\mathbf{i} + a_2\mathbf{j} + a_3\mathbf{k}) \times (b_1\mathbf{i} + b_2\mathbf{j} + b_3\mathbf{k})$$

$$= a_1 b_1 \mathbf{i} \times \mathbf{i} + a_1 b_2 \mathbf{i} \times \mathbf{j}$$

$$+ a_1 b_3 \mathbf{i} \times \mathbf{k} + a_2 b_1 \mathbf{j} \times \mathbf{i} + a_2 b_2 \mathbf{j} \times \mathbf{j}$$

$$+ a_2 b_3 \mathbf{j} \times \mathbf{k} + a_3 b_1 \mathbf{k} \times \mathbf{i} + a_3 b_2 \mathbf{k} \times \mathbf{j}$$

$$+ a_3 b_3 \mathbf{k} \times \mathbf{k}$$

But by the definition of a vector product,

$$i \times j = k, \ j \times k = i \text{ and } k \times i = j$$

Also $i \times i = j \times j = k \times k = (1)(1)\sin 0° = 0$

Remembering that $a \times b = -b \times a$ gives:

$$a \times b = a_1 b_2 k - a_1 b_3 j - a_2 b_1 k + a_2 b_3 i$$
$$+ a_3 b_1 j - a_3 b_2 i$$

Grouping the i, j and k terms together, gives:

$$a \times b = (a_2 b_3 - a_3 b_2)i + (a_3 b_1 - a_1 b_3)j$$
$$+ (a_1 b_2 - a_2 b_1)k$$

The vector product can be written in determinant form as:

$$a \times b = \begin{vmatrix} i & j & k \\ a_1 & a_2 & a_3 \\ b_1 & b_2 & b_3 \end{vmatrix} \quad (5)$$

The 3×3 determinant $\begin{vmatrix} i & j & k \\ a_1 & a_2 & a_3 \\ b_1 & b_2 & b_3 \end{vmatrix}$ is evaluated as:

$$i \begin{vmatrix} a_2 & a_3 \\ b_2 & b_3 \end{vmatrix} - j \begin{vmatrix} a_1 & a_3 \\ b_1 & b_3 \end{vmatrix} + k \begin{vmatrix} a_1 & a_2 \\ b_1 & b_2 \end{vmatrix}$$

where

$$\begin{vmatrix} a_2 & a_3 \\ b_2 & b_3 \end{vmatrix} = a_2 b_3 - a_3 b_2,$$

$$\begin{vmatrix} a_1 & a_3 \\ b_1 & b_3 \end{vmatrix} = a_1 b_3 - a_3 b_1 \text{ and}$$

$$\begin{vmatrix} a_1 & a_2 \\ b_1 & b_2 \end{vmatrix} = a_1 b_2 - a_2 b_1$$

The magnitude of the vector product of two vectors can be found by expressing it in scalar product form and then using the relationship

$$a \cdot b = a_1 b_1 + a_2 b_2 + a_3 b_3$$

Squaring both sides of a vector product equation gives:

$$(|a \times b|)^2 = a^2 b^2 \sin^2 \theta = a^2 b^2 (1 - \cos^2 \theta)$$
$$= a^2 b^2 - a^2 b^2 \cos^2 \theta \quad (6)$$

It is stated in Section 28.2 that $a \cdot b = ab\cos\theta$, hence

$$a \cdot a = a^2 \cos\theta$$

But $\theta = 0°$, thus $a \cdot a = a^2$

Also, $\cos\theta = \dfrac{a \cdot b}{ab}$

Multiplying both sides of this equation by $a^2 b^2$ and squaring gives:

$$a^2 b^2 \cos^2 \theta = \frac{a^2 b^2 (a \cdot b)^2}{a^2 b^2} = (a \cdot b)^2$$

Substituting in equation (6) above for $a^2 = a \cdot a$, $b^2 = b \cdot b$ and $a^2 b^2 \cos^2 \theta = (a \cdot b)^2$ gives:

$$(|a \times b|)^2 = (a \cdot a)(b \cdot b) - (a \cdot b)^2$$

That is,

$$|a \times b| = \sqrt{[(a \cdot a)(b \cdot b) - (a \cdot b)^2]} \quad (7)$$

Problem 7. For the vectors $a = i + 4j - 2k$ and $b = 2i - j + 3k$ find (a) $a \times b$ and (b) $|a \times b|$

(a) From equation (5),

$$a \times b = \begin{vmatrix} i & j & k \\ a_1 & a_2 & a_3 \\ b_1 & b_2 & b_3 \end{vmatrix}$$

$$= i \begin{vmatrix} a_2 & a_3 \\ b_2 & b_3 \end{vmatrix} - j \begin{vmatrix} a_1 & a_3 \\ b_1 & b_3 \end{vmatrix} + k \begin{vmatrix} a_1 & a_2 \\ b_1 & b_2 \end{vmatrix}$$

Hence

$$a \times b = \begin{vmatrix} i & j & k \\ 1 & 4 & -2 \\ 2 & -1 & 3 \end{vmatrix}$$

$$= i \begin{vmatrix} 4 & -2 \\ -1 & 3 \end{vmatrix} - j \begin{vmatrix} 1 & -2 \\ 2 & 3 \end{vmatrix} + k \begin{vmatrix} 1 & 4 \\ 2 & -1 \end{vmatrix}$$

$$= i(12 - 2) - j(3 + 4) + k(-1 - 8)$$

$$= 10i - 7j - 9k$$

(b) From equation (7)

$$|\mathbf{a} \times \mathbf{b}| = \sqrt{[(\mathbf{a} \cdot \mathbf{a})(\mathbf{b} \cdot \mathbf{b}) - (\mathbf{a} \cdot \mathbf{b})^2]}$$

Now $\quad \mathbf{a} \cdot \mathbf{a} = (1)(1) + (4 \times 4) + (-2)(-2)$

$$= 21$$

$$\mathbf{b} \cdot \mathbf{b} = (2)(2) + (-1)(-1) + (3)(3)$$

$$= 14$$

and $\quad \mathbf{a} \cdot \mathbf{b} = (1)(2) + (4)(-1) + (-2)(3)$

$$= -8$$

Thus $\quad |\mathbf{a} \times \mathbf{b}| = \sqrt{(21 \times 14 - 64)}$

$$= \sqrt{230} = \mathbf{15.17}$$

Problem 8. If $p = 4i + j - 2k$, $q = 3i - 2j + k$ and $r = i - 2k$ find (a) $(p - 2q) \times r$ (b) $p \times (2r \times 3q)$

(a) $(p - 2q) \times r = [4i + j - 2k$

$$- 2(3i - 2j + k)] \times (i - 2k)$$

$$= (-2i + 5j - 4k) \times (i - 2k)$$

$$= \begin{vmatrix} i & j & k \\ -2 & 5 & -4 \\ 1 & 0 & -2 \end{vmatrix}$$

from equation (5)

$$= i \begin{vmatrix} 5 & -4 \\ 0 & -2 \end{vmatrix} - j \begin{vmatrix} -2 & -4 \\ 1 & -2 \end{vmatrix}$$

$$+ k \begin{vmatrix} -2 & 5 \\ 1 & 0 \end{vmatrix}$$

$$= i(-10 - 0) - j(4 + 4)$$

$$+ k(0 - 5), \text{ i.e.}$$

$$(p - 2q) \times r = \mathbf{-10i - 8j - 5k}$$

(b) $(2r \times 3q) = (2i - 4k) \times (9i - 6j + 3k)$

$$= \begin{vmatrix} i & j & k \\ 2 & 0 & -4 \\ 9 & -6 & 3 \end{vmatrix}$$

$$= i(0 - 24) - j(6 + 36)$$

$$+ k(-12 - 0)$$

$$= \mathbf{-24i - 42j - 12k}$$

Hence

$$\mathbf{p} \times (2\mathbf{r} \times 3\mathbf{q}) = (4i + j - 2k)$$

$$\times (-24i - 42j - 12k)$$

$$= \begin{vmatrix} i & j & k \\ 4 & 1 & -2 \\ -24 & -42 & -12 \end{vmatrix}$$

$$= i(-12 - 84) - j(-48 - 48)$$

$$+ k(-168 + 24)$$

$$= \mathbf{-96i + 96j - 144k}$$

or $\mathbf{-48(2i - 2j + 3k)}$

Practical applications of vector products

Problem 9. Find the moment and the magnitude of the moment of a force of $(i + 2j - 3k)$ newtons about point B having co-ordinates (0, 1, 1), when the force acts on a line through A whose co-ordinates are (1, 3, 4)

The moment M about point B of a force vector F which has a position vector of r from A is given by:

$$M = r \times F$$

r is the vector from B to A, i.e. $r = BA$
But $BA = BO + OA = OA - OB$ (see Problem 13, page 317), that is:

$$r = (i + 3j + 4k) - (j + k)$$

$$= i + 2j + 3k$$

Moment,

$$M = r \times F = (i + 2j + 3k) \times (i + 2j - 3k)$$

$$= \begin{vmatrix} i & j & k \\ 1 & 2 & 3 \\ 1 & 2 & -3 \end{vmatrix}$$

$$= i(-6 - 6) - j(-3 - 3)$$

$$+ k(2 - 2)$$

$$= \mathbf{-12i + 6j \, N\,m}$$

The magnitude of M,

$$|M| = |r \times F|$$

$$= \sqrt{[(r \cdot r)(F \cdot F) - (r \cdot F)^2]}$$

$$r \cdot r = (1)(1) + (2)(2) + (3)(3) = 14$$

$$F \cdot F = (1)(1) + (2)(2) + (-3)(-3) = 14$$

$$r \cdot F = (1)(1) + (2)(2) + (3)(-3) = -4$$

$$|M| = \sqrt{[14 \times 14 - (-4)^2]}$$

$$= \sqrt{180}\, \text{Nm} = \mathbf{13.42\, N\, m}$$

Problem 10. The axis of a circular cylinder coincides with the z-axis and it rotates with an angular velocity of $(2i - 5j + 7k)$ rad/s. Determine the tangential velocity at a point P on the cylinder, whose co-ordinates are $(j + 3k)$ metres, and also determine the magnitude of the tangential velocity.

The velocity v of point P on a body rotating with angular velocity ω about a fixed axis is given by:

$$v = \omega \times r$$

where r is the point on vector P

Thus $\qquad v = (2i - 5j + 7k) \times (j + 3k)$

$$= \begin{vmatrix} i & j & k \\ 2 & -5 & 7 \\ 0 & 1 & 3 \end{vmatrix}$$

$$= i(-15 - 7) - j(6 - 0) + k(2 - 0)$$

$$= (-22i - 6j + 2k)\, \text{m/s}$$

The magnitude of v,

$$|v| = \sqrt{[(\omega \cdot \omega)(r \cdot r) - (r \cdot \omega)^2]}$$

$$\omega \cdot \omega = (2)(2) + (-5)(-5) + (7)(7) = 78$$

$$r \cdot r = (0)(0) + (1)(1) + (3)(3) = 10$$

$$\omega \cdot r = (2)(0) + (-5)(1) + (7)(3) = 16$$

Hence,

$$|v| = \sqrt{(78 \times 10 - 16^2)}$$

$$= \sqrt{524}\, \text{m/s} = \mathbf{22.89\, m/s}$$

Now try the following Practice Exercise

Practice Exercise 126   Vector products (Answers on page 875)

In Problems 1 to 4, determine the quantities stated when

$p = 3i + 2k$, $q = i - 2j + 3k$ and

$r = -4i + 3j - k$

1. (a) $p \times q$ (b) $q \times p$

2. (a) $|p \times r|$ (b) $|r \times q|$

3. (a) $2p \times 3r$ (b) $(p + r) \times q$

4. (a) $p \times (r \times q)$ (b) $(3p \times 2r) \times q$

5. For vectors $p = 4i - j + 2k$ and $q = -2i + 3j - 2k$ determine: (a) $p \cdot q$ (b) $p \times q$ (c) $|p \times q|$ (d) $q \times p$ and (e) the angle between the vectors.

6. For vectors $a = -7i + 4j + \frac{1}{2}k$ and $b = 6i - 5j - k$ find (a) $a \cdot b$ (b) $a \times b$ (c) $|a \times b|$ (d) $b \times a$ and (e) the angle between the vectors.

7. Forces of $(i + 3j)$, $(-2i - j)$, $(i - 2j)$ newtons act at three points having position vectors of $(2i + 5j)$, $4j$ and $(-i + j)$ metres respectively. Calculate the magnitude of the moment.

8. A force of $(2i - j + k)$ newtons acts on a line through point P having co-ordinates $(0, 3, 1)$ metres. Determine the moment vector and its magnitude about point Q having co-ordinates $(4, 0, -1)$ metres.

9. A sphere is rotating with angular velocity ω about the z-axis of a system, the axis coinciding with the axis of the sphere. Determine the velocity vector and its magnitude at position $(-5i + 2j - 7k)$ m, when the angular velocity is $(i + 2j)$ rad/s.

10. Calculate the velocity vector and its magnitude for a particle rotating about the z-axis at an angular velocity of $(3i - j + 2k)$ rad/s when the position vector of the particle is at $(i - 5j + 4k)$ m.

28.4   Vector equation of a line

The equation of a straight line may be determined, given that it passes through point A with position vector $\boldsymbol{a}$ relative to O, and is parallel to vector $\boldsymbol{b}$. Let r be the position vector of a point P on the line, as shown in Fig. 28.10.

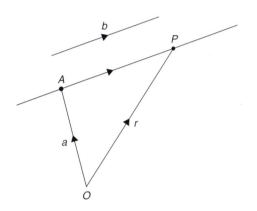

Figure 28.10

By vector addition, $\boldsymbol{OP} = \boldsymbol{OA} + \boldsymbol{AP}$,
i.e. $\boldsymbol{r} = \boldsymbol{a} + \boldsymbol{AP}$
However, as the straight line through A is parallel to the free vector $\boldsymbol{b}$ (**free vector** means one that has the same magnitude, direction and sense), then $\boldsymbol{AP} = \lambda \boldsymbol{b}$, where λ is a scalar quantity. Hence, from above,

$$\boldsymbol{r} = \boldsymbol{a} + \lambda \boldsymbol{b} \tag{8}$$

If, say, $\boldsymbol{r} = x\boldsymbol{i} + y\boldsymbol{j} + z\boldsymbol{k}$, $\boldsymbol{a} = a_1\boldsymbol{i} + a_2\boldsymbol{j} + a_3\boldsymbol{k}$ and $\boldsymbol{b} = b_1\boldsymbol{i} + b_2\boldsymbol{j} + b_3\boldsymbol{k}$, then from equation (8),

$$x\boldsymbol{i} + y\boldsymbol{j} + z\boldsymbol{k} = (a_1\boldsymbol{i} + a_2\boldsymbol{j} + a_3\boldsymbol{k})$$
$$+ \lambda(b_1\boldsymbol{i} + b_2\boldsymbol{j} + b_3\boldsymbol{k})$$

Hence $x = a_1 + \lambda b_1$, $y = a_2 + \lambda b_2$ and $z = a_3 + \lambda b_3$.
Solving for λ gives:

$$\frac{x - a_1}{b_1} = \frac{y - a_2}{b_2} = \frac{z - a_3}{b_3} = \lambda \tag{9}$$

Equation (9) is the standard Cartesian form for the vector equation of a straight line.

Problem 11. (a) Determine the vector equation of the line through the point with position vector $2\boldsymbol{i} + 3\boldsymbol{j} - \boldsymbol{k}$ which is parallel to the vector $\boldsymbol{i} - 2\boldsymbol{j} + 3\boldsymbol{k}$.
(b) Find the point on the line corresponding to $\lambda = 3$ in the resulting equation of part (a).
(c) Express the vector equation of the line in standard Cartesian form.

(a)   From equation (8),

$$\boldsymbol{r} = \boldsymbol{a} + \lambda \boldsymbol{b}$$

i.e. $\boldsymbol{r} = (2\boldsymbol{i} + 3\boldsymbol{j} - \boldsymbol{k}) + \lambda(\boldsymbol{i} - 2\boldsymbol{j} + 3\boldsymbol{k})$

or $\boldsymbol{r} = (2 + \lambda)\boldsymbol{i} + (3 - 2\lambda)\boldsymbol{j} + (3\lambda - 1)\boldsymbol{k}$

which is the vector equation of the line.

(b)   When $\lambda = 3$, $\boldsymbol{r} = 5\boldsymbol{i} - 3\boldsymbol{j} + 8\boldsymbol{k}$

(c)   From equation (9),

$$\frac{x - a_1}{b_1} = \frac{y - a_2}{b_2} = \frac{z - a_3}{b_3} = \lambda$$

Since $\boldsymbol{a} = 2\boldsymbol{i} + 3\boldsymbol{j} - \boldsymbol{k}$, then $a_1 = 2$,

$a_2 = 3$ and $a_3 = -1$ and

$\boldsymbol{b} = \boldsymbol{i} - 2\boldsymbol{j} + 3\boldsymbol{k}$, then

$b_1 = 1, b_2 = -2$ and $b_3 = 3$

Hence, the Cartesian equations are:

$$\frac{x - 2}{1} = \frac{y - 3}{-2} = \frac{z - (-1)}{3} = \lambda$$

i.e. $x - 2 = \dfrac{3 - y}{2} = \dfrac{z + 1}{3} = \lambda$

Problem 12. The equation

$$\frac{2x - 1}{3} = \frac{y + 4}{3} = \frac{-z + 5}{2}$$

represents a straight line. Express this in vector form.

Comparing the given equation with equation (9) shows that the coefficients of x, y and z need to be equal to unity.

Thus $\dfrac{2x-1}{3} = \dfrac{y+4}{3} = \dfrac{-z+5}{2}$ becomes:

$$\frac{x-\frac{1}{2}}{\frac{3}{2}} = \frac{y+4}{3} = \frac{z-5}{-2}$$

Again, comparing with equation (9), shows that

$a_1 = \dfrac{1}{2}, a_2 = -4$ and $a_3 = 5$ and

$b_1 = \dfrac{3}{2}, b_2 = 3$ and $b_3 = -2$

In vector form the equation is:

$$\boldsymbol{r} = (a_1 + \lambda b_1)\boldsymbol{i} + (a_2 + \lambda b_2)\boldsymbol{j} + (a_3 + \lambda b_3)\boldsymbol{k},$$
from equation (8)

i.e. $\boldsymbol{r} = \left(\dfrac{1}{2} + \dfrac{3}{2}\lambda\right)\boldsymbol{i} + (-4 + 3\lambda)\boldsymbol{j} + (5 - 2\lambda)\boldsymbol{k}$

or $\boldsymbol{r} = \dfrac{1}{2}(1 + 3\lambda)\boldsymbol{i} + (3\lambda - 4)\boldsymbol{j} + (5 - 2\lambda)\boldsymbol{k}$

Now try the following Practice Exercise

Practice Exercise 127 The vector equation of a line (Answers on page 875)

1. Find the vector equation of the line through the point with position vector $5\boldsymbol{i} - 2\boldsymbol{j} + 3\boldsymbol{k}$ which is parallel to the vector $2\boldsymbol{i} + 7\boldsymbol{j} - 4\boldsymbol{k}$. Determine the point on the line corresponding to $\lambda = 2$ in the resulting equation.

2. Express the vector equation of the line in problem 1 in standard Cartesian form.

In problems 3 and 4, express the given straight line equations in vector form.

3. $\dfrac{3x-1}{4} = \dfrac{5y+1}{2} = \dfrac{4-z}{3}$

4. $2x + 1 = \dfrac{1-4y}{5} = \dfrac{3z-1}{4}$

For fully worked solutions to each of the problems in Practice Exercises 125 to 127 in this chapter, go to the website:
www.routledge.com/cw/bird

Revision Test 8 Vectors

This Revision Test covers the material contained in Chapters 26 to 28. *The marks for each question are shown in brackets at the end of each question.*

1. State whether the following are scalar or vector quantities:
   (a) A temperature of $50°C$
   (b) A downward force of $80\,N$
   (c) $300\,J$ of work
   (d) A south-westerly wind of 15 knots
   (e) $70\,m$ distance
   (f) An acceleration of $25\,m/s^2$ at $30°$ to the horizontal (6)

2. Calculate the resultant and direction of the force vectors shown in Fig. RT8.1, correct to 2 decimal places. (7)

Figure RT8.1

3. Four coplanar forces act at a point A as shown in Fig. RT8.2 Determine the value and direction of the resultant force by (a) drawing (b) by calculation using horizontal and vertical components. (10)

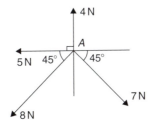

Figure RT8.2

4. The instantaneous values of two alternating voltages are given by:

$$v_1 = 150\sin(\omega t + \pi/3) \text{ volts and}$$

$$v_2 = 90\sin(\omega t - \pi/6) \text{ volts}$$

Plot the two voltages on the same axes to scales of $1\,cm = 50$ volts and $1\,cm = \dfrac{\pi}{6}\,rad$.

Obtain a sinusoidal expression for the resultant $v_1 + v_2$ in the form $R\sin(\omega t + \alpha)$: (a) by adding ordinates at intervals and (b) by calculation. (13)

5. If velocity $v_1 = 26\,m/s$ at $52°$ and $v_2 = 17\,m/s$ at $-28°$ calculate the magnitude and direction of $v_1 + v_2$, correct to 2 decimal places, using complex numbers. (10)

6. Given $a = -3i + 3j + 5k, b = 2i - 5j + 7k$ and $c = 3i + 6j - 4k$, determine the following:
   (a) $-4b$ (b) $a + b - c$ (c) $5b - 3c$ (8)

7. If $a = 2i + 4j - 5k$ and $b = 3i - 2j + 6k$ determine:
   (a) $a \cdot b$ (b) $|a + b|$ (c) $a \times b$ (d) the angle between a and b (14)

8. Determine the work done by a force of F newtons acting at a point A on a body, when A is displaced to point B, the co-ordinates of A and B being $(2, 5, -3)$ and $(1, -3, 0)$ metres respectively, and when $F = 2i - 5j + 4k$ newtons. (4)

9. A force of $F = 3i - 4j + k$ newtons acts on a line passing through a point P. Determine moment M and its magnitude of the force F about a point Q when P has co-ordinates $(4, -1, 5)$ metres and Q has co-ordinates $(4, 0, -3)$ metres. (8)

Section F

Introduction to calculus

Chapter 29

Methods of differentiation

Why it is important to understand: Methods of differentiation

There are many practical situations engineers have to analyse which involve quantities that are varying. Typical examples include the stress in a loaded beam, the temperature of an industrial chemical, the rate at which the speed of a vehicle is increasing or decreasing, the current in an electrical circuit or the torque on a turbine blade. Further examples include the voltage on a transmission line, the rate of growth of a bacteriological culture, and the rate at which the charge on a capacitor is changing. Differential calculus, or differentiation, is a mathematical technique for analysing the way in which functions change. There are many methods and rules of differentiation which are individually covered in the following chapters. A good knowledge of algebra, in particular, laws of indices, is essential. Calculus is one of the most powerful mathematical tools used by engineers. This chapter explains how to differentiate common functions, products, quotients and function of a function – all important methods providing a basis for further study in later chapters.

At the end of this chapter, you should be able to:

- differentiate common functions
- differentiate a product using the product rule
- differentiate a quotient using the quotient rule
- differentiate a function of a function
- differentiate successively

29.1 Introduction to calculus

Calculus is a branch of mathematics involving or leading to calculations dealing with continuously varying functions – such as velocity and acceleration, rates of change and maximum and minimum values of curves.

Calculus has widespread applications in science and engineering and is used to solve complicated problems for which algebra alone is insufficient.

Calculus is a subject that falls into two parts:

(i) **differential calculus**, or **differentiation**, which is covered in Chapters 29, 30 and 34 to 41, and

(ii) **integral calculus**, or **integration**, which is covered in Chapters 31, 32 and 42 to 49.

29.2 The gradient of a curve

If a tangent is drawn at a point P on a curve, then the gradient of this tangent is said to be the **gradient of the curve** at P. In Fig. 29.1, the gradient of the curve at P is equal to the gradient of the tangent PQ.

Figure 29.1

Figure 29.2

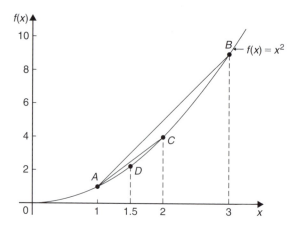

Figure 29.3

For the curve shown in Fig. 29.2, let the points A and B have co-ordinates (x_1, y_1) and (x_2, y_2), respectively. In functional notation, $y_1 = f(x_1)$ and $y_2 = f(x_2)$ as shown.

The gradient of the chord AB

$$= \frac{BC}{AC} = \frac{BD - CD}{ED} = \frac{f(x_2) - f(x_1)}{(x_2 - x_1)}$$

For the curve $f(x) = x^2$ shown in Fig. 29.3.

(i) the gradient of chord AB

$$= \frac{f(3) - f(1)}{3 - 1} = \frac{9 - 1}{2} = 4$$

(ii) the gradient of chord AC

$$= \frac{f(2) - f(1)}{2 - 1} = \frac{4 - 1}{1} = 3$$

(iii) the gradient of chord AD

$$= \frac{f(1.5) - f(1)}{1.5 - 1} = \frac{2.25 - 1}{0.5} = \mathbf{2.5}$$

(iv) if E is the point on the curve $(1.1, f(1.1))$ then the gradient of chord AE

$$= \frac{f(1.1) - f(1)}{1.1 - 1} = \frac{1.21 - 1}{0.1} = \mathbf{2.1}$$

(v) if F is the point on the curve $(1.01, f(1.01))$ then the gradient of chord AF

$$= \frac{f(1.01) - f(1)}{1.01 - 1} = \frac{1.0201 - 1}{0.01} = \mathbf{2.01}$$

Thus as point B moves closer and closer to point A the gradient of the chord approaches nearer and nearer to the value **2**. This is called the **limiting value** of the gradient of the chord AB and when B coincides with A the chord becomes the tangent to the curve.

29.3 Differentiation from first principles

In Fig. 29.4, A and B are two points very close together on a curve, δx (delta x) and δy (delta y) representing small increments in the x and y directions, respectively.

Gradient of chord $AB = \dfrac{\delta y}{\delta x}$; however,

$$\delta y = f(x + \delta x) - f(x)$$

Hence $\dfrac{\delta y}{\delta x} = \dfrac{f(x + \delta x) - f(x)}{\delta x}$

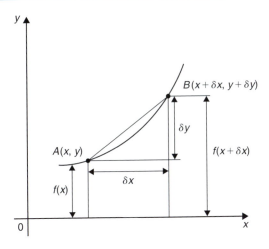

Figure 29.4

As δx approaches zero, $\dfrac{\delta y}{\delta x}$ approaches a limiting value and the gradient of the chord approaches the gradient of the tangent at A.

When determining the gradient of a tangent to a curve there are two notations used. The gradient of the curve at A in Fig. 29.4 can either be written as

$$\lim_{\delta x \to 0} \frac{\delta y}{\delta x} \quad \text{or} \quad \lim_{\delta x \to 0} \left\{ \frac{f(x+\delta x) - f(x)}{\delta x} \right\}$$

In **Leibniz**[*] notation, $\dfrac{dy}{dx} = \lim_{\delta x \to 0} \dfrac{\delta y}{\delta x}$

In **functional notation**,

$$f'(x) = \lim_{\delta x \to 0} \left\{ \frac{f(x+\delta x) - f(x)}{\delta x} \right\}$$

$\dfrac{dy}{dx}$ is the same as $f'(x)$ and is called the **differential coefficient** or the **derivative**. The process of finding the differential coefficient is called **differentiation**.

> **Problem 1.** Differentiate from first principle $f(x) = x^2$ and determine the value of the gradient of the curve at $x = 2$

To 'differentiate from first principles' means 'to find $f'(x)$' by using the expression

$$f'(x) = \lim_{\delta x \to 0} \left\{ \frac{f(x+\delta x) - f(x)}{\delta x} \right\}$$

$$f(x) = x^2$$

[*] **Who was Leibniz**? For image and resume, see page 91. For more information, go to **www.routledge.com/cw/bird**

Substituting $(x + \delta x)$ for x gives
$f(x + \delta x) = (x + \delta x)^2 = x^2 + 2x\delta x + \delta x^2$, hence

$$f'(x) = \lim_{\delta x \to 0} \left\{ \frac{(x^2 + 2x\delta x + \delta x^2) - (x^2)}{\delta x} \right\}$$

$$= \lim_{\delta x \to 0} \left\{ \frac{(2x\delta x + \delta x^2)}{\delta x} \right\}$$

$$= \lim_{\delta x \to 0} \left[2x + \delta x \right]$$

As $\delta x \to 0$, $[2x + \delta x] \to [2x + 0]$. Thus $f'(x) = 2x$, i.e. the differential coefficient of x^2 is $2x$. At $x = 2$, the gradient of the curve, $f'(x) = 2(2) = 4$

Differentiation from first principles can be a lengthy process and it would not be convenient to go through this procedure every time we want to differentiate a function. In reality we do not have to because a set of general rules have evolved from the above procedure, which we consider in the following section.

29.4 Differentiation of common functions

From differentiation by first principles of a number of examples such as in Problem 1 above, a general rule for differentiating $y = ax^n$ emerges, where a and n are constants.

The rule is: **if $y = ax^n$ then $\dfrac{dy}{dx} = anx^{n-1}$**

(or, **if $f(x) = ax^n$ then $f'(x) = anx^{n-1}$**) and is true for all real values of a and n.

For example, if $y = 4x^3$ then $a = 4$ and $n = 3$, and

$$\frac{dy}{dx} = anx^{n-1} = (4)(3)x^{3-1} = 12x^2$$

If $y = ax^n$ and $n = 0$ then $y = ax^0$ and

$$\frac{dy}{dx} = (a)(0)x^{0-1} = 0,$$

i.e. **the differential coefficient of a constant is zero**.

Fig. 29.5(a) shows a graph of $y = \sin x$. The gradient is continually changing as the curve moves from 0 to A to B to C to D. The gradient, given by $\dfrac{dy}{dx}$, may be plotted in a corresponding position below $y = \sin x$, as shown in Fig. 29.5(b).

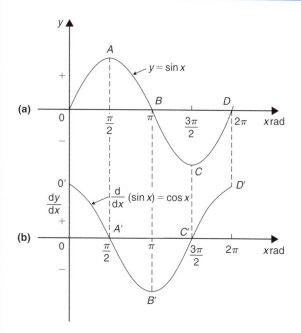

Figure 29.5

(i) At 0, the gradient is positive and is at its steepest. Hence $0'$ is a maximum positive value.

(ii) Between 0 and A the gradient is positive but is decreasing in value until at A the gradient is zero, shown as A'.

(iii) Between A and B the gradient is negative but is increasing in value until at B the gradient is at its steepest negative value. Hence B' is a maximum negative value.

(iv) If the gradient of $y = \sin x$ is further investigated between B and D then the resulting graph of $\dfrac{dy}{dx}$ is seen to be a cosine wave. Hence the rate of change of $\sin x$ is $\cos x$,

i.e. **if $y = \sin x$ then $\dfrac{dy}{dx} = \cos x$**

By a similar construction to that shown in Fig. 29.5 it may be shown that:

if $y = \sin ax$ then $\dfrac{dy}{dx} = a\cos ax$

If graphs of $y = \cos x$, $y = e^x$ and $y = \ln x$ are plotted and their gradients investigated, their differential coefficients may be determined in a similar manner to that shown for $y = \sin x$. The rate of change of a function is a measure of the derivative.

The **standard derivatives** summarised below may be proved theoretically and are true for all real values of x

y or $f(x)$	$\dfrac{dy}{dx}$ or $f'(x)$
ax^n	anx^{n-1}
$\sin ax$	$a\cos ax$
$\cos ax$	$-a\sin ax$
e^{ax}	ae^{ax}
$\ln ax$	$\dfrac{1}{x}$

The **differential coefficient of a sum or difference** is the sum or difference of the differential coefficients of the separate terms.

Thus, if $f(x) = p(x) + q(x) - r(x)$,

(where f, p, q and r are functions),

then $f'(x) = p'(x) + q'(x) - r'(x)$

Differentiation of common functions is demonstrated in the following worked Problems.

Problem 2. Find the differential coefficients of (a) $y = 12x^3$ (b) $y = \dfrac{12}{x^3}$

If $y = ax^n$ then $\dfrac{dy}{dx} = anx^{n-1}$

(a) Since $y = 12x^3$, $a = 12$ and $n = 3$ thus $\dfrac{dy}{dx} = (12)(3)x^{3-1} = \mathbf{36x^2}$

(b) $y = \dfrac{12}{x^3}$ is rewritten in the standard ax^n form as $y = 12x^{-3}$ and in the general rule $a = 12$ and $n = -3$

Thus $\dfrac{dy}{dx} = (12)(-3)x^{-3-1} = -36x^{-4} = -\dfrac{36}{x^4}$

Problem 3. Differentiate (a) $y = 6$ (b) $y = 6x$

(a) $y = 6$ may be written as $y = 6x^0$, i.e. in the general rule $a = 6$ and $n = 0$

Hence $\dfrac{dy}{dx} = (6)(0)x^{0-1} = \mathbf{0}$

In general, **the differential coefficient of a constant is always zero**.

(b) Since $y = 6x$, in the general rule $a = 6$ and $n = 1$

Hence $\dfrac{dy}{dx} = (6)(1)x^{1-1} = 6x^0 = \mathbf{6}$

In general, the differential coefficient of kx, where k is a constant, is always k.

Problem 4. Find the derivatives of

(a) $y = 3\sqrt{x}$ (b) $y = \dfrac{5}{\sqrt[3]{x^4}}$

(a) $y = 3\sqrt{x}$ is rewritten in the standard differential form as $y = 3x^{\frac{1}{2}}$

In the general rule, $a = 3$ and $n = \dfrac{1}{2}$

Thus $\dfrac{dy}{dx} = (3)\left(\dfrac{1}{2}\right)x^{\frac{1}{2}-1} = \dfrac{3}{2}x^{-\frac{1}{2}}$

$= \dfrac{3}{2x^{\frac{1}{2}}} = \dfrac{\mathbf{3}}{\mathbf{2\sqrt{x}}}$

(b) $y = \dfrac{5}{\sqrt[3]{x^4}} = \dfrac{5}{x^{\frac{4}{3}}} = 5x^{-\frac{4}{3}}$ in the standard differential form.

In the general rule, $a = 5$ and $n = -\dfrac{4}{3}$

Thus $\dfrac{dy}{dx} = (5)\left(-\dfrac{4}{3}\right)x^{-\frac{4}{3}-1} = \dfrac{-20}{3}x^{-\frac{7}{3}}$

$= \dfrac{-20}{3x^{\frac{7}{3}}} = \dfrac{\mathbf{-20}}{\mathbf{3\sqrt[3]{x^7}}}$

Problem 5. Differentiate, with respect to x,

$y = 5x^4 + 4x - \dfrac{1}{2x^2} + \dfrac{1}{\sqrt{x}} - 3$

$y = 5x^4 + 4x - \dfrac{1}{2x^2} + \dfrac{1}{\sqrt{x}} - 3$ is rewritten as

$y = 5x^4 + 4x - \dfrac{1}{2}x^{-2} + x^{-\frac{1}{2}} - 3$

When differentiating a sum, each term is differentiated in turn.

Thus $\dfrac{dy}{dx} = (5)(4)x^{4-1} + (4)(1)x^{1-1} - \dfrac{1}{2}(-2)x^{-2-1}$

$+ (1)\left(-\dfrac{1}{2}\right)x^{-\frac{1}{2}-1} - 0$

$= 20x^3 + 4 + x^{-3} - \dfrac{1}{2}x^{-\frac{3}{2}}$

i.e. $\dfrac{dy}{dx} = \mathbf{20x^3 + 4 + \dfrac{1}{x^3} - \dfrac{1}{2\sqrt{x^3}}}$

Problem 6. Find the differential coefficients of
(a) $y = 3\sin 4x$ (b) $f(t) = 2\cos 3t$ with respect to the variable.

(a) When $y = 3\sin 4x$ then $\dfrac{dy}{dx} = (3)(4\cos 4x)$

$= \mathbf{12\cos 4x}$

(b) When $f(t) = 2\cos 3t$ then
$f'(t) = (2)(-3\sin 3t) = \mathbf{-6\sin 3t}$

Problem 7. Determine the derivatives of
(a) $y = 3e^{5x}$ (b) $f(\theta) = \dfrac{2}{e^{3\theta}}$ (c) $y = 6\ln 2x$

(a) When $y = 3e^{5x}$ then $\dfrac{dy}{dx} = (3)(5)e^{5x} = \mathbf{15e^{5x}}$

(b) $f(\theta) = \dfrac{2}{e^{3\theta}} = 2e^{-3\theta}$, thus

$f'(\theta) = (2)(-3)e^{-30} = -6e^{-3\theta} = \dfrac{\mathbf{-6}}{\mathbf{e^{3\theta}}}$

(c) When $y = 6\ln 2x$ then $\dfrac{dy}{dx} = 6\left(\dfrac{1}{x}\right) = \dfrac{\mathbf{6}}{\mathbf{x}}$

Problem 8. Find the gradient of the curve
$y = 3x^4 - 2x^2 + 5x - 2$ at the points $(0, -2)$ and $(1, 4)$

The gradient of a curve at a given point is given by the corresponding value of the derivative. Thus, since $y = 3x^4 - 2x^2 + 5x - 2$

Then the gradient $= \dfrac{dy}{dx} = 12x^3 - 4x + 5$

At the point $(0, -2)$, $x = 0$
Thus the gradient $= 12(0)^3 - 4(0) + 5 = \mathbf{5}$

At the point $(1, 4)$, $x = 1$
Thus the gradient $= 12(1)^3 - 4(1) + 5 = \mathbf{13}$

Problem 9. Determine the co-ordinates of the point on the graph $y = 3x^2 - 7x + 2$ where the gradient is -1

The gradient of the curve is given by the derivative.

When $y = 3x^2 - 7x + 2$ then $\dfrac{dy}{dx} = 6x - 7$

Since the gradient is -1 then $6x - 7 = -1$, from which, $x = 1$

When $x = 1$, $y = 3(1)^2 - 7(1) + 2 = -2$

Hence the gradient is -1 at the point $(1, -2)$

Now try the following Practice Exercise

Practice Exercise 128   Differentiating common functions (Answers on page 875)

In Problems 1 to 6 find the differential coefficients of the given functions with respect to the variable.

1. (a) $5x^5$ (b) $2.4x^{3.5}$ (c) $\dfrac{1}{x}$

2. (a) $\dfrac{-4}{x^2}$ (b) 6 (c) $2x$

3. (a) $2\sqrt{x}$ (b) $3\sqrt[3]{x^5}$ (c) $\dfrac{4}{\sqrt{x}}$

4. (a) $\dfrac{-3}{\sqrt[3]{x}}$ (b) $(x - 1)^2$ (c) $2\sin 3x$

5. (a) $-4\cos 2x$ (b) $2e^{6x}$ (c) $\dfrac{3}{e^{5x}}$

6. (a) $4\ln 9x$ (b) $\dfrac{e^x - e^{-x}}{2}$ (c) $\dfrac{1 - \sqrt{x}}{x}$

7. Find the gradient of the curve $y = 2t^4 + 3t^3 - t + 4$ at the points $(0, 4)$ and $(1, 8)$

8. Find the co-ordinates of the point on the graph $y = 5x^2 - 3x + 1$ where the gradient is 2

9. (a) Differentiate $y = \dfrac{2}{\theta^2} + 2\ln 2\theta -$
    $2(\cos 5\theta + 3\sin 2\theta) - \dfrac{2}{e^{3\theta}}$

    (b) Evaluate $\dfrac{dy}{d\theta}$ in part (a) when $\theta = \dfrac{\pi}{2}$, correct to 4 significant figures.

10. Evaluate $\dfrac{ds}{dt}$, correct to 3 significant figures, when $t = \dfrac{\pi}{6}$ given $s = 3\sin t - 3 + \sqrt{t}$

11. A mass, m, is held by a spring with a stiffness constant k. The potential energy, p, of the system is given by: $p = \dfrac{1}{2}kx^2 - mgx$ where x is the displacement and g is acceleration due to gravity.

    The system is in equilibrium if $\dfrac{dp}{dx} = 0$. Determine the expression for x for system equilibrium.

12. The current i flowing in an inductor of inductance 100 mH is given by: $i = 5\sin 100t$ amperes, where t is the time t in seconds. The voltage v across the inductor is given by: $v = L\dfrac{di}{dt}$ volts.
    Determine the voltage when $t = 10$ ms.

29.5   Differentiation of a product

When $y = uv$, and u and v are both functions of x,

then
$$\dfrac{dy}{dx} = u\dfrac{dv}{dx} + v\dfrac{du}{dx}$$

This is known as the **product rule**.

Problem 10. Find the differential coefficient of $y = 3x^2 \sin 2x$

$3x^2 \sin 2x$ is a product of two terms $3x^2$ and $\sin 2x$
Let $u = 3x^2$ and $v = \sin 2x$
Using the product rule:

$$\dfrac{dy}{dx} = u\quad\dfrac{dv}{dx}\quad + \quad v\quad\dfrac{du}{dx}$$
$$\downarrow\quad\downarrow\qquad\qquad\downarrow\quad\downarrow$$

gives: $\dfrac{dy}{dx} = (3x^2)(2\cos 2x) + (\sin 2x)(6x)$

i.e. $\dfrac{dy}{dx} = 6x^2\cos 2x + 6x\sin 2x$

$$= 6x(x\cos 2x + \sin 2x)$$

Note that the differential coefficient of a product is **not** obtained by merely differentiating each term and multiplying the two answers together. The product rule formula **must** be used when differentiating products.

Problem 11. Find the rate of change of y with respect to x given $y = 3\sqrt{x}\ln 2x$

The rate of change of y with respect to x is given by $\dfrac{dy}{dx}$

$y = 3\sqrt{x}\ln 2x = 3x^{\frac{1}{2}}\ln 2x$, which is a product.

Let $u = 3x^{\frac{1}{2}}$ and $v = \ln 2x$

Then $\quad \dfrac{dy}{dx} = \quad u \quad \dfrac{dv}{dx} \quad + \quad v \quad \dfrac{du}{dx}$

$\qquad\qquad\quad \downarrow \qquad \downarrow \qquad\quad \downarrow \qquad\quad \downarrow$

$\qquad = \left(3x^{\frac{1}{2}}\right)\left(\dfrac{1}{x}\right) + (\ln 2x)\left[3\left(\dfrac{1}{2}\right)x^{\frac{1}{2}-1}\right]$

$\qquad = 3x^{\frac{1}{2}-1} + (\ln 2x)\left(\dfrac{3}{2}\right)x^{-\frac{1}{2}}$

$\qquad = 3x^{-\frac{1}{2}}\left(1 + \dfrac{1}{2}\ln 2x\right)$

i.e. $\quad \dfrac{dy}{dx} = \dfrac{3}{\sqrt{x}}\left(1 + \dfrac{1}{2}\ln 2x\right)$

Problem 12. Differentiate $y = x^3\cos 3x\ln x$

Let $u = x^3\cos 3x$ (i.e. a product) and $v = \ln x$

Then $\quad \dfrac{dy}{dx} = u\dfrac{dv}{dx} + v\dfrac{du}{dx}$

where $\dfrac{du}{dx} = (x^3)(-3\sin 3x) + (\cos 3x)(3x^2)$

and $\quad \dfrac{dv}{dx} = \dfrac{1}{x}$

Hence $\dfrac{dy}{dx} = (x^3\cos 3x)\left(\dfrac{1}{x}\right) + (\ln x)[-3x^3\sin 3x$

$\qquad\qquad\qquad\qquad\qquad\qquad + 3x^2\cos 3x]$

$\qquad = x^2\cos 3x + 3x^2\ln x(\cos 3x - x\sin 3x)$

i.e. $\quad \dfrac{dy}{dx} = x^2\{\cos 3x + 3\ln x(\cos 3x - x\sin 3x)\}$

Problem 13. Determine the rate of change of voltage, given $v = 5t\sin 2t$ volts when $t = 0.2\,s$

Rate of change of voltage $= \dfrac{dv}{dt}$

$\qquad = (5t)(2\cos 2t) + (\sin 2t)(5)$

$\qquad = 10t\cos 2t + 5\sin 2t$

When $t = 0.2$, $\dfrac{dv}{dt} = 10(0.2)\cos 2(0.2) + 5\sin 2(0.2)$

$\qquad = 2\cos 0.4 + 5\sin 0.4$ (where $\cos 0.4$

$\qquad\qquad$ means the cosine of 0.4 radians)

Hence $\dfrac{dv}{dt} = 2(0.92106) + 5(0.38942)$

$\qquad = 1.8421 + 1.9471 = 3.7892$

i.e. **the rate of change of voltage when $t = 0.2\,s$ is 3.79 volts/s, correct to 3 significant figures.**

Now try the following Practice Exercise

Practice Exercise 129   Differentiating products (Answers on page 876)

In Problems 1 to 8 differentiate the given products with respect to the variable.

1. $x\sin x$

2. $x^2 e^{2x}$

3. $x^2\ln x$

4. $2x^3\cos 3x$

5. $\sqrt{x^3}\ln 3x$

6. $e^{3t}\sin 4t$

7. $e^{4\theta}\ln 3\theta$

8. $e^t\ln t\cos t$

9.   Evaluate $\dfrac{di}{dt}$, correct to 4 significant figures, when $t = 0.1$, and $i = 15t\sin 3t$

10.   Evaluate $\dfrac{dz}{dt}$, correct to 4 significant figures, when $t = 0.5$, given that $z = 2e^{3t}\sin 2t$

29.6 Differentiation of a quotient

When $y = \dfrac{u}{v}$, and u and v are both functions of x

then $\qquad \dfrac{dy}{dx} = \dfrac{v\dfrac{du}{dx} - u\dfrac{dv}{dx}}{v^2}$

This is known as the **quotient rule**.

Problem 14. Find the differential coefficient of
$$y = \frac{4\sin 5x}{5x^4}$$

$\dfrac{4\sin 5x}{5x^4}$ is a quotient. Let $u = 4\sin 5x$ and $v = 5x^4$

(Note that v is **always** the denominator and u the numerator.)

$$\frac{dy}{dx} = \frac{v\dfrac{du}{dx} - u\dfrac{dv}{dx}}{v^2}$$

where $\quad \dfrac{du}{dx} = (4)(5)\cos 5x = 20\cos 5x$

and $\quad \dfrac{dv}{dx} = (5)(4)x^3 = 20x^3$

Hence $\quad \dfrac{dy}{dx} = \dfrac{(5x^4)(20\cos 5x) - (4\sin 5x)(20x^3)}{(5x^4)^2}$

$$= \frac{100x^4\cos 5x - 80x^3\sin 5x}{25x^8}$$

$$= \frac{20x^3[5x\cos 5x - 4\sin 5x]}{25x^8}$$

i.e. $\quad \dfrac{dy}{dx} = \dfrac{4}{5x^5}(5x\cos 5x - 4\sin 5x)$

Note that the differential coefficient is **not** obtained by merely differentiating each term in turn and then dividing the numerator by the denominator. The quotient formula **must** be used when differentiating quotients.

Problem 15. Determine the differential coefficient of $y = \tan ax$

$y = \tan ax = \dfrac{\sin ax}{\cos ax}$. Differentiation of $\tan ax$ is thus treated as a quotient with $u = \sin ax$ and $v = \cos ax$

$$\frac{dy}{dx} = \frac{v\dfrac{du}{dx} - u\dfrac{dv}{dx}}{v^2}$$

$$= \frac{(\cos ax)(a\cos ax) - (\sin ax)(-a\sin ax)}{(\cos ax)^2}$$

$$= \frac{a\cos^2 ax + a\sin^2 ax}{(\cos ax)^2} = \frac{a(\cos^2 ax + \sin^2 ax)}{\cos^2 ax}$$

$$= \frac{a}{\cos^2 ax}, \text{ since } \cos^2 ax + \sin^2 ax = 1$$

(see Chapter 17)

Hence $\quad \dfrac{dy}{dx} = a\sec^2 ax \;$ since $\; \sec^2 ax = \dfrac{1}{\cos^2 ax} \;$ (see Chapter 12).

Problem 16. Find the derivative of $y = \sec ax$

$y = \sec ax = \dfrac{1}{\cos ax}$ (i.e. a quotient). Let $u = 1$ and $v = \cos ax$

$$\frac{dy}{dx} = \frac{v\dfrac{du}{dx} - u\dfrac{dv}{dx}}{v^2}$$

$$= \frac{(\cos ax)(0) - (1)(-a\sin ax)}{(\cos ax)^2}$$

$$= \frac{a\sin ax}{\cos^2 ax} = a\left(\frac{1}{\cos ax}\right)\left(\frac{\sin ax}{\cos ax}\right)$$

i.e. $\quad \dfrac{dy}{dx} = a\sec ax \tan ax$

Problem 17. Differentiate $y = \dfrac{te^{2t}}{2\cos t}$

The function $\dfrac{te^{2t}}{2\cos t}$ is a quotient, whose numerator is a product.

Let $u = te^{2t}$ and $v = 2\cos t$ then

$\dfrac{du}{dt} = (t)(2e^{2t}) + (e^{2t})(1)$ and $\dfrac{dv}{dt} = -2\sin t$

Hence $\quad \dfrac{dy}{dx} = \dfrac{v\dfrac{du}{dx} - u\dfrac{dv}{dx}}{v^2}$

$$= \frac{(2\cos t)[2te^{2t} + e^{2t}] - (te^{2t})(-2\sin t)}{(2\cos t)^2}$$

$$= \frac{4te^{2t}\cos t + 2e^{2t}\cos t + 2te^{2t}\sin t}{4\cos^2 t}$$

$$= \frac{2e^{2t}[2t\cos t + \cos t + t \sin t]}{4\cos^2 t}$$

i.e. $\dfrac{dy}{dx} = \dfrac{e^{2t}}{2\cos^2 t}(2t\cos t + \cos t + t \sin t)$

Problem 18. Determine the gradient of the curve $y = \dfrac{5x}{2x^2+4}$ at the point $\left(\sqrt{3}, \dfrac{\sqrt{3}}{2}\right)$

Let $y = 5x$ and $v = 2x^2 + 4$

$$\frac{dy}{dx} = \frac{v\dfrac{du}{dx} - u\dfrac{dv}{dx}}{v^2} = \frac{(2x^2+4)(5) - (5x)(4x)}{(2x^2+4)^2}$$

$$= \frac{10x^2 + 20 - 20x^2}{(2x^2+4)^2} = \frac{20 - 10x^2}{(2x^2+4)^2}$$

At the point $\left(\sqrt{3}, \dfrac{\sqrt{3}}{2}\right)$, $x = \sqrt{3}$,

hence the gradient $= \dfrac{dy}{dx} = \dfrac{20 - 10(\sqrt{3})^2}{[2(\sqrt{3})^2 + 4]^2}$

$$= \frac{20 - 30}{100} = -\frac{1}{10}$$

Now try the following Practice Exercise

Practice Exercise 130   Differentiating quotients (Answers on page 876)

In Problems 1 to 7, differentiate the quotients with respect to the variable.

1. $\dfrac{\sin x}{x}$

2. $\dfrac{2\cos 3x}{x^3}$

3. $\dfrac{2x}{x^2+1}$

4. $\dfrac{\sqrt{x}}{\cos x}$

5. $\dfrac{3\sqrt{\theta^3}}{2\sin 2\theta}$

6. $\dfrac{\ln 2t}{\sqrt{t}}$

7. $\dfrac{2xe^{4x}}{\sin x}$

8. Find the gradient of the curve $y = \dfrac{2x}{x^2-5}$ at the point $(2, -4)$

9. Evaluate $\dfrac{dy}{dx}$ at $x = 2.5$, correct to 3 significant figures, given $y = \dfrac{2x^2+3}{\ln 2x}$

29.7 Function of a function

It is often easier to make a substitution before differentiating.

If y is a function of x then $\qquad \dfrac{dy}{dx} = \dfrac{dy}{du} \times \dfrac{du}{dx}$

This is known as the **'function of a function'** rule (or sometimes the **chain rule**).
For example, if $y = (3x-1)^9$ then, by making the substitution $u = (3x-1)$, $y = u^9$, which is of the 'standard' form.

Hence $\dfrac{dy}{du} = 9u^8$ and $\dfrac{du}{dx} = 3$

Then $\dfrac{dy}{dx} = \dfrac{dy}{du} \times \dfrac{du}{dx} = (9u^8)(3) = 27u^8$

Rewriting u as $(3x-1)$ gives: $\dfrac{dy}{dx} = 27(3x-1)^8$

Since y is a function of u, and u is a function of x, then y is a function of a function of x.

Problem 19. Differentiate $y = 3\cos(5x^2+2)$

Let $u = 5x^2 + 2$ then $y = 3\cos u$

Hence $\dfrac{du}{dx} = 10x$ and $\dfrac{dy}{du} = -3\sin u$

Using the function of a function rule,

$$\frac{dy}{dx} = \frac{dy}{du} \times \frac{du}{dx} = (-3\sin u)(10x) = -30x\sin u$$

Rewriting u as $5x^2 + 2$ gives:

$$\frac{dy}{dx} = -30x\sin(5x^2+2)$$

Problem 20. Find the derivative of
$$y = (4t^3 - 3t)^6$$

Let $u = 4t^3 - 3t$, then $y = u^6$

Hence $\dfrac{du}{dt} = 12t^2 - 3$ and $\dfrac{dy}{du} = 6u^5$

Using the function of a function rule,

$$\frac{dy}{dx} = \frac{dy}{du} \times \frac{du}{dx} = (6u^5)(12t^2 - 3)$$

Rewriting u as $(4t^3 - 3t)$ gives:

$$\frac{dy}{dt} = 6(4t^3 - 3t)^5(12t^2 - 3)$$

$$= 18(4t^2 - 1)(4t^3 - 3t)^5$$

Problem 21. Determine the differential coefficient of $y = \sqrt{(3x^2 + 4x - 1)}$

$$y = \sqrt{(3x^2 + 4x - 1)} = (3x^2 + 4x - 1)^{\frac{1}{2}}$$

Let $u = 3x^2 + 4x - 1$ then $y = u^{\frac{1}{2}}$

Hence $\dfrac{du}{dx} = 6x + 4$ and $\dfrac{dy}{du} = \dfrac{1}{2}u^{-\frac{1}{2}} = \dfrac{1}{2\sqrt{u}}$

Using the function of a function rule,

$$\frac{dy}{dx} = \frac{dy}{du} \times \frac{du}{dx} = \left(\frac{1}{2\sqrt{u}}\right)(6x + 4) = \frac{3x + 2}{\sqrt{u}}$$

i.e. $\dfrac{dy}{dx} = \dfrac{3x + 2}{\sqrt{(3x^2 + 4x - 1)}}$

Problem 22. Differentiate $y = 3\tan^4 3x$

Let $u = \tan 3x$ then $y = 3u^4$

Hence $\dfrac{du}{dx} = 3\sec^2 3x$, (from Problem 15), and

$$\frac{dy}{du} = 12u^3$$

Then $\dfrac{dy}{dx} = \dfrac{dy}{du} \times \dfrac{du}{dx} = (12u^3)(3\sec^2 3x)$

$$= 12(\tan 3x)^3(3\sec^2 3x)$$

i.e. $\dfrac{dy}{dx} = \mathbf{36\tan^3 3x \sec^2 3x}$

Problem 23. Find the differential coefficient of
$$y = \frac{2}{(2t^3 - 5)^4}$$

$y = \dfrac{2}{(2t^3 - 5)^4} = 2(2t^3 - 5)^{-4}$. Let $u = (2t^3 - 5)$, then
$y = 2u^{-4}$

Hence $\dfrac{du}{dt} = 6t^2$ and $\dfrac{dy}{du} = -8u^{-5} = \dfrac{-8}{u^5}$

Then $\dfrac{dy}{dt} = \dfrac{dy}{du} \times \dfrac{du}{dt} = \left(\dfrac{-8}{u^5}\right)(6t^2)$

$$= \frac{-48t^2}{(2t^3 - 5)^5}$$

Now try the following Practice Exercise

Practice Exercise 131   Function of a function (Answers on page 876)

In Problems 1 to 9, find the differential coefficients with respect to the variable.

1. $(2x - 1)^6$

2. $(2x^3 - 5x)^5$

3. $2\sin(3\theta - 2)$

4. $2\cos^5 \alpha$

5. $\dfrac{1}{(x^3 - 2x + 1)^5}$

6. $5e^{2t+1}$

7. $2\cot(5t^2 + 3)$

8. $6\tan(3y + 1)$

9. $2e^{\tan\theta}$

10. Differentiate $\theta\sin\left(\theta - \dfrac{\pi}{3}\right)$ with respect to θ, and evaluate, correct to 3 significant figures, when $\theta = \dfrac{\pi}{2}$

11. The extension, x metres, of an undamped vibrating spring after t seconds is given by:

$$x = 0.54\cos(0.3t - 0.15) + 3.2$$

Calculate the speed of the spring, given by $\dfrac{dx}{dt}$, when (a) $t = 0$, (b) $t = 2s$

29.8   Successive differentiation

When a function $y = f(x)$ is differentiated with respect to x the differential coefficient is written as $\dfrac{dy}{dx}$ or $f'(x)$. If the expression is differentiated again, the second differential coefficient is obtained and is written as $\dfrac{d^2y}{dx^2}$ (pronounced dee two y by dee x squared) or $f''(x)$ (pronounced f double-dash x).

By successive differentiation further higher derivatives such as $\dfrac{d^3y}{dx^3}$ and $\dfrac{d^4y}{dx^4}$ may be obtained.

Thus if $y = 3x^4$, $\dfrac{dy}{dx} = 12x^3$, $\dfrac{d^2y}{dx^2} = 36x^2$,

$\dfrac{d^3y}{dx^3} = 72x$, $\dfrac{d^4y}{dx^4} = 72$ and $\dfrac{d^5y}{dx^5} = 0$

Problem 24.   If $f(x) = 2x^5 - 4x^3 + 3x - 5$, find $f''(x)$

$$f(x) = 2x^5 - 4x^3 + 3x - 5$$

$$f'(x) = 10x^4 - 12x^2 + 3$$

$$f''(x) = 40x^3 - 24x = \mathbf{4x(10x^2 - 6)}$$

Problem 25.   If $y = \cos x - \sin x$, evaluate x, in the range $0 \leq x \leq \dfrac{\pi}{2}$, when $\dfrac{d^2y}{dx^2}$ is zero.

Since $y = \cos x - \sin x$, $\dfrac{dy}{dx} = -\sin x - \cos x$ and

$\dfrac{d^2y}{dx^2} = -\cos x + \sin x$

When $\dfrac{d^2y}{dx^2}$ is zero, $-\cos x + \sin x = 0$,

i.e. $\sin x = \cos x$ or $\dfrac{\sin x}{\cos x} = 1$

Hence $\tan x = 1$ and $x = \arctan 1 = \mathbf{45°}$ **or** $\dfrac{\boldsymbol\pi}{\mathbf 4}$ **rads** in the range $0 \leq x \leq \dfrac{\pi}{2}$

Problem 26.   Given $y = 2xe^{-3x}$ show that

$$\dfrac{d^2y}{dx^2} + 6\dfrac{dy}{dx} + 9y = 0$$

$y = 2xe^{-3x}$ (i.e. a product)

Hence $\dfrac{dy}{dx} = (2x)(-3e^{-3x}) + (e^{-3x})(2)$

$\qquad = -6xe^{-3x} + 2e^{-3x}$

$\dfrac{d^2y}{dx^2} = [(-6x)(-3e^{-3x}) + (e^{-3x})(-6)]$

$\qquad\qquad\qquad\qquad + (-6e^{-3x})$

$\qquad = 18xe^{-3x} - 6e^{-3x} - 6e^{-3x}$

i.e. $\dfrac{d^2y}{dx^2} = 18xe^{-3x} - 12e^{-3x}$

Substituting values into $\dfrac{d^2y}{dx^2} + 6\dfrac{dy}{dx} + 9y$ gives:

$(18xe^{-3x} - 12e^{-3x}) + 6(-6xe^{-3x} + 2e^{-3x})$

$+ 9(2xe^{-3x}) = 18xe^{-3x} - 12e^{-3x} - 36xe^{-3x}$

$+ 12e^{-3x} + 18xe^{-3x} = 0$

Thus when $y = 2xe^{-3x}$, $\dfrac{d^2y}{dx^2} + 6\dfrac{dy}{dx} + 9y = 0$

Problem 27.   Evaluate $\dfrac{d^2y}{d\theta^2}$ when $\theta = 0$ given $y = 4\sec 2\theta$

Since $y = 4\sec 2\theta$,

then $\dfrac{dy}{d\theta} = (4)(2)\sec 2\theta \tan 2\theta$ (from Problem 16)

$\qquad = 8\sec 2\theta \tan 2\theta$ (i.e. a product)

$\dfrac{d^2y}{d\theta^2} = (8\sec 2\theta)(2\sec^2 2\theta)$

$\qquad\qquad + (\tan 2\theta)[(8)(2)\sec 2\theta \tan 2\theta]$

$\qquad = 16\sec^3 2\theta + 16\sec 2\theta \tan^2 2\theta$

When $\theta = 0$, $\dfrac{d^2y}{d\theta^2} = 16\sec^3 0 + 16\sec 0 \tan^2 0$

$\qquad = 16(1) + 16(1)(0) = \mathbf{16}$

Now try the following Practice Exercise

Practice Exercise 132   Successive differentiation (Answers on page 876)

1. If $y = 3x^4 + 2x^3 - 3x + 2$ find

   (a) $\dfrac{d^2 y}{dx^2}$ (b) $\dfrac{d^3 y}{dx^3}$

2. (a) Given $f(t) = \dfrac{2}{5}t^2 - \dfrac{1}{t^3} + \dfrac{3}{t} - \sqrt{t} + 1$ determine $f''(t)$

   (b) Evaluate $f''(t)$ when $t = 1$

3. The charge q on the plates of a capacitor is given by $q = CV e^{-\frac{t}{CR}}$, where t is the time, C is the capacitance and R the resistance. Determine (a) the rate of change of charge, which is given by $\dfrac{dq}{dt}$, (b) the rate of change of current, which is given by $\dfrac{d^2 q}{dt^2}$

In Problems 4 and 5, find the second differential coefficient with respect to the variable.

4. (a) $3 \sin 2t + \cos t$ (b) $2 \ln 4\theta$

5. (a) $2 \cos^2 x$ (b) $(2x - 3)^4$

6. Evaluate $f''(\theta)$ when $\theta = 0$ given $f(\theta) = 2 \sec 3\theta$

7. Show that the differential equation $\dfrac{d^2 y}{dx^2} - 4\dfrac{dy}{dx} + 4y = 0$ is satisfied when $y = x e^{2x}$

8. Show that, if P and Q are constants and $y = P \cos(\ln t) + Q \sin(\ln t)$, then

$$t^2 \frac{d^2 y}{dt^2} + t\frac{dy}{dt} + y = 0$$

9. The displacement, s, of a mass in a vibrating system is given by: $s = (1 + t)e^{-\omega t}$ where ω is the natural frequency of vibration. Show that:

$$\frac{d^2 s}{dt^2} + 2\omega\frac{ds}{dt} + \omega^2 s = 0$$

For fully worked solutions to each of the problems in Practice Exercises 128 to 132 in this chapter, go to the website:
www.routledge.com/cw/bird

Chapter 30

Some applications of differentiation

Why it is important to understand: Some applications of differentiation

In the previous chapter some basic differentiation techniques were explored, sufficient to allow us to look at some applications of differential calculus. Some practical rates of change problems are initially explained, followed by some practical velocity and acceleration problems. Determining maximum and minimum points and points of inflexion on curves, together with some practical maximum and minimum problems follow. Tangents and normals to curves and errors and approximations complete this initial look at some applications of differentiation. In general, with these applications, the differentiation tends to be straightforward.

At the end of this chapter, you should be able to:

- determine rates of change using differentiation
- solve velocity and acceleration problems
- understand turning points
- determine the turning points on a curve and determine their nature
- solve practical problems involving maximum and minimum values
- determine points of inflexion on a curve
- determine tangents and normals to a curve
- determine small changes in functions

30.1 Rates of change

If a quantity y depends on and varies with a quantity x then the rate of change of y with respect to x is $\dfrac{dy}{dx}$.

Thus, for example, the rate of change of pressure p with height h is $\dfrac{dp}{dh}$

A rate of change with respect to time is usually just called 'the rate of change', the 'with respect to time' being assumed. Thus, for example, a rate of change of current, i, is $\dfrac{di}{dt}$ and a rate of change of temperature, θ, is $\dfrac{d\theta}{dt}$, and so on.

Problem 1. The length l metres of a certain metal rod at temperature $\theta °C$ is given by $l = 1 + 0.00005\theta + 0.0000004\theta^2$. Determine the rate of change of length, in mm/°C, when the temperature is (a) 100°C and (b) 400°C.

The rate of change of length means $\dfrac{dl}{d\theta}$

Since length $l = 1 + 0.00005\theta + 0.0000004\theta^2$,

then $\dfrac{dl}{d\theta} = 0.00005 + 0.0000008\theta$

(a) When $\theta = 100°C$,

$$\frac{dl}{d\theta} = 0.00005 + (0.0000008)(100)$$

$$= 0.00013 \, \text{m/}°C$$

$$= \textbf{0.13 mm/}°\textbf{C}$$

(b) When $\theta = 400°C$,

$$\frac{dl}{d\theta} = 0.00005 + (0.0000008)(400)$$

$$= 0.00037 \, \text{m/}°C$$

$$= \textbf{0.37 mm/}°\textbf{C}$$

Problem 2. The luminous intensity I candelas of a lamp at varying voltage V is given by $I = 4 \times 10^{-4} V^2$. Determine the voltage at which the light is increasing at a rate of 0.6 candelas per volt.

The rate of change of light with respect to voltage is given by $\dfrac{dI}{dV}$

Since $I = 4 \times 10^{-4} V^2$,

$$\frac{dI}{dV} = (4 \times 10^{-4})(2)V = 8 \times 10^{-4} V$$

When the light is increasing at 0.6 candelas per volt then $+0.6 = 8 \times 10^{-4} V$, from which, voltage

$$V = \frac{0.6}{8 \times 10^{-4}} = 0.075 \times 10^{+4}$$

$$= \textbf{750 volts}$$

Problem 3. Newton's law of cooling is given by $\theta = \theta_0 e^{-kt}$, where the excess of temperature at zero time is $\theta_0°C$ and at time t seconds is $\theta°C$. Determine the rate of change of temperature after 40 s, given that $\theta_0 = 16°C$ and $k = 0.03$

The rate of change of temperature is $\dfrac{d\theta}{dt}$

Since $\theta = \theta_0 e^{-kt}$

then $\dfrac{d\theta}{dt} = (\theta_0)(-k)e^{-kt} = -k\theta_0 e^{-kt}$

When $\theta_0 = 16, k = 0.03$ and $t = 40$

then $\dfrac{d\theta}{dt} = -(0.03)(16)e^{-(0.03)(40)}$

$$= -0.48e^{-1.2} = \textbf{-0.145}°\textbf{C/s}$$

Problem 4. The displacement s cm of the end of a stiff spring at time t seconds is given by $s = ae^{-kt} \sin 2\pi ft$. Determine the velocity of the end of the spring after 1 s, if $a = 2$, $k = 0.9$ and $f = 5$

Velocity, $v = \dfrac{ds}{dt}$ where $s = ae^{-kt} \sin 2\pi ft$ (i.e. a product).

Using the product rule,

$$\frac{ds}{dt} = (ae^{-kt})(2\pi f \cos 2\pi ft)$$

$$+ (\sin 2\pi ft)(-ake^{-kt})$$

When $a = 2$, $k = 0.9$, $f = 5$ and $t = 1$,

$$\textbf{velocity, } v = (2e^{-0.9})(2\pi 5 \cos 2\pi 5)$$

$$+ (\sin 2\pi 5)(-2)(0.9)e^{-0.9}$$

$$= 25.5455 \cos 10\pi - 0.7318 \sin 10\pi$$

$$= 25.5455(1) - 0.7318(0)$$

$$= \textbf{25.55 cm/s}$$

(Note that $\cos 10\pi$ means 'the cosine of 10π radians', *not* degrees, and $\cos 10\pi \equiv \cos 2\pi = 1$)

Now try the following Practice Exercise

Practice Exercise 133   Rates of change (Answers on page 876)

1. An alternating current, i amperes, is given by $i = 10 \sin 2\pi ft$, where f is the frequency in hertz and t the time in seconds. Determine the

rate of change of current when $t = 20$ ms, given that $f = 150$ Hz.

2. The luminous intensity, I candelas, of a lamp is given by $I = 6 \times 10^{-4} V^2$, where V is the voltage. Find (a) the rate of change of luminous intensity with voltage when $V = 200$ volts, and (b) the voltage at which the light is increasing at a rate of 0.3 candelas per volt.

3. The voltage across the plates of a capacitor at any time t seconds is given by $v = V e^{-t/CR}$, where V, C and R are constants.
    Given $V = 300$ volts, $C = 0.12 \times 10^{-6}$ F and $R = 4 \times 10^6 \, \Omega$ find (a) the initial rate of change of voltage, and (b) the rate of change of voltage after 0.5 s.

4. The pressure p of the atmosphere at height h above ground level is given by $p = p_0 e^{-h/c}$, where p_0 is the pressure at ground level and c is a constant. Determine the rate of change of pressure with height when $p_0 = 1.013 \times 10^5$ pascals and $c = 6.05 \times 10^4$ at 1450 metres.

5. The volume, v cubic metres, of water in a reservoir varies with time t, in minutes. When a valve is opened the relationship between v and t is given by: $v = 2 \times 10^4 - 20t^2 - 10t^3$. Calculate the rate of change of water volume at the time when $t = 3$ minutes.

Figure 30.1

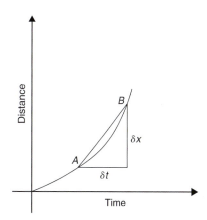

Figure 30.2

30.2 Velocity and acceleration

When a car moves a distance x metres in a time t seconds along a straight road, if the velocity v is constant then $v = \dfrac{x}{t}$ m/s, i.e. the gradient of the distance/time graph shown in Fig. 30.1 is constant.

If, however, the velocity of the car is not constant then the distance/time graph will not be a straight line. It may be as shown in Fig. 30.2.

The average velocity over a small time δt and distance δx is given by the gradient of the chord AB, i.e. the average velocity over time δt is $\dfrac{\delta x}{\delta t}$

As $\delta t \to 0$, the chord AB becomes a tangent, such that at point A, the velocity is given by:

$$v = \frac{dx}{dt}$$

Hence the velocity of the car at any instant is given by the gradient of the distance/time graph. If an expression for the distance x is known in terms of time t then the velocity is obtained by differentiating the expression.

The acceleration a of the car is defined as the rate of change of velocity. A velocity/time graph is shown in Fig. 30.3. If δv is the change in v and δt the corresponding change in time, then $a = \dfrac{\delta v}{\delta t}$

As $\delta t \to 0$, the chord CD becomes a tangent, such that at point C, the acceleration is given by:

$$a = \frac{dv}{dt}$$

Hence the acceleration of the car at any instant is given by the gradient of the velocity/time graph. If an expression for velocity is known in terms of time t then the acceleration is obtained by differentiating the expression.

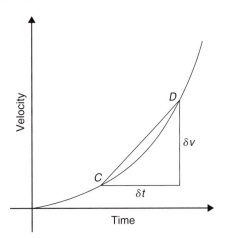

Figure 30.3

Acceleration $a = \dfrac{\mathrm{d}v}{\mathrm{d}t}$. However, $v = \dfrac{\mathrm{d}x}{\mathrm{d}t}$. Hence

$$a = \frac{\mathrm{d}}{\mathrm{d}t}\left(\frac{\mathrm{d}x}{\mathrm{d}t}\right) = \frac{\mathrm{d}^2 x}{\mathrm{d}x^2}$$

The acceleration is given by the second differential coefficient of distance x with respect to time t.

Summarising, if a body moves a distance x metres in a time t seconds then:

(i) **distance $x = f(t)$**

(ii) **velocity $v = f'(t)$ or $\dfrac{\mathrm{d}x}{\mathrm{d}t}$**, which is the gradient of the distance/time graph.

(iii) **acceleration $a = \dfrac{\mathrm{d}v}{\mathrm{d}t} = f''(t)$ or $\dfrac{\mathrm{d}^2 x}{\mathrm{d}t^2}$**, which is the gradient of the velocity/time graph.

> **Problem 5.** The distance x metres moved by a car in a time t seconds is given by $x = 3t^3 - 2t^2 + 4t - 1$. Determine the velocity and acceleration when (a) $t = 0$ and (b) $t = 1.5$ s.

Distance $\qquad x = 3t^3 - 2t^2 + 4t - 1\,\text{m}$

Velocity $\qquad v = \dfrac{\mathrm{d}x}{\mathrm{d}t} = 9t^2 - 4t + 4\,\text{m/s}$

Acceleration $\quad a = \dfrac{\mathrm{d}^2 x}{\mathrm{d}x^2} = 18t - 4\,\text{m/s}^2$

(a) When time $t = 0$,
velocity $v = 9(0)^2 - 4(0) + 4 = \mathbf{4\,m/s}$ and
acceleration $a = 18(0) - 4 = \mathbf{-4\,m/s^2}$ (i.e. a deceleration)

(b) When time $t = 1.5$ s,
velocity $v = 9(1.5)^2 - 4(1.5) + 4 = \mathbf{18.25\,m/s}$
and acceleration $a = 18(1.5) - 4 = \mathbf{23\,m/s^2}$

> **Problem 6.** Supplies are dropped from a helicoptor and the distance fallen in a time t seconds is given by $x = \frac{1}{2}gt^2$, where $g = 9.8\,\text{m/s}^2$. Determine the velocity and acceleration of the supplies after it has fallen for 2 seconds.

Distance $\qquad x = \dfrac{1}{2}gt^2 = \dfrac{1}{2}(9.8)t^2 = 4.9t^2\,\text{m}$

Velocity $\qquad v = \dfrac{\mathrm{d}v}{\mathrm{d}t} = 9.8t\,\text{m/s}$

and acceleration $\quad a = \dfrac{\mathrm{d}^2 x}{\mathrm{d}t^2} = 9.8\,\text{m/s}^2$

When time $t = 2$ s,
$\qquad$ **velocity, $v = (9.8)(2) = 19.6\,m/s$**

and **acceleration $a = 9.8\,m/s^2$**
(which is acceleration due to gravity).

> **Problem 7.** The distance x metres travelled by a vehicle in time t seconds after the brakes are applied is given by $x = 20t - \frac{5}{3}t^2$. Determine (a) the speed of the vehicle (in km/h) at the instant the brakes are applied, and (b) the distance the car travels before it stops.

(a) Distance, $x = 20t - \frac{5}{3}t^2$

$\quad$ Hence velocity $v = \dfrac{\mathrm{d}x}{\mathrm{d}t} = 20 - \dfrac{10}{3}t$

$\quad$ At the instant the brakes are applied, time $= 0$.

$\quad$ Hence **velocity, $v = 20\,m/s$**

$$= \frac{20 \times 60 \times 60}{1000}\,\text{km/h}$$

$$= \mathbf{72\,km/h}$$

$\quad$ (Note: changing from m/s to km/h merely involves multiplying by 3.6)

(b) When the car finally stops, the velocity is zero, i.e. $v = 20 - \dfrac{10}{3}t = 0$, from which, $20 = \dfrac{10}{3}t$, giving $t = 6$ s.

$\quad$ Hence the distance travelled before the car stops is given by:

$$x = 20t - \tfrac{5}{3}t^2 = 20(6) - \tfrac{5}{3}(6)^2$$

$$= 120 - 60 = \mathbf{60\,m}$$

Problem 8. The angular displacement θ radians of a flywheel varies with time t seconds and follows the equation $\theta = 9t^2 - 2t^3$. Determine (a) the angular velocity and acceleration of the flywheel when time, $t = 1\,$s, and (b) the time when the angular acceleration is zero.

(a) Angular displacement $\theta = 9t^2 - 2t^3$ rad

Angular velocity $\omega = \dfrac{d\theta}{dt} = 18t - 6t^2\,$rad/s

When time $t = 1\,$s,

$$\omega = 18(1) - 6(1)^2 = \mathbf{12\,rad/s}$$

Angular acceleration $\alpha = \dfrac{d^2\theta}{dt^2} = 18 - 12t\,$rad/s^2

When time $t = 1\,$s,

$$\alpha = 18 - 12(1) = \mathbf{6\,rad/s^2}$$

(b) When the angular acceleration is zero, $18 - 12t = 0$, from which, $18 = 12t$, giving time, $\mathbf{t = 1.5\,s}$

Problem 9. The displacement x cm of the slide valve of an engine is given by $x = 2.2\cos 5\pi t + 3.6\sin 5\pi t$. Evaluate the velocity (in m/s) when time $t = 30\,$ms.

Displacement $x = 2.2\cos 5\pi t + 3.6\sin 5\pi t$

Velocity $v = \dfrac{dx}{dt}$

$$= (2.2)(-5\pi)\sin 5\pi t + (3.6)(5\pi)\cos 5\pi t$$

$$= -11\pi\sin 5\pi t + 18\pi\cos 5\pi t\,\text{cm/s}$$

When time $t = 30\,$ms, velocity

$$= -11\pi\sin\left(5\pi\cdot\frac{30}{10^3}\right) + 18\pi\cos\left(5\pi\cdot\frac{30}{10^3}\right)$$

$$= -11\pi\sin 0.4712 + 18\pi\cos 0.4712$$

$$= -11\pi\sin 27° + 18\pi\cos 27°$$

$$= -15.69 + 50.39 = 34.7\,\text{cm/s}$$

$$= \mathbf{0.347\,m/s}$$

Now try the following Practice Exercise

Practice Exercise 134   Velocity and acceleration (Answers on page 876)

1. A missile fired from ground level rises x metres vertically upwards in t seconds and $x = 100t - \dfrac{25}{2}t^2$. Find (a) the initial velocity of the missile, (b) the time when the height of the missile is a maximum, (c) the maximum height reached, (d) the velocity with which the missile strikes the ground.

2. The distance s metres travelled by a car in t seconds after the brakes are applied is given by $s = 25t - 2.5t^2$. Find (a) the speed of the car (in km/h) when the brakes are applied, (b) the distance the car travels before it stops.

3. The equation $\theta = 10\pi + 24t - 3t^2$ gives the angle θ, in radians, through which a wheel turns in t seconds. Determine (a) the time the wheel takes to come to rest, (b) the angle turned through in the last second of movement.

4. At any time t seconds the distance x metres of a particle moving in a straight line from a fixed point is given by $x = 4t + \ln(1-t)$. Determine (a) the initial velocity and acceleration (b) the velocity and acceleration after 1.5 s (c) the time when the velocity is zero.

5. The angular displacement θ of a rotating disc is given by $\theta = 6\sin\dfrac{t}{4}$, where t is the time in seconds. Determine (a) the angular velocity of the disc when t is 1.5 s, (b) the angular acceleration when t is 5.5 s, and (c) the first time when the angular velocity is zero.

6. $x = \dfrac{20t^3}{3} - \dfrac{23t^2}{2} + 6t + 5$ represents the distance, x metres, moved by a body in t seconds. Determine (a) the velocity and acceleration at the start, (b) the velocity and acceleration when $t = 3\,$s, (c) the values of t when the body is at rest, (d) the value of t when the acceleration is $37\,$m/s^2 and (e) the distance travelled in the third second.

7. A particle has a displacement s given by $s = 30t + 27t^2 - 3t^3$ metres, where time t is in seconds. Determine the time at which the acceleration will be zero.

30.3 Turning points

In Fig. 30.4, the gradient (or rate of change) of the curve changes from positive between O and P to negative between P and Q, and then positive again between Q and R. At point P, the gradient is zero and, as x increases, the gradient of the curve changes from positive just before P to negative just after. Such a point is called a **maximum point** and appears as the 'crest of a wave'. At point Q, the gradient is also zero and, as x increases, the gradient of the curve changes from negative just before Q to positive just after. Such a point is called a **minimum point**, and appears as the 'bottom of a valley'. Points such as P and Q are given the general name of **turning points**.

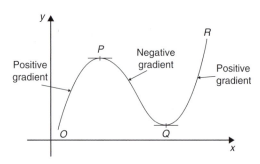

Figure 30.4

It is possible to have a turning point, the gradient on either side of which is the same. Such a point is given the special name of a **point of inflexion**, and examples are shown in Fig. 30.5.

Maximum and minimum points and points of inflexion are given the general term of **stationary points**.

Procedure for finding and distinguishing between stationary points:

(i) Given $y = f(x)$, determine $\dfrac{dy}{dx}$ (i.e. $f'(x)$)

(ii) Let $\dfrac{dy}{dx} = 0$ and solve for the values of x.

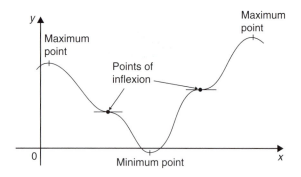

Figure 30.5

(iii) Substitute the values of x into the original equation, $y = f(x)$, to find the corresponding y-ordinate values. This establishes the co-ordinates of the stationary points.

To determine the nature of the stationary points:
Either

(iv) Find $\dfrac{d^2 y}{dx^2}$ and substitute into it the values of x found in (ii).

If the result is:
(a) positive – the point is a minimum one,
(b) negative – the point is a maximum one,
(c) zero – the point is a point of inflexion,

or

(v) Determine the sign of the gradient of the curve just before and just after the stationary points. If the sign change for the gradient of the curve is:
(a) positive to negative – the point is a maximum one,

(b) negative to positive – the point is a minimum one,

(c) positive to positive or negative to negative – the point is a point of inflexion.

For more on points of inflexion, see Section 30.5.

Problem 10. Locate the turning point on the curve $y = 3x^2 - 6x$ and determine its nature by examining the sign of the gradient on either side.

Following the above procedure:

(i) Since $y = 3x^2 - 6x$, $\dfrac{dy}{dx} = 6x - 6$

(ii) At a turning point, $\dfrac{dy}{dx} = 0$. Hence $6x - 6 = 0$, from which, $x = 1$

(iii)   When $x=1$, $y=3(1)^2-6(1)=-3$

Hence the co-ordinates of the turning point are (1, −3)

(iv)   If x is slightly less than 1, say, 0.9, then

$$\frac{dy}{dx}=6(0.9)-6=-0.6,$$

i.e. negative.
If x is slightly greater than 1, say, 1.1, then

$$\frac{dy}{dx}=6(1.1)-6=0.6,$$

i.e. positive.

Since the gradient of the curve is negative just before the turning point and positive just after (i.e. $-\vee+$), **(1, −3) is a minimum point**.

Problem 11.   Find the maximum and minimum values of the curve $y=x^3-3x+5$ by

(a)   examining the gradient on either side of the turning points, and

(b)   determining the sign of the second derivative.

Since $y=x^3-3x+5$ then $\dfrac{dy}{dx}=3x^2-3$

For a maximum or minimum value $\dfrac{dy}{dx}=0$

Hence $3x^2-3=0$, from which, $3x^2=3$ and $x=\pm1$

When $x=1$, $y=(1)^3-3(1)+5=3$

When $x=-1$, $y=(-1)^3-3(-1)+5=7$

Hence (1, 3) and $(-1,7)$ are the co-ordinates of the turning points.

(a)   Considering the point (1, 3):
If x is slightly less than 1, say 0.9, then

$$\frac{dy}{dx}=3(0.9)^2-3,$$

which is negative.
If x is slightly more than 1, say 1.1, then

$$\frac{dy}{dx}=3(1.1)^2-3,$$

which is positive.
Since the gradient changes from negative to positive, **the point (1, 3) is a minimum point**.

Considering the point $(-1, 7)$:

If x is slightly less than -1, say -1.1, then

$$\frac{dy}{dx}=3(-1.1)^2-3,$$

which is positive.

If x is slightly more than -1, say -0.9, then

$$\frac{dy}{dx}=3(-0.9)^2-3,$$

which is negative.

Since the gradient changes from positive to negative, **the point (−1, 7) is a maximum point**.

(b)   Since $\dfrac{dy}{dx}=3x^2-3$, then $\dfrac{d^2y}{dx^2}=6x$

When $x=1$, $\dfrac{d^2y}{dx^2}$ is positive, hence (1, 3) is a **minimum value**.

When $x=-1$, $\dfrac{d^2y}{dx^2}$ is negative, hence $(-1,7)$ is a **maximum value**.

Thus the maximum value is 7 and the minimum value is 3

It can be seen that the second differential method of determining the nature of the turning points is, in this case, quicker than investigating the gradient.

Problem 12.   Locate the turning point on the following curve and determine whether it is a maximum or minimum point: $y=4\theta+e^{-\theta}$

Since $y=4\theta+e^{-\theta}$

then $\dfrac{dy}{d\theta}=4-e^{-\theta}=0$

for a maximum or minimum value.

Hence $4=e^{-\theta}$, $\frac{1}{4}=e^{\theta}$, giving $\theta=\ln\frac{1}{4}=-1.3863$ (see Chapter 4).

When $\theta=-1.3863$, $y=4(-1.3863)+e^{-(-1.3863)}$
$\qquad\qquad\qquad\qquad\quad=5.5452+4.0000=-1.5452$

Thus $(-1.3863, -1.5452)$ are the co-ordinates of the turning point.

$$\frac{d^2y}{d\theta^2}=e^{-\theta}$$

When $\theta=-1.3863$,

$$\frac{d^2y}{d\theta^2}=e^{+1.3863}=4.0,$$

which is positive, hence **(−1.3863, −1.5452) is a minimum point**.

> **Problem 13.** Determine the co-ordinates of the maximum and minimum values of the graph $y = \dfrac{x^3}{3} - \dfrac{x^2}{2} - 6x + \dfrac{5}{3}$ and distinguish between them. Sketch the graph.

Following the given procedure:

(i) Since $y = \dfrac{x^3}{3} - \dfrac{x^2}{2} - 6x + \dfrac{5}{3}$ then

$$\frac{dy}{dx} = x^2 - x - 6$$

(ii) At a turning point, $\dfrac{dy}{dx} = 0$. Hence

$$x^2 - x - 6 = 0, \text{ i.e. } (x+2)(x-3) = 0,$$

from which $x = -2$ or $x = 3$

(iii) When $x = -2$,

$$y = \frac{(-2)^3}{3} - \frac{(-2)^2}{2} - 6(-2) + \frac{5}{3} = 9$$

When $x = 3$,

$$y = \frac{(3)^3}{3} - \frac{(3)^2}{2} - 6(3) + \frac{5}{3} = -11\frac{5}{6}$$

Thus the co-ordinates of the turning points are $(-2, 9)$ and $\left(3, -11\frac{5}{6}\right)$

(iv) Since $\dfrac{dy}{dx} = x^2 - x - 6$ then $\dfrac{d^2y}{dx^2} = 2x - 1$

When $x = -2$,

$$\frac{d^2y}{dx^2} = 2(-2) - 1 = -5,$$

which is negative.

Hence $(-2, 9)$ is a maximum point.

When $x = 3$,

$$\frac{d^2y}{dx^2} = 2(3) - 1 = 5,$$

which is positive.

Hence $\left(3, -11\frac{5}{6}\right)$ is a minimum point.

Knowing $(-2, 9)$ is a maximum point (i.e. crest of a wave), and $\left(3, -11\frac{5}{6}\right)$ is a minimum point (i.e. bottom of a valley) and that when $x = 0$, $y = \frac{5}{3}$, a sketch may be drawn as shown in Fig. 30.6.

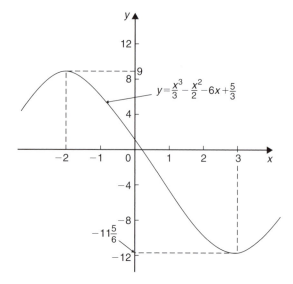

Figure 30.6

> **Problem 14.** Determine the turning points on the curve $y = 4\sin x - 3\cos x$ in the range $x = 0$ to $x = 2\pi$ radians, and distinguish between them. Sketch the curve over one cycle.

Since $y = 4\sin x - 3\cos x$

then $\dfrac{dy}{dx} = 4\cos x + 3\sin x = 0$,

for a turning point, from which,

$4\cos x = -3\sin x$ and

$$\frac{-4}{3} = \frac{\sin x}{\cos x} = \tan x$$

Hence $x = \tan^{-1}\left(\dfrac{-4}{3}\right) = 126.87°$ or $306.87°$, since tangent is negative in the second and fourth quadrants.

When $x = 126.87°$,

$$y = 4\sin 126.87° - 3\cos 126.87° = 5$$

When $x = 306.87°$,

$$y = 4\sin 306.87° - 3\cos 306.87° = -5$$

$126.87° = \left(126.87° \times \dfrac{\pi}{180}\right)$ radians

$\qquad = 2.214\,\text{rad}$

$306.87° = \left(306.87° \times \dfrac{\pi}{180}\right)$ radians

$\qquad = 5.356\,\text{rad}$

Hence (2.214, 5) and (5.356, −5) are the co-ordinates of the turning points.

$$\frac{d^2y}{dx^2} = -4\sin x + 3\cos x$$

When $x = 2.214$ rad,

$$\frac{d^2y}{dx^2} = -4\sin 2.214 + 3\cos 2.214,$$

which is negative.

Hence (2.214, 5) is a maximum point.

When $x = 5.356$ rad,

$$\frac{d^2y}{dx^2} = -4\sin 5.356 + 3\cos 5.356,$$

which is positive.

Hence (5.356, −5) is a minimum point.

A sketch of $y = 4\sin x - 3\cos x$ is shown in Fig. 30.7.

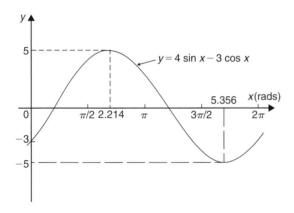

Figure 30.7

Now try the following Practice Exercise

Practice Exercise 135   Turning points (Answers on page 876)

In Problems 1 to 11, find the turning points and distinguish between them.

1. $y = x^2 - 6x$

2. $y = 8 + 2x - x^2$

3. $y = x^2 - 4x + 3$

4. $y = 3 + 3x^2 - x^3$

5. $y = 3x^2 - 4x + 2$

6. $x = \theta(6 - \theta)$

7. $y = 4x^3 + 3x^2 - 60x - 12$

8. $y = 5x - 2\ln x$

9. $y = 2x - e^x$

10. $y = t^3 - \dfrac{t^2}{2} - 2t + 4$

11. $x = 8t + \dfrac{1}{2t^2}$

12. Determine the maximum and minimum values on the graph $y = 12\cos\theta - 5\sin\theta$ in the range $\theta = 0$ to $\theta = 360°$. Sketch the graph over one cycle showing relevant points.

13. Show that the curve $y = \frac{2}{3}(t-1)^3 + 2t(t-2)$ has a maximum value of $\frac{2}{3}$ and a minimum value of -2

30.4 Practical problems involving maximum and minimum values

There are many **practical problems** involving maximum and minimum values which occur in science and engineering. Usually, an equation has to be determined from given data, and rearranged where necessary, so that it contains only one variable. Some examples are demonstrated in Problems 15 to 20.

Problem 15.   A rectangular area is formed having a perimeter of 40 cm. Determine the length and breadth of the rectangle if it is to enclose the maximum possible area.

Let the dimensions of the rectangle be x and y. Then the perimeter of the rectangle is $(2x + 2y)$. Hence

$$2x + 2y = 40$$

or $x + y = 20$ (1)

Since the rectangle is to enclose the maximum possible area, a formula for area A must be obtained in terms of one variable only.

Area $A = xy$. From equation (1), $x = 20 - y$

Hence, area $A = (20 - y)y = 20y - y^2$

$$\frac{dA}{dy} = 20 - 2y = 0$$

for a turning point, from which, $y = 10$ cm

$$\frac{d^2 A}{dy^2} = -2,$$

which is negative, giving a maximum point.
When $y = 10$ cm, $x = 10$ cm, from equation (1)
Hence the length and breadth of the rectangle are each 10 cm, i.e. a square gives the maximum possible area. When the perimeter of a rectangle is 40 cm, the maximum possible area is $10 \times 10 = \mathbf{100\, cm^2}$.

> **Problem 16.** A rectangular sheet of metal having dimensions 20 cm by 12 cm has squares removed from each of the four corners and the sides bent upwards to form an open box. Determine the maximum possible volume of the box.

The squares to be removed from each corner are shown in Fig. 30.8, having sides x cm. When the sides are bent upwards the dimensions of the box will be:
length $(20 - 2x)$ cm, breadth $(12 - 2x)$ cm and height x cm.

Figure 30.8

Volume of box,

$$V = (20 - 2x)(12 - 2x)(x)$$

$$= 240x - 64x^2 + 4x^3$$

$$\frac{dV}{dx} = 240 - 128x + 12x^2 = 0$$

for a turning point.

Hence $4(60 - 32x + 3x^2) = 0,$

i.e. $3x^2 - 32x + 60 = 0$

Using the quadratic formula,

$$x = \frac{32 \pm \sqrt{(-32)^2 - 4(3)(60)}}{2(3)}$$

$$= 8.239\, \text{cm or } 2.427\, \text{cm}.$$

Since the breadth is $(12 - 2x)$ cm then $x = 8.239$ cm is not possible and is neglected. Hence $x = 2.427$ cm

$$\frac{d^2 V}{dx^2} = -128 + 24x$$

When $x = 2.427$, $\dfrac{d^2 V}{dx^2}$ is negative, giving a maximum value.

The dimensions of the box are:

$$\text{length} = 20 - 2(2.427) = 15.146\, \text{cm},$$

$$\text{breadth} = 12 - 2(2.427) = 7.146\, \text{cm},$$

and height $= 2.427$ cm

$$\mathbf{Maximum\ volume} = (15.146)(7.146)(2.427)$$

$$= \mathbf{262.7\, cm^3}$$

> **Problem 17.** Determine the height and radius of a cylinder of volume $200\, \text{cm}^3$ which has the least surface area.

Let the cylinder have radius r and perpendicular height h.
Volume of cylinder,

$$V = \pi r^2 h = 200 \tag{1}$$

Surface area of cylinder,

$$A = 2\pi r h + 2\pi r^2$$

Least surface area means minimum surface area and a formula for the surface area in terms of one variable only is required.
From equation (1),

$$h = \frac{200}{\pi r^2} \tag{2}$$

Hence surface area,

$$A = 2\pi r \left(\frac{200}{\pi r^2} \right) + 2\pi r^2$$

$$= \frac{400}{r} + 2\pi r^2 = 400r^{-1} + 2\pi r^2$$

$$\frac{dA}{dr} = \frac{-400}{r^2} + 4\pi r = 0,$$

for a turning point.

Hence $4\pi r = \dfrac{400}{r^2}$ and $r^3 = \dfrac{400}{4\pi}$

from which,

$$r = \sqrt[3]{\left(\dfrac{100}{\pi}\right)} = 3.169\,\text{cm}$$

$$\dfrac{d^2 A}{dr^2} = \dfrac{800}{r^3} + 4\pi$$

When $r = 3.169\,\text{cm}$, $\dfrac{d^2 A}{dr^2}$ is positive, giving a minimum value.
From equation (2),

when $r = 3.169\,\text{cm}$,

$$h = \dfrac{200}{\pi(3.169)^2} = 6.339\,\text{cm}$$

Hence for the least surface area, a cylinder of volume $200\,\text{cm}^3$ has a radius of $3.169\,\text{cm}$ and height of $6.339\,\text{cm}$.

> **Problem 18.**   Determine the area of the largest piece of rectangular ground that can be enclosed by $100\,\text{m}$ of fencing, if part of an existing straight wall is used as one side.

Let the dimensions of the rectangle be x and y as shown in Fig. 30.9, where PQ represents the straight wall.

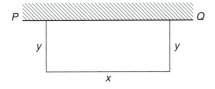

Figure 30.9

From Fig. 30.9,

$$x + 2y = 100 \tag{1}$$

Area of rectangle,

$$A = xy \tag{2}$$

Since the maximum area is required, a formula for area A is needed in terms of one variable only.
From equation (1), $x = 100 - 2y$
Hence area $A = xy = (100 - 2y)y = 100y - 2y^2$

$$\dfrac{dA}{dy} = 100 - 4y = 0,$$

for a turning point, from which, $y = 25\,\text{m}$

$$\dfrac{d^2 A}{dy^2} = -4,$$

which is negative, giving a maximum value.
When $y = 25\,\text{m}$, $x = 50\,\text{m}$ from equation (1).
Hence the **maximum possible area**
$$= xy = (50)(25) = \mathbf{1250\,m^2}$$

> **Problem 19.**   An open rectangular box with square ends is fitted with an overlapping lid which covers the top and the front face. Determine the maximum volume of the box if $6\,\text{m}^2$ of metal are used in its construction.

A rectangular box having square ends of side x and length y is shown in Fig. 30.10.

Figure 30.10

Surface area of box, A, consists of two ends and five faces (since the lid also covers the front face).
Hence

$$A = 2x^2 + 5xy = 6 \tag{1}$$

Since it is the maximum volume required, a formula for the volume in terms of one variable only is needed.
Volume of box, $V = x^2 y$
From equation (1),

$$y = \dfrac{6 - 2x^2}{5x} = \dfrac{6}{5x} - \dfrac{2x}{5} \tag{2}$$

Hence volume

$$V = x^2 y = x^2\left(\dfrac{6}{5x} - \dfrac{2x}{5}\right) = \dfrac{6x}{5} - \dfrac{2x^3}{5}$$

$$\dfrac{dV}{dx} = \dfrac{6}{5} - \dfrac{6x^2}{5} = 0$$

for a maximum or minimum value.
Hence $6 = 6x^2$, giving $x = 1\,\text{m}$ ($x = -1$ is not possible, and is thus neglected).

$$\dfrac{d^2 V}{dx^2} = \dfrac{-12x}{5}$$

When $x = 1$, $\dfrac{d^2 V}{dx^2}$ is negative, giving a maximum value.
From equation (2), when $x = 1$,

$$y = \frac{6}{5(1)} - \frac{2(1)}{5} = \frac{4}{5}$$

Hence the maximum volume of the box is given by

$$V = x^2 y = (1)^2 \left(\tfrac{4}{5}\right) = \tfrac{4}{5}\,\textbf{m}^3$$

Problem 20. Find the diameter and height of a cylinder of maximum volume which can be cut from a sphere of radius 12 cm.

A cylinder of radius r and height h is shown enclosed in a sphere of radius $R = 12$ cm in Fig. 30.11.
Volume of cylinder,

$$V = \pi r^2 h \qquad (1)$$

Using the right-angled triangle OPQ shown in Fig. 30.11,

$$r^2 + \left(\frac{h}{2}\right)^2 = R^2 \quad \text{by Pythagoras' theorem,}$$

i.e. $\qquad r^2 + \dfrac{h^2}{4} = 144 \qquad (2)$

Since the maximum volume is required, a formula for the volume V is needed in terms of one variable only. From equation (2),

$$r^2 = 144 - \frac{h^2}{4}$$

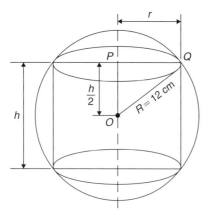

Figure 30.11

Substituting into equation (1) gives:

$$V = \pi \left(144 - \frac{h^2}{4} \right) h = 144\pi h - \frac{\pi h^3}{4}$$

$$\frac{dV}{dh} = 144\pi - \frac{3\pi h^2}{4} = 0,$$

for a maximum or minimum value.
Hence

$$144\pi = \frac{3\pi h^2}{4}$$

from which, $\qquad h = \sqrt{\dfrac{(144)(4)}{3}} = 13.86\,\text{cm}$

$$\frac{d^2 V}{dh^2} = \frac{-6\pi h}{4}$$

When $h = 13.86$, $\dfrac{d^2 V}{dh^2}$ is negative, giving a maximum value.
From equation (2),

$$r^2 = 144 - \frac{h^2}{4} = 144 - \frac{13.86^2}{4}$$

from which, radius $r = 9.80\,\text{cm}$

Diameter of cylinder $= 2r = 2(9.80) = 19.60\,\text{cm}$.

Hence the cylinder having the maximum volume that can be cut from a sphere of radius 12 cm is one in which the diameter is 19.60 cm and the height is 13.86 cm.

Now try the following Practice Exercise

Practice Exercise 136   Practical maximum and minimum problems (Answers on page 876)

1. The speed, v, of a car (in m/s) is related to time t s by the equation $v = 3 + 12t - 3t^2$. Determine the maximum speed of the car in km/h.

2. Determine the maximum area of a rectangular piece of land that can be enclosed by 1200 m of fencing.

3. A shell is fired vertically upwards and its vertical height, x metres, is given by $x = 24t - 3t^2$, where t is the time in seconds. Determine the maximum height reached.

4. A lidless box with square ends is to be made from a thin sheet of metal. Determine the least area of the metal for which the volume of the box is $3.5\,\text{m}^3$.

5. A closed cylindrical container has a surface area of $400\,\text{cm}^2$. Determine the dimensions for maximum volume.

6. Calculate the height of a cylinder of maximum volume which can be cut from a cone of height $20\,\text{cm}$ and base radius $80\,\text{cm}$.

7. The power developed in a resistor R by a battery of emf E and internal resistance r is given by $P = \dfrac{E^2 R}{(R+r)^2}$. Differentiate P with respect to R and show that the power is a maximum when $R = r$.

8. Find the height and radius of a closed cylinder of volume $125\,\text{cm}^3$ which has the least surface area.

9. Resistance to motion, F, of a moving vehicle is given by $F = \dfrac{5}{x} + 100x$. Determine the minimum value of resistance.

10. An electrical voltage E is given by
$E = (15\sin 50\pi t + 40\cos 50\pi t)$ volts,
where t is the time in seconds. Determine the maximum value of voltage.

11. The fuel economy E of a car, in miles per gallon, is given by:

$$E = 21 + 2.10 \times 10^{-2} v^2$$
$$- 3.80 \times 10^{-6} v^4$$

where v is the speed of the car in miles per hour.
Determine, correct to 3 significant figures, the most economical fuel consumption, and the speed at which it is achieved.

12. The horizontal range of a projectile, x, launched with velocity u at an angle θ to the horizontal is given by: $x = \dfrac{2u^2 \sin\theta \cos\theta}{g}$
To achieve maximum horizontal range, determine the angle the projectile should be launched at.

13. The signalling range, x, of a submarine cable is given by the formula: $x = r^2 \ln\left(\dfrac{1}{r}\right)$ where r is the ratio of the radii of the conductor and cable. Determine the value of r for maximum range.

30.5 Points of inflexion

As mentioned earlier in the chapter, it is possible to have a turning point, the gradient on either side of which is the same. This is called a **point of inflexion**.
Procedure to determine points of inflexion:

(i) Given $y = f(x)$, determine $\dfrac{dy}{dx}$ and $\dfrac{d^2 y}{dx^2}$

(ii) Solve the equation $\dfrac{d^2 y}{dx^2} = 0$

(iii) Test whether there is a change of sign occurring in $\dfrac{d^2 y}{dx^2}$. This is achieved by substituting into the expression for $\dfrac{d^2 y}{dx^2}$ firstly a value of x just less than the solution and then a value just greater than the solution.

(iv) A point of inflexion has been found if $\dfrac{d^2 y}{dx^2} = 0$ **and** there is a change of sign.

This procedure is demonstrated in the following worked Problems.

Problem 21. Determine the point(s) of inflexion (if any) on the graph of the function

$$y = x^3 - 6x^2 + 9x + 5$$

Find also any other turning points. Hence, sketch the graph.

Using the above procedure:

(i) Given $y = x^3 - 6x^2 + 9x + 5$,
$\dfrac{dy}{dx} = 3x^2 - 12x + 9$ and $\dfrac{d^2 y}{dx^2} = 6x - 12$

(ii) Solving the equation $\dfrac{d^2 y}{dx^2} = 0$ gives: $6x - 12 = 0$
from which, $6x = 12$ and $x = 2$
Hence, if there is a point of inflexion, it occurs at $x = 2$

(iii) Taking a value just less than 2, say, 1.9:
$\dfrac{d^2y}{dx^2} = 6x - 12 = 6(1.9) - 12$, which is negative.

Taking a value just greater than 2, say, 2.1:
$\dfrac{d^2y}{dx^2} = 6x - 12 = 6(2.1) - 12$, which is positive.

(iv) Since a change of sign has occurred a point of inflexion exists at $x = 2$

When $x = 2$, $y = 2^3 - 6(2)^2 + 9(2) + 5 = 7$

i.e. **a point of inflexion occurs at the co-ordinates (2, 7)**

From above, $\dfrac{dy}{dx} = 3x^2 - 12x + 9 = 0$ for a turning point

i.e. $x^2 - 4x + 3 = 0$

Using the quadratic formula or factorising (or by calculator), the solution is: $x = 1$ or $x = 3$

Since $y = x^3 - 6x^2 + 9x + 5$, then when $x = 1$, $y = 1^3 - 6(1)^2 + 9(1) + 5 = 9$

and when $x = 3$, $y = 3^3 - 6(3)^2 + 9(3) + 5 = 5$

Hence, there are turning points at $(1, 9)$ and at $(3, 5)$

Since $\dfrac{d^2y}{dx^2} = 6x - 12$, when $x = 1$, $\dfrac{d^2y}{dx^2} = 6(1) - 12$ which is negative – hence a maximum point

and when $x = 3$, $\dfrac{d^2y}{dx^2} = 6(3) - 12$ which is positive – hence a minimum point.

Thus, **(1, 9) is a maximum point and (3, 5) is a minimum point**.

A sketch of the graph $y = x^3 - 6x^2 + 9x + 5$ is shown in Figure 30.12

Problem 22. Determine the point(s) of inflexion (if any) on the graph of the function
$$y = x^4 - 24x^2 + 5x + 60$$

Using the above procedure:

(i) Given $y = x^4 - 24x^2 + 5x + 60$,
$\dfrac{dy}{dx} = 4x^3 - 48x + 5$ and $\dfrac{d^2y}{dx^2} = 12x^2 - 48$

(ii) Solving the equation $\dfrac{d^2y}{dx^2} = 0$ gives:
$12x^2 - 48 = 0$ from which, $12x^2 = 48$
and $x^2 = 4$ from which, $x = \sqrt{4} = \pm 2$
Hence, if there are points of inflexion, they occur at $x = 2$ and at $x = -2$

(iii) Taking a value just less than 2, say, 1.9:
$\dfrac{d^2y}{dx^2} = 12x^2 - 48 = 12(1.9)^2 - 48$, which is negative.

Taking a value just greater than 2, say, 2.1: $\dfrac{d^2y}{dx^2} = 12x^2 - 48 = 12(2.1)^2 - 48$, which is positive.

Taking a value just less than -2, say, -2.1:
$\dfrac{d^2y}{dx^2} = 12x^2 - 48 = 12(-2.1)^2 - 48$, which is positive.

Taking a value just greater than -2, say, -1.9:
$\dfrac{d^2y}{dx^2} = 12x^2 - 48 = 12(-1.9)^2 - 48$, which is negative.

(iv) Since changes of signs have occurred, points of inflexion exist at $x = 2$ and $x = -2$

When $x = 2$, $y = 2^4 - 24(2)^2 + 5(2) + 60 = -10$

When $x = -2$,
$y = (-2)^4 - 24(-2)^2 + 5(-2) + 60 = -30$

i.e. **points of inflexion occur at the co-ordinates (2, -10) and at (-2, -30)**

Now try the following Practice Exercise

Practice Exercise 137   Further problems on points of inflexion (Answers on page 877)

1. Find the points of inflexion (if any) on the graph of the function
$$y = \frac{1}{3}x^3 - \frac{1}{2}x^2 - 2x + \frac{1}{12}$$

2. Find the points of inflexion (if any) on the graph of the function
$$y = 4x^3 + 3x^2 - 18x - \frac{5}{8}$$

3. Find the point(s) of inflexion on the graph of the function $y = x + \sin x$ for $0 < x < 2\pi$

4. Find the point(s) of inflexion on the graph of the function $y = 3x^3 - 27x^2 + 15x + 17$

5. Find the point(s) of inflexion on the graph of the function $y = 2xe^{-x}$

6. The displacement, s, of a particle is given by: $s = 3t^3 - 9t^2 + 10$. Determine the maximum, minimum and point of inflexion of s.

Figure 30.12

30.6 Tangents and normals

Tangents

The equation of the tangent to a curve $y = f(x)$ at the point (x_1, y_1) is given by:

$$y - y_1 = m(x - x_1)$$

where $m = \dfrac{dy}{dx} = $ gradient of the curve at (x_1, y_1)

> **Problem 23.** Find the equation of the tangent to the curve $y = x^2 - x - 2$ at the point $(1, -2)$

Gradient, m

$$= \frac{dy}{dx} = 2x - 1$$

At the point $(1, -2)$, $x = 1$ and $m = 2(1) - 1 = 1$
Hence the equation of the tangent is:

$$y - y_1 = m(x - x_1)$$
i.e. $\quad y - (-2) = 1(x - 1)$
i.e. $\qquad y + 2 = x - 1$
or $\qquad\quad \boldsymbol{y = x - 3}$

The graph of $y = x^2 - x - 2$ is shown in Fig. 30.13. The line AB is the tangent to the curve at the point C, i.e. $(1, -2)$, and the equation of this line is $y = x - 3$

Normals

The normal at any point on a curve is the line which passes through the point and is at right angles to the tangent. Hence, in Fig. 30.13, the line CD is the normal.

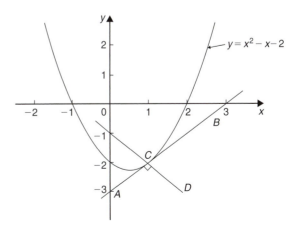

Figure 30.13

It may be shown that if two lines are at right angles then the product of their gradients is -1. Thus if m is the gradient of the tangent, then the gradient of the normal is $-\dfrac{1}{m}$

Hence the equation of the normal at the point (x_1, y_1) is given by:

$$y - y_1 = -\frac{1}{m}(x - x_1)$$

Problem 24. Find the equation of the normal to the curve $y = x^2 - x - 2$ at the point $(1, -2)$

$m = 1$ from Problem 21, hence the equation of the normal is

$$y - y_1 = -\frac{1}{m}(x - x_1)$$

i.e. $\quad y - (-2) = -\dfrac{1}{1}(x - 1)$

i.e. $\qquad y + 2 = -x + 1$

or $\qquad y = -x - 1$

Thus the line CD in Fig. 30.13 has the equation $y = -x - 1$

Problem 25. Determine the equations of the tangent and normal to the curve $y = \dfrac{x^3}{5}$ at the point $\left(-1, -\dfrac{1}{5}\right)$

Gradient m of curve $y = \dfrac{x^3}{5}$ is given by

$$m = \frac{dy}{dx} = \frac{3x^2}{5}$$

At the point $\left(-1, -\frac{1}{5}\right)$, $x = -1$ and $m = \dfrac{3(-1)^2}{5} = \dfrac{3}{5}$

Equation of the tangent is:

$$y - y_1 = m(x - x_1)$$

i.e. $\quad y - \left(-\dfrac{1}{5}\right) = \dfrac{3}{5}(x - (-1))$

i.e. $\qquad y + \dfrac{1}{5} = \dfrac{3}{5}(x + 1)$

or $\qquad 5y + 1 = 3x + 3$

or $\qquad \boldsymbol{5y - 3x = 2}$

Equation of the normal is:

$$y - y_1 = -\frac{1}{m}(x - x_1)$$

i.e. $\quad y - \left(-\dfrac{1}{5}\right) = \dfrac{-1}{(3/5)}(x - (-1))$

i.e. $\qquad y + \dfrac{1}{5} = -\dfrac{5}{3}(x + 1)$

i.e. $\qquad y + \dfrac{1}{5} = -\dfrac{5}{3}x - \dfrac{5}{3}$

Multiplying each term by 15 gives:

$$15y + 3 = -25x - 25$$

Hence **equation of the normal** is:

$$\boldsymbol{15y + 25x + 28 = 0}$$

Now try the following Practice Exercise

Practice Exercise 138 Tangents and normals (Answers on page 877)

For the curves in problems 1 to 5, at the points given, find (a) the equation of the tangent, and (b) the equation of the normal.

1. $y = 2x^2$ at the point $(1, 2)$

2. $y = 3x^2 - 2x$ at the point $(2, 8)$

3. $y = \dfrac{x^3}{2}$ at the point $\left(-1, -\dfrac{1}{2}\right)$

4. $y = 1 + x - x^2$ at the point $(-2, -5)$

5. $\theta = \dfrac{1}{t}$ at the point $\left(3, \dfrac{1}{3}\right)$

30.7 Small changes

If y is a function of x, i.e. $y = f(x)$, and the approximate change in y corresponding to a small change δx in x is required, then:

$$\frac{\delta y}{\delta x} \approx \frac{dy}{dx}$$

and $\delta y \approx \dfrac{dy}{dx} \cdot \delta x$ or $\delta y \approx f'(x) \cdot \delta x$

Problem 26.   Given $y = 4x^2 - x$, determine the approximate change in y if x changes from 1 to 1.02

Since $y = 4x^2 - x$, then

$$\frac{dy}{dx} = 8x - 1$$

Approximate change in y,

$$\delta y \approx \frac{dy}{dx} \cdot \delta x \approx (8x - 1)\delta x$$

When $x = 1$ and $\delta x = 0.02$, $\delta y \approx [8(1) - 1](0.02)$

$$\approx 0.14$$

[Obviously, in this case, the exact value of dy may be obtained by evaluating y when $x = 1.02$, i.e. $y = 4(1.02)^2 - 1.02 = 3.1416$ and then subtracting from it the value of y when $x = 1$, i.e. $y = 4(1)^2 - 1 = 3$, giving $\delta y = 3.1416 - 3 = 0.1416$

Using $\delta y = \frac{dy}{dx} \cdot \delta x$ above gave 0.14, which shows that the formula gives the approximate change in y for a small change in x]

Problem 27.   The time of swing T of a pendulum is given by $T = k\sqrt{l}$, where k is a constant. Determine the percentage change in the time of swing if the length of the pendulum l changes from 32.1 cm to 32.0 cm.

If $T = k\sqrt{l} = kl^{\frac{1}{2}}$, then

$$\frac{dT}{dl} = k\left(\frac{1}{2}l^{\frac{-1}{2}}\right) = \frac{k}{2\sqrt{l}}$$

Approximate change in T,

$$\delta t \approx \frac{dT}{dl}\delta l \approx \left(\frac{k}{2\sqrt{l}}\right)\delta l$$

$$\approx \left(\frac{k}{2\sqrt{l}}\right)(-0.1)$$

(negative since l decreases)

Percentage error

$$= \left(\frac{\text{approximate change in } T}{\text{original value of } T}\right)100\%$$

$$= \frac{\left(\dfrac{k}{2\sqrt{l}}\right)(-0.1)}{k\sqrt{l}} \times 100\%$$

$$= \left(\frac{-0.1}{2l}\right)100\% = \left(\frac{-0.1}{2(32.1)}\right)100\%$$

$$= -0.156\%$$

Hence the change in the time of swing is a decrease of 0.156%

Problem 28.   A circular template has a radius of 10 cm (± 0.02). Determine the possible error in calculating the area of the template. Find also the percentage error.

Area of circular template, $A = \pi r^2$, hence

$$\frac{dA}{dr} = 2\pi r$$

Approximate change in area,

$$\delta A \approx \frac{dA}{dr} \cdot \delta r \approx (2\pi r)\delta r$$

When $r = 10$ cm and $\delta r = 0.02$,

$$\delta A = (2\pi 10)(0.02) \approx 0.4\pi \text{ cm}^2$$

i.e. **the possible error in calculating the template area is approximately 1.257 cm^2.**

$$\textbf{Percentage error} \approx \left(\frac{0.4\pi}{\pi(10)^2}\right)100\%$$

$$= \mathbf{0.40\%}$$

Now try the following Practice Exercise

Practice Exercise 139   Small changes (Answers on page 877)

1. Determine the change in y if x changes from 2.50 to 2.51 when
    (a) $y = 2x - x^2$ (b) $y = \dfrac{5}{x}$

2. The pressure p and volume v of a mass of gas are related by the equation $pv = 50$. If the pressure increases from 25.0 to 25.4, determine the approximate change in the volume of the gas.

Find also the percentage change in the volume of the gas.

3. Determine the approximate increase in (a) the volume, and (b) the surface area of a cube of side x cm if x increases from 20.0 cm to 20.05 cm.

4. The radius of a sphere decreases from 6.0 cm to 5.96 cm. Determine the approximate change in (a) the surface area, and (b) the volume.

5. The rate of flow of a liquid through a tube is given by Poiseuilles' equation as:

$Q = \dfrac{p\pi r^4}{8\eta L}$ where Q is the rate of flow, p is the pressure difference between the ends of the tube, r is the radius of the tube, L is the length of the tube and η is the coefficient of viscosity of the liquid. η is obtained by measuring Q, p, r and L. If Q can be measured accurate to $\pm 0.5\%$, p accurate to $\pm 3\%$, r accurate to $\pm 2\%$ and L accurate to $\pm 1\%$, calculate the maximum possible percentage error in the value of η.

For fully worked solutions to each of the problems in Practice Exercises 133 to 139 in this chapter, go to the website: www.routledge.com/cw/bird

This Revision Test covers the material contained in Chapters 29 and 30. *The marks for each question are shown in brackets at the end of each question.*

1. Differentiate the following with respect to the variable:

   (a) $y = 5 + 2\sqrt{x^3} - \dfrac{1}{x^2}$ (b) $s = 4e^{2\theta}\sin 3\theta$

   (c) $y = \dfrac{3\ln 5t}{\cos 2t}$

   (d) $x = \dfrac{2}{\sqrt{(t^2 - 3t + 5)}}$ (15)

2. If $f(x) = 2.5x^2 - 6x + 2$ find the co-ordinates at the point at which the gradient is -1 (5)

3. The displacement s cm of the end of a stiff spring at time t seconds is given by:

   $s = ae^{-kt}\sin 2\pi f t$. Determine the velocity and acceleration of the end of the spring after two seconds if $a = 3$, $k = 0.75$ and $f = 20$ (10)

4. Find the co-ordinates of the turning points on the curve $y = 3x^3 + 6x^2 + 3x - 1$ and distinguish between them. Find also the point(s) of inflexion.   (14)

5. The heat capacity C of a gas varies with absolute temperature θ as shown:

   $C = 26.50 + 7.20 \times 10^{-3}\theta - 1.20 \times 10^{-6}\theta^2$

   Determine the maximum value of C and the temperature at which it occurs.   (7)

6. Determine for the curve $y = 2x^2 - 3x$ at the point $(2, 2)$: (a) the equation of the tangent (b) the equation of the normal.   (7)

7. A rectangular block of metal with a square cross-section has a total surface area of $250\,\text{cm}^2$. Find the maximum volume of the block of metal.   (7)

Chapter 31

Standard integration

Why it is important to understand: Standard integration

Engineering is all about problem solving and many problems in engineering can be solved using calculus. Physicists, chemists, engineers, and many other scientific and technical specialists use calculus in their everyday work; it is a technique of fundamental importance. Both integration and differentiation have numerous applications in engineering and science and some typical examples include determining areas, mean and rms values, volumes of solids of revolution, centroids, second moments of area, differential equations and Fourier series. Besides the standard integrals covered in this chapter, there are a number of other methods of integration covered in later chapters. For any further studies in engineering, differential and integral calculus are unavoidable.

At the end of this chapter, you should be able to:

- understand that integration is the reverse process of differentiation
- determine integrals of the form ax^n where n is fractional, zero, or a positive or negative integer
- integrate standard functions: $\cos ax$, $\sin ax$, $\sec^2 ax$, $\operatorname{cosec}^2 ax$, $\operatorname{cosec} ax \cot ax$, $\sec ax \tan ax$, e^{ax}, $\dfrac{1}{x}$
- evaluate definite integrals

31.1 The process of integration

The process of integration reverses the process of differentiation. In differentiation, if $f(x) = 2x^2$ then $f'(x) = 4x$. Thus the integral of $4x$ is $2x^2$, i.e. integration is the process of moving from $f'(x)$ to $f(x)$. By similar reasoning, the integral of $2t$ is t^2

Integration is a process of summation or adding parts together and an elongated S, shown as $\int$, is used to replace the words 'the integral of'. Hence, from above, $\int 4x = 2x^2$ and $\int 2t$ is t^2

In differentiation, the differential coefficient $\dfrac{dy}{dx}$ indicates that a function of x is being differentiated with respect to x, the dx indicating that it is 'with respect to x'. In integration the variable of integration is shown by adding d (the variable) after the function to be integrated.

Thus $\displaystyle\int 4x \, dx$ means 'the integral of $4x$ with respect to x',

and $\displaystyle\int 2t \, dt$ means 'the integral of $2t$ with respect to t'.

As stated above, the differential coefficient of $2x^2$ is $4x$, hence $\int 4x \, dx = 2x^2$. However, the differential coefficient of $2x^2 + 7$ is also $4x$. Hence $\int 4x \, dx$ is also equal to $2x^2 + 7$. To allow for the possible presence of a constant, whenever the process of integration is performed, a constant 'c' is added to the result.

Thus $\displaystyle\int 4x \, dx = 2x^2 + c$ and $\displaystyle\int 2t \, dt = t^2 + c$

'c' is called the **arbitrary constant of integration**.

31.2 The general solution of integrals of the form ax^n

The general solution of integrals of the form $\int ax^n dx$, where a and n are constants is given by:

$$\int ax^n \, dx = \frac{ax^{n+1}}{n+1} + c$$

This rule is true when n is fractional, zero, or a positive or negative integer, with the exception of $n = -1$

Using this rule gives:

(i) $\displaystyle\int 3x^4 \, dx = \frac{3x^{4+1}}{4+1} + c = \frac{3}{5}x^5 + c$

(ii) $\displaystyle\int \frac{2}{x^2} \, dx = \int 2x^{-2} \, dx = \frac{2x^{-2+1}}{-2+1} + c$

$\displaystyle = \frac{2x^{-1}}{-1} + c = \frac{-2}{x} + c$, and

(iii) $\displaystyle\int \sqrt{x} \, dx = \int x^{\frac{1}{2}} \, dx = \frac{x^{\frac{1}{2}+1}}{\frac{1}{2}+1} + c = \frac{x^{\frac{3}{2}}}{\frac{3}{2}} + c$

$\displaystyle = \frac{2}{3}\sqrt{x^3} + c$

Each of these three results may be checked by differentiation.

(a) The integral of a constant k is $kx + c$. For example,

$$\int 8 \, dx = 8x + c$$

(b) When a sum of several terms is integrated the result is the sum of the integrals of the separate terms. For example,

$$\int (3x + 2x^2 - 5) \, dx$$

$$= \int 3x \, dx + \int 2x^2 \, dx - \int 5 \, dx$$

$$= \frac{3x^2}{2} + \frac{2x^3}{3} - 5x + c$$

31.3 Standard integrals

Since integration is the reverse process of differentiation the **standard integrals** listed in Table 31.1 may be deduced and readily checked by differentiation.

Table 31.1 Standard integrals

(i) $\displaystyle\int ax^n \, dx = \frac{ax^{n+1}}{n+1} + c$

(except when $n = -1$)

(ii) $\displaystyle\int \cos ax \, dx = \frac{1}{a} \sin ax + c$

(iii) $\displaystyle\int \sin ax \, dx = -\frac{1}{a} \cos ax + c$

(iv) $\displaystyle\int \sec^2 ax \, dx = \frac{1}{a} \tan ax + c$

(v) $\displaystyle\int \operatorname{cosec}^2 ax \, dx = -\frac{1}{a} \cot ax + c$

(vi) $\displaystyle\int \operatorname{cosec} ax \cot ax \, dx = -\frac{1}{a} \operatorname{cosec} ax + c$

(vii) $\displaystyle\int \sec ax \tan ax \, dx = \frac{1}{a} \sec ax + c$

(viii) $\displaystyle\int e^{ax} \, dx = \frac{1}{a} e^{ax} + c$

(ix) $\displaystyle\int \frac{1}{x} \, dx = \ln x + c$

Problem 1. Determine (a) $\int 5x^2 \, dx$ (b) $\int 2t^3 \, dt$

The standard integral, $\displaystyle\int ax^n \, dx = \frac{ax^{n+1}}{n+1} + c$

(a) When $a = 5$ and $n = 2$ then

$$\int 5x^2 \, dx = \frac{5x^{2+1}}{2+1} + c = \frac{5x^3}{3} + c$$

(b) When $a = 2$ and $n = 3$ then

$$\int 2t^3 \, dt = \frac{2t^{3+1}}{3+1} + c = \frac{2t^4}{4} + c = \frac{1}{2}t^4 + c$$

Each of these results may be checked by differentiating them.

Problem 2. Determine

$$\int \left(4 + \frac{3}{7}x - 6x^2\right) dx$$

$\int (4 + \frac{3}{7}x - 6x^2)\,dx$ may be written as
$\int 4\,dx + \int \frac{3}{7}x\,dx - \int 6x^2\,dx$, i.e. each term is integrated
separately. (This splitting up of terms only applies,
however, for addition and subtraction.)

Hence $\displaystyle\int \left(4 + \frac{3}{7}x - 6x^2\right) dx$

$$= 4x + \left(\frac{3}{7}\right)\frac{x^{1+1}}{1+1} - (6)\frac{x^{2+1}}{2+1} + c$$

$$= 4x + \left(\frac{3}{7}\right)\frac{x^2}{2} - (6)\frac{x^3}{3} + c$$

$$= 4x + \frac{3}{14}x^2 - 2x^3 + c$$

Note that when an integral contains more than one term
there is no need to have an arbitrary constant for each;
just a single constant at the end is sufficient.

Problem 3. Determine

(a) $\displaystyle\int \frac{2x^3 - 3x}{4x}\,dx$ (b) $\displaystyle\int (1-t)^2\,dt$

(a) Rearranging into standard integral form gives:

$$\int \frac{2x^3 - 3x}{4x}\,dx$$

$$= \int \frac{2x^3}{4x} - \frac{3x}{4x}\,dx = \int \frac{x^2}{2} - \frac{3}{4}\,dx$$

$$= \left(\frac{1}{2}\right)\frac{x^{2+1}}{2+1} - \frac{3}{4}x + c$$

$$= \left(\frac{1}{2}\right)\frac{x^3}{3} - \frac{3}{4}x + c = \frac{1}{6}x^3 - \frac{3}{4}x + c$$

(b) Rearranging $\displaystyle\int (1-t)^2\,dt$ gives:

$$\int (1 - 2t + t^2)\,dt = t - \frac{2t^{1+1}}{1+1} + \frac{t^{2+1}}{2+1} + c$$

$$= t - \frac{2t^2}{2} + \frac{t^3}{3} + c$$

$$= t - t^2 + \frac{1}{3}t^3 + c$$

This problem shows that functions often have to be
rearranged into the standard form of $\int ax^n\,dx$ before
it is possible to integrate them.

Problem 4. Determine $\displaystyle\int \frac{3}{x^2}\,dx$

$\displaystyle\int \frac{3}{x^2}\,dx = \int 3x^{-2}\,dx$. Using the standard integral,
$\int ax^n\,dx$ when $a = 3$ and $n = -2$ gives:

$$\int 3x^{-2}\,dx = \frac{3x^{-2+1}}{-2+1} + c = \frac{3x^{-1}}{-1} + c$$

$$= -3x^{-1} + c = \frac{-3}{x} + c$$

Problem 5. Determine $\int 3\sqrt{x}\,dx$

For fractional powers it is necessary to appreciate
$\sqrt[n]{a^m} = a^{\frac{m}{n}}$

$$\int 3\sqrt{x}\,dx = \int 3x^{\frac{1}{2}}\,dx = \frac{3x^{\frac{1}{2}+1}}{\frac{1}{2}+1} + c$$

$$= \frac{3x^{\frac{3}{2}}}{\frac{3}{2}} + c = 2x^{\frac{3}{2}} + c = 2\sqrt{x^3} + c$$

Problem 6. Determine $\displaystyle\int \frac{-5}{9\sqrt[4]{t^3}}\,dt$

$$\int \frac{-5}{9\sqrt[4]{t^3}}\,dt = \int \frac{-5}{9t^{\frac{3}{4}}}\,dt = \int \left(-\frac{5}{9}\right)t^{-\frac{3}{4}}\,dt$$

$$= \left(-\frac{5}{9}\right)\frac{t^{-\frac{3}{4}+1}}{-\frac{3}{4}+1} + c$$

$$= \left(-\frac{5}{9}\right)\frac{t^{\frac{1}{4}}}{\frac{1}{4}} + c = \left(-\frac{5}{9}\right)\left(\frac{4}{1}\right)t^{\frac{1}{4}} + c$$

$$= -\frac{20}{9}\sqrt[4]{t} + c$$

Problem 7. Determine $\displaystyle\int \frac{(1+\theta)^2}{\sqrt{\theta}}\,d\theta$

$$\int \frac{(1+\theta)^2}{\sqrt{\theta}}\,d\theta = \int \frac{(1+2\theta+\theta^2)}{\sqrt{\theta}}\,d\theta$$

$$= \int\left(\frac{1}{\theta^{\frac{1}{2}}} + \frac{2\theta}{\theta^{\frac{1}{2}}} + \frac{\theta^2}{\theta^{\frac{1}{2}}}\right)d\theta$$

$$= \int\left(\theta^{\frac{-1}{2}} + 2\theta^{1-\left(\frac{1}{2}\right)} + \theta^{2-\left(\frac{1}{2}\right)}\right)d\theta$$

$$= \int\left(\theta^{\frac{-1}{2}} + 2\theta^{\frac{1}{2}} + \theta^{\frac{3}{2}}\right)d\theta$$

$$= \frac{\theta^{\left(\frac{-1}{2}\right)+1}}{-\frac{1}{2}+1} + \frac{2\theta^{\left(\frac{1}{2}\right)+1}}{\frac{1}{2}+1} + \frac{\theta^{\left(\frac{3}{2}\right)+1}}{\frac{3}{2}+1} + c$$

$$= \frac{\theta^{\frac{1}{2}}}{\frac{1}{2}} + \frac{2\theta^{\frac{3}{2}}}{\frac{3}{2}} + \frac{\theta^{\frac{5}{2}}}{\frac{5}{2}} + c$$

$$= 2\theta^{\frac{1}{2}} + \frac{4}{3}\theta^{\frac{3}{2}} + \frac{2}{5}\theta^{\frac{5}{2}} + c$$

$$= 2\sqrt{\theta} + \frac{4}{3}\sqrt{\theta^3} + \frac{2}{5}\sqrt{\theta^5} + c$$

Problem 8. Determine
(a) $\int 4\cos 3x\,dx$ (b) $\int 5\sin 2\theta\,d\theta$

(a)   From Table 31.1(ii),

$$\int 4\cos 3x\,dx = (4)\left(\frac{1}{3}\right)\sin 3x + c$$

$$= \frac{4}{3}\sin 3x + c$$

(b)   From Table 31.1(iii),

$$\int 5\sin 2\theta\,d\theta = (5)\left(-\frac{1}{2}\right)\cos 2\theta + c$$

$$= -\frac{5}{2}\cos 2\theta + c$$

Problem 9. Determine
(a) $\int 7\sec^2 4t\,dt$ (b) $3\int \csc^2 2\theta\,d\theta$

(a)   From Table 31.1(iv),

$$\int 7\sec^2 4t\,dt = (7)\left(\frac{1}{4}\right)\tan 4t + c$$

$$= \frac{7}{4}\tan 4t + c$$

(b)   From Table 31.1(v),

$$3\int \csc^2 2\theta\,d\theta = (3)\left(-\frac{1}{2}\right)\cot 2\theta + c$$

$$= -\frac{3}{2}\cot 2\theta + c$$

Problem 10. Determine
(a) $\displaystyle\int 5\,e^{3x}\,dx$ (b) $\displaystyle\int \frac{2}{3\,e^{4t}}\,dt$

(a)   From Table 31.1(viii),

$$\int 5\,e^{3x}\,dx = (5)\left(\frac{1}{3}\right)e^{3x} + c = \frac{5}{3}e^{3x} + c$$

(b) $\displaystyle\int \frac{2}{3\,e^{4t}}\,dt = \int \frac{2}{3}e^{-4t}\,dt = \left(\frac{2}{3}\right)\left(-\frac{1}{4}\right)e^{-4t} + c$

$$= -\frac{1}{6}e^{-4t} + c = -\frac{1}{6e^{4t}} + c$$

Problem 11. Determine
(a) $\displaystyle\int \frac{3}{5x}\,dx$ (b) $\displaystyle\int\left(\frac{2m^2+1}{m}\right)dm$

(a) $\displaystyle\int \frac{3}{5x}\,dx = \int\left(\frac{3}{5}\right)\left(\frac{1}{x}\right)dx = \frac{3}{5}\ln x + c$

(from Table 39.1(ix))

(b) $\displaystyle\int\left(\frac{2m^2+1}{m}\right)dm = \int\left(\frac{2m^2}{m} + \frac{1}{m}\right)dm$

$$= \int\left(2m + \frac{1}{m}\right)dm$$

$$= \frac{2m^2}{2} + \ln m + c$$

$$= m^2 + \ln m + c$$

Now try the following Practice Exercise

Practice Exercise 140   Standard integrals (Answers on page 877)

In Problems 1 to 12, determine the indefinite integrals.

1. (a) $\int 4\,dx$ (b) $\int 7x\,dx$

2. (a) $\int \frac{2}{5}x^2\,dx$ (b) $\int \frac{5}{6}x^3\,dx$

3. (a) $\int \left(\frac{3x^2 - 5x}{x}\right)dx$ (b) $\int (2+\theta)^2\,d\theta$

4. (a) $\int \frac{4}{3x^2}\,dx$ (b) $\int \frac{3}{4x^4}\,dx$

5. (a) $2\int \sqrt{x^3}\,dx$ (b) $\int \frac{1}{4}\sqrt[4]{x^5}\,dx$

6. (a) $\int \frac{-5}{\sqrt{t^3}}\,dt$ (b) $\int \frac{3}{7\sqrt[5]{x^4}}\,dx$

7. (a) $\int 3\cos 2x\,dx$ (b) $\int 7\sin 3\theta\,d\theta$

8. (a) $\int \frac{3}{4}\sec^2 3x\,dx$ (b) $\int 2\operatorname{cosec}^2 4\theta\,d\theta$

9. (a) $5\int \cot 2t\,\operatorname{cosec}2t\,dt$

    (b) $\int \frac{4}{3}\sec 4t\tan 4t\,dt$

10. (a) $\int \frac{3}{4}e^{2x}\,dx$ (b) $\frac{2}{3}\int \frac{dx}{e^{5x}}$

11. (a) $\int \frac{2}{3x}\,dx$ (b) $\int \left(\frac{u^2 - 1}{u}\right)du$

12. (a) $\int \frac{(2+3x)^2}{\sqrt{x}}\,dx$ (b) $\int \left(\frac{1}{t}+2t\right)^2 dt$

31.4   Definite integrals

Integrals containing an arbitrary constant c in their results are called **indefinite integrals** since their precise value cannot be determined without further information. **Definite integrals** are those in which limits are applied. If an expression is written as $[x]_a^b$, 'b' is called the **upper limit** and 'a' the **lower limit**. The operation of applying the limits is defined as $[x]_a^b = (b) - (a)$

The increase in the value of the integral x^2 as x increases from 1 to 3 is written as $\int_1^3 x^2\,dx$.

Applying the limits gives:

$$\int_1^3 x^2\,dx = \left[\frac{x^3}{3} + c\right]_1^3 = \left(\frac{3^3}{3} + c\right) - \left(\frac{1^3}{3} + c\right)$$

$$= (9 + c) - \left(\frac{1}{3} + c\right) = 8\frac{2}{3}$$

Note that the 'c' term always cancels out when limits are applied and it need not be shown with definite integrals.

Problem 12.   Evaluate

(a) $\int_1^2 3x\,dx$ (b) $\int_{-2}^3 (4 - x^2)\,dx$

(a) $\displaystyle\int_1^2 3x\,dx = \left[\frac{3x^2}{2}\right]_1^2 = \left\{\frac{3}{2}(2)^2\right\} - \left\{\frac{3}{2}(1)^2\right\}$

$$= 6 - 1\frac{1}{2} = 4\frac{1}{2}$$

(b) $\displaystyle\int_{-2}^3 (4 - x^2)\,dx = \left[4x - \frac{x^3}{3}\right]_{-2}^3$

$$= \left\{4(3) - \frac{(3)^3}{3}\right\} - \left\{4(-2) - \frac{(-2)^3}{3}\right\}$$

$$= \{12 - 9\} - \left\{-8 - \frac{-8}{3}\right\}$$

$$= \{3\} - \left\{-5\frac{1}{3}\right\} = 8\frac{1}{3}$$

Problem 13.   Evaluate $\displaystyle\int_1^4 \left(\frac{\theta + 2}{\sqrt{\theta}}\right)d\theta$, taking positive square roots only.

$$\int_1^4 \left(\frac{\theta+2}{\sqrt{\theta}}\right) d\theta = \int_1^4 \left(\frac{\theta}{\theta^{\frac{1}{2}}} + \frac{2}{\theta^{\frac{1}{2}}}\right) d\theta$$

$$= \int_1^4 \left(\theta^{\frac{1}{2}} + 2\theta^{\frac{-1}{2}}\right) d\theta$$

$$= \left[\frac{\theta^{\left(\frac{1}{2}\right)+1}}{\frac{1}{2}+1} + \frac{2\theta^{\left(\frac{-1}{2}\right)+1}}{-\frac{1}{2}+1}\right]_1^4$$

$$= \left[\frac{\theta^{\frac{3}{2}}}{\frac{3}{2}} + \frac{2\theta^{\frac{1}{2}}}{\frac{1}{2}}\right]_1^4 = \left[\frac{2}{3}\sqrt{\theta^3} + 4\sqrt{\theta}\right]_1^4$$

$$= \left\{\frac{2}{3}\sqrt{(4)^3} + 4\sqrt{4}\right\} - \left\{\frac{2}{3}\sqrt{(1)^3} + 4\sqrt{(1)}\right\}$$

$$= \left\{\frac{16}{3} + 8\right\} - \left\{\frac{2}{3} + 4\right\}$$

$$= 5\frac{1}{3} + 8 - \frac{2}{3} - 4 = \mathbf{8\frac{2}{3}}$$

Problem 14. Evaluate $\displaystyle\int_0^{\frac{\pi}{2}} 3\sin 2x \, dx$

$$\int_0^{\frac{\pi}{2}} 3\sin 2x \, dx$$

$$= \left[(3)\left(-\frac{1}{2}\right)\cos 2x\right]_0^{\frac{\pi}{2}} = \left[-\frac{3}{2}\cos 2x\right]_0^{\frac{\pi}{2}}$$

$$= \left\{-\frac{3}{2}\cos 2\left(\frac{\pi}{2}\right)\right\} - \left\{-\frac{3}{2}\cos 2(0)\right\}$$

$$= \left\{-\frac{3}{2}\cos \pi\right\} - \left\{-\frac{3}{2}\cos 0\right\}$$

$$= \left\{-\frac{3}{2}(-1)\right\} - \left\{-\frac{3}{2}(1)\right\} = \frac{3}{2} + \frac{3}{2} = \mathbf{3}$$

Problem 15. Evaluate $\displaystyle\int_1^2 4\cos 3t \, dt$

$$\int_1^2 4\cos 3t \, dt = \left[(4)\left(\frac{1}{3}\right)\sin 3t\right]_1^2 = \left[\frac{4}{3}\sin 3t\right]_1^2$$

$$= \left\{\frac{4}{3}\sin 6\right\} - \left\{\frac{4}{3}\sin 3\right\}$$

Note that limits of trigonometric functions are always expressed in radians – thus, for example, $\sin 6$ means the sine of 6 radians $= -0.279415\ldots$

Hence $\displaystyle\int_1^2 4\cos 3t \, dt$

$$= \left\{\frac{4}{3}(-0.279415\ldots)\right\} - \left\{\frac{4}{3}(0.141120\ldots)\right\}$$

$$= (-0.37255) - (0.18816) = \mathbf{-0.5607}$$

Problem 16. Evaluate

$$\text{(a)} \int_1^2 4e^{2x} \, dx \quad \text{(b)} \int_1^4 \frac{3}{4u} \, du,$$

each correct to 4 significant figures.

(a) $\displaystyle\int_1^2 4e^{2x} \, dx = \left[\frac{4}{2}e^{2x}\right]_1^2 = 2[e^{2x}]_1^2 = 2[e^4 - e^2]$

$$= 2[54.5982 - 7.3891] = \mathbf{94.42}$$

(b) $\displaystyle\int_1^4 \frac{3}{4u} \, du = \left[\frac{3}{4}\ln u\right]_1^4 = \frac{3}{4}[\ln 4 - \ln 1]$

$$= \frac{3}{4}[1.3863 - 0] = \mathbf{1.040}$$

Now try the following exercise

Practice Exercise 141   Definite integrals (Answers on page 877)

In problems 1 to 8, evaluate the definite integrals (where necessary, correct to 4 significant figures).

1.   (a) $\displaystyle\int_1^4 5x^2 \, dx$ (b) $\displaystyle\int_{-1}^1 -\frac{3}{4}t^2 \, dt$

2.   (a) $\displaystyle\int_{-1}^2 (3 - x^2) \, dx$ (b) $\displaystyle\int_1^3 (x^2 - 4x + 3) \, dx$

3.   (a) $\displaystyle\int_0^\pi \frac{3}{2}\cos\theta \, d\theta$ (b) $\displaystyle\int_0^{\frac{\pi}{2}} 4\cos\theta \, d\theta$

4.   (a) $\displaystyle\int_{\frac{\pi}{6}}^{\frac{\pi}{3}} 2\sin 2\theta \, d\theta$ (b) $\displaystyle\int_0^2 3\sin t \, dt$

5.   (a) $\displaystyle\int_0^1 5\cos 3x \, dx$ (b) $\displaystyle\int_0^{\frac{\pi}{6}} 3\sec^2 2x \, dx$

6. (a) $\displaystyle\int_1^2 \operatorname{cosec}^2 4t \, dt$

   (b) $\displaystyle\int_{\frac{\pi}{4}}^{\frac{\pi}{2}} (3\sin 2x - 2\cos 3x) \, dx$

7. (a) $\displaystyle\int_0^1 3e^{3t} \, dt$ (b) $\displaystyle\int_{-1}^2 \frac{2}{3e^{2x}} \, dx$

8. (a) $\displaystyle\int_2^3 \frac{2}{3x} \, dx$ (b) $\displaystyle\int_1^3 \frac{2x^2+1}{x} \, dx$

9. The entropy change ΔS, for an ideal gas is given by:

   $$\Delta S = \int_{T_1}^{T_2} C_v \frac{dT}{T} - R \int_{V_1}^{V_2} \frac{dV}{V}$$

   where T is the thermodynamic temperature, V is the volume and $R = 8.314$. Determine the entropy change when a gas expands from 1 litre to 3 litres for a temperature rise from 100 K to 400 K given that:

   $$C_v = 45 + 6 \times 10^{-3} T + 8 \times 10^{-6} T^2$$

10. The p.d. between boundaries a and b of an electric field is given by: $V = \displaystyle\int_a^b \frac{Q}{2\pi r \varepsilon_0 \varepsilon_r} \, dr$

    If $a = 10$, $b = 20$, $Q = 2 \times 10^{-6}$ coulombs, $\varepsilon_0 = 8.85 \times 10^{-12}$ and $\varepsilon_r = 2.77$, show that $V = 9$ kV.

11. The average value of a complex voltage waveform is given by:

    $$V_{AV} = \frac{1}{\pi} \int_0^\pi (10\sin \omega t + 3\sin 3\omega t$$

    $$+ 2\sin 5\omega t) \, d(\omega t)$$

    Evaluate V_{AV} correct to 2 decimal places.

12. The volume of liquid in a tank is given by:

    $$v = \int_{t_1}^{t_2} q \, dt.$$ Determine the volume of a chemical, given

    $q = (5 - 0.05t + 0.003t^2) \text{m}^3/\text{s}$, $t_1 = 0$ and $t_2 = 16 \text{s}$.

For fully worked solutions to each of the problems in Practice Exercises 140 and 141 in this chapter, go to the website:
www.routledge.com/cw/bird

Chapter 32

Some applications of integration

Why it is important to understand: Some applications of integration

Engineering is all about problem solving and many problems in engineering can be solved using integral calculus. One important application is to find the area bounded by a curve; often such an area can have a physical significance like the work done by a motor, or the distance travelled by a vehicle. Other examples can involve position, velocity, force, charge density, resistivity and current density. Electrical currents and voltages often vary with time and engineers may wish to know the average or mean value of such a current or voltage over some particular time interval. An associated quantity is the root mean square (rms) value of a current which is used, for example, in the calculation of the power dissipated by a resistor. Mean and rms values are required with alternating currents and voltages, pressure of sound waves, and much more. Revolving a plane figure about an axis generates a volume, called a solid of revolution, and integration may be used to calculate such a volume. There are many applications in engineering, and particularly in manufacturing. Centroids of basic shapes can be intuitive – such as the centre of a circle; centroids of more complex shapes can be found using integral calculus – as long as the area, volume or line of an object can be described by a mathematical equation. Centroids are of considerable importance in manufacturing, and in mechanical, civil and structural design engineering. The second moment of area is a property of a cross-section that can be used to predict the resistance of a beam to bending and deflection around an axis that lies in the cross-sectional plane. The stress in, and deflection of, a beam under load depends not only on the load but also on the geometry of the beam's cross-section; larger values of second moment cause smaller values of stress and deflection. This is why beams with larger second moments of area, such as I-beams, are used in building construction in preference to other beams with the same cross-sectional area. The second moment of area has applications in many scientific disciplines including fluid mechanics, engineering mechanics, and biomechanics – for example to study the structural properties of bone during bending. The static roll stability of a ship depends on the second moment of area of the waterline section – short fat ships are stable, long thin ones are not. It is clear that calculations involving areas, mean and rms values, volumes, centroids and second moment of area are very important in many areas of engineering.

At the end of this chapter, you should be able to:

- calculate areas under and between curves
- determine the mean and rms value of a function over a given range
- determine the volume of a solid of revolution between given limits
- determine the centroid of an area between a curve and given axes
- define and use the theorem of Pappus to determine the centroid of an area
- determine the second moment of area and radius of gyration of regular sections and composite areas

32.1 Introduction

There are a number of applications of integral calculus in engineering. The determination of areas, mean and rms values, volumes, centroids and second moments of area and radius of gyration are included in this chapter.

32.2 Areas under and between curves

In Fig. 32.1,

$$\text{total shaded area} = \int_a^b f(x)\,dx - \int_b^c f(x)\,dx$$
$$+ \int_c^d f(x)\,dx$$

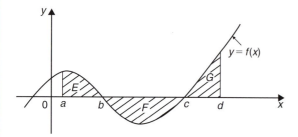

Figure 32.1

Problem 1. Determine the area between the curve $y = x^3 - 2x^2 - 8x$ and the x-axis.

$y = x^3 - 2x^2 - 8x = x(x^2 - 2x - 8) = x(x+2)(x-4)$
When $y = 0$, $x = 0$ or $(x+2) = 0$ or $(x-4) = 0$, i.e. when $y = 0$, $x = 0$ or -2 or 4, which means that the curve crosses the x-axis at 0, -2, and 4. Since the curve is a continuous function, only one other co-ordinate value needs to be calculated before a sketch of the curve can be produced. When $x = 1$, $y = -9$, showing that the part of the curve between $x = 0$ and $x = 4$ is negative. A sketch of $y = x^3 - 2x^2 - 8x$ is shown in Fig. 32.2. (Another method of sketching Fig. 32.2 would have been to draw up a table of values.)

Shaded area

$$= \int_{-2}^0 (x^3 - 2x^2 - 8x)\,dx - \int_0^4 (x^3 - 2x^2 - 8x)\,dx$$

$$= \left[\frac{x^4}{4} - \frac{2x^3}{3} - \frac{8x^2}{2}\right]_{-2}^0 - \left[\frac{x^4}{4} - \frac{2x^3}{3} - \frac{8x^2}{2}\right]_0^4$$

$$= \left(6\frac{2}{3}\right) - \left(-42\frac{2}{3}\right) = 49\frac{1}{3} \text{ square units}$$

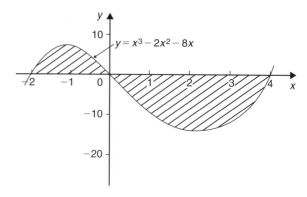

Figure 32.2

Problem 2. Determine the area enclosed between the curves $y = x^2 + 1$ and $y = 7 - x$.

At the points of intersection the curves are equal. Thus, equating the y values of each curve gives:

$$x^2 + 1 = 7 - x$$

from which, $\qquad x^2 + x - 6 = 0$

Factorising gives $\quad (x-2)(x+3) = 0$

from which $x = 2$ and $x = -3$
By firstly determining the points of intersection the range of x-values has been found. Tables of values are produced as shown below.

x	-3	-2	-1	0	1	2
$y = x^2 + 1$	10	5	2	1	2	5

x	-3	0	2
$y = 7 - x$	10	7	5

A sketch of the two curves is shown in Fig. 32.3.

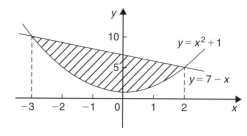

Figure 32.3

Shaded area $= \int_{-3}^{2}(7-x)\,dx - \int_{-3}^{2}(x^2+1)\,dx$

$= \int_{-3}^{2}[(7-x)-(x^2+1)]\,dx$

$= \int_{-3}^{2}(6-x-x^2)\,dx$

$= \left[6x - \dfrac{x^2}{2} - \dfrac{x^3}{3}\right]_{-3}^{2}$

$= \left(12 - 2 - \dfrac{8}{3}\right) - \left(-18 - \dfrac{9}{2} + 9\right)$

$= \left(7\dfrac{1}{3}\right) - \left(-13\dfrac{1}{2}\right)$

$= 20\dfrac{5}{6}$ square units

Problem 3. Determine by integration the area bounded by the three straight lines $y = 4 - x$, $y = 3x$ and $3y = x$

Each of the straight lines are shown sketched in Fig. 32.4.

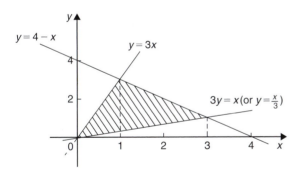

Figure 32.4

Shaded area

$= \int_{0}^{1}\left(3x - \dfrac{x}{3}\right)dx + \int_{1}^{3}\left[(4-x) - \dfrac{x}{3}\right]dx$

$= \left[\dfrac{3x^2}{2} - \dfrac{x^2}{6}\right]_{0}^{1} + \left[4x - \dfrac{x^2}{2} - \dfrac{x^2}{6}\right]_{1}^{3}$

$= \left[\left(\dfrac{3}{2} - \dfrac{1}{6}\right) - (0)\right] + \left[\left(12 - \dfrac{9}{2} - \dfrac{9}{6}\right)\right.$

$\left. - \left(4 - \dfrac{1}{2} - \dfrac{1}{6}\right)\right]$

$= \left(1\dfrac{1}{3}\right) + \left(6 - 3\dfrac{1}{3}\right) = 4$ **square units**

Now try the following Practice Exercise

Practice Exercise 142   Areas under and between curves (Answers on page 877)

1. Find the area enclosed by the curve $y = 4\cos 3x$, the x-axis and ordinates $x = 0$ and $x = \dfrac{\pi}{6}$

2. Sketch the curves $y = x^2 + 3$ and $y = 7 - 3x$ and determine the area enclosed by them.

3. Determine the area enclosed by the three straight lines $y = 3x$, $2y = x$ and $y + 2x = 5$

32.3   Mean and rms values

With reference to Fig. 32.5,

$$\text{mean value, } \bar{y} = \dfrac{1}{b-a}\int_{a}^{b} y\,dx$$

and $$\text{rms value} = \sqrt{\left\{\dfrac{1}{b-a}\int_{a}^{b} y^2\,dx\right\}}$$

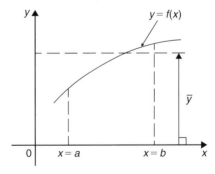

Figure 32.5

Problem 4. A sinusoidal voltage $v = 100\sin\omega t$ volts. Use integration to determine over half a cycle (a) the mean value, and (b) the rms value.

(a) Half a cycle means the limits are 0 to π radians.

Mean value, $\bar{y} = \dfrac{1}{\pi - 0}\int_{0}^{\pi} v\,d(\omega t)$

$= \dfrac{1}{\pi}\int_{0}^{\pi} 100\sin\omega t\,d(\omega t)$

$$= \frac{100}{\pi}[-\cos \omega t]_0^{\pi}$$

$$= \frac{100}{\pi}[(-\cos \pi) - (-\cos 0)]$$

$$= \frac{100}{\pi}[(+1) - (-1)] = \frac{200}{\pi}$$

$$= \textbf{63.66 volts}$$

[Note that for a sine wave,

$$\textbf{mean value} = \frac{\textbf{2}}{\pi} \times \textbf{maximum value}$$

In this case, mean value $= \frac{2}{\pi} \times 100 = 63.66 \, \text{V}]$

(b) rms value

$$= \sqrt{\left\{ \frac{1}{\pi - 0} \int_0^{\pi} v^2 \, d(\omega t) \right\}}$$

$$= \sqrt{\left\{ \frac{1}{\pi} \int_0^{\pi} (100 \sin \omega t)^2 \, d(\omega t) \right\}}$$

$$= \sqrt{\left\{ \frac{10\,000}{\pi} \int_0^{\pi} \sin^2 \omega t \, d(\omega t) \right\}}$$

which is not a 'standard' integral.

It is shown in Chapter 19 that
$\cos 2A = 1 - 2 \sin^2 A$ and this formula is used whenever $\sin^2 A$ needs to be integrated.

Rearranging $\cos 2A = 1 - 2 \sin^2 A$ gives

$$\sin^2 A = \frac{1}{2}(1 - \cos 2A)$$

Hence $\sqrt{\left\{ \frac{10\,000}{\pi} \int_0^{\pi} \sin^2 \omega t \, d(\omega t) \right\}}$

$$= \sqrt{\left\{ \frac{10\,000}{\pi} \int_0^{\pi} \frac{1}{2}(1 - \cos 2\omega t) \, d(\omega t) \right\}}$$

$$= \sqrt{\left\{ \frac{10\,000}{\pi} \frac{1}{2} \left[\omega t - \frac{\sin 2\omega t}{2} \right]_0^{\pi} \right\}}$$

$$= \sqrt{\left\{ \frac{10\,000}{\pi} \frac{1}{2} \left[\left(\pi - \frac{\sin 2\pi}{2} \right) - \left(0 - \frac{\sin 0}{2} \right) \right] \right\}}$$

$$= \sqrt{\left\{ \frac{10\,000}{\pi} \frac{1}{2}[\pi] \right\}}$$

$$= \sqrt{\left\{ \frac{10\,000}{2} \right\}} = \frac{100}{\sqrt{2}} = \textbf{70.71 volts}$$

[Note that for a sine wave,

$$\textbf{rms value} = \frac{\textbf{1}}{\sqrt{\textbf{2}}} \times \textbf{maximum value.}$$

In this case,

$$\text{rms value} = \frac{1}{\sqrt{2}} \times 100 = 70.71 \, \text{V}]$$

Now try the following Practice Exercise

Practice Exercise 143   Mean and rms values (Answers on page 877)

1. The vertical height h km of a missile varies with the horizontal distance d km, and is given by $h = 4d - d^2$. Determine the mean height of the missile from $d = 0$ to $d = 4$ km.

2. The distances of points y from the mean value of a frequency distribution are related to the variate x by the equation $y = x + \frac{1}{x}$. Determine the standard deviation (i.e. the rms value), correct to 4 significant figures for values of x from 1 to 2

3. A current $i = 25 \sin 100\pi t$ mA flows in an electrical circuit. Determine, using integral calculus, its mean and rms values each correct to 2 decimal places over the range $t = 0$ to $t = 10$ ms.

4. A wave is defined by the equation:

$$v = E_1 \sin \omega t + E_3 \sin 3\omega t$$

where E_1, E_3 and ω are constants. Determine the rms value of v over the interval $0 \leq t \leq \frac{\pi}{\omega}$

32.4   Volumes of solids of revolution

With reference to Fig. 32.6, the volume of revolution V obtained by rotating area A through one revolution

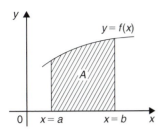

Figure 32.6

about the x-axis is given by:

$$V = \int_a^b \pi y^2 \, dx$$

If a curve $x = f(y)$ is rotated $360°$ about the y-axis between the limits $y = c$ and $y = d$ then the volume generated, V, is given by:

$$V = \int_c^d \pi x^2 \, dy$$

Problem 5. The curve $y = x^2 + 4$ is rotated one revolution about the x-axis between the limits $x = 1$ and $x = 4$. Determine the volume of solid of revolution produced.

Revolving the shaded area shown in Fig. 32.7 $360°$ about the x-axis produces a solid of revolution given by:

$$\text{Volume} = \int_1^4 \pi y^2 \, dx = \int_1^4 \pi (x^2 + 4)^2 \, dx$$

$$= \int_1^4 \pi (x^4 + 8x^2 + 16) \, dx$$

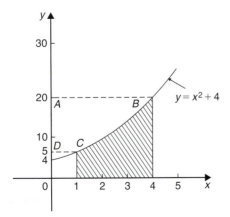

Figure 32.7

$$= \pi \left[\frac{x^5}{5} + \frac{8x^3}{3} + 16x \right]_1^4$$

$$= \pi [(204.8 + 170.67 + 64)$$

$$- (0.2 + 2.67 + 16)]$$

$$= \mathbf{420.6\pi \ cubic \ units}$$

Problem 6. Determine the area enclosed by the two curves $y = x^2$ and $y^2 = 8x$. If this area is rotated $360°$ about the x-axis determine the volume of the solid of revolution produced.

At the points of intersection the co-ordinates of the curves are equal. Since $y = x^2$ then $y^2 = x^4$. Hence equating the y^2 values at the points of intersection:

$$x^4 = 8x$$

from which, $\quad x^4 - 8x = 0$

and $\qquad x(x^3 - 8) = 0$

Hence, at the points of intersection, $x = 0$ and $x = 2$
When $x = 0$, $y = 0$ and when $x = 2$, $y = 4$. The points of intersection of the curves $y = x^2$ and $y^2 = 8x$ are therefore at (0,0) and (2,4). A sketch is shown in Fig. 32.8. If $y^2 = 8x$ then $y = \sqrt{8x}$

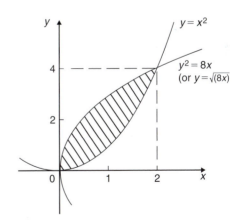

Figure 32.8

Shaded area

$$= \int_0^2 \left(\sqrt{8x} - x^2 \right) dx = \int_0^2 \left(\sqrt{8} \right) x^{\frac{1}{2}} - x^2 \, dx$$

$$= \left[\left(\sqrt{8} \right) \frac{x^{\frac{3}{2}}}{\frac{3}{2}} - \frac{x^3}{3} \right]_0^2 = \left\{ \frac{\sqrt{8}\sqrt{8}}{\frac{3}{2}} - \frac{8}{3} \right\} - \{0\}$$

$$= \frac{16}{3} - \frac{8}{3} = \frac{8}{3} = 2\frac{2}{3} \text{ square units}$$

The volume produced by revolving the shaded area about the x-axis is given by:
{(volume produced by revolving $y^2 = 8x$)
  $-$ (volume produced by revolving $y = x^2$)}

i.e. **volume** $= \int_0^2 \pi(8x)\mathrm{d}x - \int_0^2 \pi(x^4)\mathrm{d}x$

$$= \pi \int_0^2 (8x - x^4)\mathrm{d}x = \pi \left[\frac{8x^2}{2} - \frac{x^5}{5} \right]_0^2$$

$$= \pi \left[\left(16 - \frac{32}{5} \right) - (0) \right]$$

$$= \mathbf{9.6\pi} \text{ cubic units}$$

Now try the following Practice Exercise

Practice Exercise 144 Volumes (Answers on page 877)

1. The curve $xy = 3$ is revolved one revolution about the x-axis between the limits $x = 2$ and $x = 3$. Determine the volume of the solid produced.

2. The area between $\frac{y}{x^2} = 1$ and $y + x^2 = 8$ is rotated $360°$ about the x-axis. Find the volume produced.

3. The curve $y = 2x^2 + 3$ is rotated about (a) the x-axis between the limits $x = 0$ and $x = 3$, and (b) the y-axis, between the same limits. Determine the volume generated in each case.

4. The profile of a rotor blade is bounded by the lines $x = 0.2$, $y = 2x$, $y = e^{-x}$, $x = 1$ and the x-axis. The blade thickness t varies linearly with x and is given by: $t = (1.1 - x)K$, where K is a constant.

    (a) Sketch the rotor blade, labelling the limits.

    (b) Determine, using an iterative method, the value of x, correct to 3 decimal places, where $2x = e^{-x}$.

    (c) Calculate the cross-sectional area of the blade, correct to 3 decimal places.

    (d) Calculate the volume of the blade in terms of K, correct to 3 decimal places.

32.5 Centroids

A **lamina** is a thin, flat sheet having uniform thickness. The **centre of gravity** of a lamina is the point where it balances perfectly, i.e. the lamina's **centre of mass**. When dealing with an area (i.e. a lamina of negligible thickness and mass) the term **centre of area** or **centroid** is used for the point where the centre of gravity of a lamina of that shape would lie.
If $\bar{x}$ and $\bar{y}$ denote the co-ordinates of the centroid C of area A of Fig. 32.9, then:

$$\bar{x} = \frac{\int_a^b xy\,\mathrm{d}x}{\int_a^b y\,\mathrm{d}x} \quad \text{and} \quad \bar{y} = \frac{\frac{1}{2}\int_a^b y^2\,\mathrm{d}x}{\int_a^b y\,\mathrm{d}x}$$

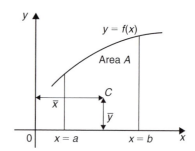

Figure 32.9

Problem 7. Find the position of the centroid of the area bounded by the curve $y = 3x^2$, the x-axis and the ordinates $x = 0$ and $x = 2$

If $(\bar{x}, \bar{y})$ are co-ordinates of the centroid of the given area then:

$$\bar{x} = \frac{\int_0^2 xy\,\mathrm{d}x}{\int_0^2 y\,\mathrm{d}x} = \frac{\int_0^2 x(3x^2)\,\mathrm{d}x}{\int_0^2 3x^2\,\mathrm{d}x}$$

$$= \frac{\int_0^2 3x^3\,\mathrm{d}x}{\int_0^2 3x^2\,\mathrm{d}x} = \frac{\left[\frac{3x^4}{4} \right]_0^2}{[x^3]_0^2}$$

$$= \frac{12}{8} = 1.5$$

$$\bar{y} = \frac{\dfrac{1}{2}\displaystyle\int_0^2 y^2\,dx}{\displaystyle\int_0^2 y\,dx} = \frac{\dfrac{1}{2}\displaystyle\int_0^2 (3x^2)^2\,dx}{8}$$

$$= \frac{\dfrac{1}{2}\displaystyle\int_0^2 9x^4\,dx}{8} = \frac{\dfrac{9}{2}\left[\dfrac{x^5}{5}\right]_0^2}{8}$$

$$= \frac{\dfrac{9}{2}\left(\dfrac{32}{5}\right)}{8} = \frac{18}{5} = 3.6$$

Hence the centroid lies at (1.5, 3.6)

Problem 8. Determine the co-ordinates of the centroid of the area lying between the curve $y = 5x - x^2$ and the x-axis.

$y = 5x - x^2 = x(5 - x)$. When $y = 0$, $x = 0$ or $x = 5$. Hence the curve cuts the x-axis at 0 and 5 as shown in Fig. 32.10. Let the co-ordinates of the centroid be $(\bar{x}, \bar{y})$ then, by integration,

$$\bar{x} = \frac{\displaystyle\int_0^5 xy\,dx}{\displaystyle\int_0^5 y\,dx} = \frac{\displaystyle\int_0^5 x(5x - x^2)\,dx}{\displaystyle\int_0^5 (5x - x^2)\,dx}$$

$$= \frac{\displaystyle\int_0^5 (5x^2 - x^3)\,dx}{\displaystyle\int_0^5 (5x - x^2)\,dx} = \frac{\left[\dfrac{5x^3}{3} - \dfrac{x^4}{4}\right]_0^5}{\left[\dfrac{5x^2}{2} - \dfrac{x^3}{3}\right]_0^5}$$

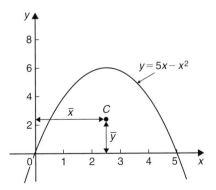

Figure 32.10

$$= \frac{\dfrac{625}{3} - \dfrac{625}{4}}{\dfrac{125}{2} - \dfrac{125}{3}} = \frac{\dfrac{625}{12}}{\dfrac{125}{6}}$$

$$= \left(\frac{625}{12}\right)\left(\frac{6}{125}\right) = \frac{5}{2} = \mathbf{2.5}$$

$$\bar{y} = \frac{\dfrac{1}{2}\displaystyle\int_0^5 y^2\,dx}{\displaystyle\int_0^5 y\,dx} = \frac{\dfrac{1}{2}\displaystyle\int_0^5 (5x - x^2)^2\,dx}{\displaystyle\int_0^5 (5x - x^2)\,dx}$$

$$= \frac{\dfrac{1}{2}\displaystyle\int_0^5 (25x^2 - 10x^3 + x^4)\,dx}{\dfrac{125}{6}}$$

$$= \frac{\dfrac{1}{2}\left[\dfrac{25x^3}{3} - \dfrac{10x^4}{4} + \dfrac{x^5}{5}\right]_0^5}{\dfrac{125}{6}}$$

$$= \frac{\dfrac{1}{2}\left(\dfrac{25(125)}{3} - \dfrac{6250}{4} + 625\right)}{\dfrac{125}{6}} = \mathbf{2.5}$$

Hence the centroid of the area lies at (2.5, 2.5)

(Note from Fig. 32.10 that the curve is symmetrical about $x = 2.5$ and thus $\bar{x}$ could have been determined 'on sight'.)

Now try the following Practice Exercise

Practice Exercise 145 Centroids (Answers on page 878)

In Problems 1 and 2, find the position of the centroids of the areas bounded by the given curves, the x-axis and the given ordinates.

1. $y = 3x + 2$ $x = 0, x = 4$

2. $y = 5x^2$ $x = 1, x = 4$

3. Determine the position of the centroid of a sheet of metal formed by the curve $y = 4x - x^2$ which lies above the x-axis.

4. Find the co-ordinates of the centroid of the area which lies between the curve $y/x = x - 2$ and the x-axis.

5. Sketch the curve $y^2 = 9x$ between the limits $x = 0$ and $x = 4$. Determine the position of the centroid of this area.

32.6 Theorem of Pappus

A theorem of Pappus* states:

'If a plane area is rotated about an axis in its own plane but not intersecting it, the volume of the solid formed is given by the product of the area and the distance moved by the centroid of the area'.

With reference to Fig. 32.11, when the curve $y = f(x)$ is rotated one revolution about the x-axis between the limits $x = a$ and $x = b$, the volume V generated is given by:

volume $V = (A)(2\pi \bar{y})$, from which, $\bar{y} = \dfrac{V}{2\pi A}$

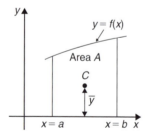

Figure 32.11

Problem 9. (a) Calculate the area bounded by the curve $y = 2x^2$, the x-axis and ordinates $x = 0$ and $x = 3$. (b) If this area is revolved (i) about the x-axis and (ii) about the y-axis, find the volumes of the solids produced. (c) Locate the position of the centroid using (i) integration, and (ii) the theorem of Pappus.

***Who was Pappus?** **Pappus of Alexandria** (c. 290–c. 350) was one of the last great Greek mathematicians of Antiquity. *Collection*, his best-known work, is a compendium of mathematics in eight volumes. It covers a wide range of topics, including geometry, recreational mathematics, doubling the cube, polygons and polyhedra. To find out more go to **www.routledge.com/cw/bird**

(a) The required area is shown shaded in Fig. 32.12.

$$\text{Area} = \int_0^3 y\,dx = \int_0^3 2x^2\,dx$$

$$= \left[\frac{2x^3}{3}\right]_0^3 = \textbf{18 square units}$$

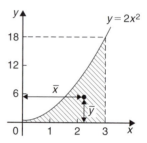

Figure 32.12

(b) (i) When the shaded area of Fig. 32.12 is revolved $360°$ about the x-axis, the volume generated

$$= \int_0^3 \pi y^2\,dx = \int_0^3 \pi (2x^2)^2\,dx$$

$$= \int_0^3 4\pi x^4\,dx = 4\pi\left[\frac{x^5}{5}\right]_0^3$$

$$= 4\pi\left(\frac{243}{5}\right) = \textbf{194.4}\pi\,\textbf{cubic units}$$

(ii) When the shaded area of Fig. 32.12 is revolved $360°$ about the y-axis, the volume generated

$$= (\text{volume generated by } x = 3)$$
$$- (\text{volume generated by } y = 2x^2)$$

$$= \int_0^{18} \pi (3)^2\,dy - \int_0^{18} \pi \left(\frac{y}{2}\right) dy$$

$$= \pi \int_0^{18} \left(9 - \frac{y}{2}\right) dy = \pi \left[9y - \frac{y^2}{4}\right]_0^{18}$$

$$= \textbf{81}\pi\,\textbf{cubic units}$$

(c) If the co-ordinates of the centroid of the shaded area in Fig. 32.12 are $(\bar{x}, \bar{y})$ then:

(i) by integration,

$$\overline{x} = \frac{\int_0^3 xy\,\mathrm{d}x}{\int_0^3 y\,\mathrm{d}x} = \frac{\int_0^3 x(2x^2)\,\mathrm{d}x}{18}$$

$$= \frac{\int_0^3 2x^3\,\mathrm{d}x}{18} = \frac{\left[\dfrac{2x^4}{4}\right]_0^3}{18}$$

$$= \frac{81}{36} = 2.25$$

$$\overline{y} = \frac{\dfrac{1}{2}\int_0^3 y^2\,\mathrm{d}x}{\int_0^3 y\,\mathrm{d}x} = \frac{\dfrac{1}{2}\int_0^3 (2x^2)^2\,\mathrm{d}x}{18}$$

$$= \frac{\dfrac{1}{2}\int_0^3 4x^4\,\mathrm{d}x}{18} = \frac{\dfrac{1}{2}\left[\dfrac{4x^5}{5}\right]_0^3}{18} = 5.4$$

(ii) using the theorem of Pappus:

Volume generated when shaded area is revolved about $OY = (\text{area})(2\pi\,\overline{x})$

i.e. $81\pi = (18)(2\pi\,\overline{x})$

from which, $\overline{x} = \dfrac{81\pi}{36\pi} = 2.25$

Volume generated when shaded area is revolved about $OX = (\text{area})(2\pi\,\overline{y})$

i.e. $194.4\pi = (18)(2\pi\,\overline{y})$

from which, $\overline{y} = \dfrac{194.4\pi}{36\pi} = 5.4$

Hence the centroid of the shaded area in Fig. 32.12 is at (2.25, 5.4)

> **Problem 10.** A metal disc has a radius of 5.0 cm and is of thickness 2.0 cm. A semicircular groove of diameter 2.0 cm is machined centrally around the rim to form a pulley. Determine, using Pappus' theorem, the volume and mass of metal removed and the volume and mass of the pulley if the density of the metal is 8000 kg m^{-3}

A side view of the rim of the disc is shown in Fig. 32.13.

Figure 32.13

When area $PQRS$ is rotated about axis XX the volume generated is that of the pulley. The centroid of the semicircular area removed is at a distance of $\dfrac{4r}{3\pi}$ from its diameter (see *Engineering Mathematics* 8th edition, Chapter 62), i.e. $\dfrac{4(1.0)}{3\pi}$, i.e. 0.424 cm from PQ. Thus the distance of the centroid from XX is $5.0 - 0.424$, i.e. 4.576 cm.

The distance moved through in one revolution by the centroid is $2\pi(4.576)$ cm.

$$\text{Area of semicircle} = \frac{\pi r^2}{2} = \frac{\pi(1.0)^2}{2} = \frac{\pi}{2}\,\text{cm}^2$$

By the theorem of Pappus,
volume generated $=$ area $\times$ distance moved by centroid $= \left(\dfrac{\pi}{2}\right)(2\pi)(4.576)$

i.e. **volume of metal removed $= 45.16$ cm^3**

Mass of metal removed $=$ density $\times$ volume

$$= 8000\,\text{kg m}^{-3} \times \frac{45.16}{10^6}\,\text{m}^3$$

$$= 0.3613\,\text{kg or } 361.3\,\text{g}$$

volume of pulley $=$ volume of cylindrical disc
$\qquad\qquad\qquad\qquad$ $-$ volume of metal removed

$$= \pi(5.0)^2(2.0) - 45.16$$

$$= \mathbf{111.9\,cm^3}$$

Mass of pulley $=$ density $\times$ volume

$$= 8000\,\text{kg m}^{-3} \times \frac{111.9}{10^6}\,\text{m}^3$$

$$= \mathbf{0.8952\,kg\ or\ 895.2\,g}$$

Now try the following Practice Exercise

Practice Exercise 146   The theorem of Pappus (Answers on page 878)

1. A right-angled isosceles triangle having a hypotenuse of 8 cm is revolved one revolution about one of its equal sides as axis. Determine the volume of the solid generated using Pappus' theorem.

2. Using (a) the theorem of Pappus, and (b) integration, determine the position of the centroid of a metal template in the form of a quadrant of a circle of radius 4 cm. (The equation of a circle, centre 0, radius r is $x^2 + y^2 = r^2$)

3. (a) Determine the area bounded by the curve $y = 5x^2$, the x-axis and the ordinates $x = 0$ and $x = 3$

   (b) If this area is revolved 360° about (i) the x-axis, and (ii) the y-axis, find the volumes of the solids of revolution produced in each case.

   (c) Determine the co-ordinates of the centroid of the area using (i) integral calculus, and (ii) the theorem of Pappus.

4. A metal disc has a radius of 7.0 cm and is of thickness 2.5 cm. A semicircular groove of diameter 2.0 cm is machined centrally around the rim to form a pulley. Determine the volume of metal removed using Pappus' theorem and express this as a percentage of the original volume of the disc. Find also the mass of metal removed if the density of the metal is 7800 kg m^{-3}.

For more on areas, mean and rms values, volumes and centroids, see *Engineering Mathematics* 8th edition, Chapters 59 to 63.

32.7 Second moments of area of regular sections

The **first moment of area** about a fixed axis of a lamina of area A, perpendicular distance y from the centroid of the lamina is defined as Ay cubic units. The **second moment of area** of the same lamina as above is given by Ay^2, i.e. the perpendicular distance from the centroid of the area to the fixed axis is squared.

Second moments of areas are usually denoted by I and have units of mm^4, cm^4, and so on.

Radius of gyration

Several areas, $a_1, a_2, a_3, \ldots$ at distances $y_1, y_2, y_3, \ldots$ from a fixed axis, may be replaced by a single area A, where $A = a_1 + a_2 + a_3 + \cdots$ at distance k from the axis, such that $Ak^2 = \sum ay^2$.

k is called the **radius of gyration** of area A about the given axis. Since $Ak^2 = \sum ay^2 = I$ then the radius of gyration,

$$k = \sqrt{\frac{I}{A}}$$

The second moment of area is a quantity much used in the theory of bending of beams, in the torsion of shafts, and in calculations involving water planes and centres of pressure.

The procedure to determine the second moment of area of regular sections about a given axis is (i) to find the second moment of area of a typical element and (ii) to sum all such second moments of area by integrating between appropriate limits.

For example, the second moment of area of the rectangle shown in Fig. 32.14 about axis PP is found by initially considering an elemental strip of width δx, parallel to and distance x from axis PP. Area of shaded strip = $b\delta x$

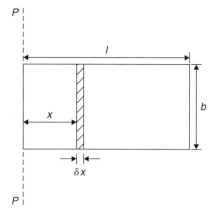

Figure 32.14

Second moment of area of the shaded strip about $PP = (x^2)(b\,\delta x)$

The second moment of area of the whole rectangle about PP is obtained by summing all such strips between $x = 0$ and $x = l$, i.e. $\sum_{x=0}^{x=l} x^2 b\,\delta x$

It is a fundamental theorem of integration that

$$\lim_{\delta x \to 0} \sum_{x=0}^{x=l} x^2 b\, \delta x = \int_0^l x^2 b\, \mathrm{d}x$$

Thus the second moment of area of the rectangle about PP

$$= b \int_0^l x^2\, \mathrm{d}x = b \left[\frac{x^3}{3} \right]_0^l = \frac{bl^3}{3}$$

Since the total area of the rectangle, $A = lb$, then

$$I_{pp} = (lb) \left(\frac{l^2}{3} \right) = \frac{Al^2}{3}$$

$$I_{pp} = Ak_{pp}^2 \quad \text{thus} \quad k_{pp}^2 = \frac{l^2}{3}$$

i.e. the radius of gyration about axes PP,

$$k_{pp} = \sqrt{\frac{l^2}{3}} = \frac{l}{\sqrt{3}}$$

Parallel axis theorem

In Fig. 32.15, axis GG passes through the centroid C of area A. Axes DD and GG are in the same plane, are parallel to each other and distance d apart. The parallel axis theorem states:

$$I_{DD} = I_{GG} + Ad^2$$

Using the parallel axis theorem the second moment of area of a rectangle about an axis through the centroid

Figure 32.15

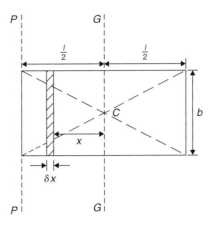

Figure 32.16

may be determined. In the rectangle shown in Fig. 32.16,
$$I_{pp} = \frac{bl^3}{3} \quad \text{(from above)}.$$

From the parallel axis theorem

$$I_{pp} = I_{GG} + (bl) \left(\frac{1}{2} \right)^2$$

i.e. $$\frac{bl^3}{3} = I_{GG} + \frac{bl^3}{4}$$

from which, $$I_{GG} = \frac{bl^3}{3} - \frac{bl^3}{4} = \frac{bl^3}{12}$$

Perpendicular axis theorem

In Fig. 32.17, axes OX, OY and OZ are mutually perpendicular. If OX and OY lie in the plane of area A then the perpendicular axis theorem states:

$$I_{OZ} = I_{OX} + I_{OY}$$

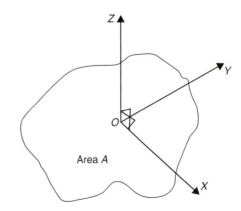

Figure 32.17

A summary of derived standard results for the second moment of area and radius of gyration of regular sections are listed in Table 32.1.

Table 32.1 Summary of standard results of the second moments of areas of regular sections

Shape	Position of axis	Second moment of area, I	Radius of gyration, k
Rectangle length l, breadth b	(1) Coinciding with b	$\dfrac{bl^3}{3}$	$\dfrac{l}{\sqrt{3}}$
	(2) Coinciding with l	$\dfrac{lb^3}{3}$	$\dfrac{b}{\sqrt{3}}$
	(3) Through centroid, parallel to b	$\dfrac{bl^3}{12}$	$\dfrac{l}{\sqrt{12}}$
	(4) Through centroid, parallel to l	$\dfrac{lb^3}{12}$	$\dfrac{b}{\sqrt{12}}$
Triangle Perpendicular height h, base b	(1) Coinciding with b	$\dfrac{bh^3}{12}$	$\dfrac{h}{\sqrt{6}}$
	(2) Through centroid, parallel to base	$\dfrac{bh^3}{36}$	$\dfrac{h}{\sqrt{18}}$
	(3) Through vertex, parallel to base	$\dfrac{bh^3}{4}$	$\dfrac{h}{\sqrt{2}}$
Circle radius r	(1) Through centre, perpendicular to plane (i.e. polar axis)	$\dfrac{\pi r^4}{2}$	$\dfrac{r}{\sqrt{2}}$
	(2) Coinciding with diameter	$\dfrac{\pi r^4}{4}$	$\dfrac{r}{2}$
	(3) About a tangent	$\dfrac{5\pi r^4}{4}$	$\dfrac{\sqrt{5}}{2}r$
Semicircle radius r	Coinciding with diameter	$\dfrac{\pi r^4}{8}$	$\dfrac{r}{2}$

Problem 11. Determine the second moment of area and the radius of gyration about axes AA, BB and CC for the rectangle shown in Fig. 32.18.

Figure 32.18

From Table 32.1, the second moment of area about axis AA,

$$I_{AA} = \frac{bl^3}{3} = \frac{(4.0)(12.0)^3}{3} = 2304\,\text{cm}^4$$

Radius of gyration, $k_{AA} = \dfrac{l}{\sqrt{3}} = \dfrac{12.0}{\sqrt{3}} = 6.93\,\text{cm}$

Similarly, $\quad I_{BB} = \dfrac{lb^3}{3} = \dfrac{(12.0)(4.0)^3}{3} = 256\,\text{cm}^4$

and $\quad k_{BB} = \dfrac{b}{\sqrt{3}} = \dfrac{4.0}{\sqrt{3}} = 2.31\,\text{cm}$

The second moment of area about the centroid of a rectangle is $\dfrac{bl^3}{12}$ when the axis through the centroid is parallel with the breadth b. In this case, the axis CC is parallel with the length l.

Hence $I_{CC} = \dfrac{lb^3}{12} = \dfrac{(12.0)(4.0)^3}{12} = 64\,\text{cm}^4$

and $k_{CC} = \dfrac{b}{\sqrt{12}} = \dfrac{4.0}{\sqrt{12}} = 1.15\,\text{cm}$

Problem 12. Find the second moment of area and the radius of gyration about axis PP for the rectangle shown in Fig. 32.19.

Figure 32.19

$I_{GG} = \dfrac{lb^3}{12}$ where $l = 40.0\,\text{mm}$ and $b = 15.0\,\text{mm}$

Hence $I_{GG} = \dfrac{(40.0)(15.0)^3}{12} = 11\,250\,\text{mm}^4$

From the parallel axis theorem, $I_{PP} = I_{GG} + Ad^2$, where $A = 40.0 \times 15.0 = 600\,\text{mm}^2$ and $d = 25.0 + 7.5 = 32.5\,\text{mm}$, the perpendicular distance between GG and PP. Hence,

$I_{PP} = 11\,250 + (600)(32.5)^2$

$\quad = 645\,000\,\text{mm}^4$

$I_{PP} = Ak_{PP}^2$, from which,

$k_{PP} = \sqrt{\dfrac{I_{PP}}{\text{area}}} = \sqrt{\left(\dfrac{645\,000}{600}\right)} = 32.79\,\text{mm}$

Problem 13. Determine the second moment of area and radius of gyration about axis QQ of the triangle BCD shown in Fig. 32.20.

Figure 32.20

Using the parallel axis theorem: $I_{QQ} = I_{GG} + Ad^2$, where I_{GG} is the second moment of area about the centroid of the triangle,

i.e. $\dfrac{bh^3}{36} = \dfrac{(8.0)(12.0)^3}{36} = 384\,\text{cm}^4$,

A is the area of the triangle,

$= \tfrac{1}{2}bh = \tfrac{1}{2}(8.0)(12.0) = 48\,\text{cm}^2$

and d is the distance between axes GG and QQ,

$= 6.0 + \tfrac{1}{3}(12.0) = 10\,\text{cm}$

Hence the second moment of area about axis QQ,

$I_{QQ} = 384 + (48)(10)^2 = 5184\,\text{cm}^4$

Radius of gyration,

$k_{QQ} = \sqrt{\dfrac{I_{QQ}}{\text{area}}} = \sqrt{\left(\dfrac{5184}{48}\right)} = 10.4\,\text{cm}$

Problem 14. Determine the second moment of area and radius of gyration of the circle shown in Fig. 32.21 about axis YY.

Figure 32.21

In Fig. 32.21, $I_{GG} = \dfrac{\pi r^4}{4} = \dfrac{\pi}{4}(2.0)^4 = 4\pi\,\text{cm}^4$.

Using the parallel axis theorem, $I_{YY} = I_{GG} + Ad^2$, where $d = 3.0 + 2.0 = 5.0\,\text{cm}$.

Hence $I_{YY} = 4\pi + [\pi(2.0)^2](5.0)^2$

$\quad = 4\pi + 100\pi = 104\pi = 327\,\text{cm}^4$

Radius of gyration,

$$k_{YY} = \sqrt{\frac{I_{YY}}{\text{area}}} = \sqrt{\left(\frac{104\pi}{\pi(2.0)^2}\right)} = \sqrt{26} = \textbf{5.10 cm}$$

Problem 15. Determine the second moment of area and radius of gyration for the semicircle shown in Fig. 32.22 about axis XX.

Figure 32.22

The centroid of a semicircle lies at $\dfrac{4r}{3\pi}$ from its diameter.

Using the parallel axis theorem:

$$I_{BB} = I_{GG} + Ad^2$$

where $\quad I_{BB} = \dfrac{\pi r^4}{8}$ (from Table 32.1)

$$= \frac{\pi(10.0)^4}{8} = 3927\,\text{mm}^4,$$

$$A = \frac{\pi r^2}{2} = \frac{\pi(10.0)^2}{2} = 157.1\,\text{mm}^2$$

and $\quad d = \dfrac{4r}{3\pi} = \dfrac{4(10.0)}{3\pi} = 4.244\,\text{mm}$

Hence $\quad 3927 = I_{GG} + (157.1)(4.244)^2$

i.e. $\quad 3927 = I_{GG} + 2830$

from which, $\quad I_{GG} = 3927 - 2830 = 1097\,\text{mm}^4$

Using the parallel axis theorem again:

$$I_{XX} = I_{GG} + A(15.0 + 4.244)^2$$

i.e. $\quad \boldsymbol{I_{XX}} = 1097 + (157.1)(19.244)^2$

$$= 1097 + 58\,179$$

$$= 59\,276\,\text{mm}^4 \text{ or } \textbf{59 280 mm}^4$$

$$\text{correct to 4 significant figures.}$$

Radius of gyration, $\boldsymbol{k_{XX}} = \sqrt{\dfrac{I_{XX}}{\text{area}}} = \sqrt{\left(\dfrac{59\,276}{157.1}\right)}$

$$= \textbf{19.42 mm}$$

Problem 16. Determine the polar second moment of area of the propeller shaft cross-section shown in Fig. 32.23.

Figure 32.23

The polar second moment of area of a circle $= \dfrac{\pi r^4}{2}$

The polar second moment of area of the shaded area is given by the polar second moment of area of the 7.0 cm diameter circle minus the polar second moment of area of the 6.0 cm diameter circle.

Hence the polar second moment of area of the cross-section shown

$$= \frac{\pi}{2}\left(\frac{7.0}{2}\right)^4 - \frac{\pi}{2}\left(\frac{6.0}{2}\right)^4$$

$$= 235.7 - 127.2 = \textbf{108.5 cm}^4$$

Problem 17. Determine the second moment of area and radius of gyration of a rectangular lamina of length 40 mm and width 15 mm about an axis through one corner, perpendicular to the plane of the lamina.

The lamina is shown in Fig. 32.24.

Figure 32.24

From the perpendicular axis theorem:

$$I_{ZZ} = I_{XX} + I_{YY}$$

$$I_{XX} = \frac{lb^3}{3} = \frac{(40)(15)^3}{3} = 45\,000\,\text{mm}^4$$

and $I_{YY} = \dfrac{bl^3}{3} = \dfrac{(15)(40)^3}{3} = 320\,000\,\text{mm}^4$

Hence $I_{ZZ} = 45\,000 + 320\,000$

$= 365\,000\,\text{mm}^4$ or $36.5\,\text{cm}^4$

Radius of gyration,

$$k_{ZZ} = \sqrt{\dfrac{I_{ZZ}}{\text{area}}} = \sqrt{\left(\dfrac{365\,000}{(40)(15)}\right)}$$

$$= 24.7\,\text{mm} \text{ or } 2.47\,\text{cm}$$

Problem 18.   Determine correct to 3 significant figures, the second moment of area about axis XX for the composite area shown in Fig. 32.25.

Figure 32.25

For the semicircle,

$$I_{XX} = \dfrac{\pi r^4}{8} = \dfrac{\pi (4.0)^4}{8} = 100.5\,\text{cm}^4$$

For the rectangle,

$$I_{XX} = \dfrac{bl^3}{3} = \dfrac{(6.0)(8.0)^3}{3} = 1024\,\text{cm}^4$$

For the triangle, about axis TT through centroid C_T,

$$I_{TT} = \dfrac{bh^3}{36} = \dfrac{(10)(6.0)^3}{36} = 60\,\text{cm}^4$$

By the parallel axis theorem, the second moment of area of the triangle about axis XX

$$= 60 + \left[\tfrac{1}{2}(10)(6.0)\right]\left[8.0 + \tfrac{1}{3}(6.0)\right]^2 = 3060\,\text{cm}^4$$

Total second moment of area about XX

$$= 100.5 + 1024 + 3060$$

$$= 4184.5$$

$$= 4180\,\text{cm}^4, \text{ correct to 3 significant figures.}$$

Problem 19.   Determine the second moment of area and the radius of gyration about axis XX for the I-section shown in Fig. 32.26.

Figure 32.26

The I-section is divided into three rectangles, D, E and F and their centroids denoted by C_D, C_E and C_F respectively.

For rectangle D:
The second moment of area about C_D (an axis through C_D parallel to XX)

$$= \dfrac{bl^3}{12} = \dfrac{(8.0)(3.0)^3}{12} = 18\,\text{cm}^4$$

Using the parallel axis theorem:

$$I_{XX} = 18 + Ad^2$$

where $A = (8.0)(3.0) = 24\,\text{cm}^2$ and $d = 12.5\,\text{cm}$

Hence $I_{XX} = 18 + 24(12.5)^2 = 3768\,\text{cm}^4$.

For rectangle E:

The second moment of area about C_E (an axis through C_E parallel to XX)

$$= \frac{bl^3}{12} = \frac{(3.0)(7.0)^3}{12} = 85.75\,\text{cm}^4$$

Using the parallel axis theorem:

$$I_{XX} = 85.75 + (7.0)(3.0)(7.5)^2 = 1267\,\text{cm}^4$$

For rectangle F:

$$I_{XX} = \frac{bl^3}{3} = \frac{(15.0)(4.0)^3}{3} = 320\,\text{cm}^4$$

Total second moment of area for the *I*-section about axis *XX*,

$$I_{XX} = 3768 + 1267 + 320 = \mathbf{5355\,\text{cm}^4}$$

Total area of *I*-section

$$= (8.0)(3.0) + (3.0)(7.0) + (15.0)(4.0)$$
$$= 105\,\text{cm}^2$$

Radius of gyration,

$$k_{XX} = \sqrt{\frac{I_{XX}}{\text{area}}} = \sqrt{\left(\frac{5355}{105}\right)} = \mathbf{7.14\,\text{cm}}$$

Now try the following Practice Exercise

Practice Exercise 147   Second moments of areas of regular sections (Answers on page 878)

1. Determine the second moment of area and radius of gyration for the rectangle shown in Fig. 32.27 about (a) axis AA (b) axis BB and (c) axis CC.

Figure 32.27

2. Determine the second moment of area and radius of gyration for the triangle shown in Fig. 32.28 about (a) axis DD (b) axis EE and

(c) an axis through the centroid of the triangle parallel to axis DD.

Figure 32.28

3. For the circle shown in Fig. 32.29, find the second moment of area and radius of gyration about (a) axis FF and (b) axis HH.

Figure 32.29

4. For the semicircle shown in Fig. 32.30, find the second moment of area and radius of gyration about axis JJ.

Figure 32.30

5. For each of the areas shown in Fig. 32.31 determine the second moment of area and radius of gyration about axis LL, by using the parallel axis theorem.

Figure 32.31

Section G

6. Calculate the radius of gyration of a rectangular door 2.0 m high by 1.5 m wide about a vertical axis through its hinge.

7. A circular door of a boiler is hinged so that it turns about a tangent. If its diameter is 1.0 m, determine its second moment of area and radius of gyration about the hinge.

8. A circular cover, centre O, has a radius of 12.0 cm. A hole of radius 4.0 cm and centre X, where $OX = 6.0$ cm, is cut in the cover. Determine the second moment of area and the radius of gyration of the remainder about a diameter through O perpendicular to OX.

9. For the sections shown in Fig. 32.32, find the second moment of area and the radius of gyration about axis XX.

Figure 32.32

10. Determine the second moments of areas about the given axes for the shapes shown in Fig. 32.33. (In Fig. 32.33(b), the circular area is removed.)

Figure 32.33

11. Find the second moment of area and radius of gyration about the axis XX for the beam section shown in Fig. 32.34.

Figure 32.34

For fully worked solutions to each of the problems in Practice Exercises 142 to 147 in this chapter, go to the website:
www.routledge.com/cw/bird

Chapter 33

Introduction to differential equations

Why it is important to understand: **Solution of first-order differential equations by separation of variables**

Differential equations play an important role in modelling virtually every physical, technical, or biological process, from celestial motion, to bridge design, to interactions between neurons. Further applications are found in fluid dynamics with the design of containers and funnels, in heat conduction analysis with the design of heat spreaders in microelectronics, in rigid-body dynamic analysis, with falling objects, and in exponential growth of current in an R–L circuit, to name but a few. This chapter introduces first-order differential equations – the subject is clearly of great importance in many different areas of engineering.

At the end of this chapter, you should be able to:

- sketch a family of curves given a simple derivative
- define a differential equation – first-order, second-order, general solution, particular solution, boundary conditions
- solve a differential equation of the form $\dfrac{dy}{dx} = f(x)$
- solve a differential equation of the form $\dfrac{dy}{dx} = f(y)$
- solve a differential equation of the form $\dfrac{dy}{dx} = f(x) \cdot f(y)$

33.1 Family of curves

Integrating both sides of the derivative $\dfrac{dy}{dx} = 3$ with respect to x gives $y = \int 3\,dx$, i.e., $y = 3x + c$, where c is an arbitrary constant.

$y = 3x + c$ represents a **family of curves**, each of the curves in the family depending on the value of c. Examples include $y = 3x + 8$, $y = 3x + 3$, $y = 3x$ and $y = 3x - 10$ and these are shown in Fig. 33.1.

Each are straight lines of gradient 3. A particular curve of a family may be determined when a point on the

Figure 33.1

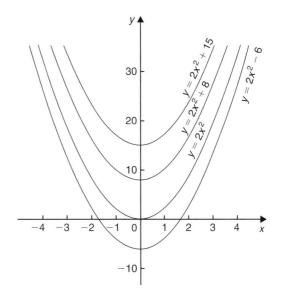

Figure 33.2

curve is specified. Thus, if $y = 3x + c$ passes through the point $(1, 2)$ then $2 = 3(1) + c$, from which, $c = -1$. The equation of the curve passing through $(1, 2)$ is therefore $y = 3x - 1$

Problem 1. Sketch the family of curves given by the equation $\dfrac{dy}{dx} = 4x$ and determine the equation of one of these curves which passes through the point $(2, 3)$

Integrating both sides of $\dfrac{dy}{dx} = 4x$ with respect to x gives:

$$\int \frac{dy}{dx}\, dx = \int 4x\, dx, \quad \text{i.e., } y = 2x^2 + c$$

Some members of the family of curves having an equation $y = 2x^2 + c$ include $y = 2x^2 + 15$, $y = 2x^2 + 8$, $y = 2x^2$ and $y = 2x^2 - 6$, and these are shown in Fig. 33.2. To determine the equation of the curve passing through the point $(2, 3)$, $x = 2$ and $y = 3$ are substituted into the equation $y = 2x^2 + c$

Thus $3 = 2(2)^2 + c$, from which $c = 3 - 8 = -5$

Hence the equation of the curve passing through the point $(2, 3)$ is $y = 2x^2 - 5$

Now try the following Practice Exercise

Practice Exercise 148   Families of curves (Answers on page 878)

1. Sketch a family of curves represented by each of the following differential equations:

    (a) $\dfrac{dy}{dx} = 6$ (b) $\dfrac{dy}{dx} = 3x$ (c) $\dfrac{dy}{dx} = x + 2$

2. Sketch the family of curves given by the equation $\dfrac{dy}{dx} = 2x + 3$ and determine the equation of one of these curves which passes through the point $(1, 3)$

33.2 Differential equations

A **differential equation** is one that contains differential coefficients.
Examples include

$$\text{(i) } \frac{dy}{dx} = 7x \quad \text{and} \quad \text{(ii) } \frac{d^2y}{dx^2} + 5\frac{dy}{dx} + 2y = 0$$

Differential equations are classified according to the highest derivative which occurs in them. Thus example (i) above is a **first-order differential equation**, and example (ii) is a **second-order differential equation**.

The **degree** of a differential equation is that of the highest power of the highest differential which the equation contains after simplification.

Thus $\left(\dfrac{d^2 x}{dt^2}\right)^3 + 2\left(\dfrac{dx}{dt}\right)^5 = 7$ is a second-order differential equation of degree three.

Starting with a differential equation it is possible, by integration and by being given sufficient data to determine unknown constants, to obtain the original function. This process is called '**solving the differential equation**'. A solution to a differential equation which contains one or more arbitrary constants of integration is called the **general solution** of the differential equation.

When additional information is given so that constants may be calculated the **particular solution** of the differential equation is obtained. The additional information is called **boundary conditions**. It was shown in Section 49.1 that $y = 3x + c$ is the general solution of the differential equation $\dfrac{dy}{dx} = 3$

Given the boundary conditions $x = 1$ and $y = 2$, produces the particular solution of $y = 3x - 1$

Equations which can be written in the form

$$\frac{dy}{dx} = f(x), \quad \frac{dy}{dx} = f(y) \text{ and } \frac{dy}{dx} = f(x) \cdot f(y)$$

can all be solved by integration. In each case it is possible to separate the y's to one side of the equation and the x's to the other. Solving such equations is therefore known as solution by **separation of variables**.

33.3 The solution of equations of the form $\dfrac{dy}{dx} = f(x)$

A differential equation of the form $\dfrac{dy}{dx} = f(x)$ is solved by direct integration,

i.e.
$$y = \int f(x)\, dx$$

Problem 2. Determine the general solution of $x\dfrac{dy}{dx} = 2 - 4x^3$

Rearranging $x\dfrac{dy}{dx} = 2 - 4x^3$ gives:

$$\frac{dy}{dx} = \frac{2 - 4x^3}{x} = \frac{2}{x} - \frac{4x^3}{x} = \frac{2}{x} - 4x^2$$

Integrating both sides gives:

$$y = \int \left(\frac{2}{x} - 4x^2\right) dx$$

i.e. $y = 2\ln x - \dfrac{4}{3}x^3 + c,$

which is the general solution.

Problem 3. Find the particular solution of the differential equation $5\dfrac{dy}{dx} + 2x = 3$, given the boundary conditions $y = 1\dfrac{2}{5}$ when $x = 2$

Since $5\dfrac{dy}{dx} + 2x = 3$ then $\dfrac{dy}{dx} = \dfrac{3 - 2x}{5} = \dfrac{3}{5} - \dfrac{2x}{5}$

Hence $y = \displaystyle\int \left(\frac{3}{5} - \frac{2x}{5}\right) dx$

i.e. $y = \dfrac{3x}{5} - \dfrac{x^2}{5} + c,$

which is the general solution.

Substituting the boundary conditions $y = 1\frac{2}{5}$ and $x = 2$ to evaluate c gives:

$1\frac{2}{5} = \frac{6}{5} - \frac{4}{5} + c$, from which, $c = 1$

Hence the particular solution is $y = \dfrac{3x}{5} - \dfrac{x^2}{5} + 1$

Problem 4. Solve the equation $2t\left(t - \dfrac{d\theta}{dt}\right) = 5$, given $\theta = 2$ when $t = 1$

Rearranging gives:

$$t - \frac{d\theta}{dt} = \frac{5}{2t} \quad \text{and} \quad \frac{d\theta}{dt} = t - \frac{5}{2t}$$

Integrating gives:

$$\theta = \int \left(t - \frac{5}{2t}\right) dt$$

i.e. $\theta = \dfrac{t^2}{2} - \dfrac{5}{2}\ln t + c,$

which is the general solution.

When $\theta = 2$, $t = 1$, thus $2 = \frac{1}{2} - \frac{5}{2}\ln 1 + c$ from which, $c = \frac{3}{2}$

Hence the particular solution is:

$$\theta = \frac{t^2}{2} - \frac{5}{2}\ln t + \frac{3}{2}$$

i.e. $\theta = \dfrac{1}{2}(t^2 - 5\ln t + 3)$

Problem 5. The bending moment M of a beam is given by $\dfrac{dM}{dx} = -w(l-x)$, where w and x are constants. Determine M in terms of x given: $M = \frac{1}{2}wl^2$ when $x = 0$

$$\frac{dM}{dx} = -w(l-x) = -wl + wx$$

Integrating with respect to x gives:

$$M = -wlx + \frac{wx^2}{2} + c$$

which is the general solution.

When $M = \frac{1}{2}wl^2, x = 0$

Thus $\dfrac{1}{2}wl^2 = -wl(0) + \dfrac{w(0)^2}{2} + c$

from which, $c = \dfrac{1}{2}wl^2$

Hence the particular solution is:

$$M = -wlx + \frac{w(x)^2}{2} + \frac{1}{2}wl^2$$

i.e. $\quad M = \dfrac{1}{2}w(l^2 - 2lx + x^2)$

or $\quad M = \dfrac{1}{2}w(l-x)^2$

Now try the following Practice Exercise

Practice Exercise 149   Solving equations of the form $\dfrac{dy}{dx} = f(x)$ (Answers on page 878)

In Problems 1 to 5, solve the differential equations.

1. $\dfrac{dy}{dx} = \cos 4x - 2x$

2. $2x\dfrac{dy}{dx} = 3 - x^3$

3. $\dfrac{dy}{dx} + x = 3$, given $y = 2$ when $x = 1$

4. $3\dfrac{dy}{d\theta} + \sin\theta = 0$, given $y = \dfrac{2}{3}$ when $\theta = \dfrac{\pi}{3}$

5. $\dfrac{1}{e^x} + 2 = x - 3\dfrac{dy}{dx}$, given $y = 1$ when $x = 0$

6. The gradient of a curve is given by:

$$\frac{dy}{dx} + \frac{x^2}{2} = 3x$$

Find the equation of the curve if it passes through the point $\left(1, \frac{1}{3}\right)$

7. The acceleration a of a body is equal to its rate of change of velocity, $\dfrac{dv}{dt}$. Find an equation for v in terms of t, given that when $t = 0$, velocity $v = u$

8. An object is thrown vertically upwards with an initial velocity u of 20 m/s. The motion of the object follows the differential equation $\dfrac{ds}{dt} = u - gt$, where s is the height of the object in metres at time t seconds and $g = 9.8\,\text{m/s}^2$. Determine the height of the object after three seconds if $s = 0$ when $t = 0$

33.4   The solution of equations of the form $\dfrac{dy}{dx} = f(y)$

A differential equation of the form $\dfrac{dy}{dx} = f(y)$ is initially rearranged to give $dx = \dfrac{dy}{f(y)}$ and then the solution is obtained by direct integration,

i.e. $\qquad \displaystyle\int dx = \int \frac{dy}{f(y)}$

Problem 6. Find the general solution of $\dfrac{dy}{dx} = 3 + 2y$

Rearranging $\dfrac{dy}{dx} = 3 + 2y$ gives:

$$dx = \frac{dy}{3 + 2y}$$

Integrating both sides gives:

$$\int dx = \int \frac{dy}{3+2y}$$

Thus, by using the substitution $u = (3+2y)$ – see Chapter 42,

$$x = \tfrac{1}{2}\ln(3+2y) + c \qquad (1)$$

It is possible to give the general solution of a differential equation in a different form. For example, if $c = \ln k$, where k is a constant, then:

$$x = \tfrac{1}{2}\ln(3+2y) + \ln k,$$

i.e. $\quad x = \ln(3+2y)^{\frac{1}{2}} + \ln k$

or $\quad x = \ln[k\sqrt{(3+2y)}] \qquad (2)$

by the laws of logarithms, from which,

$$e^x = k\sqrt{(3+2y)} \qquad (3)$$

Equations (1), (2) and (3) are all acceptable general solutions of the differential equation

$$\frac{dy}{dx} = 3+2y$$

Problem 7. Determine the particular solution of $(y^2 - 1)\dfrac{dy}{dx} = 3y$ given that $y=1$ when $x = 2\dfrac{1}{6}$

Rearranging gives:

$$dx = \left(\frac{y^2 - 1}{3y}\right) dy = \left(\frac{y}{3} - \frac{1}{3y}\right) dy$$

Integrating gives:

$$\int dx = \int \left(\frac{y}{3} - \frac{1}{3y}\right) dy$$

i.e. $\quad x = \dfrac{y^2}{6} - \dfrac{1}{3}\ln y + c,$

which is the general solution.
When $y=1$, $x=2\tfrac{1}{6}$, thus $2\tfrac{1}{6} = \tfrac{1}{6} - \tfrac{1}{3}\ln 1 + c$, from which, $c=2$

Hence the particular solution is:

$$x = \frac{y^2}{6} - \frac{1}{3}\ln y + 2$$

Problem 8. (a) The variation of resistance, R ohms, of an aluminium conductor with temperature $\theta\,°C$ is given by $\dfrac{dR}{d\theta} = \alpha R$, where α is the temperature coefficient of resistance of aluminium. If $R = R_0$ when $\theta = 0°C$, solve the equation for R. (b) If $\alpha = 38 \times 10^{-4}/°C$, determine the resistance of an aluminium conductor at $50°C$, correct to 3 significant figures, when its resistance at $0°C$ is $24.0\,\Omega$.

(a) $\quad \dfrac{dR}{d\theta} = \alpha R$ is of the form $\dfrac{dy}{dx} = f(y)$

Rearranging gives: $d\theta = \dfrac{dR}{\alpha R}$

Integrating both sides gives:

$$\int d\theta = \int \frac{dR}{\alpha R}$$

i.e. $\quad \theta = \dfrac{1}{\alpha}\ln R + c,$

which is the general solution.
Substituting the boundary conditions $R = R_0$ when $\theta = 0$ gives:

$$0 = \frac{1}{\alpha}\ln R_0 + c$$

from which $\quad c = -\dfrac{1}{\alpha}\ln R_0$

Hence the particular solution is

$$\theta = \frac{1}{\alpha}\ln R - \frac{1}{\alpha}\ln R_0 = \frac{1}{\alpha}(\ln R - \ln R_0)$$

i.e. $\quad \theta = \dfrac{1}{\alpha}\ln\left(\dfrac{R}{R_0}\right)$ or $\alpha\theta = \ln\left(\dfrac{R}{R_0}\right)$

Hence $e^{\alpha\theta} = \dfrac{R}{R_0}$ from which, $\boldsymbol{R = R_0 e^{\alpha\theta}}$

(b) Substituting $\alpha = 38 \times 10^{-4}$, $R_0 = 24.0$ and $\theta = 50$ into $R = R_0 e^{\alpha\theta}$ gives the resistance at $50°C$, i.e.
$$\boldsymbol{R_{50}} = 24.0\,e^{(38\times10^{-4}\times50)} = \boldsymbol{29.0\,ohms}$$

Now try the following Practice Exercise

Practice Exercise 150 Solving equations of the form $\dfrac{dy}{dx} = f(y)$ (Answers on page 878)

In Problems 1 to 3, solve the differential equations.

1. $\dfrac{dy}{dx} = 2 + 3y$

2. $\dfrac{dy}{dx} = 2\cos^2 y$

3. $(y^2 + 2)\dfrac{dy}{dx} = 5y$, given $y = 1$ when $x = \dfrac{1}{2}$

4. The current in an electric circuit is given by the equation

$$Ri + L\dfrac{di}{dt} = 0$$

where L and R are constants. Show that $i = Ie^{\frac{-Rt}{L}}$, given that $i = I$ when $t = 0$

5. The velocity of a chemical reaction is given by $\dfrac{dx}{dt} = k(a - x)$, where x is the amount transferred in time t, k is a constant and a is the concentration at time $t = 0$ when $x = 0$. Solve the equation and determine x in terms of t.

6. (a) Charge Q coulombs at time t seconds is given by the differential equation $R\dfrac{dQ}{dt} + \dfrac{Q}{C} = 0$, where C is the capacitance in farads and R the resistance in ohms. Solve the equation for Q given that $Q = Q_0$ when $t = 0$

   (b) A circuit possesses a resistance of $250 \times 10^3\,\Omega$ and a capacitance of $8.5 \times 10^{-6}\,F$, and after 0.32 seconds the charge falls to 8.0 C. Determine the initial charge and the charge after one second, each correct to 3 significant figures.

7. A differential equation relating the difference in tension T, pulley contact angle θ and coefficient of friction μ is $\dfrac{dT}{d\theta} = \mu T$. When $\theta = 0$, $T = 150\,N$, and $\mu = 0.30$ as slipping starts. Determine the tension at the point of slipping when $\theta = 2$ radians. Determine also the value of θ when T is 300 N.

8. The rate of cooling of a body is given by $\dfrac{d\theta}{dt} = k\theta$, where k is a constant. If $\theta = 60°C$ when $t = 2$ minutes and $\theta = 50°C$ when $t = 5$ minutes, determine the time taken for θ to fall to $40°C$, correct to the nearest second.

33.5 The solution of equations of the form $\dfrac{dy}{dx} = f(x) \cdot f(y)$

A differential equation of the form $\dfrac{dy}{dx} = f(x) \cdot f(y)$, where $f(x)$ is a function of x only and $f(y)$ is a function of y only, may be rearranged as $\dfrac{dy}{f(y)} = f(x)\,dx$, and then the solution is obtained by direct integration, i.e.

$$\int \dfrac{dy}{f(y)} = \int f(x)\,dx$$

Problem 9. Solve the equation $4xy\dfrac{dy}{dx} = y^2 - 1$

Separating the variables gives:

$$\left(\dfrac{4y}{y^2 - 1}\right)dy = \dfrac{1}{x}\,dx$$

Integrating both sides gives:

$$\int\left(\dfrac{4y}{y^2 - 1}\right)dy = \int\left(\dfrac{1}{x}\right)dx$$

Using the substitution $u = y^2 - 1$, the general solution is:

$$2\ln(y^2 - 1) = \ln x + c \tag{1}$$

or $\quad \ln(y^2 - 1)^2 - \ln x = c$

from which, $\quad \ln\left\{\dfrac{(y^2 - 1)^2}{x}\right\} = c$

and $\quad \dfrac{(y^2 - 1)^2}{x} = e^c \tag{2}$

If in equation (1), $c = \ln A$, where A is a different constant,

then $\quad \ln(y^2 - 1)^2 = \ln x + \ln A$

i.e. $\quad \ln(y^2 - 1)^2 = \ln Ax$

i.e. $(y^2 - 1)^2 = Ax$ (3)

Equations (1) to (3) are thus three valid solutions of the differential equations

$$4xy\frac{dy}{dx} = y^2 - 1$$

Problem 10. Determine the particular solution of $\frac{d\theta}{dt} = 2e^{3t-2\theta}$, given that $t = 0$ when $\theta = 0$

$$\frac{d\theta}{dt} = 2e^{3t-2\theta} = 2(e^{3t})(e^{-2\theta}),$$

by the laws of indices.

Separating the variables gives:

$$\frac{d\theta}{e^{-2\theta}} = 2e^{3t} dt,$$

i.e. $e^{2\theta} d\theta = 2e^{3t} dt$

Integrating both sides gives:

$$\int e^{2\theta} d\theta = \int 2e^{3t} dt$$

Thus the general solution is:

$$\frac{1}{2}e^{2\theta} = \frac{2}{3}e^{3t} + c$$

When $t = 0$, $\theta = 0$, thus:

$$\frac{1}{2}e^{0} = \frac{2}{3}e^{0} + c$$

from which, $c = \frac{1}{2} - \frac{2}{3} = -\frac{1}{6}$

Hence the particular solution is:

$$\frac{1}{2}e^{2\theta} = \frac{2}{3}e^{3t} - \frac{1}{6}$$

or $3e^{2\theta} = 4e^{3t} - 1$

Problem 11. Find the curve which satisfies the equation $xy = (1 + x^2)\frac{dy}{dx}$ and passes through the point $(0, 1)$

Separating the variables gives:

$$\frac{x}{(1 + x^2)}dx = \frac{dy}{y}$$

Integrating both sides gives:

$$\tfrac{1}{2}\ln(1 + x^2) = \ln y + c$$

When $x = 0$, $y = 1$ thus $\frac{1}{2}\ln 1 = \ln 1 + c$, from which, $c = 0$

Hence the particular solution is $\frac{1}{2}\ln(1 + x^2) = \ln y$

i.e. $\ln(1 + x^2)^{\frac{1}{2}} = \ln y$, from which, $(1 + x^2)^{\frac{1}{2}} = y$

Hence the equation of the curve is $y = \sqrt{(1 + x^2)}$

Problem 12. The current i in an electric circuit containing resistance R and inductance L in series with a constant voltage source E is given by the differential equation $E - L\left(\frac{di}{dt}\right) = Ri$. Solve the equation and find i in terms of time t given that when $t = 0$, $i = 0$

In the $R-L$ series circuit shown in Fig. 33.3, the supply p.d., E, is given by

$$E = V_R + V_L$$

$$V_R = iR \text{ and } V_L = L\frac{di}{dt}$$

Hence $E = iR + L\frac{di}{dt}$

from which $E - L\frac{di}{dt} = Ri$

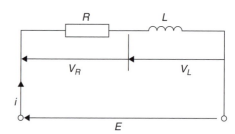

Figure 33.3

Most electrical circuits can be reduced to a differential equation.

Rearranging $E - L\frac{di}{dt} = Ri$ gives $\frac{di}{dt} = \frac{E - Ri}{L}$

and separating the variables gives:

$$\frac{di}{E - Ri} = \frac{dt}{L}$$

Integrating both sides gives:

$$\int \frac{di}{E - Ri} = \int \frac{dt}{L}$$

Hence the general solution is:

$$-\frac{1}{R}\ln(E - Ri) = \frac{t}{L} + c$$

(by making a substitution $u = E - Ri$, see Chapter 42).

When $t = 0$, $i = 0$, thus $-\dfrac{1}{R} \ln E = c$

Thus the particular solution is:

$$-\frac{1}{R} \ln (E - Ri) = \frac{t}{L} - \frac{1}{R} \ln E$$

Transposing gives:

$$-\frac{1}{R} \ln (E - Ri) + \frac{1}{R} \ln E = \frac{t}{L}$$

$$\frac{1}{R}[\ln E - \ln (E - Ri)] = \frac{t}{L}$$

$$\ln \left(\frac{E}{E - Ri} \right) = \frac{Rt}{L}$$

from which $\dfrac{E}{E - Ri} = e^{\frac{Rt}{L}}$

Hence $\dfrac{E - Ri}{E} = e^{\frac{-Rt}{L}}$ and $E - Ri = Ee^{\frac{-Rt}{L}}$ and $Ri = E - Ee^{\frac{-Rt}{L}}$

Hence current,

$$i = \frac{E}{R} \left(1 - e^{\frac{-Rt}{L}} \right)$$

which represents the law of growth of current in an inductive circuit as shown in Fig. 33.4.

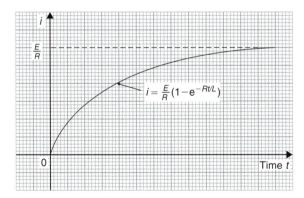

Figure 33.4

Problem 13. For an adiabatic expansion of a gas

$$C_v \frac{dp}{p} + C_p \frac{dV}{V} = 0$$

where C_p and C_v are constants. Given $n = \dfrac{C_p}{C_v}$ show that $pV^n = $ constant.

Separating the variables gives:

$$C_v \frac{dp}{p} = -C_p \frac{dV}{V}$$

Integrating both sides gives:

$$C_v \int \frac{dp}{p} = -C_p \int \frac{dV}{V}$$

i.e. $C_v \ln p = -C_p \ln V + k$

Dividing throughout by constant C_v gives:

$$\ln p = -\frac{C_p}{C_v} \ln V + \frac{k}{C_v}$$

Since $\dfrac{C_p}{C_v} = n$, then $\ln p + n \ln V = K$,

where $K = \dfrac{k}{C_v}$

i.e. $\ln p + \ln V^n = K$ or $\ln pV^n = K$, by the laws of logarithms.

Hence $pV^n = e^K$, i.e. $\boldsymbol{pV^n = \text{constant}}$.

Now try the following Practice Exercise

Practice Exercise 151   Solving equations of the form $\dfrac{dy}{dx} = f(x) \cdot f(y)$ (Answers on page 878)

In Problems 1 to 4, solve the differential equations.

1. $\dfrac{dy}{dx} = 2y \cos x$

2. $(2y - 1)\dfrac{dy}{dx} = (3x^2 + 1)$, given $x = 1$ when $y = 2$

3. $\dfrac{dy}{dx} = e^{2x - y}$, given $x = 0$ when $y = 0$

4. $2y(1 - x) + x(1 + y)\dfrac{dy}{dx} = 0$, given $x = 1$ when $y = 1$

Section G

5. Show that the solution of the equation $\dfrac{y^2+1}{x^2+1} = \dfrac{y}{x}\dfrac{dy}{dx}$ is of the form

$$\sqrt{\left(\dfrac{y^2+1}{x^2+1}\right)} = \text{constant.}$$

6. Solve $xy = (1-x^2)\dfrac{dy}{dx}$ for y, given $x=0$ when $y=1$

7. Determine the equation of the curve which satisfies the equation $xy\dfrac{dy}{dx} = x^2 - 1$, and which passes through the point $(1, 2)$

8. The p.d., V, between the plates of a capacitor C charged by a steady voltage E through a resistor R is given by the equation $CR\dfrac{dV}{dt} + V = E$

   (a) Solve the equation for V given that at $t=0$, $V=0$

   (b) Calculate V, correct to 3 significant figures, when $E=25$V, $C=20 \times 10^{-6}$F, $R=200 \times 10^3\ \Omega$ and $t=3.0$s

9. Determine the value of p, given that $x^3\dfrac{dy}{dx} = p - x$, and that $y=0$ when $x=2$ and when $x=6$

For fully worked solutions to each of the problems in Practice Exercises 148 to 151 in this chapter,
go to the website:
www.routledge.com/cw/bird

This Revision Test covers the material contained in Chapters 31 to 33. *The marks for each question are shown in brackets at the end of each question.*

1. Determine: (a) $\int 3\sqrt{t^5}\,dt$ (b) $\int \frac{2}{\sqrt[3]{x^2}}\,dx$

   (c) $\int (2+\theta)^2\,d\theta$ (9)

2. Evaluate the following integrals, each correct to 4 significant figures:

   (a) $\int_0^{\frac{\pi}{3}} 3\sin 2t\,dt$ (b) $\int_1^2 \left(\frac{2}{x^2}+\frac{1}{x}+\frac{3}{4}\right)dx$

   (c) $\int_0^1 \frac{3}{e^{2t}}\,dt$ (13)

3. Calculate the area between the curve $y=x^3-x^2-6x$ and the x-axis. (10)

4. A voltage $v=25\sin 50\pi t$ volts is applied across an electrical circuit. Determine, using integration, its mean and rms values over the range $t=0$ to $t=20$ ms, each correct to 4 significant figures. (11)

5. Sketch on the same axes the curves $x^2=2y$ and $y^2=16x$ and determine the co-ordinates of the points of intersection. Determine (a) the area enclosed by the curves, and (b) the volume of the solid produced if the area is rotated one revolution about the x-axis. (13)

6. Calculate the position of the centroid of the sheet of metal formed by the x-axis and the part of the curve $y=5x-x^2$ which lies above the x-axis. (9)

7. A cylindrical pillar of diameter 400 mm has a groove cut around its circumference as shown in Fig. RT10.1. The section of the groove is a semicircle of diameter 50 mm. Given that the centroid of a semicircle from its base is $\frac{4r}{3\pi}$, use the theorem of Pappus to determine the volume of material removed, in cm^3, correct to 3 significant figures. (8)

Figure RT10.1

8. A circular door is hinged so that it turns about a tangent. If its diameter is 1.0 m find its second moment of area and radius of gyration about the hinge. (5)

9. Solve the differential equation: $x\frac{dy}{dx}+x^2=5$ given that $y=2.5$ when $x=1$ (4)

10. Determine the equation of the curve which satisfies the differential equation $2xy\frac{dy}{dx}=x^2+1$ and which passes through the point $(1,2)$ (5)

11. A capacitor C is charged by applying a steady voltage E through a resistance R. The p.d. between the plates V is given by the differential equation:
$$CR\frac{dV}{dt}+V=E$$

   (a) Solve the equation for V given that when time $t=0$, $V=0$

   (b) Evaluate voltage V when $E=50$ V, $C=10\,\mu$F, $R=200$ kΩ and $t=1.2$ s (13)

For lecturers/instructors/teachers, fully worked solutions to each of the problems in Revision Test 10, together with a full marking scheme, are available at the website:
www.routledge.com/cw/bird

Section H

Further differential calculus

Chapter 34

Differentiation of parametric equations

Why it is important to understand: Differentiation of parametric equations

Rather than using a single equation to define two variables with respect to one another, parametric equations exist as a set that relates the two variables to one another with respect to a third variable. Some curves are easier to describe using a pair of parametric equations. The coordinates x and y of the curve are given using a third variable t, such as $x = f(t)$ and $y = g(t)$, where t is referred to as the parameter. Hence, for a given value of t, a point (x, y) is determined. For example, let t be the time while x and y are the positions of a particle; the parametric equations then describe the path of the particle at different times. Parametric equations are useful in defining three-dimensional curves and surfaces, such as determining the velocity or acceleration of a particle following a three-dimensional path. CAD systems use parametric versions of equations. Sometimes in engineering, differentiation of parametric equations is necessary, for example, when determining the radius of curvature of part of the surface when finding the surface tension of a liquid. Knowledge of standard differentials and the function of a function rule from previous chapters are needed to be able to differentiate parametric equations.

At the end of this chapter, you should be able to:

- recognise parametric equations – ellipse, parabola, hyperbola, rectangular hyperbola, cardioids, asteroid and cycloid
- differentiate parametric equations

34.1 Introduction to parametric equations

Certain mathematical functions can be expressed more simply by expressing, say, x and y separately in terms of a third variable. For example, $y = r \sin \theta$, $x = r \cos \theta$. Then, any value given to θ will produce a pair of values for x and y, which may be plotted to provide a curve of $y = f(x)$.

The third variable, θ, is called a **parameter** and the two expressions for y and x are called **parametric equations**.

The above example of $y = r \sin \theta$ and $x = r \cos \theta$ are the parametric equations for a circle. The equation of any point on a circle, centre at the origin and of radius r is given by: $x^2 + y^2 = r^2$, as shown in Chapter 14.

To show that $y = r \sin \theta$ and $x = r \cos \theta$ are suitable parametric equations for such a circle:

Left-hand side of equation

$$= x^2 + y^2$$

$$= (r\cos\theta)^2 + (r\sin\theta)^2$$

$$= r^2\cos^2\theta + r^2\sin^2\theta$$

$$= r^2\left(\cos^2\theta + \sin^2\theta\right)$$

$$= r^2 = \text{right-hand side}$$

(since $\cos^2\theta + \sin^2\theta = 1$, as shown in

Chapter 17)

34.2 Some common parametric equations

The following are some of the most common parametric equations, and Fig. 34.1 shows typical shapes of these curves.

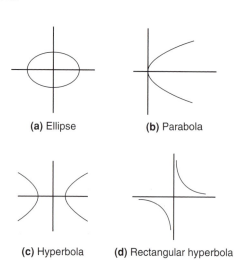

(a) Ellipse   **(b)** Parabola

(c) Hyperbola   **(d)** Rectangular hyperbola

(e) Cardioid   **(f)** Astroid

(g) Cycloid

Figure 34.1

(a) Ellipse $x = a\cos\theta,\ y = b\sin\theta$

(b) Parabola $x = a\,t^2,\ y = 2at$

(c) Hyperbola $x = a\sec\theta,\ y = b\tan\theta$

(d) Rectangular hyperbola $x = ct,\ y = \dfrac{c}{t}$

(e) Cardioid $x = a\,(2\cos\theta - \cos 2\theta),$
$y = a\,(2\sin\theta - \sin 2\theta)$

(f) Astroid $x = a\cos^3\theta,\ y = a\sin^3\theta$

(g) Cycloid $x = a\,(\theta - \sin\theta),\ y = a\,(1 - \cos\theta)$

34.3 Differentiation in parameters

When x and y are given in terms of a parameter, say θ, then by the function of a function rule of differentiation (from Chapter 29):

$$\frac{dy}{dx} = \frac{dy}{d\theta} \times \frac{d\theta}{dx}$$

It may be shown that this can be written as:

$$\frac{dy}{dx} = \frac{\dfrac{dy}{d\theta}}{\dfrac{dx}{d\theta}} \tag{1}$$

For the second differential,

$$\frac{d^2y}{dx^2} = \frac{d}{dx}\left(\frac{dy}{dx}\right) = \frac{d}{d\theta}\left(\frac{dy}{dx}\right) \cdot \frac{d\theta}{dx}$$

or

$$\frac{d^2y}{dx^2} = \frac{\dfrac{d}{d\theta}\left(\dfrac{dy}{dx}\right)}{\dfrac{dx}{d\theta}} \tag{2}$$

Problem 1. Given $x = 5\theta - 1$ and $y = 2\theta\,(\theta - 1)$, determine $\dfrac{dy}{dx}$ in terms of θ.

$x = 5\theta - 1$, hence $\dfrac{dx}{d\theta} = 5$

$y = 2\theta(\theta - 1) = 2\theta^2 - 2\theta,$

hence $\dfrac{dy}{d\theta} = 4\theta - 2 = 2\,(2\theta - 1)$

From equation (1),

$$\frac{dy}{dx} = \frac{\dfrac{dy}{d\theta}}{\dfrac{dx}{d\theta}} = \frac{2(2\theta - 1)}{5} \text{ or } \frac{2}{5}(2\theta - 1)$$

Problem 2. The parametric equations of a function are given by $y = 3\cos 2t$, $x = 2\sin t$. Determine expressions for (a) $\dfrac{dy}{dx}$ (b) $\dfrac{d^2y}{dx^2}$

(a) $y = 3\cos 2t$, hence $\dfrac{dy}{dt} = -6\sin 2t$

$x = 2\sin t$, hence $\dfrac{dx}{dt} = 2\cos t$

From equation (1),

$$\frac{dy}{dx} = \frac{\dfrac{dy}{dt}}{\dfrac{dx}{dt}} = \frac{-6\sin 2t}{2\cos t} = \frac{-6(2\sin t \cos t)}{2\cos t}$$

from double angles, Chapter 19

i.e. $\dfrac{dy}{dx} = -6\sin t$

(b) From equation (2),

$$\frac{d^2y}{dx^2} = \frac{\dfrac{d}{dt}\left(\dfrac{dy}{dx}\right)}{\dfrac{dx}{dt}} = \frac{\dfrac{d}{dt}(-6\sin t)}{2\cos t} = \frac{-6\cos t}{2\cos t}$$

i.e. $\dfrac{d^2y}{dx^2} = -3$

Problem 3. The equation of a tangent drawn to a curve at point (x_1, y_1) is given by:

$$y - y_1 = \frac{dy_1}{dx_1}(x - x_1)$$

Determine the equation of the tangent drawn to the parabola $x = 2t^2$, $y = 4t$ at the point t.

At point t, $x_1 = 2t^2$, hence $\dfrac{dx_1}{dt} = 4t$

and $y_1 = 4t$, hence $\dfrac{dy_1}{dt} = 4$

From equation (1),

$$\frac{dy}{dx} = \frac{\dfrac{dy}{dt}}{\dfrac{dx}{dt}} = \frac{4}{4t} = \frac{1}{t}$$

Hence, the equation of the tangent is:

$$y - 4t = \frac{1}{t}\left(x - 2t^2\right)$$

Problem 4. The parametric equations of a cycloid are $x = 4(\theta - \sin\theta)$, $y = 4(1 - \cos\theta)$. Determine (a) $\dfrac{dy}{dx}$ (b) $\dfrac{d^2y}{dx^2}$

(a) $x = 4(\theta - \sin\theta)$,

hence $\dfrac{dx}{d\theta} = 4 - 4\cos\theta = 4(1 - \cos\theta)$

$y = 4(1 - \cos\theta)$, hence $\dfrac{dy}{d\theta} = 4\sin\theta$

From equation (1),

$$\frac{dy}{dx} = \frac{\dfrac{dy}{d\theta}}{\dfrac{dx}{d\theta}} = \frac{4\sin\theta}{4(1 - \cos\theta)} = \frac{\sin\theta}{(1 - \cos\theta)}$$

(b) From equation (2),

$$\frac{d^2y}{dx^2} = \frac{\dfrac{d}{d\theta}\left(\dfrac{dy}{dx}\right)}{\dfrac{dx}{d\theta}} = \frac{\dfrac{d}{d\theta}\left(\dfrac{\sin\theta}{1 - \cos\theta}\right)}{4(1 - \cos\theta)}$$

$$= \frac{\dfrac{(1 - \cos\theta)(\cos\theta) - (\sin\theta)(\sin\theta)}{(1 - \cos\theta)^2}}{4(1 - \cos\theta)}$$

$$= \frac{\cos\theta - \cos^2\theta - \sin^2\theta}{4(1 - \cos\theta)^3}$$

$$= \frac{\cos\theta - \left(\cos^2\theta + \sin^2\theta\right)}{4(1 - \cos\theta)^3}$$

$$= \frac{\cos\theta - 1}{4(1 - \cos\theta)^3}$$

$$= \frac{-(1 - \cos\theta)}{4(1 - \cos\theta)^3} = \frac{-1}{4(1 - \cos\theta)^2}$$

Section H

Now try the following Practice Exercise

Practice Exercise 152   Differentiation of parametric equations (Answers on page 878)

1. Given $x = 3t - 1$ and $y = t(t-1)$, determine $\dfrac{dy}{dx}$ in terms of t

2. A parabola has parametric equations: $x = t^2$, $y = 2t$. Evaluate $\dfrac{dy}{dx}$ when $t = 0.5$

3. The parametric equations for an ellipse are $x = 4\cos\theta$, $y = \sin\theta$. Determine (a) $\dfrac{dy}{dx}$ (b) $\dfrac{d^2 y}{dx^2}$

4. Evaluate $\dfrac{dy}{dx}$ at $\theta = \dfrac{\pi}{6}$ radians for the hyperbola whose parametric equations are $x = 3\sec\theta$, $y = 6\tan\theta$

5. The parametric equations for a rectangular hyperbola are $x = 2t$, $y = \dfrac{2}{t}$. Evaluate $\dfrac{dy}{dx}$ when $t = 0.40$

The equation of a tangent drawn to a curve at point (x_1, y_1) is given by:

$$y - y_1 = \frac{dy_1}{dx_1}(x - x_1)$$

Use this in Problems 6 and 7.

6. Determine the equation of the tangent drawn to the ellipse $x = 3\cos\theta$, $y = 2\sin\theta$ at $\theta = \dfrac{\pi}{6}$

7. Determine the equation of the tangent drawn to the rectangular hyperbola $x = 5t$, $y = \dfrac{5}{t}$ at $t = 2$

34.4   Further worked problems on differentiation of parametric equations

Problem 5.   The equation of the normal drawn to a curve at point (x_1, y_1) is given by:

$$y - y_1 = -\frac{1}{\dfrac{dy_1}{dx_1}}(x - x_1)$$

Determine the equation of the normal drawn to the astroid $x = 2\cos^3\theta$, $y = 2\sin^3\theta$ at the point $\theta = \dfrac{\pi}{4}$

$x = 2\cos^3\theta$, hence $\dfrac{dx}{d\theta} = -6\cos^2\theta\ \sin\theta$

$y = 2\sin^3\theta$, hence $\dfrac{dy}{d\theta} = 6\sin^2\theta\ \cos\theta$

From equation (1),

$$\frac{dy}{dx} = \frac{\dfrac{dy}{d\theta}}{\dfrac{dx}{d\theta}} = \frac{6\sin^2\theta\cos\theta}{-6\cos^2\theta\sin\theta} = -\frac{\sin\theta}{\cos\theta} = -\tan\theta$$

When $\theta = \dfrac{\pi}{4}$, $\dfrac{dy}{dx} = -\tan\dfrac{\pi}{4} = -1$

$x_1 = 2\cos^3\dfrac{\pi}{4} = 0.7071$ and $y_1 = 2\sin^3\dfrac{\pi}{4} = 0.7071$

Hence, **the equation of the normal is**:

$$y - 0.7071 = -\frac{1}{-1}(x - 0.7071)$$

i.e. $y - 0.7071 = x - 0.7071$

i.e. $\boldsymbol{y = x}$

Problem 6.   The parametric equations for a hyperbola are $x = 2\sec\theta$, $y = 4\tan\theta$. Evaluate (a) $\dfrac{dy}{dx}$ (b) $\dfrac{d^2 y}{dx^2}$, correct to 4 significant figures, when $\theta = 1$ radian.

(a) $x = 2\sec\theta$, hence $\dfrac{dx}{d\theta} = 2\sec\theta\tan\theta$

$y = 4\tan\theta$, hence $\dfrac{dy}{d\theta} = 4\sec^2\theta$

From equation (1),

$$\frac{dy}{dx} = \frac{\dfrac{dy}{d\theta}}{\dfrac{dx}{d\theta}} = \frac{4\sec^2\theta}{2\sec\theta\tan\theta} = \frac{2\sec\theta}{\tan\theta}$$

$$= \frac{2\left(\dfrac{1}{\cos\theta}\right)}{\left(\dfrac{\sin\theta}{\cos\theta}\right)} = \frac{2}{\sin\theta} \text{ or } 2\operatorname{cosec}\theta$$

When $\theta = 1$ rad, $\dfrac{dy}{dx} = \dfrac{2}{\sin 1} = \mathbf{2.377}$, correct to 4 significant figures.

(b) From equation (2),

$$\frac{d^2y}{dx^2} = \frac{\frac{d}{d\theta}\left(\frac{dy}{dx}\right)}{\frac{dx}{d\theta}} = \frac{\frac{d}{d\theta}(2\cosec\theta)}{2\sec\theta\tan\theta}$$

$$= \frac{-2\cosec\theta\cot\theta}{2\sec\theta\tan\theta}$$

$$= \frac{-\left(\frac{1}{\sin\theta}\right)\left(\frac{\cos\theta}{\sin\theta}\right)}{\left(\frac{1}{\cos\theta}\right)\left(\frac{\sin\theta}{\cos\theta}\right)}$$

$$= -\left(\frac{\cos\theta}{\sin^2\theta}\right)\left(\frac{\cos^2\theta}{\sin\theta}\right)$$

$$= -\frac{\cos^3\theta}{\sin^3\theta} = -\cot^3\theta$$

When $\theta = 1$ rad, $\dfrac{d^2y}{dx^2} = -\cot^3 1 = -\dfrac{1}{(\tan 1)^3}$
$= -0.2647$, correct to 4 significant figures.

Problem 7. When determining the surface tension of a liquid, the radius of curvature, ρ, of part of the surface is given by:

$$\rho = \frac{\sqrt{\left[1+\left(\frac{dy}{dx}\right)^2\right]^3}}{\frac{d^2y}{dx^2}}$$

Find the radius of curvature of the part of the surface having the parametric equations $x = 3t^2$, $y = 6t$ at the point $t = 2$

$x = 3t^2$, hence $\dfrac{dx}{dt} = 6t$

$y = 6t$, hence $\dfrac{dy}{dt} = 6$

From equation (1), $\dfrac{dy}{dx} = \dfrac{\frac{dy}{dt}}{\frac{dx}{dt}} = \dfrac{6}{6t} = \dfrac{1}{t}$

From equation (2),

$$\frac{d^2y}{dx^2} = \frac{\frac{d}{dt}\left(\frac{dy}{dx}\right)}{\frac{dx}{dt}} = \frac{\frac{d}{dt}\left(\frac{1}{t}\right)}{6t} = \frac{-\frac{1}{t^2}}{6t} = -\frac{1}{6t^3}$$

Hence, radius of curvature, $\rho = \dfrac{\sqrt{\left[1+\left(\frac{dy}{dx}\right)^2\right]^3}}{\frac{d^2y}{dx^2}}$

$$= \frac{\sqrt{\left[1+\left(\frac{1}{t}\right)^2\right]^3}}{-\frac{1}{6t^3}}$$

When $t = 2$, $\rho = \dfrac{\sqrt{\left[1+\left(\frac{1}{2}\right)^2\right]^3}}{-\frac{1}{6(2)^3}} = \dfrac{\sqrt{(1.25)^3}}{-\frac{1}{48}}$

$$= -48\sqrt{(1.25)^3} = -67.08$$

Now try the following Practice Exercise

Practice Exercise 153 Differentiation of parametric equations (Answers on page 878)

1. A cycloid has parametric equations $x = 2(\theta - \sin\theta)$, $y = 2(1 - \cos\theta)$. Evaluate, at $\theta = 0.62$ rad, correct to 4 significant figures, (a) $\dfrac{dy}{dx}$ (b) $\dfrac{d^2y}{dx^2}$

    The equation of the normal drawn to a curve at point (x_1, y_1) is given by:
    $$y - y_1 = -\frac{1}{\frac{dy_1}{dx_1}}(x - x_1)$$
    Use this in Problems 2 and 3.

2. Determine the equation of the normal drawn to the parabola $x = \dfrac{1}{4}t^2$, $y = \dfrac{1}{2}t$ at $t = 2$

3. Find the equation of the normal drawn to the cycloid $x = 2(\theta - \sin\theta)$, $y = 2(1 - \cos\theta)$ at $\theta = \dfrac{\pi}{2}$ rad.

4. Determine the value of $\dfrac{d^2y}{dx^2}$, correct to 4 significant figures, at $\theta = \dfrac{\pi}{6}$ rad for the cardioid $x = 5(2\theta - \cos 2\theta)$, $y = 5(2\sin\theta - \sin 2\theta)$

Section H

Section H

5. The radius of curvature, ρ, of part of a surface when determining the surface tension of a liquid is given by:

$$\rho = \frac{\left[1 + \left(\dfrac{dy}{dx}\right)^2\right]^{3/2}}{\dfrac{d^2y}{dx^2}}$$

Find the radius of curvature (correct to 4 significant figures) of the part of the surface having parametric equations

(a) $x = 3t$, $y = \dfrac{3}{t}$ at the point $t = \dfrac{1}{2}$

(b) $x = 4\cos^3 t$, $y = 4\sin^3 t$ at $t = \dfrac{\pi}{6}$ rad.

For fully worked solutions to each of the problems in Practice Exercises 152 and 153 in this chapter, go to the website:
www.routledge.com/cw/bird

Chapter 35

Differentiation of implicit functions

Why it is important to understand: Differentiation of implicit functions

Differentiation of implicit functions is another special technique, but it occurs often enough to be important. It is needed for more complicated problems involving different rates of change. Up to this chapter we have been finding derivatives of functions of the form $y = f(x)$; unfortunately not all functions fall into this form. However, implicit differentiation is nothing more than a special case of the function of a function (or chain rule) for derivatives. Engineering applications where implicit differentiation is needed are found in optics, electronics, control, and even some thermodynamics.

At the end of this chapter, you should be able to:

- recognise implicit functions
- differentiate simple implicit functions
- differentiate implicit functions containing products and quotients

35.1 Implicit functions

When an equation can be written in the form $y = f(x)$ it is said to be an **explicit function** of x. Examples of explicit functions include

$$y = 2x^3 - 3x + 4, \quad y = 2x \ln x$$

and $y = \dfrac{3e^x}{\cos x}$

In these examples y may be differentiated with respect to x by using standard derivatives, the product rule and the quotient rule of differentiation respectively.

Sometimes with equations involving, say, y and x, it is impossible to make y the subject of the formula.

The equation is then called an **implicit function** and examples of such functions include

$$y^3 + 2x^2 = y^2 - x \text{ and } \sin y = x^2 + 2xy$$

35.2 Differentiating implicit functions

It is possible to **differentiate an implicit function** by using the **function of a function rule**, which may be stated as

$$\frac{du}{dx} = \frac{du}{dy} \times \frac{dy}{dx}$$

Thus, to differentiate y^3 with respect to x, the substitution $u = y^3$ is made, from which, $\dfrac{du}{dy} = 3y^2$

Hence, $\dfrac{\mathrm{d}}{\mathrm{d}x}(y^3) = (3y^2) \times \dfrac{\mathrm{d}y}{\mathrm{d}x}$, by the function of a function rule.

A simple rule for differentiating an implicit function is summarised as:

$$\frac{\mathrm{d}}{\mathrm{d}x}[f(y)] = \frac{\mathrm{d}}{\mathrm{d}y}[f(y)] \times \frac{\mathrm{d}y}{\mathrm{d}x} \qquad (1)$$

Problem 1. Differentiate the following functions with respect to x:

(a) $2y^4$ (b) $\sin 3t$

(a)   Let $u = 2y^4$, then, by the function of a function rule:

$$\frac{\mathrm{d}u}{\mathrm{d}x} = \frac{\mathrm{d}u}{\mathrm{d}y} \times \frac{\mathrm{d}y}{\mathrm{d}x} = \frac{\mathrm{d}}{\mathrm{d}y}(2y^4) \times \frac{\mathrm{d}y}{\mathrm{d}x}$$

$$= 8y^3 \frac{\mathrm{d}y}{\mathrm{d}x}$$

(b)   Let $u = \sin 3t$, then, by the function of a function rule:

$$\frac{\mathrm{d}u}{\mathrm{d}x} = \frac{\mathrm{d}u}{\mathrm{d}t} \times \frac{\mathrm{d}t}{\mathrm{d}x} = \frac{\mathrm{d}}{\mathrm{d}t}(\sin 3t) \times \frac{\mathrm{d}t}{\mathrm{d}x}$$

$$= 3\cos 3t \frac{\mathrm{d}t}{\mathrm{d}x}$$

Problem 2. Differentiate the following functions with respect to x:

(a) $4\ln 5y$ (b) $\dfrac{1}{5}e^{3\theta - 2}$

(a)   Let $u = 4\ln 5y$, then, by the function of a function rule:

$$\frac{\mathrm{d}u}{\mathrm{d}x} = \frac{\mathrm{d}u}{\mathrm{d}y} \times \frac{\mathrm{d}y}{\mathrm{d}x} = \frac{\mathrm{d}}{\mathrm{d}y}(4\ln 5y) \times \frac{\mathrm{d}y}{\mathrm{d}x}$$

$$= \frac{4}{y} \frac{\mathrm{d}y}{\mathrm{d}x}$$

(b)   Let $u = \dfrac{1}{5}e^{3\theta - 2}$, then, by the function of a function rule:

$$\frac{\mathrm{d}u}{\mathrm{d}x} = \frac{\mathrm{d}u}{\mathrm{d}\theta} \times \frac{\mathrm{d}\theta}{\mathrm{d}x} = \frac{\mathrm{d}}{\mathrm{d}\theta}\left(\frac{1}{5}e^{3\theta - 2}\right) \times \frac{\mathrm{d}\theta}{\mathrm{d}x}$$

$$= \frac{3}{5}e^{3\theta - 2} \frac{\mathrm{d}\theta}{\mathrm{d}x}$$

Now try the following Practice Exercise

Practice Exercise 154   Differentiating implicit functions (Answers on page 879)

In Problems 1 and 2 differentiate the given functions with respect to x.

1.   (a) $3y^5$ (b) $2\cos 4\theta$ (c) $\sqrt{k}$

2.   (a) $\dfrac{5}{2}\ln 3t$ (b) $\dfrac{3}{4}e^{2y+1}$ (c) $2\tan 3y$

3.   Differentiate the following with respect to y:
     (a) $3\sin 2\theta$ (b) $4\sqrt{x^3}$ (c) $\dfrac{2}{e^t}$

4.   Differentiate the following with respect to u:

     (a) $\dfrac{2}{(3x+1)}$ (b) $3\sec 2\theta$ (c) $\dfrac{2}{\sqrt{y}}$

35.3   Differentiating implicit functions containing products and quotients

The product and quotient rules of differentiation must be applied when differentiating functions containing products and quotients of two variables.

For example, $\dfrac{\mathrm{d}}{\mathrm{d}x}(x^2 y) = (x^2)\dfrac{\mathrm{d}}{\mathrm{d}x}(y) + (y)\dfrac{\mathrm{d}}{\mathrm{d}x}(x^2)$,

by the product rule

$$= (x^2)\left(1\frac{\mathrm{d}y}{\mathrm{d}x}\right) + y(2x),$$

by using equation (1)

$$= x^2 \frac{\mathrm{d}y}{\mathrm{d}x} + 2xy$$

Problem 3. Determine $\dfrac{\mathrm{d}}{\mathrm{d}x}(2x^3 y^2)$

In the product rule of differentiation let $u = 2x^3$ and $v = y^2$

Thus $\dfrac{\mathrm{d}}{\mathrm{d}x}(2x^3 y^2) = (2x^3)\dfrac{\mathrm{d}}{\mathrm{d}x}(y^2) + (y^2)\dfrac{\mathrm{d}}{\mathrm{d}x}(2x^3)$

$$= (2x^3)\left(2y\frac{dy}{dx}\right) + (y^2)(6x^2)$$

$$= 4x^3 y\frac{dy}{dx} + 6x^2 y^2$$

$$= 2x^2 y\left(2x\frac{dy}{dx} + 3y\right)$$

Problem 4.   Find $\dfrac{d}{dx}\left(\dfrac{3y}{2x}\right)$

In the quotient rule of differentiation let $u = 3y$ and $v = 2x$

Thus $\dfrac{d}{dx}\left(\dfrac{3y}{2x}\right) = \dfrac{(2x)\dfrac{d}{dx}(3y) - (3y)\dfrac{d}{dx}(2x)}{(2x)^2}$

$$= \frac{(2x)\left(3\dfrac{dy}{dx}\right) - (3y)(2)}{4x^2}$$

$$= \frac{6x\dfrac{dy}{dx} - 6y}{4x^2} = \frac{3}{2x^2}\left(x\frac{dy}{dx} - y\right)$$

Problem 5.   Differentiate $z = x^2 + 3x\cos 3y$ with respect to y.

$$\frac{dz}{dy} = \frac{d}{dy}(x^2) + \frac{d}{dy}(3x\cos 3y)$$

$$= 2x\frac{dx}{dy} + \left[(3x)(-3\sin 3y) + (\cos 3y)\left(3\frac{dx}{dy}\right)\right]$$

$$= 2x\frac{dx}{dy} - 9x\sin 3y + 3\cos 3y\frac{dx}{dy}$$

Now try the following Practice Exercise

Practice Exercise 155   Differentiating implicit functions involving products and quotients (Answers on page 879)

1. Determine $\dfrac{d}{dx}(3x^2 y^3)$

2. Find $\dfrac{d}{dx}\left(\dfrac{2y}{5x}\right)$

3. Determine $\dfrac{d}{du}\left(\dfrac{3u}{4v}\right)$

4. Given $z = 3\sqrt{y}\cos 3x$ find $\dfrac{dz}{dx}$

5. Determine $\dfrac{dz}{dy}$ given $z = 2x^3 \ln y$

35.4   Further implicit differentiation

An implicit function such as $3x^2 + y^2 - 5x + y = 2$, may be differentiated term by term with respect to x. This gives:

$$\frac{d}{dx}(3x^2) + \frac{d}{dx}(y^2) - \frac{d}{dx}(5x) + \frac{d}{dx}(y) = \frac{d}{dx}(2)$$

i.e. $6x + 2y\dfrac{dy}{dx} - 5 + 1\dfrac{dy}{dx} = 0$,

using equation (1) and standard derivatives.

An expression for the derivative $\dfrac{dy}{dx}$ in terms of x and y may be obtained by rearranging this latter equation. Thus:

$$(2y + 1)\frac{dy}{dx} = 5 - 6x$$

from which, $\dfrac{dy}{dx} = \dfrac{5 - 6x}{2y + 1}$

Problem 6.   Given $2y^2 - 5x^4 - 2 - 7y^3 = 0$, determine $\dfrac{dy}{dx}$

Each term in turn is differentiated with respect to x:

Hence $\dfrac{d}{dx}(2y^2) - \dfrac{d}{dx}(5x^4) - \dfrac{d}{dx}(2) - \dfrac{d}{dx}(7y^3)$

$$= \frac{d}{dx}(0)$$

i.e. $4y\dfrac{dy}{dx} - 20x^3 - 0 - 21y^2\dfrac{dy}{dx} = 0$

Rearranging gives:

$$(4y - 21y^2)\frac{dy}{dx} = 20x^3$$

i.e. $\dfrac{dy}{dx} = \dfrac{20x^3}{(4y - 21y^2)}$

Problem 7. Determine the values of $\dfrac{dy}{dx}$ when $x=4$ given that $x^2+y^2=25$

Differentiating each term in turn with respect to x gives:

$$\frac{d}{dx}(x^2)+\frac{d}{dx}(y^2)=\frac{d}{dx}(25)$$

i.e. $\qquad 2x+2y\dfrac{dy}{dx}=0$

Hence $\qquad \dfrac{dy}{dx}=-\dfrac{2x}{2y}=-\dfrac{x}{y}$

Since $x^2+y^2=25$, when $x=4$, $y=\sqrt{(25-4^2)}=\pm3$

Thus when $x=4$ and $y=\pm3$, $\dfrac{dy}{dx}=-\dfrac{4}{\pm3}=\pm\dfrac{4}{3}$

$x^2+y^2=25$ is the equation of a circle, centre at the origin and radius 5, as shown in Fig. 35.1. At $x=4$, the two gradients are shown.

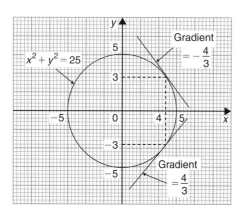

Figure 35.1

Above, $x^2+y^2=25$ was differentiated implicitly; actually, the equation could be transposed to $y=\sqrt{(25-x^2)}$ and differentiated using the function of a function rule. This gives

$$\frac{dy}{dx}=\frac{1}{2}(25-x^2)^{\frac{-1}{2}}(-2x)=-\frac{x}{\sqrt{(25-x^2)}}$$

and when $x=4$, $\dfrac{dy}{dx}=-\dfrac{4}{\sqrt{(25-4^2)}}=\pm\dfrac{4}{3}$ as obtained above.

Problem 8. (a) Find $\dfrac{dy}{dx}$ in terms of x and y given $4x^2+2xy^3-5y^2=0$

(b) Evaluate $\dfrac{dy}{dx}$ when $x=1$ and $y=2$

(a) Differentiating each term in turn with respect to x gives:

$$\frac{d}{dx}(4x^2)+\frac{d}{dx}(2xy^3)-\frac{d}{dx}(5y^2)=\frac{d}{dx}(0)$$

i.e. $8x+\left[(2x)\left(3y^2\dfrac{dy}{dx}\right)+(y^3)(2)\right]$

$$-10y\frac{dy}{dx}=0$$

i.e. $\quad 8x+6xy^2\dfrac{dy}{dx}+2y^3-10y\dfrac{dy}{dx}=0$

Rearranging gives:

$$8x+2y^3=(10y-6xy^2)\frac{dy}{dx}$$

and $\quad \dfrac{dy}{dx}=\dfrac{8x+2y^3}{10y-6xy^2}=\dfrac{4x+y^3}{y(5-3xy)}$

(b) When $x=1$ and $y=2$,

$$\frac{dy}{dx}=\frac{4(1)+(2)^3}{2[5-(3)(1)(2)]}=\frac{12}{-2}=-6$$

Problem 9. Find the gradients of the tangents drawn to the circle $x^2+y^2-2x-2y=3$ at $x=2$

The gradient of the tangent is given by $\dfrac{dy}{dx}$

Differentiating each term in turn with respect to x gives:

$$\frac{d}{dx}(x^2)+\frac{d}{dx}(y^2)-\frac{d}{dx}(2x)-\frac{d}{dx}(2y)=\frac{d}{dx}(3)$$

i.e. $\qquad 2x+2y\dfrac{dy}{dx}-2-2\dfrac{dy}{dx}=0$

Hence $\quad (2y-2)\dfrac{dy}{dx}=2-2x,$

from which $\quad \dfrac{dy}{dx}=\dfrac{2-2x}{2y-2}=\dfrac{1-x}{y-1}$

The value of y when $x=2$ is determined from the original equation.

Hence $(2)^2 + y^2 - 2(2) - 2y = 3$

i.e. $\qquad 4 + y^2 - 4 - 2y = 3$

or $\qquad y^2 - 2y - 3 = 0$

Factorising gives: $(y+1)(y-3)=0$, from which $y=-1$ or $y=3$

When $x=2$ and $y=-1$,

$$\frac{dy}{dx} = \frac{1-x}{y-1} = \frac{1-2}{-1-1} = \frac{-1}{-2} = \frac{1}{2}$$

When $x=2$ and $y=3$,

$$\frac{dy}{dx} = \frac{1-2}{3-1} = \frac{-1}{2}$$

Hence the gradients of the tangents are $\pm\dfrac{1}{2}$

The circle having the given equation has its centre at (1, 1) and radius $\sqrt{5}$ (see Chapter 14) and is shown in Fig. 35.2 with the two gradients of the tangents.

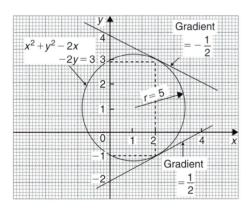

Figure 35.2

Problem 10. Pressure p and volume v of a gas are related by the law $pv^\gamma = k$, where γ and k are constants. Show that the rate of change of pressure $\dfrac{dp}{dt} = -\gamma \dfrac{p}{v} \dfrac{dv}{dt}$

Since $pv^\gamma = k$, then $p = \dfrac{k}{v^\gamma} = kv^{-\gamma}$

$$\frac{dp}{dt} = \frac{dp}{dv} \times \frac{dv}{dt}$$

by the function of a function rule

$$\frac{dp}{dv} = \frac{d}{dv}(kv^{-\gamma})$$

$$= -\gamma k v^{-\gamma-1} = \frac{-\gamma k}{v^{\gamma+1}}$$

$$\frac{dp}{dt} = \frac{-\gamma k}{v^{\gamma+1}} \times \frac{dv}{dt}$$

Since $k = pv^\gamma$,

$$\frac{dp}{dt} = \frac{-\gamma(pv^\gamma)}{v^{\gamma+1}} \frac{dv}{dt} = \frac{-\gamma pv^\gamma}{v^\gamma v^1} \frac{dv}{dt}$$

i.e. $\qquad \dfrac{dp}{dt} = -\gamma \dfrac{p}{v} \dfrac{dv}{dt}$

Now try the following Practice Exercise

Practice Exercise 156   Implicit differentiation (Answers on page 879)

In Problems 1 and 2 determine $\dfrac{dy}{dx}$

1. $x^2 + y^2 + 4x - 3y + 1 = 0$

2. $2y^3 - y + 3x - 2 = 0$

3. Given $x^2 + y^2 = 9$ evaluate $\dfrac{dy}{dx}$ when $x=\sqrt{5}$ and $y=2$

In Problems 4 to 7, determine $\dfrac{dy}{dx}$

4. $x^2 + 2x\sin 4y = 0$

5. $3y^2 + 2xy - 4x^2 = 0$

6. $2x^2y + 3x^3 = \sin y$

7. $3y + 2x\ln y = y^4 + x$

Section H

8. If $3x^2 + 2x^2y^3 - \dfrac{5}{4}y^2 = 0$ evaluate $\dfrac{dy}{dx}$ when $x = \dfrac{1}{2}$ and $y = 1$

9. Determine the gradients of the tangents drawn to the circle $x^2 + y^2 = 16$ at the point where $x = 2$. Give the answer correct to 4 significant figures.

10. Find the gradients of the tangents drawn to the ellipse $\dfrac{x^2}{4} + \dfrac{y^2}{9} = 2$ at the point where $x = 2$

11. Determine the gradient of the curve $3xy + y^2 = -2$ at the point $(1, -2)$

For fully worked solutions to each of the problems in Practice Exercises 154 to 156 in this chapter, go to the website: www.routledge.com/cw/bird

Section H

Chapter 36

Logarithmic differentiation

Why it is important to understand: **Logarithmic differentiation**

Logarithmic differentiation is a means of differentiating algebraically complicated functions or functions for which the ordinary rules of differentiation do not apply. The technique is performed in cases where it is easier to differentiate the logarithm of a function rather than the function itself. Logarithmic differentiation relies on the function of a function rule (i.e. chain rule) as well as properties of logarithms (in particular, the natural logarithm, or logarithm to the base *e*) to transform products into sums and divisions into subtractions, and can also be applied to functions raised to the power of variables of functions. Logarithmic differentiation occurs often enough in engineering calculations to make it an important technique.

At the end of this chapter, you should be able to:

- state the laws of logarithms
- differentiate simple logarithmic functions
- differentiate an implicit function involving logarithms
- differentiate more difficult logarithmic functions involving products and quotients
- differentiate functions of the form $y = [f(x)]^x$

36.1 Introduction to logarithmic differentiation

With certain functions containing more complicated products and quotients, differentiation is often made easier if the logarithm of the function is taken before differentiating. This technique, called **'logarithmic differentiation'** is achieved with a knowledge of (i) the laws of logarithms, (ii) the differential coefficients of logarithmic functions, and (iii) the differentiation of implicit functions.

36.2 Laws of logarithms

Three laws of logarithms may be expressed as:

(i) $\log(A \times B) = \log A + \log B$

(ii) $\log\left(\dfrac{A}{B}\right) = \log A - \log B$

(iii) $\log A^n = n \log A$

In calculus, Napierian logarithms (i.e. logarithms to a base of 'e') are invariably used. Thus for two functions $f(x)$ and $g(x)$ the laws of logarithms may be expressed as:

(i) $\ln[f(x) \cdot g(x)] = \ln f(x) + \ln g(x)$

(ii) $\ln\left(\dfrac{f(x)}{g(x)}\right) = \ln f(x) - \ln g(x)$

(iii) $\ln[f(x)]^n = n \ln f(x)$

Taking Napierian logarithms of both sides of the equation $y = \dfrac{f(x) \cdot g(x)}{h(x)}$ gives:

$$\ln y = \ln\left(\frac{f(x) \cdot g(x)}{h(x)}\right)$$

which may be simplified using the above laws of logarithms, giving:

$$\ln y = \ln f(x) + \ln g(x) - \ln h(x)$$

This latter form of the equation is often easier to differentiate.

36.3 Differentiation of logarithmic functions

The differential coefficient of the logarithmic function $\ln x$ is given by:

$$\frac{d}{dx}(\ln x) = \frac{1}{x}$$

More generally, it may be shown that:

$$\frac{d}{dx}[\ln f(x)] = \frac{f'(x)}{f(x)} \qquad (1)$$

For example, if $y = \ln(3x^2 + 2x - 1)$ then,

$$\frac{dy}{dx} = \frac{6x + 2}{3x^2 + 2x - 1}$$

Similarly, if $y = \ln(\sin 3x)$ then

$$\frac{dy}{dx} = \frac{3\cos 3x}{\sin 3x} = 3\cot 3x$$

Now try the following Practice Exercise

Practice Exercise 157   Differentiating logarithmic functions (Answers on page 879)

Differentiate the following:

1. $\ln(4x - 10)$

2. $\ln(\cos 3x)$

3. $\ln(3x^3 + x)$

4. $\ln(5x^2 + 10x - 7)$

5. $\ln 8x$

6. $\ln(x^2 - 1)$

7. $3\ln 4x$

8. $2\ln(\sin x)$

9. $\ln(4x^3 - 6x^2 + 3x)$

36.4 Differentiation of further logarithmic functions

As explained in Chapter 35, by using the function of a function rule:

$$\frac{d}{dx}(\ln y) = \left(\frac{1}{y}\right)\frac{dy}{dx} \qquad (2)$$

Differentiation of an expression such as

$y = \dfrac{(1+x)^2\sqrt{(x-1)}}{x\sqrt{(x+2)}}$ may be achieved by using the product and quotient rules of differentiation; however the working would be rather complicated. With logarithmic differentiation the following procedure is adopted:

(i)   Take Napierian logarithms of both sides of the equation.

Thus $\ln y = \ln\left\{\dfrac{(1+x)^2\sqrt{(x-1)}}{x\sqrt{(x+2)}}\right\}$

$$= \ln\left\{\frac{(1+x)^2(x-1)^{\frac{1}{2}}}{x(x+2)^{\frac{1}{2}}}\right\}$$

(ii)   Apply the laws of logarithms.
Thus $\ln y = \ln(1+x)^2 + \ln(x-1)^{\frac{1}{2}}$

$$-\ln x - \ln(x+2)^{\frac{1}{2}}, \text{ by laws (i)}$$
$$\text{and (ii) of Section 36.2}$$

i.e. $\ln y = 2\ln(1+x) + \tfrac{1}{2}\ln(x-1)$

$$-\ln x - \tfrac{1}{2}\ln(x+2), \text{ by law (iii)}$$
$$\text{of Section 36.2}$$

(iii)   Differentiate each term in turn with respect to x using equations (1) and (2).

Thus $\dfrac{1}{y}\dfrac{dy}{dx} = \dfrac{2}{(1+x)} + \dfrac{\frac{1}{2}}{(x-1)} - \dfrac{1}{x} - \dfrac{\frac{1}{2}}{(x+2)}$

(iv) Rearrange the equation to make $\dfrac{dy}{dx}$ the subject.

Thus $\dfrac{dy}{dx} = y\left\{\dfrac{2}{(1+x)} + \dfrac{1}{2(x-1)} - \dfrac{1}{x}\right.$

$\left. - \dfrac{1}{2(x+2)}\right\}$

(v) Substitute for y in terms of x.

Thus $\dfrac{dy}{dx} = \dfrac{(1+x)^2\sqrt{(x-1)}}{x\sqrt{(x+2)}}\left\{\dfrac{2}{(1+x)}\right.$

$\left. + \dfrac{1}{2(x-1)} - \dfrac{1}{x} - \dfrac{1}{2(x+2)}\right\}$

Problem 1.   Use logarithmic differentiation to differentiate $y = \dfrac{(x+1)(x-2)^3}{(x-3)}$.

Following the above procedure:

(i) Since $y = \dfrac{(x+1)(x-2)^3}{(x-3)}$

then $\ln y = \ln\left\{\dfrac{(x+1)(x-2)^3}{(x-3)}\right\}$

(ii) $\ln y = \ln(x+1) + \ln(x-2)^3 - \ln(x-3)$,
by laws (i) and (ii) of Section 36.2,

i.e. $\ln y = \ln(x+1) + 3\ln(x-2) - \ln(x-3)$,
by law (iii) of Section 36.2.

(iii) Differentiating with respect to x gives:

$\dfrac{1}{y}\dfrac{dy}{dx} = \dfrac{1}{(x+1)} + \dfrac{3}{(x-2)} - \dfrac{1}{(x-3)}$,

by using equations (1) and (2)

(iv) Rearranging gives:

$\dfrac{dy}{dx} = y\left\{\dfrac{1}{(x+1)} + \dfrac{3}{(x-2)} - \dfrac{1}{(x-3)}\right\}$

(v) Substituting for y gives:

$\dfrac{dy}{dx} = \dfrac{(x+1)(x-2)^3}{(x-3)}\left\{\dfrac{1}{(x+1)}\right.$

$\left. + \dfrac{3}{(x-2)} - \dfrac{1}{(x-3)}\right\}$

Problem 2.   Differentiate $y = \dfrac{\sqrt{(x-2)^3}}{(x+1)^2(2x-1)}$ with respect to x and evaluate $\dfrac{dy}{dx}$ when $x = 3$

Using logarithmic differentiation and following the above procedure:

(i) Since $y = \dfrac{\sqrt{(x-2)^3}}{(x+1)^2(2x-1)}$

then $\ln y = \ln\left\{\dfrac{\sqrt{(x-2)^3}}{(x+1)^2(2x-1)}\right\}$

$= \ln\left\{\dfrac{(x-2)^{\frac{3}{2}}}{(x+1)^2(2x-1)}\right\}$

(ii) $\ln y = \ln(x-2)^{\frac{3}{2}} - \ln(x+1)^2 - \ln(2x-1)$

i.e. $\ln y = \dfrac{3}{2}\ln(x-2) - 2\ln(x+1)$

$- \ln(2x-1)$

(iii) $\dfrac{1}{y}\dfrac{dy}{dx} = \dfrac{\frac{3}{2}}{(x-2)} - \dfrac{2}{(x+1)} - \dfrac{2}{(2x-1)}$

(iv) $\dfrac{dy}{dx} = y\left\{\dfrac{3}{2(x-2)} - \dfrac{2}{(x+1)} - \dfrac{2}{(2x-1)}\right\}$

(v) $\dfrac{dy}{dx} = \dfrac{\sqrt{(x-2)^3}}{(x+1)^2(2x-1)}\left\{\dfrac{3}{2(x-2)}\right.$

$\left. - \dfrac{2}{(x+1)} - \dfrac{2}{(2x-1)}\right\}$

When $x = 3, \dfrac{dy}{dx} = \dfrac{\sqrt{(1)^3}}{(4)^2(5)}\left(\dfrac{3}{2} - \dfrac{2}{4} - \dfrac{2}{5}\right)$

$= \pm\dfrac{1}{80}\left(\dfrac{3}{5}\right) = \pm\dfrac{3}{400}$ or ± 0.0075

Problem 3.   Given $y = \dfrac{3e^{2\theta}\sec 2\theta}{\sqrt{(\theta-2)}}$ determine $\dfrac{dy}{d\theta}$

Using logarithmic differentiation and following the procedure gives:

(i) Since $y = \dfrac{3e^{2\theta}\sec 2\theta}{\sqrt{(\theta-2)}}$

then $\quad \ln y = \ln \left\{ \dfrac{3e^{2\theta} \sec 2\theta}{\sqrt{(\theta - 2)}} \right\}$

$\qquad = \ln \left\{ \dfrac{3e^{2\theta} \sec 2\theta}{(\theta - 2)^{\frac{1}{2}}} \right\}$

(ii) $\ln y = \ln 3e^{2\theta} + \ln \sec 2\theta - \ln(\theta - 2)^{\frac{1}{2}}$

i.e. $\ln y = \ln 3 + \ln e^{2\theta} + \ln \sec 2\theta$

$\qquad\qquad\qquad\qquad - \tfrac{1}{2}\ln(\theta - 2)$

i.e. $\ln y = \ln 3 + 2\theta + \ln \sec 2\theta - \tfrac{1}{2}\ln(\theta - 2)$

(iii) Differentiating with respect to θ gives:

$\dfrac{1}{y}\dfrac{dy}{d\theta} = 0 + 2 + \dfrac{2\sec 2\theta \tan 2\theta}{\sec 2\theta} - \dfrac{\frac{1}{2}}{(\theta - 2)}$

$\qquad\qquad\qquad$ from equations (1) and (2)

(iv) Rearranging gives:

$\dfrac{dy}{d\theta} = y\left\{ 2 + 2\tan 2\theta - \dfrac{1}{2(\theta - 2)} \right\}$

(v) Substituting for y gives:

$\dfrac{dy}{d\theta} = \dfrac{3e^{2\theta}\sec 2\theta}{\sqrt{(\theta - 2)}}\left\{ 2 + 2\tan 2\theta - \dfrac{1}{2(\theta - 2)} \right\}$

Problem 4. Differentiate $y = \dfrac{x^3 \ln 2x}{e^x \sin x}$ with respect to x.

Using logarithmic differentiation and following the procedure gives:

(i) $\ln y = \ln \left\{ \dfrac{x^3 \ln 2x}{e^x \sin x} \right\}$

(ii) $\ln y = \ln x^3 + \ln(\ln 2x) - \ln(e^x) - \ln(\sin x)$

i.e. $\ln y = 3\ln x + \ln(\ln 2x) - x - \ln(\sin x)$

(iii) $\dfrac{1}{y}\dfrac{dy}{dx} = \dfrac{3}{x} + \dfrac{\frac{1}{x}}{\ln 2x} - 1 - \dfrac{\cos x}{\sin x}$

(iv) $\dfrac{dy}{dx} = y\left\{ \dfrac{3}{x} + \dfrac{1}{x\ln 2x} - 1 - \cot x \right\}$

(v) $\dfrac{dy}{dx} = \dfrac{x^3 \ln 2x}{e^x \sin x}\left\{ \dfrac{3}{x} + \dfrac{1}{x\ln 2x} - 1 - \cot x \right\}$

Now try the following Practice Exercise

Practice Exercise 158   Differentiating logarithmic functions (Answers on page 879)

In Problems 1 to 6, use logarithmic differentiation to differentiate the given functions with respect to the variable.

1. $y = \dfrac{(x - 2)(x + 1)}{(x - 1)(x + 3)}$

2. $y = \dfrac{(x + 1)(2x + 1)^3}{(x - 3)^2(x + 2)^4}$

3. $y = \dfrac{(2x - 1)\sqrt{(x + 2)}}{(x - 3)\sqrt{(x + 1)^3}}$

4. $y = \dfrac{e^{2x}\cos 3x}{\sqrt{(x - 4)}}$

5. $y = 3\theta\sin\theta\cos\theta$

6. $y = \dfrac{2x^4 \tan x}{e^{2x}\ln 2x}$

7. Evaluate $\dfrac{dy}{dx}$ when $x = 1$ given

$y = \dfrac{(x + 1)^2\sqrt{(2x - 1)}}{\sqrt{(x + 3)^3}}$

8. Evaluate $\dfrac{dy}{d\theta}$, correct to 3 significant figures, when $\theta = \dfrac{\pi}{4}$ given $y = \dfrac{2e^\theta \sin\theta}{\sqrt{\theta^5}}$

36.5   Differentiation of $[f(x)]^x$

Whenever an expression to be differentiated contains a term raised to a power which is itself a function of the variable, then logarithmic differentiation must be used. For example, the differentiation of expressions such as x^x, $(x + 2)^x$, $\sqrt[x]{(x - 1)}$ and x^{3x+2} can only be achieved using logarithmic differentiation.

Problem 5. Determine $\dfrac{dy}{dx}$ given $y = x^x$

Taking Napierian logarithms of both sides of $y = x^x$ gives:

$\ln y = \ln x^x = x \ln x$, by law (iii) of Section 36.2.
Differentiating both sides with respect to x gives:

$\dfrac{1}{y}\dfrac{dy}{dx} = (x)\left(\dfrac{1}{x}\right) + (\ln x)(1)$, using the product rule

i.e. $\dfrac{1}{y}\dfrac{dy}{dx} = 1 + \ln x$,

from which, $\dfrac{dy}{dx} = y(1 + \ln x)$

i.e. $\dfrac{dy}{dx} = x^x(1 + \ln x)$

Problem 6.   Evaluate $\dfrac{dy}{dx}$ when $x = -1$ given $y = (x+2)^x$

Taking Napierian logarithms of both sides of $y = (x+2)^x$ gives:

$\ln y = \ln(x+2)^x = x \ln(x+2)$, by law (iii) of Section 36.2

Differentiating both sides with respect to x gives:

$\dfrac{1}{y}\dfrac{dy}{dx} = (x)\left(\dfrac{1}{x+2}\right) + [\ln(x+2)](1),$

by the product rule.

Hence $\dfrac{dy}{dx} = y\left(\dfrac{x}{x+2} + \ln(x+2)\right)$

$= (x+2)^x\left\{\dfrac{x}{x+2} + \ln(x+2)\right\}$

When $x = -1$, $\dfrac{dy}{dx} = (1)^{-1}\left(\dfrac{-1}{1} + \ln 1\right)$

$= (+1)(-1) = -1$

Problem 7.   Determine (a) the differential coefficient of $y = \sqrt[x]{(x-1)}$ and (b) evaluate $\dfrac{dy}{dx}$ when $x = 2$

(a) $y = \sqrt[x]{(x-1)} = (x-1)^{\frac{1}{x}}$, since by the laws of indices $\sqrt[n]{a^m} = a^{\frac{m}{n}}$

Taking Napierian logarithms of both sides gives:

$\ln y = \ln(x-1)^{\frac{1}{x}} = \dfrac{1}{x}\ln(x-1),$

by law (iii) of Section 36.2.

Differentiating each side with respect to x gives:

$\dfrac{1}{y}\dfrac{dy}{dx} = \left(\dfrac{1}{x}\right)\left(\dfrac{1}{x-1}\right) + [\ln(x-1)]\left(\dfrac{-1}{x^2}\right),$

by the product rule.

Hence $\dfrac{dy}{dx} = y\left\{\dfrac{1}{x(x-1)} - \dfrac{\ln(x-1)}{x^2}\right\}$

i.e. $\dfrac{dy}{dx} = \sqrt[x]{(x-1)}\left\{\dfrac{1}{x(x-1)} - \dfrac{\ln(x-1)}{x^2}\right\}$

(b)   When $x = 2$, $\dfrac{dy}{dx} = \sqrt[2]{(1)}\left\{\dfrac{1}{2(1)} - \dfrac{\ln(1)}{4}\right\}$

$= \pm 1\left\{\dfrac{1}{2} - 0\right\} = \pm\dfrac{1}{2}$

Problem 8.   Differentiate x^{3x+2} with respect to x.

Let $y = x^{3x+2}$

Taking Napierian logarithms of both sides gives:

$\ln y = \ln x^{3x+2}$

i.e. $\ln y = (3x+2)\ln x$, by law (iii) of Section 36.2.
Differentiating each term with respect to x gives:

$\dfrac{1}{y}\dfrac{dy}{dx} = (3x+2)\left(\dfrac{1}{x}\right) + (\ln x)(3),$

by the product rule.

Hence $\dfrac{dy}{dx} = y\left\{\dfrac{3x+2}{x} + 3\ln x\right\}$

$= x^{3x+2}\left\{\dfrac{3x+2}{x} + 3\ln x\right\}$

$= x^{3x+2}\left\{3 + \dfrac{2}{x} + 3\ln x\right\}$

Now try the following Practice Exercise

Practice Exercise 159   Differentiating $[f(x)]^x$ type functions (Answers on page 879)

In Problems 1 to 4, differentiate with respect to x.

1. $y = x^{2x}$

2. $y = (2x-1)^x$

3. $y = \sqrt[x]{(x+3)}$

4. $y = 3x^{4x+1}$

5. Show that when $y = 2x^x$ and $x = 1$, $\dfrac{dy}{dx} = 2$

6. Evaluate $\dfrac{d}{dx}\left\{\sqrt[x]{(x-2)}\right\}$ when $x = 3$

7. Show that if $y = \theta^\theta$ and $\theta = 2$, $\dfrac{dy}{d\theta} = 6.77$, correct to 3 significant figures.

For fully worked solutions to each of the problems in Practice Exercises 157 to 159 in this chapter, go to the website:
www.routledge.com/cw/bird

This Revision Test covers the material contained in Chapters 34 to 36. *The marks for each question are shown in brackets at the end of each question.*

1. A cycloid has parametric equations given by: $x = 5(\theta - \sin\theta)$ and $y = 5(1 - \cos\theta)$. Evaluate

   (a) $\dfrac{dy}{dx}$ (b) $\dfrac{d^2y}{dx^2}$ when $\theta = 1.5$ radians. Give answers correct to 3 decimal places. (8)

2. Determine the equation of (a) the tangent, and (b) the normal, drawn to an ellipse $x = 4\cos\theta$, $y = \sin\theta$ at $\theta = \dfrac{\pi}{3}$ (8)

3. Determine expressions for $\dfrac{dz}{dy}$ for each of the following functions:

   (a) $z = 5y^2 \cos x$ (b) $z = x^2 + 4xy - y^2$ (5)

4. If $x^2 + y^2 + 6x + 8y + 1 = 0$, find $\dfrac{dy}{dx}$ in terms of x and y. (4)

5. Determine the gradient of the tangents drawn to the hyperbola $x^2 - y^2 = 8$ at $x = 3$ (4)

6. Use logarithmic differentiation to differentiate $y = \dfrac{(x+1)^2 \sqrt{(x-2)}}{(2x-1)\sqrt[3]{(x-3)^4}}$ with respect to x. (6)

7. Differentiate $y = \dfrac{3e^\theta \sin 2\theta}{\sqrt{\theta^5}}$ and hence evaluate $\dfrac{dy}{d\theta}$, correct to 2 decimal places, when $\theta = \dfrac{\pi}{3}$ (9)

8. Evaluate $\dfrac{d}{dt}\left[\sqrt[t]{(2t+1)}\right]$ when $t = 2$, correct to 4 significant figures. (6)

Section H

For lecturers/instructors/teachers, fully worked solutions to each of the problems in Revision Test 11, together with a full marking scheme, are available at the website:
www.routledge.com/cw/bird

Chapter 37

Differentiation of hyperbolic functions

Why it is important to understand: Differentiation of hyperbolic functions

Hyperbolic functions have applications in many areas of engineering. For example, the shape formed by a wire freely hanging between two points (known as a catenary curve) is described by the hyperbolic cosine. Hyperbolic functions are also used in electrical engineering applications and for solving differential equations; other applications of hyperbolic functions are found in fluid dynamics, optics, heat, mechanical engineering, and in astronomy when dealing with the curvature of light in the presence of black holes. Differentiation of hyperbolic functions is quite straightforward.

At the end of this chapter, you should be able to:

- derive the differential coefficients of hyperbolic functions
- differentiate hyperbolic functions

37.1 Standard differential coefficients of hyperbolic functions

From Chapter 16,

$$\frac{d}{dx}(\sinh x) = \frac{d}{dx}\left(\frac{e^x - e^{-x}}{2}\right) = \left[\frac{e^x - (-e^{-x})}{2}\right]$$

$$= \left(\frac{e^x + e^{-x}}{2}\right) = \cosh x$$

If $y = \sinh ax$, where a is a constant, then

$$\frac{dy}{dx} = a\cosh ax$$

$$\frac{d}{dx}(\cosh x) = \frac{d}{dx}\left(\frac{e^x + e^{-x}}{2}\right) = \left[\frac{e^x + (-e^{-x})}{2}\right]$$

$$= \left(\frac{e^x - e^{-x}}{2}\right) = \sinh x$$

If $y = \cosh ax$, where a is a constant, then

$$\frac{dy}{dx} = a\sinh ax$$

Using the quotient rule of differentiation the derivatives of $\tanh x$, $\operatorname{sech} x$, $\operatorname{cosech} x$ and $\coth x$ may be determined using the above results.

Problem 1. Determine the differential coefficient of: (a) $\th x$ (b) $\sech x$

(a) $\dfrac{d}{dx}(\th x) = \dfrac{d}{dx}\left(\dfrac{\sh x}{\ch x}\right)$

$= \dfrac{(\ch x)(\ch x) - (\sh x)(\sh x)}{\ch^2 x}$

using the quotient rule

$= \dfrac{\ch^2 x - \sh^2 x}{\ch^2 x} = \dfrac{1}{\ch^2 x} = \mathbf{sech^2\,x}$

(b) $\dfrac{d}{dx}(\sech x) = \dfrac{d}{dx}\left(\dfrac{1}{\ch x}\right)$

$= \dfrac{(\ch x)(0) - (1)(\sh x)}{\ch^2 x}$

$= \dfrac{-\sh x}{\ch^2 x} = -\left(\dfrac{1}{\ch x}\right)\left(\dfrac{\sh x}{\ch x}\right)$

$= \mathbf{-sech\,x\,th\,x}$

Problem 2. Determine $\dfrac{dy}{d\theta}$ given (a) $y = \cosech\theta$ (b) $y = \coth\theta$

(a) $\dfrac{d}{d\theta}(\cosec\theta) = \dfrac{d}{d\theta}\left(\dfrac{1}{\sh\theta}\right)$

$= \dfrac{(\sh\theta)(0) - (1)(\ch\theta)}{\sh^2\theta}$

$= \dfrac{-\ch\theta}{\sh^2\theta} = -\left(\dfrac{1}{\sh\theta}\right)\left(\dfrac{\ch\theta}{\sh\theta}\right)$

$= \mathbf{-cosech\,\theta\,coth\,\theta}$

(b) $\dfrac{d}{d\theta}(\coth\theta) = \dfrac{d}{d\theta}\left(\dfrac{\ch\theta}{\sh\theta}\right)$

$= \dfrac{(\sh\theta)(\sh\theta) - (\ch\theta)(\ch\theta)}{\sh^2\theta}$

$= \dfrac{\sh^2\theta - \ch^2\theta}{\sh^2\theta} = \dfrac{-(\ch^2\theta - \sh^2\theta)}{\sh^2\theta}$

$= \dfrac{-1}{\sh^2\theta} = \mathbf{-cosech^2\,\theta}$

Summary of differential coefficients

y or $f(x)$	$\dfrac{dy}{dx}$ or $f'(x)$
$\sinh ax$	$a\cosh ax$
$\cosh ax$	$a\sinh ax$
$\tanh ax$	$a\sech^2 ax$
$\sech ax$	$-a\sech ax\tanh ax$
$\cosech ax$	$-a\cosech ax\coth ax$
$\coth ax$	$-a\cosech^2 ax$

37.2 Further worked problems on differentiation of hyperbolic functions

Problem 3. Differentiate the following with respect to x:

(a) $y = 4\sh 2x - \dfrac{3}{7}\ch 3x$

(b) $y = 5\th\dfrac{x}{2} - 2\coth 4x$

(a) $y = 4\sh 2x - \dfrac{3}{7}\ch 3x$

$\dfrac{dy}{dx} = 4(2\cosh 2x) - \dfrac{3}{7}(3\sinh 3x)$

$= 8\cosh 2x - \dfrac{9}{7}\sinh 3x$

(b) $y = 5\th\dfrac{x}{2} - 2\coth 4x$

$\dfrac{dy}{dx} = 5\left(\dfrac{1}{2}\sech^2\dfrac{x}{2}\right) - 2(-4\cosech^2 4x)$

$= \dfrac{5}{2}\sech^2\dfrac{x}{2} + 8\cosech^2 4x$

Problem 4. Differentiate the following with respect to the variable: (a) $y = 4\sin 3t\,\ch 4t$ (b) $y = \ln(\sh 3\theta) - 4\ch^2 3\theta$

Section H

(a) $y = 4\sin 3t \, \text{ch} \, 4t$ (i.e. a product)

$$\frac{dy}{dx} = (4\sin 3t)(4\,\text{sh}\,4t) + (\text{ch}\,4t)(4)(3\cos 3t)$$

$$= 16\sin 3t \, \text{sh}\,4t + 12\,\text{ch}\,4t\cos 3t$$

$$= \mathbf{4(4\sin 3t\,\text{sh}\,4t + 3\cos 3t\,\text{ch}\,4t)}$$

(b) $y = \ln(\text{sh}\,3\theta) - 4\,\text{ch}^2 3\theta$

(i.e. a function of a function)

$$\frac{dy}{d\theta} = \left(\frac{1}{\text{sh}\,3\theta}\right)(3\,\text{ch}\,3\theta) - (4)(2\,\text{ch}\,3\theta)(3\,\text{sh}\,3\theta)$$

$$= 3\coth 3\theta - 24\,\text{ch}\,3\theta\,\text{sh}\,3\theta$$

$$= \mathbf{3(\coth 3\theta - 8\,\text{ch}\,3\theta\,\text{sh}\,3\theta)}$$

Problem 5. Show that the differential coefficient of

$$y = \frac{3x^2}{\text{ch}\,4x} \text{ is: } 6x\,\text{sech}\,4x\,(1 - 2x\,\text{th}\,4x).$$

$$y = \frac{3x^2}{\text{ch}\,4x} \quad \text{(i.e. a quotient)}$$

$$\frac{dy}{dx} = \frac{(\text{ch}\,4x)(6x) - (3x^2)(4\,\text{sh}\,4x)}{(\text{ch}\,4x)^2}$$

$$= \frac{6x(\text{ch}\,4x - 2x\,\text{sh}\,4x)}{(\text{ch}^2 4x)}$$

$$= 6x\left[\frac{\text{ch}\,4x}{\text{ch}^2 4x} - \frac{2x\,\text{sh}\,4x}{\text{ch}^2 4x}\right]$$

$$= 6x\left[\frac{1}{\text{ch}\,4x} - 2x\left(\frac{\text{sh}\,4x}{\text{ch}\,4x}\right)\left(\frac{1}{\text{ch}\,4x}\right)\right]$$

$$= 6x[\text{sech}\,4x - 2x\,\text{th}\,4x\,\text{sech}\,4x]$$

$$= \mathbf{6x\,\text{sech}\,4x\,(1 - 2x\,\text{th}\,4x)}$$

Now try the following Practice Exercise

Practice Exercise 160 Differentiation of hyperbolic functions (Answers on page 879)

In Problems 1 to 5 differentiate the given functions with respect to the variable:

1. (a) $3\,\text{sh}\,2x$ (b) $2\,\text{ch}\,5\theta$ (c) $4\,\text{th}\,9t$

2. (a) $\frac{2}{3}\text{sech}\,5x$ (b) $\frac{5}{8}\text{cosech}\frac{t}{2}$ (c) $2\coth 7\theta$

3. (a) $2\ln(\text{sh}\,x)$ (b) $\frac{3}{4}\ln\left(\text{th}\left(\frac{\theta}{2}\right)\right)$

4. (a) $\text{sh}\,2x\,\text{ch}\,2x$ (b) $3e^{2x}\,\text{th}\,2x$

5. (a) $\frac{3\,\text{sh}\,4x}{2x^3}$ (b) $\frac{\text{ch}\,2t}{\cos 2t}$

For fully worked solutions to each of the problems in Practice Exercise 160 in this chapter, go to the website:
www.routledge.com/cw/bird

Chapter 38

Differentiation of inverse trigonometric and hyperbolic functions

---

Why it is important to understand: **Differentiation of inverse trigonometric hyperbolic functions**

As has been mentioned earlier, hyperbolic functions have applications in many areas of engineering. For example, the hyperbolic sine arises in the gravitational potential of a cylinder, the hyperbolic cosine function is the shape of a hanging cable, the hyperbolic tangent arises in the calculation of and rapidity of special relativity, the hyperbolic secant arises in the profile of a laminar jet, and the hyperbolic cotangent arises in the Langevin function for magnetic polarisation. So there are plenty of applications for inverse functions in engineering and this chapter explains how to differentiate inverse trigonometric and hyperbolic functions.

At the end of this chapter, you should be able to:

- understand inverse functions
- differentiate inverse trigonometric functions
- evaluate inverse hyperbolic functions using logarithmic forms
- differentiate inverse hyperbolic functions

38.1 Inverse functions

If $y = 3x - 2$, then by transposition, $x = \dfrac{y+2}{3}$. The function $x = \dfrac{y+2}{3}$ is called the **inverse function** of $y = 3x - 2$ (see page 229).

Inverse trigonometric functions are denoted by prefixing the function with 'arc' or, more commonly, by using the $^{-1}$ notation. For example, if $y = \sin x$, then $x = \arcsin y$ or $x = \sin^{-1} y$. Similarly, if $y = \cos x$, then

$x = \arccos y$ or $x = \cos^{-1} y$, and so on. In this chapter the $^{-1}$ notation will be used. A sketch of each of the inverse trigonometric functions is shown in Fig. 38.1.

Inverse hyperbolic functions are denoted by prefixing the function with 'ar' or, more commonly, by using the $^{-1}$ notation. For example, if $y = \sinh x$, then $x = \operatorname{arsinh} y$ or $x = \sinh^{-1} y$. Similarly, if $y = \operatorname{sech} x$, then $x = \operatorname{arsech} y$ or $x = \operatorname{sech}^{-1} y$, and so on. In this chapter the $^{-1}$ notation will be used. A sketch of each of the inverse hyperbolic functions is shown in Fig. 38.2.

Figure 38.1

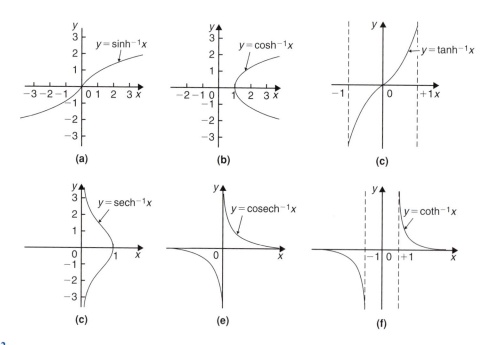

Figure 38.2

38.2 Differentiation of inverse trigonometric functions

(i) If $y = \sin^{-1} x$, then $x = \sin y$.

Differentiating both sides with respect to y gives:
$$\frac{dx}{dy} = \cos y = \sqrt{1 - \sin^2 y}$$
since $\cos^2 y + \sin^2 y = 1$, i.e. $\frac{dx}{dy} = \sqrt{1 - x^2}$

However $\frac{dy}{dx} = \frac{1}{\frac{dx}{dy}}$

Hence, when $y = \sin^{-1} x$ then
$$\frac{dy}{dx} = \frac{1}{\sqrt{1 - x^2}}$$

(ii) A sketch of part of the curve of $y = \sin^{-1} x$ is shown in Fig. 38.1(a). The principal value of $\sin^{-1} x$ is defined as the value lying between $-\pi/2$ and $\pi/2$. The gradient of the curve between points A and B is positive for all values of x and thus only the positive value is taken when evaluating $\frac{1}{\sqrt{1 - x^2}}$

(iii) Given $y = \sin^{-1} \frac{x}{a}$ then $\frac{x}{a} = \sin y$ and $x = a \sin y$

Hence $\frac{dx}{dy} = a \cos y = a\sqrt{1 - \sin^2 y}$
$$= a\sqrt{\left[1 - \left(\frac{x}{a}\right)^2\right]} = a\sqrt{\left(\frac{a^2 - x^2}{a^2}\right)}$$
$$= \frac{a\sqrt{a^2 - x^2}}{a} = \sqrt{a^2 - x^2}$$

Thus $\frac{dy}{dx} = \frac{1}{\frac{dx}{dy}} = \frac{1}{\sqrt{a^2 - x^2}}$

i.e. **when $y = \sin^{-1} \frac{x}{a}$ then $\frac{dy}{dx} = \frac{1}{\sqrt{a^2 - x^2}}$**

Since integration is the reverse process of differentiation then:
$$\int \frac{1}{\sqrt{a^2 - x^2}} dx = \sin^{-1} \frac{x}{a} + c$$

(iv) Given $y = \sin^{-1} f(x)$ the function of a function rule may be used to find $\frac{dy}{dx}$

Let $u = f(x)$ then $y = \sin^{-1} u$

Then $\frac{du}{dx} = f'(x)$ and $\frac{dy}{du} = \frac{1}{\sqrt{1 - u^2}}$
(see para. (i))

Thus $\frac{dy}{dx} = \frac{dy}{du} \times \frac{du}{dx} = \frac{1}{\sqrt{1 - u^2}} f'(x)$
$$= \frac{f'(x)}{\sqrt{1 - [f(x)]^2}}$$

(v) The differential coefficients of the remaining inverse trigonometric functions are obtained in a similar manner to that shown above and a summary of the results is shown in Table 38.1.

Table 38.1 Differential coefficients of inverse trigonometric functions

	y or $f(x)$	$\frac{dy}{dx}$ or $f'(x)$
(i)	$\sin^{-1} \frac{x}{a}$	$\frac{1}{\sqrt{a^2 - x^2}}$
	$\sin^{-1} f(x)$	$\frac{f'(x)}{\sqrt{1 - [f(x)]^2}}$
(ii)	$\cos^{-1} \frac{x}{a}$	$\frac{-1}{\sqrt{a^2 - x^2}}$
	$\cos^{-1} f(x)$	$\frac{-f'(x)}{\sqrt{1 - [f(x)]^2}}$
(iii)	$\tan^{-1} \frac{x}{a}$	$\frac{a}{a^2 + x^2}$
	$\tan^{-1} f(x)$	$\frac{f'(x)}{1 + [f(x)]^2}$
(iv)	$\sec^{-1} \frac{x}{a}$	$\frac{a}{x\sqrt{x^2 - a^2}}$
	$\sec^{-1} f(x)$	$\frac{f'(x)}{f(x)\sqrt{[f(x)]^2 - 1}}$
(v)	$\operatorname{cosec}^{-1} \frac{x}{a}$	$\frac{-a}{x\sqrt{x^2 - a^2}}$
	$\operatorname{cosec}^{-1} f(x)$	$\frac{-f'(x)}{f(x)\sqrt{[f(x)]^2 - 1}}$
(vi)	$\cot^{-1} \frac{x}{a}$	$\frac{-a}{a^2 + x^2}$
	$\cot^{-1} f(x)$	$\frac{-f'(x)}{1 + [f(x)]^2}$

Problem 1. Find $\dfrac{dy}{dx}$ given $y = \sin^{-1} 5x^2$

From Table 38.1(i), if

$$y = \sin^{-1} f(x) \text{ then } \frac{dy}{dx} = \frac{f'(x)}{\sqrt{1 - [f(x)]^2}}$$

Hence, if $y = \sin^{-1} 5x^2$ then $f(x) = 5x^2$ and $f'(x) = 10x$

Thus $\dfrac{dy}{dx} = \dfrac{10x}{\sqrt{1 - (5x^2)^2}} = \dfrac{10x}{\sqrt{1 - 25x^4}}$

Problem 2.
(a) Show that if $y = \cos^{-1} x$ then
$$\frac{dy}{dx} = \frac{1}{\sqrt{1 - x^2}}$$

(b) Hence obtain the differential coefficient of $y = \cos^{-1}(1 - 2x^2)$

(a) If $y = \cos^{-1} x$ then $x = \cos y$

Differentiating with respect to y gives:

$$\frac{dx}{dy} = -\sin y = -\sqrt{1 - \cos^2 y}$$
$$= -\sqrt{1 - x^2}$$

Hence $\dfrac{dy}{dx} = \dfrac{1}{\dfrac{dx}{dy}} = -\dfrac{1}{\sqrt{1 - x^2}}$

The principal value of $y = \cos^{-1} x$ is defined as the angle lying between 0 and π, i.e. between points C and D shown in Fig. 38.1(b). The gradient of the curve is negative between C and D and thus the differential coefficient $\dfrac{dy}{dx}$ is negative as shown above.

(b) If $y = \cos^{-1} f(x)$ then by letting $u = f(x)$, $y = \cos^{-1} u$

Then $\dfrac{dy}{du} = -\dfrac{1}{\sqrt{1 - u^2}}$ (from part (a))

and $\dfrac{du}{dx} = f'(x)$

From the function of a function rule,

$$\frac{dy}{dx} = \frac{dy}{du} \cdot \frac{du}{dx} = -\frac{1}{\sqrt{1 - u^2}} f'(x)$$

$$= \frac{-f'(x)}{\sqrt{1 - [f(x)]^2}}$$

Hence, when $y = \cos^{-1}(1 - 2x^2)$

then $\dfrac{dy}{dx} = \dfrac{-(-4x)}{\sqrt{1 - [1 - 2x^2]^2}}$

$$= \frac{4x}{\sqrt{1 - (1 - 4x^2 + 4x^4)}} = \frac{4x}{\sqrt{(4x^2 - 4x^4)}}$$

$$= \frac{4x}{\sqrt{[4x^2(1 - x^2)]}} = \frac{4x}{2x\sqrt{1 - x^2}} = \frac{2}{\sqrt{1 - x^2}}$$

Problem 3. Determine the differential coefficient of $y = \tan^{-1} \dfrac{x}{a}$ and show that the differential coefficient of $\tan^{-1} \dfrac{2x}{3}$ is $\dfrac{6}{9 + 4x^2}$

If $y = \tan^{-1} \dfrac{x}{a}$ then $\dfrac{x}{a} = \tan y$ and $x = a \tan y$

$\dfrac{dx}{dy} = a \sec^2 y = a(1 + \tan^2 y)$ since

$$\sec^2 y = 1 + \tan^2 y$$

$$= a\left[1 + \left(\frac{x}{a}\right)^2\right] = a\left(\frac{a^2 + x^2}{a^2}\right)$$

$$= \frac{a^2 + x^2}{a}$$

Hence $\dfrac{dy}{dx} = \dfrac{1}{\dfrac{dx}{dy}} = \dfrac{a}{a^2 + x^2}$

The principal value of $y = \tan^{-1} x$ is defined as the angle lying between $-\dfrac{\pi}{2}$ and $\dfrac{\pi}{2}$ and the gradient $\left(\text{i.e. } \dfrac{dy}{dx}\right)$ between these two values is always positive (see Fig. 38.1(c)).

Comparing $\tan^{-1} \dfrac{2x}{3}$ with $\tan^{-1} \dfrac{x}{a}$ shows that $a = \dfrac{3}{2}$

Hence if $y = \tan^{-1} \dfrac{2x}{3}$ then

$$\frac{dy}{dx} = \frac{\dfrac{3}{2}}{\left(\dfrac{3}{2}\right)^2 + x^2} = \frac{\dfrac{3}{2}}{\dfrac{9}{4} + x^2} = \frac{\dfrac{3}{2}}{\dfrac{9 + 4x^2}{4}}$$

$$= \frac{\dfrac{3}{2}(4)}{9 + 4x^2} = \frac{6}{9 + 4x^2}$$

Problem 4. Find the differential coefficient of $y = \ln(\cos^{-1} 3x)$

Let $u = \cos^{-1} 3x$ then $y = \ln u$

By the function of a function rule,

$$\frac{dy}{dx} = \frac{dy}{du} \cdot \frac{du}{dx} = \frac{1}{u} \times \frac{d}{dx}(\cos^{-1} 3x)$$

$$= \frac{1}{\cos^{-1} 3x}\left\{\frac{-3}{\sqrt{1-(3x)^2}}\right\}$$

i.e. $\dfrac{d}{dx}[\ln(\cos^{-1} 3x)] = \dfrac{-3}{\sqrt{1-9x^2}\,\cos^{-1} 3x}$

Problem 5.   If $y = \tan^{-1}\dfrac{3}{t^2}$ find $\dfrac{dy}{dt}$

Using the general form from Table 38.1(iii),

$$f(t) = \frac{3}{t^2} = 3t^{-2}$$

from which $f'(t) = \dfrac{-6}{t^3}$

Hence $\dfrac{d}{dt}\left(\tan^{-1}\dfrac{3}{t^2}\right) = \dfrac{f'(t)}{1+[f(t)]^2}$

$$= \frac{-\dfrac{6}{t^3}}{\left\{1+\left(\dfrac{3}{t^2}\right)^2\right\}} = \frac{-\dfrac{6}{t^3}}{\dfrac{t^4+9}{t^4}}$$

$$= \left(-\frac{6}{t^3}\right)\left(\frac{t^4}{t^4+9}\right) = -\frac{6t}{t^4+9}$$

Problem 6.   Differentiate $y = \dfrac{\cot^{-1} 2x}{1+4x^2}$

Using the quotient rule:

$$\frac{dy}{dx} = \frac{(1+4x^2)\left(\dfrac{-2}{1+(2x)^2}\right) - (\cot^{-1} 2x)(8x)}{(1+4x^2)^2}$$

from Table 38.1(vi)

$$= \frac{-2(1+4x\cot^{-1} 2x)}{(1+4x^2)^2}$$

Problem 7.   Differentiate $y = x\,\mathrm{cosec}^{-1}\,x$

Using the product rule:

$$\frac{dy}{dx} = (x)\left[\frac{-1}{x\sqrt{x^2-1}}\right] + (\mathrm{cosec}^{-1}\,x)\,(1)$$

from Table 38.1(v)

$$= \frac{-1}{\sqrt{x^2-1}} + \mathrm{cosec}^{-1} x$$

Problem 8.   Show that if
$$y = \tan^{-1}\left(\frac{\sin t}{\cos t - 1}\right) \text{ then } \frac{dy}{dt} = \frac{1}{2}$$

If $f(t) = \left(\dfrac{\sin t}{\cos t - 1}\right)$

then $f'(t) = \dfrac{(\cos t - 1)(\cos t) - (\sin t)(-\sin t)}{(\cos t - 1)^2}$

$$= \frac{\cos^2 t - \cos t + \sin^2 t}{(\cos t - 1)^2} = \frac{1-\cos t}{(\cos t - 1)^2}$$

since $\sin^2 t + \cos^2 t = 1$

$$= \frac{-(\cos t - 1)}{(\cos t - 1)^2} = \frac{-1}{\cos t - 1}$$

Using Table 38.1(iii), when
$$y = \tan^{-1}\left(\frac{\sin t}{\cos t - 1}\right)$$

then $\dfrac{dy}{dt} = \dfrac{\dfrac{-1}{\cos t - 1}}{1+\left(\dfrac{\sin t}{\cos t - 1}\right)^2}$

$$= \frac{\dfrac{-1}{\cos t - 1}}{\dfrac{(\cos t - 1)^2 + (\sin t)^2}{(\cos t - 1)^2}}$$

$$= \left(\frac{-1}{\cos t - 1}\right)\left(\frac{(\cos t - 1)^2}{\cos^2 t - 2\cos t + 1 + \sin^2 t}\right)$$

$$= \frac{-(\cos t - 1)}{2 - 2\cos t} = \frac{1-\cos t}{2(1-\cos t)} = \frac{1}{2}$$

Now try the following Practice Exercise

Practice Exercise 161   Differentiating inverse trigonometric functions (Answers on page 880)

In Problems 1 to 6, differentiate with respect to the variable.

1. (a) $\sin^{-1} 4x$ (b) $\sin^{-1}\dfrac{x}{2}$

2. (a) $\cos^{-1} 3x$ (b) $\dfrac{2}{3}\cos^{-1}\dfrac{x}{3}$

3. (a) $3\tan^{-1} 2x$ (b) $\dfrac{1}{2}\tan^{-1}\sqrt{x}$

4. (a) $2\sec^{-1} 2t$ (b) $\sec^{-1}\dfrac{3}{4}x$

5. (a) $\dfrac{5}{2}\csc^{-1}\dfrac{\theta}{2}$ (b) $\csc^{-1} x^2$

6. (a) $3\cot^{-1} 2t$ (b) $\cot^{-1}\sqrt{\theta^2-1}$

7. Show that the differential coefficient of $\tan^{-1}\left(\dfrac{x}{1-x^2}\right)$ is $\dfrac{1+x^2}{1-x^2+x^4}$

In Problems 8 to 11 differentiate with respect to the variable.

8. (a) $2x\sin^{-1} 3x$ (b) $t^2\sec^{-1} 2t$

9. (a) $\theta^2\cos^{-1}(\theta^2-1)$ (b) $(1-x^2)\tan^{-1} x$

10. (a) $2\sqrt{t}\cot^{-1} t$ (b) $x\csc^{-1}\sqrt{x}$

11. (a) $\dfrac{\sin^{-1} 3x}{x^2}$ (b) $\dfrac{\cos^{-1} x}{\sqrt{1-x^2}}$

38.3 Logarithmic forms of inverse hyperbolic functions

Inverse hyperbolic functions may be evaluated most conveniently when expressed in a **logarithmic form**.

For example, if $y=\sinh^{-1}\dfrac{x}{a}$ then $\dfrac{x}{a}=\sinh y$.

From Chapter 16, $e^y=\cosh y+\sinh y$ and $\cosh^2 y-\sinh^2 y=1$, from which,

$\cosh y=\sqrt{1+\sinh^2 y}$ which is positive since $\cosh y$ is always positive (see Fig. 16.2, page 183).

Hence $e^y=\sqrt{1+\sinh^2 y}+\sinh y$

$$=\sqrt{\left[1+\left(\dfrac{x}{a}\right)^2\right]}+\dfrac{x}{a}=\sqrt{\left(\dfrac{a^2+x^2}{a^2}\right)}+\dfrac{x}{a}$$

$$=\dfrac{\sqrt{a^2+x^2}}{a}+\dfrac{x}{a} \quad\text{or}\quad \dfrac{x+\sqrt{a^2+x^2}}{a}$$

Taking Napierian logarithms of both sides gives:

$$y=\ln\left\{\dfrac{x+\sqrt{a^2+x^2}}{a}\right\}$$

Hence, $\sinh^{-1}\dfrac{x}{a}=\ln\left\{\dfrac{x+\sqrt{a^2+x^2}}{a}\right\}$ \hfill (1)

Thus to evaluate $\sinh^{-1}\dfrac{3}{4}$, let $x=3$ and $a=4$ in equation (1).

Then $\sinh^{-1}\dfrac{3}{4}=\ln\left\{\dfrac{3+\sqrt{4^2+3^2}}{4}\right\}$

$$=\ln\left(\dfrac{3+5}{4}\right)=\ln 2=0.6931$$

By similar reasoning to the above it may be shown that:

$$\cosh^{-1}\dfrac{x}{a}=\ln\left\{\dfrac{x+\sqrt{x^2-a^2}}{a}\right\}$$

and $\quad\tanh^{-1}\dfrac{x}{a}=\dfrac{1}{2}\ln\left(\dfrac{a+x}{a-x}\right)$

Problem 9. Evaluate, correct to 4 decimal places, $\sinh^{-1} 2$

From above, $\sinh^{-1}\dfrac{x}{a}=\ln\left\{\dfrac{x+\sqrt{a^2+x^2}}{a}\right\}$

With $x=2$ and $a=1$,

$$\sinh^{-1} 2=\ln\left\{\dfrac{2+\sqrt{1^2+2^2}}{1}\right\}$$

$$=\ln(2+\sqrt{5})=\ln 4.2361$$

$$= \mathbf{1.4436},\textbf{ correct to 4 decimal places}$$

Using a calculator,

(i) press hyp

(ii) press 4 and $\sinh^{-1}($ appears

(iii) type in 2

(iv) press) to close the brackets

(v) press = and 1.443635475 appears

Hence, $\sinh^{-1} 2=\mathbf{1.4436}$, correct to 4 decimal places.

Problem 10. Show that

$$\tanh^{-1}\frac{x}{a} = \frac{1}{2}\ln\left(\frac{a+x}{a-x}\right)$$ and evaluate, correct

to 4 decimal places, $\tanh^{-1}\frac{3}{5}$

If $y = \tanh^{-1}\frac{x}{a}$ then $\frac{x}{a} = \tanh y$

From Chapter 16,

$$\tanh y = \frac{\sinh x}{\cosh x} = \frac{\frac{1}{2}(e^y - e^{-y})}{\frac{1}{2}(e^y + e^{-y})} = \frac{e^{2y} - 1}{e^{2y} + 1}$$

by dividing each term by e^{-y}

Thus, $$\frac{x}{a} = \frac{e^{2y} - 1}{e^{2y} + 1}$$

from which, $x(e^{2y} + 1) = a(e^{2y} - 1)$

Hence $x + a = ae^{2y} - xe^{2y} = e^{2y}(a - x)$

from which $e^{2y} = \left(\frac{a+x}{a-x}\right)$

Taking Napierian logarithms of both sides gives:

$$2y = \ln\left(\frac{a+x}{a-x}\right)$$

and $$y = \frac{1}{2}\ln\left(\frac{a+x}{a-x}\right)$$

Hence, $$\mathbf{\tanh^{-1}\frac{x}{a} = \frac{1}{2}\ln\left(\frac{a+x}{a-x}\right)}$$

Substituting $x = 3$ and $a = 5$ gives:

$$\tanh^{-1}\frac{3}{5} = \frac{1}{2}\ln\left(\frac{5+3}{5-3}\right) = \frac{1}{2}\ln 4$$

$$= \mathbf{0.6931}, \text{ correct to 4 decimal places.}$$

Problem 11. Prove that

$$\cosh^{-1}\frac{x}{a} = \ln\left\{\frac{x + \sqrt{x^2 - a^2}}{a}\right\}$$

and hence evaluate $\cosh^{-1}1.4$ correct to 4 decimal places.

If $y = \cosh^{-1}\frac{x}{a}$ then $\frac{x}{a} = \cos y$

$$e^y = \cosh y + \sinh y = \cosh y \pm \sqrt{\cosh^2 y - 1}$$

$$= \frac{x}{a} \pm \sqrt{\left[\left(\frac{x}{a}\right)^2 - 1\right]} = \frac{x}{a} \pm \frac{\sqrt{x^2 - a^2}}{a}$$

$$= \frac{x \pm \sqrt{x^2 - a^2}}{a}$$

Taking Napierian logarithms of both sides gives:

$$y = \ln\left\{\frac{x \pm \sqrt{x^2 - a^2}}{a}\right\}$$

Thus, assuming the principal value,

$$\mathbf{\cosh^{-1}\frac{x}{a} = \ln\left\{\frac{x + \sqrt{x^2 - a^2}}{a}\right\}}$$

$$\cosh^{-1}1.4 = \cosh^{-1}\frac{14}{10} = \cosh^{-1}\frac{7}{5}$$

In the equation for $\cosh^{-1}\frac{x}{a}$, let $x = 7$ and $a = 5$

Then $$\cosh^{-1}\frac{7}{5} = \ln\left\{\frac{7 + \sqrt{7^2 - 5^2}}{5}\right\}$$

$$= \ln 2.3798 = \mathbf{0.8670}$$

correct to 4 decimal places.

Now try the following Practice Exercise

Practice Exercise 162   Logarithmic forms of the inverse hyperbolic functions (Answers on page 880)

In Problems 1 to 3 use logarithmic equivalents of inverse hyperbolic functions to evaluate correct to 4 decimal places.

1. (a) $\sinh^{-1}\frac{1}{2}$ (b) $\sinh^{-1}4$ (c) $\sinh^{-1}0.9$

2. (a) $\cosh^{-1}\frac{5}{4}$ (b) $\cosh^{-1}3$ (c) $\cosh^{-1}4.3$

3. (a) $\tanh^{-1}\frac{1}{4}$ (b) $\tanh^{-1}\frac{5}{8}$ (c) $\tanh^{-1}0.7$

38.4 Differentiation of inverse hyperbolic functions

If $y = \sinh^{-1} \dfrac{x}{a}$ then $\dfrac{x}{a} = \sinh y$ and $x = a \sinh y$

$\dfrac{dx}{dy} = a \cosh y$ (from Chapter 37).

Also $\cosh^2 y - \sinh^2 y = 1$, from which,

$$\cosh y = \sqrt{1 + \sinh^2 y} = \sqrt{\left[1 + \left(\dfrac{x}{a}\right)^2\right]}$$

$$= \frac{\sqrt{a^2 + x^2}}{a}$$

Hence $\dfrac{dx}{dy} = a \cosh y = \dfrac{a\sqrt{a^2 + x^2}}{a} = \sqrt{a^2 + x^2}$

Then $\dfrac{dy}{dx} = \dfrac{1}{\dfrac{dx}{dy}} = \dfrac{1}{\sqrt{a^2 + x^2}}$

[An alternative method of differentiating $\sinh^{-1} \dfrac{x}{a}$ is to differentiate the logarithmic form

$\ln\left\{\dfrac{x + \sqrt{a^2 + x^2}}{a}\right\}$ with respect to x]

From the sketch of $y = \sinh^{-1} x$ shown in Fig. 38.2(a) it is seen that the gradient $\left(\text{i.e. } \dfrac{dy}{dx}\right)$ is always positive.

It follows from above that

$$\int \frac{1}{\sqrt{x^2 + a^2}}\, dx = \sinh^{-1}\frac{x}{a} + c$$

or $\ln\left\{\dfrac{x + \sqrt{a^2 + x^2}}{a}\right\} + c$

It may be shown that

$$\frac{d}{dx}(\sinh^{-1} x) = \frac{1}{\sqrt{x^2 + 1}}$$

or more generally

$$\frac{d}{dx}[\sinh^{-1} f(x)] = \frac{f'(x)}{\sqrt{[f(x)]^2 + 1}}$$

by using the function of a function rule as in Section 38.2(iv).

The remaining inverse hyperbolic functions are differentiated in a similar manner to that shown above and the results are summarised in Table 38.2.

Table 38.2 Differential coefficients of inverse hyperbolic functions

	y or $f(x)$	$\dfrac{dy}{dx}$ or $f'(x)$
(i)	$\sinh^{-1}\dfrac{x}{a}$	$\dfrac{1}{\sqrt{x^2 + a^2}}$
	$\sinh^{-1} f(x)$	$\dfrac{f'(x)}{\sqrt{[f(x)]^2 + 1}}$
(ii)	$\cosh^{-1}\dfrac{x}{a}$	$\dfrac{1}{\sqrt{x^2 - a^2}}$
	$\cosh^{-1} f(x)$	$\dfrac{f'(x)}{\sqrt{[f(x)]^2 - 1}}$
(iii)	$\tanh^{-1}\dfrac{x}{a}$	$\dfrac{a}{a^2 - x^2}$
	$\tanh^{-1} f(x)$	$\dfrac{f'(x)}{1 - [f(x)]^2}$
(iv)	$\text{sech}^{-1}\dfrac{x}{a}$	$\dfrac{-a}{x\sqrt{a^2 - x^2}}$
	$\text{sech}^{-1} f(x)$	$\dfrac{-f'(x)}{f(x)\sqrt{1 - [f(x)]^2}}$
(v)	$\text{cosech}^{-1}\dfrac{x}{a}$	$\dfrac{-a}{x\sqrt{x^2 + a^2}}$
	$\text{cosech}^{-1} f(x)$	$\dfrac{-f'(x)}{f(x)\sqrt{[f(x)]^2 + 1}}$
(vi)	$\coth^{-1}\dfrac{x}{a}$	$\dfrac{a}{a^2 - x^2}$
	$\coth^{-1} f(x)$	$\dfrac{f'(x)}{1 - [f(x)]^2}$

Problem 12. Find the differential coefficient of $y = \sinh^{-1} 2x$

From Table 38.2(i),

$$\frac{d}{dx}[\sinh^{-1} f(x)] = \frac{f'(x)}{\sqrt{[f(x)]^2 + 1}}$$

Hence $\dfrac{d}{dx}(\sinh^{-1} 2x) = \dfrac{2}{\sqrt{[(2x)^2 + 1]}}$

$$= \frac{2}{\sqrt{[4x^2 + 1]}}$$

Problem 13. Determine

$$\frac{d}{dx}\left[\cosh^{-1}\sqrt{(x^2+1)}\right]$$

If $y=\cosh^{-1} f(x)$, $\dfrac{dy}{dx}=\dfrac{f'(x)}{\sqrt{[f(x)]^2-1}}$

If $y=\cosh^{-1}\sqrt{(x^2+1)}$, then $f(x)=\sqrt{(x^2+1)}$ and

$f'(x)=\dfrac{1}{2}(x+1)^{-1/2}(2x)=\dfrac{x}{\sqrt{(x^2+1)}}$

Hence, $\dfrac{d}{dx}\left[\cosh^{-1}\sqrt{(x^2+1)}\right]$

$$=\frac{\dfrac{x}{\sqrt{(x^2+1)}}}{\sqrt{\left[\left(\sqrt{(x^2+1)}\right)^2-1\right]}}=\frac{\dfrac{x}{\sqrt{(x^2+1)}}}{\sqrt{(x^2+1-1)}}$$

$$=\frac{\dfrac{x}{\sqrt{(x^2+1)}}}{x}=\frac{1}{\sqrt{(x^2+1)}}$$

Problem 14. Show that

$\dfrac{d}{dx}\left[\tanh^{-1}\dfrac{x}{a}\right]=\dfrac{a}{a^2-x^2}$ and hence determine the

differential coefficient of $\tanh^{-1}\dfrac{4x}{3}$

If $y=\tanh^{-1}\dfrac{x}{a}$ then $\dfrac{x}{a}=\tanh y$ and $x=a\tanh y$

$\dfrac{dx}{dy}=a\,\text{sech}^2\,y=a(1-\tanh^2 y)$, since

$$1-\text{sech}^2\,y=\tanh^2 y$$

$$=a\left[1-\left(\frac{x}{a}\right)^2\right]=a\left(\frac{a^2-x^2}{a^2}\right)=\frac{a^2-x^2}{a}$$

Hence $\dfrac{dy}{dx}=\dfrac{1}{\dfrac{dx}{dy}}=\dfrac{a}{a^2-x^2}$

Comparing $\tanh^{-1}\dfrac{4x}{3}$ with $\tanh^{-1}\dfrac{x}{a}$ shows that $a=\dfrac{3}{4}$

Hence $\dfrac{d}{dx}\left[\tanh^{-1}\dfrac{4x}{3}\right]=\dfrac{\dfrac{3}{4}}{\left(\dfrac{3}{4}\right)^2-x^2}=\dfrac{\dfrac{3}{4}}{\dfrac{9}{16}-x^2}$

$$=\frac{\dfrac{3}{4}}{\dfrac{9-16x^2}{16}}=\frac{3}{4}\cdot\frac{16}{(9-16x^2)}=\frac{12}{9-16x^2}$$

Problem 15. Differentiate $\text{cosech}^{-1}(\sinh\theta)$

From Table 38.2(v),

$$\frac{d}{dx}[\text{cosech}^{-1}\,f(x)]=\frac{-f'(x)}{f(x)\sqrt{[f(x)]^2+1}}$$

Hence $\dfrac{d}{d\theta}[\text{cosech}^{-1}(\sinh\theta)]$

$$=\frac{-\cosh\theta}{\sinh\theta\sqrt{[\sinh^2\theta+1]}}$$

$$=\frac{-\cosh\theta}{\sinh\theta\sqrt{\cosh^2\theta}}\ \text{since}\ \cosh^2\theta-\sinh^2\theta=1$$

$$=\frac{-\cosh\theta}{\sinh\theta\cosh\theta}=\frac{-1}{\sinh\theta}=-\text{cosech}\,\theta$$

Problem 16. Find the differential coefficient of $y=\text{sech}^{-1}(2x-1)$

From Table 38.2(iv),

$$\frac{d}{dx}[\text{sech}^{-1}\,f(x)]=\frac{-f'(x)}{f(x)\sqrt{1-[f(x)]^2}}$$

Hence, $\dfrac{d}{dx}[\text{sech}^{-1}(2x-1)]$

$$=\frac{-2}{(2x-1)\sqrt{[1-(2x-1)^2]}}$$

$$=\frac{-2}{(2x-1)\sqrt{[1-(4x^2-4x+1)]}}$$

$$=\frac{-2}{(2x-1)\sqrt{(4x-4x^2)}}=\frac{-2}{(2x-1)\sqrt{[4x(1-x)]}}$$

$$=\frac{-2}{(2x-1)2\sqrt{[x(1-x)]}}=\frac{-1}{(2x-1)\sqrt{[x(1-x)]}}$$

Problem 17. Show that

$$\frac{d}{dx}[\coth^{-1}(\sin x)]=\sec x$$

From Table 38.2(vi),

$$\frac{d}{dx}[\coth^{-1}\,f(x)]=\frac{f'(x)}{1-[f(x)]^2}$$

Hence $\dfrac{d}{dx}[\coth^{-1}(\sin x)]=\dfrac{\cos x}{[1-(\sin x)^2]}$

$$=\frac{\cos x}{\cos^2 x}\ \text{since}\ \cos^2 x+\sin^2 x=1$$

$$=\frac{1}{\cos x}=\sec x$$

Section H

Problem 18. Differentiate
$$y = (x^2 - 1)\tanh^{-1} x$$

Using the product rule,

$$\frac{dy}{dx} = (x^2 - 1)\left(\frac{1}{1-x^2}\right) + (\tanh^{-1} x)(2x)$$

$$= \frac{-(1-x^2)}{(1-x^2)} + 2x\tanh^{-1} x = 2x\tanh^{-1} x - 1$$

Problem 19. Determine $\displaystyle\int \frac{dx}{\sqrt{(x^2+4)}}$

Since $\dfrac{d}{dx}\left(\sinh^{-1}\dfrac{x}{a}\right) = \dfrac{1}{\sqrt{(x^2+a^2)}}$

then $\displaystyle\int \frac{dx}{\sqrt{(x^2+a^2)}} = \sinh^{-1}\frac{x}{a} + c$

Hence $\displaystyle\int \frac{1}{\sqrt{(x^2+4)}}\,dx = \int \frac{1}{\sqrt{(x^2+2^2)}}\,dx$

$$= \sinh^{-1}\frac{x}{2} + c$$

Problem 20. Determine $\displaystyle\int \frac{4}{\sqrt{(x^2-3)}}\,dx$

Since $\dfrac{d}{dx}\left(\cosh^{-1}\dfrac{x}{a}\right) = \dfrac{1}{\sqrt{(x^2-a^2)}}$

then $\displaystyle\int \frac{1}{\sqrt{(x^2-a^2)}}\,dx = \cosh^{-1}\frac{x}{a} + c$

Hence $\displaystyle\int \frac{4}{\sqrt{(x^2-3)}}\,dx = 4\int \frac{1}{\sqrt{[x^2-(\sqrt{3})^2]}}\,dx$

$$= 4\cosh^{-1}\frac{x}{\sqrt{3}} + c$$

Problem 21. Find $\displaystyle\int \frac{2}{(9-4x^2)}\,dx$

Since $\tanh^{-1}\dfrac{x}{a} = \dfrac{a}{a^2-x^2}$

then $\displaystyle\int \frac{a}{a^2-x^2}\,dx = \tanh^{-1}\frac{x}{a} + c$

i.e. $\displaystyle\int \frac{1}{a^2-x^2}\,dx = \frac{1}{a}\tanh^{-1}\frac{x}{a} + c$

Hence $\displaystyle\int \frac{2}{(9-4x^2)}\,dx = 2\int \frac{1}{4\left(\frac{9}{4}-x^2\right)}\,dx$

$$= \frac{1}{2}\int \frac{1}{\left[\left(\frac{3}{2}\right)^2 - x^2\right]}\,dx$$

$$= \frac{1}{2}\left[\frac{1}{\left(\frac{3}{2}\right)}\tanh^{-1}\frac{x}{\left(\frac{3}{2}\right)} + c\right]$$

i.e. $\displaystyle\int \frac{2}{(9-4x^2)}\,dx = \frac{1}{3}\tanh^{-1}\frac{2x}{3} + c$

Now try the following Practice Exercise

Practice Exercise 163 Differentiation of inverse hyperbolic functions (Answers on page 880)

In Problems 1 to 11, differentiate with respect to the variable.

1. (a) $\sinh^{-1}\dfrac{x}{3}$ (b) $\sinh^{-1} 4x$

2. (a) $2\cosh^{-1}\dfrac{t}{3}$ (b) $\dfrac{1}{2}\cosh^{-1} 2\theta$

3. (a) $\tanh^{-1}\dfrac{2x}{5}$ (b) $3\tanh^{-1} 3x$

4. (a) $\operatorname{sech}^{-1}\dfrac{3x}{4}$ (b) $-\dfrac{1}{2}\operatorname{sech}^{-1} 2x$

5. (a) $\operatorname{cosech}^{-1}\dfrac{x}{4}$ (b) $\dfrac{1}{2}\operatorname{cosech}^{-1} 4x$

6. (a) $\coth^{-1}\dfrac{2x}{7}$ (b) $\dfrac{1}{4}\coth^{-1} 3t$

7. (a) $2\sinh^{-1}\sqrt{(x^2-1)}$

   (b) $\dfrac{1}{2}\cosh^{-1}\sqrt{(x^2+1)}$

8. (a) $\operatorname{sech}^{-1}(x-1)$ (b) $\tanh^{-1}(\tanh x)$

9. (a) $\cosh^{-1}\left(\dfrac{t}{t-1}\right)$ (b) $\coth^{-1}(\cos x)$

10. (a) $\theta\sinh^{-1}\theta$ (b) $\sqrt{x}\cosh^{-1} x$

11. (a) $\dfrac{2\operatorname{sech}^{-1}\sqrt{t}}{t^2}$ (b) $\dfrac{\tanh^{-1} x}{(1-x^2)}$

12. Show that $\dfrac{d}{dx}[x\cosh^{-1}(\cosh x)] = 2x$

In Problems 13 to 15, determine the given integrals.

13. (a) $\displaystyle\int \dfrac{1}{\sqrt{(x^2+9)}}\,dx$

   (b) $\displaystyle\int \dfrac{3}{\sqrt{(4x^2+25)}}\,dx$

14. (a) $\displaystyle\int \dfrac{1}{\sqrt{(x^2-16)}}\,dx$

   (b) $\displaystyle\int \dfrac{1}{\sqrt{(t^2-5)}}\,dt$

15. (a) $\displaystyle\int \dfrac{d\theta}{\sqrt{(36+\theta^2)}}$ (b) $\displaystyle\int \dfrac{3}{(16-2x^2)}\,dx$

For fully worked solutions to each of the problems in Practice Exercises 161 to 163 in this chapter, go to the website:
www.routledge.com/cw/bird

Chapter 39

Partial differentiation

Why it is important to understand: Partial differentiation

First-order partial derivatives can be used for numerous applications, from determining the volume of different shapes to analysing anything from water to heat flow. Second-order partial derivatives are used in many fields of engineering. One of its applications is in solving problems related to dynamics of rigid bodies and in determination of forces and strength of materials. Partial differentiation is used to estimate errors in calculated quantities that depend on more than one uncertain experimental measurement. Thermodynamic energy functions (enthalpy, Gibbs free energy, Helmholtz free energy) are functions of two or more variables. Most thermodynamic quantities (temperature, entropy, heat capacity) can be expressed as derivatives of these functions. Many laws of nature are best expressed as relations between the partial derivatives of one or more quantities. Partial differentiation is hence important in many branches of engineering.

At the end of this chapter, you should be able to:

- determine first-order partial derivatives
- determine second-order partial derivatives

39.1 Introduction to partial derivatives

In engineering, it sometimes happens that the variation of one quantity depends on changes taking place in two, or more, other quantities. For example, the volume V of a cylinder is given by $V = \pi r^2 h$. The volume will change if either radius r or height h is changed. The formula for volume may be stated mathematically as $V = f(r, h)$ which means 'V is some function of r and h'. Some other practical examples include:

(i) time of oscillation, $t = 2\pi \sqrt{\dfrac{l}{g}}$ i.e. $t = f(l, g)$

(ii) torque $T = I\alpha$, i.e. $T = f(I, \alpha)$

(iii) pressure of an ideal gas $p = \dfrac{mRT}{V}$

i.e. $p = f(T, V)$

(iv) resonant frequency $f_r = \dfrac{1}{2\pi\sqrt{LC}}$

i.e. $f_r = f(L, C)$, and so on.

When differentiating a function having two variables, **one variable is kept constant and the differential coefficient of the other variable is found with respect to that variable.** The differential coefficient obtained is called a **partial derivative** of the function.

39.2 First-order partial derivatives

A 'curly dee', ∂, is used to denote a differential coefficient in an expression containing more than one variable.

Hence if $V = \pi r^2 h$ then $\dfrac{\partial V}{\partial r}$ means 'the partial derivative of V with respect to r, with h remaining constant'. Thus,

$$\frac{\partial V}{\partial r} = (\pi h)\frac{d}{dr}(r^2) = (\pi h)(2r) = 2\pi rh$$

Similarly, $\dfrac{\partial V}{\partial h}$ means 'the partial derivative of V with respect to h, with r remaining constant'. Thus,

$$\frac{\partial V}{\partial h} = (\pi r^2)\frac{d}{dh}(h) = (\pi r^2)(1) = \pi r^2$$

$\dfrac{\partial V}{\partial r}$ and $\dfrac{\partial V}{\partial h}$ are examples of **first order partial derivatives**, since $n = 1$ when written in the form $\dfrac{\partial^n V}{\partial r^n}$

First-order partial derivatives are used when finding the total differential, rates of change and errors for functions of two or more variables (see Chapter 40), when finding maxima, minima and saddle points for functions of two variables (see Chapter 41), and with partial differential equations (see Chapter 56).

Problem 1.   If $z = 5x^4 + 2x^3 y^2 - 3y$ find
(a) $\dfrac{\partial z}{\partial x}$ and (b) $\dfrac{\partial z}{\partial y}$

(a)   To find $\dfrac{\partial z}{\partial x}$, y is kept constant.

Since $z = 5x^4 + (2y^2)x^3 - (3y)$

then,

$$\frac{\partial z}{\partial x} = \frac{d}{dx}(5x^4) + (2y^2)\frac{d}{dx}(x^3) - (3y)\frac{d}{dx}(1)$$

$$= 20x^3 + (2y^2)(3x^2) - 0$$

Hence $\dfrac{\partial z}{\partial x} = \mathbf{20x^3 + 6x^2 y^2}$

(b)   To find $\dfrac{\partial z}{\partial y}$, x is kept constant.

Since $z = (5x^4) + (2x^3)y^2 - 3y$

then,

$$\frac{\partial z}{\partial y} = (5x^4)\frac{d}{dy}(1) + (2x^3)\frac{d}{dy}(y^2) - 3\frac{d}{dy}(y)$$

$$= 0 + (2x^3)(2y) - 3$$

Hence $\dfrac{\partial z}{\partial y} = \mathbf{4x^3 y - 3}$

Problem 2.   Given $y = 4\sin 3x \cos 2t$, find $\dfrac{\partial y}{\partial x}$ and $\dfrac{\partial y}{\partial t}$

To find $\dfrac{\partial y}{\partial x}$, t is kept constant.

Hence $\dfrac{\partial y}{\partial x} = (4\cos 2t)\dfrac{d}{dx}(\sin 3x)$

$$= (4\cos 2t)(3\cos 3x)$$

i.e. $\dfrac{\partial y}{\partial x} = \mathbf{12\cos 3x \cos 2t}$

To find $\dfrac{\partial y}{\partial t}$, x is kept constant.

Hence $\dfrac{\partial y}{\partial t} = (4\sin 3x)\dfrac{d}{dt}(\cos 2t)$

$$= (4\sin 3x)(-2\sin 2t)$$

i.e. $\dfrac{\partial y}{\partial t} = \mathbf{-8\sin 3x \sin 2t}$

Problem 3.   If $z = \sin xy$ show that

$$\frac{1}{y}\frac{\partial z}{\partial x} = \frac{1}{x}\frac{\partial z}{\partial y}$$

$\dfrac{\partial z}{\partial x} = y\cos xy$, since y is kept constant.

$\dfrac{\partial z}{\partial y} = x\cos xy$, since x is kept constant.

$$\frac{1}{y}\frac{\partial z}{\partial x} = \left(\frac{1}{y}\right)(y\cos xy) = \cos xy$$

and $$\frac{1}{x}\frac{\partial z}{\partial y} = \left(\frac{1}{x}\right)(x\cos xy) = \cos xy$$

Hence $\dfrac{1}{y}\dfrac{\partial z}{\partial x} = \dfrac{1}{x}\dfrac{\partial z}{\partial y}$

Problem 4.   Determine $\dfrac{\partial z}{\partial x}$ and $\dfrac{\partial z}{\partial y}$ when

$$z = \frac{1}{\sqrt{(x^2 + y^2)}}$$

$$z = \frac{1}{\sqrt{(x^2 + y^2)}} = (x^2 + y^2)^{\frac{-1}{2}}$$

$\dfrac{\partial z}{\partial x} = -\dfrac{1}{2}(x^2 + y^2)^{\frac{-3}{2}}(2x)$, by the function of a

function of a function rule (keeping y constant)

$= \dfrac{-x}{(x^2+y^2)^{\frac{3}{2}}} = \dfrac{-x}{\sqrt{(x^2+y^2)^3}}$

$\dfrac{\partial z}{\partial y} = -\dfrac{1}{2}(x^2+y^2)^{\frac{-3}{2}}(2y)$, (keeping x constant)

$= \dfrac{-y}{\sqrt{(x^2+y^2)^3}}$

Problem 5. Pressure p of a mass of gas is given by $pV = mRT$, where m and R are constants, V is the volume and T the temperature. Find expressions for $\dfrac{\partial p}{\partial T}$ and $\dfrac{\partial p}{\partial V}$

Since $pV = mRT$ then $p = \dfrac{mRT}{V}$

To find $\dfrac{\partial p}{\partial T}$, V is kept constant.

Hence $\dfrac{\partial p}{\partial T} = \left(\dfrac{mR}{V}\right)\dfrac{\mathrm{d}}{\mathrm{d}T}(T) = \dfrac{mR}{V}$

To find $\dfrac{\partial p}{\partial V}$, T is kept constant.

Hence $\dfrac{\partial p}{\partial V} = (mRT)\dfrac{\mathrm{d}}{\mathrm{d}V}\left(\dfrac{1}{V}\right)$

$= (mRT)(-V^{-2}) = \dfrac{-mRT}{V^2}$

Problem 6. The time of oscillation, t, of a pendulum is given by $t = 2\pi\sqrt{\dfrac{l}{g}}$ where l is the length of the pendulum and g the free fall acceleration due to gravity. Determine $\dfrac{\partial t}{\partial l}$ and $\dfrac{\partial t}{\partial g}$

To find $\dfrac{\partial t}{\partial l}$, g is kept constant.

$t = 2\pi\sqrt{\dfrac{l}{g}} = \left(\dfrac{2\pi}{\sqrt{g}}\right)\sqrt{l} = \left(\dfrac{2\pi}{\sqrt{g}}\right)l^{\frac{1}{2}}$

Hence $\dfrac{\partial t}{\partial l} = \left(\dfrac{2\pi}{\sqrt{g}}\right)\dfrac{\mathrm{d}}{\mathrm{d}l}(l^{\frac{1}{2}}) = \left(\dfrac{2\pi}{\sqrt{g}}\right)\left(\dfrac{1}{2}l^{\frac{-1}{2}}\right)$

$= \left(\dfrac{2\pi}{\sqrt{g}}\right)\left(\dfrac{1}{2\sqrt{l}}\right) = \dfrac{\pi}{\sqrt{lg}}$

To find $\dfrac{\partial t}{\partial g}$, l is kept constant.

$t = 2\pi\sqrt{\dfrac{l}{g}} = (2\pi\sqrt{l})\left(\dfrac{1}{\sqrt{g}}\right)$

$= (2\pi\sqrt{l})g^{\frac{-1}{2}}$

Hence $\dfrac{\partial t}{\partial g} = (2\pi\sqrt{l})\left(-\dfrac{1}{2}g^{\frac{-3}{2}}\right)$

$= (2\pi\sqrt{l})\left(\dfrac{-1}{2\sqrt{g^3}}\right)$

$= \dfrac{-\pi\sqrt{l}}{\sqrt{g^3}} = -\pi\sqrt{\dfrac{l}{g^3}}$

Now try the following Practice Exercise

Practice Exercise 164   First order partial derivatives (Answers on page 881)

In Problems 1 to 6, find $\dfrac{\partial z}{\partial x}$ and $\dfrac{\partial z}{\partial y}$

1. $z = 2xy$

2. $z = x^3 - 2xy + y^2$

3. $z = \dfrac{x}{y}$

4. $z = \sin(4x + 3y)$

5. $z = x^3y^2 - \dfrac{y}{x^2} + \dfrac{1}{y}$

6. $z = \cos 3x \sin 4y$

7. The volume of a cone of height h and base radius r is given by $V = \frac{1}{3}\pi r^2 h$. Determine $\dfrac{\partial V}{\partial h}$ and $\dfrac{\partial V}{\partial r}$

8. The resonant frequency f_r in a series electrical circuit is given by $f_r = \dfrac{1}{2\pi\sqrt{LC}}$. Show that $\dfrac{\partial f_r}{\partial L} = \dfrac{-1}{4\pi\sqrt{CL^3}}$

9. An equation resulting from plucking a string is:

$$y = \sin\left(\frac{n\pi}{L}\right)x\left\{k\cos\left(\frac{n\pi b}{L}\right)t + c\sin\left(\frac{n\pi b}{L}\right)t\right\}$$

Determine $\dfrac{\partial y}{\partial t}$ and $\dfrac{\partial y}{\partial x}$

10. In a thermodynamic system, $k = A\text{e}^{\frac{T\Delta S - \Delta H}{RT}}$, where R, k and A are constants.

Find (a) $\dfrac{\partial k}{\partial T}$ (b) $\dfrac{\partial A}{\partial T}$ (c) $\dfrac{\partial(\Delta S)}{\partial T}$ (d) $\dfrac{\partial(\Delta H)}{\partial T}$

39.3 Second-order partial derivatives

As with ordinary differentiation, where a differential coefficient may be differentiated again, a partial derivative may be differentiated partially again to give higher-order partial derivatives.

(i) Differentiating $\dfrac{\partial V}{\partial r}$ of Section 36.2 with respect to r, keeping h constant, gives $\dfrac{\partial}{\partial r}\left(\dfrac{\partial V}{\partial r}\right)$ which is written as $\dfrac{\partial^2 V}{\partial r^2}$

Thus if $V = \pi r^2 h$,

then $\dfrac{\partial^2 V}{\partial r^2} = \dfrac{\partial}{\partial r}(2\pi rh) = 2\pi h$

(ii) Differentiating $\dfrac{\partial V}{\partial h}$ with respect to h, keeping r constant, gives $\dfrac{\partial}{\partial h}\left(\dfrac{\partial V}{\partial h}\right)$ which is written as $\dfrac{\partial^2 V}{\partial h^2}$

Thus $\dfrac{\partial^2 V}{\partial h^2} = \dfrac{\partial}{\partial h}(\pi r^2) = 0$

(iii) Differentiating $\dfrac{\partial V}{\partial h}$ with respect to r, keeping h constant, gives $\dfrac{\partial}{\partial r}\left(\dfrac{\partial V}{\partial h}\right)$ which is written as $\dfrac{\partial^2 V}{\partial r \partial h}$. Thus,

$$\dfrac{\partial^2 V}{\partial r \partial h} = \dfrac{\partial}{\partial r}\left(\dfrac{\partial V}{\partial h}\right) = \dfrac{\partial}{\partial r}(\pi r^2) = 2\pi r$$

(iv) Differentiating $\dfrac{\partial V}{\partial r}$ with respect to h, keeping r constant, gives $\dfrac{\partial}{\partial h}\left(\dfrac{\partial V}{\partial r}\right)$, which is written as $\dfrac{\partial^2 V}{\partial h \partial r}$. Thus,

$$\dfrac{\partial^2 V}{\partial h \partial r} = \dfrac{\partial}{\partial h}\left(\dfrac{\partial V}{\partial r}\right) = \dfrac{\partial}{\partial h}(2\pi rh) = 2\pi r$$

(v) $\dfrac{\partial^2 V}{\partial r^2}$, $\dfrac{\partial^2 V}{\partial h^2}$, $\dfrac{\partial^2 V}{\partial r \partial h}$ and $\dfrac{\partial^2 V}{\partial h \partial r}$ are examples of **second-order partial derivatives**.

(vi) It is seen from (iii) and (iv) that $\dfrac{\partial^2 V}{\partial r \partial h} = \dfrac{\partial^2 V}{\partial h \partial r}$ and such a result is always true for continuous functions (i.e. a graph of the function which has no sudden jumps or breaks).

Second-order partial derivatives are used in the solution of partial differential equations, in waveguide theory, in such areas of thermodynamics covering entropy and the continuity theorem, and when finding maxima, minima and saddle points for functions of two variables (see Chapter 41).

Problem 7. Given $z = 4x^2 y^3 - 2x^3 + 7y^2$ find (a) $\dfrac{\partial^2 z}{\partial x^2}$ (b) $\dfrac{\partial^2 z}{\partial y^2}$ (c) $\dfrac{\partial^2 z}{\partial x \partial y}$ (d) $\dfrac{\partial^2 z}{\partial y \partial x}$

(a) $\dfrac{\partial z}{\partial x} = 8xy^3 - 6x^2$

$\dfrac{\partial^2 z}{\partial x^2} = \dfrac{\partial}{\partial x}\left(\dfrac{\partial z}{\partial x}\right) = \dfrac{\partial}{\partial x}(8xy^3 - 6x^2)$

$= 8y^3 - 12x$

(b) $\dfrac{\partial z}{\partial y} = 12x^2 y^2 + 14y$

$\dfrac{\partial^2 z}{\partial y^2} = \dfrac{\partial}{\partial y}\left(\dfrac{\partial z}{\partial y}\right) = \dfrac{\partial}{\partial y}(12x^2 y^2 + 14y)$

$= 24x^2 y + 14$

(c) $\dfrac{\partial^2 z}{\partial x \partial y} = \dfrac{\partial}{\partial x}\left(\dfrac{\partial z}{\partial y}\right) = \dfrac{\partial}{\partial x}(12x^2 y^2 + 14y) = 24xy^2$

(d) $\dfrac{\partial^2 z}{\partial y \partial x} = \dfrac{\partial}{\partial y}\left(\dfrac{\partial z}{\partial x}\right) = \dfrac{\partial}{\partial y}(8xy^3 - 6x^2) = 24xy^2$

$\left[\text{It is noted that } \dfrac{\partial^2 z}{\partial x \partial y} = \dfrac{\partial^2 z}{\partial y \partial x}\right]$

Problem 8. Show that when $z = e^{-t}\sin\theta$,

(a) $\dfrac{\partial^2 z}{\partial t^2} = -\dfrac{\partial^2 z}{\partial \theta^2}$, and (b) $\dfrac{\partial^2 z}{\partial t \partial \theta} = \dfrac{\partial^2 z}{\partial \theta \partial t}$

(a) $\dfrac{\partial z}{\partial t} = -e^{-t}\sin\theta$ and $\dfrac{\partial^2 z}{\partial t^2} = e^{-t}\sin\theta$

$\dfrac{\partial z}{\partial \theta} = e^{-t}\cos\theta$ and $\dfrac{\partial^2 z}{\partial \theta^2} = -e^{-t}\sin\theta$

Hence $\dfrac{\partial^2 z}{\partial t^2} = -\dfrac{\partial^2 z}{\partial \theta^2}$

(b) $\dfrac{\partial^2 z}{\partial t \partial \theta} = \dfrac{\partial}{\partial t}\left(\dfrac{\partial z}{\partial \theta}\right) = \dfrac{\partial}{\partial t}(e^{-t}\cos\theta)$

$= -e^{-t}\cos\theta$

$\dfrac{\partial^2 z}{\partial \theta \partial t} = \dfrac{\partial}{\partial \theta}\left(\dfrac{\partial z}{\partial t}\right) = \dfrac{\partial}{\partial \theta}(-e^{-t}\sin\theta)$

$= -e^{-t}\cos\theta$

Hence $\dfrac{\partial^2 z}{\partial t \partial \theta} = \dfrac{\partial^2 z}{\partial \theta \partial t}$

Problem 9. Show that if $z = \dfrac{x}{y}\ln y$, then

(a) $\dfrac{\partial z}{\partial y} = x\dfrac{\partial^2 z}{\partial y \partial x}$ and (b) evaluate $\dfrac{\partial^2 z}{\partial y^2}$ when $x = -3$ and $y = 1$

(a) To find $\dfrac{\partial z}{\partial x}$, y is kept constant.

Hence $\dfrac{\partial z}{\partial x} = \left(\dfrac{1}{y}\ln y\right)\dfrac{d}{dx}(x) = \dfrac{1}{y}\ln y$

To find $\dfrac{\partial z}{\partial y}$, x is kept constant.

Hence

$\dfrac{\partial z}{\partial y} = (x)\dfrac{d}{dy}\left(\dfrac{\ln y}{y}\right)$

$= (x)\left\{\dfrac{(y)\left(\dfrac{1}{y}\right) - (\ln y)(1)}{y^2}\right\}$

using the quotient rule

$= x\left(\dfrac{1 - \ln y}{y^2}\right) = \dfrac{x}{y^2}(1 - \ln y)$

$\dfrac{\partial^2 z}{\partial y \partial x} = \dfrac{\partial}{\partial y}\left(\dfrac{\partial z}{\partial x}\right) = \dfrac{\partial}{\partial y}\left(\dfrac{\ln y}{y}\right)$

$= \dfrac{(y)\left(\dfrac{1}{y}\right) - (\ln y)(1)}{y^2}$

using the quotient rule

$= \dfrac{1}{y^2}(1 - \ln y)$

Hence $x\dfrac{\partial^2 z}{\partial y \partial x} = \dfrac{x}{y^2}(1 - \ln y) = \dfrac{\partial z}{\partial y}$

(b) $\dfrac{\partial^2 z}{\partial y^2} = \dfrac{\partial}{\partial y}\left(\dfrac{\partial z}{\partial y}\right) = \dfrac{\partial}{\partial y}\left\{\dfrac{x}{y^2}(1 - \ln y)\right\}$

$= (x)\dfrac{d}{dy}\left(\dfrac{1 - \ln y}{y^2}\right)$

$= (x)\left\{\dfrac{(y^2)\left(-\dfrac{1}{y}\right) - (1 - \ln y)(2y)}{y^4}\right\}$

using the quotient rule

$= \dfrac{x}{y^4}[-y - 2y + 2y\ln y]$

$= \dfrac{xy}{y^4}[-3 + 2\ln y] = \dfrac{x}{y^3}(2\ln y - 3)$

When $x = -3$ and $y = 1$,

$\dfrac{\partial^2 z}{\partial y^2} = \dfrac{(-3)}{(1)^3}(2\ln 1 - 3) = (-3)(-3) = \mathbf{9}$

Now try the following Practice Exercise

Practice Exercise 165   Second-order partial derivatives (Answers on page 881)

In Problems 1 to 4, find (a) $\dfrac{\partial^2 z}{\partial x^2}$ (b) $\dfrac{\partial^2 z}{\partial y^2}$

(c) $\dfrac{\partial^2 z}{\partial x \partial y}$ (d) $\dfrac{\partial^2 z}{\partial y \partial x}$

1. $z = (2x - 3y)^2$

2. $z = 2\ln xy$

3. $z = \dfrac{(x - y)}{(x + y)}$

4. $z = \sinh x \cosh 2y$

5. Given $z = x^2 \sin(x - 2y)$ find (a) $\dfrac{\partial^2 z}{\partial x^2}$ and (b) $\dfrac{\partial^2 z}{\partial y^2}$

Show also that $\dfrac{\partial^2 z}{\partial x \partial y} = \dfrac{\partial^2 z}{\partial y \partial x}$
$= 2x^2 \sin(x - 2y) - 4x \cos(x - 2y)$

6. Find $\dfrac{\partial^2 z}{\partial x^2}, \dfrac{\partial^2 z}{\partial y^2}$ and show that $\dfrac{\partial^2 z}{\partial x \partial y} = \dfrac{\partial^2 z}{\partial y \partial x}$

when $z = \cos^{-1} \dfrac{x}{y}$

7. Given $z = \sqrt{\left(\dfrac{3x}{y}\right)}$ show that

$\dfrac{\partial^2 z}{\partial x \partial y} = \dfrac{\partial^2 z}{\partial y \partial x}$ and evaluate $\dfrac{\partial^2 z}{\partial x^2}$ when

$x = \dfrac{1}{2}$ and $y = 3$

8. An equation used in thermodynamics is the Benedict–Webb–Rubine equation of state for the expansion of a gas. The equation is:

$$p = \frac{RT}{V} + \left(B_0 RT - A_0 - \frac{C_0}{T^2} \right) \frac{1}{V^2}$$
$$+ (bRT - a)\frac{1}{V^3} + \frac{A\alpha}{V^6}$$
$$+ \frac{C\left(1 + \dfrac{\gamma}{V^2}\right)}{T^2} \left(\frac{1}{V^3}\right) e^{-\frac{\gamma}{V^2}}$$

Show that $\dfrac{\partial^2 p}{\partial T^2}$

$$= \frac{6}{V^2 T^4} \left\{ \frac{C}{V}\left(1 + \frac{\gamma}{V^2}\right) e^{-\frac{\gamma}{V^2}} - C_0 \right\}$$

For fully worked solutions to each of the problems in Practice Exercises 164 and 165 in this chapter, go to the website:
www.routledge.com/cw/bird

Chapter 40

Total differential, rates of change and small changes

Why it is important to understand: Total differential, rates of change and small changes

This chapter looks at some applications of partial differentiation, i.e. the total differential for variables which may all be changing at the same time, rates of change, where different quantities have different rates of change, and small changes, where approximate errors may be calculated in situations where small changes in the variables associated with the quantity occur.

At the end of this chapter, you should be able to:

- determine the total differential of a function having more than one variable
- determine the rates of change of functions having more than one variable
- determine an approximate error in a function having more than one variable

40.1 Total differential

In Chapter 39, partial differentiation is introduced for the case where only one variable changes at a time, the other variables being kept constant. In practice, variables may all be changing at the same time.

If $z = f(u, v, w, \ldots)$, then the **total differential, dz,** is given by the sum of the separate partial differentials of z,

i.e. $\mathbf{d}z = \dfrac{\partial z}{\partial u}\,\mathbf{d}u + \dfrac{\partial z}{\partial v}\,\mathbf{d}v + \dfrac{\partial z}{\partial w}\,\mathbf{d}w + \cdots$ (1)

The total differential is the sum of the partial differentials,

i.e. $dz = \dfrac{\partial z}{\partial x}\,dx + \dfrac{\partial z}{\partial y}\,dy$

$\dfrac{\partial z}{\partial x} = 2xy^3 + \dfrac{2}{y}$ (i.e. y is kept constant)

$\dfrac{\partial z}{\partial y} = 3x^2y^2 - \dfrac{2x}{y^2}$ (i.e. x is kept constant)

Hence $dz = \left(2xy^3 + \dfrac{2}{y}\right)dx + \left(3x^2y^2 - \dfrac{2x}{y^2}\right)dy$

Problem 1. If $z = f(x, y)$ and $z = x^2y^3 + \dfrac{2x}{y} + 1$, determine the total differential, dz

Problem 2. If $z = f(u, v, w)$ and $z = 3u^2 - 2v + 4w^3v^2$ find the total differential, dz

The total differential

$$dz = \frac{\partial z}{\partial u} du + \frac{\partial z}{\partial v} dv + \frac{\partial z}{\partial w} dw$$

$\frac{\partial z}{\partial u} = 6u$ (i.e. v and w are kept constant)

$\frac{\partial z}{\partial v} = -2 + 8w^3 v$

(i.e. u and w are kept constant)

$\frac{\partial z}{\partial w} = 12w^2 v^2$ (i.e. u and v are kept constant)

Hence

$$dz = 6u\,du + (8vw^3 - 2)\,dv + (12v^2 w^2)\,dw$$

Problem 3. The pressure p, volume V and temperature T of a gas are related by $pV = kT$, where k is a constant. Determine the total differentials (a) dp and (b) dT in terms of p, V and T.

(a) Total differential $dp = \frac{\partial p}{\partial T} dT + \frac{\partial p}{\partial V} dV$

Since $pV = kT$ then $p = \frac{kT}{V}$

hence $\frac{\partial p}{\partial T} = \frac{k}{V}$ and $\frac{\partial p}{\partial V} = -\frac{kT}{V^2}$

Thus $dp = \frac{k}{V} dT - \frac{kT}{V^2} dV$

Since $pV = kT, k = \frac{pV}{T}$

Hence $dp = \frac{\left(\frac{pV}{T}\right)}{V} dT - \frac{\left(\frac{pV}{T}\right)T}{V^2} dV$

i.e. $dp = \frac{p}{T} dT - \frac{p}{V} dV$

(b) Total differential $dT = \frac{\partial T}{\partial p} dp + \frac{\partial T}{\partial V} dV$

Since $pV = kT, T = \frac{pV}{k}$

hence $\frac{\partial T}{\partial p} = \frac{V}{k}$ and $\frac{\partial T}{\partial V} = \frac{p}{k}$

Thus $dT = \frac{V}{k} dp + \frac{p}{k} dV$ and substituting $k = \frac{pV}{T}$ gives:

$$dT = \frac{V}{\left(\frac{pV}{T}\right)} dp + \frac{p}{\left(\frac{pV}{T}\right)} dV$$

i.e. $dT = \frac{T}{p} dp + \frac{T}{V} dV$

Now try the following Practice Exercise

Practice Exercise 166   Total differential (Answers on page 881)

In Problems 1 to 5, find the total differential dz.

1. $z = x^3 + y^2$

2. $z = 2xy - \cos x$

3. $z = \frac{x - y}{x + y}$

4. $z = x \ln y$

5. $z = xy + \frac{\sqrt{x}}{y} - 4$

6. If $z = f(a, b, c)$ and $z = 2ab - 3b^2 c + abc$, find the total differential, dz

7. Given $u = \ln \sin(xy)$ show that $du = \cot(xy)(y\,dx + x\,dy)$

40.2   Rates of change

Sometimes it is necessary to solve problems in which different quantities have different rates of change. From equation (1), the rate of change of z, $\frac{dz}{dt}$ is given by:

$$\frac{dz}{dt} = \frac{\partial z}{\partial u} \frac{du}{dt} + \frac{\partial z}{\partial v} \frac{dv}{dt} + \frac{\partial z}{\partial w} \frac{dw}{dt} + \cdots \qquad (2)$$

Problem 4. If $z = f(x, y)$ and $z = 2x^3 \sin 2y$ find the rate of change of z, correct to 4 significant figures, when x is 2 units and y is $\pi/6$ radians and when x is increasing at 4 units/s and y is decreasing at 0.5 units/s.

Section H

Using equation (2), the rate of change of z,

$$\frac{dz}{dt} = \frac{\partial z}{\partial x}\frac{dx}{dt} + \frac{\partial z}{\partial y}\frac{dy}{dt}$$

Since $z = 2x^3 \sin 2y$, then

$$\frac{\partial z}{\partial x} = 6x^2 \sin 2y \text{ and } \frac{\partial z}{\partial y} = 4x^3 \cos 2y$$

Since x is increasing at 4 units/s, $\dfrac{dx}{dt} = +4$

and since y is decreasing at 0.5 units/s, $\dfrac{dy}{dt} = -0.5$

Hence $\dfrac{dz}{dt} = (6x^2 \sin 2y)(+4) + (4x^3 \cos 2y)(-0.5)$

$$= 24x^2 \sin 2y - 2x^3 \cos 2y$$

When $x = 2$ units and $y = \dfrac{\pi}{6}$ radians, then

$$\frac{dz}{dt} = 24(2)^2 \sin[2(\pi/6)] - 2(2)^3 \cos[2(\pi/6)]$$

$$= 83.138 - 8.0$$

Hence the rate of change of z, $\dfrac{dz}{dt} = \textbf{75.14 units/s}$, correct to 4 significant figures.

Problem 5. The height of a right circular cone is increasing at 3 mm/s and its radius is decreasing at 2 mm/s. Determine, correct to 3 significant figures, the rate at which the volume is changing (in cm³/s) when the height is 3.2 cm and the radius is 1.5 cm.

Volume of a right circular cone, $V = \dfrac{1}{3}\pi r^2 h$

Using equation (2), the rate of change of volume,

$$\frac{dV}{dt} = \frac{\partial V}{\partial r}\frac{dr}{dt} + \frac{\partial V}{\partial h}\frac{dh}{dt}$$

$$\frac{\partial V}{\partial r} = \frac{2}{3}\pi r h \text{ and } \frac{\partial V}{\partial h} = \frac{1}{3}\pi r^2$$

Since the height is increasing at 3 mm/s,

i.e. 0.3 cm/s, then $\dfrac{dh}{dt} = +0.3$

and since the radius is decreasing at 2 mm/s,

i.e. 0.2 cm/s, then $\dfrac{dr}{dt} = -0.2$

Hence $\dfrac{dV}{dt} = \left(\dfrac{2}{3}\pi r h\right)(-0.2) + \left(\dfrac{1}{3}\pi r^2\right)(+0.3)$

$$= \frac{-0.4}{3}\pi r h + 0.1\pi r^2$$

However, $h = 3.2$ cm and $r = 1.5$ cm.

Hence $\dfrac{dV}{dt} = \dfrac{-0.4}{3}\pi(1.5)(3.2) + (0.1)\pi(1.5)^2$

$$= -2.011 + 0.707 = -1.304 \text{ cm}^3/\text{s}$$

Thus the rate of change of volume is 1.30 cm³/s decreasing.

Problem 6. The area A of a triangle is given by $A = \frac{1}{2}ac\sin B$, where B is the angle between sides a and c. If a is increasing at 0.4 units/s, c is decreasing at 0.8 units/s and B is increasing at 0.2 units/s, find the rate of change of the area of the triangle, correct to 3 significant figures, when a is 3 units, c is 4 units and B is $\pi/6$ radians.

Using equation (2), the rate of change of area,

$$\frac{dA}{dt} = \frac{\partial A}{\partial a}\frac{da}{dt} + \frac{\partial A}{\partial c}\frac{dc}{dt} + \frac{\partial A}{\partial B}\frac{dB}{dt}$$

Since $A = \dfrac{1}{2}ac\sin B, \dfrac{\partial A}{\partial a} = \dfrac{1}{2}c\sin B,$

$$\frac{\partial A}{\partial c} = \frac{1}{2}a\sin B \text{ and } \frac{\partial A}{\partial B} = \frac{1}{2}ac\cos B$$

$$\frac{da}{dt} = 0.4 \text{ units/s}, \frac{dc}{dt} = -0.8 \text{ units/s}$$

and $\dfrac{dB}{dt} = 0.2 \text{ units/s}$

Hence $\dfrac{dA}{dt} = \left(\dfrac{1}{2}c\sin B\right)(0.4) + \left(\dfrac{1}{2}a\sin B\right)(-0.8)$

$$+ \left(\frac{1}{2}ac\cos B\right)(0.2)$$

When $a = 3$, $c = 4$ and $B = \dfrac{\pi}{6}$ then:

$$\frac{dA}{dt} = \left(\frac{1}{2}(4)\sin\frac{\pi}{6}\right)(0.4) + \left(\frac{1}{2}(3)\sin\frac{\pi}{6}\right)(-0.8)$$

$$+ \left(\frac{1}{2}(3)(4)\cos\frac{\pi}{6}\right)(0.2)$$

$$= 0.4 - 0.6 + 1.039 = \textbf{0.839 units}^2/\textbf{s}, \text{ correct to 3 significant figures.}$$

Problem 7. Determine the rate of increase of diagonal AC of the rectangular solid, shown in Fig. 40.1, correct to 2 significant figures, if the sides x, y and z increase at 6 mm/s, 5 mm/s and 4 mm/s when these three sides are 5 cm, 4 cm and 3 cm respectively.

Figure 40.1

Diagonal $AB = \sqrt{(x^2 + y^2)}$

Diagonal $AC = \sqrt{(BC^2 + AB^2)}$

$$= \sqrt{[z^2 + \{\sqrt{(x^2 + y^2)}\}^2]}$$

$$= \sqrt{(z^2 + x^2 + y^2)}$$

Let $AC = b$, then $b = \sqrt{(x^2 + y^2 + z^2)}$
Using equation (2), the rate of change of diagonal b is given by:

$$\frac{db}{dt} = \frac{\partial b}{\partial x}\frac{dx}{dt} + \frac{\partial b}{\partial y}\frac{dy}{dt} + \frac{\partial b}{\partial z}\frac{dz}{dt}$$

Since $b = \sqrt{(x^2 + y^2 + z^2)}$

$$\frac{\partial b}{\partial x} = \frac{1}{2}(x^2 + y^2 + z^2)^{\frac{-1}{2}}(2x) = \frac{x}{\sqrt{(x^2 + y^2 + z^2)}}$$

Similarly, $\dfrac{\partial b}{\partial y} = \dfrac{y}{\sqrt{(x^2 + y^2 + z^2)}}$

and $\dfrac{\partial b}{\partial z} = \dfrac{z}{\sqrt{(x^2 + y^2 + z^2)}}$

$$\frac{dx}{dt} = 6\,\text{mm/s} = 0.6\,\text{cm/s},$$

$$\frac{dy}{dt} = 5\,\text{mm/s} = 0.5\,\text{cm/s},$$

and $\dfrac{dz}{dt} = 4\,\text{mm/s} = 0.4\,\text{cm/s}$

Hence $\dfrac{db}{dt} = \left[\dfrac{x}{\sqrt{(x^2 + y^2 + z^2)}}\right](0.6)$

$$+ \left[\frac{y}{\sqrt{(x^2 + y^2 + z^2)}}\right](0.5)$$

$$+ \left[\frac{z}{\sqrt{(x^2 + y^2 + z^2)}}\right](0.4)$$

When $x = 5$ cm, $y = 4$ cm and $z = 3$ cm, then:

$$\frac{db}{dt} = \left[\frac{5}{\sqrt{(5^2 + 4^2 + 3^2)}}\right](0.6)$$

$$+ \left[\frac{4}{\sqrt{(5^2 + 4^2 + 3^2)}}\right](0.5)$$

$$+ \left[\frac{3}{\sqrt{(5^2 + 4^2 + 3^2)}}\right](0.4)$$

$$= 0.4243 + 0.2828 + 0.1697 = 0.8768\,\text{cm/s}$$

Hence the rate of increase of diagonal AC is 0.88 cm/s or 8.8 mm/s, correct to 2 significant figures.

Now try the following Practice Exercise

Practice Exercise 167   Rates of change (Answers on page 881)

1. The radius of a right cylinder is increasing at a rate of 8 mm/s and the height is decreasing at a rate of 15 mm/s. Find the rate at which the volume is changing in cm^3/s when the radius is 40 mm and the height is 150 mm.

2. If $z = f(x, y)$ and $z = 3x^2y^5$, find the rate of change of z when x is 3 units and y is 2 units when x is decreasing at 5 units/s and y is increasing at 2.5 units/s.

3. Find the rate of change of k, correct to 4 significant figures, given the following data: $k = f(a, b, c)$; $k = 2b\ln a + c^2 e^a$; a is increasing at 2 cm/s; b is decreasing at 3 cm/s; c is decreasing at 1 cm/s; $a = 1.5$ cm, $b = 6$ cm and $c = 8$ cm.

4. A rectangular box has sides of length x cm, y cm and z cm. Sides x and z are expanding at rates of 3 mm/s and 5 mm/s respectively and side y is contracting at a rate of 2 mm/s. Determine the rate of change of volume when x is 3 cm, y is 1.5 cm and z is 6 cm.

5. Find the rate of change of the total surface area of a right circular cone at the instant when the base radius is 5 cm and the height is 12 cm if the radius is increasing at 5 mm/s and the height is decreasing at 15 mm/s.

40.3 Small changes

It is often useful to find an approximate value for the change (or error) of a quantity caused by small changes (or errors) in the variables associated with the quantity. If $z = f(u, v, w, \ldots)$ and $\delta u, \delta v, \delta w, \ldots$ denote **small changes** in $u, v, w, \ldots$ respectively, then the corresponding approximate change δz in z is obtained from equation (1) by replacing the differentials by the small changes.

Thus $\delta z \approx \dfrac{\partial z}{\partial u} \delta u + \dfrac{\partial z}{\partial v} \delta v + \dfrac{\partial z}{\partial w} \delta w + \cdots$ \hfill (3)

Problem 8. Pressure p and volume V of a gas are connected by the equation $pV^{1.4} = k$. Determine the approximate percentage error in k when the pressure is increased by 4% and the volume is decreased by 1.5%

Using equation (3), the approximate error in k,

$$\delta k \approx \frac{\partial k}{\partial p} \delta p + \frac{\partial k}{\partial V} \delta V$$

Let p, V and k refer to the initial values.

Since $k = pV^{1.4}$ then $\dfrac{\partial k}{\partial p} = V^{1.4}$

and $\dfrac{\partial k}{\partial V} = 1.4pV^{0.4}$

Since the pressure is increased by 4%, the change in pressure $\delta p = \dfrac{4}{100} \times p = 0.04p$

Since the volume is decreased by 1.5%, the change in volume $\delta V = \dfrac{-1.5}{100} \times V = -0.015V$

Hence the approximate error in k,

$$\delta k \approx (V)^{1.4}(0.04p) + (1.4pV^{0.4})(-0.015V)$$

$$\approx pV^{1.4}[0.04 - 1.4(0.015)]$$

$$\approx pV^{1.4}[0.019] \approx \frac{1.9}{100} pV^{1.4} \approx \frac{1.9}{100} k$$

i.e. **the approximate error in k is a 1.9% increase**.

Problem 9. Modulus of rigidity $G = (R^4\theta)/L$, where R is the radius, θ the angle of twist and L the length. Determine the approximate percentage error in G when R is increased by 2%, θ is reduced by 5% and L is increased by 4%

Using $\delta G \approx \dfrac{\partial G}{\partial R} \delta R + \dfrac{\partial G}{\partial \theta} \delta\theta + \dfrac{\partial G}{\partial L} \delta L$

Since $G = \dfrac{R^4\theta}{L}, \dfrac{\partial G}{\partial R} = \dfrac{4R^3\theta}{L}, \dfrac{\partial G}{\partial \theta} = \dfrac{R^4}{L}$

and $\dfrac{\partial G}{\partial L} = \dfrac{-R^4\theta}{L^2}$

Since R is increased by 2%, $\delta R = \dfrac{2}{100} R = 0.02R$

Similarly, $\delta\theta = -0.05\theta$ and $\delta L = 0.04L$

Hence $\delta G \approx \left(\dfrac{4R^3\theta}{L}\right)(0.02R) + \left(\dfrac{R^4}{L}\right)(-0.05\theta)$

$$+ \left(-\dfrac{R^4\theta}{L^2}\right)(0.04L)$$

$$\approx \frac{R^4\theta}{L}[0.08 - 0.05 - 0.04] \approx -0.01\frac{R^4\theta}{L},$$

i.e. $\delta G \approx -\dfrac{1}{100} G$

Hence the approximate percentage error in G is a 1% decrease.

Problem 10. The second moment of area of a rectangle is given by $I = (bl^3)/3$. If b and l are measured as 40 mm and 90 mm respectively and the measurement errors are -5 mm in b and $+8$ mm in l, find the approximate error in the calculated value of I.

Using equation (3), the approximate error in I,

$$\delta I \approx \frac{\partial I}{\partial b}\delta b + \frac{\partial I}{\partial l}\delta l$$

$$\frac{\partial I}{\partial b} = \frac{l^3}{3} \text{ and } \frac{\partial I}{\partial l} = \frac{3bl^2}{3} = bl^2$$

$$\delta b = -5 \text{ mm and } \delta l = +8 \text{ mm}$$

Hence $\delta I \approx \left(\dfrac{l^3}{3}\right)(-5) + (bl^2)(+8)$

Since $b=40$ mm and $l=90$ mm then

$$\delta I \approx \left(\frac{90^3}{3}\right)(-5) + 40(90)^2(8)$$

$$\approx -1\,215\,000 + 2\,592\,000$$

$$\approx 1\,377\,000 \text{ mm}^4 \approx 137.7 \text{ cm}^4$$

Hence the approximate error in the calculated value of I is a 137.7 cm^4 increase.

Problem 11. The time of oscillation t of a pendulum is given by $t = 2\pi\sqrt{\dfrac{l}{g}}$. Determine the approximate percentage error in t when l has an error of 0.2% too large and g 0.1% too small.

Using equation (3), the approximate change in t,

$$\delta t \approx \frac{\partial t}{\partial l}\delta l + \frac{\partial t}{\partial g}\delta g$$

Since $t = 2\pi\sqrt{\dfrac{l}{g}}$, $\dfrac{\partial t}{\partial l} = \dfrac{\pi}{\sqrt{lg}}$

and $\dfrac{\partial t}{\partial g} = -\pi\sqrt{\dfrac{l}{g^3}}$ (from Problem 6, Chapter 39)

$$\delta l = \frac{0.2}{100}l = 0.002l \text{ and } \delta g = -0.001g$$

hence $\delta t \approx \dfrac{\pi}{\sqrt{lg}}(0.002l) + -\pi\sqrt{\dfrac{l}{g^3}}(-0.001\,g)$

$$\approx 0.002\pi\sqrt{\frac{l}{g}} + 0.001\pi\sqrt{\frac{l}{g}}$$

$$\approx (0.001)\left[2\pi\sqrt{\frac{l}{g}}\right] + 0.0005\left[2\pi\sqrt{\frac{l}{g}}\right]$$

$$\approx 0.0015t \approx \frac{0.15}{100}t$$

Hence the approximate error in t is a 0.15% increase.

Now try the following Practice Exercise

Practice Exercise 168   Small changes (Answers on page 881)

1. The power P consumed in a resistor is given by $P = V^2/R$ watts. Determine the approximate change in power when V increases by 5% and R decreases by 0.5% if the original values of V and R are 50 volts and 12.5 ohms respectively.

2. An equation for heat generated H is $H = i^2 Rt$. Determine the error in the calculated value of H if the error in measuring current i is $+2\%$, the error in measuring resistance R is -3% and the error in measuring time t is $+1\%$

3. $f_r = \dfrac{1}{2\pi\sqrt{LC}}$ represents the resonant frequency of a series - connected circuit containing inductance L and capacitance C. Determine the approximate percentage change in f_r when L is decreased by 3% and C is increased by 5%

4. The second moment of area of a rectangle about its centroid parallel to side b is given by $I = bd^3/12$. If b and d are measured as 15 cm and 6 cm respectively and the measurement errors are $+12$ mm in b and -1.5 mm in d, find the error in the calculated value of I.

5. The side b of a triangle is calculated using $b^2 = a^2 + c^2 - 2ac\cos B$. If a, c and B are measured as 3 cm, 4 cm and $\pi/4$ radians respectively and the measurement errors which occur are $+0.8$ cm, -0.5 cm and $+\pi/90$ radians respectively, determine the error in the calculated value of b.

6. Q factor in a resonant electrical circuit is given by: $Q = \dfrac{1}{R}\sqrt{\dfrac{L}{C}}$. Find the percentage change in Q when L increases by 4%, R decreases by 3% and C decreases by 2%

Section H

7. The rate of flow of gas in a pipe is given by: $v = \dfrac{C\sqrt{d}}{\sqrt[6]{T^5}}$, where C is a constant, d is the diameter of the pipe and T is the thermodynamic temperature of the gas. When determining the rate of flow experimentally, d is measured and subsequently found to be in error by $+1.4\%$, and T has an error of -1.8%. Determine the percentage error in the rate of flow based on the measured values of d and T.

For fully worked solutions to each of the problems in Practice Exercises 166 to 168 in this chapter, go to the website:
www.routledge.com/cw/bird

Chapter 41

Maxima, minima and saddle points for functions of two variables

---

Why it is important to understand: Maxima, minima and saddle points for functions of two variables

The problem of finding the maximum and minimum values of functions is encountered in mechanics, physics, geometry, and in many other fields. Finding maxima, minima and saddle points for functions of two variables requires the application of partial differentiation (which was explained in the previous chapters). This is demonstrated in this chapter.

At the end of this chapter, you should be able to:

- understand functions of two independent variables
- explain a saddle point
- determine the maxima, minima and saddle points for a function of two variables
- sketch a contour map for functions of two variables

41.1 Functions of two independent variables

If a relation between two real variables, x and y, is such that when x is given, y is determined, then y is said to be a function of x and is denoted by $y = f(x)$; x is called the independent variable and y the dependent variable. If $y = f(u, v)$, then y is a function of two independent variables u and v. For example, if, say, $y = f(u, v) = 3u^2 - 2v$ then when $u = 2$ and $v = 1$, $y = 3(2)^2 - 2(1) = 10$. This may be written as $f(2, 1) = 10$. Similarly, if $u = 1$ and $v = 4$, $f(1, 4) = -5$

Consider a function of two variables x and y defined by $z = f(x, y) = 3x^2 - 2y$. If $(x, y) = (0, 0)$, then $f(0, 0) = 0$ and if $(x, y) = (2, 1)$, then $f(2, 1) = 10$. Each pair of numbers, (x, y), may be represented by a point P in the (x, y) plane of a rectangular Cartesian co-ordinate system as shown in Fig. 41.1. The corresponding value of $z = f(x, y)$ may be represented by a line PP' drawn parallel to the z-axis. Thus, if, for example, $z = 3x^2 - 2y$, as above, and P is the co-ordinate $(2, 3)$ then the length of PP' is $3(2)^2 - 2(3) = 6$. Fig. 41.2 shows that when a large number of (x, y) co-ordinates are taken for a function

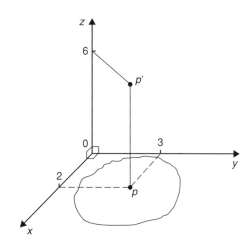

Figure 41.1

$f(x, y)$, and then $f(x, y)$ calculated for each, a large number of lines such as PP' can be constructed, and in the limit when all points in the (x, y) plane are considered, a surface is seen to result as shown in Fig. 41.2. Thus the function $z = f(x, y)$ represents a surface and not a curve.

Figure 41.2

41.2 Maxima, minima and saddle points

Partial differentiation is used when determining stationary points for functions of two variables. A function $f(x, y)$ is said to be a maximum at a point (x, y) if the value of the function there is greater than at all points in the immediate vicinity, and is a minimum if less than at all points in the immediate vicinity. Fig. 41.3 shows

Figure 41.3

geometrically a maximum value of a function of two variables and it is seen that the surface $z = f(x, y)$ is higher at $(x, y) = (a, b)$ than at any point in the immediate vicinity. Fig. 41.4 shows a minimum value of a function of two variables and it is seen that the surface $z = f(x, y)$ is lower at $(x, y) = (p, q)$ than at any point in the immediate vicinity.

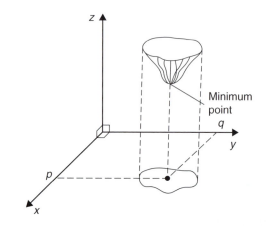

Figure 41.4

If $z = f(x, y)$ and a maximum occurs at (a, b), the curve lying in the two planes $x = a$ and $y = b$ must also have a maximum point (a, b) as shown in Fig. 41.5. Consequently, the tangents (shown as t_1 and t_2) to the curves at (a, b) must be parallel to Ox and Oy respectively. This requires that $\dfrac{\partial z}{\partial x} = 0$ and $\dfrac{\partial z}{\partial y} = 0$ at all maximum and minimum values, and the solution of these equations gives the stationary (or critical) points of z.

With functions of two variables there are three types of stationary points possible, these being a maximum

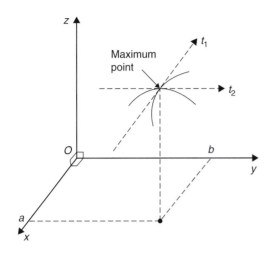

Figure 41.5

point, a minimum point, and a **saddle point**. A saddle point Q is shown in Fig. 41.6 and is such that a point Q is a maximum for curve 1 and a minimum for curve 2.

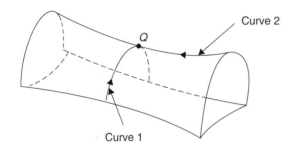

Figure 41.6

41.3 Procedure to determine maxima, minima and saddle points for functions of two variables

Given $z = f(x, y)$:

(i)   determine $\dfrac{\partial z}{\partial x}$ and $\dfrac{\partial z}{\partial y}$

(ii) for stationary points, $\dfrac{\partial z}{\partial x} = 0$ and $\dfrac{\partial z}{\partial y} = 0$,

(iii) solve the simultaneous equations $\dfrac{\partial z}{\partial x} = 0$ and $\dfrac{\partial z}{\partial y} = 0$ for x and y, which gives the co-ordinates of the stationary points,

(iv) determine $\dfrac{\partial^2 z}{\partial x^2}$, $\dfrac{\partial^2 z}{\partial y^2}$ and $\dfrac{\partial^2 z}{\partial x \partial y}$

(v)   for each of the co-ordinates of the stationary points, substitute values of x and y into $\dfrac{\partial^2 z}{\partial x^2}$, $\dfrac{\partial^2 z}{\partial y^2}$ and $\dfrac{\partial^2 z}{\partial x \partial y}$ and evaluate each,

(vi) evaluate $\left(\dfrac{\partial^2 z}{\partial x \partial y}\right)^2$ for each stationary point,

(vii) substitute the values of $\dfrac{\partial^2 z}{\partial x^2}$, $\dfrac{\partial^2 z}{\partial y^2}$ and $\dfrac{\partial^2 z}{\partial x \partial y}$ into the equation

$$\Delta = \left(\frac{\partial^2 z}{\partial x \partial y}\right)^2 - \left(\frac{\partial^2 z}{\partial x^2}\right)\left(\frac{\partial^2 z}{\partial y^2}\right)$$

and evaluate,

(viii) (a)   if $\Delta > 0$ then the stationary point is a **saddle point.**

(b)   if $\Delta < 0$ and $\dfrac{\partial^2 z}{\partial x^2} < 0$, then the stationary point is a **maximum point**,

and

(c)   if $\Delta < 0$ and $\dfrac{\partial^2 z}{\partial x^2} > 0$, then the stationary point is a **minimum point**.

41.4 Worked problems on maxima, minima and saddle points for functions of two variables

Problem 1. Show that the function $z = (x-1)^2 + (y-2)^2$ has one stationary point only and determine its nature. Sketch the surface represented by z and produce a contour map in the x–y plane.

Following the above procedure:

(i) $\dfrac{\partial z}{\partial x} = 2(x-1)$ and $\dfrac{\partial z}{\partial y} = 2(y-2)$

(ii) $2(x-1) = 0$ \hfill (1)

$2(y-2) = 0$ \hfill (2)

Section H

(iii) From equations (1) and (2), $x=1$ and $y=2$, thus the only stationary point exists at $(1, 2)$

(iv) Since $\dfrac{\partial z}{\partial x} = 2(x-1) = 2x - 2$, $\dfrac{\partial^2 z}{\partial x^2} = 2$

and since $\dfrac{\partial z}{\partial y} = 2(y-2) = 2y - 4$, $\dfrac{\partial^2 z}{\partial y^2} = 2$

and $\dfrac{\partial^2 z}{\partial x \partial y} = \dfrac{\partial}{\partial x}\left(\dfrac{\partial z}{\partial y}\right) = \dfrac{\partial}{\partial x}(2y-4) = 0$

(v) $\dfrac{\partial^2 z}{\partial x^2} = \dfrac{\partial^2 z}{\partial y^2} = 2$ and $\dfrac{\partial^2 z}{\partial x \partial y} = 0$

(vi) $\left(\dfrac{\partial^2 z}{\partial x \partial y}\right)^2 = 0$

(vii) $\Delta = (0)^2 - (2)(2) = -4$

(viii) Since $\Delta < 0$ and $\dfrac{\partial^2 z}{\partial x^2} > 0$, **the stationary point (1, 2) is a minimum**.

The surface $z = (x-1)^2 + (y-2)^2$ is shown in three dimensions in Fig. 41.7. Looking down towards the x–y plane from above, it is possible to produce a **contour map**. A contour is a line on a map which gives places having the same vertical height above a datum line (usually the mean sea-level on a geographical map). A contour map for $z = (x-1)^2 + (y-2)^2$ is shown in Fig. 41.8. The values of z are shown on the map and these give an indication of the rise and fall to a stationary point.

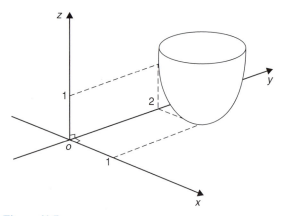

Figure 41.7

Problem 2. Find the stationary points of the surface $f(x, y) = x^3 - 6xy + y^3$ and determine their nature.

Let $z = f(x, y) = x^3 - 6xy + y^3$

Following the procedure:

(i) $\dfrac{\partial z}{\partial x} = 3x^2 - 6y$ and $\dfrac{\partial z}{\partial y} = -6x + 3y^2$

(ii) for stationary points, $3x^2 - 6y = 0$ (1)

and $\qquad -6x + 3y^2 = 0$ (2)

(iii) from equation (1), $3x^2 = 6y$

and $\qquad y = \dfrac{3x^2}{6} = \dfrac{1}{2}x^2$

and substituting in equation (2) gives:

$$-6x + 3\left(\dfrac{1}{2}x^2\right)^2 = 0$$

$$-6x + \dfrac{3}{4}x^4 = 0$$

$$3x\left(\dfrac{x^3}{4} - 2\right) = 0$$

from which, $x = 0$ or $\dfrac{x^3}{4} - 2 = 0$

i.e. $x^3 = 8$ and $x = 2$

When $x = 0$, $y = 0$ and when $x = 2$, $y = 2$ from equations (1) and (2).

Thus stationary points occur at (0, 0) and (2, 2)

(iv) $\dfrac{\partial^2 z}{\partial x^2} = 6x$, $\dfrac{\partial^2 z}{\partial y^2} = 6y$ and $\dfrac{\partial^2 z}{\partial x \partial y} = \dfrac{\partial}{\partial x}\left(\dfrac{\partial z}{\partial y}\right)$

$$= \dfrac{\partial}{\partial x}(-6x + 3y^2) = -6$$

(v) for (0, 0) $\quad \dfrac{\partial^2 z}{\partial x^2} = 0, \dfrac{\partial^2 z}{\partial y^2} = 0$

and $\qquad \dfrac{\partial^2 z}{\partial x \partial y} = -6$

for (2, 2), $\dfrac{\partial^2 z}{\partial x^2} = 12, \dfrac{\partial^2 z}{\partial y^2} = 12$

and $\qquad \dfrac{\partial^2 z}{\partial x \partial y} = -6$

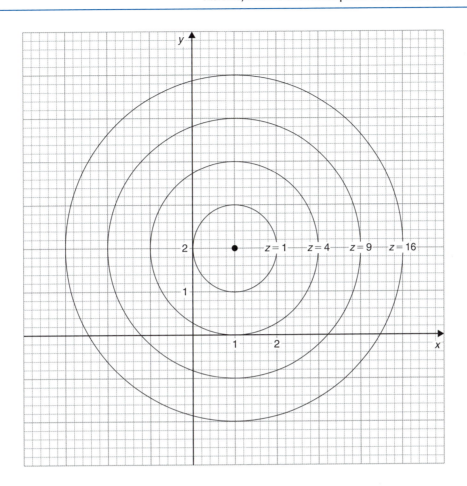

Figure 41.8

(vi)   for $(0, 0)$, $\left(\dfrac{\partial^2 z}{\partial x \partial y}\right)^2 = (-6)^2 = 36$

for $(2, 2)$, $\left(\dfrac{\partial^2 z}{\partial x \partial y}\right)^2 = (-6)^2 = 36$

(vii) $\Delta_{(0,\,0)} = \left(\dfrac{\partial^2 z}{\partial x \partial y}\right)^2 - \left(\dfrac{\partial^2 z}{\partial x^2}\right)\left(\dfrac{\partial^2 z}{\partial y^2}\right)$

$= 36 - (0)(0) = 36$

$\Delta_{(2,\,2)} = 36 - (12)(12) = -108$

(viii)   Since $\Delta_{(0,\,0)} > 0$ then **(0, 0) is a saddle point**.

Since $\Delta_{(2,\,2)} < 0$ and $\dfrac{\partial^2 z}{\partial x^2} > 0$, then **(2, 2) is a minimum point**.

Now try the following Practice Exercise

Practice Exercise 169   Maxima, minima and saddle points for functions of two variables (Answers on page 881)

1.   Find the stationary point of the surface $f(x, y) = x^2 + y^2$ and determine its nature. Sketch the contour map represented by z

2.   Find the maxima, minima and saddle points for the following functions:
(a) $f(x, y) = x^2 + y^2 - 2x + 4y + 8$
(b) $f(x, y) = x^2 - y^2 - 2x + 4y + 8$
(c) $f(x, y) = 2x + 2y - 2xy - 2x^2 - y^2 + 4$

3.   Determine the stationary values of the function $f(x, y) = x^3 - 6x^2 - 8y^2$ and distinguish between them.

Section H

Section H

4. Locate the stationary point of the function $z = 12x^2 + 6xy + 15y^2$

5. Find the stationary points of the surface $z = x^3 - xy + y^3$ and distinguish between them.

41.5 Further worked problems on maxima, minima and saddle points for functions of two variables

Problem 3. Find the co-ordinates of the stationary points on the surface

$$z = (x^2 + y^2)^2 - 8(x^2 - y^2)$$

and distinguish between them. Sketch the approximate contour map associated with z

Following the procedure:

(i) $\dfrac{\partial z}{\partial x} = 2(x^2 + y^2)2x - 16x$ and

$\dfrac{\partial z}{\partial y} = 2(x^2 + y^2)2y + 16y$

(ii) for stationary points,

$2(x^2 + y^2)2x - 16x = 0$

i.e. $4x^3 + 4xy^2 - 16x = 0$ (1)

and $2(x^2 + y^2)2y + 16y = 0$

i.e. $4y(x^2 + y^2 + 4) = 0$ (2)

(iii) From equation (1), $y^2 = \dfrac{16x - 4x^3}{4x} = 4 - x^2$

Substituting $y^2 = 4 - x^2$ in equation (2) gives

$4y(x^2 + 4 - x^2 + 4) = 0$

i.e. $32y = 0$ and $y = 0$

When $y = 0$ in equation (1), $4x^3 - 16x = 0$

i.e. $4x(x^2 - 4) = 0$

from which, $x = 0$ or $x = \pm 2$

The co-ordinates of the stationary points are (0, 0), (2, 0) and (−2, 0)

(iv) $\dfrac{\partial^2 z}{\partial x^2} = 12x^2 + 4y^2 - 16$,

$\dfrac{\partial^2 z}{\partial y^2} = 4x^2 + 12y^2 + 16$ and $\dfrac{\partial^2 z}{\partial x \partial y} = 8xy$

(v) For the point (0, 0),

$\dfrac{\partial^2 z}{\partial x^2} = -16, \dfrac{\partial^2 z}{\partial y^2} = 16$ and $\dfrac{\partial^2 z}{\partial x \partial y} = 0$

For the point (2, 0),

$\dfrac{\partial^2 z}{\partial x^2} = 32, \dfrac{\partial^2 z}{\partial y^2} = 32$ and $\dfrac{\partial^2 z}{\partial x \partial y} = 0$

For the point (−2, 0),

$\dfrac{\partial^2 z}{\partial x^2} = 32, \dfrac{\partial^2 z}{\partial y^2} = 32$ and $\dfrac{\partial^2 z}{\partial x \partial y} = 0$

(vi) $\left(\dfrac{\partial^2 z}{\partial x \partial y}\right)^2 = 0$ for each stationary point

(vii) $\Delta_{(0, 0)} = (0)^2 - (-16)(16) = 256$

$\Delta_{(2, 0)} = (0)^2 - (32)(32) = -1024$

$\Delta_{(-2, 0)} = (0)^2 - (32)(32) = -1024$

(viii) Since $\Delta_{(0, 0)} > 0$, **the point (0, 0) is a saddle point.**

Since $\Delta_{(0, 0)} < 0$ and $\left(\dfrac{\partial^2 z}{\partial x^2}\right)_{(2, 0)} > 0$, **the point (2, 0) is a minimum point**.

Since $\Delta_{(-2, 0)} < 0$ and $\left(\dfrac{\partial^2 z}{\partial x^2}\right)_{(-2, 0)} > 0$, **the point (−2, 0) is a minimum point.**

Looking down towards the x–y plane from above, an approximate contour map can be constructed to represent the value of z. Such a map is shown in Fig. 41.9. To produce a contour map requires a large number of x–y co-ordinates to be chosen and the values of z at each co-ordinate calculated. Here are a few examples of points used to construct the contour map.

When $z = 0$, $0 = (x^2 + y^2)^2 - 8(x^2 - y^2)$
In addition, when, say, $y = 0$ (i.e. on the x-axis)

$0 = x^4 - 8x^2$, i.e. $x^2(x^2 - 8) = 0$

from which, $x = 0$ or $x = \pm\sqrt{8}$

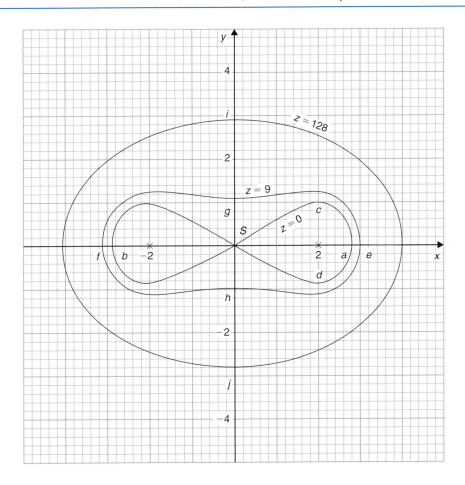

Figure 41.9

Hence the contour $z=0$ crosses the x-axis at 0 and $\pm\sqrt{8}$, i.e. at co-ordinates $(0,0),(2.83,0)$ and $(-2.83,0)$ shown as points S, a and b respectively.

When $z=0$ and $x=2$ then

$$0=(4+y^2)^2-8(4-y^2)$$

i.e. $0=16+8y^2+y^4-32+8y^2$

i.e. $0=y^4+16y^2-16$

Let $y^2=p$, then $p^2+16p-16=0$ and

$$p=\frac{-16\pm\sqrt{16^2-4(1)(-16)}}{2}$$

$$=\frac{-16\pm17.89}{2}$$

$$=0.945 \text{ or } -16.945$$

Hence $y=\sqrt{p}=\sqrt{(0.945)}$ or $\sqrt{(-16.945)}$

$$=\pm0.97 \text{ or complex roots.}$$

Hence the $z=0$ contour passes through the co-ordinates $(2,0.97)$ and $(2,-0.97)$ shown as a c and d in Fig. 41.9.

Similarly, for the **$z=9$** contour, when $y=0$,

$$9=(x^2+0^2)^2-8(x^2-0^2)$$

i.e. $9=x^4-8x^2$

i.e. $x^4-8x^2-9=0$

Hence $(x^2-9)(x^2+1)=0$

from which, $x=\pm3$ or complex roots.

Thus the $z=9$ contour passes through $(3,0)$ and $(-3,0)$, shown as e and f in Fig. 41.9.

If $z=9$ and $x=0,9\ =y^4+8y^2$

i.e. $y^4+8y^2-9=0$

i.e. $(y^2+9)(y^2-1)=0$

from which, $y=\pm1$ or complex roots.

Thus the $z=9$ contour also passes through $(0,1)$ and $(0,-1)$, shown as g and h in Fig. 41.9.

When, say, $x = 4$ and $y = 0$,

$z = (4^2)^2 - 8(4^2) = 128$

when $z = 128$ and $x = 0$, $128 = y^4 + 8y^2$

i.e. $y^4 + 8y^2 - 128 = 0$

i.e. $(y^2 + 16)(y^2 - 8) = 0$

from which, $y = \pm\sqrt{8}$ or complex roots.
Thus the $z = 128$ contour passes through $(0, 2.83)$ and $(0, -2.83)$, shown as i and j in Fig. 41.9.
In a similar manner many other points may be calculated with the resulting approximate contour map shown in Fig. 41.9. It is seen that two 'hollows' occur at the minimum points, and a 'cross-over' occurs at the saddle point S, which is typical of such contour maps.

> **Problem 4.** Show that the function
>
> $$f(x, y) = x^3 - 3x^2 - 4y^2 + 2$$
>
> has one saddle point and one maximum point. Determine the maximum value.

Let $z = f(x, y) = x^3 - 3x^2 - 4y^2 + 2$

Following the procedure:

(i) $\dfrac{\partial z}{\partial x} = 3x^2 - 6x$ and $\dfrac{\partial z}{\partial y} = -8y$

(ii) for stationary points, $3x^2 - 6x = 0$ (1)

and $\qquad\qquad\qquad\qquad -8y = 0$ (2)

(iii) From equation (1), $3x(x - 2) = 0$ from which, $x = 0$ and $x = 2$

From equation (2), $y = 0$

Hence the stationary points are $(0, 0)$ and $(2, 0)$

(iv) $\dfrac{\partial^2 z}{\partial x^2} = 6x - 6$, $\dfrac{\partial^2 z}{\partial y^2} = -8$ and $\dfrac{\partial^2 z}{\partial x \partial y} = 0$

(v) For the point $(0, 0)$,

$$\frac{\partial^2 z}{\partial x^2} = -6, \frac{\partial^2 z}{\partial y^2} = -8 \text{ and } \frac{\partial^2 z}{\partial x \partial y} = 0$$

For the point $(2, 0)$,

$$\frac{\partial^2 z}{\partial x^2} = 6, \frac{\partial^2 z}{\partial y^2} = -8 \text{ and } \frac{\partial^2 z}{\partial x \partial y} = 0$$

(vi) $\left(\dfrac{\partial^2 z}{\partial x \partial y}\right)^2 = (0)^2 = 0$

(vii) $\Delta_{(0, 0)} = 0 - (-6)(-8) = -48$

$\Delta_{(2, 0)} = 0 - (6)(-8) = 48$

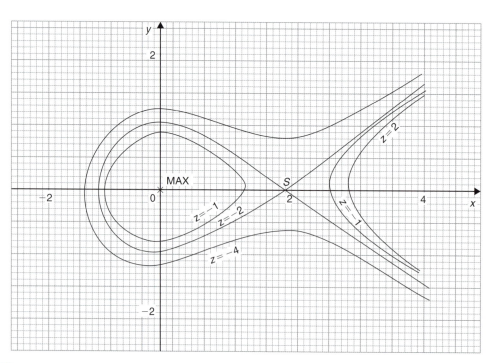

Figure 41.10

(viii)   Since $\Delta_{(0, 0)} < 0$ and $\left(\dfrac{\partial^2 z}{\partial x^2}\right)_{(0, 0)} < 0$,   **the point (0, 0) is a maximum point** and hence **the maximum value is 0**

Since $\Delta_{(2, 0)} > 0$, **the point (2, 0) is a saddle point**.

The value of z at the saddle point is $2^3 - 3(2)^2 - 4(0)^2 + 2 = -2$

An approximate contour map representing the surface $f(x, y)$ is shown in Fig. 41.10 where a 'hollow effect' is seen surrounding the maximum point and a 'cross-over' occurs at the saddle point S.

Problem 5. An open rectangular container is to have a volume of $62.5\,\text{m}^3$. Determine the least surface area of material required.

Let the dimensions of the container be x, y and z as shown in Fig. 41.11.

Volume $\qquad V = xyz = 62.5$ $\qquad\qquad$ (1)

Surface area, $\quad S = xy + 2yz + 2xz$ $\qquad$ (2)

From equation (1), $z = \dfrac{62.5}{xy}$

Substituting in equation (2) gives:

$$S = xy + 2y\left(\frac{62.5}{xy}\right) + 2x\left(\frac{62.5}{xy}\right)$$

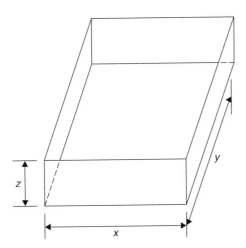

Figure 41.11

i.e. $\qquad S = xy + \dfrac{125}{x} + \dfrac{125}{y}$

which is a function of two variables

$\dfrac{\partial s}{\partial x} = y - \dfrac{125}{x^2} = 0$ for a stationary point,

hence $\quad x^2 y = 125$ $\qquad\qquad\qquad\qquad$ (3)

$\dfrac{\partial s}{\partial y} = x - \dfrac{125}{y^2} = 0$ for a stationary point,

hence $\quad xy^2 = 125$ $\qquad\qquad\qquad\qquad$ (4)

Dividing equation (3) by (4) gives:

$$\frac{x^2 y}{xy^2} = 1, \text{ i.e. } \frac{x}{y} = 1, \text{ i.e. } x = y$$

Substituting $y = x$ in equation (3) gives $x^3 = 125$, from which, $x = 5\,\text{m}$.

Hence $y = 5\,\text{m}$ also

From equation (1), $\quad (5)(5)\, z = 62.5$

from which, $\qquad\qquad\qquad z = \dfrac{62.5}{25} = 2.5\,\text{m}$

$$\frac{\partial^2 S}{\partial x^2} = \frac{250}{x^3}, \frac{\partial^2 S}{\partial y^2} = \frac{250}{y^3} \text{ and } \frac{\partial^2 S}{\partial x \partial y} = 1$$

When $x = y = 5$, $\dfrac{\partial^2 S}{\partial x^2} = 2$, $\dfrac{\partial^2 S}{\partial y^2} = 2$ and $\dfrac{\partial^2 S}{\partial x \partial y} = 1$

$$\Delta = (1)^2 - (2)(2) = -3$$

Since $\Delta < 0$ and $\dfrac{\partial^2 S}{\partial x^2} > 0$, then the surface area S is a **minimum**.

Hence the minimum dimensions of the container to have a volume of $62.5\,\text{m}^3$ are **5 m by 5 m by 2.5 m**.

From equation (2), **minimum surface area, S**

$$= (5)(5) + 2(5)(2.5) + 2(5)(2.5)$$

$$= \mathbf{75\,m^2}$$

Now try the following Practice Exercise

Practice Exercise 170   Maxima, minima and saddle points for functions of two variables (Answers on page 882)

1.   The function $z = x^2 + y^2 + xy + 4x - 4y + 3$ has one stationary value. Determine its co-ordinates and its nature.

2. An open rectangular container is to have a volume of $32\,\text{m}^3$. Determine the dimensions and the total surface area such that the total surface area is a minimum.

3. Determine the stationary values of the function

$$f(x,\,y) = x^4 + 4x^2 y^2 - 2x^2 + 2y^2 - 1$$

and distinguish between them.

4. Determine the stationary points of the surface
$$f(x,\,y) = x^3 - 6x^2 - y^2$$

5. Locate the stationary points on the surface

$$f(x,\,y) = 2x^3 + 2y^3 - 6x - 24y + 16$$

and determine their nature.

6. A large marquee is to be made in the form of a rectangular box-like shape with canvas covering on the top, back and sides. Determine the minimum surface area of canvas necessary if the volume of the marquee is to be $250\,\text{m}^3$.

For fully worked solutions to each of the problems in Practice Exercises 169 and 170 in this chapter,
go to the website:
www.routledge.com/cw/bird

Section H

Revision Test 12 Further differentiation

This Revision Test covers the material contained in Chapters 37 to 41. *The marks for each question are shown in brackets at the end of each question.*

1. Differentiate the following functions with respect to x:

   (a) $5 \ln (\text{sh} x)$ (b) $3 \text{ch}^3 2x$

   (c) $e^{2x} \text{sech} 2x$ (7)

2. Differentiate the following functions with respect to the variable:

   (a) $y = \frac{1}{5} \cos^{-1} \frac{x}{2}$

   (b) $y = 3 e^{\sin^{-1} t}$

   (c) $y = \frac{2 \sec^{-1} 5x}{x}$

   (d) $y = 3 \sinh^{-1} \sqrt{(2x^2 - 1)}$ (14)

3. Evaluate the following, each correct to 3 decimal places:

   (a) $\sinh^{-1} 3$ (b) $\cosh^{-1} 2.5$ (c) $\tanh^{-1} 0.8$ (6)

4. If $z = f(x, y)$ and $z = x \cos(x + y)$ determine

   $$\frac{\partial z}{\partial x}, \frac{\partial z}{\partial y}, \frac{\partial^2 z}{\partial x^2}, \frac{\partial^2 z}{\partial y^2}, \frac{\partial^2 z}{\partial x \partial y} \text{ and } \frac{\partial^2 z}{\partial y \partial x}$$ (12)

5. The magnetic field vector H due to a steady current I flowing around a circular wire of radius r and at a distance x from its centre is given by

   $$H = \pm \frac{I}{2} \frac{\partial}{\partial x} \left(\frac{x}{\sqrt{r^2 + x^2}} \right)$$

   Show that $H = \pm \dfrac{r^2 I}{2 \sqrt{(r^2 + x^2)^3}}$ (7)

6. If $xyz = c$, where c is constant, show that

   $$dz = -z \left(\frac{dx}{x} + \frac{dy}{y} \right)$$ (6)

7. An engineering function $z = f(x, y)$ and $z = e^{\frac{y}{2}} \ln(2x + 3y)$. Determine the rate of increase of z, correct to 4 significant figures, when $x = 2$ cm, $y = 3$ cm, x is increasing at 5 cm/s and y is increasing at 4 cm/s. (8)

8. The volume V of a liquid of viscosity coefficient η delivered after time t when passed through a tube of length L and diameter d by a pressure p is given by $V = \dfrac{pd^4 t}{128 \eta L}$. If the errors in V, p and L are 1%, 2% and 3% respectively, determine the error in η. (8)

9. Determine and distinguish between the stationary values of the function

   $$f(x, y) = x^3 - 6x^2 - 8y^2$$

   and sketch an approximate contour map to represent the surface $f(x, y)$ (20)

10. An open, rectangular fish tank is to have a volume of 13.5 m³. Determine the least surface area of glass required. (12)

Section H

For lecturers/instructors/teachers, fully worked solutions to each of the problems in Revision Test 12, together with a full marking scheme, are available at the website:
www.routledge.com/cw/bird

Section I

Further integral calculus

Chapter 42

Integration using algebraic substitutions

Why it is important to understand: Integration using algebraic substitutions

As intimated in previous chapters, most complex engineering problems cannot be solved without calculus. Calculus has widespread applications in science, economics and engineering and can solve many problems for which algebra alone is insufficient. For example, calculus is needed to calculate the force exerted on a particle a specific distance from an electrically charged wire, and is needed for computations involving arc length, centre of mass, work and pressure. Sometimes the integral is not a standard one; in these cases it may be possible to replace the variable of integration by a function of a new variable. A change in variable can reduce an integral to a standard form, and this is demonstrated in this chapter.

At the end of this chapter, you should be able to:

- appreciate when an algebraic substitution is required to determine an integral
- integrate functions which require an algebraic substitution
- determine definite integrals where an algebraic substitution is required
- appreciate that it is possible to change the limits when determining a definite integral

---

42.1 Introduction

Functions which require integrating are not always in the 'standard form' shown in Chapter 31. However, it is often possible to change a function into a form which can be integrated by using either:

- (i) an algebraic substitution (see Section 42.2),
- (ii) a trigonometric or hyperbolic substitution (see Chapter 43),
- (iii) partial fractions (see Chapter 44),
- (iv) the $t = \tan \theta/2$ substitution (see Chapter 45),
- (v) integration by parts (see Chapter 46), or
- (vi) reduction formulae (see Chapter 47).

42.2 Algebraic substitutions

With **algebraic substitutions**, the substitution usually made is to let u be equal to $f(x)$ such that $f(u)\,du$ is a standard integral. It is found that integrals of the forms,

$$k \int [f(x)]^n f'(x)\,dx \quad \text{and} \quad k \int \frac{f'(x)}{[f(x)]^n}\,dx$$

(where k and n are constants) can both be integrated by substituting u for $f(x)$.

42.3 Worked problems on integration using algebraic substitutions

Problem 1. Determine $\int \cos(3x+7)\,dx$

$\int \cos(3x+7)\,dx$ is not a standard integral of the form shown in Table 31.1, page 377, thus an algebraic substitution is made.

Let $u = 3x+7$ then $\dfrac{du}{dx} = 3$ and rearranging gives $dx = \dfrac{du}{3}$. Hence,

$$\int \cos(3x+7)\,dx = \int (\cos u)\frac{du}{3} = \int \frac{1}{3}\cos u\,du,$$

which is a standard integral

$$= \frac{1}{3}\sin u + c$$

Rewriting u as $(3x+7)$ gives:

$$\int \cos(3x+7)\,dx = \frac{1}{3}\sin(3x+7)+c,$$

which may be checked by differentiating it.

Problem 2. Find $\int(2x-5)^7\,dx$

$(2x-5)$ may be multiplied by itself seven times and then each term of the result integrated. However, this would be a lengthy process, and thus an algebraic substitution is made.

Let $u = (2x-5)$ then $\dfrac{du}{dx} = 2$ and $dx = \dfrac{du}{2}$

Hence

$$\int(2x-5)^7\,dx = \int u^7\frac{du}{2} = \frac{1}{2}\int u^7\,du$$

$$= \frac{1}{2}\left(\frac{u^8}{8}\right)+c = \frac{1}{16}u^8+c$$

Rewriting u as $(2x-5)$ gives:

$$\int(2x-5)^7\,dx = \frac{1}{16}(2x-5)^8+c$$

Problem 3. Find $\int \dfrac{4}{(5x-3)}\,dx$

Let $u = (5x-3)$ then $\dfrac{du}{dx} = 5$ and $dx = \dfrac{du}{5}$

Hence

$$\int \frac{4}{(5x-3)}\,dx = \int \frac{4}{u}\frac{du}{5} = \frac{4}{5}\int \frac{1}{u}\,du$$

$$= \frac{4}{5}\ln u + c = \frac{4}{5}\ln(5x-3)+c$$

Problem 4. Evaluate $\displaystyle\int_0^1 2e^{6x-1}\,dx$, correct to 4 significant figures.

Let $u = 6x-1$ then $\dfrac{du}{dx} = 6$ and $dx = \dfrac{du}{6}$

Hence

$$\int 2e^{6x-1}\,dx = \int 2e^u\frac{du}{6} = \frac{1}{3}\int e^u\,du$$

$$= \frac{1}{3}e^u + c = \frac{1}{3}e^{6x-1}+c$$

Thus

$$\int_0^1 2e^{6x-1}\,dx = \frac{1}{3}[e^{6x-1}]_0^1 = \frac{1}{3}[e^5 - e^{-1}] = \textbf{49.35},$$

correct to 4 significant figures.

Problem 5. Determine $\int 3x(4x^2+3)^5\,dx$

Let $u = (4x^2+3)$ then $\dfrac{du}{dx} = 8x$ and $dx = \dfrac{du}{8x}$

Hence

$$\int 3x(4x^2+3)^5\,dx = \int 3x(u)^5\frac{du}{8x}$$

$$= \frac{3}{8}\int u^5\,du, \text{ by cancelling.}$$

The original variable 'x' has been completely removed and the integral is now only in terms of u and is a standard integral.

Hence $\dfrac{3}{8}\displaystyle\int u^5\,du = \dfrac{3}{8}\left(\dfrac{u^6}{6}\right)+c$

$$= \frac{1}{16}u^6 + c = \frac{1}{16}(4x^2+3)^6+c$$

Problem 6. Evaluate $\displaystyle\int_0^{\frac{\pi}{6}} 24\sin^5\theta\cos\theta\,d\theta$

Let $u = \sin\theta$ then $\dfrac{du}{d\theta} = \cos\theta$ and $d\theta = \dfrac{du}{\cos\theta}$

Hence $\displaystyle\int 24\sin^5\theta\cos\theta\,d\theta = \int 24u^5\cos\theta\,\dfrac{du}{\cos\theta}$

$$= 24\int u^5\,du\text{, by cancelling}$$

$$= 24\dfrac{u^6}{6} + c = 4u^6 + c = 4(\sin\theta)^6 + c$$

$$= 4\sin^6\theta + c$$

Thus $\displaystyle\int_0^{\frac{\pi}{6}} 24\sin^5\theta\cos\theta\,d\theta = [4\sin^6\theta]_0^{\frac{\pi}{6}}$

$$= 4\left[\left(\sin\dfrac{\pi}{6}\right)^6 - (\sin 0)^6\right]$$

$$= 4\left[\left(\dfrac{1}{2}\right)^6 - 0\right] = \dfrac{1}{16}\text{ or }\mathbf{0.0625}$$

Now try the following Practice Exercise

Practice Exercise 171   Integration using algebraic substitutions (Answers on page 882)

In Problems 1 to 6, integrate with respect to the variable.

1. $2\sin(4x+9)$

2. $3\cos(2\theta-5)$

3. $4\sec^2(3t+1)$

4. $\dfrac{1}{2}(5x-3)^6$

5. $\dfrac{-3}{(2x-1)}$

6. $3e^{3\theta+5}$

In Problems 7 to 10, evaluate the definite integrals correct to 4 significant figures.

7. $\displaystyle\int_0^1 (3x+1)^5\,dx$

8. $\displaystyle\int_0^2 x\sqrt{(2x^2+1)}\,dx$

9. $\displaystyle\int_0^{\frac{\pi}{3}} 2\sin\left(3t+\dfrac{\pi}{4}\right)dt$

10. $\displaystyle\int_0^1 3\cos(4x-3)\,dx$

11. The mean time to failure, M years, for a set of components is given by:
$$M = \int_0^4 (1-0.25t)^{1.5}dt$$
Determine the mean time to failure.

42.4 Further worked problems on integration using algebraic substitutions

Problem 7. Find $\displaystyle\int \dfrac{x}{2+3x^2}\,dx$

Let $u = 2+3x^2$ then $\dfrac{du}{dx} = 6x$ and $dx = \dfrac{du}{6x}$

Hence
$$\int \dfrac{x}{2+3x^2}\,dx = \int \dfrac{x}{u}\dfrac{du}{6x} = \dfrac{1}{6}\int \dfrac{1}{u}\,du,$$
$$\text{by cancelling}$$

$$= \dfrac{1}{6}\ln u + c = \dfrac{1}{6}\ln(2+3x^2) + c$$

Problem 8. Determine $\displaystyle\int \dfrac{2x}{\sqrt{(4x^2-1)}}\,dx$

Let $u = 4x^2-1$ then $\dfrac{du}{dx} = 8x$ and $dx = \dfrac{du}{8x}$

Hence $\displaystyle\int \dfrac{2x}{\sqrt{(4x^2-1)}}\,dx = \int \dfrac{2x}{\sqrt{u}}\dfrac{du}{8x}$

$$= \dfrac{1}{4}\int \dfrac{1}{\sqrt{u}}\,du\text{, by cancelling}$$

$$= \dfrac{1}{4}\int u^{\frac{-1}{2}}\,du = \dfrac{1}{4}\left[\dfrac{u^{\left(\frac{-1}{2}\right)+1}}{-\dfrac{1}{2}+1}\right] + c$$

$$= \dfrac{1}{4}\left[\dfrac{u^{\frac{1}{2}}}{\dfrac{1}{2}}\right] + c = \dfrac{1}{2}\sqrt{u} + c$$

$$= \dfrac{1}{2}\sqrt{(4x^2-1)} + c$$

Problem 9. Show that

$$\int \tan\theta \, d\theta = \ln(\sec\theta) + c$$

$\int \tan\theta \, d\theta = \int \dfrac{\sin\theta}{\cos\theta} \, d\theta$. Let $u = \cos\theta$

then $\dfrac{du}{d\theta} = -\sin\theta$ and $d\theta = \dfrac{-du}{\sin\theta}$

Hence

$$\int \frac{\sin\theta}{\cos\theta} \, d\theta = \int \frac{\sin\theta}{u} \left(\frac{-du}{\sin\theta} \right)$$

$$= -\int \frac{1}{u} \, du = -\ln u + c$$

$$= -\ln(\cos\theta) + c = \ln(\cos\theta)^{-1} + c,$$

by the laws of logarithms.

Hence $\displaystyle\int \tan\theta \, d\theta = \ln(\sec\theta) + c,$

since $(\cos\theta)^{-1} = \dfrac{1}{\cos\theta} = \sec\theta$

42.5 Change of limits

When evaluating definite integrals involving substitutions it is sometimes more convenient to **change the limits** of the integral as shown in Problems 10 and 11.

Problem 10. Evaluate $\displaystyle\int_{1}^{3} 5x\sqrt{(2x^2+7)} \, dx$, taking positive values of square roots only.

Let $u = 2x^2 + 7$, then $\dfrac{du}{dx} = 4x$ and $dx = \dfrac{du}{4x}$

It is possible in this case to change the limits of integration. Thus when $x = 3$, $u = 2(3)^2 + 7 = 25$ and when $x = 1$, $u = 2(1)^2 + 7 = 9$

Hence

$$\int_{x=1}^{x=3} 5x\sqrt{(2x^2+7)} \, dx = \int_{u=9}^{u=25} 5x\sqrt{u} \frac{du}{4x}$$

$$= \frac{5}{4} \int_{9}^{25} \sqrt{u} \, du$$

$$= \frac{5}{4} \int_{9}^{25} u^{\frac{1}{2}} \, du$$

Thus the limits have been changed, and it is unnecessary to change the integral back in terms of x.

Thus $\displaystyle\int_{x=1}^{x=3} 5x\sqrt{(2x^2+7)} \, dx = \frac{5}{4} \left[\frac{u^{\frac{3}{2}}}{3/2} \right]_{9}^{25}$

$$= \frac{5}{6} \left[\sqrt{u^3} \right]_{9}^{25} = \frac{5}{6} \left[\sqrt{25^3} - \sqrt{9^3} \right]$$

$$= \frac{5}{6}(125 - 27) = \mathbf{81\frac{2}{3}}$$

Problem 11. Evaluate $\displaystyle\int_{0}^{2} \frac{3x}{\sqrt{(2x^2+1)}} \, dx$, taking positive values of square roots only.

Let $u = 2x^2 + 1$ then $\dfrac{du}{dx} = 4x$ and $dx = \dfrac{du}{4x}$

Hence $\displaystyle\int_{0}^{2} \frac{3x}{\sqrt{(2x^2+1)}} \, dx = \int_{x=0}^{x=2} \frac{3x}{\sqrt{u}} \frac{du}{4x}$

$$= \frac{3}{4} \int_{x=0}^{x=2} u^{\frac{-1}{2}} \, du$$

Since $u = 2x^2 + 1$, when $x = 2, u = 9$ and when $x = 0, u = 1$

Thus $\dfrac{3}{4} \displaystyle\int_{x=0}^{x=2} u^{\frac{-1}{2}} \, du = \frac{3}{4} \int_{u=1}^{u=9} u^{\frac{-1}{2}} \, du,$

i.e. the limits have been changed

$$= \frac{3}{4} \left[\frac{u^{\frac{1}{2}}}{\frac{1}{2}} \right]_{1}^{9} = \frac{3}{2} \left[\sqrt{9} - \sqrt{1} \right] = \mathbf{3},$$

taking positive values of square roots only.

___

Now try the following Practice Exercise

Section I

3. $3\sec^2 3x \tan 3x$

4. $2t\sqrt{(3t^2-1)}$

5. $\dfrac{\ln\theta}{\theta}$

6. $3\tan 2t$

7. $\dfrac{2e^t}{\sqrt{(e^t+4)}}$

In Problems 8 to 10, evaluate the definite integrals correct to 4 significant figures.

8. $\displaystyle\int_0^1 3x\,e^{(2x^2-1)}\,dx$

9. $\displaystyle\int_0^{\frac{\pi}{2}} 3\sin^4\theta\cos\theta\,d\theta$

10. $\displaystyle\int_0^1 \dfrac{3x}{(4x^2-1)^5}\,dx$

11. The electrostatic potential on all parts of a conducting circular disc of radius r is given by the equation:

$$V = 2\pi\sigma\int_0^9 \frac{R}{\sqrt{R^2+r^2}}\,dR$$

Solve the equation by determining the integral.

12. In the study of a rigid rotor the following integration occurs:

$$Z_r = \int_0^\infty (2J+1)e^{\frac{-J(J+1)h^2}{8\pi^2 IkT}}\,dJ$$

Determine Z_r for constant temperature T assuming h, I and k are constants.

13. In electrostatics,

$$E = \int_0^\pi \left\{ \frac{a^2\sigma\sin\theta}{2\varepsilon\sqrt{(a^2-x^2-2ax\cos\theta)}} d\theta \right\}$$

where a, σ and ε are constants, x is greater than a, and x is independent of θ. Show that

$$E = \frac{a^2\sigma}{\varepsilon x}$$

14. The time taken, t hours, for a vehicle to reach a velocity of 130 km/h with an initial speed of 60 km/h is given by: $t = \displaystyle\int_{60}^{130} \dfrac{dv}{650-3v}$

where v is the velocity in km/h. Determine t, correct to the nearest second.

For fully worked solutions to each of the problems in Practice Exercises 171 and 172 in this chapter, go to the website:
www.routledge.com/cw/bird

Chapter 43

Integration using trigonometric and hyperbolic substitutions

Why it is important to understand: Integration using trigonometric and hyperbolic substitutions

Calculus is the most powerful branch of mathematics. It is capable of computing many quantities accurately, which cannot be calculated using any other branch of mathematics. Many integrals are not 'standard' ones that we can determine from a list of results. Some need substitutions to rearrange them into a standard form. There are a number of trigonometric and hyperbolic substitutions that may be used for certain integrals to change them into a form that can be integrated. These are explained in this chapter which provides another piece of the integral calculus jigsaw.

At the end of this chapter, you should be able to:

- integrate functions of the form $\sin^2 x$, $\cos^2 x$, $\tan^2 x$ and $\cot^2 x$
- integrate functions having powers of sines and cosines
- integrate functions that are products of sines and cosines
- integrate using the $\sin\theta$ substitution
- integrate using the $\tan\theta$ substitution
- integrate using the $\sinh\theta$ substitution
- integrate using the $\cosh\theta$ substitution

43.1 Introduction

Table 43.1 gives a summary of the integrals that require the use of **trigonometric and hyperbolic substitutions** and their application is demonstrated in Problems 1 to 27.

43.2 Worked problems on integration of $\sin^2 x$, $\cos^2 x$, $\tan^2 x$ and $\cot^2 x$

Problem 1. Evaluate $\displaystyle\int_0^{\frac{\pi}{4}} 2\cos^2 4t \, dt$

Table 43.1 Integrals using trigonometric and hyperbolic substitutions

$f(x)$	$\int f(x)\,\mathrm{d}x$	Method	See problem
1. $\cos^2 x$	$\dfrac{1}{2}\left(x+\dfrac{\sin 2x}{2}\right)+c$	Use $\cos 2x = 2\cos^2 x - 1$	1
2. $\sin^2 x$	$\dfrac{1}{2}\left(x-\dfrac{\sin 2x}{2}\right)+c$	Use $\cos 2x = 1 - 2\sin^2 x$	2
3. $\tan^2 x$	$\tan x - x + c$	Use $1 + \tan^2 x = \sec^2 x$	3
4. $\cot^2 x$	$-\cot x - x + c$	Use $\cot^2 x + 1 = \operatorname{cosec}^2 x$	4
5. $\cos^m x \sin^n x$	(a) If either m or n is odd (but not both), use $$\cos^2 x + \sin^2 x = 1$$ (b) If both m and n are even, use either $$\cos 2x = 2\cos^2 x - 1 \text{ or } \cos 2x = 1 - 2\sin^2 x$$		5, 6 7, 8
6. $\sin A \cos B$		Use $\frac{1}{2}[\sin(A+B)+\sin(A-B)]$	9
7. $\cos A \sin B$		Use $\frac{1}{2}[\sin(A+B)-\sin(A-B)]$	10
8. $\cos A \cos B$		Use $\frac{1}{2}[\cos(A+B)+\cos(A-B)]$	11
9. $\sin A \sin B$		Use $-\frac{1}{2}[\cos(A+B)-\cos(A-B)]$	12
10. $\dfrac{1}{\sqrt{(a^2-x^2)}}$	$\sin^{-1}\dfrac{x}{a}+c$	Use $x = a\sin\theta$ substitution	13, 14
11. $\sqrt{(a^2-x^2)}$	$\dfrac{a^2}{2}\sin^{-1}\dfrac{x}{a}+\dfrac{x}{2}\sqrt{(a^2-x^2)}+c$	Use $x = a\sin\theta$ substitution	15, 16
12. $\dfrac{1}{a^2+x^2}$	$\dfrac{1}{a}\tan^{-1}\dfrac{x}{a}+c$	Use $x = a\tan\theta$ substitution	17–19
13. $\dfrac{1}{\sqrt{(x^2+a^2)}}$	$\sinh^{-1}\dfrac{x}{a}+c$ or $\ln\left\{\dfrac{x+\sqrt{(x^2+a^2)}}{a}\right\}+c$	Use $x = a\sinh\theta$ substitution	20–22
14. $\sqrt{(x^2+a^2)}$	$\dfrac{a^2}{2}\sinh^{-1}\dfrac{x}{a}+\dfrac{x}{2}\sqrt{(x^2+a^2)}+c$	Use $x = a\sinh\theta$ substitution	23
15. $\dfrac{1}{\sqrt{(x^2-a^2)}}$	$\cosh^{-1}\dfrac{x}{a}+c$ or $\ln\left\{\dfrac{x+\sqrt{(x^2-a^2)}}{a}\right\}+c$	Use $x = a\cosh\theta$ substitution	24, 25
16. $\sqrt{(x^2-a^2)}$	$\dfrac{x}{2}\sqrt{(x^2-a^2)}-\dfrac{a^2}{2}\cosh^{-1}\dfrac{x}{a}+c$	Use $x = a\cosh\theta$ substitution	26, 27

Since $\cos 2t = 2\cos^2 t - 1$ (from Chapter 19),

then $\cos^2 t = \dfrac{1}{2}(1 + \cos 2t)$ and

$$\cos^2 4t = \dfrac{1}{2}(1 + \cos 8t)$$

Hence $\displaystyle \int_0^{\frac{\pi}{4}} 2\cos^2 4t \, dt$

$$= 2\int_0^{\frac{\pi}{4}} \dfrac{1}{2}(1 + \cos 8t) \, dt$$

$$= \left[t + \dfrac{\sin 8t}{8} \right]_0^{\frac{\pi}{4}}$$

$$= \left[\dfrac{\pi}{4} + \dfrac{\sin 8\left(\frac{\pi}{4}\right)}{8} \right] - \left[0 + \dfrac{\sin 0}{8} \right]$$

$$= \dfrac{\pi}{4} \text{ or } \mathbf{0.7854}$$

Problem 2. Determine $\int \sin^2 3x \, dx$

Since $\cos 2x = 1 - 2\sin^2 x$ (from Chapter 19),

then $\sin^2 x = \dfrac{1}{2}(1 - \cos 2x)$ and

$$\sin^2 3x = \dfrac{1}{2}(1 - \cos 6x)$$

Hence $\displaystyle \int \sin^2 3x \, dx = \int \dfrac{1}{2}(1 - \cos 6x) \, dx$

$$= \dfrac{1}{2}\left(x - \dfrac{\sin 6x}{6} \right) + c$$

Problem 3. Find $3\int \tan^2 4x \, dx$

Since $1 + \tan^2 x = \sec^2 x$, then $\tan^2 x = \sec^2 x - 1$ and $\tan^2 4x = \sec^2 4x - 1$

Hence $3\displaystyle \int \tan^2 4x \, dx = 3\int (\sec^2 4x - 1) \, dx$

$$= 3\left(\dfrac{\tan 4x}{4} - x \right) + c$$

Problem 4. Evaluate $\displaystyle \int_{\frac{\pi}{6}}^{\frac{\pi}{3}} \dfrac{1}{2}\cot^2 2\theta \, d\theta$

Since $\cot^2 \theta + 1 = \operatorname{cosec}^2 \theta$, then $\cot^2 \theta = \operatorname{cosec}^2 \theta - 1$ and $\cot^2 2\theta = \operatorname{cosec}^2 2\theta - 1$

Hence $\displaystyle \int_{\frac{\pi}{6}}^{\frac{\pi}{3}} \dfrac{1}{2}\cot^2 2\theta \, d\theta$

$$= \dfrac{1}{2}\int_{\frac{\pi}{6}}^{\frac{\pi}{3}} (\operatorname{cosec}^2 2\theta - 1) \, d\theta = \dfrac{1}{2}\left[\dfrac{-\cot 2\theta}{2} - \theta \right]_{\frac{\pi}{6}}^{\frac{\pi}{3}}$$

$$= \dfrac{1}{2}\left[\left(\dfrac{-\cot 2\left(\frac{\pi}{3}\right)}{2} - \dfrac{\pi}{3} \right) - \left(\dfrac{-\cot 2\left(\frac{\pi}{6}\right)}{2} - \dfrac{\pi}{6} \right) \right]$$

$$= \dfrac{1}{2}[(0.2887 - 1.0472) - (-0.2887 - 0.5236)]$$

$$= \mathbf{0.0269}$$

Now try the following Practice Exercise

Practice Exercise 173   Integration of $\sin^2 x$, $\cos^2 x$, $\tan^2 x$ and $\cot^2 x$ (Answers on page 882)

In Problems 1 to 4, integrate with respect to the variable.

1. $\sin^2 2x$

2. $3\cos^2 t$

3. $5\tan^2 3\theta$

4. $2\cot^2 2t$

In Problems 5 to 8, evaluate the definite integrals, correct to 4 significant figures.

5. $\displaystyle \int_0^{\frac{\pi}{3}} 3\sin^2 3x \, dx$

6. $\displaystyle \int_0^{\frac{\pi}{4}} \cos^2 4x \, dx$

7. $\displaystyle \int_0^1 2\tan^2 2t \, dt$

8. $\displaystyle \int_{\frac{\pi}{6}}^{\frac{\pi}{3}} \cot^2 \theta \, d\theta$

Section I

43.3 Worked problems on integration of powers of sines and cosines

Problem 5. Determine $\int \sin^5 \theta \, d\theta$

Since $\cos^2\theta + \sin^2\theta = 1$ then $\sin^2\theta = (1 - \cos^2\theta)$

Hence $\int \sin^5\theta \, d\theta$

$$= \int \sin\theta (\sin^2\theta)^2 \, d\theta = \int \sin\theta (1 - \cos^2\theta)^2 \, d\theta$$

$$= \int \sin\theta (1 - 2\cos^2\theta + \cos^4\theta) \, d\theta$$

$$= \int (\sin\theta - 2\sin\theta\cos^2\theta + \sin\theta\cos^4\theta) \, d\theta$$

$$= -\cos\theta + \frac{2\cos^3\theta}{3} - \frac{\cos^5\theta}{5} + c$$

Whenever a power of a cosine is multiplied by a sine of power 1, or vice-versa, the integral may be determined by inspection as shown.

In general, $\int \cos^n\theta \sin\theta \, d\theta = \dfrac{-\cos^{n+1}\theta}{(n+1)} + c$

and $\quad \int \sin^n\theta \cos\theta \, d\theta = \dfrac{\sin^{n+1}\theta}{(n+1)} + c$

Problem 6. Evaluate $\displaystyle\int_0^{\frac{\pi}{2}} \sin^2 x \cos^3 x \, dx$

$$\int_0^{\frac{\pi}{2}} \sin^2 x \cos^3 x \, dx = \int_0^{\frac{\pi}{2}} \sin^2 x \cos^2 x \cos x \, dx$$

$$= \int_0^{\frac{\pi}{2}} (\sin^2 x)(1 - \sin^2 x)(\cos x) \, dx$$

$$= \int_0^{\frac{\pi}{2}} (\sin^2 x \cos x - \sin^4 x \cos x) \, dx$$

$$= \left[\frac{\sin^3 x}{3} - \frac{\sin^5 x}{5} \right]_0^{\frac{\pi}{2}}$$

$$= \left[\frac{\left(\sin\frac{\pi}{2}\right)^3}{3} - \frac{\left(\sin\frac{\pi}{2}\right)^5}{5} \right] - [0 - 0]$$

$$= \frac{1}{3} - \frac{1}{5} = \frac{2}{15} \text{ or } \mathbf{0.1333}$$

Problem 7. Evaluate $\displaystyle\int_0^{\frac{\pi}{4}} 4\cos^4\theta \, d\theta$, correct to 4 significant figures.

$$\int_0^{\frac{\pi}{4}} 4\cos^4\theta \, d\theta = 4 \int_0^{\frac{\pi}{4}} (\cos^2\theta)^2 \, d\theta$$

$$= 4 \int_0^{\frac{\pi}{4}} \left[\frac{1}{2}(1 + \cos 2\theta) \right]^2 \, d\theta$$

$$= \int_0^{\frac{\pi}{4}} (1 + 2\cos 2\theta + \cos^2 2\theta) \, d\theta$$

$$= \int_0^{\frac{\pi}{4}} \left[1 + 2\cos 2\theta + \frac{1}{2}(1 + \cos 4\theta) \right] \, d\theta$$

$$= \int_0^{\frac{\pi}{4}} \left(\frac{3}{2} + 2\cos 2\theta + \frac{1}{2}\cos 4\theta \right) \, d\theta$$

$$= \left[\frac{3\theta}{2} + \sin 2\theta + \frac{\sin 4\theta}{8} \right]_0^{\frac{\pi}{4}}$$

$$= \left[\frac{3}{2}\left(\frac{\pi}{4}\right) + \sin\frac{2\pi}{4} + \frac{\sin 4(\pi/4)}{8} \right] - [0]$$

$$= \frac{3\pi}{8} + 1 = \mathbf{2.178},$$

correct to 4 significant figures.

Problem 8. Find $\int \sin^2 t \cos^4 t \, dt$

$$\int \sin^2 t \cos^4 t \, dt = \int \sin^2 t (\cos^2 t)^2 \, dt$$

$$= \int \left(\frac{1 - \cos 2t}{2} \right) \left(\frac{1 + \cos 2t}{2} \right)^2 \, dt$$

$$= \frac{1}{8} \int (1 - \cos 2t)(1 + 2\cos 2t + \cos^2 2t) \, dt$$

$$= \frac{1}{8} \int (1 + 2\cos 2t + \cos^2 2t - \cos 2t$$
$$\qquad\qquad - 2\cos^2 2t - \cos^3 2t) \, dt$$

$$= \frac{1}{8} \int (1 + \cos 2t - \cos^2 2t - \cos^3 2t) \, dt$$

$$= \frac{1}{8} \int \left[1 + \cos 2t - \left(\frac{1 + \cos 4t}{2} \right) \right.$$
$$\qquad\qquad \left. - \cos 2t (1 - \sin^2 2t) \right] \, dt$$

$$= \frac{1}{8} \int \left(\frac{1}{2} - \frac{\cos 4t}{2} + \cos 2t \sin^2 2t \right) \, dt$$

$$= \frac{1}{8} \left(\frac{t}{2} - \frac{\sin 4t}{8} + \frac{\sin^3 2t}{6} \right) + c$$

Now try the following Practice Exercise

In Problems 1 to 6, integrate with respect to the variable.

1. $\sin^3 \theta$

2. $2\cos^3 2x$

3. $2\sin^3 t \cos^2 t$

4. $\sin^3 x \cos^4 x$

5. $2\sin^4 2\theta$

6. $\sin^2 t \cos^2 t$

43.4 Worked problems on integration of products of sines and cosines

Problem 9.   Determine $\int \sin 3t \cos 2t \, dt$

$$\int \sin 3t \cos 2t \, dt$$

$$= \int \frac{1}{2}[\sin(3t + 2t) + \sin(3t - 2t)] \, dt,$$

from 6 of Table 43.1, which follows from Section 19.4, page 210,

$$= \frac{1}{2} \int (\sin 5t + \sin t) \, dt$$

$$= \frac{1}{2}\left(\frac{-\cos 5t}{5} - \cos t\right) + c$$

Problem 10.   Find $\int \frac{1}{3}\cos 5x \sin 2x \, dx$

$$\int \frac{1}{3}\cos 5x \sin 2x \, dx$$

$$= \frac{1}{3}\int \frac{1}{2}[\sin(5x + 2x) - \sin(5x - 2x)] \, dx,$$

from 7 of Table 43.1

$$= \frac{1}{6}\int (\sin 7x - \sin 3x) \, dx$$

$$= \frac{1}{6}\left(\frac{-\cos 7x}{7} + \frac{\cos 3x}{3}\right) + c$$

Problem 11.   Evaluate $\int_0^1 2\cos 6\theta \cos \theta \, d\theta$, correct to 4 decimal places.

$$\int_0^1 2\cos 6\theta \cos \theta \, d\theta$$

$$= 2\int_0^1 \frac{1}{2}[\cos(6\theta + \theta) + \cos(6\theta - \theta)] \, d\theta,$$

from 8 of Table 43.1

$$= \int_0^1 (\cos 7\theta + \cos 5\theta) \, d\theta = \left[\frac{\sin 7\theta}{7} + \frac{\sin 5\theta}{5}\right]_0^1$$

$$= \left(\frac{\sin 7}{7} + \frac{\sin 5}{5}\right) - \left(\frac{\sin 0}{7} + \frac{\sin 0}{5}\right)$$

'$\sin 7$' means 'the sine of 7 radians' ($\equiv 401°4'$) and $\sin 5 \equiv 286°29'$

Hence $\int_0^1 2\cos 6\theta \cos \theta \, d\theta$

$$= (0.09386 + (-0.19178)) - (0)$$

$$= \mathbf{-0.0979}, \text{ correct to 4 decimal places.}$$

Problem 12.   Find $3\int \sin 5x \sin 3x \, dx$

$$3\int \sin 5x \sin 3x \, dx$$

$$= 3\int -\frac{1}{2}[\cos(5x + 3x) - \cos(5x - 3x)] \, dx,$$

from 9 of Table 43.1

$$= -\frac{3}{2}\int (\cos 8x - \cos 2x) \, dx$$

$$= -\frac{3}{2}\left(\frac{\sin 8x}{8} - \frac{\sin 2x}{2}\right) + c \text{ or}$$

$$\frac{3}{16}(\mathbf{4 \sin 2x - \sin 8x}) + c$$

Now try the following Practice Exercise

Practice Exercise 175   Integration of products of sines and cosines (Answers on page 882)

In Problems 1 to 4, integrate with respect to the variable.

1. $\sin 5t \cos 2t$

2. $2 \sin 3x \sin x$

3. $3 \cos 6x \cos x$

4. $\dfrac{1}{2} \cos 4\theta \sin 2\theta$

In Problems 5 to 8, evaluate the definite integrals.

5. $\displaystyle\int_0^{\frac{\pi}{2}} \cos 4x \cos 3x \, dx$

6. $\displaystyle\int_0^1 2 \sin 7t \cos 3t \, dt$

7. $-4 \displaystyle\int_0^{\frac{\pi}{3}} \sin 5\theta \sin 2\theta \, d\theta$

8. $\displaystyle\int_1^2 3 \cos 8t \sin 3t \, dt$

43.5 Worked problems on integration using the $\sin \theta$ substitution

Problem 13.   Determine $\displaystyle\int \frac{1}{\sqrt{(a^2 - x^2)}} \, dx$

Let $x = a \sin\theta$, then $\dfrac{dx}{d\theta} = a \cos\theta$ and $dx = a \cos\theta \, d\theta$

Hence $\displaystyle\int \frac{1}{\sqrt{(a^2 - x^2)}} \, dx$

$= \displaystyle\int \frac{1}{\sqrt{(a^2 - a^2 \sin^2\theta)}} \, a \cos\theta \, d\theta$

$= \displaystyle\int \frac{a \cos\theta \, d\theta}{\sqrt{[a^2(1 - \sin^2\theta)]}}$

$= \displaystyle\int \frac{a \cos\theta \, d\theta}{\sqrt{(a^2 \cos^2\theta)}}, \quad \text{since } \sin^2\theta + \cos^2\theta = 1$

$= \displaystyle\int \frac{a \cos\theta \, d\theta}{a \cos\theta} = \int d\theta = \theta + c$

Since $x = a \sin\theta$, then $\sin\theta = \dfrac{x}{a}$ and $\theta = \sin^{-1} \dfrac{x}{a}$

Hence $\displaystyle\int \frac{1}{\sqrt{(a^2 - x^2)}} \, dx = \boldsymbol{\sin^{-1} \dfrac{x}{a} + c}$

Problem 14.   Evaluate $\displaystyle\int_0^3 \frac{1}{\sqrt{(9 - x^2)}} \, dx$

From Problem 13, $\displaystyle\int_0^3 \frac{1}{\sqrt{(9 - x^2)}} \, dx$

$= \left[\sin^{-1} \dfrac{x}{3} \right]_0^3, \quad \text{since } a = 3$

$= (\sin^{-1} 1 - \sin^{-1} 0) = \dfrac{\pi}{2} \text{ or } \mathbf{1.5708}$

Problem 15.   Find $\displaystyle\int \sqrt{(a^2 - x^2)} \, dx$

Let $x = a \sin\theta$ then $\dfrac{dx}{d\theta} = a \cos\theta$ and $dx = a \cos\theta \, d\theta$

Hence $\displaystyle\int \sqrt{(a^2 - x^2)} \, dx$

$= \displaystyle\int \sqrt{(a^2 - a^2 \sin^2\theta)} \, (a \cos\theta \, d\theta)$

$= \displaystyle\int \sqrt{[a^2(1 - \sin^2\theta)]} \, (a \cos\theta \, d\theta)$

$= \displaystyle\int \sqrt{(a^2 \cos^2\theta)} \, (a \cos\theta \, d\theta)$

$= \displaystyle\int (a \cos\theta)(a \cos\theta \, d\theta)$

$= a^2 \displaystyle\int \cos^2\theta \, d\theta = a^2 \int \left(\frac{1 + \cos 2\theta}{2} \right) d\theta$

$\qquad\qquad\text{(since } \cos 2\theta = 2 \cos^2\theta - 1)$

$= \dfrac{a^2}{2} \left(\theta + \dfrac{\sin 2\theta}{2} \right) + c$

$= \dfrac{a^2}{2} \left(\theta + \dfrac{2 \sin\theta \cos\theta}{2} \right) + c$

$\qquad\qquad \text{since from Chapter 19, } \sin 2\theta = 2 \sin\theta \cos\theta$

$= \dfrac{a^2}{2} [\theta + \sin\theta \cos\theta] + c$

Section I

Since $x = a\sin\theta$, then $\sin\theta = \dfrac{x}{a}$ and $\theta = \sin^{-1}\dfrac{x}{a}$

Also, $\cos^2\theta + \sin^2\theta = 1$, from which,

$$\cos\theta = \sqrt{(1 - \sin^2\theta)} = \sqrt{\left[1 - \left(\dfrac{x}{a}\right)^2\right]}$$

$$= \sqrt{\left(\dfrac{a^2 - x^2}{a^2}\right)} = \dfrac{\sqrt{(a^2 - x^2)}}{a}$$

Thus $\displaystyle\int \sqrt{(a^2 - x^2)}\,dx = \dfrac{a^2}{2}[\theta + \sin\theta\cos\theta]$

$$= \dfrac{a^2}{2}\left[\sin^{-1}\dfrac{x}{a} + \left(\dfrac{x}{a}\right)\dfrac{\sqrt{(a^2 - x^2)}}{a}\right] + c$$

$$= \dfrac{a^2}{2}\sin^{-1}\dfrac{x}{a} + \dfrac{x}{2}\sqrt{(a^2 - x^2)} + c$$

Problem 16. Evaluate $\displaystyle\int_0^4 \sqrt{(16 - x^2)}\,dx$

From Problem 15, $\displaystyle\int_0^4 \sqrt{(16 - x^2)}\,dx$

$$= \left[\dfrac{16}{2}\sin^{-1}\dfrac{x}{4} + \dfrac{x}{2}\sqrt{(16 - x^2)}\right]_0^4$$

$$= \left[8\sin^{-1}1 + 2\sqrt{(0)}\right] - [8\sin^{-1}0 + 0]$$

$$= 8\sin^{-1}1 = 8\left(\dfrac{\pi}{2}\right) = \mathbf{4\pi} \text{ or } \mathbf{12.57}$$

Now try the following Practice Exercise

Practice Exercise 176   Integration using the $\sin\theta$ substitution (Answers on page 882)

1. Determine $\displaystyle\int \dfrac{5}{\sqrt{(4 - t^2)}}\,dt$

2. Determine $\displaystyle\int \dfrac{3}{\sqrt{(9 - x^2)}}\,dx$

3. Determine $\displaystyle\int \sqrt{(4 - x^2)}\,dx$

4. Determine $\displaystyle\int \sqrt{(16 - 9t^2)}\,dt$

5. Evaluate $\displaystyle\int_0^4 \dfrac{1}{\sqrt{(16 - x^2)}}\,dx$

6. Evaluate $\displaystyle\int_0^1 \sqrt{(9 - 4x^2)}\,dx$

43.6 Worked problems on integration using the $\tan\theta$ substitution

Problem 17. Determine $\displaystyle\int \dfrac{1}{(a^2 + x^2)}\,dx$

Let $x = a\tan\theta$ then $\dfrac{dx}{d\theta} = a\sec^2\theta$ and $dx = a\sec^2\theta\,d\theta$

Hence $\displaystyle\int \dfrac{1}{(a^2 + x^2)}\,dx$

$$= \int \dfrac{1}{(a^2 + a^2\tan^2\theta)}(a\sec^2\theta\,d\theta)$$

$$= \int \dfrac{a\sec^2\theta\,d\theta}{a^2(1 + \tan^2\theta)}$$

$$= \int \dfrac{a\sec^2\theta\,d\theta}{a^2\sec^2\theta}, \quad \text{since } 1 + \tan^2\theta = \sec^2\theta$$

$$= \int \dfrac{1}{a}\,d\theta = \dfrac{1}{a}(\theta) + c$$

Since $x = a\tan\theta$, $\theta = \tan^{-1}\dfrac{x}{a}$

Hence $\displaystyle\int \dfrac{1}{(a^2 + x^2)}\,dx = \dfrac{1}{a}\tan^{-1}\dfrac{x}{a} + c$

Problem 18. Evaluate $\displaystyle\int_0^2 \dfrac{1}{(4 + x^2)}\,dx$

From Problem 17, $\displaystyle\int_0^2 \dfrac{1}{(4 + x^2)}\,dx$

$$= \dfrac{1}{2}\left[\tan^{-1}\dfrac{x}{2}\right]_0^2 \quad \text{since } a = 2$$

$$= \dfrac{1}{2}(\tan^{-1}1 - \tan^{-1}0) = \dfrac{1}{2}\left(\dfrac{\pi}{4} - 0\right)$$

$$= \dfrac{\pi}{8} \text{ or } \mathbf{0.3927}$$

Problem 19. Evaluate $\int_0^1 \dfrac{5}{(3+2x^2)}\,dx$, correct to 4 decimal places.

$$\int_0^1 \frac{5}{(3+2x^2)}\,dx = \int_0^1 \frac{5}{2[(3/2)+x^2]}\,dx$$

$$= \frac{5}{2}\int_0^1 \frac{1}{[\sqrt{(3/2)}]^2 + x^2}\,dx$$

$$= \frac{5}{2}\left[\frac{1}{\sqrt{(3/2)}}\tan^{-1}\frac{x}{\sqrt{(3/2)}}\right]_0^1$$

$$= \frac{5}{2}\sqrt{\left(\frac{2}{3}\right)}\left[\tan^{-1}\sqrt{\left(\frac{2}{3}\right)} - \tan^{-1}0\right]$$

$$= (2.0412)[0.6847 - 0]$$

$$= \mathbf{1.3976}, \text{ correct to 4 decimal places.}$$

Now try the following Practice Exercise

Practice Exercise 177   Integration using the tan θ substitution (Answers on page 882)

1. Determine $\int \dfrac{3}{4+t^2}\,dt$

2. Determine $\int \dfrac{5}{16+9\theta^2}\,d\theta$

3. Evaluate $\int_0^1 \dfrac{3}{1+t^2}\,dt$

4. Evaluate $\int_0^3 \dfrac{5}{4+x^2}\,dx$

43.7 Worked problems on integration using the $\sinh\theta$ substitution

Problem 20. Determine $\int \dfrac{1}{\sqrt{(x^2+a^2)}}\,dx$

Let $x = a\sinh\theta$, then $\dfrac{dx}{d\theta} = a\cosh\theta$ and $dx = a\cosh\theta\,d\theta$

Hence $\int \dfrac{1}{\sqrt{(x^2+a^2)}}\,dx$

$$= \int \frac{1}{\sqrt{(a^2\sinh^2\theta + a^2)}}(a\cosh\theta\,d\theta)$$

$$= \int \frac{a\cosh\theta\,d\theta}{\sqrt{(a^2\cosh^2\theta)}}$$

since $\cosh^2\theta - \sinh^2\theta = 1$

$$= \int \frac{a\cosh\theta}{a\cosh\theta}\,d\theta = \int d\theta = \theta + c$$

$$= \sinh^{-1}\frac{x}{a} + c, \text{ since } x = a\sinh\theta$$

It is shown on page 440 that

$$\sinh^{-1}\frac{x}{a} = \ln\left\{\frac{x + \sqrt{(x^2+a^2)}}{a}\right\}$$

which provides an alternative solution to

$$\int \frac{1}{\sqrt{(x^2+a^2)}}\,dx$$

Problem 21. Evaluate $\int_0^2 \dfrac{1}{\sqrt{(x^2+4)}}\,dx$, correct to 4 decimal places.

$$\int_0^2 \frac{1}{\sqrt{(x^2+4)}}\,dx = \left[\sinh^{-1}\frac{x}{2}\right]_0^2 \text{ or}$$

$$\left[\ln\left\{\frac{x + \sqrt{(x^2+4)}}{2}\right\}\right]_0^2$$

from Problem 20, where $a = 2$

Using the logarithmic form,

$$\int_0^2 \frac{1}{\sqrt{(x^2+4)}}\,dx$$

$$= \left[\ln\left(\frac{2+\sqrt{8}}{2}\right) - \ln\left(\frac{0+\sqrt{4}}{2}\right)\right]$$

$$= \ln 2.4142 - \ln 1 = \mathbf{0.8814},$$

correct to 4 decimal places.

Problem 22. Evaluate $\int_1^2 \dfrac{2}{x^2\sqrt{(1+x^2)}}\,dx$, correct to 3 significant figures.

Section I

Since the integral contains a term of the form $\sqrt{(a^2 + x^2)}$, then let $x = \sinh\theta$, from which

$$\frac{dx}{d\theta} = \cosh\theta \text{ and } dx = \cosh\theta \, d\theta$$

Hence $\int \frac{2}{x^2\sqrt{(1+x^2)}} \, dx$

$$= \int \frac{2(\cosh\theta \, d\theta)}{\sinh^2\theta\sqrt{(1+\sinh^2\theta)}}$$

$$= 2 \int \frac{\cosh\theta \, d\theta}{\sinh^2\theta \cosh\theta}$$

since $\cosh^2\theta - \sinh^2\theta = 1$

$$= 2 \int \frac{d\theta}{\sinh^2\theta} = 2 \int \text{cosech}^2\theta \, d\theta$$

$$= -2\coth\theta + c$$

$$\coth\theta = \frac{\cosh\theta}{\sinh\theta} = \frac{\sqrt{(1+\sinh^2\theta)}}{\sinh\theta} = \frac{\sqrt{(1+x^2)}}{x}$$

Hence $\int_1^2 \frac{2}{x^2\sqrt{1+x^2}} \, dx$

$$= -[2\coth\theta]_1^2 = -2\left[\frac{\sqrt{(1+x^2)}}{x}\right]_1^2$$

$$= -2\left[\frac{\sqrt{5}}{2} - \frac{\sqrt{2}}{1}\right] = \mathbf{0.592},$$

correct to 3 significant figures

Problem 23. Find $\int \sqrt{(x^2 + a^2)} \, dx$

Let $x = a\sinh\theta$ then $\frac{dx}{d\theta} = a\cosh\theta$ and
$dx = a\cosh\theta \, d\theta$

Hence $\int \sqrt{(x^2 + a^2)} \, dx$

$$= \int \sqrt{(a^2\sinh^2\theta + a^2)}(a\cosh\theta \, d\theta)$$

$$= \int \sqrt{[a^2(\sinh^2\theta + 1)]}(a\cosh\theta \, d\theta)$$

$$= \int \sqrt{(a^2\cosh^2\theta)}(a\cosh\theta \, d\theta)$$

since $\cosh^2\theta - \sinh^2\theta = 1$

$$= \int (a\cosh\theta)(a\cosh\theta) \, d\theta = a^2 \int \cosh^2\theta \, d\theta$$

$$= a^2 \int \left(\frac{1 + \cosh 2\theta}{2}\right) d\theta$$

$$= \frac{a^2}{2}\left(\theta + \frac{\sinh 2\theta}{2}\right) + c$$

$$= \frac{a^2}{2}[\theta + \sinh\theta\cosh\theta] + c,$$

since $\sinh 2\theta = 2\sinh\theta\cosh\theta$

Since $x = a\sinh\theta$, then $\sinh\theta = \frac{x}{a}$ and $\theta = \sinh^{-1}\frac{x}{a}$

Also since $\cosh^2\theta - \sinh^2\theta = 1$

then $\cosh\theta = \sqrt{(1 + \sinh^2\theta)}$

$$= \sqrt{\left[1 + \left(\frac{x}{a}\right)^2\right]} = \sqrt{\left(\frac{a^2 + x^2}{a^2}\right)}$$

$$= \frac{\sqrt{(a^2 + x^2)}}{a}$$

Hence $\int \sqrt{(x^2 + a^2)} \, dx$

$$= \frac{a^2}{2}\left[\sinh^{-1}\frac{x}{a} + \left(\frac{x}{a}\right)\frac{\sqrt{(x^2 + a^2)}}{a}\right] + c$$

$$= \frac{a^2}{2}\sinh^{-1}\frac{x}{a} + \frac{x}{2}\sqrt{(x^2 + a^2)} + c$$

Now try the following Practice Exercise

Practice Exercise 178   Integration using the $\sinh\theta$ substitution (Answers on page 882)

1. Find $\int \frac{2}{\sqrt{(x^2 + 16)}} \, dx$

2. Find $\int \frac{3}{\sqrt{(9 + 5x^2)}} \, dx$

3. Find $\int \sqrt{(x^2 + 9)} \, dx$

4. Find $\int \sqrt{(4t^2 + 25)} \, dt$

5. Evaluate $\int_0^3 \frac{4}{\sqrt{(t^2 + 9)}} \, dt$

6. Evaluate $\int_0^1 \sqrt{(16 + 9\theta^2)} \, d\theta$

43.8 Worked problems on integration using the $\cosh\theta$ substitution

Problem 24. Determine $\displaystyle\int \frac{1}{\sqrt{(x^2-a^2)}}\,dx$

Let $x=a\cosh\theta$ then $\dfrac{dx}{d\theta}=a\sinh\theta$ and $dx=a\sinh\theta\,d\theta$

Hence $\displaystyle\int \frac{1}{\sqrt{(x^2-a^2)}}\,dx$

$$= \int \frac{1}{\sqrt{(a^2\cosh^2\theta-a^2)}}\,(a\sinh\theta\,d\theta)$$

$$= \int \frac{a\sinh\theta\,d\theta}{\sqrt{[a^2(\cosh^2\theta-1)]}}$$

$$= \int \frac{a\sinh\theta\,d\theta}{\sqrt{(a^2\sinh^2\theta)}}$$

$$\text{since } \cosh^2\theta-\sinh^2\theta=1$$

$$= \int \frac{a\sinh\theta\,d\theta}{a\sinh\theta} = \int d\theta = \theta+c$$

$$= \cosh^{-1}\frac{x}{a}+c, \quad \text{since } x=a\cosh\theta$$

It is shown on page 440 that

$$\cosh^{-1}\frac{x}{a} = \ln\left\{\frac{x+\sqrt{(x^2-a^2)}}{a}\right\}$$

which provides as alternative solution to

$$\int \frac{1}{\sqrt{(x^2-a^2)}}\,dx$$

Problem 25. Determine $\displaystyle\int \frac{2x-3}{\sqrt{(x^2-9)}}\,dx$

$$\int \frac{2x-3}{\sqrt{(x^2-9)}}\,dx = \int \frac{2x}{\sqrt{(x^2-9)}}\,dx$$

$$-\int \frac{3}{\sqrt{(x^2-9)}}\,dx$$

The first integral is determined using the algebraic substitution $u=(x^2-9)$, and the second integral is of the form $\displaystyle\int \frac{1}{\sqrt{(x^2-a^2)}}\,dx$ (see Problem 24)

Hence $\displaystyle\int \frac{2x}{\sqrt{(x^2-9)}}\,dx - \int \frac{3}{\sqrt{(x^2-9)}}\,dx$

$$= 2\sqrt{(x^2-9)}-3\cosh^{-1}\frac{x}{3}+c$$

Problem 26. $\displaystyle\int \sqrt{(x^2-a^2)}\,dx$

Let $x=a\cosh\theta$ then $\dfrac{dx}{d\theta}=a\sinh\theta$ and $dx=a\sinh\theta\,d\theta$

Hence $\displaystyle\int \sqrt{(x^2-a^2)}\,dx$

$$= \int \sqrt{(a^2\cosh^2\theta-a^2)}\,(a\sinh\theta\,d\theta)$$

$$= \int \sqrt{[a^2(\cosh^2\theta-1)]}\,(a\sinh\theta\,d\theta)$$

$$= \int \sqrt{(a^2\sinh^2\theta)}\,(a\sinh\theta\,d\theta)$$

$$= a^2\int \sinh^2\theta\,d\theta = a^2\int\left(\frac{\cosh 2\theta-1}{2}\right)d\theta$$

$$\text{since } \cosh 2\theta=1+2\sinh^2\theta$$

$$\text{from Table 16.1, page 185,}$$

$$= \frac{a^2}{2}\left[\frac{\sinh 2\theta}{2}-\theta\right]+c$$

$$= \frac{a^2}{2}[\sinh\theta\cosh\theta-\theta]+c,$$

$$\text{since } \sinh 2\theta=2\sinh\theta\cosh\theta$$

Since $x=a\cosh\theta$ then $\cosh\theta=\dfrac{x}{a}$ and

$$\theta=\cosh^{-1}\frac{x}{a}$$

Also, since $\cosh^2\theta-\sinh^2\theta=1$, then

$$\sinh\theta = \sqrt{(\cosh^2\theta-1)}$$

$$= \sqrt{\left[\left(\frac{x}{a}\right)^2-1\right]} = \frac{\sqrt{(x^2-a^2)}}{a}$$

Hence $\int \sqrt{(x^2 - a^2)} \, dx$

$$= \frac{a^2}{2} \left[\frac{\sqrt{(x^2 - a^2)}}{a} \left(\frac{x}{a} \right) - \cosh^{-1} \frac{x}{a} \right] + c$$

$$= \frac{x}{2}\sqrt{(x^2 - a^2)} - \frac{a^2}{2} \cosh^{-1} \frac{x}{a} + c$$

Problem 27. Evaluate $\int_2^3 \sqrt{(x^2 - 4)} \, dx$

$$\int_2^3 \sqrt{(x^2 - 4)} \, dx = \left[\frac{x}{2} \sqrt{(x^2 - 4)} - \frac{4}{2} \cosh^{-1} \frac{x}{2} \right]_2^3$$

from Problem 26, when $a = 2$,

$$= \left(\frac{3}{5}\sqrt{5} - 2 \cosh^{-1} \frac{3}{2} \right)$$

$$- (0 - 2 \cosh^{-1} 1)$$

Since $\cosh^{-1} \frac{x}{a} = \ln \left\{ \frac{x + \sqrt{(x^2 - a^2)}}{a} \right\}$ then

$$\cosh^{-1} \frac{3}{2} = \ln \left\{ \frac{3 + \sqrt{(3^2 - 2^2)}}{2} \right\}$$

$$= \ln 2.6180 = 0.9624$$

Similarly, $\cosh^{-1} 1 = 0$

Hence $\int_2^3 \sqrt{(x^2 - 4)} \, dx$

$$= \left[\frac{3}{2}\sqrt{5} - 2(0.9624) \right] - [0]$$

$$= \mathbf{1.429}, \text{ correct to 4 significant figures.}$$

Now try the following Practice Exercise

Practice Exercise 179   Integration using the cosh θ substitution (Answers on page 883)

1. Find $\int \frac{1}{\sqrt{(t^2 - 16)}} \, dt$

2. Find $\int \frac{3}{\sqrt{(4x^2 - 9)}} \, dx$

3. Find $\int \sqrt{(\theta^2 - 9)} \, d\theta$

4. Find $\int \sqrt{(4\theta^2 - 25)} \, d\theta$

5. Evaluate $\int_1^2 \frac{2}{\sqrt{(x^2 - 1)}} \, dx$

6. Evaluate $\int_2^3 \sqrt{(t^2 - 4)} \, dt$

For fully worked solutions to each of the problems in Practice Exercises 173 to 179 in this chapter, go to the website:
www.routledge.com/cw/bird

Chapter 44

Integration using partial fractions

> **Why it is important to understand:** Integration using partial fractions
>
> Sometimes expressions which at first sight look impossible to integrate using standard techniques may in fact be integrated by first expressing them as simpler partial fractions and then using earlier learned techniques. As explained in Chapter 2, the algebraic technique of resolving a complicated fraction into partial fractions is often needed by electrical and mechanical engineers for not only determining certain integrals in calculus, but for determining inverse Laplace transforms, and for analysing linear differential equations like resonant circuits and feedback control systems.

At the end of this chapter, you should be able to:

- integrate functions using partial fractions with linear factors
- integrate functions using partial fractions with repeated linear factors
- integrate functions using partial fractions with quadratic factors

44.1 Introduction

The process of expressing a fraction in terms of simpler fractions – called **partial fractions** – is discussed in Chapter 2, with the forms of partial fractions used being summarised in Table 2.1, page 16.

Certain functions have to be resolved into partial fractions before they can be integrated as demonstrated in the following worked problems.

44.2 Worked problems on integration using partial fractions with linear factors

Problem 1. Determine $\int \dfrac{11-3x}{x^2+2x-3}\,dx$

As shown in Problem 1, page 16:

$$\frac{11-3x}{x^2+2x-3} \equiv \frac{2}{(x-1)} - \frac{5}{(x+3)}$$

Hence $\displaystyle\int \frac{11-3x}{x^2+2x-3}\,dx$

$$= \int \left\{ \frac{2}{(x-1)} - \frac{5}{(x+3)} \right\} dx$$

$$= 2\,\ln(x-1) - 5\,\ln(x+3) + c$$

(by algebraic substitutions — see Chapter 42)

or $\ln\left\{ \dfrac{(x-1)^2}{(x+3)^5} \right\} + c$ by the laws of logarithms

Problem 2. Find

$$\int \frac{2x^2 - 9x - 35}{(x+1)(x-2)(x+3)} \, dx$$

It was shown in Problem 2, page 16:

$$\frac{2x^2 - 9x - 35}{(x+1)(x-2)(x+3)} \equiv \frac{4}{(x+1)} - \frac{3}{(x-2)} + \frac{1}{(x+3)}$$

Hence $\int \frac{2x^2 - 9x - 35}{(x+1)(x-2)(x+3)} \, dx$

$$\equiv \int \left\{ \frac{4}{(x+1)} - \frac{3}{(x-2)} + \frac{1}{(x+3)} \right\} dx$$

$$= 4\ln(x+1) - 3\ln(x-2) + \ln(x+3) + c$$

or $\ln \left\{ \frac{(x+1)^4(x+3)}{(x-2)^3} \right\} + c$

Problem 3. Determine $\int \frac{x^2 + 1}{x^2 - 3x + 2} \, dx$

By dividing out (since the numerator and denominator are of the same degree) and resolving into partial fractions it was shown in Problem 3, page 17:

$$\frac{x^2 + 1}{x^2 - 3x + 2} \equiv 1 - \frac{2}{(x-1)} + \frac{5}{(x-2)}$$

Hence $\int \frac{x^2 + 1}{x^2 - 3x + 2} \, dx$

$$\equiv \int \left\{ 1 - \frac{2}{(x-1)} + \frac{5}{(x-2)} \right\} dx$$

$$= (x-2)\ln(x-1) + 5\ln(x-2) + c$$

or $x + \ln \left\{ \frac{(x-2)^5}{(x-1)^2} \right\} + c$

Problem 4. Evaluate

$$\int_2^3 \frac{x^3 - 2x^2 - 4x - 4}{x^2 + x - 2} \, dx,$$

correct to 4 significant figures.

By dividing out and resolving into partial fractions it was shown in Problem 4, page 17:

$$\frac{x^3 - 2x^2 - 4x - 4}{x^2 + x - 2} \equiv x - 3 + \frac{4}{(x+2)} - \frac{3}{(x-1)}$$

Hence $\int_2^3 \frac{x^3 - 2x^2 - 4x - 4}{x^2 + x - 2} \, dx$

$$\equiv \int_2^3 \left\{ x - 3 + \frac{4}{(x+2)} - \frac{3}{(x-1)} \right\} dx$$

$$= \left[\frac{x^2}{2} - 3x + 4\ln(x+2) - 3\ln(x-1) \right]_2^3$$

$$= \left(\frac{9}{2} - 9 + 4\ln 5 - 3\ln 2 \right)$$

$$- (2 - 6 + 4\ln 4 - 3\ln 1)$$

$$= -1.687, \text{ correct to 4 significant figures.}$$

Now try the following Practice Exercise

Practice Exercise 180 Integration using partial fractions with linear factors (Answers on page 883)

In Problems 1 to 5, integrate with respect to x.

1. $\int \frac{12}{(x^2 - 9)} \, dx$

2. $\int \frac{4(x-4)}{(x^2 - 2x - 3)} \, dx$

3. $\int \frac{3(2x^2 - 8x - 1)}{(x+4)(x+1)(2x-1)} \, dx$

4. $\int \frac{x^2 + 9x + 8}{x^2 + x - 6} \, dx$

5. $\int \frac{3x^3 - 2x^2 - 16x + 20}{(x-2)(x+2)} \, dx$

In Problems 6 and 7, evaluate the definite integrals correct to 4 significant figures.

6. $\int_3^4 \frac{x^2 - 3x + 6}{x(x-2)(x-1)} \, dx$

7. $\int_4^6 \frac{x^2 - x - 14}{x^2 - 2x - 3} \, dx$

8. Determine the value of k, given that:

$$\int_0^1 \frac{(x-k)}{(3x+1)(x+1)}\,dx = 0$$

9. The velocity constant k of a given chemical reaction is given by:

$$kt = \int \left(\frac{1}{(3-0.4x)(2-0.6x)} \right) dx$$

where $x = 0$ when $t = 0$. Show that:

$$kt = \ln \left\{ \frac{2(3-0.4x)}{3(2-0.6x)} \right\}$$

10. The velocity v of an object in a medium at time t seconds is given by: $t = \int_{20}^{80} \frac{dv}{v(2v-1)}$
Evaluate t, in milliseconds, correct to 2 decimal places.

44.3 Worked problems on integration using partial fractions with repeated linear factors

Problem 5. Determine $\int \frac{2x+3}{(x-2)^2}\,dx$

It was shown in Problem 5, page 18:

$$\frac{2x+3}{(x-2)^2} \equiv \frac{2}{(x-2)} + \frac{7}{(x-2)^2}$$

Thus $\int \frac{2x+3}{(x-2)^2}\,dx \equiv \int \left\{ \frac{2}{(x-2)} + \frac{7}{(x-2)^2} \right\} dx$

$$= \mathbf{2\,ln(x-2) - \frac{7}{(x-2)} + c}$$

$$\left[\int \frac{7}{(x-2)^2}\,dx \text{ is determined using the algebraic substitution } u = (x-2) - \text{see Chapter 42.} \right]$$

Problem 6. Find $\int \frac{5x^2 - 2x - 19}{(x+3)(x-1)^2}\,dx$.

It was shown in Problem 6, page 18:

$$\frac{5x^2 - 2x - 19}{(x+3)(x-1)^2} \equiv \frac{2}{(x+3)} + \frac{3}{(x-1)} - \frac{4}{(x-1)^2}$$

Hence $\int \frac{5x^2 - 2x - 19}{(x+3)(x-1)^2}\,dx$

$$\equiv \int \left\{ \frac{2}{(x+3)} + \frac{3}{(x-1)} - \frac{4}{(x-1)^2} \right\} dx$$

$$= \mathbf{2\,ln\,(x+3) + 3\,ln\,(x-1) + \frac{4}{(x-1)} + c}$$

or $\mathbf{ln \left\{ (x+3)^2(x-1)^3 \right\} + \frac{4}{(x-1)} + c}$

Problem 7. Evaluate

$$\int_{-2}^1 \frac{3x^2 + 16x + 15}{(x+3)^3}\,dx,$$

correct to 4 significant figures.

It was shown in Problem 7, page 19:

$$\frac{3x^2 + 16x + 15}{(x+3)^3} \equiv \frac{3}{(x+3)} - \frac{2}{(x+3)^2} - \frac{6}{(x+3)^3}$$

Hence $\int \frac{3x^2 + 16x + 15}{(x+3)^3}\,dx$

$$\equiv \int_{-2}^1 \left\{ \frac{3}{(x+3)} - \frac{2}{(x+3)^2} - \frac{6}{(x+3)^3} \right\} dx$$

$$= \left[3\ln(x+3) + \frac{2}{(x+3)} + \frac{3}{(x+3)^2} \right]_{-2}^1$$

$$= \left(3\ln 4 + \frac{2}{4} + \frac{3}{16} \right) - \left(3\ln 1 + \frac{2}{1} + \frac{3}{1} \right)$$

$$= \mathbf{-0.1536}, \text{ correct to 4 significant figures.}$$

Section I

Now try the following Practice Exercise

Practice Exercise 181   Integration using partial fractions with repeated linear factors (Answers on page 883)

In Problems 1 and 2, integrate with respect to x.

1. $\displaystyle\int \frac{4x-3}{(x+1)^2}\, dx$

2. $\displaystyle\int \frac{5x^2-30x+44}{(x-2)^3}\, dx$

In Problems 3 and 4, evaluate the definite integrals correct to 4 significant figures.

3. $\displaystyle\int_1^2 \frac{x^2+7x+3}{x^2(x+3)}$

4. $\displaystyle\int_6^7 \frac{18+21x-x^2}{(x-5)(x+2)^2}\, dx$

5. Show that $\displaystyle\int_0^1 \left(\frac{4t^2+9t+8}{(t+2)(t+1)^2}\right) dt = 2.546$,

   correct to 4 significant figures.

44.4   Worked problems on integration using partial fractions with quadratic factors

Problem 8.   Find $\displaystyle\int \frac{3+6x+4x^2-2x^3}{x^2(x^2+3)}\, dx.$

It was shown in Problem 9, page 20:

$$\frac{3+6x+4x^2-2x^3}{x^2(x^2+3)} \equiv \frac{2}{x} + \frac{1}{x^2} + \frac{3-4x}{(x^2+3)}$$

Thus $\displaystyle\int \frac{3+6x+4x^2-2x^3}{x^2(x^2+3)}\, dx$

$\displaystyle \equiv \int \left(\frac{2}{x} + \frac{1}{x^2} + \frac{(3-4x)}{(x^2+3)}\right) dx$

$\displaystyle = \int \left\{\frac{2}{x} + \frac{1}{x^2} + \frac{3}{(x^2+3)} - \frac{4x}{(x^2+3)}\right\} dx$

$$\int \frac{3}{(x^2+3)}\, dx = 3\int \frac{1}{x^2+(\sqrt{3})^2}\, dx$$

$$= \frac{3}{\sqrt{3}} \tan^{-1}\frac{x}{\sqrt{3}}, \text{ from } 12, \text{ Table } 43.1, \text{ page } 479.$$

$\displaystyle\int \frac{4x}{x^2+3}\, dx$ is determined using the algebraic substitution $u=(x^2+3)$

Hence $\displaystyle\int \left\{\frac{2}{x} + \frac{1}{x^2} + \frac{3}{(x^2+3)} - \frac{4x}{(x^2+3)}\right\} dx$

$\displaystyle = 2\ln x - \frac{1}{x} + \frac{3}{\sqrt{3}}\tan^{-1}\frac{x}{\sqrt{3}} - 2\ln(x^2+3) + c$

$\displaystyle = \ln\left(\frac{x}{x^2+3}\right)^2 - \frac{1}{x} + \sqrt{3}\tan^{-1}\frac{x}{\sqrt{3}} + c$

Problem 9.   Determine $\displaystyle\int \frac{1}{(x^2-a^2)}\, dx$

Let $\displaystyle\frac{1}{(x^2-a^2)} \equiv \frac{A}{(x-a)} + \frac{B}{(x+a)}$

$\displaystyle \equiv \frac{A(x+a)+B(x-a)}{(x+a)(x-a)}$

Equating the numerators gives:

$$1 \equiv A(x+a)+B(x-a)$$

Let $x=a$, then $A=\dfrac{1}{2a}$, and let $x=-a$, then

$B=-\dfrac{1}{2a}$

Hence $\displaystyle\int \frac{1}{(x^2-a^2)}\, dx$

$\displaystyle \equiv \int \frac{1}{2a}\left[\frac{1}{(x-a)} - \frac{1}{(x+a)}\right] dx$

$\displaystyle = \frac{1}{2a}[\ln(x-a) - \ln(x+a)] + c$

$\displaystyle = \frac{1}{2a}\ln\left(\frac{x-a}{x+a}\right) + c$

Problem 10.   Evaluate

$$\int_3^4 \frac{3}{(x^2-4)}\, dx,$$

correct to 3 significant figures.

From Problem 9,

$$\int_3^4 \frac{3}{(x^2-4)} \, dx = 3 \left[\frac{1}{2(2)} \ln \left(\frac{x-2}{x+2} \right) \right]_3^4$$

$$= \frac{3}{4} \left[\ln \frac{2}{6} - \ln \frac{1}{5} \right]$$

$$= \frac{3}{4} \ln \frac{5}{3} = \mathbf{0.383}, \text{ correct to 3}$$

$$\text{significant figures.}$$

Problem 11. Determine $\displaystyle\int \frac{1}{(a^2-x^2)} \, dx$

Using partial fractions, let

$$\frac{1}{(a^2-x^2)} \equiv \frac{1}{(a-x)(a+x)} \equiv \frac{A}{(a-x)} + \frac{B}{(a+x)}$$

$$\equiv \frac{A(a+x)+B(a-x)}{(a-x)(a+x)}$$

Then $1 \equiv A(a+x) + B(a-x)$

Let $x = a$ then $A = \dfrac{1}{2a}$. Let $x = -a$ then $B = \dfrac{1}{2a}$

Hence $\displaystyle\int \frac{1}{(a^2-x^2)} \, dx$

$$= \int \frac{1}{2a} \left[\frac{1}{(a-x)} + \frac{1}{(a+x)} \right] dx$$

$$= \frac{1}{2a} [-\ln(a-x) + \ln(a+x)] + c$$

$$= \frac{1}{2a} \ln \left(\frac{a+x}{a-x} \right) + c$$

Problem 12. Evaluate

$$\int_0^2 \frac{5}{(9-x^2)} \, dx,$$

correct to 4 decimal places.

From Problem 11,

$$\int_0^2 \frac{5}{(9-x^2)} \, dx = 5 \left[\frac{1}{2(3)} \ln \left(\frac{3+x}{3-x} \right) \right]_0^2$$

$$= \frac{5}{6} \left[\ln \frac{5}{1} - \ln 1 \right]$$

$$= \mathbf{1.3412}, \text{ correct to 4 decimal places.}$$

Now try the following Practice Exercise

Practice Exercise 182   Integration using partial fractions with quadratic factors (Answers on page 883)

1. Determine $\displaystyle\int \frac{x^2-x-13}{(x^2+7)(x-2)} \, dx$

In Problems 2 to 4, evaluate the definite integrals correct to 4 significant figures.

2. $\displaystyle\int_5^6 \frac{6x-5}{(x-4)(x^2+3)} \, dx$

3. $\displaystyle\int_1^2 \frac{4}{(16-x^2)} \, dx$

4. $\displaystyle\int_4^5 \frac{2}{(x^2-9)} \, dx$

5. Show that $\displaystyle\int_1^2 \left(\frac{2+\theta+6\theta^2-2\theta^3}{\theta^2(\theta^2+1)} \right) d\theta$

$$= 1.606, \text{ correct to 4 significant figures.}$$

Section I

For fully worked solutions to each of the problems in Practice Exercises 180 to 182 in this chapter, go to the website:
www.routledge.com/cw/bird

The $t = \tan\dfrac{\theta}{2}$ substitution

At the end of this chapter, you should be able to:

- develop formulae for $\sin\theta$, $\cos\theta$ and $d\theta$ in terms of t, where $t = \tan\dfrac{\theta}{2}$

- integrate functions using $t = \tan\dfrac{\theta}{2}$ substitution

45.1 Introduction

Integrals of the form $\displaystyle\int \dfrac{1}{a\cos\theta + b\sin\theta + c}\, d\theta$, where a, b and c are constants, may be determined by using the substitution $t = \tan\dfrac{\theta}{2}$. The reason is explained below.

If angle A in the right-angled triangle ABC shown in Fig. 45.1 is made equal to $\dfrac{\theta}{2}$ then, since tangent $=$ $\dfrac{\text{opposite}}{\text{adjacent}}$, if $BC = t$ and $AB = 1$, then $\tan\dfrac{\theta}{2} = t$

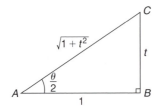

Figure 45.1

By Pythagoras' theorem, $AC = \sqrt{1 + t^2}$

Therefore $\sin\dfrac{\theta}{2} = \dfrac{t}{\sqrt{1+t^2}}$ and $\cos\dfrac{\theta}{2} = \dfrac{1}{\sqrt{1+t^2}}$ Since $\sin 2x = 2\sin x \cos x$ (from double angle formulae, Chapter 19), then

$$\sin\theta = 2\sin\dfrac{\theta}{2}\cos\dfrac{\theta}{2}$$

$$= 2\left(\dfrac{t}{\sqrt{1+t^2}}\right)\left(\dfrac{t}{\sqrt{1+t^2}}\right)$$

i.e. $$\sin\theta = \dfrac{2t}{(1+t^2)} \tag{1}$$

Since $\cos 2x = \cos^2\dfrac{\theta}{2} - \sin^2\dfrac{\theta}{2}$

$$= \left(\dfrac{1}{\sqrt{1+t^2}}\right)^2 - \left(\dfrac{t}{\sqrt{1+t^2}}\right)^2$$

Higher Engineering Mathematics. 978-1-138-67357-1, © 2017 John Bird. Published by Taylor & Francis. All rights reserved.

i.e. $$\cos\theta = \frac{1-t^2}{1+t^2} \quad (2)$$

Also, since $t = \tan\dfrac{\theta}{2}$,

$\dfrac{dt}{d\theta} = \dfrac{1}{2}\sec^2\dfrac{\theta}{2} = \dfrac{1}{2}\left(1+\tan^2\dfrac{\theta}{2}\right)$ from trigonometric identities,

i.e. $$\frac{dt}{d\theta} = \frac{1}{2}(1+t^2)$$

from which, $$d\theta = \frac{2\,dt}{1+t^2} \quad (3)$$

Equations (1), (2) and (3) are used to determine integrals of the form $\displaystyle\int \frac{1}{a\cos\theta + b\sin\theta + c}\,d\theta$ where a, b or c may be zero.

45.2 Worked problems on the $t = \tan\dfrac{\theta}{2}$ substitution

Problem 1. Determine $\displaystyle\int \frac{d\theta}{\sin\theta}$

If $t = \tan\dfrac{\theta}{2}$ then $\sin\theta = \dfrac{2t}{1+t^2}$ and $d\theta = \dfrac{2\,dt}{1+t^2}$ from equations (1) and (3).

Thus $\displaystyle\int \frac{d\theta}{\sin\theta} = \int \frac{1}{\sin\theta}\,d\theta$

$$= \int \frac{1}{\frac{2t}{1+t^2}}\left(\frac{2\,dt}{1+t^2}\right)$$

$$= \int \frac{1}{t}\,dt = \ln t + c$$

Hence $\displaystyle\int \frac{d\theta}{\sin\theta} = \ln\left(\tan\frac{\theta}{2}\right) + c$

Problem 2. Determine $\displaystyle\int \frac{dx}{\cos x}$

If $\tan\dfrac{x}{2}$ then $\cos x = \dfrac{1-t^2}{1+t^2}$ and $dx = \dfrac{2\,dt}{1+t^2}$ from equations (2) and (3).

Thus $\displaystyle\int \frac{dx}{\cos x} = \int \frac{1}{\frac{1-t^2}{1+t^2}}\left(\frac{2\,dt}{1+t^2}\right)$

$$= \int \frac{2}{1-t^2}\,dt$$

$\dfrac{2}{1-t^2}$ may be resolved into partial fractions (see Chapter 2).

Let $\dfrac{2}{1-t^2} = \dfrac{2}{(1-t)(1+t)}$

$$= \frac{A}{(1-t)} + \frac{B}{(1+t)}$$

$$= \frac{A(1+t) + B(1-t)}{(1-t)(1+t)}$$

Hence $2 = A(1+t) + B(1-t)$

When $t = 1, 2 = 2A$, from which, $A = 1$

When $t = -1, 2 = 2B$, from which, $B = 1$

Hence $\displaystyle\int \frac{2\,dt}{1-t^2} = \int \frac{1}{(1-t)} + \frac{1}{(1+t)}\,dt$

$$= -\ln(1-t) + \ln(1+t) + c$$

$$= \ln\left\{\frac{(1+t)}{(1-t)}\right\} + c$$

Thus $\displaystyle\int \frac{dx}{\cos x} = \ln\left\{\frac{1+\tan\frac{x}{2}}{1-\tan\frac{x}{2}}\right\} + c$

Note that since $\tan\dfrac{\pi}{4} = 1$, the above result may be written as:

$$\int \frac{dx}{\cos x} = \ln\left\{\frac{\tan\frac{\pi}{4} + \tan\frac{x}{2}}{1 - \tan\frac{\pi}{4}\tan\frac{x}{2}}\right\} + c$$

$$= \ln\left\{\tan\left(\frac{\pi}{4} + \frac{x}{2}\right)\right\} + c$$

from compound angles, Chapter 19.

Problem 3. Determine $\displaystyle\int \frac{dx}{1+\cos x}$

If $\tan\dfrac{x}{2}$ then $\cos x = \dfrac{1-t^2}{1+t^2}$ and $dx = \dfrac{2\,dt}{1+t^2}$ from equations (2) and (3).

Thus $\displaystyle\int \frac{dx}{1+\cos x} = \int \frac{1}{1+\cos x} dx$

$$= \int \frac{1}{1+\dfrac{1-t^2}{1+t^2}}\left(\frac{2\,dt}{1+t^2}\right)$$

$$= \int \frac{1}{\dfrac{(1+t^2)+(1-t^2)}{1+t^2}}\left(\frac{2\,dt}{1+t^2}\right)$$

$$= \int dt$$

Hence $\displaystyle\int \frac{dx}{1+\cos x} = t+c = \tan\frac{x}{2}+c$

Problem 4. Determine $\displaystyle\int \frac{d\theta}{5+4\cos\theta}$

If $t=\tan\dfrac{\theta}{2}$ then $\cos\theta = \dfrac{1-t^2}{1+t^2}$ and $dx = \dfrac{2\,dt}{1+t^2}$ from equations (2) and (3).

Thus $\displaystyle\int \frac{d\theta}{5+4\cos\theta} = \int \frac{\left(\dfrac{2\,dt}{1+t^2}\right)}{5+4\left(\dfrac{1-t^2}{1+t^2}\right)}$

$$= \int \frac{\left(\dfrac{2\,dt}{1+t^2}\right)}{\dfrac{5(1+t^2)+4(1-t^2)}{(1+t^2)}}$$

$$= 2\int \frac{dt}{t^2+9} = 2\int \frac{dt}{t^2+3^2}$$

$$= 2\left(\frac{1}{3}\tan^{-1}\frac{t}{3}\right)+c,$$

from 12 of Table 43.1, page 479. Hence

$$\int \frac{d\theta}{5+4\cos\theta} = \frac{2}{3}\tan^{-1}\left(\frac{1}{3}\tan\frac{\theta}{2}\right)+c$$

Now try the following Practice Exercise

Practice Exercise 183 The $t=\tan\dfrac{\theta}{2}$

substitution (Answers on page 883)

Integrate the following with respect to the variable:

1. $\displaystyle\int \frac{d\theta}{1+\sin\theta}$

2. $\displaystyle\int \frac{dx}{1-\cos x+\sin x}$

3. $\displaystyle\int \frac{d\alpha}{3+2\cos\alpha}$

4. $\displaystyle\int \frac{dx}{3\sin x-4\cos x}$

45.3 Further worked problems on the $t=\tan\dfrac{\theta}{2}$ substitution

Problem 5. Determine $\displaystyle\int \frac{dx}{\sin x+\cos x}$

If $\tan\dfrac{x}{2}$ then $\sin x = \dfrac{2t}{1+t^2}$, $\cos x = \dfrac{1-t^2}{1+t^2}$ and $dx = \dfrac{2\,dt}{1+t^2}$ from equations (1), (2) and (3).

Thus

$$\int \frac{dx}{\sin x+\cos x} = \int \frac{\dfrac{2\,dt}{1+t^2}}{\left(\dfrac{2t}{1+t^2}\right)+\left(\dfrac{1-t^2}{1+t^2}\right)}$$

$$= \int \frac{\dfrac{2\,dt}{1+t^2}}{\dfrac{2t+1-t^2}{1+t^2}} = \int \frac{2\,dt}{1+2t-t^2}$$

$$= \int \frac{-2\,dt}{t^2-2t-1} = \int \frac{-2\,dt}{(t-1)^2-2}$$

$$= \int \frac{2\,dt}{(\sqrt{2})^2-(t-1)^2}$$

$$= 2\left[\frac{1}{2\sqrt{2}}\ln\left\{\frac{\sqrt{2}+(t-1)}{\sqrt{2}-(t-1)}\right\}\right]+c$$

(see Problem 11, Chapter 44, page 493),

i.e. $\displaystyle\int \frac{dx}{\sin x+\cos x}$

$$= \frac{1}{\sqrt{2}}\ln\left\{\frac{\sqrt{2}-1+\tan\dfrac{x}{2}}{\sqrt{2}+1-\tan\dfrac{x}{2}}\right\}+c$$

Problem 6. Determine
$$\int \frac{dx}{7 - 3\sin x + 6\cos x}$$

From equations (1) and (3),

$$\int \frac{dx}{7 - 3\sin x + 6\cos x}$$

$$= \int \frac{\dfrac{2\,dt}{1+t^2}}{7 - 3\left(\dfrac{2t}{1+t^2}\right) + 6\left(\dfrac{1-t^2}{1+t^2}\right)}$$

$$= \int \frac{\dfrac{2\,dt}{1+t^2}}{\dfrac{7(1+t^2) - 3(2t) + 6(1-t^2)}{1+t^2}}$$

$$= \int \frac{2\,dt}{7 + 7t^2 - 6t + 6 - 6t^2}$$

$$= \int \frac{2\,dt}{t^2 - 6t + 13} = \int \frac{2\,dt}{(t-3)^2 + 2^2}$$

$$= 2\left[\frac{1}{2}\tan^{-1}\left(\frac{t-3}{2}\right)\right] + c$$

from 12, Table 43.1, page 479. Hence

$$\int \frac{dx}{7 - 3\sin x + 6\cos x}$$

$$= \tan^{-1}\left(\frac{\tan\dfrac{x}{2} - 3}{2}\right) + c$$

Problem 7. Determine $\displaystyle\int \frac{d\theta}{4\cos\theta + 3\sin\theta}$

From equations (1) to (3),

$$\int \frac{d\theta}{4\cos\theta + 3\sin\theta}$$

$$= \int \frac{\dfrac{2\,dt}{1+t^2}}{4\left(\dfrac{1-t^2}{1+t^2}\right) + 3\left(\dfrac{2t}{1+t^2}\right)}$$

$$= \int \frac{2\,dt}{4 - 4t^2 + 6t} = \int \frac{dt}{2 + 3t - 2t^2}$$

$$= -\frac{1}{2}\int \frac{dt}{t^2 - \dfrac{3}{2}t - 1}$$

$$= -\frac{1}{2}\int \frac{dt}{\left(t - \dfrac{3}{4}\right)^2 - \dfrac{25}{16}}$$

$$= \frac{1}{2}\int \frac{dt}{\left(\dfrac{5}{4}\right)^2 - \left(t - \dfrac{3}{4}\right)^2}$$

$$= \frac{1}{2}\left[\frac{1}{2\left(\dfrac{5}{4}\right)}\ln\left\{\frac{\dfrac{5}{4} + \left(t - \dfrac{3}{4}\right)}{\dfrac{5}{4} - \left(t - \dfrac{3}{4}\right)}\right\}\right] + c$$

from Problem 11, Chapter 44, page 493

$$= \frac{1}{5}\ln\left\{\frac{\dfrac{1}{2} + t}{2 - t}\right\} + c$$

Hence $\displaystyle\int \frac{d\theta}{4\cos\theta + 3\sin\theta}$

$$= \frac{1}{5}\ln\left\{\frac{\dfrac{1}{2} + \tan\dfrac{\theta}{2}}{2 - \tan\dfrac{\theta}{2}}\right\} + c$$

or $\displaystyle\quad \frac{1}{5}\ln\left\{\frac{1 + 2\tan\dfrac{\theta}{2}}{4 - 2\tan\dfrac{\theta}{2}}\right\} + c$

Now try the following Practice Exercise

Practice Exercise 184   The $t = \tan\dfrac{\theta}{2}$ substitution (Answers on page 883)

In Problems 1 to 4, integrate with respect to the variable.

1. $\displaystyle\int \frac{d\theta}{5 + 4\sin\theta}$

2. $\displaystyle\int \frac{dx}{1 + 2\sin x}$

3. $\displaystyle\int \frac{\mathrm{d}p}{3-4\sin p+2\cos p}$

4. $\displaystyle\int \frac{\mathrm{d}\theta}{3-4\sin\theta}$

5. Show that

$$\int \frac{\mathrm{d}t}{1+3\cos t}=\frac{1}{2\sqrt{2}}\ln\left\{\frac{\sqrt{2}+\tan\frac{t}{2}}{\sqrt{2}-\tan\frac{t}{2}}\right\}+c$$

6. Show that $\displaystyle\int_0^{\pi/3}\frac{3\,\mathrm{d}\theta}{\cos\theta}=3.95$, correct to 3 significant figures.

7. Show that

$$\int_0^{\pi/2}\frac{\mathrm{d}\theta}{2+\cos\theta}=\frac{\pi}{3\sqrt{3}}$$

Revision Test 13   Further integration

This Revision Test covers the material contained in Chapters 42 to 45. *The marks for each question are shown in brackets at the end of each question.*

1. Determine the following integrals:

   (a) $\displaystyle\int 5(6t+5)^7\,dt$ (b) $\displaystyle\int \frac{3\ln x}{x}\,dx$

   (c) $\displaystyle\int \frac{2}{\sqrt{(2\theta-1)}}\,d\theta$ (10)

2. Evaluate the following definite integrals:

   (a) $\displaystyle\int_0^{\frac{\pi}{2}} 2\sin\left(2t+\frac{\pi}{3}\right)\,dt$ (b) $\displaystyle\int_0^1 3x\,e^{4x^2-3}\,dx$ (10)

3. Determine the following integrals:

   (a) $\displaystyle\int \cos^3 x\sin^2 x\,dx$ (b) $\displaystyle\int \frac{2}{\sqrt{(9-4x^2)}}\,dx$

   (c) $\displaystyle\int \frac{2}{\sqrt{(4x^2-9)}}\,dx$ (14)

4. Evaluate the following definite integrals, correct to 4 significant figures:

   (a) $\displaystyle\int_0^{\frac{\pi}{2}} 3\sin^2 t\,dt$ (b) $\displaystyle\int_0^{\frac{\pi}{3}} 3\cos 5\theta\sin 3\theta\,d\theta$

   (c) $\displaystyle\int_0^2 \frac{5}{4+x^2}\,dx$ (15)

5. Determine:

   (a) $\displaystyle\int \frac{x-11}{x^2-x-2}\,dx$

   (b) $\displaystyle\int \frac{3-x}{(x^2+3)(x+3)}\,dx$ (21)

6. Evaluate $\displaystyle\int_1^2 \frac{3}{x^2(x+2)}\,dx$ correct to 4 significant figures.   (12)

7. Determine: $\displaystyle\int \frac{dx}{2\sin x+\cos x}$ (8)

8. Evaluate $\displaystyle\int_{\frac{\pi}{3}}^{\frac{\pi}{2}} \frac{dx}{3-2\sin x}$ correct to 3 decimal places.   (10)

Section I

Chapter 46

Integration by parts

> **Why it is important to understand: Integration by parts**
>
> Integration by parts is a very important technique that is used often in engineering and science. It is frequently used to change the integral of a product of functions into an ideally simpler integral. It is the foundation for the theory of differential equations and is used with Fourier series. We have looked at standard integrals followed by various techniques to change integrals into standard ones; integration by parts is a particularly important technique for integrating a product of two functions.

At the end of this chapter, you should be able to:

- appreciate when integration by parts is required
- integrate functions using integration by parts
- evaluate definite integrals using integration by parts

46.1 Introduction

From the product rule of differentiation:

$$\frac{d}{dx}(uv) = v\frac{du}{dx} + u\frac{dv}{dx},$$

where u and v are both functions of x.

Rearranging gives: $u\dfrac{dv}{dx} = \dfrac{d}{dx}(uv) - v\dfrac{du}{dx}$

Integrating both sides with respect to x gives:

$$\int u\frac{dv}{dx}\,dx = \int \frac{d}{dx}(uv)\,dx - \int v\frac{du}{dx}\,dx$$

i.e.

$$\int u\frac{dv}{dx}\,dx = uv - \int v\frac{du}{dx}\,dx$$

or

$$\int u\,dv = uv - \int v\,du$$

This is known as the **integration by parts formula** and provides a method of integrating such products of simple functions as $\int xe^x\,dx$, $\int t\sin t\,dt$, $\int e^{\theta}\cos\theta\,d\theta$ and $\int x\ln x\,dx$.

Given a product of two terms to integrate the initial choice is: 'which part to make equal to u' and 'which part to make equal to v'. The choice must be such that the 'u part' becomes a constant after successive differentiation and the 'dv part' can be integrated from standard integrals. Invariably, the following rule holds: If a product to be integrated contains an algebraic term (such as x, t^2 or 3θ) then this term is chosen as the u part. The one exception to this rule is when a '$\ln x$' term is involved; in this case $\ln x$ is chosen as the 'u part'.

46.2 Worked problems on integration by parts

Problem 1. Determine $\int x\cos x\,dx$

From the integration by parts formula,

$$\int u \, dv = uv - \int v \, du$$

Let $u = x$, from which $\dfrac{du}{dx} = 1$, i.e. $du = dx$ and let $dv = \cos x \, dx$, from which $v = \int \cos x \, dx = \sin x$.

Expressions for u, du and v are now substituted into the 'by parts' formula as shown below.

$$\int \boxed{u} \boxed{dv} = \boxed{u} \boxed{v} - \int \boxed{v} \boxed{du}$$

$$\int \boxed{x} \boxed{\cos x \, dx} = \boxed{(x)} \boxed{(\sin x)} - \int \boxed{(\sin x)} \boxed{(dx)}$$

i.e. $\displaystyle\int x \cos x \, dx = x \sin x - (-\cos x) + c$

$$= x\sin x + \cos x + c$$

[This result may be checked by differentiating the right-hand side,

i.e. $\dfrac{d}{dx}(x \sin x + \cos x + c)$

$\quad = [(x)(\cos x) + (\sin x)(1)] - \sin x + 0$

$\qquad\qquad\qquad\qquad\qquad$ using the product rule

$\quad = x \cos x$, which is the function being integrated.]

Problem 2. Find $\int 3te^{2t} \, dt$

Let $u = 3t$, from which, $\dfrac{du}{dt} = 3$, i.e. $du = 3 \, dt$ and

let $dv = e^{2t} \, dt$, from which, $v = \int e^{2t} \, dt = \dfrac{1}{2}e^{2t}$

Substituting into $\int u \, dv = uv - \int v \, du$ gives:

$$\int 3te^{2t} \, dt = (3t)\left(\frac{1}{2}e^{2t}\right) - \int \left(\frac{1}{2}e^{2t}\right)(3 \, dt)$$

$$= \frac{3}{2}te^{2t} - \frac{3}{2}\int e^{2t} \, dt$$

$$= \frac{3}{2}te^{2t} - \frac{3}{2}\left(\frac{e^{2t}}{2}\right) + c$$

Hence

$$\int 3te^{2t} \, dt = \tfrac{3}{2}e^{2t}\left(t - \tfrac{1}{2}\right) + c,$$

which may be checked by differentiating.

Problem 3. Evaluate $\displaystyle\int_0^{\frac{\pi}{2}} 2\theta \sin\theta \, d\theta$

Let $u = 2\theta$, from which, $\dfrac{du}{d\theta} = 2$, i.e. $du = 2 \, d\theta$ and let $dv = \sin\theta \, d\theta$, from which,

$$v = \int \sin\theta \, d\theta = -\cos\theta$$

Substituting into $\int u \, dv = uv - \int v \, du$ gives:

$$\int 2\theta \sin\theta \, d\theta = (2\theta)(-\cos\theta) - \int (-\cos\theta)(2 \, d\theta)$$

$$= -2\theta \cos\theta + 2\int \cos\theta \, d\theta$$

$$= -2\theta \cos\theta + 2\sin\theta + c$$

Hence $\displaystyle\int_0^{\frac{\pi}{2}} 2\theta \sin\theta \, d\theta$

$$= [-2\theta \cos\theta + 2\sin\theta]_0^{\frac{\pi}{2}}$$

$$= \left[-2\left(\frac{\pi}{2}\right)\cos\frac{\pi}{2} + 2\sin\frac{\pi}{2}\right] - [0 + 2\sin 0]$$

$$= (-0 + 2) - (0 + 0) = 2$$

$$\text{since } \cos\frac{\pi}{2} = 0 \text{ and } \sin\frac{\pi}{2} = 1$$

Problem 4. Evaluate $\displaystyle\int_0^1 5xe^{4x} \, dx$, correct to 3 significant figures.

Let $u = 5x$, from which $\dfrac{du}{dx} = 5$, i.e. $du = 5 \, dx$ and

let $dv = e^{4x} \, dx$, from which, $v = \int e^{4x} \, dx = \frac{1}{4}e^{4x}$

Substituting into $\int u \, dv = uv - \int v \, du$ gives:

$$\int 5xe^{4x} \, dx = (5x)\left(\frac{e^{4x}}{4}\right) - \int \left(\frac{e^{4x}}{4}\right)(5 \, dx)$$

$$= \frac{5}{4}xe^{4x} - \frac{5}{4}\int e^{4x} \, dx$$

$$= \frac{5}{4}xe^{4x} - \frac{5}{4}\left(\frac{e^{4x}}{4}\right) + c$$

$$= \frac{5}{4}e^{4x}\left(x - \frac{1}{4}\right) + c$$

Hence $\int_0^1 5xe^{4x}\,dx$

$$= \left[\frac{5}{4}e^{4x}\left(x - \frac{1}{4}\right)\right]_0^1$$

$$= \left[\frac{5}{4}e^4\left(1 - \frac{1}{4}\right)\right] - \left[\frac{5}{4}e^0\left(0 - \frac{1}{4}\right)\right]$$

$$= \left(\frac{15}{16}e^4\right) - \left(-\frac{5}{16}\right)$$

$$= 51.186 + 0.313 = 51.499 = \mathbf{51.5},$$

correct to 3 significant figures

Problem 5. Determine $\int x^2 \sin x\,dx$

Let $u = x^2$, from which, $\dfrac{du}{dx} = 2x$, i.e. $du = 2x\,dx$, and let $dv = \sin x\,dx$, from which,

$$v = \int \sin x\,dx = -\cos x$$

Substituting into $\int u\,dv = uv - \int v\,du$ gives:

$$\int x^2 \sin x\,dx = (x^2)(-\cos x) - \int (-\cos x)(2x\,dx)$$

$$= -x^2 \cos x + 2\left[\int x \cos x\,dx\right]$$

The integral, $\int x \cos x\,dx$, is not a 'standard integral' and it can only be determined by using the integration by parts formula again.
From Problem 1, $\int x \cos x\,dx = x \sin x + \cos x$

Hence $\int x^2 \sin x\,dx$

$$= -x^2 \cos x + 2\{x \sin x + \cos x\} + c$$

$$= -x^2 \cos x + 2x \sin x + 2\cos x + c$$

$$= \mathbf{(2 - x^2)\cos x + 2x\sin x + c}$$

In general, if the algebraic term of a product is of power n, then the integration by parts formula is applied n times.

Now try the following Practice Exercise

**Practice Exercise 185   Integration by parts
(Answers on page 883)**

Determine the integrals in Problems 1 to 5 using integration by parts.

1. $\displaystyle\int xe^{2x}\,dx$

2. $\displaystyle\int \frac{4x}{e^{3x}}\,dx$

3. $\displaystyle\int x \sin x\,dx$

4. $\displaystyle\int 5\theta \cos 2\theta\,d\theta$

5. $\displaystyle\int 3t^2 e^{2t}\,dt$

Evaluate the integrals in Problems 6 to 9, correct to 4 significant figures.

6. $\displaystyle\int_0^2 2xe^x\,dx$

7. $\displaystyle\int_0^{\frac{\pi}{4}} x \sin 2x\,dx$

8. $\displaystyle\int_0^{\frac{\pi}{2}} t^2 \cos t\,dt$

9. $\displaystyle\int_1^2 3x^2 e^{\frac{x}{2}}\,dx$

46.3 Further worked problems on integration by parts

Problem 6. Find $\int x \ln x\,dx$

The logarithmic function is chosen as the 'u part'.

Thus when $u = \ln x$, then $\dfrac{du}{dx} = \dfrac{1}{x}$, i.e. $du = \dfrac{dx}{x}$

Letting $dv = x\,dx$ gives $v = \int x\,dx = \dfrac{x^2}{2}$

Substituting into $\int u\,dv = uv - \int v\,du$ gives:

$$\int x \ln x\,dx = (\ln x)\left(\frac{x^2}{2}\right) - \int\left(\frac{x^2}{2}\right)\frac{dx}{x}$$

$$= \frac{x^2}{2}\ln x - \frac{1}{2}\int x\,dx$$

$$= \frac{x^2}{2}\ln x - \frac{1}{2}\left(\frac{x^2}{2}\right) + c$$

Hence $\displaystyle\int x \ln x\,dx = \frac{x^2}{2}\left(\ln x - \frac{1}{2}\right) + c$ or

$$\frac{x^2}{4}(2\ln x - 1) + c$$

Problem 7.   Determine $\int \ln x\,dx$

$\int \ln x\,dx$ is the same as $\int(1)\ln x\,dx$

Let $u = \ln x$, from which, $\dfrac{du}{dx} = \dfrac{1}{x}$, i.e. $du = \dfrac{dx}{x}$

and let $dv = 1\,dx$, from which, $v = \int 1\,dx = x$

Substituting into $\int u\,dv = uv - \int v\,du$ gives:

$$\int \ln x\,dx = (\ln x)(x) - \int x\,\frac{dx}{x}$$

$$= x\ln x - \int dx = x\ln x - x + c$$

Hence $\displaystyle\int \ln x\,dx = x(\ln x - 1) + c$

Problem 8.   Evaluate $\displaystyle\int_1^9 \sqrt{x}\ln x\,dx$, correct to 3 significant figures.

Let $u = \ln x$, from which $du = \dfrac{dx}{x}$

and let $dv = \sqrt{x}\,dx = x^{\frac{1}{2}}\,dx$, from which,

$$v = \int x^{\frac{1}{2}}\,dx = \frac{2}{3}x^{\frac{3}{2}}$$

Substituting into $\int u\,dv = uv - \int v\,du$ gives:

$$\int \sqrt{x}\ln x\,dx = (\ln x)\left(\frac{2}{3}x^{\frac{3}{2}}\right) - \int\left(\frac{2}{3}x^{\frac{3}{2}}\right)\left(\frac{dx}{x}\right)$$

$$= \frac{2}{3}\sqrt{x^3}\ln x - \frac{2}{3}\int x^{\frac{1}{2}}\,dx$$

$$= \frac{2}{3}\sqrt{x^3}\ln x - \frac{2}{3}\left(\frac{2}{3}x^{\frac{3}{2}}\right) + c$$

$$= \frac{2}{3}\sqrt{x^3}\left[\ln x - \frac{2}{3}\right] + c$$

Hence $\displaystyle\int_1^9 \sqrt{x}\ln x\,dx$

$$= \left[\frac{2}{3}\sqrt{x^3}\left(\ln x - \frac{2}{3}\right)\right]_1^9$$

$$= \left[\frac{2}{3}\sqrt{9^3}\left(\ln 9 - \frac{2}{3}\right)\right] - \left[\frac{2}{3}\sqrt{1^3}\left(\ln 1 - \frac{2}{3}\right)\right]$$

$$= \left[18\left(\ln 9 - \frac{2}{3}\right)\right] - \left[\frac{2}{3}\left(0 - \frac{2}{3}\right)\right]$$

$$= 27.550 + 0.444 = 27.994 = \mathbf{28.0},$$

correct to 3 significant figures.

Problem 9.   Find $\int e^{ax}\cos bx\,dx$

When integrating a product of an exponential and a sine or cosine function it is immaterial which part is made equal to 'u'

Let $u = e^{ax}$, from which $\dfrac{du}{dx} = ae^{ax}$,

i.e. $du = ae^{ax}\,dx$ and let $dv = \cos bx\,dx$, from which,

$$v = \int \cos bx\,dx = \frac{1}{b}\sin bx$$

Substituting into $\int u\,dv = uv - \int v\,du$ gives:

$$\int e^{ax}\cos bx\,dx$$

$$= (e^{ax})\left(\frac{1}{b}\sin bx\right) - \int\left(\frac{1}{b}\sin bx\right)(ae^{ax}\,dx)$$

$$= \frac{1}{b}e^{ax}\sin bx - \frac{a}{b}\left[\int e^{ax}\sin bx\,dx\right] \qquad (1)$$

$\int e^{ax}\sin bx\,dx$ is now determined separately using integration by parts again:

Let $u = e^{ax}$ then $du = ae^{ax}\,dx$, and let $dv = \sin bx\,dx$, from which

$$v = \int \sin bx\,dx = -\frac{1}{b}\cos bx$$

Substituting into the integration by parts formula gives:

$$\int e^{ax} \sin bx \, dx = (e^{ax})\left(-\frac{1}{b}\cos bx\right)$$

$$-\int\left(-\frac{1}{b}\cos bx\right)(ae^{ax}\,dx)$$

$$=-\frac{1}{b}e^{ax}\cos bx + \frac{a}{b}\int e^{ax}\cos bx\,dx$$

Substituting this result into equation (1) gives:

$$\int e^{ax}\cos bx\,dx = \frac{1}{b}e^{ax}\sin bx - \frac{a}{b}\left[-\frac{1}{b}e^{ax}\cos bx\right.$$

$$\left. + \frac{a}{b}\int e^{ax}\cos bx\,dx\right]$$

$$=\frac{1}{b}e^{ax}\sin bx + \frac{a}{b^2}e^{ax}\cos bx$$

$$-\frac{a^2}{b^2}\int e^{ax}\cos bx\,dx$$

The integral on the far right of this equation is the same as the integral on the left hand side and thus they may be combined.

$$\int e^{ax}\cos bx\,dx + \frac{a^2}{b^2}\int e^{ax}\cos bx\,dx$$

$$=\frac{1}{b}e^{ax}\sin bx + \frac{a}{b^2}e^{ax}\cos bx$$

i.e. $\left(1+\dfrac{a^2}{b^2}\right)\displaystyle\int e^{ax}\cos bx\,dx$

$$=\frac{1}{b}e^{ax}\sin bx + \frac{a}{b^2}e^{ax}\cos bx$$

i.e. $\left(\dfrac{b^2+a^2}{b^2}\right)\displaystyle\int e^{ax}\cos bx\,dx$

$$=\frac{e^{ax}}{b^2}(b\sin bx + a\cos bx)$$

Hence $\displaystyle\int e^{ax}\cos bx\,dx$

$$=\left(\frac{b^2}{b^2+a^2}\right)\left(\frac{e^{ax}}{b^2}\right)(b\sin bx + a\cos bx)$$

$$=\frac{e^{ax}}{a^2+b^2}(b\sin bx + a\cos bx)+c$$

Using a similar method to above, that is, integrating by parts twice, the following result may be proved:

$$\int e^{ax}\sin bx\,dx$$

$$=\frac{e^{ax}}{a^2+b^2}(a\sin bx - b\cos bx)+c \qquad (2)$$

Problem 10. Evaluate $\displaystyle\int_0^{\frac{\pi}{4}} e^t \sin 2t \, dt$, correct to 4 decimal places.

Comparing $\int e^t \sin 2t\,dt$ with $\int e^{ax}\sin bx\,dx$ shows that $x=t$, $a=1$ and $b=2$

Hence, substituting into equation (2) gives:

$$\int_0^{\frac{\pi}{4}} e^t \sin 2t\,dt$$

$$=\left[\frac{e^t}{1^2+2^2}(1\sin 2t - 2\cos 2t)\right]_0^{\frac{\pi}{4}}$$

$$=\left[\frac{e^{\frac{\pi}{4}}}{5}\left(\sin 2\left(\frac{\pi}{4}\right) - 2\cos 2\left(\frac{\pi}{4}\right)\right)\right]$$

$$-\left[\frac{e^0}{5}(\sin 0 - 2\cos 0)\right]$$

$$=\left[\frac{e^{\frac{\pi}{4}}}{5}(1-0)\right]-\left[\frac{1}{5}(0-2)\right]=\frac{e^{\frac{\pi}{4}}}{5}+\frac{2}{5}$$

$$=\mathbf{0.8387},\ \text{correct to 4 decimal places.}$$

Now try the following Practice Exercise

Practice Exercise 186   Integration by parts (Answers on page 884)

Determine the integrals in Problems 1 to 5 using integration by parts.

1. $\displaystyle\int 2x^2 \ln x\,dx$

2. $\displaystyle\int 2\ln 3x\,dx$

3. $\int x^2 \sin 3x \, dx$

4. $\int 2e^{5x} \cos 2x \, dx$

5. $\int 2\theta \sec^2 \theta \, d\theta$

Evaluate the integrals in Problems 6 to 9, correct to 4 significant figures.

6. $\int_1^2 x \ln x \, dx$

7. $\int_0^1 2e^{3x} \sin 2x \, dx$

8. $\int_0^{\frac{\pi}{2}} e^t \cos 3t \, dt$

9. $\int_1^4 \sqrt{x^3} \ln x \, dx$

10. In determining a Fourier series to represent $f(x) = x$ in the range $-\pi$ to π, Fourier coefficients are given by:

$$a_n = \frac{1}{\pi} \int_{-\pi}^{\pi} x \cos nx \, dx$$

and $\quad b_n = \frac{1}{\pi} \int_{-\pi}^{\pi} x \sin nx \, dx$

where n is a positive integer. Show by using integration by parts that $a_n = 0$ and $b_n = -\frac{2}{n} \cos n\pi$

11. The equations $C = \int_0^1 e^{-0.4\theta} \cos 1.2\theta \, d\theta$

and $\qquad S = \int_0^1 e^{-0.4\theta} \sin 1.2\theta \, d\theta$

are involved in the study of damped oscillations. Determine the values of C and S.

For fully worked solutions to each of the problems in Practice Exercises 185 and 186 in this chapter, go to the website:
www.routledge.com/cw/bird

Section I

Chapter 47

Reduction formulae

Why it is important to understand: Reduction formulae

When an integral contains a power of n, then it may sometimes be rewritten, using integration by parts, in terms of a similar integral containing $(n-1)$ or $(n-2)$, and so on. The result is a recurrence relation, i.e. an equation that recursively defines a sequence, once one or more initial terms are given; each further term of the sequence is defined as a function of the preceding terms. It may sound difficult but it's actually quite straightforward and is just one more technique for integrating certain functions.

At the end of this chapter, you should be able to:

- appreciate when reduction formulae might be used
- integrate functions of the form $\int x^n e^x \, dx$ using reduction formulae
- integrate functions of the form $\int x^n \cos x \, dx$ and $\int x^n \sin x \, dx$ using reduction formulae
- integrate functions of the form $\int \sin^n x \, dx$ and $\int \cos^n x \, dx$ using reduction formulae
- integrate further functions using reduction formulae

47.1 Introduction

When using integration by parts in Chapter 46, an integral such as $\int x^2 e^x \, dx$ requires integration by parts twice. Similarly, $\int x^3 e^x \, dx$ requires integration by parts three times. Thus, integrals such as $\int x^5 e^x \, dx$, $\int x^6 \cos x \, dx$ and $\int x^8 \sin 2x \, dx$ for example, would take a long time to determine using integration by parts. **Reduction formulae** provide a quicker method for determining such integrals and the method is demonstrated in the following sections.

47.2 Using reduction formulae for integrals of the form $\int x^n e^x \, dx$

To determine $\int x^n e^x \, dx$ using integration by parts,

let $\qquad u = x^n$ from which,

and $\dfrac{du}{dx} = nx^{n-1}$ and $du = nx^{n-1} \, dx$

and $\qquad dv = e^x \, dx$ from which,

$$v = \int e^x \, dx = e^x$$

Thus, $\displaystyle \int x^n e^x \, dx = x^n e^x - \int e^x n x^{n-1} \, dx$

using the integration by parts formula,

$$= x^n e^x - n \int x^{n-1} e^x \, dx$$

The integral on the far right is seen to be of the same form as the integral on the left-hand side, except that n has been replaced by $n-1$

Thus, if we let,

$$\int x^n e^x \, dx = I_n$$

then $\displaystyle\int x^{n-1}e^x\,dx = I_{n-1}$

Hence $\displaystyle\int x^n e^x\,dx = x^n e^x - n\int x^{n-1}e^x\,dx$

can be written as:

$$I_n = x^n e^x - nI_{n-1} \qquad (1)$$

Equation (1) is an example of a reduction formula since it expresses an integral in n in terms of the same integral in $n-1$

Problem 1. Determine $\int x^2 e^x\,dx$ using a reduction formula.

Using equation (1) with $n=2$ gives:

$$\int x^2 e^x\,dx = I_2 = x^2 e^x - 2I_1$$

and $\qquad I_1 = x^1 e^x - 1I_0$

$$I_0 = \int x^0 e^x\,dx = \int e^x\,dx = e^x + c_1$$

Hence $\qquad I_2 = x^2 e^x - 2[xe^x - 1I_0]$

$$= x^2 e^x - 2[xe^x - 1(e^x + c_1)]$$

i.e. $\displaystyle\int x^2 e^x\,dx = x^2 e^x - 2xe^x + 2e^x + 2c_1$

$$= e^x(x^2 - 2x + 2) + c$$

(where $c = 2c_1$)

As with integration by parts, in the following examples the constant of integration will be added at the last step with indefinite integrals.

Problem 2. Use a reduction formula to determine $\int x^3 e^x\,dx$

From equation (1), $I_n = x^n e^x - nI_{n-1}$

Hence $\displaystyle\int x^3 e^x\,dx = I_3 = x^3 e^x - 3I_2$

$$I_2 = x^2 e^x - 2I_1$$

$$I_1 = x^1 e^x - 1I_0$$

and $\qquad I_0 = \displaystyle\int x^0 e^x\,dx = \int e^x\,dx = e^x$

Thus $\displaystyle\int x^3 e^x\,dx = x^3 e^x - 3[x^2 e^x - 2I_1]$

$$= x^3 e^x - 3[x^2 e^x - 2(xe^x - I_0)]$$

$$= x^3 e^x - 3[x^2 e^x - 2(xe^x - e^x)]$$

$$= x^3 e^x - 3x^2 e^x + 6(xe^x - e^x)$$

$$= x^3 e^x - 3x^2 e^x + 6xe^x - 6e^x$$

i.e. $\displaystyle\int x^3 e^x\,dx = e^x(x^3 - 3x^2 + 6x - 6) + c$

Now try the following Practice Exercise

Practice Exercise 187 Using reduction formulae for integrals of the form $\int x^n e^x\,dx$ (Answers on page 884)

1. Use a reduction formula to determine $\int x^4 e^x\,dx$.

2. Determine $\int t^3 e^{2t}\,dt$ using a reduction formula.

3. Use the result of Problem 2 to evaluate $\int_0^1 5t^3 e^{2t}\,dt$, correct to 3 decimal places.

47.3 Using reduction formulae for integrals of the form $\int x^n \cos x\,dx$ and $\int x^n \sin x\,dx$

(a) $\int x^n \cos x\,dx$

Let $I_n = \int x^n \cos x\,dx$ then, using integration by parts:

if $\qquad u = x^n$ then $\dfrac{du}{dx} = nx^{n-1}$

and if $\quad dv = \cos x\,dx$ then

$$v = \int \cos x\,dx = \sin x$$

Hence $\quad I_n = x^n \sin x - \displaystyle\int (\sin x)nx^{n-1}\,dx$

$$= x^n \sin x - n\int x^{n-1}\sin x\,dx$$

Using integration by parts again, this time with $u = x^{n-1}$:

$$\frac{du}{dx} = (n-1)x^{n-2}, \text{ and } dv = \sin x\,dx,$$

from which,

$$v = \int \sin x\,dx = -\cos x$$

Hence $\quad I_n = x^n \sin x - n\left[x^{n-1}(-\cos x)\right.$

$$\left. - \int(-\cos x)(n-1)x^{n-2}\,dx\right]$$

$$= x^n \sin x + nx^{n-1}\cos x$$

$$- n(n-1)\int x^{n-2}\cos x\,dx$$

i.e. $I_n = x^n \sin x + nx^{n-1} \cos x$
$$- n(n-1)I_{n-2} \qquad (2)$$

Problem 3. Use a reduction formula to determine $\int x^2 \cos x \, dx$

Using the reduction formula of equation (2):

$$\int x^2 \cos x \, dx = I_2$$
$$= x^2 \sin x + 2x^1 \cos x - 2(1)I_0$$

and
$$I_0 = \int x^0 \cos x \, dx$$
$$= \int \cos x \, dx = \sin x$$

Hence

$$\int x^2 \cos x \, dx = x^2 \sin x + 2x \cos x - 2\sin x + c$$

Problem 4. Evaluate $\int_1^2 4t^3 \cos t \, dt$, correct to 4 significant figures.

Let us firstly find a reduction formula for $\int t^3 \cos t \, dt$.

From equation (2),

$$\int t^3 \cos t \, dt = I_3 = t^3 \sin t + 3t^2 \cos t - 3(2)I_1$$

and

$$I_1 = t^1 \sin t + 1t^0 \cos t - 1(0)I_{n-2}$$
$$= t \sin t + \cos t$$

Hence

$$\int t^3 \cos t \, dt = t^3 \sin t + 3t^2 \cos t$$
$$- 3(2)[t \sin t + \cos t]$$
$$= t^3 \sin t + 3t^2 \cos t - 6t \sin t - 6\cos t$$

Thus

$$\int_1^2 4t^3 \cos t \, dt$$
$$= [4(t^3 \sin t + 3t^2 \cos t - 6t \sin t - 6\cos t)]_1^2$$
$$= [4(8\sin 2 + 12\cos 2 - 12\sin 2 - 6\cos 2)]$$
$$- [4(\sin 1 + 3\cos 1 - 6\sin 1 - 6\cos 1)]$$
$$= (-24.53628) - (-23.31305)$$
$$= -1.223$$

Problem 5. Determine a reduction formula for $\int_0^\pi x^n \cos x \, dx$ and hence evaluate

$\int_0^\pi x^4 \cos x \, dx$, correct to 2 decimal places.

From equation (2),

$$I_n = x^n \sin x + nx^{n-1} \cos x - n(n-1)I_{n-2}$$

hence $\int_0^\pi x^n \cos x \, dx = [x^n \sin x + nx^{n-1} \cos x]_0^\pi$
$$- n(n-1)I_{n-2}$$
$$= [(\pi^n \sin \pi + n\pi^{n-1} \cos \pi)$$
$$- (0 + 0)] - n(n-1)I_{n-2}$$
$$= -n\pi^{n-1} - n(n-1)I_{n-2}$$

Hence

$$\int_0^\pi x^4 \cos x \, dx = I_4$$
$$= -4\pi^3 - 4(3)I_2 \text{ since } n = 4$$

When $n = 2$,

$$\int_0^\pi x^2 \cos x \, dx = I_2 = -2\pi^1 - 2(1)I_0$$

and
$$I_0 = \int_0^\pi x^0 \cos x \, dx$$
$$= \int_0^\pi \cos x \, dx$$
$$= [\sin x]_0^\pi = 0$$

Hence

$$\int_0^\pi x^4 \cos x \, dx = -4\pi^3 - 4(3)[-2\pi - 2(1)(0)]$$
$$= -4\pi^3 + 24\pi \text{ or } -48.63,$$

correct to 2 decimal places.

(b) $\int x^n \sin x \, dx$

Let $I_n = \int x^n \sin x \, dx$
Using integration by parts, if $u = x^n$ then
$\dfrac{du}{dx} = nx^{n-1}$ and if $dv = \sin x \, dx$ then
$v = \int \sin x \, dx = -\cos x$. Hence

$$\int x^n \sin x \, dx$$
$$= I_n = x^n(-\cos x) - \int (-\cos x)nx^{n-1} \, dx$$
$$= -x^n \cos x + n \int x^{n-1} \cos x \, dx$$

Using integration by parts again, with $u = x^{n-1}$, from which, $\dfrac{du}{dx} = (n-1)x^{n-2}$ and $dv = \cos x$, from which, $v = \int \cos x\, dx = \sin x$. Hence

$$I_n = -x^n \cos x + n\left[x^{n-1}(\sin x)\right.$$
$$\left. - \int (\sin x)(n-1)x^{n-2}\, dx\right]$$
$$= -x^n \cos x + nx^{n-1}(\sin x)$$
$$- n(n-1)\int x^{n-2}\sin x\, dx$$

i.e. $\quad \boldsymbol{I_n = -x^n\cos x + nx^{n-1}\sin x - n(n-1)I_{n-2}}$ (3)

Problem 6. Use a reduction formula to determine $\int x^3 \sin x\, dx$

Using equation (3),

$$\int x^3 \sin x\, dx = I_3$$
$$= -x^3 \cos x + 3x^2 \sin x - 3(2)I_1$$
and $\qquad I_1 = -x^1 \cos x + 1x^0 \sin x$
$$= -x\cos x + \sin x$$

Hence

$$\int x^3 \sin x\, dx = -x^3 \cos x + 3x^2 \sin x$$
$$- 6[-x\cos x + \sin x]$$
$$= -x^3\cos x + 3x^2\sin x$$
$$+ 6x\cos x - 6\sin x + c$$

Problem 7. Evaluate $\displaystyle\int_0^{\frac{\pi}{2}} 3\theta^4 \sin\theta\, d\theta$, correct to 2 decimal places.

From equation (3),

$$I_n = [-x^n \cos x + nx^{n-1}(\sin x)]_0^{\frac{\pi}{2}} - n(n-1)I_{n-2}$$
$$= \left[\left(-\left(\frac{\pi}{2}\right)^n \cos\frac{\pi}{2} + n\left(\frac{\pi}{2}\right)^{n-1}\sin\frac{\pi}{2}\right) - (0)\right]$$
$$- n(n-1)I_{n-2}$$
$$= n\left(\frac{\pi}{2}\right)^{n-1} - n(n-1)I_{n-2}$$

Hence

$$\int_0^{\frac{\pi}{2}} 3\theta^4 \sin\theta\, d\theta = 3\int_0^{\frac{\pi}{2}} \theta^4 \sin\theta\, d\theta$$
$$= 3I_4$$
$$= 3\left[4\left(\frac{\pi}{2}\right)^3 - 4(3)I_2\right]$$
$$I_2 = 2\left(\frac{\pi}{2}\right)^1 - 2(1)I_0$$
and $I_0 = \displaystyle\int_0^{\frac{\pi}{2}} \theta^0 \sin\theta\, d\theta = [-\cos x]_0^{\frac{\pi}{2}}$
$$= [-0 - (-1)] = 1$$

Hence

$$3\int_0^{\frac{\pi}{2}} \theta^4 \sin\theta\, d\theta$$
$$= 3I_4$$
$$= 3\left[4\left(\frac{\pi}{2}\right)^3 - 4(3)\left\{2\left(\frac{\pi}{2}\right)^1 - 2(1)I_0\right\}\right]$$
$$= 3\left[4\left(\frac{\pi}{2}\right)^3 - 4(3)\left\{2\left(\frac{\pi}{2}\right)^1 - 2(1)(1)\right\}\right]$$
$$= 3\left[4\left(\frac{\pi}{2}\right)^3 - 24\left(\frac{\pi}{2}\right)^1 + 24\right]$$
$$= 3(15.503 - 37.699 + 24)$$
$$= 3(1.8039) = \mathbf{5.41}$$

Now try the following Practice Exercise

Practice Exercise 188 Using reduction formulae for integrals of the form $\int x^n \cos x\, dx$ and $\int x^n \sin x\, dx$ (Answers on page 884)

1. Use a reduction formula to determine $\int x^5 \cos x\, dx$

2. Evaluate $\displaystyle\int_0^{\pi} x^5 \cos x\, dx$, correct to 2 decimal places.

3. Use a reduction formula to determine $\int x^5 \sin x\, dx$

4. Evaluate $\displaystyle\int_0^{\pi} x^5 \sin x\, dx$, correct to 2 decimal places.

47.4 Using reduction formulae for integrals of the form $\int \sin^n x\,dx$ and $\int \cos^n x\,dx$

(a) $\int \sin^n x\,dx$

Let $I_n = \int \sin^n x\,dx \equiv \int \sin^{n-1} x \sin x\,dx$ from laws of indices.

Using integration by parts, let $u = \sin^{n-1} x$, from which,

$$\frac{du}{dx} = (n-1)\sin^{n-2} x \cos x \quad \text{and}$$

$$du = (n-1)\sin^{n-2} x \cos x\,dx$$

and let $dv = \sin x\,dx$, from which,

$v = \int \sin x\,dx = -\cos x$. Hence,

$$I_n = \int \sin^{n-1} x \sin x\,dx$$

$$= (\sin^{n-1} x)(-\cos x)$$

$$\quad - \int (-\cos x)(n-1)\sin^{n-2} x \cos x\,dx$$

$$= -\sin^{n-1} x \cos x$$

$$\quad + (n-1)\int \cos^2 x \sin^{n-2} x\,dx$$

$$= -\sin^{n-1} x \cos x$$

$$\quad + (n-1)\int (1-\sin^2 x)\sin^{n-2} x\,dx$$

$$= -\sin^{n-1} x \cos x$$

$$\quad + (n-1)\left\{ \int \sin^{n-2} x\,dx - \int \sin^n x\,dx \right\}$$

i.e. $\quad I_n = -\sin^{n-1} x \cos x$

$$\quad + (n-1)I_{n-2} - (n-1)I_n$$

i.e. $\quad I_n + (n-1)I_n$

$$\quad = -\sin^{n-1} x \cos x + (n-1)I_{n-2}$$

and $\quad nI_n = -\sin^{n-1} x \cos x + (n-1)I_{n-2}$

from which,

$$\int \sin^n x\,dx =$$

$$I_n = -\frac{1}{n}\sin^{n-1} x\cos x + \frac{n-1}{n}I_{n-2} \qquad (4)$$

Problem 8. Use a reduction formula to determine $\int \sin^4 x\,dx$

Using equation (4),

$$\int \sin^4 x\,dx = I_4 = -\frac{1}{4}\sin^3 x \cos x + \frac{3}{4}I_2$$

$$I_2 = -\frac{1}{2}\sin^1 x \cos x + \frac{1}{2}I_0$$

and $\quad I_0 = \int \sin^0 x\,dx = \int 1\,dx = x$

Hence

$$\int \sin^4 x\,dx = I_4 = -\frac{1}{4}\sin^3 x \cos x$$

$$\quad + \frac{3}{4}\left[-\frac{1}{2}\sin x \cos x + \frac{1}{2}(x) \right]$$

$$\quad = -\frac{1}{4}\sin^3 x \cos x - \frac{3}{8}\sin x \cos x$$

$$\quad + \frac{3}{8}x + c$$

Problem 9. Evaluate $\int_0^1 4\sin^5 t\,dt$, correct to 3 significant figures.

Using equation (4),

$$\int \sin^5 t\,dt = I_5 = -\frac{1}{5}\sin^4 t \cos t + \frac{4}{5}I_3$$

$$I_3 = -\frac{1}{3}\sin^2 t \cos t + \frac{2}{3}I_1$$

and $\quad I_1 = -\frac{1}{1}\sin^0 t \cos t + 0 = -\cos t$

Hence

$$\int \sin^5 t\,dt = -\frac{1}{5}\sin^4 t \cos t$$

$$\quad + \frac{4}{5}\left[-\frac{1}{3}\sin^2 t \cos t + \frac{2}{3}(-\cos t) \right]$$

$$\quad = -\frac{1}{5}\sin^4 t \cos t - \frac{4}{15}\sin^2 t \cos t$$

$$\quad - \frac{8}{15}\cos t + c$$

and $\displaystyle\int_0^t 4\sin^5 t\,\mathrm{d}t$

$= 4\left[-\dfrac{1}{5}\sin^4 t\cos t\right.$

$\left.\qquad -\dfrac{4}{15}\sin^2 t\cos t - \dfrac{8}{15}\cos t\right]_0^1$

$= 4\left[\left(-\dfrac{1}{5}\sin^4 1\cos 1 - \dfrac{4}{15}\sin^2 1\cos 1\right.\right.$

$\left.\left.\qquad -\dfrac{8}{15}\cos 1\right) - \left(-0 - 0 - \dfrac{8}{15}\right)\right]$

$= 4[(-0.054178 - 0.1020196$

$\qquad\qquad -0.2881612) - (-0.533333)]$

$= 4(0.0889745) = \mathbf{0.356}$

Problem 10. Determine a reduction formula for $\displaystyle\int_0^{\frac{\pi}{2}} \sin^n x\,\mathrm{d}x$ and hence evaluate $\displaystyle\int_0^{\frac{\pi}{2}} \sin^6 x\,\mathrm{d}x$

From equation (4),

$\displaystyle\int \sin^n x\,\mathrm{d}x$

$\qquad = I_n = -\dfrac{1}{n}\sin^{n-1} x\cos x + \dfrac{n-1}{n}I_{n-2}$

hence

$\displaystyle\int_0^{\frac{\pi}{2}} \sin^n x\,\mathrm{d}x = \left[-\dfrac{1}{n}\sin^{n-1} x\cos x\right]_0^{\frac{\pi}{2}} + \dfrac{n-1}{n}I_{n-2}$

$\qquad\qquad = [0-0] + \dfrac{n-1}{n}I_{n-2}$

i.e. $\qquad \boldsymbol{I_n = \dfrac{n-1}{n}I_{n-2}}$

Hence

$\displaystyle\int_0^{\frac{\pi}{2}} \sin^6 x\,\mathrm{d}x = I_6 = \dfrac{5}{6}I_4$

$\qquad I_4 = \dfrac{3}{4}I_2, \quad I_2 = \dfrac{1}{2}I_0$

and $\qquad I_0 = \displaystyle\int_0^{\frac{\pi}{2}} \sin^0 x\,\mathrm{d}x = \int_0^{\frac{\pi}{2}} 1\,\mathrm{d}x = \dfrac{\pi}{2}$

Thus

$\displaystyle\int_0^{\frac{\pi}{2}} \sin^6 x\,\mathrm{d}x = I_6 = \dfrac{5}{6}I_4 = \dfrac{5}{6}\left[\dfrac{3}{4}I_2\right]$

$\qquad\qquad = \dfrac{5}{6}\left[\dfrac{3}{4}\left\{\dfrac{1}{2}I_0\right\}\right]$

$\qquad\qquad = \dfrac{5}{6}\left[\dfrac{3}{4}\left\{\dfrac{1}{2}\left[\dfrac{\pi}{2}\right]\right\}\right] = \dfrac{\mathbf{15}}{\mathbf{96}}\boldsymbol{\pi}$

(b) $\int \cos^n x\,\mathrm{d}x$

Let $I_n = \int \cos^n x\,\mathrm{d}x \equiv \int \cos^{n-1} x\cos x\,\mathrm{d}x$ from laws of indices.

Using integration by parts, let $u = \cos^{n-1} x$ from which,

$\qquad \dfrac{\mathrm{d}u}{\mathrm{d}x} = (n-1)\cos^{n-2} x(-\sin x)$

and $\qquad \mathrm{d}u = (n-1)\cos^{n-2} x(-\sin x)\,\mathrm{d}x$

and let $\quad \mathrm{d}v = \cos x\,\mathrm{d}x$

from which, $v = \displaystyle\int \cos x\,\mathrm{d}x = \sin x$

Then

$I_n = (\cos^{n-1} x)(\sin x)$

$\qquad - \displaystyle\int (\sin x)(n-1)\cos^{n-2} x(-\sin x)\,\mathrm{d}x$

$\quad = (\cos^{n-1} x)(\sin x) + (n-1)\displaystyle\int \sin^2 x\cos^{n-2} x\,\mathrm{d}x$

$\quad = (\cos^{n-1} x)(\sin x) + (n-1)\displaystyle\int (1-\cos^2 x)\cos^{n-2} x\,\mathrm{d}x$

$\quad = (\cos^{n-1} x)(\sin x)$

$\qquad + (n-1)\left\{\displaystyle\int \cos^{n-2} x\,\mathrm{d}x - \int \cos^n x\,\mathrm{d}x\right\}$

i.e. $I_n = (\cos^{n-1} x)(\sin x) + (n-1)I_{n-2} - (n-1)I_n$

i.e. $I_n + (n-1)I_n = (\cos^{n-1} x)(\sin x) + (n-1)I_{n-2}$

i.e. $n I_n = (\cos^{n-1} x)(\sin x) + (n-1)I_{n-2}$

Thus $\quad \boldsymbol{I_n = \dfrac{1}{n}\cos^{n-1} x\sin x + \dfrac{n-1}{n}I_{n-2}}$ $\qquad$ (5)

Problem 11. Use a reduction formula to determine $\int \cos^4 x\,\mathrm{d}x$

Using equation (5),

$\displaystyle\int \cos^4 x\,\mathrm{d}x = I_4 = \dfrac{1}{4}\cos^3 x\sin x + \dfrac{3}{4}I_2$

and $\qquad I_2 = \dfrac{1}{2}\cos x\sin x + \dfrac{1}{2}I_0$

and $\qquad I_0 = \displaystyle\int \cos^0 x\,\mathrm{d}x$

$\qquad\qquad = \displaystyle\int 1\,\mathrm{d}x = x$

Hence $\int \cos^4 x \, dx$

$$= \frac{1}{4} \cos^3 x \sin x + \frac{3}{4} \left(\frac{1}{2} \cos x \sin x + \frac{1}{2} x \right)$$

$$= \frac{1}{4} \cos^3 x \sin x + \frac{3}{8} \cos x \sin x + \frac{3}{8} x + c$$

Problem 12. Determine a reduction formula for $\int_0^{\frac{\pi}{2}} \cos^n x \, dx$ and hence evaluate $\int_0^{\frac{\pi}{2}} \cos^5 x \, dx$

From equation (5),

$$\int \cos^n x \, dx = \frac{1}{n} \cos^{n-1} x \sin x + \frac{n-1}{n} I_{n-2}$$

and hence

$$\int_0^{\frac{\pi}{2}} \cos^n x \, dx = \left[\frac{1}{n} \cos^{n-1} x \sin x \right]_0^{\frac{\pi}{2}}$$

$$+ \frac{n-1}{n} I_{n-2}$$

$$= [0 - 0] + \frac{n-1}{n} I_{n-2}$$

i.e. $\int_0^{\frac{\pi}{2}} \cos^n x \, dx = I_n = \frac{n-1}{n} I_{n-2}$ (6)

(Note that this is the same reduction formula as for $\int_0^{\frac{\pi}{2}} \sin^n x \, dx$ (in Problem 10) and the result is usually known as **Wallis'* formula.**)

* **Who was Wallis?** **John Wallis** (23 November 1616–28 October 1703) was an English mathematician partially credited for the development of infinitesimal calculus, and is also credited with introducing the symbol ∞ for infinity. To find out more go to **www.routledge.com/cw/bird**

Thus, from equation (6),

$$\int_0^{\frac{\pi}{2}} \cos^5 x \, dx = \frac{4}{5} I_3, \qquad I_3 = \frac{2}{3} I_1$$

and $\qquad I_1 = \int_0^{\frac{\pi}{2}} \cos^1 x \, dx$

$$= [\sin x]_0^{\frac{\pi}{2}} = (1 - 0) = 1$$

Hence $\int_0^{\frac{\pi}{2}} \cos^5 x \, dx = \frac{4}{5} I_3 = \frac{4}{5} \left[\frac{2}{3} I_1 \right]$

$$= \frac{4}{5} \left[\frac{2}{3} (1) \right] = \frac{8}{15}$$

Now try the following Practice Exercise

47.5 Further reduction formulae

The following worked problems demonstrate further examples where integrals can be determined using reduction formulae.

Problem 13. Determine a reduction formula for $\int \tan^n x \, dx$ and hence find $\int \tan^7 x \, dx$

Let $I_n = \int \tan^n x \, dx \equiv \int \tan^{n-2} x \tan^2 x \, dx$

by the laws of indices

$$= \int \tan^{n-2} x (\sec^2 x - 1) \, dx$$

since $1 + \tan^2 x = \sec^2 x$

$$= \int \tan^{n-2} x \sec^2 x \, dx - \int \tan^{n-2} x \, dx$$

$$= \int \tan^{n-2} x \sec^2 x \, dx - I_{n-2}$$

i.e. $I_n = \dfrac{\tan^{n-1} x}{n-1} - I_{n-2}$

When $n = 7$,

$$I_7 = \int \tan^7 x \, dx = \frac{\tan^6 x}{6} - I_5$$

$$I_5 = \frac{\tan^4 x}{4} - I_3 \quad \text{and} \quad I_3 = \frac{\tan^2 x}{2} - I_1$$

$$I_1 = \int \tan x \, dx = \ln(\sec x)$$

from Problem 9, Chapter 42, page 476

Thus

$$\int \tan^7 x \, dx = \frac{\tan^6 x}{6} - \left[\frac{\tan^4 x}{4} \right.$$

$$\left. - \left(\frac{\tan^2 x}{2} - \ln(\sec x) \right) \right]$$

Hence $\int \tan^7 x \, dx$

$$= \frac{1}{6} \tan^6 x - \frac{1}{4} \tan^4 x + \frac{1}{2} \tan^2 x$$

$$- \ln(\sec x) + c$$

Problem 14. Evaluate, using a reduction formula,
$\displaystyle \int_0^{\frac{\pi}{2}} \sin^2 t \cos^6 t \, dt$

$$\int_0^{\frac{\pi}{2}} \sin^2 t \cos^6 t \, dt = \int_0^{\frac{\pi}{2}} (1 - \cos^2 t) \cos^6 t \, dt$$

$$= \int_0^{\frac{\pi}{2}} \cos^6 t \, dt - \int_0^{\frac{\pi}{2}} \cos^8 t \, dt$$

If $\qquad I_n = \displaystyle \int_0^{\frac{\pi}{2}} \cos^n t \, dt$

then

$$\int_0^{\frac{\pi}{2}} \sin^2 t \cos^6 t \, dt = I_6 - I_8$$

and from equation (6),

$$I_6 = \frac{5}{6} I_4 = \frac{5}{6} \left[\frac{3}{4} I_2 \right]$$

$$= \frac{5}{6} \left[\frac{3}{4} \left(\frac{1}{2} I_0 \right) \right]$$

and $\qquad I_0 = \displaystyle \int_0^{\frac{\pi}{2}} \cos^0 t \, dt$

$$= \int_0^{\frac{\pi}{2}} 1 \, dt = [x]_0^{\frac{\pi}{2}} = \frac{\pi}{2}$$

Hence $\qquad I_6 = \dfrac{5}{6} \cdot \dfrac{3}{4} \cdot \dfrac{1}{2} \cdot \dfrac{\pi}{2}$

$$= \frac{15\pi}{96} \text{ or } \frac{5\pi}{32}$$

Similarly, $I_8 = \dfrac{7}{8} I_6 = \dfrac{7}{8} \cdot \dfrac{5\pi}{32}$

Thus

$$\int_0^{\frac{\pi}{2}} \sin^2 t \cos^6 t \, dt = I_6 - I_8$$

$$= \frac{5\pi}{32} - \frac{7}{8} \cdot \frac{5\pi}{32}$$

$$= \frac{1}{8} \cdot \frac{5\pi}{32} = \frac{5\pi}{256}$$

Problem 15. Use integration by parts to determine a reduction formula for $\int (\ln x)^n \, dx$. Hence determine $\int (\ln x)^3 \, dx$.

Let $I_n = \int (\ln x)^n \, dx$.
Using integration by parts, let $u = (\ln x)^n$, from which,

$$\frac{du}{dx} = n (\ln x)^{n-1} \left(\frac{1}{x} \right)$$

and $\quad du = n (\ln x)^{n-1} \left(\dfrac{1}{x} \right) dx$

and let $dv = dx$, from which, $v = \int dx = x$

Then $\quad I_n = \displaystyle \int (\ln x)^n \, dx$

$$= (\ln x)^n (x) - \int (x) n (\ln x)^{n-1} \left(\frac{1}{x} \right) dx$$

$$= x (\ln x)^n - n \int (\ln x)^{n-1} \, dx$$

i.e. $I_n = x(\ln x)^n - nI_{n-1}$

When $n = 3$,

$$\int (\ln x)^3 \, dx = I_3 = x(\ln x)^3 - 3I_2$$

$I_2 = x(\ln x)^2 - 2I_1$ and $I_1 = \int \ln x \, dx = x(\ln x - 1)$ from Problem 7, page 503.

Hence

$$\int (\ln x)^3 \, dx = x(\ln x)^3 - 3[x(\ln x)^2 - 2I_1] + c$$

$$= x(\ln x)^3 - 3[x(\ln x)^2$$
$$- 2[x(\ln x - 1)]] + c$$

$$= x(\ln x)^3 - 3[x(\ln x)^2$$
$$- 2x \ln x + 2x] + c$$

$$= x(\ln x)^3 - 3x(\ln x)^2$$
$$+ 6x \ln x - 6x + c$$

$$= x[(\ln x)^3 - 3(\ln x)^2$$
$$+ 6 \ln x - 6] + c$$

Now try the following Practice Exercise

Practice Exercise 190   Reduction formulae (Answers on page 884)

1. Evaluate $\displaystyle\int_0^{\frac{\pi}{2}} \cos^2 x \sin^5 x \, dx$

2. Determine $\displaystyle\int \tan^6 x \, dx$ by using reduction formulae and hence evaluate $\displaystyle\int_0^{\frac{\pi}{4}} \tan^6 x \, dx$

3. Evaluate $\displaystyle\int_0^{\frac{\pi}{2}} \cos^5 x \sin^4 x \, dx$

4. Use a reduction formula to determine $\displaystyle\int (\ln x)^4 \, dx$

5. Show that $\displaystyle\int_0^{\frac{\pi}{2}} \sin^3 \theta \cos^4 \theta \, d\theta = \frac{2}{35}$

For fully worked solutions to each of the problems in Practice Exercises 187 to 190 in this chapter, go to the website:
www.routledge.com/cw/bird

Chapter 48

Double and triple integrals

> **Why it is important to understand: Double and triple integrals**
>
> Double and triple integrals have engineering applications in finding areas, masses and forces of two-dimensional regions, and in determining volumes, average values of functions, centres of mass, moments of inertia, and surface areas. A multiple integral is a type of definite integral extended to functions of more than one real variable. This chapter explains how to evaluate double and triple integrals and completes the many techniques of integral calculus explained in the preceding chapters.

At the end of this chapter, you should be able to:

- evaluate a double integral
- evaluate a triple integral

48.1 Double integrals

The procedure to determine a double integral of the form: $\int_{y_1}^{y_2} \int_{x_1}^{x_2} f(x,y) \, dx \, dy$ is:

(i) integrate $f(x, y)$ with respect to x between the limits of $x = x_1$ and $x = x_2$ (where y is regarded as being a constant), and

(ii) integrate the result in (i) with respect to y between the limits of $y = y_1$ and $y = y_2$

It is seen from this procedure that to determine a double integral we start with the innermost integral and then work outwards.

Double integrals may be used to determine areas under curves, second moments of area, centroids and moments of inertia.

(Sometimes $\int_{y_1}^{y_2} \int_{x_1}^{x_2} f(x, y) \, dx \, dy$ is written as: $\int_{y_1}^{y_2} dy \int_{x_1}^{x_2} f(x, y) \, dx$. All this means is that the right hand side integral is determined first).

Determining double integrals is demonstrated in the following worked problems.

Problem 1. Evaluate $\int_{1}^{3} \int_{2}^{5} (2x - 3y) \, dx \, dy$

Following the above procedure:

(i) $(2x - 3y)$ is integrated with respect to x between $x = 2$ and $x = 5$, with y regarded as a constant

i.e. $\int_{2}^{5} (2x - 3y) \, dx$

$= \left[\frac{2x^2}{2} - (3y)x \right]_{2}^{5} = \left[x^2 - 3xy \right]_{2}^{5}$

$= \left[\left(5^2 - 3(5)y \right) - \left(2^2 - 3(2)y \right) \right]$

$= (25 - 15y) - (4 - 6y) = 25 - 15y - 4 + 6y$

$= 21 - 9y$

(ii) $\displaystyle\int_1^3 \int_2^5 (2x - 3y)\, dx\, dy$

$\displaystyle = \int_1^3 (21 - 9y)\, dy$

$\displaystyle = \left[21y - \frac{9y^2}{2} \right]_1^3$

$\displaystyle = \left[\left(21(3) - \frac{9(3)^2}{2} \right) - \left(21(1) - \frac{9(1)^2}{2} \right) \right]$

$= (63 - 40.5) - (21 - 4.5)$

$= 63 - 40.5 - 21 + 4.5 = 6$

Hence, $\displaystyle\int_1^3 \int_2^5 (2x - 3y)\, dx\, dy = 6$

Problem 2. Evaluate $\displaystyle\int_0^4 \int_1^2 (3x^2 - 2)\, dx\, dy$

Following the above procedure:

(i) $(3x^2 - 2)$ is integrated with respect to x between $x = 1$ and $x = 2$,

i.e. $\displaystyle\int_1^2 (3x^2 - 2)\, dx$

$\displaystyle = \left[\frac{3x^3}{3} - 2x \right]_1^2 = \left[\left(2^3 - 2(2) \right) - \left(1^3 - 2(1) \right) \right]$

$= (8 - 4) - (1 - 2) = 8 - 4 - 1 + 2 = 5$

(ii) $\displaystyle\int_0^4 \int_1^2 (3x^2 - 2)\, dx\, dy$

$\displaystyle = \int_0^4 (5)\, dy = [5y]_0^4 = [(5(4)) - (5(0))]$

$= 20 - 0 = 20$

Hence, $\displaystyle\int_0^4 \int_1^2 (3x^2 - 2)\, dx\, dy = 20$

Problem 3. Evaluate $\displaystyle\int_1^3 \int_0^2 (2x^2 y)\, dx\, dy$

Following the above procedure:

(i) $(2x^2 y)$ is integrated with respect to x between $x = 0$ and $x = 2$,

i.e. $\displaystyle\int_0^2 (2x^2 y)\, dx = \left[\frac{2x^3 y}{3} \right]_0^2$

$\displaystyle = \left[\left(\frac{2(2)^3 y}{3} \right) - (0) \right] = \frac{16}{3} y$

(ii) $\displaystyle\int_1^3 \int_0^2 (2x^2 y)\, dx\, dy = \int_1^3 \left(\frac{16}{3} y \right) dy$

$\displaystyle = \left[\frac{16y^2}{6} \right]_1^3 = \left[\left(\frac{16(3)^2}{6} \right) - \left(\frac{16(1)^2}{6} \right) \right]$

$= 24 - 2.67 = 21.33$

Hence, $\displaystyle\int_1^3 \int_0^2 (2x^2 y)\, dx\, dy = \mathbf{21.33}$

Problem 4. Evaluate $\displaystyle\int_1^3 dy \int_0^2 (2x^2 y)\, dx$

With this configuration:

(i) $(2x^2 y)$ is integrated with respect to x between $x = 0$ and $x = 2$,

i.e. $\displaystyle\int_0^2 (2x^2 y)\, dx = \left[\frac{2x^3 y}{3} \right]_0^2$

$\displaystyle = \left[\left(\frac{2(2)^3 y}{3} \right) - (0) \right] = \frac{16}{3} y$

(ii) $\displaystyle\int_1^3 dy \int_0^2 (2x^2 y)\, dx = \int_1^3 dy \left(\frac{16}{3} y \right)$

$\displaystyle = \int_1^3 \left(\frac{16}{3} y \right) dy = \left[\frac{16y^2}{6} \right]_1^3$

$\displaystyle = \left[\left(\frac{16(3)^2}{6} \right) - \left(\frac{16(1)^2}{6} \right) \right]$

$= 24 - 2.67 = 21.33$

Hence, $\displaystyle\int_1^3 dy \int_0^2 (2x^2 y)\, dx = \mathbf{21.33}$

The last two worked problems show that

$\displaystyle\int_1^3 \int_0^2 (2x^2 y)\, dx\, dy$ gives the same answer as

$\displaystyle\int_1^3 dy \int_0^2 2x^2 y\, dx$

Problem 5. Evaluate $\displaystyle\int_1^4 \int_0^\pi (2 + \sin 2\theta)\, d\theta\, dr$

Following the above procedure:

(i) $(2 + \sin 2\theta)$ is integrated with respect to θ between $\theta = 0$ and $\theta = \pi$,

i.e. $\displaystyle\int_0^\pi (2 + \sin 2\theta)\,dx = \left[2\theta - \frac{1}{2}\cos 2\theta\right]_0^\pi$

$= \left[\left(2\pi - \frac{1}{2}\cos 2\pi\right) - \left(0 - \frac{1}{2}\cos 0\right)\right]$

$= (2\pi - 0.5) - (0 - 0.5) = 2\pi$

(ii) $\displaystyle\int_1^4 \int_0^\pi (2 + \sin 2\theta)\,d\theta\,dr$

$= \displaystyle\int_1^4 (2\pi)\,dr = [2\pi r]_1^4 = [(2\pi(4)) - (2\pi(1))]$

$= 8\pi - 2\pi = 6\pi \text{ or } \mathbf{18.85}$

Hence, $\displaystyle\int_1^4 \int_0^\pi (2 + \sin 2\theta)\,d\theta\,dr = \mathbf{18.85}$

Now try the following Practice Exercise

Practice Exercise 191   Double integrals (Answers on page 884)

Evaluate the double integrals in Problems 1 to 8.

1. $\displaystyle\int_0^3 \int_2^4 2\,dx\,dy$

2. $\displaystyle\int_1^2 \int_2^3 (2x - y)\,dx\,dy$

3. $\displaystyle\int_1^2 \int_2^3 (2x - y)\,dy\,dx$

4. $\displaystyle\int_1^5 \int_{-1}^2 (x - 5y)\,dx\,dy$

5. $\displaystyle\int_1^6 \int_2^5 (x^2 + 4y)\,dx\,dy$

6. $\displaystyle\int_1^4 \int_{-3}^2 (3xy^2)\,dx\,dy$

7. $\displaystyle\int_{-1}^3 \int_0^\pi (3 + \sin 2\theta)\,d\theta\,dr$

8. $\displaystyle\int_1^3 dx \int_2^4 (40 - 2xy)\,dy$

9.   The volume of a solid, V, bounded by the curve $4 - x - y$ between the limits $x = 0$ to $x = 1$ and $y = 0$ to $y = 2$ is given by:

$$V = \int_0^2 \int_0^1 (4 - x - y)\,dx\,dy$$

Evaluate V.

10.   The second moment of area, I, of a 5 cm by 3 cm rectangle about an axis through one corner perpendicular to the plane of the figure is given by:

$$I = \int_0^5 \int_0^3 (x^2 + y^2)\,dy\,dx$$

Evaluate I.

48.2   Triple integrals

The procedure to determine a triple integral of the form:

$$\int_{z_1}^{z_2} \int_{y_1}^{y_2} \int_{x_1}^{x_2} f(x, y, z)\,dx\,dy\,dz \text{ is:}$$

(i)   integrate $f(x, y, z)$ with respect to x between the limits of $x = x_1$ and $x = x_2$ (where y and z are regarded as being constants),

(ii)   integrate the result in (i) with respect to y between the limits of $y = y_1$ and $y = y_2$, and

(iii)   integrate the result in (ii) with respect to z between the limits of $z = z_1$ and $z = z_2$

It is seen from this procedure that to determine a triple integral we start with the innermost integral and then work outwards.

Determining triple integrals is demonstrated in the following worked problems.

Problem 6.   Evaluate

$$\int_1^2 \int_{-1}^3 \int_0^2 (x - 3y + z)\,dx\,dy\,dz$$

Following the above procedure:

(i) $(x - 3y + z)$ is integrated with respect to x between $x = 0$ and $x = 2$, with y and z regarded as constants, i.e. $\displaystyle\int_0^2 (x - 3y + z)\,dx = \left[\frac{x^2}{2} - (3y)x + (z)x\right]_0^2$

$$= \left[\left(\frac{2^2}{2} - (3y)(2) + (z)(2)\right) - (0)\right]$$

$$= 2 - 6y + 2z$$

(ii) $(2 - 6y + 2z)$ is integrated with respect to y between $y = -1$ and $y = 3$, with z regarded as a constant, i.e.

$$\int_{-1}^3 (2 - 6y + 2z)\,dy$$

$$= \left[2y - \frac{6y^2}{2} + (2z)y\right]_{-1}^3$$

$$= \left[\left(2(3) - \frac{6(3)^2}{2} + (2z)(3)\right)\right.$$

$$\left. - \left(2(-1) - \frac{6(-1)^2}{2} + (2z)(-1)\right)\right]$$

$$= [(6 - 27 + 6z) - (-2 - 3 - 2z)]$$

$$= 6 - 27 + 6z + 2 + 3 + 2z = 8z - 16$$

(iii) $(8z - 16)$ is integrated with respect to z between $z = 1$ and $z = 2$

i.e. $\displaystyle\int_1^2 (8z - 16)\,dz = \left[\frac{8z^2}{2} - 16z\right]_1^2$

$$= \left[\left(\frac{8(2)^2}{2} - 16(2)\right) - \left(\frac{8(1)^2}{2} - 16(1)\right)\right]$$

$$= [(16 - 32) - (4 - 16)]$$

$$= 16 - 32 - 4 + 16 = -4$$

Hence, $\displaystyle\int_1^2 \int_{-1}^3 \int_0^2 (x - 3y + z)\,dx\,dy\,dz = \mathbf{-4}$

Problem 7. Evaluate
$\displaystyle\int_1^3 \int_0^2 \int_0^1 (2a^2 - b^2 + 3c^2)\,da\,db\,dc$

Following the above procedure:

(i) $(2a^2 - b^2 + 3c^2)$ is integrated with respect to a between $a = 0$ and $a = 1$, with b and c regarded as constants,

i.e. $\displaystyle\int_0^1 (2a^2 - b^2 + 3c^2)\,da$

$$= \left[\frac{2a^3}{3} - (b^2)a + (3c^2)a\right]_0^1$$

$$= \left[\left(\frac{2}{3} - (b^2) + (3c^2)\right) - (0)\right]$$

$$= \frac{2}{3} - b^2 + 3c^2$$

(ii) $\left(\dfrac{2}{3} - b^2 + 3c^2\right)$ is integrated with respect to b between $b = 0$ and $b = 2$, with c regarded as a constant, i.e.

$$\int_0^2 \left(\frac{2}{3} - b^2 + 3c^2\right)db = \left[\frac{2}{3}b - \frac{b^3}{3} + (3c^2)b\right]_0^2$$

$$= \left[\left(\frac{2}{3}(2) - \frac{(2)^3}{3} + (3c^2)(2)\right) - (0)\right]$$

$$= \left(\frac{4}{3} - \frac{8}{3} + 6c^2\right) - (0) = 6c^2 - \frac{4}{3}$$

(iii) $\left(6c^2 - \dfrac{4}{3}\right)$ is integrated with respect to c between $c = 1$ and $c = 3$

i.e. $\displaystyle\int_1^3 \left(6c^2 - \frac{4}{3}\right)dc = \left[\frac{6c^3}{3} - \frac{4}{3}c\right]_1^3$

$$= \left[(54 - 4) - \left(2 - \frac{4}{3}\right)\right]$$

$$= [(50) - (0.67)] = \mathbf{49.33}$$

Hence, $\displaystyle\int_1^3 \int_0^2 \int_0^1 (2a^2 - b^2 + 3c^2)\,da\,db\,dc = 49.33$

Now try the following Practice Exercise

Practice Exercise 192   Triple integrals (Answers on page 884)

Evaluate the triple integrals in Problems 1 to 7.

1. $\displaystyle\int_1^2 \int_2^3 \int_0^1 (8xyz)\,dz\,dx\,dy$

2. $\displaystyle\int_1^2 \int_2^3 \int_0^1 (8xyz)\,dx\,dy\,dz$

3. $\displaystyle\int_0^2 \int_{-1}^2 \int_1^3 \left(x + y^2 + z^3\right)dx\,dy\,dz$

4. $\int_2^3 \int_{-1}^1 \int_0^2 (x^2 + 5y^2 - 2z)\,dx\,dy\,dz$

5. $\int_0^\pi \int_0^\pi \int_0^\pi (xy \sin z)\,dx\,dy\,dz$

6. $\int_0^4 \int_{-2}^{-1} \int_1^2 (xy)\,dx\,dy\,dz$

7. $\int_1^3 \int_0^2 \int_{-1}^1 (xz + y)\,dx\,dy\,dz$

8. A box shape X is described by the triple integral: $X = \int_0^3 \int_0^2 \int_0^1 (x + y + z)\,dz\,dy\,dx$
Evaluate X.

Section I

For fully worked solutions to each of the problems in Practice Exercises 191 and 192 in this chapter,
go to the website:
www.routledge.com/cw/bird

Chapter 49

Numerical integration

Why it is important to understand: **Numerical integration**

There are two main reasons for why there is a need to do numerical integration – analytical integration may be impossible or infeasible, or it may be necessary to integrate tabulated data rather than known functions. As has been mentioned before, there are many applications for integration. For example, Maxwell's equations can be written in integral form; numerical solutions of Maxwell's equations can be directly used for a huge number of engineering applications. Integration is involved in practically every physical theory in some way – vibration, distortion under weight, or one of many types of fluid flow – be it heat flow, air flow (over a wing), or water flow (over a ship's hull, through a pipe, or perhaps even groundwater flow regarding a contaminant), and so on; all these things can be either directly solved by integration (for simple systems), or some type of numerical integration (for complex systems). Numerical integration is also essential for the evaluation of integrals of functions available only at discrete points; such functions often arise in the numerical solution of differential equations or from experimental data taken at discrete intervals. Engineers therefore often require numerical integration and this chapter explains the procedures available.

At the end of this chapter, you should be able to:

- appreciate the need for numerical integration
- evaluate integrals using the trapezoidal rule
- evaluate integrals using the mid-ordinate rule
- evaluate integrals using Simpson's rule
- apply numerical integration to practical situations

49.1 Introduction

Even with advanced methods of integration there are many mathematical functions which cannot be integrated by analytical methods and thus approximate methods have then to be used. In many cases, such as in modelling airflow around a car, an exact answer may not even be strictly necessary. Approximate methods of definite integrals may be determined by what is termed **numerical integration**.

It may be shown that determining the value of a definite integral is, in fact, finding the area between a curve, the horizontal axis and the specified ordinates. Three methods of finding approximate areas under curves are the trapezoidal rule, the mid-ordinate rule and Simpson's rule, and these rules are used as a basis for numerical integration.

49.2 The trapezoidal rule

Let a required definite integral be denoted by $\int_a^b y \, dx$ and be represented by the area under the graph of $y = f(x)$ between the limits $x = a$ and $x = b$ as shown in Fig. 49.1.

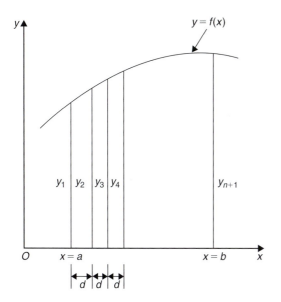

Figure 49.1

Let the range of integration be divided into n equal intervals each of width d, such that $nd = b - a$, i.e.
$$d = \frac{b-a}{n}$$

The ordinates are labelled $y_1, y_2, y_3, \ldots, y_{n+1}$ as shown. An approximation to the area under the curve may be determined by joining the tops of the ordinates by straight lines. Each interval is thus a trapezium, and since the area of a trapezium is given by:

$$\text{area} = \frac{1}{2}(\text{sum of parallel sides})\,(\text{perpendicular}$$
$$\text{distance between them)} \text{ then}$$

$$\int_a^b y\,dx \approx \frac{1}{2}(y_1 + y_2)d + \frac{1}{2}(y_2 + y_3)d$$
$$+ \frac{1}{2}(y_3 + y_4)d + \cdots \frac{1}{2}(y_n + y_{n+1})d$$

$$\approx d\left[\frac{1}{2}y_1 + y_2 + y_3 + y_4 + \cdots + y_n\right.$$
$$\left. + \frac{1}{2}y_{n+1}\right]$$

i.e. **the trapezoidal rule states:**

$$\int_a^b y\,dx \approx \binom{\text{width of}}{\text{interval}}\left\{\frac{1}{2}\binom{\text{first}+\text{last}}{\text{ordinate}}\right.$$
$$\left. + \binom{\text{sum of remaining}}{\text{ordinates}}\right\} \quad (1)$$

Problem 1. (a) Use integration to evaluate, correct to 3 decimal places, $\int_1^3 \frac{2}{\sqrt{x}}\,dx$ (b) Use the trapezoidal rule with four intervals to evaluate the integral in part (a), correct to 3 decimal places.

(a) $\displaystyle\int_1^3 \frac{2}{\sqrt{x}}\,dx = \int_1^3 2x^{-\frac{1}{2}}\,dx$

$$= \left[\frac{2x^{\left(\frac{-1}{2}\right)+1}}{-\frac{1}{2}+1}\right]_1^3 = \left[4x^{\frac{1}{2}}\right]_1^3$$

$$= 4\left[\sqrt{x}\right]_1^3 = 4\left[\sqrt{3} - \sqrt{1}\right]$$

$$= \mathbf{2.928}, \text{ correct to 3 decimal places.}$$

(b) The range of integration is the difference between the upper and lower limits, i.e. $3 - 1 = 2$. Using the trapezoidal rule with four intervals gives an interval width $d = \dfrac{3-1}{4} = 0.5$ and ordinates situated at 1.0, 1.5, 2.0, 2.5 and 3.0. Corresponding values of $\dfrac{2}{\sqrt{x}}$ are shown in the table below, each correct to 4 decimal places (which is one more decimal place than required in the problem).

x	$\dfrac{2}{\sqrt{x}}$
1.0	2.0000
1.5	1.6330
2.0	1.4142
2.5	1.2649
3.0	1.1547

From equation (1):

$$\int_1^3 \frac{2}{\sqrt{x}}\,dx \approx (0.5)\left\{\frac{1}{2}(2.0000 + 1.1547)\right.$$
$$\left. + 1.6330 + 1.4142 + 1.2649\right\}$$

$$= \mathbf{2.945}, \text{ correct to 3 decimal places.}$$

This problem demonstrates that even with just 4 intervals a close approximation to the true value of 2.928 (correct to 3 decimal places) is obtained using the trapezoidal rule.

Section I

Problem 2. Use the trapezoidal rule with eight intervals to evaluate $\int_1^3 \dfrac{2}{\sqrt{x}}\,dx$ correct to 3 decimal places.

With eight intervals, the width of each is $\dfrac{3-1}{8}$ i.e. 0.25 giving ordinates at 1.00, 1.25, 1.50, 1.75, 2.00, 2.25, 2.50, 2.75 and 3.00. Corresponding values of $\dfrac{2}{\sqrt{x}}$ are shown in the table below.

x	$\dfrac{2}{\sqrt{x}}$
1.00	2.0000
1.25	1.7889
1.50	1.6330
1.75	1.5119
2.00	1.4142
2.25	1.3333
2.50	1.2649
2.75	1.2060
3.00	1.1547

From equation (1):

$$\int_1^3 \frac{2}{\sqrt{x}}\,dx \approx (0.25)\left\{\frac{1}{2}(2.000+1.1547)+1.7889\right.$$
$$+1.6330+1.5119+1.4142$$
$$\left.+1.3333+1.2649+1.2060\right\}$$

$$= \mathbf{2.932}, \text{ correct to 3 decimal places.}$$

This problem demonstrates that the greater the number of intervals chosen (i.e. the smaller the interval width) the more accurate will be the value of the definite integral. The exact value is found when the number of intervals is infinite, which is, of course, what the process of integration is based upon.

Problem 3. Use the trapezoidal rule to evaluate $\int_0^{\frac{\pi}{2}} \dfrac{1}{1+\sin x}\,dx$ using six intervals. Give the answer correct to 4 significant figures.

With six intervals, each will have a width of $\dfrac{\frac{\pi}{2}-0}{6}$ i.e. $\dfrac{\pi}{12}$ rad (or 15°) and the ordinates occur at $0, \dfrac{\pi}{12}, \dfrac{\pi}{6}, \dfrac{\pi}{4}, \dfrac{\pi}{3}, \dfrac{5\pi}{12}$ and $\dfrac{\pi}{2}$

Corresponding values of $\dfrac{1}{1+\sin x}$ are shown in the table below.

x	$\dfrac{1}{1+\sin x}$
0	1.0000
$\dfrac{\pi}{12}$ (or 15°)	0.79440
$\dfrac{\pi}{6}$ (or 30°)	0.66667
$\dfrac{\pi}{4}$ (or 45°)	0.58579
$\dfrac{\pi}{3}$ (or 60°)	0.53590
$\dfrac{5\pi}{12}$ (or 75°)	0.50867
$\dfrac{\pi}{2}$ (or 90°)	0.50000

From equation (1):

$$\int_0^{\frac{\pi}{2}} \frac{1}{1+\sin x}\,dx \approx \left(\frac{\pi}{12}\right)\left\{\frac{1}{2}(1.00000+0.50000)\right.$$
$$+0.79440+0.66667$$
$$+0.58579+0.53590$$
$$\left.+0.50867\right\}$$

$$= \mathbf{1.006}, \text{ correct to 4}$$
$$\text{significant figures.}$$

Now try the following Practice Exercise

Practice Exercise 193   Trapezoidal rule (Answers on page 884)

In Problems 1 to 4, evaluate the definite integrals using the **trapezoidal rule**, giving the answers correct to 3 decimal places.

1. $\displaystyle\int_0^1 \frac{2}{1+x^2}\,dx$ (use eight intervals)

2. $\displaystyle\int_1^3 2\ln 3x\,dx$ (use eight intervals)

3. $\displaystyle\int_0^{\frac{\pi}{3}} \sqrt{(\sin\theta)}\,d\theta$ (use six intervals)

4. $\displaystyle\int_0^{1.4} e^{-x^2}\,dx$ (use seven intervals)

49.3   The mid-ordinate rule

Let a required definite integral be denoted again by $\int_a^b y\,dx$ and represented by the area under the graph of $y = f(x)$ between the limits $x = a$ and $x = b$, as shown in Fig. 49.2.

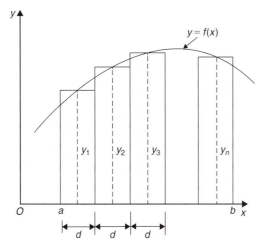

Figure 49.2

With the mid-ordinate rule each interval of width d is assumed to be replaced by a rectangle of height equal to the ordinate at the middle point of each interval, shown as $y_1, y_2, y_3, \ldots, y_n$ in Fig. 49.2.

Thus $\displaystyle\int_a^b y\,dx \approx dy_1 + dy_2 + dy_3 + \cdots + dy_n$

$\qquad\qquad \approx d(y_1 + y_2 + y_3 + \cdots + y_n)$

i.e. **the mid-ordinate rule states:**

$$\int_a^b y\,dx \approx \text{(width of interval) (sum of mid-ordinates)} \qquad (2)$$

Problem 4.   Use the mid-ordinate rule with (a) four intervals, (b) eight intervals, to evaluate $\displaystyle\int_1^3 \frac{2}{\sqrt{x}}\,dx$, correct to 3 decimal places.

(a) With four intervals, each will have a width of $\dfrac{3-1}{4}$, i.e. 0.5 and the ordinates will occur at 1.0, 1.5, 2.0, 2.5 and 3.0. Hence the mid-ordinates y_1, y_2, y_3 and y_4 occur at 1.25, 1.75, 2.25 and 2.75. Corresponding values of $\dfrac{2}{\sqrt{x}}$ are shown in the following table.

x	$\dfrac{2}{\sqrt{x}}$
1.25	1.7889
1.75	1.5119
2.25	1.3333
2.75	1.2060

From equation (2):

$$\int_1^3 \frac{2}{\sqrt{x}}\,dx \approx (0.5)[1.7889 + 1.5119 + 1.3333 + 1.2060]$$
$$= \mathbf{2.920}, \text{ correct to 3 decimal places.}$$

(b) With eight intervals, each will have a width of 0.25 and the ordinates will occur at 1.00, 1.25, 1.50, 1.75, ... and thus mid-ordinates at 1.125, 1.375, 1.625, 1.875 ...

Corresponding values of $\dfrac{2}{\sqrt{x}}$ are shown in the following table.

x	$\dfrac{2}{\sqrt{x}}$
1.125	1.8856
1.375	1.7056
1.625	1.5689
1.875	1.4606
2.125	1.3720
2.375	1.2978
2.625	1.2344
2.875	1.1795

Section I

From equation (2):

$$\int_1^3 \frac{2}{\sqrt{x}} dx \approx (0.25)[1.8856 + 1.7056$$
$$+ 1.5689 + 1.4606 + 1.3720$$
$$+ 1.2978 + 1.2344 + 1.1795]$$
$$= \mathbf{2.926}, \text{ correct to 3 decimal places.}$$

As previously, the greater the number of intervals the nearer the result is to the true value (of 2.928, correct to 3 decimal places).

Problem 5. Evaluate $\int_0^{2.4} e^{\frac{-x^2}{3}} dx$, correct to 4 significant figures, using the mid-ordinate rule with six intervals.

With six intervals each will have a width of $\dfrac{2.4 - 0}{6}$, i.e. 0.40 and the ordinates will occur at 0, 0.40, 0.80, 1.20, 1.60, 2.00 and 2.40 and thus mid-ordinates at 0.20, 0.60, 1.00, 1.40, 1.80 and 2.20. Corresponding values of $e^{\frac{-x^2}{3}}$ are shown in the following table.

x	$e^{\frac{-x^2}{3}}$
0.20	0.98676
0.60	0.88692
1.00	0.71653
1.40	0.52031
1.80	0.33960
2.20	0.19922

From equation (2):

$$\int_0^{2.4} e^{\frac{-x^2}{3}} dx \approx (0.40)[0.98676 + 0.88692$$
$$+ 0.71653 + 0.52031$$
$$+ 0.33960 + 0.19922]$$
$$= \mathbf{1.460}, \text{ correct to 4 significant figures.}$$

Now try the following Practice Exercise

Practice Exercise 194   Mid-ordinate rule (Answers on page 884)

In Problems 1 to 4, evaluate the definite integrals using the **mid-ordinate rule**, giving the answers correct to 3 decimal places.

1. $\int_0^2 \dfrac{3}{1 + t^2} dt$ (use eight intervals)

2. $\int_0^{\frac{\pi}{2}} \dfrac{1}{1 + \sin\theta} d\theta$ (use six intervals)

3. $\int_1^3 \dfrac{\ln x}{x} dx$ (use ten intervals)

4. $\int_0^{\frac{\pi}{3}} \sqrt{(\cos^3 x)} dx$ (use six intervals)

49.4 Simpson's rule

The approximation made with the trapezoidal rule is to join the top of two successive ordinates by a straight line, i.e. by using a linear approximation of the form $a + bx$. With **Simpson's* rule**, the approximation made is to join the tops of three successive ordinates by a parabola, i.e. by using a quadratic approximation of the form $a + bx + cx^2$.

Fig. 49.3 shows a parabola $y = a + bx + cx^2$ with ordinates y_1, y_2 and y_3 at $x = -d$, $x = 0$ and $x = d$ respectively.

Thus the width of each of the two intervals is d. The area enclosed by the parabola, the x-axis and ordinates $x = -d$ and $x = d$ is given by:

$$\int_{-d}^d (a + bx + cx^2) dx = \left[ax + \frac{bx^2}{2} + \frac{cx^3}{3} \right]_{-d}^d$$
$$= \left(ad + \frac{bd^2}{2} + \frac{cd^3}{3} \right)$$

---

* **Who was Simpson?** **Thomas Simpson** FRS (20 August 1710–14 May 1761) was the British mathematician who invented Simpson's rule to approximate definite integrals. For an image of Simpson, see page 242. To find out more go to **www.routledge.com/cw/bird**

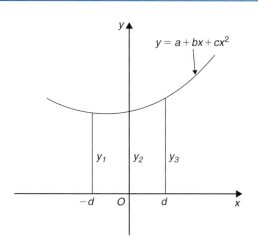

Figure 49.3

$$-\left(-ad + \frac{bd^2}{2} - \frac{cd^3}{3}\right)$$

$$= 2ad + \frac{2}{3}cd^3 \text{ or}$$

$$\frac{1}{3}d(6a + 2cd^2) \qquad (3)$$

Since $y = a + bx + cx^2$,

at $\quad x = -d, y_1 = a - bd + cd^2$

at $\quad x = 0, \ y_2 = a$

and at $\quad x = d, \ y_3 = a + bd + cd^2$

Hence $\quad y_1 + y_3 = 2a + 2cd^2$

and $\quad y_1 + 4y_2 + y_3 = 6a + 2cd^2 \qquad (4)$

Thus the area under the parabola between $x = -d$ and $x = d$ in Fig. 49.3 may be expressed as $\frac{1}{3}d(y_1 + 4y_2 + y_3)$, from equations (3) and (4), and the result is seen to be independent of the position of the origin.

Let a definite integral be denoted by $\int_a^b y\,dx$ and represented by the area under the graph of $y = f(x)$ between the limits $x = a$ and $x = b$, as shown in Fig. 49.4. The range of integration, $b - a$, is divided into an **even** number of intervals, say $2n$, each of width d.

Since an even number of intervals is specified, an odd number of ordinates, $2n + 1$, exists. Let an approximation to the curve over the first two intervals be a parabola of the form $y = a + bx + cx^2$ which passes through the tops of the three ordinates y_1, y_2 and y_3. Similarly, let an approximation to the curve over the next two intervals be the parabola which passes through the tops of the ordinates y_3, y_4 and y_5, and so on.

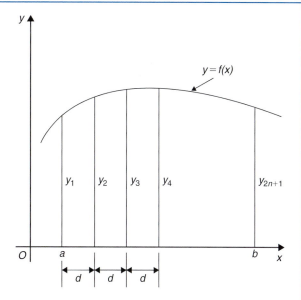

Figure 49.4

Then $\int_a^b y\,dx$

$$\approx \frac{1}{3}d(y_1 + 4y_2 + y_3) + \frac{1}{3}d(y_3 + 4y_4 + y_5)$$

$$+ \frac{1}{3}d(y_{2n-1} + 4y_{2n} + y_{2n+1})$$

$$\approx \frac{1}{3}d[(y_1 + y_{2n+1}) + 4(y_2 + y_4 + \cdots + y_{2n})$$

$$+ 2(y_3 + y_5 + \cdots + y_{2n-1})]$$

i.e. **Simpson's rule states:**

$$\int_a^b y\,dx \approx \frac{1}{3}\begin{pmatrix}\text{width of}\\\text{interval}\end{pmatrix}\left\{\begin{pmatrix}\text{first + last}\\\text{ordinate}\end{pmatrix}\right.$$

$$+ 4\begin{pmatrix}\text{sum of even}\\\text{ordinates}\end{pmatrix} \qquad (5)$$

$$\left.+ 2\begin{pmatrix}\text{sum of remaining}\\\text{odd ordinates}\end{pmatrix}\right\}$$

Note that Simpson's rule can only be applied when an even number of intervals is chosen, i.e. an odd number of ordinates.

Problem 6. Use Simpson's rule with (a) four intervals, (b) eight intervals, to evaluate $\int_1^3 \frac{2}{\sqrt{x}}\,dx$, correct to 3 decimal places.

(a) With four intervals, each will have a width of $\frac{3-1}{4}$, i.e. 0.5 and the ordinates will occur at 1.0, 1.5, 2.0, 2.5 and 3.0. The values

of the ordinates are as shown in the table of Problem 1(b), page 522.
Thus, from equation (5):

$$\int_1^3 \frac{2}{\sqrt{x}}\, dx \approx \frac{1}{3}(0.5)\,[(2.0000 + 1.1547)$$

$$+ 4(1.6330 + 1.2649) + 2(1.4142)]$$

$$= \frac{1}{3}(0.5)[3.1547 + 11.5916$$

$$+ 2.8284]$$

$$= \mathbf{2.929}, \text{ correct to 3 decimal places.}$$

(b) With eight intervals, each will have a width of $\dfrac{3-1}{8}$, i.e. 0.25 and the ordinates occur at 1.00, 1.25, 1.50, 1.75, ..., 3.0. The values of the ordinates are as shown in the table in Problem 2, page 522.
Thus, from equation (5):

$$\int_1^3 \frac{2}{\sqrt{x}}\, dx \approx \frac{1}{3}(0.25)\,[(2.0000 + 1.1547)$$

$$+ 4(1.7889 + 1.5119 + 1.3333$$

$$+ 1.2060) + 2(1.6330 + 1.4142$$

$$+ 1.2649)]$$

$$= \frac{1}{3}(0.25)[3.1547 + 23.3604$$

$$+ 8.6242]$$

$$= \mathbf{2.928}, \text{ correct to 3 decimal places.}$$

It is noted that the latter answer is exactly the same as that obtained by integration. In general, Simpson's rule is regarded as the most accurate of the three approximate methods used in numerical integration.

Problem 7. Evaluate

$$\int_0^{\frac{\pi}{3}} \sqrt{\left(1 - \frac{1}{3}\sin^2\theta\right)}\, d\theta$$

correct to 3 decimal places, using Simpson's rule with six intervals.

With six intervals, each will have a width of $\dfrac{\frac{\pi}{3} - 0}{6}$ i.e.

$\dfrac{\pi}{18}$ rad (or 10°), and the ordinates will occur at

$0, \dfrac{\pi}{18}, \dfrac{\pi}{9}, \dfrac{\pi}{6}, \dfrac{2\pi}{9}, \dfrac{5\pi}{18}$ and $\dfrac{\pi}{3}$

Corresponding values of $\sqrt{\left(1 - \dfrac{1}{3}\sin^2\theta\right)}$ are shown in the table below.

θ	0	$\dfrac{\pi}{18}$	$\dfrac{\pi}{9}$	$\dfrac{\pi}{6}$
		(or 10°)	(or 20°)	(or 30°)
$\sqrt{\left(1 - \dfrac{1}{3}\sin^2\theta\right)}$	1.0000	0.9950	0.9803	0.9574

θ	$\dfrac{2\pi}{9}$	$\dfrac{5\pi}{18}$	$\dfrac{\pi}{3}$
	(or 40°)	(or 50°)	(or 60°)
$\sqrt{\left(1 - \dfrac{1}{3}\sin^2\theta\right)}$	0.9286	0.8969	0.8660

From equation (5)

$$\int_0^{\frac{\pi}{3}} \sqrt{\left(1 - \frac{1}{3}\sin^2\theta\right)}\, d\theta$$

$$\approx \frac{1}{3}\left(\frac{\pi}{18}\right)[(1.0000 + 0.8660) + 4(0.9950$$

$$+ 0.9574 + 0.8969)$$

$$+ 2(0.9803 + 0.9286)]$$

$$= \frac{1}{3}\left(\frac{\pi}{18}\right)[1.8660 + 11.3972 + 3.8178]$$

$$= \mathbf{0.994}, \text{ correct to 3 decimal places.}$$

Problem 8. An alternating current i has the following values at equal intervals of 2.0 milliseconds:

Time (ms)	Current i (A)
0	0
2.0	3.5
4.0	8.2
6.0	10.0
8.0	7.3
10.0	2.0
12.0	0

Charge q, in millicoulombs, is given by
$q = \int_0^{12.0} i \, dt$.

Use Simpson's rule to determine the approximate charge in the 12 millisecond period.

From equation (5):

$$\text{Charge, } q = \int_0^{12.0} i \, dt \approx \frac{1}{3}(2.0)\,[(0+0) + 4(3.5$$

$$+ 10.0 + 2.0) + 2(8.2 + 7.3)]$$

$$= \mathbf{62 \, mC}$$

Now try the following Practice Exercise

**Practice Exercise 195   Simpsons rule
(Answers on page 884)**

In Problems 1 to 5, evaluate the definite integrals using **Simpson's rule**, giving the answers correct to 3 decimal places.

1. $\int_0^{\frac{\pi}{2}} \sqrt{(\sin x)} \, dx$ (use six intervals)

2. $\int_0^{1.6} \frac{1}{1 + \theta^4} \, d\theta$ (use eight intervals)

3. $\int_{0.2}^{1.0} \frac{\sin \theta}{\theta} \, d\theta$ (use eight intervals)

4. $\int_0^{\frac{\pi}{2}} x \cos x \, dx$ (use six intervals)

5. $\int_0^{\frac{\pi}{3}} e^{x^2} \sin 2x \, dx$ (use ten intervals)

In Problems 6 and 7 evaluate the definite integrals using (a) integration, (b) the trapezoidal rule, (c) the mid-ordinate rule, (d) Simpson's rule. Give answers correct to 3 decimal places.

6. $\int_1^4 \frac{4}{x^3} \, dx$ (Use 6 intervals)

7. $\int_2^6 \frac{1}{\sqrt{(2x-1)}} \, dx$ (Use 8 intervals)

In Problems 8 and 9 evaluate the definite integrals using (a) the trapezoidal rule, (b) the mid-ordinate rule, (c) Simpson's rule. Use 6 intervals in each case and give answers correct to 3 decimal places.

8. $\int_0^3 \sqrt{(1 + x^4)} \, dx$

9. $\int_{0.1}^{0.7} \frac{1}{\sqrt{(1 - y^2)}} \, dy$

10. A vehicle starts from rest and its velocity is measured every second for 8 s, with values as follows:

time t (s)	velocity v (ms^{-1})
0	0
1.0	0.4
2.0	1.0
3.0	1.7
4.0	2.9
5.0	4.1
6.0	6.2
7.0	8.0
8.0	9.4

The distance travelled in 8.0 s is given by $\int_0^{8.0} v \, dt$

Estimate this distance using Simpson's rule, giving the answer correct to 3 significant figures.

11. A pin moves along a straight guide so that its velocity v (m/s) when it is a distance x (m) from the beginning of the guide at time t (s) is given in the table on the right.

Use Simpson's rule with eight intervals to determine the approximate total distance travelled by the pin in the 4.0 s period.

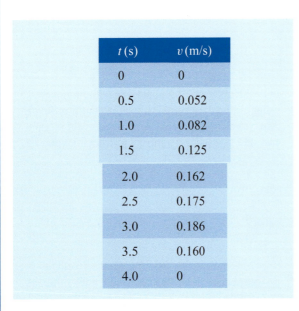

t (s)	v (m/s)
0	0
0.5	0.052
1.0	0.082
1.5	0.125
2.0	0.162
2.5	0.175
3.0	0.186
3.5	0.160
4.0	0

49.5 Accuracy of numerical integration

For a function with an increasing gradient, the trapezoidal rule will tend to over-estimate and the mid-ordinate rule will tend to under-estimate (but by half as much). The appropriate combination of the two in Simpson's rule eliminates this error term, giving a rule which will perfectly model anything up to a cubic, and have a proportionately lower error for any function of greater complexity.

In general, for a given number of strips, Simpson's rule is considered the most accurate of the three numerical methods.

For fully worked solutions to each of the problems in Practice Exercises 193 to 195 in this chapter,
go to the website:
www.routledge.com/cw/bird

Revision Test 14   Further integration

This Revision Test covers the material contained in Chapters 46 to 49. *The marks for each question are shown in brackets at the end of each question.*

1. Determine the following integrals:

   (a) $\int 5x\, e^{2x}\, \mathrm{d}x$ (b) $\int t^2 \sin 2t\, \mathrm{d}t$ (14)

2. Evaluate correct to 3 decimal places:

   $$\int_1^4 \sqrt{x} \ln x\, \mathrm{d}x \qquad (11)$$

3. Use reduction formulae to determine:

   (a) $\int x^3 e^{3x}\, \mathrm{d}x$ (b) $\int t^4 \sin t\, \mathrm{d}t$ (14)

4. Evaluate $\displaystyle\int_0^{\frac{\pi}{2}} \cos^6 x\, \mathrm{d}x$ using a reduction formula. (6)

5. Evaluate $\displaystyle\int_1^3 \int_0^{\pi} (1 + \sin 3\theta)\mathrm{d}\theta\, \mathrm{d}r$ correct to 4 significant figures. (7)

6. Evaluate $\displaystyle\int_0^3 \int_{-1}^1 \int_1^2 (2x + y^2 + 4z^3)\, \mathrm{d}x\, \mathrm{d}y\, \mathrm{d}z$ (9)

7. Evaluate $\displaystyle\int_1^3 \frac{5}{x^2}\, \mathrm{d}x$ using (a) integration (b) the trapezoidal rule (c) the mid-ordinate rule (d) Simpson's rule. In each of the approximate methods use eight intervals and give the answers correct to 3 decimal places. (19)

8. An alternating current i has the following values at equal intervals of 5 ms:

Time t (ms)	0	5	10	15	20	25	30
Current i (A)	0	4.8	9.1	12.7	8.8	3.5	0

   Charge q, in coulombs, is given by

   $$q = \int_0^{30 \times 10^{-3}} i\, \mathrm{d}t$$

   Use Simpson's rule to determine the approximate charge in the 30 ms period. (5)

Section I

For lecturers/instructors/teachers, fully worked solutions to each of the problems in Revision Test 14, together with a full marking scheme, are available at the website:
www.routledge.com/cw/bird

Further differential equations

Chapter 50

Homogeneous first-order differential equations

Why it is important to understand: Homogeneous first-order differential equations

As was previously stated, differential equations play a prominent role in engineering, physics, economics, and other disciplines. Applications are many and varied; for example, they are involved in combined heat conduction and convection with the design of heating and cooling chambers, in fluid mechanics analysis, in heat transfer analysis, in kinematic analysis of rigid body dynamics, with exponential decay in radioactive material, Newton's law of cooling and in mechanical oscillations. This chapter explains how to solve a particular type of differential equation – the homogeneous first-order type.

At the end of this chapter, you should be able to:

- Recognise a homogeneous differential equation
- solve a differential equations of the form $P\dfrac{dy}{dx} = Q$

50.1 Introduction

Certain first-order differential equations are not of the 'variable-separable' type, but can be made separable by changing the variable.

An equation of the form $P\dfrac{dy}{dx} = Q$, where P and Q are functions of both x and y of the same degree throughout, is said to be **homogeneous** in y and x. For example, $f(x, y) = x^2 + 3xy + y^2$ is a homogeneous function since each of the three terms are of degree 2. However, $f(x, y) = \dfrac{x^2 - y}{2x^2 + y^2}$ is not homogeneous since the term in y in the numerator is of degree 1 and the other three terms are of degree 2.

50.2 Procedure to solve differential equations of the form $P\dfrac{dy}{dx} = Q$

(i) Rearrange $P\dfrac{dy}{dx} = Q$ into the form $\dfrac{dy}{dx} = \dfrac{Q}{P}$

(ii) Make the substitution $y = vx$ (where v is a function of x), from which, $\dfrac{dy}{dx} = v(1) + x\dfrac{dv}{dx}$, by the product rule.

(iii) Substitute for both y and $\dfrac{dy}{dx}$ in the equation $\dfrac{dy}{dx} = \dfrac{Q}{P}$. Simplify, by cancelling, and an equation results in which the variables are separable.

(iv) Separate the variables and solve using the method shown in Chapter 33.

(v) Substitute $v = \dfrac{y}{x}$ to solve in terms of the original variables.

50.3 Worked problems on homogeneous first-order differential equations

Problem 1. Solve the differential equation: $y - x = x\dfrac{dy}{dx}$, given $x = 1$ when $y = 2$

Using the above procedure:

(i) Rearranging $y - x = x\dfrac{dy}{dx}$ gives:

$$\frac{dy}{dx} = \frac{y - x}{x}$$

which is homogeneous in x and y

(ii) Let $y = vx$, then $\dfrac{dy}{dx} = v + x\dfrac{dv}{dx}$

(iii) Substituting for y and $\dfrac{dy}{dx}$ gives:

$$v + x\frac{dv}{dx} = \frac{vx - x}{x} = \frac{x(v-1)}{x} = v - 1$$

(iv) Separating the variables gives:

$$x\frac{dv}{dx} = v - 1 - v = -1, \text{ i.e. } dv = -\frac{1}{x}dx$$

Integrating both sides gives:

$$\int dv = \int -\frac{1}{x}dx$$

Hence, $v = -\ln x + c$

(v) Replacing v by $\dfrac{y}{x}$ gives: $\dfrac{y}{x} = -\ln x + c$, which is the general solution.

When $x = 1, y = 2$, thus: $\dfrac{2}{1} = -\ln 1 + c$ from which, $c = 2$

Thus, the particular solution is: $\dfrac{y}{x} = -\ln x + 2$

or $y = -x(\ln x - 2)$ or $y = x(2 - \ln x)$

Problem 2. Find the particular solution of the equation: $x\dfrac{dy}{dx} = \dfrac{x^2 + y^2}{y}$, given the boundary conditions that $y = 4$ when $x = 1$

Using the procedure of section 50.2:

(i) Rearranging $x\dfrac{dy}{dx} = \dfrac{x^2 + y^2}{y}$ gives:

$$\frac{dy}{dx} = \frac{x^2 + y^2}{xy}$$ which is homogeneous in x and y since each of the three terms on the right-hand side are of the same degree (i.e. degree 2).

(ii) Let $y = vx$ then $\dfrac{dy}{dx} = v + x\dfrac{dv}{dx}$

(iii) Substituting for y and $\dfrac{dy}{dx}$ in the equation $\dfrac{dy}{dx} = \dfrac{x^2 + y^2}{xy}$ gives:

$$v + x\frac{dv}{dx} = \frac{x^2 + v^2 x^2}{x(vx)} = \frac{x^2 + v^2 x^2}{vx^2} = \frac{1 + v^2}{v}$$

(iv) Separating the variables gives:

$$x\frac{dv}{dx} = \frac{1 + v^2}{v} - v = \frac{1 + v^2 - v^2}{v} = \frac{1}{v}$$

Hence, $v\,dv = \dfrac{1}{x}dx$

Integrating both sides gives:

$$\int v\,dv = \int \frac{1}{x}dx \text{ i.e. } \frac{v^2}{2} = \ln x + c$$

(v) Replacing v by $\dfrac{y}{x}$ gives: $\dfrac{y^2}{2x^2} = \ln x + c$, which is the general solution.

When $x = 1, y = 4$, thus: $\dfrac{16}{2} = \ln 1 + c$ from which, $c = 8$

Hence, the particular solution is: $\dfrac{y^2}{2x^2} = \ln x + 8$

or $y^2 = 2x^2(8 + \ln x)$

Now try the following Practice Exercise

1. Find the general solution of: $x^2 = y^2 \dfrac{dy}{dx}$

2. Find the general solution of:
$$x - y + x\frac{dy}{dx} = 0$$

3. Find the particular solution of the differential equation: $(x^2 + y^2)dy = xy\,dx$, given that $x = 1$ when $y = 1$

4. Solve the differential equation: $\dfrac{x+y}{y-x} = \dfrac{dy}{dx}$

5. Find the particular solution of the differential equation: $\left(\dfrac{2y-x}{y+2x}\right)\dfrac{dy}{dx} = 1$ given that $y = 3$ when $x = 2$

50.4 Further worked problems on homogeneous first-order differential equations

Problem 3. Solve the equation:
$7x(x-y)dy = 2(x^2 + 6xy - 5y^2)dx$
given that $x = 1$ when $y = 0$

Using the procedure of Section 50.2:

(i) Rearranging gives: $\dfrac{dy}{dx} = \dfrac{2x^2 + 12xy - 10y^2}{7x^2 - 7xy}$
which is homogeneous in x and y since each of the terms on the right-hand side is of degree 2.

(ii) Let $y = vx$ then $\dfrac{dy}{dx} = v + x\dfrac{dv}{dx}$

(iii) Substituting for y and $\dfrac{dy}{dx}$ gives:
$$v + x\frac{dv}{dx} = \frac{2x^2 + 12x(vx) - 10(vx)^2}{7x^2 - 7x(vx)}$$
$$= \frac{2 + 12v - 10v^2}{7 - 7v}$$

(iv) Separating the variables gives:
$$x\frac{dv}{dx} = \frac{2 + 12v - 10v^2}{7 - 7v} - v$$
$$= \frac{(2 + 12v - 10v^2) - v(7 - 7v)}{7 - 7v}$$
$$= \frac{2 + 5v - 3v^2}{7 - 7v}$$

Hence, $\dfrac{7 - 7v}{2 + 5v - 3v^2}dv = \dfrac{dx}{x}$

Integrating both sides gives:
$$\int\left(\frac{7 - 7v}{2 + 5v - 3v^2}\right)dv = \int\frac{1}{x}dx$$

Resolving $\dfrac{7 - 7v}{2 + 5v - 3v^2}$ into partial fractions gives: $\dfrac{4}{(1+3v)} - \dfrac{1}{(2-v)}$ (see Chapter 2)

Hence, $\displaystyle\int\left(\frac{4}{(1+3v)} - \frac{1}{(2-v)}\right)dv = \int\frac{1}{x}dx$

i.e. $\dfrac{4}{3}\ln(1+3v) + \ln(2-v) = \ln x + c$

(v) Replacing v by $\dfrac{y}{x}$ gives:
$$\frac{4}{3}\ln\left(1 + \frac{3y}{x}\right) + \ln\left(2 - \frac{y}{x}\right) = \ln + c$$
or $\dfrac{4}{3}\ln\left(\dfrac{x+3y}{x}\right) + \ln\left(\dfrac{2x-y}{x}\right) = \ln + c$

When $x = 1$, $y = 0$, thus: $\dfrac{4}{3}\ln 1 + \ln 2 = \ln 1 + c$
from which, $c = \ln 2$

Hence, the particular solution is:
$$\frac{4}{3}\ln\left(\frac{x+3y}{x}\right) + \ln\left(\frac{2x-y}{x}\right) = \ln + \ln 2$$

i.e. $\ln\left(\dfrac{x+3y}{x}\right)^{\frac{4}{3}}\left(\dfrac{2x-y}{x}\right) = \ln(2x)$
from the laws of logarithms

i.e. $\left(\dfrac{x+3y}{x}\right)^{\frac{4}{3}}\left(\dfrac{2x-y}{x}\right) = 2x$

Problem 4. Show that the solution of the differential equation: $x^2 - 3y^2 + 2xy\dfrac{dy}{dx} = 0$ is: $y = x\sqrt{(8x+1)}$, given that $y = 3$ when $x = 1$

Using the procedure of Section 50.2:

(i) Rearranging gives:

$$2xy\frac{dy}{dx} = 3y^2 - x^2 \quad \text{and} \quad \frac{dy}{dx} = \frac{3y^2 - x^2}{2xy}$$

(ii) Let $y = vx$ then $\dfrac{dy}{dx} = v + x\dfrac{dv}{dx}$

(iii) Substituting for y and $\dfrac{dy}{dx}$ gives:

$$v + x\frac{dv}{dx} = \frac{3(vx)^2 - x^2}{2x(vx)} = \frac{3v^2 - 1}{2v}$$

(iv) Separating the variables gives:

$$x\frac{dv}{dx} = \frac{3v^2 - 1}{2v} - v = \frac{3v^2 - 1 - 2v^2}{2v} = \frac{v^2 - 1}{2v}$$

Hence, $\dfrac{2v}{v^2 - 1}\,dv = \dfrac{1}{x}\,dx$

Integrating both sides gives:

$$\int \frac{2v}{v^2 - 1}\,dv = \int \frac{1}{x}\,dx$$

i.e. $\ln(v^2 - 1) = \ln x + c$

(v) Replacing v by $\dfrac{y}{x}$ gives:

$$\ln\left(\frac{y^2}{x^2} - 1\right) = \ln x + c,$$

which is the general solution.

When $y = 3, x = 1$, thus: $\ln\left(\dfrac{9}{1} - 1\right) = \ln 1 + c$ from which, $c = \ln 8$

Hence, the particular solution is:

$$\ln\left(\frac{y^2}{x^2} - 1\right) = \ln x + \ln 8 = \ln 8x$$

by the laws of logarithms

Hence, $\left(\dfrac{y^2}{x^2} - 1\right) = 8x$ i.e. $\dfrac{y^2}{x^2} = 8x + 1$ and

$$y^2 = x^2(8x + 1)$$

i.e. $\qquad\qquad \boldsymbol{y = x\sqrt{(8x+1)}}$

Now try the following Practice Exercise

Practice Exercise 197   Homogeneous first-order differential equations (Answers on page 885)

1. Solve the differential equation: $xy^3\,dy = (x^4 + y^4)dx$

2. Solve: $(9xy - 11xy)\dfrac{dy}{dx} = 11y^2 - 16xy + 3x^2$

3. Solve the differential equation: $2x\dfrac{dy}{dx} = x + 3y$, given that when $x = 1$, $y = 1$

4. Show that the solution of the differential equation: $2xy\dfrac{dy}{dx} = x^2 + y^2$ can be expressed as: $x = K(x^2 - y^2)$, where K is a constant.

5. Determine the particular solution of $\dfrac{dy}{dx} = \dfrac{x^3 + y^3}{xy^2}$, given that $x = 1$ when $y = 4$

6. Show that the solution of the differential equation $\dfrac{dy}{dx} = \dfrac{y^3 - xy^2 - x^2y - 5x^3}{xy^2 - x^2y - 2x^3}$ is of the form: $\dfrac{y^2}{2x^2} + \dfrac{4y}{x} + 18\ln\left(\dfrac{y - 5x}{x}\right) = \ln x + 42$, when $x = 1$ and $y = 6$

For fully worked solutions to each of the problems in Practice Exercises 196 and 197 in this chapter, go to the website:
www.routledge.com/cw/bird

Chapter 51

Linear first-order differential equations

Why it is important to understand: Linear first-order differential equations

As has been stated in previous chapters, differential equations have many applications in engineering and science. For example, first-order differential equations model phenomena of cooling, population growth, radioactive decay, mixture of salt solutions, series circuits, survivability with AIDS, draining a tank, economics and finance, drug distribution, pursuit problem and harvesting of renewable natural resources. This chapter explains how to solve another specific type of equation, the linear first-order differential equation.

At the end of this chapter, you should be able to:

- recognise a linear differential equation
- solve a differential equation of the form $\dfrac{dy}{dx} + Py = Q$ where P and Q are functions of x only

---

51.1 Introduction

An equation of the form $\dfrac{dy}{dx} + Py = Q$, where P and Q are functions of x only is called a **linear differential equation** since y and its derivatives are of the first degree.

(i) The solution of $\dfrac{dy}{dx} + Py = Q$ is obtained by multiplying throughout by what is termed an **integrating factor**.

(ii) Multiplying $\dfrac{dy}{dx} + Py = Q$ by say R, a function of x only, gives:

$$R\frac{dy}{dx} + RPy = RQ \qquad (1)$$

(iii) The differential coefficient of a product Ry is obtained using the product rule,

i.e. $\dfrac{d}{dx}(Ry) = R\dfrac{dy}{dx} + y\dfrac{dR}{dx}$

which is the same as the left-hand side of equation (1), when R is chosen such that

$$RP = \frac{dR}{dx}$$

(iv) If $\dfrac{dR}{dx} = RP$, then separating the variables gives $\dfrac{dR}{R} = P\,dx$

Integrating both sides gives:

$$\int \frac{dR}{R} = \int P\, dx \text{ i.e. } \ln R = \int P\, dx + c$$

from which,

$$R = e^{\int P\, dx + c} = e^{\int P\, dx} e^c$$

i.e. $R = Ae^{\int P\, dx}$, where $A = e^c = $ a constant.

(v) Substituting $R = Ae^{\int P\, dx}$ in equation (1) gives:

$$Ae^{\int P\, dx}\left(\frac{dy}{dx}\right) + Ae^{\int P\, dx} Py = Ae^{\int P\, dx} Q$$

i.e. $e^{\int P\, dx}\left(\dfrac{dy}{dx}\right) + e^{\int P\, dx} Py = e^{\int P\, dx} Q$ (2)

(vi) The left-hand side of equation (2) is

$$\frac{d}{dx}\left(ye^{\int P\, dx}\right)$$

which may be checked by differentiating $ye^{\int P\, dx}$ with respect to x, using the product rule.

(vii) From equation (2),

$$\frac{d}{dx}\left(ye^{\int P\, dx}\right) = e^{\int P\, dx} Q$$

Integrating both sides gives:

$$ye^{\int P\, dx} = \int e^{\int P\, dx} Q\, dx \qquad (3)$$

(viii) $e^{\int P\, dx}$ is the **integrating factor**.

51.2 Procedure to solve differential equations of the form $\dfrac{dy}{dx} + Py = Q$

(i) Rearrange the differential equation into the form $\dfrac{dy}{dx} + Py = Q$, where P and Q are functions of x

(ii) Determine $\int P\, dx$

(iii) Determine the integrating factor $e^{\int P\, dx}$

(iv) Substitute $e^{\int P\, dx}$ into equation (3)

(v) Integrate the right-hand side of equation (3) to give the general solution of the differential equation. Given boundary conditions, the particular solution may be determined.

51.3 Worked problems on linear first-order differential equations

Problem 1. Solve $\dfrac{1}{x}\dfrac{dy}{dx} + 4y = 2$ given the boundary conditions $x = 0$ when $y = 4$

Using the above procedure:

(i) Rearranging gives $\dfrac{dy}{dx} + 4xy = 2x$, which is of the form $\dfrac{dy}{dx} + Py = Q$ where $P = 4x$ and $Q = 2x$

(ii) $\int P\, dx = \int 4x\, dx = 2x^2$

(iii) Integrating factor $e^{\int P\, dx} = e^{2x^2}$

(iv) Substituting into equation (3) gives:

$$ye^{2x^2} = \int e^{2x^2}(2x)\, dx$$

(v) Hence the general solution is:

$$ye^{2x^2} = \tfrac{1}{2}e^{2x^2} + c,$$

by using the substitution $u = 2x^2$ When $x = 0$, $y = 4$, thus $4e^0 = \tfrac{1}{2}e^0 + c$, from which, $c = \tfrac{7}{2}$

Hence the particular solution is

$$ye^{2x^2} = \tfrac{1}{2}e^{2x^2} + \tfrac{7}{2}$$

or $y = \tfrac{1}{2} + \tfrac{7}{2}e^{-2x^2}$ or $y = \tfrac{1}{2}\left(1 + 7e^{-2x^2}\right)$

Problem 2. Show that the solution of the equation $\dfrac{dy}{dx} + 1 = -\dfrac{y}{x}$ is given by $y = \dfrac{3 - x^2}{2x}$, given $x = 1$ when $y = 1$

Using the procedure of Section 51.2:

(i) Rearranging gives: $\dfrac{dy}{dx} + \left(\dfrac{1}{x}\right)y = -1$, which is of the form $\dfrac{dy}{dx} + Py = Q$, where $P = \dfrac{1}{x}$ and $Q = -1$. (Note that Q can be considered to be $-1x^0$, i.e. a function of x)

(ii) $\int P\, dx = \int \dfrac{1}{x}\, dx = \ln x$

(iii) Integrating factor $e^{\int P\, dx} = e^{\ln x} = x$ (from the definition of logarithm).

(iv) Substituting into equation (3) gives:

$$yx = \int x(-1)\,dx$$

(v) Hence the general solution is:

$$yx = \frac{-x^2}{2} + c$$

When $x=1$, $y=1$, thus $1 = \frac{-1}{2} + c$, from which, $c = \frac{3}{2}$

Hence the particular solution is:

$$yx = \frac{-x^2}{2} + \frac{3}{2}$$

i.e. $2yx = 3 - x^2$ and $\boldsymbol{y = \dfrac{3 - x^2}{2x}}$

Problem 3. Determine the particular solution of $\dfrac{dy}{dx} - x + y = 0$, given that $x=0$ when $y=2$

Using the procedure of Section 51.2:

(i) Rearranging gives $\dfrac{dy}{dx} + y = x$, which is of the form $\dfrac{dy}{dx} + P = Q$, where $P=1$ and $Q=x$. (In this case P can be considered to be $1x^0$, i.e. a function of x).

(ii) $\int P\,dx = \int 1\,dx = x$

(iii) Integrating factor $e^{\int P\,dx} = e^x$

(iv) Substituting in equation (3) gives:

$$ye^x = \int e^x(x)\,dx \qquad (4)$$

(v) $\int e^x(x)\,dx$ is determined using integration by parts (see Chapter 46).

$$\int xe^x\,dx = xe^x - e^x + c$$

Hence from equation (4): $ye^x = xe^x - e^x + c$, which is the general solution.

When $x=0$, $y=2$ thus $2e^0 = 0 - e^0 + c$, from which, $c=3$

Hence the particular solution is:

$$\boldsymbol{ye^x = xe^x - e^x + 3} \text{ or } \boldsymbol{y = x - 1 + 3e^{-x}}$$

Now try the following Practice Exercise

Practice Exercise 198   Linear first-order differential equations (Answers on page 885)

Solve the following differential equations.

1. $x\dfrac{dy}{dx} = 3 - y$

2. $\dfrac{dy}{dx} = x(1 - 2y)$

3. $t\dfrac{dy}{dt} - 5t = -y$

4. $x\left(\dfrac{dy}{dx} + 1\right) = x^3 - 2y$, given $x=1$ when $y=3$

5. $\dfrac{1}{x}\dfrac{dy}{dx} + y = 1$

6. $\dfrac{dy}{dx} + x = 2y$

51.4 Further worked problems on linear first-order differential equations

---

Problem 4. Solve the differential equation $\dfrac{dy}{d\theta} = \sec\theta + y\tan\theta$ given the boundary conditions $y=1$ when $\theta=0$

Using the procedure of Section 51.2:

(i) Rearranging gives $\dfrac{dy}{d\theta} - (\tan\theta)y = \sec\theta$, which is of the form $\dfrac{dy}{d\theta} + Py = Q$ where $P = -\tan\theta$ and $Q = \sec\theta$

(ii) $\int P\,d\theta = \int -\tan\theta\,d\theta = -\ln(\sec\theta)$
$$= \ln(\sec\theta)^{-1} = \ln(\cos\theta)$$

(iii) Integrating factor $e^{\int P\,d\theta} = e^{\ln(\cos\theta)} = \cos\theta$ (from the definition of a logarithm).

(iv) Substituting in equation (3) gives:

$$y\cos\theta = \int \cos\theta(\sec\theta)\,d\theta$$

i.e. $y\cos\theta = \int d\theta$

(v) Integrating gives: $y\cos\theta=\theta+c$, which is the general solution. When $\theta=0$, $y=1$, thus $1\cos 0=0+c$, from which, $c=1$

Hence the particular solution is:

$$y\cos\theta = \theta + 1 \text{ or } y = (\theta+1)\sec\theta$$

Problem 5.

(a) Find the general solution of the equation

$$(x-2)\frac{dy}{dx}+\frac{3(x-1)}{(x+1)}y=1$$

(b) Given the boundary conditions that $y=5$ when $x=-1$, find the particular solution of the equation given in (a).

(a) Using the procedure of Section 51.2:

(i) Rearranging gives:

$$\frac{dy}{dx}+\frac{3(x-1)}{(x+1)(x-2)}y=\frac{1}{(x-2)}$$

which is of the form

$$\frac{dy}{dx}+Py=Q, \text{ where } P=\frac{3(x-1)}{(x+1)(x-2)}$$

and $Q=\dfrac{1}{(x-2)}$

(ii) $\displaystyle\int P\,dx=\int\frac{3(x-1)}{(x+1)(x-2)}\,dx$, which is integrated using partial fractions.

Let $\dfrac{3x-3}{(x+1)(x-2)}$

$$\equiv\frac{A}{(x+1)}+\frac{B}{(x-2)}$$

$$\equiv\frac{A(x-2)+B(x+1)}{(x+1)(x-2)}$$

from which, $3x-3=A(x-2)+B(x+1)$
When $x=-1$,
$$-6=-3A, \text{ from which, } A=2$$

When $x=2$,

$$3=3B, \text{ from which, } B=1$$

Hence $\displaystyle\int\frac{3x-3}{(x+1)(x-2)}\,dx$

$$=\int\left[\frac{2}{x+1}+\frac{1}{x-2}\right]dx$$

$$=2\ln(x+1)+\ln(x-2)$$

$$=\ln[(x+1)^2(x-2)]$$

(iii) Integrating factor

$$e^{\int P\,dx}=e^{\ln[(x+1)^2(x-2)]}=(x+1)^2(x-2)$$

(iv) Substituting in equation (3) gives:

$$y(x+1)^2(x-2)$$

$$=\int(x+1)^2(x-2)\frac{1}{x-2}\,dx$$

$$=\int(x+1)^2\,dx$$

(v) **Hence the general solution is:**

$$y(x+1)^2(x-2)=\tfrac{1}{3}(x+1)^3+c$$

(b) When $x=-1$, $y=5$ thus $5(0)(-3)=0+c$, from which, $c=0$

Hence $y(x+1)^2(x-2)=\tfrac{1}{3}(x+1)^3$

i.e. $y=\dfrac{(x+1)^3}{3(x+1)^2(x-2)}$

and hence **the particular solution is**

$$y=\frac{(x+1)}{3(x-2)}$$

Now try the following Practice Exercise

Practice Exercise 199   Linear first-order differential equations (Answers on page 885)

In problems 1 and 2, solve the differential equations.

1. $\cot x\dfrac{dy}{dx}=1-2y$, given $y=1$ when $x=\dfrac{\pi}{4}$

2. $t\dfrac{d\theta}{dt}+\sec t(t\sin t+\cos t)\theta=\sec t$, given $t=\pi$ when $\theta=1$

3. Given the equation $x\dfrac{dy}{dx}=\dfrac{2}{x+2}-y$ show that the particular solution is $y=\dfrac{2}{x}\ln(x+2)$, given the boundary conditions that $x=-1$ when $y=0$

4. Show that the solution of the differential equation

$$\frac{dy}{dx}-2(x+1)^3=\frac{4}{(x+1)}y$$

is $y = (x+1)^4 \ln(x+1)^2$, given that $x=0$ when $y=0$

5. Show that the solution of the differential equation

$$\frac{dy}{dx} + ky = a \sin bx$$

is given by:

$$y = \left(\frac{a}{k^2+b^2}\right)(k \sin bx - b \cos bx)$$

$$+ \left(\frac{k^2+b^2+ab}{k^2+b^2}\right)e^{-kx},$$

given $y=1$ when $x=0$

6. The equation $\dfrac{dv}{dt} = -(av+bt)$, where a and b are constants, represents an equation of motion when a particle moves in a resisting medium. Solve the equation for v given that $v=u$ when $t=0$

7. In an alternating current circuit containing resistance R and inductance L the current i is given by: $Ri + L\dfrac{di}{dt} = E_0 \sin \omega t$. Given $i=0$ when $t=0$, show that the solution of the equation is given by:

$$i = \left(\frac{E_0}{R^2+\omega^2 L^2}\right)(R \sin \omega t - \omega L \cos \omega t)$$

$$+ \left(\frac{E_0 \omega L}{R^2+\omega^2 L^2}\right)e^{-Rt/L}$$

8. The concentration C of impurities of an oil purifier varies with time t and is described by the equation $a\dfrac{dC}{dt} = b + dm - Cm$, where a, b, d and m are constants. Given $C=c_0$ when $t=0$, solve the equation and show that:

$$C = \left(\frac{b}{m} + d\right)\left(1 - e^{-mt/\alpha}\right) + c_0 e^{-mt/\alpha}$$

9. The equation of motion of a train is given by: $m\dfrac{dv}{dt} = mk(1 - e^{-t}) - mcv$, where v is the speed, t is the time and m, k and c are constants. Determine the speed v given $v=0$ at $t=0$.

For fully worked solutions to each of the problems in Practice Exercises 198 and 199 in this chapter,
go to the website:
www.routledge.com/cw/bird

Chapter 52

Numerical methods for first-order differential equations

Why it is important to understand: Numerical methods for first-order differential equations

Most physical systems can be described in mathematical terms through differential equations. Specific types of differential equation have been solved in the preceding chapters, i.e. the separable-variable type, the homogeneous type and the linear type. However, differential equations such as those used to solve real-life problems may not necessarily be directly solvable, i.e. do not have closed form solutions. Instead, solutions can be approximated using numerical methods and in science and engineering, a numeric approximation to the solution is often good enough to solve a problem. Various numerical methods are explained in this chapter.

At the end of this chapter, you should be able to:

- state the reason for solving differential equations using numerical methods
- obtain a numerical solution to a first-order differential equation using Euler's method
- obtain a numerical solution to a first-order differential equation using the Euler–Cauchy method
- obtain a numerical solution to a first-order differential equation using the Runge–Kutta method

---

52.1 Introduction

Not all first-order differential equations may be solved by separating the variables (as in Chapter 33) or by the integrating factor method (as in Chapter 51). A number of other analytical methods of solving differential equations exist. However the differential equations that can be solved by such analytical methods is fairly restricted. Where a differential equation and known boundary conditions are given, an approximate solution may be obtained by applying a **numerical method**. There are a number of such numerical methods available and the simplest of these is called **Euler's**[*] **method**.

52.2 Euler's method

From Chapter 8, Maclaurin's series may be stated as:

$$f(x) = f(0) + x f'(0) + \frac{x^2}{2!} f''(0) + \cdots$$

Hence at some point $f(h)$ in Fig. 52.1:

$$f(h) = f(0) + h f'(0) + \frac{h^2}{2!} f''(0) + \cdots$$

If the y-axis and origin are moved a units to the left, as shown in Fig. 52.2, the equation of the same curve relative to the new axis becomes $y = f(a + x)$ and the function value at P is $f(a)$.

At point Q in Fig. 52.2:

$$f(a + h) = f(a) + h f'(a) + \frac{h^2}{2!} f''(a) + \cdots \quad (1)$$

Figure 52.1

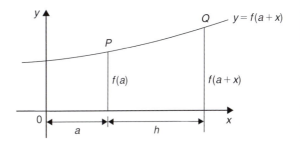

Figure 52.2

which is a statement called **Taylor's**[*] **series**.

[*] **Who was Euler?** **Leonhard Euler** (15 April 1707–18 September 1783) was a pioneering Swiss mathematician and physicist who made important discoveries in infinitesimal calculus and graph theory. He also introduced much of the modern mathematical terminology and notation. To find out more go to **www.routledge.com/cw/bird**

[*] **Who was Taylor?** **Brook Taylor** (18 August 1685–29 December 1731) solved the problem of the centre of oscillation of a body. In 1712 Taylor was elected to the Royal Society. Taylor added a new branch to mathematics now called the 'calculus of finite differences', invented integration by parts, and discovered the celebrated series known as Taylor's expansion. To find out more go to **www.routledge.com/cw/bird**

If h is the interval between two new ordinates y_0 and y_1, as shown in Fig. 52.3, and if $f(a) = y_0$ and $y_1 = f(a+h)$, then Euler's method states:

$$f(a+h) = f(a) + hf'(a)$$

i.e. $\qquad y_1 = y_0 + h(y')_0 \qquad\qquad (2)$

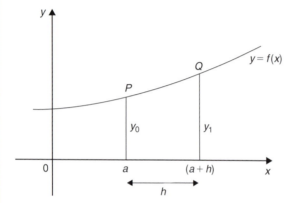

Figure 52.3

The approximation used with Euler's method is to take only the first two terms of Taylor's series shown in equation (1).

Hence if y_0, h and $(y')_0$ are known, y_1, which is an approximate value for the function at Q in Fig. 52.3, can be calculated.

Euler's method is demonstrated in the worked problems following.

52.3 Worked problems on Euler's method

Problem 1. Obtain a numerical solution of the differential equation

$$\frac{dy}{dx} = 3(1+x) - y$$

given the initial conditions that $x = 1$ when $y = 4$, for the range $x = 1.0$ to $x = 2.0$ with intervals of 0.2

Draw the graph of the solution.

$$\frac{dy}{dx} = y' = 3(1+x) - y$$

With $x_0 = 1$ and $y_0 = 4$, $(y')_0 = 3(1+1) - 4 = \mathbf{2}$

By Euler's method:

$$y_1 = y_0 + h(y')_0, \text{ from equation (2)}$$

Hence $\quad y_1 = 4 + (0.2)(2) = \mathbf{4.4}$, since $h = 0.2$

At point Q in Fig. 52.4, $x_1 = 1.2$, $y_1 = 4.4$

and $(y')_1 = 3(1 + x_1) - y_1$

i.e. $(y')_1 = 3(1 + 1.2) - 4.4 = \mathbf{2.2}$

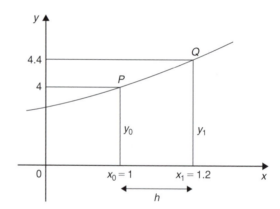

Figure 52.4

If the values of x, y and y' found for point Q are regarded as new starting values of x_0, y_0 and $(y')_0$, the above process can be repeated and values found for the point R shown in Fig. 52.5.

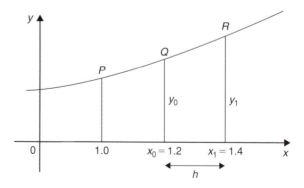

Figure 52.5

Thus at point R,

$$y_1 = y_0 + h(y')_0 \quad \text{from equation (2)}$$

$$= 4.4 + (0.2)(2.2) = \mathbf{4.84}$$

When $x_1 = 1.4$ and $y_1 = 4.84$,

$$(y')_1 = 3(1 + 1.4) - 4.84 = \mathbf{2.36}$$

This step-by-step Euler's method can be continued and it is easiest to list the results in a table, as shown in Table 52.1. The results for lines 1 to 3 have been produced above.

Table 52.1

	x_0	y_0	$(y')_0$
1.	1	4	2
2.	1.2	4.4	2.2
3.	1.4	4.84	2.36
4.	1.6	5.312	2.488
5.	1.8	5.8096	2.5904
6.	2.0	6.32768	

For line 4, where $x_0 = 1.6$:

$$y_1 = y_0 + h(y')_0$$

$$= 4.84 + (0.2)(2.36) = \mathbf{5.312}$$

and $(y')_0 = 3(1 + 1.6) - 5.312 = \mathbf{2.488}$

For line 5, where $x_0 = 1.8$:

$$y_1 = y_0 + h(y')_0$$

$$= 5.312 + (0.2)(2.488) = \mathbf{5.8096}$$

and $(y')_0 = 3(1 + 1.8) - 5.8096 = \mathbf{2.5904}$

For line 6, where $x_0 = 2.0$:

$$y_1 = y_0 + h(y')_0$$

$$= 5.8096 + (0.2)(2.5904)$$

$$= \mathbf{6.32768}$$

(As the range is 1.0 to 2.0 there is no need to calculate $(y')_0$ in line 6.) The particular solution is given by the value of y against x.

A graph of the solution of $\dfrac{dy}{dx} = 3(1 + x) - y$ with initial conditions $x = 1$ and $y = 4$ is shown in Fig. 52.6. In practice it is probably best to plot the graph as each calculation is made, which checks that there is a smooth progression and that no calculation errors have occurred.

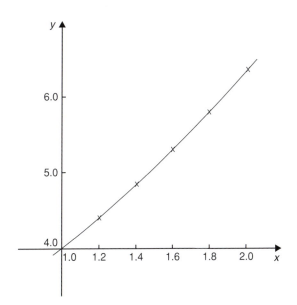

Figure 52.6

Problem 2. Use Euler's method to obtain a numerical solution of the differential equation $\dfrac{dy}{dx} + y = 2x$, given the initial conditions that at $x = 0$, $y = 1$, for the range $x = 0(0.2)1.0$. Draw the graph of the solution in this range.

$x = 0(0.2)1.0$ means that x ranges from 0 to 1.0 in equal intervals of 0.2 (i.e. $h = 0.2$ in Euler's method).

$$\frac{dy}{dx} + y = 2x,$$

hence $\dfrac{dy}{dx} = 2x - y$, i.e. $y' = 2x - y$

If initially $x_0 = 0$ and $y_0 = 1$, then

$$(y')_0 = 2(0) - 1 = -1$$

Hence line 1 in Table 52.2 can be completed with $x = 0$, $y = 1$ and $y'(0) = -1$.

Table 52.2

	x_0	y_0	$(y')_0$
1.	0	1	-1
2.	0.2	0.8	-0.4
3.	0.4	0.72	0.08
4.	0.6	0.736	0.464
5.	0.8	0.8288	0.7712
6.	1.0	0.98304	

For line 2, where $x_0 = 0.2$ and $h = 0.2$:

$$y_1 = y_0 + h(y'), \quad \text{from equation (2)}$$

$$= 1 + (0.2)(-1) = \mathbf{0.8}$$

and $(y')_0 = 2x_0 - y_0 = 2(0.2) - 0.8 = \mathbf{-0.4}$

For line 3, where $x_0 = 0.4$:

$$y_1 = y_0 + h(y')_0$$

$$= 0.8 + (0.2)(-0.4) = \mathbf{0.72}$$

and $(y')_0 = 2x_0 - y_0 = 2(0.4) - 0.72 = \mathbf{0.08}$

For line 4, where $x_0 = 0.6$:

$$y_1 = y_0 + h(y')_0$$

$$= 0.72 + (0.2)(0.08) = \mathbf{0.736}$$

and $(y')_0 = 2x_0 - y_0 = 2(0.6) - 0.736 = \mathbf{0.464}$

For line 5, where $x_0 = 0.8$:

$$y_1 = y_0 + h(y')_0$$

$$= 0.736 + (0.2)(0.464) = \mathbf{0.8288}$$

and $(y')_0 = 2x_0 - y_0 = 2(0.8) - 0.8288 = \mathbf{0.7712}$

For line 6, where $x_0 = 1.0$:

$$y_1 = y_0 + h(y')_0$$

$$= 0.8288 + (0.2)(0.7712) = \mathbf{0.98304}$$

As the range is 0 to 1.0, $(y')_0$ in line 6 is not needed. A graph of the solution of $\dfrac{dy}{dx} + y = 2x$, with initial conditions $x = 0$ and $y = 1$ is shown in Fig. 52.7.

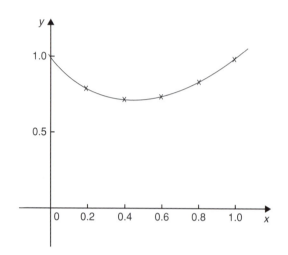

Figure 52.7

Problem 3.

(a) Obtain a numerical solution, using Euler's method, of the differential equation $\dfrac{dy}{dx} = y - x$, with the initial conditions that at $x = 0$, $y = 2$, for the range $x = 0(0.1)0.5$. Draw the graph of the solution.

(b) By an analytical method (using the integrating factor method of Chapter 51), the solution of the above differential equation is given by $y = x + 1 + e^x$

Determine the percentage error at $x = 0.3$

(a) $\dfrac{dy}{dx} = y' = y - x$

If initially $x_0 = 0$ and $y_0 = 2$
then $(y')_0 = y_0 - x_0 = 2 - 0 = 2$
Hence line 1 of Table 52.3 is completed.

For line 2, where $x_0 = 0.1$:

$$y_1 = y_0 + h(y')_0, \quad \text{from equation (2)},$$

$$= 2 + (0.1)(2) = \mathbf{2.2}$$

Table 52.3

	x_0	y_0	$(y')_0$
1.	0	2	2
2.	0.1	2.2	2.1
3.	0.2	2.41	2.21
4.	0.3	2.631	2.331
5.	0.4	2.8641	2.4641
6.	0.5	3.11051	

and $(y')_0 = y_0 - x_0$

$$= 2.2 - 0.1 = \mathbf{2.1}$$

For line 3, where $x_0 = 0.2$:

$$\mathbf{y_1} = y_0 + h(y')_0$$

$$= 2.2 + (0.1)(2.1) = \mathbf{2.41}$$

and $(y')_0 = y_0 - x_0 = 2.41 - 0.2 = \mathbf{2.21}$

For line 4, where $x_0 = 0.3$:

$$\mathbf{y_1} = y_0 + h(y')_0$$

$$= 2.41 + (0.1)(2.21) = \mathbf{2.631}$$

and $(y')_0 = y_0 - x_0$

$$= 2.631 - 0.3 = \mathbf{2.331}$$

For line 5, where $x_0 = 0.4$:

$$\mathbf{y_1} = y_0 + h(y')_0$$

$$= 2.631 + (0.1)(2.331) = \mathbf{2.8641}$$

and $(y')_0 = y_0 - x_0$

$$= 2.8641 - 0.4 = \mathbf{2.4641}$$

For line 6, where $x_0 = 0.5$:

$$\mathbf{y_1} = y_0 + h(y')_0$$

$$= 2.8641 + (0.1)(2.4641) = \mathbf{3.11051}$$

A graph of the solution of $\dfrac{dy}{dx} = y - x$ with $x = 0$, $y = 2$ is shown in Fig. 52.8.

(b) If the solution of the differential equation $\dfrac{dy}{dx} = y - x$ is given by $y = x + 1 + e^x$, then when $x = 0.3$, $y = 0.3 + 1 + e^{0.3} = 2.649859$

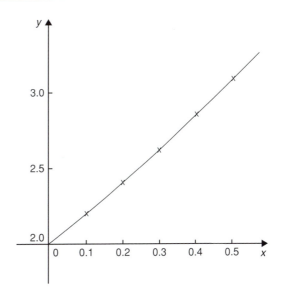

Figure 52.8

By Euler's method, when $x = 0.3$ (i.e. line 4 in Table 52.3), $y = 2.631$

Percentage error

$$= \left(\frac{\text{actual} - \text{estimated}}{\text{actual}} \right) \times 100\%$$

$$= \left(\frac{2.649859 - 2.631}{2.649859} \right) \times 100\%$$

$$= \mathbf{0.712\%}$$

Euler's method of numerical solution of differential equations is simple, but approximate. The method is most useful when the interval h is small.

Now try the following Practice Exercise

Practice Exercise 200   Eulers method (Answers on page 885)

1. Use Euler's method to obtain a numerical solution of the differential equation $\dfrac{dy}{dx} = 3 - \dfrac{y}{x}$, with the initial conditions that $x = 1$ when $y = 2$, for the range $x = 1.0$ to $x = 1.5$ with intervals of 0.1. Draw the graph of the solution in this range.

2. Obtain a numerical solution of the differential equation $\dfrac{1}{x}\dfrac{dy}{dx} + 2y = 1$, given the initial conditions that $x = 0$ when $y = 1$, in the range $x = 0(0.2)1.0$

3. (a) The differential equation $\dfrac{dy}{dx} + 1 = -\dfrac{y}{x}$ has the initial conditions that $y = 1$ at $x = 2$. Produce a numerical solution of the differential equation in the range $x = 2.0(0.1)2.5$

   (b) If the solution of the differential equation by an analytical method is given by $y = \dfrac{4}{x} - \dfrac{x}{2}$, determine the percentage error at $x = 2.2$

4. Use Euler's method to obtain a numerical solution of the differential equation $\dfrac{dy}{dx} = x - \dfrac{2y}{x}$, given the initial conditions that $y = 1$ when $x = 2$, in the range $x = 2.0(0.2)3.0$.

   If the solution of the differential equation is given by $y = \dfrac{x^2}{4}$, determine the percentage error by using Euler's method when $x = 2.8$

52.4 The Euler–Cauchy method

In Euler's method of Section 52.2, the gradient $(y')_0$ at $P_{(x_0, y_0)}$ in Fig. 52.9 across the whole interval h is used

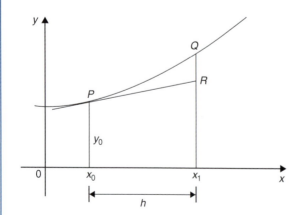

Figure 52.9

to obtain an approximate value of y_1 at point Q. QR in Fig. 52.9 is the resulting error in the result.

In an improved Euler method, called the **Euler–Cauchy* method**, the gradient at $P_{(x_0, y_0)}$ across half the interval is used and then continues with a line whose gradient approximates to the gradient of the curve at x_1, shown in Fig. 52.10.

Let y_{P_1} be the predicted value at point R using Euler's method, i.e. length RZ, where

$$y_{P_1} = y_0 + h(y')_0 \tag{3}$$

The error shown as QT in Fig. 52.10 is now less than the error QR used in the basic Euler method and the calculated results will be of greater accuracy. The corrected value, y_{C_1} in the improved Euler method is given by:

$$y_{C_1} = y_0 + \tfrac{1}{2}h[(y')_0 + f(x_1, y_{P_1})] \tag{4}$$

* **Who was Cauchy**? **Baron Augustin-Louis Cauchy** (21 August 1789–23 May 1857) was a French mathematician who became an early pioneer of analysis. A prolific writer; he wrote approximately 800 research articles. To find out more go to **www.routledge.com/cw/bird**
* **Who was Euler**? For image and resumé of Leonhard Euler, see page 543. To find out more go to **www.routledge.com/cw/bird**

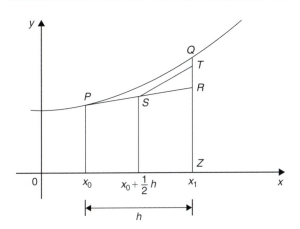

Figure 52.10

The following worked problems demonstrate how equations (3) and (4) are used in the Euler–Cauchy method.

Problem 4. Apply the Euler–Cauchy method to solve the differential equation

$$\frac{dy}{dx} = y - x$$

in the range 0(0.1)0.5, given the initial conditions that at $x = 0$, $y = 2$

$$\frac{dy}{dx} = y' = y - x$$

Since the initial conditions are $x_0 = 0$ and $y_0 = 2$ then $(y')_0 = 2 - 0 = 2$. Interval $h = 0.1$, hence $x_1 = x_0 + h = 0 + 0.1 = 0.1$
From equation (3),

$$y_{P_1} = y_0 + h(y')_0 = 2 + (0.1)(2) = 2.2$$

From equation (4),

$$y_{C_1} = y_0 + \tfrac{1}{2}h[(y')_0 + f(x_1, y_{P_1})]$$
$$= y_0 + \tfrac{1}{2}h[(y')_0 + (y_{P_1} - x_1)],$$

in this case

$$= 2 + \tfrac{1}{2}(0.1)[2 + (2.2 - 0.1)] = \mathbf{2.205}$$

$$(y')_1 = y_{C_1} - x_1 = 2.205 - 0.1 = 2.105$$

If we produce a table of values, as in Euler's method, we have so far determined lines 1 and 2 of Table 52.4. The results in line 2 are now taken as x_0, y_0 and $(y')_0$ for the next interval and the process is repeated.

Table 52.4

	x	y	y'
1.	0	2	2
2.	0.1	2.205	2.105
3.	0.2	2.421025	2.221025
4.	0.3	2.649232625	2.349232625
5.	0.4	2.89090205	2.49090205
6.	0.5	3.147446765	

For line 3, $x_1 = 0.2$

$$y_{P_1} = y_0 + h(y')_0 = 2.205 + (0.1)(2.105)$$
$$= 2.4155$$

$$y_{C_1} = y_0 + \tfrac{1}{2}h[(y')_0 + f(x_1, y_{P_1})]$$
$$= 2.205 + \tfrac{1}{2}(0.1)[2.105 + (2.4155 - 0.2)]$$
$$= \mathbf{2.421025}$$

$$(y')_0 = y_{C_1} - x_1 = 2.421025 - 0.2 = 2.221025$$

For line 4, $x_1 = 0.3$

$$y_{P_1} = y_0 + h(y')_0$$
$$= 2.421025 + (0.1)(2.221025)$$
$$= 2.6431275$$

$$y_{C_1} = y_0 + \tfrac{1}{2}h[(y')_0 + f(x_1, y_{P_1})]$$
$$= 2.421025 + \tfrac{1}{2}(0.1)[2.221025$$
$$+ (2.6431275 - 0.3)]$$
$$= \mathbf{2.649232625}$$

$$(y')_0 = y_{C_1} - x_1 = 2.649232625 - 0.3$$
$$= 2.349232625$$

For line 5, $x_1 = 0.4$

$$y_{P_1} = y_0 + h(y')_0$$
$$= 2.649232625 + (0.1)(2.349232625)$$
$$= 2.884155887$$

$$y_{C_1} = y_0 + \tfrac{1}{2} h[(y')_0 + f(x_1, y_{P_1})]$$

$$= 2.649232625 + \tfrac{1}{2}(0.1)[2.349232625$$

$$+ (2.884155887 - 0.4)]$$

$$= \mathbf{2.89090205}$$

$$(y')_0 = y_{C_1} - x_1 = 2.89090205 - 0.4$$

$$= 2.49090205$$

For line 6, $x_1 = 0.5$

$$y_{P_1} = y_0 + h(y')_0$$

$$= 2.89090205 + (0.1)(2.49090205)$$

$$= 3.139992255$$

$$y_{C_1} = y_0 + \tfrac{1}{2} h[(y')_0 + f(x_1, y_{P_1})]$$

$$= 2.89090205 + \tfrac{1}{2}(0.1)[2.49090205$$

$$+ (3.139992255 - 0.5)]$$

$$= \mathbf{3.147446765}$$

Problem 4 is the same example as Problem 3 and Table 52.5 shows a comparison of the results, i.e. it compares the results of Tables 52.3 and 52.4. $\dfrac{dy}{dx} = y - x$ may be solved analytically by the integrating factor method of Chapter 51 with the solution $y = x + 1 + e^x$. Substituting values of x of $0, 0.1, 0.2, \ldots$ give the exact values shown in Table 52.5.

The percentage error for each method for each value of x is shown in Table 52.6. For example when $x = 0.3$,

Table 52.6

x	Error in Euler method	Error in Euler–Cauchy method
0	0	0
0.1	0.234%	0.00775%
0.2	0.472%	0.0156%
0.3	0.712%	0.0236%
0.4	0.959%	0.0319%
0.5	1.214%	0.0405%

% error with Euler method

$$= \left(\frac{\text{actual} - \text{estimated}}{\text{actual}} \right) \times 100\%$$

$$= \left(\frac{2.649858808 - 2.631}{2.649858808} \right) \times 100\%$$

$$= \mathbf{0.712\%}$$

% error with Euler–Cauchy method

$$= \left(\frac{2.649858808 - 2.649232625}{2.649858808} \right) \times 100\%$$

$$= \mathbf{0.0236\%}$$

This calculation and the others listed in Table 52.6 show the Euler–Cauchy method to be more accurate than the Euler method.

Table 52.5

	x	Euler method y	Euler–Cauchy method y	Exact value $y = x + 1 + e^x$
1.	0	2	2	2
2.	0.1	2.2	2.205	2.205170918
3.	0.2	2.41	2.421025	2.421402758
4.	0.3	2.631	2.649232625	2.649858808
5.	0.4	2.8641	2.89090205	2.891824698
6.	0.5	3.11051	3.147446765	3.148721271

Problem 5. Obtain a numerical solution of the differential equation

$$\frac{dy}{dx} = 3(1+x) - y$$

in the range 1.0(0.2)2.0, using the Euler–Cauchy method, given the initial conditions that $x = 1$ when $y = 4$

This is the same as Problem 1 on page 544, and a comparison of values may be made.

$$\frac{dy}{dx} = y' = 3(1+x) - y \quad \text{i.e. } y' = 3 + 3x - y$$

$$x_0 = 1.0, \ y_0 = 4 \text{ and } h = 0.2$$

$$(y')_0 = 3 + 3x_0 - y_0 = 3 + 3(1.0) - 4 = 2$$

$x_1 = 1.2$ and from equation (3),

$$y_{P_1} = y_0 + h(y')_0 = 4 + 0.2(2) = 4.4$$

$$y_{C_1} = y_0 + \tfrac{1}{2}h[(y')_0 + f(x_1, y_{P_1})]$$

$$= y_0 + \tfrac{1}{2}h[(y')_0 + (3 + 3x_1 - y_{P_1})]$$

$$= 4 + \tfrac{1}{2}(0.2)[2 + (3 + 3(1.2) - 4.4)]$$

$$= \mathbf{4.42}$$

$$(y')_1 = 3 + 3x_1 - y_{P_1} = 3 + 3(1.2) - 4.42 = 2.18$$

Thus the first two lines of Table 52.7 have been completed.

Table 52.7

	x_0	y_0	y_0'
1.	1.0	4	2
2.	1.2	4.42	2.18
3.	1.4	4.8724	2.3276
4.	1.6	5.351368	2.448632
5.	1.8	5.85212176	2.54787824
6.	2.0	6.370739847	

For line 3, $x_1 = 1.4$

$$y_{P_1} = y_0 + h(y')_0 = 4.42 + 0.2(2.18) = 4.856$$

$$y_{C_1} = y_0 + \tfrac{1}{2}h[(y')_0 + (3 + 3x_1 - y_{P_1})]$$

$$= 4.42 + \tfrac{1}{2}(0.2)[2.18$$

$$+ (3 + 3(1.4) - 4.856)]$$

$$= \mathbf{4.8724}$$

$$(y')_1 = 3 + 3x_1 - y_{P_1} = 3 + 3(1.4) - 4.8724$$

$$= 2.3276$$

For line 4, $x_1 = 1.6$

$$y_{P_1} = y_0 + h(y')_0 = 4.8724 + 0.2(2.3276)$$

$$= 5.33792$$

$$y_{C_1} = y_0 + \tfrac{1}{2}h[(y')_0 + (3 + 3x_1 - y_{P_1})]$$

$$= 4.8724 + \tfrac{1}{2}(0.2)[2.3276$$

$$+ (3 + 3(1.6) - 5.33792)]$$

$$= \mathbf{5.351368}$$

$$(y')_1 = 3 + 3x_1 - y_{P_1}$$

$$= 3 + 3(1.6) - 5.351368$$

$$= 2.448632$$

For line 5, $x_1 = 1.8$

$$y_{P_1} = y_0 + h(y')_0 = 5.351368 + 0.2(2.448632)$$

$$= 5.8410944$$

$$y_{C_1} = y_0 + \tfrac{1}{2}h[(y')_0 + (3 + 3x_1 - y_{P_1})]$$

$$= 5.351368 + \tfrac{1}{2}(0.2)[2.448632$$

$$+ (3 + 3(1.8) - 5.8410944)]$$

$$= \mathbf{5.85212176}$$

$$(y')_1 = 3 + 3x_1 - y_{P_1}$$

$$= 3 + 3(1.8) - 5.85212176$$

$$= 2.54787824$$

For line 6, $x_1 = 2.0$

$$y_{P_1} = y_0 + h(y')_0$$

$$= 5.85212176 + 0.2(2.54787824)$$

$$= 6.361697408$$

$$y_{C_1} = y_0 + \tfrac{1}{2}h[(y')_0 + (3 + 3x_1 - y_{P_1})]$$

$$= 5.85212176 + \tfrac{1}{2}(0.2)[2.54787824$$

$$+ (3 + 3(2.0) - 6.361697408)]$$

$$= \mathbf{6.370739843}$$

Problem 6. Using the integrating factor method the solution of the differential equation $\frac{dy}{dx} = 3(1+x) - y$ of Problem 5 is $y = 3x + e^{1-x}$. When $x = 1.6$, compare the accuracy, correct to 3 decimal places, of the Euler and the Euler–Cauchy methods.

When $x = 1.6$, $y = 3x + e^{1-x} = 3(1.6) + e^{1-1.6} = 4.8 + e^{-0.6} = 5.348811636$

From Table 52.1, page 545, by Euler's method, when $x = 1.6$, $y = 5.312$

% error in the Euler method

$$= \left(\frac{5.348811636 - 5.312}{5.348811636}\right) \times 100\%$$

$$= \mathbf{0.688\%}$$

From Table 52.7 of Problem 5, by the Euler–Cauchy method, when $x = 1.6$, $y = 5.351368$

% error in the Euler–Cauchy method

$$= \left(\frac{5.348811636 - 5.351368}{5.348811636}\right) \times 100\%$$

$$= \mathbf{-0.048\%}$$

The Euler–Cauchy method is seen to be more accurate than the Euler method when $x = 1.6$.

Now try the following Practice Exercise

Practice Exercise 201   Euler–Cauchy method (Answers on page 886)

1. Apply the Euler–Cauchy method to solve the differential equation

   $$\frac{dy}{dx} = 3 - \frac{y}{x}$$

   for the range $1.0(0.1)1.5$, given the initial conditions that $x = 1$ when $y = 2$

2. Solving the differential equation in Problem 1 by the integrating factor method gives $y = \frac{3}{2}x + \frac{1}{2x}$. Determine the percentage error, correct to 3 significant figures, when $x = 1.3$ using (a) Euler's method and (b) the Euler–Cauchy method.

3. (a) Apply the Euler-Cauchy method to solve the differential equation

   $$\frac{dy}{dx} - x = y$$

   for the range $x = 0$ to $x = 0.5$ in increments of 0.1, given the initial conditions that when $x = 0$, $y = 1$

   (b) The solution of the differential equation in part (a) is given by $y = 2e^x - x - 1$. Determine the percentage error, correct to 3 decimal places, when $x = 0.4$

4. Obtain a numerical solution of the differential equation

   $$\frac{1}{x}\frac{dy}{dx} + 2y = 1$$

   using the Euler–Cauchy method in the range $x = 0(0.2)1.0$, given the initial conditions that $x = 0$ when $y = 1$

52.5 The Runge–Kutta method

The **Runge–Kutta*** method for solving first-order differential equations is widely used and provides a high degree of accuracy. Again, as with the two previous methods, the Runge–Kutta method is a step-by-step process where results are tabulated for a range of values of x. Although several intermediate calculations are needed at each stage, the method is fairly straightforward.

The seven-step **procedure for the Runge–Kutta method**, without proof, is as follows:

To solve the differential equation $\dfrac{dy}{dx} = f(x, y)$ given the initial condition $y = y_0$ at $x = x_0$ for a range of values of $x = x_0(h)x_n$:

1. Identify x_0, y_0 and h, and values of $x_1, x_2, x_3, \ldots$.

2. Evaluate $k_1 = f(x_n, y_n)$ starting with $n = 0$

3. Evaluate $k_2 = f\left(x_n + \dfrac{h}{2}, y_n + \dfrac{h}{2}k_1\right)$

4. Evaluate $k_3 = f\left(x_n + \dfrac{h}{2}, y_n + \dfrac{h}{2}k_2\right)$

5. Evaluate $k_4 = f(x_n + h, y_n + hk_3)$

6. Use the values determined from steps 2 to 5 to evaluate:

$$y_{n+1} = y_n + \frac{h}{6}\{k_1 + 2k_2 + 2k_3 + k_4\}$$

7. Repeat steps 2 to 6 for $n = 1, 2, 3, \ldots$

* **Who was Runge**? **Carl David Tolmé Runge** (1856–1927) was a German mathematician, physicist, and spectroscopist. He was co-developer of the Runge–Kutta method in the field of numerical analysis. The Runge crater on the Moon is named after him. To find out more go to **www.routledge.com/cw/bird**

* **Who was Kutta**? **Martin Wilhelm Kutta** (3 November 1867–25 December 1944) wrote a thesis that contains the now famous Runge–Kutta method for solving ordinary differential equations. To find out more go to **www.routledge.com/cw/bird**

Thus, step 1 is given, and steps 2 to 5 are intermediate steps leading to step 6. It is usually most convenient to construct a table of values.

The Runge–Kutta method is demonstrated in the following worked problems.

> **Problem 7.** Use the Runge–Kutta method to solve the differential equation:
>
> $$\frac{dy}{dx} = y - x$$
>
> in the range $0(0.1)0.5$, given the initial conditions that at $x = 0$, $y = 2$

Using the above procedure:

1. $x_0 = 0$, $y_0 = 2$ and since $h = 0.1$, and the range is from $x = 0$ to $x = 0.5$, then $x_1 = 0.1$, $x_2 = 0.2$, $x_3 = 0.3$, $x_4 = 0.4$, and $x_5 = 0.5$

Let $n = 0$ to determine y_1:

2. $k_1 = f(x_0, y_0) = f(0, 2)$;

   since $\frac{dy}{dx} = y - x$, $f(0, 2) = 2 - 0 = \mathbf{2}$

3. $k_2 = f\left(x_0 + \frac{h}{2}, y_0 + \frac{h}{2}k_1\right)$

   $= f\left(0 + \frac{0.1}{2}, 2 + \frac{0.1}{2}(2)\right)$

   $= f(0.05, 2.1) = 2.1 - 0.05 = \mathbf{2.05}$

4. $k_3 = f\left(x_0 + \frac{h}{2}, y_0 + \frac{h}{2}k_2\right)$

   $= f\left(0 + \frac{0.1}{2}, 2 + \frac{0.1}{2}(2.05)\right)$

   $= f(0.05, 2.1025)$

   $= 2.1025 - 0.05 = \mathbf{2.0525}$

5. $k_4 = f(x_0 + h, y_0 + hk_3)$

   $= f(0 + 0.1, 2 + 0.1(2.0525))$

   $= f(0.1, 2.20525)$

   $= 2.20525 - 0.1 = \mathbf{2.10525}$

6. $y_{n+1} = y_n + \frac{h}{6}\{k_1 + 2k_2 + 2k_3 + k_4\}$ and when $n = 0$:

   $$y_1 = y_0 + \frac{h}{6}\{k_1 + 2k_2 + 2k_3 + k_4\}$$

   $$= 2 + \frac{0.1}{6}\{2 + 2(2.05) + 2(2.0525)$$

   $$+ 2.10525\}$$

   $$= 2 + \frac{0.1}{6}\{12.31025\} = \mathbf{2.205171}$$

A table of values may be constructed as shown in Table 52.8. The working has been shown for the first two rows.

Let $n = 1$ to determine y_2:

2. $k_1 = f(x_1, y_1) = f(0.1, 2.205171)$; since

   $\frac{dy}{dx} = y - x$, $f(0.1, 2.205171)$

   $= 2.205171 - 0.1 = \mathbf{2.105171}$

3. $k_2 = f\left(x_1 + \frac{h}{2}, y_1 + \frac{h}{2}k_1\right)$

   $= f\left(0.1 + \frac{0.1}{2}, 2.205171 + \frac{0.1}{2}(2.105171)\right)$

   $= f(0.15, 2.31042955)$

   $= 2.31042955 - 0.15 = \mathbf{2.160430}$

4. $k_3 = f\left(x_1 + \frac{h}{2}, y_1 + \frac{h}{2}k_2\right)$

   $= f\left(0.1 + \frac{0.1}{2}, 2.205171 + \frac{0.1}{2}(2.160430)\right)$

   $= f(0.15, 2.3131925) = 2.3131925 - 0.15$

   $= \mathbf{2.163193}$

5. $k_4 = f(x_1 + h, y_1 + hk_3)$

   $= f(0.1 + 0.1, 2.205171 + 0.1(2.163193))$

   $= f(0.2, 2.421490)$

   $= 2.421490 - 0.2 = \mathbf{2.221490}$

6. $y_{n+1} = y_n + \frac{h}{6}\{k_1 + 2k_2 + 2k_3 + k_4\}$

   and when $n = 1$:

Table 52.8

n	x_n	k_1	k_2	k_3	k_4	y_n
0	0					2
1	0.1	2.0	2.05	2.0525	2.10525	2.205171
2	0.2	2.105171	2.160430	2.163193	2.221490	2.421403
3	0.3	2.221403	2.282473	2.285527	2.349956	2.649859
4	0.4	2.349859	2.417339	2.420726	2.491932	2.891824
5	0.5	2.491824	2.566415	2.570145	2.648838	3.148720

$$y_2 = y_1 + \frac{h}{6}\{k_1 + 2k_2 + 2k_3 + k_4\}$$

$$= 2.205171 + \frac{0.1}{6}\{2.105171 + 2(2.160430)$$

$$+ 2(2.163193) + 2.221490\}$$

$$= 2.205171 + \frac{0.1}{6}\{12.973907\} = \mathbf{2.421403}$$

This completes the third row of Table 52.8. In a similar manner y_3, y_4 and y_5 can be calculated and the results are as shown in Table 52.8. Such a table is best produced by using a **spreadsheet**, such as Microsoft Excel.
This problem is the same as Problem 3, page 546 which used Euler's method, and Problem 4, page 549 which used the improved Euler's method, and a comparison of results can be made.
The differential equation $\frac{dy}{dx} = y - x$ may be solved analytically using the integrating factor method of Chapter 51, with the solution:

$$y = x + 1 + e^x$$

Substituting values of x of 0, 0.1, 0.2,..., 0.5 will give the exact values. A comparison of the results obtained by Euler's method, the Euler–Cauchy method and the Runga–Kutta method, together with the exact values is shown in Table 52.9.
It is seen from Table 52.9 that **the Runge–Kutta method is exact, correct to 5 decimal places**.

Problem 8. Obtain a numerical solution of the differential equation: $\frac{dy}{dx} = 3(1+x) - y$ in the range 1.0(0.2)2.0, using the Runge–Kutta method, given the initial conditions that $x = 1.0$ when $y = 4.0$

Using the above procedure:

1. $x_0 = 1.0$, $y_0 = 4.0$ and since $h = 0.2$, and the range is from $x = 1.0$ to $x = 2.0$, then $x_1 = 1.2$, $x_2 = 1.4$, $x_3 = 1.6$, $x_4 = 1.8$, and $x_5 = 2.0$

Let $n = 0$ to determine y_1:

2. $k_1 = f(x_0, y_0) = f(1.0, 4.0)$; since

$$\frac{dy}{dx} = 3(1+x) - y,$$

$$f(1.0, 4.0) = 3(1 + 1.0) - 4.0 = \mathbf{2.0}$$

3. $k_2 = f\left(x_0 + \frac{h}{2}, y_0 + \frac{h}{2}k_1\right)$

$$= f\left(1.0 + \frac{0.2}{2}, 4.0 + \frac{0.2}{2}(2)\right)$$

$$= f(1.1, 4.2) = 3(1 + 1.1) - 4.2 = \mathbf{2.1}$$

4. $k_3 = f\left(x_0 + \frac{h}{2}, y_0 + \frac{h}{2}k_2\right)$

$$= f\left(1.0 + \frac{0.2}{2}, 4.0 + \frac{0.2}{2}(2.1)\right)$$

$$= f(1.1, 4.21)$$

$$= 3(1 + 1.1) - 4.21 = \mathbf{2.09}$$

5. $k_4 = f(x_0 + h, y_0 + hk_3)$

$$= f(1.0 + 0.2, 4.1 + 0.2(2.09))$$

$$= f(1.2, 4.418)$$

$$= 3(1 + 1.2) - 4.418 = \mathbf{2.182}$$

Table 52.9

x	Euler's method y	Euler–Cauchy method y	Runge–Kutta method y	Exact value $y = x + 1 + e^x$
0	2	2	2	2
0.1	2.2	2.205	2.205171	2.205170918
0.2	2.41	2.421025	2.421403	2.421402758
0.3	2.631	2.649232625	2.649859	2.649858808
0.4	2.8641	2.89090205	2.891824	2.891824698
0.5	3.11051	3.147446765	3.148720	3.148721271

6. $y_{n+1} = y_n + \dfrac{h}{6}\{k_1 + 2k_2 + 2k_3 + k_4\}$ and when $n = 0$:

$$y_1 = y_0 + \frac{h}{6}\{k_1 + 2k_2 + 2k_3 + k_4\}$$

$$= 4.0 + \frac{0.2}{6}\{2.0 + 2(2.1) + 2(2.09) + 2.182\}$$

$$= 4.0 + \frac{0.2}{6}\{12.562\} = \mathbf{4.418733}$$

A table of values is compiled in Table 52.10. The working has been shown for the first two rows.

Let $n = 1$ to determine y_2:

2.
$$k_1 = f(x_1, y_1) = f(1.2, 4.418733); \text{ since}$$

$$\frac{\mathrm{d}y}{\mathrm{d}x} = 3(1 + x) - y, \quad f(1.2, 4.418733)$$

$$= 3(1 + 1.2) - 4.418733 = \mathbf{2.181267}$$

3. $k_2 = f\left(x_1 + \dfrac{h}{2}, y_1 + \dfrac{h}{2}k_1\right)$

$$= f\left(1.2 + \frac{0.2}{2}, 4.418733 + \frac{0.2}{2}(2.181267)\right)$$

$$= f(1.3, 4.636860)$$

$$= 3(1 + 1.3) - 4.636860 = \mathbf{2.263140}$$

4. $k_3 = f\left(x_1 + \dfrac{h}{2}, y_1 + \dfrac{h}{2}k_2\right)$

$$= f\left(1.2 + \frac{0.2}{2}, 4.418733 + \frac{0.2}{2}(2.263140)\right)$$

$$= f(1.3, 4.645047) = 3(1 + 1.3) - 4.645047$$

$$= \mathbf{2.254953}$$

5. $k_4 = f(x_1 + h, y_1 + hk_3)$

$$= f(1.2 + 0.2, 4.418733 + 0.2(2.254953))$$

Table 52.10

n	x_n	k_1	k_2	k_3	k_4	y_n
0	1.0					4.0
1	1.2	2.0	2.1	2.09	2.182	4.418733
2	1.4	2.181267	2.263140	2.254953	2.330276	4.870324
3	1.6	2.329676	2.396708	2.390005	2.451675	5.348817
4	1.8	2.451183	2.506065	2.500577	2.551068	5.849335
5	2.0	2.550665	2.595599	2.591105	2.632444	6.367886

$$= f(1.4, 4.869724) = 3(1+1.4) - 4.869724$$

$$= \mathbf{2.330276}$$

6. $y_{n+1} = y_n + \dfrac{h}{6}\{k_1 + 2k_2 + 2k_3 + k_4\}$ and   when $n = 1$:

$$y_2 = y_1 + \frac{h}{6}\{k_1 + 2k_2 + 2k_3 + k_4\}$$

$$= 4.418733 + \frac{0.2}{6}\{2.181267 + 2(2.263140)$$

$$+ 2(2.254953) + 2.330276\}$$

$$= 4.418733 + \frac{0.2}{6}\{13.547729\} = \mathbf{4.870324}$$

This completes the third row of Table 52.10. In a similar manner y_3, y_4 and y_5 can be calculated and the results are as shown in Table 52.10. As in the previous problem such a table is best produced by using a **spreadsheet**.

This problem is the same as Problem 1, page 544 which used Euler's method, and Problem 5, page 551 which used the Euler–Cauchy method, and a comparison of results can be made.

The differential equation $\dfrac{dy}{dx} = 3(1+x) - y$ may be solved analytically using the integrating factor method of Chapter 51, with the solution:

$$y = 3x + e^{1-x}$$

Substituting values of x of 1.0, 1.2, 1.4, ..., 2.0 will give the exact values. A comparison of the results obtained by Euler's method, the Euler–Cauchy method and the

Runga–Kutta method, together with the exact values is shown in Table 52.11.

It is seen from Table 52.11 that **the Runge–Kutta method is exact, correct to 4 decimal places**.

The percentage error in the Runge–Kutta method when, say, $x = 1.6$ is:

$$\left(\frac{5.348811636 - 5.348817}{5.348811636}\right) \times 100\% = \mathbf{-0.0001\%}$$

From Problem 6, page 552, when $x = 1.6$, the percentage error for the Euler method was 0.688%, and for the Euler–Cauchy method -0.048%. Clearly, the Runge–Kutta method is the most accurate of the three methods.

Now try the following Practice Exercise

Practice Exercise 202   Runge–Kutta method (Answers on page 886)

1.   Apply the Runge–Kutta method to solve the differential equation: $\dfrac{dy}{dx} = 3 - \dfrac{y}{x}$ for the range $1.0(0.1)1.5$, given the initial conditions $x = 1$ when $y = 2$

2.   Obtain a numerical solution of the differential equation: $\dfrac{1}{x}\dfrac{dy}{dx} + 2y = 1$ using the Runge–Kutta method in the range $x = 0(0.2)1.0$, given the initial conditions that $x = 0$ when $y = 1$

Table 52.11

x	Euler's method y	Euler–Cauchy method y	Runge–Kutta method y	Exact value $y = 3x + e^{1-x}$
1.0	4	4	4	4
1.2	4.4	4.42	4.418733	4.418730753
1.4	4.84	4.8724	4.870324	4.870320046
1.6	5.312	5.351368	5.348817	5.348811636
1.8	5.8096	5.85212176	5.849335	5.849328964
2.0	6.32768	6.370739847	6.367886	6.367879441

3. (a) The differential equation: $\dfrac{dy}{dx} + 1 = -\dfrac{y}{x}$ has the initial conditions that $y = 1$ at $x = 2$. Produce a numerical solution of the differential equation, correct to 6 decimal places, using the Runge–Kutta method in the range $x = 2.0(0.1)2.5$

(b) If the solution of the differential equation by an analytical method is given by: $y = \dfrac{4}{x} - \dfrac{x}{2}$ determine the percentage error at $x = 2.2$

For fully worked solutions to each of the problems in Practice Exercises 200 to 202 in this chapter, go to the website:
www.routledge.com/cw/bird

Revision Test 15 Further first-order differential equations

This Revision Test covers the material contained in Chapters 50 to 52. *The marks for each question are shown in brackets at the end of each question.*

1. Show that the solution to the differential equation:
$$4x\frac{dy}{dx} = \frac{x^2 + y^2}{y} \text{ is of the form}$$
$$3y^2 = \sqrt{x}\left(\sqrt{x^3} - 1\right) \text{ given that } y = 0 \text{ when } x = 1 \tag{12}$$

2. Show that the solution to the differential equation
$$x\cos x\frac{dy}{dx} + (x\sin x + \cos x)y = 1$$
is given by: $xy = \sin x + k\cos x$ where k is a constant. $\tag{11}$

3. (a) Use Euler's method to obtain a numerical solution of the differential equation:
$$\frac{dy}{dx} = \frac{y}{x} + x^2 - 2$$
given the initial conditions that $x = 1$ when $y = 3$, for the range $x = 1.0\,(0.1)\,1.5$

(b) Apply the Euler–Cauchy method to the differential equation given in part (a) over the same range.

(c) Apply the integrating factor method to solve the differential equation in part (a) analytically.

(d) Determine the percentage error, correct to 3 significant figures, in each of the two numerical methods when $x = 1.2$ $\tag{30}$

4. Use the Runge–Kutta method to solve the differential equation: $\dfrac{dy}{dx} = \dfrac{y}{x} + x^2 - 2$ in the range $1.0(0.1)1.5$, given the initial conditions that at $x = 1$, $y = 3$. Work to an accuracy of 6 decimal places. $\tag{27}$

For lecturers/instructors/teachers, fully worked solutions to each of the problems in Revision Test 15, together with a full marking scheme, are available at the website: www.routledge.com/cw/bird

Second-order differential equations of the form $a\dfrac{d^2y}{dx^2} + b\dfrac{dy}{dx} + cy = 0$

Why it is important to understand: **Second-order differential equations of the form** $a\dfrac{d^2y}{dx^2} + b\dfrac{dy}{dx} + cy = 0$

Second-order differential equations have many engineering applications. These include free vibration analysis with simple and damped mass-spring systems, resonant and non-resonant vibration analysis, with modal analysis, time-varying mechanical forces or pressure, fluid-induced vibration such as intermittent wind, forced electrical and mechanical oscillations, tidal waves, acoustics, ultrasonic and random movements of support. This chapter explains the procedure to solve second-order differential equations of the form $a\dfrac{d^2y}{dx^2} + b\dfrac{dy}{dx} + cy = 0$

At the end of this chapter, you should be able to:

- identify and solve the auxiliary equation of a second-order differential equation
- solve a second-order differential equation of the form $a\dfrac{d^2y}{dx^2} + b\dfrac{dy}{dx} + cy = 0$

53.1 Introduction

An equation of the form $a\dfrac{d^2y}{dx^2} + b\dfrac{dy}{dx} + cy = 0$, where a, b and c are constants, is called a **linear second-order differential equation with constant coefficients.** When

the right-hand side of the differential equation is zero, it is referred to as a **homogeneous differential equation.** When the right-hand side is not equal to zero (as in Chapter 54) it is referred to as a **non-homogeneous differential equation.**

There are numerous engineering examples of second-order differential equations. Three examples are:

(i) $L\frac{d^2q}{dt^2} + R\frac{dq}{dt} + \frac{1}{C}q = 0$, representing an equation for charge q in an electrical circuit containing resistance R, inductance L and capacitance C in series.

(ii) $m\frac{d^2s}{dt^2} + a\frac{ds}{dt} + ks = 0$, defining a mechanical system, where s is the distance from a fixed point after t seconds, m is a mass, a the damping factor and k the spring stiffness.

(iii) $\frac{d^2y}{dx^2} + \frac{P}{EI}y = 0$, representing an equation for the deflected profile y of a pin-ended uniform strut of length l subjected to a load P. E is Young's modulus and I is the second moment of area.

If D represents $\frac{d}{dx}$ and D^2 represents $\frac{d^2}{dx^2}$ then the above equation may be stated as $(aD^2 + bD + c)y = 0$. This equation is said to be in **'D-operator'** form.

If $y = Ae^{mx}$ then $\frac{dy}{dx} = Ame^{mx}$ and $\frac{d^2y}{dx^2} = Am^2e^{mx}$

Substituting these values into $a\frac{d^2y}{dx^2} + b\frac{dy}{dx} + cy = 0$ gives:

$$a(Am^2e^{mx}) + b(Ame^{mx}) + c(Ae^{mx}) = 0$$

i.e. $Ae^{mx}(am^2 + bm + c) = 0$

Thus $y = Ae^{mx}$ is a solution of the given equation provided that $(am^2 + bm + c) = 0$. $am^2 + bm + c = 0$ is called the **auxiliary equation**, and since the equation is a quadratic, m may be obtained either by factorising or by using the quadratic formula. Since, in the auxiliary equation, a, b and c are real values, then the equation may have either

(i) two different real roots (when $b^2 > 4ac$) or

(ii) two equal real roots (when $b^2 = 4ac$) or

(iii) two complex roots (when $b^2 < 4ac$)

53.2 Procedure to solve differential equations of the form $a\frac{d^2y}{dx^2} + b\frac{dy}{dx} + cy = 0$

(a) Rewrite the differential equation

$$a\frac{d^2y}{dx^2} + b\frac{dy}{dx} + cy = 0$$

as $(aD^2 + bD + c)y = 0$

(b) Substitute m for D and solve the auxiliary equation $am^2 + bm + c = 0$ for m

(c) If the roots of the auxiliary equation are:

(i) **real and different**, say $m = \alpha$ and $m = \beta$, then the general solution is

$$y = Ae^{\alpha x} + Be^{\beta x}$$

(ii) **real and equal**, say $m = \alpha$ twice, then the general solution is

$$y = (Ax + B)e^{\alpha x}$$

(iii) **complex**, say $m = \alpha \pm j\beta$, then the general solution is

$$y = e^{\alpha x}\{A\cos\beta x + B\sin\beta x\}$$

(d) Given boundary conditions, constants A and B, may be determined and the **particular solution** of the differential equation obtained.

The particular solutions obtained in the worked problems of Section 53.3 may each be verified by substituting expressions for y, $\frac{dy}{dx}$ and $\frac{d^2y}{dx^2}$ into the original equation.

53.3 Worked problems on differential equations of the form $a\frac{d^2y}{dx^2} + b\frac{dy}{dx} + cy = 0$

Problem 1. By applying Kirchhoff's voltage law to a circuit the following differential equation is obtained: $2\frac{d^2y}{dx^2} + 5\frac{dy}{dx} - 3y = 0$. Determine the general solution. Find also the particular solution given that when $x = 0$, $y = 4$ and $\frac{dy}{dx} = 9$

Using the above procedure:

(a) $2\frac{d^2y}{dx^2} + 5\frac{dy}{dx} - 3y = 0$ in D-operator form is $(2D^2 + 5D - 3)y = 0$, where $D \equiv \frac{d}{dx}$

(b) Substituting m for D gives the auxiliary equation

$$2m^2 + 5m - 3 = 0$$

Factorising gives: $(2m-1)(m+3)=0$, from which, $m = \frac{1}{2}$ or $m = -3$

(c) Since the roots are real and different the **general solution is $y = Ae^{\frac{1}{2}x} + Be^{-3x}$**

(d) When $x = 0$, $y = 4$,

hence $\quad 4 = A + B \qquad (1)$

Since $\quad y = Ae^{\frac{1}{2}x} + Be^{-3x}$

then $\quad \dfrac{dy}{dx} = \dfrac{1}{2}Ae^{\frac{1}{2}x} - 3Be^{-3x}$

When $\quad x = 0$, $\dfrac{dy}{dx} = 9$

thus $\quad 9 = \dfrac{1}{2}A - 3B \qquad (2)$

Solving the simultaneous equations (1) and (2) gives $A = 6$ and $B = -2$

Hence the particular solution is

$$y = 6e^{\frac{1}{2}x} - 2e^{-3x}$$

Problem 2. Find the general solution of
$9\dfrac{d^2 y}{dt^2} - 24\dfrac{dy}{dt} + 16y = 0$ and also the particular solution given the boundary conditions that when
$t = 0$, $y = \dfrac{dy}{dt} = 3$

Using the procedure of Section 53.2:

(a) $9\dfrac{d^2 y}{dt^2} - 24\dfrac{dy}{dt} + 16y = 0$ in D-operator form is
$(9D^2 - 24D + 16)y = 0$ where $D \equiv \dfrac{d}{dt}$

(b) Substituting m for D gives the auxiliary equation
$9m^2 - 24m + 16 = 0$

Factorising gives: $(3m-4)(3m-4)=0$, i.e. $m = \frac{4}{3}$ twice.

(c) Since the roots are real and equal, **the general solution is $y = (At + B)e^{\frac{4}{3}t}$**

(d) When $t = 0$, $y = 3$ hence $3 = (0+B)e^0$, i.e. $B = 3$.

Since $y = (At + B)e^{\frac{4}{3}t}$

then $\quad \dfrac{dy}{dt} = (At + B)\left(\dfrac{4}{3}e^{\frac{4}{3}t}\right) + Ae^{\frac{4}{3}t}$, by the product rule.

When $\quad t = 0$, $\dfrac{dy}{dt} = 3$

thus $\quad 3 = (0 + B)\dfrac{4}{3}e^0 + Ae^0$

i.e. $3 = \dfrac{4}{3}B + A$ from which, $A = -1$, since $B = 3$

Hence the particular solution is

$$y = (-t + 3)e^{\frac{4}{3}t} \text{ or } y = (3 - t)e^{\frac{4}{3}t}$$

Problem 3. Solve the differential equation
$\dfrac{d^2 y}{dx^2} + 6\dfrac{dy}{dx} + 13y = 0$, given that when $x = 0$, $y = 3$
and $\dfrac{dy}{dx} = 7$

Using the procedure of Section 53.2:

(a) $\dfrac{d^2 y}{dx^2} + 6\dfrac{dy}{dx} + 13y = 0$ in D-operator form is
$(D^2 + 6D + 13)y = 0$, where $D \equiv \dfrac{d}{dx}$

(b) Substituting m for D gives the auxiliary equation
$m^2 + 6m + 13 = 0$

Using the quadratic formula:

$$m = \dfrac{-6 \pm \sqrt{[(6)^2 - 4(1)(13)]}}{2(1)}$$

$$= \dfrac{-6 \pm \sqrt{(-16)}}{2}$$

i.e. $m = \dfrac{-6 \pm j4}{2} = -3 \pm j2$

(c) Since the roots are complex, **the general solution is**

$$y = e^{-3x}(A\cos 2x + B\sin 2x)$$

(d) When $x = 0$, $y = 3$, hence
$3 = e^0(A\cos 0 + B\sin 0)$, i.e. $A = 3$

Since $y = e^{-3x}(A\cos 2x + B\sin 2x)$

then $\quad \dfrac{dy}{dx} = e^{-3x}(-2A\sin 2x + 2B\cos 2x)$

$$- 3e^{-3x}(A\cos 2x + B\sin 2x),$$

by the product rule,

$$= e^{-3x}[(2B - 3A)\cos 2x$$

$$- (2A + 3B)\sin 2x]$$

When $x = 0$, $\frac{\mathrm{d}y}{\mathrm{d}x} = 7$,

hence $7 = e^0[(2B - 3A)\cos 0 - (2A + 3B)\sin 0]$

i.e. $7 = 2B - 3A$, from which, $B = 8$, since $A = 3$

Hence the particular solution is

$$y = e^{-3x}(3\cos 2x + 8\sin 2x)$$

Since, from Chapter 19, page 204,

$a\cos\omega t + b\sin\omega t = R\sin(\omega t + \alpha)$, where

$R = \sqrt{(a^2 + b^2)}$ and $\alpha = \tan^{-1}\frac{a}{b}$ then

$3\cos 2x + 8\sin 2x$

$\qquad = \sqrt{(3^2 + 8^2)}\sin(2x + \tan^{-1}\frac{3}{8})$

$\qquad = \sqrt{73}\sin(2x + 20.56°)$

$\qquad = \sqrt{73}\sin(2x + 0.359)$

Thus the particular solution may also be expressed as

$$y = \sqrt{73}\,e^{-3x}\sin(2x + 0.359)$$

Now try the following Practice Exercise

Practice Exercise 203   Second-order differential equations of the form

$a\dfrac{\mathrm{d}^2 y}{\mathrm{d}x^2} + b\dfrac{\mathrm{d}y}{\mathrm{d}x} + cy = 0$ **(Answers on page 886)**

In Problems 1 to 3, determine the general solution of the given differential equations.

1. $6\dfrac{\mathrm{d}^2 y}{\mathrm{d}t^2} - \dfrac{\mathrm{d}y}{\mathrm{d}t} - 2y = 0$

2. $4\dfrac{\mathrm{d}^2\theta}{\mathrm{d}t^2} + 4\dfrac{\mathrm{d}\theta}{\mathrm{d}t} + \theta = 0$

3. $\dfrac{\mathrm{d}^2 y}{\mathrm{d}x^2} + 2\dfrac{\mathrm{d}y}{\mathrm{d}x} + 5y = 0$

In Problems 4 to 9, find the particular solution of the given differential equations for the stated boundary conditions.

4. $6\dfrac{\mathrm{d}^2 y}{\mathrm{d}x^2} + 5\dfrac{\mathrm{d}y}{\mathrm{d}x} - 6y = 0$; when $x = 0$, $y = 5$ and $\dfrac{\mathrm{d}y}{\mathrm{d}x} = -1$

5. $4\dfrac{\mathrm{d}^2 y}{\mathrm{d}t^2} - 5\dfrac{\mathrm{d}y}{\mathrm{d}t} + y = 0$; when $t = 0$, $y = 1$ and $\dfrac{\mathrm{d}y}{\mathrm{d}t} = -2$

6. $(9D^2 + 30D + 25)y = 0$, where $D \equiv \dfrac{\mathrm{d}}{\mathrm{d}x}$; when $x = 0$, $y = 0$ and $\dfrac{\mathrm{d}y}{\mathrm{d}x} = 2$

7. $\dfrac{\mathrm{d}^2 x}{\mathrm{d}t^2} - 6\dfrac{\mathrm{d}x}{\mathrm{d}t} + 9x = 0$; when $t = 0$, $x = 2$ and $\dfrac{\mathrm{d}x}{\mathrm{d}t} = 0$

8. $\dfrac{\mathrm{d}^2 y}{\mathrm{d}x^2} + 6\dfrac{\mathrm{d}y}{\mathrm{d}x} + 13y = 0$; when $x = 0$, $y = 4$ and $\dfrac{\mathrm{d}y}{\mathrm{d}x} = 0$

9. $(4D^2 + 20D + 125)\theta = 0$, where $D \equiv \dfrac{\mathrm{d}}{\mathrm{d}t}$; when $t = 0$, $\theta = 3$ and $\dfrac{\mathrm{d}\theta}{\mathrm{d}t} = 2.5$

53.4   Further worked problems on practical differential equations of the form $a\dfrac{\mathrm{d}^2 y}{\mathrm{d}x^2} + b\dfrac{\mathrm{d}y}{\mathrm{d}x} + cy = 0$

Problem 4.   The equation of motion of a body oscillating on the end of a spring is

$$\frac{\mathrm{d}^2 x}{\mathrm{d}t^2} + 100x = 0$$

where x is the displacement in metres of the body from its equilibrium position after time t seconds. Determine x in terms of t given that at time $t = 0$, $x = 2m$ and $\dfrac{\mathrm{d}x}{\mathrm{d}t} = 0$

An equation of the form $\dfrac{\mathrm{d}^2 x}{\mathrm{d}t^2} + m^2 x = 0$ is a differential equation representing simple harmonic motion (SHM). Using the procedure of Section 53.2:

(a) $\dfrac{\mathrm{d}^2 x}{\mathrm{d}t^2} + 100x = 0$ in D-operator form is $(D^2 + 100)x = 0$

(b) The auxiliary equation is $m^2 + 100 = 0$, i.e. $m^2 = -100$ and $m = \sqrt{(-100)}$, i.e. $m = \pm j10$

Section J

(c) Since the roots are complex, the general solution is $x = e^0 (A \cos 10t + B \sin 10t)$,

i.e. $x = (A \cos 10t + B \sin 10t)$ **metres**

(d) When $t = 0$, $x = 2$, thus $2 = A$

$$\frac{dx}{dt} = -10A \sin 10t + 10B \cos 10t$$

When $t = 0$, $\frac{dx}{dt} = 0$

thus $0 = -10A \sin 0 + 10B \cos 0$, i.e. $B = 0$

Hence the particular solution is

$$x = 2 \cos 10t \text{ metres}$$

Problem 5. Given the differential equation $\frac{d^2 V}{dt^2} = \omega^2 V$, where ω is a constant, show that its solution may be expressed as:

$$V = 7 \cosh \omega t + 3 \sinh \omega t$$

given the boundary conditions that when

$t = 0$, $V = 7$ and $\frac{dV}{dt} = 3\omega$

Using the procedure of Section 53.2:

(a) $\frac{d^2 V}{dt^2} = \omega^2 V$, i.e. $\frac{d^2 V}{dt^2} - \omega^2 V = 0$ in D-operator form is $(D^2 - \omega^2) v = 0$, where $D \equiv \frac{d}{dx}$

(b) The auxiliary equation is $m^2 - \omega^2 = 0$, from which, $m^2 = \omega^2$ and $m = \pm \omega$

(c) Since the roots are real and different, **the general solution is**

$$V = Ae^{\omega t} + Be^{-\omega t}$$

(d) When $t = 0$, $V = 7$ hence $7 = A + B$ (1)

$$\frac{dV}{dt} = A\omega e^{\omega t} - B\omega e^{-\omega t}$$

When $t = 0$, $\frac{dV}{dt} = 3\omega$,

thus $3\omega = A\omega - B\omega$,

i.e. $3 = A - B$ (2)

From equations (1) and (2), $A = 5$ and $B = 2$

Hence **the particular solution is**

$$V = 5e^{\omega t} + 2e^{-\omega t}$$

Since $\sinh \omega t = \frac{1}{2}(e^{\omega t} - e^{-\omega t})$

and $\cosh \omega t = \frac{1}{2}(e^{\omega t} + e^{-\omega t})$

then $\sinh \omega t + \cosh \omega t = e^{\omega t}$

and $\cosh \omega t - \sinh \omega t = e^{-\omega t}$ from Chapter 16.

Hence the particular solution may also be written as

$$V = 5(\sinh \omega t + \cosh \omega t)$$
$$+ 2(\cosh \omega t - \sinh \omega t)$$

i.e. $V = (5 + 2) \cosh \omega t + (5 - 2) \sinh \omega t$

i.e. $V = 7 \cosh \omega t + 3 \sinh \omega t$

Problem 6. The equation

$$\frac{d^2 i}{dt^2} + \frac{R}{L}\frac{di}{dt} + \frac{1}{LC} i = 0$$

represents a current i flowing in an electrical circuit containing resistance R, inductance L and capacitance C connected in series. If $R = 200$ ohms, $L = 0.20$ henry and $C = 20 \times 10^{-6}$ farads, solve the equation for i given the boundary conditions that when $t = 0$, $i = 0$ and $\frac{di}{dt} = 100$

Using the procedure of Section 53.2:

(a) $\frac{d^2 i}{dt^2} + \frac{R}{L}\frac{di}{dt} + \frac{1}{LC} i = 0$ in D-operator form is

$$\left(D^2 + \frac{R}{L}D + \frac{1}{LC}\right) i = 0 \text{ where } D \equiv \frac{d}{dt}$$

(b) The auxiliary equation is $m^2 + \frac{R}{L} m + \frac{1}{LC} = 0$

Hence $m = \dfrac{-\dfrac{R}{L} \pm \sqrt{\left[\left(\dfrac{R}{L}\right)^2 - 4(1)\left(\dfrac{1}{LC}\right)\right]}}{2}$

When $R = 200$, $L = 0.20$ and $C = 20 \times 10^{-6}$, then

$$m = \frac{-\dfrac{200}{0.20} \pm \sqrt{\left[\left(\dfrac{200}{0.20}\right)^2 - \dfrac{4}{(0.20)(20 \times 10^{-6})}\right]}}{2}$$

$$= \frac{-1000 \pm \sqrt{0}}{2} = -500$$

(c) Since the two roots are real and equal (i.e. -500 twice, since for a second-order differential equation there must be two solutions), **the general solution is $i=(At+B)e^{-500t}$**

(d) When $t=0$, $i=0$, hence $B=0$

$$\frac{di}{dt} = (At+B)(-500e^{-500t}) + (e^{-500t})(A),$$

by the product rule

When $t=0$, $\dfrac{di}{dt}=100$, thus $100=-500B+A$

i.e. $A=100$, since $B=0$

Hence the particular solution is

$$i = 100te^{-500t}$$

Problem 7. The oscillations of a heavily damped pendulum satisfy the differential equation $\dfrac{d^2x}{dt^2}+6\dfrac{dx}{dt}+8x=0$, where x cm is the displacement of the bob at time t seconds. The initial displacement is equal to $+4$ cm and the initial velocity $\left(\text{i.e. } \dfrac{dx}{dt}\right)$ is 8 cm/s. Solve the equation for x.

Using the procedure of Section 53.2:

(a) $\dfrac{d^2x}{dt^2}+6\dfrac{dx}{dt}+8x=0$ in D-operator form is

$(D^2+6D+8)x=0$, where $D \equiv \dfrac{d}{dt}$

(b) The auxiliary equation is $m^2+6m+8=0$

Factorising gives: $(m+2)(m+4)=0$, from which, $m=-2$ or $m=-4$

(c) Since the roots are real and different, **the general solution is $x=Ae^{-2t}+Be^{-4t}$**

(d) Initial displacement means that time $t=0$. At this instant, $x=4$

Thus $4=A+B$ (1)

Velocity,

$$\frac{dx}{dt} = -2Ae^{-2t} - 4Be^{-4t}$$

$$\frac{dx}{dt} = 8 \text{ cm/s when } t = 0,$$

thus $8=-2A-4B$ (2)

From equations (1) and (2),

$$A = 12 \text{ and } B = -8$$

Hence the particular solution is

$$x = 12e^{-2t} - 8e^{-4t}$$

i.e. **displacement**, $x = 4(3e^{-2t} - 2e^{-4t})\,\text{cm}$

Now try the following Practice Exercise

Practice Exercise 204   Second-order differential equations of the form $a\dfrac{d^2y}{dx^2}+b\dfrac{dy}{dx}+cy=0$ (Answers on page 886)

1. The charge q on a capacitor in a certain electrical circuit satisfies the differential equation $\dfrac{d^2q}{dt^2}+4\dfrac{dq}{dt}+5q=0$. Initially (i.e. when $t=0$), $q=Q$ and $\dfrac{dq}{dt}=0$. Show that the charge in the circuit can be expressed as: $q=\sqrt{5}\,Qe^{-2t}\sin(t+0.464)$

2. A body moves in a straight line so that its distance s metres from the origin after time t seconds is given by $\dfrac{d^2s}{dt^2}+a^2s=0$, where a is a constant. Solve the equation for s given that $s=c$ and $\dfrac{ds}{dt}=0$ when $t=\dfrac{2\pi}{a}$

3. The motion of the pointer of a galvanometer about its position of equilibrium is represented by the equation

$$I\frac{d^2\theta}{dt^2} + K\frac{d\theta}{dt} + F\theta = 0$$

If I, the moment of inertia of the pointer about its pivot, is 5×10^{-3}, K, the resistance due to friction at unit angular velocity, is 2×10^{-2} and F, the force on the spring necessary to produce unit displacement, is 0.20, solve the equation for θ in terms of t given that when $t=0$, $\theta=0.3$ and $\dfrac{d\theta}{dt}=0$

4. Determine an expression for x for a differential equation $\dfrac{d^2x}{dt^2}+2n\dfrac{dx}{dt}+n^2x=0$ which repre-

sents a critically damped oscillator, given that at time $t=0$, $x=s$ and $\dfrac{dx}{dt}=u$

5. $L\dfrac{d^2i}{dt^2}+R\dfrac{di}{dt}+\dfrac{1}{C}i=0$ is an equation representing current i in an electric circuit. If inductance L is 0.25 henry, capacitance C is 29.76×10^{-6} farads and R is 250 ohms, solve the equation for i given the boundary conditions that when $t=0$, $i=0$ and $\dfrac{di}{dt}=34$

6. The displacement s of a body in a damped mechanical system, with no external forces, satisfies the following differential equation:

$$2\dfrac{d^2s}{dt^2}+6\dfrac{ds}{dt}+4.5s=0$$

where t represents time. If initially, when $t=0$, $s=0$ and $\dfrac{ds}{dt}=4$, solve the differential equation for s in terms of t.

For fully worked solutions to each of the problems in Practice Exercises 203 and 204 in this chapter, go to the website:
www.routledge.com/cw/bird

Second-order differential equations of the form

$$a\frac{d^2y}{dx^2} + b\frac{dy}{dx} + cy = f(x)$$

Why it is important to understand: **Second-order differential equations of the form** $a\dfrac{d^2y}{dx^2} + b\dfrac{dy}{dx} + cy = f(x)$

Second-order differential equations have many engineering applications. Differential equations govern the fundamental operation of important areas such as automobile dynamics, tyre dynamics, aerodynamics, acoustics, active control systems, including speed control, engine performance and emissions control, climate control, ABS control systems, airbag deployment systems, structural dynamics of buildings, bridges and dams, for example, earthquake and wind engineering, industrial process control, control and operation of automation (robotic) systems, the operation of the electric power grid, electric power generation, orbital dynamics of satellite systems, heat transfer from electrical equipment (including computer chips), economic systems, biological systems, chemical systems, and so on. This chapter explains the procedure to solve second-order differential equations of the form $a\dfrac{d^2y}{dx^2} + b\dfrac{dy}{dx} + cy = f(x)$ for different functions f(x).

At the end of this chapter, you should be able to:

- identify the complementary function and particular integral of a second-order differential equation

- solve a second order differential equation of the form $a\dfrac{d^2y}{dx^2} + b\dfrac{dy}{dx} + cy = f(x)$ where $f(x)$ is a constant or a polynomial function, an exponential function, a sine or cosine function, or a sum or product

---

54.1 Complementary function and particular integral

If in the differential equation

$$a\frac{d^2y}{dx^2} + b\frac{dy}{dx} + cy = f(x) \qquad (1)$$

the substitution $y = u + v$ is made then:

$$a\frac{d^2(u+v)}{dx^2} + b\frac{d(u+v)}{dx} + c(u+v) = f(x)$$

Rearranging gives:

$$\left(a\frac{d^2u}{dx^2} + b\frac{du}{dx} + cu\right) + \left(a\frac{d^2v}{dx^2} + b\frac{dv}{dx} + cv\right) = f(x)$$

If we let

$$a\frac{d^2v}{dx^2} + b\frac{dv}{dx} + cv = f(x) \qquad (2)$$

then

$$a\frac{d^2u}{dx^2} + b\frac{du}{dx} + cu = 0 \qquad (3)$$

The general solution, u, of equation (3) will contain two unknown constants, as required for the general solution of equation (1). The method of solution of equation (3) is shown in Chapter 53. The function u is called the **complementary function (C.F.)**.

If the particular solution, v, of equation (2) can be determined without containing any unknown constants then $y = u + v$ will give the general solution of equation (1). The function v is called the **particular integral (P.I.)**. Hence the general solution of equation (1) is given by:

$$y = \textbf{C.F.} + \textbf{P.I.}$$

Table 54.1 Form of particular integral for different functions

Type	Straightforward cases Try as particular integral:	'Snag' cases Try as particular integral:	See problem
(a) $f(x) = $ a constant	$v = k$	$v = kx$ (used when C.F. contains a constant)	1, 2
(b) $f(x) = $ polynomial (i.e. $f(x) = L + Mx + Nx^2 + \cdots$ where any of the coefficients may be zero)	$v = a + bx + cx^2 + \cdots$		3
(c) $f(x) = $ an exponential function (i.e. $f(x) = Ae^{ax}$)	$v = ke^{ax}$	(i) $v = kxe^{ax}$ (used when e^{ax} appears in the C.F.)	4, 5
		(ii) $v = kx^2e^{ax}$ (used when e^{ax} **and** xe^{ax} both appear in the C.F.)	6
(d) $f(x) = $ a sine or cosine function (i.e. $f(x) = a\sin px + b\cos px$, where a or b may be zero)	$v = A\sin px + B\cos px$	$v = x(A\sin px + B\cos px)$ (used when $\sin px$ and/or $\cos px$ appears in the C.F.)	7, 8
(e) $f(x) = $ a sum e.g. (i) $f(x) = 4x^2 - 3\sin 2x$	(i) $v = ax^2 + bx + c$ $+ d\sin 2x + e\cos 2x$		9
(ii) $f(x) = 2 - x + e^{3x}$	(ii) $v = ax + b + ce^{3x}$		
(f) $f(x) = $ a product e.g. $f(x) = 2e^x\cos 2x$	$v = e^x(A\sin 2x + B\cos 2x)$		10

54.2 Procedure to solve differential equations of the form $a\frac{d^2y}{dx^2} + b\frac{dy}{dx} + cy = f(x)$

(i) Rewrite the given differential equation as $(aD^2 + bD + c)y = f(x)$.

(ii) Substitute m for D, and solve the auxiliary equation $am^2 + bm + c = 0$ for m.

(iii) Obtain the complementary function, u, which is achieved using the same procedure as in Section 53.2(c), page 561.

(iv) To determine the particular integral, v, firstly assume a particular integral which is suggested by $f(x)$, but which contains undetermined coefficients. Table 54.1 gives some suggested substitutions for different functions $f(x)$.

(v) Substitute the suggested P.I. into the differential equation $(aD^2 + bD + c)v = f(x)$ and equate relevant coefficients to find the constants introduced.

(vi) The general solution is given by
$y = \text{C.F.} + \text{P.I.}$, i.e. $y = u + v$

(vii) Given boundary conditions, arbitrary constants in the C.F. may be determined and the particular solution of the differential equation obtained.

54.3 Worked problems on differential equations of the form $a\frac{d^2y}{dx^2} + b\frac{dy}{dx} + cy = f(x)$ where $f(x)$ is a constant or polynomial

Problem 1. Solve the differential equation $\frac{d^2y}{dx^2} + \frac{dy}{dx} - 2y = 4$

Using the procedure of Section 54.2:

(i) $\frac{d^2y}{dx^2} + \frac{dy}{dx} - 2y = 4$ in D-operator form is $(D^2 + D - 2)y = 4$

(ii) Substituting m for D gives the auxiliary equation $m^2 + m - 2 = 0$. Factorising gives: $(m-1)(m+2) = 0$, from which $m = 1$ or $m = -2$

(iii) Since the roots are real and different, the C.F., $u = Ae^x + Be^{-2x}$

(iv) Since the term on the right-hand side of the given equation is a constant, i.e. $f(x) = 4$, let the P.I. also be a constant, say $v = k$ (see Table 54.1(a)).

(v) Substituting $v = k$ into $(D^2 + D - 2)v = 4$ gives $(D^2 + D - 2)k = 4$. Since $D(k) = 0$ and $D^2(k) = 0$ then $-2k = 4$, from which, $k = -2$. Hence the P.I., $v = -2$

(vi) The general solution is given by $y = u + v$, i.e. $y = Ae^x + Be^{-2x} - 2$

Problem 2. Determine the particular solution of the equation $\frac{d^2y}{dx^2} - 3\frac{dy}{dx} = 9$, given the boundary conditions that when $x = 0$, $y = 0$ and $\frac{dy}{dx} = 0$

Using the procedure of Section 54.2:

(i) $\frac{d^2y}{dx^2} - 3\frac{dy}{dx} = 9$ in D-operator form is $(D^2 - 3D)y = 9$

(ii) Substituting m for D gives the auxiliary equation $m^2 - 3m = 0$. Factorising gives: $m(m-3) = 0$, from which, $m = 0$ or $m = 3$

(iii) Since the roots are real and different, the C.F., $u = Ae^0 + Be^{3x}$, i.e. $u = A + Be^{3x}$

(iv) Since the C.F. contains a constant (i.e. A) then let the P.I., $v = kx$ (see Table 54.1(a)).

(v) Substituting $v = kx$ into $(D^2 - 3D)v = 9$ gives $(D^2 - 3D)kx = 9$
$D(kx) = k$ and $D^2(kx) = 0$
Hence $(D^2 - 3D)kx = 0 - 3k = 9$, from which, $k = -3$
Hence the P.I., $v = -3x$

(vi) The general solution is given by $y = u + v$, i.e. $y = A + Be^{3x} - 3x$

(vii) When $x = 0$, $y = 0$, thus $0 = A + Be^0 - 0$, i.e.
$0 = A + B$ (1)
$\frac{dy}{dx} = 3Be^{3x} - 3$; $\frac{dy}{dx} = 0$ when $x = 0$, thus
$0 = 3Be^0 - 3$ from which, $B = 1$. From equation (1), $A = -1$

Hence the particular solution is

$$y = -1 + 1e^{3x} - 3x,$$

i.e. $y = e^{3x} - 3x - 1$

Problem 3. Solve the differential equation

$$2\frac{d^2y}{dx^2} - 11\frac{dy}{dx} + 12y = 3x - 2$$

Using the procedure of Section 54.2:

(i) $2\frac{d^2y}{dx^2} - 11\frac{dy}{dx} + 12y = 3x - 2$ in D-operator form is

$$(2D^2 - 11D + 12)y = 3x - 2$$

(ii) Substituting m for D gives the auxiliary equation $2m^2 - 11m + 12 = 0$. Factorising gives: $(2m - 3)(m - 4) = 0$, from which, $m = \frac{3}{2}$ or $m = 4$

(iii) Since the roots are real and different, the C.F.,

$$u = Ae^{\frac{3}{2}x} + Be^{4x}$$

(iv) Since $f(x) = 3x - 2$ is a polynomial, let the P.I., $v = ax + b$ (see Table 54.1(b)).

(v) Substituting $v = ax + b$ into $(2D^2 - 11D + 12)v = 3x - 2$ gives:

$$(2D^2 - 11D + 12)(ax + b) = 3x - 2,$$

i.e. $2D^2(ax + b) - 11D(ax + b)$

$$+ 12(ax + b) = 3x - 2$$

i.e. $0 - 11a + 12ax + 12b = 3x - 2$

Equating the coefficients of x gives: $12a = 3$, from which, $a = \frac{1}{4}$

Equating the constant terms gives:

$$-11a + 12b = -2$$

i.e. $-11\left(\frac{1}{4}\right) + 12b = -2$ from which,

$$12b = -2 + \frac{11}{4} = \frac{3}{4} \text{ i.e. } b = \frac{1}{16}$$

Hence the P.I., $v = ax + b = \frac{1}{4}x + \frac{1}{16}$

(vi) The general solution is given by $y = u + v$, i.e.

$$y = Ae^{\frac{3}{2}x} + Be^{4x} + \frac{1}{4}x + \frac{1}{16}$$

Now try the following Practice Exercise

Practice Exercise 205 Second-order differential equations of the form $a\frac{d^2y}{dx^2} + b\frac{dy}{dx} + cy = f(x)$ **where** $f(x)$ **is a constant or polynomial (Answers on page 887)**

In Problems 1 and 2, find the general solutions of the given differential equations.

1. $2\frac{d^2y}{dx^2} + 5\frac{dy}{dx} - 3y = 6$

2. $6\frac{d^2y}{dx^2} + 4\frac{dy}{dx} - 2y = 3x - 2$

In Problems 3 and 4 find the particular solutions of the given differential equations.

3. $3\frac{d^2y}{dx^2} + \frac{dy}{dx} - 4y = 8$; when $x = 0$, $y = 0$ and $\frac{dy}{dx} = 0$

4. $9\frac{d^2y}{dx^2} - 12\frac{dy}{dx} + 4y = 3x - 1$; when $x = 0$, $y = 0$ and $\frac{dy}{dx} = -\frac{4}{3}$

5. The charge q in an electric circuit at time t satisfies the equation $L\frac{d^2q}{dt^2} + R\frac{dq}{dt} + \frac{1}{C}q = E$, where L, R, C and E are constants. Solve the equation given $L = 2H$, $C = 200 \times 10^{-6}$ F and $E = 250$ V, when (a) $R = 200\,\Omega$ and (b) R is negligible. Assume that when $t = 0$, $q = 0$ and $\frac{dq}{dt} = 0$

6. In a galvanometer the deflection θ satisfies the differential equation $\frac{d^2\theta}{dt^2} + 4\frac{d\theta}{dt} + 4\theta = 8$. Solve the equation for θ given that when $t = 0$, $\theta = \frac{d\theta}{dt} = 2$

54.4 Worked problems on differential equations of the form $a\frac{d^2y}{dx^2} + b\frac{dy}{dx} + cy = f(x)$ **where** $f(x)$ **is an exponential function**

Problem 4. Solve the equation $\dfrac{d^2y}{dx^2} - 2\dfrac{dy}{dx} + y = 3e^{4x}$ given the boundary conditions that when $x=0$, $y = -\frac{2}{3}$ and $\dfrac{dy}{dx} = 4\frac{1}{3}$

Using the procedure of Section 54.2:

(i) $\dfrac{d^2y}{dx^2} - 2\dfrac{dy}{dx} + y = 3e^{4x}$ in D-operator form is $(D^2 - 2D + 1)y = 3e^{4x}$

(ii) Substituting m for D gives the auxiliary equation $m^2 - 2m + 1 = 0$. Factorising gives: $(m-1)(m-1) = 0$, from which, $m=1$ twice.

(iii) Since the roots are real and equal the C.F., $u = (Ax + B)e^x$

(iv) Let the particular integral, $v = ke^{4x}$ (see Table 54.1(c)).

(v) Substituting $v = ke^{4x}$ into $(D^2 - 2D + 1)v = 3e^{4x}$ gives:

$$(D^2 - 2D + 1)ke^{4x} = 3e^{4x}$$

i.e. $D^2(ke^{4x}) - 2D(ke^{4x}) + 1(ke^{4x}) = 3e^{4x}$

i.e. $16ke^{4x} - 8ke^{4x} + ke^{4x} = 3e^{4x}$

Hence $9ke^{4x} = 3e^{4x}$, from which, $k = \frac{1}{3}$

Hence the P.I., $v = ke^{4x} = \frac{1}{3}e^{4x}$

(vi) The general solution is given by $y = u + v$, i.e. $y = (Ax + B)e^x + \frac{1}{3}e^{4x}$

(vii) When $x=0$, $y=-\frac{2}{3}$ thus
$-\frac{2}{3} = (0 + B)e^0 + \frac{1}{3}e^0$, from which, $B = -1$
$\dfrac{dy}{dx} = (Ax + B)e^x + e^x(A) + \frac{4}{3}e^{4x}$
When $x=0$, $\dfrac{dy}{dx} = 4\frac{1}{3}$, thus $\frac{13}{3} = B + A + \frac{4}{3}$
from which, $A = 4$, since $B = -1$

Hence the particular solution is:

$$y = (4x - 1)e^x + \frac{1}{3}e^{4x}$$

Problem 5. Solve the differential equation $2\dfrac{d^2y}{dx^2} - \dfrac{dy}{dx} - 3y = 5e^{\frac{3}{2}x}$

Using the procedure of Section 54.2:

(i) $2\dfrac{d^2y}{dx^2} - \dfrac{dy}{dx} - 3y = 5e^{\frac{3}{2}x}$ in D-operator form is $(2D^2 - D - 3)y = 5e^{\frac{3}{2}x}$

(ii) Substituting m for D gives the auxiliary equation $2m^2 - m - 3 = 0$. Factorising gives: $(2m - 3)(m + 1) = 0$, from which, $m = \frac{3}{2}$ or $m = -1$. Since the roots are real and different then the C.F., $u = Ae^{\frac{3}{2}x} + Be^{-x}$

(iii) Since $e^{\frac{3}{2}x}$ appears in the C.F. **and** in the right-hand side of the differential equation, let the P.I., $v = kxe^{\frac{3}{2}x}$ (see Table 54.1(c), snag case (i)).

(iv) Substituting $v = kxe^{\frac{3}{2}x}$ into $(2D^2 - D - 3)v = 5e^{\frac{3}{2}x}$ gives: $(2D^2 - D - 3)kxe^{\frac{3}{2}x} = 5e^{\frac{3}{2}x}$

$$D\left(kxe^{\frac{3}{2}x}\right) = (kx)\left(\tfrac{3}{2}e^{\frac{3}{2}x}\right) + \left(e^{\frac{3}{2}x}\right)(k),$$

by the product rule,

$$= ke^{\frac{3}{2}x}\left(\tfrac{3}{2}x + 1\right)$$

$$D^2\left(kxe^{\frac{3}{2}x}\right) = D\left[ke^{\frac{3}{2}x}\left(\tfrac{3}{2}x + 1\right)\right]$$

$$= \left(ke^{\frac{3}{2}x}\right)\left(\tfrac{3}{2}\right)$$
$$+ \left(\tfrac{3}{2}x + 1\right)\left(\tfrac{3}{2}ke^{\frac{3}{2}x}\right)$$

$$= ke^{\frac{3}{2}x}\left(\tfrac{9}{4}x + 3\right)$$

Hence $(2D^2 - D - 3)\left(kxe^{\frac{3}{2}x}\right)$

$$= 2\left[ke^{\frac{3}{2}x}\left(\tfrac{9}{4}x + 3\right)\right] - \left[ke^{\frac{3}{2}x}\left(\tfrac{3}{2}x + 1\right)\right]$$
$$- 3\left[kxe^{\frac{3}{2}x}\right] = 5e^{\frac{3}{2}x}$$

i.e. $\frac{9}{2}kxe^{\frac{3}{2}x} + 6ke^{\frac{3}{2}x} - \frac{3}{2}xke^{\frac{3}{2}x} - ke^{\frac{3}{2}x}$
$- 3kxe^{\frac{3}{2}x} = 5e^{\frac{3}{2}x}$

Equating coefficients of $e^{\frac{3}{2}x}$ gives: $5k = 5$, from which, $k = 1$

Hence the P.I., $v = kxe^{\frac{3}{2}x} = xe^{\frac{3}{2}x}$

(v) The general solution is $y = u + v$, i.e.
$y = Ae^{\frac{3}{2}x} + Be^{-x} + xe^{\frac{3}{2}x}$

Problem 6. Solve $\dfrac{d^2y}{dx^2} - 4\dfrac{dy}{dx} + 4y = 3e^{2x}$

Using the procedure of Section 54.2:

(i) $\dfrac{d^2y}{dx^2} - 4\dfrac{dy}{dx} + 4y = 3e^{2x}$ in D-operator form is $(D^2 - 4D + 4)y = 3e^{2x}$

(ii) Substituting m for D gives the auxiliary equation $m^2 - 4m + 4 = 0$. Factorising gives: $(m - 2)(m - 2) = 0$, from which, $m = 2$ twice.

(iii) Since the roots are real and equal, the C.F., $u = (Ax + B)e^{2x}$

(iv) Since e^{2x} **and** xe^{2x} both appear in the C.F. let the P.I., $v = kx^2e^{2x}$ (see Table 54.1(c), snag case (ii)).

(v) Substituting $v = kx^2e^{2x}$ into $(D^2 - 4D + 4)v = 3e^{2x}$ gives: $(D^2 - 4D + 4)(kx^2e^{2x}) = 3e^{2x}$

$$D(kx^2e^{2x}) = (kx^2)(2e^{2x}) + (e^{2x})(2kx)$$
$$= 2ke^{2x}(x^2 + x)$$
$$D^2(kx^2e^{2x}) = D[2ke^{2x}(x^2 + x)]$$
$$= (2ke^{2x})(2x + 1) + (x^2 + x)(4ke^{2x})$$
$$= 2ke^{2x}(4x + 1 + 2x^2)$$

Hence $(D^2 - 4D + 4)(kx^2e^{2x})$
$$= [2ke^{2x}(4x + 1 + 2x^2)]$$
$$\quad - 4[2ke^{2x}(x^2 + x)] + 4[kx^2e^{2x}]$$
$$= 3e^{2x}$$

from which, $2ke^{2x} = 3e^{2x}$ and $k = \frac{3}{2}$

Hence the P.I., $v = kx^2e^{2x} = \frac{3}{2}x^2e^{2x}$

(vi) The general solution, $y = u + v$, i.e.

$$y = (Ax + B)e^{2x} + \tfrac{3}{2}x^2e^{2x}$$

Now try the following Practice Exercise

Practice Exercise 206 Second-order differential equations of the form $a\dfrac{d^2y}{dx^2} + b\dfrac{dy}{dx} + cy = f(x)$ **where** $f(x)$ **is an exponential function (Answers on page 887)**

In Problems 1 to 4, find the general solutions of the given differential equations.

1. $\dfrac{d^2y}{dx^2} - \dfrac{dy}{dx} - 6y = 2e^x$

2. $\dfrac{d^2y}{dx^2} - 3\dfrac{dy}{dx} - 4y = 3e^{-x}$

3. $\dfrac{d^2y}{dx^2} + 9y = 26e^{2x}$

4. $9\dfrac{d^2y}{dt^2} - 6\dfrac{dy}{dt} + y = 12e^{\frac{t}{3}}$

In problems 5 and 6 find the particular solutions of the given differential equations.

5. $5\dfrac{d^2y}{dx^2} + 9\dfrac{dy}{dx} - 2y = 3e^x$; when $x = 0$, $y = \dfrac{1}{4}$ and $\dfrac{dy}{dx} = 0$

6. $\dfrac{d^2y}{dt^2} - 6\dfrac{dy}{dt} + 9y = 4e^{3t}$; when $t = 0$, $y = 2$ and $\dfrac{dy}{dt} = 0$

54.5 Worked problems on differential equations of the form $a\dfrac{d^2y}{dx^2} + b\dfrac{dy}{dx} + cy = f(x)$ where $f(x)$ is a sine or cosine function

Problem 7. Solve the differential equation $2\dfrac{d^2y}{dx^2} + 3\dfrac{dy}{dx} - 5y = 6\sin 2x$

Using the procedure of Section 54.2:

(i) $2\dfrac{d^2y}{dx^2} + 3\dfrac{dy}{dx} - 5y = 6\sin 2x$ in D-operator form is $(2D^2 + 3D - 5)y = 6\sin 2x$

(ii) The auxiliary equation is $2m^2 + 3m - 5 = 0$, from which,
$$(m - 1)(2m + 5) = 0,$$
i.e. $m = 1$ or $m = -\frac{5}{2}$

(iii) Since the roots are real and different the C.F., $u = Ae^x + Be^{-\frac{5}{2}x}$

(iv) Let the P.I., $v = A\sin 2x + B\cos 2x$ (see Table 54.1(d)).

(v)   Substituting $v = A\sin 2x + B\cos 2x$ into
$(2D^2 + 3D - 5)v = 6\sin 2x$ gives:
$(2D^2 + 3D - 5)(A\sin 2x + B\cos 2x) = 6\sin 2x$

$D(A\sin 2x + B\cos 2x)$

$$= 2A\cos 2x - 2B\sin 2x$$

$D^2(A\sin 2x + B\cos 2x)$

$$= D(2A\cos 2x - 2B\sin 2x)$$

$$= -4A\sin 2x - 4B\cos 2x$$

Hence $(2D^2 + 3D - 5)(A\sin 2x + B\cos 2x)$

$$= -8A\sin 2x - 8B\cos 2x + 6A\cos 2x$$

$$- 6B\sin 2x - 5A\sin 2x - 5B\cos 2x$$

$$= 6\sin 2x$$

Equating coefficient of $\sin 2x$ gives:

$$-13A - 6B = 6 \qquad (1)$$

Equating coefficients of $\cos 2x$ gives:

$$6A - 13B = 0 \qquad (2)$$

$$6 \times (1)\text{gives}: \quad -78A - 36B = 36 \qquad (3)$$

$$13 \times (2)\text{gives}: \quad 78A - 169B = 0 \qquad (4)$$

$$(3) + (4)\text{gives}: \quad -205B = 36$$

$$\text{from which,} \qquad B = \dfrac{-36}{205}$$

Substituting $B = \dfrac{-36}{205}$ into equation (1) or (2)
gives $A = \dfrac{-78}{205}$

Hence the P.I., $\mathbf{\mathit{v} = \dfrac{-78}{205}\sin 2x - \dfrac{36}{205}\cos 2x}$

(vi)   The general solution, $y = u + v$, i.e.

$$\mathbf{\mathit{y} = A\mathit{e}^x + B\mathit{e}^{-\frac{5}{2}x}}$$

$$\mathbf{-\dfrac{2}{205}(39\sin 2x + 18\cos 2x)}$$

Problem 8.   Solve $\dfrac{d^2y}{dx^2} + 16y = 10\cos 4x$ given
$y = 3$ and $\dfrac{dy}{dx} = 4$ when $x = 0$

Using the procedure of Section 54.2:

(i) $\dfrac{d^2y}{dx^2} + 16y = 10\cos 4x$ in D-operator form is

$$(D^2 + 16)y = 10\cos 4x$$

(ii)   The auxiliary equation is $m^2 + 16 = 0$, from
which $m = \sqrt{-16} = \pm j4$

(iii)   Since the roots are complex the C.F.,
$u = e^0(A\cos 4x + B\sin 4x)$

i.e. $\mathbf{\mathit{u} = A\cos 4x + B\sin 4x}$

(iv)   Since $\sin 4x$ occurs in the C.F. **and** in the
right-hand side of the given differential equation, let the P.I., $v = x(C\sin 4x + D\cos 4x)$ (see
Table 54.1(d), snag case – constants C and D are
used since A and B have already been used in the
C.F.).

(v)   Substituting $v = x(C\sin 4x + D\cos 4x)$ into
$(D^2 + 16)v = 10\cos 4x$ gives:

$$(D^2 + 16)[x(C\sin 4x + D\cos 4x)]$$

$$= 10\cos 4x$$

$D[x(C\sin 4x + D\cos 4x)]$

$$= x(4C\cos 4x - 4D\sin 4x)$$

$$+ (C\sin 4x + D\cos 4x)(1),$$

$$\text{by the product rule}$$

$D^2[x(C\sin 4x + D\cos 4x)]$

$$= x(-16C\sin 4x - 16D\cos 4x)$$

$$+ (4C\cos 4x - 4D\sin 4x)$$

$$+ (4C\cos 4x - 4D\sin 4x)$$

Hence $(D^2 + 16)[x(C\sin 4x + D\cos 4x)]$

$$= -16Cx\sin 4x - 16Dx\cos 4x + 4C\cos 4x$$

$$- 4D\sin 4x + 4C\cos 4x - 4D\sin 4x$$

$$+ 16Cx\sin 4x + 16Dx\cos 4x$$

$$= 10\cos 4x,$$

i.e. $-8D\sin 4x + 8C\cos 4x = 10\cos 4x$

Equating coefficients of $\cos 4x$ gives:
$8C = 10$, from which, $C = \dfrac{10}{8} = \dfrac{5}{4}$

Equating coefficients of $\sin 4x$ gives:
$-8D = 0$, from which, $D = 0$

Hence the P.I., $v = x\left(\frac{5}{4}\sin 4x\right)$

(vi) The general solution, $y = u + v$, i.e.

$$y = A\cos 4x + B\sin 4x + \tfrac{5}{4}x\sin 4x$$

(vii) When $x = 0$, $y = 3$, thus

$3 = A\cos 0 + B\sin 0 + 0$, i.e. $A = 3$

$$\frac{dy}{dx} = -4A\sin 4x + 4B\cos 4x$$

$$+ \tfrac{5}{4}x(4\cos 4x) + \tfrac{5}{4}\sin 4x$$

When $x = 0$, $\dfrac{dy}{dx} = 4$, thus

$$4 = -4A\sin 0 + 4B\cos 0 + 0 + \tfrac{5}{4}\sin 0$$

i.e. $4 = 4B$, from which, $B = 1$

Hence the particular solution is

$$y = 3\cos 4x + \sin 4x + \tfrac{5}{4}x\sin 4x$$

Now try the following Practice Exercise

Practice Exercise 207 Second order differential equations of the form $a\dfrac{d^2y}{dx^2} + b\dfrac{dy}{dx} + cy = f(x)$ where $f(x)$ is a sine or cosine function (Answers on page 887)

In Problems 1 to 3, find the general solutions of the given differential equations.

1. $2\dfrac{d^2y}{dx^2} - \dfrac{dy}{dx} - 3y = 25\sin 2x$

2. $\dfrac{d^2y}{dx^2} - 4\dfrac{dy}{dx} + 4y = 5\cos x$

3. $\dfrac{d^2y}{dx^2} + y = 4\cos x$

4. Find the particular solution of the differential equation $\dfrac{d^2y}{dx^2} - 3\dfrac{dy}{dx} - 4y = 3\sin x$; when $x = 0$, $y = 0$ and $\dfrac{dy}{dx} = 0$

5. A differential equation representing the motion of a body is $\dfrac{d^2y}{dt^2} + n^2y = k\sin pt$,

where k, n and p are constants. Solve the equation (given $n \neq 0$ and $p^2 \neq n^2$) given that when $t = 0$, $y = \dfrac{dy}{dt} = 0$

6. The motion of a vibrating mass is given by $\dfrac{d^2y}{dt^2} + 8\dfrac{dy}{dt} + 20y = 300\sin 4t$. Show that the general solution of the differential equation is given by:

$$y = e^{-4t}(A\cos 2t + B\sin 2t)$$

$$+ \frac{15}{13}(\sin 4t - 8\cos 4t)$$

7. $L\dfrac{d^2q}{dt^2} + R\dfrac{dq}{dt} + \dfrac{1}{C}q = V_0\sin\omega t$ represents the variation of capacitor charge in an electric circuit. Determine an expression for q at time t seconds given that $R = 40\,\Omega$, $L = 0.02\,\text{H}$, $C = 50 \times 10^{-6}\,\text{F}$, $V_0 = 540.8\,\text{V}$ and $\omega = 200\,\text{rad/s}$ and given the boundary conditions that when $t = 0$, $q = 0$ and $\dfrac{dq}{dt} = 4.8$

54.6 Worked problems on differential equations of the form $a\dfrac{d^2y}{dx^2} + b\dfrac{dy}{dx} + cy = f(x)$ where $f(x)$ is a sum or a product

Problem 9. Solve

$$\dfrac{d^2y}{dx^2} + \dfrac{dy}{dx} - 6y = 12x - 50\sin x$$

Using the procedure of Section 54.2:

(i) $\dfrac{d^2y}{dx^2} + \dfrac{dy}{dx} - 6y = 12x - 50\sin x$ in D-operator form is

$$(D^2 + D - 6)y = 12x - 50\sin x$$

(ii) The auxiliary equation is $(m^2 + m - 6) = 0$, from which,

$$(m - 2)(m + 3) = 0,$$

i.e. $m = 2$ or $m = -3$

(iii) Since the roots are real and different, the C.F.,
$$u=Ae^{2x}+Be^{-3x}$$

(iv) Since the right-hand side of the given differential equation is the sum of a polynomial and a sine function let the P.I. $v=ax+b+c\sin x+d\cos x$ (see Table 54.1(e)).

(v) Substituting v into
$$(D^2+D-6)v=12x-50\sin x \text{ gives:}$$

$$(D^2+D-6)(ax+b+c\sin x+d\cos x)$$
$$=12x-50\sin x$$

$$D(ax+b+c\sin x+d\cos x)$$
$$=a+c\cos x-d\sin x$$

$$D^2(ax+b+c\sin x+d\cos x)$$
$$=-c\sin x-d\cos x$$

Hence $(D^2+D-6)(v)$

$$=(-c\sin x-d\cos x)+(a+c\cos x$$
$$-d\sin x)-6(ax+b+c\sin x+d\cos x)$$
$$=12x-50\sin x$$

Equating constant terms gives:

$$a-6b=0 \tag{1}$$

Equating coefficients of x gives: $-6a=12$, from which, $a=-2$

Hence, from (1), $b=-\frac{1}{3}$

Equating the coefficients of $\cos x$ gives:

$$-d+c-6d=0$$
i.e. $c-7d=0 \tag{2}$

Equating the coefficients of $\sin x$ gives:

$$-c-d-6c=-50$$
i.e. $-7c-d=-50 \tag{3}$

Solving equations (2) and (3) gives: $c=7$ and $d=1$
Hence the P.I.,

$$v=-2x-\tfrac{1}{3}+7\sin x+\cos x$$

(vi) The general solution, $y=u+v$,

i.e. $y=Ae^{2x}+Be^{-3x}-2x$
$$-\tfrac{1}{3}+7\sin x+\cos x$$

Problem 10.   Solve the differential equation $\frac{d^2y}{dx^2}-2\frac{dy}{dx}+2y=3e^x\cos 2x$, given that when $x=0$, $y=2$ and $\frac{dy}{dx}=3$

Using the procedure of Section 54.2:

(i) $\frac{d^2y}{dx^2}-2\frac{dy}{dx}+2y=3e^x\cos 2x$ in   D-operator form is

$$(D^2-2D+2)y=3e^x\cos 2x$$

(ii) The auxiliary equation is $m^2-2m+2=0$ Using the quadratic formula,

$$m=\frac{2\pm\sqrt{[4-4(1)(2)]}}{2}$$
$$=\frac{2\pm\sqrt{-4}}{2}=\frac{2\pm j2}{2} \text{ i.e. } m=1\pm j1$$

(iii) Since the roots are complex, the C.F.,
$$u=e^x(A\cos x+B\sin x)$$

(iv) Since the right-hand side of the given differential equation is a product of an exponential and a cosine function, let the P.I., $v=e^x(C\sin 2x+D\cos 2x)$ (see Table 54.1(f) – again, constants C and D are used since A and B have already been used for the C.F.).

(v) Substituting v into $(D^2-2D+2)v=3e^x\cos 2x$ gives:

$$(D^2-2D+2)[e^x(C\sin 2x+D\cos 2x)]$$
$$=3e^x\cos 2x$$

$$D(v)=e^x(2C\cos 2x-2D\sin 2x)$$
$$+e^x(C\sin 2x+D\cos 2x)$$
$$(\equiv e^x\{(2C+D)\cos 2x$$
$$+(C-2D)\sin 2x\})$$

$$D^2(v)=e^x(-4C\sin 2x-4D\cos 2x)$$
$$+e^x(2C\cos 2x-2D\sin 2x)$$
$$+e^x(2C\cos 2x-2D\sin 2x)$$
$$+e^x(C\sin 2x+D\cos 2x)$$

$$\equiv e^x\{(-3C - 4D)\sin 2x + (4C - 3D)\cos 2x\}$$

Hence $(D^2 - 2D + 2)v$

$$= e^x\{(-3C - 4D)\sin 2x$$
$$+ (4C - 3D)\cos 2x\}$$
$$- 2e^x\{(2C + D)\cos 2x$$
$$+ (C - 2D)\sin 2x\}$$
$$+ 2e^x(C\sin 2x + D\cos 2x)$$
$$= 3e^x\cos 2x$$

Equating coefficients of $e^x\sin 2x$ gives:

$$-3C - 4D - 2C + 4D + 2C = 0$$

i.e. $-3C = 0$, from which, $C = 0$

Equating coefficients of $e^x\cos 2x$ gives:

$$4C - 3D - 4C - 2D + 2D = 3$$

i.e. $-3D = 3$, from which, $D = -1$

Hence the P.I., $v = e^x(-\cos 2x)$

(vi) The general solution, $y = u + v$, i.e.

$$y = e^x(A\cos x + B\sin x) - e^x\cos 2x$$

(vii) When $x = 0$, $y = 2$ thus

$$2 = e^0(A\cos 0 + B\sin 0) - e^0\cos 0$$

i.e. $\quad 2 = A - 1$, from which, $A = 3$

$$\frac{dy}{dx} = e^x(-A\sin x + B\cos x)$$
$$+ e^x(A\cos x + B\sin x)$$
$$- [e^x(-2\sin 2x) + e^x\cos 2x]$$

When $\quad x = 0, \dfrac{dy}{dx} = 3$

thus $\quad 3 = e^0(-A\sin 0 + B\cos 0)$

$$+ e^0(A\cos 0 + B\sin 0)$$
$$- e^0(-2\sin 0) - e^0\cos 0$$

i.e. $\quad 3 = B + A - 1$, from which,

$$B = 1, \text{ since } A = 3$$

Hence the particular solution is

$$y = e^x(3\cos x + \sin x) - e^x\cos 2x$$

Now try the following Practice Exercise

Practice Exercise 208 Second-order differential equations of the form

$$a\frac{d^2y}{dx^2} + b\frac{dy}{dx} + cy = f(x) \text{ where } f(x) \text{ is a sum}$$

or product (Answers on page 887)

In Problems 1 to 4, find the general solutions of the given differential equations.

1. $8\dfrac{d^2y}{dx^2} - 6\dfrac{dy}{dx} + y = 2x + 40\sin x$

2. $\dfrac{d^2y}{d\theta^2} - 3\dfrac{dy}{d\theta} + 2y = 2\sin 2\theta - 4\cos 2\theta$

3. $\dfrac{d^2y}{dx^2} + \dfrac{dy}{dx} - 2y = x^2 + e^{2x}$

4. $\dfrac{d^2y}{dt^2} - 2\dfrac{dy}{dt} + 2y = e^t\sin t$

In Problems 5 to 6 find the particular solutions of the given differential equations.

5. $\dfrac{d^2y}{dx^2} - 7\dfrac{dy}{dx} + 10y = e^{2x} + 20$; when $x = 0$,
   $y = 0$ and $\dfrac{dy}{dx} = -\dfrac{1}{3}$

6. $2\dfrac{d^2y}{dx^2} - \dfrac{dy}{dx} - 6y = 6e^x\cos x$; when $x = 0$,
   $y = -\dfrac{21}{29}$ and $\dfrac{dy}{dx} = -6\dfrac{20}{29}$

For fully worked solutions to each of the problems in Practice Exercises 205 to 208 in this chapter,
go to the website:
www.routledge.com/cw/bird

Chapter 55

Power series methods of solving ordinary differential equations

Why it is important to understand: Power series methods of solving ordinary differential equations

The differential equations studied so far have all had closed form solutions, that is, their solutions could be expressed in terms of elementary functions, such as exponential, trigonometric, polynomial, and logarithmic functions, and most such elementary functions have expansions in terms of power series. However, there are a whole class of functions which are not elementary functions and which occur frequently in mathematical physics and engineering. These equations can sometimes be solved by discovering a power series that satisfies the differential equation, but the solution series may not be summable to an elementary function. In this chapter the methods of solution to such equations are explained.

At the end of this chapter, you should be able to:

- appreciate the reason for using power series methods to solve differential equations
- determine higher order differential coefficients as a series
- use Leibniz's theorem to obtain the nth derivative of a given function
- obtain a power series solution of a differential equation by the Leibniz–Maclaurin method
- obtain a power series solution of a differential equation by the Frobenius method
- determine the general power series solution of Bessel's equation
- express Bessel's equation in terms of gamma functions
- determine the general power series solution of Legendre's equation
- determine Legendre polynomials
- determine Legendre polynomials using Rodrigues' formula

55.1 Introduction

Second-order ordinary differential equations that cannot be solved by analytical methods (as shown in Chapters 53 and 54), i.e. those involving variable coefficients, can often be solved in the form of an infinite series of powers of the variable. This chapter looks at some of the methods that make this possible – by the Leibniz–Maclaurin and Frobinius methods, involving Bessel's and Legendre's equations, Bessel and gamma functions and Legendre's polynomials. Before introducing Leibniz's theorem, some trends with higher differential coefficients are considered. To better understand this chapter it is necessary to be able to:

(i) differentiate standard functions (as explained in Chapters 29 and 37),

(ii) appreciate the binomial theorem (as explained in Chapter 7), and

(iii) use Maclaurin's theorem (as explained in Chapter 8).

55.2 Higher order differential coefficients as series

The following is an extension of successive differentiation (see page 355), but looking for trends, or series, as the differential coefficient of common functions rises.

(i) If $y = e^{ax}$, then $\dfrac{dy}{dx} = ae^{ax}$, $\dfrac{d^2 y}{dx^2} = a^2 e^{ax}$, and so on.

If we abbreviate $\dfrac{dy}{dx}$ as y', $\dfrac{d^2 y}{dx^2}$ as y'', ... and $\dfrac{d^n y}{dx^n}$ as $y^{(n)}$, then $y' = ae^{ax}$, $y'' = a^2 e^{ax}$, and the

emerging pattern gives: $\quad y^{(n)} = a^n e^{ax}$ (1)

For example, if $y = 3e^{2x}$, then
$$\frac{d^7 y}{dx^7} = y^{(7)} = 3(2^7) e^{2x} = 384 e^{2x}$$

(ii) If $y = \sin ax$,

$$y' = a\cos ax = a\sin\left(ax + \frac{\pi}{2}\right)$$

$$y'' = -a^2 \sin ax = a^2 \sin(ax + \pi)$$

$$= a^2 \sin\left(ax + \frac{2\pi}{2}\right)$$

$$y''' = -a^3 \cos x$$

$$= a^3 \sin\left(ax + \frac{3\pi}{2}\right) \text{ and so on.}$$

In general, $\quad y^{(n)} = a^n \sin\left(ax + \dfrac{n\pi}{2}\right)$ (2)

For example, if

$$y = \sin 3x, \text{ then } \frac{d^5 y}{dx^5} = y^{(5)}$$

$$= 3^5 \sin\left(3x + \frac{5\pi}{2}\right) = 3^5 \sin\left(3x + \frac{\pi}{2}\right)$$

$$= 243 \cos 3x$$

(iii) If $y = \cos ax$,

$$y' = -a\sin ax = a\cos\left(ax + \frac{\pi}{2}\right)$$

$$y'' = -a^2 \cos ax = a^2 \cos\left(ax + \frac{2\pi}{2}\right)$$

$$y''' = a^3 \sin ax = a^3 \cos\left(ax + \frac{3\pi}{2}\right) \text{ and so on.}$$

In general, $\quad y^{(n)} = a^n \cos\left(ax + \dfrac{n\pi}{2}\right)$ (3)

For example, if $y = 4\cos 2x$,

$$\text{then } \frac{d^6 y}{dx^6} = y^{(6)} = 4(2^6) \cos\left(2x + \frac{6\pi}{2}\right)$$

$$= 4(2^6) \cos(2x + 3\pi)$$

$$= 4(2^6) \cos(2x + \pi)$$

$$= -256 \cos 2x$$

(iv) If $y = x^a$, $y' = ax^{a-1}$, $y'' = a(a-1)x^{a-2}$,

$$y''' = a(a-1)(a-2)x^{a-3},$$

and $y^{(n)} = a(a-1)(a-2)\ldots(a-n+1)x^{a-n}$

or $y^{(n)} = \dfrac{a!}{(a-n)!} x^{a-n}$ (4)

where a is a positive integer.

For example, if $y = 2x^6$, then $\dfrac{d^4 y}{dx^4} = y^{(4)}$

$$= (2)\frac{6!}{(6-4)!} x^{6-4}$$

$$= (2)\frac{6 \times 5 \times 4 \times 3 \times 2 \times 1}{2 \times 1} x^2$$

$$= 720 x^2$$

(v) If $y = \sinh ax$, $y' = a \cosh ax$

$$y'' = a^2 \sinh ax$$

$$y''' = a^3 \cosh ax, \text{ and so on}$$

Since $\sinh ax$ is not periodic (see graph on page 43), it is more difficult to find a general statement for $y^{(n)}$. However, this is achieved with the following general series:

$$y^{(n)} = \frac{a^n}{2}\{[1 + (-1)^n]\sinh ax$$

$$+ [1 - (-1)^n]\cosh ax\} \qquad (5)$$

For example, if

$$y = \sinh 2x, \text{ then } \frac{d^5 y}{dx^5} = y^{(5)}$$

$$= \frac{2^5}{2}\{[1 + (-1)^5]\sinh 2x$$

$$+ [1 - (-1)^5]\cosh 2x\}$$

$$= \frac{2^5}{2}\{[0]\sinh 2x + [2]\cosh 2x\}$$

$$= 32 \cosh 2x$$

(vi) If $y = \cosh ax$,

$$y' = a \sinh ax$$

$$y'' = a^2 \cosh ax$$

$$y''' = a^3 \sinh ax, \text{ and so on}$$

Since $\cosh ax$ is not periodic (see graph on page 181), again it is more difficult to find a general statement for $y^{(n)}$. However, this is achieved with the following general series:

$$y^{(n)} = \frac{a^n}{2}\{[1 - (-1)^n]\sinh ax$$

$$+ [1 + (-1)^n]\cosh ax\} \qquad (6)$$

For example, if $y = \frac{1}{9}\cosh 3x$,

then $\dfrac{d^7 y}{dx^7} = y^{(7)} = \left(\dfrac{1}{9}\right)\dfrac{3^7}{2}(2\sinh 3x)$

$$= 243 \sinh 3x$$

(vii) If $y = \ln ax$, $y' = \dfrac{1}{x}$, $y'' = -\dfrac{1}{x^2}$, $y''' = \dfrac{2}{x^3}$, and so on.

In general, $y^{(n)} = (-1)^{n-1}\dfrac{(n-1)!}{x^n} \qquad (7)$

For example, if $y = \ln 5x$, then

$$\frac{d^6 y}{dx^6} = y^{(6)} = (-1)^{6-1}\left(\frac{5!}{x^6}\right) = -\frac{120}{x^6}$$

Note that if $y = \ln x$, $y' = \dfrac{1}{x}$; if in equation (7),

$n = 1$ then $y' = (-1)^0 \dfrac{(0)!}{x^1}$

$(-1)^0 = 1$ and if $y' = \dfrac{1}{x}$ then $(0)! = 1$ (check that $(-1)^0 = 1$ and $(0)! = 1$ on a calculator).

Now try the following Practice Exercise

Exercise 209   Higher-order differential coefficients as series (Answers on page 887)

Determine the following derivatives:

1. (a) $y^{(4)}$ when $y = e^{2x}$ (b) $y^{(5)}$ when $y = 8e^{\frac{t}{2}}$

2. (a) $y^{(4)}$ when $y = \sin 3t$
    (b) $y^{(7)}$ when $y = \dfrac{1}{50}\sin 5\theta$

3. (a) $y^{(8)}$ when $y = \cos 2x$
    (b) $y^{(9)}$ when $y = 3\cos\dfrac{2}{3}t$

4. (a) $y^{(7)}$ when $y = 2x^9$ (b) $y^{(6)}$ when $y = \dfrac{t^7}{8}$

5. (a) $y^{(7)}$ when $y = \dfrac{1}{4}\sinh 2x$
    (b) $y^{(6)}$ when $y = 2\sinh 3x$

6. (a) $y^{(7)}$ when $y = \cosh 2x$
    (b) $y^{(8)}$ when $y = \dfrac{1}{9}\cosh 3x$

7. (a) $y^{(4)}$ when $y = 2\ln 3\theta$
    (b) $y^{(7)}$ when $y = \dfrac{1}{3}\ln 2t$

55.3   Leibniz's theorem

If $y = uv \qquad (8)$

where u and v are each functions of x, then by using the product rule,

$$y' = uv' + vu' \qquad (9)$$

$$y'' = uv'' + v'u' + vu'' + u'v'$$

$$= u''v + 2u'v' + uv'' \qquad (10)$$

$$y''' = u''v' + vu''' + 2u'v'' + 2v'u'' + uv''' + v''u'$$

$$= u'''v + 3u''v' + 3u'v'' + uv''' \qquad (11)$$

$$y^{(4)} = u^{(4)}v + 4u^{(3)}v^{(1)} + 6u^{(2)}v^{(2)}$$

$$+ 4u^{(1)}v^{(3)} + uv^{(4)} \qquad (12)$$

From equations (8) to (12) it is seen that

(a) the nth derivative of u decreases by 1 moving from left to right,

(b) the nth derivative of v increases by 1 moving from left to right,

(c) the coefficients $1, 4, 6, 4, 1$ are the normal binomial coefficients (see page 59).

In fact, $(uv)^{(n)}$ may be obtained by expanding $(u+v)^{(n)}$ using the binomial theorem (see page 60), where the 'powers' are interpreted as derivatives. Thus, expanding $(u+v)^{(n)}$ gives:

$$y^{(n)} = (uv)^{(n)} = u^{(n)}v + nu^{(n-1)}v^{(1)}$$

$$+ \frac{n(n-1)}{2!}u^{(n-2)}v^{(2)}$$

$$+ \frac{n(n-1)(n-2)}{3!}u^{(n-3)}v^{(3)} + \cdots \qquad (13)$$

Equation (13) is a statement of **Leibniz's theorem**[*] which can be used to differentiate a product n times. The theorem is demonstrated in the following worked problems.

> **Problem 1.** Determine $y^{(n)}$ when $y = x^2 e^{3x}$

For a product $y = uv$, the function taken as

(i) u is the one whose nth derivative can readily be determined (from equations (1) to (7)),

---

[*] **Who was Leibniz?** For image and resume see page 91. For more information, go to **www.routledge.com/cw/bird**

(ii) v is the one whose derivative reduces to zero after a few stages of differentiation.

Thus, when $y = x^2 e^{3x}$, $v = x^2$, since its third derivative is zero, and $u = e^{3x}$ since the nth derivative is known from equation (1), i.e. $3^n e^{ax}$

Using Leinbiz's theorem (equation (13),

$$y^{(n)} = u^{(n)}v + nu^{(n-1)}v^{(1)} + \frac{n(n-1)}{2!}u^{(n-2)}v^{(2)}$$

$$+ \frac{n(n-1)(n-2)}{3!}u^{(n-3)}v^{(3)} + \cdots$$

where in this case $v = x^2$, $v^{(1)} = 2x$, $v^{(2)} = 2$ and $v^{(3)} = 0$

Hence, $\quad y^{(n)} = (3^n e^{3x})(x^2) + n(3^{n-1}e^{3x})(2x)$

$$+ \frac{n(n-1)}{2!}(3^{n-2}e^{3x})(2)$$

$$+ \frac{n(n-1)(n-2)}{3!}(3^{n-3}e^{3x})(0)$$

$$= 3^{n-2}e^{3x}(3^2 x^2 + n(3)(2x)$$

$$+ n(n-1) + 0)$$

i.e. $\quad y^{(n)} = e^{3x}3^{n-2}(9x^2 + 6nx + n(n-1))$

> **Problem 2.** If $x^2 y'' + 2xy' + y = 0$ show that:
> $xy^{(n+2)} + 2(n+1)xy^{(n+1)} + (n^2 + n + 1)y^{(n)} = 0$

Differentiating each term of $x^2 y'' + 2xy' + y = 0$ n times, using Leibniz's theorem of equation (13), gives:

$$\left\{ y^{(n+2)}x^2 + n\, y^{(n+1)}(2x) + \frac{n(n-1)}{2!}y^{(n)}(2) + 0 \right\}$$

$$+ \{ y^{(n+1)}(2x) + n\, y^{(n)}(2) + 0 \} + \{ y^{(n)} \} = 0$$

i.e. $\quad x^2 y^{(n+2)} + 2nxy^{(n+1)} + n(n-1)y^{(n)}$

$$+ 2xy^{(n+1)} + 2n\, y^{(n)} + y^{(n)} = 0$$

i.e. $\quad x^2 y^{(n+2)} + 2(n+1)xy^{(n+1)}$

$$+ (n^2 - n + 2n + 1)y^{(n)} = 0$$

or $x^2 y^{(n+2)} + 2(n+1)xy^{(n+1)}$

$$+ (n^2 + n + 1)y^{(n)} = 0$$

Problem 3. Differentiate the following differential equation n times:
$$(1+x^2)y'' + 2xy' - 3y = 0$$

By Leibniz's equation, equation (13),

$$\left\{ y^{(n+2)}(1+x^2) + n\,y^{(n+1)}(2x) + \frac{n(n-1)}{2!}y^{(n)}(2) + 0 \right\}$$

$$+ 2\{y^{(n+1)}(x) + n\,y^{(n)}(1) + 0\} - 3\{y^{(n)}\} = 0$$

i.e. $(1+x^2)y^{(n+2)} + 2nxy^{(n+1)} + n(n-1)y^{(n)}$

$$+ 2xy^{(n+1)} + 2ny^{(n)} - 3y^{(n)} = 0$$

or $(1+x^2)y^{(n+2)} + 2(n+1)xy^{(n+1)}$

$$+ (n^2 - n + 2n - 3)y^{(n)} = 0$$

i.e. $(1+x^2)y^{(n+2)} + 2(n+1)xy^{(n+1)}$

$$+ (n^2 + n - 3)y^{(n)} = 0$$

Problem 4. Find the fifth derivative of $y = x^4 \sin x$

If $y = x^4 \sin x$, then using Leibniz's equation with $u = \sin x$ and $v = x^4$ gives:

$$y^{(n)} = \left[\sin\left(x + \frac{n\pi}{2}\right)x^4 \right]$$

$$+ n\left[\sin\left(x + \frac{(n-1)\pi}{2}\right)4x^3 \right]$$

$$+ \frac{n(n-1)}{2!}\left[\sin\left(x + \frac{(n-2)\pi}{2}\right)12x^2 \right]$$

$$+ \frac{n(n-1)(n-2)}{3!}\left[\sin\left(x + \frac{(n-3)\pi}{2}\right)24x \right]$$

$$+ \frac{n(n-1)(n-2)(n-3)}{4!}\left[\sin\left(x \right.\right.$$

$$\left.\left. + \frac{(n-4)\pi}{2}\right)24 \right]$$

and $y^{(5)} = x^4 \sin\left(x + \frac{5\pi}{2}\right) + 20x^3 \sin(x + 2\pi)$

$$+ \frac{(5)(4)}{2}(12x^2)\sin\left(x + \frac{3\pi}{2}\right)$$

$$+ \frac{(5)(4)(3)}{(3)(2)}(24x)\sin(x + \pi)$$

$$+ \frac{(5)(4)(3)(2)}{(4)(3)(2)}(24)\sin\left(x + \frac{\pi}{2}\right)$$

Since $\sin\left(x + \frac{5\pi}{2}\right) \equiv \sin\left(x + \frac{\pi}{2}\right) \equiv \cos x,$

$$\sin(x + 2\pi) \equiv \sin x, \sin\left(x + \frac{3\pi}{2}\right) \equiv -\cos x,$$

and $\sin(x + \pi) \equiv -\sin x,$

then $y^{(5)} = x^4 \cos x + 20x^3 \sin x + 120x^2(-\cos x)$

$$+ 240x(-\sin x) + 120\cos x$$

i.e. $y^{(5)} = (x^4 - 120x^2 + 120)\cos x$

$$+ (20x^3 - 240x)\sin x$$

Now try the following Practice Exercise

**Practice Exercise 210   Leibniz's theorem
(Answers on page 887)**

Use the theorem of Leibniz in the following problems:

1. Obtain the nth derivative of: $x^2 y$

2. If $y = x^3 e^{2x}$ find $y^{(n)}$ and hence $y^{(3)}$

3. Determine the fourth derivative of: $y = 2x^3 e^{-x}$

4. If $y = x^3 \cos x$ determine the fifth derivative

5. Find an expression for $y^{(4)}$ if $y = e^{-t}\sin t$

6. If $y = x^5 \ln 2x$ find $y^{(3)}$

7. Given $2x^2 y'' + xy' + 3y = 0$ show that $2x^2 y^{(n+2)} + (4n + 1)xy^{(n+1)}$
$$+ (2n^2 - n + 3)y^{(n)} = 0$$

8. If $y = (x^3 + 2x^2)e^{2x}$ determine an expansion for $y^{(5)}$

55.4 Power series solution by the Leibniz–Maclaurin method

For second-order differential equations that cannot be solved by algebraic methods, the **Leibniz–Maclaurin**[*] **method** produces a solution in the form of infinite series of powers of the unknown variable. The following simple **five-step procedure** may be used in the Leibniz–Maclaurin method:

(i) Differentiate the given equation n times, using the Leibniz theorem of equation (13),

(ii) rearrange the result to obtain the recurrence relation at $x = 0$,

(iii) determine the values of the derivatives at $x = 0$, i.e. find $(y)_0$ and $(y')_0$,

(iv) substitute in the Maclaurin expansion for $y = f(x)$ (see page 70, equation (5)),

(v) simplify the result where possible and apply boundary condition (if given).

The Leibniz–Maclaurin method is demonstrated, using the above procedure, in the following worked problems.

Problem 5. Determine the power series solution of the differential equation:
$\dfrac{d^2 y}{dx^2} + x\dfrac{dy}{dx} + 2y = 0$ using the Leibniz–Maclaurin method, given the boundary conditions that at $x = 0$, $y = 1$ and $\dfrac{dy}{dx} = 2$

Following the above procedure:

(i) The differential equation is rewritten as: $y'' + xy' + 2y = 0$ and from the Leibniz theorem of equation (13), each term is differentiated n times, which gives:

$$y^{(n+2)} + \{y^{(n+1)}(x) + n\, y^{(n)}(1) + 0\} + 2\, y^{(n)} = 0$$

i.e. $\qquad y^{(n+2)} + xy^{(n+1)} + (n+2)\, y^{(n)} = 0$

$$(14)$$

* **Who was Maclaurin**? For image and resume of Maclaurin, see page 69. To find out more go to **www.routledge.com/cw/bird**

(ii) At $x = 0$, equation (14) becomes:

$$y^{(n+2)} + (n+2)\, y^{(n)} = 0$$

from which, $\quad y^{(n+2)} = -(n+2)\, y^{(n)}$

This equation is called a **recurrence relation** or **recurrence formula**, because each recurring term depends on a previous term.

(iii) Substituting $n = 0, 1, 2, 3, \ldots$ will produce a set of relationships between the various coefficients.

For $n = 0$, $\quad (y'')_0 = -2(y)_0$

$n = 1$, $\quad (y''')_0 = -3(y')_0$

$n = 2$, $\quad (y^{(4)})_0 = -4(y'')_0 = -4\{-2(y)_0\}$
$$= 2 \times 4(y)_0$$

$n = 3$, $\quad (y^{(5)})_0 = -5(y''')_0 = -5\{-3(y')_0\}$
$$= 3 \times 5(y')_0$$

$n = 4$, $(y^{(6)})_0 = -6(y^{(4)})_0 = -6\{2 \times 4(y)_0\}$
$$= -2 \times 4 \times 6(y)_0$$

$n = 5$, $(y^{(7)})_0 = -7(y^{(5)})_0 = -7\{3 \times 5(y')_0\}$
$$= -3 \times 5 \times 7(y')_0$$

$n = 6$, $(y^{(8)})_0 = -8(y^{(6)})_0 =$
$$-8\{-2 \times 4 \times 6(y)_0\} = 2 \times 4 \times 6 \times 8(y)_0$$

(iv) Maclaurin's theorem from page 70 may be written as:c

$$y = (y)_0 + x(y')_0 + \frac{x^2}{2!}(y'')_0 + \frac{x^3}{3!}(y''')_0$$
$$+ \frac{x^4}{4!}(y^{(4)})_0 + \cdots$$

Substituting the above values into Maclaurin's theorem gives:

$$y = (y)_0 + x(y')_0 + \frac{x^2}{2!}\{-2(y)_0\}$$
$$+ \frac{x^3}{3!}\{-3(y')_0\} + \frac{x^4}{4!}\{2 \times 4(y)_0\}$$
$$+ \frac{x^5}{5!}\{3 \times 5(y')_0\} + \frac{x^6}{6!}\{-2 \times 4 \times 6(y)_0\}$$
$$+ \frac{x^7}{7!}\{-3 \times 5 \times 7(y')_0\}$$
$$+ \frac{x^8}{8!}\{2 \times 4 \times 6 \times 8(y)_0\}$$

(v)   Collecting similar terms together gives:

$$y = (y)_0 \left\{ 1 - \frac{2x^2}{2!} + \frac{2 \times 4x^4}{4!} \right.$$

$$- \frac{2 \times 4 \times 6x^6}{6!} + \frac{2 \times 4 \times 6 \times 8x^8}{8!}$$

$$\left. - \cdots \right\} + (y')_0 \left\{ x - \frac{3x^3}{3!} + \frac{3 \times 5x^5}{5!} \right.$$

$$\left. - \frac{3 \times 5 \times 7x^7}{7!} + \cdots \right\}$$

i.e. $y = (y)_0 \left\{ 1 - \frac{x^2}{1} + \frac{x^4}{1 \times 3} - \frac{x^6}{3 \times 5} \right.$

$$\left. + \frac{x^8}{3 \times 5 \times 7} - \cdots \right\}$$

$$+ (y')_0 \times \left\{ \frac{x}{1} - \frac{x^3}{1 \times 2} + \frac{x^5}{2 \times 4} \right.$$

$$\left. - \frac{x^7}{2 \times 4 \times 6} + \cdots \right\}$$

The boundary conditions are that at $x = 0$, $y = 1$ and $\frac{dy}{dx} = 2$, i.e. $(y)_0 = 1$ and $(y')_0 = 2$

Hence, the power series solution of the differential equation: $\frac{d^2y}{dx^2} + x\frac{dy}{dx} + 2y = 0$ is:

$$y = \left\{ 1 - \frac{x^2}{1} + \frac{x^4}{1 \times 3} - \frac{x^6}{3 \times 5} \right.$$

$$\left. + \frac{x^8}{3 \times 5 \times 7} - \cdots \right\} + 2 \left\{ \frac{x}{1} - \frac{x^3}{1 \times 2} \right.$$

$$\left. + \frac{x^5}{2 \times 4} - \frac{x^7}{2 \times 4 \times 6} + \cdots \right\}$$

Problem 6.   Determine the power series solution of the differential equation:
$\frac{d^2y}{dx^2} + \frac{dy}{dx} + xy = 0$ given the boundary conditions that at $x = 0$, $y = 0$ and $\frac{dy}{dx} = 1$, using the Leibniz–Maclaurin method.

Following the above procedure:

(i)   The differential equation is rewritten as: $y'' + y' + xy = 0$ and from the Leibniz theorem of equation (13), each term is differentiated n times, which gives:

$$y^{(n+2)} + y^{(n+1)} + y^{(n)}(x) + n\,y^{(n-1)}(1) + 0 = 0$$

i.e. $y^{(n+2)} + y^{(n+1)} + xy^{(n)} + n\,y^{(n-1)} = 0$
(15)

(ii)   At $x = 0$, equation (15) becomes:

$$y^{(n+2)} + y^{(n+1)} + n\,y^{(n-1)} = 0$$

from which, $y^{(n+2)} = -\{y^{(n+1)} + n\,y^{(n-1)}\}$

This is the **recurrence relation** and applies for $n \geq 1$

(iii)   Substituting $n = 1, 2, 3, \ldots$ will produce a set of relationships between the various coefficients.

For $n = 1$, $(y''')_0 = -\{(y'')_0 + (y)_0\}$

$$n = 2,\quad (y^{(4)})_0 = -\{(y''')_0 + 2(y')_0\}$$

$$n = 3,\quad (y^{(5)})_0 = -\{(y^{(4)})_0 + 3(y'')_0\}$$

$$n = 4,\quad (y^{(6)})_0 = -\{(y^{(5)})_0 + 4(y''')_0\}$$

$$n = 5,\quad (y^{(7)})_0 = -\{(y^{(6)})_0 + 5(y^{(4)})_0\}$$

$$n = 6,\quad (y^{(8)})_0 = -\{(y^{(7)})_0 + 6(y^{(5)})_0\}$$

From the given boundary conditions, at $x = 0$, $y = 0$, thus $(y)_0 = 0$, and at $x = 0$, $\frac{dy}{dx} = 1$, thus $(y')_0 = 1$

From the given differential equation, $y'' + y' + xy = 0$, and, at $x = 0$, $(y'')_0 + (y')_0 + (0)y = 0$ from which, $(y'')_0 = -(y')_0 = -1$

Thus, $(y)_0 = \mathbf{0}$, $(y')_0 = \mathbf{1}$, $(y'')_0 = \mathbf{-1}$,

$$(y''')_0 = -\{(y'')_0 + (y)_0\} = -(-1 + 0) = \mathbf{1}$$

$$(y^{(4)})_0 = -\{(y''')_0 + 2(y')_0\}$$

$$= -[1 + 2(1)] = \mathbf{-3}$$

$$(y^{(5)})_0 = -\{(y^{(4)})_0 + 3(y'')_0\}$$

$$= -[-3 + 3(-1)] = \mathbf{6}$$

$$(y^{(6)})_0 = -\{(y^{(5)})_0 + 4(y''')_0\}$$

$$= -[6 + 4(1)] = \mathbf{-10}$$

$$(y^{(7)})_0 = -\{(y^{(6)})_0 + 5(y^{(4)})_0\}$$

$$= -[-10 + 5(-3)] = \mathbf{25}$$

$$(y^{(8)})_0 = -\{(y^{(7)})_0 + 6(y^{(5)})_0\}$$

$$= -[25 + 6(6)] = \mathbf{-61}$$

(iv) Maclaurin's theorem states:

$$y = (y)_0 + x(y')_0 + \frac{x^2}{2!}(y'')_0 + \frac{x^3}{3!}(y''')_0$$
$$+ \frac{x^4}{4!}(y^{(4)})_0 + \cdots$$

and substituting the above values into Maclaurin's theorem gives:

$$y = 0 + x(1) + \frac{x^2}{2!}\{-1\} + \frac{x^3}{3!}\{1\} + \frac{x^4}{4!}\{-3\}$$
$$+ \frac{x^5}{5!}\{6\} + \frac{x^6}{6!}\{-10\} + \frac{x^7}{7!}\{25\}$$
$$+ \frac{x^8}{8!}\{-61\} + \cdots$$

(v) Simplifying, the power series solution of the differential equation: $\frac{d^2y}{dx^2} + \frac{dy}{dx} + xy = 0$ is given by:

$$y = x - \frac{x^2}{2!} + \frac{x^3}{3!} - \frac{3x^4}{4!} + \frac{6x^5}{5!} - \frac{10x^6}{6!}$$
$$+ \frac{25x^7}{7!} - \frac{61x^8}{8!} + \cdots$$

Now try the following Practice Exercise

Practice Exercise 211   Power series solutions by the Leibniz-Maclaurin method (Answers on page 887)

1. Determine the power series solution of the differential equation: $\frac{d^2y}{dx^2} + 2x\frac{dy}{dx} + y = 0$ using the Leibniz–Maclaurin method, given that at $x = 0$, $y = 1$ and $\frac{dy}{dx} = 2$

2. Show that the power series solution of the differential equation: $(x+1)\frac{d^2y}{dx^2} + (x-1)\frac{dy}{dx} - 2y = 0$, using the Leibniz–Maclaurin method, is given by: $y = 1 + x^2 + e^{-x}$ given the boundary conditions that at $x = 0$, $y = 2$ and $\frac{dy}{dx} = -1$

3. Find the particular solution of the differential equation: $(x^2+1)\frac{d^2y}{dx^2} + x\frac{dy}{dx} - 4y = 0$ using the Leibniz–Maclaurin method, given

the boundary conditions that at $x = 0$, $y = 1$ and $\frac{dy}{dx} = 1$

4. Use the Leibniz–Maclaurin method to determine the power series solution for the differential equation: $x\frac{d^2y}{dx^2} + \frac{dy}{dx} + xy = 1$ given that at $x = 0$, $y = 1$ and $\frac{dy}{dx} = 2$

55.5 Power series solution by the Frobenius method

A differential equation of the form $y'' + Py' + Qy = 0$, where P and Q are both functions of x, such that the equation can be represented by a power series, may be solved by the **Frobenius method**.*

* **Who was Frobenius?** **Ferdinand Georg Frobenius** (26 October 1849–3 August 1917) was a German mathematician best known for his contributions to the theory of elliptic functions, differential equations and to group theory. He is also known for determinantal identities, known as Frobenius–Stickelberger formulae. To find out more go to **www.routledge.com/cw/bird**

The following **four-step procedure** may be used in the Frobenius method:

(i) Assume a trial solution of the form $y = x^c\{a_0 + a_1x + a_2x^2 + a_3x^3 + \cdots + a_rx^r + \cdots\}$

(ii) differentiate the trial series,

(iii) substitute the results in the given differential equation,

(iv) equate coefficients of corresponding powers of the variable on each side of the equation; this enables index c and coefficients a_1, a_2, a_3, ... from the trial solution, to be determined.

This introductory treatment of the Frobenius method covering the simplest cases is demonstrated, using the above procedure, in the following worked problems.

Problem 7. Determine, using the Frobenius method, the general power series solution of the differential equation: $3x\dfrac{d^2y}{dx^2} + \dfrac{dy}{dx} - y = 0$

The differential equation may be rewritten as: $3xy'' + y' - y = 0$

(i) Let a trial solution be of the form

$$y = x^c\left\{a_0 + a_1x + a_2x^2 + a_3x^3 + \cdots \right.$$
$$\left. + a_rx^r + \cdots\right\} \qquad (16)$$

where $a_0 \neq 0$,

i.e. $y = a_0x^c + a_1x^{c+1} + a_2x^{c+2} + a_3x^{c+3}$
$$+ \cdots + a_rx^{c+r} + \cdots \qquad (17)$$

(ii) Differentiating equation (17) gives:

$$y' = a_0cx^{c-1} + a_1(c+1)x^c$$
$$+ a_2(c+2)x^{c+1} + \cdots$$
$$+ a_r(c+r)x^{c+r-1} + \cdots$$

and $y'' = a_0c(c-1)x^{c-2} + a_1c(c+1)x^{c-1}$
$$+ a_2(c+1)(c+2)x^c + \cdots$$
$$+ a_r(c+r-1)(c+r)x^{c+r-2} + \cdots$$

(iii) Substituting y, y' and y'' into each term of the given equation $3xy'' + y' - y = 0$ gives:

$$3xy'' = 3a_0c(c-1)x^{c-1} + 3a_1c(c+1)x^c$$
$$+ 3a_2(c+1)(c+2)x^{c+1} + \cdots$$
$$+ 3a_r(c+r-1)(c+r)x^{c+r-1} + \cdots \quad (a)$$

$$y' = a_0cx^{c-1} + a_1(c+1)x^c + a_2(c+2)x^{c+1}$$
$$+ \cdots + a_r(c+r)x^{c+r-1} + \cdots \qquad (b)$$

$$-y = -a_0x^c - a_1x^{c+1} - a_2x^{c+2} - a_3x^{c+3}$$
$$- \cdots - a_rx^{c+r} - \cdots \qquad (c)$$

(iv) The sum of these three terms forms the left-hand side of the equation. Since the right-hand side is zero, the coefficients of each power of x can be equated to zero.
For example, the coefficient of x^{c-1} is equated to zero giving: $3a_0c(c-1) + a_0c = 0$

or $a_0c[3c - 3 + 1] = a_0c(3c - 2) = 0 \qquad (18)$

The coefficient of x^c is equated to zero giving: $3a_1c(c+1) + a_1(c+1) - a_0 = 0$

i.e. $a_1(3c^2 + 3c + c + 1) - a_0$
$$= a_1(3c^2 + 4c + 1) - a_0 = 0$$

or $a_1(3c+1)(c+1) - a_0 = 0 \qquad (19)$

In each of series (a), (b) and (c) an x^c term is involved, after which, a general relationship can be obtained for x^{c+r}, where $r \geq 0$
In series (a) and (b), terms in x^{c+r-1} are present; replacing r by $(r+1)$ will give the corresponding terms in x^{c+r}, which occurs in all three equations, i.e.

in series (a), $3a_{r+1}(c+r)(c+r+1)x^{c+r}$

in series (b), $a_{r+1}(c+r+1)x^{c+r}$

in series (c), $-a_rx^{c+r}$

Equating the total coefficients of x^{c+r} to zero gives:

$$3a_{r+1}(c+r)(c+r+1) + a_{r+1}(c+r+1)$$
$$- a_r = 0$$

which simplifies to:

$$a_{r+1}\{(c+r+1)(3c+3r+1)\} - a_r = 0 \quad (20)$$

Equation (18), which was formed from the coefficients of the lowest power of x, i.e. x^{c-1}, is called the **indicial equation**, from which the value of c is obtained. From equation (18), since $a_0 \neq 0$, then $c = 0$ or $c = \dfrac{2}{3}$

(a) When $c = 0$

From equation (19), if $c = 0$, $a_1(1 \times 1) - a_0 = 0$, i.e. $a_1 = a_0$

From equation (20), if $c = 0$,
$a_{r+1}(r+1)(3r+1) - a_r = 0$,
i.e. $a_{r+1} = \dfrac{a_r}{(r+1)(3r+1)} \quad r \geq 0$

Thus, when $r = 1$, $a_2 = \dfrac{a_1}{(2 \times 4)} = \dfrac{a_0}{(2 \times 4)}$
$\qquad$ since $a_1 = a_0$

when $r = 2$, $a_3 = \dfrac{a_2}{(3 \times 7)} = \dfrac{a_0}{(2 \times 4)(3 \times 7)}$
$\qquad$ or $\dfrac{a_0}{(2 \times 3)(4 \times 7)}$

when $r = 3$, $a_4 = \dfrac{a_3}{(4 \times 10)}$
$\qquad = \dfrac{a_0}{(2 \times 3 \times 4)(4 \times 7 \times 10)}$

$\qquad$ and so on.

From equation (16), the trial solution was:

$$y = x^c\{a_0 + a_1 x + a_2 x^2 + a_3 x^3 + \cdots + a_r x^r + \cdots\}$$

Substituting $c = 0$ and the above values of $a_1, a_2, a_3, \ldots$ into the trial solution gives:

$$y = x^0 \left\{ a_0 + a_0 x + \left(\frac{a_0}{(2 \times 4)} \right) x^2 \right.$$

$$+ \left(\frac{a_0}{(2 \times 3)(4 \times 7)} \right) x^3$$

$$\left. + \left(\frac{a_0}{(2 \times 3 \times 4)(4 \times 7 \times 10)} \right) x^4 + \cdots \right\}$$

i.e. $y = a_0 \left\{ 1 + x + \dfrac{x^2}{(2 \times 4)} + \dfrac{x^3}{(2 \times 3)(4 \times 7)} \right.$

$$\left. + \frac{x^4}{(2 \times 3 \times 4)(4 \times 7 \times 10)} + \cdots \right\} \quad (21)$$

(b) When $c = \dfrac{2}{3}$

From equation (19), if $c = \dfrac{2}{3}$, $a_1(3)\left(\dfrac{5}{3}\right) - a_0 = 0$, i.e.
$$a_1 = \frac{a_0}{5}$$

From equation (20), if $c = \dfrac{2}{3}$

$$a_{r+1}\left(\frac{2}{3} + r + 1\right)(2 + 3r + 1) - a_r = 0,$$

i.e. $a_{r+1}\left(r + \dfrac{5}{3}\right)(3r + 3) - a_r$

$$= a_{r+1}(3r^2 + 8r + 5) - a_r = 0,$$

i.e. $a_{r+1} = \dfrac{a_r}{(r+1)(3r+5)} \quad r \geq 0$

Thus, when $r = 1$, $a_2 = \dfrac{a_1}{(2 \times 8)} = \dfrac{a_0}{(2 \times 5 \times 8)}$
$\qquad$ since $a_1 = \dfrac{a_0}{5}$

when $r = 2$, $a_3 = \dfrac{a_2}{(3 \times 11)}$

$$= \frac{a_0}{(2 \times 3)(5 \times 8 \times 11)}$$

when $r = 3$, $a_4 = \dfrac{a_3}{(4 \times 14)}$

$$= \frac{a_0}{(2 \times 3 \times 4)(5 \times 8 \times 11 \times 14)}$$

$\qquad$ and so on.

From equation (16), the trial solution was:

$$y = x^c\{a_0 + a_1 x + a_2 x^2 + a_3 x^3 + \cdots + a_r x^r + \cdots\}$$

Substituting $c = \dfrac{2}{3}$ and the above values of $a_1, a_2, a_3, \ldots$ into the trial solution gives:

$$y = x^{\frac{2}{3}} \left\{ a_0 + \left(\frac{a_0}{5} \right) x + \left(\frac{a_0}{2 \times 5 \times 8} \right) x^2 \right.$$

$$+ \left(\frac{a_0}{(2 \times 3)(5 \times 8 \times 11)} \right) x^3$$

$$\left. + \left(\frac{a_0}{(2 \times 3 \times 4)(5 \times 8 \times 11 \times 14)} \right) x^4 + \cdots \right\}$$

i.e. $y = a_0 x^{\frac{2}{3}} \left\{ 1 + \dfrac{x}{5} + \dfrac{x^2}{(2 \times 5 \times 8)} \right.$

$+ \dfrac{x^3}{(2 \times 3)(5 \times 8 \times 11)}$

$\left. + \dfrac{x^4}{(2 \times 3 \times 4)(5 \times 8 \times 11 \times 14)} + \cdots \right\}$ (22)

Since a_0 is an arbitrary (non-zero) constant in each solution, its value could well be different.
Let $a_0 = A$ in equation (21), and $a_0 = B$ in equation (22).
Also, if the first solution is denoted by $u(x)$ and the second by $v(x)$, then the general solution of the given differential equation is $y = u(x) + v(x)$. Hence,

$y = A \left\{ 1 + x + \dfrac{x^2}{(2 \times 4)} + \dfrac{x^3}{(2 \times 3)(4 \times 7)} \right.$

$\left. + \dfrac{x^4}{(2 \times 3 \times 4)(4 \times 7 \times 10)} + \cdots \right\}$

$+ B x^{\frac{2}{3}} \left\{ 1 + \dfrac{x}{5} + \dfrac{x^2}{(2 \times 5 \times 8)} \right.$

$+ \dfrac{x^3}{(2 \times 3)(5 \times 8 \times 11)}$

$\left. + \dfrac{x^4}{(2 \times 3 \times 4)(5 \times 8 \times 11 \times 14)} + \cdots \right\}$

Problem 8. Use the Frobenius method to determine the general power series solution of the differential equation:

$$2x^2 \dfrac{d^2 y}{dx^2} - x \dfrac{dy}{dx} + (1 - x)y = 0$$

The differential equation may be rewritten as: $2x^2 y'' - xy' + (1 - x)y = 0$

(i) Let a trial solution be of the form

$y = x^c \{ a_0 + a_1 x + a_2 x^2 + a_3 x^3 + \cdots$

$+ a_r x^r + \cdots \}$ (23)

where $a_0 \neq 0$,

i.e. $y = a_0 x^c + a_1 x^{c+1} + a_2 x^{c+2} + a_3 x^{c+3}$

$+ \cdots + a_r x^{c+r} + \cdots$ (24)

(ii) Differentiating equation (24) gives:

$y' = a_0 c x^{c-1} + a_1 (c+1) x^c + a_2 (c+2) x^{c+1}$

$+ \cdots + a_r (c+r) x^{c+r-1} + \cdots$

and $y'' = a_0 c(c-1) x^{c-2} + a_1 c(c+1) x^{c-1}$

$+ a_2 (c+1)(c+2) x^c + \cdots$

$+ a_r (c+r-1)(c+r) x^{c+r-2} + \cdots$

(iii) Substituting y, y' and y'' into each term of the given equation $2x^2 y'' - xy' + (1-x)y = 0$ gives:

$2x^2 y'' = 2a_0 c(c-1) x^c + 2a_1 c(c+1) x^{c+1}$

$+ 2a_2 (c+1)(c+2) x^{c+2} + \cdots$

$+ 2a_r (c+r-1)(c+r) x^{c+r} + \cdots$ (a)

$-xy' = -a_0 c x^c - a_1 (c+1) x^{c+1}$

$- a_2 (c+2) x^{c+2} - \cdots$

$- a_r (c+r) x^{c+r} - \cdots$ (b)

$(1-x)y = (1-x)(a_0 x^c + a_1 x^{c+1} + a_2 x^{c+2}$

$+ a_3 x^{c+3} + \cdots + a_r x^{c+r} + \cdots)$

$= a_0 x^c + a_1 x^{c+1} + a_2 x^{c+2} + a_3 x^{c+3}$

$+ \cdots + a_r x^{c+r} + \cdots$

$- a_0 x^{c+1} - a_1 x^{c+2} - a_2 x^{c+3}$

$- a_3 x^{c+4} - \cdots - a_r x^{c+r+1} - \cdots$ (c)

(iv) The **indicial equation**, which is obtained by equating the coefficient of the lowest power of x to zero, gives the value(s) of c. Equating the total coefficients of x^c (from equations (a) to (c)) to zero gives:

$2a_0 c(c-1) - a_0 c + a_0 = 0$

i.e. $a_0 [2c(c-1) - c + 1] = 0$

i.e. $a_0 [2c^2 - 2c - c + 1] = 0$

i.e. $a_0 [2c^2 - 3c + 1] = 0$

i.e. $a_0 [(2c-1)(c-1)] = 0$

from which, $c = 1$ or $c = \dfrac{1}{2}$

The coefficient of the general term, i.e. x^{c+r}, gives (from equations (a) to (c)):

$2a_r (c+r-1)(c+r) - a_r (c+r)$

$+ a_r - a_{r-1} = 0$

from which,

$a_r [2(c+r-1)(c+r) - (c+r) + 1] = a_{r-1}$

and $a_r = \dfrac{a_{r-1}}{2(c+r-1)(c+r)-(c+r)+1}$ (25)

(a) With $c=1$, $a_r = \dfrac{a_{r-1}}{2(r)(1+r)-(1+r)+1}$

$= \dfrac{a_{r-1}}{2r+2r^2-1-r+1}$

$= \dfrac{a_{r-1}}{2r^2+r} = \dfrac{\boldsymbol{a_{r-1}}}{\boldsymbol{r(2r+1)}}$

Thus, when $r=1$,

$a_1 = \dfrac{a_0}{1(2+1)} = \dfrac{a_0}{1\times3}$

when $r=2$,

$a_2 = \dfrac{a_1}{2(4+1)} = \dfrac{a_1}{(2\times5)}$

$= \dfrac{a_0}{(1\times3)(2\times5)}$ or $\dfrac{a_0}{(1\times2)\times(3\times5)}$

when $r=3$,

$a_3 = \dfrac{a_2}{3(6+1)} = \dfrac{a_2}{3\times7}$

$= \dfrac{a_0}{(1\times2\times3)\times(3\times5\times7)}$

when $r=4$,

$a_4 = \dfrac{a_3}{4(8+1)} = \dfrac{a_3}{4\times9}$

$= \dfrac{a_0}{(1\times2\times3\times4)\times(3\times5\times7\times9)}$

and so on.

From equation (23), the trial solution was:

$y = x^c \left\{ a_0 + a_1 x + a_2 x^2 + a_3 x^3 + \cdots \right.$

$\left. + a_r x^r + \cdots \right\}$

Substituting $c=1$ and the above values of $a_1, a_2,$ $a_3, \ldots$ into the trial solution gives:

$y = x^1 \left\{ a_0 + \dfrac{a_0}{(1\times3)}x + \dfrac{a_0}{(1\times2)\times(3\times5)}x^2 \right.$

$+ \dfrac{a_0}{(1\times2\times3)\times(3\times5\times7)}x^3$

$\left. + \dfrac{a_0}{(1\times2\times3\times4)\times(3\times5\times7\times9)}x^4 \right.$

$\left. + \cdots \right\}$

i.e. $y = a_0 x^1 \left\{ 1 + \dfrac{x}{(1\times3)} + \dfrac{x^2}{(1\times2)\times(3\times5)} \right.$

$+ \dfrac{x^3}{(1\times2\times3)\times(3\times5\times7)}$

$\left. + \dfrac{x^4}{(1\times2\times3\times4)\times(3\times5\times7\times9)} \right.$

$\left. + \cdots \right\}$ (26)

(b) With $c = \dfrac{1}{2}$

$a_r = \dfrac{a_{r-1}}{2(c+r-1)(c+r)-(c+r)+1}$

from equation (25)

i.e. $a_r = \dfrac{a_{r-1}}{2\left(\dfrac{1}{2}+r-1\right)\left(\dfrac{1}{2}+r\right)-\left(\dfrac{1}{2}+r\right)+1}$

$= \dfrac{a_{r-1}}{2\left(r-\dfrac{1}{2}\right)\left(r+\dfrac{1}{2}\right)-\dfrac{1}{2}-r+1}$

$= \dfrac{a_{r-1}}{2\left(r^2-\dfrac{1}{4}\right)-\dfrac{1}{2}-r+1}$

$= \dfrac{a_{r-1}}{2r^2-\dfrac{1}{2}-\dfrac{1}{2}-r+1} = \dfrac{a_{r-1}}{2r^2-r}$

$= \dfrac{\boldsymbol{a_{r-1}}}{\boldsymbol{r(2r-1)}}$

Thus, when $r=1$, $a_1 = \dfrac{a_0}{1(2-1)} = \dfrac{a_0}{1\times1}$

when $r=2$, $a_2 = \dfrac{a_1}{2(4-1)} = \dfrac{a_1}{(2\times3)}$

$= \dfrac{a_0}{(2\times3)}$

when $r=3$, $a_3 = \dfrac{a_2}{3(6-1)} = \dfrac{a_2}{3\times5}$

$= \dfrac{a_0}{(2\times3)\times(3\times5)}$

when $r=4$, $a_4 = \dfrac{a_3}{4(8-1)} = \dfrac{a_3}{4\times7}$

$= \dfrac{a_0}{(2\times3\times4)\times(3\times5\times7)}$

and so on.

From equation (23), the trial solution was:

$$y = x^c \left\{ a_0 + a_1 x + a_2 x^2 + a_3 x^3 + \cdots \right.$$

$$\left. + a_r x^r + \cdots \right\}$$

Substituting $c = \dfrac{1}{2}$ and the above values of $a_1, a_2, a_3, \ldots$ into the trial solution gives:

$$y = x^{\frac{1}{2}} \left\{ a_0 + a_0 x + \frac{a_0}{(2 \times 3)} x^2 + \frac{a_0}{(2 \times 3) \times (3 \times 5)} x^3 \right.$$

$$\left. + \frac{a_0}{(2 \times 3 \times 4) \times (3 \times 5 \times 7)} x^4 + \cdots \right\}$$

i.e. $y = a_0 x^{\frac{1}{2}} \left\{ 1 + x + \dfrac{x^2}{(2 \times 3)} \right.$

$$+ \frac{x^3}{(2 \times 3) \times (3 \times 5)}$$

$$+ \frac{x^4}{(2 \times 3 \times 4) \times (3 \times 5 \times 7)}$$

$$\left. + \cdots \right\} \qquad (27)$$

Since a_0 is an arbitrary (non-zero) constant in each solution, its value could well be different. Let $a_0 = A$ in equation (26), and $a_0 = B$ in equation (27). Also, if the first solution is denoted by $u(x)$ and the second by $v(x)$, then the general solution of the given differential equation is $y = u(x) + v(x)$,

i.e. $y = Ax \left\{ 1 + \dfrac{x}{(1 \times 3)} + \dfrac{x^2}{(1 \times 2) \times (3 \times 5)} \right.$

$$+ \frac{x^3}{(1 \times 2 \times 3) \times (3 \times 5 \times 7)}$$

$$+ \frac{x^4}{(1 \times 2 \times 3 \times 4) \times (3 \times 5 \times 7 \times 9)}$$

$$\left. + \cdots \right\} + B x^{\frac{1}{2}} \left\{ 1 + x + \frac{x^2}{(2 \times 3)} \right.$$

$$+ \frac{x^3}{(2 \times 3) \times (3 \times 5)}$$

$$\left. + \frac{x^4}{(2 \times 3 \times 4) \times (3 \times 5 \times 7)} + \cdots \right\}$$

Problem 9. Use the Frobenius method to determine the general power series solution of the differential equation: $\dfrac{d^2 y}{dx^2} - 2y = 0$

The differential equation may be rewritten as: $y'' - 2y = 0$

(i) Let a trial solution be of the form

$$y = x^c \left\{ a_0 + a_1 x + a_2 x^2 + a_3 x^3 + \cdots \right.$$

$$\left. + a_r x^r + \cdots \right\} \qquad (28)$$

where $a_0 \neq 0$,

i.e. $y = a_0 x^c + a_1 x^{c+1} + a_2 x^{c+2} + a_3 x^{c+3}$

$$+ \cdots + a_r x^{c+r} + \cdots \qquad (29)$$

(ii) Differentiating equation (29) gives:

$$y' = a_0 c x^{c-1} + a_1(c+1) x^c + a_2(c+2) x^{c+1}$$

$$+ \cdots + a_r(c+r) x^{c+r-1} + \cdots$$

and $y'' = a_0 c(c-1) x^{c-2} + a_1 c(c+1) x^{c-1}$

$$+ a_2(c+1)(c+2) x^c + \cdots$$

$$+ a_r(c+r-1)(c+r) x^{c+r-2} + \cdots$$

(iii) Replacing r by $(r+2)$ in $a_r(c+r-1)(c+r) \, x^{c+r-2}$ gives: $a_{r+2}(c+r+1)(c+r+2) x^{c+r}$
Substituting y and y'' into each term of the given equation $y'' - 2y = 0$ gives:

$$y'' - 2y = a_0 c(c-1) x^{c-2} + a_1 c(c+1) x^{c-1}$$

$$+ [a_2(c+1)(c+2) - 2a_0] x^c + \cdots$$

$$+ [a_{r+2}(c+r+1)(c+r+2)$$

$$- 2a_r] x^{c+r} + \cdots = 0 \qquad (30)$$

(iv) The **indicial equation** is obtained by equating the coefficient of the lowest power of x to zero.

Hence, $a_0 c(c-1) = 0$ from which, $\boldsymbol{c = 0}$ or $\boldsymbol{c = 1}$ since $a_0 \neq 0$

For the term in x^{c-1}, i.e. $a_1 c(c+1) = 0$

With $\boldsymbol{c = 1}$, $\boldsymbol{a_1 = 0}$; however, when $\boldsymbol{c = 0}$, $\boldsymbol{a_1}$ **is indeterminate**, since any value of a_1 combined with the zero value of c would make the product zero.

For the term in x^c,

$$a_2(c+1)(c+2) - 2a_0 = 0 \text{ from which,}$$

$$a_2 = \frac{2a_0}{(c+1)(c+2)} \tag{31}$$

For the term in x^{c+r},

$$a_{r+2}(c+r+1)(c+r+2) - 2a_r = 0$$

from which,

$$a_{r+2} = \frac{2a_r}{(c+r+1)(c+r+2)} \tag{32}$$

(a) **When $c=0$:** a_1 is indeterminate, and from equation (31)

$$a_2 = \frac{2a_0}{(1 \times 2)} = \frac{2a_0}{2!}$$

In general, $a_{r+2} = \dfrac{2a_r}{(r+1)(r+2)}$ and

when $r=1$, $a_3 = \dfrac{2a_1}{(2 \times 3)} = \dfrac{2a_1}{(1 \times 2 \times 3)} = \dfrac{2a_1}{3!}$

when $r=2$, $a_4 = \dfrac{2a_2}{3 \times 4} = \dfrac{4a_0}{4!}$

Hence, $y = x^0 \left\{ a_0 + a_1 x + \dfrac{2a_0}{2!} x^2 + \dfrac{2a_1}{3!} x^3 \right.$

$$\left. + \frac{4a_0}{4!} x^4 + \cdots \right\}$$

from equation (28)

$$= a_0 \left\{ 1 + \frac{2x^2}{2!} + \frac{4x^4}{4!} + \cdots \right\}$$

$$+ a_1 \left\{ x + \frac{2x^3}{3!} + \frac{4x^5}{5!} + \cdots \right\}$$

Since a_0 and a_1 are arbitrary constants depending on boundary conditions, let $a_0 = P$ and $a_1 = Q$, then:

$$y = P \left\{ 1 + \frac{2x^2}{2!} + \frac{4x^4}{4!} + \cdots \right\}$$

$$+ Q \left\{ x + \frac{2x^3}{3!} + \frac{4x^5}{5!} + \cdots \right\} \tag{33}$$

(b) **When $c=1$:** $a_1 = 0$, and from equation (31),

$$a_2 = \frac{2a_0}{(2 \times 3)} = \frac{2a_0}{3!}$$

Since $c=1$, $a_{r+2} = \dfrac{2a_r}{(c+r+1)(c+r+2)}$

$$= \frac{2a_r}{(r+2)(r+3)}$$

from equation (32) and when $r=1$,

$$a_3 = \frac{2a_1}{(3 \times 4)} = 0 \text{ since } a_1 = 0$$

when $r=2$,

$$a_4 = \frac{2a_2}{(4 \times 5)} = \frac{2}{(4 \times 5)} \times \frac{2a_0}{3!} = \frac{4a_0}{5!}$$

when $r=3$,

$$a_5 = \frac{2a_3}{(5 \times 6)} = 0$$

Hence, when $c=1$,

$$y = x^1 \left\{ a_0 + \frac{2a_0}{3!} x^2 + \frac{4a_0}{5!} x^4 + \cdots \right\}$$

from equation (28)

i.e. $y = a_0 \left\{ x + \dfrac{2x^3}{3!} + \dfrac{4x^5}{5!} + \cdots \right\}$

Again, a_0 is an arbitrary constant; let $a_0 = K$,

then $\quad y = K \left\{ x + \dfrac{2x^3}{3!} + \dfrac{4x^5}{5!} + \cdots \right\}$

However, this latter solution is not a separate solution, for it is the same form as the second series in equation (33). Hence, equation (33) with its two arbitrary constants P and Q gives the general solution. This is always the case when the two values of c differ by an integer (i.e. whole number). From the above three worked problems, the following can be deduced, and in future assumed:

(i) if two solutions of the indicial equation differ by a quantity *not* an integer, then two independent solutions $y = u(x) + v(x)$ result, the general solution of which is $y = Au + Bv$ (note: Problem 7 had $c=0$ and $\dfrac{2}{3}$ and Problem 8 had $c=1$ and $\dfrac{1}{2}$; in neither case did c differ by an integer)

(ii) if two solutions of the indicial equation *do* differ by an integer, as in Problem 9 where $c=0$ and 1, and if one coefficient is indeterminate, as with when $c=0$, then the complete solution is always given by using this value of c. Using the second value of c, i.e. $c=1$ in Problem 9, always gives a series which is one of the series in the first solution.

Now try the following Practice Exercise

Practice Exercise 212   Power series solutions by the Frobenius method (Answers on page 888)

1. Produce, using the Frobenius method, a power series solution for the differential equation:
    $$2x\frac{d^2y}{dx^2} + \frac{dy}{dx} - y = 0$$

2. Use the Frobenius method to determine the general power series solution of the differential equation: $\frac{d^2y}{dx^2} + y = 0$

3. Determine the power series solution of the differential equation: $3x\frac{d^2y}{dx^2} + 4\frac{dy}{dx} - y = 0$ using the Frobenius method.

4. Show, using the Frobenius method, that the power series solution of the differential equation: $\frac{d^2y}{dx^2} - y = 0$ may be expressed as $y = P\cosh x + Q\sinh x$, where P and Q are constants. [Hint: check the series expansions for $\cosh x$ and $\sinh x$ on page 188].

55.6 Bessel's equation and Bessel's functions

One of the most important differential equations in applied mathematics is **Bessel's*equation** and is of the form:

$$x^2\frac{d^2y}{dx^2} + x\frac{dy}{dx} + (x^2 - v^2)y = 0$$

where v is a real constant. The equation, which has applications in electric fields, vibrations and heat conduction, may be solved using Frobenius' method of the previous section.

Problem 10.   Determine the general power series solution of Bessel's equation.

Bessel's equation $x^2\frac{d^2y}{dx^2} + x\frac{dy}{dx} + (x^2 - v^2)y = 0$ may be rewritten as: $x^2y'' + xy' + (x^2 - v^2)y = 0$

Using the Frobenius method from page 585:

(i)   Let a trial solution be of the form

$$y = x^c\{a_0 + a_1x + a_2x^2 + a_3x^3 + \cdots + a_rx^r + \cdots\} \tag{34}$$

$$\text{where } a_0 \neq 0,$$

i.e. $y = a_0x^c + a_1x^{c+1} + a_2x^{c+2} + a_3x^{c+3}$
$$+ \cdots + a_rx^{c+r} + \cdots \tag{35}$$

(ii)   Differentiating equation (35) gives:

$$y' = a_0cx^{c-1} + a_1(c+1)x^c$$
$$+ a_2(c+2)x^{c+1} + \cdots$$
$$+ a_r(c+r)x^{c+r-1} + \cdots$$

* **Who was Bessel?** **Friedrich Wilhelm Bessel** (22 July 1784–17 March 1846) was a German mathematician, astronomer, and the systematiser of the Bessel functions. Bessel produced a refinement on the orbital calculations for Halley's Comet and produced precise positions for 3222 stars. To find out more go to **www.routledge.com/cw/bird**

and $y'' = a_0c(c-1)x^{c-2} + a_1c(c+1)x^{c-1}$

$$+ a_2(c+1)(c+2)x^c + \cdots$$

$$+ a_r(c+r-1)(c+r)x^{c+r-2} + \cdots$$

(iii) Substituting y, y' and y'' into each term of the given equation: $x^2y'' + xy' + (x^2 - v^2)y = 0$ gives:

$$a_0c(c-1)x^c + a_1c(c+1)x^{c+1}$$

$$+ a_2(c+1)(c+2)x^{c+2} + \cdots$$

$$+ a_r(c+r-1)(c+r)x^{c+r} + \cdots + a_0cx^c$$

$$+ a_1(c+1)x^{c+1} + a_2(c+2)x^{c+2} + \cdots$$

$$+ a_r(c+r)x^{c+r} + \cdots + a_0x^{c+2} + a_1x^{c+3}$$

$$+ a_2x^{c+4} + \cdots + a_rx^{c+r+2} + \cdots - a_0v^2x^c$$

$$- a_1v^2x^{c+1} - \cdots - a_rv^2x^{c+r} + \cdots = 0$$

(36)

(iv) The **indicial equation** is obtained by equating the coefficient of the lowest power of x to zero.

Hence, $\quad a_0c(c-1) + a_0c - a_0v^2 = 0$

from which, $\quad a_0[c^2 - c + c - v^2] = 0$

i.e. $\qquad\qquad a_0[c^2 - v^2] = 0$

from which, $\quad c = +v$ or $c = -v$ since $a_0 \neq 0$

For the term in x^{c+r},

$$a_r(c+r-1)(c+r) + a_r(c+r) + a_{r-2}$$
$$- a_rv^2 = 0$$

$$a_r[(c+r-1)(c+r) + (c+r) - v^2] = -a_{r-2}$$

i.e. $\quad a_r[(c+r)(c+r-1+1) - v^2] = -a_{r-2}$

i.e. $\qquad\qquad a_r[(c+r)^2 - v^2] = -a_{r-2}$

i.e. the **recurrence relation** is:

$$a_r = \frac{a_{r-2}}{v^2 - (c+r)^2} \quad \text{for} \quad r \geq 2 \qquad (37)$$

For the term in x^{c+1},

$$a_1[c(c+1) + (c+1) - v^2] = 0$$

i.e. $\qquad\qquad a_1[(c+1)^2 - v^2] = 0$

but if $c = v$ $\quad a_1[(v+1)^2 - v^2] = 0$

i.e. $\qquad\qquad a_1[2v+1] = 0$

Similarly, if $c = -v \quad a_1[1 - 2v] = 0$

The terms $(2v+1)$ *and* $(1-2v)$ cannot both be zero since v is a real constant, hence $a_1 = 0$

Since $\quad a_1 = 0$, then from equation (37) $a_3 = a_5 = a_7 = \ldots = 0$

and

$$a_2 = \frac{a_0}{v^2 - (c+2)^2}$$

$$a_4 = \frac{a_0}{[v^2 - (c+2)^2][v^2 - (c+4)^2]}$$

$$a_6 = \frac{a_0}{[v^2 - (c+2)^2][v^2 - (c+4)^2][v^2 - (c+6)^2]}$$

and so on.

When $c = +v$,

$$a_2 = \frac{a_0}{v^2 - (v+2)^2} = \frac{a_0}{v^2 - v^2 - 4v - 4}$$

$$= \frac{-a_0}{4 + 4v} = \frac{-a_0}{2^2(v+1)}$$

$$a_4 = \frac{a_0}{[v^2 - (v+2)^2][v^2 - (v+4)^2]}$$

$$= \frac{a_0}{[-2^2(v+1)][-2^3(v+2)]}$$

$$= \frac{a_0}{2^5(v+1)(v+2)}$$

$$= \frac{a_0}{2^4 \times 2(v+1)(v+2)}$$

$$a_6 = \frac{a_0}{[v^2 - (v+2)^2][v^2 - (v+4)^2][v^2 - (v+6)^2]}$$

$$= \frac{a_0}{[2^4 \times 2(v+1)(v+2)][-12(v+3)]}$$

$$= \frac{-a_0}{2^4 \times 2(v+1)(v+2) \times 2^2 \times 3(v+3)}$$

$$= \frac{-a_0}{2^6 \times 3!(v+1)(v+2)(v+3)} \text{ and so on.}$$

The resulting solution for $c = +v$ is given by:

$y = u =$

$$A x^v \left\{ 1 - \frac{x^2}{2^2(v+1)} + \frac{x^4}{2^4 \times 2!(v+1)(v+2)} \right.$$

$$\left. - \frac{x^6}{2^6 \times 3!(v+1)(v+2)(v+3)} + \cdots \right\}$$

$$(38)$$

which is valid provided v is not a negative integer and where A is an arbitrary constant.

When $c = -v$,

$$a_2 = \frac{a_0}{v^2 - (-v+2)^2} = \frac{a_0}{v^2 - (v^2 - 4v + 4)}$$

$$= \frac{-a_0}{4 - 4v} = \frac{-a_0}{2^2(v-1)}$$

$$a_4 = \frac{a_0}{[2^2(v-1)][v^2 - (-v+4)^2]}$$

$$= \frac{a_0}{[2^2(v-1)][2^3(v-2)]}$$

$$= \frac{a_0}{2^4 \times 2(v-1)(v-2)}$$

Similarly, $a_6 = \dfrac{a_0}{2^6 \times 3!(v-1)(v-2)(v-3)}$

Hence,

$y = w =$

$$B x^{-v} \left\{ 1 + \frac{x^2}{2^2(v-1)} + \frac{x^4}{2^4 \times 2!(v-1)(v-2)} \right.$$

$$\left. + \frac{x^6}{2^6 \times 3!(v-1)(v-2)(v-3)} + \cdots \right\}$$

which is valid provided v is not a positive integer and where B is an arbitrary constant.

The complete solution of Bessel's equation:

$$x^2 \frac{d^2 y}{dx^2} + x \frac{dy}{dx} + \left(x^2 - v^2 \right) y = 0 \text{ is:}$$

$y = u + w =$

$$A x^v \left\{ 1 - \frac{x^2}{2^2(v+1)} + \frac{x^4}{2^4 \times 2!(v+1)(v+2)} \right.$$

$$\left. - \frac{x^6}{2^6 \times 3!(v+1)(v+2)(v+3)} + \cdots \right\}$$

$$+ B x^{-v} \left\{ 1 + \frac{x^2}{2^2(v-1)} \right.$$

$$+ \frac{x^4}{2^4 \times 2!(v-1)(v-2)}$$

$$\left. + \frac{x^6}{2^6 \times 3!(v-1)(v-2)(v-3)} + \cdots \right\} \quad (39)$$

The gamma function

The solution of the Bessel equation of Problem 10 may be expressed in terms of **gamma functions**. Γ is the upper case Greek letter gamma, and the gamma function $\Gamma(x)$ is defined by the integral

$$\Gamma(x) = \int_0^\infty t^{x-1} e^{-t} dt \quad (40)$$

and is convergent for $x > 0$

From equation (40), $\Gamma(x+1) = \displaystyle\int_0^\infty t^x e^{-t} dt$

and by using integration by parts (see page 500):

$$\Gamma(x+1) = \left[\left(t^x \right) \left(\frac{e^{-t}}{-1} \right) \right]_0^\infty$$

$$- \int_0^\infty \left(\frac{e^{-t}}{-1} \right) x\, t^{x-1} dx$$

$$= (0 - 0) + x \int_0^\infty e^{-t} t^{x-1} dt$$

$$= x \Gamma(x) \quad \text{from equation (40)}$$

This is an important recurrence relation for gamma functions.

Thus, since $\Gamma(x+1) = x \Gamma(x)$

then similarly, $\Gamma(x+2) = (x+1)\Gamma(x+1)$

$$= (x+1)x\Gamma(x) \quad (41)$$

and $\Gamma(x+3) = (x+2)\Gamma(x+2)$

$$= (x+2)(x+1)x\Gamma(x),$$

and so on.

These relationships involving gamma functions are used with Bessel functions.

Bessel functions

The power series solution of the Bessel equation may be written in terms of gamma functions as shown in Problem 11 below.

Problem 11. Show that the power series solution of the Bessel equation of Problem 10 may be written in terms of the Bessel functions $J_v(x)$ and $J_{-v}(x)$ as:

$$AJ_v(x) + BJ_{-v}(x)$$

$$= \left(\frac{x}{2}\right)^v \left\{ \frac{1}{\Gamma(v+1)} - \frac{x^2}{2^2(1!)\Gamma(v+2)} \right.$$

$$\left. + \frac{x^4}{2^4(2!)\Gamma(v+4)} - \cdots \right\}$$

$$+ \left(\frac{x}{2}\right)^{-v} \left\{ \frac{1}{\Gamma(1-v)} - \frac{x^2}{2^2(1!)\Gamma(2-v)} \right.$$

$$\left. + \frac{x^4}{2^4(2!)\Gamma(3-v)} - \cdots \right\}$$

From Problem 10 above, **when $c = +v$,**

$$a_2 = \frac{-a_0}{2^2(v+1)}$$

If we let $a_0 = \dfrac{1}{2^v \Gamma(v+1)}$

then

$$a_2 = \frac{-1}{2^2(v+1)\,2^v\Gamma(v+1)} = \frac{-1}{2^{v+2}(v+1)\Gamma(v+1)}$$

$$= \frac{-1}{2^{v+2}\Gamma(v+2)} \quad \text{from equation (41)}$$

Similarly, $a_4 = \dfrac{a_2}{v^2 - (c+4)^2}$ from equation (37)

$$= \frac{a_2}{(v-c-4)(v+c+4)} = \frac{a_2}{-4(2v+4)}$$

$$\text{since } c = v$$

$$= \frac{-a_2}{2^3(v+2)} = \frac{-1}{2^3(v+2)} \frac{-1}{2^{v+2}\Gamma(v+2)}$$

$$= \frac{1}{2^{v+4}(2!)\Gamma(v+3)}$$

$$\text{since } (v+2)\Gamma(v+2) = \Gamma(v+3)$$

and $a_6 = \dfrac{-1}{2^{v+6}(3!)\Gamma(v+4)}$ and so on.

The **recurrence relation** is:

$$a_r = \frac{(-1)^{r/2}}{2^{v+r}\left(\dfrac{r}{2}!\right)\Gamma\left(v + \dfrac{r}{2} + 1\right)}$$

And if we let $r = 2k$, then

$$a_{2k} = \frac{(-1)^k}{2^{v+2k}(k!)\Gamma(v+k+1)} \tag{42}$$

$$\text{for } k = 1, 2, 3, \ldots$$

Hence, it is possible to write the new form for equation (38) as:

$$y = Ax^v \left\{ \frac{1}{2^v\Gamma(v+1)} - \frac{x^2}{2^{v+2}(1!)\Gamma(v+2)} \right.$$

$$\left. + \frac{x^4}{2^{v+4}(2!)\Gamma(v+3)} - \cdots \right\}$$

This is called *the Bessel function of the first-order kind, of order v,* and is denoted by $J_v(x)$,

i.e. $J_v(x) = \left(\dfrac{x}{2}\right)^v \left\{ \dfrac{1}{\Gamma(v+1)} - \dfrac{x^2}{2^2(1!)\Gamma(v+2)} \right.$

$$\left. + \frac{x^4}{2^4(2!)\Gamma(v+3)} - \cdots \right\}$$

provided v is not a negative integer.

For the second solution, **when $c = -v$,** replacing v by $-v$ in equation (42) above gives:

$$a_{2k} = \frac{(-1)^k}{2^{2k-v}(k!)\,\Gamma(k-v+1)}$$

from which, when $k = 0, a_0 = \dfrac{(-1)^0}{2^{-v}(0!)\Gamma(1-v)}$

$$= \frac{1}{2^{-v}\Gamma(1-v)} \quad \text{since } 0! = 1 \text{ (see page 579)}$$

when $k = 1$, $a_2 = \dfrac{(-1)^1}{2^{2-v}(1!)\Gamma(1-v+1)}$

$$= \frac{-1}{2^{2-v}(1!)\Gamma(2-v)}$$

when $k = 2$, $a_4 = \dfrac{(-1)^2}{2^{4-v}(2!)\Gamma(2-v+1)}$

$$= \frac{1}{2^{4-v}(2!)\Gamma(3-v)}$$

when $k = 3$, $a_6 = \dfrac{(-1)^3}{2^{6-v}(3!)\Gamma(3-v+1)}$

$$= \frac{1}{2^{6-v}(3!)\Gamma(4-v)} \quad \text{and so on.}$$

Hence, $y = Bx^{-v}\left\{\dfrac{1}{2^{-v}\Gamma(1-v)} - \dfrac{x^2}{2^{2-v}(1!)\Gamma(2-v)}\right.$

$$\left. + \dfrac{x^4}{2^{4-v}(2!)\Gamma(3-v)} - \cdots \right\}$$

i.e. $J_{-v}(x) = \left(\dfrac{x}{2}\right)^{-v}\left\{\dfrac{1}{\Gamma(1-v)} - \dfrac{x^2}{2^2(1!)\Gamma(2-v)}\right.$

$$\left. + \dfrac{x^4}{2^4(2!)\Gamma(3-v)} - \cdots \right\}$$

provided v is not a positive integer.

$J_v(x)$ and $J_{-v}(x)$ are two independent solutions of the Bessel equation; the complete solution is:

$y = AJ_v(x) + BJ_{-v}(x)$ where A and B are constants

i.e. $y = AJ_v(x) + BJ_{-v}(x)$

$$= A\left(\dfrac{x}{2}\right)^v\left\{\dfrac{1}{\Gamma(v+1)} - \dfrac{x^2}{2^2(1!)\Gamma(v+2)}\right.$$

$$\left. + \dfrac{x^4}{2^4(2!)\Gamma(v+4)} - \cdots \right\}$$

$$+ B\left(\dfrac{x}{2}\right)^{-v}\left\{\dfrac{1}{\Gamma(1-v)} - \dfrac{x^2}{2^2(1!)\Gamma(2-v)}\right.$$

$$\left. + \dfrac{x^4}{2^4(2!)\Gamma(3-v)} - \cdots \right\}$$

In general terms: $J_v(x) = \left(\dfrac{x}{2}\right)^v \displaystyle\sum_{k=0}^{\infty} \dfrac{(-1)^k x^{2k}}{2^{2k}(k!)\Gamma(v+k+1)}$

and $J_{-v}(x) = \left(\dfrac{x}{2}\right)^{-v} \displaystyle\sum_{k=0}^{\infty} \dfrac{(-1)^k x^{2k}}{2^{2k}(k!)\Gamma(k-v+1)}$

Another Bessel function

It may be shown that another series for $J_n(x)$ is given by:

$$J_n(x) = \left(\dfrac{x}{2}\right)^n\left\{\dfrac{1}{n!} - \dfrac{1}{(n+1)!}\left(\dfrac{x}{2}\right)^2\right.$$

$$\left. + \dfrac{1}{(2!)(n+2)!}\left(\dfrac{x}{2}\right)^4 - \cdots \right\}$$

From this series two commonly used function are derived,

i.e. $J_0(x) = \dfrac{1}{(0!)} - \dfrac{1}{(1!)^2}\left(\dfrac{x}{2}\right)^2 + \dfrac{1}{(2!)^2}\left(\dfrac{x}{2}\right)^4$

$$- \dfrac{1}{(3!)^2}\left(\dfrac{x}{2}\right)^6 + \cdots$$

$$= 1 - \dfrac{x^2}{2^2(1!)^2} + \dfrac{x^4}{2^4(2!)^2} - \dfrac{x^6}{2^6(3!)^2} + \cdots$$

and $J_1(x) = \dfrac{x}{2}\left\{\dfrac{1}{(1!)} - \dfrac{1}{(1!)(2!)}\left(\dfrac{x}{2}\right)^2\right.$

$$\left. + \dfrac{1}{(2!)(3!)}\left(\dfrac{x}{2}\right)^4 - \cdots \right\}$$

$$= \dfrac{x}{2} - \dfrac{x^3}{2^3(1!)(2!)} + \dfrac{x^5}{2^5(2!)(3!)}$$

$$- \dfrac{x^7}{2^7(3!)(4!)} + \cdots$$

Tables of Bessel functions are available for a range of values of n and x, and in these, $J_0(x)$ and $J_1(x)$ are most commonly used.

Graphs of $J_0(x)$, which looks similar to a cosine, and $J_1(x)$, which looks similar to a sine, are shown in Figure 55.1.

Now try the following Practice Exercise

Practice Exercise 213   Bessel's equation and Bessel's functions (Answers on page 888)

1. Determine the power series solution of Bessel's equation: $x^2\dfrac{d^2y}{dx^2} + x\dfrac{dy}{dx} + (x^2 - v^2)y = 0$ when $v = 2$, up to and including the term in x^6.

2. Find the power series solution of the Bessel function: $x^2y'' + xy' + (x^2 - v^2)y = 0$ in terms of the Bessel function $J_3(x)$ when $v = 3$. Give the answer up to and including the term in x^7.

3. Evaluate the Bessel functions $J_0(x)$ and $J_1(x)$ when $x = 1$, correct to 3 decimal places.

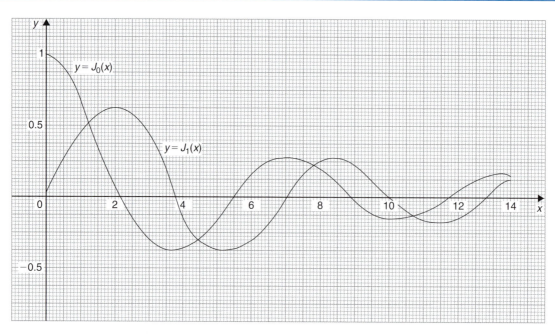

Figure 55.1

55.7 Legendre's equation and Legendre polynomials

Another important differential equation in physics and engineering applications is **Legendre's**[*] equation of the form: $(1-x^2)\dfrac{d^2 y}{dx^2} - 2x\dfrac{dy}{dx} + k(k+1)y = 0$ or $(1-x^2)y'' - 2xy' + k(k+1)y = 0$ where k is a real constant.

> **Problem 12.** Determine the general power series solution of Legendre's equation.

To solve Legendre's equation $(1-x^2)y'' - 2xy' + k(k+1)y = 0$ using the Frobenius method:

(i) Let a trial solution be of the form

$$y = x^c\left\{a_0 + a_1 x + a_2 x^2 + a_3 x^3 + \cdots + a_r x^r + \cdots\right\} \quad (43)$$

where $a_0 \neq 0$,

i.e. $y = a_0 x^c + a_1 x^{c+1} + a_2 x^{c+2} + a_3 x^{c+3} + \cdots + a_r x^{c+r} + \cdots \quad (44)$

[*] **Who was Legendre?** **Adrien-Marie Legendre** (18 September 1752–10 January 1833) was a French mathematician. Legendre developed the least squares method, which is used in linear regression, signal processing, statistics, and curve fitting. To find out more go to **www.routledge.com/cw/bird**

(ii) Differentiating equation (44) gives:

$$y' = a_0 c x^{c-1} + a_1(c+1)x^c$$

$$+ a_2(c+2)x^{c+1} + \cdots$$

$$+ a_r(c+r)x^{c+r-1} + \cdots$$

and $y'' = a_0 c(c-1)x^{c-2} + a_1 c(c+1)x^{c-1}$

$$+ a_2(c+1)(c+2)x^c + \cdots$$

$$+ a_r(c+r-1)(c+r)x^{c+r-2} + \cdots$$

(iii) Substituting y, y' and y'' into each term of the given equation:

$$\left(1 - x^2\right) y'' - 2xy' + k(k+1)y = 0 \text{ gives:}$$

$$a_0 c(c-1)x^{c-2} + a_1 c(c+1)x^{c-1}$$

$$+ a_2(c+1)(c+2)x^c + \cdots$$

$$+ a_r(c+r-1)(c+r)x^{c+r-2} + \cdots$$

$$- a_0 c(c-1)x^c - a_1 c(c+1)x^{c+1}$$

$$- a_2(c+1)(c+2)x^{c+2} - \cdots$$

$$- a_r(c+r-1)(c+r)x^{c+r} - \cdots - 2a_0 cx^c$$

$$- 2a_1(c+1)x^{c+1} - 2a_2(c+2)x^{c+2} - \cdots$$

$$- 2a_r(c+r)x^{c+r} - \cdots + k^2 a_0 x^c$$

$$+ k^2 a_1 x^{c+1} + k^2 a_2 x^{c+2} + \cdots + k^2 a_r x^{c+r}$$

$$+ \cdots + k a_0 x^c + k a_1 x^{c+1} + \cdots$$

$$+ k a_r x^{c+r} + \cdots = 0 \tag{45}$$

(iv) The **indicial equation** is obtained by equating the coefficient of the lowest power of x (i.e. x^{c-2}) to zero. Hence, $a_0 c(c-1)=0$ from which, **$c=0$** or **$c=1$** since $a_0 \neq 0$

For the term in x^{c-1}, i.e. $a_1 c(c+1)=0$, with $c=1$, $a_1=0$; however, when $c=0$, a_1 is indeterminate, since any value of a_1 combined with the zero value of c would make the product zero. For the term in x^{c+r},

$$a_{r+2}(c+r+1)(c+r+2) - a_r(c+r-1)$$
$$(c+r) - 2a_r(c+r) + k^2 a_r + k a_r = 0$$

from which,

$$a_{r+2} = \frac{a_r\left[(c+r-1)(c+r)+2(c+r)-k^2-k\right]}{(c+r+1)(c+r+2)}$$

$$= \frac{a_r[(c+r)(c+r+1) - k(k+1)]}{(c+r+1)(c+r+2)} \tag{46}$$

When $c=0$,

$$a_{r+2} = \frac{a_r[r(r+1) - k(k+1)]}{(r+1)(r+2)}$$

For $r=0$,

$$a_2 = \frac{a_0[-k(k+1)]}{(1)(2)}$$

For $r=1$,

$$a_3 = \frac{a_1[(1)(2) - k(k+1)]}{(2)(3)}$$

$$= \frac{-a_1[k^2 + k - 2]}{3!} = \frac{-a_1(k-1)(k+2)}{3!}$$

For $r=2$,

$$a_4 = \frac{a_2[(2)(3) - k(k+1)]}{(3)(4)} = \frac{-a_2\left[k^2 + k - 6\right]}{(3)(4)}$$

$$= \frac{-a_2(k+3)(k-2)}{(3)(4)}$$

$$= \frac{-(k+3)(k-2)}{(3)(4)} \cdot \frac{a_0[-k(k+1)]}{(1)(2)}$$

$$= \frac{a_0 k(k+1)(k+3)(k-2)}{4!}$$

For $r=3$,

$$a_5 = \frac{a_3[(3)(4) - k(k+1)]}{(4)(5)} = \frac{-a_3[k^2 + k - 12]}{(4)(5)}$$

$$= \frac{-a_3(k+4)(k-3)}{(4)(5)}$$

$$= \frac{-(k+4)(k-3)}{(4)(5)} \cdot \frac{-a_1(k-1)(k+2)}{(2)(3)}$$

$$= \frac{a_1(k-1)(k-3)(k+2)(k+4)}{5!} \text{ and so on.}$$

Substituting values into equation (43) gives:

$$y = x^0 \left\{ a_0 + a_1 x - \frac{a_0 k(k+1)}{2!} x^2 \right.$$

$$- \frac{a_1(k-1)(k+2)}{3!} x^3$$

$$+ \frac{a_0 k(k+1)(k-2)(k+3)}{4!} x^4$$

$$+ \frac{a_1(k-1)(k-3)(k+2)(k+4)}{5!} x^5$$

$$\left. + \cdots \right\}$$

i.e. $y = a_0 \left\{ 1 - \frac{k(k+1)}{2!} x^2 \right.$

$$\left. + \frac{k(k+1)(k-2)(k+3)}{4!} x^4 - \cdots \right\}$$

$$+ a_1 \left\{ x - \frac{(k-1)(k+2)}{3!} x^3 \right.$$

$$\left. + \frac{(k-1)(k-3)(k+2)(k+4)}{5!} x^5 - \cdots \right\}$$

(47)

From page 591, it was stated that if two solutions of the indicial equation differ by an integer, as in this case, where $c = 0$ and 1, and if one coefficient is indeterminate, as with when $c = 0$, then the complete solution is always given by using this value of c. Using the second value of c, i.e. $c = 1$ in this problem, will give a series which is one of the series in the first solution. (This may be checked for $c = 1$ and where $a_1 = 0$; the result will be the first part of equation (47) above.)

Legendre's polynomials

(A **polynomial** is an expression of the form: $f(x) = a + bx + cx^2 + dx^3 + \cdots$.) When k in equation (47) above is an integer, say, n, one of the solution series terminates after a finite number of terms. For example, if $k = 2$, then the first series terminates after the term in x^2. The resulting polynomial in x, denoted by $P_n(x)$, is called a **Legendre polynomial**. Constants a_0 and a_1 are chosen so that $y = 1$ when $x = 1$. This is demonstrated in the following worked problems.

Problem 13. Determine the Legendre polynomial $P_2(x)$

Since in $P_2(x)$, $n = k = 2$, then from the first part of equation (47), i.e. the even powers of x:

$$y = a_0 \left\{ 1 - \frac{2(3)}{2!} x^2 + 0 \right\} = a_0 \{ 1 - 3x^2 \}$$

a_0 is chosen to make $y = 1$ when $x = 1$

i.e. $1 = a_0 \{ 1 - 3(1)^2 \} = -2a_0$, from which, $a_0 = -\dfrac{1}{2}$

Hence, $\mathbf{P_2(x)} = -\dfrac{1}{2} \left(1 - 3x^2 \right) = \dfrac{1}{2} (3x^2 - 1)$

Problem 14. Determine the Legendre polynomial $P_3(x)$

Since in $P_3(x)$, $n = k = 3$, then from the second part of equation (47), i.e. the odd powers of x:

$$y = a_1 \left\{ x - \frac{(k-1)(k+2)}{3!} x^3 \right.$$

$$\left. + \frac{(k-1)(k-3)(k+2)(k+4)}{5!} x^5 - \cdots \right\}$$

i.e. $y = a_1 \left\{ x - \dfrac{(2)(5)}{3!} x^3 + \dfrac{(2)(0)(5)(7)}{5!} x^5 \right\}$

$$= a_1 \left\{ x - \frac{5}{3} x^3 + 0 \right\}$$

a_1 is chosen to make $y = 1$ when $x = 1$.

i.e. $1 = a_1 \left\{ 1 - \dfrac{5}{3} \right\} = a_1 \left(-\dfrac{2}{3} \right)$ from which, $a_1 = -\dfrac{3}{2}$

Hence, $P_3(x) = -\dfrac{3}{2} \left(x - \dfrac{5}{3} x^3 \right)$ or $\mathbf{P_3(x)} = \dfrac{1}{2} (5x^3 - 3x)$

Rodrigues' formula

An alternative method of determining Legendre polynomials is by using **Rodrigues'*** **formula**, which states:

$$P_n(x) = \frac{1}{2^n n!} \frac{d^n (x^2 - 1)^n}{dx^n} \qquad (48)$$

This is demonstrated in the following worked problems.

Problem 15.   Determine the Legendre polynomial $P_2(x)$ using Rodrigues' formula.

In Rodrigues' formula, $P_n(x) = \dfrac{1}{2^n n!} \dfrac{d^n (x^2 - 1)^n}{dx^n}$ and when $n = 2$,

$$P_2(x) = \frac{1}{2^2 2!} \frac{d^2 (x^2 - 1)^2}{dx^2}$$

* **Who was Rodrigues?** **Benjamin Olinde Rodrigues** (1795–1851), was a French banker, mathematician, and social reformer. Rodrigues is remembered for three results: Rodrigues' rotation formula for vectors; the Rodrigues formula about series of orthogonal polynomials; and the Euler–Rodrigues parameters. To find out more go to **www.routledge.com/cw/bird**

$$= \frac{1}{2^3} \frac{d^2 (x^4 - 2x^2 + 1)}{dx^2}$$

$$\frac{d}{dx}(x^4 - 2x^2 + 1)$$

$$= 4x^3 - 4x$$

and $\dfrac{d^2 (x^4 - 2x^2 + 1)}{dx^2} = \dfrac{d(4x^3 - 4x)}{dx} = 12x^2 - 4$

Hence, $P_2(x) = \dfrac{1}{2^3} \dfrac{d^2 (x^4 - 2x^2 + 1)}{dx^2} = \dfrac{1}{8}(12x^2 - 4)$

i.e. $\boldsymbol{P_2(x) = \dfrac{1}{2}(3x^2 - 1)}$, the same as in Problem 13.

Problem 16.   Determine the Legendre polynomial $P_3(x)$ using Rodrigues' formula.

In Rodrigues' formula, $P_n(x) = \dfrac{1}{2^n n!} \dfrac{d^n (x^2 - 1)^n}{dx^n}$ and when $n = 3$,

$$P_3(x) = \frac{1}{2^3 3!} \frac{d^3 (x^2 - 1)^3}{dx^3}$$

$$= \frac{1}{2^3 (6)} \frac{d^3 (x^2 - 1)(x^4 - 2x^2 + 1)}{dx^3}$$

$$= \frac{1}{(8)(6)} \frac{d^3 (x^6 - 3x^4 + 3x^2 - 1)}{dx^3}$$

$$\frac{d(x^6 - 3x^4 + 3x^2 - 1)}{dx} = 6x^5 - 12x^3 + 6x$$

$$\frac{d(6x^5 - 12x^3 + 6x)}{dx} = 30x^4 - 36x^2 + 6$$

and $\dfrac{d(30x^4 - 36x^2 + 6)}{dx} = 120x^3 - 72x$

Hence, $P_3(x) = \dfrac{1}{(8)(6)} \dfrac{d^3 (x^6 - 3x^4 + 3x^2 - 1)}{dx^3}$

$$= \frac{1}{(8)(6)}(120x^3 - 72x) = \frac{1}{8}(20x^3 - 12x)$$

i.e. $\boldsymbol{P_3(x) = \dfrac{1}{2}(5x^3 - 3x)}$, the same as in Problem 14.

Now try the following Practice Exercise

Practice Exercise 214   Legendre's equation and Legendre polynomials (Answers on page 888)

1. Determine the power series solution of the Legendre equation:

$(1 - x^2) y'' - 2xy' + k(k + 1)y = 0$ when
(a) $k = 0$ (b) $k = 2$, up to and including the term in x^5.

2. Find the following Legendre polynomials:
(a) $P_1(x)$ (b) $P_4(x)$ (c) $P_5(x)$

For fully worked solutions to each of the problems in Practice Exercises 209 to 214 in this chapter, go to the website:
www.routledge.com/cw/bird

Chapter 56

An introduction to partial differential equations

Why it is important to understand: **An introduction to partial differential equations**

In engineering, physics and economics, quantities are frequently encountered – for example energy – that depend on many variables, such as position, velocity and temperature. Usually this dependency is expressed through a partial differential equation, and solving these equations is important for understanding these complex relationships. Solving ordinary differential equations involves finding a function (or a set of functions) of one independent variable, but partial differential equations are for functions of two or more variables. Examples of physical models using partial differential equations are the heat equation for the evolution of the temperature distribution in a body, the wave equation for the motion of a wave front, the flow equation for the flow of fluids and Laplace's equation for an electrostatic potential or elastic strain field. In such cases, not only are the initial conditions needed, but also boundary conditions for the region in which the model applies; thus boundary value problems have to be solved. This chapter provides an introduction to the often complex subject of partial differential equations.

At the end of this chapter, you should be able to:

- recognise some important engineering partial differential equations

- solve a partial differential equation by direct partial integration

- solve differential equations by separating the variables

- solve the wave equation $\dfrac{\partial^2 u}{\partial x^2} = \dfrac{1}{c^2} \dfrac{\partial^2 u}{\partial t^2}$

- solve the heat conduction equation $\dfrac{\partial^2 u}{\partial x^2} = \dfrac{1}{c^2} \dfrac{\partial u}{\partial t}$

- solve Laplace's equation $\dfrac{\partial^2 u}{\partial x^2} + \dfrac{\partial^2 u}{\partial y^2} = 0$

56.1 Introduction

A partial differential equation is an equation that contains one or more partial derivatives. Examples include:

(i) $\quad a\dfrac{\partial u}{\partial x} + b\dfrac{\partial u}{\partial y} = c$

(ii) $\quad \dfrac{\partial^2 u}{\partial x^2} = \dfrac{1}{c^2}\dfrac{\partial u}{\partial t}$

(known as the heat conduction equation)

(iii) $\quad \dfrac{\partial^2 u}{\partial x^2} + \dfrac{\partial^2 u}{\partial y^2} = 0$

(known as Laplace's equation)

Equation (i) is a **first-order partial differential equation**, and equations (ii) and (iii) are **second-order partial differential equations** since the highest power of the differential is 2.

Partial differential equations occur in many areas of engineering and technology; electrostatics, heat conduction, magnetism, wave motion, hydrodynamics and aerodynamics all use models that involve partial differential equations. Such equations are difficult to solve, but techniques have been developed for the simpler types. In fact, for all but the simplest cases, there are a number of numerical methods of solutions of partial differential equations available.

To be able to solve simple partial differential equations knowledge of the following is required:

(a) partial integration,

(b) first- and second-order partial differentiation – as explained in Chapter 39, and

(c) the solution of ordinary differential equations – as explained in Chapters 33 and 50–54.

It should be appreciated that whole books have been written on partial differential equations and their solutions. This chapter does no more than introduce the topic.

56.2 Partial integration

Integration is the reverse process of differentiation.

Thus, if, for example, $\dfrac{\partial u}{\partial t} = 5\cos x \sin t$ is integrated partially with respect to t, then the $5\cos x$ term is considered as a constant,

and $\quad u = \displaystyle\int 5\cos x \sin t\, dt = (5\cos x)\int \sin t\, dt$

$\qquad\qquad = (5\cos x)(-\cos t) + c$

$\qquad\qquad = \boldsymbol{-5\cos x \cos t + f(x)}$

Similarly, if $\dfrac{\partial^2 u}{\partial x \partial y} = 6x^2 \cos 2y$ is integrated partially with respect to y,

then $\quad \dfrac{\partial u}{\partial x} = \displaystyle\int 6x^2 \cos 2y\, dy = \left(6x^2\right)\int \cos 2y\, dy$

$\qquad\qquad = \left(6x^2\right)\left(\dfrac{1}{2}\sin 2y\right) + f(x)$

$\qquad\qquad = 3x^2 \sin 2y + f(x)$

and integrating $\dfrac{\partial u}{\partial x}$ partially with respect to x gives:

$\quad u = \displaystyle\int [3x^2 \sin 2y + f(x)]\, dx$

$\qquad = x^3 \sin 2y + (x)f(x) + g(y)$

$f(x)$ and $g(y)$ are functions that may be determined if extra information, called **boundary conditions** or **initial conditions**, are known.

56.3 Solution of partial differential equations by direct partial integration

The simplest form of partial differential equations occurs when a solution can be determined by direct partial integration. This is demonstrated in the following worked problems.

Problem 1. Solve the differential equation $\dfrac{\partial^2 u}{\partial x^2} = 6x^2(2y - 1)$ given the boundary conditions that at $x = 0$, $\dfrac{\partial u}{\partial x} = \sin 2y$ and $u = \cos y$

Since $\dfrac{\partial^2 u}{\partial x^2} = 6x^2(2y - 1)$ then integrating partially with respect to x gives:

$\dfrac{\partial u}{\partial x} = \displaystyle\int 6x^2(2y - 1)dx = (2y - 1)\int 6x^2 dx$

$\qquad = (2y - 1)\dfrac{6x^3}{3} + f(y)$

$\qquad = 2x^3(2y - 1) + f(y)$

where $f(y)$ is an arbitrary function.

From the boundary conditions, when $x = 0$,

$$\frac{\partial u}{\partial x} = \sin 2y$$

Hence, $\sin 2y = 2(0)^3(2y - 1) + f(y)$

from which, $f(y) = \sin 2y$

Now $\frac{\partial u}{\partial x} = 2x^3(2y - 1) + \sin 2y$

Integrating partially with respect to x gives:

$$u = \int [2x^3(2y - 1) + \sin 2y]dx$$

$$= \frac{2x^4}{4}(2y - 1) + x(\sin 2y) + F(y)$$

From the boundary conditions, when $x = 0$, $u = \cos y$, hence

$$\cos y = \frac{(0)^4}{2}(2y - 1) + (0)\sin 2y + F(y)$$

from which, $F(y) = \cos y$

Hence, the solution of $\dfrac{\partial^2 u}{\partial x^2} = 6x^2(2y - 1)$ for the given boundary conditions is:

$$\boxed{u = \frac{x^4}{2}(2y - 1) + x\sin y + \cos y}$$

Problem 2. Solve the differential equation:
$\dfrac{\partial^2 u}{\partial x \partial y} = \cos(x + y)$ given that $\dfrac{\partial u}{\partial x} = 2$ when $y = 0$, and $u = y^2$ when $x = 0$

Since $\dfrac{\partial^2 u}{\partial x \partial y} = \cos(x + y)$ then integrating partially with respect to y gives:

$$\frac{\partial u}{\partial x} = \int \cos(x + y)dy = \sin(x + y) + f(x)$$

From the boundary conditions, $\dfrac{\partial u}{\partial x} = 2$ when $y = 0$, hence

$$2 = \sin x + f(x)$$

from which, $f(x) = 2 - \sin x$

i.e. $\dfrac{\partial u}{\partial x} = \sin(x + y) + 2 - \sin x$

Integrating partially with respect to x gives:

$$u = \int [\sin(x + y) + 2 - \sin x]dx$$

$$= -\cos(x + y) + 2x + \cos x + f(y)$$

From the boundary conditions, $u = y^2$ when $x = 0$, hence

$$y^2 = -\cos y + 0 + \cos 0 + f(y)$$

$$= 1 - \cos y + f(y)$$

from which, $f(y) = y^2 - 1 + \cos y$

Hence, the solution of $\dfrac{\partial^2 u}{\partial x \partial y} = \cos(x + y)$ is given by:

$$\boxed{u = -\cos(x + y) + 2x + \cos x + y^2 - 1 + \cos y}$$

Problem 3. Verify that
$\phi(x, y, z) = \dfrac{1}{\sqrt{x^2 + y^2 + z^2}}$ satisfies the partial
differential equation: $\dfrac{\partial^2 \phi}{\partial x^2} + \dfrac{\partial^2 \phi}{\partial y^2} + \dfrac{\partial^2 \phi}{\partial z^2} = 0$

The partial differential equation

$\dfrac{\partial^2 \phi}{\partial x^2} + \dfrac{\partial^2 \phi}{\partial y^2} + \dfrac{\partial^2 \phi}{\partial z^2} = 0$ is called **Laplace's equation**.

If $\phi(x, y, z) = \dfrac{1}{\sqrt{x^2 + y^2 + z^2}} = (x^2 + y^2 + z^2)^{-\frac{1}{2}}$

then differentiating partially with respect to x gives:

$$\frac{\partial \phi}{\partial x} = -\frac{1}{2}(x^2 + y^2 + z^2)^{-\frac{3}{2}}(2x)$$

$$= -x(x^2 + y^2 + z^2)^{-\frac{3}{2}}$$

and $\dfrac{\partial^2 \phi}{\partial x^2} = (-x)\left[-\dfrac{3}{2}(x^2 + y^2 + z^2)^{-\frac{5}{2}}(2x) \right]$

$$+ (x^2 + y^2 + z^2)^{-\frac{3}{2}}(-1)$$

by the product rule

$$= \frac{3x^2}{(x^2 + y^2 + z^2)^{\frac{5}{2}}} - \frac{1}{(x^2 + y^2 + z^2)^{\frac{3}{2}}}$$

$$= \frac{(3x^2) - (x^2 + y^2 + z^2)}{(x^2 + y^2 + z^2)^{\frac{5}{2}}}$$

Similarly, it may be shown that

$$\frac{\partial^2 \phi}{\partial y^2} = \frac{(3y^2) - (x^2 + y^2 + z^2)}{(x^2 + y^2 + z^2)^{\frac{5}{2}}}$$

and $\frac{\partial^2 \phi}{\partial z^2} = \frac{(3z^2) - (x^2 + y^2 + z^2)}{(x^2 + y^2 + z^2)^{\frac{5}{2}}}$

Thus,

$$\frac{\partial^2 \phi}{\partial x^2} + \frac{\partial^2 \phi}{\partial y^2} + \frac{\partial^2 \phi}{\partial z^2} = \frac{(3x^2) - (x^2 + y^2 + z^2)}{(x^2 + y^2 + z^2)^{\frac{5}{2}}}$$

$$+ \frac{(3y^2) - (x^2 + y^2 + z^2)}{(x^2 + y^2 + z^2)^{\frac{5}{2}}}$$

$$+ \frac{(3z^2) - (x^2 + y^2 + z^2)}{(x^2 + y^2 + z^2)^{\frac{5}{2}}}$$

$$= \frac{\begin{pmatrix} 3x^2 - (x^2 + y^2 + z^2) \\ +3y^2 - (x^2 + y^2 + z^2) \\ +3z^2 - (x^2 + y^2 + z^2) \end{pmatrix}}{(x^2 + y^2 + z^2)^{\frac{5}{2}}} = 0$$

Thus, $\dfrac{1}{\sqrt{x^2 + y^2 + z^2}}$ satisfies the Laplace equation

$$\frac{\partial^2 \phi}{\partial x^2} + \frac{\partial^2 \phi}{\partial y^2} + \frac{\partial^2 \phi}{\partial z^2} = 0$$

Now try the following Practice Exercise

Practice Exercise 215   The solution of partial differential equations by direct partial integration (Answers on page 888)

1. Determine the general solution of $\dfrac{\partial u}{\partial y} = 4ty$

2. Solve $\dfrac{\partial u}{\partial t} = 2t \cos\theta$ given that $u = 2t$ when $\theta = 0$

3. Verify that $u(\theta, t) = \theta^2 + \theta t$ is a solution of $\dfrac{\partial u}{\partial \theta} - 2\dfrac{\partial u}{\partial t} = t$

4. Verify that $u = e^{-y} \cos x$ is a solution of $\dfrac{\partial^2 u}{\partial x^2} + \dfrac{\partial^2 u}{\partial y^2} = 0$

5. Solve $\dfrac{\partial^2 u}{\partial x \partial y} = 8e^y \sin 2x$ given that at $y = 0$, $\dfrac{\partial u}{\partial x} = \sin x$, and at $x = \dfrac{\pi}{2}, u = 2y^2$

6. Solve $\dfrac{\partial^2 u}{\partial x^2} = y(4x^2 - 1)$ given that at $x = 0$, $u = \sin y$ and $\dfrac{\partial u}{\partial x} = \cos 2y$

7. Solve $\dfrac{\partial^2 u}{\partial x \partial t} = \sin(x + t)$ given that $\dfrac{\partial u}{\partial x} = 1$ when $t = 0$, and when $u = 2t$ when $x = 0$

8. Show that $u(x, y) = xy + \dfrac{x}{y}$ is a solution of $2x\dfrac{\partial^2 u}{\partial x \partial y} + y\dfrac{\partial^2 u}{\partial y^2} = 2x$

9. Find the particular solution of the differential equation $\dfrac{\partial^2 u}{\partial x \partial y} = \cos x \cos y$ given the initial conditions that when $y = \pi$, $\dfrac{\partial u}{\partial x} = x$, and when $x = \pi, u = 2\cos y$

10. Verify that $\phi(x, y) = x \cos y + e^x \sin y$ satisfies the differential equation $\dfrac{\partial^2 \phi}{\partial x^2} + \dfrac{\partial^2 \phi}{\partial y^2} + x \cos y = 0$

56.4 Some important engineering partial differential equations

There are many types of partial differential equations. Some typically found in engineering and science include:

(a) The **wave equation**, where the equation of motion is given by:

$$\frac{\partial^2 u}{\partial x^2} = \frac{1}{c^2}\frac{\partial^2 u}{\partial t^2}$$

where $c^2 = \dfrac{T}{\rho}$, with T being the tension in a string and ρ being the mass/unit length of the string.

(b) The **heat conduction equation** is of the form:

$$\frac{\partial^2 u}{\partial x^2} = \frac{1}{c^2}\frac{\partial u}{\partial t}$$

where $c^2 = \dfrac{h}{\sigma\rho}$, with h being the thermal conductivity of the material, σ the specific heat of the material, and ρ the mass/unit length of material.

(c) **Laplace's*** **equation**, used extensively with electrostatic fields is of the form:

$$\frac{\partial^2 u}{\partial x^2} + \frac{\partial^2 u}{\partial y^2} + \frac{\partial^2 u}{\partial z^2} = 0$$

(d) The **transmission equation**, where the potential u in a transmission cable is of the form:

$$\frac{\partial^2 u}{\partial x^2} = A\frac{\partial^2 u}{\partial t^2} + B\frac{\partial u}{\partial t} + Cu \text{ where } A, B \text{ and } C \text{ are}$$
constants.

Some of these equations are used in the next sections.

* **Who was Laplace? Pierre-Simon, marquis de Laplace** (23 March 1749–5 March 1827) was a French mathematician and astronomer who formulated Laplace's equation, and pioneered the Laplace transform which appears in many branches of mathematical physics. To find out more go to **www.routledge.com/cw/bird**

56.5   Separating the variables

Let $u(x,t) = X(x)T(t)$, where $X(x)$ is a function of x only and $T(t)$ is a function of t only, be a trial solution to the wave equation $\dfrac{\partial^2 u}{\partial x^2} = \dfrac{1}{c^2}\dfrac{\partial^2 u}{\partial t^2}$. If the trial solution is simplified to $u = XT$, then $\dfrac{\partial u}{\partial x} = X'T$ and $\dfrac{\partial^2 u}{\partial x^2} = X''T$.

Also $\dfrac{\partial u}{\partial t} = XT'$ and $\dfrac{\partial^2 u}{\partial t^2} = XT''$

Substituting into the partial differential equation $\dfrac{\partial^2 u}{\partial x^2} = \dfrac{1}{c^2}\dfrac{\partial^2 u}{\partial t^2}$ gives:

$$X''T = \frac{1}{c^2}XT''$$

Separating the variables gives:

$$\frac{X''}{X} = \frac{1}{c^2}\frac{T''}{T}$$

Let $\mu = \dfrac{X''}{X} = \dfrac{1}{c^2}\dfrac{T''}{T}$ where μ is a constant.

Thus, since $\mu = \dfrac{X''}{X}$ (a function of x only), it must be independent of t; and, since $\mu = \dfrac{1}{c^2}\dfrac{T''}{T}$ (a function of t only), it must be independent of x.

If μ is independent of x *and* t, it can only be a constant. If $\mu = \dfrac{X''}{X}$ then $X'' = \mu X$ or $X'' - \mu X = 0$ and if $\mu = \dfrac{1}{c^2}\dfrac{T''}{T}$ then $T'' = c^2\mu T$ or $T'' - c^2\mu T = 0$

Such ordinary differential equations are of the form found in Chapter 53, and their solutions will depend on whether $\mu > 0$, $\mu = 0$ or $\mu < 0$

Problem 4 will be a reminder of solving ordinary differential equations of this type.

> **Problem 4.**   Find the general solution of the following differential equations:
>
> (a) $X'' - 4X = 0$ (b) $T'' + 4T = 0$

(a) If $X'' - 4X = 0$ then the auxiliary equation (see Chapter 53) is:

$$m^2 - 4 = 0 \text{ i.e. } m^2 = 4 \text{ from which,}$$

$$m = +2 \text{ or } m = -2$$

Thus, the general solution is:

$$X = Ae^{2x} + Be^{-2x}$$

(b) If $T'' + 4T = 0$ then the auxiliary equation is:

$$m^2 + 4 = 0 \text{ i.e. } m^2 = -4 \text{ from which,}$$

$$m = \sqrt{-4} = \pm j2$$

Thus, the general solution is:

$$T = e^0\{A\cos 2t + B\sin 2t\} = A\cos 2t + B\sin 2t$$

Now try the following Practice Exercise

> **Practice Exercise 216   Revising the solution of ordinary differential equations (Answers on page 888)**
>
> 1. Solve $T'' = c^2\mu T$ given $c = 3$ and $\mu = 1$
>
> 2. Solve $T'' - c^2\mu T = 0$ given $c = 3$ and $\mu = -1$
>
> 3. Solve $X'' = \mu X$ given $\mu = 1$
>
> 4. Solve $X'' - \mu X = 0$ given $\mu = -1$

56.6   The wave equation

An **elastic string** is a string with elastic properties, i.e. the string satisfies Hooke's law. Fig. 56.1 shows a flexible elastic string stretched between two points at $x = 0$ and $x = L$ with uniform tension T. The string will vibrate if the string is displaced slightly from its initial position of rest and released, the end points remaining fixed. The position of any point P on the string depends on its distance from one end, and on the instant in time. Its displacement u at any time t can be expressed as $u = f(x,t)$, where x is its distance from 0

The equation of motion is as stated in Section 56.4 (a),

i.e. $\dfrac{\partial^2 u}{\partial x^2} = \dfrac{1}{c^2}\dfrac{\partial^2 u}{\partial t^2}$

The boundary and initial conditions are:

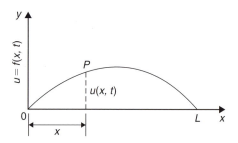

Figure 56.1

(i)   The string is fixed at both ends, i.e. $x = 0$ and $x = L$ for all values of time t.

Hence, $u(x,t)$ becomes:

$$\left.\begin{array}{l} u(0,t) = 0 \\ u(L,t) = 0 \end{array}\right\} \text{ for all values of } t \geq 0$$

(ii) If the initial deflection of P at $t = 0$ is denoted by $f(x)$ then $u(x,0) = f(x)$

(iii) Let the initial velocity of P be $g(x)$, then

$$\left[\frac{\partial u}{\partial t}\right]_{t=0} = g(x)$$

Initially a **trial solution** of the form $u(x,t) = X(x)T(t)$ is assumed, where $X(x)$ is a function of x only and $T(t)$ is a function of t only. The trial solution may be simplified to $u = XT$ and the variables separated as explained in the previous section to give:

$$\frac{X''}{X} = \frac{1}{c^2}\frac{T''}{T}$$

When both sides are equated to a constant μ this results in two ordinary differential equations:

$$T'' - c^2\mu T = 0 \quad \text{and} \quad X'' - \mu X = 0$$

Three cases are possible, depending on the value of μ.

Case 1: $\mu > 0$

For convenience, let $\mu = p^2$, where p is a real constant. Then the equations

$$X'' - p^2 X = 0 \quad \text{and} \quad T'' - c^2 p^2 T = 0$$

have solutions: $X = Ae^{px} + Be^{-px}$ and $T = Ce^{cpt} + De^{-cpt}$ where A, B, C and D are constants. But $X = 0$ at $x = 0$, hence $0 = A + B$ i.e. $B = -A$ and $X = 0$ at $x = L$, hence
$0 = Ae^{pL} + Be^{-pL} = A(e^{pL} - e^{-pL})$
Assuming $(e^{pL} - e^{-pL})$ is not zero, then $A = 0$ and since $B = -A$, then $B = 0$ also.
This corresponds to the string being stationary; since it is non-oscillatory, this solution will be disregarded.

Case 2: $\mu = 0$

In this case, since $\mu = p^2 = 0$, $T'' = 0$ and $X'' = 0$. We will assume that $T(t) \neq 0$. Since $X'' = 0$, $X' = a$ and $X = ax + b$ where a and b are constants. But $X = 0$ at $x = 0$, hence $b = 0$ and $X = ax$ and $X = 0$ at $x = L$, hence $a = 0$. Thus, again, the solution is non-oscillatory and is also disregarded.

Case 3: $\mu < 0$

For convenience,
let $\mu = -p^2$ then $X'' + p^2 X = 0$ from which,

$$X = A\cos px + B\sin px \qquad (1)$$

and $T'' + c^2 p^2 T = 0$ from which,

$$T = C\cos cpt + D\sin cpt \qquad (2)$$

(see Problem 4 above).

Thus, the suggested solution $u = XT$ now becomes:

$$u = \{A\cos px + B\sin px\}\{C\cos cpt + D\sin cpt\} \qquad (3)$$

Applying the boundary conditions:

(i) $u = 0$ when $x = 0$ for all values of t,

thus $0 = \{A\cos 0 + B\sin 0\}\{C\cos cpt + D\sin cpt\}$

i.e. $\qquad 0 = A\{C\cos cpt + D\sin cpt\}$

from which, $A = 0$ (since $\{C\cos cpt + D\sin cpt\} \neq 0$)

Hence, $\qquad u = \{B\sin px\}\{C\cos cpt + D\sin cpt\} \qquad (4)$

(ii) $u = 0$ when $x = L$ for all values of t

Hence, $\quad 0 = \{B\sin pL\}\{C\cos cpt + D\sin cpt\}$

Now $B \neq 0$ or $u(x, t)$ would be identically zero.
Thus $\sin pL = 0$ i.e. $pL = n\pi$ or $p = \dfrac{n\pi}{L}$ for integer values of n.
Substituting in equation (4) gives:

$$u = \left\{B\sin\frac{n\pi x}{L}\right\}\left\{C\cos\frac{cn\pi t}{L} + D\sin\frac{cn\pi t}{L}\right\}$$

i.e. $u = \sin\dfrac{n\pi x}{L}\left\{A_n\cos\dfrac{cn\pi t}{L} + B_n\sin\dfrac{cn\pi t}{L}\right\}$

(where constant $A_n = BC$ and $B_n = BD$). There will be many solutions, depending on the value of n. Thus, more generally,

$$u_n(x, t) = \sum_{n=1}^{\infty}\left\{\sin\frac{n\pi x}{L}\left(A_n\cos\frac{cn\pi t}{L}\right.\right.$$
$$\left.\left. + B_n\sin\frac{cn\pi t}{L}\right)\right\} \qquad (5)$$

To find A_n and B_n we put in the initial conditions not yet taken into account.

(i) At $t = 0$, $u(x, 0) = f(x)$ for $0 \leq x \leq L$
Hence, from equation (5),

$$u(x, 0) = f(x) = \sum_{n=1}^{\infty}\left\{A_n\sin\frac{n\pi x}{L}\right\} \qquad (6)$$

(ii) Also at $t = 0$, $\left[\dfrac{\partial u}{\partial t}\right]_{t=0} = g(x)$ for $0 \leq x \leq L$

Differentiating equation (5) with respect to t gives:

$$\frac{\partial u}{\partial t} = \sum_{n=1}^{\infty}\left\{\sin\frac{n\pi x}{L}\left(A_n\left(-\frac{cn\pi}{L}\sin\frac{cn\pi t}{L}\right)\right.\right.$$
$$\left.\left. + B_n\left(\frac{cn\pi}{L}\cos\frac{cn\pi t}{L}\right)\right)\right\}$$

and when $t = 0$,

$$g(x) = \sum_{n=1}^{\infty}\left\{\sin\frac{n\pi x}{L}B_n\frac{cn\pi}{L}\right\}$$

i.e. $g(x) = \dfrac{c\pi}{L}\sum_{n=1}^{\infty}\left\{B_n n\sin\dfrac{n\pi x}{L}\right\} \qquad (7)$

From Fourier series (see page 802) it may be shown that: A_n is twice the mean value of $f(x)\sin\dfrac{n\pi x}{L}$ between $x = 0$ and $x = L$

i.e. $A_n = \dfrac{2}{L}\displaystyle\int_0^L f(x)\sin\dfrac{n\pi x}{L}\,\mathrm{d}x$ for n = 1, 2, 3, ... (8)

and $B_n\left(\dfrac{cn\pi}{L}\right)$ is twice the mean value of

$g(x)\sin\dfrac{n\pi x}{L}$ between $x = 0$ and $x = L$

i.e. $\qquad B_n = \dfrac{L}{cn\pi}\left(\dfrac{2}{L}\right)\displaystyle\int_0^L g(x)\sin\dfrac{n\pi x}{L}\,\mathrm{d}x$

or $\qquad B_n = \dfrac{2}{cn\pi}\displaystyle\int_0^L g(x)\sin\dfrac{n\pi x}{L}\,\mathrm{d}x \qquad (9)$

Summary of solution of the wave equation

The above may seem complicated; however a practical problem may be solved using the following **eight-point procedure**:

1. Identify clearly the initial and boundary conditions.

2. Assume a solution of the form $u = XT$ and express the equations in terms of X and T and their derivatives.

3. Separate the variables by transposing the equation and equate each side to a constant, say, μ; two separate equations are obtained, one in x and the other in t.

4. Let $\mu = -p^2$ to give an oscillatory solution.

5. The two solutions are of the form:

$$X = A \cos px + B \sin px$$

and $\quad T = C \cos cpt + D \sin cpt$.

Then $u(x,t) = \{A \cos px + B \sin px\}\{C \cos cpt + D \sin cpt\}$

6. Apply the boundary conditions to determine constants A and B.

7. Determine the general solution as an infinite sum.

8. Apply the remaining initial and boundary conditions and determine the coefficients A_n and B_n from equations (8) and (9), using Fourier series techniques.

Problem 5. Fig. 56.2 shows a stretched string of length 50 cm which is set oscillating by displacing its mid-point a distance of 2 cm from its rest position and releasing it with zero velocity. Solve the wave equation: $\dfrac{\partial^2 u}{\partial x^2} = \dfrac{1}{c^2}\dfrac{\partial^2 u}{\partial t^2}$ where $c^2 = 1$, to determine the resulting motion $u(x,t)$

Figure 56.2

Following the above procedure,

1. The boundary and initial conditions given are:

$$\left.\begin{array}{l} u(0,t) = 0 \\ u(50,t) = 0 \end{array}\right\} \quad \text{i.e. fixed end points}$$

$$u(x,0) = f(x) = \frac{2}{25}x \quad 0 \le x \le 25$$

$$= -\frac{2}{25}x + 4 = \frac{100 - 2x}{25}$$

$$25 \le x \le 50$$

(Note: $y = mx + c$ is a straight line graph, so the gradient, m, between 0 and 25 is 2/25 and the y-axis intercept is zero, thus $y = f(x) = \dfrac{2}{25}x + 0$; between 25 and 50, the gradient $= -2/25$ and the y-axis intercept is at 4, thus $f(x) = -\dfrac{2}{25}x + 4$)

$$\left[\frac{\partial u}{\partial t}\right]_{t=0} = 0 \quad \text{i.e. zero initial velocity.}$$

2. Assuming a solution $u = XT$, where X is a function of x only, and T is a function of t only, then $\dfrac{\partial u}{\partial x} = X'T$ and $\dfrac{\partial^2 u}{\partial x^2} = X''T$ and $\dfrac{\partial u}{\partial y} = XT'$ and $\dfrac{\partial^2 u}{\partial y^2} = XT''$. Substituting into the partial differential equation, $\dfrac{\partial^2 u}{\partial x^2} = \dfrac{1}{c^2}\dfrac{\partial^2 u}{\partial t^2}$ gives:

$$X''T = \frac{1}{c^2}XT'' \quad \text{i.e.} \quad X''T = XT'' \text{ since } c^2 = 1$$

3. Separating the variables gives: $\dfrac{X''}{X} = \dfrac{T''}{T}$

Let constant,

$$\mu = \frac{X''}{X} = \frac{T''}{T} \quad \text{then } \mu = \frac{X''}{X} \text{ and } \mu = \frac{T''}{T}$$

from which,

$$X'' - \mu X = 0 \quad \text{and} \quad T'' - \mu T = 0$$

4. Letting $\mu = -p^2$ to give an oscillatory solution gives:

$$X'' + p^2 X = 0 \quad \text{and} \quad T'' + p^2 T = 0$$

The auxiliary equation for each is: $m^2 + p^2 = 0$ from which, $m = \sqrt{-p^2} = \pm jp$

5. Solving each equation gives:
$X = A \cos px + B \sin px$, and
$T = C \cos pt + D \sin pt$
Thus,
$u(x,t) = \{A \cos px + B \sin px\}\{C \cos pt + D \sin pt\}$

6. Applying the boundary conditions to determine constants A and B gives:

(i) $u(0,t)=0$, hence $0=A\{C\cos pt+D\sin pt\}$ from which we conclude that $A=0$

Therefore,

$$u(x,t)=B\sin px\{C\cos pt+D\sin pt\} \quad \text{(a)}$$

(ii) $u(50,t)=0$, hence

$0=B\sin 50p\{C\cos pt+D\sin pt\}. \quad B\neq 0,$
hence $\sin 50p=0$ from which, $50p=n\pi$ and
$p=\dfrac{n\pi}{50}$

7. Substituting in equation (a) gives:

$$u(x,t)=B\sin\frac{n\pi x}{50}\left\{C\cos\frac{n\pi t}{50}+D\sin\frac{n\pi t}{50}\right\}$$

or, more generally,

$$u_n(x,t)=\sum_{n=1}^{\infty}\sin\frac{n\pi x}{50}\left\{A_n\cos\frac{n\pi t}{50}\right.$$
$$\left.+B_n\sin\frac{n\pi t}{50}\right\} \quad \text{(b)}$$

where $A_n=BC$ and $B_n=BD$

8. From equation (8),

$$A_n=\frac{2}{L}\int_0^L f(x)\sin\frac{n\pi x}{L}\,dx$$

$$=\frac{2}{50}\left[\int_0^{25}\left(\frac{2}{25}x\right)\sin\frac{n\pi x}{50}\,dx\right.$$
$$\left.+\int_{25}^{50}\left(\frac{100-2x}{25}\right)\sin\frac{n\pi x}{50}\,dx\right]$$

Each integral is determined using integration by parts (see Chapter 46, page 500) with the result:

$$A_n=\frac{16}{n^2\pi^2}\sin\frac{n\pi}{2}$$

From equation (9),

$$B_n=\frac{2}{cn\pi}\int_0^L g(x)\sin\frac{n\pi x}{L}\,dx$$

$$\left[\frac{\partial u}{\partial t}\right]_{t=0}=0=g(x)\text{ thus, }B_n=0$$

Substituting into equation (b) gives:

$$u_n(x,t)=\sum_{n=1}^{\infty}\sin\frac{n\pi x}{50}\left\{A_n\cos\frac{n\pi t}{50}\right.$$
$$\left.+B_n\sin\frac{n\pi t}{50}\right\}$$

$$=\sum_{n=1}^{\infty}\sin\frac{n\pi x}{50}\left\{\frac{16}{n^2\pi^2}\sin\frac{n\pi}{2}\cos\frac{n\pi t}{50}\right.$$
$$\left.+(0)\sin\frac{n\pi t}{50}\right\}$$

Hence,

$$u(x,t)=\frac{16}{\pi^2}\sum_{n=1}^{\infty}\frac{1}{n^2}\sin\frac{n\pi x}{50}\sin\frac{n\pi}{2}\cos\frac{n\pi t}{50}$$

For stretched string problems as in Problem 5 above, the main parts of the procedure are:

1. Determine A_n from equation (8).

Note that $\dfrac{2}{L}\displaystyle\int_0^L f(x)\sin\dfrac{n\pi x}{L}\,dx$ is **always** equal to $\dfrac{8d}{n^2\pi^2}\sin\dfrac{n\pi}{2}$ (see Fig. 56.3)

2. Determine B_n from equation (9)

3. Substitute in equation (5) to determine $u(x,t)$

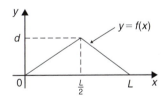

Figure 56.3

Now try the following Practice Exercise

Practice Exercise 217   The wave equation (Answers on page 888)

1. An elastic string is stretched between two points 40 cm apart. Its centre point is displaced 1.5 cm from its position of rest at right angles to the original direction of the string and then released with zero velocity. Determine the subsequent motion $u(x,t)$ by applying the wave equation $\dfrac{\partial^2 u}{\partial x^2}=\dfrac{1}{c^2}\dfrac{\partial^2 u}{\partial t^2}$ with $c^2=9$

2. The centre point of an elastic string between two points P and Q, 80 cm apart, is deflected a distance of 1 cm from its position of rest perpendicular to PQ and released initially with zero velocity. Apply the wave equation $\dfrac{\partial^2 u}{\partial x^2} = \dfrac{1}{c^2} \dfrac{\partial^2 u}{\partial t^2}$ where $c = 8$, to determine the motion of a point distance x from P at time t.

56.7 The heat conduction equation

The heat conduction equation $\dfrac{\partial^2 u}{\partial x^2} = \dfrac{1}{c^2} \dfrac{\partial u}{\partial t}$ is solved in a similar manner to that for the wave equation; the equation differs only in that the right-hand side contains a first partial derivative instead of the second.

The conduction of heat in a uniform bar depends on the initial distribution of temperature and on the physical properties of the bar, i.e. the thermal conductivity, h, the specific heat of the material, σ, and the mass per unit length, ρ, of the bar. In the above equation, $c^2 = \dfrac{h}{\sigma \rho}$

With a uniform bar insulated, except at its ends, any heat flow is along the bar and, at any instant, the temperature u at a point P is a function of its distance x from one end, and of the time t. Consider such a bar, shown in Fig. 56.4, where the bar extends from $x = 0$ to $x = L$, the temperature of the ends of the bar is maintained at zero, and the initial temperature distribution along the bar is defined by $f(x)$.

Thus, the boundary conditions can be expressed as:

$$\left. \begin{array}{l} u(0, t) = 0 \\ u(L, t) = 0 \end{array} \right\} \quad \text{for all } t \geq 0$$

and $\quad u(x, 0) = f(x)$ for $0 \leq x \leq L$

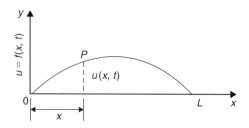

Figure 56.4

As with the wave equation, a solution of the form $u(x, t) = X(x)T(t)$ is assumed, where X is a function of x only and T is a function of t only. If the trial solution is simplified to $u = XT$, then

$$\frac{\partial u}{\partial x} = X'T \quad \frac{\partial^2 u}{\partial x^2} = X''T \text{ and } \frac{\partial u}{\partial t} = XT'$$

Substituting into the partial differential equation, $\dfrac{\partial^2 u}{\partial x^2} = \dfrac{1}{c^2} \dfrac{\partial u}{\partial t}$ gives:

$$X''T = \frac{1}{c^2} XT'$$

Separating the variables gives:

$$\frac{X''}{X} = \frac{1}{c^2} \frac{T'}{T}$$

Let $\quad -p^2 = \dfrac{X''}{X} = \dfrac{1}{c^2} \dfrac{T'}{T}$ where $-p^2$ is a constant.

If $-p^2 = \dfrac{X''}{X}$ then $X'' = -p^2 X$ or $X'' + p^2 X = 0$, giving $X = A \cos px + B \sin px$

and if $-p^2 = \dfrac{1}{c^2} \dfrac{T'}{T}$ then $\dfrac{T'}{T} = -p^2 c^2$ and integrating with respect to t gives:

$$\int \frac{T'}{T} \, dt = \int -p^2 c^2 \, dt$$

from which, $\ln T = -p^2 c^2 t + c_1$
The left-hand integral is obtained by an algebraic substitution (see Chapter 42).

If $\ln T = -p^2 c^2 t + c_1$ then $T = e^{-p^2 c^2 t + c_1} = e^{-p^2 c^2 t} e^{c_1}$ i.e. $\boldsymbol{T = k e^{-p^2 c^2 t}}$ (where constant $k = e^{c_1}$).

Hence, $u(x, t) = XT = \{A \cos px + B \sin px\} k e^{-p^2 c^2 t}$
i.e. $u(x, t) = \{P \cos px + Q \sin px\} e^{-p^2 c^2 t}$ where $P = Ak$ and $Q = Bk$

Applying the boundary conditions $u(0, t) = 0$ gives:
$0 = \{P \cos 0 + Q \sin 0\} e^{-p^2 c^2 t} = P e^{-p^2 c^2 t}$ from which, $P = 0$ and $u(x, t) = Q \sin px \, e^{-p^2 c^2 t}$

Also, $u(L, t) = 0$ thus, $0 = Q \sin pL \, e^{-p^2 c^2 t}$ and since $Q \neq 0$ then $\sin pL = 0$ from which, $pL = n\pi$ or $p = \dfrac{n\pi}{L}$ where $n = 1, 2, 3, \ldots$

There are therefore many values of $u(x, t)$.
Thus, in general,

$$u(x, t) = \sum_{n=1}^{\infty} \left\{ Q_n e^{-p^2 c^2 t} \sin \frac{n\pi x}{L} \right\}$$

Applying the remaining boundary condition, that when $t = 0, u(x, t) = f(x)$ for $0 \leq x \leq L$, gives:

$$f(x) = \sum_{n=1}^{\infty} \left\{ Q_n \sin \frac{n\pi x}{L} \right\}$$

From Fourier series, $Q_n = 2 \times$ mean value of $f(x) \sin \frac{n\pi x}{L}$ from x to L

Hence, $\quad Q_n = \frac{2}{L} \int_0^L f(x) \sin \frac{n\pi x}{L} \, dx$

Thus, $\quad u(x, t) =$

$$\frac{2}{L} \sum_{n=1}^{\infty} \left\{ \left(\int_0^L f(x) \sin \frac{n\pi x}{L} \, dx \right) e^{-p^2 c^2 t} \sin \frac{n\pi x}{L} \right\}$$

This method of solution is demonstrated in the following worked problem.

> **Problem 6.** A metal bar, insulated along its sides, is 1 m long. It is initially at room temperature of $15°C$ and at time $t = 0$, the ends are placed into ice at $0°C$. Find an expression for the temperature at a point P at a distance x m from one end at any time t seconds after $t = 0$.

The temperature u along the length of bar is shown in Fig. 56.5.

The heat conduction equation is $\dfrac{\partial^2 u}{\partial x^2} = \dfrac{1}{c^2} \dfrac{\partial u}{\partial t}$ and the given boundary conditions are:

$$u(0, t) = 0, \ u(1, t) = 0 \ \text{ and } \ u(x, 0) = 15$$

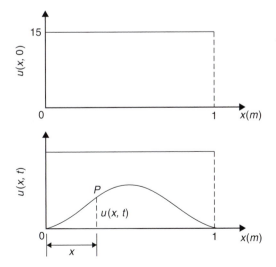

Figure 56.5

Assuming a solution of the form $u = XT$, then, from above,

$$X = A \cos px + B \sin px$$

and $\quad T = k\,e^{-p^2 c^2 t}$

Thus, the general solution is given by:

$$u(x, t) = \{P \cos px + Q \sin px\} e^{-p^2 c^2 t}$$

$$u(0, t) = 0 \text{ thus } 0 = P\,e^{-p^2 c^2 t}$$

from which, $P = 0$ and $u(x, t) = \{Q \sin px\} e^{-p^2 c^2 t}$

Also, $u(1, t) = 0$ thus $0 = \{Q \sin p\} e^{-p^2 c^2 t}$

Since $Q \neq 0$, $\sin p = 0$ from which, $p = n\pi$ where $n = 1, 2, 3, \ldots$

Hence, $u(x, t) = \sum_{n=1}^{\infty} \left\{ Q_n\,e^{-p^2 c^2 t} \sin n\pi x \right\}$

The final initial condition given was that at $t = 0$, $u = 15$, i.e. $u(x, 0) = f(x) = 15$

Hence, $15 = \sum_{n=1}^{\infty} \{Q_n \sin n\pi x\}$ where, from Fourier coefficients, $Q_n = 2 \times$ mean value of $15 \sin n\pi x$ from $x = 0$ to $x = 1$,

i.e. $\quad Q_n = \dfrac{2}{1} \displaystyle\int_0^1 15 \sin n\pi x \, dx = 30 \left[-\dfrac{\cos n\pi x}{n\pi} \right]_0^1$

$$= -\frac{30}{n\pi} [\cos n\pi - \cos 0]$$

$$= \frac{30}{n\pi} (1 - \cos n\pi)$$

$$= 0 \ (\text{when } n \text{ is even}) \text{ and } \frac{60}{n\pi} (\text{when } n \text{ is odd})$$

Hence, the required solution is:

$$\boldsymbol{u(x, t)} = \sum_{n=1}^{\infty} \left\{ Q_n\,e^{-p^2 c^2 t} \sin n\pi x \right\}$$

$$= \frac{60}{\pi} \sum_{n(\text{odd})=1}^{\infty} \frac{1}{n} (\sin n\pi x)\,e^{-n^2 \pi^2 c^2 t}$$

Now try the following Practice Exercise

> **Practice Exercise 218   The heat conduction equation (Answers on page 888)**
>
> 1. A metal bar, insulated along its sides, is 4 m long. It is initially at a temperature of $10°C$

and at time $t = 0$, the ends are placed into ice at $0°C$. Find an expression for the temperature at a point P at a distance x m from one end at any time t seconds after $t = 0$

2. An insulated uniform metal bar, 8 m long, has the temperature of its ends maintained at $0°C$, and at time $t = 0$ the temperature distribution $f(x)$ along the bar is defined by $f(x) = x(8 - x)$. If $c^2 = 1$, solve the heat conduction equation $\dfrac{\partial^2 u}{\partial x^2} = \dfrac{1}{c^2} \dfrac{\partial u}{\partial t}$ to determine the temperature u at any point in the bar at time t.

3. The ends of an insulated rod PQ, 20 units long, are maintained at $0°C$. At time $t = 0$, the temperature within the rod rises uniformly from each end, reaching $4°C$ at the mid-point of PQ. Find an expression for the temperature $u(x, t)$ at any point in the rod, distance x from P at any time t after $t = 0$. Assume the heat conduction equation to be $\dfrac{\partial^2 u}{\partial x^2} = \dfrac{1}{c^2} \dfrac{\partial u}{\partial t}$ and take $c^2 = 1$

56.8 Laplace's equation

The distribution of electrical potential, or temperature, over a plane area subject to certain boundary conditions, can be described by **Laplace's**[*] equation. The potential at a point P in a plane (see Fig. 56.6) can be indicated by an ordinate axis and is a function of its position, i.e. $z = u(x, y)$, where $u(x, y)$ is the solution of the Laplace two-dimensional equation $\dfrac{\partial^2 u}{\partial x^2} + \dfrac{\partial^2 u}{\partial y^2} = 0$

The method of solution of Laplace's equation is similar to the previous examples, as shown below.
Fig. 56.7 shows a rectangle $OPQR$ bounded by the lines $x = 0$, $y = 0$, $x = a$, and $y = b$, for which we are required to find a solution of the equation $\dfrac{\partial^2 u}{\partial x^2} + \dfrac{\partial^2 u}{\partial y^2} = 0$. The solution $z = (x, y)$ will give, say, the potential at any point within the rectangle $OPQR$. The boundary conditions are:

---

[*] Who was Laplace? See page 605 for image and resume of Pierre-Simon, marquis de Laplace. To find out more go to **www.routledge.com/cw/bird**

Figure 56.6

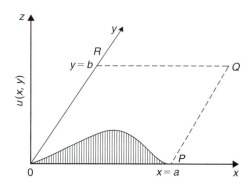

Figure 56.7

$u = 0$ when $x = 0$ i.e. $u(0, y) = 0$ for $0 \le y \le b$

$u = 0$ when $x = a$ i.e. $u(a, y) = 0$ for $0 \le y \le b$

$u = 0$ when $y = b$ i.e. $u(x, b) = 0$ for $0 \le x \le a$

$u = f(x)$ when $y = 0$ i.e. $u(x, 0) = f(x)$
$\qquad\qquad\qquad\qquad\qquad\qquad$ for $0 \le x \le a$

As with previous partial differential equations, a solution of the form $u(x, y) = X(x)Y(y)$ is assumed, where X is a function of x only, and Y is a function of y only. Simplifying to $u = XY$, determining partial derivatives, and substituting into $\dfrac{\partial^2 u}{\partial x^2} + \dfrac{\partial^2 u}{\partial y^2} = 0$ gives:

$X''Y + XY'' = 0$

Separating the variables gives: $\dfrac{X''}{X} = -\dfrac{Y''}{Y}$

Letting each side equal a constant, $-p^2$, gives the two equations:

$$X'' + p^2 X = 0 \quad \text{and} \quad Y'' - p^2 Y = 0$$

from which, $X = A \cos px + B \sin px$ and
$Y = C\, e^{py} + D\, e^{-py}$ or $Y = C \cosh py + D \sinh py$ (see Problem 5, page 564 for this conversion).
This latter form can also be expressed as:
$Y = E \sinh p(y + \phi)$ by using compound angles.
Hence $u(x, y) = XY$

$$= \{A \cos px + B \sin px\}\{E \sinh p(y + \phi)\}$$

or $u(x, y)$

$$= \{P \cos px + Q \sin px\}\{\sinh p(y + \phi)\}$$

where $P = AE$ and $Q = BE$

The first boundary condition is: $u(0, y) = 0$, hence $0 = P \sinh p(y + \phi)$ from which, $P = 0$. Hence, $u(x, y) = Q \sin px \sinh p(y + \phi)$

The second boundary condition is: $u(a, y) = 0$, hence $0 = Q \sin pa \sinh p(y + \phi)$ from which, $\sin pa = 0$, hence, $pa = n\pi$ or $p = \dfrac{n\pi}{a}$ for $n = 1, 2, 3, \ldots$

The third boundary condition is: $u(x, b) = 0$, hence, $0 = Q \sin px \sinh p(b + \phi)$ from which, $\sinh p(b + \phi) = 0$ and $\phi = -b$

Hence, $u(x, y) = Q \sin px \sinh p(y - b) = Q_1 \sin px \sinh p(b - y)$ where $Q_1 = -Q$

Since there are many solutions for integer values of n,

$$u(x, y) = \sum_{n=1}^{\infty} Q_n \sin px \sinh p(b - y)$$

$$= \sum_{n=1}^{\infty} Q_n \sin \frac{n\pi x}{a} \sinh \frac{n\pi}{a}(b - y)$$

The fourth boundary condition is: $u(x, 0) = f(x)$,

hence, $f(x) = \sum_{n=1}^{\infty} Q_n \sin \dfrac{n\pi x}{a} \sinh \dfrac{n\pi b}{a}$

i.e. $f(x) = \sum_{n=1}^{\infty} \left(Q_n \sinh \dfrac{n\pi b}{a} \right) \sin \dfrac{n\pi x}{a}$

From Fourier series coefficients,

$$\left(Q_n \sinh \frac{n\pi b}{a} \right) = 2 \times \text{the mean value of}$$

$$f(x) \sin \frac{n\pi x}{a} \text{ from } x = 0 \text{ to } x = a$$

$$= \int_0^a f(x) \sin \frac{n\pi x}{a} \, dx \text{ from which,}$$

Q_n may be determined.

This is demonstrated in the following worked problem.

Problem 7. A square plate is bounded by the lines $x = 0, y = 0, x = 1$ and $y = 1$. Apply the Laplace equation $\dfrac{\partial^2 u}{\partial x^2} + \dfrac{\partial^2 u}{\partial y^2} = 0$ to determine the

---

potential distribution $u(x, y)$ over the plate, subject to the following boundary conditions:

$u = 0$ when $x = 0$ $0 \le y \le 1$

$u = 0$ when $x = 1$ $0 \le y \le 1$

$u = 0$ when $y = 0$ $0 \le x \le 1$

$u = 4$ when $y = 1$ $0 \le x \le 1$

Initially a solution of the form $u(x, y) = X(x)Y(y)$ is assumed, where X is a function of x only, and Y is a function of y only. Simplifying to $u = XY$, determining partial derivatives, and substituting into $\dfrac{\partial^2 u}{\partial x^2} + \dfrac{\partial^2 u}{\partial y^2} = 0$

gives: $X''Y + XY'' = 0$

Separating the variables gives: $\dfrac{X''}{X} = -\dfrac{Y''}{Y}$

Letting each side equal a constant, $-p^2$, gives the two equations:

$$X'' + p^2 X = 0 \quad \text{and} \quad Y'' - p^2 Y = 0$$

from which, $X = A \cos px + B \sin px$

and $Y = Ce^{py} + De^{-py}$

or $Y = C \cosh py + D \sinh py$

or $Y = E \sinh p(y + \phi)$

Hence $u(x, y) = XY$

$$= \{A \cos px + B \sin px\}\{E \sinh p(y + \phi)\}$$

or $u(x, y)$

$$= \{P \cos px + Q \sin px\}\{\sinh p(y + \phi)\}$$

where $P = AE$ and $Q = BE$

The first boundary condition is: $u(0, y) = 0$, hence $0 = P \sinh p(y + \phi)$ from which, $P = 0$

Hence, $u(x, y) = Q \sin px \sinh p(y + \phi)$

The second boundary condition is: $u(1, y) = 0$, hence $0 = Q \sin p(1) \sinh p(y + \phi)$ from which, $\sin p = 0$, hence, $p = n\pi$ for $n = 1, 2, 3, \ldots$

The third boundary condition is: $u(x, 0) = 0$, hence, $0 = Q \sin px \sinh p(\phi)$ from which, $\sinh p(\phi) = 0$ and $\phi = 0$

Hence, $u(x, y) = Q \sin px \sinh py$

Since there are many solutions for integer values of n,

$$u(x, y) = \sum_{n=1}^{\infty} Q_n \sin px \sinh py$$

$$= \sum_{n=1}^{\infty} Q_n \sin n\pi x \sinh n\pi y \qquad \text{(a)}$$

The fourth boundary condition is: $u(x, 1) = 4 = f(x)$,

hence, $f(x) = \sum_{n=1}^{\infty} Q_n \sin n\pi x \sinh n\pi (1)$.

From Fourier series coefficients,

$Q_n \sinh n\pi = 2 \times$ the mean value of

$$f(x) \sin n\pi x \text{ from } x = 0 \text{ to } x = 1$$

i.e. $= \dfrac{2}{1} \displaystyle\int_0^1 4 \sin n\pi x \, dx$

$= 8 \left[-\dfrac{\cos n\pi x}{n\pi} \right]_0^1$

$= -\dfrac{8}{n\pi} (\cos n\pi - \cos 0)$

$= \dfrac{8}{n\pi} (1 - \cos n\pi)$

$= 0$ (for even values of n),

$\qquad = \dfrac{16}{n\pi}$ (for odd values of n)

Hence, $Q_n = \dfrac{16}{n\pi(\sinh n\pi)} = \dfrac{16}{n\pi} \operatorname{cosech} n\pi$

Hence, from equation (a),

$$u(x,y) = \sum_{n=1}^{\infty} Q_n \sin n\pi x \sinh n\pi y$$

$$= \dfrac{16}{\pi} \sum_{n(\text{odd})=1}^{\infty} \dfrac{1}{n} (\operatorname{cosech} n\pi \sin n\pi x \sinh n\pi y)$$

Now try the following Practice Exercise

Practice Exercise 219   The Laplace equation (Answers on page 889)

1. A rectangular plate is bounded by the lines $x = 0, y = 0, x = 1$ and $y = 3$. Apply the Laplace equation $\dfrac{\partial^2 u}{\partial x^2} + \dfrac{\partial^2 u}{\partial y^2} = 0$ to determine the potential distribution $u(x, y)$ over the plate, subject to the following boundary conditions:

   $u = 0$ when $x = 0$ $0 \le y \le 2$
   $u = 0$ when $x = 1$ $0 \le y \le 2$
   $u = 0$ when $y = 2$ $0 \le x \le 1$
   $u = 5$ when $y = 3$ $0 \le x \le 1$

2. A rectangular plate is bounded by the lines $x = 0, y = 0, x = 3, y = 2$. Determine the potential distribution $u(x, y)$ over the rectangle using the Laplace equation $\dfrac{\partial^2 u}{\partial x^2} + \dfrac{\partial^2 u}{\partial y^2} = 0$, subject to the following boundary conditions:

   $u(0, y) = 0$ $0 \le y \le 2$
   $u(3, y) = 0$ $0 \le y \le 2$
   $u(x, 2) = 0$ $0 \le x \le 3$
   $u(x, 0) = x(3 - x)$ $0 \le x \le 3$

For fully worked solutions to each of the problems in Practice Exercises 215 to 219 in this chapter, go to the website:
www.routledge.com/cw/bird

Revision Test 16 Second-order differential equations, power series methods, and partial differential equations

This Revision Test covers the material contained in Chapters 53 to 56. *The marks for each question are shown in brackets at the end of each question.*

1. Find the particular solution of the following differential equations:

   (a) $12\dfrac{d^2y}{dt^2} - 3y = 0$ given that when $t = 0$, $y = 3$ and $\dfrac{dy}{dt} = \dfrac{1}{2}$

   (b) $\dfrac{d^2y}{dx^2} + 2\dfrac{dy}{dx} + 2y = 10e^x$ given that when $x = 0$,

   $y = 0$ and $\dfrac{dy}{dx} = 1$ \hfill (20)

2. In a galvanometer the deflection θ satisfies the differential equation:

   $$\dfrac{d^2\theta}{dt^2} + 2\dfrac{d\theta}{dt} + \theta = 4$$

   Solve the equation for θ given that when $t = 0$, $\theta = 0$ and $\dfrac{d\theta}{dt} = 0$ \hfill (12)

3. Determine $y^{(n)}$ when $y = 2x^3e^{4x}$ \hfill (10)

4. Determine the power series solution of the differential equation: $\dfrac{d^2y}{dx^2} + 2x\dfrac{dy}{dx} + y = 0$ using the Leibniz–Maclaurin method, given the boundary conditions that at $x = 0$, $y = 2$ and $\dfrac{dy}{dx} = 1$ \hfill (20)

5. Use the Frobenius method to determine the general power series solution of the differential equation: $\dfrac{d^2y}{dx^2} + 4y = 0$ \hfill (21)

6. Determine the general power series solution of Bessel's equation:

   $$x^2\dfrac{d^2y}{dx^2} + x\dfrac{dy}{dx} + (x^2 - v^2)y = 0$$

   and hence state the series up to and including the term in x^6 when $v = +3$ \hfill (26)

7. Determine the general solution of $\dfrac{\partial u}{\partial x} = 5xy$ \hfill (2)

8. Solve the differential equation $\dfrac{\partial^2 u}{\partial x^2} = x^2(y-3)$ given the boundary conditions that at $x = 0$, $\dfrac{\partial u}{\partial x} = \sin y$ and $u = \cos y$ \hfill (6)

9. Figure RT15.1 shows a stretched string of length 40 cm which is set oscillating by displacing its mid-point a distance of 1 cm from its rest position and releasing it with zero velocity. Solve the wave equation: $\dfrac{\partial^2 u}{\partial x^2} = \dfrac{1}{c^2}\dfrac{\partial^2 u}{\partial t^2}$ where $c^2 = 1$, to determine the resulting motion $u(x, t)$ \hfill (23)

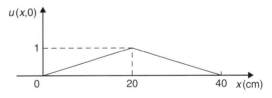

Figure RT15.1

For lecturers/instructors/teachers, fully worked solutions to each of the problems in Revision Test 16, together with a full marking scheme, are available at the website:
www.routledge.com/cw/bird

Section K

Statistics and probability

Chapter 57

Presentation of statistical data

Why it is important to understand: Presentation of statistical data

Statistics is the study of the collection, organisation, analysis, and interpretation of data. It deals with all aspects of this, including the planning of data collection in terms of the design of surveys and experiments. Statistics is applicable to a wide variety of academic disciplines, including natural and social sciences, engineering, government, and business. Statistical methods can be used for summarising or describing a collection of data. Engineering statistics combines engineering and statistics. Design of experiments is a methodology for formulating scientific and engineering problems using statistical models. Quality control and process control use statistics as a tool to manage conformance to specifications of manufacturing processes and their products. Time and methods engineering use statistics to study repetitive operations in manufacturing in order to set standards and find optimum manufacturing procedures. Reliability engineering measures the ability of a system to perform for its intended function (and time) and has tools for improving performance. Probabilistic design involves the use of probability in product and system design. System identification uses statistical methods to build mathematical models of dynamical systems from measured data. System identification also includes the optimal design of experiments for efficiently generating informative data for fitting such models. This chapter introduces the presentation of statistical data.

At the end of this chapter, you should be able to:

- distinguish between discrete and continuous data
- present data diagrammatically – pictograms, horizontal and vertical bar charts, percentage component bar charts, pie diagrams
- produce a tally diagram for a set of data
- form a frequency distribution from a tally diagram
- construct a histogram from a frequency distribution
- construct a frequency polygon from a frequency distribution
- produce a cumulative frequency distribution from a set of grouped data
- construct an ogive from a cumulative frequency distribution

57.1 Some statistical terminology

Data are obtained largely by two methods:

(a) by counting – for example, the number of stamps sold by a post office in equal periods of time, and

(b) by measurement – for example, the heights of a group of people.

When data are obtained by counting and only whole numbers are possible, the data are called **discrete**. Measured data can have any value within certain limits and are called **continuous** (see Problem 1).

A **set** is a group of data and an individual value within the set is called a **member** of the set. Thus, if the masses of five people are measured correct to the nearest 0.1 kg and are found to be 53.1 kg, 59.4 kg, 62.1 kg, 77.8 kg and 64.4 kg, then the set of masses in kilograms for these five people is:

$$\{53.1, 59.4, 62.1, 77.8, 64.4\}$$

and one of the members of the set is 59.4

A set containing all the members is called a **population**. Some members selected at random from a population are called a **sample**. Thus all car registration numbers form a population, but the registration numbers of, say, 20 cars taken at random throughout the country are a sample drawn from that population.

The number of times that the value of a member occurs in a set is called the **frequency** of that member. Thus in the set $\{2, 3, 4, 5, 4, 2, 4, 7, 9\}$, member 4 has a frequency of three, member 2 has a frequency of two and the other members have a frequency of one.

The **relative frequency** with which any member of a set occurs is given by the ratio:

$$\frac{\text{frequency of member}}{\text{total frequency of all members}}$$

For the set: $\{2, 3, 5, 4, 7, 5, 6, 2, 8\}$, the relative frequency of member 5 is $\frac{2}{9}$

Often, relative frequency is expressed as a percentage and the **percentage relative frequency** is: (relative frequency $\times 100$)%

Problem 1. Data are obtained on the topics given below. State whether they are discrete or continuous data.

(a) The number of days on which rain falls in a month for each month of the year.

(b) The mileage travelled by each of a number of salesmen.

(c) The time that each of a batch of similar batteries lasts.

(d) The amount of money spent by each of several families on food.

(a) The number of days on which rain falls in a given month must be an integer value and is obtained by **counting** the number of days. Hence, these data are **discrete**.

(b) A salesman can travel any number of miles (and parts of a mile) between certain limits and these data are **measured**. Hence the data are **continuous**.

(c) The time that a battery lasts is **measured** and can have any value between certain limits. Hence these data are **continuous**.

(d) The amount of money spent on food can only be expressed correct to the nearest pence, the amount being **counted**. Hence, these data are **discrete**.

Now try the following Practice Exercise

Practice Exercise 220   Discrete and continuous data (Answers on page 889)

In Problems 1 and 2, state whether data relating to the topics given are discrete or continuous.

1. (a) The amount of petrol produced daily, for each of 31 days, by a refinery.

    (b) The amount of coal produced daily by each of 15 miners.

    (c) The number of bottles of milk delivered daily by each of 20 milkmen.

    (d) The size of ten samples of rivets produced by a machine.

2. (a) The number of people visiting an exhibition on each of five days.

    (b) The time taken by each of 12 athletes to run 100 metres.

    (c) The value of stamps sold in a day by each of 20 post offices.

(d)   The number of defective items produced in each of ten one-hour periods by a machine.

57.2   Presentation of ungrouped data

Ungrouped data can be presented diagrammatically in several ways and these include:

(a) **pictograms**, in which pictorial symbols are used to represent quantities (see Problem 2),

(b) **horizontal bar charts**, having data represented by equally spaced horizontal rectangles (see Problem 3), and

(c) **vertical bar charts**, in which data are represented by equally spaced vertical rectangles (see Problem 4).

Trends in ungrouped data over equal periods of time can be presented diagrammatically by a **percentage component bar chart**. In such a chart, equally spaced rectangles of any width, but whose height corresponds to 100%, are constructed. The rectangles are then subdivided into values corresponding to the percentage relative frequencies of the members (see Problem 5).

A **pie diagram** is used to show diagrammatically the parts making up the whole. In a pie diagram, the area of a circle represents the whole, and the areas of the sectors of the circle are made proportional to the parts which make up the whole (see Problem 6).

Problem 2.   The number of television sets repaired in a workshop by a technician in six one-month periods is as shown below. Present these data as a pictogram.

Month	Number repaired
January	11
February	6
March	15
April	9
May	13
June	8

Each symbol shown in Fig. 57.1 represents two television sets repaired. Thus, in January, $5\frac{1}{2}$ symbols

are used to represent the 11 sets repaired, in February, three symbols are used to represent the six sets repaired, and so on.

Month	Number of TV sets repaired	⬚ ≡ 2 sets
January	⬚ ⬚ ⬚ ⬚ ⬚ ◖	
February	⬚ ⬚ ⬚	
March	⬚ ⬚ ⬚ ⬚ ⬚ ⬚ ⬚ ◖	
April	⬚ ⬚ ⬚ ⬚ ◖	
May	⬚ ⬚ ⬚ ⬚ ⬚ ⬚ ◖	
June	⬚ ⬚ ⬚ ⬚	

Figure 57.1

Problem 3.   The distances in miles travelled by four salesmen in a week are as shown below.

Salesmen	P	Q	R	S
Distance travelled miles	413	264	597	143

Use a horizontal bar chart to represent these data diagrammatically.

Equally spaced horizontal rectangles of any width, but whose length is proportional to the distance travelled, are used. Thus, the length of the rectangle for salesman P is proportional to 413 miles, and so on. The horizontal bar chart depicting these data is shown in Fig. 57.2.

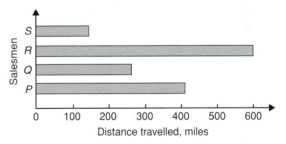

Figure 57.2

Problem 4.   The number of issues of tools or materials from a store in a factory is observed for seven one-hour periods in a day, and the results of the survey are as follows:

Period	1	2	3	4	5	6	7
Number of issues	34	17	9	5	27	13	6

Present these data on a vertical bar chart.

In a vertical bar chart, equally spaced vertical rectangles of any width, but whose height is proportional to the quantity being represented, are used. Thus the height of the rectangle for period 1 is proportional to 34 units, and so on. The vertical bar chart depicting these data is shown in Fig. 57.3.

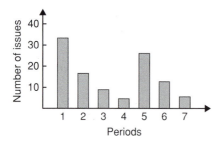

Figure 57.3

Problem 5. The numbers of various types of dwellings sold by a company annually over a three-year period are as shown below. Draw percentage component bar charts to present these data.

	Year 1	Year 2	Year 3
Four-roomed bungalows	24	17	7
Five-roomed bungalows	38	71	118
Four-roomed houses	44	50	53
Five-roomed houses	64	82	147
Six-roomed houses	30	30	25

A table of percentage relative frequency values, correct to the nearest 1%, is the first requirement. Since,

percentage relative frequency

$$= \frac{\text{frequency of member} \times 100}{\text{total frequency}}$$

then for four-roomed bungalows in year 1:
percentage relative frequency

$$= \frac{24 \times 100}{24 + 38 + 44 + 64 + 30} = 12\%$$

The percentage relative frequencies of the other types of dwellings for each of the three years are similarly calculated and the results are as shown in the table below.

	Year 1 (%)	Year 2 (%)	Year 3 (%)
Four-roomed bungalows	12	7	2
Five-roomed bungalows	19	28	34
Four-roomed houses	22	20	15
Five-roomed houses	32	33	42
Six-roomed houses	15	12	7

The percentage component bar chart is produced by constructing three equally spaced rectangles of any width, corresponding to the three years. The heights of the rectangles correspond to 100% relative frequency, and are subdivided into the values in the table of percentages shown above. A key is used (different types of shading or different colour schemes) to indicate corresponding percentage values in the rows of the table of percentages. The percentage component bar chart is shown in Fig. 57.4.

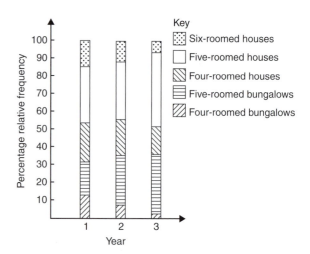

Figure 57.4

Problem 6. The retail price of a product costing £2 is made up as follows: materials 10 p, labour 20 p, research and development 40 p, overheads 70 p, profit 60 p. Present these data on a pie diagram.

A circle of any radius is drawn, and the area of the circle represents the whole, which in this case is £2. The circle is subdivided into sectors so that the areas of the sectors are proportional to the parts, i.e. the parts which make up the total retail price. For the area of a sector to be proportional to a part, the angle at the centre of the circle must be proportional to that part. The whole, £2 or 200 p, corresponds to 360°. Therefore,

$$10 \text{ p corresponds to } 360 \times \frac{10}{200} \text{ degrees, i.e. } 18°$$

$$20 \text{ p corresponds to } 360 \times \frac{20}{200} \text{ degrees, i.e. } 36°$$

and so on, giving the angles at the centre of the circle for the parts of the retail price as: 18°, 36°, 72°, 126° and 108°, respectively.

The pie diagram is shown in Fig. 57.5.

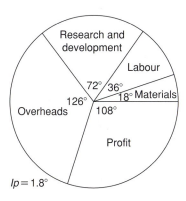

$Ip \equiv 1.8°$

Figure 57.5

Problem 7.

(a) Using the data given in Fig. 57.2 only, calculate the amount of money paid to each salesman for travelling expenses, if they are paid an allowance of 37 p per mile.

(b) Using the data presented in Fig. 57.4, comment on the housing trends over the three-year period.

(c) Determine the profit made by selling 700 units of the product shown in Fig. 57.5.

(a) By measuring the length of rectangle P the mileage covered by salesman P is equivalent to 413 miles. Hence salesman P receives a travelling allowance of

$$\frac{£413 \times 37}{100}, \text{ i.e. } \textbf{£152.81}$$

Similarly, for salesman Q, the miles travelled are 264 and his allowance is

$$\frac{£264 \times 37}{100}, \text{ i.e. } \textbf{£97.68}$$

Salesman R travels 597 miles and he receives

$$\frac{£597 \times 37}{100}, \text{ i.e. } \textbf{£220.89}$$

Finally, salesman S receives

$$\frac{£143 \times 37}{100}, \text{ i.e. } \textbf{£52.91}$$

(b) An analysis of Fig. 57.4 shows that five-roomed bungalows and five-roomed houses are becoming more popular, the greatest change in the three years being a 15% increase in the sales of five-roomed bungalows.

(c) Since 1.8° corresponds to 1 p and the profit occupies 108° of the pie diagram, then the profit per unit is

$$\frac{108 \times 1}{1.8}, \text{ that is, } 60 \text{ p}$$

The profit when selling 700 units of the product is

$$£\frac{700 \times 60}{100}, \text{ that is, } \textbf{£420}$$

Now try the following Practice Exercise

Practice Exercise 221   Presentation of ungrouped data (Answers on page 889)

1. The number of vehicles passing a stationary observer on a road in six ten-minute intervals is as shown. Draw a pictogram to represent these data.

Period of time	1	2	3	4	5	6
Number of vehicles	35	44	62	68	49	41

2. The number of components produced by a factory in a week is as shown below:

Day	Number of components
Mon	1580
Tues	2190
Wed	1840
Thur	2385
Fri	1280

Show these data on a pictogram.

3. For the data given in Problem 1 above, draw a horizontal bar chart.

4. Present the data given in Problem 2 above on a horizontal bar chart.

5. For the data given in Problem 1 above, construct a vertical bar chart.

6. Depict the data given in Problem 2 above on a vertical bar chart.

7. A factory produces three different types of components. The percentages of each of these components produced for three one-month periods are as shown below. Show this information on percentage component bar charts and comment on the changing trend in the percentages of the types of component produced.

Month	1	2	3
Component P	20	35	40
Component Q	45	40	35
Component R	35	25	25

8. A company has five distribution centres and the mass of goods in tonnes sent to each centre during four one-week periods, is as shown.

Week	1	2	3	4
Centre A	147	160	174	158
Centre B	54	63	77	69
Centre C	283	251	237	211
Centre D	97	104	117	144
Centre E	224	218	203	194

Use a percentage component bar chart to present these data and comment on any trends.

9. The employees in a company can be split into the following categories: managerial 3, supervisory 9, craftsmen 21, semi-skilled 67, others 44. Show these data on a pie diagram.

10. The way in which an apprentice spent his time over a one-month period is as follows: drawing office 44 hours, production 64 hours, training 12 hours, at college 28 hours.
Use a pie diagram to depict this information.

11. (a) With reference to Fig. 57.5, determine the amount spent on labour and materials to produce 1650 units of the product.

    (b) If in year 2 of Fig. 57.4, 1% corresponds to 2.5 dwellings, how many bungalows are sold in that year.

12. (a) If the company sell 23 500 units per annum of the product depicted in Fig. 57.5, determine the cost of their overheads per annum.

    (b) If 1% of the dwellings represented in year 1 of Fig. 57.4 corresponds to two dwellings, find the total number of houses sold in that year.

57.3 Presentation of grouped data

When the number of members in a set is small, say ten or fewer, the data can be represented diagrammatically without further analysis, by means of pictograms, bar charts, percentage components bar charts or pie diagrams (as shown in Section 57.2).

For sets having more than ten members, those members having similar values are grouped together in **classes** to form a **frequency distribution**. To assist in accurately counting members in the various classes, a **tally diagram** is used (see Problems 8 and 12).

A frequency distribution is merely a table showing classes and their corresponding frequencies (see Problems 8 and 12).

The new set of values obtained by forming a frequency distribution is called **grouped data**.

The terms used in connection with grouped data are shown in Fig. 57.6(a). The size or range of a class is given by the **upper class boundary value** minus the **lower class boundary value**, and in Fig. 57.6 is $7.65 - 7.35$, i.e. 0.30. The **class interval** for the class shown in Fig. 57.6(b) is 7.4 to 7.6 and the class mid-point value is given by

$$\frac{\left(\begin{array}{c}\text{upper class}\\\text{boundary value}\end{array}\right)+\left(\begin{array}{c}\text{lower class}\\\text{boundary value}\end{array}\right)}{2}$$

and in Fig. 57.6 is $\dfrac{7.65+7.35}{2}$, i.e. 7.5

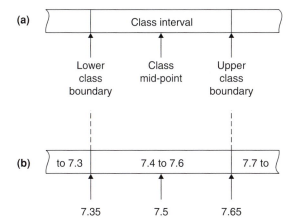

Figure 57.6

One of the principal ways of presenting grouped data diagrammatically is by using a **histogram**, in which the **areas** of vertical, adjacent rectangles are made proportional to frequencies of the classes (see Problem 9). When class intervals are equal, the heights of the rectangles of a histogram are equal to the frequencies of the classes. For histograms having unequal class intervals, the area must be proportional to the frequency. Hence, if the class interval of class A is twice the class interval of class B, then for equal frequencies, the height of the rectangle representing A is half that of B (see Problem 11). Another method of presenting grouped data diagrammatically is by using a **frequency polygon**, which is the graph produced by plotting frequency against class mid-point values and joining the co-ordinates with straight lines (see Problem 12).

A **cumulative frequency distribution** is a table showing the cumulative frequency for each value of upper class boundary. The cumulative frequency for a particular value of upper class boundary is obtained by adding the frequency of the class to the sum of the previous frequencies. A cumulative frequency distribution is formed in Problem 13.

The curve obtained by joining the co-ordinates of cumulative frequency (vertically) against upper class boundary (horizontally) is called an **ogive** or a **cumulative frequency distribution curve** (see Problem 13).

Problem 8.   The data given below refer to the gain of each of a batch of 40 transistors, expressed correct to the nearest whole number. Form a frequency distribution for these data having seven classes.

81	83	87	74	76	89	82	84
86	76	77	71	86	85	87	88
84	81	80	81	73	89	82	79
81	79	78	80	85	77	84	78
83	79	80	83	82	79	80	77

The **range** of the data is the value obtained by taking the value of the smallest member from that of the largest member. Inspection of the set of data shows that, range $= 89 - 71 = 18$. The size of each class is given approximately by range divided by the number of classes. Since seven classes are required, the size of each class is 18/7, that is, approximately 3. To achieve seven equal classes spanning a range of values from 71 to 89, the class intervals are selected as: 70–72, 73–75, and so on.

To assist with accurately determining the number in each class, a **tally diagram** is produced, as shown in Table 57.1(a). This is obtained by listing the classes in the left-hand column, and then inspecting each of the 40 members of the set in turn and allocating them to the appropriate classes by putting '1s' in the appropriate rows. Every fifth '1' allocated to the particular row is shown as an oblique line crossing the four previous '1s', to help with final counting.

A **frequency distribution** for the data is shown in Table 57.1(b) and lists classes and their corresponding frequencies, obtained from the tally diagram. (Class mid-point values are also shown in the table, since they are used for constructing the histogram for these data (see Problem 9)).

Problem 9.   Construct a histogram for the data given in Table 57.1(b).

The histogram is shown in Fig. 57.7. The width of the rectangles correspond to the upper class boundary values minus the lower class boundary values and the heights of the rectangles correspond to the class frequencies. The easiest way to draw a histogram is to mark the class mid-point values on the horizontal scale and draw

Section K

the rectangles symmetrically about the appropriate class mid-point values and touching one another.

Table 57.1(a)

Class	Tally
70–72	1
73–75	11
76–78	⋃⋃ 11
79–81	⋃⋃ ⋃⋃ 11
82–84	⋃⋃ 1111
85–87	⋃⋃ 1
88–90	111

Table 57.1(b)

Class	Class mid-point	Frequency
70–72	71	1
73–75	74	2
76–78	77	7
79–81	80	12
82–84	83	9
85–87	86	6
88–90	89	3

Figure 57.7

Problem 10. The amount of money earned weekly by 40 people working part-time in a factory, correct to the nearest £10, is shown below. Form a frequency distribution having six classes for these data.

80	90	70	110	90	160	110	80
140	30	90	50	100	110	60	100

80	90	110	80	100	90	120	70
130	170	80	120	100	110	40	110
50	100	110	90	100	70	110	80

Inspection of the set given shows that the majority of the members of the set lie between £80 and £110 and that there are a much smaller number of extreme values ranging from £30 to £170. If equal class intervals are selected, the frequency distribution obtained does not give as much information as one with unequal class intervals. Since the majority of members are between £80 and £100, the class intervals in this range are selected to be smaller than those outside of this range. There is no unique solution and one possible solution is shown in Table 57.2.

Table 57.2

Class	Frequency
20–40	2
50–70	6
80–90	12
100–110	14
120–140	4
150–170	2

Problem 11. Draw a histogram for the data given in Table 57.2.

When dealing with unequal class intervals, the histogram must be drawn so that the areas (and not the heights) of the rectangles are proportional to the frequencies of the classes. The data given are shown in columns 1 and 2 of Table 57.3. Columns 3 and 4 give the upper and lower class boundaries, respectively. In column 5, the class ranges (i.e. upper class boundary minus lower class boundary values) are listed. The heights of the rectangles are proportional to the ratio $\dfrac{\text{frequency}}{\text{class range}}$, as shown in column 6. The histogram is shown in Fig. 57.8.

Table 57.3

1 Class	2 Frequency	3 Upper class boundary	4 Lower class boundary	5 Class range	6 Height of rectangle
20–40	2	45	15	30	$\frac{2}{30} = \frac{1}{15}$
50–70	6	75	45	30	$\frac{6}{30} = \frac{3}{15}$
80–90	12	95	75	20	$\frac{12}{20} = \frac{9}{15}$
100–110	14	115	95	20	$\frac{14}{20} = \frac{10\frac{1}{2}}{15}$
120–140	4	145	115	30	$\frac{4}{30} = \frac{2}{15}$
150–170	2	175	145	30	$\frac{2}{30} = \frac{1}{15}$

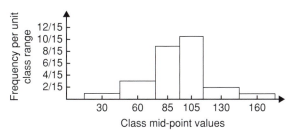

Figure 57.8

Problem 12.   The masses of 50 ingots in kilograms are measured correct to the nearest 0.1 kg and the results are as shown below. Produce a frequency distribution having about seven classes for these data and then present the grouped data as (a) a frequency polygon and (b) a histogram.

8.0	8.6	8.2	7.5	8.0	9.1	8.5	7.6	8.2	7.8
8.3	7.1	8.1	8.3	8.7	7.8	8.7	8.5	8.4	8.5
7.7	8.4	7.9	8.8	7.2	8.1	7.8	8.2	7.7	7.5
8.1	7.4	8.8	8.0	8.4	8.5	8.1	7.3	9.0	8.6
7.4	8.2	8.4	7.7	8.3	8.2	7.9	8.5	7.9	8.0

The **range** of the data is the member having the largest value minus the member having the smallest value. Inspection of the set of data shows that:

$$\text{range} = 9.1 - 7.1 = 2.0$$

The size of each class is given approximately by

$$\frac{\text{range}}{\text{number of classes}}$$

Since about seven classes are required, the size of each class is 2.0/7, that is approximately 0.3, and thus the **class limits** are selected as 7.1 to 7.3, 7.4 to 7.6, 7.7 to 7.9, and so on.

The **class mid-point** for the 7.1 to 7.3 class is $\frac{7.35 + 7.05}{2}$, i.e. 7.2, for the 7.4 to 7.6 class is $\frac{7.65 + 7.35}{2}$, i.e. 7.5, and so on.

To assist with accurately determining the number in each class, a **tally diagram** is produced as shown in Table 57.4. This is obtained by listing the classes in the left-hand column and then inspecting each of the 50 members of the set of data in turn and allocating it

Table 57.4

Class	Tally
7.1 to 7.3	111
7.4 to 7.6	1̶H̶T̶1
7.7 to 7.9	1̶H̶T̶1 1111
8.0 to 8.2	1̶H̶T̶1 1̶H̶T̶1 1111
8.3 to 8.5	1̶H̶T̶1 1̶H̶T̶1 1
8.6 to 8.8	1̶H̶T̶1 1
8.9 to 9.1	11

to the appropriate class by putting a '1' in the appropriate row. Each fifth '1' allocated to a particular row is marked as an oblique line to help with final counting.

A **frequency distribution** for the data is shown in Table 57.5 and lists classes and their corresponding frequencies. Class mid-points are also shown in this table, since they are used when constructing the frequency polygon and histogram.

Table 57.5

Class	Class mid-point	Frequency
7.1 to 7.3	7.2	3
7.4 to 7.6	7.5	5
7.5 to 7.9	7.8	9
8.0 to 8.2	8.1	14
8.1 to 8.5	8.4	11
8.2 to 8.8	8.7	6
8.9 to 9.1	9.0	2

A **frequency polygon** is shown in Fig. 57.9, the co-ordinates corresponding to the class mid-point/frequency values, given in Table 57.5. The co-ordinates are joined by straight lines and the polygon is 'anchored-down' at each end by joining to the next class mid-point value and zero frequency.

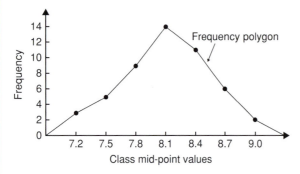

Figure 57.9

A **histogram** is shown in Fig. 57.10, the width of a rectangle corresponding to (upper class boundary value – lower class boundary value) and height corresponding to the class frequency. The easiest way to draw a histogram is to mark class mid-point values on the horizontal scale and to draw the rectangles symmetrically about the appropriate class mid-point values and touching one another. A histogram for the data given in Table 57.5 is shown in Fig. 57.10.

Figure 57.10

Problem 13. The frequency distribution for the masses in kilograms of 50 ingots is:

7.1 to 7.3   3,   7.4 to 7.6   5,   7.7 to 7.9 9,

8.0 to 8.2 14,   8.3 to 8.5   11,   8.6 to 8.8, 6,

8.9 to 9.1   2

Form a cumulative frequency distribution for these data and draw the corresponding ogive.

A **cumulative frequency distribution** is a table giving values of cumulative frequency for the value of upper class boundaries, and is shown in Table 57.6. Columns 1 and 2 show the classes and their frequencies. Column 3 lists the upper class boundary values for the classes given in column 1. Column 4 gives the cumulative frequency values for all frequencies less than the upper class boundary values given in column 3. Thus, for example, for the 7.7 to 7.9 class

Table 57.6

1 Class	2 Frequency	3 Upper class boundary	4 Cumulative frequency
		Less than	
7.1–7.3	3	7.35	3
7.4–7.6	5	7.65	8
7.7–7.9	9	7.95	17
8.0–8.2	14	8.25	31
8.3–8.5	11	8.55	42
8.6–8.8	6	8.85	48
8.9–9.1	2	9.15	50

shown in row 3, the cumulative frequency value is the sum of all frequencies having values of less than 7.95, i.e. $3+5+9=17$, and so on. The **ogive** for the cumulative frequency distribution given in Table 57.6 is shown in Fig. 57.11. The co-ordinates corresponding to each upper class boundary/cumulative frequency value are plotted and the co-ordinates are joined by straight lines (not the best curve drawn through the co-ordinates as in experimental work). The ogive is 'anchored' at its start by adding the co-ordinate $(7.05, 0)$.

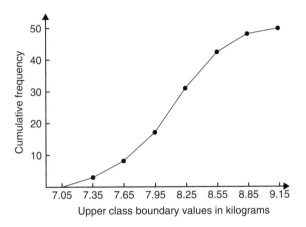

Figure 57.11

Now try the following Practice Exercise

Practice Exercise 222   Presentation of grouped data (Answers on page 889)

1. The mass in kilograms, correct to the nearest one-tenth of a kilogram, of 60 bars of metal are as shown. Form a frequency distribution of about eight classes for these data.

39.8	40.3	40.6	40.0	39.6
39.6	40.2	40.3	40.4	39.8
40.2	40.3	39.9	39.9	40.0
40.1	40.0	40.1	40.1	40.2
39.7	40.4	39.9	40.1	39.9
39.5	40.0	39.8	39.5	39.9
40.1	40.0	39.7	40.4	39.3
40.7	39.9	40.2	39.9	40.0
40.1	39.7	40.5	40.5	39.9
40.8	40.0	40.2	40.0	39.9
39.8	39.7	39.5	40.1	40.2
40.6	40.1	39.7	40.2	40.3

2. Draw a histogram for the frequency distribution given in the solution of Problem 1.

3. The information given below refers to the value of resistance in ohms of a batch of 48 resistors of similar value. Form a frequency distribution for the data, having about six classes, and draw a frequency polygon and histogram to represent these data diagramatically.

21.0	22.4	22.8	21.5	22.6	21.1	21.6	22.3
22.9	20.5	21.8	22.2	21.0	21.7	22.5	20.7
23.2	22.9	21.7	21.4	22.1	22.2	22.3	21.3
22.1	21.8	22.0	22.7	21.7	21.9	21.1	22.6
21.4	22.4	22.3	20.9	22.8	21.2	22.7	21.6
22.2	21.6	21.3	22.1	21.5	22.0	23.4	21.2

4. The time taken in hours to the failure of 50 specimens of a metal subjected to fatigue failure tests are as shown. Form a frequency distribution, having about eight classes and unequal class intervals, for these data.

28	22	23	20	12	24	37	28	21	25
21	14	30	23	27	13	23	7	26	19
24	22	26	3	21	24	28	40	27	24
20	25	23	26	47	21	29	26	22	33
27	9	13	35	20	16	20	25	18	22

5. Form a cumulative frequency distribution and hence draw the ogive for the frequency distribution given in the solution to Problem 3.

6. Draw a histogram for the frequency distribution given in the solution to Problem 4.

Section K

Section K

7. The frequency distribution for a batch of 50 capacitors of similar value, measured in microfarads, is:

10.5–10.9   2,   11.0–11.4   7,
11.5–11.9   10,   12.0–12.4 12,
12.5–12.9   11,   13.0–13.4   8

Form a cumulative frequency distribution for these data.

8. Draw an ogive for the data given in the solution of Problem 7.

9. The diameter in millimetres of a reel of wire is measured in 48 places and the results are as shown.

2.10	2.29	2.32	2.21	2.14	2.22
2.28	2.18	2.17	2.20	2.23	2.13

2.26	2.10	2.21	2.17	2.28	2.15
2.16	2.25	2.23	2.11	2.27	2.34
2.24	2.05	2.29	2.18	2.24	2.16
2.15	2.22	2.14	2.27	2.09	2.21
2.11	2.17	2.22	2.19	2.12	2.20
2.23	2.07	2.13	2.26	2.16	2.12

(a) Form a frequency distribution of diameters having about six classes.

(b) Draw a histogram depicting the data.

(c) Form a cumulative frequency distribution.

(d) Draw an ogive for the data.

For fully worked solutions to each of the problems in Practice Exercises 220 to 222 in this chapter, go to the website:
www.routledge.com/cw/bird

Chapter 58

Mean, median, mode and standard deviation

Why it is important to understand: Mean, median, mode and standard deviation

Statistics is a field of mathematics that pertains to data analysis. In many real-life situations, it is helpful to describe data by a single number that is most representative of the entire collection of numbers. Such a number is called a measure of central tendency; the most commonly used measures are mean, median, mode and standard deviation, the latter being the average distance between the actual data and the mean. Statistics is important in the field of engineering since it provides tools to analyse collected data. For example, a chemical engineer may wish to analyse temperature measurements from a mixing tank. Statistical methods can be used to determine how reliable and reproducible the temperature measurements are, how much the temperature varies within the data set, what future temperatures of the tank may be, and how confident the engineer can be in the temperature measurements made. When performing statistical analysis on a set of data, the mean, median, mode, and standard deviation are all helpful values to calculate; this chapter explains how to determine these measures of central tendency.

At the end of this chapter, you should be able to:

- determine the mean, median and mode for a set of ungrouped data
- determine the mean, median and mode for a set of grouped data
- draw a histogram from a set of grouped data
- determine the mean, median and mode from a histogram
- calculate the standard deviation from a set of ungrouped data
- calculate the standard deviation from a set of grouped data
- determine the quartile values from an ogive
- determine quartile, decile and percentile values from a set of data

58.1 Measures of central tendency

A single value, which is representative of a set of values, may be used to give an indication of the general size of the members in a set, the word 'average' often being used to indicate the single value.

The statistical term used for 'average' is the arithmetic mean or just the **mean**. Other measures of central tendency may be used and these include the **median** and the **modal** values.

58.2 Mean, median and mode for discrete data

Mean

The **arithmetic mean value** is found by adding together the values of the members of a set and dividing by the number of members in the set. Thus, the mean of the set of numbers: $\{4, 5, 6, 9\}$ is:

$$\frac{4+5+6+9}{4}, \text{ i.e. } 6$$

In general, the mean of the set: $\{x_1, x_2, x_3, \ldots, x_n\}$ is

$$\bar{x} = \frac{x_1 + x_2 + x_3 + \cdots + x_n}{n}, \text{ written as } \frac{\sum x}{n}$$

where $\sum$ is the Greek letter 'sigma' and means 'the sum of', and $\bar{x}$ (called x-bar) is used to signify a mean value.

Median

The **median value** often gives a better indication of the general size of a set containing extreme values. The set: $\{7, 5, 74, 10\}$ has a mean value of 24, which is not really representative of any of the values of the members of the set. The median value is obtained by:

(a) **ranking** the set in ascending order of magnitude, and

(b) selecting the value of the **middle member** for sets containing an odd number of members, or finding the value of the mean of the two middle members for sets containing an even number of members.

For example, the set: $\{7, 5, 74, 10\}$ is ranked as $\{5, 7, 10, 74\}$, and since it contains an even number of members (four in this case), the mean of 7 and 10 is taken, giving a median value of 8.5. Similarly, the set: $\{3, 81, 15, 7, 14\}$ is ranked as $\{3, 7, 14, 15, 81\}$ and the median value is the value of the middle member, i.e. 14.

Mode

The **modal value**, or **mode**, is the most commonly occurring value in a set. If two values occur with the same frequency, the set is 'bi-modal'. The set: $\{5, 6, 8, 2, 5, 4, 6, 5, 3\}$ has a model value of 5, since the member having a value of 5 occurs three times.

Problem 1. Determine the mean, median and mode for the set:

$$\{2, 3, 7, 5, 5, 13, 1, 7, 4, 8, 3, 4, 3\}$$

The mean value is obtained by adding together the values of the members of the set and dividing by the number of members in the set.

Thus, **mean value**,

$$\bar{x} = \frac{\begin{array}{c}2+3+7+5+5+13+1\\+7+4+8+3+4+3\end{array}}{13} = \frac{65}{13} = 5$$

To obtain the median value the set is ranked, that is, placed in ascending order of magnitude, and since the set contains an odd number of members the value of the middle member is the median value. Ranking the set gives:

$$\{1, 2, 3, 3, 3, 4, 4, 5, 5, 7, 7, 8, 13\}$$

The middle term is the seventh member, i.e. 4, thus the **median value is 4**. The **modal value** is the value of the most commonly occurring member and is **3**, which occurs three times, all other members only occurring once or twice.

Problem 2. The following set of data refers to the amount of money in pounds taken by a news vendor for six days. Determine the mean, median and modal values of the set:

$$\{27.90, 34.70, 54.40, 18.92, 47.60, 39.68\}$$

$$\textbf{Mean value} = \frac{\begin{array}{c}27.90+34.70+54.40\\+18.92+47.60+39.68\end{array}}{6} = \textbf{£37.20}$$

The ranked set is:

$$\{18.92, 27.90, 34.70, 39.68, 47.60, 54.40\}$$

Since the set has an even number of members, the mean of the middle two members is taken to give the median value, i.e.

$$\textbf{Median value} = \frac{34.70+39.68}{2} = \textbf{£37.19}$$

Since no two members have the same value, this set has **no mode**.

Now try the following Practice Exercise

Practice Exercise 223   Mean, median and mode for discrete data (Answers on page 889)

In Problems 1 to 4, determine the mean, median and modal values for the sets given.

1.   {3, 8, 10, 7, 5, 14, 2, 9, 8}

2.   {26, 31, 21, 29, 32, 26, 25, 28}

3.   {4.72, 4.71, 4.74, 4.73, 4.72, 4.71, 4.73, 4.72}

4.   {73.8, 126.4, 40.7, 141.7, 28.5, 237.4, 157.9}

58.3   Mean, median and mode for grouped data

The mean value for a set of grouped data is found by determining the sum of the (frequency × class mid-point values) and dividing by the sum of the frequencies,

i.e.   mean value $\overline{x} = \dfrac{f_1 x_1 + f_2 x_2 + \cdots + f_n x_n}{f_1 + f_2 + \cdots + f_n}$

$$= \dfrac{\sum(fx)}{\sum f}$$

where f is the frequency of the class having a mid-point value of x, and so on.

Problem 3.   The frequency distribution for the value of resistance in ohms of 48 resistors is as shown. Determine the mean value of resistance.

20.5–20.9   3,   21.0–21.4   10,
21.5–21.9   11,   22.0–22.4   13,
22.5–22.9   9,   23.0–23.4   2

The class mid-point/frequency values are:

20.7   3,   21.2   10,   21.7   11,   22.2   13,
22.7   9   and   23.2   2

For grouped data, the mean value is given by:

$$\overline{x} = \dfrac{\sum(fx)}{\sum f}$$

where f is the class frequency and x is the class mid-point value. Hence mean value,

$$\overline{x} = \dfrac{\begin{array}{c}(3 \times 20.7) + (10 \times 21.2) + (11 \times 21.7) \\ + (13 \times 22.2) + (9 \times 22.7) + (2 \times 23.2)\end{array}}{48}$$

$$= \dfrac{1052.1}{48} = 21.919$$

i.e. **the mean value is 21.9 ohms**, correct to 3 significant figures.

Histogram

The mean, median and modal values for grouped data may be determined from a **histogram**. In a histogram, frequency values are represented vertically and variable values horizontally. The mean value is given by the value of the variable corresponding to a vertical line drawn through the centroid of the histogram. The median value is obtained by selecting a variable value such that the area of the histogram to the left of a vertical line drawn through the selected variable value is equal to the area of the histogram on the right of the line. The modal value is the variable value obtained by dividing the width of the highest rectangle in the histogram in proportion to the heights of the adjacent rectangles. The method of determining the mean, median and modal values from a histogram is shown in Problem 4.

Problem 4.   The time taken in minutes to assemble a device is measured 50 times and the results are as shown. Draw a histogram depicting this data and hence determine the mean, median and modal values of the distribution.

14.5–15.5   5,   16.5–17.5   8,

18.5–19.5   16,   20.5–21.5   12,

22.5–23.5   6,   24.5–25.5   3

The histogram is shown in Fig. 58.1. The mean value lies at the centroid of the histogram. With reference to any arbitrary axis, say YY shown at a time of 14 minutes, the position of the horizontal value of the centroid can be obtained from the relationship $AM = \sum(am)$, where A is the area of the histogram, M is the horizontal distance of the centroid from the axis YY, a is the area of a rectangle of the histogram and m is the distance of the centroid of the rectangle from YY. The areas of the individual rectangles are shown circled on the histogram giving a total area of 100 square units. The positions, m, of the centroids of the individual rectangles are $1, 3, 5, \ldots$ units from YY. Thus

$$100M = (10 \times 1) + (16 \times 3) + (32 \times 5)$$

$$+ (24 \times 7) + (12 \times 9) + (6 \times 11)$$

i.e. $M = \dfrac{560}{100} = 5.6$ units from YY

Section K

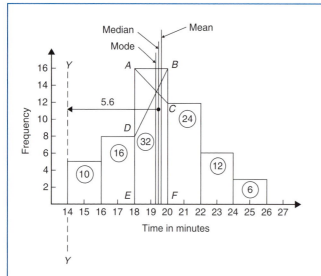

Figure 58.1

Thus the position of the **mean** with reference to the time scale is $14 + 5.6$, i.e. **19.6 minutes**.

The median is the value of time corresponding to a vertical line dividing the total area of the histogram into two equal parts. The total area is 100 square units, hence the vertical line must be drawn to give 50 units of area on each side. To achieve this with reference to Fig. 58.1, rectangle *ABFE* must be split so that $50 - (10 + 16)$ units of area lie on one side and $50 - (24 + 12 + 6)$ units of area lie on the other. This shows that the area of *ABFE* is split so that 24 units of area lie to the left of the line and 8 units of area lie to the right, i.e. the vertical line must pass through 19.5 minutes. Thus the **median value** of the distribution is **19.5 minutes**.

The mode is obtained by dividing the line *AB*, which is the height of the highest rectangle, proportionally to the heights of the adjacent rectangles. With reference to Fig. 58.1, this is done by joining *AC* and *BD* and drawing a vertical line through the point of intersection of these two lines. This gives the **mode** of the distribution and is **19.3 minutes**.

Now try the following Practice Exercise

Practice Exercise 224   Mean, median and mode for grouped data (Answers on page 890)

1. 21 bricks have a mean mass of 24.2 kg, and 29 similar bricks have a mass of 23.6 kg. Determine the mean mass of the 50 bricks.

2. The frequency distribution given below refers to the heights in centimetres of 100 people.

Determine the mean value of the distribution, correct to the nearest millimetre.

150–156   5,   157–163   18,

164–170   20,   171–177   27,

178–184   22,   185–191   8

3. The gain of 90 similar transistors is measured and the results are as shown.

83.5–85.5   6,   86.5–88.5   39,

89.5–91.5   27,   92.5–94.5   15,

95.5–97.5   3

By drawing a histogram of this frequency distribution, determine the mean, median and modal values of the distribution.

4. The diameters, in centimetres, of 60 holes bored in engine castings are measured and the results are as shown. Draw a histogram depicting these results and hence determine the mean, median and modal values of the distribution.

2.011–2.014   7,   2.016–2.019   16,

2.021–2.024   23,   2.026–2.029   9,

2.031–2.034   5

58.4   Standard deviation

(a) Discrete data

The standard deviation of a set of data gives an indication of the amount of dispersion, or the scatter, of members of the set from the measure of central tendency. Its value is the root-mean-square value of the members of the set and for discrete data is obtained as follows:

(a) determine the measure of central tendency, usually the mean value (occasionally the median or modal values are specified),

(b) calculate the deviation of each member of the set from the mean, giving

$$(x_1 - \overline{x}), (x_2 - \overline{x}), (x_3 - \overline{x}), \ldots,$$

(c) determine the squares of these deviations, i.e.

$$(x_1 - \overline{x})^2, (x_2 - \overline{x})^2, (x_3 - \overline{x})^2, \ldots,$$

(d) find the sum of the squares of the deviations, that is

$$(x_1 - \overline{x})^2 + (x_2 - \overline{x})^2 + (x_3 - \overline{x})^2, \ldots,$$

(e) divide by the number of members in the set, n, giving

$$\frac{(x_1 - \overline{x})^2 + (x_2 - \overline{x})^2 + (x_3 - \overline{x})^2 + \cdots}{n}$$

(f) determine the square root of (e).

The standard deviation is indicated by σ (the Greek letter small 'sigma') and is written mathematically as:

$$\textbf{Standard deviation}, \sigma = \sqrt{\left\{ \frac{\sum (x - \overline{x})^2}{n} \right\}}$$

where x is a member of the set, $\overline{x}$ is the mean value of the set and n is the number of members in the set. The value of standard deviation gives an indication of the distance of the members of a set from the mean value. The set: $\{1, 4, 7, 10, 13\}$ has a mean value of 7 and a standard deviation of about 4.2. The set $\{5, 6, 7, 8, 9\}$ also has a mean value of 7, but the standard deviation is about 1.4. This shows that the members of the second set are mainly much closer to the mean value than the members of the first set. The method of determining the standard deviation for a set of discrete data is shown in Problem 5.

> **Problem 5.** Determine the standard deviation from the mean of the set of numbers: $\{5, 6, 8, 4, 10, 3\}$ correct to 4 significant figures.

The arithmetic mean,

$$\overline{x} = \frac{\sum x}{n} = \frac{5 + 6 + 8 + 4 + 10 + 3}{6} = 6$$

Standard deviation, $\sigma = \sqrt{\left\{ \frac{\sum (x - \overline{x})^2}{n} \right\}}$

The $(x - \overline{x})^2$ values are: $(5 - 6)^2$, $(6 - 6)^2$, $(8 - 6)^2$, $(4 - 6)^2$, $(10 - 6)^2$ and $(3 - 6)^2$

The sum of the $(x - \overline{x})^2$ values,

i.e. $\sum (x - \overline{x})^2 = 1 + 0 + 4 + 4 + 16 + 9 = 34$

and $\dfrac{\sum (x - \overline{x})^2}{n} = \dfrac{34}{6} = 5.\dot{6}$

since there are six members in the set.

Hence, **standard deviation**,

$$\sigma = \sqrt{\left\{ \frac{\sum (x - \overline{x})^2}{n} \right\}} = \sqrt{5.\dot{6}}$$

$$= \textbf{2.380}, \text{ correct to 4 significant figures.}$$

(b) Grouped data

For **grouped data, standard deviation**

$$\sigma = \sqrt{\left\{ \frac{\sum \{f(x - \overline{x})^2\}}{\sum f} \right\}}$$

where f is the class frequency value, x is the class mid-point value and $\overline{x}$ is the mean value of the grouped data. The method of determining the standard deviation for a set of grouped data is shown in Problem 6.

> **Problem 6.** The frequency distribution for the values of resistance in ohms of 48 resistors is as shown. Calculate the standard deviation from the mean of the resistors, correct to 3 significant figures.
>
> 20.5–20.9   3,   21.0–21.4   10,
>
> 21.5–21.9   11,   22.0–22.4   13,
>
> 22.5–22.9   9,   23.0–23.4   2

The standard deviation for grouped data is given by:

$$\sigma = \sqrt{\left\{ \frac{\sum \{f(x - \overline{x})^2\}}{\sum f} \right\}}$$

From Problem 3, the distribution mean value, $\overline{x} = 21.92$, correct to 4 significant figures.

The 'x-values' are the class mid-point values, i.e. 20.7, 21.2, 21.7, ...

Thus the $(x - \overline{x})^2$ values are $(20.7 - 21.92)^2$, $(21.2 - 21.92)^2$, $(21.7 - 21.92)^2, \ldots$

and the $f(x - \overline{x})^2$ values are $3(20.7 - 21.92)^2$, $10(21.2 - 21.92)^2$, $11(21.7 - 21.92)^2, \ldots$

The $\sum f(x - \overline{x})^2$ values are

$$4.4652 + 5.1840 + 0.5324 + 1.0192 + 5.4756$$
$$+ 3.2768 = 19.9532$$

$$\frac{\sum \{f(x - \bar{x})^2\}}{\sum f} = \frac{19.9532}{48} = 0.41569$$

and **standard deviation**,

$$\sigma = \sqrt{\left\{ \frac{\sum \{f(x - \bar{x})^2\}}{\sum f} \right\}} = \sqrt{0.41569}$$

$$= \mathbf{0.645}, \text{ correct to 3 significant figures.}$$

Now try the following Practice Exercise

Practice Exercise 225   Standard deviation (Answers on page 890)

1. Determine the standard deviation from the mean of the set of numbers:

    $$\{35, 22, 25, 23, 28, 33, 30\}$$

    correct to 3 significant figures.

2. The values of capacitances, in microfarads, of ten capacitors selected at random from a large batch of similar capacitors are:

    34.3, 25.0, 30.4, 34.6, 29.6, 28.7, 33.4,

    32.7, 29.0 and 31.3

    Determine the standard deviation from the mean for these capacitors, correct to 3 significant figures.

3. The tensile strength in megapascals for 15 samples of tin were determined and found to be:

    34.61, 34.57, 34.40, 34.63, 34.63,

    34.51, 34.49, 34.61, 34.52, 34.55,

    34.58, 34.53, 34.44, 34.48 and 34.40

    Calculate the mean and standard deviation from the mean for these 15 values, correct to 4 significant figures.

4. Calculate the standard deviation from the mean for the mass of the 50 bricks given in Problem 1 of Exercise 224, page 634, correct to 3 significant figures.

5. Determine the standard deviation from the mean, correct to 4 significant figures, for the heights of the 100 people given in Problem 2 of Exercise 224, page 634.

6. Calculate the standard deviation from the mean for the data given in Problem 4 of Exercise 225, page 634, correct to 3 significant figures.

58.5 Quartiles, deciles and percentiles

Other measures of dispersion which are sometimes used are the quartile, decile and percentile values. The **quartile values** of a set of discrete data are obtained by selecting the values of members which divide the set into four equal parts. Thus for the set: $\{2, 3, 4, 5, 5, 7, 9, 11, 13, 14, 17\}$ there are 11 members and the values of the members dividing the set into four equal parts are 4, 7, and 13. These values are signified by Q_1, Q_2 and Q_3 and called the first, second and third quartile values, respectively. It can be seen that the second quartile value, Q_2, is the value of the middle member and hence is the median value of the set.

For grouped data the ogive may be used to determine the quartile values. In this case, points are selected on the vertical cumulative frequency values of the ogive, such that they divide the total value of cumulative frequency into four equal parts. Horizontal lines are drawn from these values to cut the ogive. The values of the variable corresponding to these cutting points on the ogive give the quartile values (see Problem 7).

When a set contains a large number of members, the set can be split into ten parts, each containing an equal number of members. These ten parts are then called **deciles**. For sets containing a very large number of members, the set may be split into 100 parts, each containing an equal number of members. One of these parts is called a **percentile**.

Problem 7. The frequency distribution given below refers to the overtime worked by a group of craftsmen during each of 48 working weeks in a year.

> 25–29   5,   30–34   4,   35–39   7,
>
> 40–44   11,   45–49   12,   50–54   8,
>
> 55–59   1

Draw an ogive for this data and hence determine the quartile values.

The cumulative frequency distribution (i.e. upper class boundary/cumulative frequency values) is:

> 29.5   5,   34.5   9,   39.5   16,   44.5   27,
>
> 49.5   39,   54.5   47,   59.5   48

The ogive is formed by plotting these values on a graph, as shown in Fig. 58.2. The total frequency is divided into four equal parts, each having a range of 48/4, i.e. 12. This gives cumulative frequency values of 0 to 12 corresponding to the first quartile, 12 to 24 corresponding to the second quartile, 24 to 36 corresponding to the third quartile and 36 to 48 corresponding to the fourth quartile of the distribution, i.e. the distribution is divided into four equal parts. The quartile values are those of the variable corresponding to cumulative frequency values of 12, 24 and 36, marked Q_1, Q_2 and Q_3 in Fig. 58.2. These values, correct to the nearest hour, are **37 hours, 43 hours and 48 hours**, respectively. The Q_2 value is also equal to the median value of the distribution. One measure of the dispersion of a distribution is called the **semi-interquartile range** and is given by $(Q_3 - Q_1)/2$, and is $(48 - 37)/2$ in this case, i.e. **$5\frac{1}{2}$ hours**.

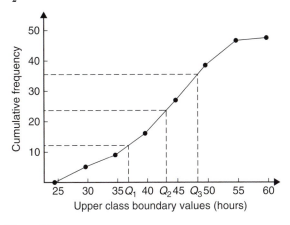

Figure 58.2

Problem 8. Determine the numbers contained in the (a) 41st to 50th percentile group, and (b) eighth decile group of the set of numbers shown below:

> 14   22   17   21   30   28   37   7   23   32
>
> 24   17   20   22   27   19   26   21   15   29

The set is ranked, giving:

> 7 14 15 17 17 19 20 21 21 22 22 23
>
> 24 26 27 28 29 30 32 37

(a) There are 20 numbers in the set, hence the first 10% will be the two numbers 7 and 14, the second 10% will be 15 and 17, and so on. Thus the 41st to 50th percentile group will be the numbers **21 and 22**

(b) The first decile group is obtained by splitting the ranked set into ten equal groups and selecting the first group, i.e. the numbers 7 and 14. The second decile group are the numbers 15 and 17, and so on. Thus the eighth decile group contains the numbers **27 and 28**

Now try the following Practice Exercise

Practice Exercise 226   Quartiles, deciles and percentiles (Answers on page 890)

1. The number of working days lost due to accidents for each of 12 one-monthly periods are as shown. Determine the median and first and third quartile values for this data.

   > 27 37 40 28 23 30 35 24 30 32 31 28

2. The number of faults occurring on a production line in a nine-week period are as shown below. Determine the median and quartile values for the data.

   > 30 27 25 24 27 37 31 27 35

3. Determine the quartile values and semi-interquartile range for the frequency distribution given in Problem 2 of Exercise 224, page 634.

4. Determine the numbers contained in the fifth decile group and in the 61st to 70th percentile groups for the set of numbers:

40 46 28 32 37 42 50 31 48 45
32 38 27 33 40 35 25 42 38 41

5. Determine the numbers in the sixth decile group and in the 81st to 90th percentile group for the set of numbers:

43 47 30 25 15 51 17 21
36 44 33 17 35 58 51 35

37 33 44 56 40 49 22
44 40 31 41 55 50 16

For fully worked solutions to each of the problems in Practice Exercises 223 to 226 in this chapter, go to the website:
www.routledge.com/cw/bird

Chapter 59

Probability

Why it is important to understand: Probability

Engineers deal with uncertainty in their work, often with precision and analysis, and probability theory is widely used to model systems in engineering and scientific applications. There are a number of examples of where probability is used in engineering. For example, with electronic circuits, scaling down the power and energy of such circuits reduces the reliability and predictability of many individual elements, but the circuits must nevertheless be engineered so that the overall circuit is reliable. Centres for disease control need to decide whether to institute massive vaccination or other preventative measures in the face of globally threatening, possibly mutating diseases in humans and animals. System designers must weigh the costs and benefits of measures for reliability and security, such as levels of backups and firewalls, in the face of uncertainty about threats from equipment failures or malicious attackers. Models incorporating probability theory have been developed and are continuously being improved for understanding the brain, gene pools within populations, weather and climate forecasts, microelectronic devices, and imaging systems such as computer aided tomography (CAT) scan and radar. The electric power grid, including power generating stations, transmission lines, and consumers, is a complex system with many redundancies; however, breakdowns occur, and guidance for investment comes from modelling the most likely sequences of events that could cause outage. Similar planning and analysis is done for communication networks, transportation networks, water, and other infrastructure. Probabilities, permutations and combinations are used daily in many different fields that range from gambling and games, to mechanical or structural failure rates, to rates of detection in medical screening. Uncertainty is clearly all around us, in our daily lives and in many professions. Use of standard deviation is widely used when results of opinion polls are described. The language of probability theory lets people break down complex problems, and to argue about pieces of them with each other, and then aggregate information about subsystems to analyse a whole system. This chapter briefly introduces the important subject of probability.

At the end of this chapter, you should be able to:

- define probability
- define expectation, dependent event, independent event and conditional probability
- state the addition and multiplication laws of probability
- use the laws of probability in simple calculations
- use the laws of probability in practical situations
- determine permutations and combinations
- understand and use Bayes' theorem

59.1 Introduction to probability

The **probability** of something happening is the likelihood or chance of it happening. Values of probability lie between 0 and 1, where 0 represents an absolute impossibility and 1 represents an absolute certainty. The probability of an event happening usually lies somewhere between these two extreme values and is expressed either as a proper or decimal fraction. Examples of probability are:

that a length of copper wire
has zero resistance at 100°C   0

that a fair, six-sided dice will
stop with a 3 upwards $\frac{1}{6}$ or 0.1667

that a fair coin will land with
a head upwards $\frac{1}{2}$ or 0.5

that a length of copper wire has
some resistance at 100°C   1

If p is the probability of an event happening and q is the probability of the same event not happening, then the total probability is $p+q$ and is equal to unity, since it is an absolute certainty that the event either does or does not occur, i.e. $p+q=1$

Expectation

The **expectation**, E, of an event happening is defined in general terms as the product of the probability p of an event happening and the number of attempts made, n, i.e. $E=pn$

Thus, since the probability of obtaining a 3 upwards when rolling a fair dice is $\frac{1}{6}$, the expectation of getting a 3 upwards within four throws of the dice is $\frac{1}{6} \times 4$, i.e. $\frac{2}{3}$

Thus expectation is the average occurrence of an event.

Dependent event

A **dependent event** is one in which the probability of an event happening affects the probability of another event happening. Let five transistors be taken at random from a batch of 100 transistors for test purposes, and the probability of there being a defective transistor, p_1, be determined. At some later time, let another five transistors be taken at random from the 95 remaining transistors in the batch and the probability of there being a defective transistor, p_2, be determined. The value of p_2 is different from p_1 since batch size has effectively altered from 100 to 95, i.e. probability p_2 is dependent on probability p_1. Since five transistors are drawn, and then another five transistors drawn without replacing the first five, the second random selection is said to be **without replacement**.

Independent event

An independent event is one in which the probability of an event happening does not affect the probability of another event happening. If five transistors are taken at random from a batch of transistors and the probability of a defective transistor p_1 is determined and the process is repeated after the original five have been replaced in the batch to give p_2, then p_1 is equal to p_2. Since the five transistors are replaced between draws, the second selection is said to be **with replacement**.

Conditional probability

Conditional probability is concerned with the probability of say event B occurring, given that event A has already taken place.

If A and B are independent events, then the fact that event A has already occurred will not affect the probability of event B.

If A and B are dependent events, then event A having occurred will effect the probability of event B.

59.2 Laws of probability

The addition law of probability

The addition law of probability is recognized by the word '**or**' joining the probabilities. If p_A is the probability of event A happening and p_B is the probability of event B happening, the probability of **event A or event B** happening is given by $p_A + p_B$ (provided events A and B are **mutually exclusive**, i.e. A and B are events which cannot occur together). Similarly, the probability of events A or B or C or ... N happening is given by

$$p_A + p_B + p_C + \cdots + p_N$$

The multiplication law of probability

The multiplication law of probability is recognized by the word '**and**' joining the probabilities. If p_A is the probability of event A happening and p_B is the probability of event B happening, the probability of **event A and event B** happening is given by $p_A \times p_B$. Similarly, the probability of events A **and** B **and** C **and** ... N happening is given by

$$p_A \times p_B \times p_C \times \cdots \times p_N$$

59.3 Worked problems on probability

Problem 1. Determine the probabilities of selecting at random (a) a man, and (b) a woman from a crowd containing 20 men and 33 women.

(a) The probability of selecting at random a man, p, is given by the ratio

$$\frac{\text{number of men}}{\text{number in crowd}}$$

i.e. $p = \dfrac{20}{20+33} = \dfrac{20}{53}$ or **0.3774**

(b) The probability of selecting at random a woman, q, is given by the ratio

$$\frac{\text{number of women}}{\text{number in crowd}}$$

i.e. $q = \dfrac{33}{20+33} = \dfrac{33}{53}$ or **0.6226**

(Check: the total probability should be equal to 1;

$$p = \frac{20}{53} \text{ and } q = \frac{33}{53}$$

thus the total probability,

$$p+q = \frac{20}{53} + \frac{33}{53} = 1$$

hence no obvious error has been made.)

Problem 2. Find the expectation of obtaining a 4 upwards with three throws of a fair dice.

Expectation is the average occurrence of an event and is defined as the probability times the number of attempts. The probability, p, of obtaining a 4 upwards for one throw of the dice is $\frac{1}{6}$
Also, three attempts are made, hence $n=3$ and the expectation, E, is pn, i.e. $E = \frac{1}{6} \times 3 = \frac{1}{2}$ or **0.50**

Problem 3. Calculate the probabilities of selecting at random:
(a) the winning horse in a race in which ten horses are running,

(b) the winning horses in both the first and second races if there are ten horses in each race.

(a) Since only one of the ten horses can win, the probability of selecting at random the winning horse is $\dfrac{\text{number of winners}}{\text{number of horses}}$, i.e. $\dfrac{1}{10}$ or **0.10**

(b) The probability of selecting the winning horse in the first race is $\frac{1}{10}$. The probability of selecting the winning horse in the second race is $\frac{1}{10}$. The probability of selecting the winning horses in the first **and** second race is given by the multiplication law of probability, i.e.

$$\textbf{probability} = \frac{1}{10} \times \frac{1}{10}$$

$$= \frac{1}{100} \text{ or } \textbf{0.01}$$

Problem 4. The probability of a component failing in one year due to excessive temperature is $\dfrac{1}{20}$, due to excessive vibration is $\dfrac{1}{25}$ and due to excessive humidity is $\dfrac{1}{50}$. Determine the probabilities that during a one-year period a component: (a) fails due to excessive temperature and excessive vibration, (b) fails due to excessive vibration or excessive humidity, and (c) will not fail because of both excessive temperature and excessive humidity.

Let p_A be the probability of failure due to excessive temperature, then

$$p_A = \frac{1}{20} \text{ and } \overline{p_A} = \frac{19}{20}$$

(where $\overline{p_A}$ is the probability of not failing).

Let p_B be the probability of failure due to excessive vibration, then

$$p_B = \frac{1}{25} \text{ and } \overline{p_B} = \frac{24}{25}$$

Let p_C be the probability of failure due to excessive humidity, then

$$p_C = \frac{1}{50} \text{ and } \overline{p_C} = \frac{49}{50}$$

(a) The probability of a component failing due to excessive temperature **and** excessive vibration is given by:

$$p_A \times p_B = \frac{1}{20} \times \frac{1}{25} = \frac{1}{500} \text{ or } \textbf{0.002}$$

(b) The probability of a component failing due to excessive vibration **or** excessive humidity is:

$$p_B + p_C = \frac{1}{25} + \frac{1}{50} = \frac{3}{50} \text{ or } \textbf{0.06}$$

(c) The probability that a component will not fail due to excessive temperature **and** will not fail due to excess humidity is:

$$\overline{p_A} \times \overline{p_C} = \frac{19}{20} \times \frac{49}{50} = \frac{931}{1000} \text{ or } \textbf{0.931}$$

Problem 5. A batch of 100 capacitors contains 73 which are within the required tolerance values, 17 which are below the required tolerance values, and the remainder are above the required tolerance values. Determine the probabilities that when randomly selecting a capacitor and then a second capacitor: (a) both are within the required tolerance values when selecting with replacement, and (b) the first one drawn is below and the second one drawn is above the required tolerance value, when selection is without replacement.

(a) The probability of selecting a capacitor within the required tolerance values is $\frac{73}{100}$. The first capacitor drawn is now replaced and a second one is drawn from the batch of 100. The probability of this capacitor being within the required tolerance values is also $\frac{73}{100}$

Thus, the probability of selecting a capacitor within the required tolerance values for both the first **and** the second draw is

$$\frac{73}{100} \times \frac{73}{100} = \frac{5329}{10\,000} \text{ or } \textbf{0.5329}$$

(b) The probability of obtaining a capacitor below the required tolerance values on the first draw is $\frac{17}{100}$. There are now only 99 capacitors left in the batch, since the first capacitor is not replaced. The probability of drawing a capacitor above the required tolerance values on the second draw is $\frac{10}{99}$, since there are $(100 - 73 - 17)$, i.e. ten capacitors above the required tolerance value. Thus, the probability of randomly selecting a capacitor below the required tolerance values and followed by randomly selecting a capacitor above the tolerance values is

$$\frac{17}{100} \times \frac{10}{99} = \frac{170}{9900} = \frac{17}{990} \text{ or } \textbf{0.0172}$$

Now try the following Practice Exercise

Practice Exercise 227   Probability (Answers on page 890)

1. In a batch of 45 lamps there are ten faulty lamps. If one lamp is drawn at random, find the probability of it being (a) faulty and (b) satisfactory.

2. A box of fuses are all of the same shape and size and comprises 23 2 A fuses, 47 5 A fuses and 69 13 A fuses. Determine the probability of selecting at random (a) a 2 A fuse, (b) a 5 A fuse and (c) a 13 A fuse.

3. (a) Find the probability of having a 2 upwards when throwing a fair six-sided dice. (b) Find the probability of having a 5 upwards when throwing a fair six-sided dice. (c) Determine the probability of having a 2 and then a 5 on two successive throws of a fair six-sided dice.

4. Determine the probability that the total score is 8 when two like dice are thrown.

5. The probability of event A happening is $\frac{3}{5}$ and the probability of event B happening is $\frac{2}{3}$. Calculate the probabilities of (a) both A and B happening, (b) only event A happening, i.e. event A happening and event B not happening, (c) only event B happening, and (d) either A, or B, or A and B happening.

6. When testing 1000 soldered joints, four failed during a vibration test and five failed due to having a high resistance. Determine the probability of a joint failing due to (a) vibration, (b) high resistance, (c) vibration or high resistance and (d) vibration and high resistance.

59.4 Further worked problems on probability

Problem 6. A batch of 40 components contains five which are defective. A component is drawn at random from the batch and tested and then a second component is drawn. Determine the probability that neither of the components is defective when drawn (a) with replacement, and (b) without replacement.

(a) With replacement

The probability that the component selected on the first draw is satisfactory is $\dfrac{35}{40}$, i.e. $\dfrac{7}{8}$. The component is now replaced and a second draw is made. The probability that this component is also satisfactory is $\dfrac{7}{8}$. Hence, the probability that both the first component drawn **and** the second component drawn are satisfactory is:

$$\frac{7}{8} \times \frac{7}{8} = \frac{49}{64} \text{ or } \mathbf{0.7656}$$

(b) Without replacement

The probability that the first component drawn is satisfactory is $\dfrac{7}{8}$. There are now only 34 satisfactory components left in the batch and the batch number is 39. Hence, the probability of drawing a satisfactory component on the second draw is $\dfrac{34}{39}$. Thus the probability that the first component drawn **and** the second component drawn are satisfactory, i.e. neither is defective, is:

$$\frac{7}{8} \times \frac{34}{39} = \frac{238}{312} \text{ or } \mathbf{0.7628}$$

Problem 7. A batch of 40 components contains five which are defective. If a component is drawn at random from the batch and tested and then a second component is drawn at random, calculate the probability of having one defective component, both with and without replacement.

The probability of having one defective component can be achieved in two ways. If p is the probability of drawing a defective component and q is the probability of drawing a satisfactory component, then the probability of having one defective component is given by drawing a satisfactory component and then a defective component **or** by drawing a defective component and then a satisfactory one, i.e. by $q \times p + p \times q$

With replacement:

$$p = \frac{5}{40} = \frac{1}{8}$$

$$\text{and} \qquad q = \frac{35}{40} = \frac{7}{8}$$

Hence, probability of having one defective component is:

$$\frac{1}{8} \times \frac{7}{8} + \frac{7}{8} \times \frac{1}{8}$$

i.e.

$$\frac{7}{64} + \frac{7}{64} = \frac{7}{32} \text{ or } \mathbf{0.2188}$$

Without replacement:

$p_1 = \dfrac{1}{8}$ and $q_1 = \dfrac{7}{8}$ on the first of the two draws. The batch number is now 39 for the second draw, thus,

$$p_2 = \frac{5}{39} \text{ and } q_2 = \frac{35}{39}$$

$$p_1 q_2 + q_1 p_2 = \frac{1}{8} \times \frac{35}{39} + \frac{7}{8} \times \frac{5}{39}$$

$$= \frac{35 + 35}{312}$$

$$= \frac{70}{312} \text{ or } \mathbf{0.2244}$$

Problem 8. A box contains 74 brass washers, 86 steel washers and 40 aluminium washers. Three washers are drawn at random from the box without replacement. Determine the probability that all three are steel washers.

Assume, for clarity of explanation, that a washer is drawn at random, then a second, then a third (although this assumption does not affect the results obtained). The total number of washers is $74 + 86 + 40$, i.e. 200. The probability of randomly selecting a steel washer on the first draw is $\dfrac{86}{200}$. There are now 85 steel washers in a batch of 199. The probability of randomly selecting a steel washer on the second draw is $\dfrac{85}{199}$. There are now 84 steel washers in a batch of 198. The probability of randomly selecting a steel washer on the third draw is $\dfrac{84}{198}$. Hence the probability of selecting a steel washer on the first draw **and** the second draw **and** the third draw is:

$$\frac{86}{200} \times \frac{85}{199} \times \frac{84}{198} = \frac{614\,040}{7\,880\,400} = \mathbf{0.0779}$$

Problem 9. For the box of washers given in Problem 8 above, determine the probability that there are no aluminium washers drawn, when three washers are drawn at random from the box without replacement.

The probability of not drawing an aluminium washer on the first draw is $1 - \left(\dfrac{40}{200}\right)$, i.e. $\dfrac{160}{200}$. There are now 199 washers in the batch of which 159 are not aluminium washers. Hence, the probability of not drawing an aluminium washer on the second draw is $\dfrac{159}{199}$. Similarly, the probability of not drawing an aluminium washer on the third draw is $\dfrac{158}{198}$. Hence the probability of not drawing an aluminium washer on the first **and** second **and** third draws is

$$\frac{160}{200} \times \frac{159}{199} \times \frac{158}{198} = \frac{4\,019\,520}{7\,880\,400} = \mathbf{0.5101}$$

Problem 10. For the box of washers in Problem 8 above, find the probability that there are two brass washers and either a steel or an aluminium washer when three are drawn at random, without replacement.

Two brass washers (A) and one steel washer (B) can be obtained in any of the following ways:

1st draw	2nd draw	3rd draw
A	A	B
A	B	A
B	A	A

Two brass washers and one aluminium washer (C) can also be obtained in any of the following ways:

1st draw	2nd draw	3rd draw
A	A	C
A	C	A
C	A	A

Thus there are six possible ways of achieving the combinations specified. If A represents a brass washer, B a steel washer and C an aluminium washer, then the combinations and their probabilities are as shown:

Draw			Probability
First	Second	Third	
A	A	B	$\dfrac{74}{200} \times \dfrac{73}{199} \times \dfrac{86}{198} = 0.0590$
A	B	A	$\dfrac{74}{200} \times \dfrac{86}{199} \times \dfrac{73}{198} = 0.0590$
B	A	A	$\dfrac{86}{200} \times \dfrac{74}{199} \times \dfrac{73}{198} = 0.0590$
A	A	C	$\dfrac{74}{200} \times \dfrac{73}{199} \times \dfrac{40}{198} = 0.0274$
A	C	A	$\dfrac{74}{200} \times \dfrac{40}{199} \times \dfrac{73}{198} = 0.0274$
C	A	A	$\dfrac{40}{200} \times \dfrac{74}{199} \times \dfrac{73}{198} = 0.0274$

The probability of having the first combination **or** the second, **or** the third, and so on, is given by the sum of the probabilities,

i.e. by $3 \times 0.0590 + 3 \times 0.0274$, that is, **0.2592**

Now try the following Practice Exercise

Practice Exercise 228   Probability (Answers on page 890)

1. The probability that component A will operate satisfactorily for five years is 0.8 and that B will operate satisfactorily over that same period of time is 0.75. Find the probabilities that in a five year period: (a) both components operate satisfactorily, (b) only component A will operate satisfactorily, and (c) only component B will operate satisfactorily.

2. In a particular street, 80% of the houses have telephones. If two houses selected at random are visited, calculate the probabilities that (a) they both have a telephone and (b) one has a telephone but the other does not have a telephone.

3. Veroboard pins are packed in packets of 20 by a machine. In a thousand packets, 40 have less than 20 pins. Find the probability that if two packets are chosen at random, one will contain less than 20 pins and the other will contain 20 pins or more.

4. A batch of 1 kW fire elements contains 16 which are within a power tolerance and four which are not. If three elements are selected at random from the batch, calculate the probabilities that (a) all three are within the power tolerance and (b) two are within but one is not within the power tolerance.

5. An amplifier is made up of three transistors, A, B and C. The probabilities of A, B or C being defective are $\dfrac{1}{20}, \dfrac{1}{25}$ and $\dfrac{1}{50}$, respectively. Calculate the percentage of amplifiers produced (a) which work satisfactorily and (b) which have just one defective transistor.

6. A box contains 14 40 W lamps, 28 60 W lamps and 58 25 W lamps, all the lamps being of the same shape and size. Three lamps are drawn at random from the box, first one, then a second, then a third. Determine the probabilities of: (a) getting one 25 W, one 40 W and one 60 W lamp, with replacement, (b) getting one 25 W, one 40 W and one 60 W lamp without replacement, and (c) getting either one 25 W and two 40 W or one 60 W and two 40 W lamps with replacement.

59.5   Permutations and combinations

Permutations

If n different objects are available, they can be arranged in different orders of selection. Each different ordered arrangement is called a **permutation**. For example, permutations of the three letters X, Y and Z taken together are:

$$XYZ, XZY, YXZ, YZX, ZXY \text{ and } ZYX$$

This can be expressed as $^3P_3 = 6$, the raised 3 denoting the number of items from which the arrangements are made, and the lowered 3 indicating the number of items used in each arrangement.

If we take the same three letters XYZ two at a time the permutations

$$XY, YX, XZ, ZX, YZ, ZY$$

can be found, and denoted by $^3P_2 = 6$
(Note that the order of the letters matter in permutations, i.e. YX is a different permutation from XY.) In general, $^nP_r = n(n-1)(n-2)\ldots(n-r+1)$ or

$$^nP_r = \frac{n!}{(n-r)!} \text{ as stated in Chapter 15}$$

For example, $^5P_4 = 5(4)(3)(2) = 120$ or
$$^5P_4 = \frac{5!}{(5-4)!} = \frac{5!}{1!} = (5)(4)(3)(2) = 120$$
Also, $^3P_3 = 6$ from above; using $^nP_r = \dfrac{n!}{(n-r)!}$ gives
$$^3P_3 = \frac{3!}{(3-3)!} = \frac{6}{0!}. \text{ Since this must equal 6, then } 0! = 1$$
(check this with your calculator).

Combinations

If selections of the three letters X, Y, Z are made without regard to the order of the letters in each group, i.e. XY is now the same as YX for example, then each group is called a **combination**. The number of possible combinations is denoted by ^nC_r, where n is the total number of items and r is the number in each selection.
In general,

$$^nC_r = \frac{n!}{r!(n-r)!}$$

For example,

$$^5C_4 = \frac{5!}{4!(5-4)!} = \frac{5!}{4!}$$
$$= \frac{5 \times 4 \times 3 \times 2 \times 1}{4 \times 3 \times 2 \times 1} = 5$$

Problem 11.   Calculate the number of permutations there are of: (a) five distinct objects taken two at a time, (b) four distinct objects taken two at a time.

(a) $\quad ^5P_2 = \dfrac{5!}{(5-2)!} = \dfrac{5!}{3!} = \dfrac{5 \times 4 \times 3 \times 2}{3 \times 2} = 20$

(b) $\quad ^4P_2 = \dfrac{4!}{(4-2)!} = \dfrac{4!}{2!} = 12$

Problem 12. Calculate the number of combinations there are of: (a) five distinct objects taken two at a time, (b) four distinct objects taken two at a time.

(a) $\quad ^5C_2 = \dfrac{5!}{2!\,(5-2)!} = \dfrac{5!}{2!\,3!}$

$\qquad = \dfrac{5 \times 4 \times 3 \times 2 \times 1}{(2 \times 1)(3 \times 2 \times 1)} = 10$

(b) $\quad ^4C_2 = \dfrac{4!}{2!\,(4-2)!} = \dfrac{4!}{2!\,2!} = 6$

Problem 13. A class has 24 students. Four can represent the class at an exam board. How many combinations are possible when choosing this group?

Number of combinations possible,

$$^nC_r = \frac{n!}{r!\,(n-r!)}$$

i.e. $\quad ^{24}C_4 = \dfrac{24!}{4!\,(24-4)!} = \dfrac{24!}{4!\,20!} = 10\,626$

Problem 14. In how many ways can a team of 11 be picked from 16 possible players?

Number of ways $= {}^nC_r = {}^{16}C_{11}$

$$= \frac{16!}{11!\,(16-11)!} = \frac{16!}{11!\,5!} = 4368$$

Now try the following Practice Exercise

Practice Exercise 229   Permutations and combinations (Answers on page 890)

1. Calculate the number of permutations there are of: (a) 15 distinct objects taken two at a time, (b) nine distinct objects taken four at a time.

2. Calculate the number of combinations there are of: (a) 12 distinct objects taken five at a time, (b) six distinct objects taken four at a time.

3. In how many ways can a team of six be picked from ten possible players?

4. 15 boxes can each hold one object. In how many ways can ten identical objects be placed in the boxes?

5. Six numbers between 1 and 49 are chosen when playing the National Lottery. Determine the probability of winning the top prize (i.e. six correct numbers!) if ten tickets were purchased and six different numbers were chosen on each ticket.

59.6   Bayes' theorem

Bayes' theorem is one of probability theory (originally stated by the Reverend Thomas Bayes), and may be seen as a way of understanding how the probability that a theory is true is affected by a new piece of evidence. The theorem has been used in a wide variety of contexts, ranging from marine biology to the development of 'Bayesian' spam blockers for email systems; in science, it has been used to try to clarify the relationship between theory and evidence. Insights in the philosophy of science involving confirmation, falsification and other topics can be made more precise, and sometimes extended or corrected, by using Bayes' theorem. Bayes' theorem may be stated mathematically as:

$P(A_1 | B)$

$$= \frac{P(B|A_1)\,P(A_1)}{P(B|A_1)\,P(A_1) + P(B|A_2)\,P(A_2) + \dots}$$

or $P(A_i | B)$

$$= \frac{P(B|A_i)\,P(A_i)}{\displaystyle\sum_{j=1}^{n} P(B|A_j)\,P(A_j)} \quad (i = 1, 2, \dots, n)$$

where $P(A|B)$ is the probability of A *given* B, i.e. *after* B is observed.

$P(A)$ and $P(B)$ are the probabilities of A and B without regard to each other, and $P(B|A)$ is the probability of observing event B given that A is true.

In the Bayes theorem formula, 'A' represents a theory or hypothesis that is to be tested, and 'B' represents a new piece of evidence that seems to confirm or disprove the theory. Bayes' theorem is demonstrated in the following worked problem.

Problem 15. An outdoor degree ceremony is taking place tomorrow, 5 July, in the hot climate of Dubai. In recent years it has rained only 2 days in the four-month period June to September. However, the weather forecaster has predicted rain for tomorrow. When it actually rains, the weatherman correctly forecasts rain 85% of the time. When it doesn't rain, he incorrectly forecasts rain 15% of the time. Determine the probability that it will rain tomorrow.

There are two possible mutually-exclusive events occurring here – it either rains or it does not rain. Also, a third event occurs when the weatherman predicts rain.

Let the notation for these events be:

Event A_1 It rains at the ceremony

Event A_2 It does not rain at the ceremony

Event B The weatherman predicts rain

The probability values are:

$$P(A_1) = \frac{2}{30+31+31+30} = \frac{1}{61}$$

(i.e. it rains 2 days in the months June to September

$$P(A_2) = \frac{120}{30+31+31+30} = \frac{60}{61}$$

(i.e. it does not rain for 120 of the 122 days

in the months June to September)

$$P(B|A_1) = 0.85$$

(i.e. when it rains, the weatherman predicts rain

85% of the time)

$$P(B|A_2) = 0.15$$

(i.e. when it does not rains, the weatherman

predicts rain 15% of the time)

Using Bayes' theorem to determine the probability that it will rain tomorrow, given the forecast of rain by the weatherman:

$$P(A_1|B) = \frac{P(B|A_1)P(A_1)}{P(B|A_1)P(A_1) + P(B|A_2)P(A_2)}$$

$$= \frac{(0.85)\left(\frac{1}{61}\right)}{0.85 \times \frac{1}{61} + 0.15 \times \frac{60}{61}} = \frac{0.0139344}{0.1614754}$$

$$= \mathbf{0.0863} \text{ or } \mathbf{8.63\%}$$

Even when the weatherman predicts rain, it rains only between 8% and 9% of the time. **Hence, there is a good chance it will not rain tomorrow in Dubai for the degree ceremony**.

Now try the following Practice Exercise

Practice Exercise 230   Bayes' theorem (Answers on page 890)

1. Machines A, B and C produce similar vehicle engine parts. Of the total output, machine A produces 35%, machine B 20% and machine C 45%. The proportions of the output from each machine that do not conform to the specification are 7% for A, 4% for B and 3% for C. Determine the proportion of those parts that do not conform to the specification that are produced by machine A.

2. A doctor is called to see a sick child. The doctor has prior information that 85% of sick children in that area have the flu, while the other 15% are sick with measles. For simplicity, assume that there are no other maladies in that area. A symptom of measles is a rash. The probability of children developing a rash and having measles is 94% and the probability of children with flu occasionally also developing a rash is 7%. Upon examining the child, the doctor finds a rash. Determine the probability that the child has measles.

3. In a study, oncologists were asked what the odds of breast cancer would be in a woman

Section K

who was initially thought to have a 1% risk of cancer but who ended up with a positive mammogram result (a mammogram accurately classifies about 80% of cancerous tumours and 90% of benign tumours). 95 out of a hundred oncologists estimated the probability of cancer to be about 75%. Use Bayes' theorem to determine the probability of cancer.

For fully worked solutions to each of the problems in Practice Exercises 227 to 230 in this chapter, go to the website:
www.routledge.com/cw/bird

Revision Test 17 Presentation of statistical data, mean, median, mode, standard deviation and probability

This Revision Test covers the material contained in Chapters 57 to 59. *The marks for each question are shown in brackets at the end of each question.*

1. A company produces five products in the following proportions:

   Product A 24   Product B 16   Product C 15
   Product D 11   Product E   6

   Present these data visually by drawing (a) a vertical bar chart, (b) a percentage component bar chart, (c) a pie diagram. (13)

2. The following lists the diameters of 40 components produced by a machine, each measured correct to the nearest hundredth of a centimetre:

1.39	1.36	1.38	1.31	1.33	1.40	1.28
1.40	1.24	1.28	1.42	1.34	1.43	1.35
1.36	1.36	1.35	1.45	1.29	1.39	1.38
1.38	1.35	1.42	1.30	1.26	1.37	1.33
1.37	1.34	1.34	1.32	1.33	1.30	1.38
1.41	1.35	1.38	1.27	1.37		

   (a) Using eight classes form a frequency distribution and a cumulative frequency distribution.
   (b) For the above data draw a histogram, a frequency polygon and an ogive. (21)

3. Determine for the ten measurements of lengths shown below:

   (a) the arithmetic mean, (b) the median, (c) the mode, and (d) the standard deviation.

   28 m, 20 m, 32 m, 44 m, 28 m, 30 m, 30 m, 26 m, 28 m and 34 m (10)

4. The heights of 100 people are measured correct to the nearest centimetre with the following results:

   150–157 cm   5   158–165 cm   18
   166–173 cm   42   174–181 cm   27
   182–189 cm   8

   Determine for the data (a) the mean height and (b) the standard deviation. (12)

5. Draw an ogive for the data of component measurements given below, and hence determine the median and the first and third quartile values for this distribution.

Class intervals (mm)	Frequency	Cumulative frequency
1.24–1.26	2	2
1.27–1.29	4	6
1.30–1.32	4	10
1.33–1.35	10	20
1.36–1.38	11	31
1.39–1.41	5	36
1.42–1.44	3	39
1.45–1.47	1	40

   (10)

6. Determine the probabilities of:

   (a) drawing a white ball from a bag containing 6 black and 14 white balls,

   (b) winning a prize in a raffle by buying six tickets when a total of 480 tickets are sold,

   (c) selecting at random a female from a group of 12 boys and 28 girls,

   (d) winning a prize in a raffle by buying eight tickets when there are five prizes and a total of 800 tickets are sold. (8)

7. The probabilities of an engine failing are given by: p_1, failure due to overheating; p_2, failure due to ignition problems; p_3, failure due to fuel blockage. When $p_1 = \frac{1}{8}$, $p_2 = \frac{1}{5}$ and $p_3 = \frac{2}{7}$, determine the probabilities of:

   (a) all three failures occurring,
   (b) the first and second but not the third failure occurring,
   (c) only the second failure occurring,
   (d) the first or the second failure occurring but not the third. (12)

8. In a box containing 120 similar transistors 70 are satisfactory, 37 give too high a gainx under normal operating conditions and the remainder give too low a gain.

   Calculate the probability that when drawing two transistors in turn, at random, **with replacement**, of having

   (a) two satisfactory,

   (b) none with low gain,

   (c) one with high gain and one satisfactory,

   (d) one with low gain and none satisfactory.

   Determine the probabilities in (a), (b) and (c) above if the transistors are drawn **without replacement**. (14)

For lecturers/instructors/teachers, fully worked solutions to each of the problems in Revision Test 17, together with a full marking scheme, are available at the website:
www.routledge.com/cw/bird

The binomial and Poisson distributions

At the end of this chapter, you should be able to:

- define the binomial distribution
- use the binomial distribution
- apply the binomial distribution to industrial inspection
- draw a histogram of probabilities
- define the Poisson distribution
- apply the Poisson distribution to practical situations

60.1 The binomial distribution

The binomial distribution deals with two numbers only, these being the probability that an event will happen, p, and the probability that an event will not happen, q. Thus, when a coin is tossed, if p is the probability of the coin landing with a head upwards, q is the probability of the coin landing with a tail upwards. $p + q$ must always be equal to unity. A binomial distribution can be used for finding, say, the probability of getting three heads in seven tosses of the coin, or in industry for determining defect rates as a result of sampling. One way of defining a binomial distribution is as follows:

'If p is the probability that an event will happen and q is the probability that the event will not happen, then the probabilities that the event will happen 0, 1, 2, 3, …, n times in n trials are given by the successive terms of the expansion of (q + p)^n, *taken from left to right.'*

The binomial expansion of $(q + p)^n$ is:

$$q^n + nq^{n-1}p + \frac{n(n-1)}{2!}q^{n-2}p^2$$
$$+ \frac{n(n-1)(n-2)}{3!}q^{n-3}p^3 + \cdots$$

from Chapter 7.
This concept of a binomial distribution is used in Problems 1 and 2.

Problem 1. Determine the probabilities of having (a) at least one girl and (b) at least one girl and one boy in a family of four children, assuming equal probability of male and female birth.

The probability of a girl being born, p, is 0.5 and the probability of a girl not being born (male birth), q, is also 0.5. The number in the family, n, is 4. From above, the probabilities of 0, 1, 2, 3, 4 girls in a family of four are given by the successive terms of the expansion of $(q + p)^4$ taken from left to right. From the binomial expansion:

$$(q + p)^4 = q^4 + 4q^3p + 6q^2p^2 + 4qp^3 + p^4$$

Hence the probability of no girls is q^4,

i.e. $\qquad\qquad\qquad 0.5^4 = 0.0625$

the probability of one girl is $4q^3p$,

i.e. $\qquad\qquad 4 \times 0.5^3 \times 0.5 = 0.2500$

the probability of two girls is $6q^2p^2$,

i.e. $\qquad\qquad 6 \times 0.5^2 \times 0.5^2 = 0.3750$

the probability of three girls is $4qp^3$,

i.e. $\qquad\qquad 4 \times 0.5 \times 0.5^3 = 0.2500$

the probability of four girls is p^4,

i.e. $\qquad\qquad\qquad 0.5^4 = 0.0625$
$$\overline{\qquad\qquad\qquad}$$
Total probability, $(q + p)^4 = 1.0000$
$$\overline{\qquad\qquad\qquad}$$

(a) The probability of having at least one girl is the sum of the probabilities of having 1, 2, 3 and 4 girls, i.e.

$$0.2500 + 0.3750 + 0.2500 + 0.0625 = \mathbf{0.9375}$$

(Alternatively, the probability of having at least one girl is: $1 - $ (the probability of having no girls), i.e. $1 - 0.0625$, giving **0.9375**, as obtained previously.)

(b) The probability of having at least one girl and one boy is given by the sum of the probabilities of having: one girl and three boys, two girls and two boys and three girls and two boys, i.e.

$$0.2500 + 0.3750 + 0.2500 = \mathbf{0.8750}$$

(Alternatively, this is also the probability of having $1 - $ (probability of having no girls + probability of having no boys), i.e.
$1 - 2 \times 0.0625 = \mathbf{0.8750}$, as obtained previously.)

Problem 2. A dice is rolled nine times. Find the probabilities of having a 4 upwards (a) three times and (b) fewer than four times.

Let p be the probability of having a 4 upwards. Then $p = 1/6$, since dice have six sides.
Let q be the probability of not having a 4 upwards. Then $q = 5/6$. The probabilities of having a 4 upwards 0, 1, 2, …, n times are given by the successive terms of the expansion of $(q + p)^n$, taken from left to right. From the binomial expansion:

$$(q + p)^9 = q^9 + 9q^8p + 36q^7p^2 + 84q^6p^3 + \cdots$$

The probability of having a 4 upwards zero times is

$$q^9 = (5/6)^9 = 0.1938$$

The probability of having a 4 upwards once is

$$9q^8p = 9(5/6)^8(1/6) = 0.3489$$

The probability of having a 4 upwards twice is

$$36q^7p^2 = 36(5/6)^7(1/6)^2 = 0.2791$$

The probability of having a 4 upwards three times is

$$84q^6p^3 = 84(5/6)^6(1/6)^3 = 0.1302$$

(a) The probability of having a 4 upwards three times is **0.1302**

(b) The probability of having a 4 upwards fewer than four times is the sum of the probabilities of having a 4 upwards zero, one, two, and three times, i.e.

$$0.1938 + 0.3489 + 0.2791 + 0.1302 = \mathbf{0.9520}$$

Industrial inspection

In industrial inspection, p is often taken as the probability that a component is defective and q is the probability that the component is satisfactory. In this case, a binomial distribution may be defined as:

'The probabilities that 0, 1, 2, 3,..., n components are defective in a sample of n components, drawn at random from a large batch of components, are given by the successive terms of the expansion of $(q+p)^n$, taken from left to right.'

This definition is used in Problems 3 and 4.

Problem 3. A machine is producing a large number of bolts automatically. In a box of these bolts, 95% are within the allowable tolerance values with respect to diameter, the remainder being outside of the diameter tolerance values. Seven bolts are drawn at random from the box. Determine the probabilities that (a) two and (b) more than two of the seven bolts are outside of the diameter tolerance values.

Let p be the probability that a bolt is outside of the allowable tolerance values, i.e. is defective, and let q be the probability that a bolt is within the tolerance values, i.e. is satisfactory. Then $p = 5\%$, i.e. 0.05 per unit and $q = 95\%$, i.e. 0.95 per unit. The sample number is 7. The probabilities of drawing $0, 1, 2, \ldots, n$ defective bolts are given by the successive terms of the expansion of $(q + p)^n$, taken from left to right. In this problem

$$(q + p)^n = (0.95 + 0.05)^7$$

$$= 0.95^7 + 7 \times 0.95^6 \times 0.05$$

$$+ 21 \times 0.95^5 \times 0.05^2 + \cdots$$

Thus the probability of no defective bolts is

$$0.95^7 = 0.6983$$

The probability of one defective bolt is

$$7 \times 0.95^6 \times 0.05 = 0.2573$$

The probability of two defective bolts is

$$21 \times 0.95^5 \times 0.05^2 = 0.0406, \text{ and so on.}$$

(a) The probability that two bolts are outside of the diameter tolerance values is **0.0406**.

(b) To determine the probability that more than two bolts are defective, the sum of the probabilities of three bolts, four bolts, five bolts, six bolts and seven bolts being defective can be determined. An easier way to find this sum is to find $1 - ($sum of zero bolts, one bolt and two bolts being defective), since the sum of all the terms is unity. Thus, the probability of there being more than two bolts outside of the tolerance values is:

$$1 - (0.6983 + 0.2573 + 0.0406), \text{ i.e. } \mathbf{0.0038}$$

Problem 4. A package contains 50 similar components and inspection shows that four have been damaged during transit. If six components are drawn at random from the contents of the package determine the probabilities that in this sample (a) one and (b) fewer than three are damaged.

The probability of a component being damaged, p, is 4 in 50, i.e. 0.08 per unit. Thus, the probability of a component not being damaged, q, is $1 - 0.08$, i.e. 0.92. The probability of there being $0, 1, 2, \ldots, 6$ damaged components is given by the successive terms of $(q + p)^6$, taken from left to right.

$$(q + p)^6 = q^6 + 6q^5 p + 15q^4 p^2 + 20q^3 p^3 + \cdots$$

(a) The probability of one damaged component is

$$6q^5 p = 6 \times 0.92^5 \times 0.08 = \mathbf{0.3164}$$

(b) The probability of fewer than three damaged components is given by the sum of the probabilities of zero, one and two damaged components.

$$q^6 + 6q^5 p + 15q^4 p^2$$

$$= 0.92^6 + 6 \times 0.92^5 \times 0.08$$

$$+ 15 \times 0.92^4 \times 0.08^2$$

$$= 0.6064 + 0.3164 + 0.0688 = \mathbf{0.9916}$$

Histogram of probabilities

The terms of a binomial distribution may be represented pictorially by drawing a histogram, as shown in Problem 5.

Problem 5. The probability of a student successfully completing a course of study in three years is 0.45. Draw a histogram showing the probabilities of $0, 1, 2, \ldots, 10$ students successfully completing the course in three years.

Let p be the probability of a student successfully completing a course of study in three years and q be the probability of not doing so. Then $p = 0.45$ and $q = 0.55$. The number of students, n, is 10

The probabilities of $0, 1, 2, \ldots, 10$ students successfully completing the course are given by the successive terms of the expansion of $(q + p)^{10}$, taken from left to right.

$$(q + p)^{10} = q^{10} + 10q^9 p + 45q^8 p^2 + 120q^7 p^3$$
$$+ 210q^6 p^4 + 252q^5 p^5 + 210q^4 p^6$$
$$+ 120q^3 p^7 + 45q^2 p^8 + 10qp^9 + p^{10}$$

Substituting $q = 0.55$ and $p = 0.45$ in this expansion gives the values of the successive terms as: 0.0025, 0.0207, 0.0763, 0.1665, 0.2384, 0.2340, 0.1596, 0.0746, 0.0229, 0.0042 and 0.0003. The histogram depicting these probabilities is shown in Fig. 60.1.

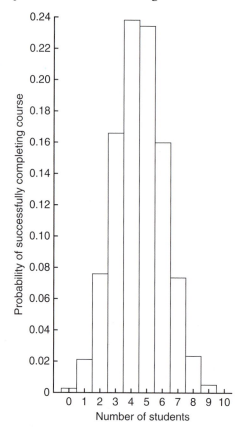

Figure 60.1

Now try the following Practice Exercise

Practice Exercise 231   The binomial distribution (Answers on page 890)

1. Concrete blocks are tested and it is found that, on average, 7% fail to meet the required specification. For a batch of nine blocks, determine the probabilities that (a) three blocks and (b) fewer than four blocks will fail to meet the specification.

2. If the failure rate of the blocks in Problem 1 rises to 15%, find the probabilities that (a) no blocks and (b) more than two blocks will fail to meet the specification in a batch of nine blocks.

3. The average number of employees absent from a firm each day is 4%. An office within the firm has seven employees. Determine the probabilities that (a) no employee and (b) three employees will be absent on a particular day.

4. A manufacturer estimates that 3% of his output of a small item is defective. Find the probabilities that in a sample of ten items (a) fewer than two and (b) more than two items will be defective.

5. Five coins are tossed simultaneously. Determine the probabilities of having 0, 1, 2, 3, 4 and 5 heads upwards, and draw a histogram depicting the results.

6. If the probability of rain falling during a particular period is 2/5, find the probabilities of having 0, 1, 2, 3, 4, 5, 6 and 7 wet days in a week. Show these results on a histogram.

7. An automatic machine produces, on average, 10% of its components outside of the tolerance required. In a sample of ten components from this machine, determine the probability of having three components outside of the tolerance required by assuming a binomial distribution.

60.2   The Poisson distribution

When the number of trials, n, in a binomial distribution becomes large (usually taken as larger than ten), the calculations associated with determining the values of the terms becomes laborious. If n is large and p is small, and the product np is less than 5, a very good approximation to a binomial distribution is given by the corresponding **Poisson*** **distribution**, in which calculations are usually simpler.

The Poisson approximation to a binomial distribution may be defined as follows:

'The probabilities that an event will happen 0, 1, 2, 3, ... , n times in n trials are given by the successive terms of the expression

$$e^{-\lambda}\left(1+\lambda+\frac{\lambda^2}{2!}+\frac{\lambda^3}{3!}+\cdots\right)$$

taken from left to right.'

The symbol λ is the expectation of an event happening and is equal to np.

Problem 6. If 3% of the gearwheels produced by a company are defective, determine the

---

probabilities that in a sample of 80 gearwheels (a) two and (b) more than two will be defective.

The sample number, n, is large, the probability of a defective gearwheel, p, is small and the product np is 80×0.03, i.e. 2.4, which is less than 5

Hence a Poisson approximation to a binomial distribution may be used. The expectation of a defective gearwheel, $\lambda = np = 2.4$

The probabilities of $0, 1, 2, \ldots$ defective gearwheels are given by the successive terms of the expression

$$e^{-\lambda}\left(1+\lambda+\frac{\lambda^2}{2!}+\frac{\lambda^3}{3!}+\cdots\right)$$

taken from left to right, i.e. by

$$e^{-\lambda}, \lambda e^{-\lambda}, \frac{\lambda^2 e^{-\lambda}}{2!}, \ldots$$

Thus probability of no defective gearwheels is

$$e^{-\lambda} = e^{-2.4} = 0.0907$$

probability of one defective gearwheel is

$$\lambda e^{-\lambda} = 2.4 e^{-2.4} = 0.2177$$

probability of two defective gearwheels is

$$\frac{\lambda^2 e^{-\lambda}}{2!} = \frac{2.4^2 e^{-2.4}}{2 \times 1} = 0.2613$$

(a) The probability of having two defective gearwheels is **0.2613**

(b) The probability of having more than two defective gearwheels is $1 -$ (the sum of the probabilities of having zero, one, and two defective gearwheels), i.e.

$$1 - (0.0907 + 0.2177 + 0.2613),$$

that is, **0.4303**

The principal use of a Poisson distribution is to determine the theoretical probabilities when p, the probability of an event happening, is known, but q, the probability of the event not happening is unknown. For example, the average number of goals scored per match by a football team can be calculated, but it is not possible to quantify the number of goals which were not scored. In this type of problem, a Poisson distribution may be defined as follows:

* **Who was Poisson**? **Siméon Denis Poisson** (21 June 1781–25 April 1840), was a French mathematician, geometer, and physicist. His work on the theory of electricity and magnetism virtually created a new branch of mathematical physics, and his study of celestial mechanics discussed the stability of the planetary orbits. To find out more go to **www.routledge.com/cw/bird**

'The probabilities of an event occurring 0, 1, 2, 3, ... times are given by the successive terms of the expression

$$e^{-\lambda}\left(1 + \lambda + \frac{\lambda^2}{2!} + \frac{\lambda^3}{3!} + \cdots\right),$$

taken from left to right.'

The symbol λ is the value of the average occurrence of the event.

Problem 7. A production department has 35 similar milling machines. The number of breakdowns on each machine averages 0.06 per week. Determine the probabilities of having (a) one, and (b) fewer than three machines breaking down in any week.

Since the average occurrence of a breakdown is known but the number of times when a machine did not break down is unknown, a Poisson distribution must be used. The expectation of a breakdown for 35 machines is 35×0.06, i.e. 2.1 breakdowns per week. The probabilities of a breakdown occurring 0, 1, 2, ... times are given by the successive terms of the expression

$$e^{-\lambda}\left(1 + \lambda + \frac{\lambda^2}{2!} + \frac{\lambda^3}{3!} + \cdots\right)$$

taken from left to right.

Hence probability of no breakdowns

$$e^{-\lambda} = e^{-2.1} = 0.1225$$

probability of one breakdown is

$$\lambda e^{-\lambda} = 2.1 e^{-2.1} = 0.2572$$

probability of two breakdowns is

$$\frac{\lambda^2 e^{-\lambda}}{2!} = \frac{2.1^2 e^{-2.1}}{2 \times 1} = 0.2700$$

(a) The probability of one breakdown per week is **0.2572**

(b) The probability of fewer than three breakdowns per week is the sum of the probabilities of zero, one, and two breakdowns per week,

   i.e. $0.1225 + 0.2572 + 0.2700$, i.e. **0.6497**

Histogram of probabilities

The terms of a Poisson distribution may be represented pictorially by drawing a histogram, as shown in Problem 8.

Problem 8. The probability of a person having an accident in a certain period of time is 0.0003. For a population of 7500 people, draw a histogram showing the probabilities of 0, 1, 2, 3, 4, 5 and 6 people having an accident in this period.

The probabilities of 0, 1, 2, ... people having an accident are given by the terms of expression

$$e^{-\lambda}\left(1 + \lambda + \frac{\lambda^2}{2!} + \frac{\lambda^3}{3!} + \cdots\right)$$

taken from left to right.
The average occurrence of the event, λ, is 7500×0.0003, i.e. 2.25

The probability of no people having an accident is

$$e^{-\lambda} = e^{-2.25} = 0.1054$$

The probability of one person having an accident is

$$\lambda e^{-\lambda} = 2.25 e^{-2.25} = 0.2371$$

The probability of two people having an accident is

$$\frac{\lambda^2 e^{-\lambda}}{2!} = \frac{2.25^2 e^{-2.25}}{2!} = 0.2668$$

and so on, giving probabilities of 0.2001, 0.1126, 0.0506 and 0.0190 for 3, 4, 5 and 6 respectively having an accident. The histogram for these probabilities is shown in Fig. 60.2.

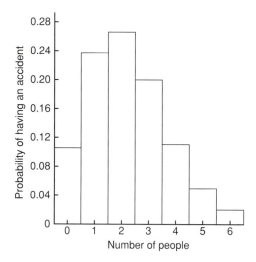

Figure 60.2

Now try the following Practice Exercise

Practice Exercise 232   The Poisson distribution (Answers on page 890)

1. In problem 7 of Exercise 231, page 654, determine the probability of having three components outside of the required tolerance using the Poisson distribution.

2. The probability that an employee will go to hospital in a certain period of time is 0.0015. Use a Poisson distribution to determine the probability of more than two employees going to hospital during this period of time if there are 2000 employees on the payroll.

3. When packaging a product, a manufacturer finds that one packet in 20 is underweight. Determine the probabilities that in a box of 72 packets (a) two and (b) fewer than four will be underweight.

4. A manufacturer estimates that 0.25% of his output of a component is defective. The components are marketed in packets of 200. Determine the probability of a packet containing fewer than three defective components.

5. The demand for a particular tool from a store is, on average, five times a day and the demand follows a Poisson distribution. How many of these tools should be kept in the stores so that the probability of there being one available when required is greater than 10%?

6. Failure of a group of particular machine tools follows a Poisson distribution with a mean value of 0.7. Determine the probabilities of 0, 1, 2, 3, 4 and 5 failures in a week and present these results on a histogram.

For fully worked solutions to each of the problems in Practice Exercises 231 and 232 in this chapter, go to the website:
www.routledge.com/cw/bird

Chapter 61

The normal distribution

Why it is important to understand: The normal distribution

A normal distribution is a very important statistical data distribution pattern occurring in many natural phenomena, such as height, blood pressure, lengths of objects produced by machines, marks in a test, errors in measurements, and so on. In general, when data is gathered, we expect to see a particular pattern to the data, called a *normal distribution*. This is a distribution where the data is evenly distributed around the mean in a very regular way, which when plotted as a histogram will result in a *bell curve*. The normal distribution is the most important of all probability distributions; it is applied directly to many practical problems in every engineering discipline. There are two principal applications of the normal distribution to engineering and reliability. One application deals with the analysis of items which exhibit failure to wear, such as mechanical devices – frequently the wear-out failure distribution is sufficiently close to normal that the use of this distribution for predicting or assessing reliability is valid. Another application is in the analysis of manufactured items and their ability to meet specifications. No two parts made to the same specification are exactly alike; the variability of parts leads to a variability in systems composed of those parts. The design must take this variability into account, otherwise the system may not meet the specification requirement due to the combined effect of part variability.

At the end of this chapter, you should be able to:

- recognise a normal curve
- use the normal distribution in calculations
- test for a normal distribution using probability paper

61.1 Introduction to the normal distribution

When data is obtained, it can frequently be considered to be a sample (i.e. a few members) drawn at random from a large population (i.e. a set having many members). If the sample number is large, it is theoretically possible to choose class intervals which are very small, but which still have a number of members falling within each class. A frequency polygon of this data then has a large number of small line segments and approximates to a continuous curve. Such a curve is called a **frequency or a distribution curve**.

An extremely important symmetrical distribution curve is called the **normal curve** and is as shown in Fig. 61.1. This curve can be described by a mathematical equation and is the basis of much of the work done in more advanced statistics. Many natural occurrences such as the heights or weights of a group of people, the sizes of components produced by a particular machine and the life length of certain components approximate to a normal distribution.

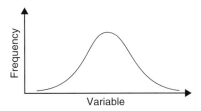

Figure 61.1

Normal distribution curves can differ from one another in the following four ways:

(a) by having different mean values

(b) by having different values of standard deviations

(c) the variables having different values and different units and

(d) by having different areas between the curve and the horizontal axis.

A normal distribution curve is **standardised** as follows:

(a) The mean value of the unstandardised curve is made the origin, thus making the mean value, $\bar{x}$, zero.

(b) The horizontal axis is scaled in standard deviations. This is done by letting $z = \dfrac{x - \bar{x}}{\sigma}$, where z is called the **normal standard variate**, x is the value of the variable, $\bar{x}$ is the mean value of the distribution and σ is the standard deviation of the distribution.

(c) The area between the normal curve and the horizontal axis is made equal to unity.

When a normal distribution curve has been standardised, the normal curve is called a **standardised normal curve** or a **normal probability curve**, and any normally distributed data may be represented by the **same** normal probability curve.

The area under part of a normal probability curve is directly proportional to probability and the value of the shaded area shown in Fig. 61.2 can be determined by evaluating:

$$\int \frac{1}{\sqrt{(2\pi)}} e^{\left(\frac{z^2}{2}\right)} dz, \quad \text{where } z = \frac{x - \bar{x}}{\sigma}$$

To save repeatedly determining the values of this function, tables of partial areas under the standardised normal curve are available in many mathematical formulae books, and such a table is shown in Table 61.1, on page 661.

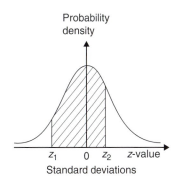

Figure 61.2

Problem 1. The mean height of 500 people is 170 cm and the standard deviation is 9 cm. Assuming the heights are normally distributed, determine the number of people likely to have heights between 150 cm and 195 cm.

The mean value, $\bar{x}$, is 170 cm and corresponds to a normal standard variate value, z, of zero on the standardised normal curve. A height of 150 cm has a z-value given by $z = \dfrac{x - \bar{x}}{\sigma}$ standard deviations, i.e. $\dfrac{150 - 170}{9}$ or -2.22 standard deviations. Using a table of partial areas beneath the standardised normal curve (see Table 61.1), a z-value of -2.22 corresponds to an area of 0.4868 between the mean value and the ordinate $z = -2.22$. The negative z-value shows that it lies to the left of the $z = 0$ ordinate.

This area is shown shaded in Fig. 61.3(a). Similarly, 195 cm has a z-value of $\dfrac{195 - 170}{9}$ that is 2.78 standard deviations. From Table 61.1, this value of z corresponds to an area of 0.4973, the positive value of z showing that it lies to the right of the $z = 0$ ordinate. This area is shown shaded in Fig. 61.3(b). The total area shaded in Figs. 61.3(a) and (b) is shown in Fig. 61.3(c) and is $0.4868 + 0.4973$, i.e. 0.9841 of the total area beneath the curve.

However, the area is directly proportional to probability. Thus, the probability that a person will have a height of between 150 and 195 cm is 0.9841. For a group of 500 people, 500×0.9841, i.e. **492 people are likely to have heights in this range**. The value of 500×0.9841 is 492.05, but since answers based on a normal probability distribution can only be approximate, results are usually given correct to the nearest whole number.

Problem 2. For the group of people given in Problem 1, find the number of people likely to have heights of less than 165 cm.

Figure 61.3

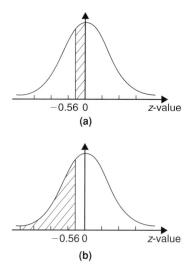

Figure 61.4

A height of 165 cm corresponds to $\dfrac{165-170}{9}$ i.e. -0.56 standard deviations.

The area between $z=0$ and $z=-0.56$ (from Table 61.1) is 0.2123, shown shaded in Fig. 61.4(a). The total area under the standardised normal curve is unity and since the curve is symmetrical, it follows that the total area to the left of the $z=0$ ordinate is 0.5000. Thus the area to the left of the $z=-0.56$ ordinate ('left' means 'less than', 'right' means 'more than') is $0.5000-0.2123$, i.e. 0.2877 of the total area, which is shown shaded in Fig 61.4(b). The area is directly proportional to probability and since the total area beneath the standardised normal curve is unity, the probability of a person's height being less than 165 cm is 0.2877. For a group of 500 people, 500×0.2877, i.e. **144 people are likely to have heights of less than 165 cm**.

> **Problem 3.** For the group of people given in Problem 1 find how many people are likely to have heights of more than 194 cm.

194 cm corresponds to a z-value of $\dfrac{194-170}{9}$ that is 2.67 standard deviations. From Table 61.1, the area

between $z=0$, $z=2.67$ and the standardised normal curve is 0.4962, shown shaded in Fig. 61.5(a). Since the standardised normal curve is symmetrical, the total area to the right of the $z=0$ ordinate is 0.5000, hence the shaded area shown in Fig. 61.5(b) is $0.5000-0.4962$, i.e. 0.0038. This area represents the probability of a person having a height of more than 194 cm, and for 500 people, the number of people likely to have a height of more than 194 cm is 0.0038×500, i.e. **two people**.

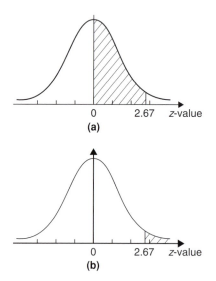

Figure 61.5

> **Problem 4.** A batch of 1500 lemonade bottles have an average contents of 753 ml and the standard deviation of the contents is 1.8 ml. If the volumes of the contents are normally distributed, find

Table 61.1 Partial areas under the standardised normal curve

$z = \dfrac{x - \bar{x}}{\sigma}$	0	1	2	3	4	5	6	7	8	9
0.0	0.0000	0.0040	0.0080	0.0120	0.0159	0.0199	0.0239	0.0279	0.0319	0.0359
0.1	0.0398	0.0438	0.0478	0.0517	0.0557	0.0596	0.0636	0.0678	0.0714	0.0753
0.2	0.0793	0.0832	0.0871	0.0910	0.0948	0.0987	0.1026	0.1064	0.1103	0.1141
0.3	0.1179	0.1217	0.1255	0.1293	0.1331	0.1388	0.1406	0.1443	0.1480	0.1517
0.4	0.1554	0.1591	0.1628	0.1664	0.1700	0.1736	0.1772	0.1808	0.1844	0.1879
0.5	0.1915	0.1950	0.1985	0.2019	0.2054	0.2086	0.2123	0.2157	0.2190	0.2224
0.6	0.2257	0.2291	0.2324	0.2357	0.2389	0.2422	0.2454	0.2486	0.2517	0.2549
0.7	0.2580	0.2611	0.2642	0.2673	0.2704	0.2734	0.2760	0.2794	0.2823	0.2852
0.8	0.2881	0.2910	0.2939	0.2967	0.2995	0.3023	0.3051	0.3078	0.3106	0.3133
0.9	0.3159	0.3186	0.3212	0.3238	0.3264	0.3289	0.3315	0.3340	0.3365	0.3389
1.0	0.3413	0.3438	0.3451	0.3485	0.3508	0.3531	0.3554	0.3577	0.3599	0.3621
1.1	0.3643	0.3665	0.3686	0.3708	0.3729	0.3749	0.3770	0.3790	0.3810	0.3830
1.2	0.3849	0.3869	0.3888	0.3907	0.3925	0.3944	0.3962	0.3980	0.3997	0.4015
1.3	0.4032	0.4049	0.4066	0.4082	0.4099	0.4115	0.4131	0.4147	0.4162	0.4177
1.4	0.4192	0.4207	0.4222	0.4236	0.4251	0.4265	0.4279	0.4292	0.4306	0.4319
1.5	0.4332	0.4345	0.4357	0.4370	0.4382	0.4394	0.4406	0.4418	0.4430	0.4441
1.6	0.4452	0.4463	0.4474	0.4484	0.4495	0.4505	0.4515	0.4525	0.4535	0.4545
1.7	0.4554	0.4564	0.4573	0.4582	0.4591	0.4599	0.4608	0.4616	0.4625	0.4633
1.8	0.4641	0.4649	0.4656	0.4664	0.4671	0.4678	0.4686	0.4693	0.4699	0.4706
1.9	0.4713	0.4719	0.4726	0.4732	0.4738	0.4744	0.4750	0.4756	0.4762	0.4767
2.0	0.4772	0.4778	0.4783	0.4785	0.4793	0.4798	0.4803	0.4808	0.4812	0.4817
2.1	0.4821	0.4826	0.4830	0.4834	0.4838	0.4842	0.4846	0.4850	0.4854	0.4857
2.2	0.4861	0.4864	0.4868	0.4871	0.4875	0.4878	0.4881	0.4884	0.4887	0.4890
2.3	0.4893	0.4896	0.4898	0.4901	0.4904	0.4906	0.4909	0.4911	0.4913	0.4916
2.4	0.4918	0.4920	0.4922	0.4925	0.4927	0.4929	0.4931	0.4932	0.4934	0.4936
2.5	0.4938	0.4940	0.4941	0.4943	0.4945	0.4946	0.4948	0.4949	0.4951	0.4952
2.6	0.4953	0.4955	0.4956	0.4957	0.4959	0.4960	0.4961	0.4962	0.4963	0.4964
2.7	0.4965	0.4966	0.4967	0.4968	0.4969	0.4970	0.4971	0.4972	0.4973	0.4974
2.8	0.4974	0.4975	0.4976	0.4977	0.4977	0.4978	0.4979	0.4980	0.4980	0.4981
2.9	0.4981	0.4982	0.4982	0.4983	0.4984	0.4984	0.4985	0.4985	0.4986	0.4986
3.0	0.4987	0.4987	0.4987	0.4988	0.4988	0.4989	0.4989	0.4989	0.4990	0.4990
3.1	0.4990	0.4991	0.4991	0.4991	0.4992	0.4992	0.4992	0.4992	0.4993	0.4993
3.2	0.4993	0.4993	0.4994	0.4994	0.4994	0.4994	0.4994	0.4995	0.4995	0.4995
3.3	0.4995	0.4995	0.4995	0.4996	0.4996	0.4996	0.4996	0.4996	0.4996	0.4997
3.4	0.4997	0.4997	0.4997	0.4997	0.4997	0.4997	0.4997	0.4997	0.4997	0.4998
3.5	0.4998	0.4998	0.4998	0.4998	0.4998	0.4998	0.4998	0.4998	0.4998	0.4998
3.6	0.4998	0.4998	0.4999	0.4999	0.4999	0.4999	0.4999	0.4999	0.4999	0.4999
3.7	0.4999	0.4999	0.4999	0.4999	0.4999	0.4999	0.4999	0.4999	0.4999	0.4999
3.8	0.4999	0.4999	0.4999	0.4999	0.4999	0.4999	0.4999	0.4999	0.4999	0.4999
3.9	0.5000	0.5000	0.5000	0.5000	0.5000	0.5000	0.5000	0.5000	0.5000	0.5000

(a) the number of bottles likely to contain less than 750 ml,

(b) the number of bottles likely to contain between 751 and 754 ml,

(c) the number of bottles likely to contain more than 757 ml, and

(d) the number of bottles likely to contain between 750 and 751 ml.

(a) The z-value corresponding to 750 ml is given by $\dfrac{x - \bar{x}}{\sigma}$ i.e. $\dfrac{750 - 753}{1.8} = -1.67$ standard deviations. From Table 61.1, the area between $z = 0$ and $z = -1.67$ is 0.4525. Thus the area to the left of the $z = -1.67$ ordinate is $0.5000 - 0.4525$ (see Problem 2), i.e. 0.0475. This is the probability of a bottle containing less than 750 ml. Thus, for a batch of 1500 bottles, it is likely that 1500×0.0475, i.e. **71 bottles will contain less than 750 ml**.

(b) The z-value corresponding to 751 and 754 ml are $\dfrac{751 - 753}{1.8}$ and $\dfrac{754 - 753}{1.8}$ i.e. -1.11 and 0.56 respectively. From Table 61.1, the areas corresponding to these values are 0.3665 and 0.2123 respectively. Thus the probability of a bottle containing between 751 and 754 ml is $0.3665 + 0.2123$ (see Problem 1), i.e. 0.5788. For 1500 bottles, it is likely that 1500×0.5788, i.e. **868 bottles will contain between 751 and 754 ml**.

(c) The z-value corresponding to 757 ml is $\dfrac{757 - 753}{1.8}$, i.e. 2.22 standard deviations. From Table 61.1, the area corresponding to a z-value of 2.22 is 0.4868. The area to the right of the $z = 2.22$ ordinate is $0.5000 - 0.4868$ (see Problem 3), i.e. 0.0132. Thus, for 1500 bottles, it is likely that 1500×0.0132, i.e. **20 bottles will have contents of more than 757 ml**.

(d) The z-value corresponding to 750 ml is -1.67 (see part (a)), and the z-value corresponding to 751 ml is -1.11 (see part (b)). The areas corresponding to these z-values are 0.4525 and 0.3665 respectively, and both these areas lie on the left of the $z = 0$ ordinate. The area between $z = -1.67$

and $z = -1.11$ is $0.4525 - 0.3665$, i.e. 0.0860 and this is the probability of a bottle having contents between 750 and 751 ml. For 1500 bottles, it is likely that 1500×0.0860, i.e. **129 bottles will be in this range**.

Now try the following Practice Exercise

Practice Exercise 233   Introduction to the normal distribution (Answers on page 890)

1. A component is classed as defective if it has a diameter of less than 69 mm. In a batch of 350 components, the mean diameter is 75 mm and the standard deviation is 2.8 mm. Assuming the diameters are normally distributed, determine how many are likely to be classed as defective.

2. The masses of 800 people are normally distributed, having a mean value of 64.7 kg and a standard deviation of 5.4 kg. Find how many people are likely to have masses of less than 54.4 kg.

3. 500 tins of paint have a mean content of 1010 ml and the standard deviation of the contents is 8.7 ml. Assuming the volumes of the contents are normally distributed, calculate the number of tins likely to have contents whose volumes are less than (a) 1025 ml (b) 1000 ml and (c) 995 ml.

4. For the 350 components in Problem 1, if those having a diameter of more than 81.5 mm are rejected, find, correct to the nearest component, the number likely to be rejected due to being oversized.

5. For the 800 people in Problem 2, determine how many are likely to have masses of more than (a) 70 kg and (b) 62 kg.

6. The mean diameter of holes produced by a drilling machine bit is 4.05 mm and the standard deviation of the diameters is 0.0028 mm. For 20 holes drilled using this machine, determine, correct to the nearest whole number, how many are likely to have diameters of between (a) 4.048 and 4.0553 mm and (b) 4.052 and 4.056 mm, assuming the diameters are normally distributed.

7. The intelligence quotients of 400 children have a mean value of 100 and a standard deviation of 14. Assuming that I.Q.s are normally distributed, determine the number of children likely to have I.Q.s of between (a) 80 and 90, (b) 90 and 110 and (c) 110 and 130

8. The mean mass of active material in tablets produced by a manufacturer is 5.00 g and the standard deviation of the masses is 0.036 g. In a bottle containing 100 tablets, find how many tablets are likely to have masses of (a) between 4.88 and 4.92 g, (b) between 4.92 and 5.04 g and (c) more than 5.04 g.

61.2   Testing for a normal distribution

It should never be assumed that because data is continuous it automatically follows that it is normally distributed. One way of checking that data is normally distributed is by using **normal probability paper**, often just called **probability paper**. This is special graph paper which has linear markings on one axis and percentage probability values from 0.01 to 99.99 on the other axis (see Figs. 61.6 and 61.7). The divisions on the probability axis are such that a straight line graph results for normally distributed data when percentage cumulative frequency values are plotted against upper class boundary values. If the points do not lie in a reasonably straight line, then the data is not normally distributed. The method used to test the normality of a distribution is shown in Problems 5 and 6. The mean value and standard deviation of normally distributed data may be determined using normal probability paper. For normally distributed data, the area beneath the standardised normal curve and a z-value of unity (i.e. one standard deviation) may be obtained from Table 61.1. For one standard deviation, this area is 0.3413, i.e. 34.13%. An area of ± 1 standard deviation is symmetrically placed on either side of the $z=0$ value, i.e. is symmetrically placed on either side of the 50% cumulative frequency value. Thus an area corresponding to ± 1 standard deviation extends from percentage cumulative frequency values of $(50+34.13)\%$ to $(50-34.13)\%$, i.e. from 84.13% to 15.87%. For most purposes, these values are taken as 84% and 16%. Thus, when using normal probability paper, the standard deviation of the distribution is given by:

Figure 61.6

$$\frac{\left(\begin{array}{l}\text{variable value for 84\% cumulative frequency} -\\ \text{variable value for 16\% cumulative frequency}\end{array}\right)}{2}$$

Problem 5.   Use normal probability paper to determine whether the data given below, which refers to the masses of 50 copper ingots, is approximately normally distributed. If the data is normally distributed, determine the mean and standard deviation of the data from the graph drawn.

Class mid-point value (kg)	Frequency
29.5	2
30.5	4
31.5	6
32.5	8
33.5	9
34.5	8

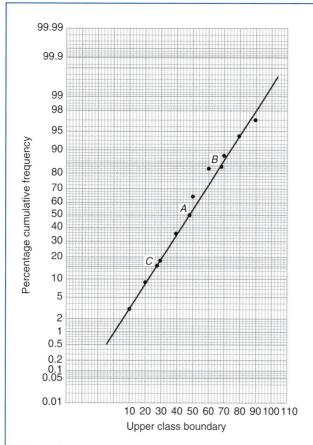

Figure 61.7

Class mid-point value (kg)	Frequency
35.5	6
36.5	4
37.5	2
38.5	1

To test the normality of a distribution, the upper class boundary/percentage cumulative frequency values are plotted on normal probability paper. The upper class boundary values are: 30, 31, 32, ..., 38, 39. The corresponding cumulative frequency values (for 'less than' the upper class boundary values) are: 2, $(4+2)=6$, $(6+4+2)=12$, 20, 29, 37, 43, 47, 49 and 50. The corresponding percentage cumulative frequency values are $\frac{2}{50} \times 100 = 4$, $\frac{6}{50} \times 100 = 12$, 24, 40, 58, 74, 86, 94, 98 and 100%

The co-ordinates of upper class boundary/percentage cumulative frequency values are plotted as shown

in Fig. 61.6. When plotting these values, it will always be found that the co-ordinate for the 100% cumulative frequency value cannot be plotted, since the maximum value on the probability scale is 99.99. **Since the points plotted in Fig. 61.6 lie very nearly in a straight line, the data is approximately normally distributed.**

The mean value and standard deviation can be determined from Fig. 61.6. Since a normal curve is symmetrical, the mean value is the value of the variable corresponding to a 50% cumulative frequency value, shown as point P on the graph. This shows that **the mean value is 33.6 kg**. The standard deviation is determined using the 84% and 16% cumulative frequency values, shown as Q and R in Fig. 61.6. The variable values for Q and R are 35.7 and 31.4 respectively; thus two standard deviations correspond to $35.7 - 31.4$, i.e. 4.3, showing that the standard deviation of the distribution is approximately $\frac{4.3}{2}$ i.e. **2.15 standard deviations**.

The mean value and standard deviation of the distribution can be calculated using

$$\text{mean}, \bar{x} = \frac{(\sum fx)}{(\sum f)}$$

and standard deviation,

$$\sigma = \sqrt{\left\{ \frac{\left(\sum [f(x - \bar{x})^2] \right)}{(\sum f)} \right\}}$$

where f is the frequency of a class and x is the class mid-point value. Using these formulae gives a mean value of the distribution of 33.6 (as obtained graphically) and a standard deviation of 2.12, showing that the graphical method of determining the mean and standard deviation give quite realistic results.

Problem 6. Use normal probability paper to determine whether the data given below is normally distributed. Use the graph and assume a normal distribution whether this is so or not, to find approximate values of the mean and standard deviation of the distribution.

Class mid-point values	Frequency
5	1
15	2
25	3
35	6

Class mid-point values	Frequency
45	9
55	6
65	2
75	2
85	1
95	1

To test the normality of a distribution, the upper class boundary/percentage cumulative frequency values are plotted on normal probability paper. The upper class boundary values are: 10, 20, 30, ..., 90 and 100. The corresponding cumulative frequency values are 1, 1+2=3, 1+2+3=6, 12, 21, 27, 29, 31, 32 and 33. The percentage cumulative frequency values are $\frac{1}{33} \times 100 = 3$, $\frac{3}{33} \times 100 = 9$, 18, 36, 64, 82, 88, 94, 97 and 100

The co-ordinates of upper class boundary values/percentage cumulative frequency values are plotted as shown in Fig. 61.7. Although six of the points lie approximately in a straight line, three points corresponding to upper class boundary values of 50, 60 and 70 are not close to the line and indicate that **the distribution is not normally** distributed. However, if a normal distribution is assumed, the mean value corresponds to the variable value at a cumulative frequency of 50% and, from Fig. 61.7, point A is **48**. The value of the standard deviation of the distribution can be obtained from the variable values corresponding to the 84% and 16% cumulative frequency values, shown as B and C in Fig. 61.7 and give: $2\sigma = 69 - 28$, i.e. the standard deviation $\sigma = $ **20.5**. The calculated values of the mean and standard deviation of the distribution are 45.9 and 19.4 respectively, showing that errors are introduced if the graphical method of determining these values is used for data which is not normally distributed.

Now try the following Practice Exercise

Practice Exercise 234   Testing for a normal distribution (Answers on page 891)

1. A frequency distribution of 150 measurements is as shown:

Class mid-point value	Frequency
26.4	5
26.6	12
26.8	24
27.0	36
27.2	36
27.4	25
27.6	12

Use normal probability paper to show that this data approximates to a normal distribution and hence determine the approximate values of the mean and standard deviation of the distribution. Use the formula for mean and standard deviation to verify the results obtained.

2. A frequency distribution of the class mid-point values of the breaking loads for 275 similar fibres is as shown below:

Load (kN)	17	19	21	23	25	27	29	31
Frequency	9	23	55	78	64	28	14	4

Use normal probability paper to show that this distribution is approximately normally distributed and determine the mean and standard deviation of the distribution (a) from the graph and (b) by calculation.

For fully worked solutions to each of the problems in Practice Exercises 233 and 234 in this chapter, go to the website:
www.routledge.com/cw/bird

Section K

Chapter 62

Linear correlation

Why it is important to understand: Linear correlation

Correlation coefficients measure the strength of association between two variables. The most common correlation coefficient, called the product-moment correlation coefficient, measures the strength of the linear association between variables. A positive value indicates a positive correlation and the higher the value, the stronger the correlation. Similarly, a negative value indicates a negative correlation and the lower the value the stronger the correlation. This chapter explores linear correlation and the meaning of values obtained calculating the coefficient of correlation.

At the end of this chapter, you should be able to:

- recognise linear correlation
- state the product-moment formula
- appreciate the significance of a coefficient of correlation
- determine the linear coefficient of correlation between two given variables

62.1 Introduction to linear correlation

Correlation is a measure of the amount of association existing between two variables. For linear correlation, if points are plotted on a graph and all the points lie on a straight line, then **perfect linear correlation** is said to exist. When a straight line having a positive gradient can reasonably be drawn through points on a graph **positive or direct linear correlation** exists, as shown in Fig. 62.1(a). Similarly, when a straight line having a negative gradient can reasonably be drawn through points on a graph, **negative or inverse linear correlation** exists, as shown in Fig. 62.1(b). When there is no apparent relationship between co-ordinate values plotted on a graph then **no correlation** exists between the points, as shown in Fig. 62.1(c). In statistics, when two variables are being investigated, the location of the co-ordinates on a rectangular co-ordinate system is called a **scatter diagram** – as shown in Fig. 62.1.

62.2 The Pearson product-moment formula for determining the linear correlation coefficient

The Pearson product-moment correlation coefficient (or **Pearson correlation coefficient**, for short) is a measure of the strength of a linear association between two variables and is denoted by the symbol r. A Pearson product-moment correlation attempts to draw a line of best fit through the data of two variables, and the Pearson correlation coefficient, r, indicates how far away all these data points are to this line of best fit (how well the data points fit this new model/line of best fit). The **product-moment formula** states:

coefficient of correlation,

$$r = \frac{\sum xy}{\sqrt{\{(\sum x^2)(\sum y^2)\}}} \tag{1}$$

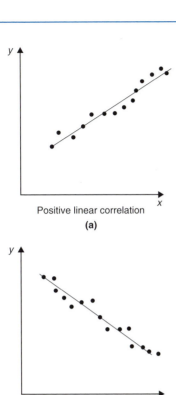

Positive linear correlation
(a)

Negative linear correlation
(b)

No correlation
(c)

Figure 62.1

where the x-values are the values of the deviations of co-ordinates X from $\overline{X}$, their mean value and the y-values are the values of the deviations of co-ordinates Y from $\overline{Y}$, their mean value. That is, $x = (X - \overline{X})$ and $y = (Y - \overline{Y})$. The results of this determination give values of r lying between $+1$ and -1, where $+1$ indicates perfect direct correlation, -1 indicates perfect inverse correlation and 0 indicates that no correlation exists. Between these values, the smaller the value of r, the less is the amount of correlation which exists. Generally, values of r in the ranges 0.7 to 1 and -0.7 to -1 show that a fair amount of correlation exists.

62.3 The significance of a coefficient of correlation

When the value of the coefficient of correlation has been obtained from the product-moment formula, some care is needed before coming to conclusions based on this result. Checks should be made to ascertain the following two points:

(a) that a 'cause and effect' relationship exists between the variables; it is relatively easy, mathematically, to show that some correlation exists between, say, the number of ice creams sold in a given period of time and the number of chimneys swept in the same period of time, although there is no relationship between these variables;

(b) that a linear relationship exists between the variables; the product-moment formula given in Section 62.2 is based on linear correlation. Perfect non-linear correlation may exist (for example, the co-ordinates exactly following the curve $y = x^3$), but this gives a low value of coefficient of correlation since the value of r is determined using the product-moment formula, based on a linear relationship.

62.4 Worked problems on linear correlation

Problem 1.   In an experiment to determine the relationship between force on a wire and the resulting extension, the following data is obtained:

Force (N)	10	20	30	40	50	60	70
Extension (mm)	0.22	0.40	0.61	0.85	1.20	1.45	1.70

Determine the linear coefficient of correlation for this data.

Let X be the variable force values and Y be the dependent variable extension values. The coefficient of correlation is given by:

$$r = \frac{\sum xy}{\sqrt{\{(\sum x^2)(\sum y^2)\}}}$$

where $x = (X - \overline{X})$ and $y = (Y - \overline{Y})$, $\overline{X}$ and $\overline{Y}$ being the mean values of the X and Y values respectively.

Section K

Using a tabular method to determine the quantities of this formula gives:

X	Y	$x=(X-\overline{X})$	$y=(Y-\overline{Y})$
10	0.22	−30	−0.699
20	0.40	−20	−0.519
30	0.61	−10	−0.309
40	0.85	0	−0.069
50	1.20	10	0.281
60	1.45	20	0.531
70	1.70	30	0.781

$$\sum X=280, \quad \overline{X}=\frac{280}{7}=40$$

$$\sum Y=6.43, \quad \overline{Y}=\frac{6.43}{7}=0.919$$

xy	x^2	y^2
20.97	900	0.489
10.38	400	0.269
3.09	100	0.095
0	0	0.005
2.81	100	0.079
10.62	400	0.282
23.43	900	0.610

$$\sum xy=71.30 \quad \sum x^2=2800 \quad \sum y^2=1.829$$

Thus $\quad r=\dfrac{71.3}{\sqrt{[2800\times1.829]}}=\mathbf{0.996}$

This shows that a **very good direct correlation exists** between the values of force and extension.

Problem 2. The relationship between expenditure on welfare services and absenteeism for similar periods of time is shown below for a small company.

Expenditure (£'000)	3.5	5.0	7.0	10	12	15	18
Days lost	241	318	174	110	147	122	86

Determine the coefficient of linear correlation for this data.

Let X be the expenditure in thousands of pounds and Y be the days lost.

The coefficient of correlation,

$$r=\frac{\sum xy}{\sqrt{\{(\sum x^2)(\sum y^2)\}}}$$

where $x=(X-\overline{X})$ and $y=(Y-\overline{Y})$, $\overline{X}$ and $\overline{Y}$ being the mean values of X and Y respectively. Using a tabular approach:

X	Y	$x=(X-\overline{X})$	$y=(Y-\overline{Y})$
3.5	241	−6.57	69.9
5.0	318	−5.07	146.9
7.0	174	−3.07	2.9
10	110	−0.07	−61.1
12	147	1.93	−24.1
15	122	4.93	−49.1
18	86	7.93	−85.1

$$\sum X=70.5, \quad \overline{X}=\frac{70.5}{7}=10.07$$

$$\sum Y=1198, \quad \overline{Y}=\frac{1198}{7}=171.1$$

xy	x^2	y^2
−459.2	43.2	4886
−744.8	25.7	21 580
−8.9	9.4	8
4.3	0	3733
−46.5	3.7	581
−242.1	24.3	2411
−674.8	62.9	7242

$$\sum xy=-2172 \quad \sum x^2=169.2 \quad \sum y^2=40441$$

Thus

$$r=\frac{-2172}{\sqrt{[169.2\times40441]}}=\mathbf{-0.830}$$

This shows that there is **fairly good inverse correlation** between the expenditure on welfare and days lost due to absenteeism.

Problem 3. The relationship between monthly car sales and income from the sale of petrol for a garage is as shown:

Cars sold	2 5 3 12 14 7 3 28 14 7 3 13
Income from petrol sales (£′000)	12 9 13 21 17 22 31 47 17 10 9 11

Determine the linear coefficient of correlation between these quantities.

Let X represent the number of cars sold and Y the income, in thousands of pounds, from petrol sales. Using the tabular approach:

X	Y	$x=(X-\overline{X})$	$y=(Y-\overline{Y})$
2	12	−7.25	−6.25
5	9	−4.25	−9.25
3	13	−6.25	−5.25
12	21	2.75	2.75
14	17	4.75	−1.25
7	22	−2.25	3.75
3	31	−6.25	12.75
28	47	18.75	28.75
14	17	4.75	−1.25
7	10	−2.25	−8.25
3	9	−6.25	−9.25
13	11	3.75	−7.25

$$\sum X=111, \quad \overline{X}=\frac{111}{12}=9.25$$

$$\sum Y=219, \quad \overline{Y}=\frac{219}{12}=18.25$$

xy	x^2	y^2
45.3	52.6	39.1
39.3	18.1	85.6
32.8	39.1	27.6
7.6	7.6	7.6
−5.9	22.6	1.6
−8.4	5.1	14.1
−79.7	39.1	162.6
539.1	351.6	826.6
−5.9	22.6	1.6
18.6	5.1	68.1
57.8	39.1	85.6
−27.2	14.1	52.6
$\sum xy=613.4$	$\sum x^2=616.7$	$\sum y^2=1372.7$

The coefficient of correlation,

$$r = \frac{\sum xy}{\sqrt{\{(\sum x^2)(\sum y^2)\}}}$$

$$= \frac{613.4}{\sqrt{\{(616.7)(1372.7)\}}} = \mathbf{0.667}$$

Thus, there is **no appreciable correlation** between petrol and car sales.

Now try the following Practice Exercise

Practice Exercise 235 Linear correlation (Answers on page 891)

In Problems 1 to 3, determine the coefficient of correlation for the data given, correct to 3 decimal places.

1. | X | 14 | 18 | 23 | 30 | 50 |
   |---|---|---|---|---|---|
   | Y | 900 | 1200 | 1600 | 2100 | 3800 |

2. | X | 2.7 | 4.3 | 1.2 | 1.4 | 4.9 |
   |---|---|---|---|---|---|
   | Y | 11.9 | 7.10 | 33.8 | 25.0 | 7.50 |

3. | X | 24 | 41 | 9 | 18 | 73 |
   |---|---|---|---|---|---|
   | Y | 39 | 46 | 90 | 30 | 98 |

4. In an experiment to determine the relationship between the current flowing in an electrical circuit and the applied voltage, the results obtained are:

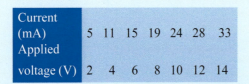

Current (mA)	5	11	15	19	24	28	33
Applied voltage (V)	2	4	6	8	10	12	14

Determine, using the product-moment formula, the coefficient of correlation for these results.

5. A gas is being compressed in a closed cylinder and the values of pressures and corresponding volumes at constant temperature are as shown:

Pressure (kPa)	Volume (m^3)
160	0.034
180	0.036
200	0.030
220	0.027
240	0.024
260	0.025
280	0.020
300	0.019

Find the coefficient of correlation for these values.

6. The relationship between the number of miles travelled by a group of engineering salesmen in ten equal time periods and the corresponding value of orders taken is given below. Calculate the coefficient of correlation using the product-moment formula for these values.

Miles travelled	Orders taken (£'000)
1370	23
1050	17
980	19
1770	22
1340	27
1560	23
2110	30
1540	23
1480	25
1670	19

7. The data shown below refers to the number of times machine tools had to be taken out of service, in equal time periods, due to faults occurring and the number of hours worked by maintenance teams. Calculate the coefficient of correlation for this data.

Machines out of service	4	13	2	9	16	8	7
Maintenance hours	400	515	360	440	570	380	415

For fully worked solutions to each of the problems in Practice Exercise 235 in this chapter,
go to the website:
www.routledge.com/cw/bird

Chapter 63

Linear regression

Why it is important to understand: Linear regression

The general process of fitting data to a linear combination of basic functions is termed linear regression. Linear least squares regression is by far the most widely used modelling method; it is what most people mean when they say they have used 'regression', 'linear regression' or 'least squares' to fit a model to their data. Not only is linear least squares regression the most widely used modelling method, but it has been adapted to a broad range of situations that are outside its direct scope. It plays a strong underlying role in many other modelling methods. This chapter explains how regression lines are determined.

At the end of this chapter, you should be able to:

- explain linear regression
- understand least-squares regression lines
- determine, for two variables X and Y, the equations of the regression lines of X on Y and Y on X

63.1 Introduction to linear regression

Regression analysis, usually termed **regression**, is used to draw the line of 'best fit' through co-ordinates on a graph. The techniques used enable a mathematical equation of the straight line form $y = mx + c$ to be deduced for a given set of co-ordinate values, the line being such that the sum of the deviations of the co-ordinate values from the line is a minimum, i.e. it is the line of 'best fit'. When a regression analysis is made, it is possible to obtain two lines of best fit, depending on which variable is selected as the dependent variable and which variable is the independent variable. For example, in a resistive electrical circuit, the current flowing is directly proportional to the voltage applied to the circuit. There are two ways of obtaining experimental values relating the current and voltage. Either, certain voltages are applied to the circuit and the current values are measured, in which case the voltage is the independent variable and the current is the dependent variable; or, the voltage can be adjusted until a desired value of current is flowing and the value of voltage is measured, in which case the current is the independent value and the voltage is the dependent value.

63.2 The least-squares regression lines

For a given set of co-ordinate values, (X_1, Y_1), $(X_2, Y_2), \ldots, (X_n, Y_n)$ let the X values be the independent variables and the Y-values be the dependent values. Also let $D_1, \ldots, D_n$ be the vertical distances between the line shown as PQ in Fig. 63.1 and the points representing the co-ordinate values. The least-squares regression line, i.e. the line of best fit, is the line which makes the value of $D_1^2 + D_2^2 + \cdots + D_n^2$ a minimum value. The equation of the least-squares regression line is usually written as $Y = a_0 + a_1 X$, where a_0 is the y-axis intercept value and a_1 is the gradient of the line (analogous to c and m in the equation $y = mx + c$). The

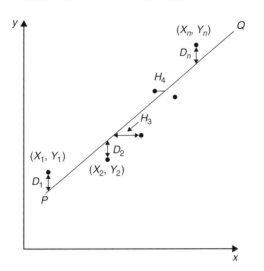

Figure 63.1

values of a_0 and a_1 to make the sum of the 'deviations squared' a minimum can be obtained from the two equations:

$$\sum Y = a_0 N + a_1 \sum X \qquad (1)$$

$$\sum (XY) = a_0 \sum X + a_1 \sum X^2 \qquad (2)$$

where X and Y are the co-ordinate values, N is the number of co-ordinates and a_0 and a_1 are called the **regression coefficients** of Y on X. Equations (1) and (2) are called the **normal equations** of the regression lines of Y on X. The regression line of Y on X is used to estimate values of Y for given values of X. If the Y-values (vertical axis) are selected as the independent variables, the horizontal distances between the line shown as PQ in Fig. 63.1 and the co-ordinate values (H_3, H_4, etc.) are taken as the deviations. The equation of the regression line is of the form: $X = b_0 + b_1 Y$ and the normal equations become:

$$\sum X = b_0 N + b_1 \sum Y \qquad (3)$$

$$\sum (XY) = b_0 \sum Y + b_1 \sum Y^2 \qquad (4)$$

where X and Y are the co-ordinate values, b_0 and b_1 are the regression coefficients of X on Y and N is the number of co-ordinates. These normal equations are of the regression line of X on Y, which is slightly different to the regression line of Y on X. The regression line of X on Y is used to estimated values of X for given values of Y. The regression line of Y on X is used to determine any value of Y corresponding to a given value of X. If the value of Y lies within the range of Y-values of the extreme co-ordinates, the process of finding the corresponding value of X is called **linear interpolation**. If it lies outside of the range of Y-values of the extreme co-ordinates then the process is called **linear extrapolation** and the assumption must be made that the line of best fit extends outside of the range of the co-ordinate values given.

By using the regression line of X on Y, values of X corresponding to given values of Y may be found by either interpolation or extrapolation.

63.3 Worked problems on linear regression

Problem 1. In an experiment to determine the relationship between frequency and the inductive reactance of an electrical circuit, the following results were obtained:

Frequency (Hz)	Inductive reactance (ohms)
50	30
100	65
150	90
200	130
250	150
300	190
350	200

Determine the equation of the regression line of inductive reactance on frequency, assuming a linear relationship.

Since the regression line of inductive reactance on frequency is required, the frequency is the independent variable, X, and the inductive reactance is the dependent variable, Y. The equation of the regression line of Y on X is:

$$Y = a_0 + a_1 X$$

and the regression coefficients a_0 and a_1 are obtained by using the normal equations

$$\sum Y = a_0 N + a_1 \sum X$$

and $\quad \sum XY = a_0 \sum X + a_1 \sum X^2$

(from equations (1) and (2))

A tabular approach is used to determine the summed quantities.

Frequency, X	Inductive reactance, Y	X^2
50	30	2500
100	65	10 000
150	90	22 500
200	130	40 000
250	150	62 500
300	190	90 000
350	200	122 500

$$\sum X = 1400 \quad \sum Y = 855 \quad \sum X^2 = 350\,000$$

XY	Y^2
1500	900
6500	4225
13 500	8100
26 000	16 900
37 500	22 500
57 000	36 100
70 000	40 000

$$\sum XY = 212\,000 \quad \sum Y^2 = 128\,725$$

The number of co-ordinate values given, N is 7. Substituting in the normal equations gives:

$$855 = 7a_0 + 1400a_1 \tag{1}$$

$$212\,000 = 1400a_0 + 350\,000a_1 \tag{2}$$

$1400 \times (1)$ gives:

$$1\,197\,000 = 9800a_0 + 1\,960\,000a_1 \tag{3}$$

$7 \times (2)$ gives:

$$1\,484\,000 = 9800a_0 + 2\,450\,000a_1 \tag{4}$$

$(4) - (3)$ gives:

$$287\,000 = 0 + 490\,000a_1$$

from which, $a_1 = \dfrac{287\,000}{490\,000} = 0.586$

Substituting $a_1 = 0.586$ in equation (1) gives:

$$855 = 7a_0 + 1400(0.586)$$

i.e. $a_0 = \dfrac{855 - 820.4}{7} = 4.94$

Thus the equation of the regression line of inductive reactance on frequency is:

$$Y = 4.94 + 0.586X$$

Problem 2.   For the data given in Problem 1, determine the equation of the regression line of frequency on inductive reactance, assuming a linear relationship.

In this case, the inductive reactance is the independent variable X and the frequency is the dependent variable Y. From equations (3) and (4), the equation of the regression line of X on Y is:

$$X = b_0 + b_1 Y$$

and the normal equations are

$$\sum X = b_0 N + b_1 \sum Y$$

and $$\sum XY = b_0 \sum Y + b_1 \sum Y^2$$

From the table shown in Problem 1, the simultaneous equations are:

$$1400 = 7b_0 + 855b_1$$

$$212\,000 = 855b_0 + 128\,725b_1$$

Solving these equations in a similar way to that in Problem 1 gives:

$$b_0 = -6.15$$

and $b_1 = 1.69$,   correct to 3 significant figures.

Thus the equation of the regression line of frequency on inductive reactance is:

$$X = -6.15 + 1.69\,Y$$

Problem 3.   Use the regression equations calculated in Problems 1 and 2 to find (a) the value of inductive reactance when the frequency is 175 Hz and (b) the value of frequency when the inductive reactance is 250 ohms, assuming the line of best fit extends outside of the given co-ordinate values. Draw a graph showing the two regression lines.

(a) From Problem 1, the regression equation of inductive reactance on frequency is
$Y = 4.94 + 0.586X$. When the frequency, X, is 175 Hz, $Y = 4.94 + 0.586(175) = 107.5$, correct to 4 significant figures, i.e. the inductive reactance is **107.5 ohms** when the frequency is 175 Hz.

(b) From Problem 2, the regression equation of frequency on inductive reactance is
$X = -6.15 + 1.69Y$. When the inductive reactance, Y, is 250 ohms,
$X = -6.15 + 1.69(250) = 416.4$ Hz, correct to 4 significant figures, i.e. the frequency is **416.4 Hz** when the inductive reactance is 250 ohms.

The graph depicting the two regression lines is shown in Fig. 63.2. To obtain the regression line of inductive reactance on frequency the regression line equation $Y = 4.94 + 0.586X$ is used, and X (frequency) values of 100 and 300 have been selected in order to find the corresponding Y values. These values gave the co-ordinates as (100, 63.5) and (300, 180.7), shown as points A and B in Fig. 63.2. Two co-ordinates for the regression line of frequency on inductive reactance are calculated using the equation $X = -6.15 + 1.69Y$, the values of inductive reactance of 50 and 150 being used to obtain the co-ordinate values. These values gave co-ordinates (78.4, 50) and (247.4, 150), shown as points C and D in Fig. 63.2.

It can be seen from Fig. 63.2 that to the scale drawn, the two regression lines coincide. Although it is not necessary to do so, the co-ordinate values are also shown

to indicate that the regression lines do appear to be the lines of best fit. A graph showing co-ordinate values is called a **scatter diagram** in statistics.

Problem 4. The experimental values relating centripetal force and radius for a mass travelling at constant velocity in a circle are as shown:

Force (N)	5	10	15	20	25	30	35	40
Radius (cm)	55	30	16	12	11	9	7	5

Determine the equations of (a) the regression line of force on radius and (b) the regression line of radius on force. Hence, calculate the force at a radius of 40 cm and the radius corresponding to a force of 32 newtons.

Let the radius be the independent variable X, and the force be the dependent variable Y. (This decision is usually based on a 'cause' corresponding to X and an 'effect' corresponding to Y.)

(a) The equation of the regression line of force on radius is of the form $Y = a_0 + a_1 X$ and the constants a_0 and a_1 are determined from the normal equations:

$$\sum Y = a_0 N + a_1 \sum X$$
and $$\sum XY = a_0 \sum X + a_1 \sum X^2$$
(from equations (1) and (2))

Using a tabular approach to determine the values of the summations gives:

Radius, X	Force, Y	X^2
55	5	3025
30	10	900
16	15	256
12	20	144
11	25	121
9	30	81
7	35	49
5	40	25
$\sum X = 145$	$\sum Y = 180$	$\sum X^2 = 4601$

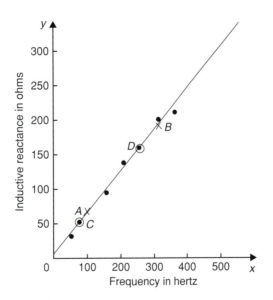

Figure 63.2

XY	Y^2
275	25
300	100
240	225
240	400
275	625
270	900
245	1225
200	1600
$\sum XY = 2045$	$\sum Y^2 = 5100$

Thus $180 = 8a_0 + 145a_1$

and $2045 = 145a_0 + 4601a_1$

Solving these simultaneous equations gives $a_0 = 33.7$ and $a_1 = -0.617$, correct to 3 significant figures. Thus the equation of the regression line of force on radius is:

$$Y = 33.7 - 0.617X$$

(b) The equation of the regression line of radius on force is of the form $X = b_0 + b_1 Y$ and the constants b_0 and b_1 are determined from the normal equations:

$$\sum X \ = b_0 N + b_1 \sum Y$$

and $\sum XY = b_0 \sum Y + b_1 \sum Y^2$

(from equations (3) and (4))

The values of the summations have been obtained in part (a) giving:

$$145 = 8b_0 + 180b_1$$

and $2045 = 180b_0 + 5100b_1$

Solving these simultaneous equations gives $b_0 = 44.2$ and $b_1 = -1.16$, correct to 3 significant figures. Thus the equation of the regression line of radius on force is:

$$X = 44.2 - 1.16Y$$

The force, Y, at a radius of 40 cm, is obtained from the regression line of force on radius, i.e. $y = 33.7 - 0.617(40) = 9.02$,

i.e. **the force at a radius of 40 cm is 9.02 N**

The radius, X, when the force is 32 newtons is obtained from the regression line of radius on force, i.e. $X = 44.2 - 1.16(32) = 7.08$,

i.e. **the radius when the force is 32 N is 7.08 cm**.

Now try the following Practice Exercise

Practice Exercise 236   Linear regression (Answers on page 891)

In Problems 1 and 2, determine the equation of the regression line of Y on X, correct to 3 significant figures.

1.

X	14	18	23	30	50
Y	900	1200	1600	2100	3800

2.

X	6	3	9	15	2	14	21	13
Y	1.3	0.7	2.0	3.7	0.5	2.9	4.5	2.7

In Problems 3 and 4, determine the equations of the regression lines of X on Y for the data stated, correct to 3 significant figures.

3. The data given in Problem 1

4. The data given in Problem 2

5. The relationship between the voltage applied to an electrical circuit and the current flowing is as shown:

Current (mA)	Applied voltage (V)
2	5
4	11
6	15
8	19
10	24
12	28
14	33

Assuming a linear relationship, determine the equation of the regression line of applied voltage, Y, on current, X, correct to 4 significant figures.

6. For the data given in Problem 5, determine the equation of the regression line of current on applied voltage, correct to 3 significant figures.

7. Draw the scatter diagram for the data given in Problem 5 and show the regression lines of applied voltage on current and current on applied voltage. Hence determine the values of (a) the applied voltage needed to give a current of 3 mA and (b) the current flowing when the applied voltage is 40 volts, assuming the regression lines are still true outside of the range of values given.

8. In an experiment to determine the relationship between force and momentum, a force X, is applied to a mass, by placing the mass on an inclined plane, and the time, Y, for the velocity to change from u m/s to v m/s is measured. The results obtained are as follows:

Force (N)	Time (s)
11.4	0.56
18.7	0.35
11.7	0.55
12.3	0.52
14.7	0.43
18.8	0.34
19.6	0.31

Determine the equation of the regression line of time on force, assuming a linear relationship between the quantities, correct to 3 significant figures.

9. Find the equation for the regression line of force on time for the data given in Problem 8, correct to 3 decimal places.

10. Draw a scatter diagram for the data given in Problem 8 and show the regression lines of time on force and force on time. Hence find (a) the time corresponding to a force of 16 N, and (b) the force at a time of 0.25 s, assuming the relationship is linear outside of the range of values given.

For fully worked solutions to each of the problems in Practice Exercise 236 in this chapter, go to the website:
www.routledge.com/cw/bird

Revision Test 18 Binomial, Poisson and normal distributions, correlation and regression

This Revision Test covers the material contained in chapters 60 to 63. *The marks for each question are shown in brackets at the end of each question.*

1. A machine produces 15% defective components. In a sample of five, drawn at random, calculate, using the binomial distribution, the probability that:

   (a) there will be four defective items,

   (b) there will be not more than three defective items,

   (c) all the items will be non-defective.

   Draw a histogram showing the probabilities of 0, 1, 2,..., 5 defective items. (20)

2. 2% of the light bulbs produced by a company are defective. Determine, using the Poisson distribution, the probability that in a sample of 80 bulbs: (a) three bulbs will be defective, (b) not more than three bulbs will be defective, (c) at least two bulbs will be defective. (13)

3. Some engineering components have a mean length of 20 mm and a standard deviation of 0.25 mm. Assume that the data on the lengths of the components is normally distributed.

   In a batch of 500 components, determine the number of components likely to:

   (a) have a length of less than 19.95 mm,

   (b) be between 19.95 mm and 20.15 mm,

   (c) be longer than 20.54 mm. (15)

4. In a factory, cans are packed with an average of 1.0 kg of a compound and the masses are normally distributed about the average value. The standard deviation of a sample of the contents of the cans is 12 g. Determine the percentage of cans containing (a) less than 985 g, (b) more than 1030 g, (c) between 985 g and 1030 g. (10)

5. The data given below gives the experimental values obtained for the torque output, X, from an electric motor and the current, Y, taken from the supply.

Torque X	Current Y
0	3
1	5
2	6
3	6
4	9
5	11
6	12
7	12
8	14
9	13

Determine the linear coefficient of correlation for this data. (18)

6. Some results obtained from a tensile test on a steel specimen are shown below:

Tensile force (kN)	Extension (mm)
4.8	3.5
9.3	8.2
12.8	10.1
17.7	15.6
21.6	18.4
26.0	20.8

Assuming a linear relationship:

   (a) determine the equation of the regression line of extension on force,

   (b) determine the equation of the regression line of force on extension,

   (c) estimate (i) the value of extension when the force is 16 kN, and (ii) the value of force when the extension is 17 mm. (24)

For lecturers/instructors/teachers, fully worked solutions to each of the problems in Revision Test 18, together with a full marking scheme, are available at the website:

www.routledge.com/cw/bird

Chapter 64

Sampling and estimation theories

Why it is important to understand: Sampling and estimation theories

Estimation theory is a branch of statistics and signal processing that deals with estimating the values of parameters based on measured/empirical data that has a random component. Estimation theory can be found at the heart of many electronic signal processing systems designed to extract information; these systems include radar, sonar, speech, image, communications, control and seismology. This chapter introduces some of the principles involved with sampling and estimation theories.

At the end of this chapter, you should be able to:

- understand sampling distributions
- determine the standard error of the means
- understand point and interval estimates and confidence intervals
- calculate confidence limits
- estimate the mean and standard deviation of a population from sample data
- estimate the mean of a population based on a small sample data using a Student's t distribution

---

64.1 Introduction

The concepts of elementary sampling theory and estimation theories introduced in this chapter will provide the basis for a more detailed study of inspection, control and quality control techniques used in industry. Such theories can be quite complicated; in this chapter a full treatment of the theories and the derivation of formulae have been omitted for clarity – basic concepts only have been developed.

64.2 Sampling distributions

In statistics, it is not always possible to take into account all the members of a set and in these circumstances, a **sample**, or many samples, are drawn from a population. Usually when the word sample is used, it means that a **random sample** is taken. If each member of a population has the same chance of being selected, then a sample taken from that population is called random. A sample which is not random is said to be **biased** and this usually occurs when some influence affects the selection.

When it is necessary to make predictions about a population based on random sampling, often many samples of, say, N members are taken, before the predictions are made. If the mean value and standard deviation of each of the samples is calculated, it is found that the results vary from sample to sample, even though the samples are all taken from the same population. In the theories introduced in the following sections, it is important to know whether the differences in the values obtained are due to chance or whether the differences obtained are related in some way. If M samples of N members are drawn at random from a population, the mean values for the M samples together form a set of data. Similarly, the standard deviations of the M samples collectively form a set of data. Sets of data based on many samples drawn from a population are called **sampling distributions**. They are often used to describe the chance fluctuations of mean values and standard deviations based on random sampling.

64.3   The sampling distribution of the means

Suppose that it is required to obtain a sample of two items from a set containing five items. If the set is the five letters A, B, C, D and E, then the different samples which are possible are:

$$AB, \ AC, \ AD, \ AE, \ BC, \ BD, \ BE,$$

$$CD, \ CE \text{ and } DE,$$

that is, ten different samples. The number of possible different samples in this case is given by $\dfrac{5 \times 4}{2 \times 1}$ i.e. 10. Similarly, the number of different ways in which a sample of three items can be drawn from a set having ten members can be shown to be $\dfrac{10 \times 9 \times 8}{3 \times 2 \times 1}$ i.e. 120. It follows that when a small sample is drawn from a large population, there are very many different combinations of members possible. With so many different samples possible, quite a large variation can occur in the mean values of various samples taken from the same population.

Usually, the greater the number of members in a sample, the closer will be the mean value of the sample to that of the population. Consider the set of numbers 3, 4, 5, 6, and 7. For a sample of two members, the lowest value of the mean is $\dfrac{3+4}{2}$, i.e. 3.5; the highest is $\dfrac{6+7}{2}$, i.e. 6.5, giving a range of mean values of $6.5 - 3.5 = 3$. For a sample of three members, the range is $\dfrac{3+4+5}{3}$ to $\dfrac{5+6+7}{3}$ that is, 2. As the number in the sample increases, the range decreases until, in the limit, if the sample contains all the members of the set, the range of mean values is zero. When many samples are drawn from a population and a sample distribution of the mean values of the sample is formed, the range of the mean values is small provided the number in the sample is large. Because the range is small it follows that the standard deviation of all the mean values will also be small, since it depends on the distance of the mean values from the distribution mean. The relationship between the standard deviation of the mean values of a sampling distribution and the number in each sample can be expressed as follows:

Theorem 1 'If all possible samples of size N are drawn from a finite population, N_p, without replacement, and the standard deviation of the mean values of the sampling distribution of means is determined then:

$$\sigma_{\bar{x}} = \frac{\sigma}{\sqrt{N}} \sqrt{\left(\frac{N_p - N}{N_p - 1} \right)}$$

where $\sigma_{\bar{x}}$ is the standard deviation of the sampling distribution of means and σ is the standard deviation of the population.'

The standard deviation of a sampling distribution of mean values is called the **standard error of the means**, thus **standard error of the means**,

$$\sigma_{\bar{x}} = \frac{\sigma}{\sqrt{N}} \sqrt{\left(\frac{N_p - N}{N_p - 1} \right)} \tag{1}$$

Equation (1) is used for a finite population of size N_p and/or for sampling without replacement. The word error in the 'standard error of the means' does not mean that a mistake has been made but rather that there is a degree of uncertainty in predicting the mean value of a population based on the mean values of the samples. The formula for the standard error of the means is true for all values of the number in the sample, N. When N_p is very large compared with N or when the population is infinite (this can be considered to be the case when sampling is done with replacement, the correction factor $\sqrt{\left(\frac{N_p - N}{N_p - 1} \right)}$ approaches unit and equation (1) becomes

$$\sigma_{\bar{x}} = \frac{\sigma}{\sqrt{N}} \tag{2}$$

Equation (2) is used for an infinite population and/or for sampling with replacement.

Problem 1.   Verify Theorem 1 above for the set of numbers {3, 4, 5, 6, 7} when the sample size is 2

The only possible different samples of size 2 which can be drawn from this set without replacement are:

$$(3, 4), \ (3, 5), \ (3, 6), \ (3, 7), \ (4, 5),$$

$$(4, 6), \ (4, 7), \ (5, 6), \ (5, 7) \text{ and } (6, 7)$$

The mean values of these samples form the following sampling distribution of means:

$$3.5, \ 4, \ 4.5, \ 5, \ 4.5, \ 5, \ 5.5, \ 5.5, \ 6 \text{ and } 6.5$$

The mean of the sampling distributions of means,

$$\mu_{\bar{x}} = \frac{\left(\begin{array}{c} 3.5 + 4 + 4.5 + 5 + 4.5 + 5 \\ +5.5 + 5.5 + 6 + 6.5 \end{array} \right)}{10} = \frac{50}{10} = 5$$

The standard deviation of the sampling distribution of means,

$$\sigma_{\bar{x}} = \sqrt{\left[\frac{\begin{array}{c} (3.5-5)^2 + (4-5)^2 + (4.5-5)^2 \\ +(5-5)^2 + \cdots + (6.5-5)^2 \end{array}}{10} \right]}$$

$$= \sqrt{\frac{7.5}{10}} = \pm 0.866$$

Thus, **the standard error of the means is 0.866**. The standard deviation of the population,

$$\sigma = \sqrt{\left[\frac{\begin{array}{c} (3-5)^2 + (4-5)^2 + (5-5)^2 \\ +(6-5)^2 + (7-5) \end{array}}{5} \right]}$$

$$= \sqrt{2} = \pm 1.414$$

But from Theorem 1:

$$\sigma_{\bar{x}} = \frac{\sigma}{\sqrt{N}} \sqrt{\left(\frac{N_p - N}{N_p - 1} \right)}$$

and substituting for N_p, N and σ in equation (1) gives:

$$\sigma_{\bar{x}} = \frac{\pm 1.414}{\sqrt{2}} \sqrt{\left(\frac{5-2}{5-1} \right)} = \sqrt{\frac{3}{4}} = \pm 0.866,$$

as obtained by considering all samples from the population. Thus Theorem 1 is verified.

In Problem 1 above, it can be seen that the mean of the population,

$$\left(\frac{3+4+5+6+7}{5} \right)$$

is 5 and also that the mean of the sampling distribution of means, $\mu_{\bar{x}}$ is 5. This result is generalised in Theorem 2.

Theorem 2 'If all possible samples of size N are drawn from a population of size N_p and the mean value of the sampling distribution of means $\mu_{\bar{x}}$ is determined then

$$\mu_{\bar{x}} = \mu \tag{3}$$

where μ is the mean value of the population'.

In practice, all possible samples of size N are not drawn from the population. However, if the sample size is large (usually taken as 30 or more), then the relationship between the mean of the sampling distribution of means and the mean of the population is very near to that shown in equation (3). Similarly, the relationship between the standard error of the means and the standard deviation of the population is very near to that shown in equation (2). Another important property of a sampling distribution is that when the sample size, N, is large, **the sampling distribution of means approximates to a normal distribution**, of mean value $\mu_{\bar{x}}$ and standard deviation $\sigma_{\bar{x}}$. This is true for all normally distributed populations and also for populations which are not normally distributed provided the population size is at least twice as large as the sample size. This property of normality of a sampling distribution is based on a special case of the 'central limit theorem', an important theorem relating to sampling theory. Because the sampling distribution of means and standard deviations is normally distributed, the table of the partial areas under the standardised normal curve (shown in Table 61.1 on page 661) can be used to determine the probabilities of a particular sample lying between, say, ± 1 standard deviation, and so on. This point is expanded in Problem 3.

Problem 2. The heights of 3000 people are normally distributed with a mean of 175 cm and a standard deviation of 8 cm. If random samples are taken of 40 people, predict the standard deviation and the mean of the sampling distribution of means if sampling is done (a) with replacement, and (b) without replacement.

For the population: number of members, $N_p = 3000$ standard deviation, $\sigma = 8$ cm; mean, $\mu = 175$ cm.

For the samples: number in each sample, $N = 40$

(a) When sampling is done **with replacement**, the total number of possible samples (two or more can be the same) is infinite. Hence, from equation (2) the **standard error of the mean (i.e. the**

standard deviation of the sampling distribution of means)

$$\sigma_{\bar{x}} = \frac{\sigma}{\sqrt{N}} = \frac{8}{\sqrt{40}} = 1.265\,\text{cm}$$

From equation (3), **the mean of the sampling distribution**

$$\mu_{\bar{x}} = \mu = 175\,\text{cm}$$

(b) When sampling is done **without replacement**, the total number of possible samples is finite and hence equation (1) applies. Thus **the standard error of the means**

$$\sigma_{\bar{x}} = \frac{\sigma}{\sqrt{N}}\sqrt{\left(\frac{N_p - N}{N_p - 1}\right)}$$

$$= \frac{8}{\sqrt{40}}\sqrt{\left(\frac{3000 - 40}{3000 - 1}\right)}$$

$$= (1.265)(0.9935) = \mathbf{1.257\,cm}$$

As stated, following equation (3), provided the sample size is large, the mean of the sampling distribution of means is the same for both finite and infinite populations. Hence, from equation (3),

$$\mu_{\bar{x}} = 175\,\text{cm}$$

Problem 3. 1500 ingots of a metal have a mean mass of 6.5 kg and a standard deviation of 0.5 kg. Find the probability that a sample of 60 ingots chosen at random from the group, without replacement, will have a combined mass of (a) between 378 and 396 kg, and (b) more than 399 kg.

For the population: numbers of members, $N_p = 1500$; standard deviation, $\sigma = 0.5$ kg; mean $\mu = 6.5$ kg
For the sample: number in sample, $N = 60$
If many samples of 60 ingots had been drawn from the group, then the mean of the sampling distribution of means, $\mu_{\bar{x}}$ would be equal to the mean of the population. Also, the standard error of means is given by

$$\sigma_{\bar{x}} = \frac{\sigma}{\sqrt{N}}\sqrt{\left(\frac{N_p - N}{N_p - 1}\right)}$$

In addition, the sample distribution would have been approximately normal. Assume that the sample given in the problem is one of many samples. For many (theoretical) samples:

the mean of the sampling distribution of means, $\mu_{\bar{x}} = \mu = 6.5$ kg.

Also, the standard error of the means,

$$\sigma_{\bar{x}} = \frac{\sigma}{\sqrt{N}}\sqrt{\left(\frac{N_p - N}{N_p - 1}\right)} = \frac{0.5}{\sqrt{60}}\sqrt{\left(\frac{1500 - 60}{1500 - 1}\right)}$$

$$= 0.0633\,\text{kg}$$

Thus, the sample under consideration is part of a normal distribution of mean value 6.5 kg and a standard error of the means of 0.0633 kg.

(a) If the combined mass of 60 ingots is between 378 and 396 kg, then the mean mass of each of the 60 ingots lies between $\frac{378}{60}$ and $\frac{396}{60}$ kg, i.e. between 6.3 kg and 6.6 kg.

Since the masses are normally distributed, it is possible to use the techniques of the normal distribution to determine the probability of the mean mass lying between 6.3 and 6.6 kg. The normal standard variate value, z, is given by

$$z = \frac{x - \bar{x}}{\sigma}$$

hence for the sampling distribution of means, this becomes,

$$z = \frac{x - \mu_{\bar{x}}}{\sigma_{\bar{x}}}$$

Thus, 6.3 kg corresponds to a z-value of $\frac{6.3 - 6.5}{0.0633} = -3.16$ standard deviations.

Similarly, 6.6 kg corresponds to a z-value of $\frac{6.6 - 6.5}{0.0633} = 1.58$ standard deviations.

Using Table 61.1 (page 661), the areas corresponding to these values of standard deviations are 0.4992 and 0.4430 respectively. Hence **the probability of the mean mass lying between 6.3 kg and 6.6kg is 0.4992 + 0.4430 = 0.9422**. (This means that if 10 000 samples are drawn, 9422 of these samples will have a combined mass of between 378 and 396kg.)

(b) If the combined mass of 60 ingots is 399 kg, the mean mass of each ingot is $\frac{399}{60}$, that is, 6.65 kg.

The z-value for 6.65 kg is $\frac{6.65 - 6.5}{0.0633}$, i.e. 2.37 standard deviations. From Table 61.1 (page 661), the area corresponding to this z-value is 0.4911. But this is the area between the ordinate $z = 0$ and ordinate $z = 2.37$. The 'more than' value required is the total area to the right of the $z = 0$ ordinate, less the value between $z = 0$ and $z = 2.37$, i.e. $0.5000 - 0.4911$. Thus, since areas are proportional to probabilities for the standardised normal curve, **the probability of the mean mass being more than 6.65 kg** is $0.5000 - 0.4911$, i.e. **0.0089**. (This means that only 89 samples in 10 000, for example, will have a combined mass exceeding 399 kg.)

Now try the following Practice Exercise

Practice Exercise 237 Sampling distribution of means (Answers on page 891)

1. The lengths of 1500 bolts are normally distributed with a mean of 22.4 cm and a standard deviation of 0.0438 cm. If 30 samples are drawn at random from this population, each sample being 36 bolts, determine the mean of the sampling distribution and standard error of the means when sampling is done with replacement.

2. Determine the standard error of the means in Problem 1 if sampling is done without replacement, correct to four decimal places.

3. A power punch produces 1800 washers per hour. The mean inside diameter of the washers is 1.70 cm and the standard deviation is 0.013 cm. Random samples of 20 washers are drawn every five minutes. Determine the mean of the sampling distribution of means and the standard error of the means for one hour's output from the punch (a) with replacement and (b) without replacement, correct to three significant figures.

    A large batch of electric light bulbs have a mean time to failure of 800 hours and the standard deviation of the batch is 60 hours. Use this data and also Table 61.1 on page 661 to solve Problems 4 to 6.

4. If a random sample of 64 light bulbs is drawn from the batch, determine the probability that the mean time to failure will be less than 785 hours, correct to three decimal places.

5. Determine the probability that the mean time to failure of a random sample of 16 light bulbs will be between 790 hours and 810 hours, correct to three decimal places.

6. For a random sample of 64 light bulbs, determine the probability that the mean time to failure will exceed 820 hours, correct to two significant figures.

7. The contents of a consignment of 1200 tins of a product have a mean mass of 0.504 kg and a standard deviation of 92 g. Determine the probability that a random sample of 40 tins drawn from the consignment will have a combined mass of (a) less than 20.13 kg, (b) between 20.13 kg and 20.17 kg, and (c) more than 20.17 kg, correct to three significant figures.

64.4 The estimation of population parameters based on a large sample size

When a population is large, it is not practical to determine its mean and standard deviation by using the basic formulae for these parameters. In fact, when a population is infinite, it is impossible to determine these values. For large and infinite populations the values of the mean and standard deviation may be estimated by using the data obtained from samples drawn from the population.

Point and interval estimates

An estimate of a population parameter, such as mean or standard deviation, based on a single number is called a **point estimate**. An estimate of a population parameter given by two numbers between which the parameter may be considered to lie is called an **interval estimate**. Thus if an estimate is made of the length of an object and the result is quoted as 150 cm, this is a point estimate. If the result is quoted as 150 ± 10 cm, this is an interval estimate and indicates that the length lies between 140 and 160 cm. Generally, a point estimate does not indicate how close the value is to the true value of the quantity and should be accompanied by additional information on which its merits may be judged. A statement of the error or the precision of an estimate is often called its **reliability**. In statistics, when estimates are made of population parameters based on samples, usually interval estimates are used. The word estimate does not suggest that we adopt the approach 'let's guess that the mean value is about ...' but rather that a value is carefully selected and the degree of confidence which can be placed in the estimate is given in addition.

Confidence intervals

It is stated in Section 64.3 that when samples are taken from a population, the mean values of these samples are approximately normally distributed, that is, the mean values forming the sampling distribution of means is approximately normally distributed. It is also true that if the standard deviations of each of the samples is found, then the standard deviations of all the samples are approximately normally distributed, that is, the standard deviations of the sampling distribution of standard deviations are approximately normally distributed. Parameters such as the mean or the standard deviation of a sampling distribution are called **sampling statistics**, S. Let μ_s be the mean value of a sampling statistic of the sampling distribution, that is, the mean value of the means of the samples or the mean value of the standard deviations of the samples. Also let σ_s be the standard deviation of a sampling statistic of the sampling distribution, that is, the standard deviation of the means of the samples or the standard deviation of the standard deviations of the samples. Because the sampling distribution of the means and of the standard deviations are normally distributed, it is possible to predict the probability of the sampling statistic lying in the intervals:

mean ± 1 standard deviation,

mean ± 2 standard deviations,

or mean ± 3 standard deviations,

by using tables of the partial areas under the standardised normal curve given in Table 61.1 on page 661. from this table, the area corresponding to a z-value of $+1$ standard deviation is 0.3413, thus the area corresponding to ± 1 standard deviation is 2×0.3413, that is, 0.6826. Thus the percentage probability of a sampling statistic lying between the mean ± 1 standard deviation is 68.26%. Similarly, the probability of a sampling statistic lying between the mean ± 2 standard deviations is 95.44% and of lying between the mean ± 3 standard deviations is 99.74%

The values 68.26%, 95.44% and 99.74% are called the **confidence levels** for estimating a sampling statistic. A confidence level of 68.26% is associated with two distinct values, these being $S - (1$ standard deviation), i.e. $S - \sigma_s$ and $S + (1$ standard deviation), i.e. $S + \sigma_s$. These two values are called the **confidence limits** of the estimate and the distance between the confidence limits is called the **confidence interval**. A confidence interval indicates the expectation or confidence of finding an estimate of the population statistic in that interval, based on a sampling statistic. The list in

Table 64.1 is based on values given in Table 61.1, and gives some of the confidence levels used in practice and their associated z-values (some of the values given are based on interpolation). When the table is used in this context, z-values are usually indicated by 'z_c' and are called the **confidence coefficients**.

Table 64.1

Confidence level, %	Confidence coefficient, z_c
99	2.58
98	2.33
96	2.05
95	1.96
90	1.645
80	1.28
50	0.6745

Any other values of confidence levels and their associated confidence coefficients can be obtained using Table 61.1.

Problem 4. Determine the confidence coefficient corresponding to a confidence level of 98.5%

98.5% is equivalent to a per unit value of 0.9850. This indicates that the area under the standardised normal curve between $-z_c$ and $+z_c$, i.e. corresponding to $2z_c$, is 0.9850 of the total area. Hence the area between the mean value and z_c is $\dfrac{0.9850}{2}$ i.e. 0.4925 of the total area.

The z-value corresponding to a partial area of 0.4925 is 2.43 standard deviations from Table 61.1. Thus, **the confidence coefficient corresponding to a confidence limit of 98.5% is 2.43**

(a) Estimating the mean of a population when the standard deviation of the population is known

When a sample is drawn from a large population whose standard deviation is known, the mean value of the sample, $\overline{x}$, can be determined. This mean value can be used to make an estimate of the mean value of the population, μ. When this is done, the estimated mean value of the population is given as lying between two values, that is, lying in the confidence interval between the confidence limits. If a high level of confidence is required in the estimated value of μ, then the range of the confidence interval will be large. For example, if the required confidence level is 96%, then from Table 64.1 the confidence interval is from $-z_c$ to $+z_c$, that is, $2 \times 2.05 = 4.10$ standard deviations wide. Conversely, a low level of confidence has a narrow confidence interval and a confidence level of, say, 50%, has a confidence interval of 2×0.6745, that is 1.3490 standard deviations. The 68.26% confidence level for an estimate of the population mean is given by estimating that the population mean, μ, is equal to the same mean, $\bar{x}$, and then stating the confidence interval of the estimate. Since the 68.26% confidence level is associated with '± 1 standard deviation of the means of the sampling distribution', then the 68.26% confidence level for the estimate of the population mean is given by:

$$\bar{x} \pm 1\sigma_{\bar{x}}$$

In general, any particular confidence level can be obtained in the estimate by using $\bar{x} \pm z_c \sigma_{\bar{x}}$, where z_c is the confidence coefficient corresponding to the particular confidence level required. Thus for a 96% confidence level, the confidence limits of the population mean are given by $\bar{x} \pm 2.05\sigma_{\bar{x}}$. Since only one sample has been drawn, the standard error of the means, $\sigma_{\bar{x}}$, is not known. However, it is shown in Section 64.3 that

$$\sigma_{\bar{x}} = \frac{\sigma}{\sqrt{N}}\sqrt{\left(\frac{N_p - N}{N_p - 1}\right)}$$

Thus, **the confidence limits of the mean of the population are**:

$$\bar{x} \pm \frac{z_c \sigma}{\sqrt{N}}\sqrt{\left(\frac{N_p - N}{N_p - 1}\right)} \qquad (4)$$

for a finite population of size N_p
The confidence limits for the mean of the population are:

$$\bar{x} \pm \frac{z_c \sigma}{\sqrt{N}} \qquad (5)$$

for an infinite population.
Thus for a sample of size N and mean $\bar{x}$, drawn from an infinite population having a standard deviation of σ, the mean value of the population is estimated to be, for example,

$$\bar{x} \pm \frac{2.33\sigma}{\sqrt{N}}$$

for a confidence level of 98%. This indicates that the mean value of the population lies between

$$\bar{x} - \frac{2.33\sigma}{\sqrt{N}} \text{ and } \bar{x} + \frac{2.33\sigma}{\sqrt{N}}$$

with 98% confidence in this prediction.

Problem 5. It is found that the standard deviation of the diameters of rivets produced by a certain machine over a long period of time is 0.018 cm. The diameters of a random sample of 100 rivets produced by this machine in a day have a mean value of 0.476 cm. If the machine produces 2500 rivets a day, determine (a) the 90% confidence limits, and (b) the 97% confidence limits for an estimate of the mean diameter of all the rivets produced by the machine in a day.

For the population:

standard deviation, $\sigma = 0.018$ cm

number in the population, $N_p = 2500$

For the sample:

number in the sample, $N = 100$

mean, $\bar{x} = 0.476$ cm

There is a finite population and the standard deviation of the population is known, hence expression (4) is used for determining an estimate of the confidence limits of the population mean, i.e.

$$\bar{x} \pm \frac{z_c \sigma}{\sqrt{N}}\sqrt{\left(\frac{N_p - N}{N_p - 1}\right)}$$

(a) For a 90% confidence level, the value of z_c, the confidence coefficient, is 1.645 from Table 64.1. Hence, the estimate of the confidence limits of the population mean, μ, is

$$0.476 \pm \left(\frac{(1.645)(0.018)}{\sqrt{100}}\right)\sqrt{\left(\frac{2500 - 100}{2500 - 1}\right)}$$

i.e. $0.476 \pm (0.00296)(0.9800)$
$= 0.476 \pm 0.0029$ cm

Thus, **the 90% confidence limits are 0.473 cm and 0.479 cm**.
This indicates that if the mean diameter of a sample of 100 rivets is 0.476 cm, then it is predicted that the mean diameter of all the rivets will be between 0.476 cm and 0.479 cm and this prediction is made with confidence that it will be correct nine times out of ten.

(b) For a 97% confidence level, the value of z_c has to be determined from a table of partial areas under the standardised normal curve given in Table 61.1, as it is not one of the values given in Table 64.1. The total area between ordinates drawn at $-z_c$ and $+z_c$ has to be 0.9700. Because the standardised normal curve is symmetrical, the area between $z_c = 0$ and z_c is $\dfrac{0.9700}{2}$, i.e. 0.4850 from Table 61.1 an area of 0.4850 corresponds to a z_c value of 2.17. Hence, the estimated value of the confidence limits of the population mean is between

$$\bar{x} \pm \frac{z_c \sigma}{\sqrt{N}} \sqrt{\left(\frac{N_p - N}{N_p - 1} \right)}$$

$$= 0.476 \pm \left(\frac{(2.17)(0.018)}{\sqrt{100}} \right) \sqrt{\left(\frac{2500 - 100}{2500 - 1} \right)}$$

$$= 0.476 \pm (0.0039)(0.9800)$$

$$= 0.476 \pm 0.0038$$

Thus, **the 97% confidence limits are 0.472 cm and 0.030 cm**.

It can be seen that the higher value of confidence level required in part (b) results in a larger confidence interval.

Problem 6. The mean diameter of a long length of wire is to be determined. The diameter of the wire is measured in 25 places selected at random throughout its length and the mean of these values is 0.425 mm. If the standard deviation of the diameter of the wire is given by the manufacturers as 0.030 mm, determine (a) the 80% confidence interval of the estimated mean diameter of the wire, and (b) with what degree of confidence it can be said that 'the mean diameter is 0.425 ± 0.012 mm'.

For the population: $\sigma = 0.030$ mm
For the sample: $N = 25, \bar{x} = 0.425$ mm
Since an infinite number of measurements can be obtained for the diameter of the wire, the population is infinite and the estimated value of the confidence interval of the population mean is given by expression (5).

(a) For an 80% confidence level, the value of z_c is obtained from Table 64.1 and is 1.28
The 80% confidence level estimate of the confidence interval of

$$\mu = \bar{x} \pm \frac{z_c \sigma}{\sqrt{N}} = 0.425 \pm \frac{(1.28)(0.030)}{\sqrt{25}}$$

$$= 0.425 \pm 0.0077 \, \text{mm}$$

i.e. **the 80% confidence interval is from 0.417 mm to 0.433 mm**.

This indicates that the estimated mean diameter of the wire is between 0.417 mm and 0.433 mm and that this prediction is likely to be correct 80 times out of 100

(b) To determine the confidence level, the given data is equated to expression (5), giving

$$0.425 \pm 0.012 = \bar{x} \pm z_c \frac{\sigma}{\sqrt{N}}$$

But $\bar{x} = 0.425$ therefore

$$\pm z_c \frac{\sigma}{\sqrt{N}} = \pm 0.012$$

i.e. $z_c = \dfrac{0.012\sqrt{N}}{\sigma} = \pm \dfrac{(0.012)(5)}{0.030} = \pm 2$

Using Table 61.1 of partial areas under the standardised normal curve, a z_c value of 2 standard deviations corresponds to an area of 0.4772 between the mean value ($z_c = 0$) and $+2$ standard deviations. Because the standardised normal curve is symmetrical, the area between the mean and ± 2 standard deviations is

$$0.4772 \times 2, \text{ i.e. } 0.9544$$

Thus the confidence level corresponding to 0.425 ± 0.012 mm is 95.44%

(b) Estimating the mean and standard deviation of a population from sample data

The standard deviation of a large population is not known and, in this case, several samples are drawn from the population. The mean of the sampling distribution of means, $\mu_{\bar{x}}$ and the standard deviation of the sampling distribution of means (i.e. the standard error of the means), $\sigma_{\bar{x}}$, may be determined. The confidence limits of the mean value of the population, μ, are given by

$$\mu_{\bar{x}} \pm z_c \sigma_{\bar{x}} \tag{6}$$

where z_c is the confidence coefficient corresponding to the confidence level required.
To make an estimate of the standard deviation, σ, of a normally distributed population:

(i) a sampling distribution of the standard deviations of the samples is formed, and

(ii) the standard deviation of the sampling distribution is determined by using the basic standard deviation formula.

This standard deviation is called the standard error of the standard deviations and is usually signified by σ_s. If s is the standard deviation of a sample, then the confidence limits of the standard deviation of the population are given by:

$$s \pm z_c \sigma_s \qquad (7)$$

where z_c is the confidence coefficient corresponding to the required confidence level.

Problem 7. Several samples of 50 fuses selected at random from a large batch are tested when operating at a 10% overload current and the mean time of the sampling distribution before the fuses failed is 16.50 minutes. The standard error of the means is 1.4 minutes. Determine the estimated mean time to failure of the batch of fuses for a confidence level of 90%

For the sampling distribution: the mean, $\mu_{\bar{x}} = 16.50$, the standard error of the means, $\sigma_{\bar{x}} = 1.4$

The estimated mean of the population is based on sampling distribution data only and so expression (6) is used i.e. the confidence limits of the estimated mean of the population are $\mu_{\bar{x}} \pm z_c \sigma_{\bar{x}}$

For a 90% confidence level, $z_c = 1.645$ (from Table 64.1), thus,

$$\mu_{\bar{x}} \pm z_c \sigma_{\bar{x}} = 16.50 \pm (1.645)(1.4)$$
$$= 16.50 \pm 2.30 \text{m}.$$

Thus, **the 90% confidence level of the mean time to failure is from 14.20 minutes to 18.80 minutes**.

Problem 8. The sampling distribution of random samples of capacitors drawn from a large batch is found to have a standard error of the standard deviations of 0.12 μF. Determine the 92% confidence interval for the estimate of the standard deviation of the whole batch, if in a particular sample, the standard deviation is 0.60 μF. It can be assumed that the values of capacitance of the batch are normally distributed.

For the sample: the standard deviation, $s = 0.60 \mu F$. For the sampling distribution: the standard error of the standard deviations,

$$\sigma_s = 0.12 \mu F$$

When the confidence level is 92%, then by using Table 58.1 of partial areas order the standardised normal curve,

$$\text{area} = \frac{0.9200}{2} = 0.4600,$$

giving z_c as ± 1.751 standard deviations (by interpolation).

Since the population is normally distributed, the confidence limits of the standard deviation of the population may be estimated by using expression (7), i.e. $s \pm z_c \sigma_s = 0.60 \pm (1.751)(0.12) = 0.60 \pm 0.21 \mu F$.

Thus, **the 92% confidence interval for the estimate of the standard deviation for the batch is from 0.39 μF to 0.81 μF**.

Now try the following Practice Exercise

Practice Exercise 238 Estimation of population parameters based on a large sample size (Answers on page 891)

1. Measurements are made on a random sample of 100 components drawn from a population of size 1546 and having a standard deviation of 2.93 mm. The mean measurement of the components in the sample is 67.45 mm. Determine the 95% and 99% confidence limits for an estimate of the mean of the population.

2. The standard deviation of the masses of 500 blocks is 150 kg. A random sample of 40 blocks has a mean mass of 2.40 Mg.
   (a) Determine the 95% and 99% confidence intervals for estimating the mean mass of the remaining 460 blocks.
   (b) With what degree of confidence can it be said that the mean mass of the remaining 460 blocks is 2.40 ± 0.035 Mg?

3. In order to estimate the thermal expansion of a metal, measurements of the change of length for a known change of temperature are taken by a group of students. The sampling distribution of the results has a mean of $12.89 \times 10^{-4} \text{m}^\circ \text{C}^{-1}$ and a standard error of the means of $0.04 \times 10^{-4} \text{m}^\circ \text{C}^{-1}$. Determine the 95% confidence interval for an estimate of the true value of the thermal expansion of the metal, correct to two decimal places.

4. The standard deviation of the time to failure of an electronic component is estimated as 100 hours. Determine how large a sample of these components must be, in order to be 90% confident that the error in the estimated time to failure will not exceed (a) 20 hours and (b) 10 hours.

5. A sample of 60 slings of a certain diameter, used for lifting purposes, are tested to destruction (that is, loaded until they snapped). The mean and standard deviation of the breaking loads are 11.09 tonnes and 0.73 tonnes respectively. Find the 95% confidence interval for the mean of the snapping loads of all the slings of this diameter produced by this company.

6. The time taken to assemble a servomechanism is measured for 40 operatives and the mean time is 14.63 minutes with a standard deviation of 2.45 minutes. Determine the maximum error in estimating the true mean time to assemble the servomechanism for all operatives, based on a 95% confidence level.

64.5 Estimating the mean of a population based on a small sample size

The methods used in Section 64.4 to estimate the population mean and standard deviation rely on a relatively large sample size, usually taken as 30 or more. This is because when the sample size is large the sampling distribution of a parameter is approximately normally distributed. When the sample size is small, usually taken as less than 30, the techniques used for estimating the population parameters in Section 64.4 become more and more inaccurate as the sample size becomes smaller, since the sampling distribution no longer approximates to a normal distribution. Investigations were carried out into the effect of small sample sizes on the estimation theory by W. S. Gosset in the early twentieth century and, as a result of his work, tables are available which enable a realistic estimate to be made when sample sizes are small. In these tables, the t-value is determined from the relationship

$$t = \frac{(\bar{x} - \mu)}{s} \sqrt{(N - 1)}$$

where $\bar{x}$ is the mean value of a sample, μ is the mean value of the population from which the sample is drawn, s is the standard deviation of the sample and N is the number of independent observations in the sample. He published his findings under the pen name of 'Student', and these tables are often referred to as the **'Student's t distribution'**.

The confidence limits of the mean value of a population based on a small sample drawn at random from the population are given by

$$\bar{x} \pm \frac{t_c s}{\sqrt{(N - 1)}} \qquad (8)$$

In this estimate, t_c is called the confidence coefficient for small samples, analogous to z_c for large samples, s is the standard deviation of the sample, $\bar{x}$ is the mean value of the sample and N is the number of members in the sample. Table 64.2 is called 'percentile values for Student's t distribution. The columns are headed t_p where p is equal to 0.995, 0.99, 0.975, $\cdots$, 0.55. For a confidence level of, say, 95%, the column headed $t_{0.95}$ is selected and so on. The rows are headed with the Greek letter 'nu', v, and are numbered from 1 to 30 in steps of 1, together with the numbers 40, 60, 120 and ∞. These numbers represent a quantity called the **degrees of freedom**, which is defined as follows:

'The sample number, N, minus the number of population parameters which must be estimated for the sample.'

When determining the t-value, given by

$$t = \frac{(\bar{x} - \mu)}{s} \sqrt{(N - 1)}$$

it is necessary to know the sample parameters $\bar{x}$ and s and the population parameter μ. $\bar{x}$ and s can be calculated for the sample, but usually an estimate has to be made of the population mean μ, based on the sample mean value. The number of degrees of freedom, v, is given by the number of independent observations in the sample, N, minus the number of population parameters which have to be estimated, k, i.e. $v = N - k$. For the equation

$$t = \frac{(\bar{x} - \mu)}{s} \sqrt{(N - 1)}$$

only μ has to be estimated, hence $k = 1$, and $v = N - 1$. When determining the mean of a population based on a small sample size, only one population parameter is to be estimated, and hence v can always be taken as $(N - 1)$. The method used to estimate the mean of a population based on a small sample is shown in Problems 9 to 11.

Problem 9. A sample of 12 measurements of the diameter of a bar are made and the mean of the sample is 1.850 cm. The standard deviation of the sample is 0.16 mm. Determine (a) the 90% confidence limits and (b) the 70% confidence limits for an estimate of the actual diameter of the bar.

For the sample: the sample size, $N = 12$; mean, $\bar{x} = 1.850$ cm; standard deviation $s = 0.16$ mm $= 0.016$ cm. Since the sample number is less than 30, the small sample estimate as given in expression (8) must be used. The number of degrees of freedom, i.e. sample size minus the number of estimations of population parameters to be made, is $12 - 1$, i.e. 11

(a) The percentile value corresponding to a confidence coefficient value of $t_{0.90}$ and a degree of freedom value of $v = 11$ can be found by using Table 64.2, and is 1.36, that is, $t_c = 1.36$. The estimated value of the mean of the population is given by

$$\bar{x} \pm \frac{t_c s}{\sqrt{(N-1)}} = 1.850 \pm \frac{(1.36)(0.016)}{\sqrt{11}}$$

$$= 1.850 \pm 0.0066 \, \text{cm}$$

Thus, **the 90% confidence limits are 1.843 cm and 1.857 cm**.
This indicates that the actual diameter is likely to lie between 1.843 cm and 1.857 cm and that this prediction stands a 90% chance of being correct.

(b) The percentile value corresponding to $t_{0.70}$ and to $v = 11$ is obtained from Table 64.2, and is 0.540, that is, $t_c = 0.540$
The estimated value of the 70% confidence limits is given by:

$$\bar{x} \pm \frac{t_c s}{\sqrt{(N-1)}} = 1.850 \pm \frac{(0.540)(0.016)}{\sqrt{11}}$$

$$= 1.850 \pm 0.0026 \, \text{cm}$$

Thus, **the 70% confidence limits are 1.847 cm and 1.850 cm**, i.e. the actual diameter of the bar is between 1.847 cm and 1.850 cm and this result has a 70% probability of being correct.

Problem 10. A sample of nine electric lamps are selected randomly from a large batch and are tested until they fail. The mean and standard deviations of the time to failure are 1210 hours and 26 hours respectively. Determine the confidence level based on an estimated failure time of 1210 ± 6.5 hours.

For the sample: sample size, $N = 9$; standard deviation, $s = 26$ hours; mean, $\bar{x} = 1210$ hours. The confidence limits are given by:

$$\bar{x} \pm \frac{t_c s}{\sqrt{(N-1)}}$$

and these are equal to 1210 ± 6.5

Since $\bar{x} = 1210$ hours,

then $\qquad \pm \dfrac{t_c s}{\sqrt{(N-1)}} = \pm 6.5$

i.e. $\quad t_c = \pm \dfrac{6.5\sqrt{(N-1)}}{s} = \pm \dfrac{(6.5)\sqrt{8}}{26}$

$$= \pm 0.707$$

From Table 64.2, a t_c value of 0.707, having a v value of $N - 1$, i.e. 8, gives a t_p value of $t_{0.75}$
Hence, **the confidence level of an estimated failure time of 1210 ± 6.5 hours is 75%**, i.e. it is likely that 75% of all of the lamps will fail between 1203.5 and 1216.5 hours.

Problem 11. The specific resistance of some copper wire of nominal diameter 1 mm is estimated by determining the resistance of six samples of the wire. The resistance values found (in ohms per metre) were:

2.16, 2.14, 2.17, 2.15, 2.16 and 2.18

Determine the 95% confidence interval for the true specific resistance of the wire.

For the sample: sample size, $N = 6$, and mean,

$$\bar{x} = \frac{2.16 + 2.14 + 2.17 + 2.15 + 2.16 + 2.18}{6}$$

$$= 2.16 \, \Omega \, \text{m}^{-1}$$

standard deviation,

$$s = \sqrt{\left\{ \frac{\begin{array}{c}(2.16 - 2.16)^2 + (2.14 - 2.16)^2 \\ +(2.17 - 2.16)^2 + (2.15 - 2.16)^2 \\ +(2.16 - 2.16)^2 + (2.18 - 2.16)^2\end{array}}{6} \right\}}$$

$$= \sqrt{\frac{0001}{6}} = 0.0129 \, \Omega \, \text{m}^{-1}$$

The percentile value corresponding to a confidence coefficient value of $t_{0.95}$ and a degree of freedom value of $N - 1$, i.e. $6 - 1 = 5$ is 2.02 from Table 64.2. The estimated value of the 95% confidence limits is given by:

$$\bar{x} \pm \frac{t_c s}{\sqrt{(N-1)}} = 2.16 \pm \frac{(2.02)(0.0129)}{\sqrt{5}}$$

$$= 2.16 \pm 0.01165 \, \Omega \, \text{m}^{-1}$$

Thus, **the 95% confidence limits are 2.148 Ωm^{-1} and 2.172 Ωm^{-1}** which indicates that there is a 95% chance

Table 64.2 Percentile values (t_p) for Student's t distribution with v degrees of freedom (shaded area $= p$)

v	$t_{0.995}$	$t_{0.99}$	$t_{0.975}$	$t_{0.95}$	$t_{0.90}$	$t_{0.80}$	$t_{0.75}$	$t_{0.70}$	$t_{0.60}$	$t_{0.55}$
1	63.66	31.82	12.71	6.31	3.08	1.376	1.000	0.727	0.325	0.158
2	9.92	6.96	4.30	2.92	1.89	1.061	0.816	0.617	0.289	0.142
3	5.84	4.54	3.18	2.35	1.64	0.978	0.765	0.584	0.277	0.137
4	4.60	3.75	2.78	2.13	1.53	0.941	0.741	0.569	0.271	0.134
5	4.03	3.36	2.57	2.02	1.48	0.920	0.727	0.559	0.267	0.132
6	3.71	3.14	2.45	1.94	1.44	0.906	0.718	0.553	0.265	0.131
7	3.50	3.00	2.36	1.90	1.42	0.896	0.711	0.549	0.263	0.130
8	3.36	2.90	2.31	1.86	1.40	0.889	0.706	0.546	0.262	0.130
9	3.25	2.82	2.26	1.83	1.38	0.883	0.703	0.543	0.261	0.129
10	3.17	2.76	2.23	1.81	1.37	0.879	0.700	0.542	0.260	0.129
11	3.11	2.72	2.20	1.80	1.36	0.876	0.697	0.540	0.260	0.129
12	3.06	2.68	2.18	1.78	1.36	0.873	0.695	0.539	0.259	0.128
13	3.01	2.65	2.16	1.77	1.35	0.870	0.694	0.538	0.259	0.128
14	2.98	2.62	2.14	1.76	1.34	0.868	0.692	0.537	0.258	0.128
15	2.95	2.60	2.13	1.75	1.34	0.866	0.691	0.536	0.258	0.128
16	2.92	2.58	2.12	1.75	1.34	0.865	0.690	0.535	0.258	0.128
17	2.90	2.57	2.11	1.74	1.33	0.863	0.689	0.534	0.257	0.128
18	2.88	2.55	2.10	1.73	1.33	0.862	0.688	0.534	0.257	0.127
19	2.86	2.54	2.09	1.73	1.33	0.861	0.688	0.533	0.257	0.127
20	2.84	2.53	2.09	1.72	1.32	0.860	0.687	0.533	0.257	0.127
21	2.83	2.52	2.08	1.72	1.32	0.859	0.686	0.532	0.257	0.127
22	2.82	2.51	2.07	1.72	1.32	0.858	0.686	0.532	0.256	0.127
23	2.81	2.50	2.07	1.71	1.32	0.858	0.685	0.532	0.256	0.127
24	2.80	2.49	2.06	1.71	1.32	0.857	0.685	0.531	0.256	0.127
25	2.79	2.48	2.06	1.71	1.32	0.856	0.684	0.531	0.256	0.127
26	2.78	2.48	2.06	1.71	1.32	0.856	0.684	0.531	0.256	0.127
27	2.77	2.47	2.05	1.70	1.31	0.855	0.684	0.531	0.256	0.127
28	2.76	2.47	2.05	1.70	1.31	0.855	0.683	0.530	0.256	0.127
29	2.76	2.46	2.04	1.70	1.31	0.854	0.683	0.530	0.256	0.127
30	2.75	2.46	2.04	1.70	1.31	0.854	0.683	0.530	0.256	0.127
40	2.70	2.42	2.02	1.68	1.30	0.851	0.681	0.529	0.255	0.126
60	2.66	2.39	2.00	1.67	1.30	0.848	0.679	0.527	0.254	0.126
120	2.62	2.36	1.98	1.66	1.29	0.845	0.677	0.526	0.254	0.126
∞	2.58	2.33	1.96	1.645	1.28	0.842	0.674	0.524	0.253	0.126

that the true specific resistance of the wire lies between $2.148 \, \Omega\text{m}^{-1}$ and $2.172 \, \Omega\text{m}^{-1}$

Now try the following Practice Exercise

Practice Exercise 239 Estimating the mean of a population based on a small sample size (Answers on page 891)

1. The value of the ultimate tensile strength of a material is determined by measurements on ten samples of the materials. The mean and standard deviation of the results are found to be 5.17 MPa and 0.06 MPa respectively. Determine the 95% confidence interval for the mean of the ultimate tensile strength of the material.

2. Use the data given in Problem 1 above to determine the 97.5% confidence interval for the mean of the ultimate tensile strength of the material.

3. The specific resistance of a reel of German silver wire of nominal diameter 0.5 mm is estimated by determining the resistance of seven samples of the wire. These were found to have resistance values (in ohms per metre) of:
   1.12,   1.15,   1.10,   1.14,   1.15,   1.10 and 1.11
   Determine the 99% confidence interval for the true specific resistance of the reel of wire.

4. In determining the melting point of a metal, five determinations of the melting point are made. The mean and standard deviation of the five results are 132.27°C and 0.742°C. Calculate the confidence with which the prediction 'the melting point of the metal is between 131.48°C and 133.06°C' can be made.

Chapter 65

Significance testing

Why it is important to understand: Significance testing

In statistical testing, a result is called statistically significant if it is unlikely to have occurred by chance, and hence provides enough evidence to reject the hypothesis of 'no effect'. The tests involve comparing the observed values with theoretical values. The tests establish whether there is a relationship between the variables, or whether pure chance could produce the observed results. For most scientific research, a statistical significance test eliminates the possibility that the results arose by chance, allowing a rejection of the null hypothesis. This chapter introduces the principles of significance testing.

At the end of this chapter, you should be able to:

- understand hypotheses
- appreciate type I and type II errors
- calculate type I and type II errors using binomial and Poisson approximations
- appreciate significance tests for population means
- determine hypotheses using significance testing
- compare two sample means given a level of significance

65.1 Hypotheses

Industrial applications of statistics is often concerned with making decisions about populations and population parameters. For example, decisions about which is the better of two processes or decisions about whether to discontinue production on a particular machine because it is producing an economically unacceptable number of defective components are often based on deciding the mean or standard deviation of a population, calculated using sample data drawn from the population. In reaching these decisions, certain assumptions are made, which may or may not be true. The assumptions made are called **statistical hypotheses** or just **hypotheses** and are usually concerned with statements about probability distributions of populations.

For example, in order to decide whether a dice is fair, that is, unbiased, a hypothesis can be made that a particular number, say 5, should occur with a probability of one in six, since there are six numbers on a dice. Such a hypothesis is called a **null hypothesis** and is an initial statement. The symbol H_0 is used to indicate a null hypothesis. Thus, if p is the probability of throwing a 5, then $H_0 : p = \frac{1}{6}$ means, 'the null hypothesis that the probability of throwing a 5 is $\frac{1}{6}$'. Any hypothesis which differs from a given hypothesis is called an **alternative hypothesis**, and is indicated by the symbol H_1. Thus, if after many trials, it is found that the dice is biased and that a 5 only occurs, on average, one in every seven throws, then several alternative hypotheses may be formulated. For example: $H_1 : p = \frac{1}{7}$ or $H_1 : p < \frac{1}{6}$ or $H_1 : p > \frac{1}{8}$ or $H_1 : p \neq \frac{1}{6}$ are all possible alternative hypotheses to the null hypothesis that $p = \frac{1}{6}$

Hypotheses may also be used when comparisons are being made. If we wish to compare, say, the strength of two metals, a null hypothesis may be formulated that there is **no difference** between the strengths of the two metals. If the forces that the two metals can withstand are F_1 and F_2, then the null hypothesis is $H_0 : F_1 = F_2$. If it is found that the null hypothesis has to be rejected, that is, that the strengths of the two metals are not the same, then the alternative hypotheses could be of several forms. For example, $H_1 : F_1 > F_2$ or $H_1 : F_2 > F_1$ or $H_1 : F_1 \neq F_2$. These are all alternative hypotheses to the original null hypothesis.

65.2 Type I and type II errors

To illustrate what is meant by type I and type II errors, let us consider an automatic machine producing, say, small bolts. These are stamped out of a length of metal and various faults may occur. For example, the heads or the threads may be incorrectly formed, the length might be incorrect, and so on. Assume that, say, three bolts out of every 100 produced are defective in some way. If a sample of 200 bolts is drawn at random, then the manufacturer might be satisfied that his defect rate is still 3% provided there are six defective bolts in the sample. Also, the manufacturer might be satisfied that his defect rate is 3% or less provided that there are six or fewer bolts defective in the sample. He might then formulate the following hypotheses:

$H_0 : p = 0.03$ (the null hypothesis that
the defect rate is 3%)

The null hypothesis indicates that a 3% defect rate is acceptable to the manufacturer. Suppose that he also makes a decision that should the defect rate rise to 5% or more, he will take some action. Then the alternative hypothesis is:

$H_1 : p \geq 0.05$ (the alternative hypothesis that
the defect rate is equal to or
greater than 5%)

The manufacturer's decisions, which are related to these hypotheses, might well be:

(i) a null hypothesis that a 3% defect rate is acceptable, on the assumption that the associated number of defective bolts is insufficient to endanger his firm's good name;

(ii) if the null hypothesis is rejected and the defect rate rises to 5% or over, stop the machine and adjust or renew parts as necessary; since the machine is not then producing bolts, this will reduce his profit.

These decisions may seem logical at first sight, but by applying the statistical concepts introduced in previous chapters it can be shown that the manufacturer is not necessarily making very sound decisions. This is shown as follows.

When drawing a random sample of 200 bolts from the machine with a defect rate of 3%, by the laws of probability, some samples will contain no defective bolts, some samples will contain one defective bolt, and so on.

A **binomial distribution** can be used to determine the probabilities of getting $0, 1, 2, \ldots, 9$ defective bolts in the sample. Thus the probability of getting ten or more defective bolts in a sample, **even with a 3% defect rate**, is given by: $1-$ (the sum of probabilities of getting $0, 1, 2, \ldots, 9$ defective bolts). This is an extremely large calculation, given by:

$$1 - \left(0.97^{200} + 200 \times 0.97^{199} \times 0.03 \right.$$
$$\left. + \frac{200 \times 199}{2} \times 0.97^{198} \times 0.03^2 \text{ to 10 terms} \right)$$

An alternative way of calculating the required probability is to use the **normal approximation** to the binomial distribution. This may be stated as follows:

'If the probability of a defective item is p *and a non-defective item is* q*, then if a sample of* N *items is drawn at random from a large population, provided both* Np *and* Nq *are greater than 5, the binomial distribution approximates to a normal distribution of mean* Np *and standard deviation* $\sqrt{(Npq)}$*.'*

The defect rate is 3%, thus $p = 0.03$. Since $q = 1 - p$, $q = 0.97$. Sample size $N = 200$. Since Np and Nq are greater than 5, a normal approximation to the binomial distribution can be used.

The mean of the normal distribution,

$$\bar{x} = Np = 200 \times 0.03 = 6$$

The standard deviation of the normal distribution

$$\sigma = \sqrt{(Npq)}$$
$$= \sqrt{[(200)(0.03)(0.97)]} = 2.41$$

The normal standard variate for ten bolts is

$$z = \frac{\text{variate} - \text{mean}}{\text{standard deviation}}$$
$$= \frac{10 - 6}{2.41} = 1.66$$

Table 61.1 on page 661 is used to determine the area between the mean and a z-value of 1.66, and is 0.4515

The probability of having ten or more defective bolts is the total area under the standardised normal curve minus the area to the left of the $z = 1.66$ ordinate, i.e. $1-(0.5+0.4515)$, i.e. $1 - 0.9515 = 0.0485 \approx 5\%$. Thus the probability of getting ten or more defective bolts in a sample of 200 bolts, **even though the defect rate is still 3%**, is 5%. It follows that as a result of the manufacturer's decisions, for five times in every 100 the number of defects in the sample will exceed ten, the alternative hypothesis will be adopted and the machine will be stopped (and profit lost) unnecessarily. In general terms:

> **'A hypothesis has been rejected when it should have been accepted.'**

When this occurs, it is called a **type I error**, and, in this example, the type I error is 5%

Assume now that the defect rate has risen to 5%, i.e. the expectancy of a defective bolt is now ten. A second error resulting from this decision occurs due to the probability of getting fewer than ten defective bolts in a random sample, even though the defect rate has risen to 5%. Using the normal approximation to a binomial distribution: $N = 200$, $p = 0.05$, $q = 0.95$. Np and Nq are greater than 5, hence a normal approximation to a binomial distribution is a satisfactory method. The normal distribution has:

mean, $\overline{x} = Np = (200)(0.05) = 10$

standard deviation,

$$\sigma = \sqrt{(Npq)}$$
$$= \sqrt{[(200)(0.05)(0.95)]} = 3.08$$

The normal standard variate for nine defective bolts,

$$z = \frac{\text{variate} - \text{mean}}{\text{standard deviation}}$$
$$= \frac{9 - 10}{3.08} = -0.32$$

Using Table 61.1 of partial areas under the standardised normal curve given on page 661, a z-value of -0.32 corresponds to an area between the mean and the ordinate at $z = -0.32$ to 0.1255. Thus, the probability of there being nine or fewer defective bolts in the sample is given by the area to the left of the $z = 0.32$ ordinate, i.e. $0.5000 - 0.1255$, that is, 0.3745. Thus, the probability of getting nine or fewer defective bolts in a sample of 200 bolts, **even though the defect rate has risen to 5%**, is 37%. It follows that as a result of the manufacturer's decisions, for 37 samples in every 100, the

machine will be left running even though the defect rate has risen to 5%. In general terms:

> **'A hypothesis has been accepted when it should have been rejected.'**

When this occurs, it is called a **type II error**, and, in this example, the type II error is 37%

Tests of hypotheses and rules of decisions should be designed to minimise the errors of decision. This is achieved largely by trial and error for a particular set of circumstances. Type I errors can be reduced by increasing the number of defective items allowable in a sample, but this is at the expense of allowing a larger percentage of defective items to leave the factory, increasing the criticism from customers. Type II errors can be reduced by increasing the percentage defect rate in the alternative hypothesis. If a higher percentage defect rate is given in the alternative hypothesis, the type II errors are reduced very effectively, as shown in the second of the two tables below, relating the decision rule to the magnitude of the type II errors. Some examples of the magnitude of type I errors are given below, for a sample of 1000 components being produced by a machine with a mean defect rate of 5%

Decision rule Stop production if the number of defective components is equal to or greater than:	Type I error (%)
52	38.6
56	19.2
60	7.35
64	2.12
68	0.45

Decision rule Stop production when the number of defective components is 60, when the defect rate is (%):	Type II error (%)
5.5	75.49
7	10.75
8.5	0.23
10	0.00

The magnitude of the type II errors for the output of the same machine, again based on a random sample of 1000 components and a mean defect rate of 5%, is given on page 693.

When testing a hypothesis, the largest value of probability which is acceptable for a type I error is called the **level of significance** of the test. The level of significance is indicated by the symbol α (alpha) and the levels commonly adopted are 0.1, 0.05, 0.01, 0.005 and 0.002. A level of significance of, say, 0.05 means that five times in 100 the hypothesis has been rejected when it should have been accepted.

In significance tests, the following terminology is frequently adopted:

(i)   if the level of significance is 0.01 or less, i.e. the confidence level is 99% or more, the results are considered to be **highly significant**, i.e. the results are considered likely to be correct,

(ii)   if the level of significance is 0.05 or between 0.05 and 0.01, i.e. the confidence level is 95% or between 95% and 99%, the results are considered to be **probably significant**, i.e. the results are probably correct,

(iii)   if the level of significance is greater than 0.05, i.e. the confidence level is less than 95%, the results are considered to be **not significant**, that is, there are doubts about the correctness of the results obtained.

This terminology indicates that the use of a level of significance of 0.05 for 'probably significant' is, in effect, a rule of thumb. Situations can arise when the probability changes with the nature of the test being done and the use being made of the results.

The example of a machine producing bolts, used to illustrate type I and type II errors, is based on a single random sample being drawn from the output of the machine. In practice, sampling is a continuous process and using the data obtained from several samples, sampling distributions are formed. From the concepts introduced in Chapter 64, the means and standard deviations of samples are normally distributed, thus for a particular sample its mean and standard deviation are part of a normal distribution. For a set of results to be probably significant a confidence level of 95% is required for a particular hypothesis being probably correct. This is equivalent to the hypothesis being rejected when the level of significance is greater than 0.05. For this to occur, the z-value of the mean of the samples will lie between -1.96 and $+1.96$ (since the area under the standardised normal distribution curve between these

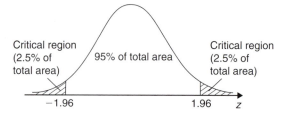

Figure 65.1

z-values is 95%). The shaded area in Fig. 65.1 is based on results which are probably significant, i.e. having a level of significance of 0.05, and represents the probability of rejecting a hypothesis when it is correct. The z-values of less than -1.96 and more than 1.96 are called **critical values** and the shaded areas in Fig. 65.1 are called the **critical regions** or regions for which the hypothesis is rejected. Having formulated hypotheses, the rules of decision and a level of significance, the magnitude of the type I error is given. Nothing can now be done about type II errors and in most cases they are accepted in the hope that they are not too large.

When critical regions occur on both sides of the mean of a normal distribution, as shown in Fig. 65.1, they are as a result of **two-tailed** or **two-sided tests**. In such tests, consideration has to be given to values on both sides of the mean. For example, if it is required to show that the percentage of metal, p, in a particular alloy is x%, then a two-tailed test is used, since the null hypothesis is incorrect if the percentage of metal is either less than x or more than x. The hypothesis is then of the form:

$$H_0 : p = x\% \quad H_1 : p \neq x\%$$

However, for the machine producing bolts, the manufacturer's decision is not affected by the fact that a sample contains say one or two defective bolts. He is only concerned with the sample containing, say, ten or more defective bolts. Thus a 'tail' on the left of the mean is not required. In this case a **one-tailed test** or a **one-sided test** is really required. If the defect rate is, say, d and the per unit values economically acceptable to the manufacturer are u_1 and u_2, where u_1 is an acceptable defect rate and u_2 is the maximum acceptable defect rate, then the hypotheses in this case are of the form:

$$H_0 : d = u_1 \quad H_1 : d > u_2$$

and the critical region lies on the right-hand side of the mean, as shown in Fig. 65.2(a). A one-tailed test can have its critical region either on the right-hand side or on the left-hand side of the mean. For example, if lamps are being tested and the manufacturer is only interested

Figure 65.2

in those lamps whose life length does not meet a certain minimum requirement, then the hypotheses are of the form:

$$H_0 : l = h \quad H_1 : l < h$$

where l is the life length and h is the number of hours to failure. In this case the critical region lies on the left-hand side of the mean, as shown in Fig. 65.2(b).

The z-values for various levels of confidence are given in Table 64.1 on page 683. The corresponding levels of significance (a confidence level of 95% is equivalent to a level of significance of 0.05 in a two-tailed test) and their z-values for both one-tailed and two-tailed tests are given in Table 65.1. It can be seen that two values of z are given for one-tailed tests, the negative value for critical regions lying to the left of the mean and a positive value for critical regions lying to the right of the mean.

The problem of the machine producing 3% defective bolts can now be reconsidered from a significance testing point of view. A random sample of 200 bolts is drawn, and the manufacturer is interested in a change in the defect rate in a specified direction (i.e. an increase),

hence the hypotheses tests are designed accordingly. If the manufacturer is willing to accept a defect rate of 3%, but wants adjustments made to the machine if the defect rate exceeds 3%, then the hypotheses will be:

(i) a null hypothesis such that the defect rate, p, is equal to 3%,

i.e. $H_0 : p = 0.03$, and

(ii) an alternative hypothesis such that the defect rate is greater than 3%,

i.e. $H_1 : p > 0.03$

The first rule of decision is as follows: let the level of significance, α, be 0.05; this will limit the type I error, that is, the error due to rejecting the hypothesis when it should be accepted, to 5%, which means that the results are probably correct. The second rule of decision is to decide the number of defective bolts in a sample for which the machine is stopped and adjustments are made. For a one-tailed test, a level of significance of 0.05 and the critical region lying to the right of the mean of the standardised normal distribution, the z-value from Table 65.1 is 1.645. If the defect rate p is 0.03%, the mean of the normal distribution is given by $Np = 200 \times 0.03 = 6$ and the standard deviation is $\sqrt{(Npq)} = \sqrt{(200 \times 0.03 \times 0.97)} = 2.41$, using the normal approximation to a binomial distribution. Since the z-value is $\dfrac{\text{variate} - \text{mean}}{\text{standard deviation}}$, then

$$1.645 = \frac{\text{variate} - 6}{2.41}$$ giving a variate value of 9.96. This variate is the number of defective bolts in a sample such that when this number is reached or exceeded the null hypothesis is rejected. For 95 times out of 100 this will be the correct thing to do. The second rule of decision will thus be 'reject H_0 if the number of defective bolts in a random sample is equal to or exceeds ten, otherwise accept H_0'. That is, the machine is adjusted when the number of defective bolts in a random sample reaches ten and this will be the correct decision 95% of the time. The type II error can now be calculated, but there is little point, since having fixed the sample number and the level of significance, there is nothing that can be done about it.

Table 65.1

Level of significance, α	0.1	0.05	0.01	0.005	0.002
z-value, one-tailed test	-1.28 or 1.28	-1.645 or 1.645	-2.33 or 2.33	-2.58 or 2.58	-2.88 or 2.88
z-value, two-tailed text	-1.645 and 1.645	-1.96 and 1.96	-2.58 and 2.58	-2.81 and 2.81	-3.08 and 3.08

Section K

A two-tailed test is used when it is required to test for changes in an **unspecified direction**. For example, if the manufacturer of bolts, used in the previous example, is inspecting the diameter of the bolts, he will want to know whether the diameters are too large or too small. Let the nominal diameter of the bolts be 2 mm. In this case the hypotheses will be:

$$H_0 : d = 2.00\,\text{mm} \quad H_1 : d \neq 2.00\,\text{mm},$$

where d is the mean diameter of the bolts. His first decision is to set the level of significance, to limit his type I error. A two-tailed test is used, since adjustments must be made to the machine if the diameter does not lie within specified limits. The method of using such a significance test is given in Section 65.3.

When determining the magnitude of type I and type II errors, it is often possible to reduce the amount of work involved by using a normal or a Poisson distribution rather than binomial distribution. A summary of the criteria for the use of these distributions and their form is given below, for a sample of size N, a probability of defective components p and a probability of non-defective components q

Binomial distribution

From Chapter 60, the probability of having $0, 1, 2, 3, \ldots$ defective components in a random sample of N components is given by the successive terms of the expansion of $(q + p)^N$, taken from the left. Thus:

Number of defective components	Probability
0	q^N
1	$Nq^{N-1}p$
2	$\dfrac{N(N-1)}{2!}q^{N-2}p^2$
3	$\dfrac{N(N-1)(N-2)}{3!}q^{N-3}p^3 \ldots$

Poisson approximation to a binomial distribution

When $N \geq 50$ and $Np < 5$, the Poisson distribution is approximately the same as the binomial distribution. In the Poisson distribution, the expectation $\lambda = Np$ and from Chapter 60, the probability of $0, 1, 2, 3, \ldots$ defective components in a random sample of N components is given by the successive terms of

$$e^{-\lambda}\left(1 + \lambda + \frac{\lambda^2}{2!} + \frac{\lambda^3}{3!} + \cdots\right)$$

taken from the left. Thus,

Number of defective components	0	1	2	3
Probability	$e^{-\lambda}$	$\lambda e^{-\lambda}$	$\dfrac{\lambda^2 e^{-\lambda}}{2!}$	$\dfrac{\lambda^3 e^{-\lambda}}{3!}$

Normal approximation to a binomial distribution

When both Np and Nq are greater than 5, the normal distribution is approximately the same as the binomial distribution, The normal distribution has a mean of Np and a standard deviation of $\sqrt{(Npq)}$

Problem 1. Wood screws are produced by an automatic machine and it is found over a period of time that 7% of all the screws produced are defective. Random samples of 80 screws are drawn periodically from the output of the machine. If a decision is made that production continues until a sample contains more than seven defective screws, determine the type I error based on this decision for a defect rate of 7%. Also determine the magnitude of the type II error when the defect rate has risen to 10%

$$N = 80, \quad p = 0.07, \quad q = 0.93$$

Since both Np and Nq are greater than 5, a normal approximation to the binomial distribution is used.

Mean of the normal distribution,

$$Np = 80 \times 0.07 = 5.6$$

Standard deviation of the normal distribution,

$$\sqrt{(Npq)} = \sqrt{(80 \times 0.07 \times 0.93)} = 2.28$$

A **type I error** is the probability of rejecting a hypothesis when it is correct, hence, the type I error in this problem is the probability of stopping the machine, that is, the probability of getting more than seven defective screws in a sample, even though the defect rate is still seven%. The z-value corresponding to seven defective screws is given by:

$$\frac{\text{variate} - \text{mean}}{\text{standard deviation}} = \frac{7 - 5.6}{2.28} = 0.61$$

Using Table 61.1 of partial areas under the standardised normal curve given on page 661, the area between the mean and a z-value of 0.61 is 0.2291. Thus, the probability of more than seven defective screws is the area to the right of the z-ordinate at 0.61, that is,

[total area − (area to the left of mean

+area between mean and $z = 0.61$)]

i.e. $1 - (0.5 + 0.2291)$. This gives a probability of 0.2709. It is usual to express type I errors as a percentage, giving

type I error = 27.1%

A **type II error** is the probability of accepting a hypothesis when it should be rejected. The type II error in this problem is the probability of a sample containing fewer than seven defective screws, even though the defect rate has risen to 10%. The values are now:

$$N = 80, \quad p = 0.1, \quad q = 0.9$$

As Np and Nq are both greater than 5, a normal approximation to a binomial distribution is used, in which the mean Np is $80 \times 0.1 = 8$ and the standard deviation $\sqrt{(Npq)} = \sqrt{(80 \times 0.1 \times 0.9)} = 2.68$

The z-value for a variate of seven defective screws is $\dfrac{7 - 8}{2.68} = -0.37$

Using Table 61.1 of partial areas given on page 661, the area between the mean and $z = -0.37$ is 0.1443. Hence, the probability of getting fewer than seven defective screws, even though the defect rate is 10% is (area to the left of mean − area between mean and a z-value of -0.37), i.e. $0.5 - 0.1443 = 0.3557$. It is usual to express type II errors as a percentage, giving

type II error = 35.6%

Problem 2. The sample size in Problem 1 is reduced to 50. Determine the type I error if the defect rate remains at 7% and the type II error when the defect rate rises to 9%. The decision is now to stop the machine for adjustment if a sample contains four or more defective screws.

$$N = 50, \quad p = 0.07$$

When $N \geq 50$ and $Np < 5$, the Poisson approximation to a binomial distribution is used. The expectation $\lambda = Np = 3.5$. The probabilities of $0, 1, 2, 3, \ldots$ defective

screws are given by $e^{-\lambda}, \lambda e^{-\lambda}, \dfrac{\lambda^2 e^{-\lambda}}{2!}, \dfrac{\lambda^3 e^{-\lambda}}{3!}, \ldots$ Thus,

probability of a sample containing no defective screws, $\qquad e^{-\lambda} = 0.0302$

probability of a sample containing one defective screw, $\qquad \lambda e^{-\lambda} = 0.1057$

probability of a sample containing two defective screws, $\qquad \dfrac{\lambda^2 e^{-\lambda}}{2!} = 0.1850$

probability of a sample containing three defective screws, $\qquad \dfrac{\lambda^3 e^{-\lambda}}{3} = 0.2158$

probability of a sample containing zero, one, two, or three defective screws is $\underline{\;0.5367\;}$

Hence, the probability of a sample containing four or more defective screws is $1 - 0.5367 = 0.4633$. Thus the **type I error**, that is, rejecting the hypothesis when it should be accepted or stopping the machine for adjustment when it should continue running, is **46.3%**. When the defect rate has risen to 9%, $p = 0.09$ and $Np = \lambda = 4.5$. Since $N \geq 50$ and $Np < 5$, the Poisson approximation to a binomial distribution can still be used. Thus,

probability of a sample containing no defective screws, $\qquad e^{-\lambda} = 0.0111$

probability of a sample containing one defective screw, $\qquad \lambda e^{-\lambda} = 0.0500$

probability of a sample containing two defective screws, $\qquad \dfrac{\lambda^2 e^{-\lambda}}{2!} = 0.1125$

probability of a sample containing three defective screws, $\qquad \dfrac{\lambda^3 e^{-\lambda}}{3!} = 0.1687$

probability of a sample containing zero, one, two, or three defective screws is $\underline{\;0.3423\;}$

That is, the probability of a sample containing less than four defective screws is 0.3423. Thus, the **type II error**, that is, accepting the hypothesis when it should have been rejected or leaving the machine running when it should be stopped, is **34.2%**

Problem 3. The sample size in Problem 1 is now reduced to 25. Determine the type I error if the defect rate remains at 7%, and the type II error when the defect rate rises to 10%. The decision is now to stop the machine for adjustment if a sample contains three or more defective screws.

$$N = 25, \quad p = 0.07, \quad q = 0.93$$

The criteria for a normal approximation to a binomial distribution and for a Poisson approximation to a binomial distribution are not met, hence the binomial distribution is applied.

Probability of no defective screws in a sample,
$$q^N = 0.93^{25} \qquad\qquad = 0.1630$$

Probability of one defective screw in a sample,
$$Nq^{N-1}p = 25 \times 0.93^{24} \times 0.07 \qquad = 0.3066$$

Probability of two defective screws in a sample,
$$\frac{N(N-1)}{2}q^{N-2}p^2$$
$$= \frac{25 \times 24}{2} \times 0.93^{23} \times 0.07^2 \qquad = 0.2770$$

Probability of zero, one, or two defective screws
in a sample $\qquad\qquad\qquad = 0.7466$

Thus, the probability of a type I error, i.e. stopping the machine even though the defect rate is still 7%, is $1 - 0.7466 = 0.2534$. Hence, the **type I error is 25.3%**
When the defect rate has risen to 10%:

$$N = 25, \quad p = 0.1, \quad q = 0.9$$

Probability of no defective screws in a sample,
$$q^N = 0.9^{25} \qquad\qquad = 0.0718$$

Probability of one defective screw in a sample,
$$Nq^{N-1}P = 25 \times 0.9^{24} \times 0.1 \qquad = 0.1994$$

Probability of two defective screws in a sample,
$$\frac{N(N-1)}{2}q^{N-2}p^2$$
$$= \frac{25 \times 24}{2} \times 0.9^{23} \times 0.1^2 \qquad = 0.2659$$

Probability of zero, one, or two defective screws
in a sample $\qquad\qquad\qquad = 0.5371$

That is, the probability of a **type II error**, i.e. leaving the machine running even though the defect rate has risen to 10%, is **53.7%**

Now try the following Practice Exercise

Practice Exercise 240   Type I and type II (Answers on page 891)

Problems 1 and 2 refer to an automatic machine producing piston rings for car engines. Random samples of 1000 rings are drawn from the output of the machine periodically for inspection purposes. A defect rate of 5% is acceptable to the manufacturer, but if the defect rate is believed to have exceeded this value, the machine producing the rings is stopped and adjusted.

In Problem 1, determine the type I errors which occur for the decision rules stated.

1. Stop production and adjust the machine if a sample contains (a) 54 (b) 62 and (c) 70 or more defective rings.

In Problem 2, determine the type II errors which are made if the decision rule is to stop production if there are more than 60 defective components in the sample.

2. When the actual defect rate has risen to (a) 6% (b) 7.5% and (c) 9%

3. A random sample of 100 components is drawn from the output of a machine whose defect rate is 3%. Determine the type I error if the decision rule is to stop production when the sample contains: (a) four or more defective components, (b) five or more defective components, and (c) six or more defective components.

4. If there are four or more defective components in a sample drawn from the machine given in Problem 3 above, determine the type II error when the actual defect rate is: (a) 5% (b) 6% (c) 7%

65.3 Significance tests for population means

When carrying out tests or measurements, it is often possible to form a hypothesis as a result of these tests. For example, the boiling point of water is found to be: 101.7°C, 99.8°C, 100.4°C, 100.3°C, 99.5°C and 98.9°C, as a result of six tests. The mean of these six results is 100.1°C. Based on these results, how confidently can it be predicted that at this particular height above sea level and at this particular barometric pressure, water boils at 100.1°C? In other words, are the results based on sampling **significantly different** from the true result? There are a variety of ways of testing significance, but only one or two of these in common use are introduced in this section. Usually, in significance tests, some predictions about population parameters, based on sample data, are required. In significance tests for population means, a random sample is drawn from the population and the mean value of the sample, $\bar{x}$, is determined. The testing procedure depends on whether or not the standard deviation of the population is known.

(a) When the standard deviation of the population is known

A null hypothesis is made that there is no difference between the value of a sample mean $\bar{x}$ and that of the population mean, μ, i.e. H_0: $x = \mu$. If many samples had been drawn from a population and a sampling distribution of means had been formed, then, provided N is large (usually taken as $N \geq 30$) the mean value would form a normal distribution, having a mean value of $\mu_{\bar{x}}$ and a standard deviation or standard error of the means (see Section 64.3).

The particular value of $\bar{x}$ of a large sample drawn for a significance test is therefore part of a normal distribution and it is possible to determine by how much $\bar{x}$ is likely to differ from $\mu_{\bar{x}}$ in terms of the normal standard variate z.

The relationship is $z = \dfrac{\bar{x} - \mu_{\bar{x}}}{\sigma_{\bar{x}}}$.

However, with reference to Chapter 64, page 679,

$\sigma_{\bar{x}} = \dfrac{\sigma}{\sqrt{N}} \sqrt{\left(\dfrac{N_p - N}{N_p - 1}\right)}$ for finite populations,

$= \dfrac{\sigma}{\sqrt{N}}$ for infinite populations, and $\mu_{\bar{x}} = \mu$

where N is the sample size, N_p is the size of the population, μ is the mean of the population and σ the standard deviation of the population.

Substituting for $\mu_{\bar{x}}$ and $\sigma_{\bar{x}}$ in the equation for z gives:

$$z = \dfrac{\bar{x} - \mu}{\dfrac{\sigma}{\sqrt{N}}} \quad \text{for infinite populations,} \tag{1}$$

$$z = \dfrac{\bar{x} - \mu}{\dfrac{\sigma}{\sqrt{N}} \sqrt{\left(\dfrac{N_p - N}{N_p - 1}\right)}} \tag{2}$$

for populations of size N_p

In Table 65.1 on page 695, the relationship between z-values and levels of significance for both one-tailed and two-tailed tests are given. It can be seen from this table for a level of significance of, say, 0.05 and a two-tailed test, the z-value is $+1.96$, and z-values outside of this range are not significant. Thus, for a given level

of significance (i.e. a known value of z), the mean of the population, μ, can be predicted by using equations (1) and (2) above, based on the mean of a sample $\bar{x}$. Alternatively, if the mean of the population is known, the significance of a particular value of z, based on sample data, can be established. If the z-value based on the mean of a random sample for a two-tailed test is found to be, say, 2.01, then at a level of significance of 0.05, that is, the results being probably significant, the mean of the sampling distribution is said to differ significantly from what would be expected as a result of the null hypothesis (i.e. that $\bar{x} = \mu$), due to the result of the test being classed as 'not significant' (see page 694). The hypothesis would then be rejected and an alternative hypothesis formed, i.e. H_1: $\bar{x} \neq \mu$. The rules of decision for such a test would be:

(i) reject the hypothesis at a 0.05 level of significance, i.e. if the z-value of the sample mean is outside of the range -1.96 to $+1.96$

(ii) accept the hypothesis otherwise.

For small sample sizes (usually taken as $N < 30$), the sampling distribution is not normally distributed, but approximates to Student's t-distributions (see Section 64.5). In this case, t-values rather than z-values are used and the equations analogous to equations (1) and (2) are:

$$|t| = \dfrac{\bar{x} - \mu}{\dfrac{\sigma}{\sqrt{N}}} \quad \text{for infinite populations} \tag{3}$$

$$|t| = \dfrac{\bar{x} - \mu}{\dfrac{\sigma}{\sqrt{N}} \sqrt{\left(\dfrac{N_p - N}{N_p - 1}\right)}} \tag{4}$$

for populations of size N_p

where $|t|$ means the modulus of t, i.e. the positive value of t.

(b) When the standard deviation of the population is not known

It is found, in practice, that if the standard deviation of a sample is determined, its value is less than the value of the standard deviation of the population from which it is drawn. This is as expected, since the range of a sample is likely to be less than the range of the population. The difference between the two standard deviations becomes more pronounced when the sample size is small. Investigations have shown that the variance, s^2, of a sample of N items is approximately related to the variance, σ^2, of the population from which it is drawn by:

$$s^2 = \left(\frac{N-1}{N}\right)\sigma^2$$

The factor $\left(\frac{N-1}{N}\right)$ is known as **Bessel's correction**. This relationship may be used to find the relationship between the standard deviation of a sample, s, and an estimate of the standard deviation of a population, $\hat{\sigma}$, and is:

$$\hat{\sigma}^2 = s^2\left(\frac{N}{N-1}\right) \text{ i.e. } \hat{\sigma} = s\sqrt{\left(\frac{N}{N-1}\right)}$$

For large samples, say, a minimum of N being 30, the factor $\sqrt{\left(\frac{N}{N-1}\right)}$ is $\sqrt{\frac{30}{29}}$ which is approximately equal to 1.017. Thus, for large samples s is very nearly equal to $\hat{\sigma}$ and the factor $\sqrt{\left(\frac{N}{N-1}\right)}$ can be omitted without introducing any appreciable error. In equations (1) and (2), s can be written for σ, giving:

$$z = \frac{\bar{x}-\mu}{\dfrac{s}{\sqrt{N}}} \quad \text{for infinite populations} \tag{5}$$

and

$$z = \frac{\bar{x}-\mu}{\dfrac{s}{\sqrt{N}}\sqrt{\left(\dfrac{N_p-N}{N_p-1}\right)}} \tag{6}$$

for populations of size N_p

For small samples, the factor $\sqrt{\left(\frac{N}{N-1}\right)}$ cannot be disregarded and substituting $\sigma = s\sqrt{\left(\frac{N}{N-1}\right)}$ in equations (3) and (4) gives:

$$|t| = \frac{\bar{x}-\mu}{\dfrac{s\sqrt{\left(\dfrac{N}{N-1}\right)}}{\sqrt{N}}} = \frac{(\bar{x}-\mu)\sqrt{(N-1)}}{s} \tag{7}$$

for infinite populations, and

$$|t| = \frac{\bar{x}-\mu}{\dfrac{s\sqrt{\left(\dfrac{N}{N-1}\right)}}{\sqrt{N}}\sqrt{\left(\dfrac{N_p-N}{N_p-1}\right)}}$$

$$= \frac{(\bar{x}-\mu)\sqrt{(N-1)}}{s\sqrt{\left(\dfrac{N_p-N}{N_p-1}\right)}} \tag{8}$$

for populations of size N_p

The equations given in this section are parts of tests which are applied to determine population means. The way in which some of them are used is shown in the following worked problems.

Problem 4. Sugar is packed in bags by an automatic machine. The mean mass of the contents of a bag is 1.000 kg. Random samples of 36 bags are selected throughout the day and the mean mass of a particular sample is found to be 1.003 kg. If the manufacturer is willing to accept a standard deviation on all bags packed of 0.01 kg and a level of significance of 0.05, above which values the machine must be stopped and adjustments made, determine if, as a result of the sample under test, the machine should be adjusted.

Population mean $\mu = 1.000$ kg, sample mean $\bar{x} = 1.003$ kg, population standard deviation $\sigma = 0.01$ kg and sample size, $N = 36$

A null hypothesis for this problem is that the sample mean and the mean of the population are equal, i.e. $H_0: \bar{x} = \mu$

Since the manufacturer is interested in deviations on both sides of the mean, the alternative hypothesis is that the sample mean is not equal to the population mean, i.e. $H_1: \bar{x} \neq \mu$

The decision rules associated with these hypotheses are:

(i) reject H_0 if the z-value of the sample mean is outside of the range of the z-values corresponding to a level of significance of 0.05 for a two-tailed test, i.e. stop machine and adjust, and

(ii) accept H_0 otherwise, i.e. keep the machine running.

The sample size is over 30 so this is a 'large sample' problem and the population can be considered to be infinite. Because values of $\bar{x}$, μ, σ and N are all known, equation (1) can be used to determine the z-value of the sample mean,

i.e. $z = \dfrac{\bar{x} - \mu}{\dfrac{\sigma}{\sqrt{N}}} = \dfrac{1.003 - 1.000}{\dfrac{0.01}{\sqrt{36}}} = \pm\dfrac{0.003}{0.0016}$

$$= \pm 1.8$$

The z-value corresponding to a level of significance of 0.05 for a two-tailed test is given in Table 65.1 on page 695 and is ± 1.96. Since the z-value of the sample is within this range, **the null hypothesis is accepted and the machine should not be adjusted**.

Problem 5. The mean lifetime of a random sample of 50 similar torch bulbs drawn from a batch of 500 bulbs is 72 hours. The standard deviation of the lifetime of the sample is 10.4 hours. The batch is classed as inferior if the mean lifetime of the batch is less than the population mean of 75 hours. Determine whether, as a result of the sample data, the batch is considered to be inferior at a level of significance of (a) 0.05 and (b) 0.01.

Population size, $N_p = 500$, population mean, $\mu = 75$ hours, mean of sample, $\bar{x} = 72$ hours, standard deviation of sample, $s = 10.4$ hours, size of sample, $N = 50$

The null hypothesis is that the mean of the sample is equal to the mean of the population, i.e. $H_0: \bar{x} = \mu$

The alternative hypothesis is that the mean of the sample is less than the mean of the population, i.e. $H_1: \bar{x} < \mu$

(The fact that $\bar{x} = 72$ should not lead to the conclusion that the batch is necessarily inferior. At a level of significance of 0.05, the result is 'probably significant', but since this corresponds to a confidence level of 95%, there are still five times in every 100 when the result can be significantly different, that is, be outside of the range of z-values for this data. This particular sample result may be one of these five times.)

The decision rules associated with the hypotheses are:

(i) reject H_0 if the z-value (or t-value) of the sample mean is less than the z-value (or t-value) corresponding to a level of significance of (a) 0.05 and (b) 0.01, i.e. the batch is inferior,

(ii) accept H_0 otherwise, i.e. the batch is not inferior.

The data given is N, N_p, $\bar{x}$, s and μ. The alternative hypothesis indicates a one-tailed distribution and since $N > 30$ the 'large sample' theory applies.

From equation (6),

$$z = \dfrac{\bar{x} - \mu}{\dfrac{s}{\sqrt{N}}\sqrt{\left(\dfrac{N_p - N}{N_p - 1}\right)}} = \dfrac{72 - 75}{\dfrac{10.4}{\sqrt{50}}\sqrt{\left(\dfrac{500 - 50}{500 - 1}\right)}}$$

$$= \dfrac{-3}{(1.471)(0.9496)} = -2.15$$

(a) For a level of significance of 0.05 and a one-tailed test, all values to the left of the z-ordinate at -1.645 (see Table 65.1 on page 695) indicate that the results are 'not significant', that is, they differ significantly from the null hypothesis. Since the z-value of the sample mean is -2.15, i.e. less than -1.645, **the batch is considered to be inferior at a level of significance of 0.05**

(b) The z-value for a level of significance of 0.01 for a one-tailed test is -2.33 and in this case, z-values of the sample means lying to the left of the z-ordinate at -2.33 are 'not significant'. Since the z-value of the sample lies to the right of this ordinate, it does not differ significantly from the null hypothesis and **the batch is not considered to be inferior at a level of significance of 0.01**

(At first sight, for a mean value to be significant at a level of significance of 0.05, but not at 0.01, appears to be incorrect. However, it is stated earlier in the chapter that for a result to be probably significant, i.e. at a level of significance of between 0.01 and 0.05, the range of z-values is less than the range for the result to be highly significant, that is, having a level of significance of 0.01 or better. Hence the results of the problem are logical.)

Problem 6. An analysis of the mass of carbon in six similar specimens of cast iron, each of mass 425.0 g, yielded the following results:

  17.1 g, 17.3 g, 16.8 g, 16.9 g, 17.8 g, and 17.4 g

Test the hypothesis that the percentage of carbon is 4.00% assuming an arbitrary level of significance of (a) 0.2 and (b) 0.1.

The sample mean,

$$\bar{x} = \dfrac{17.1 + 17.3 + 16.8 + 16.9 + 17.8 + 17.4}{6}$$

$$= 17.22$$

The sample standard deviation,

$$s = \sqrt{\left\{ \frac{\begin{array}{c}(17.1 - 17.22)^2 + (17.3 - 17.22)^2 \\ +(16.8 - 17.22)^2 + \cdots + (17.4 - 17.22)^2\end{array}}{6} \right\}}$$

$$= 0.334$$

The null hypothesis is that the sample and population means are equal, i.e. $H_0 : \bar{x} = \mu$

The alternative hypothesis is that the sample and population means are not equal, i.e. $H_1 : \bar{x} \neq \mu$

The decision rules are:

(i) reject H_0 if the z- or t-value of the sample mean is outside of the range of the z- or t-value corresponding to a level of significance of (a) 0.2 and (b) 0.1, i.e. the mass of carbon is not 4.00%,

(ii) accept H_0 otherwise, i.e. the mass of carbon is 4.00%

The number of tests taken, N, is 6 and an infinite number of tests could have been taken, hence the population is considered to be infinite. Because $N < 30$, a t-distribution is used.

If the mean mass of carbon in the bulk of the metal is 4.00%, the mean mass of carbon in a specimen is 4.00% of 425.0, i.e. 17.00 g, thus $\mu = 17.00$

From equation (7),

$$|t| = \frac{(\bar{x} - \mu)\sqrt{(N-1)}}{s}$$

$$= \frac{(17.22 - 17.00)\sqrt{(6-1)}}{0.334}$$

$$= \mathbf{1.473}$$

In general, for any two-tailed distribution there is a critical region both to the left and to the right of the mean of the distribution. For a level of significance of 0.2, 0.1 of the percentile value of a t-distribution lies to the left of the mean and 0.1 of the percentile value lies to the right of the mean. Thus, for a level of significance of α, a value $t_{(1-\frac{\alpha}{2})}$, is required for a two-tailed distribution when using Table 64.2 on page 689. This conversion is necessary because the t-distribution is given in terms of levels of confidence and for a one-tailed distribution. The row t-value for a value of α of 0.2 is $t_{(1-\frac{0.2}{2})}$, i.e. $t_{0.90}$. The degrees of freedom v are $N - 1$, i.e. 5. From Table 64.2 on page 689, the percentile value corresponding to $(t_{0.90}, v = 5)$ is 1.48, and for a two-tailed test, ± 1.48. Since the mean value of the sample is within this range, the hypothesis is accepted at a level of significance of 0.2

The t-value for $\alpha = 0.1$ is $t_{(1-\frac{0.1}{2})}$, i.e. $t_{0.95}$. The percentile value corresponding to $t_{0.95}$, $v = 5$ is 2.02 and since the mean value of the sample is within the range ± 2.02, the hypothesis is also accepted at this level of significance. **Thus, it is probable that the mass of metal contains 4% carbon at levels of significance of 0.2 and 0.1**

Now try the following Practice Exercise

Practice Exercise 241   Significance tests for population means (Answers on page 891)

1. A batch of cables produced by a manufacturer have a mean breaking strength of 2000 kN and a standard deviation of 100 kN. A sample of 50 cables is found to have a mean breaking strength of 2050 kN. Test the hypothesis that the breaking strength of the sample is greater than the breaking strength of the population from which it is drawn at a level of significance of 0.01

2. Nine estimations of the percentage of copper in a bronze alloy have a mean of 80.8% and standard deviation of 1.2%. Assuming that the percentage of copper in samples is normally distributed, test the null hypothesis that the true percentage of copper is 80% against an alternative hypothesis that it exceeds 80%, at a level of significance of 0.05

3. The internal diameter of a pipe has a mean diameter of 3.0000 cm with a standard deviation of 0.015 cm. A random sample of 30 measurements are taken and the mean of the samples is 3.0078 cm. Test the hypothesis that the mean diameter of the pipe is 3.0000 cm at a level of significance of 0.01

4. A fishing line has a mean breaking strength of 10.25 kN. Following a special treatment on the line, the following results are obtained for 20 specimens taken from the line

Breaking strength (kN)	Frequency
9.8	1
10	1
10.1	4

Breaking strength (kN)	Frequency
10.2	5
10.5	3
10.7	2
10.8	2
10.9	1
11.0	1

Test the hypothesis that the special treatment has improved the breaking strength at a level of significance of 0.05

5. A machine produces ball bearings having a mean diameter of 0.50 cm. A sample of ten ball bearings is drawn at random and the sample mean is 0.53 cm with a standard deviation of 0.03 cm. Test the hypothesis that the mean diameter is 0.50 cm at a level of significance of (a) 0.05 and (b) 0.01

6. Six similar switches are tested to destruction at an overload of 20% of their normal maximum current rating. The mean number of operations before failure is 8200 with a standard deviation of 145. The manufacturer of the switches claims that they can be operated at least 8000 times at a 20% overload current. Can the manufacturer's claim be supported at a level of significance of (a) 0.1 and (b) 0.02

65.4   Comparing two sample means

The techniques introduced in Section 65.3 can be used for comparison purposes. For example, it may be necessary to compare the performance of, say, two similar lamps produced by different manufacturers or different operators carrying out a test or tests on the same items using different equipment. The null hypothesis adopted for tests involving two different populations is that there is **no difference** between the mean values of the populations.

The technique is based on the following theorem:

If x_1 and x_2 are the means of random samples of size N_1 and N_2 drawn from populations having means

of μ_1 and μ_2 and standard deviations of σ_1 and σ_2, then the sampling distribution of the differences of the means, $(\overline{x}_1 - \overline{x}_2)$, is a close approximation to a normal distribution, having a mean of zero and a standard deviation of

$$\sqrt{\left(\frac{\sigma_1^2}{N_1} + \frac{\sigma_2^2}{N_2}\right)}$$

For large samples, when comparing the mean values of two samples, the variate is the difference in the means of the two samples, $\overline{x}_1 - \overline{x}_2$; the mean of sampling distribution (and hence the difference in population means is zero and the standard error of the sampling distribution $\sigma_{\overline{x}}$ is

$$\sqrt{\left(\frac{\sigma_1^2}{N_1} + \frac{\sigma_2^2}{N_2}\right)}$$

Hence, the z-value is

$$\frac{(\overline{x}_1 - \overline{x}_2) - 0}{\sqrt{\left(\frac{\sigma_1^2}{N_1} + \frac{\sigma_2^2}{N_2}\right)}} = \frac{\overline{x}_1 - \overline{x}_2}{\sqrt{\left(\frac{\sigma_1^2}{N_1} + \frac{\sigma_2^2}{N_2}\right)}} \quad (9)$$

For small samples, Student's t-distribution values are used and in this case:

$$|t| = \frac{\overline{x}_1 - \overline{x}_2}{\sqrt{\left(\frac{\sigma_1^2}{N_1} + \frac{\sigma_2^2}{N_2}\right)}} \quad (10)$$

where $|t|$ means the modulus of t, i.e. the positive value of t.

When the standard deviation of the population is not known, then Bessel's correction is applied to estimate it from the sample standard deviation (i.e. the estimate of the population variance, $\sigma^2 = s^2\left(\frac{N}{N-1}\right)$ (see page 690). For large populations, the factor $\left(\frac{N}{N-1}\right)$ is small and may be neglected. However, when $N < 30$, this correction factor should be included. Also, since estimates of both σ_1 and σ_2 are being made, the k factor in the degrees of freedom in Student's t-distribution tables becomes 2 and v is given by $(N_1 + N_2 - 2)$. With these factors taken into account, when testing the hypotheses that samples come from the same population, or that there is no difference between the mean values of two populations, the t-value is given by:

$$|t| = \frac{\overline{x}_1 - \overline{x}_2}{\sigma\sqrt{\left(\frac{1}{N_1} + \frac{1}{N_2}\right)}} \quad (11)$$

An estimate of the standard deviation σ is based on a concept called 'pooling'. This states that if one estimate of the variance of a population is based on a sample, giving a result of $\sigma_1^2 = \dfrac{N_1 s_1^2}{N_1 - 1}$ and another estimate is based on a second sample, giving $\sigma_2^2 = \dfrac{N_2 s_2^2}{N_2 - 1}$, then a better estimate of the population variance, σ^2, is given by:

$$\sigma^2 = \frac{N_1 s_1^2 + N_2 s_2^2}{(N_1 - 1) + (N_2 - 1)}$$

i.e.
$$\sigma = \sqrt{\left(\frac{N_1 s_1^2 + N_2 s_2^2}{N_1 + N_2 - 2} \right)} \qquad (12)$$

Problem 7. An automatic machine is producing components, and as a result of many tests the standard deviation of their size is 0.02 cm. Two samples of 40 components are taken, the mean size of the first sample being 1.51 cm and the second 1.52 cm. Determine whether the size has altered appreciably if a level of significance of 0.05 is adopted, i.e. that the results are probably significant.

Since both samples are drawn from the same population, $\sigma_1 = \sigma_2 = \sigma = 0.02$ cm. Also $N_1 = N_2 = 40$ and $\overline{x}_1 = 1.51$ cm, $\overline{x}_2 = 1.52$ cm.

The level of significance, $\alpha = 0.05$

The null hypothesis is that the size of the component has not altered, i.e. $\overline{x}_1 = \overline{x}_2$, hence it is H_0: $\overline{x}_1 - \overline{x}_2 = 0$. The alternative hypothesis is that the size of the components has altered, i.e. that $\overline{x}_1 \neq \overline{x}_2$, hence it is $H_1 : \overline{x}_1 - \overline{x}_2 \neq 0$

For a large sample having a known standard deviation of the population, the z-value of the difference of means of two samples is given by equation (9), i.e.

$$Z = \frac{\overline{x}_1 - \overline{x}_2}{\sqrt{\left(\dfrac{\sigma_1^2}{N_1} + \dfrac{\sigma_2^2}{N_2} \right)}}$$

Since $N_1 = N_2 = $ say, N, and $\sigma_1 = \sigma_2 = \sigma$, this equation becomes

$$z = \frac{\overline{x}_1 - \overline{x}_2}{\sigma \sqrt{\left(\dfrac{2}{N} \right)}} = \frac{1.51 - 1.52}{0.02 \sqrt{\left(\dfrac{2}{40} \right)}} = -2.236$$

Since the difference between $\overline{x}_1$ and $\overline{x}_2$ has no specified direction, a two-tailed test is indicated. The z-value corresponding to a level of significance of 0.05 and a two-tailed test is $+1.96$ (see Table 65.1, page 695). The result for the z-value for the difference of means is outside of the range $+1.96$, that is, **it is probable that the size has altered appreciably at a level of significance of 0.05**

Problem 8. The electrical resistances of two products are being compared. The parameters of product 1 are:

> sample size 40, mean value of sample 74 ohms, standard deviation of whole of product 1 batch is 8 ohms

Those of product 2 are:

> sample size 50, mean value of sample 78 ohms, standard deviation of whole of product 2 batch is 7 ohms

Determine if there is any significant difference between the two products at a level of significance of (a) 0.05 and (b) 0.01

Let the mean of the batch of product 1 be μ_1, and that of product 2 be μ_2

The null hypothesis is that the means are the same, i.e. $H_0 : \mu_1 - \mu_2 = 0$

The alternative hypothesis is that the means are not the same, i.e. $H_1 : \mu_1 - \mu_2 \neq 0$

The population standard deviations are known, i.e. $\sigma_1 = 8$ ohms and $\sigma_2 = 7$ ohms, the sample means are known, i.e. $\overline{x}_1 = 74$ ohms and $\overline{x}_2 = 78$ ohms. Also the sample sizes are known, i.e. $N_1 = 40$ and $N_2 = 50$. Hence, equation (9) can be used to determine the z-value of the difference of the sample means. From equation (9),

$$z = \frac{\overline{x}_1 - \overline{x}_2}{\sqrt{\left(\dfrac{\sigma_1^2}{N_1} + \dfrac{\sigma_2^2}{N_2} \right)}} = \frac{74 - 78}{\sqrt{\left(\dfrac{8^2}{40} + \dfrac{7^2}{50} \right)}}$$

$$= \frac{-4}{1.606} = -2.49$$

(a) For a two-tailed test, the results are probably significant at a 0.05 level of significance when z lies between -1.96 and $+1.96$. Hence the z-value of the difference of means shows there is 'no significance', i.e. that **product 1 is significantly different from product 2 at a level of significance of 0.05**

(b) For a two-tailed test, the results are highly significant at a 0.01 level of significance when z lies between -2.58 and $+2.58$. Hence there is **no significant difference between product 1 and product 2 at a level of significance of 0.01**

Problem 9. The reaction time in seconds of two people, A and B, are measured by electrodermal responses and the results of the tests are as shown below.

Person A(s)	0.243	0.243	0.239
Person B(s)	0.238	0.239	0.225
Person A(s)	0.232	0.229	0.241
Person B(s)	0.236	0.235	0.234

Find if there is any significant difference between the reaction times of the two people at a level of significance of 0.1

The mean, $\bar{x}$, and standard deviation, s, of the response times of the two people are determined.

$$\bar{x}_A = \frac{0.243+0.243+0.239+0.232+0.229+0.241}{6}$$

$$= 0.2378\,s$$

$$\bar{x}_B = \frac{0.238+0.239+0.225+0.236+0.235+0.234}{6}$$

$$= 0.2345\,s$$

$$s_A = \sqrt{\left[\frac{(0.234-0.2378)^2+(0.243-0.2378)^2+\cdots+(0.241-0.2378)^2}{6}\right]}$$

$$= 0.00543\,s$$

$$s_B = \sqrt{\left[\frac{(0.238-0.2345)^2+(0.239-0.2345)^2+\cdots+(0.234-0.2345)^2}{6}\right]}$$

$$= 0.00457\,s$$

The null hypothesis is that there is no difference between the reaction times of the two people, i.e. $H_0 \colon \bar{x}_A - \bar{x}_B = 0$

The alternative hypothesis is that the reaction times are different, i.e. $H_1 \colon \bar{x}_A - \bar{x}_B \neq 0$ indicating a two-tailed test.

The sample numbers (combined) are less than 30 and a t-distribution is used. The standard deviation of all the reaction times of the two people is not known, so an estimate based on the standard deviations of the samples is used. Applying Bessel's correction, the estimate of the standard deviation of the population,

$$\sigma^2 = s^2\left(\frac{N}{N-1}\right)$$

gives $\quad \sigma_A = (0.00543)\sqrt{\left(\frac{6}{5}\right)} = 0.00595$

and $\quad \sigma_B = (0.00457)\sqrt{\left(\frac{6}{5}\right)} = 0.00501$

From equation (10), the t-value of the difference of the means is given by:

$$|t| = \frac{\bar{x}_A - \bar{x}_B}{\sqrt{\left(\frac{\sigma_A^2}{N_A}+\frac{\sigma_B^2}{N_B}\right)}}$$

$$= \frac{0.2378-0.2345}{\sqrt{\left(\frac{0.00595^2}{6}+\frac{0.00501^2}{6}\right)}}$$

$$= \mathbf{1.039}$$

For a two-tailed test and a level of significance of 0.1, the column heading in the t-distribution of Table 64.2 (on page 689) is $t_{0.95}$ (refer to Problem 6). The degrees of freedom due to k being 2 is $v = N_1 + N_2 - 2$, i.e. $6 + 6 - 2 = 10$. The corresponding t-value from Table 64.2 is 1.81. Since the t-value of the difference of the means is within the range ± 1.81, there is **no significant difference between the reaction times at a level of significance of 0.1**

Problem 10. An analyst carries out ten analyses on equal masses of a substance which is found to contain a mean of 49.20 g of a metal, with a standard deviation of 0.41 g. A trainee operator carries out 12 analyses on equal masses of the same substance which is found to contain a mean of 49.30 g, with a standard deviation of 0.32 g. Is there any significance between the results of the operators?

Let μ_1 and μ_2 be the mean values of the amounts of metal found by the two operators.

The null hypothesis is that there is no difference between the results obtained by the two operators, i.e. $H_0: \mu_1 = \mu_2$

The alternative hypothesis is that there is a difference between the results of the two operators, i.e. $H_1: \mu_1 \neq \mu_2$

Under the hypothesis H_0 the standard deviations of the amount of metal, σ, will be the same, and from equation (12)

$$\sigma = \sqrt{\left(\frac{N_1 s_1^2 + N_2 s_2^2}{N_1 + N_2 - 2} \right)}$$

$$= \sqrt{\left(\frac{(10)(0.41)^2 + (12)(0.32)^2}{10 + 12 - 2} \right)}$$

$$= 0.3814$$

The t-value of the results obtained is given by equation (11), i.e.,

$$|t| = \frac{\overline{x}_1 - \overline{x}_2}{\sigma \sqrt{\left(\frac{1}{N_1} + \frac{1}{N_2} \right)}} = \frac{49.20 - 49.30}{(0.3814)\sqrt{\left(\frac{1}{10} + \frac{1}{12} \right)}}$$

$$= -0.612$$

For the results to be probably significant, a two-tailed test and a level of significance of 0.05 is taken. H_0 is rejected outside of the range $t_{-0.975}$ and $t_{0.975}$. The number of degrees of freedom is $N_1 + N_2 - 2$. For $t_{0.975}$, $v = 20$, from Table 64.2 on page 689, the range is from -2.09 to $+2.09$. Since the t-value based on the sample data is within this range, **there is no significant difference between the results of the two operators at a level of significance of 0.05**

Now try the following Practice Exercise

Practice Exercise 242   Comparing two sample means (Answers on page 891)

1. A comparison is being made between batteries used in calculators. Batteries of type A have a mean lifetime of 24 hours with a standard deviation of 4 hours, this data being calculated from a sample of 100 of the batteries. A sample of 80 of the type B batteries has a mean lifetime of 40 hours with a standard deviation of 6 hours. Test the hypothesis that the type B batteries have a mean lifetime of at least 15 hours more than those of type A, at a level of significance of 0.05

2. Two randomly selected groups of 50 operatives in a factory are timed during an assembly operation. The first group take a mean time of 112 minutes with a standard deviation of 12 minutes. The second group take a mean time of 117 minutes with a standard deviation of 9 minutes. Test the hypothesis that the mean time for the assembly operation is the same for both groups of employees at a level of significance of 0.05

3. Capacitors having a nominal capacitance of 24 μF but produced by two different companies are tested. The values of actual capacitance are:

Company 1	21.4 23.6 24.8 22.4 26.3
Company 2	22.4 27.7 23.5 29.1 25.8

Test the hypothesis that the mean capacitance of capacitors produced by company 2 are higher than those produced by company 1 at a level of significance of 0.01

$$\left(\text{Bessel's correction is } \hat{\sigma}^2 = \frac{s^2 N}{N - 1} \right)$$

4. A sample of 100 relays produced by manufacturer A operated on average 1190 times before failure occurred, with a standard deviation of 90.75. Relays produced by manufacturer B, operated on average 1220 times before failure with a standard deviation of 120. Determine if the number of operations before failure are significantly different for the two manufacturers at a level of significance of (a) 0.05 and (b) 0.01

5. A sample of 12 car engines produced by manufacturer A showed that the mean petrol consumption over a measured distance was 4.8 litres with a standard deviation of 0.40 litres. Twelve similar engines for manufacturer B were tested over the same distance and the mean petrol consumption was 5.1 litres with a standard deviation of 0.36 litres. Test the hypothesis that the engines produced by manufacturer A are more economical than those produced by manufacturer B at a level of significance of (a) 0.01 and (b) 0.05

6. Four-star and unleaded petrol is tested in five similar cars under identical conditions. For four-star petrol, the cars covered a mean distance of 21.4 kilometres with a standard deviation of 0.54 kilometres for a given mass of petrol. For the same mass of unleaded petrol, the mean distance covered was 22.6 kilometres with a standard deviation of 0.48 kilometres. Test the hypothesis that unleaded petrol gives more kilometres per litre than four-star petrol at a level of significance of 0.05

Chapter 66

Chi-square and distribution-free tests

Why it is important to understand: Chi-square and distribution-free tests

Chi-square and distribution-free tests are used in science and engineering. Chi-square is a statistical test commonly used to compare observed data with data we would expect to obtain according to a specific hypothesis. Distribution-free methods do not rely on assumptions that the data are drawn from a given probability distribution. Non-parametric methods are widely used for studying populations that take on a ranked order. These test are explained in this chapter.

At the end of this chapter, you should be able to:

- calculate a Chi-square value for a given distribution
- test hypotheses on fitting data to theoretical distribution using the Chi-square distribution
- recognise distribution-free test
- use the sign test for two samples
- use the Wilcoxon signed-rank test for two samples
- use the Mann–Whitney test for two samples

66.1 Chi-square values

The significance tests introduced in Chapter 65 rely very largely on the normal distribution. For large sample numbers where z-values are used, the mean of the samples and the standard error of the means of the samples are assumed to be normally distributed (central limit theorem). For small sample numbers where t-values are used, the population from which samples are taken should be approximately normally distributed for the t-values to be meaningful. **Chi-square tests** (pronounced KY and denoted by the Greek letter χ), which are introduced in this chapter, do not rely on the population or a sampling statistic such as the mean or standard error of the means being normally distributed. Significance tests based on z- and t-values are concerned with the parameters of a distribution, such as the mean and the standard deviation, whereas Chi-square tests are concerned with the individual members of a set and are associated with **non-parametric tests**.

Observed and expected frequencies

The results obtained from trials are rarely exactly the same as the results predicted by statistical theories. For example, if a coin is tossed 100 times, it is unlikely that the result will be exactly 50 heads and 50 tails. Let us assume that, say, five people each toss a coin 100 times and note the number of, say, heads obtained. Let the results obtained be as shown below.

Person	A	B	C	D	E
Observed frequency	43	54	60	48	57
Expected frequency	50	50	50	50	50

A measure of the discrepancy existing between the observed frequencies shown in row 2 and the expected frequencies shown in row 3 can be determined by calculating the Chi-square value. The Chi-square value is defined as follows:

$$\chi^2 = \sum \left\{ \frac{(o-e)^2}{e} \right\}$$

where o and e are the observed and expected frequencies respectively.

Problem 1. Determine the Chi-square value for the coin-tossing data given above.

The χ^2 value for the given data may be calculated by using a tabular approach as shown below.

Person	Observed frequency, o	Expected frequency, e
A	43	50
B	54	50
C	60	50
D	48	50
E	57	50

$o - e$	$(o-e)^2$	$\dfrac{(o-e)^2}{e}$
−7	49	0.98
4	16	0.32
10	100	2.00
−2	4	0.08
7	49	0.98

$$\chi^2 = \sum \left\{ \frac{(o-e)^2}{e} \right\} = \overline{4.36}$$

Hence the Chi-square value $\chi^2 = 4.36$

If the value of χ^2 is zero, then the observed and expected frequencies agree exactly. The greater the difference between the χ^2-value and zero, the greater the discrepancy between the observed and expected frequencies.

Now try the following Practice Exercise

Practice Exercise 243   Determining Chi-square values (Answers on page 892)

1.   A dice is rolled 240 times and the observed and expected frequencies are as shown.

Face	Observed frequency	Expected frequency
1	49	40
2	35	40
3	32	40
4	46	40
5	49	40
6	29	40

Determine the χ^2-value for this distribution.

2.   The numbers of telephone calls received by the switchboard of a company in 200 five-minute intervals are shown in the distribution below.

Number of calls	Observed frequency	Expected frequency
0	11	16
1	44	42
2	53	52
3	46	42
4	24	26
5	12	14
6	7	6
7	3	2

Calculate the χ^2-value for this data.

Section K

66.2 Fitting data to theoretical distributions

For theoretical distributions such as the binomial, Poisson and normal distributions, expected frequencies can be calculated. For example, from the theory of the binomial distribution, the probability of having $0, 1, 2, \ldots, n$ defective items in a sample of n items can be determined from the successive terms of $(q + p)^n$, where p is the defect rate and $q = 1 - p$. These probabilities can be used to determine the expected frequencies of having $0, 1, 2, \ldots, n$ defective items. As a result of counting the number of defective items when sampling, the observed frequencies are obtained. The expected and observed frequencies can be compared by means of a Chi-square test and predictions can be made as to whether the differences are due to random errors, due to some fault in the method of sampling, or due to the assumptions made. As for normal and t distributions, a table is available for relating various calculated values of χ^2 to those likely because of random variations, at various levels of confidence. Such a table is shown in Table 66.1. In Table 66.1, the column on the left denotes the number of degrees of freedom, v, and when the χ^2-values refer to fitting data to theoretical distributions, the number of degrees of freedom is usually $(N - 1)$, where N is the number of rows in the table from which χ^2 is calculated. However, when the population parameters such as the mean and standard deviation are based on sample data, the number of degrees of freedom is given by $v = N - 1 - M$, where M is the number of estimated population parameters. An application of this is shown in Problem 4.
The columns of the table headed $\chi^2_{0.995}$, $\chi^2_{0.99}, \ldots$ give the percentile of χ^2-values corresponding to levels of confidence of 99.5%, 99%, ... (i.e. levels of significance of $0.005, 0.01, \ldots$). On the far right of the table, the columns headed $\ldots, \chi^2_{0.01}, \chi^2_{0.005}$ also correspond to levels of confidence of $\ldots 99\%$, 99.5%, and are used to predict the 'too good to be true' type results, where the fit obtained is so good that the method of sampling must be suspect. The method in which χ^2-values are used to test the goodness of fit of data to probability distributions is shown in the following problems.

> **Problem 2.** As a result of a survey carried out of 200 families, each with five children, the distribution shown below was produced. Test the null hypothesis that the observed frequencies are consistent with male and female births being equally probable, assuming a binomial distribution, a level of significance of 0.05 and a 'too good to be true' fit at a confidence level of 95%

Number of boys (B) and girls (G)	Number of families
5B, 0G	11
4B, 1G	35
3B, 2G	69
2B, 3G	55
1B, 4G	25
0B, 5G	5

To determine the expected frequencies

Using the usual binomial distribution symbols, let p be the probability of a male birth and $q = 1 - p$ be the probability of a female birth. The probabilities of having 5 boys, 4 boys, $\ldots$, 0 boys are given by the successive terms of the expansion of $(q + p)^n$. Since there are five children in each family, $n = 5$, and

$$(q + p)^5 = q^5 + 5q^4 p + 10q^3 p^2 + 10q^2 p^3 + 5qp^4 + p^5$$

When $q = p = 0.5$, the probabilities of 5 boys, 4 boys, $\ldots$, 0 boys are

$$0.03125, \ 0.15625, \ 0.3125, \ 0.3125,$$
$$0.15625 \text{ and } 0.3125$$

For 200 families, the expected frequencies, rounded off to the nearest whole number are: $6, 31, 63, 63, 31$ and 6 respectively.

To determine the χ^2-value

Using a tabular approach, the χ^2-value is calculated using $\chi^2 = \sum \left\{ \frac{(0-e)^2}{e} \right\}$

Number of boys (B) and girls (G)	Observed frequency, o	Expected frequency, e
5B, 0G	11	6
4B, 1G	35	31
3B, 2G	69	63
2B, 3G	55	63
1B, 4G	25	31
0B, 5G	5	6

Table 66.1 Chi-square distribution

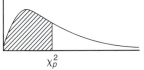

χ_p^2

Percentile values (χ_p^2) for the Chi-square distribution with v degrees of freedom

v	$\chi_{0.995}^2$	$\chi_{0.99}^2$	$\chi_{0.975}^2$	$\chi_{0.95}^2$	$\chi_{0.90}^2$	$\chi_{0.75}^2$	$\chi_{0.50}^2$	$\chi_{0.25}^2$	$\chi_{0.10}^2$	$\chi_{0.05}^2$	$\chi_{0.025}^2$	$\chi_{0.01}^2$	$\chi_{0.005}^2$
1	7.88	6.63	5.02	3.84	2.71	1.32	0.455	0.102	0.0158	0.0039	0.0010	0.0002	0.0000
2	10.6	9.21	7.38	5.99	4.61	2.77	1.39	0.575	0.211	0.103	0.0506	0.0201	0.0100
3	12.8	11.3	9.35	7.81	6.25	4.11	2.37	1.21	0.584	0.352	0.216	0.115	0.072
4	14.9	13.3	11.1	9.49	7.78	5.39	3.36	1.92	1.06	0.711	0.484	0.297	0.207
5	16.7	15.1	12.8	11.1	9.24	6.63	4.35	2.67	1.61	1.15	0.831	0.554	0.412
6	18.5	16.8	14.4	12.6	10.6	7.84	5.35	3.45	2.20	1.64	1.24	0.872	0.676
7	20.3	18.5	16.0	14.1	12.0	9.04	6.35	4.25	2.83	2.17	1.69	1.24	0.989
8	22.0	20.1	17.5	15.5	13.4	10.2	7.34	5.07	3.49	2.73	2.18	1.65	1.34
9	23.6	21.7	19.0	16.9	14.7	11.4	8.34	5.90	4.17	3.33	2.70	2.09	1.73
10	25.2	23.2	20.5	18.3	16.0	12.5	9.34	6.74	4.87	3.94	3.25	2.56	2.16
11	26.8	24.7	21.9	19.7	17.3	13.7	10.3	7.58	5.58	4.57	3.82	3.05	2.60
12	28.3	26.2	23.3	21.0	18.5	14.8	11.3	8.44	6.30	5.23	4.40	3.57	3.07
13	29.8	27.7	24.7	22.4	19.8	16.0	12.3	9.30	7.04	5.89	5.01	4.11	3.57
14	31.3	29.1	26.1	23.7	21.1	17.1	13.3	10.2	7.79	6.57	5.63	4.66	4.07
15	32.8	30.6	27.5	25.0	22.3	18.2	14.3	11.0	8.55	7.26	6.26	5.23	4.60
16	34.3	32.0	28.8	26.3	23.5	19.4	15.3	11.9	9.31	7.96	6.91	5.81	5.14
17	35.7	33.4	30.2	27.6	24.8	20.5	16.3	12.8	10.1	8.67	7.56	6.41	5.70
18	37.2	34.8	31.5	28.9	26.0	21.6	17.3	13.7	10.9	9.39	8.23	7.01	6.26
19	38.6	36.2	32.9	30.1	27.2	22.7	18.3	14.6	11.7	10.1	8.91	7.63	6.84
20	40.0	37.6	34.4	31.4	28.4	23.8	19.3	15.5	12.4	10.9	9.59	8.26	7.43
21	41.4	38.9	35.5	32.7	29.6	24.9	20.3	16.3	13.2	11.6	10.3	8.90	8.03
22	42.8	40.3	36.8	33.9	30.8	26.0	21.3	17.2	14.0	12.3	11.0	9.54	8.64
23	44.2	41.6	38.1	35.2	32.0	27.1	22.3	18.1	14.8	13.1	11.7	10.2	9.26
24	45.6	43.0	39.4	36.4	33.2	28.2	23.3	19.0	15.7	13.8	12.4	10.9	9.89
25	46.9	44.3	40.6	37.7	34.4	29.3	24.3	19.9	16.5	14.6	13.1	11.5	10.5
26	48.3	45.9	41.9	38.9	35.6	30.4	25.3	20.8	17.3	15.4	13.8	12.2	11.2
27	49.6	47.0	43.2	40.1	36.7	31.5	26.3	21.7	18.1	16.2	14.6	12.9	11.8

Table 66.1 (*Continued*)

v	$\chi^2_{0.995}$	$\chi^2_{0.99}$	$\chi^2_{0.975}$	$\chi^2_{0.95}$	$\chi^2_{0.90}$	$\chi^2_{0.75}$	$\chi^2_{0.50}$	$\chi^2_{0.25}$	$\chi^2_{0.10}$	$\chi^2_{0.05}$	$\chi^2_{0.025}$	$\chi^2_{0.01}$	$\chi^2_{0.005}$
28	51.0	48.3	44.5	41.3	37.9	32.6	27.3	22.7	18.9	16.9	15.3	13.6	12.5
29	52.3	49.6	45.7	42.6	39.1	33.7	28.3	23.6	19.8	17.7	16.0	14.3	13.1
30	53.7	50.9	47.7	43.8	40.3	34.8	29.3	24.5	20.6	18.5	16.8	15.0	13.8
40	66.8	63.7	59.3	55.8	51.8	45.6	39.3	33.7	29.1	26.5	24.4	22.2	20.7
50	79.5	76.2	71.4	67.5	63.2	56.3	49.3	42.9	37.7	34.8	32.4	29.7	28.0
60	92.0	88.4	83.3	79.1	74.4	67.0	59.3	52.3	46.5	43.2	40.5	37.5	35.5
70	104.2	100.4	95.0	90.5	85.5	77.6	69.3	61.7	55.3	51.7	48.8	45.4	43.3
80	116.3	112.3	106.6	101.9	96.6	88.1	79.3	71.1	64.3	60.4	57.2	53.5	51.2
90	128.3	124.1	118.1	113.1	107.6	98.6	89.3	80.6	73.3	69.1	65.6	61.8	59.2
100	140.2	135.8	129.6	124.3	118.5	109.1	99.3	90.1	82.4	77.9	74.2	70.1	67.3

$0-e$	$(o-e)^2$	$\dfrac{(o-e)^2}{e}$
5	25	4.167
4	16	0.516
6	36	0.571
−8	64	1.016
−6	36	1.161
−1	1	0.167

$$\chi^2 = \sum \left\{ \frac{(o-e)^2}{e} \right\} = \underline{\underline{7.598}}$$

To test the significance of the χ^2-value

The number of degrees of freedom is given by $v = N - 1$ where N is the number of rows in the table above, thus $v = 6 - 1 = 5$. For level of significance of 0.05, the confidence level is 95%, i.e. 0.95 per unit. From Table 66.1 for the $\chi^2_{0.95}$, $v = 5$ value, the percentile value χ^2_p is 11.1. Since the calculated value of χ^2 is less than χ^2_p **the null hypothesis that the observed frequencies are consistent with male and female births being equally probable is accepted**.

For a confidence level of 95%, the $\chi^2_{0.05}$, $v = 5$ value from Table 66.1 is 1.15 and because the calculated value of χ^2 (i.e. 7.598) is greater than this value, **the fit is not so good as to be unbelievable**.

Problem 3. The deposition of grit particles from the atmosphere is measured by counting the number of particles on 200 prepared cards in a specified time. The following distribution was obtained.

Number of particles	0	1	2	3	4	5	6
Number of cards		41	69	44	27	12	6	1

Test the null hypothesis that the deposition of grit particles is according to a Poisson distribution at a level of significance of 0.01 and determine if the data is 'too good to be true' at a confidence level of 99%

To determine the expected frequency

The expectation or average occurrence is given by:

$$\lambda = \frac{\text{total number of particles deposited}}{\text{total number of cards}}$$

$$= \frac{69 + 88 + 81 + 48 + 30 + 6}{200} = 1.61$$

The expected frequencies are calculated using a Poisson distribution, where the probabilities of there being $0, 1, 2, \ldots, 6$ particles deposited are given by the successive terms of $e^{-\lambda}\left(1 + \lambda + \dfrac{\lambda^2}{2!} + \dfrac{\lambda^3}{3!} + \cdots\right)$ taken from left to right,

i.e. $e^{-\lambda}, \; \lambda e^{-\lambda}, \; \dfrac{\lambda^2 e^{-\lambda}}{2!}, \; \dfrac{\lambda^3 e^{-\lambda}}{3!} \cdots$

Calculating these terms for $\lambda = 1.61$ gives:

Number of particles deposited	Probability	Expected frequency
0	0.1999	40
1	0.3218	64
2	0.2591	52
3	0.1390	28
4	0.0560	11
5	0.0180	4
6	0.0048	1

To determine the χ^2-value

The χ^2-value is calculated using a tabular method as shown below.

Number of grit particles	Observed frequency, o	Expected frequency, e
0	41	40
1	69	64
2	44	52
3	27	28
4	12	11
5	6	4
6	1	1

$o-e$	$(o-e)^2$	$\dfrac{(o-e)^2}{e}$
1	1	0.0250
5	25	0.3906
−8	64	1.2308
−1	1	0.0357
1	1	0.0909
2	4	1.0000
0	0	0.0000

$$\chi^2 = \sum \left\{ \frac{(o-e)^2}{e} \right\} = \underline{2.773}$$

To test the significance of the χ^2-value

The number of degrees of freedom is $v = N - 1$, where N is the number of rows in the table above, giving $v = 7 - 1 = 6$. The percentile *value* of χ^2 is determined from Table 66.1, for $(\chi^2_{0.99}, v = 6)$, and is 16.8. Since the calculated value of χ^2 (i.e. 2.773 is smaller than the percentile value, **the hypothesis that the grit deposition is according to a Poisson distribution is accepted**. For a confidence level of 99%, the $(\chi^2_{0.01}, v = 6)$ value is obtained from Table 66.1, and is 0.872. Since the calculated value of χ^2 is greater than this value, **the fit is not 'too good to be true'**.

Problem 4. The diameters of a sample of 500 rivets produced by an automatic process have the following size distribution.

Diameter (mm)	Frequency
4.011	12
4.015	47
4.019	86
4.023	123
4.027	107
4.031	97
4.035	28

Test the null hypothesis that the diameters of the rivets are normally distributed at a level of significance of 0.05 and also determine if the distribution gives a 'too good' fit at a level of confidence of 90%

To determine the expected frequencies

In order to determine the expected frequencies, the mean and standard deviation of the distribution are required. These population parameters, μ and σ, are based on sample data, $\bar{x}$ and s, and an allowance is made in the number of degrees of freedom used for estimating the population parameters from sample data.

The sample mean,

$$\bar{x} = \frac{\substack{12(4.011) + 47(4.015) + 86(4.019) + 123(4.023) \\ + 107(4.027) + 97(4.031) + 28(4.035)}}{500}$$

$$= \frac{2012.176}{500} = \mathbf{4.024}$$

The sample standard deviation s is given by:

$$s = \sqrt{\left[\frac{\begin{array}{c}12(4.011-4.024)^2+47(4.015-4.024)^2\\ +\cdots+28(4.035-4.024)^2\end{array}}{500}\right]}$$

$$= \sqrt{\frac{0.017212}{500}} = \mathbf{0.00587}$$

The class boundaries for the diameters are 4.009 to 4.013, 4.013 to 4.017, and so on, and are shown in column 2 of Table 66.2. Using the theory of the normal probability distribution, the probability for each class and hence the expected frequency is calculated as shown in Table 66.2.

In column 3, the z-values corresponding to the class boundaries are determined using $z = \dfrac{x - \bar{x}}{s}$ which in this case is $z = \dfrac{x - 4.024}{0.00587}$. The area between a z-value in column 3 and the mean of the distribution at $z = 0$ is determined using the table of partial areas under the standardised normal distribution curve given in Table 61.1 on page 661, and is shown in column 4. By subtracting the area between the mean and the z-value of the lower class boundary from that of the upper class boundary, the area and hence the probability of a particular class is obtained, and is shown in column 5. There is one exception in column 5, corresponding to class boundaries of 4.021 and 4.025, where the areas are added to give the probability of the 4.023 class. This is because these areas lie immediately to the left and right of the mean value. Column 6 is obtained by multiplying the probabilities in column 5 by the sample number. 500 The sum of column 6 is not equal to 500 because the area under the standardized normal curve for z-values of less than -2.56 and more than 2.21 are neglected. The error introduced by doing this is 10 in 500, i.e. 2%, and is acceptable in most problems of this type. If it is not acceptable, each expected frequency can be increased by the percentage error.

Table 66.2

1 Class mid-point	2 Class boundaries, x	3 z-value for class boundary	4 Area from 0 to z	5 Area for class	6 Expected frequency
	4.009	−2.56	0.4948		
4.011				0.0255	13
	4.013	−1.87	0.4693		
4.015				0.0863	43
	4.017	−1.19	0.3830		
4.019				0.1880	94
	4.021	−0.51	0.1950		
4.023				0.2628	131
	4.025	0.17	0.0678		
4.027				0.2345	117
	4.029	0.85	0.3023		
4.031				0.1347	67
	4.033	1.53	0.4370		
4.035				0.0494	25
	4.037	2.21	0.4864		
				Total	490

To determine the χ^2-value

The χ^2-value is calculated using a tabular method as shown below.

Diameter of rivets	Observed frequency, o	Expected frequency, e
4.011	12	13
4.015	47	43
4.019	86	94
4.023	123	131
4.027	107	117
4.031	97	67
4.035	28	25

$o-e$	$(o-e)^2$	$\dfrac{(o-e)^2}{e}$
-1	1	0.0769
4	16	0.3721
-8	64	0.6809
-8	64	0.4885
-10	100	0.8547
30	900	13.4328
3	9	0.3600

$$\chi^2 = \Sigma\left\{\frac{(o-e)^2}{e}\right\} = \underline{16.2659}$$

To text the significance of the χ^2-value

The number of degrees of freedom is given by $N-1-M$, where M is the number of estimated parameters in the population. Both the mean and the standard deviation of the population are based on the sample value, $M=2$, hence $v=7-1-2=4$. From Table 66.1, the χ_p^2-value corresponding to $\chi_{0.95}^2$ and v_4 is 9.49. **Hence the null hypothesis that the diameters of the rivets are normally distributed is rejected**. For $\chi_{0.10}^2$, v_4, the χ_p^2-value is 1.06, hence **the fit is not 'too good'**. Since the null hypothesis is rejected, the second significance test need not be carried out.

Now try the following Practice Exercise

Practice Exercise 244   Fitting data to theoretical distributions (Answers on page 892)

1. Test the null hypothesis that the observed data given below fits a binomial distribution of the form $250(0.6+0.4)^7$ at a level of significance of 0.05

Observed frequency	8	27	62	79	45	24	5	0

   Is the fit of the data 'too good' at a level of confidence of 90%?

2. The data given below refers to the number of people injured in a city by accidents for weekly periods throughout a year. It is believed that the data fits a Poisson distribution. Test the goodness of fit at a level of significance of 0.05

Number of people injured in the week	Number of weeks
0	5
1	12
2	13
3	9
4	7
5	4
6	2

3. The resistances of a sample of carbon resistors are as shown below.

Resistance (MΩ)	Frequency
1.28	7
1.29	19
1.30	41
1.31	50
1.32	73

Section K

Resistance (MΩ)	Frequency
1.33	52
1.34	28
1.35	17
1.36	9

Test the null hypothesis that this data corresponds to a normal distribution at a level of significance of 0.05

4. The quality assurance department of a firm selects 250 capacitors at random from a large quantity of them and carries out various tests on them. The results obtained are as follows:

Number of tests failed	Number of capacitors
0	113
1	77
2	39
3	16
4	4
5	1
6 and over	0

Test the goodness of fit of this distribution to a Poisson distribution at a level of significance of 0.05

5. Test the hypothesis that the maximum load before breaking supported by certain cables produced by a company follows a normal distribution at a level of significance of 0.05, based on the experimental data given below. Also test to see if the data is 'too good' at a level of significance of 0.05

Maximum load (MN)	Number of cables
8.5	2
9.0	5
9.5	12

Maximum load (MN)	Number of cables
10.0	17
10.5	14
11.0	6
11.5	3
12.0	1

66.3 Introduction to distribution-free tests

Sometimes, sampling distributions arise from populations with unknown parameters. Tests that deal with such distributions are called **distribution-free tests**; since they do not involve the use of parameters, they are known as **non-parametric tests**. Three such tests are explained in this chapter – the **sign test** in Section 66.4 following, the **Wilcoxon signed-rank test** in Section 66.5 and the **Mann–Whitney test** in Section 66.6.

66.4 The sign test

The sign test is the simplest, quickest and oldest of all non-parametric tests.

Procedure

(i) State for the data the null and alternative hypotheses, H_0 and H_1

(ii) Know whether the stated significance level, α, is for a one-tailed or a two-tailed test. Let, for example, $H_0 : x = \phi$, then if $H_1 : x \neq \phi$ then a two-tailed test is suggested because x could be less than or more than ϕ (thus use α_2 in Table 66.3 on page 718), but if say $H_1 : x < \phi$ or $H_1 : x > \phi$ then a one-tailed test is suggested (thus use α_1 in Table 66.3).

(iii) Assign plus or minus signs to each piece of data – compared with ϕ (see Problems 5 and 6) or assign plus and minus signs to the difference for paired observations (see Problem 7).

(iv) Sum either the number of plus signs or the number of minus signs. For the two-tailed test, whichever is the smallest is taken; for a one-tailed test, the one which would be expected to have the smaller value when H_1 is true is used. The sum decided upon is denoted by S.

(v) Use Table 66.3 for given values of n and α_1 or α_2 to read the critical region of S. For example, if, say, $n = 16$ and $\alpha_1 = 5\%$, then from Table 66.3, $S \leq 4$. Thus if S in part (iv) is greater than 4 we accept the null hypothesis H_0 and if S is less than or equal to 4 we accept the alternative hypothesis H_1

This procedure for the sign test is demonstrated in the following Problems.

Problem 5. A manager of a manufacturer is concerned about suspected slow progress in dealing with orders. He wants at least half of the orders received to be processed within a working day (i.e. seven hours). A little later he decides to time 17 orders selected at random, to check if his request had been met. The times spent by the 17 orders being processed were as follows:

$4\frac{3}{4}$ h	$9\frac{1}{4}$ h	$15\frac{1}{2}$ h	11 h	$8\frac{1}{4}$ h	$6\frac{1}{2}$ h
9 h	$8\frac{3}{4}$ h	$10\frac{3}{4}$ h	$3\frac{1}{2}$ h	$8\frac{1}{2}$ h	$9\frac{1}{2}$ h
$15\frac{1}{4}$ h	13 h	8 h	$7\frac{3}{4}$ h	$6\frac{3}{4}$ h	

Use the sign test at a significance level of 5% to check if the manager's request for quicker processing is being met.

Using the above procedure:

(i) The hypotheses are $H_0: t = 7$ h and $H: t > 7$ h, where t is time.

(ii) Since H_1 is $t > 7$ h, a one-tail test is assumed, i.e. $\alpha_1 = 5\%$

(iii) In the sign test each value of data is assigned a $+$ or $-$ sign. For the above data let us assign a $+$ for times greater than seven hours and a $-$ for less than seven hours. This gives the following pattern:

$$- \; + \; + \; + \; + \; - \; + \; + \; +$$
$$- \; + \; + \; + \; + \; + \; + \; -$$

(iv) The test statistic, S in this case, is the number of minus signs ($-$ if H_0 were true there would be an equal number of $+$ and $-$ signs). Table 66.3 gives critical values for the sign test and is given in terms of small values; hence in this case S is the number of $-$ signs, i.e. $S = 4$

(v) From Table 66.3, with a sample size $n = 17$, for a significance level of $\alpha_1 = 5\%$, $S \leq 4$.
Since $S = 4$ in our data, the result **is significant at $\alpha_1 = 5\%$, i.e. the alternative hypothesis is accepted – it appears that the manager's request for quicker processing of orders is not being met**.

Problem 6. The following data represents the number of hours that a portable car vacuum cleaner operates before recharging is required.

Operating time (h)	1.4	2.3	0.8	1.4	1.8	1.5
	1.9	1.4	2.1	1.1	1.6	

Use the sign test to test the hypothesis, at a 5% level of significance, that this particular vacuum cleaner operates, on average, 1.7 hours before needing a recharge.

Using the procedure:

(i) Null hypothesis $H_0: t = 1.7$ h
Alternative hypothesis $H_1: t \neq 1.7$ h

(ii) Significance level, $\alpha_2 = 5\%$ (since this is a two-tailed test).

(iii) Assuming a $+$ sign for times >1.7 and a $-$ sign for times <1.7 gives:

$$- \; + \; - \; - \; + \; - \; + \; - \; + \; - \; -$$

(iv) There are four plus signs and seven minus signs; taking the smallest number, $S = 4$

(v) From Table 66.3, where $n = 11$ and $\alpha_2 = 5\%$, $S \leq 1$

Since $S = 4$ falls in the acceptance region (i.e. in this case is greater than 1), **the null hypothesis is accepted, i.e. the average operating time is not significantly different from 1.7 h**.

Problem 7. An engineer is investigating two different types of metering devices, A and B, for an electronic fuel injection system to determine if they differ in their fuel mileage performance. The system is installed on 12 different cars, and a test is run with each metering system in turn on each car.

Table 66.3 Critical values for the sign test

n	$\alpha_1 = 5\%$ $\alpha_2 = 10\%$	$2\frac{1}{2}\%$ 5%	1% 2%	$\frac{1}{2}\%$ 1%	n	$\alpha_1 = 5\%$ $\alpha_2 = 10\%$	$2\frac{1}{2}\%$ 5%	1% 2%	$\frac{1}{2}\%$ 1%
1	—	—	—	—	26	8	7	6	6
2	—	—	—	—	27	8	7	7	6
3	—	—	—	—	28	9	8	7	6
4	—	—	—	—	29	9	8	7	7
5	0	—	—	—	30	10	9	8	7
6	0	0	—	—	31	10	9	8	7
7	0	0	0	—	32	10	9	8	8
8	1	0	0	0	33	11	10	9	8
9	1	1	0	0	34	11	10	9	9
10	1	1	0	0	35	12	11	10	9
11	2	1	1	0	36	12	11	10	9
12	2	2	1	1	37	13	12	10	10
13	3	2	1	1	38	13	12	11	10
14	3	2	2	1	39	13	12	11	11
15	3	3	2	2	40	14	13	12	11
16	4	3	2	2	41	14	13	12	11
17	4	4	3	2	42	15	14	13	12
18	5	4	3	3	43	15	14	13	12
19	5	4	4	3	44	16	15	13	13
20	5	5	4	3	45	16	15	14	13
21	6	5	4	4	46	16	15	14	13
22	6	5	5	4	47	17	16	15	14
23	7	6	5	4	48	17	16	15	14
24	7	6	5	5	49	18	17	15	15
25	7	7	6	5	50	18	17	16	15

The observed fuel mileage data (in miles/gallon) is shown on the right.
Use the sign test at a level of significance of 5% to determine whether there is any difference between the two systems.

A	18.7	20.3	20.8	18.3	16.4	16.8
B	17.6	21.2	19.1	17.5	16.9	16.4

A	17.2	19.1	17.9	19.8	18.2	19.1
B	17.7	19.2	17.5	21.4	17.6	18.8

Using the procedure:

(i) $H_0: F_A = F_B$ and $H_1: F_A \neq F_B$ where F_A and F_B are the fuels in miles/gallon for systems A and B respectively.

(ii) $\alpha_2 = 5\%$ (since it is a two-tailed test).

(iii) The difference between the observations is determined and a $+$ or a $-$ sign assigned to each as shown below:

$$(A-B) \quad +1.1 \quad -0.9 \quad +1.7 \quad +0.8$$
$$-0.5 \quad +0.4 \quad -0.5 \quad -0.1$$
$$+0.4 \quad -1.6 \quad +0.6 \quad +0.3$$

(iv) There are seven '+ signs and five '− signs'. Taking the smallest number, **S = 5**

(v) From Table 66.3, with $n = 12$ and $\alpha_2 = 5\%$, **$S \leq 2$**

Since from (iv), S is not equal or less than 2, **the null hypothesis cannot be rejected, i.e. the two metering devices produce the same fuel mileage performance**.

Now try the following Practice Exercise

Practice Exercise 245   The sign test (Answers on page 892)

1. The following data represent the number of hours of flight training received by 16 trainee pilots prior to their first solo flight:

11.5 h	20 h	9 h	12.5 h	15 h	19 h
11 h	10.5 h	13 h	22 h	14.5 h	16.5 h
17 h	18 h	14 h	12 h		

Use the sign test at a significance level of 2% to test the claim that, on average, the trainees solo after 15 hours of flight training.

2. In a laboratory experiment, 18 measurements of the coefficient of friction, μ, between metal and leather gave the following results:

0.60	0.57	0.51	0.55	0.66	0.56
0.52	0.59	0.58	0.48	0.59	0.63
0.61	0.69	0.57	0.51	0.58	0.54

Use the sign test at a level of significance of 5% to test the null hypothesis $\mu = 0.56$ against an alternative hypothesis $\mu \neq 0.56$

3. 18 random samples of two types of 9 V batteries are taken and the mean lifetime (in hours) of each are:

Type A	8.2	7.0	11.3	13.9	9.0
	13.8	16.2	8.6	9.4	3.6
	7.5	6.5	18.0	11.5	13.4
	6.9	14.2	12.4		
Type B	15.3	15.4	11.2	16.1	18.1
	17.1	17.7	8.4	13.5	7.8
	9.8	10.6	16.4	12.7	16.8
	9.9	12.9	14.7		

Use the sign test, at a level of significance of 5%, to test the null hypothesis that the two samples come from the same population.

66.5 Wilcoxon signed-rank test

The sign test represents data by using only plus and minus signs, all other information being ignored. The Wilcoxon signed-rank test does make some use of the sizes of the differences between the observed values and the hypothesised median. However, the distribution needs to be continuous and reasonably symmetric.

Procedure

(i) State for the data the null and alternative hypotheses, H_0 and H_1

(ii) Know whether the stated significance level, α, is for a one-tailed or a two-tailed test (see (ii) in the procedure for the sign test on page 716).

(iii) Find the difference of each piece of data compared with the null hypothesis (see Problems 8 and 9) or assign plus and minus signs to the difference for paired observations (see Problem 10).

(iv) Rank the differences, ignoring whether they are positive or negative.

(v) The Wilcoxon signed-rank statistic T is calculated as the sum of the ranks of either the positive differences or the negative differences – whichever is the smaller for a two-tailed test,

Section K

and the one which would be expected to have the smaller value when H_1 is true for a one-tailed test.

(vi) Use Table 66.4 for given values of n, and α_1 or α_2 to read the critical region of T. For example, if, say, $n = 16$ and $\alpha_1 = 5\%$, then from Table 66.4, $T \leq 35$. Thus if T in part (v) is greater than 35

we accept the null hypothesis H_0 and if T is less than or equal to 35 we accept the alternative hypothesis H_1

This procedure for the Wilcoxon signed-rank test is demonstrated in the following Problems.

Table 66.4 Critical values for the Wilcoxon signed-rank test

n	$\alpha_1 = 5\%$ $\alpha_2 = 10\%$	$2\frac{1}{2}\%$ 5%	1% 2%	$\frac{1}{2}\%$ 1%	n	$\alpha_1 = 5\%$ $\alpha_2 = 10\%$	$2\frac{1}{2}\%$ 5%	1% 2%	$\frac{1}{2}\%$ 1%
1	—	—	—	—	26	110	98	84	75
2	—	—	—	—	27	119	107	92	83
3	—	—	—	—	28	130	116	101	91
4	—	—	—	—	29	140	126	110	100
5	0	—	—	—	30	151	137	120	109
6	2	0	—	—	31	163	147	130	118
7	3	2	0	—	32	175	159	140	128
8	5	3	1	0	33	187	170	151	138
9	8	5	3	1	34	200	182	162	148
10	10	8	5	3	35	213	195	173	159
11	13	10	7	5	36	227	208	185	171
12	17	13	9	7	37	241	221	198	182
13	21	17	12	9	38	256	235	211	194
14	25	21	15	12	39	271	249	224	207
15	30	25	19	15	40	286	264	238	220
16	35	29	23	19	41	302	279	252	233
17	41	34	27	23	42	319	294	266	247
18	47	40	32	27	43	336	310	281	261
19	53	46	37	32	44	353	327	296	276
20	60	52	43	37	45	371	343	312	291
21	67	58	49	42	46	389	361	328	307
22	75	65	55	48	47	407	378	345	322
23	83	73	62	54	48	426	396	362	339
24	91	81	69	61	49	446	415	379	355
25	100	89	76	68	50	466	434	397	373

Problem 8. A manager of a manufacturer is concerned about suspected slow progress in dealing with orders. He wants at least half of the orders received to be processed within a working day (i.e. seven hours). A little later he decides to time 17 orders selected at random, to check if his request had been met. The times spent by the 17 orders being processed were as follows:

$4\frac{3}{4}$ h	$9\frac{3}{4}$ h	$15\frac{1}{2}$ h	11 h	$8\frac{1}{4}$ h	$6\frac{1}{2}$ h
9 h	$8\frac{3}{4}$ h	$10\frac{3}{4}$ h	$3\frac{1}{2}$ h	$8\frac{1}{2}$ h	$9\frac{1}{2}$ h
$15\frac{1}{4}$ h	13 h	8 h	$7\frac{3}{4}$ h	$6\frac{3}{4}$ h	

Use the Wilcoxon signed-rank test at a significance level of 5% to check if the manager's request for quicker processing is being met.

(This is the same as Problem 5 where the sign test was used.)

Using the procedure:

(i) The hypotheses are $H_0 : t = 7\,\text{h}$ and $H_1 : t > 7\,\text{h}$, where t is time.

(ii) Since H_1 is $t > 7$ h, a one-tail test is assumed, i.e. $\alpha_1 = 5\%$

(iii) Taking the difference between the time taken for each order and 7 h gives:

$-2\frac{1}{4}$ h	$+2\frac{3}{4}$ h	$+8\frac{1}{2}$ h	$+4$ h	$+1\frac{1}{4}$ h
$-\frac{1}{2}$ h	$+2$ h	$+1\frac{3}{4}$ h	$+3\frac{3}{4}$ h	$-3\frac{1}{2}$ h
$+1\frac{1}{2}$ h	$+2\frac{1}{2}$ h	$+8\frac{1}{4}$ h	$+6$ h	$+1$ h
$+\frac{3}{4}$ h	$-\frac{1}{4}$ h			

(iv) These differences may now be ranked from 1 to 17, ignoring whether they are positive or negative:

Rank	1	2	3	4	5	6
Difference	$-\frac{1}{4}$	$-\frac{1}{2}$	$\frac{3}{4}$	1	$1\frac{1}{4}$	$1\frac{1}{2}$

Rank	7	8	9	10	11	12
Difference	$1\frac{3}{4}$	2	$-2\frac{1}{4}$	$2\frac{1}{2}$	$2\frac{3}{4}$	$-3\frac{1}{2}$

Rank	13	14	15	16	17
Difference	$3\frac{3}{4}$	4	6	$8\frac{1}{4}$	$8\frac{1}{2}$

(v) The Wilcoxon signed-rank statistic T is calculated as **the sum of the ranks** of the negative differences for a one-tailed test.
The sum of the ranks for the negative values is:

$$T = 1 + 2 + 9 + 12 = \mathbf{24}$$

(vi) Table 66.4 gives the critical values of T for the Wilcoxon signed-rank test. For $n = 17$ and a significance level $\alpha_1 = 5\%$, $T \leq 41$

Hence the conclusion is that since $T = 24$ the result is within the 5% critical region. **There is therefore strong evidence to support H_1, the alternative hypothesis, that the median processing time is greater than seven hours.**

Problem 9. The following data represents the number of hours that a portable car vacuum cleaner operates before recharging is required.

Operating time (h)	1.4	2.3	0.8	1.4	1.8	1.5
	1.9	1.4	2.1	1.1	1.6	

Use the Wilcoxon signed-rank test to test the hypothesis, at a 5% level of significance, that this particular vacuum cleaner operates, on average, 1.7 hours before needing a recharge.

(This is the same as Problem 6 where the sign test was used.)

Using the procedure:

(i) $H_0 : t = 1.7\,\text{h}$ and $H_1 : t \neq 1.7\,\text{h}$.

(ii) Significance level, $\alpha_2 = 5\%$ (since this is a two-tailed test).

(iii) Taking the difference between each operating time and 1.7 h gives:

$$-0.3\,\text{h} \quad +0.6\,\text{h} \quad -0.9\,\text{h} \quad -0.3\,\text{h}$$
$$+0.1\,\text{h} \quad -0.2\,\text{h} \quad +0.2\,\text{h} \quad -0.3\,\text{h}$$
$$+0.4\,\text{h} \quad -0.6\,\text{h} \quad -0.1\,\text{h}$$

(iv) These differences may now be ranked from 1 to 11 (ignoring whether they are positive or negative). Some of the differences are equal to each other. For example, there are two 0.1s (ignoring signs) that would occupy positions 1 and 2 when ordered. We average these as far as rankings are concerned, i.e. each is assigned a ranking of $\dfrac{1+2}{2}$ i.e. 1.5. Similarly, the two 0.2 values in positions 3 and 4 when ordered are each as signed rankings of $\frac{3+4}{2}$ i.e. 3.5, and the three 0.3 values in positions 5, 6, and 7 are each assigned rankings of $\frac{5+6+7}{3}$ i.e. 6, and so on. The rankings are therefore:

Rank	1.5	1.5	3.5	3.5
Difference	+0.1	−0.1	−0.2	+0.2

Rank	6	6	6	8
Difference	−0.3	−0.3	−0.3	+0.4

Rank	9.5	9.5	11
Difference	+0.6	−0.6	−0.9

(v) There are four positive terms and seven negative terms. Taking the smaller number, the four positive terms have rankings of 1.5, 3.5, 8 and 9.5. Summing the positive ranks gives: $T = 1.5 + 3.5 + 8 + 9.5 = \mathbf{22.5}$

(vi) From Table 66.4, when $n = 11$ and $\alpha_2 = 5\%$, $\mathbf{T \le 10}$

Since $T = 22.5$ falls in the acceptance region (i.e. in this case is greater than 10), **the null hypothesis is accepted, i.e. the average operating time is not significantly different from 1.7 h.**

[Note that if, say, a piece of the given data was 1.7 h, such that the difference was zero, that data is ignored and n would be 10 instead of 11 in this case.]

Problem 10. An engineer is investigating two different types of metering devices, A and B, for an electronic fuel injection system to determine if they differ in their fuel mileage performance. The system is installed on 12 different cars, and a test is run with each metering system in turn on each car. The observed fuel mileage data (in miles/gallon) is shown below:

A	18.7	20.3	20.8	18.3	16.4	16.8
B	17.6	21.2	19.1	17.5	16.9	16.4

A	17.2	19.1	17.9	19.8	18.2	19.1
B	17.7	19.2	17.5	21.4	17.6	18.8

Use the Wilcoxon signed-rank test, at a level of significance of 5%, to determine whether there is any difference between the two systems.

(This is the same as Problem 7 where the sign test was used.)

Using the procedure:

(i) $H_0 : F_A = F_B$ and $H_1 : F_A \ne F_B$ where F_A and F_B are the fuels in miles/gallon for systems A and B respectively.

(ii) $\alpha_2 = \mathbf{5\%}$ (since it is a two-tailed test).

(iii) The difference between the observations is determined and a + or a − sign assigned to each as shown below:

$$(A - B) \quad \begin{array}{cccc} +1.1 & -0.9 & +1.7 & +0.8 \\ -0.5 & +0.4 & -0.5 & -0.1 \\ +0.4 & -1.6 & +0.6 & +0.3 \end{array}$$

(iv) The differences are now ranked from 1 to 12 (ignoring whether they are positive or negative). When ordered, 0.4 occupies positions 3 and 4; their average is 3.5 and both are assigned this value when ranked. Similarly, 0.5 occupies positions 5 and 6 and their average of 5.5 is assigned to each when ranked.

Rank	1	2	3.5	3.5
Difference	−0.1	+0.3	+0.4	+0.4

Rank	5.5	5.5	7	8
Difference	−0.5	−0.5	+0.6	+0.8

Rank	9	10	11	12
Difference	−0.9	+1.1	−1.6	+1.7

(v) There are seven + signs and five − signs. Taking the smaller number, the negative signs have rankings of 1, 5.5, 5.5, 9 and 11
Summing the negative ranks gives:

$$T = 1 + 5.5 + 5.5 + 9 + 11 = \mathbf{32}$$

(vi) From Table 66.4, when $n = 12$ and $\alpha_2 = 5\%$, $\mathbf{T \le 13}$
Since from (iv), T is not equal or less than 13, **the null hypothesis cannot be rejected, i.e. the two metering devices produce the same fuel mileage performance.**

Now try the following Practice Exercise

Practice Exercise 246 Wilcoxon signed-rank test (Answers on page 892)

1. The time to repair an electronic instrument is a random variable. The repair times (in hours) for 16 instruments are as follows:

218	275	264	210	161	374	178	265
150	360	185	171	215	100	474	248

Use the Wilcoxon signed-rank test, at a 5% level of significance, to test the hypothesis that the mean repair time is 220 hours.

2. 18 samples of serum are analysed for their sodium content. The results, expressed as ppm are as follows:

169	151	166	155	149	154
164	151	147	142	168	152
149	129	153	154	149	143

At a level of significance of 5%, use the Wilcoxon signed-rank test to test the null hypothesis that the average value for the method of analysis used is 150 ppm.

3. A paint supplier claims that a new additive will reduce the drying time of their acrylic paint. To test his claim, 12 pieces of wood are painted, one half of each piece with paint containing the regular additive and the other half with paint containing the new additive. The drying times (in hours) were measured as follows:

New additive	4.5	5.5	3.9	3.6	4.1	6.3
Regular additive	4.7	5.9	3.9	3.8	4.4	6.5

New additive	5.9	6.7	5.1	3.6	4.0	3.0
Regular additive	6.9	6.5	5.3	3.6	3.9	3.9

Use the Wilcoxon signed-rank test at a significance level of 5% to test the hypothesis that there is no difference, on average, in the drying times of the new and regular additive paints.

66.6 The Mann–Whitney test

As long as the sample sizes are not too large, for tests involving two samples, the Mann–Whitney test is easy to apply, is powerful and is widely used.

Procedure

(i) State for the data the null and alternative hypotheses, H_0 and H_1

(ii) Know whether the stated significance level, α, is for a one-tailed or a two-tailed test (see (ii) in the procedure for the sign test on page 716).

(iii) Arrange all the data in ascending order whilst retaining their separate identities.

(iv) If the data is now a mixture of, say, As and Bs, write under each letter A the number of Bs that precede it in the sequence (or vice-versa).

(v) Add together the numbers obtained from (iv) and denote total by U. U is defined as whichever type of count would be expected to be smallest when H_1 is true.

(vi) Use Table 66.5 on pages 724 to 727 for given values of n_1 and n_2, and α_1 or α_2 to read the critical region of U. For example, if, say, $n_1 = 10$ and $n_2 = 16$ and $\alpha_2 = 5\%$, then from Table 66.5, $U \le 42$. If U in part (v) is greater than 42 we accept the null hypothesis H_0, and if U is equal or less than 42, we accept the alternative hypothesis H_1

The procedure for the Mann–Whitney test is demonstrated in the following problems.

Problem 11. Ten British cars and eight non-British cars are compared for faults during their first 10 000 miles of use. The percentage of cars of each type developing faults were as follows:

Non-British cars, P	5	8	14	10	15
British cars, Q	18	9	25	6	21

Non-British cars, P	7	12	4		
British cars, Q	20	28	11	16	34

Use the Mann–Whitney test, at a level of significance of 1%, to test whether non-British cars have better average reliability than British models.

Table 66.5 Critical values for the Mann–Whitney test

n_1	n_2	$\alpha_1 = 5\%$ $\alpha_2 = 10\%$	$2\frac{1}{2}\%$ 5%	1% 2%	$\frac{1}{2}\%$ 1%	n_1	n_2	$\alpha_1 = 5\%$ $\alpha_2 = 10\%$	$2\frac{1}{2}\%$ 5%	1% 2%	$\frac{1}{2}\%$ 1%
2	2	—	—	—	—	3	13	6	4	2	1
2	3	—	—	—	—	3	14	7	5	2	1
2	4	—	—	—	—	3	15	7	5	3	2
2	5	0	—	—	—	3	16	8	6	3	2
2	6	0	—	—	—	3	17	9	6	4	2
2	7	0	—	—	—	3	18	9	7	4	2
2	8	1	0	—	—	3	19	10	7	4	3
2	9	1	0	—	—	3	20	11	8	5	3
2	10	1	0	—	—						
2	11	1	0	—	—	4	4	1	0	—	—
2	12	2	1	—	—	4	5	2	1	0	—
2	13	2	1	0	—	4	6	3	2	1	0
2	14	3	1	0	—	4	7	4	3	1	0
2	15	3	1	0	—	4	8	5	4	2	1
2	16	3	1	0	—	4	9	6	4	3	1
2	17	3	2	0	—	4	10	7	5	3	2
2	18	4	2	0	—	4	11	8	6	4	2
2	19	4	2	1	0	4	12	9	7	5	3
2	20	4	2	1	0	4	13	10	8	5	3
						4	14	11	9	6	4
3	3	0	—	—	—	4	15	12	10	7	5
3	4	0	—	—	—	4	16	14	11	7	5
3	5	1	0	—	—	4	17	15	11	8	6
3	6	2	1	—	—	4	18	16	12	9	6
3	7	2	1	0	—	4	19	17	13	9	7
3	8	3	2	0	—	4	20	18	14	10	8
3	9	4	2	1	0						
3	10	4	3	1	0	5	5	4	2	1	0
3	11	5	3	1	0	5	6	5	3	2	1
3	12	5	4	2	1	5	7	6	5	3	1

Table 66.5 (*Continued*)

n_1	n_2	$\alpha_1 = 5\%$ / $\alpha_2 = 10\%$	$2\frac{1}{2}\%$ / 5%	1% / 2%	$\frac{1}{2}\%$ / 1%	n_1	n_2	$\alpha_1 = 5\%$ / $\alpha_2 = 10\%$	$2\frac{1}{2}\%$ / 5%	1% / 2%	$\frac{1}{2}\%$ / 1%
5	8	8	6	4	2	7	7	11	8	6	4
5	9	9	7	5	3	7	8	13	10	7	6
5	10	11	8	6	4	7	9	15	12	9	7
5	11	12	9	7	5	7	10	17	14	11	9
5	12	13	11	8	6	7	11	19	16	12	10
5	13	15	12	9	7	7	12	21	18	14	12
5	14	16	13	10	7	7	13	24	20	16	13
5	15	18	14	11	8	7	14	26	22	17	15
5	16	19	15	12	9	7	15	28	24	19	16
5	17	20	17	13	10	7	16	30	26	21	18
5	18	22	18	14	11	7	17	33	28	23	19
5	19	23	19	15	12	7	18	35	30	24	21
5	20	25	20	16	13	7	19	37	32	26	22
						7	20	39	34	28	24
6	6	7	5	3	2						
6	7	8	6	4	3	8	8	15	13	9	7
6	8	10	8	6	4	8	9	18	15	11	9
6	9	12	10	7	5	8	10	20	17	13	11
6	10	14	11	8	6	8	11	23	19	15	13
6	11	16	13	9	7	8	12	26	22	17	15
6	12	17	14	11	9	8	13	28	24	20	17
6	13	19	16	12	10	8	14	31	26	22	18
6	14	21	17	13	11	8	15	33	29	24	20
6	15	23	19	15	12	8	16	36	31	26	22
6	16	25	21	16	13	8	17	39	34	28	24
6	17	26	22	18	15	8	18	41	36	30	26
6	18	28	24	19	16	8	19	44	38	32	28
6	19	30	25	20	17	8	20	47	41	34	30
6	20	32	27	22	18						

Table 66.5 (*Continued*)

n_1	n_2	$\alpha_1 = 5\%$ $\alpha_2 = 10\%$	$2\frac{1}{2}\%$ 5%	1% 2%	$\frac{1}{2}\%$ 1%	n_1	n_2	$\alpha_1 = 5\%$ $\alpha_2 = 10\%$	$2\frac{1}{2}\%$ 5%	1% 2%	$\frac{1}{2}\%$ 1%
9	9	21	17	14	11	11	16	54	47	41	36
9	10	24	20	16	13	11	17	57	51	44	39
9	11	27	23	18	16	11	18	61	55	47	42
9	12	30	26	21	18	11	19	65	58	50	45
9	13	33	28	23	20	11	20	69	62	53	48
9	14	36	31	26	22						
9	15	39	34	28	24	12	12	42	37	31	27
9	16	42	37	31	27	12	13	47	41	35	31
9	17	45	39	33	29	12	14	51	45	38	34
9	18	48	42	36	31	12	15	55	49	42	37
9	19	51	45	38	33	12	16	60	53	46	41
9	20	54	48	40	36	12	17	64	57	49	44
						12	18	68	61	53	47
10	10	27	23	19	16	12	19	72	65	56	51
10	11	31	26	22	18	12	20	77	69	60	54
10	12	34	29	24	21						
10	13	37	33	27	24	13	13	51	45	39	34
10	14	41	36	30	26	13	14	56	50	43	38
10	15	44	39	33	29	13	15	61	54	47	42
10	16	48	42	36	31	13	16	65	59	51	45
10	17	51	45	38	34	13	17	70	63	55	49
10	18	55	48	41	37	13	18	75	67	59	53
10	19	58	52	44	39	13	19	80	72	63	57
10	20	62	55	47	42	13	20	84	76	67	60
11	11	34	30	25	21	14	14	61	55	47	42
11	12	38	33	28	24	14	15	66	59	51	46
11	13	42	37	31	27	14	16	71	64	56	50
11	14	46	40	34	30	14	17	77	69	60	54
11	15	50	44	37	33	14	18	82	74	65	58

Table 66.5 (*Continued*)

n_1	n_2	$\alpha_1 = 5\%$ $\alpha_2 = 10\%$	$2\frac{1}{2}\%$ 5%	1% 2%	$\frac{1}{2}\%$ 1%	n_1	n_2	$\alpha_1 = 5\%$ $\alpha_2 = 10\%$	$2\frac{1}{2}\%$ 5%	1% 2%	$\frac{1}{2}\%$ 1%
14	19	87	78	69	63	17	17	96	87	77	70
14	20	92	83	73	67	17	18	102	92	82	75
						17	19	109	99	88	81
15	15	72	64	56	51	17	20	115	105	93	86
15	16	77	70	61	55						
15	17	83	75	66	60	18	18	109	99	88	81
15	18	88	80	70	64	18	19	116	106	94	87
15	19	94	85	75	69	18	20	123	112	100	92
15	20	100	90	80	73						
						19	19	123	112	101	93
16	16	83	75	66	60	19	20	130	119	107	99
16	17	89	81	71	65						
16	18	95	86	76	70	20	20	138	127	114	105
16	19	101	92	82	74						
16	20	107	98	87	79						

(cont. from page 723)

Using the above procedure:

(i) The hypotheses are:

H_0: Equal proportions of British and non-British cars have breakdowns.

H_1: A higher proportion of British cars have breakdowns.

(ii) Level of significance $\alpha_1 = 1\%$

(iii) Let the sizes of the samples be n_P and n_Q, where $n_P = 8$ and $n_Q = 10$. The Mann–Whitney test compares every item in sample P in turn with every item in sample Q, a record being kept of the number of times, say, that the item from P is greater than Q, or vice-versa. In this case there are $n_P n_Q$, i.e. $(8)(10) = 80$ comparisons to be made. All the data is arranged into ascending order whilst retaining their separate identities – an easy way is to arrange a linear scale as shown in Fig. 66.1, on page 728.

From Fig. 66.1, a list of Ps and Qs can be ranked giving:

P	P	Q	P	P	Q	P	Q	P	P	P	Q	Q	Q

Q	Q	Q	Q

(iv) Write under each letter P the number of Qs that precede it in the sequence, giving:

P	P	Q	P	P	Q	P	Q	P	P	P	Q
0	0		1	1		2		3	3	3	

Q	Q	Q	Q	Q	Q

(v) Add together these eight numbers, denoting the sum by U, i.e.

$$U = 0 + 0 + 1 + 1 + 2 + 3 + 3 + 3 = \mathbf{13}$$

Figure 66.1

(vi) The critical regions are of the form $U \le$ critical region.

From Table 66.5, for a sample size 8 and 10 at significance level $\alpha 1 = 1\%$ the critical regions is $U \le 13$

The value of U in our case, from (v), is 13 which is significant at 1% significance level.

The Mann–Whitney test has therefore confirmed that **there is evidence that the non-British cars have better reliability than the British cars in the first 10 000 miles, i.e. the alternative hypothesis applies.**

Problem 12. Two machines, A and B, are used to measure vibration in a particular rubber product. The data given below are the vibrational forces, in kilograms, of random samples from each machine:

A	9.7	10.2	11.2	12.4	14.1	22.3
	29.6	31.7	33.0	33.2	33.4	46.2
	50.7	52.5	55.4			
B	20.6	25.3	29.2	35.2	41.9	48.5
	54.1	57.1	59.8	63.2	68.5	

Use the Mann–Whitney test at a significance level of 5% to determine if there is any evidence of the two machines producing different results.

Using the procedure:

(i) H_0: There is no difference in results from the machines, on average.

H_1: The results from the two machines are different, on average.

(ii) $\alpha_2 = 5\%$

(iii) Arranging the data in order gives:

9.7	10.2	11.2	12.4	14.1	20.6	22.3
A	A	A	A	A	B	A

25.3	29.2	29.6	31.7	33.0	33.2	33.4
B	B	A	A	A	A	A

35.2	41.9	46.2	48.5	50.7	52.5	54.1
B	B	A	B	A	A	B

55.4	57.1	59.8	63.2	68.5
A	B	B	B	B

(iv) The number of Bs preceding the As in the sequence is as follows:

A	A	A	A	A	B	A	B	B
0	0	0	0	0		1		

A	A	A	A	A	B	B	A	B
3	3	3	3	3			5	

A	A	B	A	B	B	B	B
6	6		7				

(v) Adding the number from (iv) gives:

$$U = 0+0+0+0+0+1+3+3+3+3$$
$$+3+5+6+6+7 = \mathbf{40}$$

(vi) From Table 66.5, for $n_1 = 11$ and $n_2 = 15$, and $\alpha_2 = 5\%$, $U \le \mathbf{44}$

Since our value of U from (v) is less than 44, H_0 is rejected and H_1 accepted, i.e. **the results from the two machines are different**.

Now try the following Practice Exercise

Practice Exercise 247   The Mann–Whitney test (Answers on page 892)

1. The tar content of two brands of cigarettes (in mg) was measured as follows:

Brand P	22.6	4.1	3.9	0.7	3.2
Brand Q	3.4	6.2	3.5	4.7	6.3

Brand P	6.1	1.7	2.3	5.6	2.0
Brand Q	5.5	3.8	2.1		

Use the Mann–Whitney test at a 0.05 level of significance to determine if the tar contents of the two brands are equal.

2. A component is manufactured by two processes. Some components from each process are selected at random and tested for breaking strength to determine if there is a difference between the processes. The results are:

Process A	9.7	10.5	10.1	11.6	9.8
Process B	11.3	8.6	9.6	10.2	10.9

Process A	8.9	11.2	12.0	9.2
Process B	9.4	10.8		

At a level of significance of 10%, use the Mann–Whitney test to determine if there is a difference between the mean breaking strengths of the components manufactured by the two processes.

3. An experiment, designed to compare two preventive methods against corrosion, gave the following results for the maximum depths of pits (in mm) in metal strands:

Method A	143	106	135	147	139	132	153	140
Method B	98	105	137	94	112	103		

Use the Mann–Whitney test, at a level of significance of 0.05, to determine whether the two tests are equally effective.

4. Repeat Problem 3 of Exercise 245, page 719 using the Mann–Whitney test.

For fully worked solutions to each of the problems in Practice Exercises 243 to 247 in this chapter, go to the website:
www.routledge.com/cw/bird

This Revision Test covers the material contained in Chapters 64 to 66. *The marks for each question are shown in brackets at the end of each question.*

1. 1200 metal bolts have a mean mass of 7.2 g and a standard deviation of 0.3 g. Determine the standard error of the means. Calculate also the probability that a sample of 60 bolts chosen at random, without replacement, will have a mass of (a) between 7.1 g and 7.25 g, and (b) more than 7.3 g.   (12)

2. A sample of ten measurements of the length of a component are made and the mean of the sample is 3.650 cm. The standard deviation of the samples is 0.030 cm. Determine (a) the 99% confidence limits, and (b) the 90% confidence limits for an estimate of the actual length of the component.   (10)

3. An automated machine produces metal screws and over a period of time it is found that eight% are defective. Random samples of 75 screws are drawn periodically.

    (a) If a decision is made that production continues until a sample contains more than eight defective screws, determine the type I error based on this decision for a defect rate of 8%.

    (b) Determine the magnitude of the type II error when the defect rate has risen to 12%

    The above sample size is now reduced to 55 screws. The decision now is to stop the machine for adjustment if a sample contains four or more defective screws.

    (c) Determine the type I error if the defect rate remains at 8%

    (d) Determine the type II error when the defect rate rises to 9%   (22)

4. In a random sample of 40 similar light bulbs drawn from a batch of 400 the mean lifetime is found to be 252 hours. The standard deviation of the lifetime of the sample is 25 hours. The batch is classed as inferior if the mean lifetime of the batch is less than the population mean of 260 hours. As a result of the sample data, determine whether the batch is considered to be inferior at a level of significance of (a) 0.05 and (b) 0.01   (9)

5. The lengths of two products are being compared.

    Product 1: sample size $= 50$, mean value of sample $= 6.5$ cm, standard deviation of whole of batch $= 0.40$ cm.

    Product 2: sample size $= 60$, mean value of sample $= 6.65$ cm, standard deviation of whole of batch $= 0.35$ cm.

    Determine if there is any significant difference between the two products at a level of significance of (a) 0.05 and (b) 0.01   (7)

6. The resistance of a sample of 400 resistors produced by an automatic process have the following resistance distribution.

Resistance (Ω)	Frequency
50.11	9
50.15	35
50.19	61
50.23	102
50.27	89
50.31	83
50.35	21

    Calculate for the sample: (a) the mean and (b) the standard deviation. (c) Test the null hypothesis that the resistances of the resistors are normally distributed at a level of significance of 0.05, and determine if the distribution gives a 'too good' fit at a level of confidence of 90%   (25)

7. A fishing line is manufactured by two processes, A and B. To determine if there is a difference in the mean breaking strengths of the lines, eight lines by each process are selected and tested for breaking strength. The results are as follows:

Process A	8.6	7.1	6.9	6.5	7.9	6.3	7.8	8.1
Process B	6.8	7.6	8.2	6.2	7.5	8.9	8.0	8.7

Determine if there is a difference between the mean breaking strengths of the lines manufactured by the two processes, at a significance level of 0.10, using (a) the sign test, (b) the Wilcoxon signed-rank test, (c) the Mann–Whitney test.   (15)

Section L

Laplace transforms

Chapter 67

Introduction to Laplace transforms

Why it is important to understand: **Introduction to Laplace transforms**

The Laplace transform is a very powerful mathematical tool applied in various areas of engineering and science. With the increasing complexity of engineering problems, Laplace transforms help in solving complex problems with a very simple approach; the transform is an integral transform method which is particularly useful in solving linear ordinary differential equations. It has very wide applications in various areas of physics, electrical engineering, control engineering, optics, mathematics and signal processing. This chapter just gets us started in understanding some standard Laplace transforms.

At the end of this chapter, you should be able to:

- define a Laplace transform
- recognise common notations used for the Laplace transform
- derive Laplace transforms of elementary functions
- use a standard list of Laplace transforms to determine the transform of common functions

67.1 Introduction

The solution of most electrical circuit problems can be reduced ultimately to the solution of differential equations. The use of **Laplace*** **transforms** provides an alternative method to those discussed in Chapters 33 and 50 to 54 for solving linear differential equations.

67.2 Definition of a Laplace transform

The Laplace transform of the function $f(t)$ is defined by the integral $\int_0^\infty e^{-st} f(t) \, dt$, where s is a parameter assumed to be a real number.

Common notations used for the Laplace transform

There are various commonly used notations for the Laplace transform of $f(t)$ and these include:

(i) $\mathcal{L}\{f(t)\}$ or $L\{f(t)\}$

(ii) $\mathcal{L}(f)$ or Lf

(iii) $\overline{f}(s)$ or $f(s)$

Also, the letter p is sometimes used instead of s as the parameter. The notation adopted in this book will be $f(t)$ for the original function and $\mathcal{L}\{f(t)\}$ for its Laplace transform.

Hence, from above:

$$\mathcal{L}\{f(t)\} = \int_0^\infty e^{-st} f(t) \, dt \qquad (1)$$

67.3 Linearity property of the Laplace transform

From equation (1),

$$\mathcal{L}\{kf(t)\} = \int_0^\infty e^{-st} k \, f(t) \, dt$$

$$= k \int_0^\infty e^{-st} f(t) \, dt$$

---

* **Who was Laplace**? See page 605 for image and resume of Pierre-Simon, marquis de Laplace. To find out more go to **www.routledge.com/cw/bird**

i.e. $\mathcal{L}\{k \, f(t)\} = k\mathcal{L}\{f(t)\} \qquad (2)$

where k is any constant.

Similarly,

$$\mathcal{L}\{a \, f(t) + bg(t)\} = \int_0^\infty e^{-st} (a \, f(t) + bg(t)) \, dt$$

$$= a \int_0^\infty e^{-st} f(t) \, dt$$

$$+ b \int_0^\infty e^{-st} g(t) \, dt$$

i.e. $\mathcal{L}\{af(t) + bg(t)\} = a\mathcal{L}\{f(t)\} + b\mathcal{L}\{g(t)\}, \qquad (3)$

where a and b are any real constants.
The Laplace transform is termed a **linear operator** because of the properties shown in equations (2) and (3).

67.4 Laplace transforms of elementary functions

Using the definition of the Laplace transform in equation (1) a number of elementary functions may be transformed. For example:

(a) $f(t) = 1$. From equation (1),

$$\mathcal{L}\{1\} = \int_0^\infty e^{-st} (1) \, dt = \left[\frac{e^{-st}}{-s} \right]_0^\infty$$

$$= -\frac{1}{s} [e^{-s(\infty)} - e^0] = -\frac{1}{s} [0 - 1]$$

$$= \frac{1}{s} \text{ (provided } s > 0)$$

(b) $f(t) = k$. From equation (2),

$$\mathcal{L}\{k\} = k\mathcal{L}\{1\}$$

Hence $\mathcal{L}\{k\} = k\left(\dfrac{1}{s}\right) = \dfrac{k}{s}$, from (a) above.

(c) $f(t) = e^{at}$ (where a is a real constant $\neq 0$).
From equation (1),

$$\mathcal{L}\{e^{at}\} = \int_0^\infty e^{-st} (e^{at}) \, dt = \int_0^\infty e^{-(s-a)t} \, dt,$$

from the laws of indices,

$$= \left[\frac{e^{-(s-a)t}}{-(s-a)}\right]_0^{\infty} = \frac{1}{-(s-a)}(0-1)$$

$$= \frac{1}{s-a}$$

(provided $(s-a) > 0$, i.e. $s > a$)

(d) $f(t) = \cos at$ (where a is a real constant).
From equation (1),

$$\mathcal{L}\{\cos at\} = \int_0^{\infty} e^{-st} \cos at \, dt$$

$$= \left[\frac{e^{-st}}{s^2+a^2}(a\sin at - s\cos at)\right]_0^{\infty}$$

by integration by parts twice (see page 500),

$$= \left[\frac{e^{-s(\infty)}}{s^2+a^2}(a\sin a(\infty) - s\cos a(\infty))\right.$$

$$\left. - \frac{e^0}{s^2+a^2}(a\sin 0 - s\cos 0)\right]$$

$$= \frac{s}{s^2+a^2} \text{ (provided } s > 0)$$

(e) $f(t) = t$. From equation (1),

$$\mathcal{L}\{t\} = \int_0^{\infty} e^{-st} t \, dt = \left[\frac{te^{-st}}{-s} - \int \frac{e^{-st}}{-s} \, dt\right]_0^{\infty}$$

$$= \left[\frac{te^{-st}}{-s} - \frac{e^{-st}}{s^2}\right]_0^{\infty}$$

by integration by parts

$$= \left[\frac{\infty e^{-s(\infty)}}{-s} - \frac{e^{-s(\infty)}}{s^2}\right] - \left[0 - \frac{e^0}{s^2}\right]$$

$$= (0-0) - \left(0 - \frac{1}{s^2}\right)$$

since $(\infty \times 0) = 0$

$$= \frac{1}{s^2} \text{ (provided } s > 0)$$

(f) $f(t) = t^n$ (where $n = 0, 1, 2, 3, \ldots$).
By a similar method to (e) it may be shown that $\mathcal{L}\{t^2\} = \frac{2}{s^3}$ and $\mathcal{L}\{t^3\} = \frac{(3)(2)}{s^4} = \frac{3!}{s^4}$. These results can be extended to n being any positive integer.
Thus $\mathcal{L}\{t^n\} = \dfrac{n!}{s^{n+1}}$ provided $s > 0$)

(g) $f(t) = \sinh at$. From Chapter 16,
$\sinh at = \dfrac{1}{2}(e^{at} - e^{-at})$. Hence,

$$\mathcal{L}\{\sinh at\} = \mathcal{L}\left\{\frac{1}{2}e^{at} - \frac{1}{2}e^{-at}\right\}$$

$$= \frac{1}{2}\mathcal{L}\{e^{at}\} - \frac{1}{2}\mathcal{L}\{e^{-at}\}$$

from equations (2) and (3)

$$= \frac{1}{2}\left[\frac{1}{s-a}\right] - \frac{1}{2}\left[\frac{1}{s+a}\right]$$

from (c) above

$$= \frac{1}{2}\left[\frac{1}{s-a} - \frac{1}{s+a}\right]$$

$$= \frac{a}{s^2-a^2} \text{ (provided } s > a)$$

A list of elementary standard Laplace transforms are summarised in Table 67.1.

Table 67.1 Elementary standard Laplace transforms

	Function $f(t)$	Laplace transforms $\mathcal{L}\{f(t)\} = \int_0^{\infty} e^{-st} f(t)\,dt$
(i)	1	$\dfrac{1}{s}$
(ii)	k	$\dfrac{k}{s}$
(iii)	e^{at}	$\dfrac{1}{s-a}$
(iv)	$\sin at$	$\dfrac{a}{s^2+a^2}$
(v)	$\cos at$	$\dfrac{s}{s^2+a^2}$
(vi)	t	$\dfrac{1}{s^2}$
(vii)	t^2	$\dfrac{2!}{s^3}$
(viii)	t^n $(n = 1, 2, 3, \ldots)$	$\dfrac{n!}{s^{n+1}}$
(ix)	$\cosh at$	$\dfrac{s}{s^2-a^2}$
(x)	$\sinh at$	$\dfrac{a}{s^2-a^2}$

67.5 Worked problems on standard Laplace transforms

Problem 1. Using a standard list of Laplace transforms, determine the following:

(a) $\mathcal{L}\left\{1+2t-\dfrac{1}{3}t^4\right\}$ (b) $\mathcal{L}\{5e^{2t}-3e^{-t}\}$

(a) $\mathcal{L}\left\{1+2t-\dfrac{1}{3}t^4\right\}$

$$= \mathcal{L}\{1\} + 2\mathcal{L}\{t\} - \frac{1}{3}\mathcal{L}\{t^4\}$$

from equations (2) and (3)

$$= \frac{1}{s} + 2\left(\frac{1}{s^2}\right) - \frac{1}{3}\left(\frac{4!}{s^{4+1}}\right)$$

from (i), (vi) and (viii) of Table 67.1

$$= \frac{1}{s} + \frac{2}{s^2} - \frac{1}{3}\left(\frac{4.3.2.1}{s^5}\right)$$

$$= \frac{1}{s} + \frac{2}{s^2} - \frac{8}{s^5}$$

(b) $\mathcal{L}\{5e^{2t}-3e^{-t}\} = 5\mathcal{L}(e^{2t}) - 3\mathcal{L}\{e^{-t}\}$

from equations (2) and (3)

$$= 5\left(\frac{1}{s-2}\right) - 3\left(\frac{1}{s-(-1)}\right)$$

from (iii) of Table 67.1

$$= \frac{5}{s-2} - \frac{3}{s+1}$$

$$= \frac{5(s+1) - 3(s-2)}{(s-2)(s+1)}$$

$$= \frac{2s+11}{s^2-s-2}$$

Problem 2. Find the Laplace transforms of:
(a) $6\sin 3t - 4\cos 5t$ (b) $2\cosh 2\theta - \sinh 3\theta$

(a) $\mathcal{L}\{6\sin 3t - 4\cos 5t\}$

$$= 6\mathcal{L}\{\sin 3t\} - 4\mathcal{L}\{\cos 5t\}$$

$$= 6\left(\frac{3}{s^2+3^2}\right) - 4\left(\frac{s}{s^2+5^2}\right)$$

from (iv) and (v) of Table 67.1

$$= \frac{18}{s^2+9} - \frac{4s}{s^2+25}$$

(b) $\mathcal{L}\{2\cosh 2\theta - \sinh 3\theta\}$

$$= 2\mathcal{L}\{\cosh 2\theta\} - \mathcal{L}\{\sinh 3\theta\}$$

$$= 2\left(\frac{s}{s^2-2^2}\right) - \left(\frac{3}{s^2-3^2}\right)$$

from (ix) and (x) of Table 67.1

$$= \frac{2s}{s^2-4} - \frac{3}{s^2-9}$$

Problem 3. Prove that
(a) $\mathcal{L}\{\sin at\} = \dfrac{a}{s^2+a^2}$ (b) $\mathcal{L}\{t^2\} = \dfrac{2}{s^3}$

(c) $\mathcal{L}\{\cosh at\} = \dfrac{s}{s^2-a^2}$

(a) From equation (1),

$$\mathcal{L}\{\sin at\} = \int_0^\infty e^{-st} \sin at \, dt$$

$$= \left[\frac{e^{-st}}{s^2+a^2}(-s\sin at - a\cos at)\right]_0^\infty$$

by integration by parts

$$= \frac{1}{s^2+a^2}[e^{-s(\infty)}(-s\sin a(\infty)$$

$$- a\cos a(\infty)) - e^0(-s\sin 0$$

$$- a\cos 0)]$$

$$= \frac{1}{s^2+a^2}[(0) - 1(0-a)]$$

$$= \frac{a}{s^2+a^2} \quad \text{(provided } s > 0)$$

(b) From equation (1),

$$\mathcal{L}\{t^2\} = \int_0^\infty e^{-st} t^2 \, dt$$

$$= \left[\frac{t^2 e^{-st}}{-s} - \frac{2t e^{-st}}{s^2} - \frac{2e^{-st}}{s^3}\right]_0^\infty$$

by integration by parts twice

$$= \left[(0 - 0 - 0) - \left(0 - 0 - \frac{2}{s^3} \right) \right]$$

$$= \frac{2}{s^3} \text{ (provided } s > 0)$$

(c) From equation (1),

$$\mathcal{L}\{\cosh at\} = \mathcal{L}\left\{ \frac{1}{2}(e^{at} + e^{-at}) \right\}$$

from Chapter 16

$$= \frac{1}{2}\mathcal{L}\{e^{at}\} + \frac{1}{2}\mathcal{L}\{e^{-at}\} \text{ from}$$

equations (2) and (3)

$$= \frac{1}{2}\left(\frac{1}{s-a} \right) + \frac{1}{2}\left(\frac{1}{s-(-a)} \right)$$

from (iii) of Table 67.1

$$= \frac{1}{2}\left[\frac{1}{s-a} + \frac{1}{s+a} \right]$$

$$= \frac{1}{2}\left[\frac{(s+a)+(s-a)}{(s-a)(s+a)} \right]$$

$$= \frac{s}{s^2 - a^2} \text{ (provided } s > a)$$

Problem 4. Determine the Laplace transforms of:
(a) $\sin^2 t$ (b) $\cosh^2 3x$

(a) Since $\cos 2t = 1 - 2\sin^2 t$ then
$$\sin^2 t = \frac{1}{2}(1 - \cos 2t). \text{ Hence,}$$

$$\mathcal{L}\{\sin^2 t\} = \mathcal{L}\left\{ \frac{1}{2}(1 - \cos 2t) \right\}$$

$$= \frac{1}{2}\mathcal{L}\{1\} - \frac{1}{2}\mathcal{L}\{\cos 2t\}$$

$$= \frac{1}{2}\left(\frac{1}{s} \right) - \frac{1}{2}\left(\frac{s}{s^2 + 2^2} \right)$$

from (i) and (v) of Table 67.1

$$= \frac{(s^2 + 4) - s^2}{2s(s^2 + 4)} = \frac{4}{2s(s^2 + 4)}$$

$$= \frac{2}{s(s^2 + 4)}$$

(b) Since $\cosh 2x = 2\cosh^2 x - 1$ then
$$\cosh^2 x = \frac{1}{2}(1 + \cosh 2x) \text{ from Chapter 16.}$$

Hence $\cosh^2 3x = \frac{1}{2}(1 + \cosh 6x)$

Thus $\mathcal{L}\{\cosh^2 3x\} = \mathcal{L}\left\{ \frac{1}{2}(1 + \cosh 6x) \right\}$

$$= \frac{1}{2}\mathcal{L}\{1\} + \frac{1}{2}\mathcal{L}\{\cosh 6x\}$$

$$= \frac{1}{2}\left(\frac{1}{s} \right) + \frac{1}{2}\left(\frac{s}{s^2 - 6^2} \right)$$

$$= \frac{2s^2 - 36}{2s(s^2 - 36)} = \frac{s^2 - 18}{s(s^2 - 36)}$$

Problem 5. Find the Laplace transform of $3\sin(\omega t + \alpha)$, where ω and α are constants.

Using the compound angle formula for $\sin(A + B)$, from Chapter 19, $\sin(\omega t + \alpha)$ may be expanded to $(\sin \omega t \cos \alpha + \cos \omega t \sin \alpha)$. Hence,

$$\mathcal{L}\{3\sin(\omega t + \alpha)\}$$

$$= \mathcal{L}\{3(\sin \omega t \cos \alpha + \cos \omega t \sin \alpha)\}$$

$$= 3\cos \alpha \mathcal{L}\{\sin \omega t\} + 3\sin \alpha \mathcal{L}\{\cos \omega t\},$$

since α is a constant

$$= 3\cos \alpha \left(\frac{\omega}{s^2 + \omega^2} \right) + 3\sin \alpha \left(\frac{s}{s^2 + \omega^2} \right)$$

from (iv) and (v) of Table 67.1

$$= \frac{3}{(s^2 + \omega^2)}(\omega \cos \alpha + s \sin \alpha)$$

Now try the following Practice Exercise

Practice Exercise 248   Introduction to Laplace transforms (Answers on page 892)

Determine the Laplace transforms in Problems 1 to 9.

1. (a) $2t - 3$ (b) $5t^2 + 4t - 3$

2. (a) $\dfrac{t^3}{24} - 3t + 2$ (b) $\dfrac{t^5}{15} - 2t^4 + \dfrac{t^2}{2}$

3. (a) $5e^{3t}$ (b) $2e^{-2t}$

4. (a) $4\sin 3t$ (b) $3\cos 2t$

5. (a) $7\cosh 2x$ (b) $\dfrac{1}{3}\sinh 3t$

6. (a) $2\cos^2 t$ (b) $3\sin^2 2x$

7. (a) $\cosh^2 t$ (b) $2\sinh^2 2\theta$

8. $4\sin(at+b)$, where a and b are constants.

9. $3\cos(\omega t-\alpha)$, where ω and α are constants.

10. Show that $\mathcal{L}(\cos^2 3t - \sin^2 3t) = \dfrac{s}{s^2+36}$

For fully worked solutions to each of the problems in Practice Exercise 248 in this chapter, go to the website:
www.routledge.com/cw/bird

Properties of Laplace transforms

At the end of this chapter, you should be able to:

- derive the Laplace transform of $e^{at} f(t)$
- use a standard list of Laplace transforms to determine transforms of the form $e^{at} f(t)$
- derive the Laplace transforms of derivatives
- state and use the initial and final value theorems

68.1 The Laplace transform of $e^{at} f(t)$

From Chapter 67, the definition of the Laplace transform of $f(t)$ is:

$$\mathcal{L}\{f(t)\} = \int_0^\infty e^{-st} f(t)\, dt \qquad (1)$$

Thus $\mathcal{L}\{e^{at} f(t)\} = \int_0^\infty e^{-st}(e^{at} f(t))\, dt$

$$= \int_0^\infty e^{-(s-a)} f(t)\, dt \qquad (2)$$

(where a is a real constant)

Hence the substitution of $(s-a)$ for s in the transform shown in equation (1) corresponds to the multiplication of the original function $f(t)$ by e^{at}. This is known as a shift theorem.

68.2 Laplace transforms of the form $e^{at} f(t)$

From equation (2), Laplace transforms of the form $e^{at} f(t)$ may be deduced. For example:

(i) $\mathcal{L}\{e^{at} t^n\}$

Since $\mathcal{L}\{t^n\} = \dfrac{n!}{s^{n+1}}$ from (viii) of Table 67.1, page 737.

then $\mathcal{L}\{e^{at}\ t^n\} = \dfrac{n!}{(s-a)^{n+1}}$ from equation (2) above (provided $s > a$)

(ii) $\mathcal{L}\{e^{at}\sin\omega t\}$

Since $\mathcal{L}\{\sin\omega t\} = \dfrac{\omega}{s^2+\omega^2}$ from (iv) of Table 67.1, page 737.

then $\mathcal{L}\{e^{at}\sin\omega t\} = \dfrac{\omega}{(s-a)^2+\omega^2}$ from equation (2) (provided $s > a$)

(iii) $\mathcal{L}\{e^{at}\cosh\omega t\}$

Since $\mathcal{L}\{\cosh\omega t\} = \dfrac{s}{s^2-\omega^2}$ from (ix) of Table 67.1, page 737.

then $\mathcal{L}\{e^{at}\cosh\omega t\} = \dfrac{s-a}{(s-a)^2-\omega^2}$ from equation (2) (provided $s > a$)

A summary of Laplace transforms of the form $e^{at}f(t)$ is shown in Table 68.1.

Problem 1. Determine (a) $\mathcal{L}\{2t^4 e^{3t}\}$ (b) $\mathcal{L}\{4e^{3t}\cos 5t\}$

(a) From (i) of Table 68.1,

$$\mathcal{L}\{2t^4 e^{3t}\} = 2\mathcal{L}\{t^4 e^{3t}\} = 2\left(\dfrac{4!}{(s-3)^{4+1}}\right)$$

$$= \dfrac{2(4)(3)(2)}{(s-3)^5} = \dfrac{48}{(s-3)^5}$$

Table 68.1 Laplace transforms of the form $e^{at}f(t)$

Function $e^{at}f(t)$ (a is a real constant)	Laplace transform $\mathcal{L}\{e^{at}f(t)\}$
(i) $e^{at}t^n$	$\dfrac{n!}{(s-a)^{n+1}}$
(ii) $e^{at}\sin\omega t$	$\dfrac{\omega}{(s-a)^2+\omega^2}$
(iii) $e^{at}\cos\omega t$	$\dfrac{s-a}{(s-a)^2+\omega^2}$
(iv) $e^{at}\sinh\omega t$	$\dfrac{\omega}{(s-a)^2-\omega^2}$
(v) $e^{at}\cosh\omega t$	$\dfrac{s-a}{(s-a)^2-\omega^2}$

(b) From (iii) of Table 68.1,

$$\mathcal{L}\{4e^{3t}\cos 5t\} = 4\mathcal{L}\{e^{3t}\cos 5t\}$$

$$= 4\left(\dfrac{s-3}{(s-3)^2+5^2}\right)$$

$$= \dfrac{4(s-3)}{s^2-6s+9+25}$$

$$= \dfrac{4(s-3)}{s^2-6s+34}$$

Problem 2. Determine (a) $\mathcal{L}\{e^{-2t}\sin 3t\}$ (b) $\mathcal{L}\{3e^\theta\cosh 4\theta\}$

(a) From (ii) of Table 68.1,

$$\mathcal{L}\{e^{-2t}\sin 3t\} = \dfrac{3}{(s-(-2))^2+3^2} = \dfrac{3}{(s+2)^2+9}$$

$$= \dfrac{3}{s^2+4s+4+9} = \dfrac{3}{s^2+4s+13}$$

(b) From (v) of Table 68.1,

$$\mathcal{L}\{3e^\theta\cosh 4\theta\} = 3\mathcal{L}\{e^\theta\cosh 4\theta\} = \dfrac{3(s-1)}{(s-1)^2-4^2}$$

$$= \dfrac{3(s-1)}{s^2-2s+1-16} = \dfrac{3(s-1)}{s^2-2s-15}$$

Problem 3. Determine the Laplace transforms of (a) $5e^{-3t}\sinh 2t$ (b) $2e^{3t}(4\cos 2t - 5\sin 2t)$

(a) From (iv) of Table 68.1,

$$\mathcal{L}\{5e^{-3t}\sinh 2t\} = 5\mathcal{L}\{e^{-3t}\sinh 2t\}$$

$$= 5\left(\dfrac{2}{(s-(-3))^2-2^2}\right)$$

$$= \dfrac{10}{(s+3)^2-2^2} = \dfrac{10}{s^2+6s+9-4}$$

$$= \dfrac{10}{s^2+6s+5}$$

(b) $\mathcal{L}\{2e^{3t}(4\cos 2t - 5\sin 2t)\}$

$$= 8\mathcal{L}\{e^{3t}\cos 2t\} - 10\mathcal{L}\{e^{3t}\sin 2t\}$$

$$= \frac{8(s-3)}{(s-3)^2+2^2} - \frac{10(2)}{(s-3)^2+2^2}$$

from (iii) and (ii) of Table 68.1

$$= \frac{8(s-3)-10(2)}{(s-3)^2+2^2} = \frac{8s-44}{s^2-6s+13}$$

Problem 4. Show that

$$\mathcal{L}\left\{3e^{-\frac{1}{2}x}\sin^2 x\right\} = \frac{48}{(2s+1)(4s^2+4s+17)}$$

Since $\cos 2x = 1-2\sin^2 x$, $\sin^2 x = \frac{1}{2}(1-\cos 2x)$

Hence,

$$\mathcal{L}\left\{3e^{-\frac{1}{2}x}\sin^2 x\right\}$$

$$= \mathcal{L}\left\{3e^{-\frac{1}{2}x}\frac{1}{2}(1-\cos 2x)\right\}$$

$$= \frac{3}{2}\mathcal{L}\left\{e^{-\frac{1}{2}x}\right\} - \frac{3}{2}\mathcal{L}\left\{e^{-\frac{1}{2}x}\cos 2x\right\}$$

$$= \frac{3}{2}\left(\frac{1}{s-\left(-\frac{1}{2}\right)}\right) - \frac{3}{2}\left(\frac{\left(s-\left(-\frac{1}{2}\right)\right)}{\left(s-\left(-\frac{1}{2}\right)\right)^2+2^2}\right)$$

from (iii) of Table 67.1 (page 737) and (iii) of Table 68.1 above,

$$= \frac{3}{2\left(s+\frac{1}{2}\right)} - \frac{3\left(s+\frac{1}{2}\right)}{2\left[\left(s+\frac{1}{2}\right)^2+2^2\right]}$$

$$= \frac{3}{2s+1} - \frac{6s+3}{4\left(s^2+s+\frac{1}{4}+4\right)}$$

$$= \frac{3}{2s+1} - \frac{6s+3}{4s^2+4s+17}$$

$$= \frac{3(4s^2+4s+17)-(6s+3)(2s+1)}{(2s+1)(4s^2+4s+17)}$$

$$= \frac{12s^2+12s+51-12s^2-6s-6s-3}{(2s+1)(4s^2+4s+17)}$$

$$= \frac{48}{(2s+1)(4s^2+4s+17)}$$

Now try the following Practice Exercise

Practice Exercise 249 Laplace transforms of the form $e^{at}f(t)$ (Answers on page 893)

Determine the Laplace transforms of the following functions:

1. (a) $2te^{2t}$ (b) t^2e^t

2. (a) $4t^3e^{-2t}$ (b) $\frac{1}{2}t^4e^{-3t}$

3. (a) $e^t\cos t$ (b) $3e^{2t}\sin 2t$

4. (a) $5e^{-2t}\cos 3t$ (b) $4e^{-5t}\sin t$

5. (a) $2e^t\sin^2 t$ (b) $\frac{1}{2}e^{3t}\cos^2 t$

6. (a) $e^t\sinh t$ (b) $3e^{2t}\cosh 4t$

7. (a) $2e^{-t}\sinh 3t$ (b) $\frac{1}{4}e^{-3t}\cosh 2t$

8. (a) $2e^t(\cos 3t - 3\sin 3t)$

   (b) $3e^{-2t}(\sinh 2t - 2\cosh 2t)$

68.3 The Laplace transforms of derivatives

(a) First derivative

Let the first derivative of $f(t)$ be $f'(t)$ then, from equation (1),

$$\mathcal{L}\{f'(t)\} = \int_0^\infty e^{-st}f'(t)\,dt$$

From Chapter 46, when integrating by parts

$$\int u\frac{dv}{dt}\,dt = uv - \int v\frac{du}{dt}\,dt$$

When evaluating $\int_0^\infty e^{-st}f'(t)\,dt$,

let $u = e^{-st}$ and $\frac{dv}{dt} = f'(t)$

from which,

$$\frac{du}{dt} = -se^{-st} \text{ and } v = \int f'(t)\,dt = f(t)$$

Section L

Hence $\displaystyle\int_0^\infty e^{-st} f'(t)\,dt$

$$= \left[e^{-st} f(t) \right]_0^\infty - \int_0^\infty f(t)(-se^{-st})\,dt$$

$$= [0 - f(0)] + s \int_0^\infty e^{-st} f(t)\,dt$$

$$= -f(0) + s\mathcal{L}\{f(t)\}$$

assuming $e^{-st} f(t) \to 0$ as $t \to \infty$, and $f(0)$ is the value of $f(t)$ at $t = 0$. Hence,

$$\left. \begin{array}{c} \mathcal{L}\{f'(t)\} = s\mathcal{L}\{f(t)\} - f(0) \\[2mm] \text{or}\quad \mathcal{L}\left\{\dfrac{dy}{dx}\right\} = s\mathcal{L}\{y\} - y(0) \end{array} \right\} \tag{3}$$

where $y(0)$ is the value of y at $x = 0$

(b) Second derivative

Let the second derivative of $f(t)$ be $f''(t)$, then from equation (1),

$$\mathcal{L}\{f''(t)\} = \int_0^\infty e^{-st} f''(t)\,dt$$

Integrating by parts gives:

$$\int_0^\infty e^{-st} f''(t)\,dt = \left[e^{-st} f'(t) \right]_0^\infty + s\int_0^\infty e^{-st} f'(t)\,dt$$

$$= [0 - f'(0)] + s\mathcal{L}\{f'(t)\}$$

assuming $e^{-st} f'(t) \to 0$ as $t \to \infty$, and $f'(0)$ is the value of $f'(t)$ at $t = 0$. Hence $\{f''(t)\} = -f'(0) + s[s(f(t)) - f(0)]$, from equation (3),

i.e.

$$\left. \begin{array}{c} \mathcal{L}\{f''(t)\} \\[1mm] = s^2\mathcal{L}\{f(t)\} - sf(0) - f'(0) \\[3mm] \text{or}\quad \mathcal{L}\left\{\dfrac{d^2y}{dx^2}\right\} \\[3mm] = s^2\mathcal{L}\{y\} - sy(0) - y'(0) \end{array} \right\} \tag{4}$$

where $y'(0)$ is the value of $\dfrac{dy}{dx}$ at $x = 0$

Equations (3) and (4) are important and are used in the solution of differential equations (see Chapter 71) and simultaneous differential equations (Chapter 72).

Problem 5. Use the Laplace transform of the first derivative to derive:

(a) $\mathcal{L}\{k\} = \dfrac{k}{s}$ (b) $\mathcal{L}\{2t\} = \dfrac{2}{s^2}$

(c) $\mathcal{L}\{e^{-at}\} = \dfrac{1}{s+a}$

From equation (3), $\mathcal{L}\{f'(t)\} = s\mathcal{L}\{f(t)\} - f(0)$

(a) Let $f(t) = k$, then $f'(t) = 0$ and $f(0) = k$

Substituting into equation (3) gives:

$$\mathcal{L}\{0\} = s\mathcal{L}\{k\} - k$$

i.e. $k = s\mathcal{L}\{k\}$

Hence $\mathcal{L}\{k\} = \dfrac{k}{s}$

(b) Let $f(t) = 2t$ then $f'(t) = 2$ and $f(0) = 0$

Substituting into equation (3) gives:

$$\mathcal{L}\{2\} = s\mathcal{L}\{2t\} - 0$$

i.e. $\dfrac{2}{s} = s\mathcal{L}\{2t\}$

Hence $\mathcal{L}\{2t\} = \dfrac{2}{s^2}$

(c) Let $f(t) = e^{-at}$ then $f'(t) = -ae^{-at}$ and $f(0) = 1$

Substituting into equation (3) gives:

$$\mathcal{L}\{-ae^{-at}\} = s\mathcal{L}\{e^{-at}\} - 1$$

$$-a\mathcal{L}\{e^{-at}\} = s\mathcal{L}\{e^{-at}\} - 1$$

$$1 = s\mathcal{L}\{e^{-at}\} + a\mathcal{L}\{e^{-at}\}$$

$$1 = (s+a)\mathcal{L}\{e^{-at}\}$$

Hence $\mathcal{L}\{e^{-at}\} = \dfrac{1}{s+a}$

Problem 6. Use the Laplace transform of the second derivative to derive

$$\mathcal{L}\{\cos at\} = \dfrac{s}{s^2 + a^2}$$

From equation (4),

$$\mathcal{L}\{f''(t)\} = s^2\mathcal{L}\{f(t)\} - sf(0) - f'(0)$$

Let $f(t) = \cos at$, then $f'(t) = -a\sin at$ and $f''(t) = -a^2\cos at$, $f(0) = 1$ and $f'(0) = 0$

Substituting into equation (4) gives:

$$\mathcal{L}\{-a^2\cos at\} = s^2\{\cos at\} - s(1) - 0$$

i.e. $-a^2\mathcal{L}\{\cos at\} = s^2\mathcal{L}\{\cos at\} - s$

Hence $s = (s^2 + a^2)\mathcal{L}\{\cos at\}$

from which, $\mathcal{L}\{\cos at\} = \dfrac{s}{s^2+a^2}$

Now try the following Practice Exercise

Practice Exercise 250   Laplace transforms of derivatives (Answers on page 893)

1. Derive the Laplace transform of the first derivative from the definition of a Laplace transform. Hence derive the transform

$$\mathcal{L}\{1\} = \dfrac{1}{s}$$

2. Use the Laplace transform of the first derivative to derive the transforms:

   (a) $\mathcal{L}\{e^{at}\} = \dfrac{1}{s-a}$ (b) $\mathcal{L}\{3t^2\} = \dfrac{6}{s^3}$

3. Derive the Laplace transform of the second derivative from the definition of a Laplace transform. Hence derive the transform

$$\mathcal{L}\{\sin at\} = \dfrac{a}{s^2+a^2}$$

4. Use the Laplace transform of the second derivative to derive the transforms:

   (a) $\mathcal{L}\{\sinh at\} = \dfrac{a}{s^2 - a^2}$

   (b) $\mathcal{L}\{\cosh at\} = \dfrac{s}{s^2 - a^2}$

68.4 The initial and final value theorems

There are several Laplace transform theorems used to simplify and interpret the solution of certain problems. Two such theorems are the initial value theorem and the final value theorem.

(a) The initial value theorem states:

$$\lim_{t \to 0}\ [f(t)] = \lim_{s \to \infty}\ [s\mathcal{L}\{f(t)\}]$$

For example, if $f(t) = 3e^{4t}$ then

$$\mathcal{L}\{3e^{4t}\} = \dfrac{3}{s-4}$$

from (iii) of Table 67.1, page 737.

By the initial value theorem,

$$\lim_{t \to 0}[3e^{4t}] = \lim_{s \to \infty}\left[s\left(\dfrac{3}{s-4}\right)\right]$$

i.e. $3e^0 = \infty\left(\dfrac{3}{\infty - 4}\right)$

i.e.   **3 = 3**, which illustrates the theorem.

Problem 7. Verify the initial value theorem for the voltage function $(5 + 2\cos 3t)$ volts, and state its initial value.

Let $f(t) = 5 + 2\cos 3t$

$$\mathcal{L}\{f(t)\} = \mathcal{L}\{5 + 2\cos 3t\} = \dfrac{5}{s} + \dfrac{2s}{s^2+9}$$

from (ii) and (v) of Table 67.1, page 737.

By the initial value theorem,

$$\lim_{t \to 0}[f(t)] = \lim_{s \to \infty}[s\mathcal{L}\{f(t)\}]$$

i.e. $\lim_{t \to 0}[5 + 2\cos 3t] = \lim_{s \to \infty}\left[s\left(\dfrac{5}{s} + \dfrac{2s}{s^2+9}\right)\right]$

$$= \lim_{s \to \infty}\left[5 + \dfrac{2s^2}{s^2+9}\right]$$

i.e. $5 + 2(1) = 5 + \dfrac{2\infty^2}{\infty^2 + 9} = 5 + 2$

i.e. **7 = 7**, which verifies the theorem in this case.

The initial value of the voltage is thus **7 V**

Problem 8. Verify the initial value theorem for the function $(2t - 3)^2$ and state its initial value.

Let $f(t) = (2t - 3)^2 = 4t^2 - 12t + 9$

Let $\mathcal{L}\{f(t)\} = \mathcal{L}(4t^2 - 12t + 9)$

$$= 4\left(\dfrac{2}{s^3}\right) - \dfrac{12}{s^2} + \dfrac{9}{s}$$

from (vii), (vi) and (ii) of Table 67.1, page 737.

By the initial value theorem,

$$\text{limit}_{t\to 0}[(2t-3)^2] = \text{limit}_{s\to\infty}\left[s\left(\frac{8}{s^3} - \frac{12}{s^2} + \frac{9}{s}\right)\right]$$

$$= \text{limit}_{s\to\infty}\left[\frac{8}{s^2} - \frac{12}{s} + 9\right]$$

i.e. $\qquad (0-3)^2 = \dfrac{8}{\infty^2} - \dfrac{12}{\infty} + 9$

i.e. **9 = 9**, which verifies the theorem in this case.

The initial value of the given function is thus **9**

(b) The final value theorem states:

$$\text{limit}_{t\to\infty}\ [f(t)] = \text{limit}_{s\to 0}\ [s\mathcal{L}\{f(t)\}]$$

For example, if $f(t) = 3e^{-4t}$ then:

$$\text{limit}_{t\to\infty}[3e^{-4t}] = \text{limit}_{s\to 0}\left[s\left(\frac{3}{s+4}\right)\right]$$

i.e. $\qquad 3e^{-\infty} = (0)\left(\dfrac{3}{0+4}\right)$

i.e. **0 = 0**, which illustrates the theorem.

> **Problem 9.** Verify the final value theorem for the function $(2 + 3e^{-2t}\sin 4t)$ cm, which represents the displacement of a particle. State its final steady value.

Let $\qquad f(t) = 2 + 3e^{-2t}\sin 4t$

$$\mathcal{L}\{f(t)\} = \mathcal{L}\{2 + 3e^{-2t}\sin 4t\}$$

$$= \frac{2}{s} + 3\left(\frac{4}{(s-(-2))^2 + 4^2}\right)$$

$$= \frac{2}{s} + \frac{12}{(s+2)^2 + 16}$$

from (ii) of Table 67.1, page 737 and (ii) of Table 68.1 on page 742.

By the final value theorem,

$$\text{limit}_{t\to\infty}[f(t)] = \text{limit}_{s\to 0}[s\mathcal{L}\{f(t)\}]$$

i.e. $\text{limit}_{t\to\infty}[2 + 3e^{-2t}\sin 4t]$

$$= \text{limit}_{s\to 0}\left[s\left(\frac{2}{s} + \frac{12}{(s+2)^2 + 16}\right)\right]$$

$$= \text{limit}_{s\to 0}\left[2 + \frac{12s}{(s+2)^2 + 16}\right]$$

i.e. $2 + 0 = 2 + 0$

i.e. **2 = 2**, which verifies the theorem in this case.

The final value of the displacement is thus 2 cm.

The initial and final value theorems are used in pulse circuit applications where the response of the circuit for small periods of time, or the behaviour immediately after the switch is closed, are of interest. The final value theorem is particularly useful in investigating the stability of systems (such as in automatic aircraft-landing systems) and is concerned with the steady state response for large values of time t, i.e. after all transient effects have died away.

Now try the following Practice Exercise

> **Practice Exercise 251   Initial and final value theorems (Answers on page 893)**
>
> 1. State the initial value theorem. Verify the theorem for the functions (a) $3 - 4\sin t$ (b) $(t-4)^2$ and state their initial values.
>
> 2. Verify the initial value theorem for the voltage functions: (a) $4 + 2\cos t$ (b) $t - \cos 3t$ and state their initial values.
>
> 3. State the final value theorem and state a practical application where it is of use. Verify the theorem for the function $4 + e^{-2t}(\sin t + \cos t)$ representing a displacement and state its final value.
>
> 4. Verify the final value theorem for the function $3t^2 e^{-4t}$ and determine its steady state value.

For fully worked solutions to each of the problems in Practice Exercises 249 to 251 in this chapter,
go to the website:
www.routledge.com/cw/bird

Inverse Laplace transforms

Why it is important to understand: Inverse Laplace transforms

Laplace transforms and their inverses are a mathematical technique which allows us to solve differential equations, by primarily using algebraic methods. This simplification in the solving of equations, coupled with the ability to directly implement electrical components in their transformed form, makes the use of Laplace transforms widespread in both electrical engineering and control systems engineering. Laplace transforms have many further applications in mathematics, physics, optics, signal processing, and probability. This chapter specifically explains how the inverse Laplace transform is determined, which can also involve the use of partial fractions. In addition, poles and zeros of transfer functions are briefly explained; these are of importance in stability and control systems.

At the end of this chapter, you should be able to:

- define the inverse Laplace transform
- use a standard list to determine the inverse Laplace transforms of simple functions
- determine inverse Laplace transforms using partial fractions
- define a pole and a zero
- determine poles and zeros for transfer functions, showing them on a pole–zero diagram

---

69.1 Definition of the inverse Laplace transform

If the Laplace transform of a function $f(t)$ is $F(s)$, i.e. $\mathcal{L}\{f(t)\} = F(s)$, then $f(t)$ is called the **inverse Laplace transform** of $F(s)$ and is written as $f(t) = \mathcal{L}^{-1}\{F(s)\}$

For example, since $\mathcal{L}\{1\} = \dfrac{1}{s}$ then $\mathcal{L}^{-1}\left\{\dfrac{1}{s}\right\} = 1$

Similarly, since $\mathcal{L}\{\sin at\} = \dfrac{a}{s^2 + a^2}$ then

$$\mathcal{L}^{-1}\left\{\frac{a}{s^2 + a^2}\right\} = \sin at, \text{ and so on.}$$

69.2 Inverse Laplace transforms of simple functions

Tables of Laplace transforms, such as the tables in Chapters 67 and 68 (see pages 737 and 742) may be used to find inverse Laplace transforms.

However, for convenience, a summary of inverse Laplace transforms is shown in Table 69.1.

Problem 1. Find the following inverse Laplace transforms:

(a) $\mathcal{L}^{-1}\left\{\dfrac{1}{s^2 + 9}\right\}$ (b) $\mathcal{L}^{-1}\left\{\dfrac{5}{3s - 1}\right\}$

Table 69.1 Inverse Laplace transforms

	$F(s) = \mathcal{L}\{f(t)\}$	$\mathcal{L}^{-1}\{F(s)\} = f(t)$
(i)	$\dfrac{1}{s}$	1
(ii)	$\dfrac{k}{s}$	k
(iii)	$\dfrac{1}{s-a}$	e^{at}
(iv)	$\dfrac{a}{s^2+a^2}$	$\sin at$
(v)	$\dfrac{s}{s^2+a^2}$	$\cos at$
(vi)	$\dfrac{1}{s^2}$	t
(vii)	$\dfrac{2!}{s^3}$	t^2
(viii)	$\dfrac{n!}{s^{n+1}}$	t^n
(ix)	$\dfrac{a}{s^2-a^2}$	$\sinh at$
(x)	$\dfrac{s}{s^2-a^2}$	$\cosh at$
(xi)	$\dfrac{n!}{(s-a)^{n+1}}$	$e^{at} t^n$
(xii)	$\dfrac{\omega}{(s-a)^2+\omega^2}$	$e^{at} \sin \omega t$
(xiii)	$\dfrac{s-a}{(s-a)^2+\omega^2}$	$e^{at} \cos \omega t$
(xiv)	$\dfrac{\omega}{(s-a)^2-\omega^2}$	$e^{at} \sinh \omega t$
(xv)	$\dfrac{s-a}{(s-a)^2-\omega^2}$	$e^{at} \cosh \omega t$

(a) From (iv) of Table 69.1,

$$\mathcal{L}^{-1}\left\{\frac{a}{s^2+a^2}\right\} = \sin at,$$

Hence $\mathcal{L}^{-1}\left\{\dfrac{1}{s^2+9}\right\} = \mathcal{L}^{-1}\left\{\dfrac{1}{s^2+3^2}\right\}$

$$= \frac{1}{3}\mathcal{L}^{-1}\left\{\frac{3}{s^2+3^2}\right\}$$

$$= \frac{1}{3}\sin 3t$$

(b) $\mathcal{L}^{-1}\left\{\dfrac{5}{3s-1}\right\} = \mathcal{L}^{-1}\left\{\dfrac{5}{3\left(s-\dfrac{1}{3}\right)}\right\}$

$$= \frac{5}{3}\mathcal{L}^{-1}\left\{\frac{1}{\left(s-\dfrac{1}{3}\right)}\right\} = \frac{5}{3}e^{\frac{1}{3}t}$$

from (iii) of Table 69.1

Problem 2. Find the following inverse Laplace transforms:

(a) $\mathcal{L}^{-1}\left\{\dfrac{6}{s^3}\right\}$ (b) $\mathcal{L}^{-1}\left\{\dfrac{3}{s^4}\right\}$

(a) From (vii) of Table 69.1, $\mathcal{L}^{-1}\left\{\dfrac{2}{s^3}\right\} = t^2$

Hence $\mathcal{L}^{-1}\left\{\dfrac{6}{s^3}\right\} = 3\mathcal{L}^{-1}\left\{\dfrac{2}{s^3}\right\} = \mathbf{3t^2}$

(b) From (viii) of Table 69.1, if s is to have a power of 4 then $n = 3$

Thus $\mathcal{L}^{-1}\left\{\dfrac{3!}{s^4}\right\} = t^3$ i.e. $\mathcal{L}^{-1}\left\{\dfrac{6}{s^4}\right\} = t^3$

Hence $\mathcal{L}^{-1}\left\{\dfrac{3}{s^4}\right\} = \dfrac{1}{2}\mathcal{L}^{-1}\left\{\dfrac{6}{s^4}\right\} = \dfrac{1}{2}t^3$

Problem 3. Determine

(a) $\mathcal{L}^{-1}\left\{\dfrac{7s}{s^2+4}\right\}$ (b) $\mathcal{L}^{-1}\left\{\dfrac{4s}{s^2-16}\right\}$

(a) $\mathcal{L}^{-1}\left\{\dfrac{7s}{s^2+4}\right\} = 7\mathcal{L}^{-1}\left\{\dfrac{s}{s^2+2^2}\right\} = \mathbf{7\cos 2t}$

from (v) of Table 69.1

(b) $\mathcal{L}^{-1}\left\{\dfrac{4s}{s^2-16}\right\} = 4\mathcal{L}^{-1}\left\{\dfrac{s}{s^2-4^2}\right\}$

$$= \mathbf{4\cosh 4t}$$

from (x) of Table 69.1

Problem 4. Find

(a) $\mathcal{L}^{-1}\left\{\dfrac{3}{s^2-7}\right\}$ (b) $\mathcal{L}^{-1}\left\{\dfrac{2}{(s-3)^5}\right\}$

(a) From (ix) of Table 69.1,

$$\mathcal{L}^{-1}\left\{\frac{a}{s^2-a^2}\right\}=\sinh at$$

Thus

$$\mathcal{L}^{-1}\left\{\frac{3}{s^2-7}\right\}=3\mathcal{L}^{-1}\left\{\frac{1}{s^2-(\sqrt{7})^2}\right\}$$

$$=\frac{3}{\sqrt{7}}\mathcal{L}^{-1}\left\{\frac{\sqrt{7}}{s^2-(\sqrt{7})^2}\right\}$$

$$=\frac{3}{\sqrt{7}}\sinh\sqrt{7}t$$

(b) From (xi) of Table 69.1,

$$\mathcal{L}^{-1}\left\{\frac{n!}{(s-a)^{n+1}}\right\}=e^{at}t^n$$

Thus $\mathcal{L}^{-1}\left\{\dfrac{1}{(s-a)^{n+1}}\right\}=\dfrac{1}{n!}e^{at}t^n$

and comparing with $\mathcal{L}^{-1}\left\{\dfrac{2}{(s-3)^5}\right\}$ shows that $n=4$ and $a=3$

Hence

$$\mathcal{L}^{-1}\left\{\frac{2}{(s-3)^5}\right\}=2\mathcal{L}^{-1}\left\{\frac{1}{(s-3)^5}\right\}$$

$$=2\left(\frac{1}{4!}e^{3t}t^4\right)=\frac{1}{12}e^{3t}t^4$$

Problem 5. Determine

(a) $\mathcal{L}^{-1}\left\{\dfrac{3}{s^2-4s+13}\right\}$

(b) $\mathcal{L}^{-1}\left\{\dfrac{2(s+1)}{s^2+2s+10}\right\}$

(a) $\mathcal{L}^{-1}\left\{\dfrac{3}{s^2-4s+13}\right\}=\mathcal{L}^{-1}\left\{\dfrac{3}{(s-2)^2+3^2}\right\}$

$$=e^{2t}\sin 3t$$

from (xii) of Table 69.1

(b) $\mathcal{L}^{-1}\left\{\dfrac{2(s+1)}{s^2+2s+10}\right\}=\mathcal{L}^{-1}\left\{\dfrac{2(s+1)}{(s+1)^2+3^2}\right\}$

$$=2e^{-t}\cos 3t$$

from (xiii) of Table 69.1

Problem 6. Determine

(a) $\mathcal{L}^{-1}\left\{\dfrac{5}{s^2+2s-3}\right\}$

(b) $\mathcal{L}^{-1}\left\{\dfrac{4s-3}{s^2-4s-5}\right\}$

(a) $\mathcal{L}^{-1}\left\{\dfrac{5}{s^2+2s-3}\right\}=\mathcal{L}^{-1}\left\{\dfrac{5}{(s+1)^2-2^2}\right\}$

$$=\mathcal{L}^{-1}\left\{\frac{\frac{5}{2}(2)}{(s+1)^2-2^2}\right\}$$

$$=\frac{5}{2}e^{-t}\sinh 2t$$

from (xiv) of Table 69.1

(b) $\mathcal{L}^{-1}\left\{\dfrac{4s-3}{s^2-4s-5}\right\}=\mathcal{L}^{-1}\left\{\dfrac{4s-3}{(s-2)^2-3^2}\right\}$

$$=\mathcal{L}^{-1}\left\{\frac{4(s-2)+5}{(s-2)^2-3^2}\right\}$$

$$=\mathcal{L}^{-1}\left\{\frac{4(s-2)}{(s-2)^2-3^2}\right\}$$

$$+\mathcal{L}^{-1}\left\{\frac{5}{(s-2)^2-3^2}\right\}$$

$$=4e^{2t}\cosh 3t+\mathcal{L}^{-1}\left\{\frac{\frac{5}{3}(3)}{(s-2)^2-3^2}\right\}$$

from (xv) of Table 69.1

$$=4e^{2t}\cosh 3t+\frac{5}{3}e^{2t}\sinh 3t$$

from (xiv) of Table 69.1

Now try the following Practice Exercise

Determine the inverse Laplace transforms of the following:

1. (a) $\dfrac{7}{s}$ (b) $\dfrac{2}{s-5}$

2. (a) $\dfrac{3}{2s+1}$ (b) $\dfrac{2s}{s^2+4}$

3. (a) $\dfrac{1}{s^2+25}$ (b) $\dfrac{4}{s^2+9}$

4. (a) $\dfrac{5s}{2s^2+18}$ (b) $\dfrac{6}{s^2}$

5. (a) $\dfrac{5}{s^3}$ (b) $\dfrac{8}{s^4}$

6. (a) $\dfrac{3s}{\frac{1}{2}s^2-8}$ (b) $\dfrac{7}{s^2-16}$

7. (a) $\dfrac{15}{3s^2-27}$ (b) $\dfrac{4}{(s-1)^3}$

8. (a) $\dfrac{1}{(s+2)^4}$ (b) $\dfrac{3}{(s-3)^5}$

9. (a) $\dfrac{s+1}{s^2+2s+10}$ (b) $\dfrac{3}{s^2+6s+13}$

10. (a) $\dfrac{2(s-3)}{s^2-6s+13}$ (b) $\dfrac{7}{s^2-8s+12}$

11. (a) $\dfrac{2s+5}{s^2+4s-5}$ (b) $\dfrac{3s+2}{s^2-8s+25}$

69.3 Inverse Laplace transforms using partial fractions

Sometimes the function whose inverse is required is not recognisable as a standard type, such as those listed in Table 69.1. In such cases it may be possible, by using partial fractions, to resolve the function into simpler fractions which may be inverted on sight. For example, the function,

$$F(s) = \frac{2s-3}{s(s-3)}$$

cannot be inverted on sight from Table 69.1. However, by using partial fractions, $\dfrac{2s-3}{s(s-3)} \equiv \dfrac{1}{s}+\dfrac{1}{s-3}$ which may be inverted as $1+e^{3t}$ from (i) and (iii) of Table 67.1. Partial fractions are discussed in Chapter 2, and a summary of the forms of partial fractions is given in Table 2.1 on page 16.

Problem 7. Determine $\mathcal{L}^{-1}\left\{\dfrac{4s-5}{s^2-s-2}\right\}$

$$\frac{4s-5}{s^2-s-2} \equiv \frac{4s-5}{(s-2)(s+1)} \equiv \frac{A}{(s-2)}+\frac{B}{(s+1)}$$

$$\equiv \frac{A(s+1)+B(s-2)}{(s-2)(s+1)}$$

Hence $4s-5 \equiv A(s+1)+B(s-2)$

When $s=2$, $3=3A$, from which, $A=1$

When $s=-1$, $-9=-3B$, from which, $B=3$

Hence $\mathcal{L}^{-1}\left\{\dfrac{4s-5}{s^2-s-2}\right\}$

$$\equiv \mathcal{L}^{-1}\left\{\frac{1}{s-2}+\frac{3}{s+1}\right\}$$

$$= \mathcal{L}^{-1}\left\{\frac{1}{s-2}\right\}+\mathcal{L}^{-1}\left\{\frac{3}{s+1}\right\}$$

$$= e^{2t}+3e^{-t}, \text{ from (iii) of Table 69.1}$$

Problem 8. Find $\mathcal{L}^{-1}\left\{\dfrac{3s^3+s^2+12s+2}{(s-3)(s+1)^3}\right\}$

$$\frac{3s^3+s^2+12s+2}{(s-3)(s+1)^3}$$

$$\equiv \frac{A}{s-3}+\frac{B}{s+1}+\frac{C}{(s+1)^2}+\frac{D}{(s+1)^3}$$

$$\equiv \frac{\left(\begin{array}{c}A(s+1)^3+B(s-3)(s+1)^2\\+C(s-3)(s+1)+D(s-3)\end{array}\right)}{(s-3)(s+1)^3}$$

Hence

$$3s^3+s^2+12s+2 \equiv A(s+1)^3+B(s-3)(s+1)^2$$

$$+C(s-3)(s+1)+D(s-3)$$

When $s=3$, $128=64A$, from which, $A=2$

When $s=-1, -12=-4D$, from which, $D=3$

Equating s^3 terms gives: $3=A+B$ from which, $B=1$

Equating constant terms gives:

$$2=A-3B-3C-3D$$

i.e. $2=2-3-3C-9$

from which, $3C=-12$ and $C=-4$

Hence

$$\mathcal{L}^{-1}\left\{\frac{3s^3+s^2+12s+2}{(s-3)(s+1)^3}\right\}$$

$$\equiv \mathcal{L}^{-1}\left\{\frac{2}{s-3}+\frac{1}{s+1}-\frac{4}{(s+1)^2}+\frac{3}{(s+1)^3}\right\}$$

$$=2e^{3t}+e^{-t}-4e^{-t}t+\frac{3}{2}e^{-t}t^2$$

from (iii) and (xi) of Table 69.1

Problem 9.   Determine
$$\mathcal{L}^{-1}\left\{\frac{5s^2+8s-1}{(s+3)(s^2+1)}\right\}$$

$$\frac{5s^2+8s-1}{(s+3)(s^2+1)}\equiv \frac{A}{s+3}+\frac{Bs+C}{(s^2+1)}$$

$$\equiv \frac{A(s^2+1)+(Bs+C)(s+3)}{(s+3)(s^2+1)}$$

Hence $5s^2+8s-1\equiv A(s^2+1)+(Bs+C)(s+3)$

When $s=-3, 20=10A$, from which, $A=2$

Equating s^2 terms gives: $5=A+B$, from which, $B=3$, since $A=2$

Equating s terms gives: $8=3B+C$, from which, $C=-1$, since $B=3$

Hence $\mathcal{L}^{-1}\left\{\frac{5s^2+8s-1}{(s+3)(s^2+1)}\right\}$

$$\equiv \mathcal{L}^{-1}\left\{\frac{2}{s+3}+\frac{3s-1}{s^2+1}\right\}$$

$$\equiv \mathcal{L}^{-1}\left\{\frac{2}{s+3}\right\}+\mathcal{L}^{-1}\left\{\frac{3s}{s^2+1}\right\}$$

$$-\mathcal{L}^{-1}\left\{\frac{1}{s^2+1}\right\}$$

$$=2e^{-3t}+3\cos t-\sin t$$

from (iii), (v) and (iv) of Table 69.1

Problem 10.   Find $\mathcal{L}^{-1}\left\{\frac{7s+13}{s(s^2+4s+13)}\right\}$

$$\frac{7s+13}{s(s^2+4s+13)}\equiv \frac{A}{s}+\frac{Bs+C}{s^2+4s+13}$$

$$\equiv \frac{A(s^2+4s+13)+(Bs+C)(s)}{s(s^2+4s+13)}$$

Hence $7s+13\equiv A(s^2+4s+13)+(Bs+C)(s)$.

When $s=0, 13=13A$, from which, $A=1$

Equating s^2 terms gives: $0=A+B$, from which, $B=-1$

Equating s terms gives: $7=4A+C$, from which, $C=3$

Hence $\mathcal{L}^{-1}\left\{\frac{7s+13}{s(s^2+4s+13)}\right\}$

$$\equiv \mathcal{L}^{-1}\left\{\frac{1}{s}+\frac{-s+3}{s^2+4s+13}\right\}$$

$$\equiv \mathcal{L}^{-1}\left\{\frac{1}{s}\right\}+\mathcal{L}^{-1}\left\{\frac{-s+3}{(s+2)^2+3^2}\right\}$$

$$\equiv \mathcal{L}^{-1}\left\{\frac{1}{s}\right\}+\mathcal{L}^{-1}\left\{\frac{-(s+2)+5}{(s+2)^2+3^2}\right\}$$

$$\equiv \mathcal{L}^{-1}\left\{\frac{1}{s}\right\}-\mathcal{L}^{-1}\left\{\frac{s+2}{(s+2)^2+3^2}\right\}$$

$$+\mathcal{L}^{-1}\left\{\frac{5}{(s+2)^2+3^2}\right\}$$

$$\equiv 1-e^{-2t}\cos 3t+\frac{5}{3}e^{-2t}\sin 3t$$

from (i), (xiii) and (xii) of Table 69.1

Now try the following Practice Exercise

Practice Exercise 253   Inverse Laplace transforms using partial fractions (Answers on page 893)

Use partial fractions to find the inverse Laplace transforms of the following functions:

1. $\dfrac{11-3s}{s^2+2s-3}$

2. $\dfrac{2s^2 - 9s - 35}{(s+1)(s-2)(s+3)}$

3. $\dfrac{5s^2 - 2s - 19}{(s+3)(s-1)^2}$

4. $\dfrac{3s^2 + 16s + 15}{(s+3)^3}$

5. $\dfrac{7s^2 + 5s + 13}{(s^2+2)(s+1)}$

6. $\dfrac{3 + 6s + 4s^2 - 2s^3}{s^2(s^2+3)}$

7. $\dfrac{26 - s^2}{s(s^2+4s+13)}$

69.4 Poles and zeros

It was seen in the previous section that Laplace transforms, in general, have the form $f(s) = \dfrac{\phi(s)}{\theta(s)}$. This is the same form as most transfer functions for engineering systems, a **transfer function** being one that relates the response at a given pair of terminals to a source or stimulus at another pair of terminals.

Let a function in the s domain be given by: $f(s) = \dfrac{\phi(s)}{(s-a)(s-b)(s-c)}$ where $\phi(s)$ is of less degree than the denominator.

Poles: The values $a, b, c, \dots$ that makes the denominator zero, and hence $f(s)$ infinite, are called the system poles of $f(s)$.
If there are no repeated factors, the poles are **simple poles**.
If there are repeated factors, the poles are **multiple poles**.

Zeros: Values of s that make the numerator $\phi(s)$ zero, and hence $f(s)$ zero, are called the system zeros of $f(s)$.

For example: $\dfrac{s-4}{(s+1)(s-2)}$ has simple poles at $s = -1$ and $s = +2$, and a zero at $s = 4$ $\dfrac{s+3}{(s+1)^2(2s+5)}$ has a simple pole at $s = -\dfrac{5}{2}$ and double poles at $s = -1$, and

a zero at $s = -3$ and $\dfrac{s+2}{s(s-1)(s+4)(2s+1)}$ has simple poles at $s = 0, +1, -4$, and $-\dfrac{1}{2}$ and a zero at $s = -2$

Pole–zero diagram

The poles and zeros of a function are values of complex frequency s and can therefore be plotted on the complex frequency or s-plane. The resulting plot is the **pole–zero diagram** or **pole–zero map**. On the rectangular axes, the real part is labelled the σ-**axis** and the imaginary part the $j\omega$-axis.

The location of a pole in the s-plane is denoted by a cross ($\times$) and the location of a zero by a small circle (o). This is demonstrated in the following examples. From the pole–zero diagram it may be determined that the magnitude of the transfer function will be larger when it is closer to the poles and smaller when it is close to the zeros. This is important in understanding what the system does at various frequencies and is crucial in the study of **stability** and **control theory** in general.

Problem 11. Determine for the transfer function:

$$R(s) = \frac{400(s+10)}{s(s+25)(s^2+10s+125)}$$

(a) the zero and (b) the poles. Show the poles and zero on a pole–zero diagram.

(a) For the numerator to be zero, $(s+10) = 0$

Hence, $s = -10$ **is a zero** of $R(s)$

(b) For the denominator to be zero, $s = 0$ or $s = -25$ or $s^2 + 10s + 125 = 0$

Using the quadratic formula.

$$s = \frac{-10 \pm \sqrt{10^2 - 4(1)(125)}}{2} = \frac{-10 \pm \sqrt{-400}}{2}$$

$$= \frac{-10 \pm j20}{2}$$

$$= (-5 \pm j10)$$

Hence, **poles occur at $s = 0$, $s = -25$, $(-5+j10)$ and $(-5-j10)$**

The pole–zero diagram is shown in Figure 69.1.

Figure 69.1

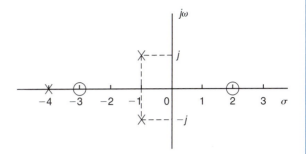

Figure 69.2

It is seen from these problems that poles and zeros are always real or complex conjugate.

Now try the following Practice Exercise

Problem 12.   Determine the poles and zeros for

the function: $F(s) = \dfrac{(s+3)(s-2)}{(s+4)(s^2+2s+2)}$

and plot them on a pole–zero map.

For the numerator to be zero, $(s+3)=0$ and $(s-2)=0$, hence **zeros occur at $s=-3$ and at $s=+2$** Poles occur when the denominator is zero, i.e. when $(s+4)=0$, i.e. $s=-4$, and when $s^2+2s+2=0$

i.e. $s = \dfrac{-2 \pm \sqrt{2^2 - 4(1)(2)}}{2} = \dfrac{-2 \pm \sqrt{-4}}{2}$

$= \dfrac{-2 \pm j2}{2} = (-1+j) \text{ or } (-1-j)$

The poles and zeros are shown on the pole–zero map of $F(s)$ in Figure 69.2.

Practice Exercise 254   Poles and zeros (Answers on page 893)

1. Determine for the transfer function:

   $R(s) = \dfrac{50\,(s+4)}{s\,(s+2)(s^2 - 8s + 25)}$

   (a) the zero and (b) the poles. Show the poles and zeros on a pole–zero diagram.

2. Determine the poles and zeros for the function: $F(s) = \dfrac{(s-1)(s+2)}{(s+3)(s^2 - 2s + 5)}$ and plot them on a pole–zero map.

3. For the function $G(s) = \dfrac{s-1}{(s+2)(s^2 + 2s + 5)}$ determine the poles and zeros and show them on a pole–zero diagram.

4. Find the poles and zeros for the transfer function: $H(s) = \dfrac{s^2 - 5s - 6}{s(s^2 + 4)}$ and plot the results in the s-plane.

Section L

For fully worked solutions to each of the problems in Practice Exercises 252 to 254 in this chapter, go to the website:
www.routledge.com/cw/bird

Chapter 70

The Laplace transform of the Heaviside function

Why it is important to understand: **The Laplace transform of the Heaviside function**

The Heaviside unit step function is used in the mathematics of control theory and signal processing to represent a signal that switches on at a specified time and stays switched on indefinitely. It is also used in structural mechanics to describe different types of structural loads. The Heaviside function has applications in engineering where periodic functions are represented. In many physical situations things change suddenly; brakes are applied, a switch is thrown, collisions occur. The Heaviside unit function is very useful for representing sudden change.

At the end of this chapter, you should be able to:

- define the Heaviside unit step function
- use a standard list to determine the Laplace transform of $H(t-c)$
- use a standard list to determine the Laplace transform of $H(t-c) \cdot f(t-c)$
- determine the inverse transforms of Heaviside functions

70.1 Heaviside unit step function

In engineering applications, functions are frequently encountered whose values change abruptly at specified values of time t. One common example is when a voltage is switched on or off in an electrical circuit at a specified value of time t.

The switching process can be described mathematically by the function called the **Unit Step Function** – otherwise known as the **Heaviside unit step function**. Fig. 70.1 shows a function that maintains a zero value for all values of t up to $t = c$ and a value of 1 for all values of $t \geq c$. This is the Heaviside unit step function and is denoted by:

$$f(t) = H(t-c) \quad \text{or} \quad u(t-c)$$

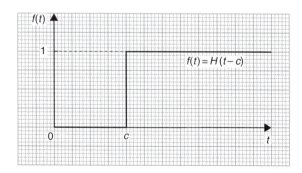

Figure 70.1

where the c indicates the value of t at which the function changes from a value of zero to a value of unity (i.e. 1).

It follows that $f(t)=H(t-5)$ is as shown in Fig. 70.2 and $f(t)=3H(t-4)$ is as shown in Fig. 70.3.

Figure 70.2

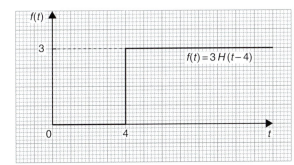

Figure 70.3

If the unit step occurs at the origin, then $c=0$ and $f(t)=H(t-0)$, i.e. **H(t)** as shown in Fig. 70.4.

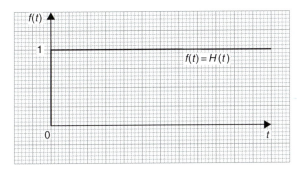

Figure 70.4

Fig. 70.5(a) shows a graph of $f(t)=t^2$; the graph shown in Fig. 70.5(b) is $f(t)=H(t-2)\cdot t^2$ where for $t<2, H(t-2)t^2=0$ and when $t\geq 2$, $H(t-2)\cdot t^2=t^2$. The function $H(t-2)$ suppresses the function t^2 for all values of t up to $t=2$ and then 'switches on' the function t^2 at $t=2$.

A common situation in an electrical circuit is for a voltage V to be applied at a particular time, say, $t=a$,

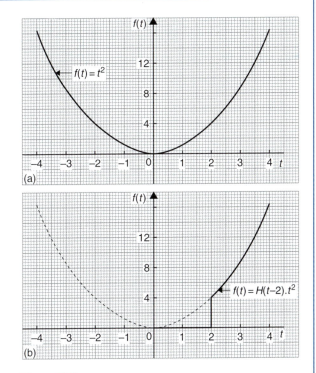

Figure 70.5

and removed later, say at $t=b$. Such a situation is written using step functions as:

$$V(t) = H(t-a) - H(t-b)$$

For example, Fig. 70.6 shows the function $f(t)=H(t-2)-H(t-5)$

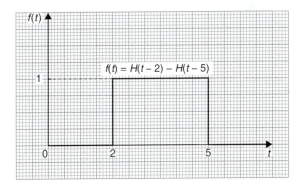

Figure 70.6

Representing the Heaviside unit step function is further explored in the following worked problems.

Problem 1. A 12 V source is switched on at time $t=3$ s. Sketch the waveform and write the function in terms of the Heaviside step function.

The function is shown sketched in Fig. 70.7.

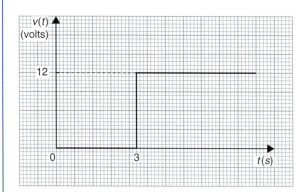

Figure 70.7

The Heaviside step function is:

$$V(t) = 12H(t-3)$$

Problem 2. Write the function
$$V(t) = \begin{cases} 1 & \text{for } 0 < t < a \\ 0 & \text{for } t > a \end{cases}$$
in terms of the Heaviside step function and sketch the waveform.

The voltage has a value of 1 up until time $t = a$; then it is turned off.
The function is shown sketched in Fig. 70.8.

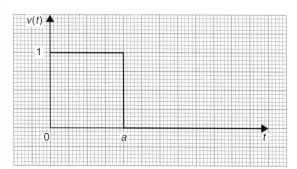

Figure 70.8

The Heaviside step function is:

$$V(t) = H(t) - H(t-a)$$

Problem 3. Sketch the graph of $f(t) = 5H(t-2)$

A function $H(t-2)$ has a maximum value of 1 and starts when $t = 2$.

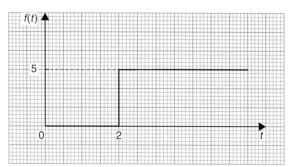

Figure 70.9

A function $5H(t-2)$ has a maximum value of 5 and starts when $t = 2$, as shown in Fig. 70.9.

Problem 4. Sketch the graph of
$$f(t) = H(t - \pi/3).\sin t$$

Fig. 70.10(a) shows a graph of $f(t) = \sin t$; the graph shown in Fig. 70.10(b) is $f(t) = H(t - \pi/3) \cdot \sin t$ where the graph of $\sin t$ does not 'switch on' until $t = \pi/3$

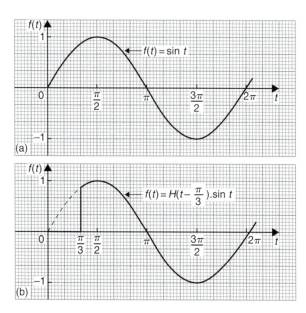

Figure 70.10

Problem 5. Sketch the graph of
$$f(t) = 2H(t - 2\pi/3) \cdot \sin(t - \pi/6)$$

Fig. 70.11(a) shows a graph of $f(t) = 2\sin(t - \pi/6)$; the graph shown in Fig. 70.11(b) is

$$f(t) = 2H(t - 2\pi/3) \cdot \sin(t - \pi/6)$$

where the graph of $2\sin(t - \pi/6)$ does not 'switch on' until $t = 2\pi/3$

Figure 70.11

Problem 6.   Sketch the graphs of
(a) $f(t) = H(t-1).e^{-t}$
(b) $f(t) = [H(t-1) - H(t-3)].e^{-t}$

Fig. 70.12(a) shows a graph of $f(t) = e^{-t}$

(a)   The graph shown in Fig. 70.12(b) is $f(t) = H(t-1) \cdot e^{-t}$ where the graph of e^{-t} does not 'switch on' until $t = 1$.

(b)   Fig. 70.12(c) shows the graph of $f(t) = [H(t-1) - H(t-3)] \cdot e^{-t}$ where the graph of e^{-t} does not 'switch on' until $t = 1$, but then 'switches off' at $t = 3$.

Figure 70.12

Figure 70.12 (*Continued*)

Now try the following Practice Exercise

Practice Exercise 255   Heaviside unit step function (Answers on page 893)

1.   A 6 V source is switched on at time $t = 4$ s. Write the function in terms of the Heaviside step function and sketch the waveform.

2.   Write the function $V(t) = \begin{cases} 2 & \text{for } 0 < t < 5 \\ 0 & \text{for } t > 5 \end{cases}$ in terms of the Heaviside step function and sketch the waveform.

In problems 3 to 12, sketch graphs of the given functions.

3. $f(t) = H(t-2)$

4. $f(t) = H(t)$

5. $f(t) = 4H(t-1)$

Section L

6. $f(t) = 7H(t-5)$

7. $f(t) = H\left(t - \dfrac{\pi}{4}\right) \cdot \cos t$

8. $f(t) = 3H\left(t - \dfrac{\pi}{2}\right) \cdot \cos\left(t - \dfrac{\pi}{6}\right)$

9. $f(t) = H(t-1) \cdot t^2$

10. $f(t) = H(t-2) \cdot e^{-\frac{t}{2}}$

11. $f(t) = [H(t-2) - H(t-5)] \cdot e^{-\frac{t}{4}}$

12. $f(t) = 5H\left(t - \dfrac{\pi}{3}\right) \cdot \sin\left(t + \dfrac{\pi}{4}\right)$

70.2 Laplace transform of $H(t-c)$

From the definition of a Laplace transform,

$$\mathcal{L}\{H(t-c)\} = \int_0^\infty e^{-st} H(t-c)\,dt$$

However, $e^{-st}H(t-c) = \begin{cases} 0 & \text{for } 0 < t < c \\ e^{-st} & \text{for } t \geq c \end{cases}$

Hence, $\mathcal{L}\{H(t-c)\} = \int_0^\infty e^{-st} H(t-c)\,dt$

$$= \int_c^\infty e^{-st}\,dt = \left[\frac{e^{-st}}{-s}\right]_c^\infty = \left[\frac{e^{-s(\infty)}}{-s} - \frac{e^{-sc}}{-s}\right]$$

$$= \left[0 - \frac{e^{-sc}}{-s}\right] = \frac{e^{-sc}}{s}$$

When $c = 0$ (i.e. a unit step at the origin), $\mathcal{L}\{H(t)\} = \dfrac{e^{-s(0)}}{s} = \dfrac{1}{s}$

Summarising, $\mathcal{L}\{H(t)\} = \dfrac{1}{s}$ and $\mathcal{L}\{H(t-c)\} = \dfrac{e^{-cs}}{s}$

From the definition of $H(t)$: $\mathcal{L}\{1\} = \{1 \cdot H(t)\}$

$$\mathcal{L}\{t\} = \{t \cdot H(t)\}$$

and $$\mathcal{L}\{f(t)\} = \{f(t) \cdot H(t)\}$$

70.3 Laplace transform of $H(t-c) \cdot f(t-c)$

It may be shown that:

$$\mathcal{L}\{H(t-c) \cdot f(t-c)\} = e^{-cs}\mathcal{L}\{f(t)\} = e^{-cs}F(s)$$

where $F(s) = \mathcal{L}\{f(t)\}$

This is demonstrated in the following worked problems.

Problem 7. Determine $\mathcal{L}\{4H(t-5)\}$

From above, $\mathcal{L}\{H(t-c) \cdot f(t-c)\} = e^{-cs}F(s)$ where in this case, $F(s) = \mathcal{L}\{4\}$ and $c = 5$

Hence, $\mathcal{L}\{4H(t-5)\} = e^{-5s}\left(\dfrac{4}{s}\right)$ from (ii) of Table 67.1, page 737

$$= \frac{4e^{-5s}}{s}$$

Problem 8. Determine $\mathcal{L}\{H(t-3) \cdot (t-3)^2\}$

From above, $\mathcal{L}\{H(t-c) \cdot f(t-c)\} = e^{-cs}F(s)$ where in this case, $F(s) = \mathcal{L}\{t^2\}$ and $c = 3$

Note that $F(s)$ is the transform of t^2 and not of $(t-3)^2$

Hence, $\mathcal{L}\{H(t-3) \cdot f(t-3)^2\} = e^{-3s}\left(\dfrac{2!}{s^3}\right)$ from (vii) of Table 67.1, page 737

$$= \frac{2e^{-3s}}{s^3}$$

Problem 9. Determine $\mathcal{L}\{H(t-2).\sin(t-2)\}$

From above, $\mathcal{L}\{H(t-c) \cdot f(t-c)\} = e^{-cs}F(s)$ where in this case, $F(s) = \mathcal{L}\{\sin t\}$ and $c = 2$

Hence, $\mathcal{L}\{H(t-2) \cdot \sin(t-2)\} = e^{-2s}\left(\dfrac{1}{s^2 + 1^2}\right)$ from (iv) of Table 67.1, page 737

$$= \frac{e^{-2s}}{s^2 + 1}$$

Problem 10. Determine $\mathcal{L}\{H(t-1) \cdot \sin 4(t-1)\}$

From above, $\mathcal{L}\{H(t-c) \cdot f(t-c)\} = e^{-cs}F(s)$ where in this case, $F(s) = \mathcal{L}\{\sin 4t\}$ and $c = 1$

Hence, $\mathcal{L}\{H(t-1) \cdot \sin 4(t-1)\} = e^{-s}\left(\dfrac{4}{s^2 + 4^2}\right)$ from (iv) of Table 67.1, page 737

$$= \frac{4e^{-s}}{s^2 + 16}$$

Problem 11. Determine $\mathcal{L}\{H(t-3) \cdot e^{t-3}\}$

From above, $\mathcal{L}\{H(t-c) \cdot f(t-c)\} = e^{-cs}F(s)$ where in this case, $F(s) = \mathcal{L}\{e^t\}$ and $c = 3$

Hence, $\mathcal{L}\{H(t-3) \cdot e^{t-3}\} = e^{-3s}\left(\dfrac{1}{s-1}\right)$ from (iii) of Table 67.1, page 737

$$= \dfrac{e^{-3s}}{s-1}$$

Problem 12. Determine

$$\mathcal{L}\left\{H\left(t - \dfrac{\pi}{2}\right) \cdot \cos 3\left(t - \dfrac{\pi}{2}\right)\right\}$$

From above, $\mathcal{L}\{H(t-c) \cdot f(t-c)\} = e^{-cs}F(s)$ where in this case, $F(s) = \mathcal{L}\{\cos 3t\}$ and $c = \dfrac{\pi}{2}$

Hence, $\mathcal{L}\left\{\left(t - \dfrac{\pi}{2}\right) \cdot \cos 3\left(1 - \dfrac{\pi}{2}\right)\right\} =$

$e^{-\frac{\pi}{2}s}\left(\dfrac{s}{s^2 + 3^2}\right)$ from (v) of Table 67.1, page 737

$$= \dfrac{s\,e^{-\frac{\pi}{2}s}}{s^2 + 9}$$

Now try the following Practice Exercise

Practice Exercise 256   Laplace transform of $H(t-c) \cdot f(t-c)$ (Answers on page 895)

1. Determine $\mathcal{L}\{H(t-1)\}$

2. Determine $\mathcal{L}\{7H(t-3)\}$

3. Determine $\mathcal{L}\{H(t-2) \cdot (t-2)^2\}$

4. Determine $\mathcal{L}\{H(t-3) \cdot \sin(t-3)\}$

5. Determine $\mathcal{L}\{H(t-4) \cdot e^{t-4}\}$

6. Determine $\mathcal{L}\{H(t-5) \cdot \sin 3(t-5)\}$

7. Determine $\mathcal{L}\{H(t-1) \cdot (t-1)^3\}$

8. Determine $\mathcal{L}\{H(t-6) \cdot \cos 3(t-6)\}$

9. Determine $\mathcal{L}\{5H(t-5) \cdot \sinh 2(t-5)\}$

10. Determine $\mathcal{L}\left\{H\left(t - \dfrac{\pi}{3}\right) \cdot \cos 2\left(t - \dfrac{\pi}{3}\right)\right\}$

11. Determine $\mathcal{L}\{2H(t-3) \cdot e^{t-3}\}$

12. Determine $\mathcal{L}\{3H(t-2) \cdot \cosh(t-2)\}$

70.4 Inverse Laplace transforms of Heaviside functions

In the previous section it was stated that: $\mathcal{L}\{H(t-c) . f(t-c)\} = e^{-cs}F(s)$ where $F(s) = \mathcal{L}\{f(t)\}$

Written in reverse, this becomes:

if $F(s) = \mathcal{L}\{f(t)\}$, then $e^{-cs}F(s) = \mathcal{L}\{H(t-c) \cdot f(t-c)\}$

This is known as the **second shift theorem** and is used when finding **inverse Laplace transforms**, as demonstrated in the following worked problems.

Problem 13. Determine $\mathcal{L}^{-1}\left\{\dfrac{3e^{-2s}}{s}\right\}$

Part of the numerator corresponds to e^{-cs} where $c = 2$ This indicates $H(t-2)$

Then $\dfrac{3}{s} = F(s) = \mathcal{L}\{3\}$ from (ii) of Table 67.1, page 737

Hence, $\mathcal{L}^{-1}\left\{\dfrac{3e^{-2s}}{s}\right\} = 3H(t-2)$

Problem 14. Determine the inverse of $\dfrac{e^{-3s}}{s^2}$

The numerator corresponds to e^{-cs} where $c = 3$. This indicates $H(t-3)$

$\dfrac{1}{s^2} = F(s) = \mathcal{L}\{t\}$ from (vi) of Table 67.1, page 737

Then $\mathcal{L}^{-1}\left\{\dfrac{e^{-3s}}{s^2}\right\} = H(t-3) \cdot (t-3)$

Problem 15. Determine $\mathcal{L}^{-1}\left\{\dfrac{8e^{-4s}}{s^2 + 4}\right\}$

Part of the numerator corresponds to e^{-cs} where $c = 4$. This indicates $H(t-4)$

$\dfrac{8}{s^2 + 4}$ may be written as: $4\left(\dfrac{2}{s^2 + 2^2}\right)$

Then $4\left(\dfrac{2}{s^2 + 2^2}\right) = F(s) = \mathcal{L}\{4 \sin 2t\}$ from (iv) of Table 67.1, page 737

Hence, $\mathcal{L}^{-1}\left\{\dfrac{8e^{-4s}}{s^2 + 4}\right\} = H(t-4) \cdot 4\sin 2(t-4)$

$$= 4H(t-4) \cdot \sin 2(t-4)$$

Section L

Problem 16. Determine $\mathcal{L}^{-1}\left\{\dfrac{5s\,e^{-2s}}{s^2+9}\right\}$

Part of the numerator corresponds to e^{-cs} where $c = 2$. This indicates $H(t-2)$

$\dfrac{5s}{s^2+9}$ may be written as: $5\left(\dfrac{s}{s^2+3^2}\right)$

Then $5\left(\dfrac{s}{s^2+3^2}\right) = F(s) = \mathcal{L}\{5\cos 3t\}$ from (v) of Table 67.1, page 737

Hence, $\mathcal{L}^{-1}\left\{\dfrac{5s\,e^{-2s}}{s^2+3^2}\right\} = H(t-2)\cdot 5\cos 3(t-2)$

$= 5H(t-2)\cdot\cos 3(t-2)$

Problem 17. Determine $\mathcal{L}^{-1}\left\{\dfrac{7e^{-3s}}{s^2-1}\right\}$

Part of the numerator corresponds to e^{-cs} where $c = 3$. This indicates $H(t-3)$

$\dfrac{7}{s^2-1}$ may be written as: $7\left(\dfrac{1}{s^2-1^2}\right)$

Then $7\left(\dfrac{1}{s^2-1^2}\right) = F(s) = \mathcal{L}\{7\sinh t\}$ from (x) of Table 67.1, page 737

Hence, $\mathcal{L}^{-1}\left\{\dfrac{7e^{-3s}}{s^2-1}\right\} = H(t-3)\cdot 7\sinh(t-3)$

$= 7H(t-3)\cdot\sinh(t-3)$

Now try the following Practice Exercise

Practice Exercise 257 Inverse Laplace transformer of Heaviside functions (Answers on page 896)

1. Determine $\mathcal{L}^{-1}\left\{\dfrac{e^{-9s}}{s}\right\}$

2. Determine $\mathcal{L}^{-1}\left\{\dfrac{4e^{-3s}}{s}\right\}$

3. Determine $\mathcal{L}^{-1}\left\{\dfrac{2e^{-2s}}{s^2}\right\}$

4. Determine $\mathcal{L}^{-1}\left\{\dfrac{5e^{-2s}}{s^2+1}\right\}$

5. Determine $\mathcal{L}^{-1}\left\{\dfrac{3s\,e^{-4s}}{s^2+16}\right\}$

6. Determine $\mathcal{L}^{-1}\left\{\dfrac{6e^{-2s}}{s^2-1}\right\}$

7. Determine $\mathcal{L}^{-1}\left\{\dfrac{3e^{-6s}}{s^3}\right\}$

8. Determine $\mathcal{L}^{-1}\left\{\dfrac{2s\,e^{-4s}}{s^2-16}\right\}$

9. Determine $\mathcal{L}^{-1}\left\{\dfrac{2s\,e^{-\frac{1}{2}s}}{s^2+5}\right\}$

10. Determine $\mathcal{L}^{-1}\left\{\dfrac{4e^{-7s}}{s-1}\right\}$

For fully worked solutions to each of the problems in Practice Exercises 255 to 257 in this chapter, go to the website:
www.routledge.com/cw/bird

Section L

Chapter 71

The solution of differential equations using Laplace transforms

Why it is important to understand: The solution of differential equations using Laplace transforms

Laplace transforms and their inverses are a mathematical technique which allows us to solve differential equations, by primarily using algebraic methods. This simplification in the solving of equations, coupled with the ability to directly implement electrical components in their transformed form, makes the use of Laplace transforms widespread in both electrical engineering and control systems engineering. The procedures explained in previous chapters are used in this chapter which demonstrates how differential equations are solved using Laplace transforms.

At the end of this chapter, you should be able to:

- understand the procedure to solve differential equations using Laplace transforms
- solve differential equations using Laplace transforms

71.1 Introduction

An alternative method of solving differential equations to that used in Chapters 33 and 50 to 54 is possible by using Laplace transforms.

71.2 Procedure to solve differential equations by using Laplace transforms

(i) Take the Laplace transform of both sides of the differential equation by applying the formulae for the Laplace transforms of derivatives (i.e. equations (3) and (4) of Chapter 68) and, where necessary, using a list of standard Laplace transforms, such as Tables 67.1 and 68.1 on pages 737 and 742.

(ii) Put in the given initial conditions, i.e. $y(0)$ and $y'(0)$.

(iii) Rearrange the equation to make $\mathcal{L}\{y\}$ the subject.

(iv) Determine y by using, where necessary, partial fractions, and taking the inverse of each term by using Table 69.1 on page 748.

71.3 Worked problems on solving differential equations using Laplace transforms

Problem 1. Use Laplace transforms to solve the differential equation

$$2\frac{d^2y}{dx^2} + 5\frac{dy}{dx} - 3y = 0, \text{ given that when } x = 0,$$

$$y = 4 \text{ and } \frac{dy}{dx} = 9$$

This is the same problem as Problem 1 of Chapter 53, page 581 and a comparison of methods can be made. Using the above procedure:

(i) $\quad 2\mathcal{L}\left\{\dfrac{d^2y}{dx^2}\right\} + 5\mathcal{L}\left\{\dfrac{dy}{dx}\right\} - 3\mathcal{L}\{y\} = \mathcal{L}\{0\}$

$$2[s^2\mathcal{L}\{y\} - sy(0) - y'(0)] + 5[s\mathcal{L}\{y\}$$
$$- y(0)] - 3\mathcal{L}\{y\} = 0,$$

from equations (3) and (4) of Chapter 68.

(ii) $\quad y(0) = 4$ and $y'(0) = 9$

Thus $\quad 2[s^2\mathcal{L}\{y\} - 4s - 9] + 5[s\mathcal{L}\{y\} - 4]$
$$- 3\mathcal{L}\{y\} = 0$$

i.e. $\quad 2s^2\mathcal{L}\{y\} - 8s - 18 + 5s\mathcal{L}\{y\} - 20$
$$- 3\mathcal{L}\{y\} = 0$$

(iii) Rearranging gives:

$$(2s^2 + 5s - 3)\mathcal{L}\{y\} = 8s + 38$$

i.e. $\quad \mathcal{L}\{y\} = \dfrac{8s + 38}{2s^2 + 5s - 3}$

(iv) $\quad y = \mathcal{L}^{-1}\left\{\dfrac{8s + 38}{2s^2 + 5s - 3}\right\}$

$$\frac{8s + 38}{2s^2 + 5s - 3} \equiv \frac{8s + 38}{(2s - 1)(s + 3)}$$

$$\equiv \frac{A}{2s - 1} + \frac{B}{s + 3}$$

$$\equiv \frac{A(s + 3) + B(2s - 1)}{(2s - 1)(s + 3)}$$

Hence $8s + 38 = A(s + 3) + B(2s - 1)$

When $s = \dfrac{1}{2}$, $42 = 3\dfrac{1}{2}A$, from which, $A = 12$

When $s = -3$, $14 = -7B$, from which, $B = -2$

Hence $y = \mathcal{L}^{-1}\left\{\dfrac{8s + 38}{2s^2 + 5s - 3}\right\}$

$$= \mathcal{L}^{-1}\left\{\frac{12}{2s - 1} - \frac{2}{s + 3}\right\}$$

$$= \mathcal{L}^{-1}\left\{\frac{12}{2\left(s - \frac{1}{2}\right)}\right\} - \mathcal{L}^{-1}\left\{\frac{2}{s + 3}\right\}$$

Hence $y = 6e^{\frac{1}{2}x} - 2e^{-3x}$, from (iii) of

Table 69.1.

Problem 2. Use Laplace transforms to solve the differential equation:

$$\frac{d^2y}{dx^2} + 6\frac{dy}{dx} + 13y = 0, \text{ given that when } x = 0, y = 3$$

and $\dfrac{dy}{dx} = 7$

This is the same as Problem 3 of Chapter 53, page 562. Using the above procedure:

(i) $\quad \mathcal{L}\left\{\dfrac{d^2x}{dy^2}\right\} + 6\mathcal{L}\left\{\dfrac{dy}{dx}\right\} + 13\mathcal{L}\{y\} = \mathcal{L}\{0\}$

Hence $\quad [s^2\mathcal{L}\{y\} - sy(0) - y'(0)]$
$$+ 6[s\mathcal{L}\{y\} - y(0)] + 13\mathcal{L}\{y\} = 0,$$

from equations (3) and (4) of Chapter 68.

(ii) $\quad y(0) = 3$ and $y'(0) = 7$

Thus $\quad s^2\mathcal{L}\{y\} - 3s - 7 + 6s\mathcal{L}\{y\}$
$$- 18 + 13\mathcal{L}\{y\} = 0$$

(iii)   Rearranging gives:

$$(s^2 + 6s + 13)\mathcal{L}\{y\} = 3s + 25$$

i.e. $$\mathcal{L}\{y\} = \frac{3s + 25}{s^2 + 6s + 13}$$

(iv) $$y = \mathcal{L}^{-1}\left\{ \frac{3s + 25}{s^2 + 6s + 13} \right\}$$

$$= \mathcal{L}^{-1}\left\{ \frac{3s + 25}{(s + 3)^2 + 2^2} \right\}$$

$$= \mathcal{L}^{-1}\left\{ \frac{3(s + 3) + 16}{(s + 3)^2 + 2^2} \right\}$$

$$= \mathcal{L}^{-1}\left\{ \frac{3(s + 3)}{(s + 3)^2 + 2^2} \right\}$$

$$+ \mathcal{L}^{-1}\left\{ \frac{8(2)}{(s + 3)^2 + 2^2} \right\}$$

$$= 3e^{-3t}\cos 2t + 8e^{-3t}\sin 2t, \text{ from (xiii)}$$

and (xii) of Table 69.1

Hence $y = e^{-3t}(3\cos 2t + 8\sin 2t)$

Problem 3. Use Laplace transforms to solve the differential equation:
$$\frac{d^2 y}{dx^2} - 3\frac{dy}{dx} = 9, \text{ given that when } x = 0, y = 0 \text{ and}$$
$$\frac{dy}{dx} = 0$$

This is the same problem as Problem 2 of Chapter 54, page 569. Using the procedure:

(i) $$\mathcal{L}\left\{ \frac{d^2 y}{dx^2} \right\} - 3\mathcal{L}\left\{ \frac{dy}{dx} \right\} = \mathcal{L}\{9\}$$

Hence $[s^2\mathcal{L}\{y\} - sy(0) - y'(0)]$
$$- 3[s\mathcal{L}\{y\} - y(0)] = \frac{9}{s}$$

(ii) $y(0) = 0$ and $y'(0) = 0$

Hence $s^2\mathcal{L}\{y\} - 3s\mathcal{L}\{y\} = \dfrac{9}{s}$

(iii)   Rearranging gives:

$$(s^2 - 3s)\mathcal{L}\{y\} = \frac{9}{s}$$

i.e. $$\mathcal{L}\{y\} = \frac{9}{s(s^2 - 3s)} = \frac{9}{s^2(s - 3)}$$

(iv) $$y = \mathcal{L}^{-1}\left\{ \frac{9}{s^2(s - 3)} \right\}$$

$$\frac{9}{s^2(s - 3)} \equiv \frac{A}{s} + \frac{B}{s^2} + \frac{C}{s - 3}$$

$$\equiv \frac{A(s)(s - 3) + B(s - 3) + Cs^2}{s^2(s - 3)}$$

Hence $9 \equiv A(s)(s - 3) + B(s - 3) + Cs^2$

When $s = 0, 9 = -3B$, from which, $B = -3$

When $s = 3, 9 = 9C$, from which, $C = 1$

Equating s^2 terms gives: $0 = A + C$, from which, $A = -1$, since $C = 1$. Hence,

$$\mathcal{L}^{-1}\left\{ \frac{9}{s^2(s - 3)} \right\} = \mathcal{L}^{-1}\left\{ -\frac{1}{s} - \frac{3}{s^2} + \frac{1}{s - 3} \right\}$$

$$= -1 - 3x + e^{3x}, \text{ from (i),}$$

(vi) and (iii) of Table 69.1.

i.e. $\boldsymbol{y = e^{3x} - 3x - 1}$

Problem 4. Use Laplace transforms to solve the differential equation:
$$\frac{d^2 y}{dx^2} - 7\frac{dy}{dx} + 10y = e^{2x} + 20, \text{ given that when}$$
$$x = 0, y = 0 \text{ and } \frac{dy}{dx} = -\frac{1}{3}$$

Using the procedure:

(i) $$\mathcal{L}\left\{ \frac{d^2 y}{dx^2} \right\} - 7\mathcal{L}\left\{ \frac{dy}{dx} \right\} + 10\mathcal{L}\{y\} = \mathcal{L}\{e^{2x} + 20\}$$

Hence $[s^2\mathcal{L}\{y\} - sy(0) - y'(0)] - 7[s\mathcal{L}\{y\}$
$$- y(0)] + 10\mathcal{L}\{y\} = \frac{1}{s - 2} + \frac{20}{s}$$

(ii) $y(0) = 0$ and $y'(0) = -\dfrac{1}{3}$

Hence $s^2\mathcal{L}\{y\} - 0 - \left(-\dfrac{1}{3} \right) - 7s\mathcal{L}\{y\} + 0$
$$+ 10\mathcal{L}\{y\} = \frac{21s - 40}{s(s - 2)}$$

(iii) $(s^2 - 7s + 10)\mathcal{L}\{y\} = \dfrac{21s - 40}{s(s - 2)} - \dfrac{1}{3}$

$$= \frac{3(21s - 40) - s(s - 2)}{3s(s - 2)}$$

$$= \frac{-s^2 + 65s - 120}{3s(s-2)}$$

Hence $\mathcal{L}\{y\} = \dfrac{-s^2 + 65s - 120}{3s(s-2)(s^2 - 7s + 10)}$

$$= \frac{1}{3}\left[\frac{-s^2 + 65s - 120}{s(s-2)(s-2)(s-5)}\right]$$

$$= \frac{1}{3}\left[\frac{-s^2 + 65s - 120}{s(s-5)(s-2)^2}\right]$$

(iv) $y = \dfrac{1}{3}\mathcal{L}^{-1}\left\{\dfrac{-s^2 + 65s - 120}{s(s-5)(s-2)^2}\right\}$

$$\frac{-s^2 + 65s - 120}{s(s-5)(s-2)^2}$$

$$\equiv \frac{A}{s} + \frac{B}{s-5} + \frac{C}{s-2} + \frac{D}{(s-2)^2}$$

$$\equiv \frac{\left(\begin{array}{c} A(s-5)(s-2)^2 + B(s)(s-2)^2 \\ + C(s)(s-5)(s-2) + D(s)(s-5) \end{array}\right)}{s(s-5)(s-2)^2}$$

Hence

$-s^2 + 65s - 120$

$\equiv A(s-5)(s-2)^2 + B(s)(s-2)^2$

$\qquad + C(s)(s-5)(s-2) + D(s)(s-5)$

When $s = 0, -120 = -20A$, from which, $A = 6$

When $s = 5, 180 = 45B$, from which, $B = 4$

When $s = 2, 6 = -6D$, from which, $D = -1$

Equating s^3 terms gives: $0 = A + B + C$, from which, $C = -10$

Hence $\dfrac{1}{3}\mathcal{L}^{-1}\left\{\dfrac{-s^2 + 65s - 120}{s(s-5)(s-2)^2}\right\}$

$$= \frac{1}{3}\mathcal{L}^{-1}\left\{\frac{6}{s} + \frac{4}{s-5} - \frac{10}{s-2} - \frac{1}{(s-2)^2}\right\}$$

$$= \frac{1}{3}[6 + 4e^{5x} - 10e^{2x} - xe^{2x}]$$

Thus $y = 2 + \dfrac{4}{3}e^{5x} - \dfrac{10}{3}e^{2x} - \dfrac{x}{3}e^{2x}$

Problem 5. The current flowing in an electrical circuit is given by the differential equation $Ri + L(di/dt) = E$, where E, L and R are constants. Use Laplace transforms to solve the equation for current i given that when $t = 0$, $i = 0$

Using the procedure:

(i) $\mathcal{L}\{Ri\} + \mathcal{L}\left\{L\dfrac{di}{dt}\right\} = \mathcal{L}\{E\}$

i.e. $R\mathcal{L}\{i\} + L[s\mathcal{L}\{i\} - i(0)] = \dfrac{E}{s}$

(ii) $i(0) = 0$, hence $R\mathcal{L}\{i\} + Ls\mathcal{L}\{i\} = \dfrac{E}{s}$

(iii) Rearranging gives:

$$(R + Ls)\mathcal{L}\{i\} = \frac{E}{s}$$

i.e. $\mathcal{L}\{i\} = \dfrac{E}{s(R + Ls)}$

(iv) $i = \mathcal{L}^{-1}\left\{\dfrac{E}{s(R + Ls)}\right\}$

$$\frac{E}{s(R + Ls)} \equiv \frac{A}{s} + \frac{B}{R + Ls}$$

$$\equiv \frac{A(R + Ls) + Bs}{s(R + Ls)}$$

Hence $E = A(R + Ls) + Bs$

When $s = 0, E = AR$,

from which, $A = \dfrac{E}{R}$

When $s = -\dfrac{R}{L}, E = B\left(-\dfrac{R}{L}\right)$

from which, $B = -\dfrac{EL}{R}$

Hence $\mathcal{L}^{-1}\left\{\dfrac{E}{s(R + Ls)}\right\}$

$$= \mathcal{L}^{-1}\left\{\frac{E/R}{s} + \frac{-EL/R}{R + Ls}\right\}$$

$$= \mathcal{L}^{-1}\left\{\frac{E}{Rs} - \frac{EL}{R(R + Ls)}\right\}$$

$$= \mathcal{L}^{-1}\left\{\frac{E}{R}\left(\frac{1}{s}\right) - \frac{E}{R}\left(\frac{1}{\dfrac{R}{L} + s}\right)\right\}$$

$$= \frac{E}{R} \mathcal{L}^{-1} \left\{ \frac{1}{s} - \frac{1}{\left(s + \frac{R}{L}\right)} \right\}$$

Hence **current** $i = \frac{E}{R} \left(1 - e^{-\frac{Rt}{L}} \right)$

Now try the following Practice Exercise

Practice Exercise 258   Solving differential equations using Laplace transforms (Answers on page 896)

1. A first-order differential equation involving current i in a series $R-L$ circuit is given by:
    $\frac{di}{dt} + 5i = \frac{E}{2}$ and $i = 0$ at time $t = 0$

    Use Laplace transforms to solve for i when (a) $E = 20$ (b) $E = 40\,e^{-3t}$ and (c) $E = 50\sin 5t$

In Problems 2 to 9, use Laplace transforms to solve the given differential equations.

2. $9\frac{d^2 y}{dt^2} - 24\frac{dy}{dt} + 16y = 0$, given $y(0) = 3$ and $y'(0) = 3$

3. $\frac{d^2 x}{dt^2} + 100x = 0$, given $x(0) = 2$ and $x'(0) = 0$

4. $\frac{d^2 i}{dt^2} + 1000\frac{di}{dt} + 250\,000i = 0$, given $i(0) = 0$ and $i'(0) = 100$

5. $\frac{d^2 x}{dt^2} + 6\frac{dx}{dt} + 8x = 0$, given $x(0) = 4$ and $x'(0) = 8$

6. $\frac{d^2 y}{dx^2} - 2\frac{dy}{dx} + y = 3\,e^{4x}$, given $y(0) = -\frac{2}{3}$ and $y'(0) = 4\frac{1}{3}$

7. $\frac{d^2 y}{dx^2} - 3\frac{dy}{dx} - 4y = 3\sin x$, given $y(0) = 0$ and $y'(0) = 0$

8. $\frac{d^2 y}{dx^2} + \frac{dy}{dx} - 2y = 3\cos 3x - 11\sin 3x$, given $y(0) = 0$ and $y'(0) = 6$

9. $\frac{d^2 y}{dx^2} - 2\frac{dy}{dx} + 2y = 3\,e^x \cos 2x$, given $y(0) = 2$ and $y'(0) = 5$

10. The free oscillations of a lightly damped elastic system are given by the equation:

    $$\frac{d^2 y}{dt^2} + 2\frac{dy}{dt} + 5y = 0$$

    where y is the displacement from the equilibrium position. If when time $t = 0$, $y = 2$ and $\frac{dy}{dt} = 0$, determine an expression for the displacement.

11. Solve, using Laplace transforms, Problems 4 to 9 of Exercise 203, page 563 and Problems 1 to 5 of Exercise 204, page 565.

12. Solve, using Laplace transforms, Problems 3 to 6 of Exercise 205, page 570, Problems 5 and 6 of Exercise 206, page 572, Problems 4 and 7 of Exercise 207, page 574 and Problems 5 and 6 of Exercise 208, page 576.

For fully worked solutions to each of the problems in Practice Exercise 258 in this chapter, go to the website: www.routledge.com/cw/bird

Section L

Chapter 72

The solution of simultaneous differential equations using Laplace transforms

Why it is important to understand: The solution of simultaneous differential equations using Laplace transforms

As stated in previous chapters, Laplace transforms have many applications in mathematics, physics, optics, electrical engineering, control engineering, signal processing, and probability and Laplace transforms and their inverses are a mathematical technique which allows us to solve differential equations, by primarily using algebraic methods. Specifically, this chapter explains the procedure for solving simultaneous differential equations; this requires all of the knowledge gained in the preceding chapters.

At the end of this chapter, you should be able to:

- understand the procedure to solve simultaneous differential equations using Laplace transforms
- solve simultaneous differential equations using Laplace transforms

72.1 Introduction

It is sometimes necessary to solve simultaneous differential equations. An example occurs when two electrical circuits are coupled magnetically where the equations relating the two currents i_1 and i_2 are typically:

$$L_1 \frac{di_1}{dt} + M \frac{di_2}{dt} + R_1 i_1 = E_1$$

$$L_2 \frac{di_2}{dt} + M \frac{di_1}{dt} + R_2 i_2 = 0$$

where L represents inductance, R resistance, M mutual inductance and E_1 the p.d. applied to one of the circuits.

72.2 Procedure to solve simultaneous differential equations using Laplace transforms

(i) Take the Laplace transform of both sides of each simultaneous equation by applying the formulae for the Laplace transforms of derivatives (i.e. equations (3) and (4) of Chapter 68, page 744) and using a list of standard Laplace transforms, as in Table 67.1, page 738 and Table 68.1, page 742.

(ii) Put in the initial conditions, i.e. $x(0)$, $y(0)$, $x'(0)$, $y'(0)$

(iii) Solve the simultaneous equations for $\mathcal{L}\{y\}$ and $\mathcal{L}\{x\}$ by the normal algebraic method.

(iv) Determine y and x by using, where necessary, partial fractions, and taking the inverse of each term.

72.3 Worked problems on solving simultaneous differential equations by using Laplace transforms

Problem 1. Solve the following pair of simultaneous differential equations

$$\frac{dy}{dt} + x = 1$$

$$\frac{dx}{dt} - y + 4e^t = 0$$

given that at $t = 0$, $x = 0$ and $y = 0$

Using the above procedure:

(i) $\mathcal{L}\left\{\dfrac{dy}{dt}\right\} + \mathcal{L}\{x\} = \mathcal{L}\{1\}$ (1)

$\mathcal{L}\left\{\dfrac{dx}{dt}\right\} - \mathcal{L}\{y\} + 4\mathcal{L}\{e^t\} = 0$ (2)

Equation (1) becomes:

$$[s\mathcal{L}\{y\} - y(0)] + \mathcal{L}\{x\} = \frac{1}{s} \qquad (1')$$

from equation (3), page 744 and Table 67.1, page 738.

Equation (2) becomes:

$$[s\mathcal{L}\{x\} - x(0)] - \mathcal{L}\{y\} = -\frac{4}{s-1} \qquad (2')$$

(ii) $x(0) = 0$ and $y(0) = 0$ hence

Equation $(1')$ becomes:

$$s\mathcal{L}\{y\} + \mathcal{L}\{x\} = \frac{1}{s} \qquad (1'')$$

and equation $(2')$ becomes:

$$s\mathcal{L}\{x\} - \mathcal{L}\{y\} = -\frac{4}{s-1}$$

$$\text{or } -\mathcal{L}\{y\} + s\mathcal{L}\{x\} = -\frac{4}{s-1} \qquad (2'')$$

(iii) $1 \times$ equation $(1'')$ and $s \times$ equation $(2'')$ gives:

$$s\mathcal{L}\{y\} + \mathcal{L}\{x\} = \frac{1}{s} \qquad (3)$$

$$-s\mathcal{L}\{y\} + s^2\mathcal{L}\{x\} = -\frac{4s}{s-1} \qquad (4)$$

Adding equations (3) and (4) gives:

$$(s^2 + 1)\mathcal{L}\{x\} = \frac{1}{s} - \frac{4s}{s-1}$$

$$= \frac{(s-1) - s(4s)}{s(s-1)}$$

$$= \frac{-4s^2 + s - 1}{s(s-1)}$$

from which, $\mathcal{L}\{x\} = \dfrac{-4s^2 + s - 1}{s(s-1)(s^2+1)} \qquad (5)$

Using partial fractions

$$\frac{-4s^2 + s - 1}{s(s-1)(s^2+1)} \equiv \frac{A}{s} + \frac{B}{(s-1)} + \frac{Cs+D}{(s^2+1)}$$

$$= \frac{\left(\begin{array}{c} A(s-1)(s^2+1) + Bs(s^2+1) \\ +(Cs+D)s(s-1) \end{array}\right)}{s(s-1)(s^2+1)}$$

Hence

$$-4s^2 + s - 1 = A(s-1)(s^2+1) + Bs(s^2+1)$$
$$+ (Cs+D)s(s-1)$$

When $s = 0$, $-1 = -A$ hence $A = 1$

When $s = 1$, $-4 = 2B$ hence $B = -2$

Equating s^3 coefficients:

$$0 = A + B + C \quad \text{hence} \quad C = 1$$
$$(\text{since } A = 1 \text{ and } B = -2)$$

Equating s^2 coefficients:

$$-4 = -A + D - C \quad \text{hence} \quad D = -2$$
$$(\text{since } A = 1 \text{ and } C = 1)$$

Thus $\mathcal{L}\{x\} = \dfrac{-4s^2 + s - 1}{s(s-1)(s^2+1)}$

$= \dfrac{1}{s} - \dfrac{2}{(s-1)} + \dfrac{s-2}{(s^2+1)}$

(iv)   Hence

$x = \mathcal{L}^{-1}\left\{\dfrac{1}{s} - \dfrac{2}{(s-1)} + \dfrac{s-2}{(s^2+1)}\right\}$

$= \mathcal{L}^{-1}\left\{\dfrac{1}{s} - \dfrac{2}{(s-1)} + \dfrac{s}{(s^2+1)} - \dfrac{2}{(s^2+1)}\right\}$

i.e. $x = 1 - 2e^t + \cos t - 2\sin t,$

from Table 69.1, page 748

From the second equation given in the question,

$\dfrac{dx}{dt} - y + 4e^t = 0$

from which,

$y = \dfrac{dx}{dt} + 4e^t$

$= \dfrac{d}{dt}(1 - 2e^t + \cos t - 2\sin t) + 4e^t$

$= -2e^t - \sin t - 2\cos t + 4e^t$

i.e. $y = 2e^t - \sin t - 2\cos t$

[Alternatively, to determine y, return to equations $(1'')$ and $(2'')$]

Problem 2. Solve the following pair of simultaneous differential equations

$3\dfrac{dx}{dt} - 5\dfrac{dy}{dt} + 2x = 6$

$2\dfrac{dy}{dt} - \dfrac{dx}{dt} - y = -1$

given that at $t = 0$, $x = 8$ and $y = 3$

Using the above procedure:

(i) $3\mathcal{L}\left\{\dfrac{dx}{dt}\right\} - 5\mathcal{L}\left\{\dfrac{dy}{dt}\right\} + 2\mathcal{L}\{x\} = \mathcal{L}\{6\}$ (1)

$2\mathcal{L}\left\{\dfrac{dy}{dt}\right\} - \mathcal{L}\left\{\dfrac{dx}{dt}\right\} - \mathcal{L}\{y\} = \mathcal{L}\{-1\}$ (2)

Equation (1) becomes:

$3[s\mathcal{L}\{x\} - x(0)] - 5[s\mathcal{L}\{y\} - y(0)]$

$+ 2\mathcal{L}\{x\} = \dfrac{6}{s}$

from equation (3), page 744, and Table 67.1, page 738.

i.e. $3s\mathcal{L}\{x\} - 3x(0) - 5s\mathcal{L}\{y\}$

$+ 5y(0) + 2\mathcal{L}\{x\} = \dfrac{6}{s}$

i.e. $(3s+2)\mathcal{L}\{x\} - 3x(0) - 5s\mathcal{L}\{y\}$

$+ 5y(0) = \dfrac{6}{s}$ (1')

Equation (2) becomes:

$2[s\mathcal{L}\{y\} - y(0)] - [s\mathcal{L}\{x\} - x(0)]$

$- \mathcal{L}\{y\} = -\dfrac{1}{s}$

from equation (3), page 744, and Table 67.1, page 738,

i.e. $2s\mathcal{L}\{y\} - 2y(0) - s\mathcal{L}\{x\}$

$+ x(0) - \mathcal{L}\{y\} = -\dfrac{1}{s}$

i.e. $(2s-1)\mathcal{L}\{y\} - 2y(0) - s\mathcal{L}\{x\}$

$+ x(0) = -\dfrac{1}{s}$ (2')

(ii) $x(0) = 8$ and $y(0) = 3$, hence equation $(1')$ becomes

$(3s+2)\mathcal{L}\{x\} - 3(8) - 5s\mathcal{L}\{y\}$

$+ 5(3) = \dfrac{6}{s}$ (1'')

and equation $(2')$ becomes

$(2s-1)\mathcal{L}\{y\} - 2(3) - s\mathcal{L}\{x\}$

$+ 8 = -\dfrac{1}{s}$ (2'')

i.e. $(3s+2)\mathcal{L}\{x\} - 5s\mathcal{L}\{y\} = \dfrac{6}{s} + 9$ (1'')

$(3s+2)\mathcal{L}\{x\} - 5s\mathcal{L}\{y\}$

$= \dfrac{6}{s} + 9$ (1''')

$-s\mathcal{L}\{x\} + (2s-1)\mathcal{L}\{y\}$ (A)

$= -\dfrac{1}{s} - 2$ (2''')

(iii) $s \times$ equation $(1''')$ and $(3s + 2) \times$ equation $(2''')$ gives:

$$s(3s+2)\mathcal{L}\{x\} - 5s^2\mathcal{L}\{y\} = s\left(\frac{6}{s}+9\right) \quad (3)$$

$$-s(3s+2)\mathcal{L}\{x\} + (3s+2)(2s-1)\mathcal{L}\{y\}$$

$$= (3s+2)\left(-\frac{1}{s}-2\right) \quad (4)$$

i.e. $s(3s+2)\mathcal{L}\{x\} - 5s^2\mathcal{L}\{y\} = 6 + 9s \quad (3')$

$$-s(3s+2)\mathcal{L}\{x\} + (6s^2+s-2)\mathcal{L}\{y\}$$

$$= -6s - \frac{2}{s} - 7 \quad (4')$$

Adding equations $(3')$ and $(4')$ gives:

$$(s^2+s-2)\mathcal{L}\{y\} = -1 + 3s - \frac{2}{s}$$

$$= \frac{-s+3s^2-2}{s}$$

from which, $\mathcal{L}\{y\} = \dfrac{3s^2-s-2}{s(s^2+s-2)}$

Using partial fractions

$$\frac{3s^2-s-2}{s(s^2+s-2)}$$

$$\equiv \frac{A}{s} + \frac{B}{(s+2)} + \frac{C}{(s-1)}$$

$$= \frac{A(s+2)(s-1) + Bs(s-1) + Cs(s+2)}{s(s+2)(s-1)}$$

i.e. $3s^2 - s - 2 = A(s+2)(s-1)$

$$+ Bs(s-1) + Cs(s+2)$$

When $s=0, -2=-2A$, hence $A=1$

When $s=1, 0=3C$, hence $C=0$

When $s=-2, 12=6B$, hence $B=2$

Thus $\mathcal{L}\{y\} = \dfrac{3s^2-s-2}{s(s^2+s-2)} = \dfrac{1}{s} + \dfrac{2}{(s+2)}$

(iv) Hence $y = \mathcal{L}^{-1}\left\{\dfrac{1}{s} + \dfrac{2}{s+2}\right\} = \mathbf{1 + 2e^{-2t}}$

Returning to equations (A) to determine $\mathcal{L}\{x\}$ and hence x:

$(2s-1) \times$ equation $(1''')$ and $5s \times (2''')$ gives:

$$(2s-1)(3s+2)\mathcal{L}\{x\} - 5s(2s-1)\mathcal{L}\{y\}$$

$$= (2s-1)\left(\frac{6}{s}+9\right) \quad (5)$$

and $-s(5s)\mathcal{L}\{x\} + 5s(2s-1)\mathcal{L}\{y\}$

$$= 5s\left(-\frac{1}{s}-2\right) \quad (6)$$

i.e. $(6s^2+s-2)\mathcal{L}\{x\} - 5s(2s-1)\mathcal{L}\{y\}$

$$= 12 + 18s - \frac{6}{s} - 9 \quad (5')$$

and $-5s^2\mathcal{L}\{x\} + 5s(2s-1)\mathcal{L}\{y\}$

$$= -5 - 10s \quad (6')$$

Adding equations $(5')$ and $(6')$ gives:

$$(s^2+s-2)\mathcal{L}\{x\} = -2 + 8s - \frac{6}{s}$$

$$= \frac{-2s+8s^2-6}{s}$$

from which, $\mathcal{L}\{x\} = \dfrac{8s^2-2s-6}{s(s^2+s-2)}$

$$= \frac{8s^2-2s-6}{s(s+2)(s-1)}$$

Using partial fractions

$$\frac{8s^2-2s-6}{s(s+2)(s-1)}$$

$$\equiv \frac{A}{s} + \frac{B}{(s+2)} + \frac{C}{(s-1)}$$

$$= \frac{A(s+2)(s-1) + Bs(s-1) + Cs(s+2)}{s(s+2)(s-1)}$$

i.e. $8s^2 - 2s - 6 = A(s+2)(s-1)$

$$+ Bs(s-1) + Cs(s+2)$$

Section L

When $s=0$, $-6=-2A$, hence $A=3$

When $s=1$, $0=3C$, hence $C=0$

When $s=-2$, $30=6B$, hence $B=5$

Thus $\mathcal{L}\{x\} = \dfrac{8s^2-2s-6}{s(s+2)(s-1)} = \dfrac{3}{s} + \dfrac{5}{(s+2)}$

Hence $x = \mathcal{L}^{-1}\left\{\dfrac{3}{s} + \dfrac{5}{s+2}\right\} = 3+5e^{-2t}$

Therefore the solutions of the given simultaneous differential equations are

$$y = 1+2e^{-2t} \quad \text{and} \quad x = 3+5e^{-2t}$$

(These solutions may be checked by substituting the expressions for x and y into the original equations.)

> **Problem 3.** Solve the following pair of simultaneous differential equations
>
> $$\dfrac{d^2x}{dt^2} - x = y$$
>
> $$\dfrac{d^2y}{dt^2} + y = -x$$
>
> given that at $t=0$, $x=2$, $y=-1$, $\dfrac{dx}{dt}=0$
>
> and $\dfrac{dy}{dt}=0$

Using the procedure:

(i) $\quad [s^2\mathcal{L}\{x\} - sx(0) - x'(0)] - \mathcal{L}\{x\} = \mathcal{L}\{y\}$ (1)

$\quad [s^2\mathcal{L}\{y\} - sy(0) - y'(0)] + \mathcal{L}\{y\} = -\mathcal{L}\{x\}$ (2)

from equation (4), page 744

(ii) $\quad x(0)=2$, $y(0)=-1$, $x'(0)=0$ and $y'(0)=0$

hence $\quad s^2\mathcal{L}\{x\} - 2s - \mathcal{L}\{x\} = \mathcal{L}\{y\}$ (1′)

$\quad s^2\mathcal{L}\{y\} + s + \mathcal{L}\{y\} = -\mathcal{L}\{x\}$ (2′)

(iii) Rearranging gives:

$$(s^2-1)\mathcal{L}\{x\} - \mathcal{L}\{y\} = 2s \quad (3)$$

$$\mathcal{L}\{x\} + (s^2+1)\mathcal{L}\{y\} = -s \quad (4)$$

Equation $(3) \times (s^2+1)$ and equation $(4) \times 1$ gives:

$$(s^2+1)(s^2-1)\mathcal{L}\{x\} - (s^2+1)\mathcal{L}\{y\}$$
$$= (s^2+1)2s \quad (5)$$

$$\mathcal{L}\{x\} + (s^2+1)\mathcal{L}\{y\} = -s \quad (6)$$

Adding equations (5) and (6) gives:

$$[(s^2+1)(s^2-1) + 1]\mathcal{L}\{x\} = (s^2+1)2s - s$$

i.e. $\quad s^4\mathcal{L}\{x\} = 2s^3 + s = s(2s^2+1)$

from which, $\quad \mathcal{L}\{x\} = \dfrac{s(2s^2+1)}{s^4} = \dfrac{2s^2+1}{s^3}$

$$= \dfrac{2s^2}{s^3} + \dfrac{1}{s^3} = \dfrac{2}{s} + \dfrac{1}{s^3}$$

(iv) Hence $\quad x = \mathcal{L}^{-1}\left\{\dfrac{2}{s} + \dfrac{1}{s^3}\right\}$

i.e. $\qquad x = 2 + \dfrac{1}{2}t^2$

Returning to equations (3) and (4) to determine y:

$1 \times$ equation (3) and $(s^2-1) \times$ equation (4) gives:

$$(s^2-1)\mathcal{L}\{x\} - \mathcal{L}\{y\} = 2s \quad (7)$$

$$(s^2-1)\mathcal{L}\{x\} + (s^2-1)(s^2+1)\mathcal{L}\{y\}$$
$$= -s(s^2-1) \quad (8)$$

Equation $(7)-$ equation (8) gives:

$$[-1 - (s^2-1)(s^2+1)]\mathcal{L}\{y\}$$
$$= 2s + s(s^2-1)$$

i.e. $\qquad -s^4\mathcal{L}\{y\} = s^3 + s$

and $\qquad \mathcal{L}\{y\} = \dfrac{s^3+s}{-s^4} = -\dfrac{1}{s} - \dfrac{1}{s^3}$

from which, $\qquad y = \mathcal{L}^{-1}\left\{-\dfrac{1}{s} - \dfrac{1}{s^3}\right\}$

i.e. $\qquad y = -1 - \dfrac{1}{2}t^2$

Now try the following Practice Exercise

Practice Exercise 259   Solving simultaneous differential equations using Laplace transforms (Answers on page 896)

Solve the following pairs of simultaneous differential equations:

1. $2\dfrac{dx}{dt} + \dfrac{dy}{dt} = 5\,e^t$

$\dfrac{dy}{dt} - 3\dfrac{dx}{dt} = 5$

given that when $t=0$, $x=0$ and $y=0$

2. $2\dfrac{dy}{dt} - y + x + \dfrac{dx}{dt} - 5\sin t = 0$

$3\dfrac{dy}{dt} + x - y + 2\dfrac{dx}{dt} - e^t = 0$

given that at $t=0$, $x=0$ and $y=0$

3. $\dfrac{d^2x}{dt^2} + 2x = y$

$\dfrac{d^2y}{dt^2} + 2y = x$

given that at $t=0$, $x=4$, $y=2$, $\dfrac{dx}{dt}=0$

and $\dfrac{dy}{dt}=0$

For fully worked solutions to each of the problems in Practice Exercise 259 in this chapter, go to the website:
www.routledge.com/cw/bird

Section L

Revision Test 20   Laplace transforms

This Revision Test covers the material contained in Chapters 67 to 72. *The marks for each question are shown in brackets at the end of each question.*

1. Find the Laplace transforms of the following functions:

    (a) $2t^3 - 4t + 5$ (b) $3e^{-2t} - 4\sin 2t$

    (c) $3\cosh 2t$ (d) $2t^4 e^{-3t}$

    (e) $5e^{2t}\cos 3t$ (f) $2e^{3t}\sinh 4t$ (16)

2. Find the inverse Laplace transforms of the following functions:

    (a) $\dfrac{5}{2s+1}$ (b) $\dfrac{12}{s^5}$

    (c) $\dfrac{4s}{s^2+9}$ (d) $\dfrac{5}{s^2-9}$

    (e) $\dfrac{3}{(s+2)^4}$ (f) $\dfrac{s-4}{s^2-8s-20}$

    (g) $\dfrac{8}{s^2-4s+3}$ (17)

3. Use partial fractions to determine the following:

    (a) $\mathcal{L}^{-1}\left\{\dfrac{5s-1}{s^2-s-2}\right\}$

    (b) $\mathcal{L}^{-1}\left\{\dfrac{2s^2+11s-9}{s(s-1)(s+3)}\right\}$

    (c) $\mathcal{L}^{-1}\left\{\dfrac{13-s^2}{s(s^2+4s+13)}\right\}$ (24)

4. Determine the poles and zeros for the transfer function: $F(s) = \dfrac{(s+2)(s-3)}{(s+3)(s^2+2s+5)}$ and plot them on a pole–zero diagram.   (10)

5. Sketch the graphs of (a) $f(t) = 4H(t-3)$ (b) $f(t) = 3[H(t-2) - H(t-5)]$ (5)

6. Determine (a) $\mathcal{L}\{H(t-2).e^{t-2}\}$ (b) $\mathcal{L}\{5H(t-1).\sin(t-1)\}$ (6)

7. Determine (a) $\mathcal{L}^{-1}\left\{\dfrac{3e^{-2s}}{s^2}\right\}$ (b) $\mathcal{L}^{-1}\left\{\dfrac{3se^{-5s}}{s^2+4}\right\}$ (9)

8. In a galvanometer the deflection θ satisfies the differential equation:

    $$\frac{d^2\theta}{dt^2} + 2\frac{d\theta}{dt} + \theta = 4$$

    Use Laplace transforms to solve the equation for θ given that when $t=0$, $\theta=0$ and $\dfrac{d\theta}{dt}=0$ (13)

9. Solve the following pair of simultaneous differential equations:

    $$3\frac{dx}{dt} = 3x + 2y$$

    $$2\frac{dy}{dt} + 3x = 6y$$

    given that when $t=0$, $x=1$ and $y=3$ (20)

Section M

Fourier series

Chapter 73

Fourier series for periodic functions of period 2π

Why it is important to understand: **Fourier series for periodic functions of periodic 2π**

A Fourier series changes a periodic function into an infinite expansion of a function in terms of sines and cosines. In engineering and physics, expanding functions in terms of sines and cosines is useful because it makes it possible to more easily manipulate functions that are just too difficult to represent analytically. The fields of electronics, quantum mechanics and electrodynamics all make great use of Fourier series. The Fourier series has become one of the most widely used and useful mathematical tools available to any scientist. This chapter introduces and explains Fourier series.

At the end of this chapter, you should be able to:

- describe a Fourier series
- understand periodic functions
- state the formula for a Fourier series and Fourier coefficients
- obtain Fourier series for given functions

73.1 Introduction

Fourier* series provides a method of analysing periodic functions into their constituent components. Alternating currents and voltages, displacement, velocity and acceleration of slider-crank mechanisms and acoustic waves are typical practical examples in engineering and science where periodic functions are involved and often require analysis.

73.2 Periodic functions

A function $f(x)$ is said to be **periodic** if $f(x+T)=f(x)$ for all values of x, where T is some positive number. T is the interval between two successive repetitions and is called the **period** of the functions $f(x)$. For example, $y=\sin x$ is periodic in x with period 2π since $\sin x = \sin(x+2\pi)$ $= \sin(x+4\pi)$, and so on. In general, if $y=\sin \omega t$ then the period of the waveform is $2\pi/\omega$. The function shown in Fig. 73.1 is also periodic of period 2π and is

* **Who was Fourier?** **Jean Baptiste Joseph Fourier** (21 March 1768–16 May 1830) was a French mathematician and physicist best known for initiating the investigation of Fourier series and their applications to problems of heat transfer and vibrations. Fourier is also credited with the discovery of the greenhouse effect. To find out more go to **www.routledge.com/cw/bird**

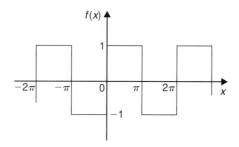

Figure 73.1

defined by:

$$f(x) = \begin{cases} -1, & \text{when} \quad -\pi < x < 0 \\ 1, & \text{when} \quad 0 < x < \pi \end{cases}$$

If a graph of a function has no sudden jumps or breaks it is called a **continuous function**, examples being the graphs of sine and cosine functions. However, other graphs make finite jumps at a point or points in the interval. The square wave shown in Fig. 73.1 has finite discontinuities at $x=\pi$, 2π, 3π, and so on. A great advantage of Fourier series over other series is that it can be applied to functions which are discontinuous as well as those which are continuous.

73.3 Fourier series

(i) The basis of a Fourier series is that all functions of practical significance which are defined in the interval $-\pi \leq x \leq \pi$ can be expressed in terms of a convergent trigonometric series of the form:

$$f(x) = a_0 + a_1 \cos x + a_2 \cos 2x$$
$$+ a_3 \cos 3x + \cdots + b_1 \sin x$$
$$+ b_2 \sin 2x + b_3 \sin 3x + \cdots$$

when $a_0, a_1, a_2, \ldots b_1, b_2, \ldots$ are real constants, i.e.

$$f(x) = a_0 + \sum_{n=1}^{\infty} (a_n \cos nx + b_n \sin nx) \quad (1)$$

where for the range $-\pi$ to π:

$$a_0 = \frac{1}{2\pi} \int_{-\pi}^{\pi} f(x)\,dx$$

$$a_n = \frac{1}{\pi} \int_{-\pi}^{\pi} f(x) \cos nx\,dx$$

$$(n = 1, 2, 3, \ldots)$$

Fourier series for periodic functions of period 2π **777**

and $b_n = \frac{1}{\pi} \int_{-\pi}^{\pi} f(x) \sin nx \, dx$

$(n = 1, 2, 3, \ldots)$

(ii) a_0, a_n and b_n are called the **Fourier coefficients** of the series and if these can be determined, the series of equation (1) is called the **Fourier series** corresponding to $f(x)$.

(iii) An alternative way of writing the series is by using the $a \cos x + b \sin x = c \sin(x + \alpha)$ relationship introduced in Chapter 19, i.e.

$f(x) = a_0 + c_1 \sin(x + \alpha_1) + c_2 \sin(2x + \alpha_2)$

$+ \cdots + c_n \sin(nx + \alpha_n),$

where a_0 is a constant,

$c_1 = \sqrt{(a_1^2 + b_1^2)}, \ldots c_n = \sqrt{(a_n^2 + b_n^2)}$

are the amplitudes of the various components, and phase angle

$\alpha_n = \tan^{-1} \frac{a_n}{b_n}$

(iv) For the series of equation (1): the term $(a_1 \cos x + b_1 \sin x)$ or $c_1 \sin(x + \alpha_1)$ is called the **first harmonic** or the **fundamental**, the term $(a_2 \cos 2x + b_2 \sin 2x)$ or $c_2 \sin(2x + \alpha_2)$ is called the **second harmonic**, and so on.

For an exact representation of a complex wave, an infinite number of terms are, in general, required. In many practical cases, however, it is sufficient to take the first few terms only (see Problem 2).

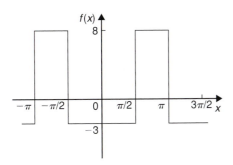

Figure 73.2

The sum of a Fourier series at a point of discontinuity is given by the arithmetic mean of the two limiting values of $f(x)$ as x approaches the point of discontinuity from the two sides. For example, for the waveform shown in

Fig. 73.2, the sum of the Fourier series at the points of discontinuity (i.e. at $\frac{\pi}{2}, \pi, \ldots$ is given by:

$\frac{8 + (-3)}{2} = \frac{5}{2}$ or $2\frac{1}{2}$

73.4 Worked problems on Fourier series of periodic functions of period 2π

Problem 1. Obtain a Fourier series for the periodic function $f(x)$ defined as:

$f(x) = \begin{cases} -k, & \text{when } -\pi < x < 0 \\ +k, & \text{when } 0 < x < \pi \end{cases}$

The function is periodic outside of this range with period 2π.

The square wave function defined is shown in Fig. 73.3. Since $f(x)$ is given by two different expressions in the two halves of the range the integration is performed in two parts, one from $-\pi$ to 0 and the other from 0 to π.

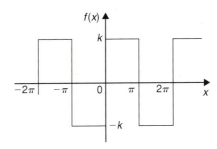

Figure 73.3

From Section 73.3(i):

$a_0 = \frac{1}{2\pi} \int_{-\pi}^{\pi} f(x) \, dx$

$= \frac{1}{2\pi} \left[\int_{-\pi}^{0} -k \, dx + \int_{0}^{\pi} k \, dx \right]$

$= \frac{1}{2\pi} \{[-kx]_{-\pi}^{0} + [kx]_{0}^{\pi}\} = 0$

[a_0 is in fact the **mean value** of the waveform over a complete period of 2π and this could have been deduced on sight from Fig. 73.3]

From Section 73.3(i):

$$a_n = \frac{1}{\pi} \int_{-\pi}^{\pi} f(x) \cos nx \, dx$$

$$= \frac{1}{\pi} \left\{ \int_{-\pi}^{0} -k \cos nx \, dx + \int_{0}^{\pi} k \cos nx \, dx \right\}$$

$$= \frac{1}{\pi} \left\{ \left[\frac{-k \sin nx}{n} \right]_{-\pi}^{0} + \left[\frac{k \sin nx}{n} \right]_{0}^{\pi} \right\} = 0$$

Hence a_1, a_2, a_3, ... are all zero (since $\sin 0 = \sin(-n\pi) = \sin n\pi = 0$), and therefore no cosine terms will appear in the Fourier series.

From Section 73.3(i):

$$b_n = \frac{1}{\pi} \int_{-\pi}^{\pi} f(x) \sin nx \, dx$$

$$= \frac{1}{\pi} \left\{ \int_{-\pi}^{0} -k \sin nx \, dx + \int_{0}^{\pi} k \sin nx \, dx \right\}$$

$$= \frac{1}{\pi} \left\{ \left[\frac{k \cos nx}{n} \right]_{-\pi}^{0} + \left[\frac{-k \cos nx}{n} \right]_{0}^{\pi} \right\}$$

When n is odd:

$$b_n = \frac{k}{\pi} \left\{ \left[\left(\frac{1}{n} \right) - \left(-\frac{1}{n} \right) \right] \right.$$

$$\left. + \left[-\left(-\frac{1}{n} \right) - \left(-\frac{1}{n} \right) \right] \right\}$$

$$= \frac{k}{\pi} \left\{ \frac{2}{n} + \frac{2}{n} \right\} = \frac{4k}{n\pi}$$

Hence $b_1 = \dfrac{4k}{\pi}$, $b_3 = \dfrac{4k}{3\pi}$, $b_5 = \dfrac{4k}{5\pi}$, and so on.

When n is even:

$$b_n = \frac{k}{\pi} \left\{ \left[\frac{1}{n} - \frac{1}{n} \right] + \left[-\frac{1}{n} - \left(-\frac{1}{n} \right) \right] \right\} = 0$$

Hence, from equation (1), the Fourier series for the function shown in Fig. 73.3 is given by:

$$f(x) = a_0 + \sum_{n=1}^{\infty} (a_n \cos nx + b_n \sin nx)$$

$$= 0 + \sum_{n=1}^{\infty} (0 + b_n \sin nx)$$

i.e. $\quad f(x) = \dfrac{4k}{\pi} \sin x + \dfrac{4k}{3\pi} \sin 3x + \dfrac{4k}{5\pi} \sin 5x + \cdots$

i.e. $\quad \boldsymbol{f(x) = \dfrac{4k}{\pi} \left(\sin x + \dfrac{1}{3} \sin 3x + \dfrac{1}{5} \sin 5x + \cdots \right)}$

Problem 2. For the Fourier series of Problem 1 let $k = \pi$. Show by plotting the first three partial sums of this Fourier series that as the series is added together term by term the result approximates more and more closely to the function it represents.

If $k = \pi$ in the Fourier series of Problem 1 then:

$$f(x) = 4(\sin x + \tfrac{1}{3} \sin 3x + \tfrac{1}{5} \sin 5x + \cdots)$$

$4 \sin x$ is termed the first partial sum of the Fourier series of $f(x)$, $(4 \sin x + \tfrac{4}{3} \sin 3x)$ is termed the second partial sum of the Fourier series, and $(4 \sin x + \tfrac{4}{3} \sin 3x + \tfrac{4}{5} \sin 5x)$ is termed the third partial sum, and so on.

Let $\quad P_1 = 4 \sin x$,

$$P_2 = \left(4 \sin x + \tfrac{4}{3} \sin 3x \right)$$

and $\quad P_3 = \left(4 \sin x + \tfrac{4}{3} \sin 3x + \tfrac{4}{5} \sin 5x \right)$

Graphs of P_1, P_2 and P_3, obtained by drawing up tables of values, and adding waveforms are shown in Figs. 73.4(a) to (c) and they show that the series is convergent, i.e. continually approximating towards a definite limit as more and more partial sums are taken, and in the limit will have the sum $f(x) = \pi$.
Even with just three partial sums, the waveform is starting to approach the rectangular wave the Fourier series is representing.

Problem 3. If in the Fourier series of Problem 1, $k = 1$, deduce a series for $\dfrac{\pi}{4}$ at the point $x = \dfrac{\pi}{2}$

If $k = 1$ in the Fourier series of Problem 1:

$$f(x) = \frac{4}{\pi} \left(\sin x + \frac{1}{3} \sin 3x + \frac{1}{5} \sin 5x + \cdots \right)$$

When $x = \dfrac{\pi}{2}$, $f(x) = 1$,

$$\sin x = \sin \frac{\pi}{2} = 1,$$

$$\sin 3x = \sin \frac{3\pi}{2} = -1,$$

$$\sin 5x = \sin \frac{5\pi}{2} = 1, \text{ and so on.}$$

Hence $\quad 1 = \dfrac{4}{\pi} \left[1 + \dfrac{1}{3}(-1) + \dfrac{1}{5}(1) + \dfrac{1}{7}(-1) + \cdots \right]$

i.e. $\quad \dfrac{\pi}{4} = 1 - \dfrac{1}{3} + \dfrac{1}{5} - \dfrac{1}{7} + \cdots$

Figure 73.4

Problem 4.   Determine the Fourier series for the full wave rectified sine wave $i = 5\sin\dfrac{\theta}{2}$ shown in Fig. 73.5.

Figure 73.5

$i = 5\sin\dfrac{\theta}{2}$ is a periodic function of period 2π.
Thus

$$i = f(\theta) = a_0 + \sum_{n=1}^{\infty}(a_n\cos n\theta + b_n\sin n\theta)$$

In this case it is better to take the range 0 to 2π instead of $-\pi$ to $+\pi$ since the waveform is continuous between 0 and 2π.

$$a_0 = \frac{1}{2\pi}\int_0^{2\pi} f(\theta)\,\mathrm{d}\theta = \frac{1}{2\pi}\int_0^{2\pi} 5\sin\frac{\theta}{2}\,\mathrm{d}\theta$$

$$= \frac{5}{2\pi}\left[-2\cos\frac{\theta}{2}\right]_0^{2\pi}$$

$$= \frac{5}{\pi}\left[\left(-\cos\frac{2\pi}{2}\right) - (-\cos 0)\right]$$

$$= \frac{5}{\pi}[(1)-(-1)] = \frac{10}{\pi}$$

$$a_n = \frac{1}{\pi}\int_0^{2\pi} 5\sin\frac{\theta}{2}\cos n\theta\,\mathrm{d}\theta$$

$$= \frac{5}{\pi}\int_0^{2\pi}\frac{1}{2}\left\{\sin\left(\frac{\theta}{2}+n\theta\right)\right.$$

$$\left. + \sin\left(\frac{\theta}{2}-n\theta\right)\right\}\mathrm{d}\theta$$
(see Chapter 43, page 482)

$$= \frac{5}{2\pi}\left[\frac{-\cos\left[\theta\left(\frac{1}{2}+n\right)\right]}{\left(\frac{1}{2}+n\right)}\right.$$

$$\left. - \frac{\cos\left[\theta\left(\frac{1}{2}-n\right)\right]}{\left(\frac{1}{2}-n\right)}\right]_0^{2\pi}$$

$$= \frac{5}{2\pi}\left\{\left[\frac{-\cos\left[2\pi\left(\frac{1}{2}+n\right)\right]}{\left(\frac{1}{2}+n\right)}\right.\right.$$

$$\left. - \frac{\cos\left[2\pi\left(\frac{1}{2}-n\right)\right]}{\left(\frac{1}{2}-n\right)}\right]$$

$$\left. - \left[\frac{-\cos 0}{\left(\frac{1}{2}+n\right)} - \frac{\cos 0}{\left(\frac{1}{2}-n\right)}\right]\right\}$$

When n is both odd and even,

$$a_n = \frac{5}{2\pi}\left\{\left[\frac{1}{\left(\frac{1}{2}+n\right)} + \frac{1}{\left(\frac{1}{2}-n\right)}\right]\right.$$

$$-\left[\frac{-1}{\left(\frac{1}{2}+n\right)}-\frac{1}{\left(\frac{1}{2}-n\right)}\right]\Bigg\}$$

$$=\frac{5}{2\pi}\left\{\frac{2}{\left(\frac{1}{2}+n\right)}+\frac{2}{\left(\frac{1}{2}-n\right)}\right\}$$

$$=\frac{5}{\pi}\left\{\frac{1}{\left(\frac{1}{2}+n\right)}+\frac{1}{\left(\frac{1}{2}-n\right)}\right\}$$

Hence

$$a_1=\frac{5}{\pi}\left[\frac{1}{\frac{3}{2}}+\frac{1}{-\frac{1}{2}}\right]=\frac{5}{\pi}\left[\frac{2}{3}-\frac{2}{1}\right]=\frac{-20}{3\pi}$$

$$a_2=\frac{5}{\pi}\left[\frac{1}{\frac{5}{2}}+\frac{1}{-\frac{3}{2}}\right]=\frac{5}{\pi}\left[\frac{2}{5}-\frac{2}{3}\right]=\frac{-20}{(3)(5)\pi}$$

$$a_3=\frac{5}{\pi}\left[\frac{1}{\frac{7}{2}}+\frac{1}{-\frac{5}{2}}\right]=\frac{5}{\pi}\left[\frac{2}{7}-\frac{2}{5}\right]=\frac{-20}{(5)(7)\pi}$$

and so on

$$b_n=\frac{1}{\pi}\int_0^{2\pi}5\sin\frac{\theta}{2}\sin n\theta\,d\theta$$

$$=\frac{5}{\pi}\int_0^{2\pi}-\frac{1}{2}\left\{\cos\left[\theta\left(\frac{1}{2}+n\right)\right]\right.$$

$$\left.-\cos\left[\theta\left(\frac{1}{2}-n\right)\right]\right\}d\theta$$

from Chapter 43

$$=\frac{5}{2\pi}\left[\frac{\sin\left[\theta\left(\frac{1}{2}-n\right)\right]}{\left(\frac{1}{2}-n\right)}-\frac{\sin\left[\theta\left(\frac{1}{2}+n\right)\right]}{\left(\frac{1}{2}+n\right)}\right]_0^{2\pi}$$

$$=\frac{5}{2\pi}\left\{\left[\frac{\sin 2\pi\left(\frac{1}{2}-n\right)}{\left(\frac{1}{2}-n\right)}-\frac{\sin 2\pi\left(\frac{1}{2}+n\right)}{\left(\frac{1}{2}+n\right)}\right]\right.$$

$$\left.-\left[\frac{\sin 0}{\left(\frac{1}{2}-n\right)}-\frac{\sin 0}{\left(\frac{1}{2}+n\right)}\right]\right\}$$

When n is both odd and even, $b_n=0$ since $\sin(-\pi)$, $\sin 0$, $\sin\pi$, $\sin 3\pi$, ... are all zero. Hence the Fourier series for the rectified sine wave,

$i=5\sin\dfrac{\theta}{2}$ is given by:

$$f(\theta)=a_0+\sum_{n=1}^{\infty}(a_n\cos n\theta+b_n\sin n\theta)$$

i.e. $i=f(\theta)=\dfrac{10}{\pi}-\dfrac{20}{3\pi}\cos\theta-\dfrac{20}{(3)(5)\pi}\cos 2\theta$

$$-\frac{20}{(5)(7)\pi}\cos 3\theta-\cdots$$

i.e. $\boldsymbol{i=\dfrac{20}{\pi}\left(\dfrac{1}{2}-\dfrac{\cos\theta}{(3)}-\dfrac{\cos 2\theta}{(3)(5)}-\dfrac{\cos 3\theta}{(5)(7)}-\cdots\right)}$

Now try the following Practice Exercise

Practice Exercise 260   Fourier series of periodic functions of period 2π (Answers on page 896)

1. Determine the Fourier series for the periodic function:

$$f(x)=\begin{cases}-2,&\text{when}\quad-\pi<x<0\\+2,&\text{when}\quad\;\;0<x<\pi\end{cases}$$

which is periodic outside this range of period 2π.

2. For the Fourier series in Problem 1, deduce a series for $\dfrac{\pi}{4}$ at the point where $x=\dfrac{\pi}{2}$

3. For the waveform shown in Fig. 73.6 determine (a) the Fourier series for the function and (b) the sum of the Fourier series at the points of discontinuity.

Figure 73.6

4. For Problem 3, draw graphs of the first three partial sums of the Fourier series and show that as the series is added together term by term the result approximates more and more closely to the function it represents.

5. Find the term representing the third harmonic for the periodic function of period 2π given by:

$$f(x)=\begin{cases}0,&\text{when}\quad-\pi<x<0\\1,&\text{when}\quad\;\;0<x<\pi\end{cases}$$

6. Determine the Fourier series for the periodic function of period 2π defined by:

$$f(t) = \begin{cases} 0, & \text{when} \quad -\pi < t < 0 \\ 1, & \text{when} \quad 0 < t < \dfrac{\pi}{2} \\ -1, & \text{when} \quad \dfrac{\pi}{2} < t < \pi \end{cases}$$

The function has a period of 2π.

7. Show that the Fourier series for the periodic function of period 2π defined by

$$f(\theta) = \begin{cases} 0, & \text{when} \quad -\pi < \theta < 0 \\ \sin\theta, & \text{when} \quad 0 < \theta < \pi \end{cases}$$

is given by:

$$f(\theta) = \frac{2}{\pi}\left(\frac{1}{2} - \frac{\cos 2\theta}{(3)} - \frac{\cos 4\theta}{(3)(5)}\right.$$

$$\left. - \frac{\cos 6\theta}{(5)(7)} - \cdots \right)$$

For fully worked solutions to each of the problems in Practice Exercise 260 in this chapter, go to the website:
www.routledge.com/cw/bird

Chapter 74

Fourier series for a non-periodic function over range 2π

Why it is important to understand: **Fourier series for a non-periodic function over range 2π**

As was stated in the previous chapter, there are many practical uses of Fourier series in science and engineering. The technique has practical applications in the resolution of sound waves into their different frequencies, for example, in an MP3 player, in telecommunications and Wi-Fi, in computer graphics and image processing, in climate variation, in water waves, and much more. In this chapter, the Fourier series for non-periodic functions is explained.

At the end of this chapter, you should be able to:

- appreciate that Fourier expansions of non-periodic functions have limited range
- determine Fourier series for non-periodic functions over a range of 2π

74.1 Expansion of non-periodic functions

If a function $f(x)$ is not periodic then it cannot be expanded in a Fourier series for **all** values of x. However, it is possible to determine a Fourier series to represent the function over any range of width 2π.

Given a non-periodic function, a new function may be constructed by taking the values of $f(x)$ in the given range and then repeating them outside of the given range at intervals of 2π. Since this new function is, by construction, periodic with period 2π, it may then be expanded in a Fourier series for

all values of x. For example, the function $f(x) = x$ is not a periodic function. However, if a Fourier series for $f(x) = x$ is required then the function is constructed outside of this range so that it is periodic with period 2π as shown by the broken lines in Fig. 74.1.

For non-periodic functions, such as $f(x) = x$, the sum of the Fourier series is equal to $f(x)$ at all points in the given range but it is not equal to $f(x)$ at points outside of the range.

For determining a Fourier series of a non-periodic function over a range 2π, exactly the same formulae for the Fourier coefficients are used as in Section 73.3(i).

Figure 74.1

74.2 Worked problems on Fourier series of non-periodic functions over a range of 2π

Problem 1. Determine the Fourier series to represent the function $f(x) = 2x$ in the range $-\pi$ to $+\pi$.

The function $f(x) = 2x$ is not periodic. The function is shown in the range $-\pi$ to π in Fig. 74.2 and is then constructed outside of that range so that it is periodic of period 2π (see broken lines) with the resulting sawtooth waveform.

For a Fourier series:

$$f(x) = a_0 + \sum_{n=1}^{\infty}(a_n \cos nx + b_n \sin nx)$$

From Section 73.3(i),

$$a_0 = \frac{1}{2\pi}\int_{-\pi}^{\pi} f(x)\, dx$$

$$= \frac{1}{2\pi}\int_{-\pi}^{\pi} 2x\, dx = \frac{2}{2\pi}\left[\frac{x^2}{2}\right]_{-\pi}^{\pi} = 0$$

$$a_n = \frac{1}{\pi}\int_{-\pi}^{\pi} f(x)\cos nx\, dx = \frac{1}{\pi}\int_{-\pi}^{\pi} 2x\cos nx\, dx$$

$$= \frac{2}{\pi}\left[\frac{x\sin nx}{n} - \int \frac{\sin nx}{n}\, dx\right]_{-\pi}^{\pi}$$

by parts (see Chapter 46)

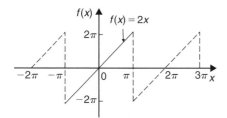

Figure 74.2

$$= \frac{2}{\pi}\left[\frac{x\sin nx}{n} + \frac{\cos nx}{n^2}\right]_{-\pi}^{\pi}$$

$$= \frac{2}{\pi}\left[\left(0 + \frac{\cos n\pi}{n^2}\right) - \left(0 + \frac{\cos n(-\pi)}{n^2}\right)\right] = 0$$

$$b_n = \frac{1}{\pi}\int_{-\pi}^{\pi} f(x)\sin nx\, dx = \frac{1}{\pi}\int_{-\pi}^{\pi} 2x\sin nx\, dx$$

$$= \frac{2}{\pi}\left[\frac{-x\cos nx}{n} - \int\left(\frac{-\cos nx}{n}\right) dx\right]_{-\pi}^{\pi}$$

by parts

$$= \frac{2}{\pi}\left[\frac{-x\cos nx}{n} + \frac{\sin nx}{n^2}\right]_{-\pi}^{\pi}$$

$$= \frac{2}{\pi}\left[\left(\frac{-\pi\cos n\pi}{n} + \frac{\sin n\pi}{n^2}\right)\right.$$

$$\left. - \left(\frac{-(-\pi)\cos n(-\pi)}{n} + \frac{\sin n(-\pi)}{n^2}\right)\right]$$

$$= \frac{2}{\pi}\left[\frac{-\pi\cos n\pi}{n} - \frac{\pi\cos(-n\pi)}{n}\right] = \frac{-4}{n}\cos n\pi$$

When n is odd, $b_n = \frac{4}{n}$. Thus $b_1 = 4$, $b_3 = \frac{4}{3}$, $b_5 = \frac{4}{5}$, and so on.

When n is even, $b_n = \frac{-4}{n}$. Thus $b_2 = -\frac{4}{2}$, $b_4 = -\frac{4}{4}$, $b_6 = -\frac{4}{6}$, and so on.

Thus $f(x) = 2x = 4\sin x - \frac{4}{2}\sin 2x + \frac{4}{3}\sin 3x$

$$- \frac{4}{4}\sin 4x + \frac{4}{5}\sin 5x - \frac{4}{6}\sin 6x + \cdots$$

i.e. $2x = 4\left(\sin x - \frac{1}{2}\sin 2x + \frac{1}{3}\sin 3x - \frac{1}{4}\sin 4x\right.$

$$\left. + \frac{1}{5}\sin 5x - \frac{1}{6}\sin 6x + \cdots\right) \quad (1)$$

for values of $f(x)$ between $-\pi$ and π. For values of $f(x)$ outside the range $-\pi$ to $+\pi$ the sum of the series is not equal to $f(x)$.

Problem 2. In the Fourier series of Problem 1, by letting $x = \pi/2$, deduce a series for $\pi/4$.

When $x = \pi/2$, $f(x) = \pi$ from Fig. 74.2.

Thus, from the Fourier series of equation (1):

$$2\left(\frac{\pi}{2}\right) = 4\left(\sin\frac{\pi}{2} - \frac{1}{2}\sin\frac{2\pi}{2} + \frac{1}{3}\sin\frac{3\pi}{2}\right.$$

$$-\frac{1}{4}\sin\frac{4\pi}{2} + \frac{1}{5}\sin\frac{5\pi}{2}$$

$$\left. -\frac{1}{6}\sin\frac{6\pi}{2} + \cdots\right)$$

$$\pi = 4\left(1 - 0 - \frac{1}{3} - 0 + \frac{1}{5} - 0 - \frac{1}{7} - \cdots\right)$$

i.e. $\dfrac{\pi}{4} = 1 - \dfrac{1}{3} + \dfrac{1}{5} - \dfrac{1}{7} + \cdots$

Problem 3. Obtain a Fourier series for the function defined by:

$$f(x) = \begin{cases} x, & \text{when } 0 < x < \pi \\ 0, & \text{when } \pi < x < 2\pi \end{cases}$$

The defined function is shown in Fig. 74.3 between 0 and 2π. The function is constructed outside of this range so that it is periodic of period 2π, as shown by the broken line in Fig. 74.3.

Figure 74.3

For a Fourier series:

$$f(x) = a_0 + \sum_{n=1}^{\infty}(a_n\cos nx + b_n\sin nx)$$

It is more convenient in this case to take the limits from 0 to 2π instead of from $-\pi$ to $+\pi$. The value of the Fourier coefficients are unaltered by this change of limits. Hence

$$a_0 = \frac{1}{2\pi}\int_0^{2\pi} f(x)\,dx = \frac{1}{2\pi}\left[\int_0^{\pi} x\,dx + \int_{\pi}^{2\pi} 0\,dx\right]$$

$$= \frac{1}{2\pi}\left[\frac{x^2}{2}\right]_0^{\pi} = \frac{1}{2\pi}\left(\frac{\pi^2}{2}\right) = \frac{\pi}{4}$$

$$a_n = \frac{1}{\pi}\int_0^{2\pi} f(x)\cos nx\,dx$$

$$= \frac{1}{\pi}\left[\int_0^{\pi} x\cos nx\,dx + \int_{\pi}^{2\pi} 0\,dx\right]$$

$$= \frac{1}{\pi}\left[\frac{x\sin nx}{n} + \frac{\cos nx}{n^2}\right]_0^{\pi}$$

(from Problem 1, by parts)

$$= \frac{1}{\pi}\left\{\left[\frac{\pi\sin n\pi}{n} + \frac{\cos n\pi}{n^2}\right] - \left[0 + \frac{\cos 0}{n^2}\right]\right\}$$

$$= \frac{1}{\pi n^2}(\cos n\pi - 1)$$

When n is even, $a_n = 0$.

When n is odd, $a_n = \dfrac{-2}{\pi n^2}$

Hence $a_1 = \dfrac{-2}{\pi}, a_3 = \dfrac{-2}{3^2\pi}, a_5 = \dfrac{-2}{5^2\pi}$, and so on

$$b_n = \frac{1}{\pi}\int_0^{2\pi} f(x)\sin nx\,dx$$

$$= \frac{1}{\pi}\left[\int_0^{\pi} x\sin nx\,dx - \int_{\pi}^{2\pi} 0\,dx\right]$$

$$= \frac{1}{\pi}\left[\frac{-x\cos nx}{n} + \frac{\sin nx}{n^2}\right]_0^{\pi}$$

(from Problem 1, by parts)

$$= \frac{1}{\pi}\left\{\left[\frac{-\pi\cos n\pi}{n} + \frac{\sin n\pi}{n^2}\right] - \left[0 + \frac{\sin 0}{n^2}\right]\right\}$$

$$= \frac{1}{\pi}\left[\frac{-\pi\cos n\pi}{n}\right] = \frac{-\cos n\pi}{n}$$

Hence $b_1 = -\cos\pi = 1$, $b_2 = -\dfrac{1}{2}$, $b_3 = \dfrac{1}{3}$, and so on.

Thus the Fourier series is:

$$f(x) = a_0 + \sum_{n=1}^{\infty}(a_n\cos nx + b_n\sin nx)$$

i.e. $\quad f(x) = \dfrac{\pi}{4} - \dfrac{2}{\pi}\cos x - \dfrac{2}{3^2\pi}\cos 3x$

$$-\frac{2}{5^2\pi}\cos 5x - \cdots + \sin x$$

$$-\frac{1}{2}\sin 2x + \frac{1}{3}\sin 3x - \cdots$$

i.e. $f(x)$

$$= \frac{\pi}{4} - \frac{2}{\pi}\left(\cos x + \frac{\cos 3x}{3^2} + \frac{\cos 5x}{5^2} + \cdots\right)$$

$$+ \left(\sin x - \frac{1}{2}\sin 2x + \frac{1}{3}\sin 3x - \cdots\right)$$

Problem 4. For the Fourier series of Problem 3: (a) What is the sum of the series at the point of discontinuity (i.e. at $x = \pi$)? (b) What is the amplitude and phase angle of the third harmonic? and (c) Let $x = 0$, and deduce a series for $\pi^2/8$

(a) The sum of the Fourier series at the point of discontinuity is given by the arithmetic mean of the two limiting values of $f(x)$ as x approaches the point of discontinuity from the two sides.

Hence sum of the series at $x = \pi$ is

$$\frac{\pi - 0}{2} = \frac{\pi}{2}$$

(b) The third harmonic term of the Fourier series is

$$\left(-\frac{2}{3^2 \pi} \cos 3x + \frac{1}{3} \sin 3x\right)$$

This may also be written in the form $c \sin(3x + \alpha)$,

where amplitude, $c = \sqrt{\left[\left(\frac{-2}{3^2\pi}\right)^2 + \left(\frac{1}{3}\right)^2\right]}$

$$= 0.341$$

and phase angle,

$$\alpha = \tan^{-1}\left(\frac{\frac{-2}{3^2\pi}}{\frac{1}{3}}\right)$$

$$= -11.98° \quad \text{or} \quad -0.209 \, \text{radians}$$

Hence the third harmonic is given by

$0.341 \sin(3x - 0.209)$

(c) When $x = 0$, $f(x) = 0$ (see Fig. 74.3).

Hence, from the Fourier series:

$$0 = \frac{\pi}{4} - \frac{2}{\pi}\left(\cos 0 + \frac{1}{3^2}\cos 0 + \frac{1}{5^2}\cos 0 + \cdots\right) + (0)$$

i.e. $-\frac{\pi}{4} = -\frac{2}{\pi}\left(1 + \frac{1}{3^2} + \frac{1}{5^2} + \frac{1}{7^2} + \cdots\right)$

Hence $\quad \frac{\pi^2}{8} = 1 + \frac{1}{3^2} + \frac{1}{5^2} + \frac{1}{7^2} + \cdots$

Problem 5. Deduce the Fourier series for the function $f(\theta) = \theta^2$ in the range 0 to 2π.

$f(\theta) = \theta^2$ is shown in Fig. 74.4 in the range 0 to 2π. The function is not periodic but is constructed outside of this range so that it is periodic of period 2π, as shown by the broken lines.

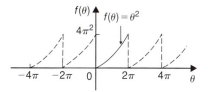

Figure 74.4

For a Fourier series:

$$f(x) = a_0 + \sum_{n=1}^{\infty}(a_n \cos nx + b_n \sin nx)$$

$$a_0 = \frac{1}{2\pi}\int_0^{2\pi} f(\theta)d\theta = \frac{1}{2\pi}\int_0^{2\pi}\theta^2 \, d\theta$$

$$= \frac{1}{2\pi}\left[\frac{\theta^3}{3}\right]_0^{2\pi} = \frac{1}{2\pi}\left[\frac{8\pi^3}{3} - 0\right] = \frac{4\pi^2}{3}$$

$$a_n = \frac{1}{\pi}\int_0^{2\pi} f(\theta)\cos n\theta \, d\theta$$

$$= \frac{1}{\pi}\int_0^{2\pi}\theta^2 \cos n\theta \, d\theta$$

$$= \frac{1}{\pi}\left[\frac{\theta^2 \sin n\theta}{n} + \frac{2\theta \cos n\theta}{n^2} - \frac{2\sin n\theta}{n^3}\right]_0^{2\pi}$$

by parts

$$= \frac{1}{\pi}\left[\left(0 + \frac{4\pi \cos 2\pi n}{n^2} - 0\right) - (0)\right]$$

$$= \frac{4}{n^2}\cos 2\pi n = \frac{4}{n^2} \text{ when } n = 1, 2, 3, \ldots$$

Hence $a_1 = \frac{4}{1^2}$, $a_2 = \frac{4}{2^2}$, $a_3 = \frac{4}{3^2}$ and so on

$$b_n = \frac{1}{\pi}\int_0^{2\pi} f(\theta)\sin n\theta \, d\theta = \frac{1}{\pi}\int_0^{2\pi}\theta^2 \sin n\theta \, d\theta$$

$$= \frac{1}{\pi}\left[\frac{-\theta^2 \cos n\theta}{n} + \frac{2\theta \sin n\theta}{n^2} + \frac{2\cos n\theta}{n^3}\right]_0^{2\pi}$$

by parts

$$= \frac{1}{\pi}\left[\left(\frac{-4\pi^2\cos 2\pi n}{n} + 0 + \frac{2\cos 2\pi n}{n^3}\right)\right.$$

$$\left. - \left(0 + 0 + \frac{2\cos 0}{n^3}\right)\right]$$

$$= \frac{1}{\pi}\left[\frac{-4\pi^2}{n} + \frac{2}{n^3} - \frac{2}{n^3}\right] = \frac{-4\pi}{n}$$

Hence $b_1 = \dfrac{-4\pi}{1}, b_2 = \dfrac{-4\pi}{2}, b_3 = \dfrac{-4\pi}{3}$, and so on.

Thus $f(\theta) = \theta^2$

$$= \frac{4\pi^2}{3} + \sum_{n=1}^{\infty}\left(\frac{4}{n^2}\cos n\theta - \frac{4\pi}{n}\sin n\theta\right)$$

i.e. $\boldsymbol{\theta^2 =}$

$$\frac{4\pi^2}{3} + 4\left(\cos\theta + \frac{1}{2^2}\cos 2\theta + \frac{1}{3^2}\cos 3\theta + \cdots\right)$$

$$- 4\pi\left(\sin\theta + \frac{1}{2}\sin 2\theta + \frac{1}{3}\sin 3\theta + \cdots\right)$$

for values of θ between 0 and 2π.

> **Problem 6.** In the Fourier series of Problem 5, let $\theta = \pi$ and determine a series for $\dfrac{\pi^2}{12}$

When $\theta = \pi$, $f(\theta) = \pi^2$

Hence $\pi^2 = \dfrac{4\pi^2}{3} + 4\left(\cos\pi + \dfrac{1}{4}\cos 2\pi\right.$

$$\left. + \frac{1}{9}\cos 3\pi + \frac{1}{16}\cos 4\pi + \cdots\right)$$

$$- 4\pi\left(\sin\pi + \frac{1}{2}\sin 2\pi\right.$$

$$\left. + \frac{1}{3}\sin 3\pi + \cdots\right)$$

i.e. $\pi^2 - \dfrac{4\pi^2}{3} = 4\left(-1 + \dfrac{1}{4} - \dfrac{1}{9}\right.$

$$\left. + \frac{1}{16} - \cdots\right) - 4\pi(0)$$

$$-\frac{\pi^2}{3} = 4\left(-1 + \frac{1}{4} - \frac{1}{9} + \frac{1}{16} - \cdots\right)$$

$$\frac{\pi^2}{3} = 4\left(1 - \frac{1}{4} + \frac{1}{9} - \frac{1}{16} + \cdots\right)$$

Hence $\dfrac{\pi^2}{12} = 1 - \dfrac{1}{4} + \dfrac{1}{9} - \dfrac{1}{16} + \cdots$

or $\dfrac{\pi^2}{12} = 1 - \dfrac{1}{2^2} + \dfrac{1}{3^2} - \dfrac{1}{4^2} + \cdots$

Now try the following Practice Exercise

Practice Exercise 261   Fourier series of non-periodic functions of period 2π (Answers on page 896)

1. Show that the Fourier series for the function $f(x) = x$ over the range $x = 0$ to $x = 2\pi$ is given by:

$$f(x) = \pi - 2\left(\sin x + \frac{1}{2}\sin 2x\right.$$

$$\left. + \frac{1}{3}\sin 3x + \frac{1}{4}\sin 4x + \ldots\right)$$

2. Determine the Fourier series for the function defined by:

$$f(t) = \begin{cases} 1-t, & \text{when } -\pi < t < 0 \\ 1+t, & \text{when } 0 < t < \pi \end{cases}$$

Draw a graph of the function within and outside of the given range.

3. Find the Fourier series for the function $f(x) = x + \pi$ within the range $-\pi < x < \pi$

4. Determine the Fourier series up to and including the third harmonic for the function defined by:

$$f(x) = \begin{cases} x, & \text{when } 0 < x < \pi \\ 2\pi - x, & \text{when } \pi < x < 2\pi \end{cases}$$

Sketch a graph of the function within and outside of the given range, assuming the period is 2π.

5. Expand the function $f(\theta) = \theta^2$ in a Fourier series in the range $-\pi < \theta < \pi$

Sketch the function within and outside of the given range.

6. For the Fourier series obtained in Problem 5, let $\theta = \pi$ and deduce the series for $\displaystyle\sum_{n=1}^{\infty}\frac{1}{n^2}$

7. Sketch the waveform defined by:

$$f(x) = \begin{cases} 1 + \dfrac{2x}{\pi}, & \text{when } -\pi < x < 0 \\[2mm] 1 - \dfrac{2x}{\pi}, & \text{when } 0 < x < \pi \end{cases}$$

Determine the Fourier series in this range.

8. For the Fourier series of Problem 8, deduce a series for $\dfrac{\pi^2}{8}$

For fully worked solutions to each of the problems in Practice Exercise 261 in this chapter, go to the website:
www.routledge.com/cw/bird

COMPANION
@
WEBSITE

Chapter 75

Even and odd functions and half-range Fourier series

> **Why it is important to understand: Even and odd functions and half-range Fourier series**
>
> It has already been noted in previous chapters that the Fourier series is a very useful tool. The Fourier series has many applications; in fact, any field of physical science that uses sinusoidal signals, such as engineering, applied mathematics and chemistry will make use of the Fourier series. Applications are found in electrical engineering, such as in determining the harmonic components in a.c. waveforms, in vibration analysis, acoustics, optics, signal processing, image processing and in quantum mechanics. If it can be found 'on sight' that a function is even or odd, then determining the Fourier series becomes an easier exercise. This is explained in this chapter, together with half-range Fourier series.

At the end of this chapter, you should be able to:

- define even and odd functions
- determine Fourier cosine series and Fourier sine series
- determine Fourier half-range cosine series and Fourier half-range sine series

---

75.1 Even and odd functions

Even functions

A function $y = f(x)$ is said to be **even** if $f(-x) = f(x)$ for all values of x. Graphs of even functions are always **symmetrical about the y-axis** (i.e. a mirror image). Two examples of even functions are $y = x^2$ and $y = \cos x$ as shown in Fig. 20.25, page 228.

Odd functions

A function $y = f(x)$ is said to be **odd** if $f(-x) = -f(x)$ for all values of x. Graphs of odd functions are always **symmetrical about the origin**. Two examples

of odd functions are $y = x^3$ and $y = \sin x$ as shown in Fig. 20.26, page 228.

Many functions are neither even nor odd, two such examples being shown in Fig. 20.27, page 228.

See also Problems 3 and 4, pages 228 and 229.

75.2 Fourier cosine and Fourier sine series

(a) Fourier cosine series

The Fourier series of an **even periodic** function $f(x)$ having period 2π contains **cosine terms only** (i.e.

contains no sine terms) and may contain a constant term.

Hence $f(x) = a_0 + \sum\limits_{n=1}^{\infty} a_n \cos nx$

where $\quad a_0 = \dfrac{1}{2\pi} \int_{-\pi}^{\pi} f(x)\, dx$

$\qquad = \dfrac{1}{\pi} \int_{0}^{\pi} f(x)\, dx$

(due to symmetry)

and $\quad a_n = \dfrac{1}{\pi} \int_{-\pi}^{\pi} f(x) \cos nx\, dx$

$\qquad = \dfrac{2}{\pi} \int_{0}^{\pi} f(x) \cos nx\, dx$

(b) Fourier sine series

The Fourier series of an **odd** periodic function $f(x)$ having period 2π contains sine terms only (i.e. contains no constant term and no cosine terms).

Hence $\quad f(x) = \sum\limits_{n=1}^{\infty} b_n \sin nx$

where $\quad b_n = \dfrac{1}{\pi} \int_{-\pi}^{\pi} f(x) \sin nx\, dx$

$\qquad = \dfrac{2}{\pi} \int_{0}^{\pi} f(x) \sin nx\, dx$

Problem 1. Determine the Fourier series for the periodic function defined by:

$$f(x) = \begin{cases} -2, & \text{when } -\pi < x < -\dfrac{\pi}{2} \\ 2, & \text{when } -\dfrac{\pi}{2} < x < \dfrac{\pi}{2} \\ -2, & \text{when } \dfrac{\pi}{2} < x < \pi \end{cases}$$

and has a period of 2π

The square wave shown in Fig. 75.1 is an even function since it is symmetrical about the $f(x)$ axis.

Hence from para. (a) the Fourier series is given by:

$$f(x) = a_0 + \sum_{n=1}^{\infty} a_n \cos nx$$

(i.e. the series contains no sine terms.)

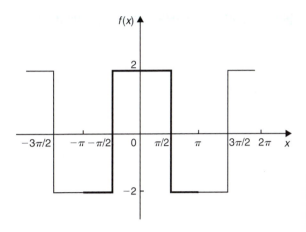

Figure 75.1

From para. (a),

$a_0 = \dfrac{1}{\pi} \int_{0}^{\pi} f(x)\, dx$

$\quad = \dfrac{1}{\pi} \left\{ \int_{0}^{\pi/2} 2\, dx + \int_{\pi/2}^{\pi} -2\, dx \right\}$

$\quad = \dfrac{1}{\pi} \left\{ [2x]_{0}^{\pi/2} + [-2x]_{\pi/2}^{\pi} \right\}$

$\quad = \dfrac{1}{\pi} [(\pi) + [(-2\pi) - (-\pi)]] = 0$

$a_n = \dfrac{2}{\pi} \int_{0}^{\pi} f(x) \cos nx\, dx$

$\quad = \dfrac{2}{\pi} \left\{ \int_{0}^{\pi/2} 2 \cos nx\, dx + \int_{\pi/2}^{\pi} -2 \cos nx\, dx \right\}$

$\quad = \dfrac{4}{\pi} \left\{ \left[\dfrac{\sin nx}{n} \right]_{0}^{\pi/2} + \left[\dfrac{-\sin nx}{n} \right]_{\pi/2}^{\pi} \right\}$

$\quad = \dfrac{4}{\pi} \left\{ \left(\dfrac{\sin(\pi/2)n}{n} - 0 \right) \right.$

$\qquad \left. + \left(0 - \dfrac{-\sin(\pi/2)n}{n} \right) \right\}$

$\quad = \dfrac{4}{\pi} \left(\dfrac{2\sin(\pi/2)n}{n} \right) = \dfrac{8}{\pi n} \left(\sin \dfrac{n\pi}{2} \right)$

When n is even, $a_n = 0$

When n is odd, $a_n = \dfrac{8}{\pi n}$ for $n = 1, 5, 9, \ldots$

and $$a_n = \frac{-8}{\pi n} \quad \text{for } n = 3, 7, 11, \ldots$$

Hence $a_1 = \frac{8}{\pi}$, $a_3 = \frac{-8}{3\pi}$, $a_5 = \frac{8}{5\pi}$, and so on.
Hence the Fourier series for the waveform of Fig. 75.1 is given by:

$$f(x) = \frac{8}{\pi} \left(\cos x - \frac{1}{3} \cos 3x + \frac{1}{5} \cos 5x \right.$$
$$\left. - \frac{1}{7} \cos 7x + \cdots \right)$$

Problem 2. In the Fourier series of Problem 1 let $x = 0$ and deduce a series for $\pi/4$.

When $x = 0$, $f(x) = 2$ (from Fig. 75.1).

Thus, from the Fourier series,

$$2 = \frac{8}{\pi} \left(\cos 0 - \frac{1}{3} \cos 0 + \frac{1}{5} \cos 0 \right.$$
$$\left. - \frac{1}{7} \cos 0 + \cdots \right)$$

Hence $$\frac{2\pi}{8} = 1 - \frac{1}{3} + \frac{1}{5} - \frac{1}{7} + \cdots$$

i.e. $$\frac{\pi}{4} = 1 - \frac{1}{3} + \frac{1}{5} - \frac{1}{7} + \cdots$$

Problem 3. Obtain the Fourier series for the square wave shown in Fig. 75.2.

Figure 75.2

The square wave shown in Fig. 75.2 is an odd function since it is symmetrical about the origin.

Hence, from para. (b), the Fourier series is given by:

$$f(x) = \sum_{n=1}^{\infty} b_n \sin nx$$

The function is defined by:

$$f(x) = \begin{cases} -2, & \text{when } -\pi < x < 0 \\ 2, & \text{when } 0 < x < \pi \end{cases}$$

From para. (b), $$b_n = \frac{2}{\pi} \int_0^{\pi} f(x) \sin nx \, dx$$

$$= \frac{2}{\pi} \int_0^{\pi} 2 \sin nx \, dx$$

$$= \frac{4}{\pi} \left[\frac{-\cos nx}{n} \right]_0^{\pi}$$

$$= \frac{4}{\pi} \left[\left(\frac{-\cos n\pi}{n} \right) - \left(-\frac{1}{n} \right) \right]$$

$$= \frac{4}{\pi n} (1 - \cos n\pi)$$

When n is even, $b_n = 0$

When n is odd, $$b_n = \frac{4}{\pi n} (1 - (-1)) = \frac{8}{\pi n}$$

Hence $$b_1 = \frac{8}{\pi}, \ b_3 = \frac{8}{3\pi}, \ b_5 = \frac{8}{5\pi},$$

and so on.

Hence the Fourier series is:

$$f(x) = \frac{8}{\pi} \left(\sin x + \frac{1}{3} \sin 3x + \frac{1}{5} \sin 5x \right.$$
$$\left. + \frac{1}{7} \sin 7x + \cdots \right)$$

Problem 4. Determine the Fourier series for the function $f(\theta) = \theta^2$ in the range $-\pi < \theta < \pi$. The function has a period of 2π.

A graph of $f(\theta) = \theta^2$ is shown in Fig. 75.3 in the range $-\pi$ to π with period 2π. The function is symmetrical about the $f(\theta)$ axis and is thus an even function. Thus a Fourier cosine series will result of the form:

$$f(\theta) = a_0 + \sum_{n=1}^{\infty} a_n \cos n\theta$$

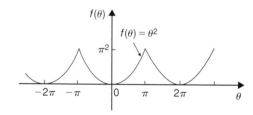

Figure 75.3

From para. (a),

$$a_0 = \frac{1}{\pi} \int_0^\pi f(\theta) d\theta = \frac{1}{\pi} \int_0^\pi \theta^2 d\theta$$

$$= \frac{1}{\pi} \left[\frac{\theta^3}{3} \right]_0^\pi = \frac{\pi^2}{3}$$

and $a_n = \frac{2}{\pi} \int_0^\pi f(\theta) \cos n\theta \, d\theta$

$$= \frac{2}{\pi} \int_0^\pi \theta^2 \cos n\theta \, d\theta$$

$$= \frac{2}{\pi} \left[\frac{\theta^2 \sin n\theta}{n} + \frac{2\theta \cos n\theta}{n^2} - \frac{2 \sin n\theta}{n^3} \right]_0^\pi$$

by parts

$$= \frac{2}{\pi} \left[\left(0 + \frac{2\pi \cos n\pi}{n^2} - 0 \right) - (0) \right]$$

$$= \frac{4}{n^2} \cos n\pi$$

When n is odd, $a_n = \frac{-4}{n^2}$. Hence $a_1 = \frac{-4}{1^2}$,
$a_3 = \frac{-4}{3^2}$, $a_5 = \frac{-4}{5^2}$, and so on.

When n is even, $a_n = \frac{4}{n^2}$. Hence $a_2 = \frac{4}{2^2}$, $a_4 = \frac{4}{4^2}$, and so on.

Hence the Fourier series is:

$$f(\theta) = \theta^2 = \frac{\pi^2}{3} - 4 \left(\cos\theta - \frac{1}{2^2} \cos 2\theta + \frac{1}{3^2} \cos 3\theta \right.$$
$$\left. - \frac{1}{4^2} \cos 4\theta + \frac{1}{5^2} \cos 5\theta - \cdots \right)$$

Problem 5. For the Fourier series of Problem 4, let $\theta = \pi$ and show that $\displaystyle\sum_{n=1}^\infty \frac{1}{n^2} = \frac{\pi^2}{6}$

When $\theta = \pi$, $f(\theta) = \pi^2$ (see Fig. 75.3). Hence from the Fourier series:

$$\pi^2 = \frac{\pi^2}{3} - 4 \left(\cos\pi - \frac{1}{2^2} \cos 2\pi + \frac{1}{3^2} \cos 3\pi \right.$$
$$\left. - \frac{1}{4^2} \cos 4\pi + \frac{1}{5^2} \cos 5\pi - \cdots \right)$$

i.e.

$$\pi^2 - \frac{\pi^2}{3} = -4 \left(-1 - \frac{1}{2^2} - \frac{1}{3^2} - \frac{1}{4^2} - \frac{1}{5^2} - \cdots \right)$$

$$\frac{2\pi^2}{3} = 4 \left(1 + \frac{1}{2^2} + \frac{1}{3^2} + \frac{1}{4^2} + \frac{1}{5^2} + \cdots \right)$$

i.e. $\dfrac{2\pi^2}{(3)(4)} = 1 + \dfrac{1}{2^2} + \dfrac{1}{3^2} + \dfrac{1}{4^2} + \dfrac{1}{5^2} + \cdots$

i.e. $\dfrac{\pi^2}{6} = \dfrac{1}{1^2} + \dfrac{1}{2^2} + \dfrac{1}{3^2} + \dfrac{1}{4^2} + \dfrac{1}{5^2} + \cdots$

Hence

$$\sum_{n=1}^\infty \frac{1}{n^2} = \frac{\pi^2}{6}$$

Now try the following Practice Exercise

Practice Exercise 262   Fourier cosine and Fourier sine series (Answers on page 897)

1. Determine the Fourier series for the function defined by:

$$f(x) = \begin{cases} -1, & -\pi < x < -\dfrac{\pi}{2} \\ 1, & -\dfrac{\pi}{2} < x < \dfrac{\pi}{2} \\ -1, & \dfrac{\pi}{2} < x < \pi \end{cases}$$

which is periodic outside of this range of period 2π.

2. Obtain the Fourier series of the function defined by:

$$f(t) = \begin{cases} t + \pi, & -\pi < t < 0 \\ t - \pi, & 0 < t < \pi \end{cases}$$

which is periodic of period 2π. Sketch the given function.

3. Determine the Fourier series defined by

$$f(x) = \begin{cases} 1 - x, & -\pi < x < 0 \\ 1 + x, & 0 < x < \pi \end{cases}$$

which is periodic of period 2π

4. In the Fourier series of Problem 3, let $x = 0$ and deduce a series for $\pi^2/8$

5. Show that the Fourier series for the triangular waveform shown in Fig. 75.4 is given by:

$$y = \frac{8}{\pi^2} \left(\sin\theta - \frac{1}{3^2} \sin 3\theta + \frac{1}{5^2} \sin 5\theta \right.$$
$$\left. - \frac{1}{7^2} \sin 7\theta + \cdots \right)$$

The function is periodic of period 2π.

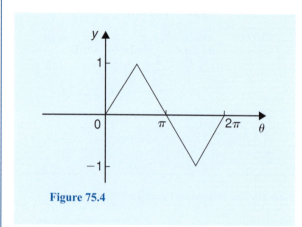

Figure 75.4

75.3 Half-range Fourier series

(a) When a function is defined over the range say 0 to π instead of from 0 to 2π it may be expanded in a series of sine terms only or of cosine terms only. The series produced is called a **half-range Fourier series**.

(b) If a **half-range cosine series** is required for the function $f(x)=x$ in the range 0 to π then an **even** periodic function is required. In Fig. 75.5, $f(x)=x$ is shown plotted from $x=0$ to $x=\pi$. Since an even function is symmetrical about the $f(x)$ axis the line AB is constructed as shown. If the triangular waveform produced is assumed to be periodic of period 2π outside of this range then the waveform is as shown in Fig. 75.5. When a half-range cosine series is required then the Fourier coefficients a_0 and a_n are calculated as in Section 75.2(a), i.e.

$$f(x) = a_0 + \sum_{n=1}^{\infty} a_n \cos nx$$

where $a_0 = \dfrac{1}{\pi}\displaystyle\int_0^{\pi} f(x)\,dx$

and $a_n = \dfrac{2}{\pi}\displaystyle\int_0^{\pi} f(x)\cos nx\,dx$

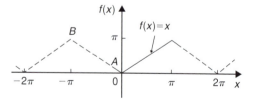

Figure 75.5

(c) If a **half-range sine series** is required for the function $f(x)=x$ in the range 0 to π then an odd periodic function is required. In Fig. 75.6, $f(x)=x$ is shown plotted from $x=0$ to $x=\pi$. Since an odd function is symmetrical about the origin the line CD is constructed as shown. If the sawtooth waveform produced is assumed to be periodic of period 2π outside of this range, then the waveform is as shown in Fig. 75.6. When a half-range sine series is required then the Fourier coefficient b_n is calculated as in Section 75.2(b), i.e.

$$f(x) = \sum_{n=1}^{\infty} b_n \sin nx$$

where $b_n = \dfrac{2}{\pi}\displaystyle\int_0^{\pi} f(x)\sin nx\,dx$

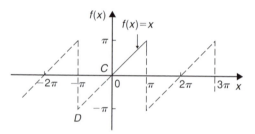

Figure 75.6

> **Problem 6.** Determine the half-range Fourier cosine series to represent the function $f(x)=3x$ in the range $0 \le x \le \pi$

From para. (b), for a half-range cosine series:

$$f(x) = a_0 + \sum_{n=1}^{\infty} a_n \cos nx$$

When $f(x)=3x$,

$$a_0 = \frac{1}{\pi}\int_0^{\pi} f(x)\,dx = \frac{1}{\pi}\int_0^{\pi} 3x\,dx$$

$$= \frac{3}{\pi}\left[\frac{x^2}{2}\right]_0^{\pi} = \frac{3\pi}{2}$$

$$a_n = \frac{2}{\pi}\int_0^{\pi} f(x)\cos nx\,dx$$

$$= \frac{2}{\pi}\int_0^{\pi} 3x\cos nx\,dx$$

$$= \frac{6}{\pi}\left[\frac{x\sin nx}{n} + \frac{\cos nx}{n^2}\right]_0^{\pi} \quad \text{by parts}$$

$$= \frac{6}{\pi}\left[\left(\frac{\pi \sin n\pi}{n} + \frac{\cos n\pi}{n^2}\right) - \left(0 + \frac{\cos 0}{n^2}\right)\right]$$

$$= \frac{6}{\pi}\left(0 + \frac{\cos n\pi}{n^2} - \frac{\cos 0}{n^2}\right)$$

$$= \frac{6}{\pi n^2}(\cos n\pi - 1)$$

When n is even, $a_n = 0$

When n is odd, $a_n = \frac{6}{\pi n^2}(-1-1) = \frac{-12}{\pi n^2}$

Hence $a_1 = \frac{-12}{\pi}$, $a_3 = \frac{-12}{\pi 3^2}$, $a_5 = \frac{-12}{\pi 5^2}$, and so on.

Hence the half-range Fourier cosine series is given by:

$$f(x) = 3x = \frac{3\pi}{2} - \frac{12}{\pi}\left(\cos x + \frac{1}{3^2}\cos 3x\right.$$

$$\left. + \frac{1}{5^2}\cos 5x + \cdots\right)$$

Problem 7. Find the half-range Fourier sine series to represent the function $f(x) = 3x$ in the range $0 \le x \le \pi$

From para. (c), for a half-range sine series:

$$f(x) = \sum_{n=1}^{\infty} b_n \sin nx$$

When $f(x) = 3x$,

$$b_n = \frac{2}{\pi}\int_0^{\pi} f(x)\sin nx \, dx = \frac{2}{\pi}\int_0^{\pi} 3x \sin nx \, dx$$

$$= \frac{6}{\pi}\left[\frac{-x\cos nx}{n} + \frac{\sin nx}{n^2}\right]_0^{\pi} \quad \text{by parts}$$

$$= \frac{6}{\pi}\left[\left(\frac{-\pi \cos n\pi}{n} + \frac{\sin n\pi}{n^2}\right) - (0+0)\right]$$

$$= -\frac{6}{n}\cos n\pi$$

When n is odd, $b_n = \frac{6}{n}$

Hence $b_1 = \frac{6}{1}$, $b_3 = \frac{6}{3}$, $b_5 = \frac{6}{5}$ and so on.

When n is even, $b_n = -\frac{6}{n}$

Hence $b_2 = -\frac{6}{2}$, $b_4 = -\frac{6}{4}$, $b_6 = -\frac{6}{6}$ and so on.

Hence the half-range Fourier sine series is given by:

$$f(x) = 3x = 6\left(\sin x - \frac{1}{2}\sin 2x + \frac{1}{3}\sin 3x\right.$$

$$\left. - \frac{1}{4}\sin 4x + \frac{1}{5}\sin 5x - \cdots\right)$$

Problem 8. Expand $f(x) = \cos x$ as a half-range Fourier sine series in the range $0 \le x \le \pi$, and sketch the function within and outside of the given range.

When a half-range sine series is required then an odd function is implied, i.e. a function symmetrical about the origin. A graph of $y = \cos x$ is shown in Fig. 75.7 in the range 0 to π. For $\cos x$ to be symmetrical about the origin the function is as shown by the broken lines in Fig. 75.7 outside of the given range.

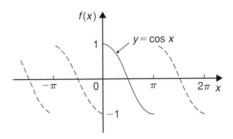

Figure 75.7

From para. (c), for a half-range Fourier sine series:

$$f(x) = \sum_{n=1}^{\infty} b_n \sin nx \, dx$$

$$b_n = \frac{2}{\pi}\int_0^{\pi} f(x)\sin nx \, dx$$

$$= \frac{2}{\pi}\int_0^{\pi} \cos x \sin nx \, dx$$

$$= \frac{2}{\pi}\int_0^{\pi} \frac{1}{2}[\sin(x+nx) - \sin(x-nx)]\,dx$$

$$= \frac{1}{\pi}\left[\frac{-\cos[x(1+n)]}{(1+n)} + \frac{\cos[x(1-n)]}{(1-n)}\right]_0^{\pi}$$

$$= \frac{1}{\pi}\left[\left(\frac{-\cos[\pi(1+n)]}{(1+n)} + \frac{\cos[\pi(1-n)]}{(1-n)}\right)\right.$$

$$\left. - \left(\frac{-\cos 0}{(1+n)} + \frac{\cos 0}{(1-n)}\right)\right]$$

When n is odd,

$$b_n = \frac{1}{\pi}\left[\left(\frac{-1}{(1+n)}+\frac{1}{(1-n)}\right)\right.$$

$$\left.-\left(\frac{-1}{(1+n)}+\frac{1}{(1-n)}\right)\right] = 0$$

When n is even,

$$b_n = \frac{1}{\pi}\left[\left(\frac{1}{(1+n)}-\frac{1}{(1-n)}\right)\right.$$

$$\left.-\left(\frac{-1}{(1+n)}+\frac{1}{(1-n)}\right)\right]$$

$$= \frac{1}{\pi}\left(\frac{2}{(1+n)}-\frac{2}{(1-n)}\right)$$

$$= \frac{1}{\pi}\left(\frac{2(1-n)-2(1+n)}{1-n^2}\right)$$

$$= \frac{1}{\pi}\left(\frac{-4n}{1-n^2}\right) = \frac{4n}{\pi(n^2-1)}$$

Hence $b_2 = \frac{8}{3\pi}$, $b_4 = \frac{16}{15\pi}$, $b_6 = \frac{24}{35\pi}$ and so on.

Hence the half-range Fourier sine series for $f(x)$ in the range 0 to π is given by:

$$f(x) = \frac{8}{3\pi}\sin 2x + \frac{16}{15\pi}\sin 4x$$

$$+ \frac{24}{35\pi}\sin 6x + \cdots$$

or $f(x) = \frac{8}{\pi}\left(\frac{1}{3}\sin 2x + \frac{2}{(3)(5)}\sin 4x\right.$

$$\left.+ \frac{3}{(5)(7)}\sin 6x + \cdots\right)$$

Now try the following Practice Exercise

Practice Exercise 263   Half-range Fourier series (Answers on page 897)

1. Determine the half-range sine series for the function defined by:

$$f(x) = \begin{cases} x, & 0 < x < \dfrac{\pi}{2} \\ 0, & \dfrac{\pi}{2} < x < \pi \end{cases}$$

2. Obtain (a) the half-range cosine series and (b) the half-range sine series for the function

$$f(t) = \begin{cases} 0, & 0 < t < \dfrac{\pi}{2} \\ 1, & \dfrac{\pi}{2} < t < \pi \end{cases}$$

3. Find the half-range Fourier sine series for the function $f(x) = \sin^2 x$ in the range $0 \le x \le \pi$. Sketch the function within and outside of the given range.

4. Determine the half-range Fourier cosine series in the range $x = 0$ to $x = \pi$ for the function defined by:

$$f(x) = \begin{cases} x, & 0 < x < \dfrac{\pi}{2} \\ (\pi - x), & \dfrac{\pi}{2} < x < \pi \end{cases}$$

For fully worked solutions to each of the problems in Practice Exercises 262 and 263 in this chapter, go to the website:
www.routledge.com/cw/bird

Chapter 76

Fourier series over any range

> **Why it is important to understand: Fourier series over any range**
>
> As has been mentioned in preceding chapters, the Fourier series has many applications; in fact, any field of physical science that uses sinusoidal signals, such as engineering, applied mathematics and chemistry, will make use of the Fourier series. In communications, the Fourier series is essential to understanding how a signal behaves when it passes through filters, amplifiers and communications channels. In astronomy, radar and digital signal processing Fourier analysis is used to map the planet. In geology, seismic research uses Fourier analysis, and in optics, Fourier analysis is used in light diffraction. This chapter explains how to determine the Fourier series of a periodic function over any range.

At the end of this chapter, you should be able to:

- understand the Fourier series of a periodic function of period L
- determine the Fourier series of a periodic function of period L
- determine the half-range Fourier series for functions of period L

76.1 Expansion of a periodic function of period L

(a) A periodic function $f(x)$ of period L repeats itself when x increases by L, i.e. $f(x+L)=f(x)$. The change from functions dealt with previously having period 2π to functions having period L is not difficult since it may be achieved by a change of variable.

(b) To find a Fourier series for a function $f(x)$ in the range $-\dfrac{L}{2}\le x\le\dfrac{L}{2}$ a new variable u is introduced such that $f(x)$, as a function of u, has period 2π. If $u=\dfrac{2\pi x}{L}$ then, when $x=-\dfrac{L}{2}$, $u=-\pi$ and when $x=\dfrac{L}{2}$, $u=+\pi$. Also, let $f(x)=f\left(\dfrac{Lu}{2\pi}\right)=F(u)$. The Fourier series for

$F(u)$ is given by:

$$F(u)=a_0+\sum_{n=1}^{\infty}(a_n\cos nu+b_n\sin nu),$$

where $a_0=\dfrac{1}{2\pi}\displaystyle\int_{-\pi}^{\pi}F(u)\,du$,

$$a_n=\frac{1}{\pi}\int_{-\pi}^{\pi}F(u)\cos nu\,du$$

and $b_n=\dfrac{1}{\pi}\displaystyle\int_{-\pi}^{\pi}F(u)\sin nu\,du$

(c) It is however more usual to change the formula of para. (b) to terms of x. Since $u=\dfrac{2\pi x}{L}$, then

$$du=\frac{2\pi}{L}\,dx,$$

and the limits of integration are $-\dfrac{L}{2}$ to $+\dfrac{L}{2}$ instead of from $-\pi$ to $+\pi$. Hence the Fourier series expressed in terms of x is given by:

$$f(x) = a_0 + \sum_{n=1}^{\infty}\left[a_n \cos\left(\frac{2\pi nx}{L}\right)\right.$$

$$\left. + b_n \sin\left(\frac{2\pi nx}{L}\right)\right]$$

where, in the range $-\dfrac{L}{2}$ to $+\dfrac{L}{2}$:

$$a_0 = \frac{1}{L}\int_{\frac{-L}{2}}^{\frac{L}{2}} f(x)\,dx,$$

$$a_n = \frac{2}{L}\int_{\frac{-L}{2}}^{\frac{L}{2}} f(x)\cos\left(\frac{2\pi nx}{L}\right)dx$$

and

$$b_n = \frac{2}{L}\int_{\frac{-L}{2}}^{\frac{L}{2}} f(x)\sin\left(\frac{2\pi nx}{L}\right)dx$$

The limits of integration may be replaced by any interval of length L, such as from 0 to L.

Problem 1. The voltage from a square wave generator is of the form:

$$v(t) = \begin{cases} 0, & -4 < t < 0 \\ 10, & 0 < t < 4 \end{cases}$$

and has a period of 8 ms.

Find the Fourier series for this periodic function.

The square wave is shown in Fig. 76.1. From para. (c), the Fourier series is of the form:

$$v(t) = a_0 + \sum_{n=1}^{\infty}\left[a_n \cos\left(\frac{2\pi nt}{L}\right) + b_n \sin\left(\frac{2\pi nt}{L}\right)\right]$$

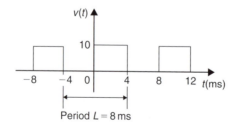

Period $L = 8$ ms

Figure 76.1

$$a_0 = \frac{1}{L}\int_{\frac{-L}{2}}^{\frac{L}{2}} v(t)\,dt = \frac{1}{8}\int_{-4}^{4} v(t)\,dt$$

$$= \frac{1}{8}\left\{\int_{-4}^{0} 0\,dt + \int_{0}^{4} 10\,dt\right\} = \frac{1}{8}[10t]_0^4 = 5$$

$$a_n = \frac{2}{L}\int_{\frac{-L}{2}}^{\frac{L}{2}} v(t)\cos\left(\frac{2\pi nt}{L}\right)dt$$

$$= \frac{2}{8}\int_{-4}^{4} v(t)\cos\left(\frac{2\pi nt}{8}\right)dt$$

$$= \frac{1}{4}\left\{\int_{-4}^{0} 0\cos\left(\frac{\pi nt}{4}\right)dt + \int_{0}^{4} 10\cos\left(\frac{\pi nt}{4}\right)dt\right\}$$

$$= \frac{1}{4}\left[\frac{10\sin\left(\dfrac{\pi nt}{4}\right)}{\left(\dfrac{\pi n}{4}\right)}\right]_0^4 = \frac{10}{\pi n}[\sin \pi n - \sin 0]$$

$$= 0 \text{ for } n = 1, 2, 3, \ldots$$

$$b_n = \frac{2}{L}\int_{\frac{-L}{2}}^{\frac{L}{2}} v(t)\sin\left(\frac{2\pi nt}{L}\right)dt$$

$$= \frac{2}{8}\int_{-4}^{4} v(t)\sin\left(\frac{2\pi nt}{8}\right)dt$$

$$= \frac{1}{4}\left\{\int_{-4}^{0} 0\sin\left(\frac{\pi nt}{4}\right)dt + \int_{0}^{4} 10\sin\left(\frac{\pi nt}{4}\right)dt\right\}$$

$$= \frac{1}{4}\left[\frac{-10\cos\left(\dfrac{\pi nt}{4}\right)}{\left(\dfrac{\pi n}{4}\right)}\right]_0^4$$

$$= \frac{-10}{\pi n}[\cos \pi n - \cos 0]$$

When n is even, $b_n = 0$

When n is odd, $b_1 = \dfrac{-10}{\pi}(-1-1) = \dfrac{20}{\pi}$,

$$b_3 = \frac{-10}{3\pi}(-1-1) = \frac{20}{3\pi},$$

$$b_5 = \frac{20}{5\pi}, \text{ and so on.}$$

Thus the Fourier series for the function $v(t)$ is given by:

$$v(t) = 5 + \frac{20}{\pi}\left[\sin\left(\frac{\pi t}{4}\right) + \frac{1}{3}\sin\left(\frac{3\pi t}{4}\right)\right.$$
$$\left. + \frac{1}{5}\sin\left(\frac{5\pi t}{4}\right) + \cdots\right]$$

Problem 2. Obtain the Fourier series for the function defined by:

$$f(x) = \begin{cases} 0, & \text{when} & -2 < x < -1 \\ 5, & \text{when} & -1 < x < 1 \\ 0, & \text{when} & 1 < x < 2 \end{cases}$$

The function is periodic outside of this range of period 4

The function $f(x)$ is shown in Fig. 76.2 where period, $L = 4$. Since the function is symmetrical about the $f(x)$ axis it is an even function and the Fourier series contains no sine terms (i.e. $b_n = 0$)

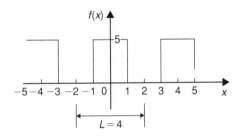

Figure 76.2

Thus, from para. (c),

$$f(x) = a_0 + \sum_{n=1}^{\infty} a_n \cos\left(\frac{2\pi nx}{L}\right)$$

$$a_0 = \frac{1}{L}\int_{\frac{-L}{2}}^{\frac{L}{2}} f(x)\,dx = \frac{1}{4}\int_{-2}^{2} f(x)\,dx$$

$$= \frac{1}{4}\left\{\int_{-2}^{-1} 0\,dx + \int_{-1}^{1} 5\,dx + \int_{1}^{2} 0\,dx\right\}$$

$$= \frac{1}{4}[5x]_{-1}^{1} = \frac{1}{4}[(5) - (-5)] = \frac{10}{4} = \frac{5}{2}$$

$$a_n = \frac{2}{L}\int_{\frac{-L}{2}}^{\frac{L}{2}} f(x)\cos\left(\frac{2\pi nx}{L}\right)dx$$

$$= \frac{2}{4}\int_{-2}^{2} f(x)\cos\left(\frac{2\pi nx}{4}\right)dx$$

$$= \frac{1}{2}\left\{\int_{-2}^{-1} 0\cos\left(\frac{\pi nx}{2}\right)dx\right.$$

$$+ \int_{-1}^{1} 5\cos\left(\frac{\pi nx}{2}\right)dx$$

$$\left. + \int_{1}^{2} 0\cos\left(\frac{\pi nx}{2}\right)dx\right\}$$

$$= \frac{5}{2}\left[\frac{\sin\frac{\pi nx}{2}}{\frac{\pi n}{2}}\right]_{-1}^{1}$$

$$= \frac{5}{\pi n}\left[\sin\left(\frac{\pi n}{2}\right) - \sin\left(\frac{-\pi n}{2}\right)\right]$$

When n is even, $a_n = 0$
When n is odd,

$$a_1 = \frac{5}{\pi}(1 - (-1)) = \frac{10}{\pi}$$

$$a_3 = \frac{5}{3\pi}(-1 - 1) = \frac{-10}{3\pi}$$

$$a_5 = \frac{5}{5\pi}(1 - (-1)) = \frac{10}{5\pi} \text{ and so on.}$$

Hence the Fourier series for the function $f(x)$ is given by:

$$f(x) = \frac{5}{2} + \frac{10}{\pi}\left[\cos\left(\frac{\pi x}{2}\right) - \frac{1}{3}\cos\left(\frac{3\pi x}{2}\right)\right.$$

$$\left. + \frac{1}{5}\cos\left(\frac{5\pi x}{2}\right) - \frac{1}{7}\cos\left(\frac{7\pi x}{2}\right) + \cdots\right]$$

Problem 3. Determine the Fourier series for the function $f(t) = t$ in the range $t = 0$ to $t = 3$

The function $f(t) = t$ in the interval 0 to 3 is shown in Fig. 76.3. Although the function is not periodic it may be constructed outside of this range so that it is periodic of period 3, as shown by the broken lines in Fig. 76.3.

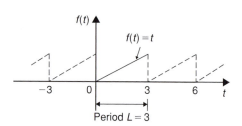

Figure 76.3

From para. (c), the Fourier series is given by:

$$f(t) = a_0 + \sum_{n=1}^{\infty}\left[a_n \cos\left(\frac{2\pi nt}{L}\right) + b_n \sin\left(\frac{2\pi nt}{L}\right)\right]$$

$$a_0 = \frac{1}{L}\int_{-\frac{L}{2}}^{\frac{L}{2}} f(t)\,dx = \frac{1}{L}\int_0^L f(t)\,dx$$

$$= \frac{1}{3}\int_0^3 t\,dt = \frac{1}{3}\left[\frac{t^2}{2}\right]_0^3 = \frac{3}{2}$$

$$a_n = \frac{2}{L}\int_{-\frac{L}{2}}^{\frac{L}{2}} f(t)\cos\left(\frac{2\pi nt}{L}\right)dt$$

$$= \frac{2}{L}\int_0^L t\cos\left(\frac{2\pi nt}{L}\right)dt$$

$$= \frac{2}{3}\int_0^3 t\cos\left(\frac{2\pi nt}{3}\right)dt$$

$$= \frac{2}{3}\left[\frac{t\sin\left(\frac{2\pi nt}{3}\right)}{\left(\frac{2\pi n}{3}\right)} + \frac{\cos\left(\frac{2\pi nt}{3}\right)}{\left(\frac{2\pi n}{3}\right)^2}\right]_0^3$$

by parts

$$= \frac{2}{3}\left[\left\{\frac{3\sin 2\pi n}{\left(\frac{2\pi n}{3}\right)} + \frac{\cos 2\pi n}{\left(\frac{2\pi n}{3}\right)^2}\right\}\right.$$

$$\left. - \left\{0 + \frac{\cos 0}{\left(\frac{2\pi n}{3}\right)^2}\right\}\right] = 0$$

$$b_n = \frac{2}{L}\int_{-\frac{L}{2}}^{\frac{L}{2}} f(t)\sin\left(\frac{2\pi nt}{L}\right)dt$$

$$= \frac{2}{L}\int_0^L t\sin\left(\frac{2\pi nt}{L}\right)dt$$

$$= \frac{2}{3}\int_0^3 t\sin\left(\frac{2\pi nt}{3}\right)dt$$

$$= \frac{2}{3}\left[\frac{-t\cos\left(\frac{2\pi nt}{3}\right)}{\left(\frac{2\pi n}{3}\right)} + \frac{\sin\left(\frac{2\pi nt}{3}\right)}{\left(\frac{2\pi n}{3}\right)^2}\right]_0^3$$

by parts

$$= \frac{2}{3}\left[\left\{\frac{-3\cos 2\pi n}{\left(\frac{2\pi n}{3}\right)} + \frac{\sin 2\pi n}{\left(\frac{2\pi n}{3}\right)^2}\right\}\right.$$

$$\left. - \left\{0 + \frac{\sin 0}{\left(\frac{2\pi n}{3}\right)^2}\right\}\right]$$

$$= \frac{2}{3}\left[\frac{-3\cos 2\pi n}{\left(\frac{2\pi n}{3}\right)}\right] = \frac{-3}{\pi n}\cos 2\pi n = \frac{-3}{\pi n}$$

Hence $b_1 = \dfrac{-3}{\pi}$, $b_2 = \dfrac{-3}{2\pi}$, $b_3 = \dfrac{-3}{3\pi}$ and so on.

Thus the Fourier series for the function $f(t)$ in the range 0 to 3 is given by:

$$f(t) = \frac{3}{2} - \frac{3}{\pi}\left[\sin\left(\frac{2\pi t}{3}\right) + \frac{1}{2}\sin\left(\frac{4\pi t}{3}\right)\right.$$

$$\left. + \frac{1}{3}\sin\left(\frac{6\pi t}{3}\right) + \cdots\right]$$

Now try the following Practice Exercise

Practice Exercise 264   Fourier series over any range L (Answers on page 897)

1. The voltage from a square wave generator is of the form:

$$v(t) = \begin{cases} 0, & -10 < t < 0 \\ 5, & 0 < t < 10 \end{cases}$$

and is periodic of period 20. Show that the Fourier series for the function is given by:

$$v(t) = \frac{5}{2} + \frac{10}{\pi}\left[\sin\left(\frac{\pi t}{10}\right) + \frac{1}{3}\sin\left(\frac{3\pi t}{10}\right)\right.$$

$$\left. + \frac{1}{5}\sin\left(\frac{5\pi t}{10}\right) + \cdots\right]$$

2. Find the Fourier series for $f(x) = x$ in the range $x = 0$ to $x = 5$

3. A periodic function of period 4 is defined by:

$$f(x) = \begin{cases} -3, & -2 < x < 0 \\ +3, & 0 < x < 2 \end{cases}$$

Sketch the function and obtain the Fourier series for the function.

4. Determine the Fourier series for the half-wave rectified sinusoidal voltage $V \sin t$ defined by:

$$f(t) = \begin{cases} V\sin t, & 0 < t < \pi \\ 0, & \pi < t < 2\pi \end{cases}$$

which is periodic of period 2π

76.2 Half-range Fourier series for functions defined over range L

(a) By making the substitution $u = \dfrac{\pi x}{L}$ (see Section 76.1), the range $x = 0$ to $x = L$ corresponds to the range $u = 0$ to $u = \pi$. Hence a function may be expanded in a series of either cosine terms or sine terms only, i.e. a **half-range Fourier series**.

(b) A **half-range cosine series** in the range 0 to L can be expanded as:

$$f(x) = a_0 + \sum_{n=1}^{\infty} a_n \cos\left(\frac{n\pi x}{L}\right)$$

where
$$a_0 = \frac{1}{L}\int_0^L f(x)\,dx \quad\text{and}$$

$$a_n = \frac{2}{L}\int_0^L f(x)\cos\left(\frac{n\pi x}{L}\right)dx$$

(c) A **half-range sine series** in the range 0 to L can be expanded as:

$$f(x) = \sum_{n=1}^{\infty} b_n \sin\left(\frac{n\pi x}{L}\right)$$

where $b_n = \dfrac{2}{L}\displaystyle\int_0^L f(x)\sin\left(\frac{n\pi x}{L}\right)dx$

Problem 4. Determine the half-range Fourier cosine series for the function $f(x) = x$ in the range $0 \le x \le 2$. Sketch the function within and outside of the given range.

A half-range Fourier cosine series indicates an even function. Thus the graph of $f(x) = x$ in the range 0 to 2 is shown in Fig. 76.4 and is extended outside of this range so as to be symmetrical about the $f(x)$ axis as shown by the broken lines.

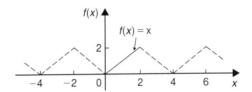

Figure 76.4

From para. (b), for a half-range cosine series:

$$f(x) = a_0 + \sum_{n=1}^{\infty} a_n \cos\left(\frac{n\pi x}{L}\right)$$

$$a_0 = \frac{1}{L}\int_0^L f(x)\,dx = \frac{1}{2}\int_0^2 x\,dx$$

$$= \frac{1}{2}\left[\frac{x^2}{2}\right]_0^2 = 1$$

$$a_n = \frac{2}{L}\int_0^L f(x)\cos\left(\frac{n\pi x}{L}\right)dx$$

$$= \frac{2}{2}\int_0^2 x\cos\left(\frac{n\pi x}{2}\right)dx$$

$$= \left[\frac{x\sin\left(\dfrac{n\pi x}{2}\right)}{\left(\dfrac{n\pi}{2}\right)} + \frac{\cos\left(\dfrac{n\pi x}{2}\right)}{\left(\dfrac{n\pi}{2}\right)^2}\right]_0^2$$

$$= \left[\left(\frac{2\sin n\pi}{\left(\dfrac{n\pi}{2}\right)} + \frac{\cos n\pi}{\left(\dfrac{n\pi}{2}\right)^2}\right) - \left(0 + \frac{\cos 0}{\left(\dfrac{n\pi}{2}\right)^2}\right)\right]$$

$$= \left[\frac{\cos n\pi}{\left(\dfrac{n\pi}{2}\right)^2} - \frac{1}{\left(\dfrac{n\pi}{2}\right)^2}\right]$$

$$= \left(\frac{2}{\pi n}\right)^2 (\cos n\pi - 1)$$

When n is even, $a_n = 0$

$$a_1 = \frac{-8}{\pi^2}, \quad a_3 = \frac{-8}{\pi^2 3^2}, \quad a_5 = \frac{-8}{\pi^2 5^2} \text{ and so on.}$$

Hence the half-range Fourier cosine series for $f(x)$ in the range 0 to 2 is given by:

$$f(x) = 1 - \frac{8}{\pi^2}\left[\cos\left(\frac{\pi x}{2}\right) + \frac{1}{3^2}\cos\left(\frac{3\pi x}{2}\right) + \frac{1}{5^2}\cos\left(\frac{5\pi x}{2}\right) + \cdots\right]$$

Problem 5. Find the half-range Fourier sine series for the function $f(x) = x$ in the range $0 \le x \le 2$. Sketch the function within and outside of the given range.

A half-range Fourier sine series indicates an odd function. Thus the graph of $f(x) = x$ in the range 0 to 2 is shown in Fig. 76.5 and is extended outside of this range so as to be symmetrical about the origin, as shown by the broken lines.

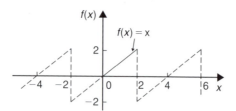

Figure 76.5

From para. (c), for a half-range sine series:

$$f(x) = \sum_{n=1}^{\infty} b_n \sin\left(\frac{n\pi x}{L}\right)$$

$$b_n = \frac{2}{L}\int_0^L f(x)\sin\left(\frac{n\pi x}{L}\right)dx$$

$$= \frac{2}{2}\int_0^2 x\sin\left(\frac{n\pi x}{L}\right)dx$$

$$= \left[\frac{-x\cos\left(\frac{n\pi x}{2}\right)}{\left(\frac{n\pi}{2}\right)} + \frac{\sin\left(\frac{n\pi x}{2}\right)}{\left(\frac{n\pi}{2}\right)^2}\right]_0^2$$

$$= \left[\left(\frac{-2\cos n\pi}{\left(\frac{n\pi}{2}\right)} + \frac{\sin n\pi}{\left(\frac{n\pi}{2}\right)^2}\right) - \left(0 + \frac{\sin 0}{\left(\frac{n\pi}{2}\right)^2}\right)\right]$$

$$= \frac{-2\cos n\pi}{\frac{n\pi}{2}} = \frac{-4}{n\pi}\cos n\pi$$

Hence $b_1 = \frac{-4}{\pi}(-1) = \frac{4}{\pi}$

$$b_2 = \frac{-4}{2\pi}(1) = \frac{-4}{2\pi}$$

$$b_3 = \frac{-4}{3\pi}(-1) = \frac{4}{3\pi} \quad \text{and so on.}$$

Thus the half-range Fourier sine series in the range 0 to 2 is given by:

$$f(x) = \frac{4}{\pi}\left[\sin\left(\frac{\pi x}{2}\right) - \frac{1}{2}\sin\left(\frac{2\pi x}{2}\right) + \frac{1}{3}\sin\left(\frac{3\pi x}{2}\right) - \frac{1}{4}\sin\left(\frac{4\pi x}{2}\right) + \cdots\right]$$

Now try the following Practice Exercise

Practice Exercise 265 Half-range Fourier series over range L (Answers on page 897)

1. Determine the half-range Fourier cosine series for the function $f(x) = x$ in the range $0 \le x \le 3$. Sketch the function within and outside of the given range.

2. Find the half-range Fourier sine series for the function $f(x) = x$ in the range $0 \le x \le 3$. Sketch the function within and outside of the given range.

3. Determine the half-range Fourier sine series for the function defined by:

$$f(t) = \begin{cases} t, & 0 < t < 1 \\ (2-t), & 1 < t < 2 \end{cases}$$

4. Show that the half-range Fourier cosine series for the function $f(\theta) = \theta^2$ in the range 0 to 4 is given by:

$$f(\theta) = \frac{16}{3} - \frac{64}{\pi^2}\left(\cos\left(\frac{\pi\theta}{4}\right) - \frac{1}{2^2}\cos\left(\frac{2\pi\theta}{4}\right) + \frac{1}{3^2}\cos\left(\frac{3\pi\theta}{4}\right) - \cdots\right)$$

Sketch the function within and outside of the given range.

For fully worked solutions to each of the problems in Practice Exercises 264 and 265 in this chapter, go to the website:
www.routledge.com/cw/bird

A numerical method of harmonic analysis

At the end of this chapter, you should be able to:

- define harmonic analysis
- perform a harmonic analysis on data in tabular or graphical form
- consider complex waveform considerations to reduce working of harmonic analysis

77.1 Introduction

Many practical waveforms can be represented by simple mathematical expressions, and, by using Fourier series, the magnitude of their harmonic components determined, as shown in Chapters 73 to 76. For waveforms not in this category, analysis may be achieved by numerical methods. **Harmonic analysis** is the process of resolving a periodic, non-sinusoidal quantity into a series of sinusoidal components of ascending order of frequency.

77.2 Harmonic analysis on data given in tabular or graphical form

The Fourier coefficients a_0, a_n and b_n used in Chapters 73 to 76 all require functions to be integrated, i.e.

$$a_0 = \frac{1}{2\pi} \int_{-\pi}^{\pi} f(x)\,dx = \frac{1}{2\pi} \int_{0}^{2\pi} f(x)\,dx$$

$$= \text{mean value of } f(x)$$

in the range $-\pi$ to π or 0 to 2π

$$a_n = \frac{1}{\pi} \int_{-\pi}^{\pi} f(x) \cos nx \, dx$$

$$= \frac{1}{\pi} \int_0^{2\pi} f(x) \cos nx \, dx$$

$$= \text{twice the mean value of } f(x) \cos nx$$
$$\text{in the range } 0 \text{ to } 2\pi$$

$$b_n = \frac{1}{\pi} \int_{-\pi}^{\pi} f(x) \sin nx \, dx$$

$$= \frac{1}{\pi} \int_0^{2\pi} f(x) \sin nx \, dx$$

$$= \text{twice the mean value of } f(x) \sin nx$$
$$\text{in the range } 0 \text{ to } 2\pi$$

However, irregular waveforms are not usually defined by mathematical expressions and thus the Fourier coefficients cannot be determined by using calculus. In these cases, approximate methods, such as the **trapezoidal rule**, can be used to evaluate the Fourier coefficients.

Most practical waveforms to be analysed are periodic. Let the period of a waveform be 2π and be divided into p equal parts as shown in Fig. 77.1. The width of each interval is thus $\frac{2\pi}{p}$. Let the ordinates be labelled y_0, $y_1, y_2, \dots y_p$ (note that $y_0 = y_p$). The trapezoidal rule states:

$$\text{Area} = (\text{width of interval}) \left[\frac{1}{2}(\text{first} + \text{last ordinate}) \right.$$
$$\left. + \text{sum of remaining ordinates} \right]$$

$$\approx \frac{2\pi}{p} \left[\frac{1}{2}(y_0 + y_p) + y_1 + y_2 + y_3 + \cdots \right]$$

Since $y_0 = y_p$, then $\frac{1}{2}(y_0 + y_p) = y_0 = y_p$

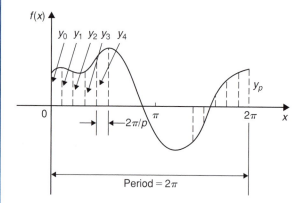

Figure 77.1

Hence area $\approx \dfrac{2\pi}{p} \sum_{k=1}^{p} y_k$

$$\text{Mean value} = \frac{\text{area}}{\text{length of base}}$$

$$\approx \frac{1}{2\pi} \left(\frac{2\pi}{p} \right) \sum_{k=1}^{p} y_k \approx \frac{1}{p} \sum_{k=1}^{p} y_k$$

However, $a_0 = $ mean value of $f(x)$ in the range 0 to 2π.

Thus $\quad a_0 \approx \dfrac{1}{p} \sum_{k=1}^{p} y_k \qquad (1)$

Similarly, $a_n = $ twice the mean value of $f(x) \cos nx$ in the range 0 to 2π,

thus $\quad a_n \approx \dfrac{2}{p} \sum_{k=1}^{p} y_k \cos nx_k \qquad (2)$

and $b_n = $ twice the mean value of $f(x) \sin nx$ in the range 0 to 2π,

thus $\quad b_n \approx \dfrac{2}{p} \sum_{k=1}^{p} y_k \sin nx_k \qquad (3)$

Problem 1. The values of the voltage V volts at different moments in a cycle are given by:

$\theta°$ (degrees)	V (volts)	$\theta°$ (degrees)	V (volts)
30	62	210	−28
60	35	240	24
90	−38	270	80
120	−64	300	96
150	−63	330	90
180	−52	360	70

Draw the graph of voltage V against angle θ and analyse the voltage into its first three constituent harmonics, each coefficient correct to 2 decimal places.

The graph of voltage V against angle θ is shown in Fig. 77.2. The range 0 to 2π is divided into 12 equal intervals giving an interval width of $\frac{2\pi}{12}$, i.e. $\frac{\pi}{6}$ rad or $30°$. The values of the ordinates $y_1, y_2, y_3, \dots$ are $62, 35, -38, \dots$ from the given table of values. If a larger number of intervals are used, results having

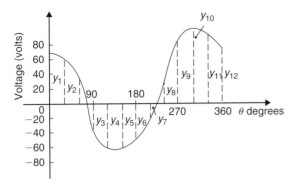

Figure 77.2

a greater accuracy are achieved. The data is tabulated in the proforma shown in Table 77.1, on page 804.

From equation (1), $a_0 \approx \dfrac{1}{p} \displaystyle\sum_{k=1}^{p} y_k = \dfrac{1}{12}(212)$

$$= 17.67 \text{ (since } p = 12)$$

From equation (2), $a_n \approx \dfrac{2}{p} \displaystyle\sum_{k=1}^{p} y_k \cos n x_k$

hence $\qquad a_1 \approx \dfrac{2}{12}(417.94) = 69.66$

$$a_2 \approx \dfrac{2}{12}(-39) = -6.50$$

and $\qquad a_3 \approx \dfrac{2}{12}(-49) = -8.17$

From equation (3), $b_n \approx \dfrac{2}{p} \displaystyle\sum_{k=1}^{p} y_k \sin n x_k$

hence $\qquad b_1 \approx \dfrac{2}{12}(-278.53) = -46.42$

$$b_2 \approx \dfrac{2}{12}(29.43) = 4.91$$

and $\qquad b_3 \approx \dfrac{2}{12}(55) = 9.17$

Substituting these values into the Fourier series:

$$f(x) = a_0 + \sum_{n=1}^{\infty}(a_n \cos nx + b_n \sin nx)$$

gives: $V = 17.67 + 69.66 \cos\theta - 6.50 \cos 2\theta$

$$- 8.17 \cos 3\theta + \cdots - 46.42 \sin\theta$$

$$+ 4.91 \sin 2\theta + 9.17 \sin 3\theta + \cdots \qquad (4)$$

Note that in equation (4), $(-46.42 \sin\theta + 69.66 \cos\theta)$ comprises the fundamental, $(4.91 \sin 2\theta - 6.50 \cos 2\theta)$ comprises the second harmonic and $(9.17 \sin 3\theta - 8.17 \cos 3\theta)$ comprises the third harmonic.
It is shown in Chapter 19 that:

$$a \sin\omega t + b \cos\omega t = R \sin(\omega t + \alpha)$$

where $\quad a = R\cos\alpha, \quad b = R\sin\alpha, \quad R = \sqrt{a^2 + b^2} \quad$ and $\alpha = \tan^{-1}\dfrac{b}{a}$

For the fundamental, $R = \sqrt{(-46.42)^2 + (69.66)^2}$

$$= 83.71$$

If $\qquad a = R\cos\alpha, \text{ then } \cos\alpha = \dfrac{a}{R} = \dfrac{-46.42}{83.71}$

which is negative,

and if $\quad b = R\sin\alpha, \text{ then } \sin\alpha = \dfrac{b}{R} = \dfrac{69.66}{83.71}$

which is positive.

The only quadrant where $\cos\alpha$ is negative *and* $\sin\alpha$ is positive is the second quadrant.

Hence $\alpha = \tan^{-1}\dfrac{b}{a} = \tan^{-1}\dfrac{69.66}{-46.42}$

$$= 123.68° \text{ or } 2.16\,\text{rad}$$

Thus $(-46.42 \sin\theta + 69.66 \cos\theta)$

$$= 83.71 \sin(\theta + 2.16)$$

By a similar method it may be shown that the second harmonic

$$(4.91 \sin 2\theta - 6.50 \cos 2\theta) = 8.15 \sin(2\theta - 0.92)$$

and the third harmonic

$$(9.17 \sin 3\theta - 8.17 \cos 3\theta) = 12.28 \sin(3\theta - 0.73)$$

Hence equation (4) may be re-written as:

$$V = 17.67 + 83.71\,\textbf{sin}(\theta + 2.16)$$

$$+ 8.15\,\textbf{sin}(2\theta - 0.92)$$

$$+ 12.28\,\textbf{sin}(3\theta - 0.73)\,\textbf{volts}$$

which is the form used in Chapter 27 with complex waveforms.

Table 77.1

Ordinates	$\theta°$	V	$\cos\theta$	$V\cos\theta$	$\sin\theta$	$V\sin\theta$	$\cos2\theta$	$V\cos2\theta$	$\sin2\theta$	$V\sin2\theta$	$\cos3\theta$	$V\cos3\theta$	$\sin3\theta$	$V\sin3\theta$
y_1	30	62	0.866	53.69	0.5	31	0.5	31	0.866	53.69	0	0	1	62
y_2	60	35	0.5	17.5	0.866	30.31	−0.5	−17.5	0.866	30.31	−1	−35	0	0
y_3	90	−38	0	0	1	−38	−1	38	0	0	0	0	−1	38
y_4	120	−64	−0.5	32	0.866	−55.42	−0.5	32	−0.866	55.42	1	−64	0	0
y_5	150	−63	−0.866	54.56	0.5	−31.5	0.5	−31.5	−0.866	54.56	0	0	1	−63
y_6	180	−52	−1	52	0	0	1	−52	0	0	−1	52	0	0
y_7	210	−28	−0.866	24.25	−0.5	14	0.5	−14	0.866	−24.25	0	0	−1	28
y_8	240	24	−0.5	−12	−0.866	−20.78	−0.5	−12	0.866	20.78	1	24	0	0
y_9	270	80	0	0	−1	−80	−1	−80	0	0	0	0	1	80
y_{10}	300	96	0.5	48	−0.866	−83.14	−0.5	−48	−0.866	−83.14	−1	−96	0	0
y_{11}	330	90	0.866	77.94	−0.5	−45	0.5	45	−0.866	−77.94	0	0	−1	−90
y_{12}	360	70	1	70	0	0	1	70	0	0	1	70	0	0
		$\displaystyle\sum_{k=1}^{12} y_k = (212)$		$\displaystyle\sum_{k=1}^{12} y_k\cos\theta_k$ $= 417.94$		$\displaystyle\sum_{k=1}^{12} y_k\sin\theta_k$ $= -278.53$		$\displaystyle\sum_{k=1}^{12} y_k\cos2\theta_k$ $= -39$		$\displaystyle\sum_{k=1}^{12} y_k\sin2\theta_k$ $= 29.43$		$\displaystyle\sum_{k=1}^{12} y_k\cos3\theta_k$ $= -49$		$\displaystyle\sum_{k=1}^{12} y_k\sin3\theta_k$ $= 55$

Now try the following Practice Exercise

Determine the Fourier series to represent the periodic functions given by the tables of values in Problems 1 to 3, up to and including the third harmonic and each coefficient correct to 2 decimal places. Use 12 ordinates in each case.

1.

Angle $\theta°$	30	60	90	120	150	180
Displacement y	40	43	38	30	23	17

Angle $\theta°$	210	240	270	300	330	360
Displacement y	11	9	10	13	21	32

2.

Angle $\theta°$	0	30	60	90	120	150
Voltage v	−5.0	−1.5	6.0	12.5	16.0	16.5

Angle $\theta°$	180	210	240	270	300	330
Voltage v	15.0	12.5	6.5	−4.0	−7.0	−7.5

3.

Angle $\theta°$	30	60	90	120	150	180
Current i	0	−1.4	−1.8	−1.9	−1.8	−1.3

Angle $\theta°$	210	240	270	300	330	360
Current i	0	2.2	3.8	3.9	3.5	2.5

77.3 Complex waveform considerations

It is sometimes possible to predict the harmonic content of a waveform on inspection of particular waveform characteristics.

(i) If a periodic waveform is such that the area above the horizontal axis is equal to the area below then the mean value is zero. Hence $a_0 = 0$ (see Fig. 77.3(a)).

(ii) An **even function** is symmetrical about the vertical axis and contains **no sine terms** (see Fig. 77.3(b)).

(iii) An **odd function** is symmetrical about the origin and contains **no cosine terms** (see Fig. 77.3(c)).

(a) $a_0 = 0$ **(b)** Contains no sine terms

(c) Contains no cosine terms   **(d)** Contains only even harmonics

(e) Contains only odd harmonics

Figure 77.3

(iv) $f(x) = f(x + \pi)$ represents a waveform which repeats after half a cycle and **only even harmonics** are present (see Fig. 77.3(d)).

(v) $f(x) = -f(x + \pi)$ represents a waveform for which the positive and negative cycles are identical in shape and **only odd harmonics** are present (see Fig. 77.3(e)).

Problem 2. Without calculating Fourier coefficients state which harmonics will be present in the waveforms shown in Fig. 77.4.

(a)

(b)

Figure 77.4

(a) The waveform shown in Fig. 77.4(a) is symmetrical about the origin and is thus an odd function. An odd function contains no cosine terms. Also, the waveform has the characteristic $f(x) = -f(x + \pi)$, i.e. the positive and negative half cycles are identical in shape. Only odd harmonics can be present in such a waveform. Thus the waveform shown in Fig. 77.4(a) contains **only odd sine terms**. Since the area above the x-axis is equal to the area below, $a_0 = 0$.

(b) The waveform shown in Fig. 77.4(b) is symmetrical about the $f(x)$ axis and is thus an even function. An even function contains no sine terms. Also, the waveform has the characteristic $f(x) = f(x + \pi)$, i.e. the waveform repeats itself after half a cycle. Only even harmonics can be present in such a waveform. Thus the waveform shown in Fig. 77.4(b) contains **only even cosine terms** (together with a constant term, a_0).

Problem 3. An alternating current i amperes is shown in Fig. 77.5. Analyse the waveform into its constituent harmonics as far as and including the fifth harmonic, correct to 2 decimal places, by taking $30°$ intervals.

Figure 77.5

With reference to Fig. 77.5, the following characteristics are noted:

(i) The mean value is zero since the area above the θ axis is equal to the area below it. Thus the constant term, or d.c. component, $a_0 = 0$

(ii) Since the waveform is symmetrical about the origin the function i is odd, which means

that there are no cosine terms present in the Fourier series.

(iii) The waveform is of the form $f(\theta) = -f(\theta + \pi)$ which means that only odd harmonics are present.

Investigating waveform characteristics has thus saved unnecessary calculations and in this case the Fourier series has only odd sine terms present, i.e.

$$i = b_1 \sin \theta + b_3 \sin 3\theta + b_5 \sin 5\theta + \cdots$$

A proforma, similar to Table 77.1, but without the 'cosine terms' columns and without the 'even sine terms' columns is shown in Table 77.2 up to and including the fifth harmonic, from which the Fourier coefficients b_1, b_3 and b_5 can be determined. Twelve co-ordinates are chosen and labelled y_1, y_2, y_3, ... y_{12} as shown in Fig. 77.5

From equation (3),

$$b_n = \frac{2}{p} \sum_{k=1}^{p} i_k \sin n\theta_k, \text{ where } p = 12$$

Hence $b_1 \approx \frac{2}{12}(48.24) = 8.04$,

$$b_3 \approx \frac{2}{12}(-12) = -2.00,$$

and $b_5 \approx \frac{2}{12}(-0.24) = -0.04$

Thus the Fourier series for current i is given by:

$$i = 8.04 \sin \theta - 2.00 \sin 3\theta - 0.04 \sin 5\theta$$

Now try the following Practice Exercise

Practice Exercise 267   A numerical method of harmonic analysis (Answers on page 898)

1. Without performing calculations, state which harmonics will be present in the waveforms shown in Fig. 77.6.

2. Analyse the periodic waveform of displacement y against angle θ in Fig. 77.7(a) into its constituent harmonics as far as and including the third harmonic, by taking $30°$ intervals.

Table 77.2

Ordinate	θ	i	$\sin\theta$	$i\sin\theta$	$\sin 3\theta$	$i\sin 3\theta$	$\sin 5\theta$	$i\sin 5\theta$
y_1	30	2	0.5	1	1	2	0.5	1
y_2	60	7	0.866	6.06	0	0	−0.866	−6.06
y_3	90	10	1	10	−1	−10	1	10
y_4	120	7	0.866	6.06	0	0	−0.866	−6.06
y_5	150	2	0.5	1	1	2	0.5	1
y_6	180	0	0	0	0	0	0	0
y_7	210	−2	−0.5	1	−1	2	−0.5	1
y_8	240	−7	−0.866	6.06	0	0	0.866	−6.06
y_9	270	−10	−1	10	1	−10	−1	10
y_{10}	300	−7	−0.866	6.06	0	0	0.866	−6.06
y_{11}	330	−2	−0.5	1	−1	2	−0.5	1
y_{12}	360	0	0	0	0	0	0	0
				$\sum_{k=1}^{12} y_k \sin\theta_k = 48.24$		$\sum_{k=1}^{12} y_k \sin 3\theta_k = -12$		$\sum_{k=1}^{12} y_k \sin 5\theta_k = -0.24$

(a)

(b)

Figure 77.6

3. For the waveform of current shown in Fig. 77.7(b) state why only a d.c.

component and even cosine terms will appear in the Fourier series and determine the series, using $\pi/6$ rad intervals, up to and including the sixth harmonic.

(a)

(b)

Figure 77.7

Section M

4. Determine the Fourier series as far as the third harmonic to represent the periodic function y given by the waveform in Fig. 77.8. Take 12 intervals when analysing the waveform.

Figure 77.8

For fully worked solutions to each of the problems in Practice Exercises 266 and 267 in this chapter, go to the website:
www.routledge.com/cw/bird

Chapter 78

The complex or exponential form of a Fourier series

Why it is important to understand: **The complex or exponential form of a Fourier series**

A Fourier series may be represented not only as a sum of sines and cosines, as in previous chapters, but as a sum of complex exponentials. The complex exponentials provide a more convenient and compact way of expressing the Fourier series than the trigonometric form. It also allows the magnitude and phase spectra to be easily calculated. This form is widely used by engineers, for example, in circuit theory and control theory. This chapter explains how the trigonometric and exponential forms are equivalent.

At the end of this chapter, you should be able to:

- derive the exponential or complex form of a Fourier series
- derive the complex coefficients for a Fourier series
- determine the complex Fourier series for a given function
- deduce the complex coefficient symmetry relationships
- understand the frequency spectrum of a waveform
- determine phasors in exponential form for various sinusoidal voltages

78.1 Introduction

The form used for the Fourier series in Chapters 73 to 77 consisted of cosine and sine terms. However, there is another form that is commonly used – one that directly gives the amplitude terms in the frequency spectrum and relates to phasor notation. This form involves the use of complex numbers (see Chapters 22 and 23). It is called the **exponential** or **complex form** of a Fourier series.

78.2 Exponential or complex notation

It was shown on page 270, equations (4) and (5) that:

$$e^{j\theta} = \cos\theta + j\sin\theta \tag{1}$$

$$\text{and} \quad e^{-j\theta} = \cos\theta - j\sin\theta \tag{2}$$

Adding equations (1) and (2) gives:

$$e^{j\theta} + e^{-j\theta} = 2\cos\theta$$

from which, $\quad \cos\theta = \dfrac{e^{j\theta} + e^{-j\theta}}{2} \qquad (3)$

Similarly, equation (1) – equation (2) gives:

$$e^{j\theta} - e^{-j\theta} = 2j\sin\theta$$

from which, $\qquad \sin\theta = \dfrac{e^{j\theta} - e^{-j\theta}}{2j} \qquad (4)$

Thus, from page 796, the Fourier series $f(x)$ over any range L,

$$f(x) = a_0 + \sum_{n=1}^{\infty}\left[a_n \cos\left(\frac{2\pi nx}{L}\right) + b_n \sin\left(\frac{2\pi nx}{L}\right) \right]$$

may be written as:

$$f(x) = a_0 + \sum_{n=1}^{\infty}\left[a_n\left(\frac{e^{j\frac{2\pi nx}{L}} + e^{-j\frac{2\pi nx}{L}}}{2}\right) \right.$$
$$\left. + b_n\left(\frac{e^{j\frac{2\pi nx}{L}} - e^{-j\frac{2\pi nx}{L}}}{2j}\right) \right]$$

Multiplying top and bottom of the b_n term by $-j$ (and remembering that $j^2 = -1$) gives:

$$f(x) = a_0 + \sum_{n=1}^{\infty}\left[a_n\left(\frac{e^{j\frac{2\pi nx}{L}} + e^{-j\frac{2\pi nx}{L}}}{2}\right) \right.$$
$$\left. - jb_n\left(\frac{e^{j\frac{2\pi nx}{L}} - e^{-j\frac{2\pi nx}{L}}}{2}\right) \right]$$

Rearranging gives:

$$f(x) = a_0 + \sum_{n=1}^{\infty}\left[\left(\frac{a_n - jb_n}{2}\right) e^{j\frac{2\pi nx}{L}} \right.$$
$$\left. + \left(\frac{a_n + jb_n}{2}\right) e^{-j\frac{2\pi nx}{L}} \right] \qquad (5)$$

The Fourier coefficients a_0, a_n and b_n may be replaced by complex coefficients c_0, c_n and c_{-n} such that

$$c_0 = a_0 \qquad (6)$$

$$c_n = \frac{a_n - jb_n}{2} \qquad (7)$$

and $\quad c_{-n} = \dfrac{a_n + jb_n}{2} \qquad (8)$

where c_{-n} represents the complex conjugate of c_n (see page 257).

Thus, equation (5) may be rewritten as:

$$f(x) = c_0 + \sum_{n=1}^{\infty} c_n e^{j\frac{2\pi nx}{L}} + \sum_{n=1}^{\infty} c_{-n} e^{-j\frac{2\pi nx}{L}} \qquad (9)$$

Since $e^0 = 1$, the c_0 term can be absorbed into the summation since it is just another term to be added to the summation of the c_n term when $n = 0$. Thus,

$$f(x) = \sum_{n=0}^{\infty} c_n e^{j\frac{2\pi nx}{L}} + \sum_{n=1}^{\infty} c_{-n} e^{-j\frac{2\pi nx}{L}} \qquad (10)$$

The c_{-n} term may be rewritten by changing the limits $n = 1$ to $n = \infty$ to $n = -1$ to $n = -\infty$. Since n has been made negative, the exponential term becomes $e^{j\frac{2\pi nx}{L}}$ and c_{-n} becomes c_n. Thus,

$$f(x) = \sum_{n=0}^{\infty} c_n e^{j\frac{2\pi nx}{L}} + \sum_{n=-1}^{-\infty} c_n e^{j\frac{2\pi nx}{L}}$$

Since the summations now extend from $-\infty$ to -1 and from 0 to $+\infty$, equation (10) may be written as:

$$\boldsymbol{f(x) = \sum_{n=-\infty}^{\infty} c_n e^{j\frac{2\pi nx}{L}}} \qquad (11)$$

Equation (11) is the **complex** or **exponential form** of the Fourier series.

From equation (7), the complex coefficient c_n was defined as: $c_n = \dfrac{a_n - jb_n}{2}$

However, a_n and b_n are defined (from page 796) by:

$$a_n = \frac{2}{L}\int_{-\frac{L}{2}}^{\frac{L}{2}} f(x)\cos\left(\frac{2\pi nx}{L}\right)dx \quad \text{and}$$

$$b_n = \frac{2}{L}\int_{-\frac{L}{2}}^{\frac{L}{2}} f(x)\sin\left(\frac{2\pi nx}{L}\right)dx$$

Thus, $\quad c_n = \dfrac{\left(\begin{array}{c} \frac{2}{L}\int_{-\frac{L}{2}}^{\frac{L}{2}} f(x)\cos\left(\frac{2\pi nx}{L}\right)dx \\[6pt] - j\frac{2}{L}\int_{-\frac{L}{2}}^{\frac{L}{2}} f(x)\sin\left(\frac{2\pi nx}{L}\right)dx \end{array}\right)}{2}$

$$= \frac{1}{L} \int_{-\frac{L}{2}}^{\frac{L}{2}} f(x) \cos\left(\frac{2\pi nx}{L}\right) dx$$

$$- j\frac{1}{L} \int_{-\frac{L}{2}}^{\frac{L}{2}} f(x) \sin\left(\frac{2\pi nx}{L}\right) dx$$

From equations (3) and (4),

$$c_n = \frac{1}{L} \int_{-\frac{L}{2}}^{\frac{L}{2}} f(x) \left(\frac{e^{j\frac{2\pi nx}{L}} + e^{-j\frac{2\pi nx}{L}}}{2}\right) dx$$

$$- j\frac{1}{L} \int_{-\frac{L}{2}}^{\frac{L}{2}} f(x) \left(\frac{e^{j\frac{2\pi nx}{L}} - e^{-j\frac{2\pi nx}{L}}}{2j}\right) dx$$

from which,

$$c_n = \frac{1}{L} \int_{-\frac{L}{2}}^{\frac{L}{2}} f(x) \left(\frac{e^{j\frac{2\pi nx}{L}} + e^{-j\frac{2\pi nx}{L}}}{2}\right) dx$$

$$- \frac{1}{L} \int_{-\frac{L}{2}}^{\frac{L}{2}} f(x) \left(\frac{e^{j\frac{2\pi nx}{L}} - e^{-j\frac{2\pi nx}{L}}}{2}\right) dx$$

i.e. $$c_n = \frac{1}{L} \int_{-\frac{L}{2}}^{\frac{L}{2}} f(x) e^{-j\frac{2\pi nx}{L}} dx \qquad (12)$$

Care needs to be taken when determining c_0. If n appears in the denominator of an expression the expansion can be invalid when $n = 0$. In such circumstances it is usually simpler to evaluate c_0 by using the relationship:

$$c_0 = a_0 = \frac{1}{L} \int_{-\frac{L}{2}}^{\frac{L}{2}} f(x) dx \quad \text{(from page 796).} \qquad (13)$$

Problem 1. Determine the complex Fourier series for the function defined by:

$$f(x) = \begin{cases} 0, & \text{when } -2 \le x \le -1 \\ 5, & \text{when } -1 \le x \le 1 \\ 0, & \text{when } 1 \le x \le 2 \end{cases}$$

The function is periodic outside this range of period 4

This is the same Problem as Problem 2 on page 797 and we can use this to demonstrate that the two forms of Fourier series are equivalent.
The function $f(x)$ is shown in Fig. 78.1, where the period, $L = 4$
From equation (11), the complex Fourier series is given by:

$$f(x) = \sum_{n=-\infty}^{\infty} c_n e^{j\frac{2\pi nx}{L}}$$

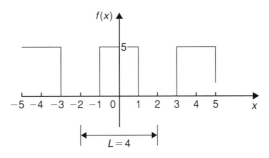

Figure 78.1

where c_n is given by:

$$c_n = \frac{1}{L} \int_{-\frac{L}{2}}^{\frac{L}{2}} f(x) e^{-j\frac{2\pi nx}{L}} dx \quad \text{(from equation 12).}$$

With reference to Fig. 78.1, when $L = 4$,

$$c_n = \frac{1}{4}\left\{\int_{-2}^{-1} 0 \, dx + \int_{-1}^{1} 5 e^{-j\frac{2\pi nx}{4}} dx + \int_{1}^{2} 0 \, dx\right\}$$

$$= \frac{1}{4}\int_{-1}^{1} 5 e^{-j\frac{\pi nx}{2}} dx = \frac{5}{4}\left[\frac{e^{-j\frac{\pi nx}{2}}}{-\frac{j\pi n}{2}}\right]_{-1}^{1}$$

$$= \frac{-5}{j2\pi n}\left[e^{-j\frac{\pi nx}{2}}\right]_{-1}^{1} = \frac{-5}{j2\pi n}\left(e^{-j\frac{\pi n}{2}} - e^{j\frac{\pi n}{2}}\right)$$

$$= \frac{5}{\pi n}\left(\frac{e^{j\frac{\pi n}{2}} - e^{-j\frac{\pi n}{2}}}{2j}\right)$$

$$= \frac{5}{\pi n}\sin\frac{\pi n}{2} \quad \text{(from equation (4)).}$$

Hence, from equation (11), **the complex form of the Fourier series** is given by:

$$f(x) = \sum_{n=-\infty}^{\infty} c_n e^{j\frac{2\pi nx}{L}} = \sum_{n=-\infty}^{\infty} \frac{5}{\pi n}\sin\frac{\pi n}{2} e^{j\frac{\pi nx}{2}} \qquad (14)$$

Let us show how this result is equivalent to the result involving sine and cosine terms determined on page 797. From equation (13),

$$c_0 = a_0 = \frac{1}{L} \int_{-\frac{L}{2}}^{\frac{L}{2}} f(x) dx = \frac{1}{4}\int_{-1}^{1} 5 \, dx$$

$$= \frac{5}{4}[x]_{-1}^{1} = \frac{5}{4}[1 - (-1)] = \frac{5}{2}$$

Since $c_n = \frac{5}{\pi n}\sin\frac{\pi n}{2}$, then

$$c_1 = \frac{5}{\pi}\sin\frac{\pi}{2} = \frac{5}{\pi}$$

$$c_2 = \frac{5}{2\pi}\sin\pi = 0$$

(in fact, **all even terms will be zero** since $\sin n\pi = 0$)

$$c_3 = \frac{5}{\pi n}\sin\frac{\pi n}{2} = \frac{5}{3\pi}\sin\frac{3\pi}{2} = -\frac{5}{3\pi}$$

By similar substitution,

$$c_5 = \frac{5}{5\pi} \qquad c_7 = -\frac{5}{7\pi}, \text{ and so on.}$$

Similarly,

$$c_{-1} = \frac{5}{-\pi}\sin\frac{-\pi}{2} = \frac{5}{\pi}$$

$$c_{-2} = -\frac{5}{2\pi}\sin\frac{-2\pi}{2} = 0 = c_{-4} = c_{-6}, \text{ and so on.}$$

$$c_{-3} = -\frac{5}{3\pi}\sin\frac{-3\pi}{2} = -\frac{5}{3\pi}$$

$$c_{-5} = -\frac{5}{5\pi}\sin\frac{-5\pi}{2} = \frac{5}{5\pi}, \text{ and so on.}$$

Hence, the extended complex form of the Fourier series shown in equation (14) becomes:

$$f(x) = \frac{5}{2} + \frac{5}{\pi}e^{j\frac{\pi x}{2}} - \frac{5}{3\pi}e^{j\frac{3\pi x}{2}} + \frac{5}{5\pi}e^{j\frac{5\pi x}{2}}$$

$$- \frac{5}{7\pi}e^{j\frac{7\pi x}{2}} + \cdots + \frac{5}{\pi}e^{-j\frac{\pi x}{2}}$$

$$- \frac{5}{3\pi}e^{-j\frac{3\pi x}{2}} + \frac{5}{5\pi}e^{-j\frac{5\pi x}{2}}$$

$$- \frac{5}{7\pi}e^{-j\frac{7\pi x}{2}} + \cdots$$

$$= \frac{5}{2} + \frac{5}{\pi}\left(e^{j\frac{\pi x}{2}} + e^{-j\frac{\pi x}{2}}\right)$$

$$- \frac{5}{3\pi}\left(e^{j\frac{3\pi x}{2}} + e^{-j\frac{3\pi x}{2}}\right)$$

$$+ \frac{5}{5\pi}\left(e^{\frac{5\pi x}{2}} + e^{-j\frac{5\pi x}{2}}\right) - \cdots$$

$$= \frac{5}{2} + \frac{5}{\pi}(2)\left(\frac{e^{j\frac{\pi x}{2}} + e^{-j\frac{\pi x}{2}}}{2}\right)$$

$$- \frac{5}{3\pi}(2)\left(\frac{e^{j\frac{3\pi x}{2}} + e^{-j\frac{3\pi x}{2}}}{2}\right)$$

$$+ \frac{5}{5\pi}(2)\left(\frac{e^{j\frac{5\pi x}{2}} + e^{-j\frac{5\pi x}{2}}}{2}\right) - \cdots$$

$$= \frac{5}{2} + \frac{10}{\pi}\cos\left(\frac{\pi x}{2}\right) - \frac{10}{3\pi}\cos\left(\frac{3\pi x}{2}\right)$$

$$+ \frac{10}{5\pi}\cos\left(\frac{5\pi x}{2}\right) - \cdots$$

(from equation (3))

i.e. $f(x) = \dfrac{5}{2} + \dfrac{10}{\pi}\left[\cos\left(\dfrac{\pi x}{2}\right) - \dfrac{1}{3}\cos\left(\dfrac{3\pi x}{2}\right)\right.$

$$\left. + \frac{1}{5}\cos\left(\frac{5\pi x}{2}\right) - \cdots\right]$$

which is the same as obtained on page 797.

Hence, $\displaystyle\sum_{n=-\infty}^{\infty} \frac{5}{\pi n}\sin\frac{n\pi}{2}\, e^{j\frac{\pi nx}{2}}$ is equivalent to

$$\frac{5}{2} + \frac{10}{\pi}\left[\cos\left(\frac{\pi x}{2}\right) - \frac{1}{3}\cos\left(\frac{3\pi x}{2}\right)\right.$$

$$\left. + \frac{1}{5}\cos\left(\frac{5\pi x}{2}\right) - \cdots\right]$$

Problem 2. Show that the complex Fourier series for the function $f(t) = t$ in the range $t = 0$ to $t = 1$, and of period 1, may be expressed as:

$$f(t) = \frac{1}{2} + \frac{j}{2\pi}\sum_{n=-\infty}^{\infty}\frac{e^{j2\pi nt}}{n}$$

The sawtooth waveform is shown in Fig. 78.2. From equation (11), the complex Fourier series is given by:

$$f(t) = \sum_{n=-\infty}^{\infty} c_n\, e^{j\frac{2\pi nt}{L}}$$

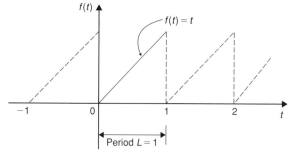

Figure 78.2

and when the period $L=1$, then:

$$f(t) = \sum_{n=-\infty}^{\infty} c_n\, e^{j2\pi nt}$$

where, from equation (12),

$$c_n = \frac{1}{L}\int_{-\frac{L}{2}}^{\frac{L}{2}} f(t)\,e^{-j\frac{2\pi nt}{L}}\,dt = \frac{1}{L}\int_0^L f(t)\,e^{-j\frac{2\pi nt}{L}}\,dt$$

and when $L=1$ and $f(t)=t$, then:

$$c_n = \frac{1}{1}\int_0^1 t\,e^{-j\frac{2\pi nt}{1}}\,dt = \int_0^1 t\,e^{-j2\pi nt}\,dt$$

Using integration by parts (see Chapter 46), let $u=t$, from which, $\dfrac{du}{dt}=1$, and $dt=du$, and

let $dv=e^{-j2\pi nt}$, from which,

$$v=\int e^{-j2\pi nt}\,dt = \frac{e^{-j2\pi nt}}{-j2\pi n}$$

Hence, $c_n = \displaystyle\int_0^1 t\,e^{-j2\pi nt} = uv - \int v\,du$

$$= \left[t\,\frac{e^{-j2\pi nt}}{-j2\pi n}\right]_0^1 - \int_0^1 \frac{e^{-j2\pi nt}}{-j2\pi n}\,dt$$

$$= \left[t\,\frac{e^{-j2\pi nt}}{-j2\pi n} - \frac{e^{-j2\pi nt}}{(-j2\pi n)^2}\right]_0^1$$

$$= \left(\frac{e^{-j2\pi n}}{-j2\pi n} - \frac{e^{-j2\pi n}}{(-j2\pi n)^2}\right)$$

$$- \left(0 - \frac{e^0}{(-j2\pi n)^2}\right)$$

From equation (2),

$$c_n = \left(\frac{\cos 2\pi n - j\sin 2\pi n}{-j2\pi n} - \frac{\cos 2\pi n - j\sin 2\pi n}{(-j2\pi n)^2}\right.$$
$$\left. + \frac{1}{(-j2\pi n)^2}\right)$$

However, $\cos 2\pi n = 1$ and $\sin 2\pi n = 0$ for all positive and negative integer values of n.

Thus, $c_n = \dfrac{1}{-j2\pi n} - \dfrac{1}{(-j2\pi n)^2} + \dfrac{1}{(-j2\pi n)^2}$

$$= \frac{1}{-j2\pi n} = \frac{1(j)}{-j2\pi n(j)}$$

i.e. $c_n = \dfrac{j}{2\pi n}$

From equation (13),

$$c_0 = a_0 = \frac{1}{L}\int_{-\frac{L}{2}}^{\frac{L}{2}} f(t)\,dt$$

$$= \frac{1}{L}\int_0^L f(t)\,dt = \frac{1}{1}\int_0^1 t\,dt$$

$$= \left[\frac{t^2}{2}\right]_0^1 = \left[\frac{1}{2}-0\right] = \frac{1}{2}$$

Hence, the complex Fourier series is given by:

$$f(t) = \sum_{n=-\infty}^{\infty} c_n\, e^{j\frac{2\pi nt}{L}} \quad \text{from equation (11)}$$

i.e. $f(t) = \dfrac{1}{2} + \displaystyle\sum_{n=-\infty}^{\infty} \frac{j}{2\pi n}\,e^{j2\pi nt}$

$$= \frac{1}{2} + \frac{j}{2\pi}\sum_{n=-\infty}^{\infty} \frac{e^{j2\pi nt}}{n}$$

Problem 3. Show that the exponential form of the Fourier series for the waveform described by:

$$f(x) = \begin{cases} 0 \text{ when } -4 \le x \le 0 \\ 10 \text{ when } \ \ 0 \le x \le 4 \end{cases}$$

and has a period of 8, is given by:

$$f(x) = \sum_{n=-\infty}^{\infty} \frac{5j}{n\pi}(\cos n\pi - 1)\,e^{j\frac{n\pi x}{4}}$$

From equation (12),

$$c_n = \frac{1}{L}\int_{-\frac{L}{2}}^{\frac{L}{2}} f(x)\,e^{-j\frac{2\pi nx}{L}}\,dx$$

$$= \frac{1}{8}\left[\int_{-4}^0 0\,e^{-j\frac{\pi nx}{4}}\,dx + \int_0^4 10\,e^{-j\frac{\pi nx}{4}}\,dx\right]$$

$$= \frac{10}{8}\left[\frac{e^{-j\frac{\pi nt}{4}}}{-j\frac{\pi n}{4}}\right]_0^4 = \frac{10}{8}\left(\frac{4}{-j\pi n}\right)\left[e^{-j\pi n}-1\right]$$

$$= \frac{5j}{-j^2\pi n}\left(e^{-j\pi n}-1\right) = \frac{5j}{\pi n}\left(e^{-j\pi n}-1\right)$$

From equation (2), $e^{-j\theta} = \cos\theta - j\sin\theta$, thus $e^{-j\pi n} = \cos\pi n - j\sin\pi n = \cos\pi n$ for all integer

values of n. Hence,

$$c_n = \frac{5j}{\pi n}\left(e^{-j\pi n} - 1\right) = \frac{5j}{\pi n}(\cos n\pi - 1)$$

From equation (11), the exponential Fourier series is given by:

$$f(x) = \sum_{n=-\infty}^{\infty} c_n e^{j\frac{2\pi nx}{L}}$$

$$= \sum_{n=-\infty}^{\infty} \frac{5j}{n\pi}(\cos n\pi - 1)e^{j\frac{n\pi x}{4}}$$

Now try the following Practice Exercise

Practice Exercise 268   The complex form of a Fourier series (Answers on page 898)

1. Determine the complex Fourier series for the function defined by:
$$f(t) = \begin{cases} 0, & \text{when } -\pi \le t \le 0 \\ 2, & \text{when } 0 \le t \le \pi \end{cases}$$
The function is periodic outside of this range of period 2π.

2. Show that the complex Fourier series for the waveform shown in Fig. 78.3, that has period 2, may be represented by:
$$f(t) = 2 + \sum_{\substack{n=-\infty \\ (n \ne 0)}}^{\infty} \frac{j2}{\pi n}(\cos n\pi - 1)e^{j\pi nt}$$

Figure 78.3

3. Show that the complex Fourier series of Problem 2 is equivalent to:

$$f(t) = 2 + \frac{8}{\pi}\left(\sin \pi t + \frac{1}{3}\sin 3\pi t + \frac{1}{5}\sin 5\pi t + \cdots\right)$$

4. Determine the exponential form of the Fourier series for the function defined by: $f(t) = e^{2t}$ when $-1 < t < 1$ and has period 2.

78.4   Symmetry relationships

If even or odd symmetry is noted in a function, then time can be saved in determining coefficients.

The Fourier coefficients present in the complex Fourier series form are affected by symmetry. Summarising from previous chapters:

An **even function** is symmetrical about the vertical axis and contains no sine terms, i.e. $b_n = 0$

For even symmetry,

$$a_0 = \frac{1}{L}\int_0^L f(x)\,dx \quad \text{and}$$

$$a_n = \frac{2}{L}\int_0^L f(x)\cos\left(\frac{2\pi nx}{L}\right)dx$$

$$= \frac{4}{L}\int_0^{\frac{L}{2}} f(x)\cos\left(\frac{2\pi nx}{L}\right)dx$$

An **odd function** is symmetrical about the origin and contains no cosine terms, $a_0 = a_n = 0$

For odd symmetry,

$$b_n = \frac{2}{L}\int_0^L f(x)\sin\left(\frac{2\pi nx}{L}\right)dx$$

$$= \frac{4}{L}\int_0^{\frac{L}{2}} f(x)\sin\left(\frac{2\pi nx}{L}\right)dx$$

From equation (7), page 810, $c_n = \dfrac{a_n - jb_n}{2}$

Thus, for **even symmetry**, $b_n = 0$ and

$$c_n = \frac{a_n}{2} = \frac{2}{L}\int_0^{\frac{L}{2}} f(x)\cos\left(\frac{2\pi nx}{L}\right)dx \qquad (15)$$

For **odd symmetry**, $a_n = 0$ and

$$c_n = \frac{-jb_n}{2} = -j\frac{2}{L}\int_0^{\frac{L}{2}} f(x)\sin\left(\frac{2\pi nx}{L}\right)dx \qquad (16)$$

For example, in Problem 1 on page 811, the function $f(x)$ is even, since the waveform is symmetrical about the $f(x)$ axis. Thus equation (15) could have been used, giving:

$$c_n = \frac{2}{L} \int_0^{\frac{L}{2}} f(x) \cos\left(\frac{2\pi nx}{L}\right) dx$$

$$= \frac{2}{4} \int_0^2 f(x) \cos\left(\frac{2\pi nx}{4}\right) dx$$

$$= \frac{1}{2} \left\{ \int_0^1 5 \cos\left(\frac{\pi nx}{2}\right) dx + \int_1^2 0 \, dx \right\}$$

$$= \frac{5}{2} \left[\frac{\sin\left(\frac{\pi nx}{2}\right)}{\frac{\pi n}{2}} \right]_0^1 = \frac{5}{2}\left(\frac{2}{\pi n}\right)\left(\sin\frac{n\pi}{2} - 0\right)$$

$$= \frac{5}{\pi n} \sin\frac{n\pi}{2}$$

which is the same answer as in Problem 1; however, a knowledge of even functions has produced the coefficient more quickly.

Problem 4. Obtain the Fourier series, in complex form, for the square wave shown in Fig. 78.4.

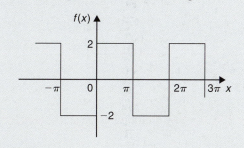

Figure 78.4

Method A

The square wave shown in Figure 78.4 is an **odd function** since it is symmetrical about the origin. The period of the waveform, $L = 2\pi$.

Thus, using equation (16):

$$c_n = -j\frac{2}{L} \int_0^{\frac{L}{2}} f(x) \sin\left(\frac{2\pi nx}{L}\right) dx$$

$$= -j\frac{2}{2\pi} \int_0^\pi 2 \sin\left(\frac{2\pi nx}{2\pi}\right) dx$$

$$= -j\frac{2}{\pi} \int_0^\pi \sin nx \, dx = -j\frac{2}{\pi}\left[\frac{-\cos nx}{n}\right]_0^\pi$$

$$= -j\frac{2}{\pi n}\left((-\cos\pi n) - (-\cos 0)\right)$$

i.e. $c_n = -j\dfrac{2}{\pi n}[1 - \cos\pi n]$ (17)

Method B

If it had **not** been noted that the function was odd, equation (12) would have been used, i.e.

$$c_n = \frac{1}{L} \int_{-\frac{L}{2}}^{\frac{L}{2}} f(x) e^{-j\frac{2\pi nx}{L}} dx$$

$$= \frac{1}{2\pi} \int_{-\pi}^{\pi} f(x) e^{-j\frac{2\pi nx}{2\pi}} dx$$

$$= \frac{1}{2\pi} \left\{ \int_{-\pi}^0 -2e^{-jnx} dx + \int_0^\pi 2e^{-jnx} dx \right\}$$

$$= \frac{1}{2\pi} \left\{ \left[\frac{-2e^{-jnx}}{-jn}\right]_{-\pi}^0 + \left[\frac{2e^{-jnx}}{-jn}\right]_0^\pi \right\}$$

$$= \frac{1}{2\pi}\left(\frac{2}{jn}\right)\left\{ \left[e^{-jnx}\right]_{-\pi}^0 - \left[e^{-jnx}\right]_0^\pi \right\}$$

$$= \frac{1}{2\pi}\left(\frac{2}{jn}\right)\left\{ \left[e^0 - e^{+jn\pi}\right] - \left[e^{-jn\pi} - e^0\right] \right\}$$

$$= \frac{1}{j\pi n}\left\{ 1 - e^{jn\pi} - e^{-jn\pi} + 1 \right\}$$

$$= \frac{1}{jn\pi}\left\{ 2 - 2\left(\frac{e^{jn\pi} + e^{-jn\pi}}{2}\right) \right\}$$

by rearranging

$$= \frac{2}{jn\pi}\left\{ 1 - \left(\frac{e^{jn\pi} + e^{-jn\pi}}{2}\right) \right\}$$

$$= \frac{2}{jn\pi}\{1 - \cos n\pi\}$$ from equation (3)

$$= \frac{-j2}{-j(jn\pi)}\{1 - \cos n\pi\}$$

by multiplying top and bottom by $-j$

i.e. $c_n = -j\dfrac{2}{n\pi}(1 - \cos n\pi)$ (17)

It is clear that method A is by far the shorter of the two methods.

From equation (11), the complex Fourier series is given by:

$$f(x) = \sum_{n=-\infty}^{\infty} c_n \, e^{j\frac{2\pi nx}{L}}$$

$$= \sum_{n=-\infty}^{\infty} -j\frac{2}{n\pi}(1 - \cos n\pi)\,e^{jnx} \qquad (18)$$

> **Problem 5.** Show that the complex Fourier series obtained in Problem 4 above is equivalent to
>
> $$f(x) = \frac{8}{\pi}\left(\sin x + \frac{1}{3}\sin 3x + \frac{1}{5}\sin 5x\right.$$
>
> $$\left. + \frac{1}{7}\sin 7x + \cdots\right)$$
>
> (which was the Fourier series obtained in terms of sines and cosines in Problem 3 on page 790).

From equation (17) above, $c_n = -j\dfrac{2}{n\pi}(1 - \cos n\pi)$

When $n = 1$,

$$c_1 = -j\frac{2}{(1)\pi}(1 - \cos\pi)$$

$$= -j\frac{2}{\pi}\left(1 - (-1)\right) = -\frac{j4}{\pi}$$

When $n = 2$,

$$c_2 = -j\frac{2}{2\pi}(1 - \cos 2\pi) = 0;$$

in fact, all even values of c_n will be zero.

When $n = 3$,

$$c_3 = -j\frac{2}{3\pi}(1 - \cos 3\pi)$$

$$= -j\frac{2}{3\pi}(1 - (-1)) = -\frac{j4}{3\pi}$$

By similar reasoning,

$$c_5 = -\frac{j4}{5\pi}, \quad c_7 = -\frac{j4}{7\pi}, \text{ and so on.}$$

When $n = -1$,

$$c_{-1} = -j\frac{2}{(-1)\pi}(1 - \cos(-\pi))$$

$$= +j\frac{2}{\pi}(1 - (-1)) = +\frac{j4}{\pi}$$

When $n = -3$,

$$c_{-3} = -j\frac{2}{(-3)\pi}(1 - \cos(-3\pi))$$

$$= +j\frac{2}{3\pi}(1 - (-1)) = +\frac{j4}{3\pi}$$

By similar reasoning,

$$c_{-5} = +\frac{j4}{5\pi}, \quad c_{-7} = +\frac{j4}{7\pi}, \text{ and so on.}$$

Since the waveform is odd, $c_0 = a_0 = 0$
From equation (18) above,

$$f(x) = \sum_{n=-\infty}^{\infty} -j\frac{2}{n\pi}(1 - \cos n\pi)\,e^{jnx}$$

Hence,

$$f(x) = -\frac{j4}{\pi}e^{jx} - \frac{j4}{3\pi}e^{j3x} - \frac{j4}{5\pi}e^{j5x}$$

$$- \frac{j4}{7\pi}e^{j7x} - \cdots + \frac{j4}{\pi}e^{-jx} + \frac{j4}{3\pi}e^{-j3x}$$

$$+ \frac{j4}{5\pi}e^{-j5x} + \frac{j4}{7\pi}e^{-j7x} + \cdots$$

$$= \left(-\frac{j4}{\pi}e^{jx} + \frac{j4}{\pi}e^{-jx}\right)$$

$$+ \left(-\frac{j4}{3\pi}e^{3x} + \frac{j4}{3\pi}e^{-3x}\right)$$

$$+ \left(-\frac{j4}{5\pi}e^{5x} + \frac{j4}{5\pi}e^{-5x}\right) + \cdots$$

$$= -\frac{j4}{\pi}\left(e^{jx} - e^{-jx}\right) - \frac{j4}{3\pi}\left(e^{3x} - e^{-3x}\right)$$

$$- \frac{j4}{5\pi}\left(e^{5x} - e^{-5x}\right) + \cdots$$

$$= \frac{4}{j\pi}\left(e^{jx} - e^{-jx}\right) + \frac{4}{j3\pi}\left(e^{3x} - e^{-3x}\right)$$

$$+ \frac{4}{j5\pi}\left(e^{5x} - e^{-5x}\right) + \cdots$$

by multiplying top and bottom by j

$$= \frac{8}{\pi}\left(\frac{e^{jx} - e^{-jx}}{2j}\right) + \frac{8}{3\pi}\left(\frac{e^{j3x} - e^{-j3}}{2j}\right)$$

$$+ \frac{8}{5\pi}\left(\frac{e^{j5x} - e^{-j5x}}{2j}\right) + \cdots$$

by rearranging

$$= \frac{8}{\pi}\sin x + \frac{8}{3\pi}\sin 3x + \frac{8}{3x}\sin 5x + \cdots$$

from equation (4), page 810

i.e.

$$f(x) = \frac{8}{\pi}\left(\sin x + \frac{1}{3}\sin 3x + \frac{1}{5}\sin 5x\right.$$

$$\left. + \frac{1}{7}\sin 7x + \cdots\right)$$

Hence,

$$f(x) = \sum_{n=-\infty}^{\infty} -j\frac{2}{n\pi}(1-\cos n\pi)\,e^{jnx}$$

$$\equiv \frac{8}{\pi}\left(\sin x + \frac{1}{3}\sin 3x + \frac{1}{5}\sin 5x\right.$$

$$\left. + \frac{1}{7}\sin 7x + \cdots\right)$$

Now try the following Practice Exercise

Practice Exercise 269   Symmetry relationships (Answers on page 898)

1. Determine the exponential form of the Fourier series for the periodic function defined by:

$$f(x) = \begin{cases} -2, & \text{when } -\pi \le x \le -\dfrac{\pi}{2} \\ 2, & \text{when } -\dfrac{\pi}{2} \le x \le +\dfrac{\pi}{2} \\ -2, & \text{when } +\dfrac{\pi}{2} \le x \le +\pi \end{cases}$$

   and has a period of 2π.

2. Show that the exponential form of the Fourier series in problem 1 above is equivalent to:

$$f(x) = \frac{8}{\pi}\left(\cos x - \frac{1}{3}\cos 3x + \frac{1}{5}\cos 5x\right.$$

$$\left. - \frac{1}{7}\cos 7x + \cdots\right)$$

3. Determine the complex Fourier series to represent the function $f(t) = 2t$ in the range $-\pi$ to $+\pi$.

4. Show that the complex Fourier series in problem 3 above is equivalent to:

$$f(t) = 4\left(\sin t - \frac{1}{2}\sin 2t + \frac{1}{3}\sin 3t\right.$$

$$\left. - \frac{1}{4}\sin 4t + \cdots\right)$$

78.5   The frequency spectrum

In the Fourier analysis of periodic waveforms seen in previous chapters, although waveforms physically exist in the time domain, they can be regarded as comprising components with a variety of frequencies. The amplitude and phase of these components are obtained from the Fourier coefficients a_n and b_n; this is known as a **frequency domain**. Plots of amplitude/frequency and phase/frequency are together known as the **spectrum** of a waveform. A simple example is demonstrated in Problem 6, following.

Problem 6.   A pulse of height 20 and width 2 has a period of 10. Sketch the spectrum of the waveform.

The pulse is shown in Fig. 78.5.
The complex coefficient is given by equation (12):

$$c_n = \frac{1}{L}\int_{-\frac{L}{2}}^{\frac{L}{2}} f(t)e^{-j\frac{2\pi nt}{L}}\,dt$$

$$= \frac{1}{10}\int_{-1}^{1} 20e^{-j\frac{2\pi nt}{10}}\,dt = \frac{20}{10}\left[\frac{e^{-j\frac{\pi nt}{5}}}{-\frac{j\pi n}{5}}\right]_{-1}^{1}$$

$$= \frac{20}{10}\left(\frac{5}{-j\pi n}\right)\left[e^{-j\frac{\pi n}{5}} - e^{j\frac{\pi n}{5}}\right]$$

$$= \frac{20}{\pi n}\left[\frac{e^{j\frac{\pi n}{5}} - e^{-j\frac{\pi n}{5}}}{2j}\right]$$

i.e. $c_n = \dfrac{20}{\pi n}\sin\dfrac{n\pi}{5}$

from equation (4), page 810.
From equation (13),

$$c_0 = \frac{1}{L}\int_{-\frac{L}{2}}^{\frac{L}{2}} f(x)\,dx = \frac{1}{10}\int_{-1}^{1} 20\,dt$$

$$= \frac{1}{10}[20t]_{-1}^{1} = \frac{1}{10}[20 - (-20)] = 4$$

$$c_1 = \frac{20}{\pi}\sin\frac{\pi}{5} = 3.74 \quad \text{and}$$

$$c_{-1} = -\frac{20}{\pi}\sin\left(-\frac{\pi}{5}\right) = 3.74$$

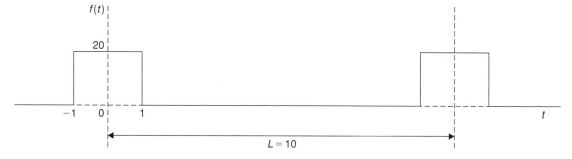

f(t)

20

−1   0   1

L = 10

t

Figure 78.5

Further values of c_n and c_{-n}, up to $n = 10$, are calculated and are shown in the following table.

n	c_n	c_{-n}
0	4	4
1	3.74	3.74
2	3.03	3.03
3	2.02	2.02
4	0.94	0.94
5	0	0
6	−0.62	−0.62
7	−0.86	−0.86
8	−0.76	−0.76
9	−0.42	−0.42
10	0	0

A graph of $|c_n|$ plotted against the number of the harmonic, n, is shown in Figure 78.6.

Figure 78.7 shows the corresponding plot of c_n against n.

Since c_n is real (i.e. no j terms) then the phase must be either $0°$ or $\pm 180°$, depending on the sign of the sine, as shown in Figure 78.8.

When c_n is positive, i.e. between $n = -4$ and $n = +4$, angle $\alpha_n = 0°$

When c_n is negative, then $\alpha_n = \pm 180°$; between $n = +6$ and $n = +9$, α_n is taken as $+180°$, and between $n = -6$ and $n = -9$, α_n is taken as $-180°$

Fig. 78.6 to 78.8 together form the spectrum of the waveform shown in Fig. 78.5.

78.6 Phasors

Electrical engineers in particular often need to analyse alternating current circuits, i.e. circuits containing a

Figure 78.6

Figure 78.7

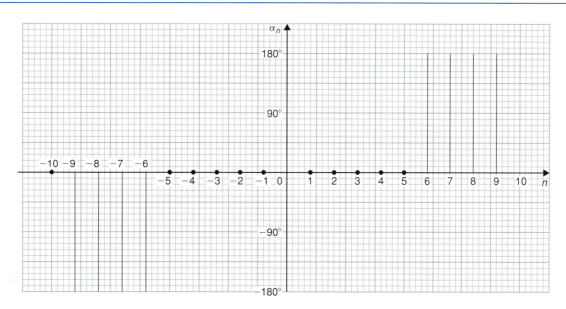

Figure 78.8

sinusoidal input and resulting sinusoidal currents and voltages within the circuit.

It was shown in Chapter 15, page 171, that a general sinusoidal voltage function can be represented by:

$$v = V_m \sin(\omega t + \alpha) \text{ volts} \qquad (19)$$

where V_m is the maximum voltage or amplitude of the voltage v, ω is the angular velocity ($=2\pi f$, where f is the frequency), and α is the phase angle compared with $v = V_m \sin \omega t$

Similarly, a sinusoidal expression may also be expressed in terms of cosine as:

$$v = V_m \cos(\omega t + \alpha) \text{ volts} \qquad (20)$$

It is quite complicated to add, subtract, multiply and divide quantities in the time domain form of equations (19) and (20). As an alternative method of analysis a waveform representation called a **phasor** is used. A phasor has two distinct parts – a magnitude and an angle; for example, the polar form of a complex number, say $5\angle\pi/6$, can represent a phasor, where 5 is the magnitude or modulus, and $\pi/6$ radians is the angle or argument. Also, it was shown on page 270 that $5\angle\pi/6$ may be written as $5\,e^{j\pi/6}$ in exponential form.

In Chapter 23, equation (4), page 270, it is shown that:

$$e^{j\theta} = \cos\theta + j\sin\theta \qquad (21)$$

which is known as **Euler's* formula**.

From equation (21),

$$e^{j(\omega t + \alpha)} = \cos(\omega t + \alpha) + j\sin(\omega t + \alpha)$$

and $\quad V_m e^{j(\omega t + \alpha)} = V_m \cos(\omega t + \alpha)$

$$+ j V_m \sin(\omega t + \alpha)$$

Thus a sinusoidal varying voltage such as in equation (19) or equation (20) can be considered to be either the real or the imaginary part of $V_m\, e^{j(\omega t + \alpha)}$, depending on whether the cosine or sine function is being considered.

$V_m\, e^{j(\omega t + \alpha)}$ may be rewritten as $V_m e^{j\omega t} e^{j\alpha}$ since $a^{m+n} = a^m \times a^n$ from the laws of indices, page 1.

The $e^{j\omega t}$ term can be considered to arise from the fact that a radius is rotated with an angular velocity ω, and α is the angle at which the radius starts to rotate at time $t = 0$ (see Chapter 15, page 171).

Thus, $V_m e^{j\omega t} e^{j\alpha}$ defines a **phasor**. In a particular circuit the angular velocity ω is the same for all the elements thus the phasor can be adequately described by $V_m \angle \alpha$, as suggested above.

---

* **Who was Euler?** For image and resume of Euler, see page 543. To find out more go to **www.routledge.com/cw/bird**

Alternatively, if

$$v = V_m \cos(\omega t + \alpha) \text{ volts}$$

and

$$\cos \theta = \frac{1}{2} \left(e^{j\vartheta} + e^{-j\theta} \right)$$

from equation (3), page 810

then

$$v = V_m \left[\frac{1}{2} \left(e^{j(\omega t + \alpha)} + e^{-j(\omega t + \alpha)} \right) \right]$$

i.e.

$$v = \frac{1}{2} V_m e^{j\omega t} e^{j\alpha} + \frac{1}{2} V_m e^{-j\omega t} e^{-j\alpha}$$

Thus, v is the sum of two phasors, each with half the amplitude, with one having a positive value of angular velocity (i.e. rotating anticlockwise) and a positive value of α, and the other having a negative value of angular velocity (i.e. rotating clockwise) and a negative value of α, as shown in Figure 78.9.

The two phasors are $\frac{1}{2} V_m \angle \alpha$ and $\frac{1}{2} V_m \angle -\alpha$.

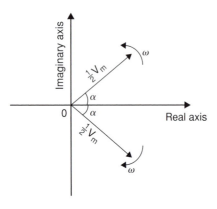

Figure 78.9

From equation (11), page 810, the Fourier representation of a waveform in complex form is:

$$c_n e^{j\frac{2\pi nt}{L}} = c_n e^{j\omega nt} \quad \text{for positive values of } n$$

$$\left(\text{since } \omega = \frac{2\pi}{L} \right)$$

and $\qquad c_n e^{-j\omega nt}$ for negative values of n.

It can thus be considered that these terms represent phasors, those with positives powers being phasors rotating with a positive angular velocity (i.e. anticlockwise), and those with negative powers being phasors rotating with a negative angular velocity (i.e. clockwise).

In the above equations,

$n = 0$ represents a non-rotating component, since $e^0 = 1$,
$n = 1$ represents a rotating component with angular velocity of 1ω,
$n = 2$ represents a rotating component with angular velocity of 2ω, and so on.

Thus we have a set of phasors, the algebraic sum of which at some instant of time gives the magnitude of the waveform at that time.

Problem 7. Determine the pair of phasors that can be used to represent the following voltages:
(a) $v = 8 \cos 2t$ \qquad (b) $v = 8 \cos(2t - 1.5)$

(a) From equation (3), page 810,

$$\cos \theta = \frac{1}{2} (e^{j\theta} + e^{-j\theta})$$

Hence,

$$v = 8 \cos 2t = 8 \left[\frac{1}{2} \left(e^{j2t} + e^{-j2t} \right) \right]$$

$$= 4e^{j2t} + 4e^{-j2t}$$

This represents a phasor of length 4 rotating anticlockwise (i.e. in the positive direction) with an angular velocity of 2 rad/s, and another phasor of length 4 and rotating clockwise (i.e. in the negative direction) with an angular velocity of 2 rad/s. Both phasors have zero phase angle. Fig. 78.10 shows the two phasors.

(b) From equation (3), page 810,

$$\cos \theta = \frac{1}{2} \left(e^{j\theta} + e^{-j\theta} \right)$$

Hence, $v = 8 \cos(2t - 1.5)$

$$= 8 \left[\frac{1}{2} \left(e^{j(2t-1.5)} + e^{-j(2t-1.5)} \right) \right]$$

$$= 4e^{j(2t-1.5)} + 4e^{-j(2t-1.5)}$$

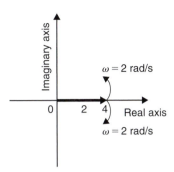

Figure 78.10

i.e. $v = 4e^{2t}\,e^{-j1.5} + 4e^{-j2t}\,e^{j1.5}$

This represents a phasor of length 4 and phase angle -1.5 radians rotating anticlockwise (i.e. in the positive direction) with an angular velocity of 2 rad/s, and another phasor of length 4 and phase angle $+1.5$ radians and rotating clockwise (i.e. in the negative direction) with an angular velocity of 2 rad/s. Fig. 78.11 shows the two phasors.

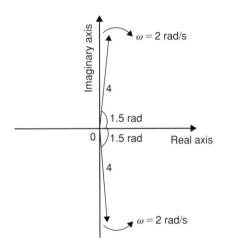

Figure 78.11

Problem 8.   Determine the pair of phasors that can be used to represent the third harmonic

$$v = 8\cos 3t - 20\sin 3t$$

Using $\quad \cos t = \dfrac{1}{2}\left(e^{jt} + e^{-jt}\right)$

and $\quad \sin t = \dfrac{1}{2j}\left(e^{jt} - e^{-jt}\right)\quad$ from page 810

gives: $\ v = 8\cos 3t - 20\sin 3t$

$$= 8\left[\frac{1}{2}\left(e^{j3t} + e^{-j3t}\right)\right]$$

$$- 20\left[\frac{1}{2j}\left(e^{j3t} - e^{-j3t}\right)\right]$$

$$= 4e^{j3t} + 4e^{-j3t} - \frac{10}{j}e^{j3t} + \frac{10}{j}e^{-j3t}$$

$$= 4e^{j3t} + 4e^{-j3t} - \frac{10(j)}{j(j)}e^{j3t} + \frac{10(j)}{j(j)}e^{-j3t}$$

$$= 4e^{j3t} + 4e^{-j3t} + 10je^{j3t} - 10je^{-j3t}$$

$$\text{since } j^2 = -1$$

$$= (4 + j10)e^{j3t} + (4 - j10)e^{-j3t}$$

$$(4 + j10) = \sqrt{4^2 + 10^2}\angle\tan^{-1}\left(\frac{10}{4}\right)$$

$$= 10.77\angle 1.19$$

and $(4 - j10)$

$$= 10.77\angle -1.19$$

Hence, $\boldsymbol{v = 10.77\angle 1.19 + 10.77\angle -1.19}$

Thus v comprises a phasor $10.77\angle 1.19$ rotating anti-clockwise with an angular velocity if 3 rad/s, and a phasor $10.77\angle -1.19$ rotating clockwise with an angular velocity of 3 rad/s.

Section M

Now try the following Practice Exercise

Practice Exercise 270   **Phasors (Answers on page 898)**

1. Determine the pair of phasors that can be used to represent the following voltages:

    (a) $v = 4\cos 4t$ (b) $v = 4\cos(4t + \pi/2)$

2. Determine the pair of phasors that can represent the harmonic given by:
    $$v = 10\cos 2t - 12\sin 2t$$

3. Find the pair of phasors that can represent the fundamental current: $i = 6\sin t + 4\cos t$

For fully worked solutions to each of the problems in Practice Exercises 268 to 270 in this chapter, go to the website:
www.routledge.com/cw/bird

Section N

Z-Transforms

Chapter 79

An introduction to z-transforms

Why it is important to understand: **An introduction to z-transforms**

In mathematics and signal processing, the z-transform converts a discrete-time signal, which is a sequence of real or complex numbers, into a complex frequency domain representation. It can be considered as a discrete-time equivalent of the Laplace transform.

Laplace transform methods are widely used for analysis in linear systems and are used when a system is described by a linear differential equation, with constant coefficients. However, there are numerous systems that are described by difference equations - not differential equations - and these systems are common and different from those described by differential equations.

Systems that satisfy difference equations include computer controlled systems - systems that take measurements with digital input/output boards or GPIB instruments (digital 8-bit parallel communications interface with data transfer rates up to 1 Mbyte/s), calculate an output voltage and output that voltage digitally. Frequently these systems run a program loop that executes in a fixed interval of time. Other systems that satisfy difference equations are those systems with digital filters - which are found anywhere digital signal processing/digital filtering is undertaken - that includes digital signal transmission systems like the telephone system or systems that process audio signals. A CD contains digital signal information, and when it is read off the CD, it is initially a digital signal that can be processed with a digital filter. There are an incredible number of systems used every day that have digital components which satisfy difference equations. In continuous systems Laplace transforms play a unique role. They allow system and circuit designers to analyse systems and predict performance, and to think in different terms - like frequency responses - to help understand linear continuous systems. They are a very powerful tool that shapes how engineers think about those systems. Z-transforms play the role in sampled systems that Laplace transforms play in continuous systems. In sampled systems, inputs and outputs are related by difference equations and z-transform techniques are used to solve those difference equations. In continuous systems, Laplace transforms are used to represent systems with transfer functions, while in sampled systems, z-transforms are used to represent systems with transfer functions.

This chapter merely provides an introduction to z-transforms – how to read z-transforms from a table, to appreciate various properties of z-transforms, how to determine inverse z-transforms and how to solve difference equations using z-transforms.

At the end of this chapter, you should be able to:

- define the z-transform
- prove z-transforms for simple sequences
- use a table of z-transforms to determine simple transforms
- understand the properties of z-transforms – linearity, first and second shift theorems, translations, final and initial value theorems, and derivatives of transforms
- use a table of z-transforms to determine inverse transforms
- use z-transforms to solve difference equations

79.1 Sequences

The sequence $\ldots, 2^{-3}, 2^{-2}, 2^{-1}, 2^0, 2^1, 2^2, 2^3, \ldots$ has a general term of the form 2^k and this series can be written as $\{2^k\}_{-\infty}^{\infty}$ indicating that the power or index of the number 2 has a range from $-\infty$ to ∞.

The sum $\displaystyle\sum_{k=-\infty}^{\infty} \left(\frac{2}{z}\right)^k = \ldots + \left(\frac{2}{z}\right)^{-3} + \left(\frac{2}{z}\right)^{-2} +$

$$\left(\frac{2}{z}\right)^{-1} + \left(\frac{2}{z}\right)^{0} + \left(\frac{2}{z}\right)^{1} + \left(\frac{2}{z}\right)^{2} + \ldots$$

is called the **z-transform** of the sequence $Z\{2^k\}_{-\infty}^{\infty}$ and is denoted by F(z), where the complex number z is chosen to ensure that the sum is finite.

Thus, $\{2^k\}_{-\infty}^{\infty}$ and $Z\{2^k\}_{-\infty}^{\infty} = F(z) = \displaystyle\sum_{k=-\infty}^{\infty} \left(\frac{2}{z}\right)^k$

forming what is referred to as a '**z-transform pair**'. For simplicity, the range will be restricted to $z\{x_k\}_0^{\infty}$ where $x_k = 0$ for $k < 0$ and denote it by $\{x_k\}$

i.e. $\qquad \mathbf{Z\{x_k\} = F(z) = \displaystyle\sum_{k=0}^{\infty} \frac{x_k}{z^k}}$

This is the **definition of the z-transform** of the sequence $\{x_k\}$ and is used in the following worked problems.

Problem 1. Determine the z-transform for the unit impulse $\{\delta_k\} = \{1, 0, 0, 0, \ldots\}$

The z-transform of $\{\delta_k\}$ is given by:

$$Z\{\delta_k\} = F(z) = \sum_{k=0}^{\infty} \frac{\delta_k}{z^k} = \frac{1}{z^0} + \frac{0}{z^1} + \frac{0}{z^2} + \ldots = 1$$

i.e. $\qquad \mathbf{Z\{\delta_k\} = 1}$ valid for all values of z

Problem 2. Determine the z-transform for the unit step sequence $\{u_k\} = \{1, 1, 1, 1, \ldots\} = \{1\}$

The z-transform of $\{u_k\}$ is given by:

$$Z\{u_k\} = F(z) = \sum_{k=0}^{\infty} \frac{u_k}{z^k} = \sum_{k=0}^{\infty} \frac{1}{z^k}$$

$$= \frac{1}{z^0} + \frac{1}{z^1} + \frac{1}{z^2} + \frac{1}{z^3} + \frac{1}{z^4} + \ldots$$

i.e. $\qquad \mathbf{Z\{u_k\} = 1 + \dfrac{1}{z} + \dfrac{1}{z^2} + \dfrac{1}{z^3} + \dfrac{1}{z^4} + \ldots}$ (1)

Using the binomial theorem for $(1+x)^n$, the series expansion of $\dfrac{1}{1-x}$ may be determined:

$$\frac{1}{1-x} = (1-x)^{-1}$$

$$= 1 + (-1)(-x) + \frac{(-1)(-2)}{2!}(-x)^2$$

$$+ \frac{(-1)(-2)(-3)}{3!}(-x)^3 + \ldots$$

$$= 1 + x + x^2 + x^3 + \ldots \text{ valid for } |x| < 1 \quad (2)$$

Comparing equations (1) and (2) gives: $F(z) = \dfrac{1}{1 - \dfrac{1}{z}}$

provided $\left|\dfrac{1}{z}\right| < 1$

$\dfrac{1}{1 - \dfrac{1}{z}} = \dfrac{1}{\dfrac{z-1}{z}} = \dfrac{z}{z-1}$ hence, $\mathbf{z\{u_k\} = \dfrac{z}{z-1}}$ pro-

vided $|z| > 1$

Problem 3. Show that the z-transform for the unit step sequence

$\{x_k\} = \{1, a, a^2, a^3, a^4, \ldots\} = \{a^k\}$ is given by $\dfrac{z}{z-a}$

The z-transform of $\{x_k\}$ is given by:

$$z\{x_k\} = z\{a^k\} = \sum_{k=0}^{\infty} \frac{a^k}{z^k} = \sum_{k=0}^{\infty} \left(\frac{a}{z}\right)^k$$

$$= \left(\frac{a}{z}\right)^0 + \left(\frac{a}{z}\right)^1 + \left(\frac{a}{z}\right)^2 + \left(\frac{a}{z}\right)^3 + \left(\frac{a}{z}\right)^4 + \ldots$$

$$= 1 + \frac{a}{z} + \left(\frac{a}{z}\right)^2 + \left(\frac{a}{z}\right)^3 + \left(\frac{a}{z}\right)^4 + \ldots$$

Comparing this with the series expansion of

$\dfrac{1}{1-x} = 1 + x + x^2 + x^3 + \ldots$ which is valid for $|x| < 1$

i.e. equation (2) above, shows that:

$$F(z) = 1 + \frac{a}{z} + \left(\frac{a}{z}\right)^2 + \left(\frac{a}{z}\right)^3 + \left(\frac{a}{z}\right)^4 + \ldots$$

$$= \frac{1}{1 - \dfrac{a}{z}} \text{ provided } \left|\frac{a}{z}\right| < 1$$

Hence, $\dfrac{1}{1 - \dfrac{a}{z}} = \dfrac{1}{\dfrac{z-a}{z}} = \dfrac{z}{z-a}$ and $\mathbf{F(z) = \dfrac{z}{z-a}}$

provided $|z| > |a|$

Problem 4. Determine the z-transform for the unit step sequence $\{x_k\} = \{0, 1, 2, 3, 4, ...\} = \{k\}$ and show that it is equivalent to $\dfrac{z}{(z-1)^2}$

The z-transform of $\{u_k\}$ is given by:

$$Z\{x_k\} = F(z) = \sum_{k=0}^{\infty} \frac{x_k}{z^k} = \sum_{k=0}^{\infty} \frac{k}{z^k}$$

$$= \frac{0}{z^0} + \frac{1}{z^1} + \frac{2}{z^2} + \frac{3}{z^3} + \frac{4}{z^4} + \cdots$$

$$= 0 + \frac{1}{z} + \frac{2}{z^2} + \frac{3}{z^3} + \frac{4}{z^4} + \cdots \qquad (3)$$

It was shown earlier, equation (2), that

$$\frac{1}{1-x} = (1-x)^{-1} = 1 + x + x^2 + x^3 + \cdots$$

Now $\dfrac{d}{dx}\left(1 + x + x^2 + x^3 + x^4 + \cdots\right)$

$$= 1 + 2x + 3x^2 + 4x^3 + \cdots \qquad (4)$$

and $\dfrac{d}{dx}\left[(1-x)^{-1}\right] = -(1-x)^{-2}(-1) = \dfrac{1}{(1-x)^2}$

using the function of a function rule,

i.e. $\qquad 1 + 2x + 3x^2 + 4x^3 + \cdots = \dfrac{1}{(1-x)^2}$

Comparing equations (3) and (4) shows that by multiplying F(z) by z,

then $zF(z) = z\left(0 + \dfrac{1}{z} + \dfrac{2}{z^2} + \dfrac{3}{z^3} + \dfrac{4}{z^4} + \cdots\right)$

$$= 1 + \frac{2}{z} + \frac{3}{z^2} + \frac{4}{z^3} + \cdots = \frac{1}{\left(1 - \dfrac{1}{z}\right)^2}$$

Dividing both sides by z gives:

$$F(z) = \frac{1}{z\left(1 - \dfrac{1}{z}\right)^2} = \frac{1}{z\left(\dfrac{z-1}{z}\right)^2} = \frac{1}{z\dfrac{(z-1)^2}{z^2}}$$

$$= \frac{1}{\dfrac{(z-1)^2}{z}} = \frac{z}{(z-1)^2}$$

Hence, $Z\{x_k\} = F(z) = 0 + \dfrac{1}{z} + \dfrac{2}{z^2} + \dfrac{3}{z^3} + \dfrac{4}{z^4} + \cdots$

$$= \frac{z}{(z-1)^2}$$

From the results obtained in Problems 1 to 4, together with some additional results, a summary of some z-transforms is shown in Table 79.1, on page 828, which may now be accepted and used.

Here are some further problems using Table 79.1.

Problem 5. Determine the z-transform of $5k^2$

From 4 in Table 79.1, $Z\{k^2\} = \dfrac{z(z+1)}{(z-1)^3}$

Hence, $\qquad \mathbf{Z\{5k^2\} = 5Z\{k^2\} = \dfrac{5z(z+1)}{(z-1)^3}}$

Problem 6. Determine the z-transform of
(a) 3^k (b) $(-3)^k$

(a) From 6 in Table 79.1, $Z\{a^k\} = \dfrac{z}{z-a}$

If a = 3, then $\qquad \mathbf{Z\{3^k\} = \dfrac{z}{z-3}}$

(b) From 6 in Table 79.1, $Z\{a^k\} = \dfrac{z}{z-a}$

If $a = -3$, then $\mathbf{Z\{(-3)^k\} = \dfrac{z}{z--3} = \dfrac{z}{z+3}}$

Problem 7. Determine the z-transform of $2e^{-3k}$

From 9 in Table 79.1, $Z\{e^{-ak}\} = \dfrac{z}{z-e^{-a}}$

Hence, $\mathbf{Z\{2e^{-3k}\} = 2Z\{e^{-3k}\} = \dfrac{2z}{z-e^{-3}}}$

Problem 8. Determine $Z\{\cos 3k\}$

From 11 in Table 79.1, $Z\{\cos ak\} = \dfrac{z(z-\cos a)}{z^2 - 2z\cos a + 1}$

Hence, since a = 3, $\mathbf{Z\{\cos 3k\} = \dfrac{z(z-\cos 3)}{z^2 - 2z\cos 3 + 1}}$

Problem 9. Determine the z-transform of $3\sin 2k$

From 10 in Table 79.1, $Z\{\sin ak\} = \dfrac{z\sin a}{z^2 - 2z\cos a + 1}$

Hence, since a = 2, $\mathbf{Z\{3\sin 2k\} = 3Z\{\sin 2k\}}$

$$= 3\left(\frac{z\sin 2}{z^2 - 2z\cos 2 + 1}\right)$$

$$= \frac{3z\sin 2}{z^2 - 2z\cos 2 + 1}$$

Problem 10. Determine $Z\{e^{-2k}\cos 4k\}$

From 13 in Table 79.1,

$$Z\{e^{-ak}\cos bk\} = \frac{z^2 - ze^{-a}\cos b}{z^2 - 2ze^{-a}\cos b + e^{-2a}}$$

Table 79.1

Sequence	Transform F(z)	
1. $\{\delta_k\} = \{1, 0, 0, \ldots\}$	1	for all values of z
2. $\{u_k\} = \{1, 1, 1, \ldots\}$	$\dfrac{z}{z-1}$	for $|z| > 1$
3. $\{k\} = \{0, 1, 2, 3, \ldots\}$	$\dfrac{z}{(z-1)^2}$	for $|z| > 1$
4. $\{k^2\} = \{0, 1, 4, 9, \ldots\}$	$\dfrac{z(z+1)}{(z-1)^3}$	for $|z| > 1$
5. $\{k^3\} = \{0, 1, 8, 27, \ldots\}$	$\dfrac{z(z^2+4z+1)}{(z-1)^4}$	for $|z| > 1$
6. $\{a^k\} = \{1, a, a^2, a^3, \ldots\}$	$\dfrac{z}{z-a}$	for $|z| > |a|$
7. $\{ka^k\} = \{0, a, 2a^2, 3a^3, \ldots\}$	$\dfrac{az}{(z-a)^2}$	for $|z| > |a|$
8. $\{k^2 a^k\} = \{0, a, 4a^2, 9a^3, \ldots\}$	$\dfrac{az(z+a)}{(z-a)^3}$	for $|z| > |a|$
9. $\{e^{-ak}\} = \{e^{-a}, e^{-2a}, e^{-3a}, \ldots\}$	$\dfrac{z}{z-e^{-a}}$	
10. $\sin ak = \{\sin a, \sin 2a, \ldots\}$	$\dfrac{z\sin a}{z^2 - 2z\cos a + 1}$	
11. $\cos ak = \{\cos a, \cos 2a, \ldots\}$	$\dfrac{z(z - \cos a)}{z^2 - 2z\cos a + 1}$	
12. $e^{-ak}\sin bk = \{e^{-a}\sin b, e^{-2a}\sin 2b, \ldots\}$	$\dfrac{ze^{-a}\sin b}{z^2 - 2ze^{-a}\cos b + e^{-2a}}$	
13. $e^{-ak}\cos bk = \{e^{-a}\cos b, e^{-2a}\cos 2b, \ldots\}$	$\dfrac{z^2 - ze^{-a}\cos b}{z^2 - 2ze^{-a}\cos b + e^{-2a}}$	

Hence, since $a = 2$ and $b = 4$,

$$Z\{e^{-2k}\cos 4k\} = \frac{z^2 - ze^{-2}\cos 4}{z^2 - 2ze^{-2}\cos 4 + e^{-4}}$$

Now try the following Practice Exercise

Practice Exercise 271   Determining z-transforms (Answers on page 898)

Use Table 79.1 to find the z-transforms of the following:

1. 2k	2. $3k^2$	3. $4k^3$
4. sin 3k	5. 2^k	6. $k(2)^k$
7. 5 cos 2k	8. $3e^{-2k}$	9. $\sin\frac{1}{2}k$
10. $4e^{3k}\cos 2k$	11. $(-4)^k$	12. $3k^2(2)^k$
13. $3(-5)^k$	14. $k(-3)^k$	15. $3e^{5k}$
16. $2e^{-4k}\sin 2k$		

79.2 Some properties of z-transforms

(a) Linearity property

The z-transform is a **linear transform**,

i.e. $Z(a\{x_k\} + b\{y_k\}) = aZ\{x_k\} + bZ\{y_k\}$ (4)

where a and b are constants

> **Problem 11.** Determine the z-transform of
>
> $$2\{k\} - 3\left\{e^{-2k}\right\}$$

Now, $2Z\{k\} = 2\left\{\dfrac{z}{(z-1)^2}\right\}$ from 4 in Table 79.1

and since $Z\{e^{-ak}\} = \dfrac{z}{z - e^{-a}}$ from 9 in Table 78.1,

then $Z\{e^{-2k}\} = \dfrac{z}{z - e^{-2}}$

Hence, $Z\left(2\{k\} - 3\left\{e^{-2k}\right\}\right) = 2Z\{k\} - 3Z\left\{e^{-2k}\right\}$

$= 2\left\{\dfrac{z}{(z-1)^2}\right\} - 3\left\{\dfrac{z}{z - e^{-2}}\right\}$ from equation (5)

$= \left\{\dfrac{2z}{(z-1)^2}\right\} - \left\{\dfrac{3z}{z - e^{-2}}\right\}$

$= \dfrac{2z(z - e^{-2}) - 3z(z-1)^2}{(z-1)^2(z - e^{-2})}$

$= \dfrac{2z^2 - 2ze^{-2} - 3z(z^2 - 2z + 1)}{(z-1)^2(z - e^{-2})}$

$= \dfrac{2z^2 - 2ze^{-2} - 3z^3 + 6z^2 - 3z}{(z-1)^2(z - e^{-2})}$

i.e. $2Z\{k\} - 3Z\left\{e^{-2k}\right\}$

$= \dfrac{-3z^3 + 8z^2 - z(2e^{-2} + 3)}{(z-1)^2(z - e^{-2})}$

Now try the following Practice Exercise

Practice Exercise 272 Linearity property of z-transforms (Answers on page 898)

Use Table 79.1 to find the z-transforms of the following:

1. $5\{k\} - 4\left\{e^{-3k}\right\}$

2. $4\left\{k^2\right\} + 3\{k\}$

3. $3\{\sin 2k - \cos 2k\}$

4. $2\left\{(3)^k\right\} + 4\left\{(-3)^k\right\}$

5. $\left\{k + e^{-k}\right\}$

(b) First shift theorem (shifting to the left)

It may be shown by the **first shift theorem (shifting to the left)**, that

if $Z\{x_k\} = F(z)$

then $Z\{x_{k+m}\}$

$$= z^m F(z) - \left[z^m x_0 + z^{m-1} x_1 + \ldots + z x_{m-1}\right]$$
(6)

is the z-transform of the sequence that has been shifted by m places to the left.

This theorem is often needed when solving difference equations (see Section 79.4).

> **Problem 12.** Determine $Z\{3^{k+2}\}$

Since from equation (6),

$Z\{x_{k+m}\} = z^m F(z) - \left[z^m x_0 + z^{m-1} x_1 + \ldots + z x_{m-1}\right]$

then $\quad Z\left\{3^{k+2}\right\} = z^2 Z\left\{3^k\right\} - \left[z^2 3^0 + z 3^1\right]$ (7)

From 6 of Table 79.1, $Z\left\{a^k\right\} = \dfrac{z}{z - a}$ thus

$$Z\{3^k\} = \dfrac{z}{z - 3}$$

Hence, substituting in equation (7),

$Z\left\{3^{k+2}\right\} = z^2\left(\dfrac{z}{z-3}\right) - \left[z^2 + 3z\right]$

$= \dfrac{z^3}{z-3} - \left[z^2 + 3z\right]$

$= \dfrac{z^3 - (z-3)\left[z^2 + 3z\right]}{z - 3}$

$= \dfrac{z^3 - \left[z^3 + 3z^2 - 3z^2 - 9z\right]}{z - 3}$

$= \dfrac{z^3 - \left[z^3 - 9z\right]}{z - 3}$

i.e. $\quad Z\left\{3^{k+2}\right\} = \dfrac{9z}{z - 3}$

This is the z-transform of the sequence $\{9, 27, 81, \ldots\}$ by shifting the sequence $\{1, 3, 9, 27, \ldots\}$ two places to the left and losing the first two terms.

Problem 13. Determine $Z\{5^{k+3}\}$

Since from equation (6),
$$Z\{x_{k+m}\} = z^m F(z) - \left[z^m x_0 + z^{m-1} x_1 + \ldots + z x_{m-1}\right]$$

then
$$Z\{5^{k+3}\} = z^3 Z\{5^k\} - \left[z^3 5^0 + z^2 5^1 + z 5^2\right] \tag{8}$$

From 6 of Table 79.1, $Z\{a^k\} = \dfrac{z}{z-a}$

thus
$$Z\{5^k\} = \dfrac{z}{z-5}$$

Hence, substituting in equation (8),

$$Z\{5^{k+3}\} = z^3 \left(\dfrac{z}{z-5}\right) - \left[z^3 + 5z^2 + 25z\right]$$

$$= \dfrac{z^4}{z-5} - \left[z^3 + 5z^2 + 25z\right]$$

$$= \dfrac{z^4 - (z-5)\left[z^3 + 5z^2 + 25z\right]}{z-5}$$

$$= \dfrac{z^4 - \left[z^4 + 5z^3 + 25z^2 - 5z^3 - 25z^2 - 125z\right]}{z-5}$$

$$= \dfrac{z^4 - \left[z^4 - 125z\right]}{z-5}$$

i.e.
$$Z\{5^{k+3}\} = \dfrac{125z}{z-5}$$

This is the z-transform of the sequence $\{125, 625, 3125, \ldots\}$ by shifting the sequence $\{1, 5, 25, 125, \ldots\}$ three places to the left and losing the first three terms.

Problem 14. Determine $Z\{k+1\}$

From 3 of Table 79.1, $Z\{k\} = \dfrac{z}{(z-1)^2}$

Since from equation (6),
$$Z\{x_{k+m}\} = z^m F(z) - \left[z^m x_0 + z^{m-1} x_1 + \ldots + z x_{m-1}\right]$$

then $Z\{k+1\} = z^1 Z(k) - \left[z^1 x_0\right]$

$$= z^1 \left(\dfrac{z}{(z-1)^2}\right) - \left[z^1 \times 0\right] \text{ from 3 of Table 79.1}$$

i.e.
$$Z\{k+1\} = \dfrac{z^2}{(z-1)^2}$$

Now try the following Practice Exercise

Practice Exercise 273   First shift theorem of z-transforms (Answers on page 899)

Use Table 79.1 to find the z-transforms of the following:

1. $\{2^{k+3}\}$ 2. $\{4^{k+1}\}$

3. $6\{k+1\}$ 4. $\{3^{k+2} - 2^{k+1}\}$

(c) Second shift theorem (shifting to the right)

It may be shown by the **second shift theorem (shifting to the right)**, that

if $Z\{x_k\} = F(z)$

then
$$\boldsymbol{Z\{x_{k-m}\} = z^{-m} F(z)} \tag{9}$$

is the z-transform of the sequence that has been shifted by m places to the right.

Problem 15. Determine $Z\{x_{k-2}\}$

Since from equation (9), $Z\{x_{k-m}\} = z^{-m} F(z)$

then
$$Z\{x_{k-2}\} = z^{-2} F(z)$$

$$= z^{-2} \left(\dfrac{z}{z-1}\right)$$

since $Z\{x_k\} = \dfrac{z}{z-1}$ from 2 of Table 79.1

i.e.
$$\boldsymbol{Z\{x_{k-2}\}} = \left(\dfrac{z^{-1}}{z-1}\right) = \dfrac{1}{z(z-1)}$$

This is the z-transform of the sequence $\{0, 0, 1, 1, 1, \ldots\}$ by shifting the sequence $\{1, 1, 1, 1, \ldots\}$ two places to the right and defining the first two terms as zeros.

Problem 16. Determine $Z\{a^{k-1}\}$

Since from equation (9), $Z\{x_{k-m}\} = z^{-m} F(z)$

then $Z\{a^{k-1}\} = z^{-1} F(z) = z^{-1} \left(\dfrac{z}{z-a}\right)$

since $Z\{a^k\} = \dfrac{z}{z-a}$ from 6 of Table 79.1

i.e.
$$\boldsymbol{Z\{a^{k-1}\}} = \left(\dfrac{z^{-1} \times z}{z-a}\right) = \dfrac{1}{(z-a)}$$

which is the z-transform of $\{a^k\}$ shifted one place to the right.

Now try the following Practice Exercise

Practice Exercise 274   Second shift theorem of z-transforms (Answers on page 899)

Use Table 79.1 to find the z-transforms of the following:

1. $\{x_{k-1}\}$

2. $\{x_{k-3}\}$

3. $\{a^{k-2}\}$

4. $\{a^{k-3}\}$

5. $\{3^{k-4}\}$

(d) Translation

If the sequence $\{x_k\}$ has the z-transform

$$Z\{a^k x_k\} = F(z),$$

then the sequence $\{a^k x_k\}$ has the z-transform

$$Z\{a^k x_k\} = F\left(a^{-1}z\right)$$

Problem 17. Determine $Z\{3^k k\}$

Since $Z(k) = \dfrac{z}{(z-1)^2}$ from 3 of Table 79.1

then by the translation property,

$$Z\{3^k k\} = F\left(3^{-1}z\right) = \frac{3^{-1}z}{(3^{-1}z - 1)^2}$$

$$= \frac{3^{-1}z}{\left[3^{-1}\left(z - \frac{1}{3^{-1}}\right)\right]^2}$$

$$= \frac{3^{-1}z}{3^{-2}(z-3)^2} = \frac{3^2 z}{3^1(z-3)^2}$$

i.e. $$Z\{3^k k\} = \frac{3z}{(z-3)^2}$$

(e) Final value theorem

For the sequence $\{x_k\}$ with the z-transform F(z),

$$\underset{k\to\infty}{Lim}\, x_k = \underset{z\to 1}{Lim}\left\{\left(\frac{z-1}{z}\right)F(z)\right\} \text{ provided that } \underset{k\to\infty}{Lim}\, x_k \text{ exists}$$

Problem 18. Determine $\underset{k\to\infty}{Lim}\left\{\left(\frac{1}{3}\right)^k\right\}$

$$F\{x^k\} = \frac{z}{z-1} \text{ and } F\left\{\left(\frac{1}{3}\right)^k\right\} = \frac{z}{z-\frac{1}{3}} = \frac{3z}{3z-1}$$

Now $\underset{k\to\infty}{Lim}\left\{\left(\frac{1}{3}\right)^k\right\} = \underset{z\to 1}{Lim}\left\{\left(\frac{z-1}{z}\right)F(z)\right\}$

$$= \underset{z\to 1}{Lim}\left\{\left(\frac{z-1}{z}\right)\frac{3z}{3z-1}\right\}$$

$$= \underset{z\to 1}{Lim}\left\{\frac{3(z-1)}{3z-1}\right\} = 0$$

i.e. $$\underset{k\to\infty}{Lim}\left\{\left(\frac{1}{3}\right)^k\right\} = 0$$

Problem 19. Determine $\underset{k\to\infty}{Lim}\left\{\dfrac{33z^2 - 25z}{(z-1)(3z-1)^2}\right\}$

By the final value theorem,

$$\underset{k\to\infty}{Lim}\left\{\frac{33z^2 - 25z}{(z-1)(3z-1)^2}\right\} = \underset{z\to 1}{Lim}\left\{\left(\frac{z-1}{z}\right)F(z)\right\}$$

$$= \underset{z\to 1}{Lim}\left\{\left(\frac{z-1}{z}\right)\frac{33z^2 - 25z}{(z-1)(3z-1)^2}\right\}$$

$$= \underset{z\to 1}{Lim}\left\{\frac{33z - 25}{(3z-1)^2}\right\}$$

$$= \frac{33 - 25}{2^2} = \frac{8}{4} = \mathbf{2}$$

(f) The initial value theorem

For the sequence $\{x_k\}$ with the z-transform F(z),

$$\text{initial value, } x_0 = \underset{z\to\infty}{Lim}\{F(z)\}$$

Problem 20. Determine $\underset{z\to\infty}{Lim}\{a^k\}$

$F(z) = F\{a^k\} = \dfrac{z}{z-a}$ from 6 of Table 79.1

and $\underset{z\to\infty}{Lim}\{F(z)\} = \underset{z\to\infty}{Lim}\left\{\dfrac{z}{z-a}\right\}$

$$= \underset{z\to\infty}{Lim}\left\{\frac{\frac{d}{dz}(z)\,dz}{\frac{d}{dz}(z-a)\,dz}\right\} = \underset{z\to\infty}{Lim}\left\{\frac{1}{1}\right\}$$

by L'Hôpital's rule (see Chapter 8)

i.e. $$\underset{z\to\infty}{Lim}\{a^k\} = 1$$

(g) The derivative of the transform

If $Z\{x_k\} = F(z)$

then $$-zF'(z) = Z\{kx_k\}$$

Problem 21. Determine the derivative of $Z\{ka^k\}$

$F(z) = F\{a^k\} = \dfrac{z}{z-a}$ from 6 of Table 79.1

Now $Z\{kx_k\} = -zF'(z)$ from above

i.e. $\quad Z\{kx_k\} = -zF'\left(\dfrac{z}{z-a}\right)$

$= -z\left[\dfrac{(z-a)(1) - z(1)}{(z-a)^2}\right]$ using the quotient rule

$= -z\left[\dfrac{z-a-z}{(z-a)^2}\right] = -z\left[\dfrac{-a}{(z-a)^2}\right] = \dfrac{az}{(z-a)^2}$

i.e. **the derivative of** $\mathbf{Z\{ka^k\}} = -zF'(z) = \dfrac{az}{(z-a)^2}$

which confirms result 7 in Table 79.1

Now try the following Practice Exercise

Practice Exercise 275   Properties of z-Transforms (Answers on page 899)

1. Determine $Z\{2^k k\}$

2. Determine $Z\{4^k k\}$

3. Determine $\underset{k\to\infty}{Lim}\left\{\left(\tfrac{1}{2}\right)^k\right\}$

4. Determine $\underset{k\to\infty}{Lim}\left\{\dfrac{15z^2 + 5z}{(z-1)(4z-1)^2}\right\}$

5. Determine $\underset{z\to\infty}{Lim}\{3^k\}$

6. Determine the derivative of $Z\{k(2)^k\}$

79.3   Inverse z-transforms

If the sequence $\{x_k\}$ has a Z transform $Z\{x_k\} = F(z)$, then the **inverse z-transform** is defined as:

$$Z^{-1}F(z) = \{x_k\}$$

In the following problems, some inverses may be determined directly from Table 79.1, albeit with a little manipulation; others sometimes require the use of partial fractions – just as with inverse Laplace transforms.

Problem 22. Determine the inverse z-transform of $F(z) = \dfrac{z}{z+5}$

From 6 in Table 79.1, $Z\{a^k\} = \dfrac{z}{z-a}$

hence $\quad Z^{-1}\left\{\dfrac{z}{z-a}\right\} = a^k$

Comparing $\dfrac{z}{z+5}$ with $\dfrac{z}{z-a}$ shows that $a = -5$

Thus, $\quad \mathbf{Z^{-1}\left\{\dfrac{z}{z+5}\right\} = (-5)^k}$

Problem 23. Determine the inverse z-transform of
$$F(z) = \dfrac{2z}{2z+1}$$

$$\dfrac{2z}{2z+1} = \dfrac{2z}{2\left(z+\tfrac{1}{2}\right)} = \dfrac{z}{z+\tfrac{1}{2}}$$

From 6 in Table 79.1, $Z\{a^k\} = \dfrac{z}{z-a}$

hence $\quad Z^{-1}\left\{\dfrac{z}{z-a}\right\} = a^k$

Comparing $\dfrac{z}{z+\tfrac{1}{2}}$ with $\dfrac{z}{z-a}$ shows that $a = -\dfrac{1}{2}$

Thus, $\quad \mathbf{Z^{-1}\left\{\dfrac{2z}{2z+1}\right\} = Z^{-1}\left\{\dfrac{z}{z+\tfrac{1}{2}}\right\} = \left(-\dfrac{1}{2}\right)^k}$

Problem 24. If $F(z) = \dfrac{3z}{3z-1}$ determine $Z^{-1}F(z)$

$$\dfrac{3z}{3z-1} = \dfrac{3z}{3\left(z-\tfrac{1}{3}\right)} = \dfrac{z}{z-\tfrac{1}{3}}$$

From 6 in Table 79.1, $Z\{a^k\} = \dfrac{z}{z-a}$

hence $\quad Z^{-1}\left\{\dfrac{z}{z-a}\right\} = a^k$

Comparing $\dfrac{z}{z-\tfrac{1}{3}}$ with $\dfrac{z}{z-a}$ shows that $a = \dfrac{1}{3}$

Thus, $\quad \mathbf{Z^{-1}\left\{\dfrac{3z}{3z-1}\right\} = Z^{-1}\left\{\dfrac{z}{z-\tfrac{1}{3}}\right\} = \left(\dfrac{1}{3}\right)^k}$

Problem 25. Determine the inverse z-transform of $F(z) = \dfrac{z}{z - e^2}$

From 9 in Table 79.1, $Z\{e^{-ak}\} = \dfrac{z}{z - e^{-a}}$

hence $\quad Z^{-1}\left\{\dfrac{z}{z - e^{-a}}\right\} = e^{-ak}$

Comparing $\dfrac{z}{z - e^2}$ with $\dfrac{z}{z - e^{-a}}$ shows that $a = -2$

Thus, $Z^{-1}\left\{\dfrac{z}{z - e^2}\right\} = e^{--2k} = e^{2k}$

Problem 26.   If $F(z) = \dfrac{z}{z^2 + 1}$ determine $Z^{-1}F(z)$

In Table 79.1, results 10 and 11 have z^2 in their denominators.

If the numerator is to be z in result 10, then sin a has to equal 1, i.e. $a = \frac{\pi}{2}$

From 10 in Table 79.1, $Z\{\sin ak\} = \dfrac{z \sin a}{z^2 - 2z \cos a + 1}$

hence $Z^{-1}\left\{\dfrac{z \sin a}{z^2 - 2z \cos a + 1}\right\} = \sin ak$

When $a = \dfrac{\pi}{2}$, $\dfrac{z \sin a}{z^2 - 2z \cos a + 1} = \dfrac{z \sin \frac{\pi}{2}}{z^2 - 2z \cos \frac{\pi}{2} + 1}$

$= \dfrac{z}{z^2 + 1}$

Thus, $Z^{-1}\left\{\dfrac{z}{z^2 + 1}\right\} = \sin \dfrac{\pi}{2} k$

Problem 27.   Determine the inverse z-transform of $F(z) = \dfrac{z}{z^2 - 7z + 12}$

Using partial fractions, let

$\dfrac{z}{z^2 - 7z + 12} = \dfrac{z}{(z - 4)(z - 3)} = \dfrac{A}{(z - 4)} + \dfrac{B}{(z - 3)}$

$= \dfrac{A(z - 3) + B(z - 4)}{(z - 4)(z - 3)}$

from which, $z = A(z - 3) + B(z - 4)$

Letting $z = 4$ gives: $4 = A$

Letting $z = 3$ gives: $3 = -B$ i.e. $B = -3$

Hence, $F(z) = \dfrac{z}{z^2 - 7z + 12} = \dfrac{4}{(z - 4)} - \dfrac{3}{(z - 3)}$

The nearest transform in Table 79.1 to either of these partial fractions is $Z\{a^k\} = \dfrac{z}{z - a}$

Rearranging gives: $F(z) = \dfrac{4}{(z - 4)} - \dfrac{3}{(z - 3)}$

$= \dfrac{4}{z} \times \dfrac{z}{(z - 4)} - \dfrac{3}{z} \times \dfrac{z}{(z - 3)}$

$= 4 \times z^{-1} Z\left\{4^k\right\} - 3 \times z^{-1} Z\left\{3^k\right\}$

Hence, $Z^{-1}F(z) = 4 \times \left\{4^{k-1}\right\} - 3 \times \left\{3^{k-1}\right\}$ by the second shift theorem

$= \left\{4^k\right\} - \left\{3^k\right\} = \left\{4^k - 3^k\right\}$

i.e.   **the sequence is $x_k = 4^k - 3^k$**

With the denominator of $F(z) = \dfrac{z}{z^2 - 7z + 12}$ being z, there is an alternative, and more straight-forward method of determining the inverse transform,

i.e. by initially rearranging as: $\dfrac{F(z)}{z} = \dfrac{1}{z^2 - 7z + 12}$

Using partial fractions,

$\dfrac{1}{z^2 - 7z + 12} = \dfrac{1}{(z - 4)(z - 3)}$

$= \dfrac{A}{(z - 4)} + \dfrac{B}{(z - 3)} = \dfrac{A(z - 3) + B(z - 4)}{(z - 4)(z - 3)}$

from which, $1 = A(z - 3) + B(z - 4)$

Letting $z = 4$ gives: $1 = A$

Letting $z = 3$ gives: $1 = -B$ i.e. $B = -1$

Hence, $\dfrac{F(z)}{z} = \dfrac{1}{z^2 - 7z + 12} = \dfrac{1}{(z - 4)} - \dfrac{1}{(z - 3)}$

and $F(z) = \dfrac{z}{(z - 4)} - \dfrac{z}{(z - 3)}$

and $z^{-1} F(z) = \left\{4^k\right\} - \left\{3^k\right\} = \left\{4^k - 3^k\right\}$ from 6 in Table 90.1

Problem 28.   Determine the inverse z-transform of $F(z) = \dfrac{z}{z^2 - 3z + 2}$

Since $F(z) = \dfrac{z}{z^2 - 3z + 2}$ then $\dfrac{F(z)}{z} = \dfrac{1}{z^2 - 3z + 2}$

Using partial fractions, let

$\dfrac{1}{z^2 - 3z + 2} = \dfrac{1}{(z - 1)(z - 2)} = \dfrac{A}{(z - 1)} + \dfrac{B}{(z - 2)}$

$= \dfrac{A(z - 2) + B(z - 1)}{(z - 1)(z - 2)}$

from which, $1 = A(z - 2) + B(z - 1)$

Letting $z = 1$ gives: $1 = -A$ i.e. $A = -1$

Letting $z = 2$ gives: $1 = B$

Hence,

$\dfrac{F(z)}{z} = \dfrac{-1}{(z - 1)} + \dfrac{1}{(z - 2)} = \dfrac{1}{(z - 2)} - \dfrac{1}{(z - 1)}$

Section N

and $F(z) = \dfrac{z}{(z-2)} - \dfrac{z}{(z-1)}$

Thus, $Z^{-1}F(z) = Z^{-1}\left\{\dfrac{z}{(z-2)} - \dfrac{z}{(z-1)}\right\}$

$= Z^{-1}\left\{\dfrac{z}{(z-2)}\right\} - Z^{-1}\left\{\dfrac{z}{(z-1)}\right\}$

From 6 in Table 79.1, $\mathbf{Z^{-1}F(z)} = (2)^k - (1)^k = \mathbf{(2)^k - 1}$

Problem 29. Determine the sequence $\{x_k\}$ which has the z-transform $F(z) = \dfrac{4z}{(z+3)\left(z^2 - 4z + 4\right)}$

$\dfrac{F(z)}{z} = \dfrac{1}{z} \times \dfrac{4z}{(z+3)\left(z^2 - 4z + 4\right)} = \dfrac{4}{(z+3)\left(z^2 - 4z + 4\right)}$

Using partial fractions,

$\dfrac{4}{(z+3)\left(z^2 - 4z + 4\right)} = \dfrac{4}{(z+3)(z-2)^2}$

$= \dfrac{A}{(z+3)} + \dfrac{B}{(z-2)} + \dfrac{C}{(z-2)^2}$

$= \dfrac{A(z-2)^2 + B(z+3)(z-2) + C(z+3)}{(z+3)(z-2)^2}$

from which, $4 = A(z-2)^2 + B(z+3)(z-2) + C(z+3)$

Letting $z = -3$ gives: $4 = 25A$ i.e. $A = 4/25$

Letting $z = 2$ gives: $4 = 5C$ i.e. $C = 4/5$

Equating z^2 coefficients gives: $0 = A + B$ from which, $B = -4/25$

Hence, $\dfrac{F(z)}{z} = \dfrac{4}{(z+3)(z-2)^2}$

$= \dfrac{4/25}{(z+3)} - \dfrac{4/25}{(z-2)} + \dfrac{4/5}{(z-2)^2}$

and $F(z) = \dfrac{(4/25)z}{(z+3)} - \dfrac{(4/25)z}{(z-2)} + \dfrac{(4/5)z}{(z-2)^2}$

$= \dfrac{4}{25} \times \dfrac{z}{(z+3)} - \dfrac{4}{25} \times \dfrac{z}{(z-2)} + \dfrac{4}{5} \times \dfrac{z}{(z-2)^2}$

$= \dfrac{4}{25} \times \dfrac{z}{(z+3)} - \dfrac{4}{25} \times \dfrac{z}{(z-2)} + \dfrac{2}{5} \times \dfrac{2z}{(z-2)^2}$

i.e. $Z^{-1}F(z) = \dfrac{4}{25}\left\{(-3)^k\right\} - \dfrac{4}{25}\left\{(2)^k\right\} + \dfrac{2}{5}\left\{k(2)^k\right\}$

from 6 and 7 of Table 79.1

i.e. $\{x_k\} = \dfrac{4}{25}\left\{(-3)^k\right\} - \dfrac{4}{25}\left\{(2)^k\right\} + \dfrac{2}{5}\left\{k(2)^k\right\}$

Now try the following Practice Exercise

Practice Exercise 276   Inverse z-Transforms (Answers on page 899)

In problems 1 to 12, use Table 79.1 to find the sequences which have the following z transforms

1. $\dfrac{z}{z-1}$ 2. $\dfrac{z}{z-2}$

3. $\dfrac{z}{z+1}$ 4. $\dfrac{z}{z+4}$

5. $\dfrac{z}{z-\dfrac{1}{3}}$ 6. $\dfrac{4z}{4z+1}$

7. $\dfrac{5z}{5z-1}$ 8. $\dfrac{z}{z-e^{-5}}$

9. $\dfrac{z}{z-e^3}$ 10. $\dfrac{3z}{(z-1)^2}$

11. $\dfrac{5z(z+1)}{(z-1)^3}$ 12. $\dfrac{z}{2(z-2)^2}$

In problems 13 to 20, use partial fractions to determine the inverse z-transforms of the following.

13. $\dfrac{z}{(z+1)(z-2)}$ 14. $\dfrac{3z}{(z-1)(z+3)}$

15. $\dfrac{z}{(z-1)(z+2)}$ 16. $\dfrac{z}{(z-3)(2z+1)}$

17. $\dfrac{z}{2-3z+z^2}$ 18. $\dfrac{z(3z+1)}{z^2-5z+6}$

19. $\dfrac{2z}{2z^2+z-1}$ 20. $\dfrac{3z}{z^2+3z-10}$

79.4 Using z-transforms to solve difference equations

In Chapter 71, Laplace transforms were used to solve differential equations; in this section, the solution of difference equations using z-transforms is demonstrated. Difference equations arise in a number of different ways – sometimes from the direct modelling of systems in discrete time, or as an approximation to a differential equation describing the behaviour of a system modelled as a continuous-time system. The z-transform method is based on the first shift theorem, (see earlier, page

829), and the method of solution is explained through the following problems.

Problem 30. Solve the difference equation $x_{k+1} - 2x_k = 0$ given the initial condition that $x_0 = 3$

Taking the z-transform of each term gives:

$$Z\{x_{k+1}\} - 2Z\{x_k\} = Z\{0\}$$

Since from equation (6),

$$Z\{x_{k+m}\} = z^m F(z) - \left[z^m x_0 + z^{m-1} x_1 + \ldots + z x_{m-1}\right]$$

then $\quad \left(z^1 Z\{k\} - \left[z^1(3)\right]\right) - 2Z\{x_k\} = 0$

i.e. $\quad z\,Z\{x_k\} - 3z - 2Z\{x_k\} = 0$

i.e. $\quad (z-2)\,Z\{x_k\} = 3z$

and $\quad Z\{x_k\} = \dfrac{3z}{z-2}$

Taking the inverse z-transform gives:

$$\{x_k\} = Z^{-1}\left(\frac{3z}{z-2}\right) = 3Z^{-1}\left(\frac{z}{z-2}\right)$$

i.e. $\quad \{x_k\} = 3(2^k)$ from 6 of Table 79.1

Problem 31. Solve the difference equation: $x_{k+2} - 3x_{k+1} + 2x_k = 1$ given that $x_0 = 0$ and $x_1 = 2$

Taking the z-transform of each term gives:

$$Z\{x_{k+2}\} - 3Z\{x_{k+1}\} + 2Z\{x_k\} = Z\{1\}$$

Since from equation (6),

$$Z\{x_{k+m}\} = z^m F(z) - \left[z^m x_0 + z^{m-1} x_1 + \ldots + z x_{m-1}\right]$$

$$\left(z^2 Z\{x_k\} - \left[z^2(0) + z^1(2)\right]\right)$$
$$-3\left(z^1 Z\{x_k\} - \left[z^1(0)\right]\right) + 2Z\{x_k\} = \frac{z}{z-1}$$

i.e. $\quad z^2 Z\{x_k\} - 2z - 3z Z\{x_k\} + 2Z\{x_k\} = \dfrac{z}{z-1}$

and $\quad (z^2 - 3z + 2)Z\{x_k\} = \dfrac{z}{z-1} + 2z$

$$= \frac{z + 2z(z-1)}{z-1} = \frac{2z^2 - z}{z-1} = \frac{z(2z-1)}{z-1}$$

from which, $\quad Z\{x_k\} = \dfrac{z(2z-1)}{(z-1)(z^2 - 3z + 2)}$

$$= \frac{z(2z-1)}{(z-1)(z-2)(z-1)}$$

or $\quad \dfrac{Z\{x_k\}}{z} = \dfrac{(2z-1)}{(z-1)(z-2)(z-1)}$

$$= \frac{(2z-1)}{(z-1)^2(z-2)}$$

Using partial fractions, let

$$\frac{(2z-1)}{(z-1)^2(z-2)} = \frac{A}{(z-1)} + \frac{B}{(z-1)^2} + \frac{C}{(z-2)}$$

$$= \frac{A(z-1)(z-2) + B(z-2) + C(z-1)^2}{(z-1)^2(z-2)}$$

and $\quad 2z - 1 = A(z-1)(z-2) + B(z-2) + C(z-1)^2$

Letting $z = 1$ gives: $\quad 1 = -B$ i.e. $B = -1$

Letting $z = 2$ gives: $\quad 3 = C$

Equating z^2 coefficients gives: $\quad 0 = A + C$ i.e. $A = -3$

Hence, $\quad \dfrac{Z\{x_k\}}{z} = \dfrac{(2z-1)}{(z-1)^2(z-2)} = \dfrac{-3}{(z-1)} + \dfrac{-1}{(z-1)^2}$
$$+ \frac{3}{(z-2)}$$

Therefore, $Z\{x_k\} = 3\left(\dfrac{z}{(z-2)}\right) - 3\left(\dfrac{z}{(z-1)}\right)$
$$- \frac{z}{(z-1)^2}$$

Taking the inverse z-transform gives:

$$\{x_k\} = 3Z^{-1}\left(\frac{z}{(z-2)}\right) - 3Z^{-1}\left(\frac{z}{(z-1)}\right)$$
$$- Z^{-1}\left(\frac{z}{(z-1)^2}\right)$$

$$= 3(2)^k - 3(1)^k - k \text{ from 6 and 3 of Table 79.1}$$

i.e. $\{x_k\} = 3(2^k) - 3 - k$

Problem 32. Solve the difference equation: $x_{k+2} - x_k = 1$ given that $x_0 = 0$ and $x_1 = -1$

Taking the z-transform of each term gives:

$$Z\{x_{k+2}\} - Z\{x_k\} = Z\{1\}$$

Since from equation (6),

$$Z\{x_{k+m}\} = z^m F(z) - \left[z^m x_0 + z^{m-1} x_1 + \ldots + z x_{m-1}\right]$$

$$z^2 Z\{x_k\} - \left[z^2(0) + z^1(-1)\right] - Z\{x_k\} = \frac{z}{z-1}$$

i.e. $\quad z^2 Z\{x_k\} + z - Z\{x_k\} = \dfrac{z}{z-1}$

and $\left(z^2 - 1\right) Z\{x_k\} = \dfrac{z}{z-1} - z = \dfrac{z - z(z-1)}{z-1}$

$$= \dfrac{2z - z^2}{z-1}$$

from which, $Z\{x_k\} = \dfrac{2z - z^2}{(z-1)\left(z^2 - 1\right)}$

$$= \dfrac{2z - z^2}{(z-1)(z-1)(z+1)} = \dfrac{2z - z^2}{(z-1)^2(z+1)}$$

and $\dfrac{Z\{x_k\}}{z} = \dfrac{2 - z}{(z-1)^2(z+1)}$

Using partial fractions, let

$$\dfrac{2-z}{(z-1)^2(z+1)} = \dfrac{A}{(z-1)} + \dfrac{B}{(z-1)^2} + \dfrac{C}{(z+1)}$$

$$= \dfrac{A(z-1)(z+1) + B(z+1) + C(z-1)^2}{(z-1)^2(z+1)}$$

and $2 - z = A(z-1)(z+1) + B(z+1) + C(z-1)^2$

Letting $z = 1$ gives: $1 = 2B$ i.e. $B = 1/2$

Letting $z = -1$ gives: $3 = 4C$ i.e. $C = 3/4$

Equating z^2 coefficients gives: $0 = A + C$

i.e. $\qquad\qquad\qquad A = -3/4$

Hence, $\dfrac{Z\{x_k\}}{z} = \dfrac{2 - z}{(z-1)^2(z+1)}$

$$= \dfrac{-3/4}{(z-1)} + \dfrac{1/2}{(z-1)^2} + \dfrac{3/4}{(z+1)}$$

Therefore, $Z\{x_k\} = -\dfrac{3}{4}\left(\dfrac{z}{z-1}\right) + \dfrac{1}{2}\left(\dfrac{z}{(z-1)^2}\right)$

$$- \dfrac{3}{4}\left(\dfrac{z}{z+1}\right)$$

Taking the inverse z-transform gives:

$$\{x_k\} = -\dfrac{3}{4} Z^{-1}\left(\dfrac{z}{(z-2)}\right) + \dfrac{1}{2} Z^{-1}\left(\dfrac{z}{(z-1)^2}\right)$$

$$- \dfrac{3}{4} Z^{-1}\left(\dfrac{z}{z+1}\right)$$

i.e. $\{x_k\} = -\dfrac{3}{4}\left(2^k\right) + \dfrac{1}{2}k - \dfrac{3}{4}(-1)^k$ from 6 and 3 of

$$\text{Table } 79.1$$

Problem 33. Solve the difference equation:
$x_{k+2} - 3x_{k+1} + 2x_k = 1$ given that $x_0 = 0$ and $x_1 = 1$

Taking the z-transform of both sides of the equation gives:

$$Z\{x_{k+2} - 3x_{k+1} + 2x_k\} = Z\{1\}$$

i.e. $\qquad Z\{x_{k+2}\} - 3Z\{x_{k+1}\} + 2Z\{x_k\} = Z\{1\}$

Using the first shift theorem and $Z\{x_k\} = F(z)$ gives:

$$\left(z^2 F(z) - z^2 x_0 - zx_1\right) - 3\left(zF(z) - zx_0\right) + 2F(z)$$

$$= \dfrac{z}{z-1}$$

$x_0 = 0$ and $x_1 = 1$, hence $\left(z^2 F(z) - z^2(0) - z(1)\right) -$

$$3\left(zF(z) - z(0)\right) + 2F(z) = \dfrac{z}{z-1}$$

i.e. $\qquad z^2 F(z) - z - 3zF(z) + 2F(z) = \dfrac{z}{z-1}$

and $\left(z^2 - 3z + 2\right) F(z) = \dfrac{z}{z-1} + z = \dfrac{z}{z-1} + \dfrac{z}{1}$

$$= \dfrac{z + z(z-1)}{z-1} = \dfrac{z + z^2 - z}{z-1} = \dfrac{z^2}{z-1}$$

Hence, $F(z) = \dfrac{z^2}{\left(z^2 - 3z + 2\right)(z-1)}$

$$= \dfrac{z^2}{(z-2)(z-1)(z-1)} = \dfrac{z^2}{(z-2)(z-1)^2}$$

and $\qquad\qquad \dfrac{F(z)}{z} = \dfrac{z}{(z-2)(z-1)^2}$

Using partial fractions, let

$$\dfrac{z}{(z-2)(z-1)^2} = \dfrac{A}{(z-2)} + \dfrac{B}{(z-1)} + \dfrac{C}{(z-1)^2}$$

$$= \dfrac{A(z-1)^2 + B(z-2)(z-1) + C(z-2)}{(z-2)(z-1)^2}$$

from which, $z = A(z-1)^2 + B(z-2)(z-1) + C(z-2)$

Letting $z = 2$ gives: $2 = A(1)^2$ i.e. $A = 2$

Letting $z = 1$ gives: $1 = C(-1)$ i.e. $C = -1$

Equating z^2 coefficients gives: $0 = A + B$ i.e. $B = -2$

Therefore, $\dfrac{F(z)}{z} = \dfrac{2}{(z-2)} - \dfrac{2}{(z-1)} - \dfrac{1}{(z-1)^2}$

or $\qquad F(z) = \dfrac{2z}{(z-2)} - \dfrac{2z}{(z-1)} - \dfrac{z}{(z-1)^2}$

Taking the inverse z-transform of F(z) gives:

$$Z^{-1} F(z) = 2Z^{-1} \left(\frac{z}{(z-2)} \right) - 2Z^{-1} \left(\frac{z}{(z-1)} \right)$$

$$- Z^{-1} \left(\frac{z}{(z-1)^2} \right)$$

$$= 2 \left(2^k \right) - 2(1) - k$$

from 2, 6 and 7 of Table 79.1

i.e. $$\{x_k\} = 2^{k+1} - 2 - k$$

Now try the following Practice Exercise

Practice Exercise 277   Solving difference equations (Answers on page 899)

1. Solve the difference equation:
   $x_{k+1} - 3x_k = 0$ given $x_0 = 4$

2. Solve the difference equation:
   $x_{k+1} - 3x_k = -6$ given $x_0 = 1$

3. Solve the difference equation:
   $2x_{k+2} - 5x_{k+1} + 4x_k = 0$ given $x_0 = 3$ and $x_1 = 2$

4. Solve the difference equation:
   $2x_{k+1} - x_k = 2^k$ given $x_0 = 2$

5. Solve the difference equation:
   $x_{k+2} + 3x_{k+1} + 2x_k = 0$ given $x_0 = 0$ and $x_1 = 1$

6. Solve the difference equation:
   $x_{k+2} - 5x_{k+1} + 6x_k = 5$ given $x_0 = 0$ and $x_1 = 1$

Section N

**For fully worked solutions to each of the problems in Practice Exercises 271 to 277
in this chapter, go to the website:
www.routledge.com/cw/bird**

Revision Test 21   Fourier series and z-transforms

This Revision Test covers the material contained in Chapters 73 to 79. *The marks for each question are shown in brackets at the end of each question.*

1. Obtain a Fourier series for the periodic function $f(x)$ defined as follows:

$$f(x) = \begin{cases} -1, & \text{when } -\pi \leq x \leq 0 \\ 1, & \text{when } \quad 0 \leq x \leq \pi \end{cases}$$

   The function is periodic outside of this range with period 2π. (13)

2. Obtain a Fourier series to represent $f(t) = t$ in the range $-\pi$ to $+\pi$. (13)

3. Expand the function $f(\theta) = \theta$ in the range $0 \leq \theta \leq \pi$ into (a) a half range cosine series, and (b) a half-range sine series. (18)

4. (a) Sketch the waveform defined by:

$$f(x) = \begin{cases} 0, & \text{when } -4 \leq x \leq -2 \\ 3, & \text{when } -2 \leq x \leq 2 \\ 0, & \text{when } \quad 2 \leq x \leq 4 \end{cases}$$

   and is periodic outside of this range of period 8.

   (b) State whether the waveform in (a) is odd, even or neither odd nor even.
   (c) Deduce the Fourier series for the function defined in (a). (15)

5. Displacement y on a point on a pulley when turned through an angle of θ degrees is given by:

θ	y	θ	y
30	3.99	210	0.27
60	4.01	240	0.13
90	3.60	270	0.45
120	2.84	300	1.25
150	1.84	330	2.37
180	0.88	360	3.41

   Sketch the waveform and construct a Fourier series for the first three harmonics (23)

6. A rectangular waveform is shown in Fig. RT20.1.

   (a) State whether the waveform is an odd or even function.

   (b) Obtain the Fourier series for the waveform in complex form.

   (c) Show that the complex Fourier series in (b) is equivalent to:

$$f(x) = \frac{20}{\pi}\left(\sin x + \frac{1}{3}\sin 3x + \frac{1}{5}\sin 5x + \frac{1}{7}\sin 7x + \cdots\right)$$ (18)

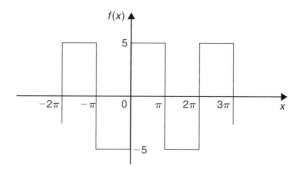

Figure RT20.1

7. Determine the z-transform of
   (a) 2^k (b) $5e^{-2k}$ (c) $2e^{-3k}\cos 5k$ (d) 4^{k+2} (10)

8. Determine the inverse z-transform of
   (a) $\dfrac{2z}{4z+1}$ (b) $\dfrac{z}{z-e^4}$ (c) $\dfrac{z}{z^2-z-2}$ (10)

9. Solve the difference equation:

   $x_{k+2} - 2x_{k+1} - 3x_k = 1$ given $x_0 = 0$ and $x_1 = 1$ (20)

Essential formulae

Number and algebra

Laws of indices:

$$a^m \times a^n = a^{m+n} \quad \frac{a^m}{a^n} = a^{m-n} \quad (a^m)^n = a^{mn}$$

$$a^{\frac{m}{n}} = \sqrt[n]{a^m} \qquad a^{-n} = \frac{1}{a^n} \qquad a^0 = 1$$

Quadratic formula:

$$\text{If } ax^2 + bx + c = 0 \text{ then } x = \frac{-b \pm \sqrt{b^2 - 4ac}}{2a}$$

Factor theorem:

If $x = a$ is a root of the equation $f(x) = 0$, then $(x - a)$ is a factor of $f(x)$.

Remainder theorem:

If $(ax^2 + bx + c)$ is divided by $(x - p)$, the remainder will be: $ap^2 + bp + c$

or if $(ax^3 + bx^2 + cx + d)$ is divided by $(x - p)$, the remainder will be: $ap^3 + bp^2 + cp + d$

Partial fractions:

Provided that the numerator $f(x)$ is of less degree than the relevant denominator, the following identities are typical examples of the form of partial fractions used:

$$\frac{f(x)}{(x+a)(x+b)(x+c)}$$
$$\equiv \frac{A}{(x+a)} + \frac{B}{(x+b)} + \frac{C}{(x+c)}$$

$$\frac{f(x)}{(x+a)^3(x+b)}$$
$$\equiv \frac{A}{(x+a)} + \frac{B}{(x+a)^2} + \frac{C}{(x+a)^3} + \frac{D}{(x+b)}$$

$$\frac{f(x)}{(ax^2 + bx + c)(x+d)}$$
$$\equiv \frac{Ax + B}{(ax^2 + bx + c)} + \frac{C}{(x+d)}$$

Definition of a logarithm:

If $y = a^x$ then $x = \log_a y$

Laws of logarithms:

$$\log(A \times B) = \log A + \log B$$

$$\log\left(\frac{A}{B}\right) = \log A - \log B$$

$$\log A^n = n \times \log A$$

Exponential series:

$$e^x = 1 + x + \frac{x^2}{2!} + \frac{x^3}{3!} + \cdots$$

(valid for all values of x)

Hyperbolic functions:

$$\sinh x = \frac{e^x - e^{-x}}{2} \qquad \operatorname{cosech} x = \frac{1}{\sinh x} = \frac{2}{e^x - e^{-x}}$$

$$\cosh x = \frac{e^x + e^{-x}}{2} \qquad \operatorname{sech} x = \frac{1}{\cosh x} = \frac{2}{e^x + e^{-x}}$$

$$\tanh x = \frac{e^x - e^{-x}}{e^x + e^{-x}} \qquad \coth x = \frac{1}{\tanh x} = \frac{e^x + e^{-x}}{e^x - e^{-x}}$$

$$\cosh^2 x - \sinh^2 x = 1 \quad 1 - \tanh^2 x = \operatorname{sech}^2 x$$

$$\coth^2 x - 1 = \operatorname{cosech}^2 x$$

Arithmetic progression:

If $a = $ first term and $d = $ common difference, then the arithmetic progression is: $a, a+d, a+2d, \ldots$

The nth term is: $a + (n-1)d$

Sum of n terms, $S_n = \dfrac{n}{2}[2a + (n-1)d]$

Geometric progression:

If $a = $ first term and $r = $ common ratio, then the geometric progression is: $a, ar, ar^2, \ldots$

The nth term is: ar^{n-1}

Sum of n terms, $S_n = \dfrac{a(1-r^n)}{(1-r)}$ or $\dfrac{a(r^n-1)}{(r-1)}$

If $-1 < r < 1$, $S_\infty = \dfrac{a}{(1-r)}$

Binomial series:

$$(a+b)^n = a^n + na^{n-1}b + \frac{n(n-1)}{2!}a^{n-2}b^2$$

$$+ \frac{n(n-1)(n-2)}{3!}a^{n-3}b^3 + \cdots$$

$$(1+x)^n = 1 + nx + \frac{n(n-1)}{2!}x^2$$

$$+ \frac{n(n-1)(n-2)}{3!}x^3 + \cdots$$

Maclaurin's series:

$$f(x) = f(0) + x\,f'(0) + \frac{x^2}{2!}\,f''(0)$$

$$+ \frac{x^3}{3!}\,f'''(0) + \cdots$$

Newton–Raphson iterative method:

If r_1 is the approximate value for a real root of the equation $f(x) = 0$, then a closer approximation to the root, r_2, is given by:

$$r_2 = r_1 - \frac{f(r_1)}{f'(r_1)}$$

Boolean algebra:

Laws and rules of Boolean algebra

Commutative laws: $A + B = B + A$

$$A \cdot B = B \cdot A$$

Associative laws: $A + B + C = (A + B) + C$

$$A \cdot B \cdot C = (A \cdot B) \cdot C$$

Distributive laws: $A \cdot (B + C) = A \cdot B + A \cdot C$

$$A + (B \cdot C) = (A + B) \cdot (A + C)$$

Sum rules: $A + \overline{A} = 1$

$$A + 1 = 1$$

$$A + 0 = A$$

$$A + A = A$$

Product rules: $A \cdot \overline{A} = 0$

$$A \cdot 0 = 0$$

$$A \cdot 1 = A$$

$$A \cdot A = A$$

Absorption rules: $A + A \cdot B = A$

$$A \cdot (A + B) = A$$

$$A + \overline{A} \cdot B = A + B$$

De Morgan's laws: $\overline{A + B} = \overline{A} \cdot \overline{B}$

$$\overline{A \cdot B} = \overline{A} + \overline{B}$$

Geometry and trigonometry

Theorem of Pythagoras:

$$b^2 = a^2 + c^2$$

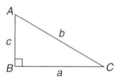

Figure FA1

Identities:

$$\sec\theta = \frac{1}{\cos\theta} \qquad \operatorname{cosec}\theta = \frac{1}{\sin\theta}$$

$$\cot\theta = \frac{1}{\tan\theta} \qquad \tan\theta = \frac{\sin\theta}{\cos\theta}$$

$$\cos^2\theta + \sin^2\theta = 1 \quad 1 + \tan^2\theta = \sec^2\theta$$

$$\cot^2\theta + 1 = \operatorname{cosec}^2\theta$$

Triangle formulae:

With reference to Fig. FA2:

Sine rule $\dfrac{a}{\sin A} = \dfrac{b}{\sin B} = \dfrac{c}{\sin C}$

Cosine rule $a^2 = b^2 + c^2 - 2bc\cos A$

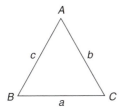

Figure FA2

Area of any triangle

(i) $\frac{1}{2} \times$ base $\times$ perpendicular height

(ii) $\frac{1}{2}ab\sin C$ or $\frac{1}{2}ac\sin B$ or $\frac{1}{2}bc\sin A$

(iii) $\sqrt{[s(s-a)(s-b)(s-c)]}$ where $s = \dfrac{a+b+c}{2}$

Compound angle formulae:

$$\sin(A \pm B) = \sin A \cos B \pm \cos A \sin B$$

$$\cos(A \pm B) = \cos A \cos B \mp \sin A \sin B$$

$$\tan(A \pm B) = \dfrac{\tan A \pm \tan B}{1 \mp \tan A \tan B}$$

If $R \sin(\omega t + \alpha) = a \sin \omega t + b \cos \omega t$,

then $a = R\cos\alpha$, $b = R\sin\alpha$,

$$R = \sqrt{(a^2 + b^2)} \text{ and } \alpha = \tan^{-1}\dfrac{b}{a}$$

Double angles:

$$\sin 2A = 2\sin A\cos A$$

$$\cos 2A = \cos^2 A - \sin^2 A = 2\cos^2 A - 1$$

$$= 1 - 2\sin^2 A$$

$$\tan 2A = \dfrac{2\tan A}{1 - \tan^2 A}$$

Products of sines and cosines into sums or differences:

$$\sin A \cos B = \tfrac{1}{2}[\sin(A+B) + \sin(A-B)]$$

$$\cos A \sin B = \tfrac{1}{2}[\sin(A+B) - \sin(A-B)]$$

$$\cos A \cos B = \tfrac{1}{2}[\cos(A+B) + \cos(A-B)]$$

$$\sin A \sin B = -\tfrac{1}{2}[\cos(A+B) - \cos(A-B)]$$

Sums or differences of sines and cosines into products:

$$\sin x + \sin y = 2\sin\left(\dfrac{x+y}{2}\right)\cos\left(\dfrac{x-y}{2}\right)$$

$$\sin x - \sin y = 2\cos\left(\dfrac{x+y}{2}\right)\sin\left(\dfrac{x-y}{2}\right)$$

$$\cos x + \cos y = 2\cos\left(\dfrac{x+y}{2}\right)\cos\left(\dfrac{x-y}{2}\right)$$

$$\cos x - \cos y = -2\sin\left(\dfrac{x+y}{2}\right)\sin\left(\dfrac{x-y}{2}\right)$$

For a **general sinusoidal function**
$y = A\sin(\omega t \pm \alpha)$, then:

A = amplitude

ω = angular velocity = $2\pi f$ rad/s

$\dfrac{2\pi}{\omega}$ = periodic time T seconds

$\dfrac{\omega}{2\pi}$ = frequency, f hertz

α = angle of lead or lag (compared with
$y = A\sin\omega t$)

Cartesian and polar co-ordinates:

If co-ordinate $(x, y) = (r, \theta)$ then $r = \sqrt{x^2 + y^2}$ and
$\theta = \tan^{-1}\dfrac{y}{x}$
If co-ordinate $(r, \theta) = (x, y)$ then $x = r\cos\theta$ and
$y = r\sin\theta$

The circle:

With reference to Fig. FA3.

$$\text{Area} = \pi r^2 \quad \text{Circumference} = 2\pi r$$

$$\pi \text{ radians} = 180°$$

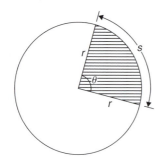

Figure FA3

For sector of circle:

$$s = r\theta \quad (\theta \text{ in rad})$$

$$\text{shaded area} = \tfrac{1}{2}r^2\theta \quad (\theta \text{ in rad})$$

Equation of a circle, centre at (a, b), radius r:

$$(x - a)^2 + (y - b)^2 = r^2$$

Linear and angular velocity:

If $v =$ linear velocity (m/s), $s =$ displacement (m), $t =$ time (s), $n =$ speed of revolution (rev/s), $\theta =$ angle (rad), $\omega =$ angular velocity (rad/s), $r =$ radius of circle (m) then:

$$v = \frac{s}{t} \quad \omega = \frac{\theta}{t} = 2\pi n \quad v = \omega r$$

$$\text{centripetal force} = \frac{mv^2}{r}$$

where $m =$ mass of rotating object.

Graphs

Equations of functions:

Equation of a straight line: $y = mx + c$

Equation of a parabola: $y = ax^2 + bx + c$

Circle, centre (a, b), radius r:

$$(x - a)^2 + (y - b)^2 = r^2$$

Equation of an ellipse, centre at origin, semi-axes a and b: $\dfrac{x^2}{a^2} + \dfrac{y^2}{b^2} = 1$

Equation of a hyperbola: $\dfrac{x^2}{a^2} - \dfrac{y^2}{b^2} = 1$

Equation of a rectangular hyperbola: $xy = c^2$

Irregular areas:

Trapezoidal rule

$$\text{Area} \approx \begin{pmatrix} \text{width of} \\ \text{interval} \end{pmatrix} \left[\frac{1}{2} \begin{pmatrix} \text{first} + \text{last} \\ \text{ordinates} \end{pmatrix} \right.$$

$$\left. + \begin{pmatrix} \text{sum of remaining} \\ \text{ordinates} \end{pmatrix} \right]$$

Mid-ordinate rule

$$\text{Area} \approx \begin{pmatrix} \text{width of} \\ \text{interval} \end{pmatrix} \begin{pmatrix} \text{sum of} \\ \text{mid-ordinates} \end{pmatrix}$$

Simpson's rule

$$\text{Area} \approx \frac{1}{3} \begin{pmatrix} \text{width of} \\ \text{interval} \end{pmatrix} \left[\begin{pmatrix} \text{first} + \text{last} \\ \text{ordinate} \end{pmatrix} \right.$$

$$+ 4 \begin{pmatrix} \text{sum of even} \\ \text{ordinates} \end{pmatrix}$$

$$\left. + 2 \begin{pmatrix} \text{sum of remaining} \\ \text{odd ordinates} \end{pmatrix} \right]$$

Vector geometry

If $\mathbf{a} = a_1\mathbf{i} + a_2\mathbf{j} + a_3\mathbf{k}$ and $b = b_1\mathbf{i} + b_2\mathbf{j} + b_3\mathbf{k}$

$$\mathbf{a} \cdot \mathbf{b} = a_1 b_1 + a_2 b_2 + a_3 b_3$$

$$|a| = \sqrt{a_1^2 + a_2^2 + a_3^2} \quad \cos\theta = \frac{a \cdot b}{|a| \, |b|}$$

$$\mathbf{a} \times \mathbf{b} = \begin{vmatrix} i & j & k \\ a_1 & a_2 & a_3 \\ b_1 & b_2 & b_3 \end{vmatrix}$$

$$|a \times b| = \sqrt{[(a \cdot a)(b \cdot b) - (a \cdot b)^2]}$$

Complex numbers

$z = a + jb = r(\cos\theta + j\sin\theta) = r\angle\theta = re^{j\theta}$ where $j^2 = -1$

Modulus $r = |z| = \sqrt{(a^2 + b^2)}$

Argument $\theta = \arg z = \tan^{-1}\dfrac{b}{a}$

Addition: $(a + jb) + (c + jd) = (a + c) + j(b + d)$

Subtraction: $(a + jb) - (c + jd) = (a - c) + j(b - d)$

Complex equations: If $m + jn = p + jq$ then $m = p$ and $n = q$

Multiplication: $z_1 z_2 = r_1 r_2 \angle(\theta_1 + \theta_2)$

Division: $\dfrac{z_1}{z_2} = \dfrac{r_1}{r_2}\angle(\theta_1 - \theta_2)$

De Moivre's theorem:

$[r\angle\theta]^n = r^n \angle n\theta = r^n(\cos n\theta + j\sin n\theta) = re^{j\theta}$

Matrices and determinants

Matrices:

If $A = \begin{pmatrix} a & b \\ c & d \end{pmatrix}$ and $B = \begin{pmatrix} e & f \\ g & h \end{pmatrix}$ then

$$A + B = \begin{pmatrix} a + e & b + f \\ c + g & d + h \end{pmatrix}$$

$$A - B = \begin{pmatrix} a - e & b - f \\ c - g & d - h \end{pmatrix}$$

$$A \times B = \begin{pmatrix} ae + bg & af + bh \\ ce + dg & cf + dh \end{pmatrix}$$

$$A^{-1} = \frac{1}{ad - bc}\begin{pmatrix} d & -b \\ -c & a \end{pmatrix}$$

If $A = \begin{pmatrix} a_1 & b_1 & c_1 \\ a_2 & b_2 & c_2 \\ a_3 & b_3 & c_3 \end{pmatrix}$ then $A^{-1} = \dfrac{B^T}{|A|}$ where

$B^T = $ transpose of cofactors of matrix A

Determinants:

$$\begin{vmatrix} a & b \\ c & d \end{vmatrix} = ad - bc$$

$$\begin{vmatrix} a_1 & b_1 & c_1 \\ a_2 & b_2 & c_2 \\ a_3 & b_3 & c_3 \end{vmatrix} = a_1\begin{vmatrix} b_2 & c_2 \\ b_3 & c_3 \end{vmatrix} - b_1\begin{vmatrix} a_2 & c_2 \\ a_3 & c_3 \end{vmatrix} + c_1\begin{vmatrix} a_2 & b_2 \\ a_3 & b_3 \end{vmatrix}$$

Differential calculus

Standard derivatives:

y or $f(x)$	$\dfrac{dy}{dx}$ or $f'(x)$
ax^n	anx^{n-1}
$\sin ax$	$a\cos ax$
$\cos ax$	$-a\sin ax$
$\tan ax$	$a\sec^2 ax$
$\sec ax$	$a\sec ax\,\tan ax$
$\operatorname{cosec} ax$	$-a\operatorname{cosec} ax\,\cot ax$
$\cot ax$	$-a\operatorname{cosec}^2 ax$
e^{ax}	ae^{ax}
$\ln ax$	$\dfrac{1}{x}$
$\sinh ax$	$a\cosh ax$
$\cosh ax$	$a\sinh ax$
$\tanh ax$	$a\operatorname{sech}^2 ax$
$\operatorname{sech} ax$	$-a\operatorname{sech} ax\,\tanh ax$
$\operatorname{cosech} ax$	$-a\operatorname{cosech} ax\,\coth ax$
$\coth ax$	$-a\operatorname{cosech}^2 ax$
$\sin^{-1}\dfrac{x}{a}$	$\dfrac{1}{\sqrt{a^2 - x^2}}$
$\sin^{-1} f(x)$	$\dfrac{f'(x)}{\sqrt{1 - [f(x)]^2}}$
$\cos^{-1}\dfrac{x}{a}$	$\dfrac{-1}{\sqrt{a^2 - x^2}}$

y or $f(x)$	$\dfrac{dy}{dx}$ or $f'(x)$
$\cos^{-1} f(x)$	$\dfrac{-f'(x)}{\sqrt{1-[f(x)]^2}}$
$\tan^{-1}\dfrac{x}{a}$	$\dfrac{a}{a^2+x^2}$
$\tan^{-1} f(x)$	$\dfrac{f'(x)}{1+[f(x)]^2}$
$\sec^{-1}\dfrac{x}{a}$	$\dfrac{a}{x\sqrt{x^2-a^2}}$
$\sec^{-1} f(x)$	$\dfrac{f'(x)}{f(x)\sqrt{[f(x)]^2-1}}$
$\csc^{-1}\dfrac{x}{a}$	$\dfrac{-a}{x\sqrt{x^2-a^2}}$
$\csc^{-1} f(x)$	$\dfrac{-f'(x)}{f(x)\sqrt{[f(x)]^2-1}}$
$\cot^{-1}\dfrac{x}{a}$	$\dfrac{-a}{a^2+x^2}$
$\cot^{-1} f(x)$	$\dfrac{-f'(x)}{1+[f(x)]^2}$
$\sinh^{-1}\dfrac{x}{a}$	$\dfrac{1}{\sqrt{x^2+a^2}}$
$\sinh^{-1} f(x)$	$\dfrac{f'(x)}{\sqrt{[f(x)]^2+1}}$
$\cosh^{-1}\dfrac{x}{a}$	$\dfrac{1}{\sqrt{x^2-a^2}}$
$\cosh^{-1} f(x)$	$\dfrac{f'(x)}{\sqrt{[f(x)]^2-1}}$
$\tanh^{-1}\dfrac{x}{a}$	$\dfrac{a}{a^2-x^2}$
$\tanh^{-1} f(x)$	$\dfrac{f'(x)}{1-[f(x)]^2}$
$\operatorname{sech}^{-1}\dfrac{x}{a}$	$\dfrac{-a}{x\sqrt{a^2-x^2}}$
$\operatorname{sech}^{-1} f(x)$	$\dfrac{-f'(x)}{f(x)\sqrt{1-[f(x)]^2}}$

y or $f(x)$	$\dfrac{dy}{dx}$ or $f'(x)$
$\operatorname{cosech}^{-1}\dfrac{x}{a}$	$\dfrac{-a}{x\sqrt{x^2+a^2}}$
$\operatorname{cosech}^{-1} f(x)$	$\dfrac{-f'(x)}{f(x)\sqrt{[f(x)]^2+1}}$
$\coth^{-1}\dfrac{x}{a}$	$\dfrac{a}{a^2-x^2}$
$\coth^{-1} f(x)$	$\dfrac{f'(x)}{1-[f(x)]^2}$

Product rule:

When $y=uv$ and u and v are functions of x then:

$$\frac{dy}{dx}=u\frac{dv}{dx}+v\frac{du}{dx}$$

Quotient rule:

When $y=\dfrac{u}{v}$ and u and v are functions of x then:

$$\frac{dy}{dx}=\frac{v\dfrac{du}{dx}-u\dfrac{dv}{dx}}{v^2}$$

Function of a function:

If u is a function of x then:

$$\frac{dy}{dx}=\frac{dy}{du}\times\frac{du}{dx}$$

Parametric differentiation:

If x and y are both functions of θ, then:

$$\frac{dy}{dx}=\frac{\dfrac{dy}{d\theta}}{\dfrac{dx}{d\theta}}\quad\text{and}\quad\frac{d^2y}{dx^2}=\frac{\dfrac{d}{d\theta}\left(\dfrac{dy}{dx}\right)}{\dfrac{dx}{d\theta}}$$

Implicit function:

$$\frac{d}{dx}[f(y)]=\frac{d}{dy}[f(y)]\times\frac{dy}{dx}$$

Maximum and minimum values:

If $y = f(x)$ then $\dfrac{dy}{dx} = 0$ for stationary points.

Let a solution of $\dfrac{dy}{dx} = 0$ be $x = a$; if the value of $\dfrac{d^2y}{dx^2}$ when $x = a$ is: *positive*, the point is a *minimum*,

negative, the point is a *maximum*.

Velocity and acceleration:

If distance $x = f(t)$, then

velocity $\quad v = f'(t)$ or $\dfrac{dx}{dt}$ and

acceleration $\quad a = f''(t)$ or $\dfrac{d^2x}{dt^2}$

Tangents and normals:

Equation of tangent to curve $y = f(x)$ at the point (x_1, y_1) is:

$$y - y_1 = m(x - x_1)$$

where $m =$ gradient of curve at (x_1, y_1).

Equation of normal to curve $y = f(x)$ at the point (x_1, y_1) is:

$$y - y_1 = -\dfrac{1}{m}(x - x_1)$$

Partial differentiation:

Total differential

If $z = f(u, v, \ldots)$, then the total differential,

$$dz = \dfrac{\partial z}{\partial u} du + \dfrac{\partial z}{\partial v} dv + \ldots$$

Rate of change

If $z = f(u, v, \ldots)$ and $\dfrac{du}{dt}, \dfrac{dv}{dt}, \ldots$ denote the rate of change of $u, v, \ldots$ respectively, then the rate of change of z,

$$\dfrac{dz}{dt} = \dfrac{\partial z}{\partial u} \cdot \dfrac{du}{dt} + \dfrac{\partial z}{\partial v} \cdot \dfrac{dv}{dt} + \ldots$$

Small changes

If $z = f(u, v, \ldots)$ and $\delta x, \delta y, \ldots$ denote small changes in $x, y, \ldots$ respectively, then the corresponding change,

$$\delta z \approx \dfrac{\partial z}{\partial x} \delta x + \dfrac{\partial z}{\partial y} \delta y + \ldots$$

To determine maxima, minima and saddle points for functions of two variables: Given $z = f(x, y)$,

(i) determine $\dfrac{\partial z}{\partial x}$ and $\dfrac{\partial z}{\partial y}$

(ii) for stationary points, $\dfrac{\partial z}{\partial x} = 0$ and $\dfrac{\partial z}{\partial y} = 0$

(iii) solve the simultaneous equations $\dfrac{\partial z}{\partial x} = 0$ and $\dfrac{\partial z}{\partial y} = 0$ for x and y, which gives the co-ordinates of the stationary points

(iv) determine $\dfrac{\partial^2 z}{\partial x^2}, \dfrac{\partial^2 z}{\partial y^2}$ and $\dfrac{\partial^2 z}{\partial x \partial y}$

(v) for each of the co-ordinates of the stationary points, substitute values of x and y into $\dfrac{\partial^2 z}{\partial x^2}, \dfrac{\partial^2 z}{\partial y^2}$ and $\dfrac{\partial^2 z}{\partial x \partial y}$ and evaluate each

(vi) evaluate $\left(\dfrac{\partial^2 z}{\partial x \partial y}\right)^2$ for each stationary point,

(vii) substitute the values of $\dfrac{\partial^2 z}{\partial x^2}, \dfrac{\partial^2 z}{\partial y^2}$ and $\dfrac{\partial^2 z}{\partial x \partial y}$ into the equation $\Delta = \left(\dfrac{\partial^2 z}{\partial x \partial y}\right)^2 - \left(\dfrac{\partial^2 z}{\partial x^2}\right)\left(\dfrac{\partial^2 z}{\partial y^2}\right)$ and evaluate

(viii) (a) if $\Delta > 0$ then the stationary point is a **saddle point**

(b) if $\Delta < 0$ and $\dfrac{\partial^2 z}{\partial x^2} < 0$, then the stationary point is a **maximum point**, and

(c) if $\Delta < 0$ and $\dfrac{\partial^2 z}{\partial x^2} > 0$, then the stationary point is a **minimum point**

Integral calculus

Standard integrals:

y	$\int y \, dx$
$a x^n$	$a\dfrac{x^{n+1}}{n+1}+c$ (except where $n=-1$)
$\cos ax$	$\dfrac{1}{a}\sin ax + c$
$\sin ax$	$-\dfrac{1}{a}\cos ax + c$
$\sec^2 ax$	$\dfrac{1}{a}\tan ax + c$
$\mathrm{cosec}^2 ax$	$-\dfrac{1}{a}\cot ax + c$
$\mathrm{cosec}\,ax\cot ax$	$-\dfrac{1}{a}\mathrm{cosec}\,ax + c$
$\sec ax\tan ax$	$\dfrac{1}{a}\sec ax + c$
e^{ax}	$\dfrac{1}{a}e^{ax} + c$
$\dfrac{1}{x}$	$\ln x + c$
$\tan ax$	$\dfrac{1}{a}\ln(\sec ax) + c$
$\cos^2 x$	$\dfrac{1}{2}\left(x+\dfrac{\sin 2x}{2}\right)+c$
$\sin^2 x$	$\dfrac{1}{2}\left(x-\dfrac{\sin 2x}{2}\right)+c$
$\tan^2 x$	$\tan x - x + c$
$\cot^2 x$	$-\cot x - x + c$
$\dfrac{1}{\sqrt{(a^2-x^2)}}$	$\sin^{-1}\dfrac{x}{a}+c$
$\sqrt{(a^2-x^2)}$	$\dfrac{a^2}{2}\sin^{-1}\dfrac{x}{a}+\dfrac{x}{2}\sqrt{(a^2-x^2)}+c$

y	$\int y \, dx$
$\dfrac{1}{(a^2+x^2)}$	$\dfrac{1}{a}\tan^{-1}\dfrac{x}{a}+c$
$\dfrac{1}{\sqrt{(x^2+a^2)}}$	$\sinh^{-1}\dfrac{x}{a}+c$ or $\ln\left[\dfrac{x+\sqrt{(x^2+a^2)}}{a}\right]+c$
$\sqrt{(x^2+a^2)}$	$\dfrac{a^2}{2}\sinh^{-1}\dfrac{x}{a}+\dfrac{x}{2}\sqrt{(x^2+a^2)}+c$
$\dfrac{1}{\sqrt{(x^2-a^2)}}$	$\cosh^{-1}\dfrac{x}{a}+c$ or $\ln\left[\dfrac{x+\sqrt{(x^2-a^2)}}{a}\right]+c$
$\sqrt{(x^2-a^2)}$	$\dfrac{x}{2}\sqrt{(x^2-a^2)}-\dfrac{a^2}{1}\cosh^{-1}\dfrac{x}{a}+c$

$t=\tan\dfrac{\theta}{2}$ substitution:

To determine $\int\dfrac{1}{a\cos\theta+b\sin\theta+c}\,d\theta$ let

$$\sin\theta=\dfrac{2t}{(1+t^2)}\quad \cos\theta=\dfrac{1-t^2}{1+t^2}\quad\text{and}$$

$$d\theta=\dfrac{2\,dt}{(1+t^2)}$$

Integration by parts:

If u and v are both functions of x then:

$$\int u\dfrac{dv}{dx}\,dx=uv-\int v\dfrac{du}{dx}\,dx$$

Reduction formulae:

$$\int x^n e^x\,dx=I_n=x^n e^x-nI_{n-1}$$

$$\int x^n\cos x\,dx=I_n=x^n\sin x+nx^{n-1}\cos x$$

$$-n(n-1)I_{n-2}$$

$$\int_0^\pi x^n \cos x \, dx = I_n = -n\pi^{n-1} - n(n-1)I_{n-2}$$

$$\int x^n \sin x \, dx = I_n = -x^n \cos x + nx^{n-1} \sin x$$
$$-n(n-1)I_{n-2}$$

$$\int \sin^n x \, dx = I_n = -\frac{1}{n} \sin^{n-1} x \cos x + \frac{n-1}{n} I_{n-2}$$

$$\int \cos^n x \, dx = I_n = \frac{1}{n} \cos^{n-1} \sin x + \frac{n-1}{n} I_{n-2}$$

$$\int_0^{\pi/2} \sin^n x \, dx = \int_0^{\pi/2} \cos^n x \, dx = I_n = \frac{n-1}{n} I_{n-2}$$

$$\int \tan^n x \, dx = I_n = \frac{\tan^{n-1} x}{n-1} - I_{n-2}$$

$$\int (\ln x)^n \, dx = I_n = x(\ln x)^n - nI_{n-1}$$

With reference to Fig. FA4.

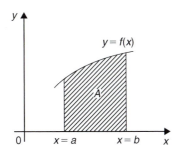

Figure FA4

Area under a curve:

$$\text{area } A = \int_a^b y \, dx$$

Mean value:

$$\text{mean value} = \frac{1}{b-a} \int_a^b y \, dx$$

Rms value:

$$\text{rms value} = \sqrt{\left\{ \frac{1}{b-a} \int_a^b y^2 \, dx \right\}}$$

Volume of solid of revolution:

$$\text{volume} = \int_a^b \pi y^2 \, dx \text{ about the } x\text{-axis}$$

Centroids:

With reference to Fig. FA5:

$$\bar{x} = \frac{\int_a^b xy \, dx}{\int_a^b y \, dx} \quad \text{and} \quad \bar{y} = \frac{\frac{1}{2}\int_a^b y^2 \, dx}{\int_a^b y \, dx}$$

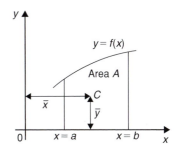

Figure FA5

Theorem of Pappus:

With reference to Fig. FA5, when the curve is rotated one revolution about the x-axis between the limits $x = a$ and $x = b$, the volume V generated is given by: $V = 2\pi A\bar{y}$

Parallel axis theorem:

If C is the centroid of area A in Fig. FA6 then

$$Ak_{BB}^2 = Ak_{GG}^2 + Ad^2 \text{ or } k_{BB}^2 = k_{GG}^2 + d^2$$

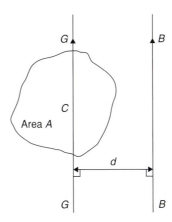

Figure FA6

Second moment of area and radius of gyration:

Shape	Position of axis	Second moment of area, I	Radius of gyration, k
Rectangle length l breadth b	(1) Coinciding with b	$\dfrac{bl^3}{3}$	$\dfrac{1}{\sqrt{3}}$
	(2) Coinciding with l	$\dfrac{lb^3}{3}$	$\dfrac{b}{\sqrt{3}}$
	(3) Through centroid, parallel to b	$\dfrac{bl^3}{12}$	$\dfrac{1}{\sqrt{12}}$
	(4) Through centroid, parallel to l	$\dfrac{lb^3}{12}$	$\dfrac{b}{\sqrt{12}}$
Triangle Perpendicular height h base b	(1) Coinciding with b	$\dfrac{bh^3}{12}$	$\dfrac{h}{\sqrt{6}}$
	(2) Through centroid, parallel to base	$\dfrac{bh^3}{36}$	$\dfrac{h}{\sqrt{18}}$
	(3) Through vertex, parallel to base	$\dfrac{bh^3}{4}$	$\dfrac{h}{\sqrt{2}}$
Circle radius r	(1) Through centre, perpendicular to plane (i.e. polar axis)	$\dfrac{\pi r^4}{2}$	$\dfrac{r}{\sqrt{2}}$
	(2) Coinciding with diameter	$\dfrac{\pi r^4}{4}$	$\dfrac{r}{2}$
	(3) About a tangent	$\dfrac{5\pi r^4}{4}$	$\dfrac{\sqrt{5}}{2}r$
Semicircle radius r	Coinciding with diameter	$\dfrac{\pi r^4}{8}$	$\dfrac{r}{2}$

Perpendicular axis theorem:

If OX and OY lie in the plane of area A in Fig. FA7,

then $Ak_{OZ}^2 = Ak_{OX}^2 + Ak_{OY}^2$ or $k_{OZ}^2 = k_{OX}^2 + k_{OY}^2$

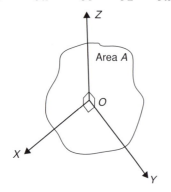

Figure FA7

Numerical integration:

Trapezoidal rule

$$\int y\,\mathrm{d}x \approx \binom{\text{width of}}{\text{interval}}\left[\frac{1}{2}\binom{\text{first} + \text{last}}{\text{ordinates}}\right.$$

$$\left. + \binom{\text{sum of remaining}}{\text{ordinates}}\right]$$

Mid-ordinate rule

$$\int y\,\mathrm{d}x \approx \binom{\text{width of}}{\text{interval}}\binom{\text{sum of}}{\text{mid-ordinates}}$$

Simpson's rule

$$\int y\,\mathrm{d}x \approx \frac{1}{3}\begin{pmatrix}\text{width of}\\\text{interval}\end{pmatrix}\left[\begin{pmatrix}\text{first}+\text{last}\\\text{ordinate}\end{pmatrix}\right.$$
$$+4\begin{pmatrix}\text{sum of even}\\\text{ordinates}\end{pmatrix}$$
$$\left.+2\begin{pmatrix}\text{sum of remaining}\\\text{odd ordinates}\end{pmatrix}\right]$$

Differential equations

First-order differential equations:

Separation of variables

If $\dfrac{\mathrm{d}y}{\mathrm{d}x}=f(x)$ then $y=\displaystyle\int f(x)\,\mathrm{d}x$

If $\dfrac{\mathrm{d}y}{\mathrm{d}x}=f(y)$ then $\displaystyle\int \mathrm{d}x=\int\dfrac{\mathrm{d}y}{f(y)}$

If $\dfrac{\mathrm{d}y}{\mathrm{d}x}=f(x)\cdot f(y)$ then $\displaystyle\int\dfrac{\mathrm{d}y}{f(y)}=\int f(x)\,\mathrm{d}x$

Homogeneous equations:

If $P\dfrac{\mathrm{d}y}{\mathrm{d}x}=Q$, where P and Q are functions of both x and y of the same degree throughout (i.e. a homogeneous first-order differential equation) then:

(i)   rearrange $P\dfrac{\mathrm{d}y}{\mathrm{d}x}=Q$ into the form $\dfrac{\mathrm{d}y}{\mathrm{d}x}=\dfrac{Q}{P}$

(ii) make the substitution $y=vx$ (where v is a function of x), from which, by the product rule,

$$\frac{\mathrm{d}y}{\mathrm{d}x}=v(1)+x\frac{\mathrm{d}v}{\mathrm{d}x}$$

(iii) substitute for both y and $\dfrac{\mathrm{d}y}{\mathrm{d}x}$ in the equation $\dfrac{\mathrm{d}y}{\mathrm{d}x}=\dfrac{Q}{P}$

(iv) simplify, by cancelling, and then separate the variables and solve using the $\dfrac{\mathrm{d}y}{\mathrm{d}x}=f(x)\cdot f(y)$ method

(v)   substitute $v=\dfrac{y}{x}$ to solve in terms of the original variables.

Linear first-order:

If $\dfrac{\mathrm{d}y}{\mathrm{d}x}+Py=Q$, where P and Q are functions of x only (i.e. a linear first-order differential equation), then

(i)   determine the integrating factor, $e^{\int P\,\mathrm{d}x}$

(ii) substitute the integrating factor (I.F.) into the equation

$$y\,(\text{I.F.})=\int (\text{I.F.})\,Q\,\mathrm{d}x$$

(iii) determine the integral $\int(\text{I.F.})Q\,\mathrm{d}x$

Numerical solutions of first-order differential equations:

Euler's method: $y_1=y_0+h(y')_0$

Euler–Cauchy method: $y_{P_1}=y_0+h(y')_0$

and $y_{C_1}=y_0+\dfrac{1}{2}h[(y')_0+f(x_1,y_{P_1})]$

Runge–Kutta method:

To solve the differential equation $\dfrac{\mathrm{d}y}{\mathrm{d}x}=f(x,y)$ given the initial condition $y=y_0$ at $x=x_0$ for a range of values of $x=x_0(h)x_n$:

1.   Identify x_0, y_0 and h, and values of $x_1, x_2, x_3,\dots$

2.   Evaluate $k_1=f(x_n,y_n)$ starting with $n=0$

3.   Evaluate $k_2=f\left(x_n+\dfrac{h}{2},y_n+\dfrac{h}{2}k_1\right)$

4.   Evaluate $k_3=f\left(x_n+\dfrac{h}{2},y_n+\dfrac{h}{2}k_2\right)$

5.   Evaluate $k_4=f(x_n+h,y_n+hk_3)$

6.   Use the values determined from steps 2 to 5 to evaluate:

$$y_{n+1}=y_n+\frac{h}{6}\{k_1+2k_2+2k_3+k_4\}$$

7.   Repeat steps 2 to 6 for $n=1,2,3,\dots$

Second-order differential equations:

If $a\dfrac{\mathrm{d}^2y}{\mathrm{d}x^2}+b\dfrac{\mathrm{d}y}{\mathrm{d}x}+cy=0$ (where a, b and c are constants) then:

(i)   rewrite the differential equation as $(a\mathrm{D}^2+b\mathrm{D}+c)y=0$

(ii) substitute m for D and solve the auxiliary equation $am^2+bm+c=0$

(iii) if the roots of the auxiliary equation are:
  (a) **real and different**, say $m=\alpha$ and $m=\beta$ then the general solution is

  $$y = Ae^{\alpha x} + Be^{\beta x}$$

  (b) **real and equal**, say $m=\alpha$ twice, then the general solution is

  $$y = (Ax + B)e^{\alpha x}$$

  (c) **complex**, say $m=\alpha \pm j\beta$, then the general solution is

  $$y = e^{\alpha x}(A\cos\beta x + B\sin\beta x)$$

(iv) given boundary conditions, constants A and B can be determined and the particular solution obtained.

If $a\dfrac{d^2y}{dx^2} + b\dfrac{dy}{dx} + cy = f(x)$ then:

(i) rewrite the differential equation as
$(aD^2 + bD + c)y = 0$

(ii) substitute m for D and solve the auxiliary equation $am^2 + bm + c = 0$

(iii) obtain the complimentary function (C.F.), u, as per (iii) above.

(iv) to find the particular integral, v, first assume a particular integral which is suggested by $f(x)$, but which contains undetermined coefficients (see Table 54.1, page 568 for guidance).

(v) substitute the suggested particular integral into the original differential equation and equate relevant coefficients to find the constants introduced.

(vi) the general solution is given by $y = u + v$

(vii) given boundary conditions, arbitrary constants in the C.F. can be determined and the particular solution obtained.

Higher derivatives:

y	$y^{(n)}$
e^{ax}	$a^n e^{ax}$
$\sin ax$	$a^n \sin\left(ax + \dfrac{n\pi}{2}\right)$
$\cos ax$	$a^n \cos\left(ax + \dfrac{n\pi}{2}\right)$
x^a	$\dfrac{a!}{(a-n)!}x^{a-n}$
$\sinh ax$	$\dfrac{a^n}{2}\left\{[1+(-1)^n]\sinh ax + [1-(-1)^n]\cosh ax\right\}$
$\cosh ax$	$\dfrac{a^n}{2}\left\{[1-(-1)^n]\sinh ax + [1+(-1)^n]\cosh ax\right\}$
$\ln ax$	$(-1)^{n-1}\dfrac{(n-1)!}{x^n}$

Leibniz's theorem:

To find the nth derivative of a product $y = uv$:

$$y^{(n)} = (uv)^{(n)} = u^{(n)}v + nu^{(n-1)}v^{(1)}$$
$$+ \frac{n(n-1)}{2!}u^{(n-2)}v^{(2)}$$
$$+ \frac{n(n-1)(n-2)}{3!}u^{(n-3)}v^{(3)} + \cdots$$

Power series solutions of second-order differential equations:

(a) **Leibniz–Maclaurin method**
  (i) Differentiate the given equation n times, using the Leibniz theorem

  (ii) rearrange the result to obtain the recurrence relation at $x = 0$

  (iii) determine the values of the derivatives at $x = 0$, i.e. find $(y)_0$ and $(y')_0$

  (iv) substitute in the Maclaurin expansion for $y = f(x)$

  (v) simplify the result where possible and apply boundary condition (if given).

(b) **Frobenius method**
  (i) Assume a trial solution of the form:
  $y = x^c\{a_0 + a_1x + a_2x^2 + a_3x^3 + \cdots + a_rx^r + \cdots\}$ $a_0 \neq 0$

(ii) differentiate the trial series to find y' and y''

(iii) substitute the results in the given differential equation

(iv) equate coefficients of corresponding powers of the variable on each side of the equation: this enables index c and coefficients $a_1, a_2, a_3, \ldots$ from the trial solution, to be determined.

Bessel's equation:

The solution of $x^2 \dfrac{d^2 y}{dx^2} + x \dfrac{dy}{dx} + (x^2 - v^2)y = 0$ is:

$$y = Ax^v \left\{ 1 - \frac{x^2}{2^2(v+1)} \right.$$

$$+ \frac{x^4}{2^4 \times 2!\,(v+1)(v+2)}$$

$$\left. - \frac{x^6}{2^6 \times 3!\,(v+1)(v+2)(v+3)} + \cdots \right\}$$

$$+ Bx^{-v} \left\{ 1 + \frac{x^2}{2^2(v-1)} + \frac{x^4}{2^4 \times 2!\,(v-1)(v-2)} \right.$$

$$\left. + \frac{x^6}{2^6 \times 3!\,(v-1)(v-2)(v-3)} + \cdots \right\}$$

or, in terms of **Bessel functions** and **gamma functions**:

$$y = A J_v(x) + B J_{-v}(x)$$

$$= A \left(\frac{x}{2}\right)^v \left\{ \frac{1}{\Gamma(v+1)} - \frac{x^2}{2^2(1!)\Gamma(v+2)} \right.$$

$$\left. + \frac{x^4}{2^4(2!)\Gamma(v+4)} - \cdots \right\}$$

$$+ B \left(\frac{x}{2}\right)^{-v} \left\{ \frac{1}{\Gamma(1-v)} - \frac{x^2}{2^2(1!)\Gamma(2-v)} \right.$$

$$\left. + \frac{x^4}{2^4(2!)\Gamma(3-v)} - \cdots \right\}$$

In general terms:

$$J_v(x) = \left(\frac{x}{2}\right)^v \sum_{k=0}^{\infty} \frac{(-1)^k x^{2k}}{2^{2k}(k!)\Gamma(v+k+1)}$$

and $J_{-v}(x) = \left(\frac{x}{2}\right)^{-v} \sum_{k=0}^{\infty} \frac{(-1)^k x^{2k}}{2^{2k}(k!)\Gamma(k-v+1)}$

and in particular:

$$J_n(x) = \left(\frac{x}{2}\right)^n \left\{ \frac{1}{n!} - \frac{1}{(n+1)!} \left(\frac{x}{2}\right)^2 \right.$$

$$\left. + \frac{1}{(2!)(n+2)!} \left(\frac{x}{2}\right)^4 - \cdots \right\}$$

$$J_0(x) = 1 - \frac{x^2}{2^2(1!)^2} + \frac{x^4}{2^4(2!)^2}$$

$$- \frac{x^6}{2^6(3!)^2} + \cdots$$

and $J_1(x) = \dfrac{x}{2} - \dfrac{x^3}{2^3(1!)(2!)} + \dfrac{x^5}{2^5(2!)(3!)}$

$$- \frac{x^7}{2^7(3!)(4!)} + \cdots$$

Legendre's equation:

The solution of $(1-x^2)\dfrac{d^2 y}{dx^2} - 2x \dfrac{dy}{dx} + k(k+1)y = 0$

is:

$$y = a_0 \left\{ 1 - \frac{k(k+1)}{2!}x^2 \right.$$

$$\left. + \frac{k(k+1)(k-2)(k+3)}{4!}x^4 - \cdots \right\}$$

$$+ a_1 \left\{ x - \frac{(k-1)(k+2)}{3!}x^3 \right.$$

$$\left. + \frac{(k-1)(k-3)(k+2)(k+4)}{5!}x^5 - \cdots \right\}$$

Rodrigues' formula:

$$P_n(x) = \frac{1}{2^n n!} \frac{d^n (x^2-1)^n}{dx^n}$$

Statistics and probability

Mean, median, mode and standard deviation:

If $x =$ variate and $f =$ frequency then:

$$\textbf{mean } \bar{x} = \frac{\sum f x}{\sum f}$$

The **median** is the middle term of a ranked set of data.

The **mode** is the most commonly occurring value in a set of data.

Standard deviation:

$$\sigma = \sqrt{\left[\frac{\sum\{f(x - \bar{x})^2\}}{\sum f}\right]} \quad \text{for a population}$$

Bayes' theorem:

$$P(A_1|B) = \frac{P(B|A_1)P(A_1)}{P(B|A_1)P(A_1) + P(B|A_2)P(A_2) + ...}$$

$$\text{or } P(A_i|B) = \frac{P(B|A_i)P(A_i)}{\sum_{j=1}^{n} P(B|A_j)P(A_j)} (i = 1, 2, ..., n)$$

Binomial probability distribution:

If n = number in sample, p = probability of the occurrence of an event and $q = 1 - p$, then the probability of $0, 1, 2, 3, ...$ occurrences is given by:

$$q^n, \quad nq^{n-1}p, \quad \frac{n(n-1)}{2!}q^{n-2}p^2,$$

$$\frac{n(n-1)(n-2)}{3!}q^{n-3}p^3, ...$$

(i.e. successive terms of the $(q + p)^n$ expansion).

Normal approximation to a binomial distribution:
Mean $= np$ Standard deviation $\sigma = \sqrt{(npq)}$

Poisson distribution:

If λ is the expectation of the occurrence of an event then the probability of $0, 1, 2, 3, ...$ occurrences is given by:

$$e^{-\lambda}, \quad \lambda e^{-\lambda}, \quad \lambda^2 \frac{e^{-\lambda}}{2!}, \quad \lambda^3 \frac{e^{-\lambda}}{3!}, ...$$

Product-moment formula for the linear correlation coefficient:

Coefficient of correlation $r = \dfrac{\sum xy}{\sqrt{[(\sum x^2)(\sum y^2)]}}$

where $x = X - \overline{X}$ and $y = Y - \overline{Y}$ and (X_1, Y_1), $(X_2, Y_2), ...$ denote a random sample from a bivariate normal distribution and $\overline{X}$ and $\overline{Y}$ are the means of the X and Y values respectively.

Normal probability distribution:

Partial areas under the standardized normal curve — see Table 61.1 on page 661.

Student's t distribution:

Percentile values (t_p) for Student's t distribution with v degrees of freedom – see Table 64.2, page 689, on the website.

Chi-square distribution:

Percentile values (χ_p^2) for the Chi-square distribution with v degrees of freedom–see Table 66.1, page 711, on the website.

$$\chi^2 = \sum\left\{\frac{(o - e)^2}{e}\right\} \quad \text{where } o \text{ and } e \text{ are the observed and}$$

expected frequencies.

Symbols:

Population
Number of members N_p, mean μ, standard deviation σ

Sample
Number of members N, mean $\bar{x}$, standard deviation s

Sampling distributions
Mean of sampling distribution of means $\mu_{\bar{x}}$
Standard error of means $\sigma_{\bar{x}}$
Standard error of the standard deviations σ_s

Standard error of the means:

Standard error of the means of a sample distribution, i.e. the standard deviation of the means of samples, is:

$$\sigma_{\bar{x}} = \frac{\sigma}{\sqrt{N}}\sqrt{\left(\frac{N_p - N}{N_p - 1}\right)}$$

for a finite population and/or for sampling without replacement, and

$$\sigma_{\bar{x}} = \frac{\sigma}{\sqrt{N}}$$

for an infinite population and/or for sampling with replacement.

The relationship between sample mean and population mean:

$\mu_{\bar{x}} = \mu$ for all possible samples of size N are drawn from a population of size N_p

Estimating the mean of a population (σ known):

The confidence coefficient for a large sample size, $(N \geq 30)$ is z_c where:

Confidence level %	Confidence coefficient z_c
99	2.58
98	2.33
96	2.05
95	1.96
90	1.645
80	1.28
50	0.6745

The confidence limits of a population mean based on sample data are given by:

$$\bar{x} \pm \frac{z_c \sigma}{\sqrt{N}} \sqrt{\left(\frac{N_p - N}{N_p - 1}\right)}$$

for a finite population of size N_p, and by

$$\bar{x} \pm \frac{z_c \sigma}{\sqrt{N}} \text{ for an infinite population}$$

Estimating the mean of a population (σ unknown):

The confidence limits of a population mean based on sample data are given by: $\mu_{\bar{x}} \pm z_c \sigma_{\bar{x}}$

Estimating the standard deviation of a population:

The confidence limits of the standard deviation of a population based on sample data are given by:

$$s \pm z_c \sigma_s$$

Estimating the mean of a population based on a small sample size:

The confidence coefficient for a small sample size $(N < 30)$ is t_c which can be determined using Table 64.1, page 683, on the website. The confidence limits of a population mean based on sample data is given by:

$$\bar{x} \pm \frac{t_c s}{\sqrt{(N-1)}}$$

Laplace transforms

Function $f(t)$	Laplace transforms $\mathcal{L}\{f(t)\} = \int_0^\infty e^{-st} f(t)\,dt$
1	$\dfrac{1}{s}$
k	$\dfrac{k}{s}$
e^{at}	$\dfrac{1}{s-a}$
$\sin at$	$\dfrac{a}{s^2 + a^2}$
$\cos at$	$\dfrac{s}{s^2 + a^2}$
t	$\dfrac{1}{s^2}$
t^n ($n = $ positve integer)	$\dfrac{n!}{s^{n+1}}$
$\cosh at$	$\dfrac{s}{s^2 - a^2}$
$\sinh at$	$\dfrac{a}{s^2 - a^2}$
$e^{-at} t^n$	$\dfrac{n!}{(s+a)^{n+1}}$
$e^{-at} \sin \omega t$	$\dfrac{\omega}{(s+a)^2 + \omega^2}$
$e^{-at} \cos \omega t$	$\dfrac{s+a}{(s+a)^2 + \omega^2}$
$e^{-at} \cosh \omega t$	$\dfrac{s+a}{(s+a)^2 - \omega^2}$
$e^{-at} \sinh \omega t$	$\dfrac{\omega}{(s+a)^2 - \omega^2}$

The Laplace transforms of derivatives:

First derivative

$$\mathcal{L}\left\{\frac{dy}{dx}\right\} = s\mathcal{L}\{y\} - y(0)$$

where $y(0)$ is the value of y at $x = 0$

Second derivative

$$\mathcal{L}\left\{\frac{dy}{dx}\right\} = s^2\mathcal{L}\{y\} - sy(0) - y'(0)$$

where $y'(0)$ is the value of $\dfrac{dy}{dx}$ at $x = 0$

Fourier series

If $f(x)$ is a periodic function of period 2π then its Fourier series is given by:

$$f(x) = a_0 + \sum_{n=1}^{\infty}(a_n\cos nx + b_n\sin nx)$$

where, for the range $-\pi$ to $+\pi$:

$$a_0 = \frac{1}{2\pi}\int_{-\pi}^{\pi} f(x)\,dx$$

$$a_n = \frac{1}{\pi}\int_{-\pi}^{\pi} f(x)\cos nx\,dx \quad (n = 1, 2, 3, \ldots)$$

$$b_n = \frac{1}{\pi}\int_{-\pi}^{\pi} f(x)\sin nx\,dx \quad (n = 1, 2, 3, \ldots)$$

If $f(x)$ is a periodic function of period L then its Fourier series is given by:

$$f(x) = a_0 + \sum_{n=1}^{\infty}\left\{a_n\cos\left(\frac{2\pi nx}{L}\right) + b_n\sin\left(\frac{2\pi nx}{L}\right)\right\}$$

where for the range $-\dfrac{L}{2}$ to $+\dfrac{L}{2}$:

$$a_0 = \frac{1}{L}\int_{-L/2}^{L/2} f(x)\,dx$$

$$a_n = \frac{2}{L}\int_{-L/2}^{L/2} f(x)\cos\left(\frac{2\pi nx}{L}\right)dx \quad (n = 1, 2, 3, \ldots)$$

$$b_n = \frac{2}{L}\int_{-L/2}^{L/2} f(x)\sin\left(\frac{2\pi nx}{L}\right)dx \quad (n = 1, 2, 3, \ldots)$$

Complex or exponential Fourier series:

$$f(x) = \sum_{n=-\infty}^{\infty} c_n e^{j\frac{2\pi nx}{L}}$$

where $c_n = \dfrac{1}{L}\displaystyle\int_{-\frac{L}{2}}^{\frac{L}{2}} f(x)e^{-j\frac{2\pi nx}{L}}\,dx$

For even symmetry,

$$c_n = \frac{2}{L}\int_0^{\frac{L}{2}} f(x)\cos\left(\frac{2\pi nx}{L}\right)dx$$

For odd symmetry,

$$c_n = -j\frac{2}{L}\int_0^{\frac{L}{2}} f(x)\sin\left(\frac{2\pi nx}{L}\right)dx$$

Sequence	Transform F(z)
1. $\{\delta_k\} = \{1,\ 0,\ 0,\ ...\}$	1 for all values of z
2. $\{u_k\} = \{1,\ 1,\ 1,\ ...\}$	$\dfrac{z}{z-1}$ for $\lvert z \rvert > 1$
3. $\{k\} = \{0,\ 1,\ 2,\ 3,\ ...\}$	$\dfrac{z}{(z-1)^2}$ for $\lvert z \rvert > 1$
4. $\{k^2\} = \{0,\ 1,\ 4,\ 9,\ ...\}$	$\dfrac{z(z+1)}{(z-1)^3}$ for $\lvert z \rvert > 1$
5. $\{k^3\} = \{0,\ 1,\ 8,\ 27,\ ...\}$	$\dfrac{z(z^2+4z+1)}{(z-1)^4}$ for $\lvert z \rvert > 1$
6. $\{a^k\} = \{1,\ a,\ a^2,\ a^3,\ ...\}$	$\dfrac{z}{z-a}$ for $\lvert z \rvert > \lvert a \rvert$
7. $\{ka^k\} = \{0,\ a,\ 2a^2,\ 3a^3,\ ...\}$	$\dfrac{az}{(z-a)^2}$ for $\lvert z \rvert > \lvert a \rvert$
8. $\{k^2 a^k\} = \{0,\ a,\ 4a^2,\ 9a^3,\ ...\}$	$\dfrac{az(z+a)}{(z-a)^3}$ for $\lvert z \rvert > \lvert a \rvert$
9. $\{e^{-ak}\} = \{e^{-a},\ e^{-2a},\ e^{-3a},\ ...\}$	$\dfrac{z}{z-e^{-a}}$
10. $\sin ak = \{\sin a,\ \sin 2a,\ ...\}$	$\dfrac{z \sin a}{z^2 - 2z \cos a + 1}$
11. $\cos ak = \{\cos a,\ \cos 2a,\ ...\}$	$\dfrac{z(z - \cos a)}{z^2 - 2z \cos a + 1}$
12. $e^{-ak} \sin bk = \{e^{-a} \sin b,\ e^{-2a} \sin 2b,\ ...\}$	$\dfrac{ze^{-a} \sin b}{z^2 - 2ze^{-a} \cos b + e^{-2a}}$
13. $e^{-ak} \cos bk = \{e^{-a} \cos b,\ e^{-2a} \cos 2b,\ ...\}$	$\dfrac{z^2 - ze^{-a} \cos b}{z^2 - 2ze^{-a} \cos b + e^{-2a}}$

These formulae are available for downloading at the website:
www.routledge.com/cw/bird

COMPANION @ WEBSITE

Answers

Answers to Practice Exercises

Chapter 1

Exercise 1 (page 4)

1. -16
2. -8
3. $3x - 5y + 5z$
4. $6a^2 - 13ab + 3ac - 5b^2 + bc$
5. $x^5y^4z^3$, $13\frac{1}{2}$
6. $\pm 4\frac{1}{2}$
7. $\dfrac{1+a}{b}$
8. $a^{\frac{11}{6}} b^{\frac{1}{3}} c^{-\frac{3}{2}}$ or $\dfrac{\sqrt[6]{a^{11}}\sqrt[3]{b}}{\sqrt{c^3}}$

Exercise 2 (page 5)

1. $-5p + 10q - 6r$
2. $x^2 - xy - 2y^2$
3. $11q - 2p$
4. $7ab(3ab - 4)$
5. $2xy(y + 3x + 4x^2)$
6. $\dfrac{2}{3y} + 12 - 3y$
7. $\dfrac{5}{y} - 1$
8. ab

Exercise 3 (page 6)

1. $\dfrac{1}{2}$
2. -3
3. $-\dfrac{1}{8}$
4. 4
5. $f = \dfrac{3F - yL}{3}$ or $f = F - \dfrac{yL}{3}$
6. $\ell = \dfrac{gt^2}{4\pi^2}$
7. $L = \dfrac{mrCR}{\mu - m}$
8. $r = \sqrt{\left(\dfrac{x-y}{x+y}\right)}$

Exercise 4 (page 8)

1. $x = 6, y = -1$
2. $a = 2, b = -3$
3. $x = 3, y = 4$
4. (a) $4, -8$ (b) $\dfrac{5}{4}, -\dfrac{3}{2}$
5. $x^2 + 3x - 10 = 0$
6. (a) $0.637, -3.137$ (b) $2.443, 0.307$

Exercise 5 (page 10)

1. $2x - y$
2. $3x - 1$
3. $5x - 2$
4. $7x + 1$
5. $x^2 + 2xy + y^2$
6. $5x + 4 + \dfrac{8}{x - 1}$
7. $3x^2 - 4x + 3 - \dfrac{2}{x + 2}$
8. $5x^3 + 18x^2 + 54x + 160 + \dfrac{481}{x - 3}$

Exercise 6 (page 12)

1. $(x - 1)(x + 3)$
2. $(x + 1)(x + 2)(x - 2)$
3. $(x + 1)\left(2x^2 + 3x - 7\right)$
4. $(x - 1)(x + 3)(2x - 5)$
5. $x^3 + 4x^2 + x - 6 = (x - 1)(x + 2)(x + 3)$
   $x = 1, x = -2$ and $x = -3$
6. $x = 1, x = 2$ and $x = -1$

Exercise 7 (page 13)

1. (a) 6 (b) 9
2. (a) -39 (b) -29
3. $(x - 1)(x - 2)(x - 3)$
4. $x = -1, x = -2$ and $x = -4$
5. $a = -3$
6. $x = 1, x = -2$ and $x = 1.5$

Chapter 2

Exercise 8 (page 18)

1. $\dfrac{2}{(x - 3)} - \dfrac{2}{(x + 3)}$
2. $\dfrac{5}{(x + 1)} - \dfrac{1}{(x - 3)}$
3. $\dfrac{3}{x} + \dfrac{2}{(x - 2)} - \dfrac{4}{(x - 1)}$
4. $\dfrac{7}{(x + 4)} - \dfrac{3}{(x + 1)} - \dfrac{2}{(2x - 1)}$

5. $1 + \dfrac{2}{(x+3)} + \dfrac{6}{(x-2)}$ **6.** $1 + \dfrac{3}{(x+1)} - \dfrac{2}{(x-3)}$

7. $3x - 2 + \dfrac{1}{(x-2)} - \dfrac{5}{(x+2)}$

Exercise 9 (page 20)

1. $\dfrac{4}{(x+1)} - \dfrac{7}{(x+1)^2}$ **2.** $\dfrac{2}{x} + \dfrac{1}{x^2} - \dfrac{1}{(x+3)}$

3. $\dfrac{5}{(x-2)} - \dfrac{10}{(x-2)^2} + \dfrac{4}{(x-2)^3}$

4. $\dfrac{2}{(x-5)} - \dfrac{3}{(x+2)} + \dfrac{4}{(x+2)^2}$

Exercise 10 (page 21)

1. $\dfrac{2x+3}{(x^2+7)} - \dfrac{1}{(x-2)}$ **2.** $\dfrac{1}{(x-4)} + \dfrac{2-x}{(x^2+3)}$

3. $\dfrac{1}{x} + \dfrac{3}{x^2} + \dfrac{2-5x}{(x^2+5)}$ **4.** $\dfrac{3}{(x-1)} + \dfrac{2}{(x-1)^2} + \dfrac{1-2x}{(x^2+8)}$

5. Proof

Chapter 3

Exercise 11 (page 24)

1. 4   **2.** 4   **3.** 3   **4.** -3

5. $\dfrac{1}{3}$ **6.** 3   **7.** 2   **8.** -2

9. $1\dfrac{1}{2}$ **10.** $\dfrac{1}{3}$ **11.** 2   **12.** 10,000

13. 100,000   **14.** 9   **15.** $\dfrac{1}{32}$ **16.** 0.01

17. $\dfrac{1}{16}$ **18.** e^3

Exercise 12 (page 26)

1. $\log 6$ **2.** $\log 15$ **3.** $\log 2$

4. $\log 3$ **5.** $\log 12$ **6.** $\log 500$

7. $\log 100$ **8.** $\log 6$ **9.** $\log 10$

10. $\log 1 = 0$ **11.** $\log 2$

12. $\log 243$ or $\log 3^5$ or $5\log 3$

13. $\log 16$ or $\log 2^4$ or $4\log 2$

14. $\log 64$ or $\log 2^6$ or $6\log 2$

15. 0.5   **16.** 1.5   **17.** $x = 2.5$

18. $t = 8$ **19.** $b = 2$ **20.** $x = 2$

21. $a = 6$ **22.** $x = 5$

Exercise 13 (page 27)

1. 1.690   **2.** 3.170   **3.** 0.2696   **4.** 6.058

5. 2.251   **6.** 3.959   **7.** 2.542   **8.** -0.3272

9. 316.2

Chapter 4

Exercise 14 (page 30)

1. (a) 0.1653   (b) 0.4584   (c) 22030

2. (a) 5.0988   (b) 0.064037   (c) 40.446

3. (a) 4.55848   (b) 2.40444   (c) 8.05124

4. (a) 48.04106   (b) 4.07482   (c) -0.08286

5. 2.739

6. 120.7 m

Exercise 15 (page 32)

1. 2.0601   **2.** (a) 7.389   (b) 0.7408

3. $1 - 2x^2 - \dfrac{8}{3}x^3 - 2x^4$

4. $2x^{1/2} + 2x^{5/2} + x^{9/2} + \dfrac{1}{3}x^{13/2} + \dfrac{1}{12}x^{17/2} + \dfrac{1}{60}x^{21/2}$

Exercise 16 (page 33)

1. 3.95, 2.05

2. 1.65, -1.30

3. (a) 28 cm^3 (b) 116 min

4. (a) 70°C   (b) 5 minutes

Exercise 17 (page 36)

1. (a) 0.55547   (b) 0.91374   (c) 8.8941

2. (a) 2.2293   (b) -0.33154 (c) 0.13087

3. 8.166   **4.** 1.522   **5.** 1.485

6. -0.4904 **7.** -0.5822 **8.** 2.197

9. 816.2   **10.** 0.8274   **11.** 1.962

12. 3   **13.** 4   **14.** 147.9

15. 4.901   **16.** 3.095

17. $t = e^{b+a\ln D} = e^b e^{a\ln D} = e^b e^{\ln D^a}$ i.e. $t = e^b D^a$

18. 500

19. $W = PV\ln\left(\dfrac{U_2}{U_1}\right)$

20. $p_2 = 348.5$ Pa

21. 992 m/s

Exercise 18 (page 39)

1. (a) $150°C$ (b) $100.5°C$
2. 99.21 kPa
3. (a) 29.32 volts   (b) 71.31×10^{-6} s
4. (a) 2.038×10^{-4} (b) 2.293 m
5. (a) $50°C$ (b) 55.45 s
6. 30.4 N, 0.807 rad
7. (a) 3.04 A   (b) 1.46 s
8. 2.45 mol/cm^3
9. (a) 7.07 A   (b) 0.966 s
10. (a) 100% (b) 67.03% (c) 1.83%
11. 2.45 mA
12. 142 ms
13. 99.752%
14. 20 min 38 s

Exercise 19 (page 42)

1. $a = 76, k = -7 \times 10^{-5}, p = 76e^{-7 \times 10^{-5}h}, 37.74$ cm
2. $\theta_0 = 152, k = -0.05$

Chapter 5

Exercise 20 (page 44)

1. (a) $t > 2$ (b) $x < 5$ 2. (a) $x > 3$ (b) $x \geq 3$
3. (a) $t \leq 1$ (b) $x \leq 6$ 4. (a) $k \geq \dfrac{3}{2}$ (b) $z > \dfrac{1}{2}$
5. (a) $y \geq -\dfrac{4}{3}$ (b) $x \geq -\dfrac{1}{2}$

Exercise 21 (page 45)

1. $-5 < t < 3$ 2. $-5 \leq y \leq -1$
3. $-\dfrac{3}{2} < x < \dfrac{5}{2}$ 4. $t > 3$ and $t < \dfrac{1}{3}$
5. $k \leq -2$ and $k \geq 4$

Exercise 22 (page 46)

1. $-4 \leq x \leq 3$ 2. $t > 5$ or $t < -9$
3. $-5 < z \leq 14$ 4. $-3 < x \leq -2$

Exercise 23 (page 47)

1. $z > 4$ or $z < -4$ 2. $-4 < z < 4$
3. $x \geq \sqrt{3}$ or $x \leq -\sqrt{3}$ 4. $-2 \leq k \leq 2$
5. $-5 \leq t \leq 7$ 6. $t \geq 7$ or $t \leq -5$
7. $y \geq 2$ or $y \leq -2$ 8. $k > -\dfrac{1}{2}$ or $k < -2$

Exercise 24 (page 48)

1. $x > 3$ or $x < -2$ 2. $-4 \leq t \leq 2$
3. $-2 < x < \dfrac{1}{2}$ 4. $y \geq 5$ or $y < -4$
5. $-4 \leq z \leq 0$
6. $\left(-\sqrt{3} - 3\right) < x < \left(\sqrt{3} - 3\right)$
7. $t \geq \left(\sqrt{2} + 11\right)$ or $t \leq \left(2 - \sqrt{11}\right)$
8. $k \geq \dfrac{1}{2}\left(1 + \sqrt{13}\right)$ or $k \leq \dfrac{1}{2}\left(1 - \sqrt{13}\right)$

Chapter 6

Exercise 25 (page 52)

1. 68 2. 6.2 3. 85.25 4. 23.5
5. 11th   6. 209 7. 346.5

Exercise 26 (page 53)

1. -0.5 2. $1.5, 3, 4.5$ 3. 7808 4. 25
5. $8.5, 12, 15.5, 19$
6. (a) 120 (b) 26070 (c) 250.5
7. £19840, £223680
8. £29820

Exercise 27 (page 55)

1. 2560 2. 273.25 3. $512, 4096$
4. 812.5 5. 8 6. $1\dfrac{2}{3}$

Exercise 28 (page 56)

1. (a) 3 (b) 2 (c) 59022 2. 10th
3. £1566, 11 years   4. 48.71 M
5. 71.53 g
6. (a) £599.14 (b) 19 years
7. $100, 139, 193, 268, 373, 518, 720, 1000$ rev/min

Chapter 7

Exercise 29 (page 60)

1. $x^7 - 7x^6y + 21x^5y^2 - 35x^4y^3 + 35x^3y^4$
$$-21x^2y^5 + 7xy^6 - y^7$$
2. $32a^5 + 240a^4b + 720a^3b^2 + 1080a^2b^3 + 810ab^4$
$$+243b^5$$

Exercise 30 (page 62)

1. $a^4 + 8a^3x + 24a^2x^2 + 32ax^3 + 16x^4$

2. $64 - 192x + 240x^2 - 160x^3 + 60x^4 - 12x^5 + x^6$

3. $16x^4 - 96x^3y + 216x^2y^2 - 216xy^3 + 81y^4$

4. $32x^5 + 160x^3 + 320x + \dfrac{320}{x} + \dfrac{160}{x^3} + \dfrac{32}{x^5}$

5. $p^{11} + 22p^{10}q + 210p^9q^2 + 1320p^8q^3 + 5280p^7q^4$

6. $34\,749p^8q^5$

7. $700\,000a^4b^4$

8. (a) 1.0243 (b) 1.3337

9. (a) 0.86813 (b) 535.51

10. 4373.88

Exercise 31 (page 64)

1. $1 + x + x^2 + x^3 + \dots,\ |x| < 1$

2. $1 - 2x + 3x^2 - 4x^3 + \dots,\ |x| < 1$

3. $\dfrac{1}{8}\left[1 - \dfrac{3}{2}x + \dfrac{3}{2}x^2 - \dfrac{5}{4}x^3 + \dots\right],\ |x| < 2$

4. $\sqrt{2}\left(1 + \dfrac{x}{4} - \dfrac{x^2}{32} + \dfrac{x^3}{128} - \dots\right),\ |x| < 2$
   or $-2 < x < 2$

5. $1 - \dfrac{3}{2}x + \dfrac{27}{8}x^2 - \dfrac{135}{16}x^3,\ |x| < \dfrac{1}{3}$

6. $\dfrac{1}{64}\left[1 - 9x + \dfrac{189}{4}x^2 + \dots\right],\ |x| < \dfrac{2}{3}$

7. Proofs

8. $4 - \dfrac{31}{15}x$

9. (a) $1 - x + \dfrac{x^2}{2},\ |x| < 1$ (b) $1 - x - \dfrac{7}{2}x^2,\ |x| < \dfrac{1}{3}$

Exercise 32 (page 66)

1. 0.6% decrease
2. 3.5% decrease
3. (a) 4.5% increase (b) 3.0% increase
4. 2.2% increase
5. 4.5% increase
6. Proof
7. 7.5% decrease
8. 2.5% increase
9. 0.9% too small
10. +7%
11. Proof
12. 5.5%

Chapter 8

Exercise 33 (page 73)

1. $\sin 2x = 2x - \dfrac{4}{3}x^3 + \dfrac{4}{15}x^5 - \dfrac{8}{315}x^7$

2. $\cosh 3x = 1 + \dfrac{9}{2}x^2 + \dfrac{27}{8}x^4 + \dfrac{81}{80}x^6$

3. $\ln 2 + \dfrac{x}{2} + \dfrac{x^2}{8}$

4. $1 - 8t^2 + \dfrac{32}{3}t^4 - \dfrac{256}{45}t^6$

5. $1 + \dfrac{3}{2}x^2 + \dfrac{9}{8}x^4 + \dfrac{9}{16}x^6$

6. $1 + 2x^2 + \dfrac{10}{3}x^4$

7. $1 + 2\theta - \dfrac{5}{2}\theta^2$

8. $x^2 - \dfrac{1}{3}x^4 + \dfrac{2}{45}x^6$

9. $81 + 216t + 216t^2 + 96t^3 + 16t^4$

Exercise 34 (page 75)

1. 1.784   2. 0.88   3. 0.53   4. 0.061

Exercise 35 (page 77)

1. $\dfrac{1}{9}$ 2. 1   3. 1   4. -1 5. $\dfrac{1}{3}$

6. $\dfrac{1}{2}$ 7. $\dfrac{1}{3}$ 8. 1   9. $\dfrac{1}{2}$

Chapter 9

Exercise 36 (page 83)

1. 1.19   2. 1.146   3. 1.20   4. 3.146   5. 1.849

Exercise 37 (page 86)

1. $-3.36, 1.69$ 2. -2.686
3. $-1.53, 1.68$ 4. $-12.01, 1.000$

Exercise 38 (page 88)

1. $-2.742, 4.742$ 2. 2.313
3. $-1.721, 2.648$ 4. $-1.386, 1.491$
5. 1.147   6. $-1.693, -0.846, 0.744$
7. 2.05   8. See above answers
9. 0.0399   10. 4.19
11. 2.9143

Chapter 10

Exercise 39 (page 92)

1. (a) 6_{10} (b) 11_{10} (c) 14_{10} (d) 9_{10}
2. (a) 21_{10} (b) 25_{10} (c) 45_{10} (d) 51_{10}
3. (a) 42_{10} (b) 56_{10} (c) 65_{10} (d) 184_{10}
4. (a) 0.8125_{10} (b) 0.78125_{10} (c) 0.21875_{10}
   (d) 0.34375_{10}

5. (a) 26.75_{10} (b) 23.375_{10} (c) 53.4375_{10}
   (d) 213.71875_{10}

Exercise 40 (page 93)

1. (a) 101_2 (b) 1111_2 (c) 10011_2 (d) 11101_2

2. (a) 11111_2 (b) 101010_2 (c) 111001_2
   (d) 111111_2

3. (a) 101111_2 (b) 111100_2 (c) 1001001_2
   (d) 1010100_2

4. (a) 0.01_2 (b) 0.00111_2 (c) 0.01001_2
   (d) 0.10011_2

5. (a) 101111.01101_2 (b) 11110.1101_2
   (c) 110101.11101_2
   (d) 111101.10101_2

Exercise 41 (page 94)

1. 101   **2.** 1011   **3.** 10100

4. 101100   **5.** 1001000   **6.** 100001010

7. 1010110111   **8.** 1001111101   **9.** 111100

10. 110111   **11.** 110011   **12.** 1101110

Exercise 42 (page 95)

1. (a) 1010101111_2 (b) 1000111100_2
   (c) 10011110001_2

2. (a) 0.01111_2 (b) 0.1011_2 (c) 0.10111_2

3. (a) 11110111.00011_2 (b) 1000000010.0111_2
   (c) 11010110100.11001_2

4. (a) 7.4375_{10} (b) 41.25_{10} (c) 7386.1875_{10}

Exercise 43 (page 98)

1. 231_{10} **2.** 44_{10} **3.** 152_{10} **4.** 753_{10}

5. 36_{16} **6.** $C8_{16}$ **7.** $5B_{16}$ **8.** EE_{16}

Exercise 44 (page 99)

1. $D7_{16}$ **2.** EA_{16}

3. $8B_{16}$ **4.** $A5_{16}$

5. 110111_2 **6.** 11101101_2

7. 100111111_2 **8.** 101000100001_2

Chapter 11

Exercise 45 (page 104)

1. $Z = C \cdot (A \cdot B + \overline{A} \cdot B)$

A	B	C	$A \cdot B$	$\overline{A}$	$\overline{A} \cdot B$	$A \cdot B + \overline{A} \cdot B$	$Z = C \cdot (A \cdot B + \overline{A} \cdot B)$
0	0	0	0	1	0	0	0
0	0	1	0	1	0	0	0
0	1	0	0	1	1	1	0
0	1	1	0	1	1	1	1
1	0	0	0	0	0	0	0
1	0	1	0	0	0	0	0
1	1	0	1	0	0	1	0
1	1	1	1	0	0	1	1

2. $Z = C \cdot (A \cdot \overline{B} + \overline{A})$

A	B	C	$\overline{A}$	$\overline{B}$	$A \cdot \overline{B}$	$A \cdot \overline{B} + \overline{A}$	$Z = C \cdot (A \cdot \overline{B} + \overline{A})$
0	0	0	1	1	0	1	0
0	0	1	1	1	0	1	1
0	1	0	1	0	0	1	0
0	1	1	1	0	0	1	1
1	0	0	0	1	1	1	0
1	0	1	0	1	1	1	1
1	1	0	0	0	0	0	0
1	1	1	0	0	0	0	0

3. $Z = A \cdot B \cdot (B \cdot \overline{C} + \overline{B} \cdot C + \overline{A} \cdot B)$

A	B	C	$\overline{A}$	$\overline{B}$	$\overline{C}$	$B \cdot \overline{C}$	$\overline{B} \cdot C$	$\overline{A} \cdot B$	$(B \cdot \overline{C} + \overline{B} \cdot C + \overline{A} \cdot B)$	$Z = A \cdot B \cdot (B \cdot \overline{C} + \overline{B} \cdot C + \overline{A} \cdot B)$
0	0	0	1	1	1	0	0	0	0	0
0	0	1	1	1	0	0	1	0	1	0
0	1	0	1	0	1	1	0	1	1	0
0	1	1	1	0	0	0	0	1	1	0
1	0	0	0	1	1	0	0	0	0	0
1	0	1	0	1	0	0	1	0	1	0
1	1	0	0	0	1	1	0	0	1	1
1	1	1	0	0	0	0	0	0	0	0

4. $Z = C.\left(B.C.\overline{A} + A.(B + \overline{C}) \right)$

A	B	C	$\overline{A}$	$\overline{C}$	$B \cdot C \cdot \overline{A}$	$B + \overline{C}$	$A \cdot (B + \overline{C})$	$B \cdot C \cdot \overline{A} + A \cdot (B + \overline{C})$	Z
0	0	0	1	1	0	1	0	0	0
0	0	1	1	0	0	0	0	0	0
0	1	0	1	1	0	1	0	0	0
0	1	1	1	0	1	1	0	1	1
1	0	0	0	1	0	1	1	1	0
1	0	1	0	0	0	0	0	0	0
1	1	0	0	1	0	1	1	1	0
1	1	1	0	0	0	1	1	1	1

5.

6.

7.

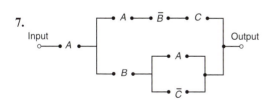

8. $\overline{A} \cdot \overline{B} \cdot C + A \cdot B \cdot \overline{C}$

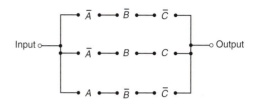

9. $\overline{A}.\overline{B}.\overline{C} + \overline{A}.B.C + A.\overline{B}.\overline{C}$

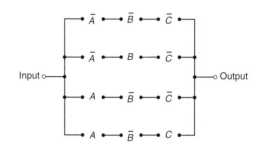

10. $\overline{A}.\overline{B}.\overline{C} + \overline{A}.B.\overline{C} + A.\overline{B}.\overline{C} + A.\overline{B}.C$

Exercise 46 (page 106)

1. $\overline{P}$ **2.** $\overline{P} + P.Q$

3. $\overline{G}$ **4.** F

5. $P.Q$ **6.** $H.\left(\overline{F} + F.\overline{G} \right)$

7. $F.\overline{G}.\overline{H} + G.H$ **8.** $\overline{Q}.\overline{R} + \overline{P}.Q.R$

9. $\overline{G}$ **10.** $F.H + G.\overline{H}$

11. $P.R + \overline{P}.\overline{R}$ **12.** $P + \overline{R}.\overline{Q}$

Exercise 47 (page 108)

1. $\overline{A}.\overline{B}$ **2.** $\overline{A} + \overline{B} + C$ **3.** $\overline{A}.\overline{B} + A.B.C$

4. 1 **5.** $\overline{P}.\left(\overline{Q} + \overline{R} \right)$

Exercise 48 (page 112)

1. Y **2.** $\overline{X} + Y$

3. $\overline{P}.\overline{Q}$ **4.** $B + A.\overline{C} + \overline{A}.C$

5. $\overline{R}.\left(\overline{P} + Q \right)$ **6.** $P.(Q + R) + \overline{P}.\overline{Q}.\overline{R}$

7. $\overline{A}.\overline{C}.(B + \overline{D})$ **8.** $\overline{B}.C.\left(\overline{A} + \overline{D} \right)$

9. $D.\left(A + B.\overline{C} \right)$ **10.** $A.\overline{D} + \overline{A}.\overline{B}.\overline{C}.D$

11. $\overline{A}.C + A.\overline{C}.D + \overline{B}.D.\left(\overline{A} + C \right)$

Exercise 49 (page 115)

1. $Z = \bar{A} + B \cdot C$

2. $Z = A \cdot \bar{B} + B \cdot \bar{C}$

3. $Z = A \cdot B \cdot \bar{C} + \bar{A} \cdot \bar{B} \cdot C$

4. $Z = (\bar{A} + B) \cdot (\bar{C} + D)$

5. $Z_1 = A.B + C$

 $Z_1 = A \cdot B + C$

6. $Z_2 = A.\bar{B} + B.C$

 $Z_2 = A \cdot \bar{B} + B \cdot C$

7. $Z_3 = A.C + B$

 $Z_3 = A \cdot C + B$

8. $Z = \bar{P} + Q$

 $Z = \bar{P} + Q$

9. $\bar{R}.\left(P + \bar{Q}\right)$

 $Z = \bar{R} \cdot (P + \bar{Q})$

10. $\bar{Q}.\left(P + \bar{R}\right)$

 $Z = \bar{Q} \cdot (P + \bar{R})$

11. $\bar{D}.\left(\bar{A}.C + \bar{B}\right)$

 $Z = \bar{D} \cdot (\bar{A} \cdot C + \bar{B})$

12. $\bar{P}.\left(Q + \bar{R}\right)$

 $Z = \bar{P} \cdot (\bar{Q} + \bar{R})$

Exercise 50 (page 118)

1. $Z = A + B \cdot C$

2. $Z = A \cdot \bar{B} + B \cdot \bar{C}$

3. $Z = A \cdot B \cdot \bar{C} + \bar{A} \cdot \bar{B} \cdot C$

4. $Z = (\bar{A} + B) \cdot (\bar{C} + D)$

5. $Z = A \cdot \bar{B} + B \cdot \bar{C} + C \cdot \bar{D}$

6.

7. $Z = \bar{P} \cdot (Q + \bar{R})$

8. $Z = A \cdot (B + C + D)$

9. $Z = A \cdot (B + C) + B \cdot C$

10. $Z = A \cdot (B + C)$

Chapter 12

Exercise 51 (page 125)

1. 24.11 mm
2. (a) 27.20 cm each (b) 45°
3. 20.81 km
4. 3.35 m, 10 cm
5. 132.7 km
6. 2.94 mm
7. 24 mm

Exercise 52 (page 127)

1. $\sin A = \dfrac{3}{5}, \cos A = \dfrac{4}{5}, \tan A = \dfrac{3}{4}, \sin B = \dfrac{4}{5},$
   $\cos B = \dfrac{3}{5}, \tan B = \dfrac{4}{3}$
2. $\sin A = \dfrac{8}{17}, \tan A = \dfrac{8}{15}$
3. (a) $\dfrac{15}{17}$ (b) $\dfrac{15}{17}$ (c) $\dfrac{8}{15}$
4. (a) 9.434 (b) −0.625 (c) 32°

Exercise 53 (page 130)

1. (a) 0.4540 (b) 0.1321 (c) −0.8399
2. (a) −0.5592 (b) 0.9307 (c) 0.2447
3. (a) −0.7002 (b) −1.1671 (c) 1.1612
4. (a) 3.4203 (b) 3.5313 (c) −1.0974
5. (a) −1.8361 (b) 3.7139 (c) −1.3421
6. (a) 0.3443 (b) −1.8510 (c) −1.2519
7. (a) 0.8660 (b) −0.1100 (c) 0.5865
8. (a) 1.0824 (b) 5.5675 (c) −1.7083
9. 13.54°, 13°32′, 0.236 rad
10. 34.20°, 34°12′, 0.597 rad
11. 39.03°, 39°2′, 0.681 rad
12. 51.92°, 51°55′, 0.906 rad
13. 23.69°, 23°41′, 0.413 rad
14. 27.01°, 27°1′, 0.471 rad
15. 29.05°
16. 20°21′
17. 1.097
18. 5.805
19. −5.325
20. 0.7199
21. 21°42′
22. 1.8258, 1.1952, 0.6546
23. 0.07448
24. 12.85
25. −1.710
26. (a) −0.8192 (b) −1.8040 (c) 0.6528
27. (a) −1.6616 (b) −0.32492 (c) 2.5985

Exercise 54 (page 132)

1. $BC = 3.50$ cm, $AB = 6.10$ cm, $\angle B = 55°$
2. $FE = 5$ cm, $\angle E = 53.13°$, $\angle F = 36.87°$
3. $GH = 9.841$ mm, $GI = 11.32$ mm, $\angle H = 49°$
4. $KL = 5.43$ cm, $JL = 8.62$ cm, $\angle J = 39°$,
   area $= 18.19$ cm^2
5. $MN = 28.86$ mm, $NO = 13.82$ mm, $\angle O = 64°25′$,
   area $= 199.4$ mm^2

6. $PR = 7.934$ m, $\angle Q = 65.06°$, $\angle R = 24.94°$,
area $= 14.64$ m^2

7. 6.54 m

Exercise 55 (page 134)

1. 48 m **2.** 110.1 m **3.** 53.0 m
4. 9.50 m **5.** 107.8 m **6.** 9.43 m, 10.56 m
7. 60 m

Exercise 56 (page 136)

1. $C = 83°$, $a = 14.1$ mm, $c = 28.9$ mm,
area $= 189$ mm^2

2. $A = 52°2'$, $c = 7.568$ cm, $a = 7.152$ cm,
area $= 25.65$ cm^2

3. $D = 19.80°$, $E = 134.20°$, $e = 36.0$ cm,
area $= 134$ cm^2

4. $E = 49°0'$, $F = 26°38'$, $f = 15.09$ mm,
area $= 185.6$ mm^2

5. $J = 44°29'$, $L = 99°31'$, $l = 5.420$ cm,
area $= 6.133$ cm^2
OR $J = 135°31'$, $L = 8°29'$, $l = 0.811$ cm,
area $= 0.917$ cm^2

6. $K = 47°8'$, $J = 97°52'$, $j = 62.2$ mm,
area $= 820.2$ mm^2
OR $K = 132°52'$, $J = 12°8'$, $j = 13.19$ mm,
area $= 174.0$ mm^2

Exercise 57 (page 137)

1. $p = 13.2$ cm, $Q = 47.34°$, $R = 78.66°$,
area $= 77.7$ cm^2

2. $p = 6.127$ m, $Q = 30.83°$, $R = 44.17°$,
area $= 6.938$ m^2

3. $X = 83.33°$, $Y = 52.62°$, $Z = 44.05°$,
area $= 27.8$ cm^2

4. $X = 29.77°$, $Y = 53.50°$, $Z = 96.73°$,
area $= 355$ mm^2

Exercise 58 (page 139)

1. 193 km
2. (a) 122.6 m (b) 94.80°, 40.66°, 44.54°
3. (a) 11.4 m (b) 17.55° **4.** 163.4 m
5. $BF = 3.9$ m, $EB = 4.0$ m **6.** 6.35 m, 5.37 m

Exercise 59 (page 141)

1. 32.48 A, 14.31° **2.** 80.42°, 59.38°, 40.20°
3. $x = 69.3$ mm, $y = 142$ mm **4.** 130°
5. 40.25 cm, 126.05° **6.** 19.8
7. 36.2 m **8.** 13.95°, 829.9 km/h
9. 13.66 mm

Chapter 13

Exercise 60 (page 146)

1. (5.83, 59.04°) or (5.83, 1.03 rad)
2. (6.61, 20.82°) or (6.61, 0.36 rad)
3. (4.47, 116.57°) or (4.47, 2.03 rad)
4. (6.55, 145.58°) or (6.55, 2.54 rad)
5. (7.62, 203.20°) or (7.62, 3.55 rad)
6. (4.33, 236.31°) or (4.33, 4.12 rad)
7. (5.83, 329.04°) or (5.83, 5.74 rad)
8. (15.68, 307.75°) or (15.68, 5.37 rad)

Exercise 61 (page 147)

1. (1.294, 4.830) **2.** (1.917, 3.960)
3. (−5.362, 4.500) **4.** (−2.884, 2.154)
5. (−9.353, −5.400) **6.** (−2.615, −3.027)
7. (0.750, −1.299) **8.** (4.252, −4.233)
9. (a) 40∠18°, 40∠90°, 40∠162°, 40∠234°, 40∠306°
(b) (38.04, 12.36), (0, 40), (−38.04, 12.36),
(−23.51, −32.36), (23.51, −32.36)
(c) 47.02 mm

Chapter 14

Exercise 62 (page 150)

1. 259.5 mm **2.** 47.68 cm **3.** 38.73 cm
4. 12 730 km **5.** 97.13 mm

Exercise 63 (page 152)

1. (a) $\dfrac{\pi}{6}$ (b) $\dfrac{5\pi}{12}$ (c) $\dfrac{5\pi}{4}$
2. (a) 0.838 (b) 1.481 (c) 4.054
3. (a) 210° (b) 80° (c) 105°
4. (a) 0°43′ (b) 154°8′ (c) 414°53′
5. 104.72 rad/s

Exercise 64 (page 154)

1. 113.1 cm^2 2. 2376 mm^2
3. 1790 mm^2 4. 802 mm^2
5. 1709 mm^2 6. 1269 m^2
7. 1548 m^2
8. 17.80 cm, 74.07 cm^2
9. (a) 59.86 mm (b) 197.8 mm
10. 26.2 cm
11. 8.67 cm, 54.48 cm   12. 82.5°
13. 748
14. (a) 0.698 rad (b) 804.2 m^2
15. (a) 396 mm^2 (b) 42.24%
16. 701.8 mm   17. 7.74 mm

Exercise 65 (page 156)

1. (a) 6 (b) (−3, 1)
2. Centre at (3, −2), radius 4
3. Circle, centre (0, 1), radius 5
4. Circle, centre (0, 0), radius 6

Exercise 66 (page 157)

1. $\omega = 90$ rad/s, $v = 13.5$ m/s
2. $v = 10$ m/s, $\omega = 40$ rad/s
3. (a) 75 rad/s, 716.2 rev/min (b) 1074 revs

Exercise 67 (page 159)

1. 2 N   2. 988 N, 5.14 km/h   3. 1.49 m/s^2

Chapter 15

Exercise 68 (page 165)

1. 227.06° and 312.94°   2. 23.27° and 156.73°
3. 122.26° and 302.26°   4. $t = 122.11°$ and 237.89°
5. $x = 64.42°$ and 295.58° 6. $\theta = 39.74°$ and 219.74°

Exercise 69 (page 171)

1. 1, 120°   2. 2, 144°   3. 3, 90°
4. 3, 720°   5. 3.5, 960°   6. 6, 360°
7. 4, 180°   8. 2, 90°   9. 5, 120°

Exercise 70 (page 174)

1. (a) 40 mA (b) 25 Hz (c) 0.04 s or 40 ms
   (d) 0.29 rad (or 16.62°) leading $40\sin 50\pi t$
2. (a) 75 cm (b) 6.37 Hz (c) 0.157 s
   (d) 0.54 rad (or 30.94°) lagging $75\sin 40t$
3. (a) 300 V (b) 100 Hz (c) 0.01 s or 10 ms
   (d) 0.412 rad (or 23.61°) lagging $300\sin 200\pi t$
4. (a) $v = 120\sin 100\pi t$ volts
   (b) $v = 120\sin(100\pi t + 0.43)$ volts
5. $i = 20\sin\left(80\pi t - \dfrac{\pi}{6}\right)$ A or
   $i = 20\sin(80\pi t - 0.524)$ A
6. $3.2\sin(100\pi t + 0.488)$ m
7. (a) 5 A, 50 Hz, 20 ms, 24.75° lagging (b) −2.093 A
   (c) 4.363 A (d) 6.375 ms (e) 3.423 ms

Exercise 71 (page 179)

1. (a) $i = (70.71\sin 628.3t + 16.97\sin 1885t)$ A
2. (a) $v = 300\sin 314.2t + 90\sin(628.3t - \pi/2)$
   $\qquad\qquad + 30\sin(1256.6t + \pi/3)$ V
3. Sketch
4. $i = (16\sin 2\pi 10^3 t + 3.2\sin 6\pi 10^3 t + 1.6\sin \pi 10^4 t)$ A
5. (a) 60 Hz, 180 Hz, 300 Hz (b) 40% (c) 10%

Chapter 16

Exercise 72 (page 182)

1. (a) 0.6846 (b) 4.376
2. (a) 1.271 (b) 5.910
3. (a) 0.5717 (b) 0.9478
4. (a) 1.754 (b) 0.08849
5. (a) 0.9285 (b) 0.1859
6. (a) 2.398 (b) 1.051
7. 56.38
8. 30.71
9. 5.042

Exercise 73 (page 186)

1–4. Proofs   5. $P = 2, Q = -4$ 6. $A = 9, B = 1$

Exercise 74 (page 188)

1. (a) 0.8814 (b) -1.6209
2. (a) ± 1.2384 (b) ± 0.9624
3. (a) -0.9962 (b) 1.0986
4. (a) ± 2.1272 (b) ± 0.4947
5. (a) 0.6442 (b) 0.9832
6. (a) 0.4162 (b) -0.6176
7. -0.8959
8. 0.6389 or -2.2484
9. 0.2554
10. (a) 67.30 (b) ± 26.42

Exercise 75 (page 189)

1. (a) 2.3524 (b) 1.3374
2. (a) 0.5211 (b) 3.6269
3. (a) $3x + \dfrac{9}{2}x^3 + \dfrac{81}{40}x^5$ (b) $1 + 2x^2 + \dfrac{2}{3}x^4$
4 – 5. Proofs

Chapter 17

Exercise 76 (page 192)

1 – 6. Proofs

Exercise 77 (page 194)

1. $\theta = 34.85°$ or $145.15°$
2. $A = 213.06°$ or $326.94°$
3. $t = 66.75°$ or $246.75°$
4. $60°, 300°$
5. $59°, 239°$
6. $41.81°, 138.19°$
7. $\pm 131.81°$
8. $39.81°, -140.19°$
9. $-30°, -150°$
10. $33.69°, 213.69°$
11. $101.31°, 281.31°$

Exercise 78 (page 195)

1. $y = 50.77°, 129.23°, 230.77°$ or $309.23°$
2. $\theta = 60°, 120°, 240°$ or $300°$
3. $\theta = 60°, 120°, 240°$ or $300°$
4. $D = 90°$ or $270°$
5. $\theta = 32.31°, 147.69°, 212.31°$ or $327.69°$

Exercise 79 (page 195)

1. $A = 19.47°, 160.53°, 203.58°$ or $336.42°$
2. $\theta = 51.34°, 123.69°, 231.34°$ or $303.69°$
3. $t = 14.48°, 165.52°, 221.81°$ or $318.19°$
4. $\theta = 60°$ or $300°$

Exercise 80 (page 196)

1. $\theta = 90°, 210°, 330°$
2. $t = 190.10°, 349.90°$
3. $\theta = 38.67°, 321.33°$
4. $\theta = 0°, 60°, 300°, 360°$
5. $\theta = 48.19°, 138.59°, 221.41°$ or $311.81°$
6. $x = 52.94°$ or $307.06°$
7. $A = 90°$
8. $t = 107.83°$ or $252.17°$
9. $a = 27.83°$ or $152.17°$
10. $\beta = 60.17°, 161.02°, 240.17°$ or $341.02°$
11. $\theta = 51.83°, 308.17°$
12. $\theta = 30°, 150°$

Chapter 18

Exercise 81 (page 199)

1 – 5. Proofs

Exercise 82 (page 201)

1. $1 - \tanh^2 \theta = \operatorname{sech}^2 \theta$
2. $\cosh(\theta + \phi) = \cosh \theta \cosh \phi + \sinh \theta \sinh \phi$
3. $\sinh(\theta - \phi) = \sinh \theta \cosh \phi - \cosh \theta \sinh \phi$
4. $\tanh 2\theta = \dfrac{2 \tanh \theta}{1 + \tanh^2 \theta}$
5. $\cosh \theta \sinh \phi = \dfrac{1}{2}[\sinh(\theta + \phi) - \sinh(\theta - \phi)]$
6. $\sinh^3 \theta = \dfrac{1}{4} \sinh 3\theta - \dfrac{3}{4} \sinh \theta$
7. $\coth^2 \theta \left(1 - \operatorname{sech}^2 \theta\right) = 1$

Chapter 19

Exercise 83 (page 204)

1. (a) $\sin 58°$ (b) $\sin 4t$ 2. (a) $\cos 104°$ (b) $\cos \dfrac{\pi}{12}$

3. Proof   4. Proof

5. (a) 0.3136 (b) 0.9495 (c) -2.4678

6. $64.72°$ or $244.72°$ 7. $67.52°$ or $247.52°$

Exercise 84 (page 208)

1. $9.434 \sin(\omega t + 1.012)$ 2. $5 \sin(\omega t - 0.644)$

3. $8.062 \sin(\omega t + 2.622)$ 4. $6.708 \sin(\omega t - 2.034)$

5. (a) $74.44°$ or $338.70°$ (b) $64.69°$ or $189.05°$

6. (a) $72.74°$ or $354.64°$ (b) $11.15°$ or $311.98°$

7. (a) $90°$ or $343.74°$ (b) $0°$ or $53.14°$

8. (a) $82.9°$ or $296°$

   (b) $32.36°, 97°, 152.36°, 217°, 272.36°$ or $337°$

9. $8.13 \sin(3\theta + 2.584)$ 10. $x = 4.0 \sin(\omega t + 0.927)$ m

11. $9.434 \sin(\omega t + 2.583)$ V

12. $x = 7.07 \sin\left(2t + \dfrac{\pi}{4}\right)$ cm

Exercise 85 (page 209)

1. $\dfrac{V^2}{2R}(1 + \cos 2t)$

2. Proofs

3. $\cos 3\theta = 4\cos^3 \theta - 3\cos \theta$

4. $-90°, 30°, 150°$

5. $-160.47°, -90°, -19.47°, 90°$

6. $-150°, -90°, -30°, 90°$

7. $-90°$

8. $45°, -135°$

Exercise 86 (page 211)

1. $\dfrac{1}{2}[\sin 9t + \sin 5t]$ 2. $\dfrac{1}{2}[\sin 10x - \sin 6x]$

3. $\cos 4t - \cos 10t$ 4. $2[\cos 4\theta + \cos 2\theta]$

5. $\dfrac{3}{2}\left[\sin \dfrac{\pi}{2} + \sin \dfrac{\pi}{6}\right]$ 6. $-\dfrac{\cos 4t}{4} - \dfrac{\cos 2t}{2} + c$

7. $-\dfrac{20}{21}$ 8. $30°, 90°$ and $150°$

Exercise 87 (page 212)

1. $2 \sin 2x \cos x$ 2. $\cos 8\theta \sin \theta$

3. $2 \cos 4t \cos t$ 4. $-\dfrac{1}{4} \sin 3t \sin 2t$

5. $\cos \dfrac{7\pi}{24} \cos \dfrac{\pi}{24}$ 6. Proofs

7. $22.5°, 45°, 67.5°, 112.5°, 135°, 157.5°$

8. $0°, 45°, 135°, 180°$ 9. $21.47°$ or $158.53°$

10. $0°, 60°, 90°, 120°, 180°, 240°, 270°, 300°, 360°$

Chapter 20

Exercise 88 (page 226)

1.

2.

3.

4.

$y = (x - 3)^2$

5.

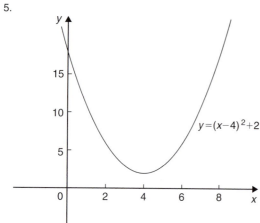

$y = (x - 4)^2 + 2$

6.

$y = x - x^2$

7.

$y = x^3 + 2$

8.

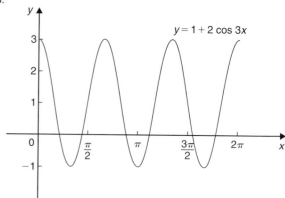

$y = 1 + 2 \cos 3x$

9.

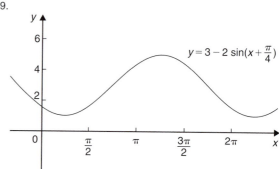

$y = 3 - 2 \sin\left(x + \dfrac{\pi}{4}\right)$

10.

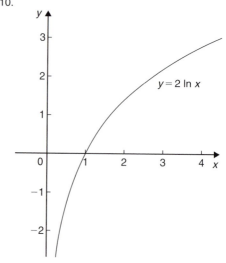

$y = 2 \ln x$

Exercise 89 (page 229)

1. (a) even (b) odd (c) neither (d) even
2. (a) odd (b) even (c) odd (d) neither
3. (a) even (b) odd

Exercise 90 (page 231)

1. $f^{-1}(x) = x - 1$ **2.** $f^{-1}(x) = \dfrac{1}{5}(x+1)$

3. $f^{-1}(x) = \sqrt[3]{x-1}$ **4.** $f^{-1}(x) = \dfrac{1}{x-2}$

5. $-\dfrac{\pi}{2}$ or -1.5708 rad   **6.** $\dfrac{\pi}{3}$ or 1.0472 rad

7. $\dfrac{\pi}{4}$ or 0.7854 rad   **8.** 0.4636 rad

9. 0.4115 rad   **10.** 0.8411 rad

11. $\dfrac{\pi}{4}$ or 0.7854 rad   **12.** 0.257

13. 1.533

Exercise 91 (page 236)

1. $y = 1, x = -1$
2. $x = 3, y = 1$ and $y = -1$
3. $x = -1, x = -2$ and $y = 1$
4. $x = 0, y = x$ and $y = -x$
5. $y = 4, y = -4$ and $x = 0$
6. $x = -1, y = x - 2$ (see Fig. 20.40 on page 870)
7. $x = 0, y = 0, y = x$ (see Fig. 20.41 on page 870)

Exercise 92 (page 240)

1. (a) Parabola with minimum value at $(-1.5, -5)$ and passing through $(0, 1.75)$
  (b) Parabola with maximum value at $(2, 70)$ and passing through $(0, 50)$
2. Circle, centre $(0, 0)$, radius 4 units
3. Parabola, symmetrical about x-axis, vertex at $(0, 0)$
4. Hyperbola, symmetrical about x- and y- axes, distance between vertices 8 units along x-axis
5. Ellipse, centre $(0, 0)$, major axis 10 units along y-axis, minor axis $2\sqrt{10}$ units along x-axis
6. Hyperbola, symmetrical about x- and y- axes, distance between vertices 6 units along x-axis
7. Rectangular hyperbola, lying in first and third quadrants only
8. Ellipse, centre $(0, 0)$, major axis 4 units along x-axis, minor axis $2\sqrt{2}$ units along y-axis
9. Circle, centre $(2, -5)$, radius 2 units
10. Ellipse, centre $(0, 0)$, major axis $2\sqrt{3}$ units along y-axis, minor axis 2 units along x-axis
11. Hyperbola, symmetrical about x- and y- axes, vertices 2 units apart along x-axis
12. Circle, centre $(0, 0)$, radius 3 units

13. Rectangular hyperbola, lying in first and third quadrants, symmetrical about x- and y- axes
14. Parabola, vertex at $(0, 0)$, symmetrical about the x-axis
15. Ellipse, centre $(0, 0)$, major axis $2\sqrt{8}$ units along y-axis, minor axis 4 units along x-axis

Chapter 21

Exercise 93 (page 243)

1. 4.5 square units   **2.** 54.7 square units
3. 63.33 m   **4.** 4.70 ha
5. 143 m^2

Exercise 94 (page 245)

1. 42.59 m^3 **2.** 147 m^3 **3.** 20.42 m^3

Exercise 95 (page 248)

1. (a) 2 A   (b) 50 V   (c) 2.5 A
2. (a) 2.5 mV   (b) 3 A
3. 0.093 As, 3.1 A
4. (a) 31.83 V   (b) 0
5. 49.13 cm^2, 368.5 kPa

Chapter 22

Exercise 96 (page 255)

1. $\pm j5$ **2.** $x = 1 \pm j$ **3.** $x = 2 \pm j$
4. $x = 3 \pm j$ **5.** $x = 0.5 \pm j0.5$ **6.** $x = 2 \pm j2$
7. $x = 0.2 \pm j0.2$
8. $x = -\dfrac{3}{4} \pm j\dfrac{\sqrt{23}}{4}$ or $x = -0.750 \pm j1.199$
9. $x = \dfrac{5}{8} \pm j\dfrac{\sqrt{87}}{8}$ or $x = 0.625 \pm j1.166$
10. (a) 1   (b) $-j$ (c) $-j2$

Exercise 97 (page 258)

1. (a) $8 + j$ (b) $-5 + j8$
2. (a) $3 - j4$ (b) $2 + j$
3. (a) 5   (b) $1 - j2$ (c) $j5$ (d) $16 + j3$
  (e) 5   (f) $3 + j4$
4. (a) $7 - j4$ (b) $-2 - j6$
5. (a) $10 + j5$ (b) $13 - j13$

Figure 20.40

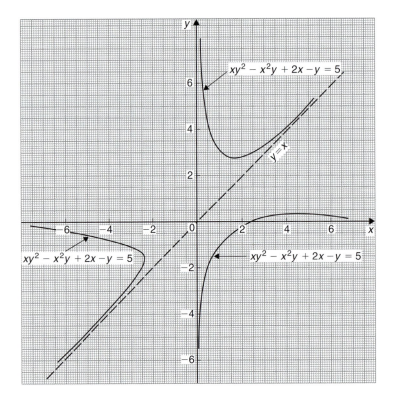

Figure 20.41

6. (a) $-13 - j2$ (b) $-35 + j20$

7. (a) $-\dfrac{2}{25} + j\dfrac{11}{25}$ (b) $-\dfrac{19}{85} + j\dfrac{43}{85}$

8. (a) $\dfrac{3}{26} + j\dfrac{41}{26}$ (b) $\dfrac{45}{26} - j\dfrac{9}{26}$

9. (a) $-j$ (b) $\dfrac{1}{2} - j\dfrac{1}{2}$

10. Proof

Exercise 98 (page 259)

1. $a = 8, b = -1$ **2.** $x = \dfrac{3}{2}, y = -\dfrac{1}{2}$

3. $a = -5, b = -12$ **4.** $x = 3, y = 1$

5. $10 + j13.75$

Exercise 99 (page 261)

1. (a) $4.472, 63.43°$ (b) $5.385, -158.20°$
   (c) $2.236, 63.43°$

2. (a) $\sqrt{13}\angle 56.31°$ (b) $4\angle 180°$ (c) $\sqrt{37}\angle 170.54°$

3. (a) $3\angle -90°$ (b) $\sqrt{125}\angle 100.30°$ (c) $\sqrt{2}\angle -135°$

4. (a) $4.330 + j2.500$ (b) $1.500 + j2.598$
   (c) $4.950 + j4.950$

5. (a) $-3.441 + j4.915$ (b) $-4.000 + j0$
   (c) $-1.750 - j3.031$

6. (a) $45\angle 65°$ (b) $10.56\angle 44°$

7. (a) $3.2\angle 42°$ (b) $2\angle 150°$

8. (a) $6.986\angle 26.79°$ (b) $7.190\angle 85.77°$

Exercise 100 (page 264)

1. (a) $R = 3\,\Omega, L = 25.5$ mH
   (b) $R = 2\,\Omega, C = 1061\,\mu$F
   (c) $R = 0, L = 44.56$ mH
   (d) $R = 4\,\Omega, C = 459.4\,\mu$F

2. 15.76 A, $23.20°$ lagging

3. 27.25 A, $3.37°$ lagging

4. 14.42 A, $43.85°$ lagging, 0.721

5. 14.6 A, $2.51°$ leading

6. 8.394 N, $208.68°$ from force A

7. $(10 + j20)\,\Omega, 22.36\angle 63.43°\,\Omega$

8. $\pm\dfrac{mh}{2\pi}$

9. (a) 922 km/h at $77.47°$ (b) 922 km/h at $-102.53°$

10. (a) $3.770\angle 8.17°$ (b) $1.488\angle 100.37°$

11. Proof

12. $353.6\angle -45°$

13. $275\angle -36.87°$ mA

Chapter 23

Exercise 101 (page 268)

1. (a) $7.594\angle 75°$ (b) $125\angle 20.61°$

2. (a) $81\angle 164°, -77.86 + j22.33$
   (b) $55.90\angle -47.18°, 38 - j41$

3. $\sqrt{10}\angle -18.43°, 3162\angle -129°$

4. $476.4\angle 119.42°, -234 + j415$

5. $45\,530\angle 12.78°, 44\,400 + j10\,070$

6. $2809\angle 63.78°, 1241 + j2520$

7. $\left(38.27 \times 10^6\right)\angle 176.15°, 10^6(-38.18 + j2.570)$

Exercise 102 (page 269)

1. (a) $\pm(1.099 + j0.455)$ (b) $\pm(0.707 + j0.707)$

2. (a) $\pm(2 - j)$ (b) $\pm(0.786 - j1.272)$

3. (a) $\pm(2.291 + j1.323)$ (b) $\pm(-2.449 + j2.449)$

4. Modulus 1.710, arguments $17.71°, 137.71°$
   and $257.71°$

5. Modulus 1.223, arguments $38.36°, 128.36°, 218.36°$
   and $308.36°$

6. Modulus 2.795, arguments $109.90°$ and $289.90°$

7. Modulus 0.3420, arguments $24.58°, 144.58°$
   and $264.58°$

8. $Z_0 = 390.2\angle -10.43°\,\Omega, \gamma = 0.1029\angle 61.92°$

Exercise 103 (page 271)

1. $5.83e^{j0.54}$

2. $4.89e^{j2.11}$

3. $-1.50 + j3.27$

4. $34.79 + j20.09$

5. $-4.52 - j3.38$

6. (a) $\ln 7 + j2.1$ (b) $2.86\angle 47.18°$ or $2.86\angle 0.82$ rad

7. $3.51\angle -0.61$ or $3.51\angle -34.72°$

8. (a) $2.06\angle 35.25°$ or $2.06\angle 0.615$ rad

(b) $4.11\angle 66.96°$ or $4.11\angle 1.17$ rad

9. $Ae^{-\frac{ht}{2m}}\cos\left(\frac{\sqrt{4mf-h^2}}{2m-a}\right)t$

Exercise 104 (page 274)

1. (a) $x^2 + y^2 = 4$

(b) a circle, centre $(0, 0)$ and radius 2

2. (a) $x^2 + y^2 = 25$

(b) a circle, centre $(0, 0)$ and radius 5

3. (a) $y = \sqrt{3}(x - 2)$ (b) a straight line

4. (a) $y = \frac{1}{\sqrt{3}}(x + 1)$ (b) a straight line

5. (a) $x^2 - 4x - 12 + y^2 = 0$ or $(x - 2)^2 + y^2 = 4^2$

(b) a circle, centre $(2, 0)$ and radius 4

6. (a) $x^2 + 6x - 16 + y^2 = 0$ or $(x + 3)^2 + y^2 = 5^2$

(b) a circle, centre $(-3, 0)$ and radius 5

7. (a) $2x^2 - 5x + 2 + 2y^2 = 0$

or $\left(x - \frac{5}{4}\right)^2 + y^2 = \left(\frac{3}{4}\right)^2$

(b) a circle, centre $\left(\frac{5}{4}, 0\right)$ and radius $\frac{3}{4}$

8. (a) $x^2 + 2x - 1 + y^2 = 0$ or $(x + 1)^2 + y^2 = 2$

(b) a circle, centre $(-1, 0)$ and radius $\sqrt{2}$

9. (a) $x^2 - x - y + y^2 = 0$

or $\left(x - \frac{1}{2}\right)^2 + \left(y - \frac{1}{2}\right)^2 = \frac{1}{2}$

(b) a circle, centre $\left(\frac{1}{2}, \frac{1}{2}\right)$ and radius $\frac{1}{\sqrt{2}}$

10. (a) $x^2 + 2x + 2y + y^2 = 0$ or $(x + 1)^2 + (y + 1)^2 = 2$

(b) a circle, centre $(-1, -1)$ and radius $\sqrt{2}$

11. (a) $y = 2x + 1.5$ (b) a straight line

12. (a) $y = 2x - 3$ (b) a straight line

13. (a) $x = \frac{1}{2}$ (b) a straight line

Chapter 24

Exercise 105 (page 281)

1. $\begin{pmatrix} 8 & 1 \\ -5 & 13 \end{pmatrix}$

2. $\begin{pmatrix} 7 & -1 & 8 \\ 3 & 1 & 7 \\ 4 & 7 & -2 \end{pmatrix}$

3. $\begin{pmatrix} -2 & -3 \\ -3 & 1 \end{pmatrix}$

4. $\begin{pmatrix} 9.3 & -6.4 \\ -7.5 & 16.9 \end{pmatrix}$

5. $\begin{pmatrix} 45 & 7 \\ -26 & 71 \end{pmatrix}$

6. $\begin{pmatrix} 4.6 & -5.6 & -7.6 \\ 17.4 & -16.2 & 28.6 \\ -14.2 & 0.4 & 17.2 \end{pmatrix}$

7. $\begin{pmatrix} -11 \\ 43 \end{pmatrix}$

8. $\begin{pmatrix} 16 & 0 \\ -27 & 34 \end{pmatrix}$

9. $\begin{pmatrix} -6.4 & 26.1 \\ 22.7 & -56.9 \end{pmatrix}$

10. $\begin{pmatrix} 135 \\ -52 \\ -85 \end{pmatrix}$

11. $\begin{pmatrix} 5 & 6 \\ 12 & -3 \\ 1 & 0 \end{pmatrix}$

12. $\begin{pmatrix} 55.4 & 3.4 & 10.1 \\ -12.6 & 10.4 & -20.4 \\ -16.9 & 25.0 & 37.9 \end{pmatrix}$

13. $A \times C = \begin{pmatrix} -6.4 & 26.1 \\ 22.7 & -56.9 \end{pmatrix}$

$C \times A = \begin{pmatrix} -33.5 & 53.1 \\ 23.1 & -29.8 \end{pmatrix}$

Hence, $A \times C \neq C \times A$

Exercise 106 (page 282)

1. 17 **2.** -3 **3.** -13.43

4. $-5 + j3$

5. $(-19.75 + j19.79)$ or $27.96\angle 134.94°$

6. $x = 6$ or $x = -1$

Exercise 107 (page 283)

1. $\begin{pmatrix} \frac{7}{17} & \frac{1}{17} \\ \frac{4}{17} & \frac{3}{17} \end{pmatrix}$

2. $\begin{pmatrix} 7\frac{5}{7} & 8\frac{4}{7} \\ -4\frac{2}{7} & -6\frac{3}{7} \end{pmatrix}$

3. $\begin{pmatrix} 0.290 & 0.551 \\ 0.186 & 0.097 \end{pmatrix}$

Exercise 108 (page 284)

1. $\begin{pmatrix} -16 & 8 & -34 \\ -14 & -46 & 63 \\ -24 & 12 & 2 \end{pmatrix}$

2. $\begin{pmatrix} -16 & -8 & -34 \\ 14 & -46 & -63 \\ -24 & -12 & 2 \end{pmatrix}$

3. -212 **4.** -328 **5.** -242.83

6. $-2 - j$

7. $26.94\angle - 139.52°$ or $(-20.49 - j17.49)$

8. (a) $\lambda = 3$ or 4 (b) $\lambda = 1$ or 2 or 3

Exercise 109 (page 285)

1. $\begin{pmatrix} 4 & -2 & 5 \\ -7 & 4 & 7 \\ 6 & 0 & -4 \end{pmatrix}$

2. $\begin{pmatrix} 3 & 5 & -1 \\ 6 & -\dfrac{2}{3} & 0 \\ \dfrac{1}{2} & 7 & \dfrac{3}{5} \end{pmatrix}$

3. $\begin{pmatrix} -16 & 14 & -24 \\ -8 & -46 & -12 \\ -34 & -63 & 2 \end{pmatrix}$

4. $\begin{pmatrix} -\dfrac{2}{5} & -3\dfrac{3}{5} & 42\dfrac{1}{3} \\ -10 & 2\dfrac{3}{10} & -18\dfrac{1}{2} \\ \dfrac{2}{3} & -6 & -32 \end{pmatrix}$

5. $-\dfrac{1}{212}\begin{pmatrix} -16 & 14 & -24 \\ -8 & -46 & -12 \\ -34 & -63 & 2 \end{pmatrix}$

6. $-\dfrac{15}{923}\begin{pmatrix} -\dfrac{2}{5} & -3\dfrac{3}{5} & 42\dfrac{1}{3} \\ -10 & 2\dfrac{3}{10} & -18\dfrac{1}{2} \\ -\dfrac{2}{3} & -6 & -32 \end{pmatrix}$

Chapter 25

Exercise 110 (page 289)

1. $x = 4, y = -3$

2. $p = 1.2, q = -3.4$

3. $x = 1, y = -1, z = 2$

4. $a = 2.5, b = 3.5, c = 6.5$

5. $p = 4.1, q = -1.9, r = -2.7$

6. $I_1 = 2, I_2 = -3$

7. $s = 2, v = -3, a = 4$

8. $\ddot{x} = 0.5, \dot{x} = 0.77, x = 1.4$

Exercise 111 (page 293)

1. $x = -1.2, y = 2.8$

2. $m = -6.4, n = -4.9$

3. $x = 1, y = 2, z = -1$

4. $p = 1.5, q = 4.5, r = 0.5$

5. $x = \dfrac{7}{20}, y = \dfrac{17}{40}, z = -\dfrac{5}{24}$

6. $F_1 = 1.5, F_2 = -4.5$

7. $I_1 = 10.77\angle 19.23°\text{A}, I_2 = 10.45\angle - 56.73°\text{A}$

8. $i_1 = -5, i_2 = -4, i_3 = 2$

9. $F_1 = 2, F_2 = -3, F_3 = 4$

10. $I_1 = 3.317\angle 22.57°\text{A}, I_2 = 1.963\angle 40.97°\text{A},$

     $I_3 = 1.010\angle - 148.32°\text{A}$

Exercise 112 (page 294)

Answers to Exercises 110 and 111
are as above

Exercise 113 (page 296)

1. $\ddot{x} = -0.30, \dot{x} = 0.60, x = 1.20$

2. $T_1 = 0.8, T_2 = 0.4, T_3 = 0.2$

3. Answers to Exercise 110 are as above

4. Answers to Exercise 111 are as above

Exercise 114 (page 301)

1. (a) $\lambda_1 = 3, \lambda_2 = -2$ (b) $\begin{pmatrix} -4 \\ 1 \end{pmatrix}, \begin{pmatrix} 1 \\ 1 \end{pmatrix}$

2. (a) $\lambda_1 = 1, \lambda_2 = 6$ (b) $\begin{pmatrix} -3 \\ 1 \end{pmatrix}, \begin{pmatrix} 2 \\ 1 \end{pmatrix}$

3. (a) $\lambda_1 = 1, \lambda_2 = 2$ (b) $\begin{pmatrix} 1 \\ -2 \end{pmatrix}, \begin{pmatrix} 1 \\ -1 \end{pmatrix}$

4. (a) $\lambda_1 = 2, \lambda_2 = 6, \lambda_3 = -2$

   (b) $\begin{pmatrix} 1 \\ -2 \\ 1 \end{pmatrix}, \begin{pmatrix} 0 \\ 1 \\ 1 \end{pmatrix}, \begin{pmatrix} 1 \\ 1 \\ 0 \end{pmatrix}$

5. (a) $\lambda_1 = 0, \lambda_2 = 1, \lambda_3 = 3$

(b) $\begin{pmatrix} 1 \\ 1 \\ 1 \end{pmatrix}, \begin{pmatrix} 1 \\ 0 \\ -1 \end{pmatrix}, \begin{pmatrix} 1 \\ -2 \\ 1 \end{pmatrix}$

6. (a) $\lambda_1 = 1, \lambda_2 = 2, \lambda_3 = 4$

(b) $\begin{pmatrix} -2 \\ 1 \\ 0 \end{pmatrix}, \begin{pmatrix} -2 \\ 1 \\ 1 \end{pmatrix}, \begin{pmatrix} 0 \\ 1 \\ 1 \end{pmatrix}$

7. (a) $\lambda_1 = 1, \lambda_2 = 2, \lambda_3 = 3$

(b) $\begin{pmatrix} 0 \\ 2 \\ -1 \end{pmatrix}, \begin{pmatrix} 1 \\ 1 \\ 0 \end{pmatrix}, \begin{pmatrix} 2 \\ 2 \\ 1 \end{pmatrix}$

Chapter 26

Exercise 115 (page 306)

1. A scalar quantity has magnitude only; a vector quantity has both magnitude and direction.
2. Scalar   **3.** Scalar   **4.** Vector   **5.** Scalar
6. Scalar   **7.** Vector   **8.** Scalar
9. Vector

Exercise 116 (page 313)

1. 17.35 N at $18.00°$ to the 12 N vector

2. 13 m/s at $22.62°$ to the 12 m/s velocity

3. 16.40 N at $37.57°$ to the 13 N force

4. 28.43 N at $129.29°$ to the 18 N force

5. 32.31 N at $21.80°$ to the 30 N displacement

6. 14.72 N at $-14.72°$ to the 5 N force

7. 29.15 m/s at $29.04°$ to the horizontal

8. 9.28 N at $16.70°$ to the horizontal

9. 6.89 m/s at $159.56°$ to the horizontal

10. 15.62 N at $26.33°$ to the 10 N force

11. 21.07 knots, E $9.22°$S

Exercise 117 (page 316)

1. (a) 54.0 N at $78.16°$ (b) 45.64 N at $4.66°$

2. (a) 31.71 m/s at $121.81°$ (b) 19.55 m/s at $8.63°$

Exercise 118 (page 317)

1. 83.5 km/h at $71.6°$ to the vertical

2. 4 minutes 55 seconds, $60°$

3. 22.79 km/h. E $9.78°$N

Exercise 119 (page 318)

1. $i - j - 4k$ **2.** $4i + j - 6k$
3. $-i + 7j - k$ **4.** $5i - 10k$
5. $-3i + 27j - 8k$ **6.** $-5i + 10k$
7. $i + 7.5j - 4k$ **8.** $20.5j - 10k$
9. $3.6i + 4.4j - 6.9k$ **10.** $2i + 40j - 43k$

Chapter 27

Exercise 120 (page 321)

1. $4.5 \sin(A + 63.5°)$

2. (a) $20.9 \sin(\omega t + 0.63)$ volts
   (b) $12.5 \sin(\omega t - 1.36)$ volts

3. $13 \sin(\omega t + 0.393)$

Exercise 121 (page 323)

1. $4.5 \sin(A + 63.5°)$

2. (a) $20.9 \sin(\omega t + 0.62)$ volts
   (b) $12.5 \sin(\omega t - 1.33)$ volts

3. $13 \sin(\omega t + 0.40)$

Exercise 122 (page 324)

1. $4.472 \sin(A + 63.44°)$

2. (a) $20.88 \sin(\omega t + 0.62)$ volts
   (b) $12.50 \sin(\omega t - 1.33)$ volts

3. $13 \sin(\omega t + 0.395)$

4. $11.11 \sin(\omega t + 0.324)$

5. $8.73 \sin(\omega t - 0.173)$

6. $1.01 \sin(\omega t - 0.698) A$

Exercise 123 (page 326)

1. $11.11 \sin(\omega t + 0.324)$

2. $8.73 \sin(\omega t - 0.173)$

3. $i = 21.79 \sin(\omega t - 0.639)$

4. $v = 5.695 \sin(\omega t + 0.695)$ V

5. $x = 14.38 \sin(\omega t + 1.444)$

6. (a) $305.3 \sin(314.2t - 0.233)$ V   (b) 50 Hz

7. (a) $10.21 \sin(628.3t + 0.818)$ V   (b) 100 Hz
   (c) 10 ms

8. (a) $79.83 \sin(300\pi t + 0.352)$ V   (b) 150 Hz
   (c) 6.667 ms

9. $150.6 \sin(\omega t - 0.247)$ V

Exercise 124 (page 328)

1. $12.07 \sin(\omega t + 0.297)$

2. $14.51 \sin(\omega t - 0.315)$

3. $9.173 \sin(\omega t + 0.396)$

4. $16.168 \sin(\omega t + 1.451)$

5. (a) $371.95 \sin(314.2t - 0.239)$ V   (b) 50 Hz

6. (a) $11.44 \sin(200\pi t + 0.715)$ V   (b) 100 Hz
   (c) 10 ms

7. (a) $79.73 \sin(300\pi t - 0.536)$ V   (b) 150 Hz
   (c) 6.667 ms   (d) 56.37 V

8. $I_N = 354.6\angle 32.41° $ A

Chapter 28

Exercise 125 (page 335)

1. (a) 7 (b) 0   2. (a) -12 (b) -4

3. (a) 11 (b) 11   4. (a) $\sqrt{13}$ (b) $\sqrt{14}$

5. (a) -16 (b) 38   6. (a) $\sqrt{19}$ (b) 7.347

7. (a) $143.82°$ (b) $44.52°$

8. (a) $0.555, \ -0.832, \ 0$ (b) $0, \ 0.097, \ -0.243$
   (c) $0.267, 0.535, -0.802$

9. $11.54°$ 10. $66.40°$

11. 53 N m

Exercise 126 (page 338)

1. (a) $4i - 7j - 6k$ (b) $-4i + 7j + 6k$

2. (a) 11.92   (b) 13.96

3. (a) $-36i - 30j + 54k$ (b) $11i + 4j - k$

4. (a) $-22i - j + 33k$ (b) $18i + 162j + 102k$

5. (i) -15 (ii) $-4i + 4j + 10k$ (iii) 11.49
   (iv) $4i - 4j - 10k$ (v) $142.55°$

6. (i) -62.5 (ii) $-1.5i - 4j + 11k$ (iii) 11.80
   (iv) $1.5i + 4j - 11k$ (v) $169.31°$

7. 10 N m

8. $M = (5i + 8j - 2k)$ N m, $|M| = 9.64$ N m

9. $v = -14i + 7j + 12k$, $[v] = 19.72$ m/s

10. $6i - 10j - 14k$, 18.22 m/s

Exercise 127 (page 340)

1. $r = (5 + 2\lambda)i + (7\lambda - 2)j + (3 - 4\lambda)k$,
   $r = 9i + 12j - 5k$

2. $\dfrac{x - 5}{2} = \dfrac{y + 2}{7} = \dfrac{3 - z}{4} = \lambda$

3. $r = \dfrac{1}{3}(1 + 4\lambda)i + \dfrac{1}{5}(2\lambda - 1)j + (4 - 3\lambda)k$

4. $r = \dfrac{1}{2}(\lambda - 1)i + \dfrac{1}{4}(1 - 5\lambda)j + \dfrac{1}{3}(1 + 4\lambda)k$

Chapter 29

Exercise 128 (page 350)

1. (a) $25x^4$ (b) $8.4x^{2.5}$ (c) $-\dfrac{1}{x^2}$

2. (a) $\dfrac{8}{x^3}$ (b) 0   (c) 2

3. (a) $\dfrac{1}{\sqrt{x}}$ (b) $5\sqrt[3]{x^2}$ (c) $-\dfrac{2}{\sqrt{x^3}}$

4. (a) $\dfrac{1}{\sqrt[3]{x^4}}$ (b) $2(x - 1)$ (c) $6\cos 3x$

5. (a) $8\sin 2x$ (b) $12e^{6x}$ (c) $-\dfrac{15}{e^{5x}}$

6. (a) $\dfrac{4}{x}$ (b) $\dfrac{e^x + e^{-x}}{2}$ (c) $-\dfrac{1}{x^2} + \dfrac{1}{2\sqrt{x^3}}$

7. $-1, 16$

8. $\left(\dfrac{1}{2}, \dfrac{3}{4}\right)$

9. (a) $-\dfrac{4}{\theta^3} + \dfrac{2}{\theta} + 10\sin 5\theta - 12\cos 2\theta + \dfrac{6}{e^{3\theta}}$
   (b) 22.30

10. 3.29

11. $x = \dfrac{mg}{k}$

12. 27.0 volts

Exercise 129 (page 351)

1. $x\cos x + \sin x$ 2. $2xe^{2x}(x+1)$

3. $x(1 + 2\ln x)$ 4. $6x^2(\cos 3x - x\sin 3x)$

5. $\sqrt{x}\left(1 + \dfrac{3}{2}\ln 3x\right)$ 6. $e^{3t}(4\cos 4t + 3\sin 4t)$

7. $e^{4\theta}\left(\dfrac{1}{\theta} + 4\ln 3\theta\right)$

8. $e^t\left\{\left(\dfrac{1}{t} + \ln t\right)\cos t - \ln t \sin t\right\}$

9. 8.732 10. 32.31

Exercise 130 (page 353)

1. $\dfrac{x\cos x - \sin x}{x^2}$ 2. $-\dfrac{6}{x^4}(x\sin 3x + \cos 3x)$

3. $\dfrac{2(1 - x^2)}{(x^2 + 1)^2}$ 4. $\dfrac{\dfrac{\cos x}{2\sqrt{x}} + \sqrt{x}\sin x}{\cos^2 x}$

5. $\dfrac{3\sqrt{\theta}\{3\sin 2\theta - 4\theta\cos 2\theta\}}{4\sin^2 2\theta}$

6. $\dfrac{1}{\sqrt{t^3}}\left(1 - \dfrac{1}{2}\ln 2t\right)$

7. $\dfrac{2e^{4x}}{\sin^2 x}\{(1 + 4x)\sin x - x\cos x\}$

8. -18 9. 3.82

Exercise 131 (page 354)

1. $12(2x - 1)^5$ 2. $5(2x^3 - 5x)^4(6x^2 - 5)$

3. $6\cos(3\theta - 2)$ 4. $-10\cos^4\alpha\sin\alpha$

5. $\dfrac{5(2 - 3x^2)}{(x^3 - 2x + 1)^6}$ 6. $10e^{2t+1}$

7. $-20t\,\mathrm{cosec}^2(5t^2 + 3)$ 8. $18\sec^2(3y + 1)$

9. $2\sec^2\theta\, e^{\tan\theta}$ 10. 1.86

11. (a) 24.21 mm/s (b) -70.46 mm/s

Exercise 132 (page 356)

1. (a) $36x^2 + 12x$ (b) $72x + 12$

2. (a) $\dfrac{4}{5} - \dfrac{12}{t^5} + \dfrac{6}{t^3} + \dfrac{1}{4\sqrt{t^3}}$ (b) -4.95

3. (a) $-\dfrac{V}{R}e^{-\frac{t}{CR}}$ (b) $\dfrac{V}{CR^2}e^{-\frac{t}{CR}}$

4. (a) $-(12\sin 2t + \cos t)$ (b) $-\dfrac{2}{\theta^2}$

5. (a) $4(\sin^2 x - \cos^2 x)$ (b) $48(2x - 3)^2$

6. 18

7. Proof
8. Proof
9. Proof

Chapter 30

Exercise 133 (page 358)

1. 3000π A/s 2. (a) 0.24 cd/V (b) 250 V

3. (a) -625 V/s (b) -220.5 V/s

4. -1.635 Pa/m 5. -390 m^3/min

Exercise 134 (page 361)

1. (a) 100 m/s (b) 4 s (c) 200 m (d) -100 m/s

2. (a) 90 km/h (b) 62.5 m

3. (a) 4 s (b) 3 rads

4. (a) 3 m/s, -1 m/s^2 (b) 6 m/s, -4 m/s^2 (c) 0.75 s

5. (a) $\omega = 1.40$ rad/s (b) $\alpha = -0.37$ rad/s^2
   (c) $t = 6.28$ s

6. (a) 6 m/s, -23 m/s^2 (b) 117 m/s, 97 m/s^2
   (c) 0.75 s or 0.4 s (d) 1.5 s (e) $75\dfrac{1}{6}$ m

7. 3 s

Exercise 135 (page 365)

1. $(3, -9)$ Minimum

2. $(1, 9)$ Maximum

3. $(2, -1)$ Minimum

4. $(0, 3)$ Minimum, $(2, 7)$ Maximum

5. Minimum at $\left(\dfrac{2}{3}, \dfrac{2}{3}\right)$

6. $(3, 9)$ Maximum

7. $(2, -88)$ Minimum, $(-2.5, 94.25)$ Maximum

8. $(0.4000, 3.8326)$ Minimum

9. $(0.6931, -0.6137)$ Maximum

10. $(1, 2.5)$ Minimum, $\left(-\dfrac{2}{3}, 4\dfrac{22}{27}\right)$ Maximum

11. $(0.5, 6)$ Minimum

12. Maximum of 13 at 337.38°,
    Minimum of -13 at 157.38°

13. Proof

Exercise 136 (page 368)

1. 54 km/h 2. 90 000 m^2 3. 48 m

4. 11.42 m^2

5. Radius = 4.607 cm, height = 9.212 cm

6. 6.67 cm   7. Proof

8. Height = 5.42 cm, radius = 2.71 cm 9. 44.72

10. 42.72 volts **11.** 50.0 miles/gallon, 52.6 miles/hour

12. 45°   **13.** 0.607

Exercise 137 (page 370)

1. $\left(\dfrac{1}{2}, -1\right)$ **2.** $\left(-\dfrac{1}{4}, 4\right)$ **3.** (π, π)

4. $(3, -100)$ **5.** $(2, 0.541)$

6. Max at $(0, 10)$, Min at $(2, -2)$, point of inflexion at $(1, 4)$

Exercise 138 (page 372)

1. (a) $y = 4x - 2$ (b) $4y + x = 9$

2. (a) $y = 10x - 12$ (b) $10y + x = 82$

3. (a) $y = \dfrac{3}{2}x + 1$ (b) $6y + 4x + 7 = 0$

4. (a) $y = 5x + 5$ (b) $5y + x + 27 = 0$

5. (a) $9\theta + t = 6$ (b) $\theta = 9t - 26\dfrac{2}{3}$ or $3\theta = 27t - 80$

Exercise 139 (page 373)

1. (a) -0.03 (b) -0.008 **2.** $-0.032, -1.6\%$

3. (a) $60\ \text{cm}^3$ (b) $12\ \text{cm}^2$

4. (a) $-6.03\ \text{cm}^2$ (b) $-18.10\ \text{cm}^3$

5. 12.5%

Chapter 31

Exercise 140 (page 380)

1. (a) $4x + c$ (b) $\dfrac{7x^2}{2} + c$

2. (a) $\dfrac{2}{15}x^3 + c$ (b) $\dfrac{5}{24}x^4 + c$

3. (a) $\dfrac{3x^2}{2} - 5x + c$ (b) $4\theta + 2\theta^2 + \dfrac{\theta^3}{3} + c$

4. (a) $-\dfrac{4}{3x} + c$ (b) $-\dfrac{1}{4x^3} + c$

5. (a) $\dfrac{4}{5}\sqrt{x^5} + c$ (b) $\dfrac{1}{9}\sqrt[4]{x^9} + c$

6. (a) $\dfrac{10}{\sqrt{t}} + c$ (b) $\dfrac{15}{7}\sqrt[5]{x} + c$

7. (a) $\dfrac{3}{2}\sin 2x + c$ (b) $-\dfrac{7}{3}\cos 3\theta + c$

8. (a) $\dfrac{1}{4}\tan 3x + c$ (b) $-\dfrac{1}{2}\cot 4\theta + c$

9. (a) $-\dfrac{5}{2}\operatorname{cosec} 2t + c$ (b) $\dfrac{1}{3}\sec 4t + c$

10. (a) $\dfrac{3}{8}e^{2x} + c$ (b) $\dfrac{-2}{15e^{5x}} + c$

11. (a) $\dfrac{2}{3}\ln x + c$ (b) $\dfrac{u^2}{2} - \ln u + c$

12. (a) $8\sqrt{x} + 8\sqrt{x^3} + \dfrac{18}{5}\sqrt{x^5} + c$

(b) $-\dfrac{1}{t} + 4t + \dfrac{4t^3}{3} + c$

Exercise 141 (page 381)

1. (a) 105 (b) -0.5 **2.** (a) 6 (b) $-1\dfrac{1}{3}$

3. (a) 0 (b) 4   **4.** (a) 1 (b) 4.248

5. (a) 0.2352 (b) 2.598   **6.** (a) 0.2527 (b) 2.638

7. (a) 19.09 (b) 2.457   **8.** (a) 0.2703 (b) 9.099

9. 55.65   **10.** Proof

11. 7.26   **12.** $77.7\ \text{m}^3$

Chapter 32

Exercise 142 (page 385)

1. $1\dfrac{1}{3}$ square units   **2.** $20\dfrac{5}{6}$ square units

3. $2\dfrac{1}{2}$ square units

Exercise 143 (page 386)

1. $2\dfrac{2}{3}$ km   **2.** 2.198

3. 15.92 mA, 17.68 mA   **4.** $\sqrt{\dfrac{E_1^2 + E_3^2}{2}}$

Exercise 144 (page 388)

1. 1.5π cubic units   **2.** $170\dfrac{2}{3}\pi$ cubic units

3. (a) 329.4π (b) 81π

4. (b) 0.352 (c) 0.419 square units (d) 0.222 K

Exercise 145 (page 389)

1. $(2.5, 4.75)$ **2.** $(3.036, 24.36)$ **3.** $(2, 1.6)$
4. $(1, -0.4)$ **5.** $(2.4, 0)$

Exercise 146 (page 392)

1. 189.6 cm^3
2. On the centre line, distance 2.40 cm from the centre, i.e. at co-ordinates $(1.70, 1.70)$
3. (a) 45 square units (b)(i) 1215π cubic units (ii) 202.5π cubic units (c) $(2.25, 13.5)$
4. 64.90 cm^3, 16.86%, 506.2 g

Exercise 147 (page 398)

1. (a) 72 cm^4, 1.73 cm (b) 128 cm^4, 2.31 cm
  (c) 512 cm^4, 4.62 cm
2. (a) 729 cm^4, 3.67 cm (b) 2187 cm^4, 6.36 cm
  (c) 243 cm^4, 2.12 cm
3. (a) 201 cm^4, 2.0 cm (b) 1005 cm^4, 4.47 cm
4. 3927 mm^4, 5.0 mm
5. (a) 335 cm^4, 4.73 cm (b) $22\,030 \text{ cm}^4$, 14.3 cm
  (c) 628 cm^4, 7.07 cm
6. 0.866 m
7. 0.245 m^4, 0.559 m
8. $14\,280 \text{ cm}^4$, 5.96 cm
9. (a) $12\,190 \text{ mm}^4$, 10.9 mm
  (b) 549.5 cm^4, 4.18 cm
10. $I_{AA} = 4224 \text{ cm}^4$, $I_{BB} = 6718 \text{ cm}^4$,
  $I_{CC} = 37\,300 \text{ cm}^4$
11. 1350 cm^4, 5.67 cm

Chapter 33

Exercise 148 (page 401)

1. Sketches   **2.** $y = x^2 + 3x - 1$

Exercise 149 (page 403)

1. $y = \dfrac{1}{4}\sin 4x - x^2 + c$ **2.** $y = \dfrac{3}{2}\ln x - \dfrac{x^3}{6} + c$
3. $y = 3x - \dfrac{x^2}{2} - \dfrac{1}{2}$ **4.** $y = \dfrac{1}{3}\cos\theta + \dfrac{1}{2}$

5. $y = \dfrac{1}{6}\left(x^2 - 4x + \dfrac{2}{e^x} + 4\right)$ **6.** $y = \dfrac{3}{2}x^2 - \dfrac{x^3}{6} - 1$
7. $v = u + at$ **8.** 15.9 m

Exercise 150 (page 404)

1. $x = \dfrac{1}{3}\ln(2 + 3y) + c$ **2.** $\tan y = 2x + c$
3. $\dfrac{y^2}{2} + 2\ln y = 5x - 2$ **4.** Proof
5. $x = a\left(1 - e^{-kt}\right)$
6. (a) $Q = Q_0 e^{-\frac{t}{CR}}$ (b) 9.30 C, 5.81 C
7. 273.3 N, 2.31 rad **8.** $8 \text{ m } 40 \text{ s}$

Exercise 151 (page 407)

1. $\ln y = 2\sin x + c$ **2.** $y^2 - y = x^3 + x$
3. $e^y = \dfrac{1}{2}e^{2x} + \dfrac{1}{2}$ **4.** $\ln(x^2 y) = 2x - y - 1$
5. Proof   **6.** $y = \dfrac{1}{\sqrt{(1 - x^2)}}$
7. $y^2 = x^2 - 2\ln x + 3$ **8.** (a) $V = E\left(1 - e^{-\frac{t}{CR}}\right)$
  (b) 13.2 V
9. 3

Chapter 34

Exercise 152 (page 416)

1. $\dfrac{1}{3}(2t - 1)$ **2.** 2
3. (a) $-\dfrac{1}{4}\cot\theta$ (b) $-\dfrac{1}{16}\operatorname{cosec}^3\theta$
4. 4 **5.** -6.25
6. $y = -1.155x + 4$ **7.** $y = -\dfrac{1}{4}x + 5$

Exercise 153 (page 417)

1. (a) 3.122 (b) -14.43 **2.** $y = -2x + 3$
3. $y = -x + \pi$ **4.** 0.02975
5. (a) 13.14 (b) 5.196

Chapter 35

Exercise 154 (page 420)

1. (a) $15y^4 \dfrac{dy}{dx}$ (b) $-8\sin 4\theta \dfrac{d\theta}{dx}$ (c) $\dfrac{1}{2\sqrt{k}} \dfrac{dk}{dx}$

2. (a) $\dfrac{5}{2t} \dfrac{dt}{dx}$ (b) $\dfrac{3}{2} e^{2y+1} \dfrac{dy}{dx}$ (c) $6\sec^2 3y \dfrac{dy}{dx}$

3. (a) $6\cos 2\theta \dfrac{d\theta}{dy}$ (b) $6\sqrt{x}\dfrac{dx}{dy}$ (c) $-\dfrac{2}{e^t}\dfrac{dt}{dy}$

4. (a) $-\dfrac{6}{(3x+1)^2}\dfrac{dx}{du}$ (b) $6\sec 2\theta \tan 2\theta \dfrac{d\theta}{du}$

   (c) $-\dfrac{1}{\sqrt{y^3}}\dfrac{dy}{du}$

Exercise 155 (page 421)

1. $3xy^2\left(3x\dfrac{dy}{dx}+2y\right)$ 2. $\dfrac{2}{5x^2}\left(x\dfrac{dy}{dx}-y\right)$

3. $\dfrac{3}{4v^2}\left(v-u\dfrac{dv}{du}\right)$ 4. $3\left(\dfrac{\cos 3x}{2\sqrt{y}}\right)\dfrac{dy}{dx}-9\sqrt{y}\sin 3x$

5. $2x^2\left(\dfrac{x}{y}+3\ln y\dfrac{dx}{dy}\right)$

Exercise 156 (page 423)

1. $\dfrac{2x+4}{3-2y}$ 2. $\dfrac{3}{1-6y^2}$ 3. $-\dfrac{\sqrt{5}}{2}$

4. $\dfrac{-(x+\sin 4y)}{4x\cos 4y}$ 5. $\dfrac{4x-y}{3y+x}$ 6. $\dfrac{x(4x+9y)}{\cos y - 2x^2}$

7. $\dfrac{1-2\ln y}{3+\dfrac{2x}{y}-4y^3}$ 8. 5 9. ± 0.5774

10. ± 1.5 11. -6

Chapter 36

Exercise 157 (page 426)

1. $\dfrac{2}{2x-5}$ 2. $-3\tan 3x$ 3. $\dfrac{9x^2+1}{3x^3+x}$

4. $\dfrac{10(x+1)}{5x^2+10x-7}$ 5. $\dfrac{1}{x}$ 6. $\dfrac{2x}{x^2-1}$

7. $\dfrac{3}{x}$ 8. $2\cot x$ 9. $\dfrac{12x^2-12x+3}{4x^3-6x^2+3x}$

Exercise 158 (page 428)

1. $\dfrac{(x-2)(x+1)}{(x-1)(x+3)}\left\{\dfrac{1}{(x-2)}+\dfrac{1}{(x+1)}\right.$
$\left.-\dfrac{1}{(x-1)}-\dfrac{1}{(x+3)}\right\}$

2. $\dfrac{(x+1)(2x+1)^3}{(x-3)^2(x+2)^4}\left\{\dfrac{1}{(x+1)}+\dfrac{6}{(2x+1)}\right.$
$\left.-\dfrac{2}{(x-3)}-\dfrac{4}{(x+2)}\right\}$

3. $\dfrac{(2x-1)\sqrt{(x+2)}}{(x-3)\sqrt{(x+1)^3}}\left\{\dfrac{2}{(2x+1)}+\dfrac{1}{2(x+2)}\right.$
$\left.-\dfrac{1}{(x-3)}-\dfrac{3}{2(x+1)}\right\}$

4. $\dfrac{e^{2x}\cos 3x}{\sqrt{(x-4)}}\left\{2-3\tan 3x-\dfrac{1}{2(x-4)}\right\}$

5. $3\theta\sin\theta\cos\theta\left\{\dfrac{1}{\theta}+\cos\theta-\tan\theta\right\}$

6. $\dfrac{2x^4\tan x}{e^{2x}\ln 2x}\left\{\dfrac{4}{x}+\dfrac{1}{\sin x\cos x}-2-\dfrac{1}{x\ln 2x}\right\}$

7. $\dfrac{13}{16}$

8. -6.71

Exercise 159 (page 429)

1. $2x^{2x}(1+\ln x)$

2. $(2x-1)^x\left\{\dfrac{2x}{2x-1}+\ln(2x-1)\right\}$

3. $\sqrt[x]{(x+3)}\left\{\dfrac{1}{x(x+3)}-\dfrac{\ln(x+3)}{x^2}\right\}$

4. $3x^{4x+1}\left(4+\dfrac{1}{x}+4\ln x\right)$

5. Proof

6. $\dfrac{1}{3}$

7. Proof

Chapter 37

Exercise 160 (page 434)

1. (a) $6\,\mathrm{ch}\,2x$ (b) $10\,\mathrm{sh}\,5\theta$ (c) $36\,\mathrm{sech}^2 9t$

2. (a) $-\dfrac{10}{3}\mathrm{sech}\,5x\tanh 5x$ (b) $-\dfrac{5}{16}\mathrm{cosech}\dfrac{t}{2}\coth\dfrac{t}{2}$
   (c) $-14\mathrm{cosech}^2 7\theta$

3. (a) $2\coth x$ (b) $\dfrac{3}{8}\mathrm{sech}\dfrac{\theta}{2}\mathrm{cosech}\dfrac{\theta}{2}$

4. (a) $2\left(\text{sh}^2 2x + \text{ch}^2 2x\right)$ (b) $6e^{2x}\left(\text{sech}^2 2x + \text{th} 2x\right)$

5. (a) $\dfrac{12x\,\text{ch}4x - 9\text{sh}4x}{2x^4}$ (b) $\dfrac{2\left(\cos 2t\,\text{sh}2t + \text{ch}2t\sin 2t\right)}{\cos^2 2t}$

Chapter 38

Exercise 161 (page 439)

1. (a) $\dfrac{4}{\sqrt{\left(1 - 16x^2\right)}}$ (b) $\dfrac{1}{\sqrt{\left(4 - x^2\right)}}$

2. (a) $\dfrac{-3}{\sqrt{\left(1 - 9x^2\right)}}$ (b) $\dfrac{-2}{3\sqrt{\left(9 - x^2\right)}}$

3. (a) $\dfrac{6}{1 + 4x^2}$ (b) $\dfrac{1}{4\sqrt{x}(1 + x)}$

4. (a) $\dfrac{2}{t\sqrt{\left(4t^2 - 1\right)}}$ (b) $\dfrac{4}{x\sqrt{\left(9x^2 - 16\right)}}$

5. (a) $\dfrac{-5}{\theta\sqrt{\left(\theta^2 - 4\right)}}$ (b) $\dfrac{-2}{x\sqrt{\left(x^4 - 1\right)}}$

6. (a) $\dfrac{-6}{1 + 4t^2}$ (b) $\dfrac{-1}{\theta\sqrt{\left(\theta^2 - 1\right)}}$

7. $\dfrac{1 + x^2}{\left(1 - x^2 + x^4\right)}$

8. (a) $\dfrac{6x}{\sqrt{\left(1 - 9x^2\right)}} + 2\sin^{-1} 3x$

  (b) $\dfrac{t}{\sqrt{\left(4t^2 - 1\right)}} + 2t\sec^{-1} 2t$

9. (a) $2\theta\cos^{-1}\left(\theta^2 - 1\right) - \dfrac{2\theta^2}{\sqrt{\left(2 - \theta^2\right)}}$

  (b) $\left(\dfrac{1 - x^2}{1 + x^2}\right) - 2x\tan^{-1}x$

10. (a) $\left(\dfrac{-2\sqrt{t}}{1 + t^2}\right) + \dfrac{1}{\sqrt{t}}\cot^{-1}t$

  (b) $\text{cosec}^{-1}\sqrt{x} - \dfrac{1}{2\sqrt{(x - 1)}}$

11. (a) $\dfrac{1}{x^3}\left\{\dfrac{3x}{\sqrt{\left(1 - 9x^2\right)}} - 2\sin^{-1} 3x\right\}$

  (b) $\dfrac{-1 + \dfrac{x}{\sqrt{1 - x^2}}\left(\cos^{-1} x\right)}{1 - x^2}$

Exercise 162 (page 441)

1. (a) 0.4812 (b) 2.0947 (c) 0.8089

2. (a) 0.6931 (b) 1.7627 (c) 2.1380

3. (a) 0.2554 (b) 0.7332 (c) 0.8673

Exercise 163 (page 444)

1. (a) $\dfrac{1}{\sqrt{\left(x^2 + 9\right)}}$ (b) $\dfrac{4}{\sqrt{\left(16x^2 + 1\right)}}$

2. (a) $\dfrac{2}{\sqrt{\left(t^2 - 9\right)}}$ (b) $\dfrac{1}{\sqrt{\left(4\theta^2 - 1\right)}}$

3. (a) $\dfrac{10}{25 - 4x^2}$ (b) $\dfrac{9}{1 - 9x^2}$

4. (a) $\dfrac{-4}{x\sqrt{\left(16 - 9x^2\right)}}$ (b) $\dfrac{-1}{2x\sqrt{\left(1 - 4x^2\right)}}$

5. (a) $\dfrac{-4}{x\sqrt{\left(x^2 + 16\right)}}$ (b) $\dfrac{-1}{2x\sqrt{\left(16x^2 + 1\right)}}$

6. (a) $\dfrac{14}{49 - 4x^2}$ (b) $\dfrac{3}{4\left(1 - 9t^2\right)}$

7. (a) $\dfrac{2}{\sqrt{\left(x^2 - 1\right)}}$ (b) $\dfrac{1}{2\sqrt{\left(x^2 + 1\right)}}$

8. (a) $\dfrac{-1}{(x - 1)\sqrt{[x(2 - x)]}}$ (b) 1

9. (a) $\dfrac{-1}{(t - 1)\sqrt{2t - 1}}$ (b) $-\text{cosec}\,x$

10. (a) $\dfrac{\theta}{\sqrt{\left(\theta^2 + 1\right)}} + \sinh^{-1}\theta$

  (b) $\dfrac{\sqrt{x}}{\sqrt{x^2 - 1}} + \dfrac{\cosh^{-1} x}{2\sqrt{x}}$

11. (a) $-\dfrac{1}{t^3}\left[\dfrac{1}{\sqrt{1 - t}} + 4\text{sech}^{-1}\sqrt{t}\right]$

  (b) $\dfrac{1 + 2x\tanh^{-1} x}{\left(1 - x^2\right)^2}$

12. $2x$

13. (a) $\sinh^{-1}\dfrac{x}{3} + c$ (b) $\dfrac{3}{2}\sinh^{-1}\dfrac{2x}{5} + c$

14. (a) $\cosh^{-1}\dfrac{x}{4} + c$ (b) $\cosh^{-1}\dfrac{t}{\sqrt{5}} + c$

15. (a) $\dfrac{1}{6}\tan^{-1}\dfrac{\theta}{6} + c$ (b) $\dfrac{3}{2\sqrt{8}}\tanh^{-1}\dfrac{x}{\sqrt{8}} + c$

Chapter 39

Exercise 164 (page 448)

1. $\dfrac{\partial z}{\partial x} = 2y, \dfrac{\partial z}{\partial y} = 2x$

2. $\dfrac{\partial z}{\partial x} = 3x^2 - 2y, \dfrac{\partial z}{\partial y} = -2x + 2y$

3. $\dfrac{\partial z}{\partial x} = \dfrac{1}{y}, \dfrac{\partial z}{\partial y} = -\dfrac{x}{y^2}$

4. $\dfrac{\partial z}{\partial x} = 4\cos(4x + 3y), \dfrac{\partial z}{\partial y} = 3\cos(4x + 3y)$

5. $\dfrac{\partial z}{\partial x} = 3x^2 y^2 + \dfrac{2y}{x^3}, \dfrac{\partial z}{\partial y} = 2x^3 y - \dfrac{1}{x^2} - \dfrac{1}{y^2}$

6. $\dfrac{\partial z}{\partial x} = -3\sin 3x \sin 4y, \dfrac{\partial z}{\partial y} = 4\cos 3x \cos 4y$

7. $\dfrac{\partial V}{\partial h} = \dfrac{1}{3}\pi r^2, \dfrac{\partial V}{\partial r} = \dfrac{2}{3}\pi rh$

8. Proof

9. $\dfrac{\partial z}{\partial x} = \left(\dfrac{n\pi b}{L}\right)\sin\left(\dfrac{n\pi}{L}\right)x\left\{c\cos\left(\dfrac{n\pi b}{L}\right)t\right.$

   $\left. -k\sin\left(\dfrac{n\pi b}{L}\right)t\right\}$

   $\dfrac{\partial z}{\partial y} = \left(\dfrac{n\pi}{L}\right)\cos\left(\dfrac{n\pi}{L}\right)x\left\{k\cos\left(\dfrac{n\pi b}{L}\right)t\right.$

   $\left. +c\sin\left(\dfrac{n\pi b}{L}\right)t\right\}$

10. (a) $\dfrac{\partial k}{\partial T} = \dfrac{A\Delta H}{RT^2}\mathrm{e}^{\frac{T\Delta S - \Delta H}{RT}}$

    (b) $\dfrac{\partial A}{\partial T} = -\dfrac{k\Delta H}{RT^2}\mathrm{e}^{\frac{\Delta H - T\Delta S}{RT}}$

    (c) $\dfrac{\partial(\Delta S)}{\partial T} = -\dfrac{\Delta H}{T^2}$

    (d) $\dfrac{\partial(\Delta H)}{\partial T} = \Delta S - R\ln\left(\dfrac{k}{A}\right)$

Exercise 165 (page 450)

1. (a) 8 (b) 18 (c) -12 (d) -12

2. (a) $-\dfrac{2}{x^2}$ (b) $-\dfrac{2}{y^2}$ (c) 0 (d) 0

3. (a) $-\dfrac{4y}{(x+3)^3}$ (b) $\dfrac{4x}{(x+3)^3}$

   (c) $\dfrac{2(x-y)}{(x+y)^3}$ (d) $\dfrac{2(x-y)}{(x+y)^3}$

4. (a) $\sinh x \cosh 2y$ (b) $4\sinh x \cosh 2y$

   (c) $2\cosh x \sinh 2y$ (d) $2\cosh x \sinh 2y$

5. (a) $\left(2 - x^2\right)\sin(x - 2y) + 4x\cos(x - 2y)$

   (b) $-4x^2 \sin(x - 2y)$

6. $\dfrac{-x}{\sqrt{(y^2 - x^2)^3}}, \dfrac{-x}{\sqrt{y^2 - x^2}}\left(\dfrac{1}{y^2} + \dfrac{x}{(y^2 - x^2)}\right)$,

   $\dfrac{\partial^2 z}{\partial x \partial y} = \dfrac{\partial^2 z}{\partial y \partial x} = \dfrac{y}{\sqrt{(y^2 - x^2)^3}}$

7. $-\dfrac{1}{\sqrt{2}}$ or -0.7071

8. Proof

Chapter 40

Exercise 166 (page 453)

1. $3x^2\,\mathrm{d}x + 2y\,\mathrm{d}y$

2. $(2y + \sin x)\,\mathrm{d}x + 2x\,\mathrm{d}y$

3. $\left(\dfrac{2y}{(x+y)^2}\right)\mathrm{d}x - \left(\dfrac{2x}{(x+y)^2}\right)\mathrm{d}y$ 4. $\ln y\,\mathrm{d}x + \dfrac{x}{y}\mathrm{d}y$

5. $\left(y + \dfrac{1}{2y\sqrt{x}}\right)\mathrm{d}x + \left(x - \dfrac{\sqrt{x}}{y^2}\right)\mathrm{d}y$

6. $b(2 + c)\,\mathrm{d}a + (2a - 6bc + ac)\,\mathrm{d}b + b(a - 3b)\,\mathrm{d}c$

7. $\mathrm{d}u = \cot(xy)[y\,\mathrm{d}x + x\,\mathrm{d}y]$

Exercise 167 (page 455)

1. $+226.2 \text{ cm}^3/\text{s}$ 2. 2520 units/s
3. 515.5 cm/s 4. $1.35 \text{ cm}^3/\text{s}$
5. $17.4 \text{ cm}^2/\text{s}$

Exercise 168 (page 457)

1. $+21$ watts 2. $+2\%$ 3. -1%
4. $+1.35 \text{ cm}^4$ 5. -0.179 cm 6. $+6\%$
7. $+2.2\%$

Chapter 41

Exercise 169 (page 463)

1. Minimum at $(0, 0)$

2. (a) Minimum at $(1, -2)$ (b) Saddle point at $(1, 2)$
   (c) Maximum at $(0, 1)$

3. Maximum point at $(0, 0)$, saddle point at $(4, 0)$

4. Minimum at $(0, 0)$

5. Saddle point at $(0, 0)$, minimum at $\left(\dfrac{1}{3}, \dfrac{1}{3}\right)$

Exercise 170 (page 467)

1. Minimum at $(-4, 4)$
2. 4 m by 4 m by 2 m, surface area $= 48$ m^2
3. Minimum at $(1, 0)$, minimum at $(-1, 0)$, saddle point at $(0, 0)$
4. Maximum at $(0, 0)$, saddle point at $(4, 0)$
5. Minimum at $(1, 2)$, maximum at $(-1, -2)$, saddle points at $(1, -2)$ and $(-1, 2)$
6. 150 m^2

Chapter 42

Exercise 171 (page 475)

1. $-\dfrac{1}{2}\cos(4x + 9) + c$
2. $\dfrac{3}{2}\sin(2\theta - 5) + c$
3. $\dfrac{4}{3}\tan(3t + 1) + c$
4. $\dfrac{1}{70}(5x - 3)^7 + c$
5. $-\dfrac{3}{2}\ln(2x - 1) + c$
6. $e^{3\theta + 5} + c$
7. 227.5
8. 4.333
9. 0.9428
10. 0.7369
11. 1.6 years

Exercise 172 (page 476)

1. $\dfrac{1}{12}\left(2x^2 - 3\right)^6 + c$
2. $-\dfrac{5}{6}\cos^6 t + c$
3. $\dfrac{1}{2}\sec^2 3x + c$ or $\dfrac{1}{2}\tan^2 3x + c$
4. $\dfrac{2}{9}\sqrt{(3t^2 - 1)^3} + c$
5. $\dfrac{1}{2}(\ln\theta)^2 + c$
6. $\dfrac{3}{2}\ln(\sec 2t) + c$
7. $4\sqrt{(e^4 + 4)} + c$
8. 1.763
9. 0.6000
10. 0.09259
11. $2\pi\sigma\left\{\sqrt{(9^2 + r^2)} - r\right\}$
12. $\dfrac{8\pi^2 I k T}{h^2}$
13. Proof
14. 11 min 50 s

Chapter 43

Exercise 173 (page 480)

1. $\dfrac{1}{2}\left(x - \dfrac{\sin 4x}{4}\right) + c$
2. $\dfrac{3}{2}\left(t + \dfrac{\sin 2t}{2}\right) + c$
3. $5\left(\dfrac{1}{3}\tan 3\theta - \theta\right) + c$
4. $-(\cot 2t + 2t) + c$

5. $\dfrac{\pi}{2}$ or 1.571
6. $\dfrac{\pi}{8}$ or 0.3927
7. -4.185
8. 0.6311

Exercise 174 (page 482)

1. $-\cos\theta + \dfrac{\cos^3\theta}{3} + c$
2. $\sin 2x - \dfrac{\sin^3 2x}{3} + c$
3. $-\dfrac{2}{3}\cos^3 t + \dfrac{2}{5}\cos^5 t + c$
4. $-\dfrac{\cos^5 x}{5} + \dfrac{\cos^7 x}{7} + c$
5. $\dfrac{3\theta}{4} - \dfrac{1}{4}\sin 4\theta + \dfrac{1}{32}\sin 8\theta + c$
6. $\dfrac{t}{8} - \dfrac{1}{32}\sin 4t + c$

Exercise 175 (page 483)

1. $-\dfrac{1}{2}\left(\dfrac{\cos 7t}{7} + \dfrac{\cos 3t}{3}\right) + c$
2. $\dfrac{\sin 2x}{2} - \dfrac{\sin 4x}{4} + c$
3. $\dfrac{3}{2}\left[\dfrac{\sin 7x}{7} + \dfrac{\sin 5x}{5}\right] + c$
4. $\dfrac{1}{4}\left(\dfrac{\cos 2\theta}{2} - \dfrac{\cos 6\theta}{6}\right) + c$
5. $\dfrac{3}{7}$ or 0.4286
6. 0.5973
7. 0.2474
8. -0.1999

Exercise 176 (page 484)

1. $5\sin^{-1}\dfrac{t}{2} + c$
2. $3\sin^{-1}\dfrac{x}{3} + c$
3. $2\sin^{-1}\dfrac{x}{2} + \dfrac{x}{2}\sqrt{(4 - x^2)} + c$
4. $\dfrac{8}{3}\sin^{-1}\dfrac{3t}{4} + \dfrac{t}{2}\sqrt{(16 - 9t^2)} + c$
5. $\dfrac{\pi}{2}$ or 1.571
6. 2.760

Exercise 177 (page 485)

1. $\dfrac{3}{2}\tan^{-1}\dfrac{t}{2} + c$
2. $\dfrac{5}{12}\tan^{-1}\dfrac{3\theta}{4} + c$
3. 2.356
4. 2.457

Exercise 178 (page 486)

1. $2\sinh^{-1}\dfrac{x}{4} + c$
2. $\dfrac{3}{\sqrt{5}}\sinh^{-1}\dfrac{\sqrt{5}}{3}x + c$
3. $\dfrac{9}{2}\sinh^{-1}\dfrac{x}{3} + \dfrac{x}{2}\sqrt{[x^2 + 9]} + c$

4. $\dfrac{25}{4}\sinh^{-1}\dfrac{2t}{5}+\dfrac{t}{2}\sqrt{[4t^2+25]}+c$

5. 3.525   **6.** 4.348

Exercise 179 (page 488)

1. $\cosh^{-1}\dfrac{t}{4}+c$ **2.** $\dfrac{3}{2}\cosh^{-1}\dfrac{2x}{3}+c$

3. $\dfrac{\theta}{2}\sqrt{(\theta^2-9)}-\dfrac{9}{2}\cosh^{-1}\dfrac{\theta}{3}+c$

4. $\theta\sqrt{\left(\theta^2-\dfrac{25}{4}\right)}-\dfrac{25}{4}\cosh^{-1}\dfrac{2\theta}{5}+c$

5. 2.634   **6.** 1.429

Chapter 44

Exercise 180 (page 490)

1. $2\ln(x-3)-2\ln(x+3)+c$ or $\ln\left(\dfrac{x-3}{x+3}\right)^2$

2. $5\ln(x+1)-\ln(x-3)+c$ or $\ln\left\{\dfrac{(x+1)^5}{(x-3)}\right\}+c$

3. $7\ln(x+4)-3\ln(x+1)-\ln(2x-1)+c$

  or $\ln\left(\dfrac{(x+4)^7}{(x+1)^3(2x-1)}\right)+c$

4. $x+2\ln(x+3)+6\ln(x-2)+c$

  or $x+\ln\left\{(x+3)^2(x-2)^6\right\}+c$

5. $\dfrac{3x^2}{2}-2x+\ln(x-2)-5\ln(x+2)+c$

6. 0.6275

7. 0.8122

8. $\dfrac{1}{3}$

9. Proof

10. 19.05 ms

Exercise 181 (page 492)

1. $4\ln(x+1)+\dfrac{7}{(x+1)}+c$

2. $5\ln(x-2)+\dfrac{10}{(x-2)}-\dfrac{2}{(x-2)^2}+c$

3. 1.663   **4.** 1.089   **5.** Proof

Exercise 182 (page 493)

1. $\ln(x^2+7)+\dfrac{3}{\sqrt7}\tan^{-1}\dfrac{x}{\sqrt7}-\ln(x-2)+c$

2. 0.5880   **3.** 0.2939   **4.** 0.1865   **5.** Proof

Chapter 45

Exercise 183 (page 496)

1. $-\dfrac{2}{1+\tan\dfrac{\theta}{2}}+c$ **2.** $\ln\left\{\dfrac{\tan\dfrac{x}{2}}{1+\tan\dfrac{x}{2}}\right\}+c$

3. $\dfrac{2}{\sqrt5}\tan^{-1}\left(\dfrac{1}{\sqrt5}\tan\dfrac{\alpha}{2}\right)+c$

4. $\dfrac{1}{5}\ln\left\{\dfrac{2\tan\dfrac{x}{2}-1}{\tan\dfrac{x}{2}+2}\right\}+c$

Exercise 184 (page 497)

1. $\dfrac{2}{3}\tan^{-1}\left(\dfrac{5\tan\dfrac{\theta}{2}+4}{3}\right)+c$

2. $\dfrac{1}{\sqrt3}\ln\left(\dfrac{\tan\dfrac{x}{2}+2-\sqrt3}{\tan\dfrac{x}{2}+2+\sqrt3}\right)+c$

3. $\dfrac{1}{\sqrt{11}}\ln\left\{\dfrac{\tan\dfrac{p}{2}-4-\sqrt{11}}{\tan\dfrac{p}{2}-4+\sqrt{11}}\right\}+c$

4. $\dfrac{1}{\sqrt7}\ln\left(\dfrac{3\tan\dfrac{\theta}{2}-4-\sqrt7}{3\tan\dfrac{\theta}{2}-4+\sqrt7}\right)+c$

5. $\dfrac{1}{2\sqrt2}\ln\left\{\dfrac{\sqrt2+\tan\dfrac{t}{2}}{\sqrt2-\tan\dfrac{t}{2}}\right\}+c$ **6.** Proof

7. Proof

Chapter 46

Exercise 185 (page 502)

1. $\dfrac{e^{2x}}{2}\left(x-\dfrac{1}{2}\right)+c$ **2.** $-\dfrac{4}{3}e^{-3x}\left(x+\dfrac{1}{3}\right)+c$

3. $-x\cos x+\sin x+c$

4. $\dfrac{5}{2}\left(\theta\sin 2\theta + \dfrac{1}{2}\cos 2\theta\right) + c$

5. $\dfrac{3}{2}e^{2t}\left(t^2 - t + \dfrac{1}{2}\right) + c$ **6.** 16.78

7. 0.2500 **8.** 0.4674

9. 15.78

Exercise 186 (page 504)

1. $\dfrac{2}{3}x^3\left(\ln x - \dfrac{1}{3}\right) + c$ **2.** $2x(\ln 3x - 1) + c$

3. $\dfrac{\cos 3x}{27}\left(2 - 9x^2\right) + \dfrac{2}{9}x\sin 3x + c$

4. $\dfrac{2}{29}e^{5x}(2\sin 2x + 5\cos 2x) + c$

5. $2[\theta\tan\theta - \ln(\sec\theta)] + c$ **6.** 0.6363

7. 11.31 **8.** -1.543

9. 12.78 **10.** Proof

11. $C = 0.66, S = 0.41$

Chapter 47

Exercise 187 (page 507)

1. $e^x(x^4 - 4x^3 + 12x^2 - 24x + 24) + c$

2. $e^{2t}\left(\dfrac{1}{2}t^3 - \dfrac{3}{4}t^2 + \dfrac{3}{4}t - \dfrac{3}{8}\right) + c$ **3.** 6.493

Exercise 188 (page 509)

1. $x^5\sin x + 5x^4\cos x - 20x^3\sin x - 60x^2\cos x$
$+ 120x\sin x + 120\cos x + c$

2. -134.87

3. $-x^5\cos x + 5x^4\sin x + 20x^3\cos x - 60x^2\sin x$
$- 120x\cos x + 120\sin x + c$

4. 62.89

Exercise 189 (page 512)

1. $-\dfrac{1}{7}\sin^6 x\cos x - \dfrac{6}{35}\sin^4 x\cos x - \dfrac{8}{35}\sin^2 x\cos x$
$- \dfrac{16}{35}\cos x + c$

2. 4

3. $\dfrac{8}{15}$

4. $\dfrac{1}{6}\cos^5 x\sin x + \dfrac{5}{24}\cos^3 x\sin x + \dfrac{5}{16}\cos x\sin x$
$+ \dfrac{5}{16}x + c$

5. $\dfrac{16}{35}$

Exercise 190 (page 514)

1. $\dfrac{8}{105}$ **2.** $\dfrac{13}{15} - \dfrac{\pi}{4}$ or 0.08127 **3.** $\dfrac{8}{315}$

4. $x(\ln x)^4 - 4x(\ln x)^3 + 12x(\ln x)^2 - 24x\ln x + 24x + c$

5. Proof

Chapter 48

Exercise 191 (page 517)

1. 12 **2.** 3.5 **3.** 0.5
4. -174 **5.** 405 **6.** -157.5
7. 15π or 47.12 **8.** 112
9. 5 **10.** 170 cm^4

Exercise 192 (page 518)

1. 15 **2.** 15 **3.** 60 **4.** -8

5. $\dfrac{\pi^4}{2}$ or 48.70 **6.** -9 **7.** 8 **8.** 18

Chapter 49

Exercise 193 (page 522)

1. 1.569 **2.** 6.979 **3.** 0.672 **4.** 0.843

Exercise 194 (page 524)

1. 3.323 **2.** 0.997 **3.** 0.605 **4.** 0.799

Exercise 195 (page 527)

1. 1.187 **2.** 1.034 **3.** 0.747
4. 0.571 **5.** 1.260
6. (a) 1.875 (b) 2.107 (c) 1.765 (d) 1.916
7. (a) 1.585 (b) 1.588 (c) 1.583 (d) 1.585
8. (a) 10.194 (b) 10.007 (c) 10.070
9. (a) 0.677 (b) 0.674 (c) 0.675
10. 28.8 m **11.** 0.485 m

Chapter 50

Exercise 196 (page 535)

1. $-\dfrac{1}{3}\ln\left(\dfrac{x^3 - y^3}{x^3}\right) = \ln x + c$

2. $y = x(c - \ln x)$

3. $x^2 = 2y^2\left(\ln y + \dfrac{1}{2}\right)$

4. $-\dfrac{1}{2}\ln\left(1 + \dfrac{2y}{x} - \dfrac{y^2}{x^2}\right) = \ln x + c$ or $x^2 + 2xy - y^2 = k$

5. $x^2 + xy - y^2 = 1$

Exercise 197 (page 536)

1. $y^4 = 4x^4(\ln x + c)$

2. $\dfrac{1}{5}\left[\dfrac{3}{13}\ln\left(\dfrac{13y - 3x}{x}\right) - \ln\left(\dfrac{y - x}{x}\right)\right] = \ln x + c$

3. $(x + y)^2 = 4x^3$ 4. Proof

5. $y^3 = x^3(3\ln x + 64)$ 6. Proof

Chapter 51

Exercise 198 (page 539)

1. $y = 3 + \dfrac{c}{x}$ 2. $y = \dfrac{1}{2} + ce^{-x^2}$

3. $y = \dfrac{5t}{2} + \dfrac{c}{t}$ 4. $y = \dfrac{x^3}{5} - \dfrac{x}{3} + \dfrac{47}{15x^2}$

5. $y = 1 + ce^{-x^2/2}$ 6. $y = \dfrac{1}{2}x + \dfrac{1}{4} + ce^{2x}$

Exercise 199 (page 540)

1. $y = \dfrac{1}{2} + \cos^2 x$ 2. $\theta = \dfrac{1}{t}(\sin t - \pi\cos t)$

3. Proof   4. Proof

5. Proof   6. $v = \dfrac{b}{a^2} - \dfrac{bt}{a} + \left(u - \dfrac{b}{a^2}\right)e^{-at}$

7. Proof

8. $C = \left(\dfrac{b}{m} + d\right)\left(1 - e^{-\frac{mt}{a}}\right) + c_0 e^{-\frac{mt}{a}}$

9. $v = k\left\{\dfrac{1}{c} - \dfrac{e^{-t}}{c - 1} + \dfrac{e^{-ct}}{c(c - 1)}\right\}$

Chapter 52

Exercise 200 (page 547)

1.

x	y
1.0	2
1.1	2.1
1.2	2.209091
1.3	2.325000
1.4	2.446154
1.5	2.571429

2.

x	y	$(y')_0$
0	1	0
0.2	1	−0.2
0.4	0.96	−0.368
0.6	0.8864	−0.46368
0.8	0.793664	−0.469824
1.0	0.699692	

3. (a)

x	y
2.0	1
2.1	0.85
2.2	0.709524
2.3	0.577273
2.4	0.452174
2.5	0.333334

(b) 1.206%

4. (a)

x	y
2.0	1
2.2	1.2
2.4	1.421818
2.6	1.664849
2.8	1.928718
3.0	2.213187

(b) 1.596%

Exercise 201 (page 552)

1.

x	y	y'
1.0	2	1
1.1	2.10454546	1.08677686
1.2	2.216666672	1.152777773
1.3	2.33461539	1.204142008
1.4	2.457142859	1.244897958
1.5	2.5883333335	

2. (a) 0.412% (b) 0.000000214%

3. (a)

x	y	y'
0	1	1
0.1	1.11	1.21
0.2	1.24205	1.44205
0.3	1.398465	1.698465
0.4	1.581804	1.981804
0.5	1.794893	

(b) 0.117%

4.

x	y	$(y')_0$
0	1	0
0.2	0.98	−0.192
0.4	0.925472	−0.3403776
0.6	0.84854666	−0.41825599
0.8	0.76433779	−0.42294046
1.0	0.68609380	

Exercise 202 (page 557)

1.

n	x_n	y_n
0	1.0	2.0
1	1.1	2.104545
2	1.2	2.216667
3	1.3	2.334615
4	1.4	2.457143
5	1.5	2.533333

2.

n	x_n	y_n
0	0	1.0
1	0.2	0.980395
2	0.4	0.926072
3	0.6	0.848838
4	0.8	0.763649
5	1.0	0.683952

3. (a)

n	x_n	y_n
0	2.0	1.0
1	2.1	0.854762
2	2.2	0.718182
3	2.3	0.589130
4	2.4	0.466667
5	2.5	0.340000

(b) No error

Chapter 53

Exercise 203 (page 563)

1. $y = Ae^{\frac{2}{3}t} + Be^{-\frac{1}{2}t}$ **2.** $\theta = (At + B)e^{-\frac{1}{2}t}$

3. $y = e^{-x}\{A\cos 2x + B\sin 2x\}$

4. $y = 3e^{\frac{2}{3}x} + 2e^{-\frac{3}{2}x}$ **5.** $y = 4e^{\frac{1}{4}t} - 3e^t$

6. $y = 2xe^{-\frac{5}{3}x}$ **7.** $x = 2(1 - 3t)e^{3t}$

8. $y = 2e^{-3x}\{2\cos 2x + 3\sin 2x\}$

9. $\theta = e^{-2.5t}\{3\cos 5t + 2\sin 5t\}$

Exercise 204 (page 565)

1. Proof   **2.** $s = c\cos at$

3. $\theta = e^{-2t}\{0.3\cos 6t + 0.1\sin 6t\}$

4. $x = \{s + (u + ns)t\}e^{-nt}$

5. $i = \dfrac{1}{20}\left(e^{-160t} - e^{-840t}\right)$

6. $s = 4te^{-\frac{3}{2}t}$

Chapter 54

Exercise 205 (page 570)

1. $y = Ae^{\frac{1}{2}x} + Be^{-3x} - 2$

2. $y = Ae^{\frac{1}{3}x} + Be^{-x} - 2 - \frac{3}{2}x$

3. $y = \frac{2}{7}\left(3e^{-\frac{4}{3}x} + 4e^x\right) - 2$

4. $y = -\left(2 + \frac{3}{4}x\right)e^{\frac{2}{3}x} + 2 + \frac{3}{4}x$

5. (a) $q = \frac{1}{20} - \left(\frac{5}{2}t + \frac{1}{20}\right)e^{-50t}$

   (b) $q = \frac{1}{20}(1 - \cos 50t)$

6. $\theta = 2\left(te^{-2t} + 1\right)$

Exercise 206 (page 572)

1. $y = Ae^{3x} + Be^{-2x} - \frac{1}{3}e^x$

2. $y = Ae^{4x} + Be^{-x} - \frac{3}{5}xe^{-x}$

3. $y = A\cos 3x + B\sin 3x + 2e^{2x}$

4. $y = (At + B)e^{\frac{t}{3}} + \frac{2}{3}t^2e^{\frac{t}{3}}$

5. $y = \frac{5}{44}\left(e^{-2x} - e^{\frac{1}{5}x}\right) + \frac{1}{4}e^x$

6. $y = 2e^{3t}\left(1 - 3t + t^2\right)$

Exercise 207 (page 574)

1. $y = Ae^{\frac{3}{2}x} + Be^{-x} - \frac{1}{5}(11\sin 2x - 2\cos 2x)$

2. $y = (Ax + B)e^{2x} - \frac{4}{5}\sin x + \frac{3}{5}\cos x$

3. $y = A\cos x + B\sin x + 2x\sin x$

4. $y = \frac{1}{170}\left(6e^{4x} - 51e^{-x}\right) - \frac{1}{34}(15\sin x - 9\cos x)$

5. $y = \frac{k}{(n^4 - p^4)}\left(p^2\left(\sin pt - \frac{p}{n}\sin nt\right)\right.$
   $\left. + n^2(\cos pt - \cos nt)\right)$

6. Proof

7. $q = (10t + 0.01)e^{-1000t} + 0.024\sin 200t$
   $- 0.010\cos 200t$

Exercise 208 (page 576)

1. $y = Ae^{\frac{1}{4}x} + Be^{\frac{1}{2}x} + 2x + 12 + \frac{8}{17}(6\cos x - 7\sin x)$

2. $y = Ae^{2\theta} + Be^{\theta} + \frac{1}{2}(\sin 2\theta + \cos 2\theta)$

3. $y = Ae^x + Be^{-2x} - \frac{3}{4} - \frac{1}{2}x - \frac{1}{2}x^2 + \frac{1}{4}e^{2x}$

4. $y = e^t(A\cos t + B\sin t) - \frac{t}{2}e^t\cos t$

5. $y = \frac{4}{3}e^{5x} - \frac{10}{3}e^{2x} - \frac{1}{3}xe^{2x} + 2$

6. $y = 2e^{-\frac{3}{2}x} - 2e^{2x} + \frac{3e^x}{29}(3\sin x - 7\cos x)$

Chapter 55

Exercise 209 (page 579)

1. (a) $16e^{2x}$ (b) $\frac{1}{4}e^{\frac{1}{2}t}$

2. (a) $81\sin 3t$ (b) $-1562.5\cos 5\theta$

3. (a) $256\cos 2x$ (b) $-\frac{2^9}{3^8}\sin\frac{2}{3}t$

4. (a) $(9!)x^2$ (b) $630\,t$

5. (a) $32\cosh 2x$ (b) $1458\sinh 3x$

6. (a) $128\sinh 2x$ (b) $729\cosh 3x$

7. (a) $-\frac{12}{\theta^4}$ (b) $\frac{240}{t^7}$

Exercise 210 (page 581)

1. $x^2y^{(n)} + 2nxy^{(n-1)} + n(n-1)y^{(n-2)}$

2. $y^{(n)} = e^{2x}2^{n-3}\{8x^3 + 12nx^2 + n(n-1)(6x)$
   $+ n(n-1)(n-2)\}$

   $y^{(3)} = e^{2x}(8x^3 + 36x^2 + 36x + 6)$

3. $y^{(4)} = 2e^{-x}(x^3 - 12x^2 + 36x - 24)$

4. $y^{(5)} = (60x - x^3)\sin x + (15x^2 - 60)\cos x$

5. $y^{(4)} = -4e^{-t}\sin t$ 6. $y^{(3)} = x^2(47 + 60\ln 2x)$

7. Proof 8. $y^{(5)} = e^{2x}2^4(2x^3 + 19x^2 + 50x + 35)$

Exercise 211 (page 584)

1. $y = \left\{1 - \frac{x^2}{2!} + \frac{5x^4}{4!} - \frac{5 \times 9x^6}{6!} + \frac{5 \times 9 \times 13x^8}{8!} - \cdots\right\}$
   $+ 2\left\{x - \frac{3x^3}{3!} + \frac{3 \times 7x^5}{5!} - \frac{3 \times 7 \times 11x^7}{7!} + \cdots\right\}$

2. Proof

3. $y = 1 + x + 2x^2 + \dfrac{x^3}{3} - \dfrac{x^5}{8} + \dfrac{x^7}{16} + \cdots\cdots$

4. $y = \left\{ 1 - \dfrac{1}{2^2}x^2 + \dfrac{1}{2^2 \times 4^2}x^4 - \dfrac{1}{2^2 \times 4^2 \times 6^2}x^6 + \cdots \right\}$

$\qquad + 2\left\{ x - \dfrac{x^3}{3^2} + \dfrac{x^5}{3^2 \times 5^2} - \dfrac{x^7}{3^2 \times 5^2 \times 7^2} \cdots \right\}$

Exercise 212 (page 591)

1. $y = A\left\{ 1 + x + \dfrac{x^2}{(2 \times 3)} + \dfrac{x^3}{(2 \times 3)(3 \times 5)} \right.$

$\qquad \left. + \dfrac{x^4}{(2 \times 3 \times 4)(3 \times 5 \times 7)} + \cdots \right\}$

$\qquad + Bx^{\frac{1}{2}}\left\{ 1 + \dfrac{x}{(1 \times 3)} + \dfrac{x^2}{(1 \times 2)(3 \times 5)} \right.$

$\qquad \cdot + \dfrac{x^3}{(1 \times 2 \times 3)(3 \times 5 \times 7)}$

$\qquad \left. + \dfrac{x^4}{(1 \times 2 \times 3 \times 4)(3 \times 5 \times 7 \times 9)} + \cdots \right\}$

2. $y = A\left\{ 1 - \dfrac{x^2}{2!} + \dfrac{x^4}{4!} - \cdots \right\} + B\left\{ x - \dfrac{x^3}{3!} + \dfrac{x^5}{5!} - \cdots \right\}$

$\qquad = P\cos x + Q\sin x$

3. $y = A\left\{ 1 + \dfrac{x}{(1 \times 4)} + \dfrac{x^2}{(1 \times 2)(4 \times 7)} \right.$

$\qquad \left. + \dfrac{x^3}{(1 \times 2 \times 3)(4 \times 7 \times 10)} + \cdots \right\}$

$\qquad + Bx^{-\frac{1}{3}}\left\{ 1 + \dfrac{x}{(1 \times 2)} + \dfrac{x^2}{(1 \times 2)(2 \times 5)} \right.$

$\qquad \left. + \dfrac{x^3}{(1 \times 2 \times 3)(2 \times 5 \times 8)} + \cdots \right\}$

4. Proof

Exercise 213 (page 595)

1. $y = Ax^2\left\{ 1 - \dfrac{x^2}{12} + \dfrac{x^4}{384} - \cdots \right\}$

$\qquad$ or $A\left\{ x^2 - \dfrac{x^4}{12} + \dfrac{x^6}{384} - \cdots \right\}$

2. $J_3(x) = \left(\dfrac{x}{2}\right)^3\left\{ \dfrac{1}{\Gamma 4} - \dfrac{x^2}{2^2\Gamma 5} + \dfrac{x^4}{2^5\Gamma 6} - \cdots \right\}$

$\qquad$ or $\dfrac{x^3}{8\Gamma 4} - \dfrac{x^5}{2^5\Gamma 5} + \dfrac{x^7}{2^8\Gamma 6} - \cdots$

3. $J_0(x) = 0.765$, $J_1(x) = 0.440$

Exercise 214 (page 600)

1. (a) $y = a_0 + a_1\left(x + \dfrac{x^3}{3} + \dfrac{x^5}{5} + \cdots \right)$

$\qquad$ (b) $y = a_0\left(1 - 3x^2\right) + a_1\left(x - \dfrac{2}{3}x^3 - \dfrac{1}{5}x^5 - \cdots \right)$

2. (a) x

$\qquad$ (b) $\dfrac{1}{8}\left(35x^4 - 30x^2 + 3\right)$ (c) $\dfrac{1}{8}\left(63x^5 - 70x^3 + 15x\right)$

Chapter 56

Exercise 215 (page 604)

1. $u = 2ty^2 + f(t)$ $\qquad$ **2.** $u = t^2(\cos\theta - 1) + 2t$

3. Proof $\qquad\qquad\quad$ **4.** Proof

5. $u = -4e^y\cos 2x - \cos x + 4\cos 2x + 2y^2 - 4e^y + 4$

6. $u = y\left(\dfrac{x^4}{3} - \dfrac{x^2}{2}\right) + x\cos 2y + \sin y$

7. $u = -\sin(x + t) + x + \sin x + 2t + \sin t$

8. Proof

9. $u = \sin x \sin y + \dfrac{x^2}{2} + 2\cos y - \dfrac{\pi^2}{2}$

10. Proof

Exercise 216 (page 606)

1. $T = Ae^{3t} + Be^{-3t}$ $\qquad$ **2.** $T = A\cos 3t + B\sin 3t$

3. $X = Ae^x + Be^{-x}$ $\qquad$ **4.** $X = A\cos x + B\sin x$

Exercise 217 (page 609)

1. $u(x,t) = \dfrac{12}{\pi^2}\displaystyle\sum_{n=1}^{\infty}\dfrac{1}{n^2}\sin\dfrac{n\pi}{2}\sin\dfrac{n\pi x}{40}\cos\dfrac{3n\pi t}{40}$

2. $u(x,t) = \dfrac{8}{\pi^2}\displaystyle\sum_{n=1}^{\infty}\dfrac{1}{n^2}\sin\dfrac{n\pi}{2}\sin\dfrac{n\pi x}{80}\cos\dfrac{n\pi t}{10}$

Exercise 218 (page 611)

1. $u(x,t) = \dfrac{40}{\pi}\displaystyle\sum_{n(\text{odd})=1}^{\infty}\dfrac{1}{n}e^{-\frac{n^2\pi^2c^2t}{16}}\sin\dfrac{n\pi x}{4}$

2. $u(x,t) = \left(\dfrac{8}{\pi}\right)^3\displaystyle\sum_{n(\text{odd})=1}^{\infty}\dfrac{1}{n^3}e^{-\frac{n^2\pi^2t}{64}}\sin\dfrac{n\pi x}{8}$

3. $u(x,t) = \dfrac{320}{\pi^2}\displaystyle\sum_{n(\text{odd})=1}^{\infty}\dfrac{1}{n^2}\sin\dfrac{n\pi}{2}\sin\dfrac{n\pi x}{20}e^{-\left(\frac{n^2\pi^2t}{400}\right)}$

Exercise 219 (page 614)

1. $u(x, y) = \dfrac{20}{\pi} \displaystyle\sum_{n(\text{odd})=1}^{\infty} \dfrac{1}{n} \text{cosech}\, n\pi \, \sin n\pi x$
$$\sinh n\pi(y - 2)$$

2. $u(x, y) = \dfrac{216}{\pi^3} \displaystyle\sum_{n(\text{odd})=1}^{\infty} \dfrac{1}{n^3} \text{cosech}\, \dfrac{2n\pi}{3} \, \sin \dfrac{n\pi x}{3}$
$$\sinh \dfrac{n\pi}{3}(2 - y)$$

Chapter 57

Exercise 220 (page 620)

1. (a) continuous (b) continuous (c) discrete
   (d) continuous
2. (a) discrete (b) continuous (c) discrete
   (d) discrete

Exercise 221 (page 623)

1. If one symbol is used to represent ten vehicles, working correct to the nearest five vehicles, gives 3.5, 4.5, 6, 7, 5 and 4 symbols respectively.
2. If one symbol represents 200 components, working correct to the nearest 100 components gives: Mon 8, Tues 11, Wed 9, Thurs 12 and Fri 6.5
3. Six equally spaced horizontal rectangles, whose lengths are proportional to 35, 44, 62, 68, 49 and 41 respectively
4. Five equally spaced horizontal rectangles, whose lengths are proportional to 1580, 2190, 1840, 2385 and 1280 units, respectively
5. Six equally spaced vertical rectangles, whose heights are proportional to 35, 44, 62, 68, 49 and 41 units, respectively
6. Six equally spaced vertical rectangles, whose heights are proportional to 1580, 2190, 1840, 2385 and 1280 units, respectively
7. Three rectangles of equal height, subdivided in the percentages shown; P increases by 20% at the expense of Q and R
8. Four rectangles of equal heights, subdivided as follows: **Week 1**: 18%, 7%, 35%, 12%, 28% **Week 2**: 20%, 8%, 32%, 13%, 27% **Week 3**: 22%, 10%, 29%, 14%, 25% **Week 4**: 20%, 9%, 27%, 19%, 25%. Little change in centres A and B, there is a reduction of around 8% in centre C, an increase of around 7% in centre D and a reduction of about 3% in centre E.

9. A circle of any radius, subdivided into sectors, having angles of 7.5°, 22.5°, 52.5°, 167.5° and 110°, respectively.
10. A circle of any radius, subdivided into sectors, having angles of 107°, 156°, 29° and 68°, respectively.
11. (a) £495 (b) 88
12. (a) £16 450 (b) 138

Exercise 222 (page 629)

1. There is no unique solution, but one solution is: 39.3 – 39.4 1; 39.5 – 39.6 5; 39.7 – 39.8 9; 39.9 – 40.0 17; 40.1 – 40.2 15; 40.3 – 40.4 7; 40.5 – 40.6 4; 40.7 – 40.8 2;
2. Rectangles, touching one another, having mid-points of 39.35, 39.55, 39.75, 39.95.... and heights of 1, 5, 9, 17,...
3. There is no unique solution, but one solution is: 20.5 – 20.9 3; 21.0 – 21.4 10; 21.5 – 21.9 11; 22.0 – 22.4 13; 22.5 – 22.9 9; 23.0 – 23.4 2
4. There is no unique solution, but one solution is: 1 – 10 3; 11 – 19 7; 20 – 22 12; 23 – 25 11; 26 – 28 10; 29 – 38 5; 39 – 48 2
5. 20.95 3; 21.45 13; 21.95 24; 22.45 37; 22.95 46; 23.45 48
6. Rectangles, touching one another, having mid-points of 5.5, 15, 21, 24, 33.5 and 43.5. The heights of the rectangles (frequency per unit class range) are 0.3, 0.78, 4, 3.67, 3.33, 0.5 and 0.2
7. (10.95 2), (11.45 9), (11.95 19), (12.45 31), (12.95 42), (13.45 50)
8. Ogive
9. (a) There is no unique solution, but one solution is: 2.05 – 2.09 3; 2.10 – 2.14 10; 2.15 – 2.19 11; 2.20 – 2.24 13; 2.25 – 2.29 9; 2.30 – 2.34 2. (b) Rectangles, touching one another, having mid-points of 2.07, 2.12 .... and heights of 3, 10 ... (c) Using the frequency distribution given in the solution to part (a) gives: 2.095 3; 2.145 13; 2.195 24; 2.245 37; 2.295 46; 2.345 48. (d) A graph of cumulative frequency against upper class boundary having the co-ordinates given in part (c).

Chapter 58

Exercise 223 (page 633)

1. Mean $7\dfrac{1}{3}$, median 8, mode 8
2. Mean 27.25, median 27, mode 26

3. Mean 4.7225, median 4.72, mode 4.72
4. Mean 115.2, median 126.4, no mode

Exercise 224 (page 634)

1. 23.85 kg   2. 171.7 cm
3. Mean 89.5, median 89, mode 88.2
4. Mean 2.02158 cm, median 2.02152 cm, Mode 2.02167 cm

Exercise 225 (page 636)

1. 4.60   2. 2.83 μF
3. Mean 34.53 MPa, standard deviation 0.07474 MPa
4. 0.296 kg   5. 9.394 cm
6. 0.00544 cm

Exercise 226 (page 637)

1. 30, 27.5, 33.5 days   2. 27, 26, 33 faults
3. $Q_1 = 164.5$ cm, $Q_2 = 172.5$ cm, $Q_3 = 179$ cm, 7.25 cm
4. 37 and 38; 40 and 41   5. 40, 40, 41; 50, 51, 51

Chapter 59

Exercise 227 (page 642)

1. (a) $\frac{2}{9}$ or 0.2222 (b) $\frac{7}{9}$ or 0.7778
2. (a) $\frac{23}{139}$ or 0.1655 (b) $\frac{47}{139}$ or 0.3381 (c) $\frac{69}{139}$ or 0.4964
3. (a) $\frac{1}{6}$ (b) $\frac{1}{6}$ (c) $\frac{1}{36}$
4. $\frac{5}{36}$
5. (a) $\frac{2}{5}$ (b) $\frac{1}{5}$ (c) $\frac{4}{15}$ (d) $\frac{13}{15}$
6. (a) $\frac{1}{250}$ (b) $\frac{1}{200}$ (c) $\frac{9}{1000}$ (d) $\frac{1}{50\,000}$

Exercise 228 (page 644)

1. (a) 0.6 (b) 0.2 (c) 0.15
2. (a) 0.64 (b) 0.32
3. 0.0768
4. (a) 0.4912 (b) 0.4211
5. (a) 89.38% (b) 10.25%
6. (a) 0.0227 (b) 0.0234 (c) 0.0169

Exercise 229 (page 646)

1. (a) 210 (b) 3024   2. (a) 792 (b) 15
3. 210   4. 3003
5. $\frac{10}{^{49}C_6} = \frac{10}{13\,983\,816} = \frac{1}{1\,398\,382}$ or 715×10^{-9}

Exercise 230 (page 647)

1. 53.26%   2. 70.32%   3. 7.48%

Chapter 60

Exercise 231 (page 654)

1. (a) 0.0186 (b) 0.9976
2. (a) 0.2316 (b) 0.1408
3. (a) 0.7514 (b) 0.0019
4. (a) 0.9655 (b) 0.0028
5. Vertical adjacent rectangles, whose heights are proportional to 0.0313, 0.1563, 0.3125, 0.3125, 0.1563 and 0.0313
6. Vertical adjacent rectangles, whose heights are proportional to 0.0280, 0.1306, 0.2613, 0.2903, 0.1935, 0.0774, 0.0172 and 0.0016
7. 0.0574

Exercise 232 (page 657)

1. 0.0613   2. 0.5768
3. (a) 0.1771 (b) 0.5153   4. 0.9856
5. The probabilities of the demand for 0, 1, 2,..... tools are 0.0067, 0.0337, 0.0842, 0.1404, 0.1755, 0.1755, 0.1462, 0.1044, 0.0653, ... This shows that the probability of wanting a tool eight times a day is 0.0653, i.e. less than 10%. Hence seven should be kept in the store
6. Vertical adjacent rectangles having heights proportional to 0.4966, 0.3476, 0.1217, 0.0284, 0.0050 and 0.0007

Chapter 61

Exercise 233 (page 662)

1. 6   2. 22
3. (a) 479 (b) 63 (c) 21   4. 4

5. (a) 131 (b) 553   **6.** (a) 15 (b) 4

7. (a) 65 (b) 209 (c) 89 **8.** (a) 1 (b) 85 (c) 13

Exercise 234 (page 665)

1. Graphically, $\bar{x} = 27.1, \sigma = 0.3$; by calculation, $\bar{x} = 27.079, \sigma = 0.3001$;

2. (a) $\bar{x} = 23.5\,\text{kN}, \sigma = 2.9\,\text{kN}$

(b) $\bar{x} = 23.364\,\text{kN}, \sigma = 2.917\,\text{kN}$

Chapter 62

Exercise 235 (page 669)

1. 0.999   **2.** -0.916 **3.** 0.422   **4.** 0.999

5. -0.962 **6.** 0.632   **7.** 0.937

Chapter 63

Exercise 236 (page 675)

1. $Y = -256 + 80.6X$ **2.** $Y = 0.0477 + 0.216X$

3. $X = 3.20 + 0.0124Y$ **4.** $X = -0.056 + 4.56Y$

5. $Y = 1.142 + 2.268X$ **6.** $X = -0.483 + 0.440Y$

7. (a) 7.95 V (b) 17.1 mA **8.** $Y = 0.881 - 0.0290X$

9. $X = 30.194 - 34.039Y$ **10.** (a) 0.417 s (b) 21.7 N

Chapter 64

Exercise 237 (page 682)

1. $\mu_{\bar{x}} = \mu = 22.4\,\text{cm}\ \sigma_{\bar{x}} = 0.0080\,\text{cm}$

2. $\sigma_{\bar{x}} = 0.0079\,\text{cm}$

3. (a) $\mu_{\bar{x}} = 1.70\,\text{cm}, \sigma_{\bar{x}} = 2.91 \times 10^{-3}\,\text{cm}$

(b) $\mu_{\bar{x}} = 1.70\,\text{cm}, \sigma_{\bar{x}} = 2.89 \times 10^{-3}\,\text{cm}$

4. 0.023

5. 0.497

6. 0.0038

7. (a) 0.0179 (b) 0.740 (c) 0.242

Exercise 238 (page 686)

1. 66.89 and 68.01 mm, 66.72 and 68.18 mm

2. (a) 2.355Mg to 2.445Mg; 2.341Mg to 2.459Mg

(b) 86%

3. $12.73 \times 10^{-4}\text{m}^\circ\text{C}^{-1}$ to $12.89 \times 10^{-4}\text{m}^\circ\text{C}^{-1}$

4. (a) at least 68 (b) at least 271

5. 10.91t to 11.27t

6. 45.6 seconds

Exercise 239 (page 690)

1. 5.133 MPa to 5.207 MPa

2. 5.125 MPa to 5.215 MPa

3. $1.10\,\Omega\,\text{m}^{-1}$ to $1.15\,\Omega\,\text{m}^{-1}$

4. 95%

Chapter 65

Exercise 240 (page 698)

1. (a) 28.1% (b) 4.09% (c) 0.19%

2. (a) 55.2% (b) 4.65% (c) 0.07%

3. (a) 35.3% (b) 18.5% (c) 8.4%

4. (a) 32.3% (b) 20.1% (c) 11.9%

Exercise 241 (page 702)

1. z (sample) $= 3.54, z_\alpha = 2.58$, hence hypothesis is rejected, where z_α is the z-value corresponding to a level of significance of α

2. $t_{0.95}, \nu_8 = 1.86, |t| = 1.89$, hence null hypothesis rejected

3. z (sample) $= 2.85, z_\alpha = \pm 2.58$, hence hypothesis is rejected

4. $\bar{x} = 10.38, s = 0.33, t_{0.95}, \nu_{19} = 1.73$, $|t| = 1.72$, hence hypothesis is accepted

5. $|t| = 3.00$, (a) $t_{0.975}, \nu_9 = 2.26$, hence hypothesis rejected,   (b) $t_{0.995}, \nu_9 = 3.25$, hence hypothesis is accepted

6. $|t| = 3.08$, (a) $t_{0.95}, \nu_5 = 2.02$, hence claim supported,   (b) $t_{0.99}, \nu_5 = 3.36$, hence claim not supported

Exercise 242 (page 706)

1. Take $\bar{x}$ as $24 + 15$, i.e. 39 hours, $z = 1.28, z_{0.05}$, one-tailed test $= 1.645$, hence hypothesis is accepted

2. $z = 2.357, z_{0.05}$, two-tailed test $= \pm 1.96$, hence hypothesis is rejected

3. $\bar{x}_1 = 23.7, s_1 = 1.73, \sigma_1 = 1.93, x_2 = 25.7$, $s_2 = 2.50, \sigma_2 = 2.80, |t| = 1.32, t_{0.995}, \nu_8 = 3.36$ hence hypothesis is accepted

4. z (sample) $= 1.99$, (a) $z_{0.05}$, two-tailed test $= \pm 1.96$, no significance, (b) $z_{0.01}$, two-tailed test $= \pm 2.58$, significant difference

5. Assuming null hypothesis of no difference, $\sigma = 0.397, |t| = 1.85$, (a) $t_{0.995}, \nu_8 = 2.82$, hypothesis rejected, (b) $t_{0.95}, \nu_{22} = 1.72$, hypothesis accepted

6. $\sigma = 0.571, |t| = 3.32, t_{0.95}, \nu_8 = 1.86$, hence hypothesis is rejected

Chapter 66

Exercise 243 (page 709)

1. 10.2   2. 3.16

Exercise 244 (page 715)

1. Expected frequencies: 7, 33, 65, 73, 48, 19, 4, 0; χ^2-value $= 3.62, \chi^2_{0.95}, \nu_7 = 14.1$, hence hypothesis accepted. $\chi^2_{0.10}, \nu_7 = 2.83$, hence data is not 'too good'

2. $\lambda = 2.404$; expected frequencies: 11, 27, 33, 26, 16, 8, 3 χ^2-value $= 42.24$; $\chi^2_{0.95}, \nu_6 = 12.6$, hence the data does not fit a Poisson distribution at a level of significance of 0.05

3. $\bar{x} = 1.32, s = 0.0180$; expected frequencies, 6, 17, 36, 55, 65, 55, 36, 17, 6; χ^2-value $= 5.98$; $\chi^2_{0.95}\nu_6 = 12.6$, hence the null hypothesis is accepted, i.e. the data does correspond to a normal distribution

4. $\lambda = 0.896$; expected frequencies are 102, 91, 41, 12, 3, 0, 0; χ^2-value $= 5.10$; $\chi^2_{0.95}, \nu_6 = 12.6$, hence this data fits a Poisson distribution at a level of significance of 0.05

5. $x = 10.09$ MN; $\sigma = 0.733$ MN; expected frequencies, 2, 5, 12, 16, 14, 8, 3, 1; χ^2-value $= 0.563$; $\chi^2_{0.95}, \nu_5 = 11.1$. Hence hypothesis accepted. $\chi^2_{0.05}, \nu_5 = 1.15$, hence the results are 'too good to be true'

Exercise 245 (page 719)

1. $H_0 : t = 15\text{h}, H_1 : t \neq 15\text{h}$ $S = 7$. From Table 66.3, $S \leq 2$, hence accept H_0

2. $S = 6$. From Table 66.3, $S \leq 4$, hence null hypothesis accepted

3. H_0: mean A = mean B, H_1: mean $A \neq$ mean B, $S = 4$. From Table 66.3, $S \leq 4$, hence H_1 is accepted

Exercise 246 (page 723)

1. $H_0 : t = 220\text{h}, H_1 : t \neq 220\text{h}, T = 74$. From Table 66.4, $T \leq 29$, hence H_0 is accepted

2. $H_0 : s = 150, H_1 : s \neq 150, T = 59.5$. From Table 66.4, $T \leq 40$, hence hypothesis H_0 is accepted

3. $H_0 : N = R, H_1 : N \neq R, T = 5$. From Table 66.4, with $n = 10$ (since two differences are zero), $T \leq 8$, Hence there is a significant difference in the drying times

Exercise 247 (page 729)

1. $H_0 : T_A = T_B, H_1 : T_A \neq T_B, U = 30$. From Table 66.5, $U \leq 17$, hence accept H_0, i.e. there is no difference between the brands

2. $H_0 : B.S._A = B.S._B, H_1 : B.S._A \neq B.S._B, \alpha_2 = 10\%, U = 28$. From Table 66.5, $U \leq 15$, hence accept H_0, i.e. there is no difference between the processes

3. $H_0 : A = B, H_1 : A \neq B, \alpha_2 = 5\%, U = 4$. From Table 66.5, $U \leq 8$, hence null hypothesis is rejected, i.e. the two methods are not equally effective

4. $H_0 : \text{mean}_A = \text{mean}_B$, $H_1 : \text{mean}_A \neq \text{mean}_B$, $\alpha_2 = 5\%, U = 90$. From Table 66.5, $U \leq 99$, hence H_0 is rejected and H_1 accepted

Chapter 67

Exercise 248 (page 739)

1. (a) $\dfrac{2}{s^2} - \dfrac{3}{s}$ (b) $\dfrac{10}{s^3} + \dfrac{4}{s^2} - \dfrac{3}{s}$

2. (a) $\dfrac{1}{4s^4} - \dfrac{3}{s^2} + \dfrac{2}{s}$ (b) $\dfrac{8}{s^6} - \dfrac{48}{s^5} + \dfrac{1}{s^3}$

3. (a) $\dfrac{5}{s-3}$ (b) $\dfrac{2}{s+2}$

4. (a) $\dfrac{12}{s^2+9}$ (b) $\dfrac{3s}{s^2+4}$

5. (a) $\dfrac{7s}{s^2-4}$ (b) $\dfrac{1}{s^3-9}$

6. (a) $\dfrac{2(s^2+2)}{s(s^2+4)}$ (b) $\dfrac{24}{s(s^2+16)}$

7. (a) $\dfrac{s^2-2}{s(s^2-4)}$ (b) $\dfrac{16}{s(s^2-16)}$

8. $\dfrac{4}{s^2+a^2}(a\cos b + s\sin b)$

9. $\dfrac{3}{s^2+\omega^2}(s\cos\alpha + \omega\sin\alpha)$

10. Proof

Chapter 68

Exercise 249 (page 743)

1. (a) $\dfrac{2}{(s-2)^2}$ (b) $\dfrac{2}{(s-1)^3}$

2. (a) $\dfrac{24}{(s+2)^4}$ (b) $\dfrac{12}{(s+3)^5}$

3. (a) $\dfrac{s-1}{s^2-2s+2}$ (b) $\dfrac{6}{s^2-4s+8}$

4. (a) $\dfrac{5(s+2)}{s^2+4s+13}$ (b) $\dfrac{4}{s^2+10s+26}$

5. (a) $\dfrac{1}{s-1} - \dfrac{s-1}{s^2-2s+5}$

   (b) $\dfrac{1}{4}\left(\dfrac{1}{s-3} - \dfrac{s-3}{s^2-6s+13}\right)$

6. (a) $\dfrac{1}{s(s-2)}$ (b) $\dfrac{3(s-2)}{s^2-4s-12}$

7. (a) $\dfrac{6}{s^2+2s-8}$ (b) $\dfrac{s+3}{4(s^2+6s+5)}$

8. (a) $\dfrac{2(s-10)}{s^2-2s+10}$ (b) $\dfrac{-6(s+1)}{s(s+4)}$

Exercise 250 (page 745)

1. Proof 2. Proof 3. Proof 4. Proof

Exercise 251 (page 746)

1. (a) 3 (b) 16 2. (a) 6 (b) –1 3. 4 4. 0

Chapter 69

Exercise 252 (page 750)

1. (a) 7 (b) $2e^{5t}$

2. (a) $\dfrac{3}{2}e^{-\frac{1}{2}t}$ (b) $2\cos 2t$

3. (a) $\dfrac{1}{5}\sin 5t$ (b) $\dfrac{4}{3}\sin 3t$

4. (a) $\dfrac{5}{2}\cos 3t$ (b) $6t$

5. (a) $\dfrac{5}{2}t^2$ (b) $\dfrac{4}{3}t^3$

6. (a) $6\cosh 4t$ (b) $\dfrac{7}{4}\sinh 4t$

7. (a) $\dfrac{5}{3}\sinh 3t$ (b) $2e^t t^2$

8. (a) $\dfrac{1}{6}e^{-2t}t^3$ (b) $\dfrac{1}{8}e^{3t}t^4$

9. (a) $e^{-t}\cos 3t$ (b) $\dfrac{3}{2}e^{-3t}\sin 2t$

10. (a) $2e^{3t}\cos 2t$ (b) $\dfrac{7}{2}e^{4t}\sinh 2t$

11. (a) $2e^{-2t}\cosh 3t + \dfrac{1}{3}e^{-2t}\sinh 3t$

    (b) $3e^{4t}\cos 3t + \dfrac{14}{3}e^{4t}\sin 3t$

Exercise 253 (page 751)

1. $2e^{-t} - 5e^{-3t}$

2. $4e^{-t} - 3e^{2t} + e^{-3t}$

3. $2e^{-3t} + 3e^t - 4e^t t$

4. $e^{-3t}(3 - 2t - 3t^2)$

5. $2\cos\sqrt{2}t + \dfrac{3}{\sqrt{2}}\sin\sqrt{2}t + 5e^{-t}$

6. $2 + t + \sqrt{3}\sin\sqrt{3}t - 4\cos\sqrt{3}t$

7. $2 - 3e^{-2t}\cos 3t - \dfrac{2}{3}e^{-2t}\sin 3t$

Exercise 254 (page 753)

1. (a) $s = -4$ (b) $s = 0$, $s = -2$, $s = 4+j3$, $s = 4-j3$

2. Poles at $s = -3, s = 1+j2, s = 1-j2$, zeros at $s = +1, s = -2$

3. Poles at $s = -2$, $s = -1+j2$, $s = -1-j2$, zero at $s = 1$

4. Poles at $s = 0$, $s = +j2$, $s = -j2$, zeros at $s = -1, s = 6$

Chapter 70

Exercise 255 (page 757)

1. $V(t) = 6(H - 4)$

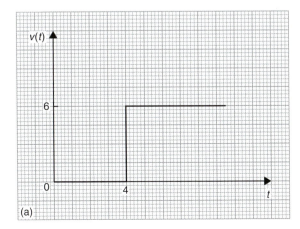

(a)

2. $2H(t) - H(t-5)$

(b)

3.

(c)

4.

(d)

5.

(e)

6.

(f)

7.

(g)

8.

(h)

9.

(i)

10.

(j)

11.

(k)

12.

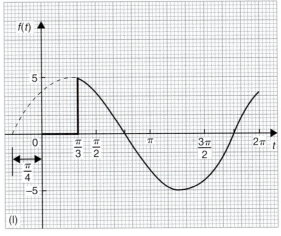

(l)

Exercise 256 (page 759)

1. $\dfrac{e^{-s}}{s}$ **2.** $\dfrac{7e^{-3s}}{s}$ **3.** $\dfrac{2e^{-2s}}{s^3}$

4. $\dfrac{e^{-3s}}{s^2+1}$ **5.** $\dfrac{e^{-4s}}{s-1}$ **6.** $\dfrac{3e^{-5s}}{s^2+9}$

7. $\dfrac{6e^{-s}}{s^4}$ **8.** $\dfrac{se^{-6s}}{s^2+9}$ **9.** $\dfrac{10e^{-5s}}{s^2-4}$

10. $\dfrac{se^{-\frac{\pi}{3}s}}{s^2+4}$ **11.** $\dfrac{2e^{-3s}}{s-1}$ **12.** $\dfrac{3se^{-2s}}{s^2-1}$

Exercise 257 (page 760)

1. $H(t-9)$

2. $4H(t-3)$

3. $2H(t-2).(t-2)$

4. $5H(t-2).\sin(t-2)$

5. $3H(t-4).\cos 4(t-4)$

6. $6H(t-2).\sinh(t-2)$

7. $1.5H(t-6).(t-6)^2$

8. $2H(t-4).\cosh 4(t-4)$

9. $2H\left(t-\dfrac{1}{2}\right).\cos\sqrt{5}\left(t-\dfrac{1}{2}\right)$

10. $4H(t-7).e^{t-7}$

Chapter 71

Exercise 258 (page 765)

1. (a) $2\left(1-e^{-5t}\right)$ (b) $10\left(e^{-3t}-e^{-5t}\right)$

   (c) $i=\dfrac{5}{2}\left(e^{-5t}+\sin 5t-\cos 5t\right)$

2. $y=(3-t)\,e^{\frac{4}{3}t}$

3. $x=2\cos 10t$

4. $i=100t e^{-500t}$

5. $y=4\left(3e^{-2t}-2e^{-4t}\right)$

6. $y=(4x-1)\,e^{x}+\dfrac{1}{3}e^{4x}$

7. $y=\dfrac{3}{85}e^{4x}-\dfrac{3}{10}e^{-x}+\dfrac{9}{34}\cos x-\dfrac{15}{34}\sin x$

8. $y=e^{x}-e^{-2x}+\sin 3x$

9. $y=3e^{x}(\cos x+\sin x)-e^{x}\cos 2x$

10. $y=e^{-t}(2\cos 2t+\sin 2t)$

11. See answers to Exercises 203 and 204 of Chapter 53

12. See answers to Exercises 205 to 208 of Chapter 54

Chapter 72

Exercise 259 (page 771)

1. $x=e^{t}-t-1$ and $y=3e^{t}+2t-3$

2. $x=5\cos t+5\sin t-e^{2t}-e^{t}-3$
   and $y=e^{2t}+2e^{t}-3-5\sin t$

3. $x=3\cos t+\cos\left(\sqrt{3}t\right)$ and $y=3\cos t-\cos\left(\sqrt{3}t\right)$

Chapter 73

Exercise 260 (page 780)

1. $f(x)=\dfrac{8}{\pi}\left(\sin x+\dfrac{1}{3}\sin 3x+\dfrac{1}{5}\sin 5x+\dots\right)$

2. $\dfrac{\pi}{4}=1-\dfrac{1}{3}+\dfrac{1}{5}-\dfrac{1}{7}+\dots.$

3. (a) $f(x)=\dfrac{1}{2}+\dfrac{2}{\pi}\left(\cos x-\dfrac{1}{3}\cos 3x+\dfrac{1}{5}\cos 5x+\dots\right)$

   (b) $\dfrac{1}{2}$

4. Graph sketching

5. $\dfrac{2}{3\pi}\sin 3x$

6. $f(x)=\dfrac{2}{\pi}\left(\cos t-\dfrac{1}{3}\cos 3t+\dfrac{1}{5}\cos 5t-\dots+\sin 2t\right.$
   $\left.+\dfrac{1}{3}\sin 6t+\dfrac{1}{5}\sin 10t+\dots\right)$

7. $f(\theta)=\dfrac{2}{\pi}\left(\dfrac{1}{2}-\dfrac{\cos 2\theta}{(3)}-\dfrac{\cos 4\theta}{(3)(5)}-\dfrac{\cos 6\theta}{(5)(7)}-\dots\right)$

Chapter 74

Exercise 261 (page 786)

1. $f(x)=\pi-2(\sin x+\dfrac{1}{2}\sin 2x+\dfrac{1}{3}\sin 3x$
   $+\dfrac{1}{4}\sin 4x+\dfrac{1}{5}\sin 5x+\dfrac{1}{6}\sin 6x+\dots)$

2. $f(t)=\dfrac{\pi}{2}+1-\dfrac{4}{\pi}\left(\cos t+\dfrac{\cos 3t}{3^2}+\dfrac{\cos 5t}{5^2}+\dots\right)$

3. $f(x)=\pi+2(\sin x-\dfrac{1}{2}\sin 2x+\dfrac{1}{3}\sin 3x$
   $-\dfrac{1}{4}\sin 4x+\dfrac{1}{5}\sin 5x-\dfrac{1}{6}\sin 6x+\dots)$

4. $f(x)=\dfrac{\pi}{2}-\dfrac{4}{\pi}\left(\cos x+\dfrac{\cos 3x}{3^2}+\dfrac{\cos 5x}{5^2}+\dots\right)$

5. $f(\theta)=\dfrac{\pi^2}{3}-4\left(\cos\theta-\dfrac{1}{2^2}\cos 2\theta+\dfrac{1}{3^2}\cos 3\theta-\dots\right)$

6. $\displaystyle\sum_{n=1}^{\infty}\dfrac{1}{n^2}=\dfrac{\pi^2}{6}$

7. $f(x) = \dfrac{8}{\pi^2}\left(\cos x + \dfrac{1}{3^2}\cos 3x + \dfrac{1}{5^2}\cos 5x\right.$

$\left. + \dfrac{1}{7^2}\cos 7x + \ldots\right)$

8. $\dfrac{\pi^2}{8} = 1 + \dfrac{1}{3^2} + \dfrac{1}{5^2} + \dfrac{1}{7^2} + \dfrac{1}{9^2} + \ldots$

Chapter 75

Exercise 262 (page 791)

1. $f(x) = \dfrac{4}{\pi}\left(\cos x - \dfrac{1}{3}\cos 3x + \dfrac{1}{5}\cos 5x\right.$

$\left. - \dfrac{1}{7}\cos 7x + \ldots\right)$

2. $f(t) = -2\left(\sin t + \dfrac{1}{2}\sin 2t + \dfrac{1}{3}\sin 3t\right.$

$\left. + \dfrac{1}{4}\sin 4t + \ldots\right)$

3. $f(x) = \dfrac{\pi}{2} + 1 - \dfrac{4}{\pi}\left(\cos x + \dfrac{1}{3^2}\cos 3x\right.$

$\left. + \dfrac{1}{5^2}\cos 5x + \ldots\right)$

4. $\dfrac{\pi^2}{8} = 1 + \dfrac{1}{3^2} + \dfrac{1}{5^2} + \dfrac{1}{7^2} + \ldots$

5. $y = \dfrac{8}{\pi^2}\left(\sin\theta - \dfrac{1}{3^2}\sin 3\theta + \dfrac{1}{5^2}\sin 5\theta\right.$

$\left. - \dfrac{1}{7^2}\sin 7\theta + \ldots\right)$

Exercise 263 (page 794)

1. $f(x) = \dfrac{2}{\pi}\left(\sin x + \dfrac{\pi}{4}\sin 2x - \dfrac{1}{9}\sin 3x\right.$

$\left. - \dfrac{\pi}{8}\sin 4x + \ldots\right)$

2. (a) $f(t) = \dfrac{1}{2} - \dfrac{2}{\pi}\left(\cos t - \dfrac{1}{3}\cos 3t + \dfrac{1}{5}\cos 5t - \ldots\right)$

(b) $f(t) = \dfrac{2}{\pi}\left(\sin t - \sin 2t + \dfrac{1}{3}\sin 3t + \dfrac{1}{5}\sin 5t\right.$

$\left. - \dfrac{1}{3}\sin 6t + \ldots\right)$

3. $f(x) = \sin^2 x = \dfrac{8}{\pi}\left(\dfrac{\sin x}{(1)(3)} - \dfrac{\sin 3x}{(1)(3)(5)} - \dfrac{\sin 5x}{(3)(5)(7)}\right.$

$\left. - \dfrac{\sin 7x}{(5)(7)(9)} - \ldots\right)$

4. $f(x) = \dfrac{\pi}{4} - \dfrac{2}{\pi}\left(\cos 2x + \dfrac{\cos 6x}{3^2} + \dfrac{\cos 10x}{5^2} + \ldots\right)$

Chapter 76

Exercise 264 (page 798)

1. $v(t) = \dfrac{5}{2} + \dfrac{10}{\pi}\left[\sin\left(\dfrac{\pi t}{10}\right) + \dfrac{1}{3}\sin\left(\dfrac{3\pi t}{10}\right)\right.$

$\left. + \dfrac{1}{5}\sin\left(\dfrac{5\pi t}{10}\right) + \ldots\right]$

2. $f(x) = \dfrac{5}{2} - \dfrac{5}{\pi}\left[\sin\left(\dfrac{2\pi x}{5}\right) + \dfrac{1}{2}\sin\left(\dfrac{4\pi x}{5}\right)\right.$

$\left. + \dfrac{1}{3}\sin\left(\dfrac{6\pi x}{5}\right) + \ldots\right]$

3. $f(x) = \dfrac{12}{\pi}\left\{\sin\left(\dfrac{\pi x}{2}\right) + \dfrac{1}{3}\sin\left(\dfrac{3\pi x}{2}\right)\right.$

$\left. + \dfrac{1}{5}\sin\left(\dfrac{5\pi x}{2}\right) + \ldots\right\}$

4. $f(t) = \dfrac{V}{\pi} + \dfrac{V}{2}\sin t - \dfrac{2V}{\pi}\left(\dfrac{\cos 2t}{(1)(3)} + \dfrac{\cos 4t}{(3)(5)}\right.$

$\left. + \dfrac{\cos 6t}{(5)(7)} + \ldots\right)$

Exercise 265 (page 800)

1. $f(x) = \dfrac{3}{2} - \dfrac{12}{\pi^2}\left\{\cos\left(\dfrac{\pi x}{3}\right) + \dfrac{1}{3^2}\cos\left(\dfrac{3\pi x}{3}\right)\right.$

$\left. + \dfrac{1}{5^2}\cos\left(\dfrac{5\pi x}{3}\right) + \ldots\right\}$

2. $f(x) = \dfrac{6}{\pi}\left\{\sin\left(\dfrac{\pi x}{3}\right) - \dfrac{1}{2}\sin\left(\dfrac{2\pi x}{3}\right) + \dfrac{1}{3}\sin\left(\dfrac{3\pi x}{3}\right)\right.$

$\left. - \dfrac{1}{4}\sin\left(\dfrac{4\pi x}{3}\right) + \ldots\right\}$

3. $f(t) = \dfrac{8}{\pi^2}\left\{\sin\left(\dfrac{\pi t}{2}\right) - \dfrac{1}{3^2}\sin\left(\dfrac{3\pi t}{2}\right)\right.$

$\left. + \dfrac{1}{5^2}\sin\left(\dfrac{5\pi t}{2}\right) - \ldots\right\}$

4. $f(\theta) = \dfrac{16}{3} - \dfrac{64}{\pi^2}\left\{\cos\left(\dfrac{\pi\theta}{4}\right) - \dfrac{1}{2^2}\cos\left(\dfrac{2\pi\theta}{4}\right)\right.$

$\left. + \dfrac{1}{3^2}\cos\left(\dfrac{3\pi\theta}{4}\right) - \ldots\right\}$

Chapter 77

Exercise 266 (page 805)

1. $y = 23.92 + 7.81\cos\theta + 14.61\sin\theta + 0.17\cos 2\theta + 2.31\sin 2\theta - 0.33\cos 3\theta + 0.50\sin 3\theta$

2. $y = 5.00 - 10.78\cos\theta + 6.83\sin\theta + 0.13\cos 2\theta + 0.79\sin 2\theta + 0.58\cos 3\theta - 1.08\sin 3\theta$

3. $i = 0.64 + 1.58\cos\theta - 2.73\sin\theta - 0.23\cos 2\theta - 0.42\sin 2\theta + 0.27\cos 3\theta + 0.05\sin 3\theta$

Exercise 267 (page 806)

1. (a) Only odd cosine terms present (b) Only even sine terms present

2. $y = 9.4 + 13.2\cos\theta - 24.1\sin\theta + 0.92\cos 2\theta - 0.14\sin 2\theta + 0.83\cos 3\theta + 0.67\sin 3\theta$

3. $i = 4.00 - 4.67\cos 2\theta + 1.00\cos 4\theta - 0.66\cos 6\theta + \dots$

4. $y = 1.83 - 25.67\cos\theta + 83.89\sin\theta$
   $+ 1.0\cos 2\theta - 0.29\sin 2\theta + 15.83\cos 3\theta$
   $+ 10.5\sin 3\theta$

Chapter 78

Exercise 268 (page 814)

1. $f(t) = \sum_{n=-\infty}^{\infty} \frac{j}{n\pi}(\cos n\pi - 1)e^{jnt}$

   $= 1 - j\frac{2}{\pi}\left(e^{jt} + \frac{1}{3}e^{j3t} + \frac{1}{5}e^{j5t} + \dots\right)$

   $+ j\frac{2}{\pi}\left(e^{-jt} + \frac{1}{3}e^{-j3t} + \frac{1}{5}e^{-j5t} + \dots\right)$

2. Proof

3. Proof

4. $f(t) = \frac{1}{2}\sum_{n=-\infty}^{\infty}\left(\frac{e^{(2-j\pi n)} - e^{-(2-j\pi n)}}{2 - j\pi n}\right)e^{j\pi nt}$

Exercise 269 (page 817)

1. $f(x) = \sum_{n=-\infty}^{\infty}\left\{\frac{4}{\pi n}\sin\left(\frac{n\pi}{2}\right)\right\}e^{jnx}$

2. $f(x) = \frac{8}{\pi}\left(\cos x - \frac{1}{3}\cos 3x + \frac{1}{5}\cos 5x\right.$
   $\left. - \frac{1}{7}\cos 7x + \dots\right)$

3. $f(t) = \sum_{n=-\infty}^{\infty}\left(\frac{j2}{n}\cos n\pi\right)e^{jnt}$

4. $f(t) = 4\left(\sin t - \frac{1}{2}\sin 2t + \frac{1}{3}\sin 3t - \frac{1}{4}\sin 4t + \dots\right)$

Exercise 270 (page 822)

1. (a) $2e^{j4t} + 2e^{-j4t}$, $2\angle 0°$ anticlockwise, $2\angle 0°$ clockwise, each with $\omega = 4$ rad/s
   (b) $2e^{j4t}e^{j\pi/2} + 2e^{-j4t}e^{-j\pi/2}$, $2\angle\pi/2$ anticlockwise, $2\angle -\pi/2$ clockwise, each with $\omega = 4$ rad/s

2. $(5 + j6)e^{j2t} + (5 - j6)e^{-j2t}$, $7.81\angle 0.88$ rotating anticlockwise, $7.81\angle -0.88$ rotating clockwise each with $\omega = 2$ rad/s

3. $(2 - j3)e^{jt} + (2 + j3)e^{-jt}$, $3.61\angle -0.98$ rotating anticlockwise, $3.61\angle 0.98$ rotating clockwise, each with $\omega = 1$ rad/s

Chapter 79

Exercise 271 (page 828)

1. $\dfrac{2z}{(z-1)^2}$

2. $\dfrac{3z(z+1)}{(z-1)^3}$

3. $\dfrac{4z(z^2 + 4z + 1)}{(z-1)^4}$

4. $\dfrac{z\sin 3}{z^2 - 2z\cos 3 + 1}$

5. $\dfrac{z}{z-2}$

6. $\dfrac{2z}{(z-2)^2}$

7. $\dfrac{5z(z - \cos 2)}{z^2 - 2z\cos 2 + 1}$

8. $\dfrac{3z}{z - e^{-2}}$

9. $\dfrac{z\sin\frac{1}{2}}{z^2 - 2z\cos\frac{1}{2} + 1}$

10. $4\left(\dfrac{z^2 - ze^3\cos 2}{z^2 - 2ze^3\cos 2 + e^6}\right)$

11. $\dfrac{z}{z+4}$

12. $\dfrac{6z(z+2)}{(z-2)^3}$

13. $\dfrac{3z}{z+5}$

14. $\dfrac{-3z}{(z+3)^2}$

15. $\dfrac{3z}{z - e^5}$

16. $\dfrac{2ze^{-4}\sin 2}{z^2 - 2ze^{-4}\cos 2 + e^{-8}}$

Exercise 272 (page 829)

1. $\left\{\dfrac{5z}{(z-1)^2}\right\} - \left\{\dfrac{4z}{z - e^{-3}}\right\}$ or $\dfrac{-4z^3 + 13z^2 - z\left(5e^{-3} + 4\right)}{(z-1)^2\left(z - e^{-3}\right)}$

2. $\dfrac{z(7z+1)}{(z-1)^3}$

3. $\dfrac{3z(\sin 2 - z + \cos 2)}{z^2 - 2z\cos 2 + 1}$

4. $\dfrac{6z(z-1)}{z^2 - 9}$

5. $\dfrac{z}{(z-1)^2} + \dfrac{z}{(z-e^{-1})}$ or $\dfrac{z^3 - z^2 + z(1-e^{-1})}{(z-1)^2(z-e^{-1})}$

Exercise 273 (page 830)

1. $\dfrac{8z}{z-2}$ **2.** $\dfrac{4z}{z-4}$ **3.** $\dfrac{6z^2}{(z-1)^2}$

4. $\dfrac{z(7z-12)}{z^2 - 5z + 6}$

Exercise 274 (page 831)

1. $\dfrac{1}{(z-1)}$ **2.** $\dfrac{1}{z^2(z-1)}$ **3.** $\dfrac{1}{z(z-a)}$

4. $\dfrac{1}{z^2(z-a)}$ **5.** $\dfrac{1}{z^3(z-3)}$

Exercise 275 (page 832)

1. $\dfrac{2z}{(z-2)^2}$ **2.** $\dfrac{4z}{(z-4)^2}$ **3.** 0

4. $20/9$ or $2\dfrac{2}{9}$ **5.** 1 **6.** $\dfrac{2z}{(z-2)^2}$

Exercise 276 (page 834)

1. 1 **2.** $(2)^k$

3. $(-1)^k$ **4.** $(-4)^k$

5. $\left(\dfrac{1}{3}\right)^k$ **6.** $\left(-\dfrac{1}{4}\right)^k$

7. $\left(\dfrac{1}{5}\right)^k$ **8.** e^{-5k}

9. e^{3k} **10.** $3k$

11. $5k^2$ **12.** $\dfrac{1}{2}k$

13. $\dfrac{1}{3}\left\{(2)^k - (-1)^k\right\}$ **14.** $\dfrac{3}{4}\left\{1 - (-3)^k\right\}$

15. $\dfrac{1}{3}\left\{1 - (-2)^k\right\}$ **16.** $\dfrac{1}{7}\left\{(3)^k - \left(-\dfrac{1}{2}\right)^k\right\}$

17. $(2)^k - 1$ **18.** $10(3)^k - 7(2)^k$

19. $\dfrac{2}{3}\left\{\left(\dfrac{1}{2}\right)^k - (-1)^k\right\}$ **20.** $\dfrac{3}{7}\left\{(2)^k - (-5)^k\right\}$

Exercise 277 (page 837)

1. $4\left(3^k\right)$ **2.** $3 - 2(3)^k$

3. $\dfrac{1}{3}\left\{10 - (4)^k\right\}$ **4.** $\dfrac{1}{3}\left\{(2)^k - 5\left(\dfrac{1}{2}\right)^k\right\}$

5. $(-1)^k - (-2)^k$ **6.** $\dfrac{5}{2} - 6(2)^k + \dfrac{7}{2}(3)^k$

Index